C_v = τ_{cr}/τ_y = web shear coefficient

C_w = torsional warping constant (Chap. 8); compression force in web of steel section (Chap. 16)

d = overall depth of section; nominal bolt diameter

d_f = distance from outside of flange to PNA (Chap. 16)

D = $Et^3/[12(1-\mu^2)]$ = flexural rigidity per unit length of plate (Chap. 6 Part II and Eq. 11.11.14; ; width of fastener hole to be deducted (Sec. 3.4); number of 1/16s of an inch in weld size

D_s = factor to account for eccentric load on stiffeners, Eqs. 11.9.28 and 11.9.35

e = eccentricity of load

E = tension—compression modulus of elasticity

E_c = modulus of elasticity of concrete = $(w^{1.5})\sqrt{f'_c}$ where w is the density of concrete in pcf (i.e., 145 pcf for normal-weight concrete) and f'_c is in ksi.

E_s = modulus of elasticity of steel, 29,000 ksi

EI_{eff} = effective stiffness of a composite section

f' = stress due to direct shear, P/A

f'' = stress due to torsional moment, Tr/I_p

f_a = required axial stress at a point using LRFD or ASD load combination

f_b = required bending stress at a point using LRFD or ASD load combination

f_{bw} = compression-tension stress due to restraint of warping (lateral bending of flanges), $(Ebh/4)(d^2\phi/dz^2)$

f'_c = specified 28-day compressive strength of concrete

f_t = required tensile stress at a point using LRFD or ASD load combination

f_{un} = factored normal (compression or tension) stress

f_v = required shear stress at a point using LRFD or ASD load combination

F_b = allowable flexural stress

F_{BM} = nominal strength of the base metal per unit area

F_{cr} = critical stress in compression; buckling stress

F_e = elastic critical buckling stress

F'_e = Eq. 12.14.11

F_{EXX} = weld metal tensile strength

F_{nt} = bolt nominal tensile strength

F_{nv} = bolt nominal shear strength

F_r = residual stress = $0.3F_y$

FS   = factor of safety

F_u = tensile strength of structural steel; tensile strength of base metal (Chap. 5)

F_u^b = tensile strength of bolt material

F'_{ut} = design *stress* limit on R_{ut}/A_b in tension on bolt subject to combined shear and tension, values in Table 4.14.1

F_y, F_{yf}, F_{yw} = yield stress; for beam flange, F_{yb}; for column web, F_{yc}; for flange, F_{yf}; for longitudinal reinforcing bars, F_{yr}; for stiffeners, F_{yst}; for tension flange, F_{yt}; for web or weld metal, F_{yw}

g = gage distance for fastener holes measured transverse to direction of load

G = shear modulus of elasticity, $E/[2(1+\mu)]$

h = unsupported web height (For unsymmetrical sections it relates to the compression side of the neutral axis.) (See AISC B4.2); overall depth (Fig. 6.13.1); distance between centers of flanges (Chap. 8); depth of web plate Chap. 11); story height (Chaps. 14 and 15)

h_c = twice the distance from the neutral axis to the inside face of the compression flange less the fillet or corner radius

h_o = distance between flange centroids

h_{sc} = hole factor (Eq. 4.9.1)

H_u = factored lateral force causing sway deflection

I, I_x, I_y = moment of inertia, about the x- or y-axis, respectively

I_f = moment of inertia of one flange about y-axis (Chap. 8)

I_g = girder moment of inertia (Chaps. 14 and 15)

I_p = polar moment of inertia about shear center

I_{st}	=	moment of inertia of the cross-sectional area of a transverse stiffener as defined following Eq. 11.11.11
I_{tr}	=	transformed cracked section moment of inertia of composite section (Chap.16)
I_{xy}	=	product of inertia, $\int xy\,dA$ (Sec. 7.10)
I_{yc}	=	moment of inertia of compression flange about y-axis
J	=	torsion constant
k	=	$\sqrt{P/EI}$ (Secs. 6.2 and 12.2); distance from outer face of flange to web toe of fillet (Sec. 7.8); plate buckling coefficient, Eq. 6.14.28; spring constant (Sec. 9.13); spring constant $= \sum P_n/h$, Eq. 14.5.9
k_c	=	coefficient in l_r for welded I-shape, $4/\sqrt{h/t_w}$ (Tables 7.4.1 and 9.6.2)
K, K_x, K_y, K_z	=	effective length factor (Sec. 6.9), with respect to the x-, y-, or z-axes, respectively
K_n	=	modified effective length factor used in design of column supporting a "leaner" column, Eq. 14.5.11
K_1, K_2	=	effective length factor in plane of bending with no lateral translation and with sidesway, respectively (Chap. 12)
L	=	length; span
L_b	=	laterally unbraced length
L_c	=	clear distance in the direction of the force (Chap. 4)
L_e	=	end distance measured in direction of line of force (Chap. 4)
L_p	=	AISC: maximum laterally unbraced length for using $M_n = M_p$, Eq. 9.6.3
L_{pd}	=	AISC: maximum laterally unbraced length for using plastic analysis, Eq.9.6.2
L_r	=	AISC: maximum laterally unbraced length for using $M_n = M_r = 0.7F_yS_x$
L_w	=	length of fillet weld
m	=	number of shear planes (Chap. 4); uniformly distributed service load torsional moment (Chap. 8)
m_u	=	factored uniformly distributed torsional moment (Chap. 8)
M_1, M_2	=	smaller moment M_1 and larger moment M_2 at the ends of a laterally unbraced segment
M_a	=	required moment using ASD load combination
M_A, M_B, M_C	=	moment (absolute value) at 1/4, 1/2, and 3/4 points, respectively, of laterally unbraced segment, Eq. 9.6.11
M_{cr}	=	elastic lateral-torsional buckling moment strength
M_E	=	equivalent constant moment, C_mM_2 for beam-column in braced frame subject to end moments M_1 and M_2 only (Sec. 12.4)
M_f	=	lateral bending moment on one flange, $(EC_w/h)(d^2\phi/dz^2)$
M_i	=	primary bending moment from first-order analysis
$M_{\ell t}$	=	first-order moment in sway analysis under H_u force
M_{max}	=	maximum moment in the unbraced segment (Sec. 9.6)
M_n	=	nominal moment strength
M'_n	=	nominal moment strength in the presence of shear
M_{nt}	=	primary factored moment for the no translation case of beam-column
M_{nx}, M_{ny}	=	nominal moment strengths about x- and y-axes, respectively (Secs. 7.11 & 9.14)
M_p, M_{px}, M_{py}	=	plastic moment strength, ZF_y, with respect to the x- and y-axes, according to second subscript
M_r	=	moment strength when extreme fiber reaches $(F_y - F_r = 0.7F_y)$
M_s	=	pure torsional moment (St. Venant torsion) (Chap. 8)
M_u, M_{ux}, M_{uy}	=	factored service load moment, about the x- or y-axes, according to second subscript
M_{uf}	=	factored moment on one flange
M_w	=	warping torsional moment (lateral bending effect) (Chap. 8)
M_x	=	moment about x-axis; bending moment from stress σ_x (Sec. 6.14)
M_{xy}	=	torsional moment from shear stress τ_{xy} (Sec. 6.14)
M_y	=	bending moment from stress σ_y (Sec. 6.14); moment about y-axis when biaxial bending is being considered (Chapters 8 and 9; Sec. 12.9); nominal moment strength M_n when extreme fiber reaches (except for biaxial bending)
M_z	=	bending moment or torsional moment at a location z along axis of member; for torsion, $M_z = M_s + M_w$
n	=	modulus of elasticity ratio, E_s/E_c; number of fasteners (Chap. 4)
N	=	bearing length (Sec. 7.8)
N_b	=	number of bolts carrying the applied tension

STEEL STRUCTURES
Design and Behavior

STEEL STRUCTURES
Design and Behavior

Emphasizing Load and Resistance Factor Design

FIFTH EDITION

CHARLES G. SALMON

University of Wisconsin-Madison

JOHN E. JOHNSON

University of Wisconsin-Madison

FARIS A. MALHAS

West Virginia University
Institute of Technology

PEARSON
Prentice
Hall

Upper Saddle River, NJ 07458

Library of Congress Cataloging-in-Publication Data on File

Vice President and Editorial Director, ECS: *Marcia J. Horton*
Acquisitions Editor: *Holly Stark*
Associate Editor: *Dee Bernhard*
Managing Editor: *Scott Disanno*
Production Editor: *Rose Kernan*
Art Director: *Kenny Beck*
Managing Editor, AV Management and Production: *Patricia Burns*
Art Editor: *Gregory Dulles*
Director, Image Resource Center: *Melinda Reo*
Manager, Rights and Permissions: *Zina Arabia*
Manager, Visual Research: *Beth Brenzel*
Manager, Cover Visual Research and Permissions: *Karen Sanatar*
Manufacturing Manager: *Alexis Heydt-Long*
Manufacturing Buyer: *Lisa McDowell*
Senior Marketing Manager: *Tim Galligan*

© 2009 Pearson Education, Inc.
Pearson Prentice Hall
Pearson Education, Inc.
Upper Saddle River, New Jersey 07458

Printed in the United States of America

10 9 8 7 6 5 4

ISBN-10: 0-13-188556-1
ISBN-13: 978-0-13-188556-1

Pearson Education Ltd., London
Pearson Education Singapore, Pte. Ltd
Pearson Education Canada, Inc.
Pearson Education-Japan
Pearson Education Australia PTY, Limited
Pearson Education North Asia, Ltd., Hong Kong
Pearson Educación de Mexico, S.A. de C.V.
Pearson Education Malaysia, Pte. Ltd.
Pearson Education Upper Saddle River, New Jersey

Contents

Preface

The publication of this fifth edition reflects the continuing changes occurring in design requirements for structural steel, particularly the significant change by the American Institute of Steel Construction (AISC) to combine the *Load and Resistance Factor Design* (LRFD) *Method with the Allowable Stress (now called Strength) Design (ASD) Method for Structural Steel Buildings.*

Design of structural steel members has developed over the past 110 years from a simple approach involving a few basic properties of steel and elementary mathematics, to a sophisticated treatment demanding a thorough knowledge of structural and material behavior. Present design practice utilizes knowledge of mechanics of materials, structural analysis, and particularly, structural stability, in combination with nationally recognized design rules for safety. The most widely used design rules are those of the American Institute of Steel Construction (AISC), given in the *Specification for Structural Steel Buildings.*

The specific occurrence dictating this fifth edition is the publication of the 2005 *AISC Specification* (effective March 9, 2005) with Commentary, along with the corresponding handbook, *Steel Construction Manual*, 13th Edition. The thirteen arises from nine editions of the Allowable Stress Design Manual and four editions of the Load and Resistance Factor Design Manual.

The fifth edition of this text follows the same philosophical approach that has gained wide acceptance of users since the first edition was published in 1971. This edition continues to strive to present in a logical manner the theoretical background needed for developing and explaining design requirements, particularly those of the 2005 *AISC Specification,* emphasizing its LRFD Method. Beginning with coverage of background material, including references to pertinent research, the development of specific formulas used in the *AISC Specification* is followed by a generous number of design examples explaining in detail the process of selecting minimum weight members to satisfy given conditions.

Emphasis throughout this fifth edition is on the 2005 AISC Specification LRFD Method. That specification is based on statistical studies of loads and the resistances of steel structures subject to various types of load effects, such as bending moment, shear, axial force, and torsional moment. The rational treatment of both loads (including load effects) and resistances results in steel structures having more uniform safety throughout.

This modern philosophy of design, discussed only briefly in one section of the second edition, is moving toward being the predominant approach to design.

Considerable emphasis has been placed on presenting for the beginner, as well as the advanced student, the necessary elastic and inelastic stability concepts, the understanding of which is essential to properly apply steel design rules. The same concepts are applicable whether design uses the Load and Resistance Factor Design Method (LRFD) or the Allowable Strength Design Method (ASD). The explanation of stability concepts is incorporated into the chapters in such a way that the reader may either study in detail the stability concepts in logical sequence, or omit or postpone study of sections containing detailed development, merely accepting qualitative explanation and proceeding directly to design.

As this fifth edition is prepared, considerable design is still done according to the traditional ASD Method (formerly often called Working Stress Design). That method focused on service (working) loads and elastically computed stresses, comparing those stresses with allowable limiting values. However, the logical trend now should be toward the more rational LRFD Method. Strength design philosophy (reflected in the 1986 and 1993 LRFD Specifications) uses factored service loads and compares the strength provided with such factored loads (or load effects). The strength in any given case depends on the "limit state", or mode of failure, such as yielding, fracture, or buckling. The traditional "plastic design", included as Chapter N of the 1989 ASD Specification, is an option integrally included as part of the AISC LRFD Method.

Throughout the text, the theory and background material, now being the same for both the LRFD and ASD philosophies of design, have been integrated. The specific design provisions and illustrative examples emphasize the Load and Resistance Factor Design Method.

The fifth edition continues the use of SI units as an addition to the primary use of Inch-Pound units. The 2005 *AISC Specification* has non-dimensionalized the formulas by including the modulus of elasticity E of steel as a variable. The inclusion of E as a variable in the formulas has made many of them cumbersome to use. In many parts of this text, the authors have used both the formulas containing E and the familiar coefficient version with F_y in ksi units. Tables and diagrams generally contain both Inch-Pound and SI units.

Depending on the proficiency required of the student, this textbook provides material for two courses of three or four semester-credit hours each. It is suggested that the beginning course in steel structures for undergraduate students might contain the material of Chapters 1 through 7, 9, 10, 12 and 16, except Sections 6.4, 6.6, 6.12 to 6.19, 7.9 to 7.11, 9.3 to 9.5, 9.8, 9.12 to 9.14, and 12.6 to 12.7. The second course would review some of the same topics of the first course, but more rapidly, emphasizing items omitted in the first course. In addition, the remaining chapters- namely Chapter 8 on torsion, Chapter 11 on plate girders, Chapter 13 on connections, Chapter 14 on frames, and Chapter 15 on frame design- are suggested for inclusion. The primary philosophy emphasized in both courses should be Load and Resistance Factor Design.

The reader will need ready access to the *AISC Manual** throughout the study of the text, particularly when working with the examples. However, it is not the objective of this text that the reader become proficient in the routine use of tables; the tables serve only as a guide to obtaining experience with variation of design parameters and as an aid in arriving at good design. The AISC Specification and Commentary are contained in the *Steel Design Manual* and are therefore not included in this book, except for various individual provisions quoted where they are explained.

Steel Construction Manual, 13th edition, 2005. Since nearly continuous reference is made to the *AISC Manual* (which also contains the 2005 Specification and Commentary), the reader will find it desirable to secure a copy of this document from the American Institute of Steel Construction, Inc., One East Wacker Drive, Suite 700, Chicago, IL 60601–2001.

The direct use of the computer is not specifically employed anywhere in the text. The authors believe the study of basic principles in the classroom is of the highest priority. However, the reader may find that acquiring the data base of standard section properties, available from AISC, will be helpful. There are many types of steel design software available to be used once the designer is knowledgable in structural steel behavior and design.

Features of this fifth edition are: (1) detailed presentation of strength-related background and design rules for the Load and Resistance Factor Design Method; (2) with some limited treatment of the Allowable Strength Design Method. The authors believe intermixing LRFD and ASD leads to confusion and mistakes. The authors also believe that teachers of structural steel design have for quite a few years now been teaching the logical LRFD Method, and should continue to do so.

Other special features of this text are (3) comprehensive treatment of design of I-shaped members subject to torsion (Chapter 8), including a simplified practical method; (4) detailed treatment of plate girder theory as it relates to Load and Resistance Factor Design (Chapter 11) and a comprehensive design example of a two-span continuous girder using two different grades of steel; (5) extensive treatment of connections (Chapter 13), including significant discussion and illustration of the design of components. All chapters of the book were extensively rewritten for the fifth edition to reflect the major changes made by the 2005 AISC Specification, and improved for better readability and understanding as a result of suggestions from users.

The authors are indebted to students, colleagues, and other users of the first four editions who have suggested improvements of wording, identified errors, and recommended items for inclusion or deletion. The suggestions have been carefully considered, resulting in this complete revision. The continued cooperation and help from the American Institute of Steel Construction is also appreciated.

Users of this fifth edition are urged to communicate with the authors regarding all aspects of this book, particularly on identification of errors and suggestions for improvement.

The senior author affectionately dedicates this book in memory of his late wife, Bette Salmon, for her patience and encouragement through the first three editions.

The third author dedicates this book to his wife, Fikrat, and his children Ameen, Sami, Tamara, and Layla for their patience, support, and encouragement.

CHARLES G. SALMON

JOHN E. JOHNSON

FARIS A. MALHAS

Conversion Factors

Some Conversion Factors, between Inch-Pound Units and SI Units, Useful in Structural Steel Design

	To Convert	To	Multiply by
Forces	kip force	kN	4.448
	lb	N	4.448
	kN	kip	0.2248
Stresses	ksi	MPa (i.e., N/mm^2)	6.895
	psi	MPa	0.006895
	MPa	ksi	0.1450
	MPa	psi	145.0
Moments	ft-kip	$kN \cdot m$	1.356
	$kN \cdot m$	ft-kip	0.7376
Uniform Loading	kip/ft	kN/m	14.59
	kN/m	kip/ft	0.06852
	kip/ft^2	kN/m^2	47.88
	psf	N/m^2	47.88
	kN/m^2	kip/ft^2	0.02089

For proper use of SI, see IEEE/ASTM SI 10-2002 Standard for Use of the International System of Units (*SI*): The Modern Metric System (Replaces ASTM E380 and ANSI/IEEE Std 268-1992), American Society for Testing and Materials, West Conshohocken, PA 2002.

Basis of Conversions (IEEE/ ASTM SI 10): 1 in. = 25.4 mm; 1 lb force = 4.448 221 615 260 5 newtons.

Basic SI units relating to structural steel design:

Quantity	Unit	Symbol
length	metre	m
mass	kilogram	kg
time	second	s

Derived SI units relating to structural design:

Quantity	Unit	Symbol	Formula
force	newton	N	$kg \cdot m/s^2$
pressure, stress	pascal	Pa	N/m^2
energy, or work	joule	J	$N \cdot m$

1

Introduction

1.1 STRUCTURAL DESIGN

Structural design may be defined as a mixture of art and science, combining the experienced engineer's intuitive feeling for the behavior of a structure with a sound knowledge of the principles of statics, dynamics, mechanics of materials, and structural analysis, to produce a safe, economical structure that will serve its intended purpose.

Until about 1850, structural design was largely an art relying on intuition to determine the size and arrangement of the structural elements. Early man-made structures essentially conformed to those which could also be observed in nature, such as beams and arches. As the principles governing the behavior of structures and structural materials have become better understood, design procedures have become more scientific.

Computations involving scientific principles should serve as a *guide* to decision making and not be followed blindly. The art or intuitive ability of the experienced engineer is utilized to make the decisions, guided by the computational results.

1.2 PRINCIPLES OF DESIGN

Design is a process by which an optimum solution is obtained. In this text the concern is with the design of structures—in particular, *steel* structures. In any design, certain criteria must be established to evaluate whether or not an optimum has been achieved. For a structure, typical criteria may be (a) minimum cost; (b) minimum weight; (c) minimum construction time; (d) minimum labor; (e) minimum cost of manufacture of owner's products; and (f) maximum efficiency of operation to owner. Usually several criteria are involved, each of which may require weighting. Observing the above possible criteria, it may be apparent that setting clearly measurable criteria (such as weight and cost) for establishing an optimum frequently will be difficult, and perhaps impossible. In most practical situations, the evaluation must be qualitative.

1

1730 ft Sears Tower, Chicago
(photo courtesy of Galen R.
Frysinger)

If a specific objective criterion can be expressed mathematically, then optimization techniques may be employed to obtain a maximum or minimum for the objective function. Optimization procedures and techniques comprise an entire subject that is outside the scope of this text. The criterion of minimum weight is emphasized throughout, under the general assumption that minimum material represents minimum cost. Other subjective criteria must be kept in mind, even though the integration of behavioral principles with design of structural steel elements in this text utilizes only simple objective criteria, such as weight or cost.

Design Procedure

The design procedure may be considered to be composed of two parts—functional design and structural framework design. Functional design ensures that intended results are achieved, such as (a) adequate working areas and clearances; (b) proper ventilation and/or air conditioning; (c) adequate transportation facilities, such as elevators, stairways, and cranes or materials handling equipment; (d) adequate lighting; and (e) aesthetics.

The structural framework design is the selection of the arrangement and sizes of structural elements so that service loads may be safely carried, and displacements are within acceptable limits.

The iterative design procedure may be outlined as follows:

1. *Planning.* Establishment of the functions which the structure must serve. Set criteria against which to measure the resulting design for being an optimum.

2. *Preliminary structural configuration.* Arrangement of the elements to serve the functions in step 1.

3. *Establishment of the loads* to be carried.

4. *Preliminary member selection.* Based on the decisions of steps 1, 2, and 3, selection of the member sizes to satisfy an objective criterion, such as least weight or cost.

5. *Analysis.* Structural analysis involving modeling the loads and the structural framework to obtain internal forces and any desired deflections.

6. *Evaluation.* Are all strength and serviceability requirements satisfied and is the result optimal? Compare the result with predetermined criteria.

7. *Redesign.* Repetition of any part of the sequence 1 through 6 found necessary or desirable as a result of evaluation. Steps 1 through 6 represent an iterative process. Usually in this text only steps 3 through 6 will be subject to this iteration, since the structural configuration and external loading will be prescribed.

8. *Final decision.* The determination of whether or not an optimum design has been achieved.

1.3   HISTORICAL BACKGROUND OF STEEL STRUCTURES

Metal as a structural material began with cast iron, used on a 100-ft (30-m) arch span which was built in England in 1777–1779 [1.1].* A number of cast-iron bridges were built during the period 1780–1820, mostly arch-shaped with main girders consisting of individual cast-iron pieces forming bars or trusses. Cast iron was also used for chain links on suspension bridges until about 1840.

Wrought iron began replacing cast iron soon after 1840, the earliest important example being the Brittania Bridge over Menai Straits in Wales, which was built in 1846–1850. This was a tubular girder bridge having spans 230–460–460–230 ft (70–140–140–70 m), which was made from wrought-iron plates and angles.

The process of rolling various shapes was developing as cast iron and wrought iron received wider usage. Bars were rolled on an industrial scale beginning about 1780. The rolling of rails began about 1820 and was extended to I-shapes by the 1870s.

The development of the Bessemer process (1855), the introduction of a basic liner in the Bessemer converter (1870), and the open-hearth furnace brought widespread use of iron ore products in building materials. Since 1890, steel has replaced wrought iron as the principal metallic building material. Currently (2008), steels having yield stresses varying from 24,000 to 100,000 pounds per square inch, or psi (165 to 690 megapascals,† or MPa), are available for structural uses. The various steels, their uses and their properties are discussed in Chapter 2.

1.4   LOADS

The determination of the loads to which a structure or structural element will be subjected is, at best, an estimate. Even if the loads are well known at one location in a structure, the distribution of load from element to element throughout the structure usually requires assumptions and approximations. Some of the most common kinds of loads are discussed in the following sections.

---

*Numbers in brackets refer to the Selected References at the end of the chapter.
†MPa, megapascals, are equivalent to Newtons per square millimeter, N/mm², in SI units.

Dead Load

Dead load is a fixed-position gravity service load, so called because it acts continuously toward the earth when the structure is in service. The weight of the structure is considered dead load, as are attachments to the structure such as pipes, electrical conduit, air-conditioning and heating ducts, lighting fixtures, floor covering, roof covering, and suspended ceilings; that is, all items that remain throughout the life of the structure.

Dead loads are usually known accurately but not until the design has been completed. Under steps 3 through 6 of the design procedure discussed in Sec. 1.2, the weight of the structure or structural element must be estimated, preliminary section selected, weight recomputed, and member selection revised if necessary. The dead load of attachments is usually known with reasonable accuracy prior to the design.

Live Load

Gravity loads acting when the structure is in service, but varying in magnitude and location, are termed *live loads*. Examples of live loads are human occupants, furniture, movable equipment, vehicles, and stored goods. Some live loads may be practically permanent, others may be highly transient. Because of the unknown nature of the magnitude, location, and density of live load items, realistic magnitudes and the positions of such loads are very difficult to determine.

Because of the public concern for adequate safety, live loads to be taken as service loads in design are usually prescribed by state and local building codes. These loads are generally empirical and conservative, based on experience and accepted practice rather than accurately computed values. Wherever local codes do not apply, or do not exist, the provisions from one of several regional and national building codes may be used. One such widely recognized code, *Minimum Design Loads for Buildings and Other Structures* ASCE 7 (formerly ANSI A58.1, published by the American National Standards Institute), for the past few years has been developed under the jurisdiction of the American Society of Civil Engineers. This code will henceforth be referred to as the ASCE 7 Standard. This Standard is updated from time to time, most recently in 2005, making ASCE 7–05 [1.2] the current specific reference. Some typical live loads from ASCE 7 are presented in Table 1.4.1.

Live load when applied to a structure should be positioned to give the maximum effect, including partial loading, alternate span loading, or full span loading as may be necessary. The simplified assumption of full uniform loading everywhere should be used only when it agrees with reality or is an appropriate approximation. The probability of having the prescribed loading applied uniformly over an entire floor, or over all floors of a building simultaneously, is almost nonexistent. Most codes recognize this by allowing for some percentage reduction from full loading. For instance, ASCE 7 allows members having an influence area of 400 sq ft or more to be designed for a reduced live load:

$$L = L_0 \left(0.25 + \frac{15}{\sqrt{K_{LL}A_T}} \right) \qquad (1.4.1)$$

where L = reduced live load per sq ft of area supported by the member
        L_0 = unreduced live load per sq ft of area supported by the member (from Table 1.4.1)
        A_T = tributary area, sq ft
      K_{LL} = live load element factor

The live load element factor K_{LL} is equal to 4 for interior columns and exterior columns without cantilevered slabs; 3 for edge columns with cantilevered slabs; 2 for corner columns

TABLE 1.4.1 Typical Minimum Uniformly Distributed Live Loads (Adapted from Ref. 1.2)

Occupancy or use	Live load	
	psf	Pa[*]

1. Hotel guest rooms
   School classrooms
   Private apartments
   Private hospital rooms
	40	1900

2. Offices	50	2400
3. Assembly halls, fixed seats Library reading rooms	60	2900
4. Corridors, above first floor in schools, libraries, and hospitals	80	3800

5. Assembly areas; theater lobbies
   Dining rooms and restaurants
   Office building lobbies
   Main floor, retail stores
   Assembly halls, movable seats
	100	4800

6. Wholesale stores, all floors
   Manufacturing, light
   Storage warehouses, light
	125	6000

7. Armories and drill halls
   Stage floors
   Library stack rooms
	150	7200

8. Manufacturing, heavy
   Sidewalks and driveways subject to trucking
   Storage warehouses, heavy
	250	12,000

[*]SI values are approximate conversions, using 1 psf (lb/sq ft) = 47.9 Pa.

with cantilevered slabs, edge beams without cantilever slabs, and interior beams; and 1 for all other members. The reduced live load L shall not be less than 50% of the live load L_0 for members supporting one floor, nor less than 40% of the live load L_0 otherwise.

Highway Live Load

Highway vehicle loading in the United States has been standardized by the American Association of State Highway and Transportation Officials (AASHTO) [1.3] into standard truck loads and lane loads that approximate a series of trucks. Three basic types of live load defined by AASHTO are used in different combinations to produce maximum effects: the HS20 Design Truck Loading is a tractor truck with a semitrailer having a total of three axles, the Design Uniform Lane Loading consists of a uniform load of 640 pounds per linear foot, or plf, and the Design Tandem Loading consists of a two-axle vehicle with each axle carrying a load of 25 kilopounds, or kips (Figs. 1.4.1a-c). These live loads are for one design lane and are referred to as HL-93.

In designing a given bridge, the three different types of live loads described above are combined in three different ways by AASHTO to obtain the maximum effect. The first live load combination is made of one HS20 truck in addition to a uniform load of 0.64 klf. The second live load combination is made of one tandem vehicle in addition to the same uniform load. The third load combination is to produce maximum negative moments between points of inflection: two HS20 trucks are placed on two adjacent spans in addition to the same uniform load as above.

Railroad bridges are designed to carry a similar semiempirical loading known as the Cooper E72 train, consisting of a series of concentrated loads a fixed distance apart followed

Figure 1.4.1
AASHTO highway loadings. (a) Uniform Lane Loading, (b) Tandem Loading, (c) HS20 Design Truck Loading [1.3]

by uniform loading. This loading is prescribed by the American Railway Engineering and Maintenance of Way Association (AREMA). [1.4].

Impact

The term *impact* as ordinarily used in structural design refers to the dynamic effect of a suddenly applied load. In the building of a structure, the materials are added slowly; people entering a building are also considered a gradual loading. Dead loads are static loads; i.e., they have no effect other than weight. Live loads may be either static or they may have a dynamic effect. Persons and furniture would be treated as static live load, but cranes and various types of machinery also have dynamic effects.

Consider the spring-mass system of Fig. 1.4.2a, where the spring may be thought of as analogous to an elastic beam. When load is gradually applied (i.e., static loading) the mass (weight) deflects an amount x and the load on the spring (beam) is equal to the weight W. In Fig. 1.4.2b the load is suddenly applied (dynamic loading), and the maximum deflection is $2x$; i.e., the maximum load on the spring (beam) is $2W$. In this case the mass vibrates in simple harmonic motion with its neutral position equal to its static deflected position. In real structures, the harmonic (vibratory) motion is damped out (reduced to zero) very rapidly. Once the motion has stopped, the force remaining in the spring is the weight W. To account for the increased force during the time the member is in motion, a load equal to twice the static load W should be used—add 100% of the static load to represent the dynamic effect. This is called a 100% impact factor.

Any live load that can have a dynamic effect should be increased by an impact factor. While a dynamic analysis of a structure could be made, such a procedure is unnecessary in ordinary design. Thus, empirical formulas and impact factors are usually used. In cases

Figure 1.4.2
Comparison of static and
dynamic loading.

(a) No vibration:
max. spring force = W

(b) Free vibration:
max. spring force = 2W

where the dynamic effect is small (say where impact would be less than about 20%), it is ordinarily accounted for by using a conservative (higher) value for the specified live load. The dynamic effects of persons in buildings and of slow-moving vehicles in parking garages are examples where ordinary design live load is conservative, and usually no explicit impact factor is added.

For highway bridge design, however, impact is always to be considered. AASHTO [1.3] prescribes empirically that the impact factor expressed as a portion of live load is

$$I = \frac{50}{L + 125} \leq 0.30 \qquad (1.4.2)$$

In Eq. 1.4.2, L (expressed in feet) is the length of the portion of the span that is loaded to give the maximum effect on the member. Since vehicles travel directly on the superstructure, all parts of it are subjected to vibration and must be designed to include impact. The substructure, including all portions not rigidly attached to the superstructure such as abutments, retaining walls, and piers, are assumed to have adequate damping or be sufficiently remote from the application point of the dynamic load so that impact need not be considered. Again, conservative static loads may account for the smaller dynamic effects.

In buildings, impact is explicitly provided for primarily in the design of supports for cranes and heavy machinery. To account for the magnification of imposed loads due to impact, ASCE 7-05 (Secs. 4.7 and 4.10) [1.2] requires an increase in the maximum live load by the following percentages:

For supports of elevators and elevator machinery	100%
For supports of light machinery, shaft- or motor-driven, not less than	20%
For supports of reciprocating machinery or power-driven units, not less than	50%
For hangers supporting floors and balconies	33%
For cab-operated traveling crane support girders and their connections	25%
For pendant-operated traveling crane support girders and their connections	10%

In the design of crane runway girders (see Fig. 1.4.3) and their connections, the horizontal forces caused by moving crane trolleys must be considered. ASCE 7-05 (Sec. 4.10) prescribes using a minimum of "20% of the sum of weights of the lifted load and of the crane trolley, but exclusive of other parts of the crane. The force shall be assumed to be applied at the top of the rails, acting in either direction normal to the runway rails, and shall be distributed with due regard for lateral stiffness of the structure supporting the rails."

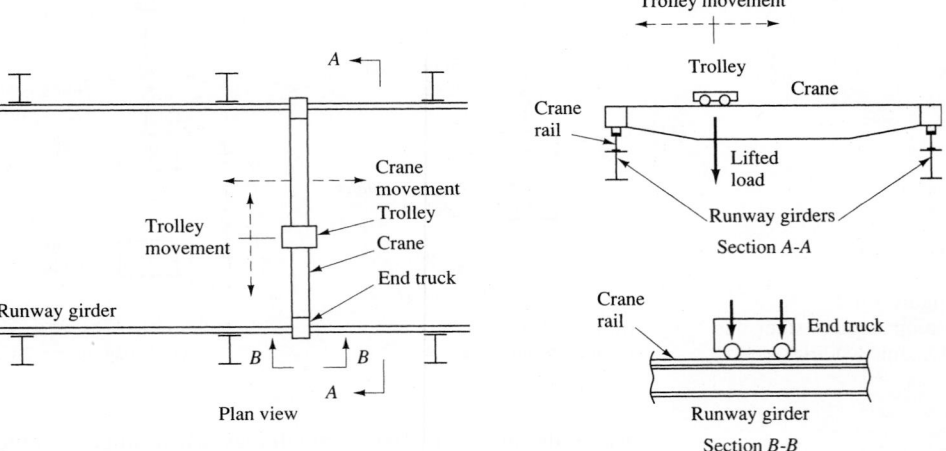

Figure 1.4.3
Crane arrangement, showing
movements that contribute
impact loading.

In addition, due to acceleration and deceleration of the entire crane, a longitudinal tractive force is transmitted to the runway girder through friction of the end truck wheels with the crane rail. ASCE 7 requires this force, unless otherwise specified, to be a minimum of 10% of the maximum wheel loads of the crane applied at the top of the rail.

Snow Load

The live loading for which roofs are designed is either totally or primarily a snow load. Since snow has a variable specific gravity, even if one knows the depth of snow for which design is to be made, the load per unit area of roof is at best only a guess.

The best procedure for establishing snow load for design is to follow ASCE 7. This Standard uses a map of the United States giving isolines of ground snow corresponding to a 50-year mean recurrence interval, for use in designing most permanent structures. The ground snow is then multiplied by a coefficient that includes the effect of roof slope, wind exposure, nonuniform accumulation on pitched or curved roofs, multiple series roofs, and multilevel roofs and roof areas adjacent to projections on a roof level.

It is apparent that the steeper the roof the less snow can accumulate. Also, partial snow loading must be considered in addition to full loading, if such loading can occur and would cause maximum effects. Wind may also act on a structure that is carrying snow load. It is unlikely, however, that maximum snow and wind loads would act simultaneously.

In general, the basic snow load used in design varies from 30 to 40 psf (1400 to 1900 Pa) in the northern and eastern states to 20 psf (960 Pa) or less in the southern states. Flat roofs in normally warm climates should be designed for 20 psf (960 Pa) even when such accumulation of snow may seem doubtful. This loading may be thought of as due to people gathered on such a roof. Furthermore, though wind is frequently ignored as a vertical force on a roof, nevertheless it may cause such an effect. For these reasons, a 20 psf (960 Pa) minimum loading, even though it may not always be snow, is reasonable. Local codes, actual weather conditions, ASCE 7, or the *Canadian Structural Commentaries Part 4* [1.6], should be used when designing for snow.

Other snow load information has been provided by Lew, Simiu, and Ellingwood in the *Building Structural Design Handbook* (Chapter 2) [1.7], and in the works of O'Rourke and Stiefel [1.36], Templin and Schriever [1.37], O'Rourke, Tobiasson, and Wood [1.38], O'Rourke, Redfield, and von Bradsky [1.39], O'Rourke, Speck, and Stiefel [1.40], Sack [1.41], O'Rourke and Galanakis [1.42], and Sack and Giever [1.43].

Wind Load

All structures are subject to wind load, but usually only those more than three or four stories high, as well as long bridges, require special consideration.

On any typical building of rectangular plan and elevation, wind exerts pressure on the windward side and suction on the leeward side, as well as either uplift or downward pressure on the roof. For most ordinary situations, vertical roof loading from wind is neglected on the assumption that snow loading will require a greater strength than wind loading. This assumption is not true for southern climates where the vertical loading due to wind must be included. Furthermore, the total lateral wind load, windward and leeward effect, is commonly assumed to be applied to the windward face of the building.

In accordance with Bernoulli's theorem for an ideal fluid striking an object, the increase in static pressure equals the decrease in dynamic pressure, given by

$$q = \frac{1}{2}\rho V^2 \tag{1.4.3}$$

where q is the dynamic pressure on the object, ρ is the mass density of air (specific weight $w = 0.07651$ pcf at sea level and 15°C), and V is the wind velocity. In terms of velocity V in miles per hour, the dynamic pressure q (psf) would be

$$q = \frac{1}{2}\left(\frac{0.07651}{32.2}\right)\left(\frac{5280V}{3600}\right)^2 = 0.00256\ V^2 \tag{1.4.4*}$$

ASCE 7 defines the velocity pressure using a modified form of the above equation:

$$q_z = 0.00256\ K_z K_{zt} K_d V^2 I \tag{1.4.5}$$

where K_d is the wind directionality factor,

K_z is the velocity pressure exposure coefficient,

K_{zt} is the topographic factor,

I is the importance factor, and

q_h is the calculated q_z at midheight of the roof.

In design, the dynamic pressure q is commonly converted into equivalent static pressure p, which may be expressed [1.2, Sec. 6]

$$p = qGC_p - q_i(GC_{pi}) \tag{1.4.6}$$

where $q = q_z$ for windward walls evaluated at height z above the ground

$q = q_h$ for leeward walls, side walls, and roofs

$q_i = q_h$ for leeward walls, side walls, and roofs of enclosed buildings

$q_i = q_z$ for positive internal pressure evaluation in partially enclosed buildings

$G = $ gust effect factor

$C_p = $ external pressure coefficient

$GC_{pi} = $ internal pressure coefficient

Typically, with wind-resisting systems only the first term is considered, because the second term cancels out as it produces equal and opposite pressures on the windward and leeward walls. Excellent details of application of wind loading to structures are available in the ASCE 7 Standard and in the *National Building Code of Canada* [1.6].

---

*In SI units, $q = 0.62V^2$, for q in MPa and V in m/sec.   (1.4.4)

For all buildings having nonplanar surfaces, plane surfaces inclined to the wind direction, or surfaces having significant openings, special determination of the wind forces should be made using sources like the ASCE 7 Standard or the *National Building Code of Canada* [1.6]. For more extensive treatment of wind loads, the reader is referred to the Report of the Task Committee on Wind Forces [1.33], Lew, Simiu, and Ellingwood in the *Building Structural Design Handbook* [1.7, Chap. 2], Mehta [1.34], and Stathopoulos, Surry, and Davenport [1.35].

Earthquake Load

An earthquake consists of horizontal and vertical ground motions, with the vertical motion usually having the much smaller magnitude. Because the horizontal motion of the ground causes the most significant effect, it is that effect which is often thought of as earthquake load. When the ground under an object (structure) having mass suddenly moves, the inertia of the mass tends to resist the movement, as shown in Fig. 1.4.4. A shear force is developed between the ground and the mass.

Most building codes having earthquake provisions require that the designer either (1) use a dynamic analysis of the structure, or (2) for normal generally rectangular, medium-height buildings, use an empirical lateral base shear force *CW*. The dynamics of earthquake action on structures is outside the scope of this text, and the reader is referred to Chopra [1.46] and Clough and Penzien [1.47].

The equivalent lateral base shear force procedure for designing earthquake-resistance has traditionally been used by most building codes to simplify the design process. For many years, a widely used source has been the Structural Engineers Association of California (SEAOC) recommendations [1.44], the latest version of which is 1999. ASCE 7 contains an equivalent lateral force procedure for "Buildings designated as regular up to 240 feet," wherein the seismic base shear *V* is given as

$$V = C_s W \tag{1.4.7}$$

where C_s = seismic response coefficient, varying from around 0.01 for a low velocity-related acceleration coefficient (say, 0.05) on good soil (say, rock) with a good seismic-resisting structural system (say, a moment-resisting frame) for the maximum 240-foot-high "regular" building, to around 0.35 for a high velocity-related acceleration coefficient (say, 0.20) on poor soil (say, soft clay or silt) with a poor moment-resisting system (say, unreinforced masonry) for a 120-foot-high "regular" building.

   W = total dead load of the building, including partitions, and portions of other loads as defined in ASCE 7.

Figure 1.4.4
Force developed by earthquake.

(a) At rest

(b) Under horizontal motion from earthquake

The seismic response coefficient is given by

$$C_s = \frac{S_{DS}}{R/I} \tag{1.4.8}$$

where S_{DS} = 5% damped design spectral response acceleration for short periods
  R = response modification factor relating to the seismic force-resisting structural system
  I = occupancy importance factor

However, the seismic response coefficient need not be greater than

$$C_s = \frac{S_{D1}}{T(R/I)} \qquad \text{for } T \leq T_L \tag{1.4.9a}$$

or

$$C_s = \frac{S_{D1}T_L}{T^2(R/I)} \qquad \text{for } T > T_L \tag{1.4.9b}$$

where S_{D1} = design spectral response parameter at a period of 1.0 s
  T = fundamental period of the structure
  T_L = long period transition period specified in ASCE7 for different regions

Also, C_s shall not be less than

$$C_s = 0.01 \qquad \text{for } S_1 < 0.6g \tag{1.4.9c}$$

$$C_s = \frac{0.5S_1}{R/I} \qquad \text{for } S_1 \geq 0.6g \tag{1.4.9d}$$

where S_1 = mapped maximum spectral response acceleration parameter

The fundamental period of the structure, T, used in Eqs. (1.4.9) shall not exceed

$$T_{\max} = C_u T_a \tag{1.4.10}$$

where C_u = coefficient relating to S_{D1},
  varying from 1.4 to 1.7 as S_{D1} varies from 0.4 to 1.7
  T_a = approximate fundamental period determined from

$$T_a = C_t h_n^x \tag{1.4.11}$$

where C_t and x are parameters related to the type of structure and h_n is the height of the structure above the base.

Note that the base shear force method is for "regular" buildings. Irregular buildings are those which contain (1) *plan structural irregularities*, such as torsional irregularity, re-entrant corners, diaphragm discontinuity, out-of-plane offsets, and nonparallel systems, and/or (2) *vertical structural irregularities*, such as stiffness irregularity (soft story), mass irregularity, geometric irregularity, in-plane discontinuity in vertical lateral force-resisting elements, and discontinuity in lateral strength (weak story). The base shear force method is also limited to buildings not exceeding 250 ft in height.

After the base shear force V is determined, the vertical distribution of seismic forces must be determined. The seismic design story shear must include direct shear as well as torsion. The building must be designed to resist overturning effects caused by seismic forces. Also, story drifts, and where required, member forces and moments due to **P**-delta effects must be determined.

ASCE 7-05 is based on the *NEHRP Recommended Provisions for the Development of Seismic Regulations for New Buildings* [1.49], which is the definitive source for seismic design. Various traditional building codes for earthquake-resistant design are compared by Chopra and Cruz [1.45]. Many states have adopted the *International Building Code (IBC)* [1.48], which contains provisions for seismic design generally based on the ASCE 7 Standard [1.2]. For steel design, AISC has recently (2006) published the *Seismic Design Manual* [1.53]. Other information on steel-related earthquake codes is provided by Popov [1.51] and Marsh [1.52].

1.5   TYPES OF STRUCTURAL STEEL MEMBERS

As discussed in Sec 1.2, the function of a structure is the principal factor determining the structural configuration. Using the structural configuration along with the design loads, individual components are selected to properly support and transmit loads throughout the structure. Steel members are selected from among the standard rolled shapes adopted by the American Institute of Steel Construction (AISC) (also given by American Society for Testing and Materials [ASTM] A6 Specification). Of course, welding permits combining plates and/or other rolled shapes to obtain any shape the designer may require.

Typical rolled shapes, the dimensions for which are found in the AISC Manual [1.15], are shown in Fig. 1.5.1. The most commonly used section is the wide-flange shape (Fig. 1.5.1a) which is formed by hot rolling in the steel mill. The wide-flange shape is designated by the nominal depth and the weight per foot, such as a W18×97 which is nominally 18 in. deep (actual depth = 18.59 in. according to *AISC Manual*) and weighs 97 pounds per foot. (In SI units the W18×97 section could be designated W460×144, meaning nominally 460 mm deep and having a mass of 144 kg/m.) Two sets of dimensions are found in the *AISC Manual*, one set stated in decimals for the designer to use in computations, and another set expressed in fractions ($\frac{1}{16}$ in. as the smallest increment) for the detailer to use on plans and shop drawings. Rolled W shapes are also designated by ASTM A6/A6M* [1.5] in accordance with web thickness as Groups 1 through 5, with the thinnest web sections in Group 1.

The American Standard beam (Fig. 1.5.1b), commonly called the I-beam, has relatively narrow and sloping flanges and a thick web compared to the wide-flange shape. Use

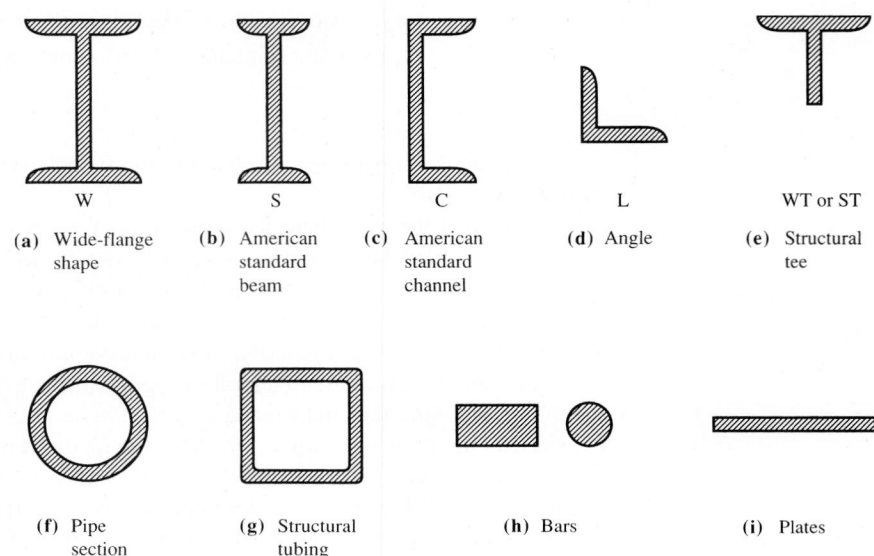

W	S	C	L	WT or ST
(a) Wide-flange shape	**(b)** American standard beam	**(c)** American standard channel	**(d)** Angle	**(e)** Structural tee

(f) Pipe section	**(g)** Structural tubing	**(h)** Bars	**(i)** Plates

Figure 1.5.1
Standard rolled shapes.

*The M refers to the SI version of the ASTM Standard.

of I-beams has become uncommon because of excessive material in the web and relative lack of lateral stiffness resulting from the narrow flanges.

The channel (Fig. 1.5.1c) and angle (Fig. 1.5.1d) are commonly used either alone or in combination with other sections. The channel is designated, for example, as C12×20.7, a nominal 12-in. deep channel having a weight of 20.7 pounds per foot. Angles are designated by their leg length (long leg first) and thickness, such as L6×4×$\frac{3}{8}$.

The structural tee (Fig. 1.5.1e) is made by cutting wide-flange or I-beams in half and is commonly used for chord members in trusses. The tee is designated, for example, as WT5×44, where the 5 is the nominal depth and 44 is the weight in pounds per foot, this tee being cut from a W10×88.

Pipe sections (Fig. 1.5.1f) are designated "standard," "extra strong," and "double-extra strong" in accordance with the thickness and are also nominally prescribed by diameter; thus 10-in.-diam. double-extra strong is an example of a particular pipe size.

Hollow structural steel (HSS) (Fig. 1.5.1g) is used where pleasing architectural appearance is desired with exposed steel. Tubing is designated by outside dimensions and thickness, such as HSS 8×6×$\frac{1}{4}$.

The sections shown in Fig. 1.5.1 are all hot-rolled; that is, they are formed from hot billet steel (blocks of steel) by passing through rolls numerous times to obtain the final shapes. Many other shapes are cold-formed from plate material having a thickness not exceeding 1 in., as shown in Fig. 1.5.2.

Regarding size and designation of cold-formed steel members, there are no truly standard shapes even though the properties of many common shapes are given in the *Cold-Formed Steel Design Manual* [1.9]. Various manufacturers produce many proprietary shapes.

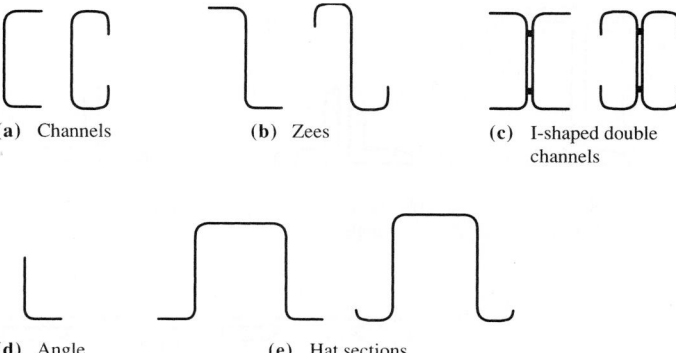

(a) Channels   (b) Zees   (c) I-shaped double channels

(d) Angle   (e) Hat sections

Figure 1.5.2
Some cold-formed shapes.

Tension Members

Tension members commonly occur as chord members in a truss, as diagonal bracing in many types of structures, as direct support for balconies, as cables in suspended roof systems, and as suspension bridge main cables and suspenders that support the roadway. Typical cross-sections of tension members are shown in Fig. 1.5.3, and their design (except for special factors relating to suspension-type cable supported structures) is treated in Chapter 3.

Compression Members

Because compression member strength is a function of the cross-sectional shape (radius of gyration), the area is generally spread out as much as is practical. Chord members in trusses, and many interior columns in buildings, are examples of members subject to axial

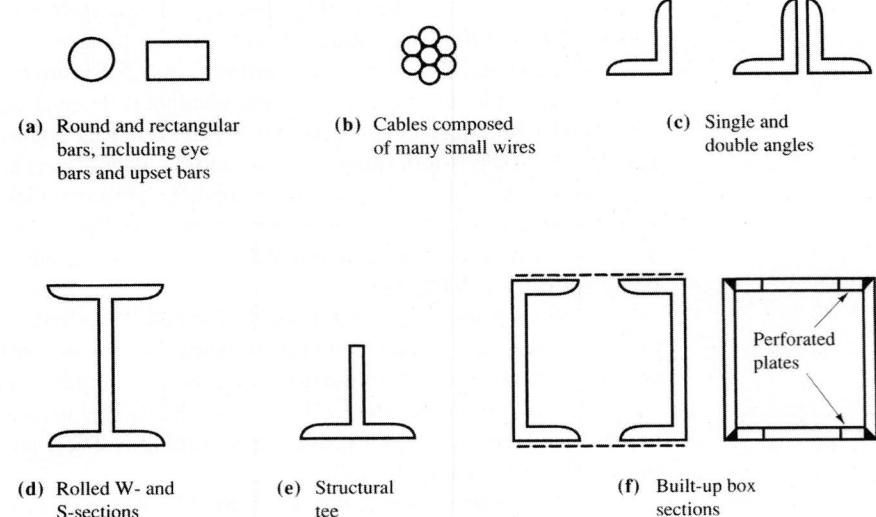

Figure 1.5.3
Typical tension members.

(a) Round and rectangular bars, including eye bars and upset bars

(b) Cables composed of many small wires

(c) Single and double angles

(d) Rolled W- and S-sections

(e) Structural tee

(f) Built-up box sections

compression. Even under the most ideal condition, pure axial compression is not attainable; so, design for "axial" loading assumes the effect of any small simultaneous bending may be neglected. Typical cross-sections of compression members are shown in Fig. 1.5.4, and their behavior and design are treated in Chapter 6.

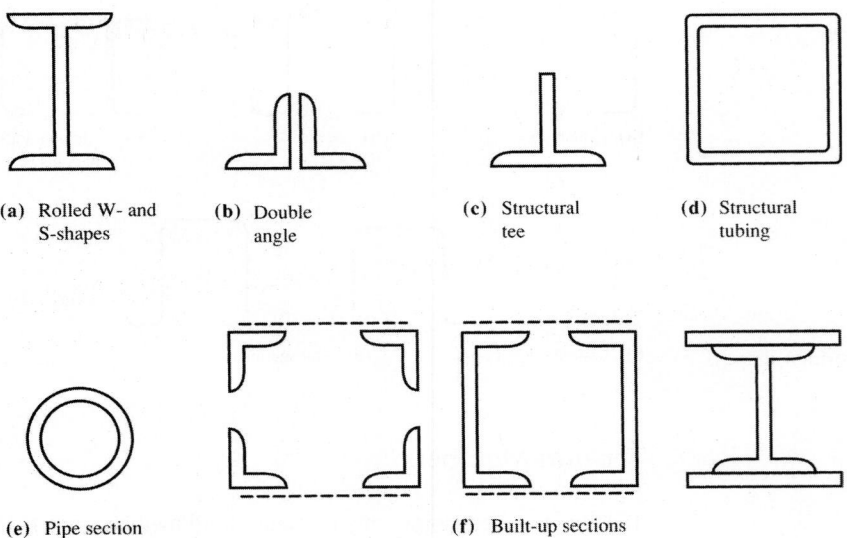

Figure 1.5.4
Typical compression members.

(a) Rolled W- and S-shapes

(b) Double angle

(c) Structural tee

(d) Structural tubing

(e) Pipe section

(f) Built-up sections

Beams

Beams are members subjected to transverse loading and are most efficient when their area is distributed so as to be located at the greatest practical distance from the neutral axis. The most common beam sections are the wide-flange (W) and I-beams (S) (Fig. 1.5.5a), as well as smaller rolled I-shaped sections designated as "miscellaneous shapes" (M).

For deeper and thinner-webbed sections than can economically be rolled, welded I-shaped sections (Fig. 1.5.5b) are used, including stiffened plate girders.

(a) Rolled W- and other I-shaped sections

(b) Welded I-shape (plate girder)

(c) Open-web joists

Figure 1.5.5
Typical beam members.

(d) Angle   (e) Channel   (f) Built-up members   (g) Composite steel-concrete

For moderate spans carrying light loads, open-web "joists" are often used (Fig. 1.5.5c). These are parallel chord truss-type members used for the support of floors and roofs. The steel may be hot-rolled or cold-formed. Such joists are designated "K-Series," "LH-Series," and "DLH-Series." The K-Series is suitable for members having the direct support of floors and roof decks in buildings. The LH-Series and DLH-Series are known as Longspan and Deep Longspan, respectively. Longspan Steel Joists are shop-fabricated trusses used "for the direct support of floor or roof slabs or decks between walls, beams, and main structural members" [1.10]. Deep Longspan Joists are used "for the direct support of roof slabs or decks between wall, beams and main structural members" [1.10]. The design of the chords for K-Series trusses is based on a yield strength* of 50 ksi (345 MPa), while the web sections may use either 36 (248 MPa) or 50 ksi (345 MPa). For the LH- and DLH-Series, the chord and web sections design must be based on a yield strength of at least 36 ksi (248 MPa) but not greater than 50 ksi (345 MPa).

The K-Series joists have depths from 8 to 30 in. for clear spans to 60 ft. The Longspan joists (LH-Series) have depths from 18 to 48 in. for clear spans to 96 ft. The Deep Longspan joists (DLH-Series) have depths from 52 to 72 in. for clear spans to 144 ft.

All of these joists are designed according to Specifications adopted by the Steel Joist Institute (SJI) [1.10], which generally agree with the *AISC Specification* [1.13] for hot-rolled steels, and with the *AISI Specification* [1.8] for coldformed steels. *Designing with Steel Joists, Joist Girders, Steel Deck* by Fisher, West, and Van de Pas [1.11] provides excellent treatment of joists and joist-related floor systems.

For beams (known as lintels) carrying loads across window and door openings, angles are frequently used; and for beams (known as girts) in wall panels, channels are frequently used.

---

*Refer to Sec. 2.4 for definition.

Bending and Axial Load

When simultaneous action of tension or compression along with bending occurs, a combined load problem arises and the type of member used will be dependent on the type of load that predominates. A member subjected to axial compression and bending is usually referred to as a *beam-column*, the behavior and design of which is dealt with in Chapter 12.

The aforementioned illustration of types of members to resist various kinds of load is intended only to show common and representative types of members and not to be all-inclusive.

1.6 STEEL STRUCTURES

Structures may be divided into three general categories: (a) framed structures, where elements may consist of tension members, columns, beams, and members under combined bending and axial load; (b) shell-type structures, where axial forces predominate; and (c) suspension-type structures, where axial tension predominates the principal support system.

Framed Structures

Most typical building construction is in this category. The multistory building usually consists of beams and columns, either rigidly connected or having simple end connections along with diagonal bracing to provide stability. Even though a multistory building is three-dimensional, it usually is designed to be much stiffer in one direction than the other; thus it may reasonably be treated as a series of plane frames. However, if the framing is

Figure 1.6.1
Floor joists (plane trusses) and steel decking. (Photo by C. G. Salmon)

Figure 1.6.2
Space truss roof erected in sections; also shows plate girder at lower level containing vertical stiffener plates and special stiffening around rectangular holes through girder web. Upjohn Office Building, Kalamazoo, Michigan. (Photo courtesy Whitehead and Kales Company)

such that the behavior of the members in one plane substantially influences the behavior in another plane, the frame must be treated as a three-dimensional space frame.

Industrial buildings and special one-story buildings such as churches, schools, and arenas, generally are either wholly or partly framed structures. Particularly the roof system may be a series of plane trusses (see Fig. 1.6.1), a space truss (see Fig. 1.6.2), a dome (see Fig. 1.6.3), or it may be part of a flat or gabled one-story rigid frame.

Figure 1.6.3
Dome roof, Brown University auditorium. (Photo courtesy Bethlehem Steel Corporation)

Figure 1.6.4
Continuous orthotropic plate
girder across Mississippi
River at St. Louis, Missouri.
(Photo by C. G. Salmon)

Bridges are mostly framed structures, such as beams and plate girders (see Fig. 1.6.4), or trusses, usually continuous (see Fig. 1.6.5).

This text is devoted to behavior and design of elements in framed structures.

Shell-Type Structures

In this type of structure, the shell serves a use function in addition to participation in carrying loads. One common type where the main stress is tension is the containment vessel used to store liquids (for both high and low temperatures), of which the elevated water tank is a notable example. Storage bins, tanks, and the hulls of ships are other examples. On many shell-type structures, a framed structure may be used in conjunction with the shell.

On walls and flat roofs, the "skin" elements may be in compression while they act together with a framework. The aircraft body is another such example.

Shell-type structures are usually designed by a specialist and are not within the scope of this text.

Figure 1.6.5
Continuous truss bridge.
Outerbridge Crossing, Staten
Island, New York. (Photo by
C. G. Salmon)

Figure 1.6.6
Cable-suspended roof for
Madison Square Garden
Sports and Entertainment
Center, New York. (Photo
courtesy Bethlehem Steel
Corporation)

Suspension-Type Structures

In the suspension-type structure, tension cables are major supporting elements. A roof may be cable-supported, as shown in Fig. 1.6.6. Probably the most common structure of this type is the suspension bridge, as shown in Fig. 1.6.7. Usually a subsystem of the structure consists of a framed structure, as in the stiffening truss for the suspension bridge. Since the tension element is the most efficient way of carrying load, structures utilizing this concept are increasingly being used.

Figure 1.6.7
Suspension bridge. Golden
Gate Bridge, San Francisco,
California. (Photo by C. G.
Salmon)

Many unusual structures utilizing various combinations of framed, shell-type, and suspension-type structures have been built. However, the typical designer must principally understand the design and behavior of framed structures.

1.7 SPECIFICATIONS AND BUILDING CODES

Structural steel design of buildings in the United States is principally based on the specification of the American Institute of Steel Construction (AISC). AISC is composed of steel fabricator and manufacturing companies, as well as individuals interested in steel design and research. The *AISC Specification* is the result of the combined judgment of researchers and practicing engineers. The research efforts have been synthesized into practical design procedures to provide a safe, economical structure. The advent of the digital computer in design practice has made feasible more elaborate design rules. The current specification which is referred to throughout this book is the 2005 *Design Specification for Structural Steel Buildings* [1.13].

A specification containing a set of rules is intended to insure safety; however, the designer must understand the behavior for which the rule applies. Otherwise, an absurd design may result, either unsafe or grossly conservative. The authors contend that it is virtually impossible to write rules that fully apply to every situation. Behavioral understanding must come first; application of rules then follows. No matter what set of rules is applicable, *the designer has the ultimate responsibility for a safe structure.*

A specification when adopted by AISC is actually a set of recommendations put forth by a highly respected group of experts in the field of steel research and design. Only when governmental bodies, such as city, state, and federal agencies, who have legal responsibility for public safety, adopt or incorporate a specification such as the AISC *Steel Design Specification* (ANSI/AISC 360–05) [1.13] into their building codes does it become legally official.

The design of steel bridges is generally in accordance with specifications of the American Association of State Highway and Transportation Officials (AASHTO) [1.3]. This is a legal set of rules because it has been adopted by the states (usually the state highway departments have this responsibility).

Railroad bridges are designed in accordance with the specifications adopted by the American Railway Engineering and Maintenance of Way Association (AREMA) [1.4]. In this case the railroads have the responsibility for safety and through their own organization adopt the rules to insure safe designs.

The term *building code* is sometimes used synonymously with specifications. More correctly a building code is a broadly based document, either a legal document such as a state or local building code, or a document widely recognized even though not legal which covers the same wide range of topics as the state or local building code. Building codes generally treat all facets relating to safety, such as structural design, architectural details, fire protection, heating and air conditioning, plumbing and sanitation, and lighting. On the other hand, specifications frequently refer to rules set forth by the architect or engineer that pertain to only one particular building while under construction. Building codes also ordinarily prescribe standard loads for which the structure is to be designed, as discussed in Sec. 1.4.

The reader should not be disturbed by the interchangeable use of building code and specification, but should clearly understand that which is legally required for design and that which could be thought of as recommended practice.

1.8 PHILOSOPHIES OF DESIGN

The 2005 AISC *Specification for Structural Steel Buildings* provides an integrated treatment of Allowable Strength Design (previously referred to by AISC as the Allowable

Stress Method) and Load and Resistance Factor Design. The new specification combines the two design methods and replaces earlier specifications that treated the two design methods separately. Allowable Stress Design has been the principal philosophy used during the past 100 years. During the past 20 years or so, structural design has been moving toward a more rational and probability-based design procedure referred to as "limit states" design. Haaijer [1.24] and Kennedy [1.25, 1.26] have presented the current status of the limit states concept and its use in design. Limit states design includes the methods commonly referred to as "ultimate strength design," "strength design," "plastic design," "load factor design," "limit design," and the more recent "load and resistance factor design (LRFD)."

Structures and structural members must have adequate strength, as well as adequate stiffness and toughness, to permit proper functioning during the service life of the structure. The design must provide some reserve strength above that which is needed to carry the service loads; that is, the structure must provide for the possibility of overload. Overloads can arise from changing the use for which a particular structure was designed, from underestimation of the effects of loads by oversimplifications in structural analysis, and from variations in construction procedures. In addition, there must be a provision for the possibility of understrength. Deviations in the dimensions of members, even though within accepted tolerances, can result in a member having less than its computed strength. The materials (steel for members, bolts, and welds) may have less strength than used in the design calculations. A steel section may occasionally have a yield stress below the minimum specified value, but still within the statistically acceptable limits.

Structural design must provide for adequate safety no matter what philosophy of design is used. Provision must be made for both *overload* and *understrength*. The study of what constitutes the proper formulation of structural safety has been continuing during the past 30 years [1.17, 1.18]. The main thrust has been to examine by various probabilistic methods the chances of "failure" occurring in a member, connection, or system.

Rather than "failure," the term "limit state" is preferred. *Limit states* means "those conditions of a structure at which it ceases to fulfill its intended function" [1.25]. Limit states are generally divided into two categories: *strength* and *serviceability*. Strength (i.e., safety) limit states are such behavioral phenomena as achieving ductile maximum strength (i.e., plastic strength), buckling, fatigue, fracture, overturning, and sliding. Serviceability limit states are those concerned with occupancy of a building, such as deflection, vibration, permanent deformation, and cracking.

Both the loads acting on the structure and its resistance (strength) to loads are variables that must be considered. In general, a thorough analysis of all uncertainties that might influence achieving a "limit state" is not practical, or perhaps even possible. The current approach to a simplified method for obtaining a probability-based assessment of structural safety uses *first-order second-moment reliability methods* [1.29]. Such methods assume that the load (or load effect) Q and the resistance R are random variables. Typical frequency distributions of these random variables are shown in Fig. 1.8.1. When the resistance R exceeds the load (or load effect) Q there will be a margin of safety. Unless R exceeds Q by a large amount, there will be some probability that R may be less than Q.

Structural "failure" (achievement of a limit state) may then be examined by comparing R with Q, or in logarithmic form observing $\ln(R/Q)$, as shown in Fig. 1.8.2. "Failure" is represented by the cross-hatched region. The distance between the failure line and the mean value of the function $[\ln(R/Q)]$ is defined as a multiple β of the standard deviation σ of the function. The multiplier β is called the *reliability index*. The larger is β the greater is the margin of safety.

As summarized by Pinkham [1.18], the reliability index β is useful in several ways:

1. It can give an indication of the consistency of safety for various components and systems using traditional design methods.

2. It can be used to establish new methods which will have consistent margins of safety.

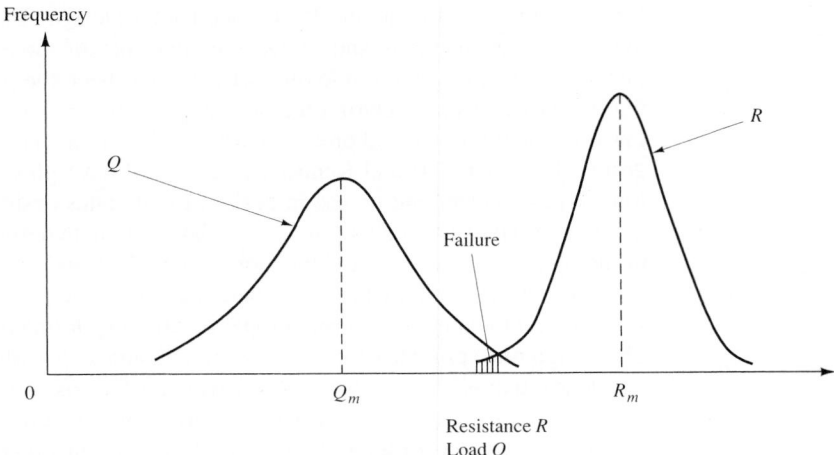

Figure 1.8.1
Frequency distributions of
load Q and resistance R.

3. It can be used to vary in a rational manner the margins of safety for those components and systems having a greater or lesser need for safety than that required in ordinary situations.

In general, the expression for the structural safety requirement may be written as

$$\phi R_n \geq [R_u = \Sigma \gamma_i Q_i] \qquad (1.8.1)$$

where the left side of Eq. 1.8.1 represents the resistance, or strength, of the component or system, and the right side represents the load expected to be carried. On the strength side, the nominal resistance R_n is multiplied by a resistance (strength reduction) factor ϕ to obtain the design strength (also called *usable strength or usable resistance*). On the load side of the equation, the various load effects Q_i (such as dead load, live load, and snow load) are multiplied by overload factors γ_i to obtain the sum $\Sigma \gamma_i Q_i$ of *factored loads*.

AISC—Load and Resistance Factor Design (LRFD)

During the past 25 years, the general "limit states design" approach has continued to gain acceptance, particularly for steel design in the United States with the adoption in 1986 of a *Load and Resistance Factor Design Specification* by AISC. The latest version of that

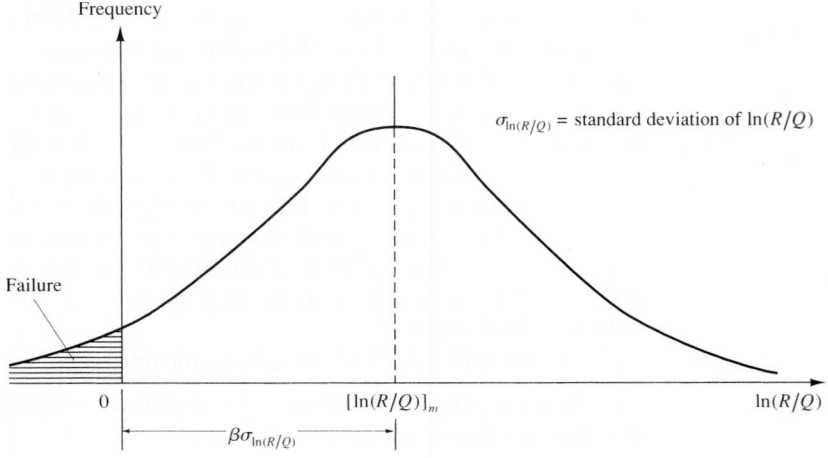

Figure 1.8.2
Reliability index β.

Specification is 2005 [1.13], referred to throughout this book as the AISC Specification LRFD Method. In Canada, limit states design for steel has been used since 1974, and since 1978 has been the only method used; the latest edition is 1989 [1.16]. The AISC Specification LRFD Method was developed under the leadership of T. V. Galambos [1.20–1.23]. The adaptation of probabilistic methods to steel design and the development of that 1986 Specification are explained by Galambos [1.20, 1.22] and by Galambos and Ravindra [1.21, 1.23].

The safety requirement of the LRFD Method is given by Eq. 1.8.1. This means the design strength ϕR_n provided by the resulting design must at least equal the sum $\Sigma \gamma_i Q_i$ of the applied factored service loads. The subscript i indicates that there are terms for each type of load Q_i acting, such as dead load D, live load L, wind load W, snow load S, and earthquake load E. The γ_i may be different for each type of load.

The AISC LRFD Method is based on the following:

1. A probability-based model [1.21, 1.29–1.31].

2. Calibration with the 1978 AISC Allowable Stress Design (ASD) Specification.

3. Evaluation using judgment and past experience, along with studies of representative structures conducted by design offices.

The development of probability-based criteria by Galambos, Ellingwood, MacGregor, and Cornell [1.29–1.31] led to the factored load combinations of the 1982 ANSI Standard, which has become ASCE 7 [1.2]. The ANSI Standard was developed for use in design with *all* structural materials. It is reasonable that the probability of overload with certain types of loads should be unrelated to the material of which a structure is built. With this concept in mind, the AISC LRFD Method adopted the ASCE 7-05 [1.2] factored load combinations as follows:

$$1.4D \tag{1.8.2}$$

$$1.2D + 1.6L + 0.5(L_r \text{ or } S \text{ or } R) \tag{1.8.3}$$

$$1.2D + 1.6(L_r \text{ or } S \text{ or } R) + (0.5L \text{ or } 0.8W) \tag{1.8.4}$$

$$1.2D + 1.6W + 0.5L + 0.5(L_r \text{ or } S \text{ or } R) \tag{1.8.5}$$

$$1.2D \pm 1.0E + 0.5L + 0.2S \tag{1.8.6}$$

$$0.9D \pm (1.6W \text{ or } 1.0E) \tag{1.8.7}$$

where the nominal *service loads* indicated by Eqs. 1.8.2 through 1.8.7 are

D = dead load (gravity load from the weight of structural elements and permanent attachments)

L = live load (gravity occupancy and movable equipment load)

L_r = roof live load

W = wind load

S = snow load

E = earthquake load

R = rainwater or ice load

Note that *D, L, W, S*, etc. are loads in a general sense, which includes bending moment, shear, axial force, and torsional moment. Sometimes these internal forces are called *load effects*. Thus, the symbol D means dead load, dead load moment, dead load shear, dead load axial force, etc. An explanation of the statistics relating to snow and wind load factors is given by Ravindra, Cornell, and Galambos [1.32]. The factors for earthquake E are reduced from 1.5 in the 1986 LRFD Specification to 1.0 in the 1993 Specification. This reflects the new earthquake loads of National Earthquake Hazards Reduction Program (NEHRP) [1.49] and the *AISC Seismic Provisions* [1.50], which specify larger earthquake forces than traditionally used.

AISC—Allowable Strength Design (ASD)

The AISC ASD Method has been included in the 2005 *AISC Specification* as an equal alternative to the LRFD Method. The term "allowable strength" is used to signify that the same strength limit states are the basis for both the ASD Method and the LRFD Method. In the past [i.e., the most recent (1978) AISC ASD Specification] ASD was based on the concept that the maximum stress due to service load shall not exceed a specified allowable stress, and the method was called "allowable stress design." The safety factors in traditional ASD were primarily based on experience and workmanship, and had remained the same for the past 75 years. While the actual level of safety provided by traditional ASD has always been variable and unknown, structures design by ASD performed satisfactorily. The ASD Method in the current AISC Specification (2005) uses the following AISC Formula (B3-2)

$$\left[\frac{\phi R_n}{\gamma} = \frac{R_n}{\Omega}\right] \geq \left[R_a = \Sigma \gamma_i Q_i\right] \tag{1.8.8}$$

The authors have reversed the order of the equation, believing that normally the designer would say "the provided strength must equal or exceed the expected loads". Note that γ is used here similarly to the LRFD Method; however, here the γ is attached to some loads (see Eqs. 1.8.9 through 1.8.14) to reflect the probability of certain service loads occurring simultaneously. The γ is 1.0 for most loads because the loads are unfactored here but are factored for overload in the LRFD Method.

The left-hand side of Eq. 1.8.8 represents the allowable strength, and the right-hand side represents the load to be carried. The safety factor Ω is explicitly given in the specification. The 2005 AISC ASD Method adopted the following service load combinations from ASCE 7-05:

$$D \tag{1.8.9}$$

$$D + L \tag{1.8.10}$$

$$D + L + (L_r \text{ or } S \text{ or } R) \tag{1.8.11}$$

$$D + (W \text{ or } 0.7E) + L + (L_r \text{ or } S \text{ or } R) \tag{1.8.12}$$

$$0.6D + W \tag{1.8.13}$$

$$0.6D + 0.7E \tag{1.8.14}$$

1.9 FACTORS OF SAFETY—ASD AND LRFD COMPARED

Allowable Strength Design (ASD)

The "safety factor" Ω used in Eq. 1.8.8 was not determined consciously by using probabilistic methods. The values used in the AISC ASD Specification have been in use for many years and are the result of experience and judgment. It is clear that the safety required must be a combination of economics and statistics. Obviously, it is not economically feasible to design a structure so that the probability of failure is zero. Prior to the development of the 1986 AISC LRFD Specification, the AISC Specifications from 1924 through 1978 did not give a rationale for the allowable stresses prescribed.

   One may state that the minimum resistance must exceed the maximum applied load by some prescribed amount. Suppose the actual load exceeds the service load by an amount ΔQ, and the actual resistance is less than the computed resistance by an amount ΔR. A structure that is just adequate would have

$$R_n - \Delta R_n = Q + \Delta Q$$
$$R_n(1 - \Delta R_n/R_n) = Q(1 + \Delta Q/Q) \tag{1.9.1}$$

The margin of safety, or "safety factor," would be the ratio of the nominal strength R_n to nominal service load Q; or

$$\Omega = \frac{R_n}{Q} = \frac{1 + \Delta Q/Q}{1 - \Delta R_n/R_n} \tag{1.9.2}$$

Equation 1.9.2 illustrates the effect of overload $(\Delta Q/Q)$ and understrength $(\Delta R_n/R_n)$; however it does not identify the factors contributing to either. If one assumes that occasional overload $(\Delta Q/Q)$ may be 40% greater than its nominal value, and that an occasional understrength $(\Delta R_n/R_n)$ may be 15% less than its nominal value, then

$$\Omega = \frac{1 + 0.4}{1 - 0.15} = \frac{1.4}{0.85} = 1.65$$

The above is an oversimplification but it shows a possible scenario for obtaining the traditional AISC value of FS = 1.67 used as the basic value in Allowable Stress Design.

Load and Resistance Factor Design (LRFD)

As discussed in Sec. 1.8, the factors for overload are variable depending upon the type of load, and the factored load combinations that must be considered are those given by the ASCE 7 Standard [1.2]. and presented as Eqs. 1.8.2 through 1.8.7. The other part of the safety-related provisions is the ϕ factor, known as the *resistance factor*. The resistance factor ϕ varies with the type of member and with the limit state being considered. Some representative resistance factors ϕ are as follows.

   Tension Members (AISC-D2)

$$\phi_t = 0.90 \quad \text{for } \textit{tensile yielding} \text{ limit state}$$
$$\phi_t = 0.75 \quad \text{for } \textit{tensile rupture} \text{ limit state}$$

Compression Members (AISC-G1)

$$\phi_c = 0.90$$

Beams (AISC-F1 and E1)

$$\phi_b = 0.90 \quad \text{for flexure}$$
$$\phi_v = 0.90 \quad \text{for shear}$$

Welds (AISC-Table J2.5)

$$\phi = \text{same as for type of action; i.e., tension, shear etc.}$$

Fasteners in Tension and Shear (AISC-J3.6 and J3.7)

$$\phi = 0.75$$

In order to establish adequate safety using probabilistic methods, the natural logarithm of the resistance R divided by the load Q, that is, $\ln(R/Q)$ as shown in Fig. 1.8.2, may be treated as a random variable. It is simpler than working with two groups (R and Q) of random variables as in Fig. 1.8.1. When $\ln(R/Q) < 0$, the limit state has been reached, and the shaded area in Fig. 1.8.2 is the probability of this event. The method used to develop LRFD uses the *mean values* R_m and Q_m and the *standard deviations* σ_R and σ_Q of the resistance and load, respectively. Frequently, the mean values and standard deviations can be estimated even though the actual distributions cannot be obtained. From these estimated quantities, the standard deviation σ of the $\ln(R/Q)$ may be approximated as

$$\sigma_{\ln(R/Q)} \approx \sqrt{V_R^2 + V_Q^2} \qquad (1.9.3)$$

where
$$V_R = \sigma_R/R_m$$
$$V_Q = \sigma_Q/Q_m$$

The margin of safety is the distance from the origin to the mean, as shown in Fig. 1.8.2, and is expressed as a multiple β of $\sigma_{\ln(R/Q)}$. The distance representing the margin of safety may be approximated [1.14] as

$$\beta\sigma_{\ln(R/Q)} \approx \beta\sqrt{V_R^2 + V_Q^2} = \ln(R_m/Q_m) \qquad (1.9.4)$$

Thus, the larger the distance the smaller the probability of reaching the limit state. The multiplier β is called the *reliability index*. The expression for β from Eq. 1.9.4 becomes

$$\beta = \frac{\ln(R_m/Q_m)}{\sqrt{V_R^2 + V_Q^2}} \qquad (1.9.5)$$

More discussion of the development of Eq. 1.9.5 is given in the AISC Commentary [1.14], by Ravindra and Galambos [1.23], and in NBS Special Publication 577 [1.29]. The treatment of the theory of probability is outside the scope of this book.

Using the factored load combinations given by the ASCE 7 Standard [1.2], the AISC Task Force and Specification Committee calibrated the 1986 LRFD Specification to generally agree with past experience. Thus, the resistance factors ϕ were set in LRFD with the objective of obtaining the following values of β:

Load combinations	Objective reliability index β
Dead load + live load (or snow load)	3.0 for members
	4.5 for connections
Dead load + live load + wind load	2.5 for members
Dead load + live load + earthquake load	1.75 for members

Because of a lower probability of wind or earthquake occurring with full gravity load, the reliability index β was made lower for those cases. The β values for connections were made higher than for members, in keeping with the tradition of making connections stronger than members.

ASCE 7 uses six factored load combinations, given as Eqs. 1.8.2 through 1.8.7. This was necessary to account for each of the separate loads (dead, live, roof, wind additive to gravity, wind opposite to gravity, and earthquake) acting at its maximum lifetime value. Loads other than dead load and the load being maximized will act at an "arbitrary point-in-time" value. The "arbitrary point-in-time" value is that value which can be expected to be on the structure at any time. The arbitrary point-in-time value of live load (L) might be as low as one-quarter of its mean maximum lifetime load, but its distribution varies widely. The arbitrary point-in-time wind (W) is the maximum daily wind. The "lifetime maximum" is taken as the 50-year recurrence value.

Thus, each factored load combination and its corresponding load occurring at its 50-year maximum are as follows:

Load combination	Load at its lifetime (50 year) maximum
1.4D	Dead load D during construction; other loads not present.
1.2D + 1.6L + 0.5S	Live load L
1.2D + 1.6S + (0.8W or 0.5L)	Roof load; i.e., snow load S or rain R other than ponding effect.
1.2D + 1.6W + 0.5L + 0.5S	Wind load W *additive* to dead load.
1.2D ± 1.0E + 0.5L + 0.2S*	Earthquake load E *additive* to dead load.
0.9D ± (1.6W or 1.0E)*	Wind load W or earthquake load E *opposite* to dead load.

*The sign following 1.2D or 0.9D is to be taken + or − so as to provide for the more severe effect.

Note: Where snow S is used in the above equations, except in Eq. 1.8.6, the meaning is snow S OR roof live load L_r OR rain R other than ponding.

Comparison of LRFD with ASD for Tension Members

The original LRFD Specification values were calibrated to the 1978 ASD Specification at a live load to dead load ratio of 3. To determine the relationship between ϕ and Ω, the nominal strengths from ASD and LRFD are equated. Using the live load and dead load combinations, with $L = 3D$, the required nominal strength can be expressed as follows:

From ASD:
$$\frac{R_n}{\Omega} = D + L = D + 3D = 4D$$

$$R_n = 4D\Omega$$

From LRFD:
$$\phi R_n = 1.2D + 1.6L = 1.2D + 1.6 \times 3D = 6D$$

$$R_n = \frac{6D}{\phi}$$

Equating and solving for Ω:
$$\Omega = \frac{6D}{\phi} \times \frac{1}{4D} = \frac{1.5}{\phi}$$

Therefore, for $\phi = 0.9$, the value of $\Omega = 1.67$ and for $\phi = 0.75$, $\Omega = 2.00$.
A similar method was used to determine all the Ω values throughout the Specification.

The comparison of safety obtained for tension members designed by the two AISC methods is indicative of the general result expected. Direct comparisons have become easy for all types of members since the nominal strength in both methods is the same.

For tension members acted upon by gravity dead and live loads, the resistance factor $\phi = 0.90$, and using Eq. 1.8.3 gives for LRFD

$$1.2D + 1.6L = 0.90R_n \qquad\qquad\qquad [1.8.3]$$

$$1.33D + 1.78L = R_n \qquad\qquad \text{LRFD}$$

In ASD the factor of safety $\Omega = 1.67$ for axial tension, which gives from Eq. 1.8.8 where $(\gamma/\phi$ is the factor of safety)

$$R_n/1.67 = \Sigma Q = D + L \qquad [1.8.8]$$

or

$$1.67D + 1.67L = R_n \qquad \text{ASD}$$

Next, dividing Eq. 1.8.3 by Eq. 1.8.8 gives

$$\frac{\text{LRFD}}{\text{ASD}} = \frac{1.33D + 1.78L}{1.67D + 1.67L} = \frac{0.8 + 1.07(L/D)}{1 + (L/D)} \qquad (1.9.6)$$

Since this is a gravity load comparison, Eq. 1.8.2 must also be used as L/D approaches zero. Thus, Eq. 1.8.2 gives

$$1.4D = 0.90R_n \qquad [1.8.2]$$

$$1.56D = R_n \qquad \text{LRFD}$$

Dividing LRFD by ASD gives

$$\frac{\text{LRFD}}{\text{ASD}} = \frac{1.56D}{1.67D + 1.67L} = \frac{0.93}{1 + (L/D)} \qquad (1.9.7)$$

Equations 1.9.6 and 1.9.7 are shown plotted in Fig. 1.9.1. The design of tension members will be about the same in both LRFD and ASD when the live load to dead load ratio (L/D) is about 3. As the L/D ratio becomes lower (that is, dead load becomes more predominant) there will be economy in using LRFD. With L/D ratio larger than 3, ASD will be slightly more economical, but rarely by more than about 3%.

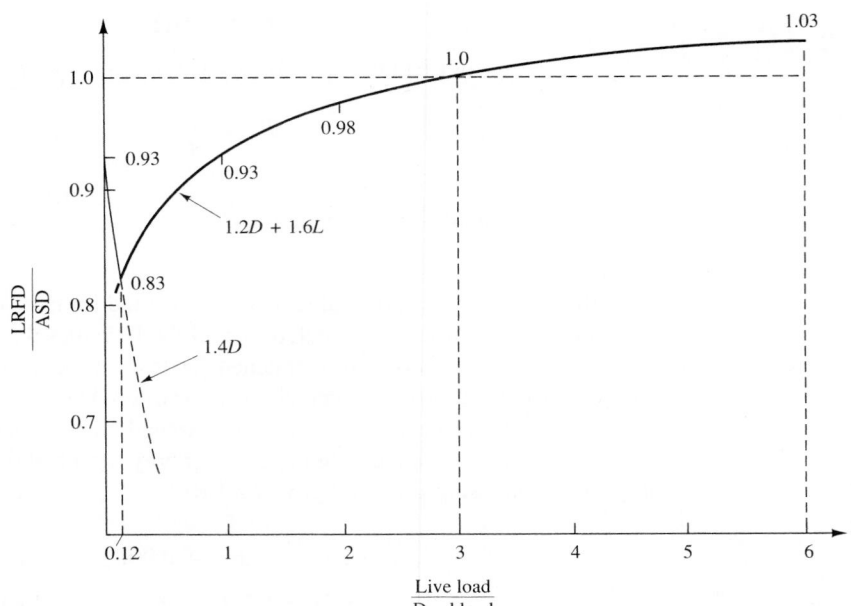

Figure 1.9.1
Comparison of load and resistance factor design with allowable stress design for tension members.

1.10   WHY SHOULD LRFD BE USED?

The many advantages of LRFD are well-expressed by Beedle [1.27], whose listing is the basis of the following:

1. LRFD is another "tool" for structural engineers to use in steel design. Why not have the same tools (variable overload factors and resistance factors) available for steel design as are available for concrete design?

2. Adoption of LRFD is not mandatory but provides a flexibility of options to the designer. The marketplace will dictate whether or not LRFD will become the sole method.

3. ASD is an approximate way to account for what LRFD does in a more rational way. The use of plastic design concepts in ASD has made ASD such that it no longer may be called an "elastic design" method.

4. The rationality of LRFD has always been attractive, and becomes an incentive permitting the better and more economical use of material for some load combinations and structural configurations. It will also likely produce safer structures in view of the arbitrary practice under ASD of combining dead and live loads and treating them the same.

5. Using multiple load factor combinations should lead to economy.

6. LRFD will facilitate the input of new information on loads and load variations as such information becomes available. Considerable knowledge of the resistance of steel structures is available. On the other hand, our knowledge of loads and their variation is much less. Separating the loading from the resistance allows one to be changed without the other if that should be desired.

7. Changes in overload factors and resistance factors ϕ are much easier to make than to change the allowable stress in ASD.

8. LRFD makes design in all materials more compatible. The variability of loads is actually unrelated to the material used in the design. Future specifications not in the limit states format for any material will put that material at a disadvantage in design.

9. LRFD provides the framework to handle unusual loads that may not be covered by the Specification. The design may have uncertainty relating to the resistance of the structure, in which case the resistance factors may be modified. On the other hand, the uncertainty may relate to the loads and different overload factors may be used.

10. Future adjustments in the calibration of the method can be made without much complication. Calibration for LRFD was done for an *average* situation but might be adjusted in the future.

11. Economy is likely to result for low live load to dead load ratios. For high live load to dead load ratios there will be slightly greater costs.

12. Safer structures may result under LRFD because the method should lead to a better awareness of structural behavior.

13. Design practice is still at the beginning with regard to serviceability limit states; however, at least LRFD provides the approach.

1.11   ANALYSIS OF THE STRUCTURE

In general, the structural analysis to obtain the service loads (or load *effects* bending moment, shear, axial force, and torsional moment) on the members is performed the same for LRFD as for ASD. Elastic methods of structural analysis are used except when the limit state is the *plastic collapse mechanism* as described in Chapter 10. A *first-order analysis* [1.12] is sufficient in usual framed structures that are braced against sway. In a first-order analysis, equilibrium equations are based on the original geometry of the structure. This means that the designer is assuming the internal forces (moments, shears, etc.) are not sufficiently affected by the change in shape of the structure to justify more complicated analysis. When the elastic displacements are small compared to the dimensions, this approximation will be satisfactory.

The most common situation where a *second-order* effect must be considered is in a multistory structure that must rely on the stiffnesses of the interacting beams and columns to resist sway under lateral loading (wind and/or earthquake). This is the so-called *unbraced frame*. In this case the lateral displacement Δ (also called *sway* or *drift*) causes additional bending moments due to the gravity loads (ΣP) acting at positions that have been displaced by an amount Δ. The analysis must include this $P\Delta$ secondary effect. There are varying degrees of sophistication that are used in analysis to include the $P\Delta$ effects. In both ASD and LRFD the second-order effects may be computed as a part of the analysis or they may be accounted for using formulas in the Specification [1.13] or Commentary [1.14].

The emphasis in this book is on designing the members to have adequate strength and proper serviceability, rather than on structural analysis. The reader is referred to Wang and Salmon [1.12] for basic structural analysis topics. Other references specifically related to analysis for unbraced frames are given in Chapter 14.

Most examples in this book use given service loads, or service load effects, acting on the structural member to be designed. These values are to be assumed the result of structural analysis. The service loads are to be factored using Eqs. 1.8.2 through 1.8.7 for design in LRFD, but used without factors in Eqs. 1.8.9 through 1.8.14 for design in ASD.

SELECTED REFERENCES

1.1. Hans Straub. *A History of Civil Engineering*. Cambridge, MA: M.I.T. Press, 1964 (pp. 173–180).

1.2. ASCE. *Minimum Design Loads for Buildings and Other Structures*, ASCE Standard 7-05. Reston, VA: American Society of Civil Engineers (1801 Alexander Bell Drive, Reston, VA 20191), 2005.

1.3. AASHTO. *Standard Specifications for Highway Bridges*, 17th ed. Washington DC: The American Association of State and Highway Transportation Officials, 01-Sep-2002.

1.4. AREMA, *Manual for Railway Engineering*, Chapter 15 (Steel Structures). Lanham, MD: The American Railway Engineering and Maintenance of Way Association (10003 Derekwood Lane, Suite 210, Lanham, MD 20706), 2006.

1.5. ASTM. *Standard Specification for General Requirements for Rolled Steel Plates, Shapes, Sheet Piling, and Bars for Structural Use*, A6/A6M-06a. West Conshohocken PA: American Society for Testing and Materials, 2006.

1.6. National Research Council of Canada. *Canadian Structural Commentaries Part 4* to the National Building Code of Canada, Ottawa, Canada, 1995.

1.7. H. S. Lew, Emil Simiu, and Bruce Ellingwood. "Loads," Chapter 2, *Building Structural Design Handbook*, Richard N. White and Charles G. Salmon, Ed. New York: John Wiley & Sons, 1987 (pp. 9–43).

1.8. AISI. *North American Specification for the Design of Cold-Formed Steel Structural Members*. Washington D.C.: American Iron and Steel Institute. 2001.

1.9. AISI. *Cold-Formed Steel Design Manual*: Part I, Dimensions and Properties; Part II, Beam Design; Part III, Column Design; Part IV, Connections; Part V, Supplementary Information; and Part VI, Test Procedures. Washington D.C.: American Iron and Steel Institute, 2002.

1.10. SJI. *Standard of Specifications, Load Tables & Weight Tables for Steel Joists & Joist Girders*, 42nd ed., Steel Joist Institute, 2005, (3127 Mr. Joe White Avenue, Myrtle Beach, SC 29577-6760)

1.11. James M. Fisher, Michael A. West, and Julius P. Van de Pas. *Designing With Steel Joists, Joist Girders, Steel Deck*. Charlotte, NC: Nucor Corporation, 1991.

1.12. Chu-Kia Wang and Charles G. Salmon. *Introductory Structural Analysis*. Englewood Cliffs, New Jersey: Prentice-Hall, Inc., 1984.

1.13. AISC. *Specification for Structural Steel Buildings* (AISC 360–05). Chicago: American Institute of Steel Construction, March 9, 2005.

1.14. AISC. *Commentary on the Design Specification for Structural Steel Buildings* (March 9, 2005). Chicago: American Institute of Steel Construction, 2005.

1.15. AISC, *Steel Construction Manual*, 13th ed. Chicago: American Institute of Steel Construction, 2005.

1.16. CAN/CSA S16-01 Consolidation (R2007), *Limit States Design of Steel Structures* (Includes Update No. 3, August 2006), and Supplement No. 1 (2005) Canadian Standards Association/National Standard of Canada.

1.17. ASCE Task Committee on Structural Safety of the Administrative Committee on Analysis and Design of the Structural Division. "Structural Safety: A Literature Review," *Journal of the Structural Division*, ASCE, **98**, ST4 (April 1972), 845–884.

1.18. Clarkson W. Pinkham. "Design Philosophies," Chapter 3, *Building Structural Design Handbook*, Richard N. White and Charles G. Salmon, Ed. New York: John Wiley & Sons, 1987 (pp. 44–54).

1.19. C. W. Pinkham and W. C. Hansell. "An Introduction to Load and Resistance Factor Design for Steel Buildings," *Engineering Journal*, AISC, **15**, 1 (First Quarter 1978), 2–7.

1.20. T. V. Galambos. "Load Factor Design of Steel Buildings," *Engineering Journal*, AISC, **9**, 3 (July, 1972), 108–113.

1.21. Theodore V. Galambos and M. K. Ravindra. "Proposed Criteria for Load and Resistance Factor Design," *Engineering Journal*, AISC, **15**, 1 (First Quarter 1978), 8–17.

1.22. Theodore V. Galambos. "Load and Resistance Factor Design," *Engineering Journal*, AISC, **18**, 3 (Third Quarter 1981), 74–82.

1.23. Mayasandra K. Ravindra and Theodore V. Galambos. "Load and Resistance Factor Design for Steel," *Journal of the Structural Division*, ASCE, **104**, ST9 (September 1978), 1337–1353.

1.24. Geerhard Haaijer. "Limit States Design: A Tool for Reducing the Complexity of Steel Structures," paper presented at AISC National Engineering Conference, March 4, 1983.

1.25. D. J. Laurie Kennedy. "Limit States Design of Steel Structures in Canada," *Journal of Structural Engineering*, ASCE, **110**, 2 (February 1984), 275–290.

1.26. D. J. Laurie Kennedy. "North American Limit States Design," *Proceedings, The 1985 International Engineering Symposium on Structural Steel*. Chicago: American Institute of Steel Construction, May 22–24, 1985.

1.27. Lynn S. Beedle. "Why LRFD?", *Modern Steel Construction*, AISC, **26**, 4 (4th Quarter 1986), 30–31.

1.28. Frank J. Heger. "Proposed AISC LRFD Design Criteria," *Journal of the Structural Division*, ASCE, **106**, ST3 (March 1980), 729–734. (Disc. **106**, ST12 (December 1980), 2576–2577)

1.29. Bruce Ellingwood, Theodore V. Galambos, James G. MacGregor, and C. Allin Cornell. *Development of a Probability Based Load Criterion for American National Standard A58*, NBS Special Publication 577. Washington, DC: US Department of Commerce, National Bureau of Standards, June 1980.

1.30. Theodore V. Galambos, Bruce Ellingwood, James G. MacGregor, and C. Allin Cornell. "Probability Based Load Criteria: Assessment of Current Design Practice," *Journal of the Structural Division*, ASCE, **108**, ST5 (May 1982), 959–977.

1.31. Bruce Ellingwood, James G. MacGregor, Theodore V. Galambos, and C. Allin Cornell. "Probability Based Load Criteria: Load Factors and Load Combinations," *Journal of the Structural Division*, ASCE, **108**, ST5 (May 1982), 978–997.

1.32. Mayasandra K. Ravindra, C. Allin Cornell, and Theodore V. Galambos. "Wind and Snow Load Factors for Use in LRFD," *Journal of the Structural Division*, ASCE, **104**, ST9 (September 1978), 1443–1457.

Wind Loads

1.33. ASCE. Task Committee on Wind Forces, Committee on Loads and Stresses, Structural Division. "Wind Forces on Structures," *Journal of the Structural Division*, ASCE, **84**, ST4 (July 1958) (Preliminary Report); and Final Report, *Transactions*, ASCE, **126**, pt. II (1961), 1124–1198.

1.34. Kishor C. Mehta. "Wind Load Provisions ANSI #A58.1–1982," *Journal of Structural Engineering*, **110**, 4 (April 1984), 769–784.

1.35. Theodore Stathopoulos, David Surry, and Alan G. Davenport. "Effective Wind Loads on Flat Roofs," *Journal of the Structural Division*, ASCE, **107**, ST2 (February 1981), 281–298.

Snow Loads

1.36. Michael J. O'Rourke and Ulrich Stiefel. "Roof Snow Loads for Structural Design," *Journal of Structural Engineering*, ASCE, **109**, 7 (July 1983), 1527–1537.

1.37. J. T. Templin and W. R. Schriever. "Loads due to Drifted Snow," *Journal of the Structural Division*, ASCE, **108**, ST8 (August 1982), 1916–1925.

1.38. Michael O'Rourke, Wayne Tobiasson, and Evelyn Wood. "Proposed Code Provisions for Drifted Snow Loads," *Journal of Structural Engineering*, **112**, 9 (September 1986), 2080–2092.

1.39. Michael J. O'Rourke, Robert Redfield, and Peter von Bradsky. "Uniform Snow Loads on Structures," *Journal of the Structural Division*, ASCE, **108**, ST12 (December 1982), 2781–2798.

1.40. Michael J. O'Rourke, Robert S. Speck, Jr., and Ulrich Stiefel. "Drift Snow Loads on Multilevel Roofs," *Journal of Structural Engineering*, ASCE, **111**, 2 (February 1985), 290–306.

1.41. R. L. Sack. "Snow Loads on Sloped Roofs," *Journal of Structural Engineering*, ASCE, **114**, 3 (March 1988), 501–517.

1.42. Michael O'Rourke and Ioannis Galanakis. "Roof Snowdrifts due to Blizzards," *Journal of Structural Engineering*, ASCE, **116**, 3 (March 1990), 641–658.

1.43. Ronald L. Sack and Paul M. Giever. "Predicting Roof Snow Loads on Gabled Structures," *Journal of Structural Engineering*, ASCE, **116**, 10 (October 1990), 2763–2779. Errata, **116**, 11 (November 1990), 3249–3250.

Earthquake Load

1.44. SEAOC. *Recommended Lateral Force Requirements and Commentary*. 7th ed. San Francisco, CA: Seismology Committee, Structural Engineers Association of California, Los Angeles, CA, 1999.

1.45. Anil K. Chopra and Ernesto F. Cruz. "Evaluation of Building Code Formulas for Earthquake Forces," *Journal of Structural Engineering*, ASCE, **112**, 8 (August 1986), 1881–1899.

1.46. Anil K. Chopra. *Dynamics of Structures, Theory and Applications to Earthquake Engineering*. (2nd ed.) Englewood Cliffs, NJ: Prentice-Hall, Inc., 2001.

1.47. R. W. Clough and Joseph Penzien. *Dynamics of Structures*. New York: McGraw-Hill, 1975.

1.48. IBC. *International Building Code*. Falls Church VA: International Code Council, 2006.

1.49. BSSC. NEHRP (National Earthquake Hazards Reduction Program) *Recommended Provisions for New Buildings and Other Structures, Part 1: Provisions* (FEMA 450). Washington, D.C.: Building Seismic Safety Council, Federal Emergency Management Agency, 2003.

1.50. AISC. *Seismic Provisions for Structural Steel Buildings*. Chicago, IL: American Institute of Steel Construction, March 9, 2006.

1.51. Egor P. Popov. "U.S. Seismic Steel Codes," *Engineering Journal*, AISC, **28**, 3 (3rd Quarter 1991), 119–128.

1.52. James W. Marsh. "Earthquakes: Steel Structures Performance and Design Code Developments," *Engineering Journal*, AISC, **30**, 2 (2nd Quarter 1993), 56–65.

2

Steels and Properties

2.1 STRUCTURAL STEELS

During most of the period from the introduction of structural steel as a major building material until about 1960, the steel used was classified as a carbon steel with the ASTM (American Society for Testing and Materials) designation A7, and had a minimum specified yield stress of 33 ksi. Most designers merely referred to "steel" without further identification, and the AISC Specification prescribed allowable stresses and procedures only for A7 steel. Other structural steels, such as a special corrosion resistant low alloy steel (A242) and a more readily weldable steel (A373), were available but they were rarely used in buildings. Bridge design made occasional use of these other steels.

Today (2008) the many steels available to the designer permit use of increased strength material in highly stressed regions rather than increase the size of members. The designer can decide whether maximum rigidity or least weight is the more desirable attribute. Corrosion resistance, hence elimination of frequent painting, may be a highly important factor. Some steels oxidize to form a dense protective coating that prevents further oxidation (corrosion), acquiring a pleasing even-textured dark red-brown appearance. Since painting is not required, it may be economical to use these "weathering steels" even though the initial cost is somewhat higher than traditional carbon steels.

Certain steels provide better weldability than others; some are more suitable than others for pressure vessels, at temperatures either well above or well below room temperatures.

Structural steels are referred to by ASTM designations, and also by many proprietary names. For design purposes the yield stress in tension is the material property that specifications, such as AISC, use to establish strength or allowable stress. The term *yield stress* is used to include either "yield point," the well-defined deviation from perfect elasticity exhibited by most of the common structural steels; or "yield strength," the unit stress at a certain offset strain for steels having no well-defined yield point. Today (2008) steels are readily available having yield stresses from 24 to 100 ksi (170 to 690 MPa).

Steels for structural use in hot-rolled applications may be classified as *carbon steels, high-strength low-alloy steels,* and *alloy steels.* The general requirements for such

John Hancock Center, Chicago, showing exterior diagonal bracing. (Photo by C. G. Salmon)

steels are covered under ASTM A6/A6M Specification [1.5]. Table 2.1.1 lists the common steels, their minimum yield stresses, and their tensile strengths. Their common uses are given in Table 2.1.2.

Carbon Steels

Carbon steels are divided into four categories based on the percentage of carbon: low carbon (less than 0.15%); mild carbon (0.15–0.29%); medium carbon (0.30–0.59%); and high carbon (0.60–1.70%). Structural carbon steels are in the mild carbon category; a steel such as A36 has maximum carbon varying from 0.25 to 0.29% depending on thickness. Structural carbon steels exhibit definite yield points as shown on curve (a) of Fig. 2.1.1. Increased carbon percent raises the yield stress but reduces ductility, making welding more difficult.

The carbon steels given in Table 2.1.1 are A36 [2.1], A53 [2.2], A500 [2.8], A501 [2.9], A529 [2.11], and A709 [2.16], Grade 36.

High-Strength Low-Alloy Steels

This category includes steels having yield stresses from 40 to 70 ksi (275 to 480 MPa), exhibiting the well-defined yield point shown in curve (b) of Fig. 2.1.1, the same as shown by

TABLE 2.1.1 Properties of Steels Used for Buildings and Bridges

ASTM[†] designation	F_y Minimum yield stress ksi (MPa)[‡]	F_u Tensile strength ksi (MPa)[‡]	Maximum thickness for plates in. (mm)	ASTM A6 groups* for shapes
A36	32 (220)	58–80 (400–550)	Over 8 (200)	—
	36 (250)	58–80 (400–550)	To 8 (200)	All
A53 Grade B	35 (240)	60 (415)		
A242	42 (290)	63 (435)	Over $1\frac{1}{2}$ to 4 (40 to 100)	4, 5
	46 (317)	67 (460)	Over $\frac{3}{4}$ to $1\frac{1}{2}$ (20 to 40)	3
	50 (345)	70 (480)	To $\frac{3}{4}$ (20)	1, 2
A500 Grade A	33 (228)	45 (310)	Round	
Grade B	42 (290)	58 (400)	Round	
Grade C	46 (317)	62 (427)	Round	
Grade A	39 (269)	45 (310)	Shaped	
Grade B	46 (317)	58 (400)	Shaped	
Grade C	50 (345)	62 (427)	Shaped	
A510	36 (250)	58 (400)		
A514	90 (620)	100–130 (690–895)	Over $2\frac{1}{2}$ to 6 (65 to 150)	
	100 (690)	110–130 (760–895)	To $2\frac{1}{2}$ (65)	
A529 Grade 50	50 (345)	65–100 (450–690)	To $\frac{1}{2}$ (13)	1
Grade 55	55 (380)	70–100 (485–690)	To 1 (25)	1, 2
A572 Grade 42	42 (290)	60 (415)	To 6 (150)	All
Grade 50	50 (345)	65 (450)	To 4 (100)	All
Grade 60	60 (415)	75 (520)	To $1\frac{1}{4}$ (32)	1, 2, 3
Grade 65	65 (450)	80 (550)	To $1\frac{1}{4}$ (32)	1, 2, 3
A588	42 (290)	63 (435)	Over 5 to 8 (125 to 200)	—
	46 (315)	67 (460)	Over 4 to 5 (100 to 125)	—
	50 (345)	70 (485)	To 4 (100)	All
A606	45 (310)	65 (450)		
	50 (345)	70 (485)		*(Continued)*

TABLE 2.1.1 Properties of Steels Used for Buildings and Bridges *(Continued)*

ASTM[†] designation	F_y Minimum yield stress ksi (MPa)[‡]	F_u Tensile strength ksi (MPa)[‡]	Maximum thickness for plates in. (mm)	ASTM A6 groups* for shapes
A618 Grades I&II	50 (345)	70 (485)	To $\frac{3}{4}$ in. walls	
Grade III	50 (345)	65 (450)		
A709 Grade 36	36 (250)	58–80 (400–550)	To 4 (100)	All
Grade 50	50 (345)	65 (450)	To 4 (100)	All
Grade 50W	50 (345)	70 (485)	To 4 (100)	All
Grade 70W	70 (485)	90–110 (620–760)	To 4 (100)	
Grade 100 & 100W	90 (620)	100–130 (690–895)	Over $2\frac{1}{2}$ to 4 (64–100)	
Grade 100 & 100W	100 (690)	110–130 (760–895)	To $2\frac{1}{2}$ (64)	
A852	70 (485)	90–110 (620–760)	To 4 (100)	
A913 Grade 50	50 (345)	60 (415)		
Grade 60	60 (415)	75 (520)		All
Grade 65	65 (450)	80 (550)		
Grade 70	70 (485)	90 (620)		
A992	50 (345)	65 (450)		All
A1008 Grade 40	40 (275)	55 (380)		
Grade 45	45 (310)	60 (415)		
Grade 50	50 (345)	65 (450)		
A1011 Grade 40	40 (275)	55 (380)		
Grade 45	45 (310)	60 (415)		
Grade 50	50 (345)	65 (450)		

[†] All steels listed are approved under the AISC Specification [1.13] A709.

*ASTM A6/A6M [1.5] places structural rolled shapes (W, M, S, HP, C, MC, and L) in Groups 1 through 5 according to size for tensile property classification. All rolled flanged sections having at least one cross-section dimension 3 in. (75 mm) or greater are included. The size basis for groups is approximately the web thickness corresponding to the maximum thickness for plates, with the thinnest web sections in Group 1. The specific sections included in each group are given in ASTM A6/A6M [1.5] and in the *AISC Manual* [1.15].

[‡] All SI values are those given in the particular ASTM Specification.

carbon steels. The addition to carbon steels of small amounts of alloy elements such as chromium, columbium, copper, manganese, molybdenum, nickel, phosphorus, vanadium, or zirconium improves some of the mechanical properties. Whereas carbon steels gain their strength by increasing carbon content, the alloy elements create increased strength from a fine rather than coarse microstructure obtained during cooling of the steel. High-strength low-alloy steels are used in the as-rolled or normalized condition; i.e., no heat treatment is used.

The high-strength low-alloy steels of Table 2.1.1 are A242 [2.3], A572 [2.12], A588 [2.13], A606 [2.14], A618 [2.15], A709 [2.16] Grades 50 and 50W, A992 [2.18], and A1011 [2.20].

TABLE 2.1.2 Uses of Various Steels

ASTM* designation	Common usage
A36 Carbon steel	General structural purposes; bolted and welded, mainly for buildings
A53 Carbon steel	Welded and seamless pipe
A242 High-strength low-alloy steel	Welded and bolted bridge construction where corrosion resistance is desired; essentially superseded by A709, Grade 50W
A500 Carbon steel	Cold-formed welded and seamless round, square, rectangular, or special shape structural tubing for bolted and welded general structural purposes
A501 Carbon steel	Hot-formed welded and seamless square, rectangular, round, or special shape structural tubing for bolted and welded general structural purposes
A514 Alloy steel, quenched and tempered	Plates in thicknesses of 6 in. (150 mm) and under, primarily for welded bridges; largely superseded by A709 for bridges
A529 Carbon steel	Plates and bars $\frac{1}{2}$ in. (13 mm) and less in thickness or diameter and Group 1 shapes [1.5] for use in bolted and welded metal building system frames and trusses
A1011/A1011M-05a Carbon steel	Hot-rolled carbon steel sheet and strip of structural quality in cut lengths or coils for cold-formed sections [maximum thickness 0.229 in. (6 mm)]
A572 High-strength, low-alloy, columbium or vanadium steel	Structural shapes, plates, sheet piling, and bars for bolted and welded buildings; welded bridges in Grades 42 and 50 only; essentially superseded by A709, Grade 50
A588 High-strength low-alloy steel	Structural shapes, plates, and bars for welded buildings and bridges where weight savings or added durability are needed; atmospheric corrosion resistance is about four times that of A36 steel; essentially superseded by A709, Grade 50 for bridges
A606 High-strength low-alloy steel	Hot- and cold-rolled sheet and strip in lengths or coils; for cold-formed sections, where enhanced durability is desired; atmospheric corrosion resistance for Type 2 at least twice that of carbon steel; and for Type 4 at least four times that of carbon steel
A607 High-strength low-alloy columbium or vanadium steel	Hot- and cold-rolled sheet and strip in lengths or coils; for cold-formed sections, where greater strength and weight savings are important; atmospheric corrosion resistance (without copper) is the same as carbon steel; with copper, corrosion resistance is twice that of carbon steel
A611 Carbon steel	Cold-rolled sheet in cut lengths or coils for making cold-formed sections
A618 High-strength low-alloy steel	Hot-formed welded and seamless square, rectangular, round, or special shape structural tubing for bolted and welded general structural purposes; Grade II has corrosion resistance about twice that of carbon steel; Grade I has corrosion resistance about four times that of carbon steel; Grade III for enhanced corrosion resistance may have copper specified
A709 Carbon; high-strength low-alloy; and quenched and tempered alloy	Structural shapes, plates, and bars in Grades 36, 50, and 50W for use in bridges; plates in Grades 100 and 100W for use in bridges; when supplementary requirements are used, requirements of A36, A572, A588, and A514 are exceeded; Grades 50W and 100W are weathering steels
A852 High-strength, low-alloy; and quenched and tempered alloy	Plates to 4 in. thick for welded and bolted construction where atmospheric corrosion resistance is desired
A913 High-strength low-alloy steel	Structural shapes for bolted and welded construction
A992 High strength low-alloy steel	Structural shapes used in building framing or bridges with a yield stress range of 50 ksi to 65 ksi, a minimum tensile stress of 65 ksi and a maximum yield-to-tensile strength ratio of 0.85, with a maximum carbon equivalent of 0.45. Supersedes A572 Grade 50.

*All steels listed are approved under the *AISC Specification* [1.15] except A611 and A709.

Figure 2.1.1
Typical stress-strain curves.

Alloy Steels

Low-alloy steels may be quenched and tempered to obtain yield strengths of 80 to 110 ksi (550 to 760 MPa). Yield strength is usually defined as the stress at 0.2% offset strain, since these steels do not exhibit a well-defined yield point. A typical stress-strain curve is shown in Fig. 2.1.1, curve (c). These steels are weldable with proper procedures, and ordinarily require no additional heat treatment after they have been welded. For special uses, stress relieving may occasionally be required. Some carbon steels, such as certain pressure vessel steels, may be quenched and tempered to give yield strengths in the 80 ksi (550 MPa) range, but most steels of this strength are low-alloy steels. These low-alloy steels generally have a maximum carbon content of about 0.20% to limit the hardness of any coarse-grain microstructure (martensite) that may form during heat treating or welding, thus reducing the danger of cracking.

The heat treatment consists of quenching [rapid cooling with water or oil from at least 1650°F (900°C) to about 300–400°F]; then tempering by reheating to at least 1150°F (620°C) and allowing to cool. Tempering, even though reducing the strength and hardness somewhat from the quenched material, greatly improves the toughness and ductility. Reduction in strength and hardness with increasing temperature is somewhat counteracted by the occurrence of a secondary hardening, resulting from precipitation of fine columbium, titanium, or vanadium carbides. This precipitation begins at about 950°F (510°C) and accelerates up to about 1250°F (680°C). Tempering at or near 1250°F to get maximum benefit from precipitating carbides may result in entering the transformation zone, thus producing the weaker microstructure that would have been obtained without quenching and tempering.

In summary, the quenching produces martensite, a very hard, strong, and brittle microstructure; reheating reduces the strength and hardness somewhat while increasing the toughness and ductility. For more detailed information concerning the metallurgy of the quenching and tempering process, the reader is referred to the *Welding Handbook* [2.21]. The quenched and tempered alloy steels of Table 2.1.1 are A514 [2.10], A709 [2.16] Grades 100 and 100W, A852 [2.17], and A913 [2.44].

2.2   FASTENER STEELS

The detailed treatment of the design of threaded fasteners appears in Chapter 4. A brief description of the materials used for bolts appears in the following paragraphs. [The headings are the ASTM specification exact titles.]

A307 [2.4], Carbon Steel Bolts and Studs, 60,000 psi Tensile Strength

This material is used for what are commonly referred to as "machine bolts." These are usually used only for temporary installations. Included are Grade A bolts for general applications, which have a *minimum* tensile strength of 60 ksi (415 MPa); and Grade B bolts for flanged joints in piping systems where one or both flanges are cast iron. The Grade B bolts have a *maximum* tensile strength limitation of 100 ksi (700 MPa). No well-defined yield point is exhibited by these bolts, and no minimum yield strength (for instance, 0.2% offset strength) is specified.

A325 [2.5], High-Strength Bolts for Structural Steel Joints

This quenched and tempered medium carbon steel is used for bolts commonly known as "high-strength structural bolts," or high-strength bolts. This material has maximum carbon of 0.30%. It is heat-treated by quenching and then by reheating (tempering) to a temperature of at least 800°F. In a tension test, this steel behaves more like heat-treated low-alloy steels than like carbon steel. It has an ultimate tensile strength of 105 ksi (724 MPa) ($1\frac{1}{8}$ to $1\frac{1}{2}$-in.-diam bolts) to 120 ksi (827 MPa) ($\frac{1}{2}$ to 1-in.-diam bolts). Its yield strength, measured at 0.2% offset, is prescribed at 81 ksi (558 MPa) minimum for $1\frac{1}{8}$ to $1\frac{1}{2}$-in.-diam bolts, and 92 ksi (634 MPa) for bolts $\frac{1}{2}$ to 1 in. diam (see Table 4.1.1).

A449 [2.6], Quenched and Tempered Steel Bolts and Studs

These bolts have tensile strengths and yield stresses (strength at 0.2% offset) the same as A325 for bolts $1\frac{1}{2}$ in. diam and smaller; however, they have the regular (instead of heavy) hexagon head and longer thread length of A307 bolts. They are also available in diameters up to 3 in. The *AISC Specification* [1.13] permits use of A449 bolts only for certain structural joints requiring diameters exceeding $1\frac{1}{2}$ in. and for high-strength anchor bolts and threaded rods.

A490 [2.7], Heat-Treated Steel Structural Bolts, 150 ksi (1035 MPa) Minimum Tensile Strength

This material has carbon content that may range up to 0.53% for $1\frac{1}{2}$ in.-diam bolts, and has alloying elements in amounts similar to the A514 [2.10] steels. After quenching in oil, the material is tempered by reheating to at least 900°F. The minimum yield strength, obtained by 0.2% offset, ranges from 115 ksi (793 MPa) (over $2\frac{1}{2}$ in. to 4 in. diam) to 130 ksi (896 MPa) (for $2\frac{1}{2}$ in. diam and under).

Galvanized High-Strength Bolts

In order to provide corrosion protection, A325 bolts may be galvanized. Hot-dip galvanizing requires the molten zinc temperature to be in the range of the heat treatment temperature;

thus, the mechanical properties obtained by heat treatment may be diminished. Whenever galvanized bolts are used, the nuts must be "oversized." If the nuts are also galvanized, they must be "double oversized."

Steels having tensile strength in the range of 200 ksi or higher are subject to hydrogen embrittlement when hydrogen is permitted to remain in the steel and high tensile stress is applied. The introduction of hydrogen occurs during the pickling operation of the galvanizing process and the subsequent "sealing-in" of the hydrogen and zinc coating [2.22]. The minimum tensile strength of A325 bolts is well below the critical 200 ksi range. On the other hand, A490 bolts have a maximum tensile strength of 170 ksi, a value considered too close to the critical range. Thus, *galvanizing of A490 bolts is not permitted.*

2.3   WELD ELECTRODE AND FILLER MATERIAL

The detailed treatment of welding and welded connections appears in Chapter 5. The electrodes used in shielded metal arc welding (SMAW) (see Sec. 5.2) also serve as the filler material and are covered by American Welding Society (AWS) A5.1 and A5.5 Specifications [2.23]. Such consumable electrodes are classified E60XX, E70XX, E80XX, E90XX, E100XX, and E110XX. The "E" denotes electrode. The first two digits indicate the tensile strength in ksi; thus the tensile strength ranges from 60 to 110 ksi (414 to 760 MPa). The "X's" represent numbers indicating the usage of the electrode.

TABLE 2.3.1 Electrodes Used for Welding[*]

Process				Minimum yield stress		Minimum tensile strength	
Shielded metal arc welding (SMAW) AWS A5.1 or A5.5	Submerged arc welding (SAW) AWS A5.17 or A5.23	Gas metal arc welding (GMAW) AWS A5.18 or A5.28	Flux cored arc welding (FCAW) AWS A5.20 or A5.29	(ksi)	(MPa)	(ksi)	(MPa)
E60XX			E6XT-X	50	345	62 min	425
	F6XX-EXXX			50	345	62–80	425–550
E70XX		ER70S-X	E7XT-X	60	415	72 min	495
	F7XX-EXXX			60	415	70–90	485–620
E80XX				67	460	80 min	550
	F8XX-EXXX		E8XT	68	470	80–100	550–690
		ER80S		65	450	80 min	550
E100XX				87	600	100 min	690
	F10XX-EXXX			88	605	100–130	690–895
		ER100S		90	620	100 min	690
			E10XT	88	605	100–120	690–830
E110X				97	670	110 min	760
	F11XX-EXXX			98	675	110–130	760–895
		ER110S		98	675	110 min	760
			E11XT	98	675	110–125	760–860

[*]Filler metal requirement given by AWS D1.1 [2.23], Table 3.1 to match the various structural steels.

For submerged arc welding (SAW) (see Sec. 5.2), the electrodes which also serve as filler material are specified under AWS A5.17 and A5.23. The weld-electrode combinations are designated F6XX-EXXX, F7XX-EXXX, etc. up to F11XX-EXXX. The "F" designates a granular flux material that shields the weld as it is made. The first one or two of the three digits following the "F" indicate the tensile strength (6 means 60 ksi, 11 means 110 ksi). The "E" stands for electrode and the other X's represent numbers relating to use. The yield stresses and tensile strengths of commonly used electrodes are given in Table 2.3.1.

2.4   STRESS-STRAIN BEHAVIOR (TENSION TEST) AT ATMOSPHERIC TEMPERATURES

Typical stress-strain curves for tension are shown in Fig 2.1.1 for the three categories of steel already discussed: carbon, high-strength low-alloy, and heat-treated high-strength low-alloy. The same behavior occurs in compression when support is provided so as to preclude buckling. The portion of each of the stress-strain curves of Fig. 2.1.1 that is utilized in ordinary design is shown enlarged in Fig. 2.4.1.

The stress-strain curves of Fig. 2.1.1 are determined using a unit stress obtained by dividing the load by the original cross-sectional area of the specimen, and the strain (inches per inch) is obtained as the elongation divided by the original length. Such curves are known as *engineering stress-strain* curves and rise to a maximum stress level (known as the tensile strength) and then fall off with increasing strain until they terminate as the specimen breaks. Insofar as the material itself is concerned, the unit stress continues to rise until failure occurs. The so-called *true-stress/true-strain* curve is obtained by using the actual cross-section even after necking down begins and using the instantaneous incremental strain.

Figure 2.4.1
Enlarged typical stress-strain curves for different yield stresses.

Stress-strain curves (as per Fig. 2.4.1) show a straight-line relationship up to a point known as the *proportional limit,* which essentially coincides with the yield point for most structural steels with yield points not exceeding 65 ksi (450 MPa). For the quenched and tempered low-alloy steels, the deviation from a straight line occurs gradually, as in curve (c), Fig. 2.4.1. Since the term yield point is not appropriate to curve (c), *yield strength* is used for the stress at an offset strain of 0.2%; or alternatively, a 0.5% extension under load, as shown in Fig 2.4.1. *Yield stress* is the general term to include the unit stress at a yield point, when one exists, or the yield strength.

The ratio of stress to strain in the initial straight line region is known as the modulus of elasticity, or Young's modulus, *E,* which for structural steels may be taken approximately as 29,000 ksi (200,000 MPa). In the straight-line region, loading and unloading results in no permanent deformation; hence it is the *elastic range.* The service load unit stress in steel design is always intended to be safely below the proportional limit, even though in order to ascertain safety factors against failure or excessive deformation, knowledge is required of the stress-strain behavior up to a strain about 15 to 20 times the maximum elastic strain.

For steels exhibiting yield points, as curves (a) and (b) of Fig. 2.4.1, the long plateau for which essentially constant stress exists is known as the *plastic range.* The load and resistance factor design method consciously uses this range. The higher strength steels typified by curve (c) of Fig. 2.4.1 also have a region that might be called the plastic range; however, in this zone the stress is continuously increasing (instead of remaining constant) as strain increases. For lack of having a region of constant stress with increasing strain, the steels whose yield stress exceeds 65 ksi are not permitted to be used for inelastic analysis (AISC-Appendix 1). Inelastic analysis (treated in Chaps. 6 and 10) relies on the ability of steel to deform (strain) at constant stress.

For strains greater than 15 to 20 times the maximum elastic strain, the stress again increases but with a much flatter slope than the original elastic slope. This increase in strength is called *strain hardening,* which continues up to the tensile strength. The slope of the stress-strain curve is known as the strain-hardening modulus, E_{st}. Average values for this modulus and the strain ϵ_{st} at which it begins have been determined [2.24] for two steels: A36 steel, E_{st} = 900 ksi (6200 MPa) at ϵ_{st} = 0.014 in. per in.; and A441, E_{st} = 700 ksi (4800 MPa) at ϵ_{st} = 0.021 in. per in. The strain-hardening range is not consciously used in design, but certain of the buckling limitations are conservatively derived to preclude buckling even at strains well beyond onset of strain hardening.

The stress-strain curve also indicates the *ductility.* Ductility is defined as the amount of permanent strain (i.e., strain exceeding proportional limit) up to the point of fracture. Measurement of ductility is obtained from the tension test by determining the percent elongation (comparing final and original lengths over a specified gage distance) of the specimen. Ductility is important because it permits yielding locally due to high stresses and thus allows the stress distribution to change. Design procedures based on inelastic behavior require large inherent ductility, particularly for treatment of stresses near holes or abrupt change in member shape, as well as for design of connections.

2.5 MATERIAL TOUGHNESS

The use of steels having higher strength than A36 without heat treatment has resulted in problems relating to lack of ductility and material fracture [2.26]; at a minimum, the use of such steels requires the structural engineer to be more conscious of material behavior.

In structural steel design, toughness is a measure of the ability of steel to resist fracture; i.e., to absorb energy. According to Rolfe [2.25], material toughness is defined as "the resistance to unstable crack propagation in the presence of a notch." Unstable crack propagation produces brittle fracture, as opposed to stable crack growth of a subcritical crack from fatigue.

For uniaxial tension, toughness can be expressed as the total area under the stress-strain curve out to the fracture point where the diagram terminates. Since uniaxial tension rarely exists in real structures, a more useful index of toughness is based on the more complex stress condition at the root of a notch.

Notch toughness is the measure of the resistance of a metal to the start and propagation of a crack at the base of a standard notch, commonly using the Charpy V-notch test. This test uses a small rectangular simply supported beam having a V-notch at midlength. The bar is fractured by a blow from a swinging pendulum. The amount of energy absorbed is calculated from the height the pendulum raises after breaking the specimen. The amount of energy absorbed will increase with increasing temperature at which the test is conducted.

Though the Charpy V-notch test has been a common means of determining notch toughness, other fracture criteria and more recently fracture mechanics have been used [2.25]. Barsom and Rolfe [2.27] and Barsom [2.28, 2.29] have excellently presented the important factors relating to fracture of steel.

Figure 2.5.1 shows the typical relationship between temperature and toughness, and also shows the transition from ductile to brittle behavior, such as one may obtain from the Charpy V-notch test. The temperature at the point where the slope is steepest (point A of Fig. 2.5.1) is the transition temperature. Since brittleness and ductility are qualitative terms, the various structural steels have different requirements for ductility at various temperatures depending on their service environment (loading, temperature, stress and strain levels, loading rate, and number of load repetitions).

For example, a moderate amount of ductility may be required for ordinary structures where very low temperatures are not expected; in such cases, 15 ft-lb has commonly been the energy absorption required. The corresponding temperature obtained from the test results shown in Fig. 2.5.1 would be about 17°F. The temperature at which marked decrease in slope begins to occur is known as the *ductility transition temperature.* This would indicate that the material may be expected to be brittle when service temperatures are below 17°F.

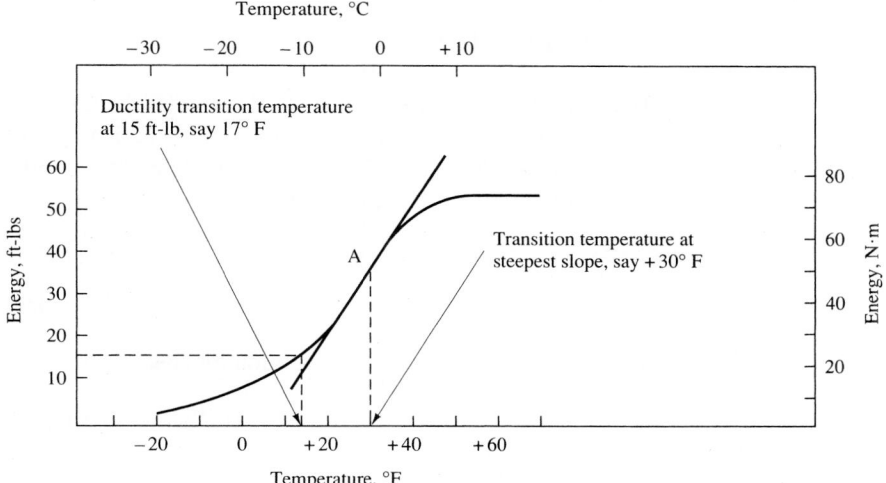

Figure 2.5.1
Transition temperature curve for carbon steel obtained from Charpy V-notch impact tests. (Adapted from Ref. 2.24)

2.6 YIELD STRENGTH FOR MULTIAXIAL STATES OF STRESS

Only when the load-carrying member is subject to uniaxial tensile stress can the properties from the tension test be expected to be identical with those of the structural member. It is easy to forget that yielding in a real structure is usually *not* the well-defined behavior observed in the tension test. Yielding is commonly assumed to be achieved when any one component of stress reaches the uniaxial value F_y.

For all states of stress other than uniaxial, a definition of yielding is needed. These definitions, and there are frequently several for a given state of stress, are called *yield conditions* (or *theories of failure*) and are equations of interaction between the stresses acting.

Energy-of-Distortion (Huber–von Mises–Hencky) Yield Criterion

This most commonly accepted theory gives the uniaxial yield stress in terms of the three principal stresses. The yield criterion* may be stated

$$\sigma_y^2 = \tfrac{1}{2}[(\sigma_1 - \sigma_2)^2 + (\sigma_2 - \sigma_3)^2 + (\sigma_3 - \sigma_1)^2] \qquad (2.6.1)$$

where $\sigma_1, \sigma_2, \sigma_3$ are the tensile or compressive stresses that act in the three principal directions; i.e., the stresses that act in the three mutually perpendicular planes of zero shear, and σ_y is the "yield stress" that may be compared with the uniaxial value F_y.

For most structural design situations, one of the principal stresses is either zero or small enough to be neglected; hence Eq. 2.6.1 reduces to the following for the case of the plane stress (all stresses considered are acting in a plane)

$$\sigma_y^2 = \sigma_1^2 + \sigma_2^2 - \sigma_1\sigma_2 \qquad (2.6.2)$$

When stresses on thin plates are involved, the principal stress acting transverse to the plane of the plate is usually zero (at least to first-order approximation). Flexural stresses on beams assume zero principal stress perpendicular to the plane of bending. Furthermore, structural shapes (Fig. 1.5.1) are comprised of thin plate elements, so that each is subject to Eq. 2.6.2. The plane stress yield criterion, Eq. 2.6.2, is the one used throughout the remaining chapters where needed, and is illustrated in Fig. 2.6.1.

Shear Yield Stress

The yield point for pure shear can be determined from a stress-strain curve with shear loading, or if the multiaxial yield criterion is known, that relationship can be used. Pure shear occurs on 45° planes to the principal planes when $\sigma_2 = -\sigma_1$, and the shear stress $\tau = \sigma_1$. Substitution of $\sigma_2 = -\sigma_1$ into Eq. 2.6.2 gives

$$\sigma_y^2 = \sigma_1^2 + \sigma_1^2 - \sigma_1(-\sigma_1) = 3\sigma_1^2 \qquad (2.6.3)$$

$$\sigma_1 = \tau = \sigma_y/\sqrt{3} = \text{shear yield} \qquad (2.6.4)$$

which indicates that the yield condition for shear stress acting alone is

$$\tau_y = \sigma_y/\sqrt{3} = 0.58\sigma_y \qquad (2.6.5)$$

---

*See Arthur P. Boresi, Richard J. Schmidt, and Omar M. Sidebottom, *Advanced Mechanics of Materials*, 5th ed. New York: John Wiley & Sons, Inc., 1993, pp. 133–134.

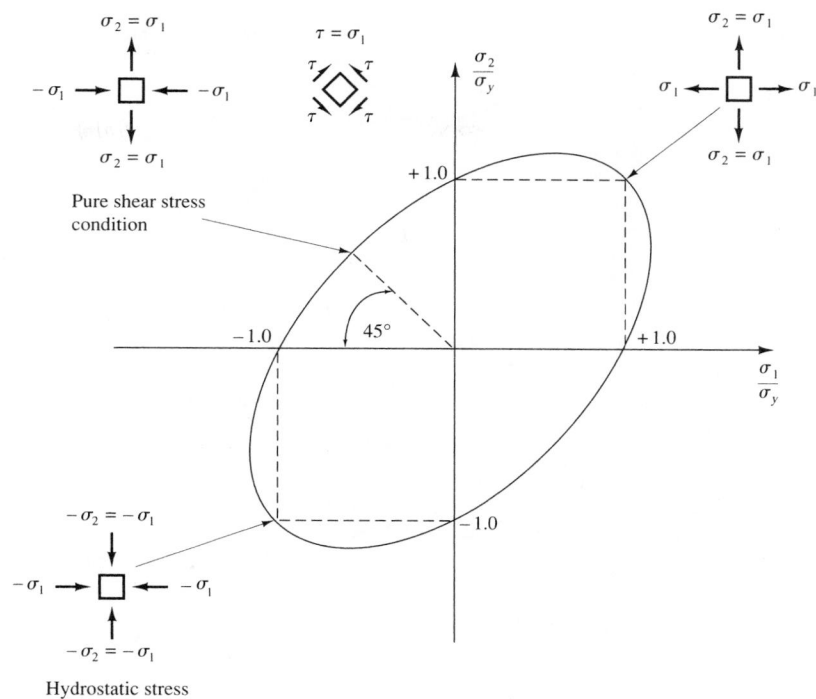

Figure 2.6.1
Energy-of-distortion yield
criterion for plane stress.

Poisson's Ratio, μ

When stress is applied in one direction, strains are induced not only in the direction of applied stress but also in the other two mutually perpendicular directions. The usual value of μ used is that obtained from the uniaxial stress condition, where it is the ratio of the transverse strain to longitudinal strain under load. For structural steels, Poisson's ratio is approximately 0.3 in the elastic range where the material is compressible and approaches 0.5 when in the plastic range where the material is essentially incompressible (i.e., constant resistance no matter what the strain).

Shear Modulus of Elasticity

Loading in pure shear produces a stress-strain curve with a straight line portion whose slope represents the shear modulus of elasticity. If Poisson's ratio μ and the tension-compression modulus of elasticity E are known, the shear modulus G is defined by the theory of elasticity as

$$G = \frac{E}{2(1 + \mu)} \tag{2.6.6}$$

which for structural steel is just over 11,000 ksi (75,800 MPa).

2.7   HIGH TEMPERATURE BEHAVIOR

The design of structures to serve under atmospheric temperature rarely involves concern about high temperature behavior. Knowledge of such behavior is desirable when specifying welding procedures, and is necessary when concerned with the effects of fire.

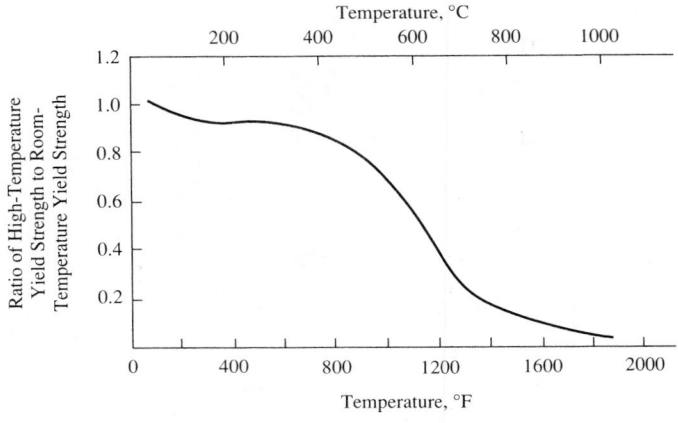

(a) Average Effect of Temperature on Yield Strength

(b) Average Effect of Temperature on Tensile Strength

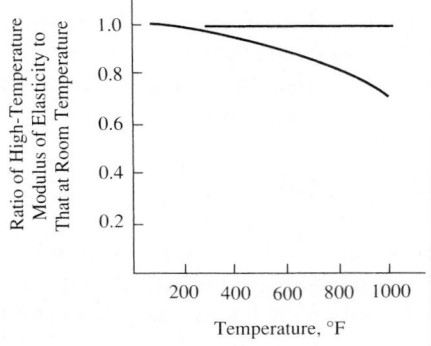

Figure 2.7.1
Typical effects of high temperature on stress-strain curve properties of structural steels. (Adapted from Ref. 2.24)

(c) Typical Effect of Temperature on Modulus of Elasticity

When temperatures exceed about 200°F (93°C), the stress-strain curve begins to become nonlinear, gradually eliminating the well-defined yield point. The modulus of elasticity, yield strength, and tensile strength all reduce as temperature increases. The range from 800 to 1000°F (430 to 540°C) is where the rate of decrease is maximum. While each steel, because of its different chemistry and microstructure, behaves somewhat differently, the general relationships are shown in Fig. 2.7.1. Steels having relatively high percentages of carbon, such as A36, exhibit "strain aging" in the range 300 to 700°F (150 to 370°C). This is evidenced by a relative rise in yield strength and tensile strength in that range. Tensile strength may rise to about 10% above that at room temperature and yield strength may recover to about its room temperature value when the temperature reaches 500 to 600°F (260 to 320°C). Strain aging results in decreased ductility.

The modulus of elasticity decrease is moderate up to 1000°F (540°C); thereafter it decreases rapidly. More importantly, at temperatures above about 500 to 600°F (260 to 320°C), steels exhibit deformation which increases with increasing time under load, a phenomenon known as *creep*. Creep is well known in concrete structures; and its effect in steel, which does not occur at atmospheric temperatures, increases with increasing temperature.

Other high temperature effects are (a) improved notch impact resistance up to about 150 to 200°F (65 to 95°C), as discussed in Sec. 2.5; (b) increased brittleness due to metallurgical changes, such as carbide precipitation discussed in Sec. 2.1, begins to occur at about 950°F (510°C); and (c) corrosion resistance of structural steels increases for temperatures up to about 1000°F (540°C). Most steels are used in applications below 1000°F, and some heat treated steels should be kept below about 800°F (430°C).

2.8 COLD WORK AND STRAIN HARDENING

After the strain $\epsilon_y = F_y/E_s$ at first yield has been exceeded appreciably, and the specimen is unloaded, reloading may give a stress-strain relationship differing from that observed during the initial loading. Elastic loading and unloading results in no residual strain; however, initial loading beyond the yield point, such as to point A of Fig. 2.8.1, results in unloading to a strain at point B. A permanent set OB has occurred. The ductility capacity has been reduced from strain OF to strain BF. Reloading exhibits behavior as if the stress-strain origin were at point B; the plastic zone prior to strain hardening is also reduced.

When loading has occurred until point C is reached, unloading follows the dashed line to point D; i.e., the origin for a new loading is now point D. The length of the line CD is greater, indicating that the yield point has increased. The increased yield point is referred to as a strain hardening effect; the ductility remaining when loading from point D is severely reduced from its original value prior to the initial loading. The process of loading

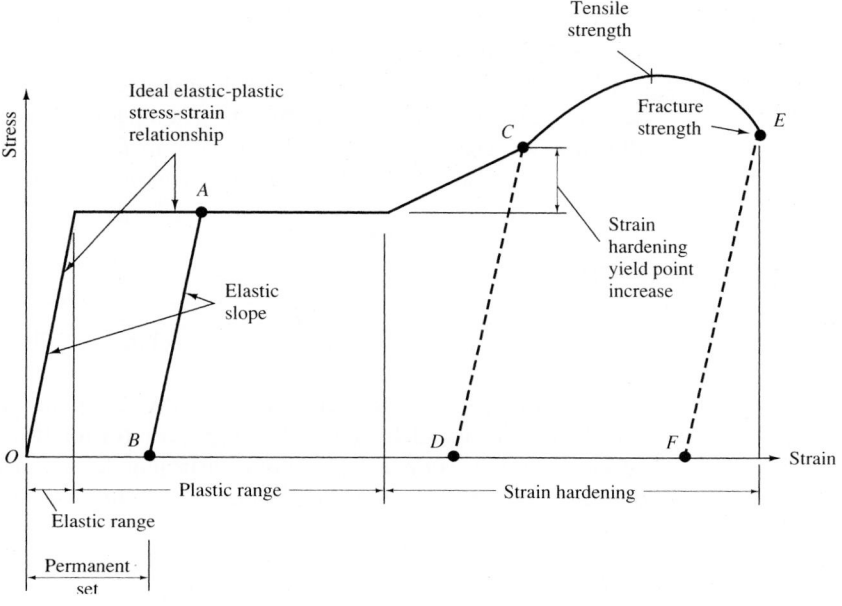

Figure 2.8.1
Effects of straining beyond
the elastic range.

Figure 2.8.2
Effect of strain aging after
straining into strain-hardening
range and unloading.

beyond the elastic range to cause a change in available ductility, when done at atmospheric temperature, is known as *cold work*. Since real structures are not loaded in uniaxial tension-compression, the cold work effect is much more complex, and any theoretical study of it is outside the scope of the text.

When structural shapes are made by cold-forming from plates at atmospheric temperature, inelastic deformations occur at the bends. Cold working into the strain hardening range at the bend locations increases the yield strength, which design specifications may permit taking into account. The *North American Specification for the Design of Cold-Formed Steel Structural Members** [1.8] has such provisions.

Upon unloading and after a period of time, the steel will have acquired different properties from those represented by points *D, C,* and *E* of Fig. 2.8.1 by a phenomenon known as *strain aging*. Strain aging, as shown in Fig. 2.8.2, produces an additional increase in yield point, restores a plastic zone of constant stress, and gives a new strain hardening zone at an elevated stress. The original shape of the stress-strain diagram is restored, but the ductility is reduced. The new stress-strain diagram may be used as if it were the original for analyzing cold-formed sections, as long as the ductility that remains is sufficient. The corner regions of cold-formed shapes generally would not require high ductility for rotational strain about the axis of the bend.

Stress relieving by annealing will eliminate the effects of cold work should that be desired. Annealing involves heating to a temperature above transformation range and allowing slow cooling; a recrystallization occurs to restore the original properties. Bittence [2.30] provides an excellent summary of the basics of heat treating.

2.9 BRITTLE FRACTURE

As has been discussed in several sections, steel that is ordinarily ductile can become brittle under various conditions.

Barsom [2.28, 2.29], Barsom and Rolfe [2.27], and Rolfe [2.25] have provided an excellent summary of fracture and fatigue control for structural engineers. Rolfe [2.25] defines *brittle fracture* as "a type of catastrophic failure that occurs without prior plastic deformation and at extremely high speeds." Fracture behavior is affected by temperature,

*Referred to henceforth as the AISI Specification.

loading rate, stress level, flaw size, plate thickness or constraint, joint geometry, and workmanship.

Effect of Temperature

Notch toughness, as determined by the Charpy-impact energy vs temperature curves (see Sec. 2.5), is an indication of the susceptibility to brittle fracture. Temperature is a vital factor in several ways: (a) the value below which notch toughness is inadequate; (b) the 600 to 800°F (320 to 430°C) range causes formation of brittle microstructure; and (c) over 1000°F (540°C) causes precipitation of carbides of alloying elements to give more brittle microstructure. The other temperature factors have already been discussed in earlier sections.

Effect of Multiaxial Stress

The complex stress condition found in usual structures, particularly at joints, is another major factor affecting brittleness. The *Primer on Brittle Fracture* [2.31] provides an excellent, rational presentation of this and forms the basis for what follows. The engineering stress-strain curve is for uniaxial stress; prior to fracture a necking down occurs, as shown in Fig. 2.9.1a. If biaxial lateral loading as shown in Fig. 2.9.1b could be applied, plastic necking down could be suppressed to the point where the bar would break in a brittle manner without elongation and without reduction in area. The fracture stress based on the unreduced cross-sectional area would be the same high value as that based on the necked-down cross section in the uniaxial tension case. The unit stress would be far above the nominal maximum tensile strength of the engineering stress-strain curve, which is always computed on the basis of original cross section.

Also the effects of notches have been alluded to in the discussion of notch toughness in Sec. 2.5. The notch serves somewhat the same purpose as the theoretical triaxial loading of Fig. 2.9.1b, in that it restrains plastic flow which otherwise would occur and thus at some higher stress may likely fail in a brittle manner. Figure 2.9.2 shows the effect of a notch in a tensile test specimen. The cross-sectional area at the base of the notch corresponds to the area of the original specimen of Fig. 2.9.1b. The reduced section tries to become narrower

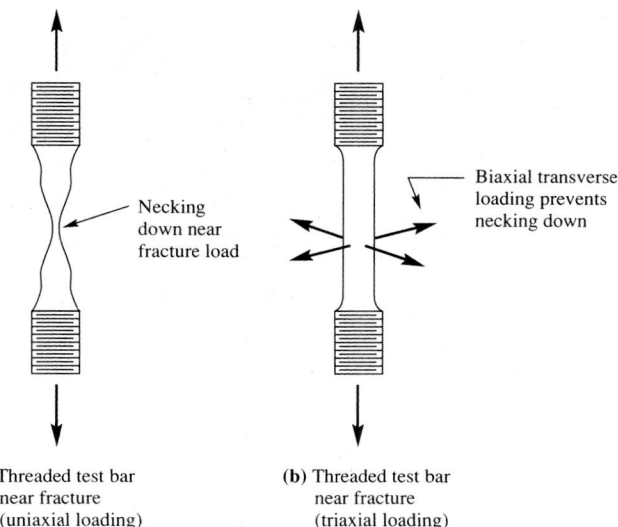

Figure 2.9.1
Uniaxial and triaxial loading.

(a) Threaded test bar
near fracture
(uniaxial loading)

(b) Threaded test bar
near fracture
(triaxial loading)

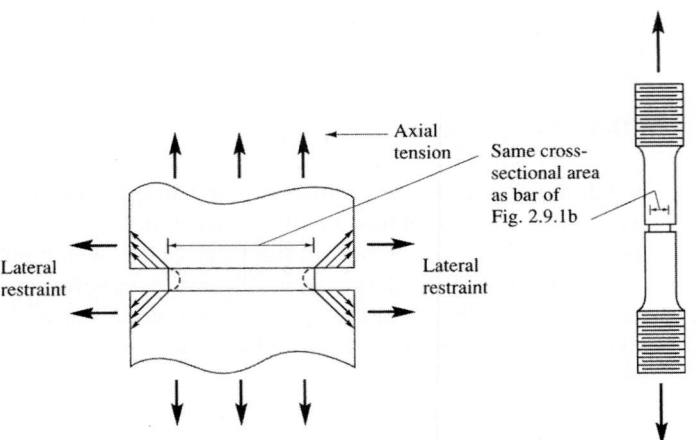

Figure 2.9.2
Effect of notch on uniaxial
tension test.

as the axial tension increases, but is resisted by the diagonal pull that develops in the corners, as shown in Fig. 2.9.2. The test bar will fail at high stress by brittle fracture.

Notches can occur in real structures by use of unfilleted corners in design or from
improperly made welds that may crack. Such occurrences can lead to brittleness. Notches
and cracked welds can, however, be minimized by good design and welding procedures.

Unusual configurations and changes in section should be made gradually so the
stress flow lines are not required to make abrupt changes. Whenever the complexity is such
as to give rise to three-dimensional stresses, the tendency for brittleness increases. Cast-
ings, for instance, have the reputation for brittleness. Primarily this is because of the built-
in three-dimensional continuity.

Multiaxial Stress Induced by Welding

In general, welding creates a built-in restraint that gives rise to biaxial and triaxial stress and
strain conditions, which result in brittle behavior. To illustrate, consider the loaded simply
supported beam of Fig. 2.9.3, which in turn supports a plate in tension. Due to flexure, the

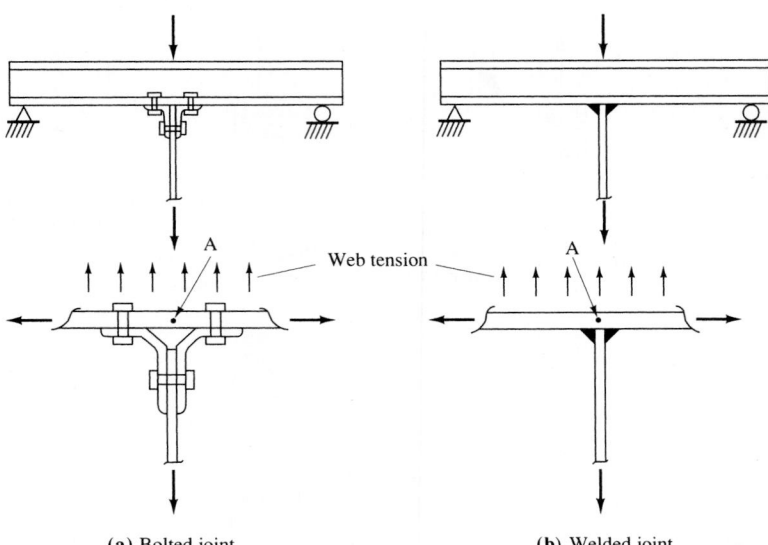

Figure 2.9.3
Comparison of stress conditions in bolted and welded
joints.

(a) Bolted joint   (b) Welded joint

bottom flange of the beam is in tension; therefore, the stress at point *A* is uniaxial tension (neglecting the small effects of beam width and attachment of flange to web). Connecting the tension plate with angles and bolts puts the flange bolts essentially in uniaxial tension, puts the bolt which passes through the suspender plate in shear, and distorts the horizontal legs of the angles in bending; so that there is no appreciable effect on the stress at point *A*. In other words, the stress conditions in the connection of Fig. 2.9.3a are approximately uniaxial in nature.

Next, consider the tensile suspender plate welded to the tension flange of the beam, as in Fig. 2.9.3b. The stress at point *A* is now biaxial because of the direct attachment to the flange at that point. The weld region, therefore, is subject to triaxial stress; biaxial from the directly applied loads, plus the resistance to deformation along the axis of the welds resulting from continuous attachment (Poisson's ratio effect). The design of welded joints should consider the possibilities of brittleness due to three-dimensional stressing. The subject of *lamellar tearing* is treated in Sec. 2.10.

Effect of Thickness

As discussed in Sec. 2.6, thin plates may usually be assumed to be in a state of plane stress in which the three dimensional stress effects may be ignored. This is not generally the case for thick plate elements for which three-dimensional stress contributes to brittleness. Brittleness in thick plates also increases due to the manufacturing process. The slower cooling rate produces a coarser microstructure, and a higher carbon content is required to achieve the strength otherwise obtained by hot working in thin plates.

The very thick rolled W-shapes (ASTM A6/A6M [1.5], Groups 4 and 5); i.e., the so-called "jumbo shapes," exhibit low fracture toughness at the core of the thick flange-to-web junction and the center of the web adjacent to it [2.32]. This low fracture toughness may cause brittle failures when these heavy W-shapes are used as tension members. For this reason their use is intended only for compression members [2.33].

When ASTM A6/A6M, Groups 4 and 5 rolled shapes are to be "used as members subject to primary tensile stresses due to tension or flexure, toughness need not be specified if splices are made by bolting. If such members are spliced using complete joint penetration welds, the steel shall be specified in the contract documents to be supplied with Charpy V-Notch testing in accordance with ASTM A6, Supplementary Requirement S5. (AISC-A3.1c) The Bethlehem Steel Technical Bulletin, *Use of Heavy Structural Shapes in Tension Applications* [2.34] provides additional guidance. Even though jumbo sections were not originally intended for tension applications, designers use them in such situations.

Effect of Dynamic Loading

The stress-strain properties referred to so far have been for static loading slowly applied. More rapid loading, such as that of forge drop-hammers, earthquake, or nuclear blast changes the stress-strain properties. Ordinarily, the increased strain rate from dynamic loading increases the yield point, tensile strength, and ductility. At about 600°F (320°C) there will be a small decrease in strength. Some increased brittleness has been noted with high strain rate, but it seems principally associated with other factors already discussed, such as notches where stress concentrations exist and the temperature effect on toughness. The more important factor relating to dynamic load application is not that a rapidly increasing strain rate occurs, but that it is combined with a rapidly *decreasing* strain rate. The effect of stress *variation* is discussed in the section on fatigue.

Table 2.9.1, from Ref. 2.31, provides a list of factors "to help determine whether or not the risk of brittle fracture is serious and requires special design considerations."

TABLE 2.9.1 The Element of Risk: Factors to Analyze in Estimating Seriousness of Brittle Fracture (Adapted From Ref. 2.31)

1. What is the minimum anticipated service temperature? The lower the temperature, the greater the susceptibility to brittle fracture.
2. Are tension stresses involved? Brittle fracture can occur only under conditions of tensile stress.
3. How thick is the material? The thicker the steel, the greater the susceptibility to brittle fracture.
4. Is there three-dimensional continuity? Three-dimensional continuity, giving rise to multiaxial states of stress, tends to restrain the steel from yielding and increases susceptibility to brittle fracture.
5. Are notches present? The presence of sharp notches increases susceptibility to brittle fracture.
6. Is loading applied at a high rate? The higher the rate of loading, the greater the susceptibility to brittle fracture.
7. Is there a changing rate of stress? Brittle fracture occurs only under conditions of increasing rate of stress.
8. Is welding involved? Weld cracks can act as severe notches.

2.10 LAMELLAR TEARING

Lamellar tearing is a form of brittle fracture occurring "in planes essentially parallel to the rolled surface of a plate under high through-thickness loading." [2.35]. Because strains resulting from service loads are well below ϵ_y, normal loads are not believed to initiate or propagate lamellar tears, and something else must be responsible. In a highly restrained welded joint "thru-thickness" strains ϵ are induced by weld metal shrinkage. The localized strains resulting from weld metal shrinkage, which can be several times larger than yield strain ϵ_y, are the source of the problem.

The subject of lamellar tearing has received considerable attention since the early 1970s, resulting in a tendency for structural engineers to blame lamellar tearing for many brittle fractures. The AISC has provided an excellent summary of the phenomenon [2.36]. Thornton [2.37] has provided design and supervision procedures to minimize lamellar tearing. For more detailed treatment, the reader is referred to Kaufman, Pense, and Stout [2.35] and Holby and Smith [2.38].

As a result of the hot rolling operation in manufacture, steel sections have different properties in the direction parallel to rolling (see Fig. 2.10.1), in the transverse direction, and in the "thru-thickness" direction. In the elastic range, both the rolling and transverse directions exhibit similar behavior, with the elastic limit for the transverse direction being only slightly below that for the rolling direction. The ductility (strain capability), however, in the "thru-thickness" direction may be well below that for the rolling direction.

Figure 2.10.1
Definition of direction terminology. (From Ref. 2.36)

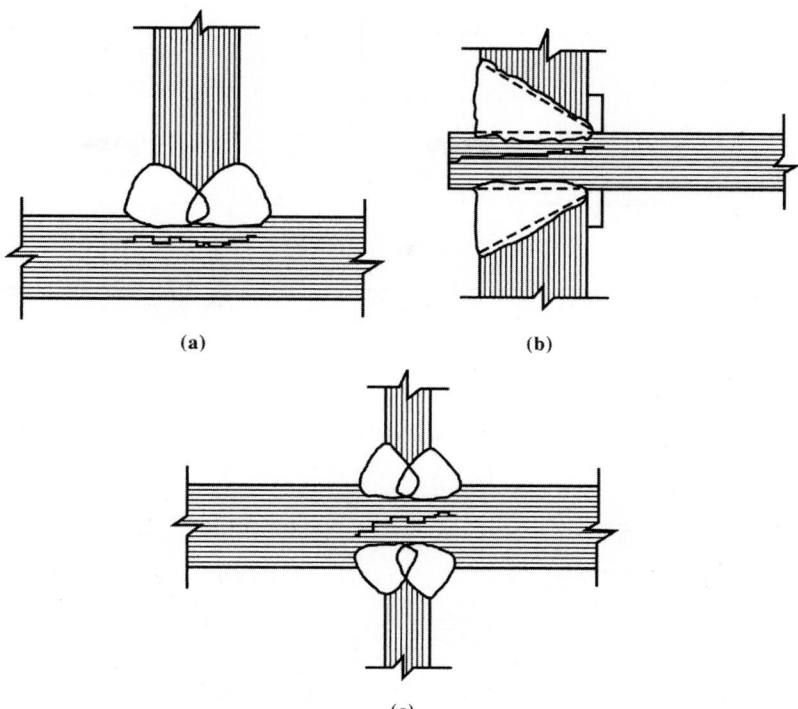

Figure 2.10.2
Joints showing typical lamellar tears resulting from
shrinkage of large welds in
thick material under high
restraint. (From Ref. 2.36)

Generally, I-shaped steel sections are adequately ductile when loaded either parallel or transverse to the rolling direction. They will deform locally to strains greater than the yield strain (F_y/E_s), carrying load with some of the material acting at the yield stress and bringing adjacent material into participation if added strength is needed. When, however, the strain is localized for instance in the "thru-thickness" direction at one thick flange of a section, a restrained situation exists because the strain cannot redistribute from the flange through the web to the opposite flange. The large localized "thru-thickness" strain may exceed the yield point strain, causing decohesion and leading to a lamellar tear.

Figure 2.10.2 shows conditions that promote lamellar tearing in welded joints. Internal joint restraint that inhibits large strains ϵ resulting from weld shrinkage can potentially cause lamellar tearing. Figure 2.10.3 indicates weld shrinkage in the "thru-thickness" direction, increasing susceptibility to lamellar tearing. The weld detail should be made such that weld shrinkage occurs in the rolling direction so that the shrinkage pulls on the fibers longitudinally in their strongest orientation. References 2.36 and 2.39 suggest ways of avoiding the problem.

2.11   FATIGUE STRENGTH

Repeated loading and unloading, primarily in *tension,* may eventually result in failure even if the yield stress is never exceeded. The term *fatigue* means failure under cyclic loading. It is a progressive failure, the final stage of which is unstable crack propagation. The fatigue strength is governed by three variables: (1) the number of cycles of loading, (2) the *range* of service load stress (the difference between the maximum and minimum stress), and (3) the initial size of a flaw. A flaw is a discontinuity, such as an extremely small crack.

In welded assemblies, a flaw could be the "notch" intersection of two elements or a "discontinuity" such as a bolt hole. Flaws may be the result of poorly made welds, rough

Susceptible detail   Improved detail

(a)

Susceptible detail   Improved detail

(b)

Susceptible detail   Improved detail

(c)

Figure 2.10.3
Susceptibility to lamellar tearing can be reduced by careful detailing of welded connections. (From Ref. 2.36)

edges resulting from shearing, punching, or flame cutting, or small holes. Such flaws may be of no concern; however, under many cycles of loading the flaw (notch effect) may give rise to a crack that increases in length with each cycle of load and reduces the section carrying the load, consequently increasing the stress intensity on the uncracked part. The fatigue strength is more dependent on the localized state of stress than is the static strength. Fatigue is always a service load consideration; the actual service load state of stress is what determines crack propagation.

The grade of steel has no apparent affect on the number of cycles to failure, and the effect of minimum stress (attributable to dead load) is considered to be negligible for design purposes. On the other hand, the specimen geometry, including the surface condition and internal soundness of the weld, have a significant effect. These factors are reflected in the *Structural Welding Code* [2.23] rules for welded structure design.

Work by Zuraski and Johnson [2.40] evaluating the remaining life in steel bridges has shown that under certain conditions repeated stressing in steel sections can actually increase their fatigue life. The phenomenon, known as *coaxing,* was first studied by Sinclair [2.41] and results from repeatedly stressing near, but below, the fatigue limit and gradually increasing the stress.

The *AISC Specification* [1.13] in Appendix 3 prescribes no fatigue effect for fewer than 20,000 cycles, which is approximately two applications a day for 25 years. Since most loadings in buildings are in that category, fatigue is generally not considered. The exceptions are crane-runway girders and structures supporting machinery. Fatigue is always considered in the design of highway bridges, which are expected to have in excess of 100,000 cycles of loading.

Volume 1 of the *Welding Handbook* [2.21] shows several good examples of the fatigue relationships for welded plate girders and cover-plated beams. Extended discussion of fatigue is given by Barsom and Rolfe [2.27].

2.12   CORROSION RESISTANCE AND WEATHERING STEELS

Since the earliest uses of steel, one of its important drawbacks was that painting was required to prevent the deterioration of the metal by corrosion (rusting). The lower-strength carbon steels were inexpensive but very vulnerable to corrosion. Corrosion resistance may be improved by the addition of copper as an alloy element. However, copper-bearing carbon steel is too expensive for general use.

High-strength low-alloy steels have several times [2.43] the corrosion resistance of structural carbon steel, with or without the addition of copper, as shown in Fig. 2.12.1. The high-strength low-alloy steels do not pit as severely as carbon steels and the rust that forms becomes a protective coating to prevent further deterioration. With certain alloy elements the high-strength low-alloy steel will develop an oxide protective coating that is pleasing in appearance and is described as follows*: "It is a very dense corrosion—actually a deeply colored brown, red, purple. . . . It has a texture and color which cannot be reproduced artificially—a character only nature can give, as with stone, marble, and granite." When steels are to be unpainted and left exposed they are called weathering steels.

As might be expected, the corrosion properties of any steel, including the weathering steels, are dependent on the chemical composition, the degree of pollution in the atmosphere, and the frequency of wetting and drying of the steel.

Since its first major use in 1958, for the Administrative Center for Deere & Company in Moline, Illinois, the use of weathering steel has received considerable attention. At first such steels were specified under ASTM A242, which as previously discussed is very general, allowing a wide variation in chemistry.

With the adoption of A588 steel in 1969, and A709 in 1975, A242 is now essentially obsolete. A588 is generally used for weathering steel in buildings and A709 Grades 50W and 100W for weathering steel in bridges (see Table 2.1.2).

Fabrication and erection of weathering steel requires care. Unsightly gouges, scratches, and dents should be avoided. Painting, even for identification, should be minimized, since all marks must be removed after the erection is completed. Scale and discoloration from welding also must be removed. The extra expense resulting from fabrication and erection is offset by the elimination of painting at intervals during the life of the structure.

The practice of using weathering steels, including the results of 30 years experience, has been summarized by Coburn [2.42], who presents the following "rules":

1. For optimum performance in the unpainted condition, the structure should be boldly exposed to the elements.

Figure 2.12.1 Comparative corrosion of steels in an industrial atmosphere. Shaded areas indicate range for individual specimens. (Adapted from Ref. 2.43)

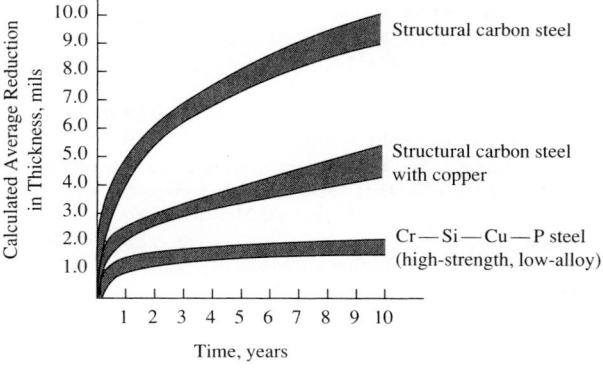

Architectural Record, August 1962.

2. The development of the protective oxide film is best achieved under normal exposure, wherein the surfaces are wet at night by dew formation and dry during daylight hours.

3. Because this wet-dry cycle cannot occur when the steel, regardless of its grade, is buried in the soil or immersed in water, the protective oxide will not form and the performance will resemble that of mild carbon steel exposed to the same conditions.

SELECTED REFERENCES

2.1. ASTM. *Standard Specification for Structural Steel* (A36/A36M-04a). West Conshohocken, PA: American Society for Testing and Materials, 2004.

2.2. ASTM. *Standard Specification for Pipe, Steel, Black and Hot-Dipped, Zinc-Coated Welded and Seamless* (A53/A53M-02). West Conshohocken, PA: American Society for Testing and Materials, 2002.

2.3. ASTM. *Standard Specification for High-Strength Low-Alloy Structural Steel* (A242/A242M-04). West Conshohocken, PA: American Society for Testing and Materials, 2004.

2.4. ASTM. *Standard Specification for Carbon Steel Bolts and Studs, 60 000 psi Tensile Strength* (A307-03). West Conshohocken, PA: American Society for Testing and Materials, 2003.

2.5. ASTM. *Standard Specification for Structural Bolts, Steel, Heat Treated, 120/105 ksi Minimum Tensile Strength* (A325-04b), West Conshohocken, PA: American Society for Testing and Materials, 2004.

2.6. ASTM. *Standard Specification for Quenched and Tempered Steel Bolts and Studs* (A449-04). West Conshohocken, PA: American Society for Testing and Materials, 2004.

2.7. ASTM. *Standard Specification for Heat-Treated Steel Structural Bolts, 150 ksi Minimum Tensile Strength* (A490-04); also *Specification for High-Strength Steel Bolts, Classes 10.9 and 10.9.3, for Structural Steel Joints [Metric]* (A490M-04). West Conshohocken, PA: American Society for Testing and Materials, 2004.

2.8. ASTM. *Standard Specification for Cold-Formed Welded and Seamless Carbon Steel Structural Tubing in Rounds and Shapes* (A500-03a). West Conshohocken, PA: American Society for Testing and Materials, 2003.

2.9. ASTM. *Standard Specification for Hot-Formed Welded and Seamless Carbon Steel Structural Tubing* (A501-01). West Conshohocken, PA: American Society for Testing and Materials, 2001.

2.10. ASTM. *Standard Specification for High-Yield Strength, Quenched and Tempered Alloy Steel Plate, Suitable for Welding* (A514/A514M-00a). West Conshohocken, PA: American Society for Testing and Materials, 2000.

2.11. ASTM. *Standard Specification for High-Strength Carbon-Manganese Steel of Structural Quality* (A529/A529M-05). West Conshohocken, PA: American Society for Testing and Materials, 2005.

2.12. ASTM. *Standard Specification for High-Strength Low-Alloy Columbium-Vanadium Steels of Structural Quality* (A572/A572M-04). West Conshohocken, PA: American Society for Testing and Materials, 2004.

2.13. ASTM. *Standard Specification for High-Strength Low-Alloy Structural Steel with 50 ksi [345 MPa] Minimum Yield Point to 4 in. [100 mm] Thick* (A588/A588M-04). West Conshohocken, PA: American Society for Testing and Materials, 2004.

2.14. ASTM. *Standard Specification for Steel, Sheet and Strip, High-Strength, Low-Alloy, Hot-Rolled and Cold-Rolled, with Improved Atmospheric Corrosion Resistance* (A606-04). West Conshohocken, PA: American Society for Testing and Materials, 2004.

2.15. ASTM. *Standard Specification for Hot-Formed Welded and Seamless High-Strength Low-Alloy Structural Tubing* (A618/A618M-04). West Conshohocken, PA: American Society for Testing and Materials, 2004.

2.16. ASTM. *Standard Specification for Structural Steel for Bridges* (A709/A709M-04). West Conshohocken, PA: American Society for Testing and Materials, 2004.

2.17. ASTM. *Standard Specification for Quenched and Tempered Low-Alloy Structural Steel Plate with 70 ksi [485 MPa] Minimum Yield Strength to 4 in. [100 mm] Thick* (A852/A852M-03). West Conshohocken, PA: American Society for Testing and Materials, 2003.

2.18. ASTM. *Standard Specification for Structural Steel Shapes* (A992/A992M-04), West Conshohocken, PA: American Society for Testing and Materials, 2004.

2.19. ASTM *Standard Specification for Steel, Sheet, Cold-Rolled, Carbon, Structural, High-Strength Low-Alloy, High-Strength Low-Alloy with Improved Formability, Solution Hardened, and Bake Hardenable* (A1008/A1008M-05b), West Conshohocken, PA: American Society for Testing and Materials, 2005.

2.20. ASTM. *Standard Specification for Steel, Sheet and Strip Hot-Rolled, Carbon, Structural, High-Strength Low-Alloy and High-Strength Low-Alloy with Improved Formability* (A1011/A1011M-04), West Conshohocken, PA: American Society for Testing and Materials, 2004.

2.21. AWS *Welding Handbook* WHB-1.8 Rev 8th ed., *Volume 1, Welding Technology* (AWS WHB-1.8), 2001, *Volume 2, Welding Processes,* 2004 revision. Miami, FL: American Welding Society, 1987, 1991. (550 N. W. LeJeune Road, P.O. Box 351040, Miami, FL 33135).

2.22. Research Council on Structural Connections. *Commentary on Specifications for Structural Joints Using ASTM A325 or A490 Bolts.* Chicago: American Institute of Steel Construction, June 8, 1988.

2.23. AWS. *Structural Welding Code—Steel,* Nineteenth Edition, Effective December 30, 2003 (ANSI/AWS D1.1–04). Miami, FL: American Welding Society, 2004.

2.24. R. L. Brockenbrough and B. G. Johnston. *Steel Design Manual.* Pittsburgh, PA: United States Steel Corporation, 1968. (Chap. 1).

2.25. S. T. Rolfe. "Fracture and Fatigue Control in Steel Structures," *Engineering Journal,* AISC, **14,** 1 (1st Quarter 1977), 2–15.

2.26. K. A. Godfrey, Jr. "High Strength Steel: Crisis or No?", *Civil Engineering,* May 1985, 50–53.

2.27. John M. Barsom and Stanley T. Rolfe. *Fracture and Fatigue Control in Structures Applications of Fracture Mechanics,* 2nd ed. Englewood Cliffs, New Jersey: Prentice-Hall, Inc., 1987.

2.28. John M. Barsom. "Material Considerations in Structural Steel Design," *Proceedings, National Engineering Conference & Conference of Operating Personnel.* Chicago: American Institute of Steel Construction, April 29–May 2, 1987, 1-1 through 1-15.

2.29. J. M. Barsom. "Material Considerations in Structural Steel Design," *Engineering Journal,* AISC, **24,** 3 (3rd Quarter 1987), 127–139.

2.30. John C. Bittence. "The Basics of Heat Treating- What it Does- How it Works- Where to Specify it," *Machine Design,* January 24, 1974, 106–111; February 7, 1974, 117–121.

2.31. *A Primer on Brittle Fracture,* Booklet 1960-A, Steel Design File, Bethlehem Steel Corporation, Bethlehem, PA.

2.32. John W. Fisher and Alan W. Pense, "Experience with Use of Heavy W Shapes in Tension," *Engineering Journal,* AISC, **24,** 2 (2nd Quarter 1987), 63–77.

2.33. "The Use of Jumbo Shapes in Non-column Applications," *Engineering Journal,* AISC, **23,** 3 (3rd Quarter 1986), 96.

2.34. "Use of Heavy Structural Shapes in Tension Applications" *Construction Marketing Technical Bulletin TB-312,* Bethlehem Steel Corporation, June 1992.

2.35. E. J. Kaufman, A. W. Pense, and R. D. Stout. "An Evaluation of Factors Significant to Lamellar Tearing," *Welding Journal,* **60,** March 1981, Research Supplement, 43s–49s.

2.36. "Commentary on Highly Restrained Welded Connections," *Engineering Journal,* AISC, **10,** 3 (3rd Quarter 1973), 61–73.

2.37. Charles H. Thornton. "Quality Control in Design and Supervision Can Eliminate Lamellar Tearing," *Engineering Journal,* AISC, **10,** 4 (4th Quarter 1973), 112–116.

2.38. E. Holby and J. F. Smith. "Lamellar Tearing: The Problem Nobody Seems to Want to Talk About," *Welding Journal,* **59,** February 1980, 37–44.

2.39. "Causes and Prevention of Lamellar Tearing," *Civil Engineering,* April, 1982, 74–75.

2.40. P. D. Zuraski and J. E. Johnson, "Research on the Remaining Life in Steel Bridges," *Proceedings, ASCE Specialty Conference on Probabilistic Mechanics and Structural Reliability.* Berkeley, CA: January 11–13, 1984, 414–418.

2.41. G. M. Sinclair. "An Investigation of the Coaxing Effect in Fatigue of Metals," *Proceedings,* ASTM, **52** (1952), 743–758.

2.42. Seymour Coburn. "Theory and Practice in Use of Weathering Steels." *Proceedings, National Engineering Conference & Conference of Operating Personnel.* Chicago: American Institute of Steel Construction, April 29–May 2, 1987, 14-1 through 14-25.

2.43. C. P. Larrabee. "Corrosion Resistance of High-Strength Low-Alloy Steels as Influenced by Composition and Environment," *Corrosion,* **9,** August 1953, 259–271.

2.44. ASTM. *Standard Specification for High-Strength Low-Alloy Steel Shapes of Structural Quality, Produced by Quenching and Self-Tempering Process (QST)* (A913/A913M-93). West Conshohocken, PA: American Society for Testing and Materials, 1993.

3

Tension Members

3.1   INTRODUCTION

Tension members are encountered in most steel structures. They occur as principal structural members in bridge and roof trusses, in truss structures such as transmission towers and wind bracing systems in multistoried buildings. They frequently appear as tie rods to stiffen a trussed floor system or to provide intermediate support for a wall girt system. Tension members may consist of a single structural shape or they may be built up from a number of structural shapes. The cross-sections of some typical tension members are shown in Fig. 3.1.1.

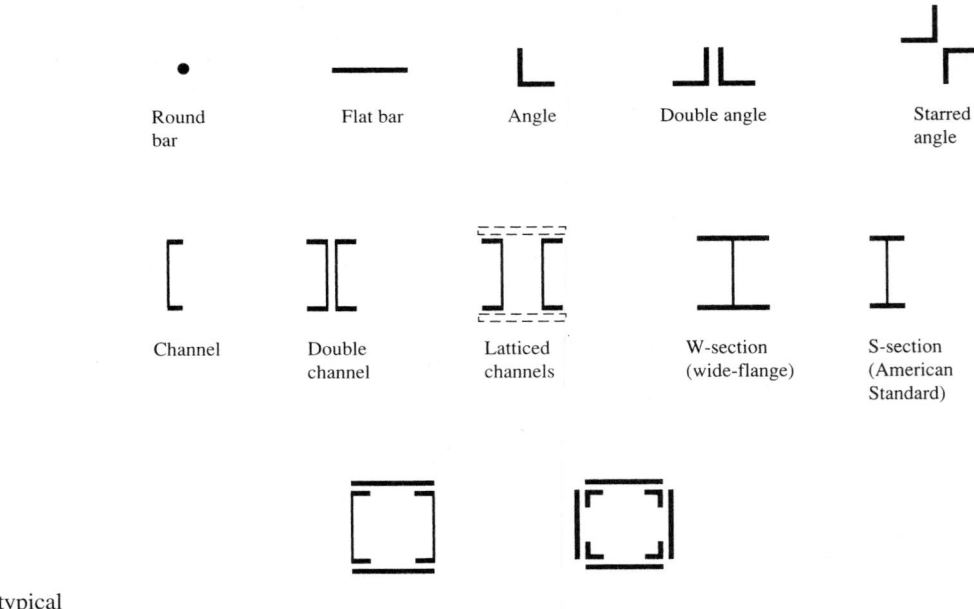

Figure 3.1.1
Cross-section of typical tension members.

Figure 3.1.2

Structural steel framework at intermediate floor level. This level suspended from roof space truss with tension rods. (Photo by C. G. Salmon)

3.2   NOMINAL STRENGTH

The strength of a tension member may be described in terms of the "limit states" that govern. The controlling strength limit state for a tension member will be one of the following: (a) yielding of the gross cross-section of the member away from the connection, (b) fracture of the effective net area (i.e., through the holes) at the connection [3.1], or (c) block shear fracture through the bolt holes at the connection.

When the limit state is general yielding of the gross section over the member length, typified by a tension member without holes (i.e., with welded connections), the nominal strength T_n may be expressed

$$T_n = F_y A_g \tag{3.2.1}$$

where F_y = yield stress

A_g = gross cross-sectional area

For tension members having holes, such as for bolts, the reduced cross-section is referred to as the *net area*. Holes in a member cause stress concentrations at service load as, for example, shown in Fig. 3.2.1a. Theory of elasticity shows that tensile stress adjacent to a hole

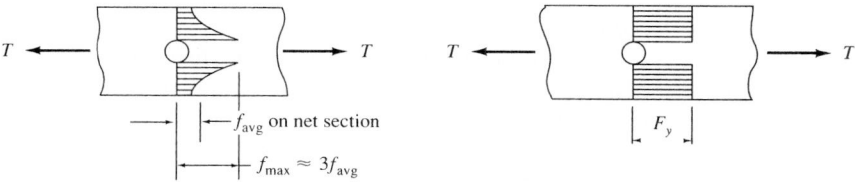

Figure 3.2.1
Stress distribution with holes present.

(a) Elastic stresses

(b) Nominal strength condition

will be about three times the average stress on the net area. However, as each fiber reaches yield strain $\epsilon_y = F_y/E_s$, its stress then becomes a constant F_y, with deformation continuing with increasing load until finally all fibers have achieved or exceeded the strain ϵ_y (Fig. 3.2.1b).

When the limit state is a localized yielding resulting in a fracture through the *effective* net area of a tension member having holes, the nominal strength T_n may be expressed

$$T_n = F_u A_e \qquad (3.2.2)$$

where　F_u = specified minimum tensile strength (see Fig. 2.1.1)

　　　　A_e = effective net area = $U A_n$ (see Secs. 3.4 and 3.5)

　　　　A_n = net area

　　　　U = reduction coefficient (an efficiency factor)

Because of *strain hardening,* that is, the rise in resistance when the tensile strain becomes large (Sec. 2.4), the actual strength of a ductile member may exceed that indicated by Eq. 3.2.1 [3.1]. However, the large elongations resulting from yielding of the member along its entire length may cause the member ends to move unacceptably far apart and distress the structure; thus, the member no longer serves its intended purpose. Either unrestrained yielding or fracture through reduced section at holes may limit the structural usefulness of the member. Traditionally, a higher margin of safety has been used in design when considering the fracture limit state than for the yielding limit state.

3.3　NET AREA

Whenever a tension member is to be fastened by means of bolts or rivets, holes must be provided at the connection. As a result, the member cross-sectional area at the connection is reduced and the strength of the member *may* also be reduced depending on the size and location of the holes.

Several methods are used to cut holes. The most common and least expensive method is to punch *standard* holes $\frac{1}{16}$ in. (1.6 mm) larger than the diameter of the rivet or bolt. In general, the plate thickness is less than the punch diameter. During the punching operation, the metal at the edge of the hole is damaged. This is accounted for in design by assuming that the extent of the damage is limited to a radial distance of $\frac{1}{32}$ in. (0.8 mm) around the hole. Therefore the total width to be deducted (AISC-D3.2) is to be taken as the nominal dimension of the *hole* normal to the direction of applied load plus $\frac{1}{16}$ in. (1.6 mm). For fasteners in standard holes, the total deduction is equal to the fastener diameter plus $\frac{1}{8}$ in. (3.2 mm).

A second method of cutting holes consists of subpunching them $\frac{3}{16}$ in. (4.8 mm) diameter undersize and then reaming the holes to the finished size after the pieces being joined are assembled. This method is more expensive than that of punching standard holes but does offer the advantage of accurate alignment. This method and the next method produce better strength, both static and fatigue, but this is ignored in design procedures.

A third method consists of drilling holes to a diameter of the bolt or rivet plus $\frac{1}{32}$ in. (0.8 mm). This method is used to join thick pieces, and is the most expensive of the common methods.

When greater latitude is needed in meeting dimensional tolerances during erection, larger than *standard* holes can be used with high strength bolts $\frac{1}{2}$ in. diameter and larger without adversely affecting the performance. The maximum sizes for *oversized, short-slotted,* and *long-slotted* holes are specified in AISC-J3.2.

EXAMPLE 3.3.1

What is the net area A_n for the tension member shown in Fig. 3.3.1?

Figure 3.3.1
Tension member for
Example 3.3.1.

Solution:

$$A_g = 4(0.25) = 1.0 \text{ sq in.}$$

$$\text{Width to be deducted for hole} = \tfrac{3}{4} + \tfrac{1}{8} = \tfrac{7}{8} \text{ in.}$$

$$A_n = A_g - (\text{width for hole})(\text{thickness of plate})$$

$$= 1.0 - 0.875 \,(0.25) = 0.78 \text{ sq in.} \qquad \blacksquare$$

3.4   EFFECT OF STAGGERED HOLES ON NET AREA

Whenever there is more than one hole and the holes are *not* lined up transverse to the loading direction, more than one potential failure line may exist. The controlling failure line is that which gives the largest stress on an effective net area. In many cases, the critical failure path is also the path that has the minimum net area.

In Fig. 3.4.1a the failure line is along the section *A–B*. In Fig. 3.4.1b showing two lines of staggered holes, the failure line might be through one hole (section *A–B*) or it might be along a diagonal path, *A–C*. At first glance one might think section *A–B* is critical since the path *A–B* is obviously shorter than path *A–C*. However, from path *A–B*, only one hole would be deducted while two holes would have to be deducted from path *A–C*. In order to determine the controlling section, both paths *A–B* and *A–C* must be investigated. Accurate checking of strength along path *A–C* is complex. However, a simplified empirical relationship proposed by Cochrane [3.2] has been adopted by AISC-D3.2 to account for the difference between the path *A–C* and the path *A–B*, expressed as a length correction,

$$\frac{s^2}{4g}$$

where *s* is the stagger, or spacing of adjacent holes parallel to the loading direction (see Fig. 3.4.1), and *g* is the gage distance transverse to the loading direction. Thus the net

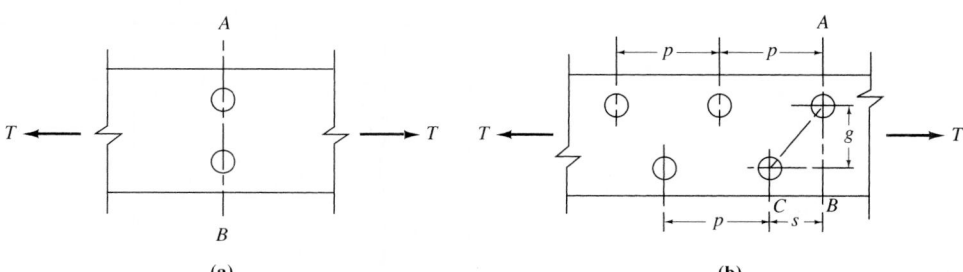

Figure 3.4.1
Paths of failure on net section.

lengths of paths $A–B$ and $A–C$ would be

$$\text{Net length of } A–B = \text{length of } (A–B) - \left(\text{width of hole} + \tfrac{1}{16} \text{ in.}\right)$$

$$\text{Net length of } A–C = \text{length of } (A–B) - 2\left(\text{width of hole} + \tfrac{1}{16} \text{ in.}\right) + \frac{s^2}{4g}$$

The minimum net area would then be determined from the minimum net length multiplied by the thickness of the plate.

In the years since Cochrane proposed the simple $s^2/4g$ expression, many investigators have proposed other rules [3.3–3.6] but none of them gives significantly better results, and all are more complicated.

Consistent with the general trend toward using strength-related design approaches, the work of Bijlaard [3.7] and others [3.8–3.10] has provided limit analysis theories to obtain net area in tension. These theories do not deviate from the $s^2/4g$ method by more than 10 to 15 percent.

The reader is referred to McGuire [3.11] for a more complete coverage of this subject of net section through staggered lines of fasteners.

EXAMPLE 3.4.1

Determine the minimum net area of the plate shown in Fig. 3.4.2, assuming $\tfrac{15}{16}$-in.-diam holes are located as shown.

Solution:
According to AISC-D3.2, the width used in deducting for holes is the hole diameter plus $\tfrac{1}{16}$ in., and the staggered length correction is $s^2/4g$.

Path AD (two holes):

$$\left[12 - 2\left(\frac{15}{16} + \frac{1}{16}\right) \right]0.25 = 2.50 \text{ sq in.}$$

Path ABD (three holes; two staggers):

$$\left[12 - 3\left(\frac{15}{16} + \frac{1}{16}\right) + \frac{(2.125)^2}{4(2.5)} + \frac{(2.125)^2}{4(4)} \right]0.25 = 2.43 \text{ sq in.}$$

Path ABC (three holes; two staggers):

$$\left[12 - 3\left(\frac{15}{16} + \frac{1}{16}\right) + \frac{(2.125)^2}{4(2.5)} + \frac{(1.875)^2}{4(4)} \right]0.25 = 2.42 \text{ sq in.}$$
$$\text{(controls)}$$

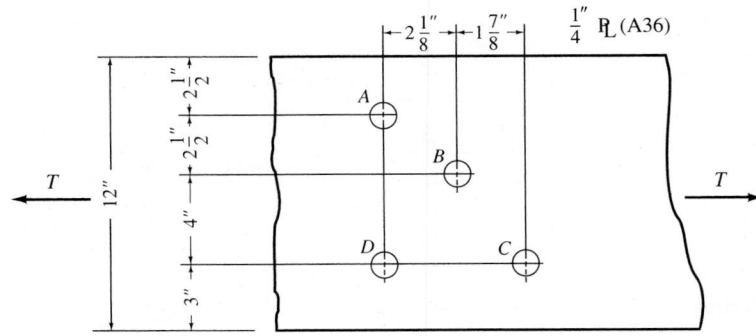

Figure 3.4.2
Example 3.4.1.

Angles

When holes are staggered on two legs of an angle, the gage length g for use in the $s^2/4g$ expression is obtained by using a length between the centers of the holes measured along the centerline of the angle thickness, i.e., the distance A–B in Fig. 3.4.3. Thus the gage distance g is

$$g = g_a - \frac{t}{2} + g_b - \frac{t}{2} = g_a + g_b - t \qquad (3.4.1)$$

Every rolled angle has a standard value for the location of holes (i.e., gage distances g_a and g_b), depending on the length of the leg and the number of lines of holes. Table 3.4.1 shows *usual gages* for angles as listed in the *AISC Manual**.

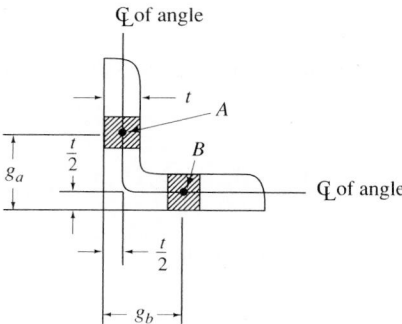

Figure 3.4.3
Gage dimensions for an angle.

TABLE 3.4.1 Usual Gages* for Angles, Inches (From AISC Manual[†])

Leg	8	7	6	5	4	$3\frac{1}{2}$	3	$2\frac{1}{2}$	2	$1\frac{3}{4}$	$1\frac{1}{2}$	$1\frac{3}{8}$	$1\frac{1}{4}$	1
g_1	$4\frac{1}{2}$	4	$3\frac{1}{2}$	3	$2\frac{1}{2}$	2	$1\frac{3}{4}$	$1\frac{3}{8}$	$1\frac{1}{8}$	1	$\frac{7}{8}$	$\frac{7}{8}$	$\frac{3}{4}$	$\frac{5}{8}$
g_2	3	$2\frac{1}{2}$	$2\frac{1}{4}$	2										
g_3	3	3	$2\frac{1}{2}$	$1\frac{3}{4}$										

*Other gages are permitted to suit specific requirements subject to clearances and edge distance limitations.
[†]*AISC Manual* [1.15], p. 1–46

EXAMPLE 3.4.2

Determine the net area A_n for the angle given in Fig. 3.4.4 if $\frac{15}{16}$-in.-diam holes are used.

L6 × 4 × $\frac{1}{2}$ (L152 × 102 × 12.7)*

$2\frac{1}{2}$

$2\frac{1}{4}$

$2\frac{1}{2}$

Figure 3.4.4
Example 3.4.2.

*legs and thickness in mm

Solution:

For net area calculation the angle may be visualized as being flattened into a plate as shown in Fig. 3.4.5:

$$A_n = A_g - \sum Dt + \sum \frac{s^2}{4g}t$$

where D is the width to be deducted for the hole.

Path *AC*:

$$4.75 - 2\left(\frac{15}{16} + \frac{1}{16}\right)0.5 = 3.75 \text{ sq in.}$$

Path *ABC*:

$$4.75 - 3\left(\frac{15}{16} + \frac{1}{16}\right)0.5 + \left[\frac{(3)^2}{4(2.5)} + \frac{(3)^2}{4(4.25)}\right]0.5 = 3.96 \text{ sq in.}$$

Since the smallest A_n is 3.75 sq in., that value governs.

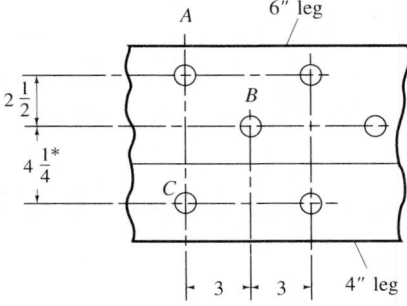

A   6" leg

$2\frac{1}{2}$

B

$4\frac{1}{4}$*

C

3   3   4" leg

Figure 3.4.5
Angle for Example 3.4.2 with
legs shown "flattened" into
one plane.

$$*g + g_1 - t = 2\frac{1}{2} + 2\frac{1}{4} - \frac{1}{2} = 4\frac{1}{4}$$

■

3.5   EFFECTIVE NET AREA

The net area as computed in Secs. 3.3 and 3.4 gives the reduced section that resists tension but still may not correctly reflect the strength. That is particularly true when the tension member has a profile consisting of elements not in a common plane and where the tensile load is transmitted at the end of the member by connection to some but not all of the elements. An angle section having connection to one leg only is an example of such a situation. For such cases the tensile force is not uniformly distributed over the net area.

AISC-D3.3 provides that the *effective net area* A_e be computed as

$$A_e = UA_n \tag{3.5.1}$$

where U = reduction coefficient

A_n = net area

The above equation logically applies for *both* fastener connections having holes and for welded connections. For welded connections, the net area equals the gross area A_g since there are no holes.

When a tensile load is applied eccentrically to a wide plate, the stress distribution across the width of the plate is nonuniform. The mechanism by which stress gets transmitted from the location of the applied load to sections distant from the load is by shear stresses acting in the plane of the plate. The fact that the stress is lower the farther the location is from the applied load means that the shear transfer "lags" or is inefficient. Thus, the nonuniformity of stress in wide plates or plate elements of rolled sections when a tensile load is applied nonuniformly, is referred to as "shear lag".

For *bolted connections,* the reduction coefficient U, the shear lag factor, relates to the eccentricity $\overline{x}$ of loading in the connection. Whenever tension is transmitted through some but not all of the cross-sectional elements, AISC-Table D3.1, Case 2, indicates the following shall be used for all tension members except plates and HSS (hollow structural sections),

$$U = 1 - \frac{\overline{x}}{L} \le 0.9 \tag{3.5.2}$$

where $\overline{x}$ = distance from centroid of element being connected eccentrically to plane of load transfer (see Fig. 3.5.1 and also AISC Commentary Figs. C-D3.1 for added guidance to determine $\overline{x}$.)

L = length of connection in the direction of loading

Equation 3.5.2 is based on the work of Munse and Chesson [3.9] and correlates within $\pm 10\%$ of tests [3.1]. A review of shear lag research for both bolted and welded connections, along with design recommendations is provided by Easterling and Giroux [3.16]. In addition to the reduction in strength due to shear lag, the efficiency of the fasteners is reduced in long (in the load direction) connections, as discussed in Chapter 4, where detailed treatment of bolted tension members occurs.

The approximate values of U (used prior to the 1993 LRFD Specification) for bolted connections to I-shaped sections, channels, and angles, are still considered acceptable, and have been moved from the Commentary to AISC Table D3.1, Cases 7 and 8. Use of the larger of the "exact" and approximate vales is permitted.

For short tension members (connecting elements), such as splice and gusset plates, where the elements of the cross-section lie essentially in a common plane, the effective net area is taken equal to A_n, but may not exceed 85% of the gross area A_g (AISC-J4.1). Tests [3.1] have shown that when any holes are present in such short elements where general

Figure 3.5.1
Eccentricity in joints;
determination of $\bar{x}$ for
computing U, Eq. 3.5.2.

yielding on the gross section cannot occur, there will be at least 15% reduction in strength from that obtained based on yielding of the gross section.

For *welded tension connections,* there are three categories:

1. Load transmitted by longitudinal welds, or by longitudinal welds in combination with transverse welds (AISC Table D3.1, Case 2)

$$A_e = UA_n = UA_g \qquad (3.5.3)$$

2. Load transmitted only to transverse welds:

$$A_e = UA_n = A_{con} \qquad (3.5.4)$$

where A_{con} = area of directly connected elements. In this case, the shear lag effect is approximated indirectly by using the reduced area A_{con}.

3. Load transmitted to a plate by longitudinal welds along both sides of the plate spaced apart such that $\ell \geq w$

$$A_e = UA_g \qquad (3.5.5)$$

where ℓ = length of weld
        w = distance between longitudinal welds (i.e., plate width)
        U = 1.00   For $\ell \geq 2w$
           = 0.87   For $2w > \ell \geq 1.5w$
           = 0.75   For $1.5w > \ell \geq w$

Welded connections for tension members are treated in Chapter 5 on welding.

EXAMPLE 3.5.1

Determine the reduction factor U to be applied in computing the effective net area for a W14×82 section connected by plates at its two flanges, as shown in Fig. 3.5.2. There are three bolts along along each connection line.

Solution:
In this case, two elements (the flanges) of the cross-section are connected but one (the web) is not connected. In accordance with Eq. 3.5.1, there is reduced efficiency of carrying load. The reduction factor U must be computed using Eq. 3.5.2. Because each flange connection can be thought of as a load on the tributary portion of the W shape, the section may be treated as two structural tees, as in Fig. 3.5.1c. The half W shape corresponds to a structural tee WT7×41, whose centroidal distance $\bar{x}$ is given by the *AISC Manual* as 1.39 in. The length ℓ of the connection is 6 in. Thus, Eq. 3.5.2 gives

$$U = 1 - \frac{1.39}{6.0} = 0.77$$

According to AISC Table D3.1, Case 7, W, M, or S shapes having flange widths not less than two-thirds of the depth, and structural tees cut from these shapes, $U = 0.90$ when there are at least three fasteners per line in the direction of stress. For this example,

$$\frac{b_f}{d} = \frac{\text{flange width}}{\text{section depth}} = \frac{10.13}{14.31} = 0.71 > 0.67 \qquad \textbf{OK}$$

Thus, the reduction factor U could be taken as 0.90.

Figure 3.5.2
Example 3.5.1.

3.6   BLOCK SHEAR STRENGTH

When thin plates are attached by bolts, a tearing limit state, known as *block* shear, may control the strength of a tension member, or the tension region at the end connection of a beam (see Chapter 13). Referring to Fig. 3.6.1a, the angle tension member attached to a gusset plate may have a tearing failure along the bolt holes, section *a-b-c*. The limit state is defined by rupture along *b-c* plane plus either yielding or rupture along plane *a-b*. Tests [3.12, 3.13] have shown it to be reasonable to add the strength in tension rupture in one plane to the shear rupture (or yielding) strength of the perpendicular plane. Welded connections could experience similar block shear failure along the periphery of the weld.

The four holes in the plate of Fig. 3.6.1b and c will contribute to a tear-out failure if the sum of the shear strengths along *a-b* and *c-d* plus the tensile strength along *b-c* is less than either of the strengths in general yielding of the member (Eq. 3.2.1) or rupture along *e-b-c-f* (Eq. 3.2.2).

Combination shear and tension tearing failures are uncommon in tension members; however, this combination mode frequently controls the design of bolted end connections to the thin webs of beams.

Tests on block shear failure in angle members [3.15] have shown that block shear controls for short connections (i.e., two or fewer connectors per gage line). These tests also indicate that shear lag is a factor in block shear failure of tension members, which is accounted for by the U_{bs} factor in the *AISC Specification*. Furthermore adding $s^2/4g$ to the net tensile area for block shear calculations on staggered bolts may be unconservative, depending on the direction of the stagger with respect to the failure path.

(a) Failure by tearing out

Figure 3.6.1
Tearing failure limit state.

(b) Large tension, small shear

(c) Large shear, small tension

AISC-J4.3. defines two block shear failure modes:

1. Rupture along the tensile plane (*b-c* in Fig. 3.6.1b) accompanied by yielding along the shear planes (*a-b* and *c-d* in Fig. 3.6.1b).

2. Rupture along the shear planes (*a-b* and *c-d* in Fig. 3.6.1c) accompanied by rupture along the tensile plane (*b-c* in Fig. 3.6.1c).

The tensile failure is defined by rupture along the net area in both modes, while the failure along the shear planes can either be rupture along the net area or yield along the gross area, whichever is smaller.

Consistent with the limit states discussed in Sec. 3.2, the gross area is used for the yielding limit state and the net area is used for the fracture limit state. In addition, following the energy-of-distortion theory (Eq. 2.6.5), the shear yield stress τ_y is taken as $0.6F_y$. Similarly, the shear strength τ_u is taken as $0.6F_u$.

The nominal strength T_n in tension is thus given by:

1. Shear yielding—tension rupture $(0.6F_yA_{gv} < 0.6F_uA_{nv})$

$$T_n = 0.6F_yA_{gv} + F_uU_{bs}A_{nt} \tag{3.6.1}$$

or

2. Shear fracture—tension rupture $(0.6F_yA_{gv} \geq 0.6F_uA_{nv})$

$$T_n = 0.6F_uA_{nv} + F_uU_{bs}A_{nt} \tag{3.6.2}$$

where A_{gv} = gross area acted upon by shear
A_{nt} = net area acted upon by tension
A_{nv} = net area acted upon by shear
F_u = specified (ASTM) minimum tensile strength
F_y = specified (ASTM) minimum yield stress

When the tension stress is uniform, use $U_{bs} = 1$, where the tension is non-uniform, use $U_{bs} = 0.5$.

Equation 3.6.1 indicates fracture (F_uA_{nt}) on the net tensile area followed by yielding $(0.6F_yA_{gv})$ along the shear planes. Equation 3.6.2 indicates fracture $(0.6F_uA_{nv})$ on the net shear area followed by rupture (F_uA_{nt}) on the net tensile area. The smaller strength along the shear planes defines the governing mode of failure. The reduction factor U_{bs} has been included in the equation to approximate the non-uniform stress distribution on the tensile plane. Such stress distribution occurs when multiple rows of bolts occur in a beam end connection.

3.7   STIFFNESS AS A DESIGN CRITERION

Even though stability is not a criterion in the design of tension members, it is still necessary to limit their length to prevent a member from becoming too flexible both during erection and final use of the structure. Tension members that are too long may sag excessively because of their own weight. In addition, they may also vibrate when subjected to wind forces as in an open truss or when supporting vibrating equipment such as fans or compressors.

To reduce the problems associated with excessive deflections and vibrations a stiffness criterion was established. This criterion is based on the slenderness ratio L/r of a member where L is the length and r the least radius of gyration $\left(r = \sqrt{I/A}\right)$. The *preferable* maximum slenderness ratio is 300 for members whose design is based on tensile force (AISC-D1). This limitation does not apply to rods in tension.

In applying the stiffness criterion to tension members, the highest slenderness ratio must be used. A symmetrical member may have two different radii of gyration, and for nonsymmetrical members one must consider the weakest principal axis. When a tension member is built up from a number of sections, the radius of gyration must be computed using the moment of inertia I and the cross-sectional area A. The value for r will be with respect to the same axis as that used to calculate the moment of inertia.

3.8   LOAD TRANSFER AT CONNECTIONS

Normally the holes in tension members are those for rivets or bolts to transfer load from one tension member into another.

Although the detailed treatment of fasteners and their behavior is in Chapter 4, the basic assumption is that each equal size fastener transfers an equal share of the load whenever the fasteners are arranged symmetrically with respect to the centroidal axis of a tension member. The following example is to illustrate the idea and its relationship to net area calculations.

EXAMPLE 3.8.1

Calculate the governing net area for plate A of the single lap joint in Fig. 3.8.1 and show free-body diagrams of portions of plate A with sections taken through each line of holes. Assume that plate B has adequate net area and does not control the strength T.

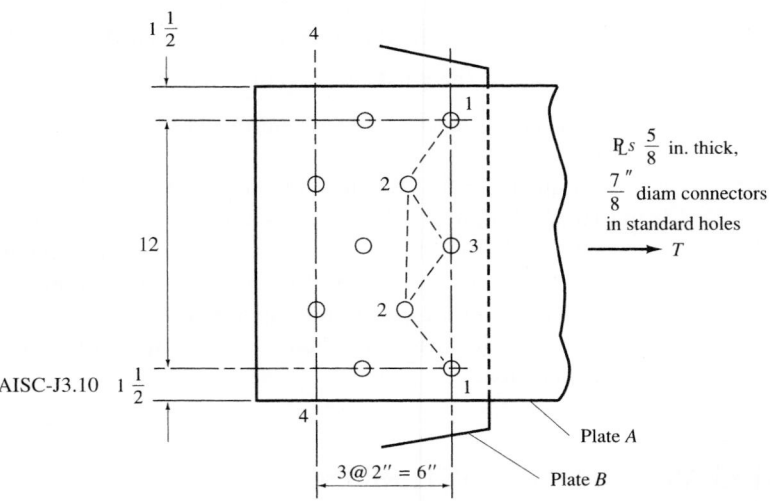

Figure 3.8.1
Single lap connection for
Example 3.8.1.

Solution:
The full tensile force T in plate A acts on section 1–1 of Fig. 3.8.1. Examination of other sections in plate A to the left of section 1–1 will involve *less than* 100% of T acting, since part of that force will have already been transferred from plate A to plate B. At section 4–4, 100% of T must now be acting in plate B while only 20% of T acts in plate A. Since there must be zero force acting on the end of plate A a short distance to the left of section 4–4, the force T must have been entirely transferred to plate B over the distance from sections 1–1 to 4–4. The free bodies of the various segments are shown in Fig. 3.8.2.

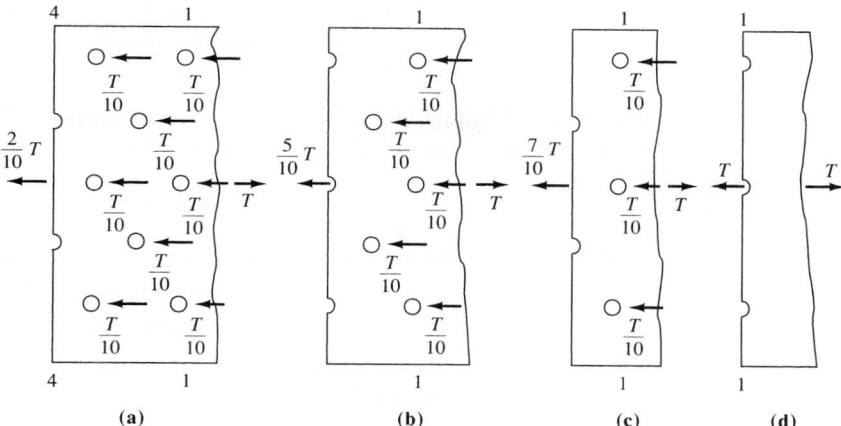

Figure 3.8.2
Load distribution in plate *A*.

(a)   (b)   (c)   (d)

Deduction in width for 1 hole = Diam of hole $+\frac{1}{16}$ in.

$\qquad$ = Diam of fastener $+\frac{1}{8}$ in.   for *standard* hole

$\qquad$ = $\frac{7}{8} + \frac{1}{8} = 1$ in.

Net area (section 1–1) $= \frac{5}{8}(15 - 3) = 7.50$ sq in.

on which 100% of *T* acts (Fig. 3.8.2d).

Net area (staggered path 1–2–3–2–1);

$$= \frac{5}{8}\left[15 - 5(1) + 4\frac{(2)^2}{4(3)}\right] = 7.08 \text{ sq in.}$$

$\qquad\qquad\qquad\qquad\qquad\qquad\qquad s^2/4g$

on which 100% of *T* also acts.

Net area (staggered path 1–2–2–1):

$$= \frac{5}{8}\left[15 - 4 + 2\frac{(2)^2}{4(3)}\right] = 7.29 \text{ sq in.}$$

on which 0.9 of *T* is presumed to act, since one connector has already transferred its share (0.10) of the load prior to reaching section 1–2–2–1. The 7.29 sq in. with 0.9*T* acting would compare with $7.29/0.9 = 8.10$ sq in. with *T* acting. A comparison of 7.50, 7.08, and 8.10 shows that section 1–2–3–2–1 governs; then $A_n = 7.08$ sq in.   ■

3.9   LOAD AND RESISTANCE FACTOR DESIGN—TENSION MEMBERS

The general philosophy of Load and Resistance Factor Design (LRFD) was described in Secs. 1.8 and 1.9. Equation 1.8.1 gives the structural safety requirement, as follows:

$$\phi R_n \geq [R_u = \Sigma \gamma_i Q_i] \qquad\qquad [1.8.1]$$

where ϕ = resistance factor (strength reduction factor)

$\qquad$ R_n = nominal strength (resistance)

$\qquad$ R_u = factored service load

$\qquad$ γ_i = overload factors (ASCE 7)

Q_i = service loads (such as dead load, live load, wind load, earthquake load) or load effects (such as bending moment, shear, axial force, and torsional moment resulting from various loads)

Equation 1.8.1 requires that the *design strength* ϕR_n equal or exceed the summation of *factored loads,* or specifically for tension members, Eq. 1.8.1 becomes

$$\phi_t T_n \geq T_u \tag{3.9.1}$$

where
ϕ_t = resistance factor relating to tension member strength
T_n = nominal strength of a tension member (see AISC-Chapter D)
$\phi_t T_n$ = design strength of a tension member
T_u = factored load on a tension member (critical load combination as per ASCE 7)

Since the loading is not related to the type of member, such as tension member or column, the various load factor combinations given in ASCE 7-05 are the same for all members in the structure. However, the resistance factor ϕ to account for the possibility that the actual strength of the member may be less than the theoretically computed strength because of variations in material properties and dimensional tolerances. These variations while individually within accepted tolerance limits may combine in the actual structure to give a strength less than the computed value. Neither the factor ϕ nor the overload factors γ are intended to account for careless errors in design or construction. The reliability of design of bolted steel tension members using LRFD has been reported by Bennett and Najem-Clarke [3.17].

According to AISC-D2, the design tension member strength $\phi_t T_n$ is the smaller of that based on *yielding in the gross section,*

$$\phi_t T_n = \phi_t F_y A_g = 0.90 F_y A_g \tag{3.9.2}$$

or *fracture in the net section,*

$$\phi_t T_n = \phi_t F_u A_e = 0.75 F_u A_e \tag{3.9.3}$$

Note that the resistance factor ϕ_t is 0.90 for the yielding limit state and 0.75 for the fracture limit state.

In addition, at the connection, the block shear strength must be evaluated. This involves evaluating the rupture strength (tension, shear, or a combination of both) along a potential tear-out path (see Sec. 3.6). The design strength requirements of AISC-J4.3 are:

1. Shear yielding—tension fracture $(0.6 F_y A_{gv} \leq 0.6 F_u A_{nv})$

$$\phi_t T_n = \phi_t (0.6 F_y A_{gv} + F_u U_{bs} A_{nt}) = 0.75(0.6 F_y A_{gv} + F_u U_{bs} A_{nt}) \tag{3.9.4}$$

or

2. Shear fracture—tension fracture $(0.6 F_y A_{gv} > 0.6 F_u A_{nv})$

$$\phi_t T_n = \phi_t (0.6 F_u A_{nv} + F_u U_{bs} A_{nt}) = 0.75(0.6 F_u A_{nv} + F_u U_{bs} A_{nt}) \tag{3.9.5}$$

where
A_{gv} = gross area acted upon by shear
A_{nt} = net area acted upon by tension
A_{nv} = net area acted upon by shear
F_u = specified (ASTM) minimum tensile strength
F_y = specified (ASTM) minimum yield stress

Note the fracture limit state ϕ_t of 0.75 is used in Eqs. 3.9.4 and 3.9.5.

TABLE 3.9.1 Tension Members—AISC Specification References

Topic	Specification section
Net area	D3.2
Effective net area	D3.3
Limiting slenderness ratio	D1
Tensile strength	D2
Built-up members	D4
Pin-connected member; eyebars	D5
Threaded rods	J3.1
Block shear	J4.3
Connecting elements	J4
Fatigue	B3.9

In the 2005 *AISC Manual* (Tables 9-1 and 9-2), the deduction for holes is computed in accordance with AISC-D3, par 2, the same as the tension fracture limit state for the member, as discussed in Sec. 3.3.

The AISC Specification [1.13] sections used in Load and Resistance Factor Design of tension members are summarized in Table 3.9.1.

EXAMPLE 3.9.1

Determine the service load capacity in tension for an L6×4×$\frac{1}{2}$ of A572 Grade 50 steel connected with $\frac{7}{8}$-in.-diam bolts in standard holes as shown in Fig. 3.9.1. Use AISC Load and Resistance Factor Design, and assume the live load to dead load ratio is 3.0.

Solution:

The angle tension member is connected to a *gusset plate,* typical of truss joints. The gusset plate is the plate at the intersection of members to which they are connected.

The maximum strength will be based on either section 1–1 with one hole deducted, or on the staggered section 1–2 through two holes. The governing section will have 100% of load *T* acting on it.

For section 1–1,

$$A_n = A_g - 1 \text{ hole} = 4.75 - \left(\frac{7}{8} + \frac{1}{8}\right)0.50 = 4.25 \text{ sq in.}$$

For section 1–2,

$$A_n = A_g - 2 \text{ holes} + (s^2/4g)t$$

$$= 4.75 - 2\left(\frac{7}{8} + \frac{1}{8}\right)0.50 + \frac{(2)^2}{4(2.5)}(0.50) = 3.95 \text{ sq in.}$$

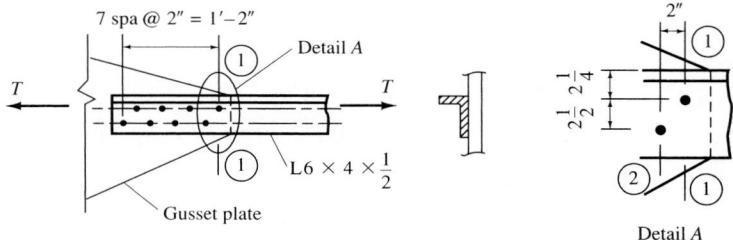

Figure 3.9.1
Tension member for
Example 3.9.1.

The two design strengths to be used in accordance with AISC-D2 are given by Eq. 3.9.2 based on *general yielding* on the gross section of the member,

$$\phi_t T_n = \phi_t F_y A_g \qquad [3.9.2]$$
$$= 0.90(50)(4.75) = 214 \text{ kips}$$

and by Eq. 3.9.3 based on *fracture* at the connection,

$$\phi_t T_n = \phi_t F_u A_e \qquad [3.9.3]$$

The angle does not have both legs connected to transfer the tensile force, so the effective net area is less than the computed net area, accounting for the eccentricity at the connection.

The effective net area A_e is

$$A_e = U A_n$$

where, according to AISC-D3.3 Case 2 the reduction coefficient U is to be computed from Eq. 3.5.2:

$$U = 1 - \frac{\bar{x}}{L} \qquad [3.5.2]$$

where the distance $\bar{x}$ from centroid of element being connected eccentrically to plane of load transfer (see Fig. 3.5.1) is for the case of the angle 0.987 in. The length L of the joint is 14 in. Equation 3.5.2 then gives

$$U = 1 - \frac{\bar{x}}{L} = 1 - \frac{0.987}{14} = 0.93$$
$$A_e = A_n U = 3.95(0.93) = 3.67 \text{ sq in.}$$

Thus, Eq. 3.9.3 gives

$$\phi_t T_n = \phi_t F_u A_e = 0.75(65)(3.67) = 179 \text{ kips}$$

Thus, the controlling $\phi_t T_n$ is the smaller of the values from Eqs. 3.9.2 (214 kips) and 3.9.3 (179 kips),

$$\phi_t T_n = 179 \text{ kips}$$

The overload factors relate the design strength to the service loads or load effects. Using the gravity load combination Eq. 1.8.3 [ASCE7-05], the factored load T_u is

$$T_u = \sum \gamma_i Q_i = 1.2D + 1.6L + 0.5(L_r \text{ or } S \text{ or } R) \qquad [1.8.3]$$

where in this example the roof loading L_r, S (snow), and R (rain) are not involved. The live load (L) is given as three times dead load (D). Thus, applying the safety requirement, Eq. 3.9.1; that is, letting $\phi_t T_n$ equal factored load T_u gives

$$\phi_t T_n = 1.2D + 1.6L = 1.2D + 1.6(3D) = 6.0D$$
$$179 = 6.0D$$
$$D = 29.8$$
$$L = 3D = 3(29.8) = 89.4 \text{ kips}$$

The total safe service load T is

$$T = D + L = 29.8 + 89.4 = 119 \text{ kips}$$

If this angle connection consisted of a very few large fasteners or if either the gusset plate or the angle were thin elements, the block shear rupture limit state of AISC-J4.3 represented by Eqs. 3.9.4 through 3.9.7, might give a lower strength than the lesser of Eqs. 3.9.2 and 3.9.3. ∎

EXAMPLE 3.9.2

Investigate the block shear failure mode on the angle L4×4×$\frac{1}{4}$ attached with three $\frac{7}{8}$-in. diam bolts to a $\frac{3}{8}$-in. gusset plate, as shown in Fig. 3.9.2. The material is A36 steel.

Solution:
The usual general yielding and fracture limit states governed by the lesser of Eqs. 3.9.2 and 3.9.3 gives

$$\phi_t T_n = \phi_t F_y A_g = 0.90(36)1.94 = 62.9 \text{ kips}$$

$$\phi_t T_n = \phi_t F_u A_e = \phi_t F_u U A_n$$

where

$$U = 1 - \frac{\overline{x}}{L} = 1 - \frac{1.09}{6.0} = 0.82$$

Thus,

$$\phi_t T_n = \phi_t F_u U A_n$$
$$= 0.75(58)(0.82)(1.94 - 0.25) = 60.3 \text{ kips}$$

The block shear potential failure along path *a-b-c* of Fig. 3.9.2 must be investigated according to AISC-J4.3. Calculating the net areas A_{nv} and A_{nt},

$$A_{nv} = (\text{length } a\text{-}b \text{ less } 2.5 \text{ holes}) \times \text{thickness}$$

$$= \left[7.5 - 2.5\left(\frac{7}{8} + \frac{1}{8}\right) \right] 0.25 = 1.25 \text{ sq in.}$$

$$A_{nt} = (\text{length } b\text{-}c \text{ less } 0.5 \text{ holes}) \times \text{thickness}$$

$$= \left[1.5 - 0.5\left(\frac{7}{8} + \frac{1}{8}\right) \right] 0.25 = 0.25 \text{ sq in.}$$

Compare $0.6F_y A_{gv}$ with $0.6F_u A_{nv}$,

$$[0.6F_y A_{gv} = 0.6(36)1.875 = 40.5] < [0.6F_u A_{nv} = 0.6(58)1.25 = 43.5]$$

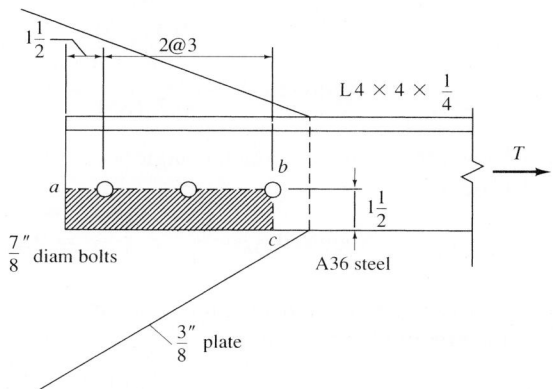

Figure 3.9.2
Tension member of
Example 3.9.2.

which means Eq. 3.6.1 controls

$$T_n(R_n \text{ as per AISC Chap. J}) = 0.6F_y A_{gv} + F_u U_{bs} A_{nt}$$

$$T_n = 40.5 + 58(1.0)0.25 = 55 \text{ kips} \qquad [3.6.1]$$

Thus, block shear strength is

$$\phi_t T_n = 0.75(55) = 41.3 \text{ kips}$$

and block shear (41.3 kips) controls over yielding on gross section (62.9 kips), or fracture on effective net section (60.3 kips). ∎

EXAMPLE 3.9.3

Select a tension diagonal member for a roof truss of A572 Grade 50 steel using AISC Load and Resistance Factor Design. The axial tension is 60 kips dead load and 6 kips live load and the member is 12 ft long. Assume $\frac{7}{8}$-in.-diam bolts are located on a single gage line in standard holes. Assume the preferable limit on slenderness ratio L/r is 240 (*not* an AISC limit).

    a. Select the lightest single angle member.
    b. Select the lightest double angle member having legs separated by $\frac{1}{4}$ in. back-to-back.

Solution:
For tension members, the design strength requirement is

$$\phi_t T_n \geq T_u$$

where the factored load T_u may be governed by Eq. 1.8.3 in general, or Eq. 1.8.2 when the proportion of dead load is large, as in this case; thus,

$$T_u = 1.2D + 1.6L = 1.2(60) + 1.6(6) = 82 \text{ kips}$$

or

$$T_u = 1.4D = 1.4(60) = 84 \text{ kips (controls)}$$

In this case, the factored load to be designed for is 84 kips.
    The strength of the members may be controlled by either

$$\phi_t T_n = \phi_t F_y A_g = 0.90(50)A_g$$

or

$$\phi_t T_n = \phi_t F_u A_e = 0.75(65)A_e$$

The design strength ϕT_n must equal the factored load T_u. The area requirements become

$$\text{Required } A_g = \frac{84}{0.90(50)} = 1.87 \text{ sq in.}$$

$$\text{Required } A_e = \frac{84}{0.75(65)} = 1.72 \text{ sq in.}$$

Estimating U for either a single or double angle section to be 0.8, in accordance with AISC Table D3.1 Case 8,

$$\text{Required } A_n = \frac{A_e}{U} = \frac{1.72}{0.8} = 2.15 \text{ sq in.}$$

The net area requirement obviously controls since it exceeds the gross area requirement. Also, the preferred minimum r to satisfy the given limitation of $L/r = 240$ is

$$\min r = \frac{L}{240} = \frac{12(12)}{240} = 0.6 \text{ in.}$$

(a)   Select single angle member. The required gross area in each case depends on the area deducted for one hole, which in turn depends on the thickness. The following tabular procedure may be found useful in making the selection:

Standard thickness t	Deduction for one hole	Required gross area	Choices from AISC manual single angle properties	
$\frac{5}{16}$	0.313^*	$2.46^\dagger$	$L5 \times 3\frac{1}{2} \times \frac{5}{16}$	$A = 2.56, r = 0.76$
$\frac{3}{8}$	0.375	2.53	$L4 \times 3\frac{1}{2} \times \frac{3}{8}$	$A = 2.67, r = 0.72$
$\frac{7}{16}$	0.438	2.59	$L3\frac{1}{2} \times 3 \times \frac{7}{16}$	$A = 2.65, r = 0.62$
$\frac{1}{2}$	0.500	2.65		

$*\ \left(\frac{7}{8} + \frac{1}{8}\right)0.3125 = 0.3125$ sq in.

$\dagger$ Required $A_g = 2.15 + 0.31 = 2.46$ sq in.

$\ddagger$ *Note:* Min $r = r_z$ for single angles.

Use $L5 \times 3\frac{1}{2} \times \frac{5}{16}$ single angle member (least area, therefore lightest).

(b)   Select double angle member. For this type of section two holes must be deducted. Selection should be made from the double angle properties given in *AISC Manual.*

Standard thickness t	Deduction for two holes	Required gross area	Choices from AISC manual double angle properties	
$\frac{1}{4}$	0.500	2.65	$L3 \times 3 \times \frac{1}{4}$	$A = 2.87, r = 0.93$
			$L3 \times 2\frac{1}{2} \times \frac{1}{4}$	$A = 2.64, r = 0.94$
$\frac{5}{16}$	0.625	2.78		
$\frac{3}{8}$	0.750	2.90	$L2\frac{1}{2} \times 2\frac{1}{2} \times \frac{5}{16}$	$A = 2.93, r = 0.76$

Use $2 - L3 \times 2\frac{1}{2} \times \frac{1}{4}$ with long legs back-to-back, with less than 1% understrength.   ■

3.10   TENSION RODS

A common and simple tension member is the threaded rod. Such rods are usually secondary members where the required strength is small, such as (a) sag rods to help support purlins in industrial buildings (Fig. 3.10.1a); (b) vertical ties to help support girts in industrial building walls (Fig. 3.10.1b); (c) hangers, such as tie rods supporting a balcony (Fig. 3.10.1c); and (d) tie rods to resist the thrust of an arch.

Tie rods are frequently used with an initial tension as diagonal wind bracing in walls, roofs, and towers. The initial tension effectively adds to the stiffness and reduces deflection and vibrational motion, which tends to cause fatigue failures in the connections. Such initial tension can be obtained by designing the member something on the order of $\frac{1}{16}$ in. short for a 20-ft length or by the use of turnbuckles which can be tightened after construction.

Figure 3.10.1
Uses of tension rods.

(a) Roof truss

(b) Wall system

(c)

EXAMPLE 3.10.1

Select the diameter for a threaded round rod of A36 steel to carry an axial tension of 6 kips live load and 3 kips dead load. Use AISC Load and Resistance Factor Design.

Solution:
The design strength of a threaded rod is given by AISC-J3.6 (Table J3.2) as

$$\phi_t T_n = 0.75 A_b F_n = 0.75 A_b (0.75 F_u) \tag{a}$$

The factored load T_u to be carried is

$$T_u = 1.2D + 1.6L = 1.2(3) + 1.6(6) = 13.2 \text{ kips}$$

Using the minimum tensile strength F_u for A36 steel as 58 ksi from Table 2.1.1 (or AISC Table 2-5), and equating the factored load T_u to the design strength $\phi_t T_n$, gives the required gross area A_b from Eq. (a) as

$$\text{Required } A_b = \frac{\text{Required } \phi_t T_n}{0.75(0.75 F_u)} = \frac{13.2}{0.75(0.75)(58)} = 0.40 \text{ sq in.}$$

Compute the diameter of the threaded rod based on the required area A_b. The area computed is the gross area A_b based on the diameter of the unthreaded body of the rod (AISC-Table J3.2).

Use $\frac{3}{4}$-in.-diam rod (10 threads per inch) ($A_b = 0.442$ sq in.). ■

EXAMPLE 3.10.2

Design sag rods to support the purlins of the industrial building roof of Fig. 3.10.2. Sag rods are spaced at the third points between roof trusses, which are spaced 24 ft apart. Use 20 psf snow load, A36 steel, and AISC LRFD Method.

Solution:

(a)   Loads. Assume cold-formed steel roofing is used, weighing 3 psf, and that the purlins have already been designed. Their weight may be approximated as a 3.5 psf roof load.

Snow load customarily is prescribed as having an intensity given in pounds per square foot (psf) of horizontal projection. Generally, a value not less than 20 psf is used, with 30 to 40 psf (1.4 to 1.9 kN/m^2) being used in northern areas (see Sec. 1.4).

The horizontal projection of the roof area is 25 cos 25° over which the snow load acts. Because the other loads on the roof are given in terms of the roof area, the snow load can be converted to a load per square foot of roof area by multiplying by cos 25°,

$$20(\cos 25°) = 18.1 \text{ psf of roof}$$

Figure 3.10.2
Roof and sag rods for
Example 3.10.2.

(b) Compute factored load on roof. The structure weight is dead load. Because there are no live loads other than snow, Eq. 1.8.4 will control,

$$1.2D + 1.6(L_r \text{ or } S \text{ or } R) \qquad\qquad [1.8.4]$$

$$1.2(3 + 3.5) + 1.6(18.1) = 36.8 \text{ psf}$$

(c) Compute factored load to be carried by a single rod. Because the rods are spaced 8 ft apart, the load carried by each rod is that load over a 4 ft strip of roof on either side of the rod. The sag rod indicated in Fig. 3.10.2 must carry the load for the entire length (i.e., 25 ft) of roof. The vertical factored load associated with a line of sag rods is

$$36.8 \text{ times tributary area (in sq ft)}$$

$$36.8(25)(8)/1000 = 7.4 \text{ kips}$$

The top rod parallel to the roof carries only that component of the load parallel to the roof, equal to

$$7.4(\sin 25°) = 7.4(0.423) = 3.1 \text{ kips}$$

The component of the vertical load perpendicular to the roof is carried by beam action in the purlins. The horizontal rod at the roof peak, indicated in Fig. 3.10.2, must carry $3.1/(\cos 25°)$,

$$T_u = 3.1/(\cos 25°) = 3.1/(0.906) = 3.4 \text{ kips}$$

(d) Select the diameter or rod. The strength requirement of AISC-J3.6 is

$$\phi_t T_n \geq T_u$$

$$\phi_t T_n = 0.75(0.75 F_u) A_b$$

Obtaining F_u from Table 2.1.1 (or AISC Table 2-4), and equating the factored load T_u to the design strength $\phi_t T_n$, gives the required area A_b as

$$\text{Required } A_b = \frac{T_u}{0.75(0.75 F_u)} = \frac{3.4}{0.75(0.75)(58)} = 0.11 \text{ sq in.}$$

Select a $\frac{3}{8}$-in. diam threaded rod having $A_b = 0.11$ sq in.

Use $\frac{3}{8}$-in.-diam rod ($A_b = 0.110$ sq in.). ∎

3.11 ALLOWABLE STRENGTH DESIGN—TENSION MEMBERS

The general philosophy of Allowable Strength Design (ASD) was described in Secs. 1.8 and 1.9. Equation 1.8.8 gives the structural safety requirement, as follows:

$$\left[\frac{\phi R_n}{\gamma} = \frac{R_n}{\Omega} \right] \geq \Sigma Q_i \qquad\qquad [1.8.8]$$

which expresses that the nominal strength R_n divided by a factor of safety Ω must exceed the sum of the factored service loads. The nominal strength divided by the factor Ω is called the "allowable strength." In the past, this procedure was called "allowable stress design." Now (2008) the equations are multiplied by the properties (area, section modulus, etc.) so that stresses are converted to forces (axial force, bending moment, torsional moment, etc.).

The nominal strength T_n for tension members may be controlled by either of Eqs. 3.2.1 or 3.2.2,

$$T_n = F_y A_g \qquad\qquad [3.2.1]$$

$$T_n = F_u A_e \qquad\qquad [3.2.2]$$

where A_g = gross cross-sectional area

A_e = effective net area = $U A_n$ (see Sec. 3.5)

U = reduction coefficient (efficiency factor)

The allowable strength in tension, according to AISC-D2, is the smaller of that based on yielding in the gross section,

$$\frac{T_n}{\Omega} = \frac{F_y A_g}{\Omega} = \frac{F_y A_g}{1.67} \qquad\qquad (3.11.1)$$

or fracture in the net section,

$$\frac{T_n}{\Omega} = \frac{F_u A_e}{\Omega} = \frac{F_u A_e}{2.00} \qquad\qquad (3.11.2)$$

or block shear strength along the tear out path,

$$\frac{T_n}{\Omega} = \frac{(0.6F_y A_{gv} + U_{bs} F_u A_{nt})}{\Omega} = \frac{(0.6F_y A_{gv} + U_{bs} F_u A_{nt})}{2.00} \qquad\qquad (3.11.3)$$

Note that the safety factor Ω is 1.67 for the yielding limit state, 2.00 for the fracture limit state, and 2.00 for the block shear limit state (which is primarily a fracture limit state).

EXAMPLE 3.11.1

Redo Example 3.9.1 using the AISC Allowable Strength Design (ASD) Method.

Solution:

The controlling allowable strength is the smaller value of

(a) $T_n/\Omega = F_y A_g/\Omega = 50(4.75)/1.67 = 142$ kips

or

(b) $T_n/\Omega = F_u A_e/\Omega = 65(3.67)/2.00 = 119$ kips   **Governs!**

The total allowable service load is then,

$$\Sigma Q = D + L = D + 3D = 4.0D$$

which means

$$119 = 4D$$

$$D = 29.8 \text{ kips}$$

$$L = 3D = 3(29.8) = 89 \text{ kips} \qquad\qquad \blacksquare$$

EXAMPLE 3.11.2

Re-do Example 3.9.2 using the Allowable Strength Method (ASD).

Solution:
The nominal strengths have already been calculated in the LRFD solution. The calculated value of T_n was not actually given in Ex. 3.9.2.

$$69.8 = 36(1.94) \qquad \textbf{OK}$$

$$80.4 = 58(0.82)(1.94 - 0.25) \qquad \textbf{OK}$$

The governing allowable strength is the smallest of the three allowable strengths.

  (a) Allowable yield strength $= T_n/\Omega = 69.8/1.67 = 41.8$ kips

  (b) Allowable fracture strength $= T_n/\Omega = 80.4/2.00 = 40.2$ kips

  (c) Allowable block shear strength $= T_n/\Omega = 55/2.00 = 27.5$ kips **Governs!**

Thus, block shear (27.5 kips) controls over yielding on the gross section (41.8 kips) or fracture on effective net section (40.2 kips). ∎

EXAMPLE 3.11.3

Re-do Example 3.9.3a using the Allowable Strength Design Method (ASD).

Solution:
The total service tension load is

$$\Sigma Q = D + L = 60 + 6 = 66 \text{ kips}$$

The allowable strength of the member may be controlled by the yield strength,

$$T_n/\Omega = F_y A_g/\Omega = 50A_g/1.67 = 29.9A_g$$

or by the fracture strength on effective net area,

$$T_n/\Omega = F_u A_e/\Omega = 65A_e/2.00 = 32.5A_e$$

The allowable strength must equal or exceed the service load ΣQ. Thus,

$$\text{Required } A_g = \frac{66}{29.9} = 2.21 \text{ sq in.}$$

$$\text{Required } A_e = \frac{66}{32.5} = 2.03 \text{ sq in.}$$

Estimating U for either the single or double angle section to be 0.85,

$$\text{Required } A_n = \frac{A_e}{U} = \frac{2.03}{0.85} = 2.39 \text{ sq in.}$$

The net area requirement obviously controls since it exceeds the gross area requirement.

Standard thickness t	Deduction for one hole	Required gross area	Choices from *AISC Manual* single angle properties	
$\frac{5}{16}$	0.313^*	$2.70^\dagger$	L5×5×$\frac{5}{16}$	$A = 3.07, r = 1.56^\ddagger$
$\frac{3}{8}$	0.375	2.77	L5×3×$\frac{3}{8}$	$A = 2.86, r = 1.02$
$\frac{7}{16}$	0.438	2.83	L4×4×$\frac{3}{8}$	$A = 2.86, r = 1.23$
$\frac{1}{2}$	0.500	2.89	L3$\frac{1}{2}$×3$\frac{1}{2}$×$\frac{7}{16}$	$A = 2.89, r = 1.06$

$^*\left(\frac{7}{8} + \frac{1}{8}\right)0.3125 = 0.3125$ sq in.

† Required $A_g = 2.39 + 0.31 = 2.70$ sq in.

‡ Min $r = r_z$ for single angles.

Use L4×4×$\frac{3}{8}$ single angle member. ∎

SELECTED REFERENCES

3.1. Geoffrey L. Kulak, John W. Fisher, and John H. A. Struik. *Guide to Design Criteria for Bolted and Riveted Joints,* 2nd ed. New York: John Wiley & Sons, 1987.

3.2. V. H. Cochrane. "Rules for Rivet Hole Deductions in Tension Members," *Engineering News-Record,* **89,** November 16, 1922, 847–848.

3.3. W. M. Wilson. Discussion of "Tension Tests of Large Riveted Joints," *Transactions,* ASCE, **105** (1940), 1268.

3.4. W. M. Wilson, W. H. Munse, and M. A. Cayci. "A Study of the Practical Efficiency under Static Loading of Riveted Joints Connecting Plates." *Bulletin 402,* U. of Illinois Engineering Experiment Station, Urbana, IL, 1952.

3.5. F. W. Schutz. "Effective Net Section of Riveted Joints," *Proceedings,* Second Illinois Structural Engineering Conference, November, 1952.

3.6. "Here's a Better Way to Design Splices," *Engineering News-Record,* **150,** Part I, January 8, 1953, 41.

3.7. P. P. Bijlaard. Discussion of "Investigation and Limit Analysis of Net Area in Tension," *Transactions,* ASCE, **120** (1955), 1156–1163.

3.8. G. W. Brady and D. C. Drucker. "Investigation and Limit Analysis of Net Area in Tension," *Transactions,* ASCE, **120** (1955), 1133–1154.

3.9. W. H. Munse and E. Chesson, Jr. "Riveted and Bolted Joints: Net Section Design," *Journal of the Structural Division,* ASCE, **89,** ST2 (February 1963), 107–126.

3.10. E. Chesson and W. H. Munse. "Behavior of Riveted Connections in Truss-Type Members," *Journal of the Structural Division,* ASCE, **83,** STI (January 1957), Paper 1150, 1–61; also *Transactions,* ASCE, **123** (1958), 1087–1128.

3.11. William McGuire. *Steel Structures.* Englewood Cliffs, NJ: Prentice-Hall, Inc., 1968 (pp. 310–328).

3.12. James M. Ricles and Joseph A. Yura. "Strength of Double-Row Bolted-Web Connections," *Journal of Structural Engineering,* ASCE, **109,** 1 (January 1983), 126–142.

3.13. Steve G. Hardash and Reidar Bjorhovde. "New Design Criteria for Gusset Plates in Tension," *Engineering Journal,* AISC, **22,** 2 (2nd Quarter 1985), 77–94.

3.14. Geoffrey L. Kulak, and Gilbert Y. Grodin. "AISC Rules for Block Shear–A Review," *Engineering Journal,* AISC, **39,** 4 (4th Quarter), 2001, 199–203.

3.15. Howard I. Epstein. "An Experimental Study of Block Shear Failure of Angles in Tension," *Engineering Journal,* AISC, **29,** 2 (2nd Quarter 1992), 75–84.

3.16. W. Samuel Easterling and Lisa Gonzales Giroux. "Shear Lag Effects in Steel Tension Members," *Engineering Journal,* AISC, **30,** 3 (3rd Quarter), 1993, 77–89.

3.17. Richard M. Bennett and F. Shima Najem-Clarke. "Reliability of Bolted Steel Tension Members," *Journal of Structural Engineering,* ASCE, **113,** 8 (August 1987), 1865–1872.

PROBLEMS

All problems are to be done according to the AISC LRFD Method or the ASD Method, as indicated by the instructor. Assume fastener strength is adequate and does not control. All holes are *standard* holes. Values of yield stress F_y and tensile strength F_u are available in Table 2.1.1. For all problems where the total number of holes is not known, assume rupture strength of AISC-J4 does not control. Where needed, assume distances from center of hole to end of piece are $1\frac{1}{2}$ in. Assume all given loads are service loads.

3.1. Compute the maximum acceptable tensile service load that may act on a single angle L6×4×$\frac{3}{4}$ that is welded along only one leg to a gusset plate; thus, there are no holes. The service live load is three times the dead load. Solve for (a) A36 steel and (b) A572 Grade 50 steel.

3.2. Compute the maximum acceptable tensile service load the angle in Prob. 3.1 may carry when connected on both legs. The 4-in. leg contains a single gage line of $\frac{7}{8}$-in.-diam bolts, and the 6-in. leg contains a double gage line of $\frac{7}{8}$-in.-diam bolts. Assume no stagger of bolts, and that all bolts participate in carrying load.

3.3. Compute the maximum acceptable service load on an A36 steel plate tension member $\frac{1}{4}$ in.×12 in. having a single line of holes parallel to the direction of loading. The load is 25% dead load and 75% live load, and $\frac{7}{8}$-in.-diam bolts are used.

3.4. Compute the net area A_n for the plate (a connecting element according to AISC-J5) shown in the accompanying figure. Then compute the maximum

value for service load T when A36 steel is used, the live load is four times the dead load, and the holes are $\frac{13}{16}$ in. diameter.

3.5. Repeat Prob. 3.4 using A572 Grade 60 steel and $\frac{15}{16}$-in.-diam holes.

3.6. Select a pair of angles to support a tensile live load (LL) and dead load (DL) for the case assigned by the instructor. Assume the angles are separated back-to-back $\frac{3}{8}$ in. by a connected gusset plate, and that the connection is welded. Assume the slenderness ratio is desired to not exceed 300.

Case	DL (kips)	LL (kips)	Steel	Length (ft)
1	70	20	A36	20
2	65	22	A36	30
3	70	20	A572 Gr 60	20
4	48	30	A36	22
5	50	30	A572 Gr 50	22
6	80	30	A36	20
7	80	100	A572 Gr 50	28

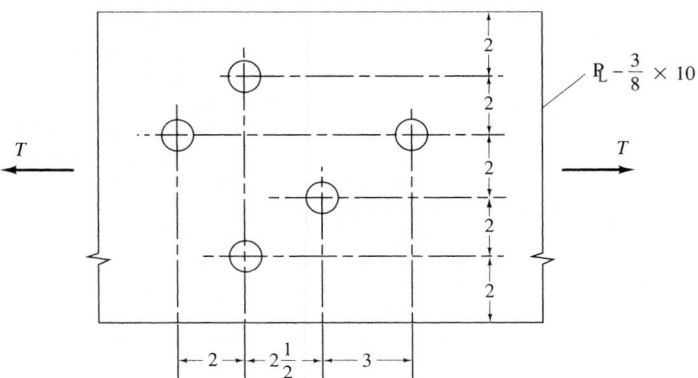

Problems 3.4 and 3.5

3.7. Select a single angle (for the case assigned by the instructor) to support a tensile load. A single gage line of at least three bolts is to be used.

Case	DL (kips)	LL (kips)	Steel	Bolt diameter (in.)	Length (ft)
1	15	40	A36	3/4	15
2	15	40	A572 Gr 50	3/4	15
3	15	40	A36	3/4	20
4	15	40	A572 Gr 60	7/8	25
5	10	30	A36	7/8	20
6	10	30	A572 Gr 50	7/8	20
7	12	35	A36	7/8	18
8	12	35	A572 Gr 60	7/8	18

3.8. Select a standard threaded rod to carry a tensile force T of 4 kips dead load and 6 kips live load. Use A572 Grade 50 steel.

3.9. Select a standard threaded rod to carry a tensile force T of 2 kips dead load and 4 kips live load. Use A36 steel.

3.10. Design sag rods to support the purlins of an industrial building roof whose span and slope are shown in the accompanying figure. Sag rods are placed at $\frac{1}{3}$ points between roof trusses, which are spaced 30 ft apart. Assume roofing and purlin weight is 9 psf of roof surface. Use standard threaded rods and A36 steel. The snow load to be carried is 20, 30 or 40 psf of horizontal projection, whichever is appropriate for your locale.

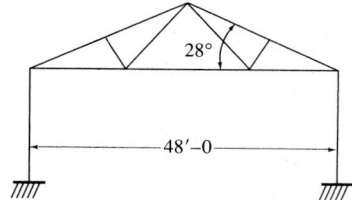

Problem 3.10

3.11. Determine the maximum allowable tensile load (20% dead load, 80% live load) for a single C15×33.9 fastened to a $\frac{1}{2}$-in. gusset plate as in the accompanying figure. Use A36 steel and assume holes are for $\frac{3}{4}$-in.-diam bolts. Base answer on tension strength of the channel, and include shear rupture strength.

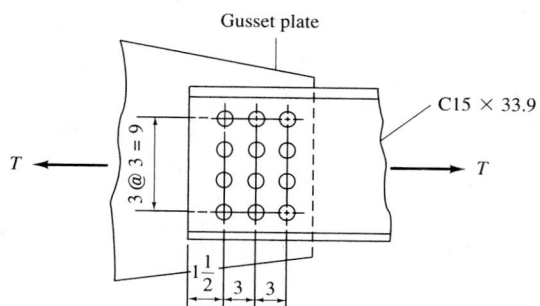

Problem 3.11

3.12. Repeat Prob. 3.11 using a C10×25 attached to a $\frac{5}{8}$-in. gusset plate. Assume the 12 bolts are in 3 lines parallel to the direction of loading, with the same 3 in. spacing.

3.13. Determine the tensile load (85% live load; 15% dead load) permitted by AISC for a pair of angles L6×4×$\frac{3}{8}$ attached to a gusset plate as shown in the accompanying figure. Use A36 steel and $\frac{3}{4}$-in.-diam bolts on standard gage lines whose distances are given in Table 3.4.1. The force T is transmitted to the gusset plate by fasteners on lines A and B; assume only open holes in the 4-in. (outstanding) legs.

Problem 3.13

3.14. Repeat Prob. 3.13 using a pair of L8×6×$\frac{3}{4}$ angles with staggered $\frac{7}{8}$-in.-diam bolts in the 8-in. leg.

3.15. Given the splice shown in the accompanying figure:
(a) Determine the maximum capacity T (25% dead load, 75% live load) based on the A36 steel plates having holes arranged as shown.

(b) What value of s should be specified to provide the maximum capacity T as computed in part (a), if the final design is to have $s_1 = s_2 = s$?

Problem 3.15

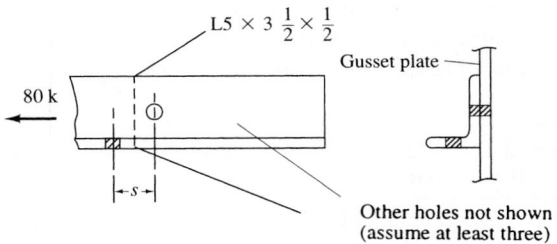

Problem 3.17

3.16. An L5×3$\frac{1}{2}$×$\frac{1}{2}$ angle, as shown in the accompanying figure, is to carry 20 kips dead load and 70 kips live load using the shortest length of connection using two gage lines of bolts in the 5-in. leg. What is the minimum acceptable stagger, theoretical and specified ($\frac{1}{2}$-in. multiples), using A572 Grade 50 steel?

3.18. Compute the minimum value of s that could theoretically be used on the angle of the accompanying figure such that the maximum factored tensile force T_u may be carried. Assume m is large enough so that a failure pattern through the open hole will not govern. Include consideration of shear rupture strength.

Problem 3.16

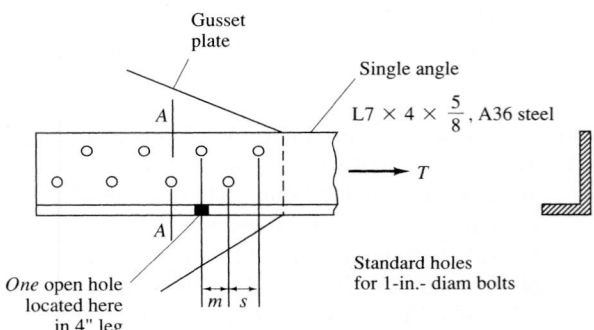

Problems 3.18 and 3.19

3.17. An L5×3$\frac{1}{2}$×$\frac{1}{2}$ angle, as shown in the accompanying figure, is to carry 20 kips dead load and 60 kips live load. Using one gage line of holes for $\frac{7}{8}$-in.-diam bolts in each leg, what would be the minimum stagger s required to accomplish this? Consider the load to be transferred by bolts in the 5-in. leg, while the holes in the 3$\frac{1}{2}$-in. leg may be considered open ones (i.e., not to transmit the tensile load). Use A36 steel.

3.19. Assuming that s for the angle of Prob. 3.18 is made as large as required in Prob. 3.18, compute the minimum distance m required so that the open hole in the 4-in. leg will not reduce the strength below its maximum possible value. If Prob. 3.18 is not solved, assume s is 3.75 in.

3.20. Design an eyebar to carry 24 kips dead load and 76 kips live load, using flame-cut A572 Grade 50 steel plate. (Refer to AISC-D6.)

4

Structural Bolts

4.1   TYPES OF BOLTS

Every structure is an assemblage of individual parts or members that must be fastened together, usually at the member ends. Welding is one method and is treated in Chapter 5. The other method is to use bolts. This chapter is concerned with bolting; in particular, high-strength bolts. High-strength bolts have replaced rivets as the means of making non-welded structural connections. However, for completeness, a brief description of the other fasteners, including rivets and unfinished machine bolts, is given.

High-Strength Bolts

The two basic types of high-strength bolts are designated as ASTM A325 [2.5] and A490 [2.7], the material properties of which are discussed in Sec. 2.2 and summarized in Table 4.1.1. These bolts are heavy hexagon-head bolts, used with heavy semifinished hexagon nuts, as shown in Fig. 4.1.1b. The threaded portion is shorter than for bolts in nonstructural applications, and may be cut or rolled. A325 bolts are of heat-treated *medium carbon* steel having an approximate yield strength of 81 to 92 ksi (560 to 630 MPa) depending on diameter. A490 bolts are also heat-treated but are of *alloy* steel having an approximate yield strength of 115 to 130 ksi (790 to 900 MPa) depending on diameter. A449 bolts are occasionally used when diameters over $1\frac{1}{2}$ in. up to 3 in. are needed, and also for anchor bolts and threaded rods.

High-strength bolts range in diameter from $\frac{1}{2}$ to $1\frac{1}{2}$ in. (3 in. for A449). The most common diameters used in building construction are $\frac{3}{4}$ in. and $\frac{7}{8}$ in., whereas the most common sizes in bridge design are $\frac{7}{8}$ in. and 1 in.

High-strength bolts are usually tightened to develop a specified tensile stress in them, which results in a predictable clamping force on the joint. The actual transfer of service loads through a joint is, therefore, due to the friction developed in the pieces being joined. Joints containing high-strength bolts are designed either as *slip-critical* (formerly called *friction-type*), where high slip resistance at service load is desired; or as *bearing-type,* where high slip resistance at service load is unnecessary.

TABLE 4.1.1 Properties of Bolts

ASTM designation	Bolt diameter in. (mm)	Proof load,[a] length measurement[b] method, ksi (MPa)	Proof load,[a] yield strength[c] method, ksi (MPa)	Minimum tensile strength, ksi (MPa)
A307 [2.4], low-carbon steel Grades A and B	$\frac{1}{4}$ to 4 (6.35 to 102)	—	—	60 (415)
A325 [2.5], high-strength steel Types 1, 2, and 3	$\frac{1}{2}$ to 1 (12.7 to 25.4)	85 (585)	92 (635)	120 (825)
Types 1, 2, and 3	$1\frac{1}{8}$ to $1\frac{1}{2}$ (28.6 to 38.1)	74 (510)	81 (560)	105 (725)
A449 [2.6], quenched and tempered steel	$\frac{1}{4}$ to 1 (6.35 to 25.4)	85 (585)	92 (635)	120 (825)
(*Note:* AISC[d] permits use only for bolts larger than $1\frac{1}{2}$ in. and for threaded rods and anchor bolts)	$1\frac{1}{8}$ to $1\frac{1}{2}$ (28.6 to 38.1) $1\frac{3}{4}$ to 3 (44.5 to 76.2)	74 (510) 55 (380)	81 (560) 58 (400)	105 (725) 90 (620)
A490 [2.7], quenched and tempered alloy steel	$\frac{1}{2}$ to $1\frac{1}{2}$ (12.7 to 38.1)	120 (825)	130 (895)	150 (1035)

[a]Actual proof load and tensile load obtained by multiplying given stress value by the tensile stress area A_s; $A_s = 0.785 \, [d_b - (0.9743/n)]^2$, where A_s = stress area in square inches, d_b = nominal diameter of bolt in inches, and n = number of threads per inch.
[b]0.5% extension under load.
[c]0.2% offset value.
[d]AISC-J3.1

Rivets

For many years rivets were the accepted means of connecting members but today (2008) they are virtually obsolete in the United States. Undriven rivets are formed from bar steel, a cylindrical shaft with a head formed on one end, as shown in Fig. 4.1.1a. Rivet steel is a mild carbon steel designated by ASTM as A502 Grade 1 ($F_y = 28$ ksi) (195 MPa) and Grade 2 ($F_y = 38$ ksi) (260 MPa), with the minimum specified yield strengths based on bar stock as rolled. The forming of undriven rivets and the driving of rivets cause changes in the mechanical properties.

Figure 4.1.1
Types of fasteners.   (a)   (b)   (c)

Installation requires heating the rivet to a light cherry-red color, inserting it into a hole and then applying pressure to the preformed head while at the same time squeezing the plain end of the rivet to form a rounded head. During this process the shank of the rivet completely or nearly fills the hole into which it had been inserted. Upon cooling, the rivet shrinks, thereby providing a clamping force. However, the amount of clamping produced by the cooling of the rivet varies from rivet to rivet and therefore cannot be counted on in design calculations.

Unfinished Bolts

Bolts of low-carbon steel designated ASTM A307 [2.4] are the least expensive bolt. They may not, however, produce the least expensive connection since more are required in a particular connection. Their primary use is in light structures, secondary or bracing members, platforms, catwalks, purlins, girts, small trusses, and similar applications in which the loads are primarily small and static in nature. Such bolts are also used as temporary fitting-up fasteners in cases where high-strength bolts, rivets, or welding may be the permanent means of connection. Unfinished bolts are sometimes called common, machine, or rough bolts and may come with square heads and square nuts.

Ribbed Bolts

Bolts of ordinary rivet steel having a rounded head and raised ribs parallel to the shank were used for many years as an alternative to rivets. The actual diameter of a ribbed bolt is slightly larger than the hole into which it is driven. In driving, the bolt actually cuts into the edges around the hole, producing a relatively tight fit. The ribbed bolt was particularly useful in bearing-type connections and in connections that had stress reversals.

A modern variation of the ribbed bolt is the *interference-body bolt* shown in Fig. 4.1.1c, which is of A325 bolt steel and, instead of longitudinal ribs, has serrations around the shank as well as parallel to the shank. Because of the serrations around the shank through the ribs, this bolt is often called an *interrupted-rib* bolt. Ribbed bolts are difficult to drive when several layers of plates are to be connected. The A325 interference-body bolt may also be more difficult to insert through several plates; however, it is used when tight fit of the bolt in the hole is desired, and it permits tightening the nut without simultaneously holding the bolt head as may be required with smooth loose-fitting, ordinary A325 bolts. These bolts are, however, rarely used in ordinary steel structures.

4.2   HISTORICAL BACKGROUND OF HIGH-STRENGTH BOLTS

The first experiments indicating the possibility of using high-strength bolts in steel-framed construction were reported by Batho and Bateman [4.1] in 1934. They concluded that bolts with a minimum yield strength of 54 ksi (370 MPa) could be relied on to prevent slip between the connected parts. Follow-up tests by Wilson and Thomas [4.2] substantiated the earlier work by reporting that high-strength bolts smaller in diameter than the holes in which they were inserted had fatigue strengths equal to that of well-driven rivets, provided that the bolts were sufficiently pretensioned.

The next major step occurred in 1947 with the formation of the Research Council on Riveted and Bolted Structural Joints. This organization began by using and extrapolating information from studies of riveted joints–in particular, the extensive annotated 1945 bibliography by De Jonge [4.3].

By 1950 the concept of using high-strength bolts and a summary of research and behavior was presented [4.4] to practicing engineers and the steel-fabrication industry. The first Specification in 1951 permitted replacement of rivets with bolts on a one-to-one basis. It was conservatively assumed that friction transfer of the load was necessary in all joints under service load conditions. The factor of safety against slip was established at a high enough level so that good fatigue resistance (i.e., no slip under varying stress or stress reversal occurring during many load cycles) was provided in every joint, similar to or better than that shown by riveted joints.

In 1954 a revision was made in the Specification to include the use of flat washers on 1:20 sloping surfaces and to allow the use of impact wrenches for installing bolts. Also, the 1954 revision permitted the surfaces in contact to be painted when the bolts were used in a *bearing-type* connection; i.e., when the strength of the connection was to be based on the bolt in bearing against the side of the hole.

In 1956 Munse [4.5] summarized bolt behavior and concluded that bolts must have as high an initial tension as practicable if high-strength bolts were to be efficient and economical. By 1960 the minimum bolt tension was increased. The *bearing-type* connection was recognized as an acceptable substitute for a riveted connection, and the connection designed on the basis of slip resistance, known then as a *friction-type* connection, would probably only be necessary when direct tension acts on the bolts or when stress reversals occur.

Also, in 1960 a simple installation procedure, known as the *turn-of-the-nut method,* was introduced as an alternative to the torque wrench method previously required. The economics of high-strength bolting further improved when only one washer [located under the element (head or nut) being turned], instead of the previously required two, could be used if the turn-of-the-nut method was used.

In 1962 the requirement for washers was eliminated except in special situations [4.6]. In 1964 the higher strength A490 bolt was introduced.

The 1985 and 1988 Specifications of the Research Council on Structural Connections (RCSC) [4.7, 4.8] changed the philosophy of design for bearing-type connections and for friction-type connections, now called *slip-critical* connections. A detailed review of the historical background is provided in the RCSC 2004 Commentary [4.9]. The *Guide to Design Criteria for Bolted and Riveted Joints* by Kulak, Fisher, and Struik [3.1] (hereinafter referred to as the *Guide*) summarizes research and makes recommendations which generally form the basis for current design of bolted connections.

4.3 CAUSES OF RIVET OBSOLESCENCE

Riveting is a method of connecting members at a joint by inserting ductile metal pins into holes in the pieces being joined and forming a head at each end to prevent the joint from coming apart. Typical types of rivets are shown in Fig. 4.3.1 (see also Fig. 4.1.1a).

Riveting required a crew of four or five experienced persons. On the other hand, the crews required for high-strength bolt installation do not need to be highly skilled. Inspection was difficult, and cutting out and replacing bad rivets was an expensive procedure. Even the preheating immediately prior to driving is critical in developing the necessary tightness after cooling.

The principal factor that delayed immediate acceptance of high-strength bolts was the high cost of the material including two hardened washers. In the early 1950s

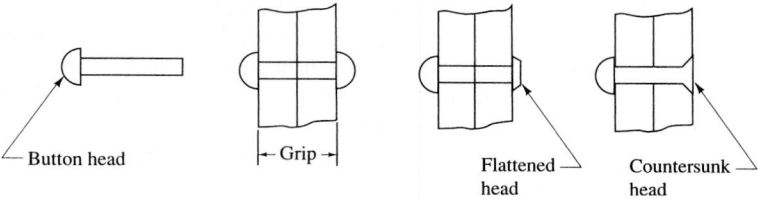

Figure 4.3.1
Types of rivets.

the reduced labor cost for installing bolts did not offset the higher bolt material cost. After the washers could be reduced to one or eliminated and the greater strength of a bolt over that of a rivet could be utilized in design, high-strength bolts became economical. Now (2008) with even higher labor cost and connection design generally requiring fewer bolts than would be required for rivets, the economy is clearly with high-strength bolts.

Welding, as treated in Chapter 5, has played an important role in reducing the use of all fasteners, both rivets and bolts.

4.4   DETAILS OF HIGH-STRENGTH BOLTS

Both the most commonly used A325 bolt and the occasionally used A490 bolt are heavy hexagon (hex) head bolts with heavy hexagon nuts, identified by the ASTM designation on the top of the head as shown in Fig. 4.4.1.

Heavy hex bolts have shorter threaded portions than other bolts; this reduces the probability of having threads occur where maximum strength is required across the shank of the bolt. The heavy hex bolt and nut are shown in Fig. 4.4.1. Requirements for marking bolts and nuts, including manufacturer's identification, are given in Fig. 4.4.2.

The A325 and A490 bolts are available as Types 1, 2, or 3. The A325 Type 1 is the medium-carbon steel bolt that has been available for many years and is the one that would be supplied if not otherwise specified. Similarly for A490, Type 1 is the usual alloy steel bolt. Type 2 for both A325 and A490 is a low-carbon martensite steel alternative to Type 1 for atmospheric temperature applications. For elevated temperature applications, Type 1 must be used. Type 3 for both A325 and A490 is a weathering steel bolt having corrosion resistance comparable to A588 weathering steel (see Table 2.1.1).

Occasionally, ASTM A449 bolts are used where diameters larger than $1\frac{1}{2}$ in. are required.

Proof Load and Bolt Tension—Slip-Critical Connections

Until the 1985 RCSC Specification [4.8], all high-strength bolts were required to be installed with a sufficient *pretension* force to create as high a compression force as practical between the pieces being connected, such that shear forces were transmitted through connections by friction between the connected pieces. That Specification relaxed the *pretension* requirement when bolts are not subject to direct tension and slip resistance between connected pieces is not required [4.8, Sec. 5].

When slip resistance is required, the pretensioning should be as high as possible without chancing permanent deformation or failure of the bolt. Bolt material exhibits a stress-strain (load-deformation) behavior that has no well-defined yield point, as shown in

Figure 4.4.1
Heavy hex structural bolt and
heavy hex nut. (From Ref. 4.9)

Nut may be chamfered
on both faces

Bolt / Nut	Type 1		Type 3
ASTM A325 bolt	XYZ A325 — Three radial lines 120° apart are optional		XYZ A325
ASTM F1852 bolt	XYZ A325 C — Three radial lines 120° apart are optional		XYZ A325 C
ASTM A490 bolt	XYZ A490		XYZ A490
ASTM A563 nut	XYZ — Arcs indicate grade C	XYZ 3 — Arcs with "3" indicate grade C3	XYZ D — Grade mark D
	XYZ DH — Grade mark DH		XYZ DH3 — Grade mark DH3

Figure 4.4.2
Required marks for
acceptable bolt and nut
assemblies. (From Ref. 4.9)

1. XYZ represents the manufacturer's identification mark.
2. ASTM F1852 twist-off-type tension-control bolt assemblies are
   also produced with heavy-hex head that has similar markings.

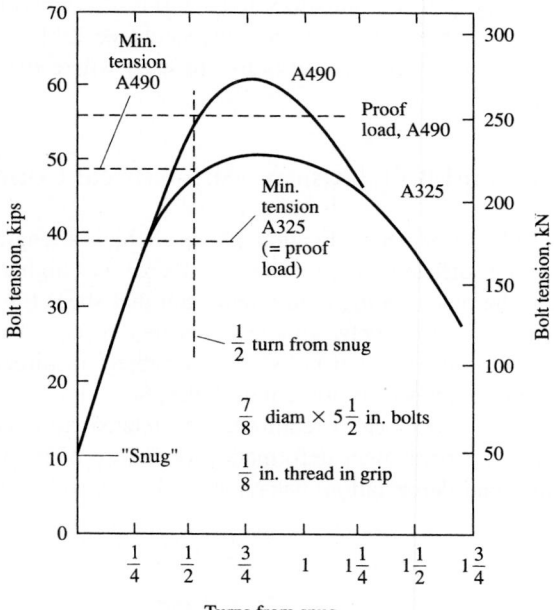

Figure 4.4.3
Typical load vs nut rotation
relationships for A325 and
A490 bolts. (From Kulak,
Fisher, and Struik [3.1])

Fig. 4.4.3. Instead of directly using a yield stress, a so-called *proof load* is used. The *proof load* is the load obtained by multiplying the tensile stress area* by a yield stress established by using either a 0.2% offset strain or a 0.5% extension under load (see Sec. 2.4). The ASTM tabulates this proof load for each diameter fastener using, for example, for $\frac{1}{2}$- to

---

*Tensile stress area $= 0.785\left(d_b - \dfrac{0.9743}{n}\right)^2$ where n = number of threads per inch

TABLE 4.4.1 Minimum Bolt Pretension* for
Fully-Tightened Bolts
(AISC Table J3.)

Bolt size		A325 bolts		A490 bolts	
(in.)	(mm)	(kips)	(kN)	(kips)	(kN)
1/2	12.7	12	53	15	67
5/8	15.9	19	85	24	107
3/4	19.1	28	125	35	156
7/8	22.2	39	173	49	218
1	25.4	51	227	64	285
1 1/8	28.6	56	249	80	356
1 1/4	31.8	71	316	102	454
1 3/8	34.9	85	378	121	538
1 1/2	38.1	103	458	148	658

*Equal to 70% of minimum tensile strength of bolts, rounded off
to the nearest kip, as specified in ASTM Specifications for A325
and A490 bolts for UNC (unified standard *coarse* screw threads
under ANSI B1.1; see *AISC Manual* [1.15], J3.1).

1-in.-diam bolts a strain offset value of 92 ksi (630 MPa) and a length measurement value of 85 ksi (590 MPa). The proof load stress is approximately a minimum of 70% and 80% of the minimum tensile strengths for A325 and A490 bolts, respectively.

Since the early 1950s, the minimum required pretension equals the proof load for A325 bolts. Using the turn-of-the-nut installation method (discussed in Sec. 4.5) no difficulty is encountered in obtaining proof load for these bolts with $\frac{1}{2}$ turn of the nut from snug position, as shown in Fig. 4.4.3. With the A490 bolt, however, the $\frac{1}{2}$ turn from snug may not achieve the proof load. Also for long bolts, more than $\frac{1}{2}$ turn from snug will be required to achieve the same tension as for short bolts.

AISC (Table J3.1) requires slip-critical connections to be pretensioned to 70% of the minimum tensile strength, as given in Table 4.4.1. This equals the proof load for A325 bolts and about 85 to 90% of proof load for A490 bolts.

The magnitude of pretension that is desirable and necessary has been the subject of considerable study by researchers [3.1].

4.5   INSTALLATION PROCEDURES

Connections Not Requiring Full Pretensioning

When slip-resistant connections are not required [4.9] and when the bolts are not subject to direct tension, bolts are permitted to be tightened "snug tight". This is a tightness that exists when all plies in a joint are in firm contact, which is further defined as the result of "a few impacts of an impact wrench or the full effort of an ironworker using an ordinary spud wrench".

Connections Requiring Full Precompression

There are four general methods for installing high-strength bolts to obtain the pretension indicated in Table 4.4.1. These are the *turn-of-the-nut tightening, calibrated wrench tightening, installation of alternative design bolts,* and *direct tension indicator tightening* [4.9].

The *turn-of-the-nut method* is the simplest. Developed in the 1950s and 1960s, this method obtains the specified pretension by a *specified rotation* of the nut from the "snug

Figure 4.5.1
Bolt elongations in a typical test joint. (From Fisher, Ramseier, and Beedle [4.11])

tight" condition, which causes a specified strain in the bolt. Although snugness or initial tightness can vary due to the surface condition of the pieces being tightened, this variation does not significantly affect the clamping force, as may be seen from Fig. 4.5.1. The clamping force of 48.6 kips (220 kN) corresponding to a $\frac{1}{2}$ turn-of-the-nut occurs at a sufficiently large bolt elongation (i.e., along the horizontal portion of the curve) that any variation in the initial snug tightness has an insignificant effect on the clamping force.

Calibrated wrench tightening uses manual torque wrenches and power wrenches adjusted to stall at a specified torque. For a given torque, variations in bolt tension have been found [3.1, p. 52] to range as high as ±30% with an average variation of ±10%. The RCSC Specification [4.9] therefore requires that calibrated wrenches be set to produce a bolt tension at least 5% in excess of the values specified in Table 4.4.2. Furthermore, calibrated wrenches must be calibrated at least *daily* and a hardened washer must be used under the element (head or nut) being tightened.

One may wonder whether there is danger of having inadequate reserve strength if the pretension exceeds the proof load; i.e., when it approaches 90% of tensile strength. Figure 4.5.2 shows the effect of various turns of the nut with the margin of safety indicated. If the calibrated wrench method is used, *strength* is the critical factor, with the typical safety margin shown in Fig. 4.5.2. The possibility of overtorquing the bolts with

Figure 4.5.2
A325 bolt behavior. (Adapted from Rumpf and Fisher [4.12])

TABLE 4.5.1 Nut Rotation[a] from Snug Tight Condition (From Ref. 4.9, Table 8.2)

Bolt Length[c]	Disposition of Outer Face of Bolted Parts		
	Both faces normal to bolt axis	One face normal to bolt axis, other sloped not more than 1:20[d]	Both faces sloped not more than 1:20 from normal to bolt axis[d]
Not more than $4d_n$	$\frac{1}{2}$ turn	$\frac{1}{2}$ turn	$\frac{2}{3}$ turn
More than $4d_b$ but not more than $8d_b$	$\frac{1}{2}$ turn	$\frac{2}{3}$ turn	$\frac{5}{6}$ turn
More than $8d_b$ but not more than $12d_b$	$\frac{2}{3}$ turn	$\frac{5}{6}$ turn	1 turn

[a] Nut rotation is relative to bolt regardless of the element (nut or bolt) being turned. For required nut rotations of $\frac{1}{2}$ turn and less, the tolerance is plus or minus 30 degrees; for required nut rotations of $\frac{2}{3}$ turn and more, the tolerance is plus or minus 45 degrees.

[b] Applicable only to joints in which all material within the grip is steel.

[c] When the bolt length exceeds $12d_b$, the required nut rotation shall be determined by actual testing in a suitable tension calibrator that simulates the conditions of solidly fitting steel.

[d] Beveled washer not used.

power wrenches is not considered a problem since such overtorquing usually fractures the bolts and they are replaced during installation. In the turn-of-the-nut method, *deformation* is the critical factor with the typical safety margin shown in Fig. 4.5.2. For either installation process, one can expect a minimum of $2\frac{1}{4}$ turns from snug to fracture. When the turn-of-the-nut method is used and bolts are tensioned using $\frac{1}{8}$ turn increments, frequently as many as four turns may be obtained from snug to fracture. The turn-of-the-nut method is the cheapest, is more reliable, and is generally preferred over the calibrated wrench method. The approved [4.9] nut rotations are indicated in Table 4.5.1.

The third general category of installation technique, the *installation of alternative design bolts,* uses proprietary fasteners designed to indirectly indicate the bolt tension or automatically provide the required tension. A procedure for qualifying such bolts is prescribed [4.9]. Sometimes the alternative design feature is a twist-off or yielding-type element.

The fourth category is the *direct tension indicator tightening.* Again, a procedure is specified [4.9] to qualify such devices. Typically, a hardened washer is used containing a number of protrusions on one face. The washer is inserted between the element being turned (head or nut) and the gripped material with the protrusions bearing against the underside of the element leaving a gap maintained by the protrusions. Upon tightening the bolt, the protrusions are flattened and the gap is reduced. The bolt tension is determined by measuring (using a feeler gage) the remaining gap, which for properly tensioned bolts will be about 0.015 in. (0.38 mm) or less [4.13].

In all cases, installation must begin at the most rigid part of the connection and progress systematically toward the least rigid areas.

4.6   NOMINAL STRENGTH OF INDIVIDUAL FASTENERS

Loads are transferred from one member to another by means of the connection between them. A few typical connections are shown in Fig. 4.6.1.

(a) Shear connections

Lap joint
(single shear)

Butt joint
(double shear)

(b) Eccentric shear
connection

(c) Tension connections

W section

Structural tee

(d) Combined shear and tension connections

Figure 4.6.1
Typical bolted connections.

The simplest device for transferring load from one steel piece to another is with a pin (a cylindrical piece of steel) inserted in holes that are aligned in the two pieces as shown in Fig. 4.6.2. The cotter pins shown would prevent the pin from sliding out. Load would be transferred by bearing of the shank of the pin against the side of the hole. From the free bodies of the pin it may be noted the transfer between plate *A* and *B* is actually made via shear on the pin (the slight rotation of the pin due to unbalanced moment would be relatively negligible). There would be negligible friction between plates *A* and *B*. The earliest steel structures, particularly trusses, were actually connected by pins.

Figure 4.6.2
Transfer of load in pin
connections.

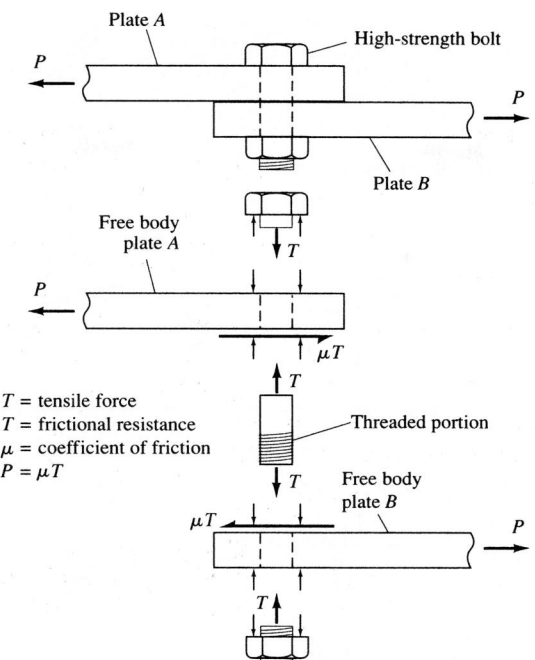

Figure 4.6.3
Transfer of load in
pretensioned high-strength
bolted connection.

When a high-strength bolt is installed to have a specified initial tension, there will be an initial precompression between the pieces being joined, as shown in Fig. 4.6.3. A transfer of plate tension loads P as shown in Fig. 4.6.3 may then occur entirely via friction at service-load levels, and there may be no bearing of the bolt shank against the side of the hole. Until the friction force μT is overcome, the shear strength of the bolt and the bearing strength of the plate will not affect the ability to transfer load across the shear plane between plates.

The *AISC Specification* [1.13] recognizes two general categories of performance requirement for high-strength bolted connections, for many years known as *bearing-type* and *friction-type*. The 1986 LRFD and 1989 ASD Specifications renamed the friction-type connection as a *slip-critical* connection.

The strength of all high-strength bolted connections in transmitting shear forces across a shear plane between steel elements is the same whether the connection is a *bearing-type* or a *slip-critical* connection. The slip-critical connection with standard holes or slots transverse to the direction of the load is to be designed for slip as a serviceability limit state, and will be designed for slip at the required strength level for a connection with oversized holes or slots parallel to the direction of the load.

The possible "limit states," or failure modes, that may control the strength of a bolted connection are shown in Fig. 4.6.4.

Tensile Strength of Fasteners

In accordance with the fracture limit state in tension discussed in Sec. 3.2 and the failure mode shown in Fig. 4.6.4e, the nominal strength R_n (here using R_n instead of T_n) of one fastener in tension is

$$R_n = F_u^b A_n \qquad [3.2.2]$$

where F_u^b = tensile strength of the bolt material. The net area A_n should be the area through the threaded portion of the bolt, known as the "tensile stress area"*. The ratio of the

---

*Tensile stress area $= 0.785\left(d_b - \dfrac{0.9743}{n} \right)^2$ where n = number of threads per inch.

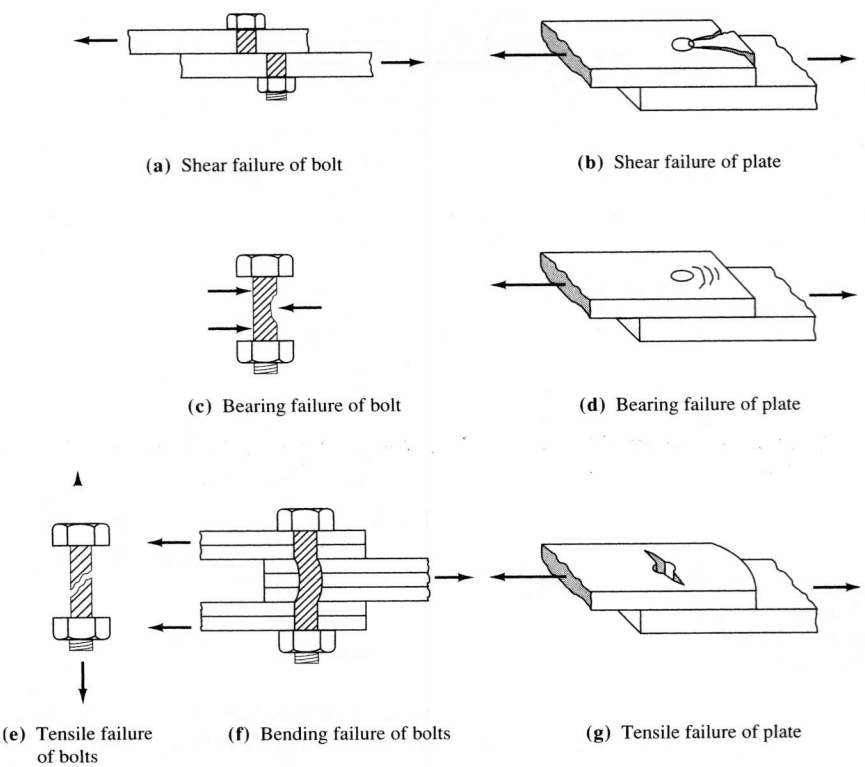

(a) Shear failure of bolt

(b) Shear failure of plate

(c) Bearing failure of bolt

(d) Bearing failure of plate

(e) Tensile failure of bolts

(f) Bending failure of bolts

(g) Tensile failure of plate

Figure 4.6.4
Possible modes of failure of bolted connections.

tensile stress area to the gross area A_b ranges from 0.75 to 0.79. Thus, in terms of the gross area A_b of one bolt, Eq. 3.2.2 becomes

$$R_n = F_u^b(0.75A_b) \qquad (4.6.1a)$$

which can be written as:

$$R_n = F_{nt}A_b \qquad (4.6.1b)$$

where F_u^b = tensile strength of the bolt material
  A_b = gross cross-sectional area of one bolt
  F_{nt} = $0.75F_u^b$, nominal tensile stress of the bolt as given in AISC Table J3.2

Shear Strength of Fasteners

In accordance with the fracture limit state as the basis for fastener strength and the failure mode shown in Fig. 4.6.4a, the nominal strength R_n for one fastener will be the ultimate shear stress τ_u across the gross area A_b of the bolt times the number m of shear planes; thus,

$$R_n = mA_b\left(0.62F_u^b\right) \qquad (4.6.2)$$

which can be written as:

$$R_n = mA_bF_{nv} \qquad (4.6.3)$$

where F_{nv} is the nominal shear stress of the bolt as given in AISC-Table J3.2.
  Note that the ultimate (nominal) shear strength was found experimentally [3.1] to be about 62% of ultimate tensile strength, about the same ratio as for the yield strengths between shear and tension (see Sec. 2.6).

Figure 4.6.5
Bearing strength related to
end distance.

Bearing Strength

The bearing limit state relates to excessive deformation around a bolt hole, as shown in Fig. 4.6.4d. A shear tear-out failure as shown in Fig. 4.6.4b is closely related to a bearing failure.

The bearing strength R_n is the force applied against the side of a hole to split or tear the plate. The larger the end clear distance L_c, measured from the edge of the hole to the edge of the plate, the smaller the possibility of having a splitting failure.

Referring to Fig. 4.6.5, the actual tearing would occur along lines 1-1 and 2-2. As a lower bound for strength, the angle α could be taken as zero, and the edge distance could be taken as the clear distance, giving the nominal strength R_n as

$$R_n = 2tL_c\tau_u^p \tag{4.6.4}$$

where τ_u^p = shear strength of plate material $\approx 0.62F_u$
  F_u = tensile strength of plate material
  L_c = clear distance along the line of force, from edge of connected part to the edge of the hole or the clear distance between holes.

Thus,

$$R_n = 2tL_c(0.62F_u) \tag{4.6.5}$$

$$R_n = 1.24F_utL_c \tag{4.6.6}$$

which applies for a single bolt hole or for the bolt hole nearest the edge when there are two or more bolt holes in the line of force.

Experience and tradition recommends that the preferred minimum center-to-center spacing of bolts be 3 bolt-diameters [AISC-J3.3]. When $L_c = 2.5d$ is used in Eq. 4.6.6, the nominal bearing strength becomes

$$R_n = 3.1F_udt \tag{4.6.7}$$

which is the basic expression when tear-out is prevented. When a bearing strength represented by Eq. 4.6.7 is achieved, and no rupture occurs, the elongation of the hole may be excessive. Thus, Eq. 4.6.7. should be used only when deformation around the hole is *not* of concern. Preferably lower strengths should be used.

4.7   LOAD AND RESISTANCE FACTOR DESIGN—BOLTS

The general philosophy of Load and Resistance Factor Design (LRFD) was described in Secs. 1.8. and 1.9. Equation 1.8.1 gives the structural safety requirement, as follows:

$$\phi R_n \geq \sum \gamma_i Q_i \tag{[1.8.1]}$$

where ϕ = resistance factor (strength reduction factor)
  R_n = nominal resistance (strength)

$$\gamma_i = \text{overload factors (ASCE 7-05)}$$

Q_i = loads (such as dead load, live load, wind load, earthquake load) or load effects (such as bending moment, shear, axial force, and torsional moment resulting from the various loads)

Equation 1.8.1 requires the *design strength* ϕR_n to equal or exceed the summation of *factored loads,* or specifically for fasteners, Eq. 1.8.1 becomes

$$\phi R_n \geq P_u \tag{4.7.1}$$

where ϕ = resistance factor, 0.75 for fracture in tension, shear on high-strength bolts, and bearing of bolt against side of hole

R_n = nominal strength of one fastener

P_u = factored load on one fastener

Since the loading is not related to the type of member, such as tension member or column, the various load factor combinations given in ASCE 7-05 are the same for all members in the structure. However, the resistance factor ϕ accounts for the possibility that the actual strength of the fastener (or member) may be less than the theoretically computed strength because of variations in material properties and dimensional tolerances. These variations, while individually within accepted tolerance limits, may combine in the actual structure to give a strength less than the computed value. Neither the ϕ factor nor the overload factors γ are intended to account for careless errors in design or construction.

The strength of a fastener may be based on (1) shear, (2) bearing, (3) tension, or (4) combined shear and tension. The nominal strengths in tension, shear, and bearing were presented in Sec. 4.6.

Design Shear Strength—No Threads in Shear Planes

The nominal shear strength R_n for a single fastener, given by Eq. 4.6.2, was used by the 1986 LRFD Specification along with a ϕ factor of 0.65 to obtain the design shear strength for connection design. The 0.65 is lower than the ϕ factor normally used for shear-related behavior because of the calibration of the method to experimental results and past design practice. Particularly, the strength of long connections, up to 50 in., is lower than indicated by the sum of the strengths of individual fasteners. AISC indicates no explicit adjustment for long connections unless they exceed 50 in.

Since the 1993 LRFD Specification a value of $\phi = 0.75$ has been adopted, the commonly used ϕ factor for shear. This change requires the nominal shear strength per fastener be reduced to account for the approximately 20% reduction in strength of connections up to 50 in. long (see Ref. 3.1, p. 100).

Thus, the nominal strength in shear is taken at 0.8 of that given by Eq. 4.6.2,

$$R_n = [\text{reduction for connection length}](0.62F_u^b)mA_b$$
$$= (0.8)(0.62F_u^b)mA_b = (0.50F_u^b)mA_b \tag{4.7.2}$$

where $0.50F_u^b$ defines the values of F_{nv} as given in the Table of AISC-J3.2 when threads are excluded from the shear planes and the design strength is

$$\phi R_n = 0.75F_{nv}mA_b \tag{4.7.3}$$

where $\phi = 0.75$, the standard ϕ value for shear

F_u^b = tensile strength of the bolt material (120 ksi for A325 bolts; 150 ksi for A490 bolts)

m = the number of shear planes participating [usually one (*single shear*) or two (*double shear*) as in Fig. 4.6.1a]

A_b = gross cross-sectional area across the unthreaded shank of the bolt

F_{nv} = nominal shear stress of fasteners (AISC Table J3.2)

Design Shear Strength—Threads in Shear Planes

When threads are in the shear plane, the area at the root of the threads should be used in place of A_b. The thread root area is approximately 70% to 75% of the gross (shank) area. The nominal shear strength then becomes

$$R_n = [\text{reduction for connection length}](0.62F_u^b)m(0.75A_b)$$
$$= (0.8)(0.62F_u^b)m(0.75A_b) = (0.37F_u^b)mA_b \tag{4.7.4}$$

Because both the 0.8 connection length effect and the 0.75 for area through the threads are approximate, the 2005 *AISC Specification* uses

$$R_n = (0.40F_u^b)mA_b \tag{4.7.5}$$

where $0.40F_u^b$ defines the value of F_{nv} as given in the Table of AISC-J3.2 when threads are included and the design strength is

$$\phi R_n = 0.75F_{nv}mA_b \tag{4.7.6}$$

Design Tension Strength

The design strength ϕR_n *based on the tension strength of the fastener,* according to AISC-J3.6 and developed as Eqs. 4.6.1, is

$$\phi R_n = \phi F_u^b(0.75A_b) = 0.75F_u^b(0.75A_b) \tag{4.7.7}$$

or

$$\phi R_n = 0.75F_{nt}A_b \tag{4.7.8}$$

where ϕ = 0.75, a value for the tensile fracture mode as discussed in Chap. 3

F_u^b = tensile strength of the bolt material (120 ksi for A325 bolts; 150 ksi for A490 bolts)

A_b = gross cross-sectional area across the unthreaded shank of the bolt

F_{nt} = nominal tensile stress of fasteners (AISC Table J3.2)

Note that the $0.75A_b$ represents the area through the threaded portion of the bolt. AISC-Table J3.2 indicates that the "Tensile Strength" is 90 ksi and 113 ksi, for A325 and A490 bolts, respectively; that is, $0.75F_u^b$. As is apparent from Eq. 4.7.7, multiplying the gross area A_b by $0.75F_u^b$ gives the correct value for R_n. Using the reduced stress on the gross area gives the same result as using the correct stress on the reduced area.

The design strengths for tension and shear on A325 and A490 bolts are summarized in Table 4.7.1.

Design Bearing Strength

The hole bearing strength is controlled either by tear-out (a bolt to bolt block shear rupture) or by bearing deformation [4.9]. The bearing strength limit state is defined based on type

TABLE 4.7.1 Design Strength* of A325 and A490
High-Strength Bolts

Fastener	F_u^b (ksi)	Tension strength (ksi) $\phi = 0.75$	Shear strength (ksi) $\phi = 0.75$
A325 bolts, when threads are *not* excluded from shear planes	120	$\phi(0.75\,F_u^b)$ $0.75(90.0) = 67.5$	$\phi(0.40\,F_u^b)$ $0.75(48) = 36.0$
A325 bolts, when threads *are* excluded from shear planes	120	$\phi(0.75\,F_u^b)$ $0.75(90.0) = 67.5$	$\phi(0.50\,F_u^b)$ $0.75(60) = 45.0$
A490 bolts, when threads are *not* excluded from shear planes	150	$\phi(0.75\,F_u^b)$ $0.75(113) = 84.8$	$\phi(0.40\,F_u^b)$ $0.75(60) = 45.0$
A490 bolts, when threads *are* excluded from shear planes	150	$\phi(0.75\,F_u^b)$ $0.75(113) = 84.8$	$\phi(0.50\,F_u^b)$ $0.75(75) = 56.3$

* Design strength ϕR_n equals stress in table times gross bolt cross-sectional area A_b.

of hole and/or acceptability of hole ovalization. AISC-J3.10 defines the various categories of nominal bearing strength as follows:

1. For a bolt in a connection with standard, oversized, and short-slotted holes, independent of the direction of loading, or a hole with a slot parallel to the direction of the bearing force:

   (a) When deformation at the bolt hole at service load is a design consideration, according to AISC Formula (J3-6a)

   $$R_n = 1.2L_c t F_u \leq 2.4dt F_u \qquad (4.7.9)$$

   (b) When deformation at the bolt hole at service load is not a design consideration, according to AISC Formula (J3-6b)

   $$R_n = 1.5L_c t F_u \leq 3.0dt F_u \qquad (4.7.10)$$

2. For a bolt in a connection with long-slotted holes with the slot perpendicular to the direction of the force, according to AISC Formula (J3-6c)

   $$R_n = 1.0L_c t F_u \leq 2.0dt F_u \qquad (4.7.11)$$

d = nominal diameter of bolt at unthreaded area

t = thickness of part against which bolt bears

F_u = tensile strength of connected part against which bolt bears

L_c = clear distance, in the direction of the force, between the edge of the hole and the edge of the adjacent hole

The design bearing strength is ϕR_n where $\phi = 0.75$.

The design equations are based on the models presented in the *Guide* [3.1], where tests have shown that a total elongation equal to diameter d is produced when the nominal strength equals $3.0dt F_u$. The lower limit value of $2.4dt F_u$ in Eq. 4.7.9 provides a bearing strength limit state that is attainable at a reasonable deformation of $\frac{1}{4}$-in. When long-slotted holes are oriented perpendicular to the direction of load, the bending component of the deformation in the material between adjacent holes or between the hole and the edge of the plate is increased. Hence the maximum nominal bearing strength is $2.0dtF_u$ in Eq. 4.7.11, which again provides a bearing strength limit state that is attainable at a reasonable deformation.

Minimum Spacing of Bolts in Line of Transmitted Force

When the spacing of bolts in the direction of the transmitted force is at least 3 bolt-diameters, the defined upper limit for all three conditions dictates the nominal bearing strength for each case. Expressing Eqs. 4.7.9, 4.7.10, and 4.7.11 as:

$$R_n = KL_c tF_u \tag{4.7.12}$$

where $K = 1.2$, 1.5, or 1.0. Solving Eq. 4.7.12 for L_c gives the minimum distance from the edge of one fastener to the edge of the adjacent hole,

$$L_c \geq \frac{R_n}{KF_u t} \tag{4.7.13}$$

Then, adding the radius $d_h/2$ of the hole to Eq. 4.7.13 gives the minimum center-to-center spacing,

$$\text{Spacing} = \frac{R_n}{KF_u t} + d_h \tag{4.7.14}$$

Because R_n in Eq. 4.7.14 is the *required* nominal strength, which equals the factored load P acting on one bolt divided by the resistance factor ϕ, Eq. 4.7.14 [AISC-J3.10] becomes

$$\text{Spacing} = \frac{P}{\phi KF_u t} + d_h \tag{4.7.15}$$

where $\phi = 0.75$
$\quad\quad P = $ factored load acting on one bolt
$\quad\quad F_u = $ tensile strength of plate material
$\quad\quad t = $ thickness of plate material
$\quad\quad d_h = $ diameter of the hole

The minimum spacing of bolts in a line is *preferably* [AISC-J3.3] 3 bolt diameters and shall not be less than $2\frac{2}{3}$ diameters.

Minimum End Distance in Direction of Transmitted Force

The edge distance is the clear distance plus half the diameter of the hole. Using Eq. 4.7.13, the minimum end distance L_e can be expressed as follows

$$L_e \geq \frac{R_n}{KF_u t} + \frac{d_h}{2} \tag{4.7.16}$$

where the required nominal strength is the factored load divided by the resistance factor ϕ. Thus, Eq. 4.7.16 becomes

$$L_e \geq \frac{P}{\phi KF_u t} + \frac{d_h}{2} \tag{4.7.17}$$

where $\phi = 0.75$

P = factored load acting on one bolt

F_u = tensile strength of plate material

t = thickness of plate material

d_h = diameter of the hole

K = 1.2, 1.5, or 1.0, according to AISC-J3.10

The end distance actually used must be the larger of that computed from a strength requirement and the minimum prescribed by Table 4.7.2 [AISC Table J3.4].

Equations 4.7.12 through 4.7.17 are design equations giving required spacing and end distance; AISC-J3.10 gives those equations as R_n equations assuming the spacing and end distance is known. The designer will usually want to obtain the spacing and end distance necessary to carry the given factored loads P.

Maximum Edge Distance

In accordance with AISC-J3.5, the maximum distance from the center of a bolt to the nearest edge is $12t$, where t is the thickness of the connected part, and this edge distance may not exceed 6 in. The purpose of this requirement is to prevent corrosion resulting from moisture entering the joint. The two contact surfaces of a joint may not be perfectly flat, and the clamping action will be lower when the bolts are far apart (or far from an edge).

TABLE 4.7.2 Minimum Edge Distances* (Center of Standard Hole** to Edge of Connected Part) (Adapted from AISC Table J3.4)

| Nominal rivet or bolt diameter | | Minimum edge distance | | | |
| | | At sheared edges | | At rolled edges of plates, shapes or bars, or gas cut edges[†] | |
(in.)	(mm)	(in.)	(mm)	(in.)	(mm)
1/2	12.7	7/8	22.2	3/4	19.1
5/8	15.9	1 1/8	28.6	7/8	22.2
3/4	19.1	1 1/4	31.8	1	25.4
7/8	22.2	1 1/2[‡]	38.1	1 1/8	28.6
1	25.4	1 3/4[‡]	44.5	1 1/4	31.8
1 1/8	28.6	2	50.8	1 1/2	38.1
1 1/4	31.8	2 1/4	57.2	1 5/8	41.3
Over 1 1/4	Over 31.8	1 3/4 × diam		1 1/4 × diam	

* Lesser edge distances are permitted to be used provided formulas from AISC-J3.10, as appropriate, are satisfied.

** When oversize or slotted holes are used, see AISC-J3.5.

[†] All edge distances in this column may be reduced $\frac{1}{8}$ in. (3.2 mm) when the hole is at a point where factored load does not exceed 25% of the maximum design strength in the element.

[‡] These are permitted to be $1\frac{1}{4}$ in. at the ends of beam connection angles and shear end plates.

Maximum Spacing of Bolts

The maximum longitudinal spacing of bolts between elements in continuous contact when the elements consist of a plate and a shape, or two plates, is given by AISC-J3.5.

a. For painted members or unpainted members not subject to corrosion,

$$s \leq 24t \leq 12 \text{ in.} \tag{4.7.18}$$

b. For unpainted members of weathering steel subject to atmospheric corrosion

$$s \leq 14t \leq 7 \text{ in.} \tag{4.7.19}$$

where t is the thickness of the thinner element.

4.8   EXAMPLES—TENSION MEMBER BEARING-TYPE CONNECTIONS—LRFD

The following examples illustrate the AISC Load and Resistance Design Method for connections of tension members. The principles related to member strength are discussed in Sec. 3.9 and those that are fastener related are in Sec. 4.7.

EXAMPLE 4.8.1

Compute the tensile service load capacity for the bearing-type connection of two members in Fig. 4.8.1 if (a) the bolt threads are excluded from the shear plane and (b) the bolt threads are included in the shear plane. Use *AISC Specification*-LRFD Method with $\frac{7}{8}$-in.-diam A325 bolts in standard holes and A572 Grade 50 steel plates. The service live load is three times the service dead load.

Solution:

(a) Threads excluded from shear plane. First compute the strength of the plates as tension members (Chapter 3):

$$A_g = 6(0.625) = 3.75 \text{ sq in.}$$

$$A_n = \left[6 - 2\left(\tfrac{7}{8} + \tfrac{1}{8}\right)\right]0.625 = 2.50 \text{ sq in.}$$

$$A_e = A_n = 2.50 \text{ sq in.}$$

Figure 4.8.1
Example 4.8.1.

Then, using Eqs. 3.9.2 and 3.9.3 gives

$$\phi T_n = \phi F_y A_g = 0.90(50)(3.75) = 169 \text{ kips}$$
$$\phi T_n = \phi F_u A_e = 0.75(65)(2.50) = 122 \text{ kips (controls)}$$

The action of the entire connection is a shear transfer of load between the two plates. The plane of contact may be thought of as the shear plane. When there is a single plane of contact involved in the load transfer, it is referred to as "single shear."

The design strength of the bolts in shear and bearing must be computed using Eqs. 4.7.3 and 4.7.9.

The design strength ϕR_n per bolt in single shear for A325-X* is

$$\phi R_n = \phi F_{nv} m A_b \qquad [4.7.3]$$
$$= 0.75(60)(1)0.6013 = 27.1 \text{ kips/bolt}$$

$$\left[L_C = 1.5 - \left(\frac{1}{2}\right)\left(\frac{7}{8} + \frac{1}{16}\right) = 1.03 \right] < [2d = 1.75]$$

Since $L_c \not> 2d$ the design bearing strength for each bolt is

$$\phi R_n = \phi(1.2 L_c t \, F_u)$$
$$= 0.75(1.2)(1.63)(0.625)(65) = 37.7 \text{ kips/bolt}$$

Note that for bearing strength, F_u is the tensile strength of the plate material, i.e., 65 ksi for A572 Grade 50. Thus, shear controls and the design strength per bolt is 27.1 kips. The total design strength based on the bolts then becomes

$$\phi T_n = (\text{number of fasteners}) R_n = 4(27.1) = 108 \text{ kips}$$

Because the single shear bolt design strength $\left(\phi T_n = 108 \text{ kips}\right)$ is lower than the plate design strength as a tension member $\left(\phi T_n = 122 \text{ kips}\right)$,

$$\phi T_n = 108 \text{ kips (for A325-X)}$$

The 1.5 in. end distance provided satisfies the minimum for sheared edge $\left(1\frac{1}{2} \text{ in.}\right)$ given in Table 4.7.2.

Block shear strength is not checked here (see Sec. 3.6); it is not likely to control when $\frac{5}{8}$-in. thickness of pieces is used.

To relate the design strength to the service load, Eq. 3.9.1 must be used

$$\phi T_n \geq T_u \qquad [3.9.1]$$

Setting the factored service load T_u equal to the design strength ϕT_n will give the maximum acceptable service load. Using the gravity dead and live load factored load

$$T_u = 1.2D + 1.6L \qquad [1.8.3]$$

Substituting into Eq. 1.8.3 the live load L as three times the dead load D, and using ϕT_n as 108 kips in Eq. 3.9.1 gives

$$\phi T_n = T_u$$
$$108 = 1.2D + 1.6(3D) = 6.0D$$

---

*Standard designation for an A325 bolt in a bearing-type connection with threads *eXcluded from* the shear planes.

$$D = 18 \text{ kips}$$

$$L = 3D = 54 \text{ kips}$$

Thus, the maximum total service load T permitted is

$$T = D + L = 18 + 54 = 72 \text{ kips (for A325-X)}$$

(b) Threads included in shear plane. The bolt design strength in single shear is less than computed in part (a). The design strength ϕR_n per bolt in single shear for A325-N* is

$$\phi R_n = \phi F_{nv} m A_b$$

$$= 0.75(48)(1)0.6013 = 21.6 \text{ kips/bolt}$$

Strength in single shear again governs, since there are no changes in the net section tension member strength, or in bearing strength. Since ϕR_n is 0.8 of that in part (a), the service load capacity T is

$$T = 0.8(72) = 57.6 \text{ kips (for A325-N)} \qquad \blacksquare$$

EXAMPLE 4.8.2

Determine the number of $\frac{3}{4}$-in.-diam A325 bolts required to develop the full strength of A572 Grade 65 steel plates in Fig. 4.8.2 (a portion of a double lap splice connection) for a bearing-type connection with threads excluded from the shear planes. Use AISC-LRFD Method and assume a double row of bolts with standard size holes.

Solution:
It is determined by inspection that the cross-sectional area of the center plate is less than the sum of the areas of the two outer plates. Therefore, in this case one need check only the strength of the center plate:

$$A_n = \left[6 - 2\left(\tfrac{3}{4} + \tfrac{1}{8}\right) \right] 0.375 = 1.59 \text{ sq in.}$$

$$\text{Max } A_n = 0.85 A_g = 0.85(6)0.375 = 1.91 \text{ sq in.}$$

$$A_e = A_n = 1.59 \text{ sq in.}$$

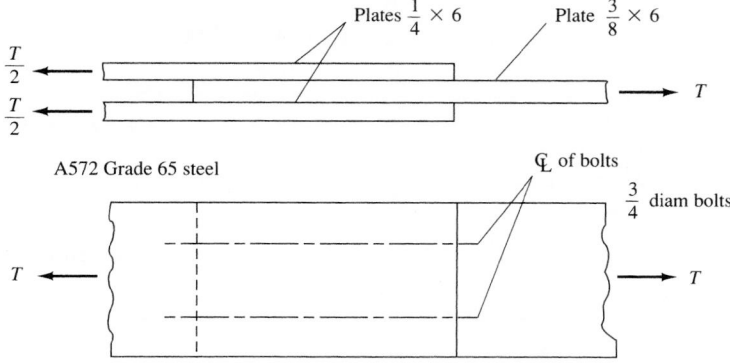

Figure 4.8.2
Example 4.8.2.

---

*Standard designation for an A325 bolt in a bearing-type connection with threads *iNcluded* in the shear planes.

Then using Eqs. 3.9.2 and 3.9.3 gives

$$\phi T_n = \phi F_y A_g = 0.90(65)(6)0.375 = 132 \text{ kips}$$

$$\phi T_n = \phi F_u A_e = 0.75(80)1.59 = 96 \text{ kips (controls)}$$

Thus, the design strength in double shear (per bolt) is

$$\phi R_n = \phi F_{nv} m A_b = 0.75(60)(2)0.4418 = 39.8 \text{ kips/bolt}$$

Assume bearing strength Eq. 4.7.9 controls, i.e., the clear edge distance or spacing is equal to or greater than $2d$. Then the design strength in bearing on the $\frac{3}{8}$-in. plate is

$$\phi R_n = \phi(2.4 F_u dt) = 0.75(2.4)(80)(0.75)0.375 = 40.5 \text{ kips/bolt}$$

$$\text{Number of bolts} = \frac{96}{39.8} = 2.4$$

The end distance must be at least 1.25 in. in accordance with Table 4.7.2.
Use 4 − $\frac{3}{4}$-in.-diam A325 bolts (A325-X).
The block shear limit state should also be checked; however, since bearing does not control in other regards, the shear rupture mode is unlikely to control. ∎

EXAMPLE 4.8.3

Determine the number of $\frac{3}{4}$-in.-diam A325 bolts in standard size holes required to carry 7 kips dead load and 43 kips live load on the plates in Fig. 4.8.3 if A36 steel is used. Assume the portion of the double lap splice is a bearing-type connection with threads excluded from the shear planes, and a double row of bolts is used. Use the AISC-LRFD Method.

Solution:
The factored load T_n that must be carried is

$$T_n = 1.2D + 1.6L = 1.3(7) + 1.6(43) = 77 \text{ kips}$$

In this example, the center plate will control the tension strength:

$$A_g = 10(0.25) = 2.5 \text{ sq in.}$$

$$A_n = \left[10 - 2\left(\tfrac{3}{4} + \tfrac{1}{8}\right)\right]0.25 = 2.06 \text{ sq in.}$$

$$\text{Max } A_n = 0.85 A_g = 0.85(2.5) = 2.13 \text{ sq in.}$$

$$\therefore A_e = A_n = 2.06 \text{ sq in}$$

Figure 4.8.3
Example 4.8.3.

$$\phi T_n = \phi F_y A_g = 0.90(36)2.50 = 81 \text{ kips (controls)}$$
$$\phi T_n = \phi F_u A_e = 0.75(58)2.06 = 90 \text{ kips}$$

and this exceeds the factored load on the members (i.e., $\phi T_n \geq T_u$).
For shear, using A325-X,

$$\phi R_n = \phi F_{nv} m A_b = 0.75(60)(2)0.4418 = 39.8 \text{ kips/bolt}$$

For bearing on the $\frac{1}{4}$-in. plate, when $L_c > 2d$.

$$\phi R_n = \phi(2.4 F_u dt) = 0.75(2.4)(58)(0.75)0.25 = 19.6 \text{ kips/bolt}$$

$$\text{Number of bolts} = \frac{77}{19.6} = 3.9 \quad \underline{Use \text{ 4 bolts}}$$

The block shear rupture limit state should also be checked; particularly here, since the plate $\left(\frac{1}{4}\text{-in.}\right)$ is thin and heavily loaded. Referring to Fig. 4.8.4, the preliminary arrangement for four bolts is to use 2 in. edge distances at the sides and end, and 3 in. longitudinal spacing.

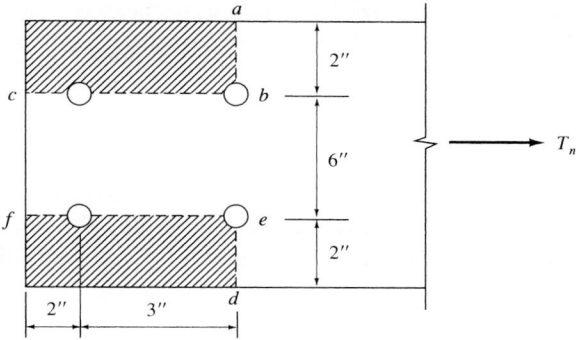

Figure 4.8.4
Block shear failure on plate
of Example 4.8.3.

Investigate the block shear strength of the resulting arrangement using AISC-J4.3.
The block shear potential failure paths *a-b-c* and *d-e-f* of Fig. 4.8.4 must be investigated according to AISC-J4. Calculating the net areas A_{nv} and A_{nt},

$$A_{nv} = (\text{length } b\text{-}c \text{ and } e\text{-}f \text{ less } 1.5 \text{ holes}) \text{ thickness}$$
$$= 2\left[5 - 1.5\left(\tfrac{3}{4} + \tfrac{1}{8}\right)\right]0.25 = 1.84 \text{ sq in.}$$
$$A_{nt} = (\text{length } a\text{-}b \text{ and } d\text{-}e \text{ less } 0.5 \text{ holes}) \text{ thickness}$$
$$= 2\left[2 - 0.5\left(\tfrac{3}{4} + \tfrac{1}{8}\right)\right]0.25 = 0.78 \text{ sq in.}$$

Compare

$$\left[R_n = 0.6 F_u A_{nv} + F_u U_{bs} A_{nt}\right] \leq \left[R_n = 0.6 F_y A_{gv} + F_u U_{bs} A_{nt}\right]$$
$$R_n = 0.6 F_u A_{nv} + F_u U_{bs} A_{nt} = 0.6(58)1.84 + 58(1.0)0.78 = 109.3 \text{ kips}$$
$$R_n = 0.6 F_y A_{gv} + F_u U_{bs} A_{nt} = 0.6(36)[2(5)0.25] + 58(1.0)0.78 = 99.2 \text{ kips}$$
$$\therefore R_n = 99.2 \text{ kips}$$
$$\phi R_n = 0.75(109.2) = 74 \text{ kips}$$

Thus, block shear (74 kips) does control over tension yielding on gross section (81 kips). The design strength is less than the factored load (77 kips) that must be carried, and the design is unsatisfactory.   ∎

4.9 SLIP-CRITICAL JOINTS

When slip resistance at service load is desired, the joint is referred to as a *slip-critical joint.* *Slip-critical joints* are normally used when *joints* are subjected to reversing fatigue load, *joints* that use oversized or slotted holes parallel to the applied load, or *joints* in which slip would be detrimental to the performance of the structure. In the 1978 AISC Specification, such connections were termed *friction-type connections.* All tensioned high-strength bolted connections actually resist load by friction. Referring to Fig. 4.6.3, the pretension force *T* in the bolt equals the clamping force between the pieces being joined. The resistance to shear is a frictional force μT, where μ is the coefficient of friction.

The coefficient of friction, or more properly the *slip coefficient,* depends on the surface condition, with such items as mill scale, oil, paint, or special surface treatments determining the value of μ.

Slip is defined as occurring when "the friction bond is definitely broken and the two surfaces slip with respect to one another by a relatively large amount" [4.14]. The range of μ varies from 0.2 to 0.6 depending on the surface condition [3.1].

To avoid directly using the coefficient of friction and to permit design of slip-critical joints using the same general approach as for bearing-type connections, the friction force μT is divided by the bolt gross cross-sectional area A_b to obtain a so-called "shear stress" on the bolt.

EXAMPLE 4.9.1

Determine the amount of force *P* required to cause slip of a $\frac{7}{8}$-in.-diam A325 bolt loaded as in Fig. 4.9.1, if the slip coefficient μ is 0.33 (a typical value for the usual "clean mill scale" surface condition). Using the service load force *P*, compute the "shear stress", $f_v = P/A_b$.

Figure 4.9.1
Example 4.9.1.

Solution:
Using the pretension load from Table 4.4.1,

$$T_b = 39 \text{ kips}$$
$$P = \mu T_b = 0.33(39) = 12.9 \text{ kips}$$

Since the overall action of the connection is a shearing effect, the "shear stress" f_v in the $\frac{7}{8}$-in.-diam bolt at the load causing slip to begin is

$$f_v = \frac{P}{A_b} = \frac{12.9}{0.6013} = 21.4 \text{ ksi} \qquad \blacksquare$$

The *AISC Specification* [1.16] uses the slip limit state (a serviceability limit state) as the criterion for design. Adequate slip resistance must be provided in joints where slip at *service load* cannot be tolerated. The actual failure of bolts in a joint will be as discussed in Sec. 4.6; that is, a shear failure of the bolts, or bearing failure in the connected material.

Slip is considered critical when it causes excessive distortion, or anywhere the loss of clamping action is considered undesirable. The strength ϕR_n of the connection is usually sufficient to resist design loads, though it must be checked.

AISC-J3.8 permits high strength bolts in slip-critical joints to be designed to prevent slip either as a serviceability limit state or at the required strength limit state. Regardless of the design limit state used, the intent is the prevention of slip at service loads. Both limit states were calibrated to give the same reliability and safety. AISC-J3.8 recommends that slip-critical connections with standard holes or slotted holes perpendicular to the direction of the force be designed based on serviceability. Any slip that may occur in such connections will cause minor changes in the geometry and, therefore, will not compromise the basic assumption used for analysis. Connections with oversized holes or with slotted holes parallel to the load are required to be designed to prevent slip at the required strength level. In such cases it is presumed that slip will lead to a significant increase in load on the bolts.

The required reliability when designing a slip-critical connection is not as obvious as that for the cases of welded and bearing-type connections. In the latter two cases, connection failure is sudden and nonductile, hence, the reliability index β value of 4.0. In slip-critical connections the exact reliability index is not critical to a connections performance. The *AISC Specification* for the slip-critical connections is intended to achieve the same level of reliability as other connections. However, the bearing and shear strengths of a slip-critical connection must always be evaluated, because of the complexity of calculating the reliability index and the absence of adequate statistical data on slip resistance.

The slip resistance of the connection is affected by three factors:

1. Slip coefficient of the faying surface which depends on the type of surface between the connected parts.

2. The pretensioning method. AISC recognizes four different methods to pretension high strength bolts: turn-of-the-nut, calibrated wrench, twist-off type tension-control bolt assemblies, and direct tension indicator assemblies.

3. Hole size and shape; oversize holes and slotted holes in the direction of the load make bolts more susceptible to slip than standard holes.

Design of Slip-Critical Connections—AISC Specification-LRFD Method

The design of slip-critical connections requires full consideration of the strength limit states. The strength of the bolts in shear, bearing, and direct tension, as discussed in Secs. 4.6 and 4.7 must be investigated. In addition, slip must be resisted at service load, such as $D + L$, and factored service load, such as $1.2D + 1.6L$, must not exceed the design strength ϕR_n, according to AISC-J3.8.

Connections having standard holes or slotted holes transverse to the direction of the load are required to be designed to resist slip at service load. Connections having oversized holes or slotted holes parallel to the direction of the load are required to resist slip at the required strength R_n (i.e., factored service load) level.

AISC-J3.8 defines the slip limit state (nominal resistance) as follows:

$$R_n = \mu D_u h_{sc} N_s T_b \tag{4.9.1}$$

where μ = the mean slip coefficient for Class A or Class B surfaces, as applicable, or as established by tests

   = 0.35 for Class A surface condition

   = 0.50 for Class B surface condition

   D_u = the ratio of the mean installed bolt pretension to the specified minimum bolt pretension = 1.13

   h_{sc} = hole size factor (reflecting the relative contact area to resist slip)

(a) Standard holes, 1.00

(b) Oversized and short-slotted holes, 0.85

(c) Long-slotted holes, 0.70

N_s = number of slip planes

T_b = minimum bolt tension as given in AISC Table J3.1

The design slip resistance ϕR_n uses $\phi = 1.0$ for the serviceability limit state, and $\phi = 0.85$ for the required strength R_u (i.e., factored service loads) limit state.

EXAMPLE 4.9.2

Determine the tensile capacity of the connection previously investigated in Example 4.8.1 as a bearing-type connection (Fig. 4.8.1); however, consider it as a slip-critical connection with the usual clean mill scale (Class A) surface condition. Use $\frac{7}{8}$-in.-diam A325 bolts in standard holes with A572 Grade 50 plate material. Use the AISC-LRFD Method.

Solution:

The tension member (plates) design strength ϕT_n was determined in Example 4.8.1 to be

$$\phi T_n = 122 \text{ kips (based on } \phi F_u A_e)$$

The nominal slip resistance R_n is

$$R_n = \mu D_u h_{sc} N_s T_b \qquad \text{[4.9.1]}$$

where
$\mu = 0.35$
$D_u = 1.13$
$h_{sc} = 1.00$
$N_s = 1$
$T_b = 39 \text{ kips}$

$$R_n = \mu D_u h_{sc} N_s T_b = 0.35(1.13)(1.00)(1)39 = 15.4 \text{ kips}$$

The resistance factor ϕ is taken as 1.0 for the serviceability limit state with standard holes. Thus,

$$\phi R_n = 1.0(15.4) = 15.4 \text{ kips}$$

The tension member design strength (with four bolts) is

$$\phi T_n = 15.4(4) = 61.7 \text{ kips}$$

Thus, the slip-critical design strength of 61.7 kips controls.

The design strength can be equated to the factored service load,

$$\phi T_n = 1.2D + 1.6L = 1.2D + 1.6(3D) = 6D$$

$$D = 61.7/6 = 10.3 \text{ kips}$$

$$\therefore T = D + L = 10.3 + 3(10.3) = 41.1 \text{ kips}$$

This service load capacity T based on slip resistance is lower than the bolt-related strength determined service load $T = 72$ kips computed in Example 4.8.1 for A325-X; thus $T = 41.1$ kips ■

The requirements for spacing, end, and edge distances are all identical whether the joint is a bearing-type connection or a slip-critical connection. Since the factored load per bolt will be lower when slip-resistance governs the number of fasteners used, the bearing-related equations for spacing of fasteners and end distance will permit smaller spacing and end distances. Of course, the minimums are then more likely to control. Generally, slip-resistance controls in slip-critical connections, rather than strength in shear or bearing.

EXAMPLE 4.9.3

Redesign the connection for Fig. 4.8.2 as a slip-critical connection using (a) oversized holes, and (b) standard holes and a Class A surface condition. Use $\frac{5}{8}$-in.-diam A325 bolts in a double row and the *AISC Specification* LRFD Method. The live load is four times the dead load.

Solution:
The design strength ϕT_n based on plate strength in tension was computed in Example 4.8.2 to be 96 kips.

For the slip-critical connection, the nominal resistance R_n is

$$R_n = \mu D_u h_{sc} N_s T_b \qquad [4.9.1]$$

For case (a) with oversized holes,
$$\mu = 0.35$$
$$D_u = 1.13$$
$$h_{sc} = 0.85$$
$$N_s = 2$$
$$T_b = 19 \text{ kips}$$

$$R_n = \mu D_u h_{sc} N_s T_b = 0.35(1.13)(0.85)(2)19 = 12.8 \text{ kips}$$

The resistance factor ϕ is taken as 0.85 for the required strength T_u limit state with over-sized holes. Thus,
$$\phi R_n = 0.85(12.8) = 10.9 \text{ kips/bolt}$$

$$\text{number of bolts required} = \frac{T_u}{\phi R_n} = \frac{96}{10.9} = 8.8 \text{ bolts}$$

Use $9-\frac{3}{4}$-in.-diam bolts (A325-SC)

For the case (b) with standard holes for the serviceability limit state, the resistance factor ϕ is taken as 1.00 for the serviceability limit state with standard holes. Thus,
$$\phi R_n = 1.0(12.8) = 12.8 \text{ kips/bolt}$$

$$\text{number of bolts required} = \frac{T_u}{\phi R_n} = \frac{96}{12.8} = 7.5 \text{ bolts}$$

Use $8-\frac{5}{8}$-in.-diam bolts (A325-SC) ■

4.10 ALLOWABLE STRENGTH DESIGN—BOLTS

The general philosophy of Allowable Strength Design Method (ASD) was described in Secs. 1.8 and 1.9. Equation 1.8.8 gives the structural design criteria, as follows:

$$\left[\frac{\phi R_n}{\gamma} = \frac{R_n}{\Omega}\right] \geq [R_a = \Sigma\gamma_iQ_i] \qquad [1.8.8]$$

where $\dfrac{R_n}{\Omega}$ = allowable strength

Ω = safety factor; 2.0 for fracture in tension, shear on high-strength bolts, and bearing of bolts against the sides of holes

R_n = nominal strength of one bolt

γ_i = ASD load combination multipliers, according to ASCE 7

R_a = required strength from the critical ASCE 7 load combination

Allowable Shear Strength of Bolts

The allowable strength in shear of a high-strength bolt when threads are excluded from the shear planes is determined by dividing the bolt nominal shear strength R_n by the safety factor Ω, as follows

$$\frac{R_n}{\Omega} = \frac{F_{nv}mA_b}{\Omega} = \frac{F_{nv}mA_b}{2.0} \qquad (4.10.1)$$

Allowable Bearing Strength of Bolts

As expressed in Eq. 1.8.8, the allowable bearing strength must exceed the required bearing strength. The allowable bearing strength is obtained by dividing the nominal strength R_n by a safety factor Ω. Thus

$$\frac{R_n}{\Omega} = \frac{kL_ctF_u}{\Omega} = \frac{kL_ctF_u}{2.0} \qquad (4.10.2)$$

The ASD design criterion becomes

$$\left[\frac{R_n}{\Omega} = \frac{kL_ctF_u}{\Omega} = \frac{kL_ctF_u}{2.0}\right] \geq [R_n = \Sigma\gamma_iQ_i] \qquad (4.10.3)$$

where k is a coefficient 1.0, 1.2, or 1.5 depending on (a) whether there are long-slotted holes with the slot direction perpendicular to the line of the force, (b) deformation at the bolt hole at service load *is* a design consideration, or (c) deformation at the bolt hole at service load *is not* a design consideration, respectively.

Slip-Critical Connections

The nominal resistance R_n for the slip limit state, as defined in Eq. 4.9.1, is also the basis for the ASD design of slip-critical connections. The safety factor Ω is as follows:

1. For slip as a serviceability limit state $\qquad \Omega = 1.50$
2. For slip as a required strength R_a limit state $\qquad \Omega = 1.76$

Based on the performance criterion selected for the connection, the appropriate Ω value is used in the AISC ASD Method for the slip-critical connection, given as

$$\left[\frac{R_n}{\Omega} = \frac{\mu D_u h_{sc} N_s T_b}{\Omega}\right] \geq R_a \qquad (4.10.4)$$

The required strength R_a is the critical ASD load combination according to ASCE 7.

Minimum Spacing and Minimum End Distance of Bolts in Line of Transmitted Force

These requirements are essentially the same for the AISC ASD Method as for the AISC LRFD Method.

4.11   EXAMPLES—TENSION MEMBERS USING ALLOWABLE STRENGTH DESIGN

EXAMPLE 4.11.1

Investigate the tension member connection of Example 4.8.1 (Fig 4.8.1) to carry a total service load of 75 kips. The connection is a bearing-type connection, with threads excluded from the shear planes, using $\frac{7}{8}$-in.-diam A325 bolts in standard holes. The plates are A572 Grade 50 steel. Use the *AISC Specification* ASD Method.

Solution:

(a) Tension allowable strength T_n/Ω_t. The areas as computed in Example 4.8.1 are

$$A_e = A_n = 2.50 \text{ sq in.}$$
$$A_g = 3.75 \text{ sq in.}$$

Using Eqs. 3.11.1 and 3.11.2,

$$\left[\frac{T_n}{\Omega_t} = \frac{F_y A_g}{\Omega_t} = \frac{50(3.75)}{1.67} = 112 \text{ kips}\right] > 75 \text{ kips} \qquad \textbf{OK}$$

or

$$\left[\frac{T_n}{\Omega_t} = \frac{F_u A_e}{\Omega_t} = \frac{65(2.50)}{2.00} = 81 \text{ kips}\right] > 75 \text{ kips} \qquad \textbf{OK}$$

(b) Bolt strengths. The allowable strength R_n/Ω in single shear from Eq. 4.6.2 is

$$\frac{R_n}{\Omega} = \frac{F_{nv} m A_b}{\Omega} = \frac{60(1)0.6013}{2.0} = 18.0 \text{ kips/bolt}$$

The allowable bearing strength, when deformation around the hole is of concern (AISC-J3.10), and $s \geq 3d$, is given by AISC Formula (J3-6a) as

$$\frac{R_n}{\Omega} = \frac{1.2 L_c t F_u}{\Omega} \leq \frac{2.4 d t F_u}{\Omega}$$

Because $L_c = 1.03 < 2d = 1.75$

$$\frac{R_n}{\Omega} = \frac{1.2 L_c t F_u}{52} = \frac{1.2(1.03)(0.625)65}{2} = 25.1 \text{ kips/bolt}$$

The shear strength is less than the bearing strength $(18 < 25.1)$ and therefore controls:

$$\frac{T_n}{\Omega} = (\text{number of bolts}) \times \frac{R_n}{\Omega} = 4(18.0) = 72 \text{ kips}$$

Thus, a service load of 75 kips will exceed the allowable strength of 72 kips by about 4%, generally too high to accept. Note that this connection investigated by AISC LRFD Method in Example 4.8.1 also obtained a service load capacity of 72 kips, assuming that live load represents 75% of the total service load. ∎

EXAMPLE 4.11.2

Investigate the acceptability of the connection of Fig. 4.8.1 to serve as a slip-critical connection carrying a service load tension of 41 kips. The connection consists of four $\frac{7}{8}$-in.-diam A325 bolts in standard holes connecting A572 Grade 50 steel plates having Class A surface condition. Use the AISC ASD Method.

Solution:

(a) Tension member capacity. The areas as computed in Example 4.8.1 are

$$A_e = A_n = 2.50 \text{ sq in.}$$
$$A_g = 3.75 \text{ sq in.}$$

Then

$$\frac{T_n}{\Omega_t} = \frac{F_y A_g}{\Omega_t} = \frac{50(3.75)}{1.67} = 112 \text{ kips}$$

$$\frac{T_n}{\Omega_t} = \frac{F_u A_e}{\Omega_t} = \frac{65(2.50)}{2.00} = 81 \text{ kips} \quad (\text{controls!})$$

(b) Bolts—slip-critical. Since the holes are standard, use the serviceability limit state. The allowable strength in single shear is determined from Eq. 4.10.5, as follows:

$$\frac{R_n}{\Omega} = \frac{\mu D_u h_{sc} N_s T_b}{\Omega} = \frac{0.35(1.13)(1.00)(1)39}{1.50} = 10.3 \text{ kips/bolt}$$

(c) Bolts—strength (i.e., check as if a bearing-type connection). The allowable strength R_n/Ω in single shear is determined from Eq. 4.10.1. Assume threads are included in the shear plane:

$$\frac{R_n}{\Omega} = \frac{F_{nv} m A_b}{\Omega} = \frac{48(1)0.6013}{2.00} = 14.4 \text{ kips/bolt}$$

The allowable bearing strength for $L_c > 2d$ is

$$\frac{1.2 L_c t F_u}{52} = \frac{1.2(1.03)(0.625)65}{2} = 25.1 \text{ kips/bolt}$$

(d) Allowable service load capacity. The result here is the usual situation; i.e., the slip-critical shear strength governs. Since the serviceability-related value of 10.3 kips is less than strength-related values of 14.4 and 25.1 kips, the allowable load is 10.3 kips/bolt. Thus,

$$T = (\text{number of bolts})R = 4(10.3) = 41.2 \text{ kips}$$

The applied load of 41 kips is essentially the allowable service load permitted on the member with its connections. ∎

4.12 ECCENTRIC SHEAR

When the load P is applied on a line of action that does not pass through the center of gravity of a bolt group, there will be an eccentric loading effect, such as in Fig. 4.12.1. A load P at an eccentricity e, as shown in Fig. 4.12.2, is statically equivalent to a moment P times e plus a concentric load P both acting on the connection. Since both the moment and the concentric load contribute shear effects on the bolt group, the situation is referred to as *eccentric shear.*

Only in the past few years has significant experimental work [3.1, 4.17, 4.18] been done to evaluate the strength of such joints. Essentially, two approaches are available to the designer: (1) the traditional elastic (vector) analysis assuming no friction with the plates rigid and the fasteners elastic; and (2) a strength analysis (plastic analysis) wherein it is assumed the eccentrically loaded fastener group rotates about an instantaneous center of rotation, and the deformation at each fastener is proportional to its distance from the center of rotation.

Since the *AISC Specification* [1.13] prescribes the capacity (allowable strength in ASD and design strength in LRFD) for a fastener but does not specify the means of analysis, any rational approach may be used by the designer. Several common approaches are treated in the remainder of this section.

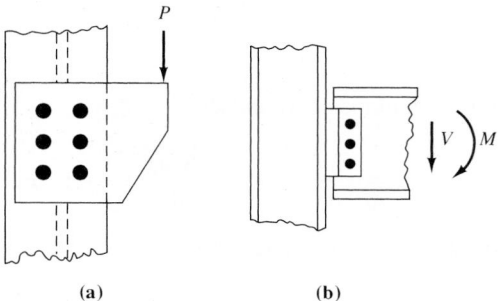

Figure 4.12.1
Typical eccentric shear connections.

(a)   (b)

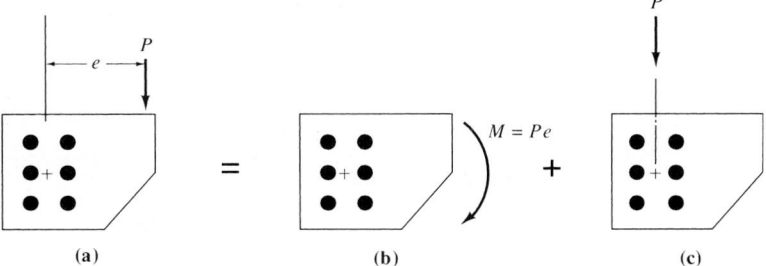

Figure 4.12.2
Combined moment and direct shear.

(a)   (b)   (c)

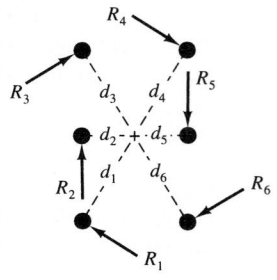

Figure 4.12.3
Pure moment connection.

(a) Connection   **(b)** Forces on connectors

Traditional Elastic (Vector) Analysis

For many years eccentrically loaded fastener groups have been analyzed by considering the fastener group areas as an elastic cross-section subjected to direct shear and torsion. The stresses resulting are nominal in the sense that they have stress units (say, psi) and provide a guide to safety but are not real stresses because the service loads are actually carried by friction. This elastic analysis method has been used because it makes use of simple mechanics of materials concepts and has been found to be a conservative procedure.

To develop the equations for use in this procedure, consider first the connection acted upon by the moment M, as shown in Fig. 4.12.3. Neglecting friction between the plates, the moment equals the sum of the forces shown in Fig. 4.12.3b times their distances to the centroid (CG) of the fastener areas:

$$M = R_1 d_1 + R_2 d_2 + \cdots + R_6 d_6 = \sum Rd \qquad (4.12.1)$$

The deformation in each fastener is assumed proportional to its distance d from the assumed center of twist. If all fasteners are considered elastic and of equal cross-sectional area A, the force R on each fastener is also proportional to its distance d from the centroid of the fastener group,

$$\frac{R_1}{d_1} = \frac{R_2}{d_2} = \cdots = \frac{R_6}{d_6} \qquad (4.12.2)$$

Rewriting the forces in terms of R_1 and d_1,

$$R_1 = \frac{R_1 d_1}{d_1}; \quad R_2 = \frac{R_1 d_2}{d_1}; \quad \cdots; \quad R_6 = \frac{R_1 d_6}{d_1} \qquad (4.12.3)$$

Substituting Eqs. 4.12.3 into Eq. 4.12.1 gives

$$M = \frac{R_1 d_1^2}{d_1} + \frac{R_1 d_2^2}{d_1} + \cdots + \frac{R_1 d_6^2}{d_1}$$

$$= \frac{R_1}{d_1} [d_1^2 + d_2^2 + d_3^2 + \cdots + d_6^2] \qquad (4.12.4a)$$

$$= \frac{R_1}{d_1} \sum d^2 \qquad (4.12.4b)$$

The force in fastener 1 is therefore

$$R_1 = \frac{M d_1}{\sum d^2} \qquad (4.12.5a)$$

and by similar reasoning, the forces on the other fasteners are

$$R_2 = \frac{Md_2}{\Sigma d^2}; \quad R_3 = \frac{Md_3}{\Sigma d^2}; \quad \cdots; \quad R_6 = \frac{Md_6}{\Sigma d^2} \tag{4.12.5b}$$

or in general,

$$R = \frac{Md}{\Sigma d^2} \tag{4.12.6}$$

which gives the force R on the fastener at the distance d from the center of rotation.

Note that if stress is desired, Eq. 4.12.6 may be divided by the bolt cross-sectional area A_b to give stress $f = R/A_b$. The denominator then would become ΣAd^2, which is the polar moment of intertia J about the center of rotation for the series of bolt cross-sectional areas. Equation 4.12.6 is essentially the familiar mechanics of materials formula for torsion on a circular shaft, Tr/J, which is discussed in Sec. 8.2. The torsional moment T is M; the radius r from the center of rotation to the point at which the stress is computed would be the distance d.

It is usually convenient to work with the horizontal and vertical components of R, R_x and R_y, respectively, obtained when the horizontal and vertical components of d, x and y, respectively, are used in Eq. 4.12.6. From Fig. 4.12.4,

$$R_x = \frac{y}{d}R \quad \text{and} \quad R_y = \frac{x}{d}R \tag{4.12.7}$$

Substituting Eq. 4.12.7 into Eq. 4.12.6 gives

$$R_x = \frac{My}{\Sigma d^2} \quad \text{and} \quad R_y = \frac{Mx}{\Sigma d^2} \tag{4.12.8}$$

Noting that $d^2 = x^2 + y^2$, Eqs. 4.12.8 may be written

$$R_x = \frac{My}{\Sigma x^2 + \Sigma y^2} \tag{4.12.9a}$$

$$R_y = \frac{Mx}{\Sigma x^2 + \Sigma y^2} \tag{4.12.9b}$$

By taking the vector sum of R_x and R_y, the total force R on the fastener becomes

$$R = \sqrt{R_x^2 + R_y^2} \tag{4.12.10}$$

To compute the total force on a fastener in an eccentric shear connection such as shown in Fig. 4.12.2a, the direct shear force R_v is

$$R_v = \frac{P}{N} \tag{4.12.11}$$

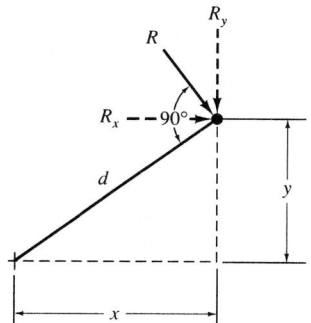

Figure 4.12.4
Horizontal and vertical
components of force R.

where N is the number of fasteners in the group. The total resultant force R then becomes

$$R = \sqrt{[R_y + R_v]^2 + R_x^2} \tag{4.12.12}$$

EXAMPLE 4.12.1

Use the elastic (vector) method to compute the maximum force R on any bolt in the eccentrically loaded bolt group of Fig. 4.12.5a. The bolts are all the same size.

Figure 4.12.5
Example 4.12.1.

(a)   (b) Forces acting on fasteners

Solution:

From Fig. 4.12.5b it can be seen that the upper and lower right fasteners are the most highly stressed. Since these two fasteners are equally stressed, only one need be investigated; check upper-right fastener. The eccentricity e, as measured from the centroid (assumed center of rotation), is

$$e = 3 + 2 = 5 \text{ in.}$$

$$M = 24(5) = 120 \text{ in.-kips}$$

$$\sum x^2 + \sum y^2 = 6(2)^2 + 4(3)^2 = 60 \text{ in.}^2$$

$$R_x = \frac{My}{\sum x^2 + \sum y^2} = \frac{120(3)}{60} = 6.0 \text{ kips} \rightarrow$$

$$R_y = \frac{Mx}{\sum x^2 + \sum y^2} = \frac{120(2)}{60} = 4.0 \text{ kips} \downarrow$$

$$R_v = \frac{P}{N} = \frac{24}{6} = 4.0 \text{ kips} \downarrow$$

$$R = \sqrt{[R_y + R_v]^2 + R_x^2}$$

$$R = \sqrt{(4.0 + 4.0)^2 + (6.0)^2} = 10.0 \text{ kips} \qquad \blacksquare$$

EXAMPLE 4.12.2

Use the elastic (vector) method to compute the force R on the top right bolt in the eccentrically loaded bolt group of Fig. 4.12.6a. The bolts are all the same size.

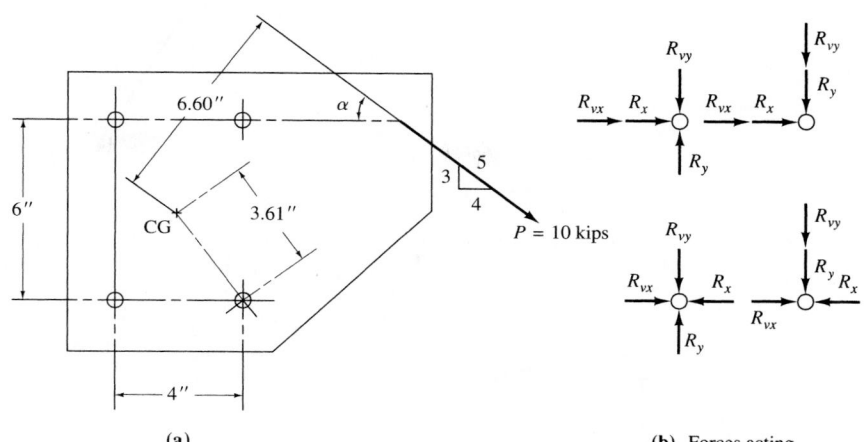

Figure 4.12.6
Example 4.12.2.

(a)

(b) Forces acting
on fasteners

Solution:

$$e = 6.6 \text{ in.}$$

$$M = 10(6.60) = 66.0 \text{ in.-kips}$$

$$\sum d^2 = 4(3.61)^2 = 52 \text{ in.}^2$$

$$R_x = \frac{My}{\sum d^2} = \frac{66.0(3)}{52} = 3.81 \text{ kips} \rightarrow$$

$$R_y = \frac{Mx}{\sum d^2} = \frac{66.0(2)}{52} = 2.54 \text{ kips} \downarrow$$

$$R_{vx} = \frac{P \cos \alpha}{N} = \frac{10(0.8)}{4} = 2.00 \text{ kips} \rightarrow$$

$$R_{vy} = \frac{P \sin \alpha}{N} = \frac{10(0.6)}{4} = 1.50 \text{ kips} \downarrow$$

$$R = \sqrt{[R_y + R_{vy}]^2 + [R_x + R_{vx}]^2}$$

$$R = \sqrt{(2.54 + 1.50)^2 + (3.81 + 2.00)^2} = 7.08 \text{ kips}$$ ∎

Ultimate Strength Analysis

This method, also called plastic analysis, currently is recognized by the *Guide* [3.1] as the most rational one.

The application of the load P causes both a translation and a rotation of the fastener group. The translation and rotation can be reduced to a pure rotation about a point defined as the *instantaneous center of rotation* (see Fig. 4.12.7).

The requirements for equilibrium are as follows:

$$\sum F_H = 0; \qquad \sum_{i=1}^{N} R_i \sin \theta_i - P \sin \delta = 0 \qquad (4.12.13)$$

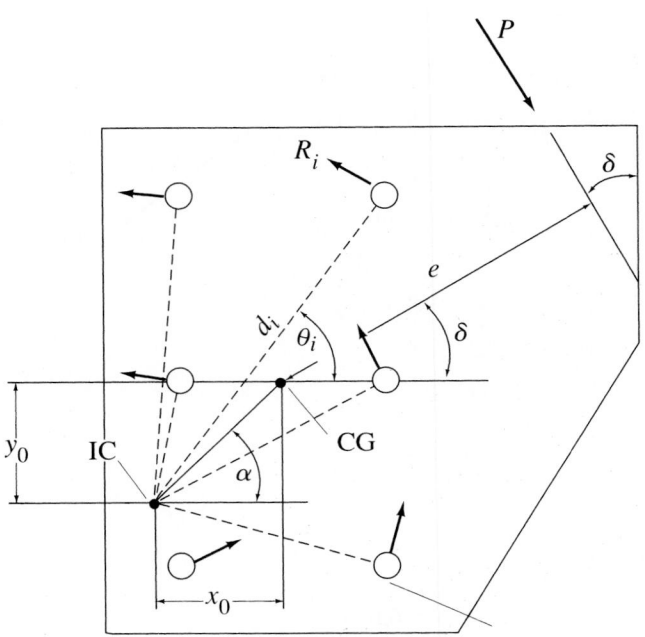

Figure 4.12.7
Instantaneous center of
rotation.

$$\sum F_V = 0; \qquad \sum_{i=1}^{N} R_i \cos \theta_i - P \cos \delta = 0 \qquad (4.12.14)$$

$$\sum M = 0; \qquad \sum_{i=1}^{N} R_i d_i - P(e + x_0 \cos \delta + y_0 \sin \delta) = 0 \qquad (4.12.15)$$

These three equations (Eqs. 4.12.13 to 4.12.15) contain three unknowns (P, x_0, and y_0) and thus exactly determine the location (x_0, y_0) of the instantaneous center and the magnitude of the applied force P. When either the resistance R_i is proportional to the deformation, or when the angle δ is equal to zero or ninety degrees, the angle α is equal to the angle δ and Eq. 4.12.15 reduces to

$$\sum M = 0; \qquad \sum_{i=1}^{N} R_i d_i - P(e + r_0) = 0 \qquad (4.12.16)$$

where r_0 is the distance between the instantaneous center and the centroid (CG) of the connection (r_0 shown in Fig. 4.12.8).

Actually the concept of instantaneous center is identical to the elastic (vector) method when the resistance R_i is assumed proportional to the deformation (i.e., stress is proportional to strain). For either the elastic or the strength method, the deformation is proportional to the distance d_i from the instantaneous center of rotation.

For the strength analysis, two approaches have been used [3.1]. For a *bearing-type* connection, slip is neglected so that the *deformation* of each fastener is proportional to its distance from the instantaneous center. The resistance of each fastener is related to its deformation according to its load-deformation relationship. An expression proposed by Fisher [4.21] and used by Crawford and Kulak [4.17] for this load R vs deformation Δ response is

$$R_i = R_{\text{ult}}(1 - e^{-10\Delta})^{0.55} \qquad (4.12.17)$$

where $R_{\text{ult}} = \tau_u A_b$. The coefficients 10 and 0.55 were experimentally determined and the maximum Δ at failure was about 0.34 in. [3.1]. Note that e in Eq. 4.12.17 is the Naperian base (2.718) and not the eccentricity. For A325 bolts, the ultimate shear strength τ_u is approximately 70% of the tensile strength (120 ksi minimum). Actually, the experimental work directly obtained $\tau_u A_b$ to be 74 kips for $\frac{3}{4}$-in.-diam A325 bolts in double shear, making τ_u equal to 0.7 of 120 ksi.

The experimental work relating to Eq. 4.12.17 used bolts loaded symmetrically (that is, in double shear); however, the general strength method could use *any* appropriate load *R* vs deformation Δ relationship, not necessarily Eq. 4.12.17.

EXAMPLE 4.12.3

Illustrate the general strength method by determining the nominal strength load P_n that may be applied to the fastener group of Fig. 4.12.5. Use Eq. 4.12.17 as the load-deformation expression, and assume that the maximum deformation Δ_{max} at failure is 0.34 in. Assume full shank cross-sections of the bolts resist shear.

Solution:
For $\frac{7}{8}$-in.-diam A325 bolts, Eq. 4.12.17 becomes

$$R_i = 0.7(120)(0.6013)(1 - e^{-10\Delta})^{0.55}$$

$$= 50.5(1 - e^{-10\Delta})^{0.55}$$

The load is applied in the *y*-direction; therefore $\delta = 0$ (Fig. 4.12.7). Using y_i/d_i for $\sin \theta_i$ and x_i/d_i for $\cos \theta_i$, Eqs. 4.12.13 through 4.12.15 become

$$\sum R_i \frac{y_i}{d_i} = 0 \tag{4.12.18}$$

$$\sum R_i \frac{x_i}{d_i} = P_n \tag{4.12.19}$$

$$\sum R_i d_i = P_n(e + r_0) \tag{4.12.20}$$

Also, a basic deformation assumption is

$$\Delta_i = \frac{d_i}{d_{max}}\Delta_{max} = \frac{d_i}{d_{max}}(0.34)$$

(a) Since an iterative process will be required to solve Eqs. (4.12.18 through 4.12.20), let the first trial $r_0 = 3$ in. (see Fig. 4.12.8).

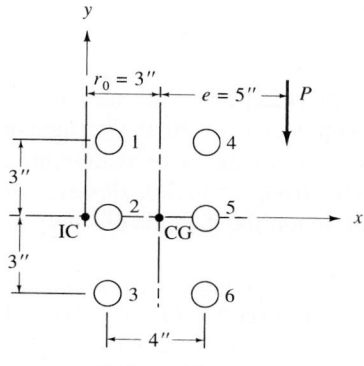

Figure 4.12.8
Strength method for Example 4.12.3.

Fasteners	x_i	y_i	d_i	Δ_i	R_i	$\dfrac{R_i x_i}{d_i}$	$R_i d_i$
1	1	3	3.162	0.184	45.9	14.53	145.3
2	1	0	1.0	0.058	32.2	32.23	32.2
3	1	−3	3.162	0.184	45.9	14.53	145.3
4	5	3	5.831	0.340	49.6	42.51	289.1
5	5	0	5.0	0.292	49.0	48.99	244.9
6	5	−3	5.831	0.340	49.6	42.51	289.1
						195.30	1145.9

$$\text{Eq. 4.12.19:} \qquad P_n = 195 \text{ kips}$$

$$\text{Eq. 4.12.20:} \qquad P_n = \frac{1145.9}{8} = 143 \text{ kips}$$

Since the values are not identical, further trials are required. One will generally find that P_n from Eq. 4.12.20 is relatively close to the correct value even though Eqs. 4.12.19 and 4.12.20 give values that are not very close. A trial value of r_0 that will give P_n between the two values but close to the value from Eq. 4.12.20 will make the calculation converge rather rapidly. The correct value is $r_0 = 2.06$ in., as shown in the following table:

Fasteners	x_i	y_i	d_i	Δ_i	R_i	$\dfrac{R_i x_i}{d_i}$	$R_i d_i$
1	0.06	3	3.001	0.202	46.7	0.93	140.2
2	0.06	0	0.060	0.004	8.6	8.55	0.5
3	0.06	−3	3.001	0.202	46.7	0.93	140.2
4	4.06	3	5.048	0.340	49.6	39.87	250.3
5	4.06	0	4.060	0.273	48.7	48.68	197.6
6	4.06	−3	5.048	0.340	49.6	39.87	250.3
						138.85	979.0

$$\text{Eq. 4.12.9:} \qquad P_n = 139 \text{ kips}$$

$$\text{Eq. 4.12.20:} \qquad P_n = \frac{979.0}{7.06} = 139 \text{ kips}$$

Thus, $P_n = 139$ kips. ∎

Studies [4.17, 4.18, 4.20, 4.22] have indicated that an ultimate strength (plastic) analysis is the most rational approach to obtain the strength of eccentric shear connections. The elastic (vector) analysis was found to be conservative, making the ratio between strength and service load range from 2.5 to 3.0. However, since that elastic method does not properly reflect actual behavior, the margins of safety are variable from case to case, though conservative.

AISC Manual Tables 7-7 to 7-14, pp. 7-32 to 7-79, provide for eccentric shear based on the ultimate strength concept described in this section. Brandt [4.19] has provided a practical procedure for making a general solution to these problems.

Slip-Critical Connections

The same strength procedure is recommended [3.1] for slip-critical connections as for bearing-type connections. Since slip-resistance is a serviceability requirement, not a strength requirement, it is logical to investigate the *strength* of either type connection by the ultimate strength approach described above.

A preferred procedure for analysis of slip-critical connections at service load is to consider the resistance R_i as constant for all fasteners, say, at whatever is the available shear strength. Since the bolts in a slip-critical joint must be installed with initial tension, there will be a fairly uniform clamping action between the pieces being joined.

EXAMPLE 4.12.4

Repeat Example 4.12.3 (Fig. 4.12.5) using $R_i = R_s$ as for a slip-critical connection using the instantaneous center approach, similar to the ultimate strength method. Assume standard holes are used and design for slip-critical serviceability limit state with $\Omega = 1.50$.

Solution:
For $R_i = R_s$ and $\delta = 0$, Eqs. 4.12.13 through 4.12.15 become

$$R_s \sum \frac{y_i}{d_i} = 0 \tag{4.12.21}$$

$$R_s \sum \frac{x_i}{d_i} - P = 0 \tag{4.12.22}$$

$$R_s \sum d_i - P(e + r_0) = 0 \tag{4.12.23}$$

Try $r_0 = 2$ in. Referring to Fig. 4.12.8,

Fastener	x_i	y_i	d_i	$\dfrac{x_i}{d_i}$
1	0	3	3.0	0
2	0	0	0	0
3	0	−3	3.0	0
4	4	3	5.0	0.8
5	4	0	4.0	1.0
6	4	−3	5.0	0.8
			20.0	2.6

From Eq. 4.12.22,

$$P = R_s \sum \frac{x_i}{d_i} = 2.6 R_s$$

From Eq. 4.12.23,

$$P = \frac{R_s \sum d_i}{e + r_0} = \frac{R_s(20)}{7} = 2.86 R_s$$

For this assumption, fastener No. 2 is at the center of rotation and therefore is not involved in Eq. 4.12.23. Also, fastener No. 2 is assumed to have *no* contribution to Eq. 4.12.22. When r_0 is assumed slightly larger than 2.0, Eq. 4.12.22 gives $P = 3.6 R_s$ because $x_i/d_i = 1.0$ for fastener No. 2. When r_0 is assumed slightly smaller than 2.0, Eq. 4.12.22 gives $P = 1.6 R_s$ because $x_i/d_i = -1$ for the same fastener. Thus the value $P = 2.86 R_s$ from Eq. 4.12.23 is accepted as the answer (i.e., $r_0 = 2.0$ in.).

The slip-critical allowable strength per bolt using the serviceability limit state can be obtained from *AISC Manual* Table 7-3, p 7-24, to be

$$R_s = r_0/\Omega_u = 10.3 \text{ kips}$$

and the service load capacity would be

$$P = 2.86(10.3) = 29.5 \text{ kips} \qquad \blacksquare$$

AISC-Load and Resistance Factor Design Method

EXAMPLE 4.12.5

Compare the service load capacities P of the eccentric shear connection of Fig. 4.12.5 when investigated by various methods. Solve assuming the connection is (a) a bearing-type connection with threads excluded from the shear plane, and (b) a slip-critical (friction-type) connection. Assume the live load is 80% and dead load is 20% of the total. Assume the plates are thick enough that bearing on the plates does not control. Use $\frac{7}{8}$-in.-diam A325 bolts and the *AISC Specification* LRFD Method.

Solution:

(a) Elastic analysis—bearing-type connection.

$$R = 10 \text{ kips (fasteners 4 and 6 in Example 4.12.1)}$$

The design strength ϕR_n in shear on the bolt for a bearing-type connection (A325-X) is

$$R_n = F_{nv} m A_b \qquad [4.7.2]$$
$$= 60(1)0.6013 = 36.1 \text{ kips/bolt}$$
$$\phi R_n = 0.75(36.1) = 27.1 \text{ kips/bolt}$$

The factored service load R_u is

$$R_u = 1.2D + 1.6L = 1.2(2) + 1.6(8) = 15.2 \text{ kips}$$

Since the factored load (15.2 kips) on the critical bolt is well below the design strength of 27.1 kips, the load P on the eccentric connection may be increased proportionally. Thus,

$$P = 24(27.1/15.2) = 42.8 \text{ kips}$$

(b) Elastic analysis—slip-critical connection (using serviceability limit state). The design strength ϕR_n is still 27.1 kips/bolt as computed in (a).

$$R_n = \mu D_u h_{sc} N_s T_b$$

Take $\mu = 0.35$, $D_u = 1.13$, $h_{sc} = 1.0$, $N = 1$, and $T_b = 39$ kips. Then

$$R_n = 0.35(1.13)(1)(1)39 = 15.4 \text{ kips/bolt}$$

Using $\phi = 1.0$,

$$\phi R_n = 1.00(15.4) = 15.4 \text{ kips/bolt}$$

which exceeds R_u of 15.2 kips. Thus,

$$P = 24(15.4/15.2) = 24.3 \text{ kips}$$

(c) Strength—bearing-type connection. The design strength ϕR_n is still 27.1 kips/bolt as computed in (a), which means the maximum nominal resistance R_n per bolt is $27.1/\phi = 36.1$ kips. Thus, P_n from Example 4.12.3 must be reduced in the ratio 36.1 to maximum R (i.e., 49.6 kips). Thus,

$$\phi P_n = 0.75(139)(36.1/49.6) = 75.9 \text{ kips}$$

The factored load on the connection is

$$P_u = 1.2(0.2P) + 1.6(0.8P) = 1.52P$$

Equating P_u to ϕR_n gives

$$P = 75.9/1.52 = 49.9 \text{ kips}$$

(d) Instantaneous center—slip-critical connection. The service load capacity from Example 4.12.4 is

$$P = 29.5 \text{ kips}$$

(e) *AISC Manual* [1.15] Tables 7-7 to 7-14 (pp 7-32 to 7-79) for $\theta = 0°$—bearing type connection. Since the horizontal spacing of bolts is 4 in., the AISC LRFD Method tables are not directly applicable. One could interpolate between Table 7-8 for 3 in. and Table 7-9 for $5\frac{1}{2}$ in.

	$e = 5$ in.	Table 7-8	Table 7-9
	$n = 3$ bolts	(p. 7-38)	(p. 7-48)
		Angle $= 0°$	Angle $= 0°$
Coefficient, $C =$		2.59	2.96

Interpolated, $C = 2.59 + (2.96 - 2.59)(1/2.5) = 2.74$

where $P_n = Cr_n =$ design strength for one bolt. Thus,

$$r_n = R_n = 36.1 \text{ kips/bolt}$$
$$P_n = Cr_n = 2.74(36.1) = 98.9 \text{ kips}$$
$$\phi P_n = 0.75(98.9) = 74.2 \text{ kips}$$
$$P = 74.2/1.52 = 48.8 \text{ kips}$$

In this case, the interpolated design strength is less than the value computed in Example 4.12.3; a conservative result but adequate within the accuracy that behavior can be predicted.

(f) *AISC Manual* [1.15] Tables 7-7 to 7–14 (pp. 7-32 to 7-79)—slip-critical connection. The interpolated coefficient $C = 2.74$ from (e) is multiplied by the service load capacity (15.4 kips) per bolt from (b). Thus,

$$\phi P_n = 2.74(15.4) = 42.2 \text{ kips}$$
$$P = 42.2/1.52 = 27.8 \text{ kips}$$

(g) Summary,

Procedure	Bearing-type load P	Slip-critical load P
Elastic (vector) method	42.8 kips	24.3 kips
Strength analysis	49.9 kips	—
Friction analysis	—	29.5 kips
AISC Manual—interpolated	48.8 kips	27.8 kips

Thus, one may note the elastic (vector) method produces the most conservative result.   ∎

Design Formula for Moment on Single Line of Fasteners

AISC Manual [1.15] provides Tables 7-7 to 7-14 that allow the designer to determine the number of connectors required for a given load and eccentricity. In unusual cases for which the tables do not apply, or when they may not be readily available, it is desirable to have a simple alternative method to use. The following development from Shedd* provides a useful simple formula.

Consider a single line of *n* equally spaced fasteners subjected to moment alone, as shown in Fig. 4.12.9. Since with uniform spacing the resistance of the fasteners is uniform from top to bottom, and according to Eq. 4.12.6, $R = Md/(\Sigma d^2)$, the force varies linearly as shown in Fig. 4.12.9.

Assuming R is the force in the outermost fastener, and that it represents the accumulation of stress that would occur on a rectangular resisting section over the distance p, one may designate the average load per inch of height at the outermost fastener as R/p.

Using similar triangles the load per inch at the extreme fiber may be determined,

$$\text{Extreme fiber value} = \frac{R}{p}\left(\frac{n}{n-1}\right) \tag{4.12.24}$$

Figure 4.12.9
Moment on a single line of fasteners.

---

*Thomas C. Shedd, *Structural Design in Steel* (John Wiley & Sons, New York 1934), p. 287.

The tensile force is the area of the triangle represented by the force per inch diagram,

$$T = \frac{1}{2}\left(\frac{np}{2}\right)\left(\frac{R}{p}\right)\left(\frac{n}{n-1}\right) = \frac{Rn^2}{4(n-1)} \tag{4.12.25}$$

The internal resisting moment is

$$M = T\left(\tfrac{2}{3}np\right) \tag{4.12.26}$$

Substitution of Eq. 4.12.25 into Eq. 4.12.26 gives

$$M = \frac{Rn^2}{4(n-1)}\left(\frac{2}{3}np\right) = \frac{Rn^3p}{6(n-1)} \tag{4.12.27}$$

Solving Eq. 4.12.27 for n^2, one obtains

$$n = \sqrt{\frac{6M}{Rp}\left(\frac{n-1}{n}\right)} \tag{4.12.28}$$

which as a first approximation becomes

$$n = \sqrt{\frac{6M}{Rp}} \tag{4.12.29}$$

which is suggested for design use.

Since Eq. 4.12.29 is for moment alone acting on a single row of fasteners, the numerical value for R to be used in it should be adjusted to account for direct shear and for more than one row of fasteners. It is suggested to use a reduced effective R for the direct-shear effect and use an increased effective R for the effect of lateral spread. For lateral spread use a multiplier on R of 1.0 for one line up to about 2.0 for a square array of connectors.

More complicated formulas have been developed to compute the maximum stress or force on a fastener, but no direct solution for the number of connectors or the number of rows is possible from such equations.

EXAMPLE 4.12.6

Determine the required number of $\frac{7}{8}$-in.-diam A325 bolts for one vertical line of bolts A-A in the bracket shown in Fig. 4.12.10. Assume it to be a bearing-type connection with threads included in the shear planes (A325-N). Use the *AISC* LRFD Method.

Solution:

(a) Factored load. Using the gravity load equation, Eq. 1.8.3,

$$P_u = 1.2(7) + 1.6(41) = 74 \text{ kips}$$

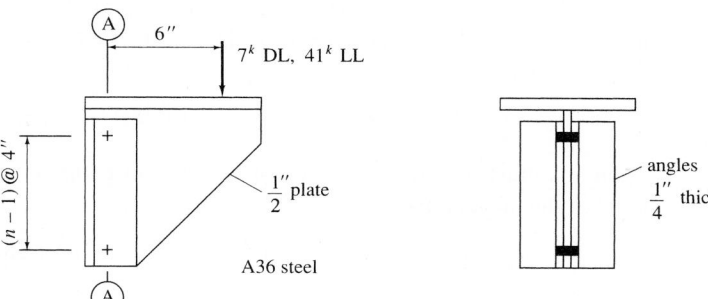

Figure 4.12.10
Example 4.12.6.

(b) Design strength of a bolt in a bearing-type connection. For double shear,

$$R_n = F_{nv}mA_b = 48(2)0.6013 = 57.7 \text{ kips}$$
$$\phi R_n = 0.75(57.7) = 43.3 \text{ kips}$$

For bearing assuming deformation at service load is a consideration [AISC-J4.10(a), Case (i)]

$$R_n = 2.4F_u dt$$
$$= 2.4(58)(0.875)0.5 = 60.9 \text{ kips}$$
$$\phi R_n = 0.75(60.9) = 45.7 \text{ kips}$$

(c) Estimate the number of bolts required, using Eq. 4.12.29.

$$n = \sqrt{\frac{6M}{Rp}} = \sqrt{\frac{6M_u}{\phi R_n p}} = \sqrt{\frac{6(74)(6)}{43.3(4)}} = 3.9$$

The R value (here ϕR_n) has *not* been adjusted for the direct shear effect; try 4 fasteners.

(d) Check the adequacy using the general ultimate strength analysis. Starting with the assumption that the instantaneous center is 3 in. to the left of the vertical line of fasteners, several iterations are required to obtain satisfaction of Eqs. 4.12.18 through 4.12.20. The final results are tabulated below. It is noted that quite a few iterations may be needed to get the values of P_n from Eqs. 4.12.19 and 4.12.20 exactly equal; however, the initial assumption of $r_0 = 3$ in. gave $P_n = 111$ kips from Eq. 4.12.20. The authors have found that even the first trial gives a reasonable approximation of the answer.

Verify that answer is $r_0 = 2.55$ in. to the left of the line of fasteners:

Fastener	x_i	y_i	d_i	Δ_i	R_i	$\dfrac{R_i x_i}{d_i}$	$R_i d_i$
1	2.55	6.00	6.519	0.340	49.6	19.39	323.2
2	2.55	2.00	3.241	0.169	45.2	35.53	146.3
3	2.55	−2.00	3.241	0.169	45.2	35.53	146.3
4	2.55	−6.00	6.519	0.340	49.6	19.39	323.2
						109.84	939.1

Eq. 4.12.19: $P_n = 110$ kips

Eq. 4.12.20: $P_n = 110$ kips (Equal proves solution)

Note that the solution above is for single shear and has assumed that no threads are in the shear planes (A325-X). Also, the maximum R_i, even though logically computed using Eq. 4.12.17, cannot exceed the AISC-Table J3.2 specified value (based on $F_{nv} = 60$ ksi),

$$R_n = F_{nv}A_b = 60(0.6013) = 36.1 \text{ kips}$$

Then, P_n must be reduced in proportion that 36.1 kips is to the maximum R_i (i.e., 49.6 kips) in the table above,

$$P_n = 110(36.1/49.6) = 80.1 \text{ kips}$$

for A325-X in *single shear.*

For A325-N, this value must be multiplied by 2 for the double shear case of this example, and multiplied by 0.8 because threads are possible in the shear planes (A325-N).

Compare ϕP_n with P_u using the shear-related ϕ value since shear controlled the fastener strength,

$$[\phi P_n = 0.75(80.1)2(0.8) = 96 \text{ kips}] > [P_u = 74 \text{ kips}] \qquad \textbf{OK}$$

The above shows that 4 fasteners in a line are more than adequate. Investigation (not shown) for 3 fasteners indicates that 3 fasteners are not adequate.

(e) Check the adequacy using the elastic (vector) method. The direct shear component R_{us} from Eq. 4.12.11 is

$$R_{us} = \frac{P_u}{N} = \frac{74}{4} = 18.5 \text{ kips} \downarrow$$

The moment component R_{ux} from Eq. 4.12.9a (noting that M equals P_u times e) is

$$R_{ux} = \frac{P_u e y}{\Sigma x^2 + \Sigma y^2} \qquad [4.12.9a]$$

$$\Sigma x^2 + \Sigma y^2 = 2[(2)^2 + (6)^2] = 80 \text{ in.}^2$$

$$R_{ux} = \frac{74(6)6}{80} = 33.3 \text{ kips} \rightarrow$$

Then, using Eq. 4.12.12, the resultant is obtained,

$$R_u = \sqrt{(18.5)^2 + (33.3)^2} = 38.1 \text{ kips} < [\phi R_n = 43.3 \text{ kips}]$$

Thus, the factored load R_u on the most heavily loaded bolt does not exceed the design strength $\phi R_n = 43.3$ kips for A325-N. Thus, 4 fasteners are acceptable.

It must be noted that the clear edge distance L_c measured in the direction of the resultant force must satisfy Eq. 4.7.13 (AISC-J3.10) requiring that $L_c \geq R_u/(\phi K F_u t)$. Note that the factored force R_u on the most heavily loaded bolt is used as P in the original equation.

Use $4—\frac{7}{8}$-*in.-diam A325-N bolts @ 4-in. pitch.* ∎

EXAMPLE 4.12.7

Determine the required number of $\frac{3}{4}$-in. diam A325 bolts in standard holes for the bracket plate of Fig. 4.12.11, assuming 4 vertical rows. Use a slip-critical connection with clean mill scale (Class A) surface condition, and use the *AISC Specification* LRFD Method.

Solution:

(a) Factored load. Since one-half the load is carried by each plate in a single shear situation, use one-half the total load in all calculations for one plate. Using the gravity load equation, Eq. 1.8.3,

$$P_u = 1.2(5) + 1.6(55) = 94 \text{ kips} \quad (\text{or } 47 \text{ kips/plate})$$

(b) Design strength ϕR_n of a bolt in a slip-critical connection. For strength, values are as for bearing-type connections. Assume no threads are to be in the shear planes (A325-X). For single shear,

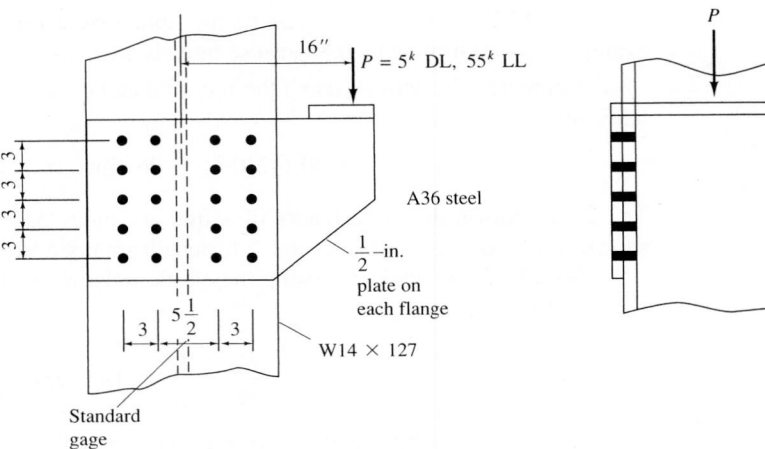

Figure 4.12.11
Eccentric shear connection of
Example 4.12.7.

$$R_n = F_{nv}mA_b$$

$$= 60(1)0.4418 = 26.5 \text{ kips} \quad \text{(nominal shear strength—single shear)}$$

$$\phi R_n = 0.75(26.5) = 19.9 \text{ kips}$$

For bearing assuming deformation at service load is a consideration [AISC-J4.10(a), Case (i)]

$$R_n = 2.4F_u dt \quad \text{(nominal bearing strength)}$$

$$= 2.4(58)(0.75)0.5 = 52.2 \text{ kips}$$

$$\phi R_n = 0.75(52.2) = 39.2 \text{ kips}$$

For slip resistance, the limit state of slip is

$$R_n = \mu D_u h_{sc} N_s T_b = 0.35(1.13)(1.0)(1)28 = 11.1 \text{ kips}$$

$$\phi R_n = 1.0(11.1) = 11.1 \text{ kips} \quad \text{(serviceability limit state with standard holes)}$$

$$R = \frac{R_n}{\Omega} = \frac{11.1}{1.5} = 7.4 \text{ kips}$$

(c) Estimate the number of bolts required, using Eq. 4.12.29. It is assumed that slip resistance at service load is the controlling limit state; strength will later be checked.

$$n = \sqrt{\frac{6M}{Rp}} = \sqrt{\frac{6(30/4)(16)}{7.4(3)}} = 5.7$$

In the above equation, the service load per plate is 30 kips, and the load per line of fasteners is 30/4, which must be used since Eq. 4.12.29 applies to one line of fasteners. No adjustment in R was made either for direct shear or for several lines of fasteners. Try 5 bolts per row.

(d) Check the adequacy using the general ultimate strength analysis. Referring to Fig. 4.12.12, and using the same method illustrated in Example 4.12.6, the value of r_0 is found to be 3.03 in. to the left of the centroid of the fastener group:

Figure 4.12.12
Fastener numbering for
Example 4.12.7.

Fastener	x_i	y_i	d_i	Δ_i	R_i	$\dfrac{R_i x_i}{d_i}$	$R_i d_i$
1	−2.72	6.00	6.588	0.211	34.6	−14.27	227.6
2	−2.72	3.00	4.049	0.129	31.1	−20.90	126.0
6	0.28	6.00	6.007	0.192	34.0	1.59	204.3
7	0.28	3.00	3.013	0.096	28.5	2.65	85.8
11	5.78	6.00	8.331	0.266	35.7	24.74	297.1
12	5.78	3.00	6.512	0.208	34.5	30.61	224.6
16	8.78	6.00	10.634	0.340	36.4	30.07	387.3
17	8.78	3.00	9.278	0.297	36.0	34.11	334.5
					$\Sigma =$	88.60	1887.2
	Multiply the above times 2 for symmetry $=$					177.20	3774.4
3	−2.72	0.00	2.720	0.087	27.5	−27.53	74.9
8	0.28	0.00	0.280	0.009	9.6	9.60	2.7
13	5.78	0.00	5.780	0.185	33.8	33.77	195.2
18	8.78	0.00	8.780	0.281	35.9	35.86	314.9
					$\Sigma =$	228.90	4362.1

Eq. 4.12.18: $P_n = 228.9$ kips

Eq. 4.12.19: $P_n = 4362.1/(16 + 3.03) = 229.2$ kips

Note that the solution above is for single shear and has assumed that no threads are in the shear planes (A325-X). In addition, consider that LRFD limits the maximum strength of a bolt in shear to

$$R_n = F_{nv}A_b = 60(0.4418) = 26.5 \text{ kips}$$

Reducing P_n from the strength analysis,

$$P_n = 229(26.5/36.4) = 167 \text{ kips}$$

$$[\phi P_n = 0.75(167) = 125 \text{ kips}] > [P_u = 47 \text{ kips}]$$ **OK**

Strength is more than adequate; it was not expected that strength would govern for this slip-resistant connection having standard holes and Class A surface condition. Only when using a surface condition having a very high slip coefficient is there a real probability of strength controlling.

(e) Using the strength analysis approach but reducing the capacity in proportion that the allowable capacity [7.4 kips from part (b)] of a slip-critical connection is to the force (36.4 kips) on the most heavily loaded fastener in the strength analysis of the bearing-type connection. Thus, the service load capacity is

$$P = 229(7.4/36.4) = 46.6 \text{ kips} > [\text{Required 30 kips}]$$

This procedure is always conservative. An alternative is to do an analysis as illustrated in Example 4.12.4 where the resistance of each fastener is the same.

(f) Check using *AISC Manual* Table 7-14 (p. 7-74) for eccentrically loaded bolt groups. The horizontal spacing of fasteners is not exactly the same as in Table 7-14; however, when the external dimensions of the array are about the same, and the vertical spacing of bolts agrees with the tables, the coefficient will be found to be about the same. If the AISC table is assumed applicable as the authors believe it is for this case, enter the table with the number ($n = 5$) of fasteners in a vertical line, and the eccentricity ($e = 16$ in.) of the load from the center of gravity of the group of fasteners,

$$\text{Find coefficient } C = 6.15$$
$$\phi P_n = C\phi R_n = 6.15(19.9) = 122 \text{ kips}$$

which compares with $\phi P_n = 125$ kips for a bearing-type connection from the analysis in (d).

For the slip-critical connection of this example,

$$P = CR = 6.15(7.4) = 46 \text{ kips}$$

which compares with $P = 46.6$ kips computed in (e).

(g) Investigate the connection using the elastic (vector) method. Since the slip-critical service load condition has already been shown to govern, the illustration of the use of the elastic (vector) method is performed for the service load (30 kips on one plate).

Check 5 bolts per row by elastic method:

$$R_s = \frac{P}{n} = \frac{30}{20} = 1.50 \text{ kips} \downarrow$$

$$\sum x^2 + \sum y^2 = 10[(2.75)^2 + (5.75)^2] + 8[(3)^2 + (6)^2] = 766 \text{ in.}^2$$

$$R_x = \frac{My}{\sum x^2 + \sum y^2} = \frac{30(16)6}{766} = 3.76 \text{ kips} \rightarrow$$

$$R_y = \frac{Mx}{\sum x^2 + \sum y^2} = \frac{30(16)5.75}{766} = 3.60 \text{ kips} \downarrow$$

$$\text{Actual } R = \sqrt{(1.50 + 3.60)^2 + (3.76)^2} = 6.34 \text{ kips} < 7.5 \qquad \textbf{OK}$$

Use 5—$\frac{3}{4}$-diam A325 bolts per row. The fasteners (3, 8, 13, and 18) located on the *x*-axis could have been omitted; however, the regular pattern of Fig. 4.12.11 is preferred.   ∎

4.13   FASTENERS ACTING IN AXIAL TENSION

Axial tension occurring without simultaneous shear exists in fasteners for tension members such as hangers (see Fig. 4.6.1c) or other members whose line of action is perpendicular to the member to which it is fastened. When such tension members are not perpendicular to their connecting members, the fasteners are subjected to both axial tension and shear. The latter, more typical case, is discussed in Sec. 4.14.

Section 4.6.1 discusses the nominal strength R_n for fasteners subject to axial tension; Eq. 4.6.1 gives the strength as

$$R_n = F_{nt}A_b \qquad [4.6.1]$$

When Eq. 4.6.1 is used in Load and Resistance Factor Design (AISC-Table J3.2), the design strength ϕR_n using the resistance factor ϕ of 0.75 is

$$\phi R_n = 0.75F_{nt}A_b \qquad (4.13.1)$$

Values of ϕF_{nt} have been given in Table 4.7.1.

When Eq. 4.6.1 is used in Allowable Strength Design, the allowable strength is the nominal strength R_n divided by the factor of safety of 2. Thus,

$$R_n/\Omega = R_n/2 = 0.50F_{nt}A_b \qquad (4.13.2)$$

and

$$R_n/\Omega \geq T_a$$

where T_a is the required tensile strength per bolt R_n/Ω must exceed the T_a caused by factored loads.

Fasteners subject to axial tension must be pretensioned according to Table 4.4.1 whether the design is for a bearing-type connection or a slip-critical connection [4.9].

Prestress Effect of High-Strength Bolts Under External Tension

To understand the effect of an externally applied load on a pretensioned high-strength bolt, consider a single bolt and the tributary portion of the connected plates as shown in Fig. 4.13.1a. The pieces being joined are of thickness t and the area of contact between the pieces is A_p. Prior to applying external load, the situation is as shown in Fig. 4.13.1b, where the bolt has been installed to have a pretension force T_b (values as in Table 4.4.1). The pieces being joined are initially compressed an amount C_i. For equilibrium,

$$C_i = T_b \qquad (4.13.3)$$

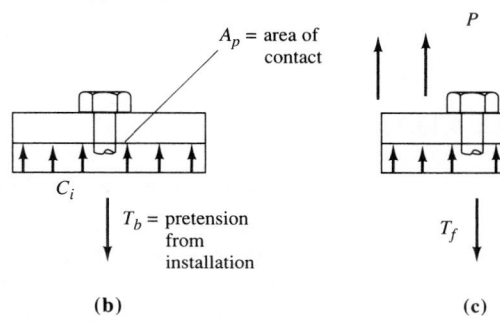

Figure 4.13.1   Prestress effect on bolted joint.

The external load P is then applied and the forces acting are shown in Fig. 4.13.1c. This time equilibrium requires

$$P + C_f = T_f \tag{4.13.4}$$

where the subscript f refers to final conditions after application of the load P.

The force P acting on the system lengthens the bolt an amount δ_b between the underside of the bolt head and the surface of contact between the two connected plates,

$$\delta_b = \frac{T_f - T_b}{A_b E_b} t \tag{4.13.5}$$

where E_b = modulus of elasticity of the bolt

T_f = final force in the bolt after external load is applied.

At the same time, the compression between the plates decreases and the plate thickness increases an amount δ_p,

$$\delta_p = \frac{C_i - C_f}{A_p E_p} t \tag{4.13.6}$$

where E_p = modulus of elasticity of the plate material

C_f = final compression force between the pieces being joined, after external load is applied

If contact is maintained, compatibility of deformation requires $\delta_b = \delta_p$; thus, Eq. 4.13.5 equated to Eq. 4.13.6 gives

$$\frac{T_f - T_b}{A_b E_b} = \frac{C_i - C_f}{A_p E_p} \tag{4.13.7}$$

Next, substitution of Eq. 4.13.3 for C_i and Eq. 4.13.4 for C_f into Eq. 4.13.7 gives

$$\frac{T_f - T_b}{A_b E_b} = \frac{T_b - T_f + P}{A_p E_p} \tag{4.13.8}$$

The moduli of elasticity E_b for the bolt and E_p for the plate are essentially the same and may be eliminated. Then solving for T_f gives

$$(T_f - T_b)\frac{A_p}{A_b} = T_b - T_f + P \tag{4.13.9}$$

$$T_f\left(1 + \frac{A_p}{A_b}\right) = T_b\left(1 + \frac{A_p}{A_b}\right) + P$$

$$T_f = T_b + \frac{P}{1 + A_p/A_b} \tag{4.13.10}$$

EXAMPLE 4.13.1

Assume $\frac{7}{8}$-in.-diam A325 bolts are used in a direct tension situation such as in Fig. 4.13.2. With bolts spaced 3 in. apart and having $1\frac{1}{2}$-in. edge distances, the tributary area of contact may reasonably be about 9 sq in. If the maximum external tensile load permitted by AISC Load and Resistance Factor Design is applied, how much does the bolt tension increase? Assume the service load is 20% dead load and 80% live load.

Figure 4.13.2
Example 4.13.1.

Solution:

(a) Design strength ϕR_n. Using Eq. 4.13.1,

$$\phi R_n = 0.75 F_{nt} A_b = 0.75(90)0.6013 = 40.6 \text{ kips}$$

(b) Permissible service load R per bolt. Equating the design strength ϕR_n to the factored load R_u gives

$$R_u = 1.2(0.2R) + 1.6(0.8R) = 1.52R$$

$$1.52R = 40.6 \text{ kips}$$

$$R = 26.7 \text{ kips}$$

(c) Initial tensile force in $\frac{7}{8}$-in.-diam A325 bolt. From Table 4.4.1,

$$T_b = 39 \text{ kips}$$

(d) Determine final tensile force in bolt. The ratio of plate contact area to bolt area is

$$\frac{A_p}{A_b} = \frac{9}{0.6013} = 15$$

This neglects subtracting the bolt area from the total tributary area, but little difference results. Using Eq. 4.13.10 with the load P per fastener equal to its maximum value R gives

$$T_f = 39 + \frac{26.7}{1 + 15} = 39 + 1.7 = 40.7 \text{ kips}$$

The increase in tension is 4.3%. The variation in actual pretension from installation may be expected to exceed this amount, so that this increase is not of concern. Furthermore the tributary area used for the example (9 sq in.) is probably the minimum one might encounter in practice, since less than a 3-in. pitch or gage is rarely used.

The important conclusion from this example is that *no significant increase in bolt tension arises until the external load equals or exceeds the pretension force,* in which case the pieces do not remain in contact and the applied force equals the bolt tension.

If the connection can distort and give rise to "prying forces" these must also be considered. (See AISC-J3.6 and the treatment in the "Split-Beam Tee Connections" part of Sec. 13.6.) In the situation of Example 4.13.1, the approximate factor of safety against overcoming initial compression between pieces is

$$\text{FS} = \frac{T_b}{P} = \frac{39}{26.7} \approx 1.5$$

In general, for A325 bolts under AISC-J3, the margin of safety against service load exceeding the proof load is approximately 1.5 for diameters up to 1 in. and approximately 1.3 for diameters over 1 in. ∎

EXAMPLE 4.13.2

Determine the required number of $\frac{3}{4}$-in.-diam A490 bolts for the connection shown in Fig. 4.13.3. Assume that the pieces making up the connection are adequate, and very stiff such that prying forces (see Sec. 13.6) can be disregarded, and that the nominal tensile stresses on the bolts govern. Assume the load is 10% dead load and 90% live load. Use the *AISC*-LRFD Method.

Solution:

(a) Design strength ϕR_n per bolt. Using Eq. 4.13.1,

$$\phi R_n = 0.75 F_{nt} A_b = 0.75(113)0.4418 = 37.4 \text{ kips}$$

(b) Calculate the factored tension force T_u.

$$T_u = 1.2(0.1)(140) + 1.6(0.9)(140) = 218 \text{ kips}$$

140 kips

140 kips

Section A - A

Figure 4.13.3
Example 4.13.2.

(c) Determine the number n of bolts required.

$$n = \frac{T_u}{\phi R_n} = \frac{218}{37.4} = 5.8, \quad \text{say } 6$$

Use 6—$\frac{3}{4}$-in.-diam A490 bolts. ∎

4.14 COMBINED SHEAR AND TENSION

In a large number of commonly used connections, both shear and tension occur and must be considered in design. Figure 4.14.1 shows a few typical connections in which the connectors are simultaneously subjected to both shear and tension. The connection shown in Fig. 4.14.1a is a common one where two angles join the beam web to the column flange. From the moment force indicated in the figure, the upper fasteners are subjected to tension in proportion to the magnitude of the applied moment. However, one may recall from structural analysis that only a small amount of end rotation on a beam is necessary to change from a fixed end to a hinged end condition. In addition, the web carries only a small part of the bending moment. Thus, one may intuitively sense that the moment shown will be relieved before a significant tension force can be developed in the fasteners. Such connections are used when little end moment is desired to be transmitted. An exception to this occurs in the case of a very deep beam such as a plate girder.

Referring next to Fig. 4.14.1b in the which the applied moment is transmitted through the flanges of the beam, the situation is different. In this case a large applied moment is intended to be transmitted so the connection is made at the flanges, the elements carrying most of the moment. Chapter 13 treats this type of connection. Figure 4.14.1c and 4.14.1d typify the two types of fastener loading in combined shear and tension which are developed in the following parts of this section.

Bearing-Type Connections—AISC LRFD Method

In earlier sections, the nominal strengths of bolts loaded separately in shear and tension were treated. When the full strength in tension is required, the full strength in shear is not

Figure 4.14.1
Typical combined shear and tension connections.

simultaneously available. Based on experimental studies [4.23, 3.1], the strength interaction equation may be represented by a circular relationship

$$\left(\frac{f_t}{F_{nt}}\right)^2 + \left(\frac{f_v}{F_{nv}}\right)^2 = 1 \tag{4.14.1}$$

where
f_t = required tensile stress due to factored service loads
f_v = required shear stress due to factored service loads
F_{nt} = nominal tensile strength of bolt material (AISC-TableJ3.2)
F_{nv} = nominal shear strength of bolt material (AISC-TableJ3.2)

AISC Specification have simplified the elliptical interaction relationship of Eq. 4.14.1 into a straight line which required reduction only in the most severe loading cases. The straight line expression is

For LRFD Method $\qquad \dfrac{f_t}{\phi F_{nt}} + \dfrac{f_v}{\phi F_{nv}} = \text{constant} = 1.3 \tag{4.14.2a}$

For ASD Method $\qquad \dfrac{\Omega f_t}{F_{nt}} + \dfrac{\Omega f_v}{F_{nv}} = \text{constant} = 1.3 \tag{4.14.2b}$

For the LRFD case, the available reduced tensile strength (F'_{nt}) in the presence of a specific level of shear stress can be determined, by multiplying both sides of Eq. 4.14.2a by F_{nt} and solving for f_t/ϕ

$$F'_{nt} = \frac{f_t}{\phi} = 1.3F_{nt} - \frac{F_{nt}}{\phi F_{nv}}f_v \le F_{nt} \tag{4.14.3}$$

which is AISC Formula J3-3a.
The tensile stress in the bolt must satisfy

$$\frac{R_{ut}}{A_b} \le \phi F'_{nt} = \phi\left[1.3F_{nt} - \frac{F_{nt}}{\phi F_{nv}}f_v\right] \le \phi F_{nt} \tag{4.14.4}$$

where R_{ut} = the factored tension load on the bolt, kips
A_b = gross cross-sectional bolt area, sq. in.
ϕ = 0.75

For A325 bolts, the value of F_{nt}/F_{nv} is equal to 1.9 and 1.5 for including and excluding threads in the shear planes, respectively. Substituting for F_{nt}/F_{nv} in Eq. 4.14.4

$$\phi F'_{nt} = \phi[1.3(90) - 1.9f_v] = \phi[117 - 1.9f_v] \le \phi F_{nt} \quad \text{for A325-N} \tag{4.14.5}$$

and

$$\phi F'_{nt} = \phi[117 - 1.5f_v] \le \phi F_{nt} \quad \text{for A325-X} \tag{4.14.6}$$

Equations 4.14.5 and 4.14.6 are shown in Figure 4.14.3 with $\phi = 0.75$ included. The equations for $\phi F'_{nt}$ are summarized for A325 and A490 bolts in Table 4.14.1.

Slip-Critical Connections—AISC LRFD Method

The presence of tension in a slip-critical connection reduces the clamping force and hence, the frictional resistance. Accordingly, the available slip resistance per bolt must be reduced. AISC-J3.9 provides a reduction factor k_s to be applied to the slip-resistance limit state computed according to AISC-J3.8. The factor k_s is as follows:

$$k_s = 1 - \frac{T_u}{D_u N_b T_b} \tag{4.14.7}$$

where $D_u = 1.13$

N_b = number of bolts carrying the applied tension

T_u = factored tension load due to LRFD combinations in ASCE 7

T_b = minimum pretension for bolt given in AISC-Table J3.1

N_b = number of bolts carrying the applied tension

Figure 4.14.2
Nondimensional shear—
tension strength interaction
curve: bearing-type
connections.

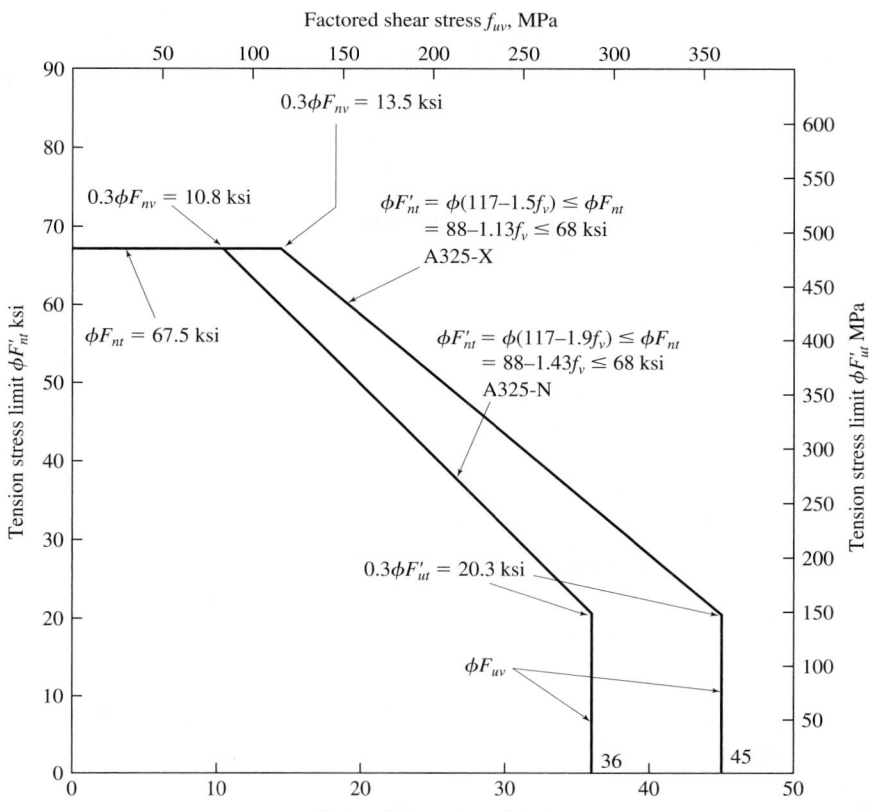

Figure 4.14.3
Interaction relationship for
combined shear and tension
in A325-X and A325-N
bearing-type connections.

TABLE 4.14.1 Design Tension Stress‡ Limit $\phi F'_{ut}$ in the Presence of Factored Shear Stress f_v (for Bearing-Type Connections) (based on Eq. 4.14.4)

Fastener	$\phi F'_{ut}$	
	(ksi)	(MPa)
A307 bolts	$\phi(59 - 1.9f_{uv}) \leq \phi(45)$	$\phi(407 - 1.9f_{uv}) \leq \phi(310)$
A325-N bolts (threads *not* excluded)	$\phi(117 - 1.9f_{uv}) \leq \phi(90)$	$\phi(807 - 1.9f_{uv}) \leq \phi(621)$
A325-X bolts (threads excluded)	$\phi(117 - 1.5f_{uv}) \leq \phi(90)$	$\phi(807 - 1.5f_{uv}) \leq \phi(621)$
A490-N bolts (threads *not* excluded)	$\phi(147 - 1.9f_{uv}) \leq \phi(113)$	$\phi(1010 - 1.9f_{uv}) \leq \phi(779)$
A490-X bolts (threads excluded)	$\phi(147 - 1.5f_{uv}) \leq \phi(113)$	$\phi(1010 - 1.5f_{uv}) \leq \phi(779)$

* Note that $\phi = 0.75$.

‡ Nominal stress due to factored load acting on gross bolt cross-sectional area, $f_t = R_{ut}/A_b$

To demonstrate the variation of the reduction factor as a function of the ratio T_u/T_b, consider a four-bolt connection and a six-bolt connection. The reduction factors will be

$$k_s = 1 - \frac{T_u}{D_u N_b T_b}$$

$$= 1 - \frac{T_u}{1.13(4)T_b} = 1 - 0.221\left(\frac{T_u}{T_b}\right) \quad \text{for a 4-bolt connection}$$

$$= 1 - \frac{T_u}{1.13(6)T_b} = 1 - 0.148\left(\frac{T_u}{T_b}\right) \quad \text{for a 6-bolt connection}$$

The equations are shown in Fig. 4.14.4.

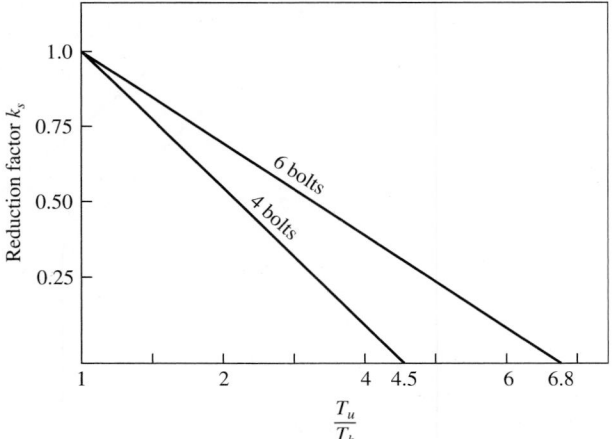

Figure 4.14.4
Reduction factor for the slip resistance limit state in the presence of tension (AISC-J3.9).

EXAMPLE 4.14.1

Using Load and Resistance Factor Design, determine the adequacy of the fasteners in Fig. 4.14.5 when $\frac{7}{8}$-in.-diam A325 bolts are used in (a) a bearing-type connection (A325-X) with threads eXcluded from the shear planes, and (b) a slip-critical connection (A325-SC) with Class A surface condition and standard holes. Assume the strength of the column flange and the ST section do not govern the answer. Neglect prying action (see Sec. 13.6). The gravity loading is 10% dead and 90% live load.

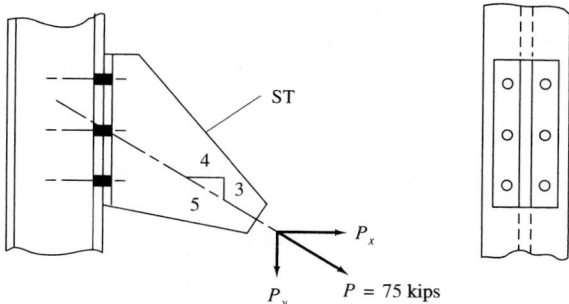

Figure 4.14.5
Example 4.14.1.

Solution:

(a) Check connection as a bearing-type connection (A325-X). Obtaining the tension and shear components of the factored applied force,

$$P_u = 1.2(0.1)(75) + 1.6(0.9)(75) = 117 \text{ kips}$$

$$\text{Tension component} = P_{ux} = 0.8(117) = 93.6 \text{ kips}$$

$$\text{Shear component} = P_{uy} = 0.6(117) = 70.2 \text{ kips}$$

The factored loads T_u and V_u per bolt are

$$\text{Tension } T_u = 93.6/6 = 15.6 \text{ kips/bolt}$$

$$\text{Shear } V_u = 70.2/6 = 11.7 \text{ kips/bolt}$$

The factored load shear stress is

$$\left[f_v = \frac{V_u}{A_b} = \frac{11.7}{0.6013} = 19.5 \text{ ksi} \right] > [0.2\phi F_{nv} = 0.2(0.75)60 = 9 \text{ ksi}]$$

Therefore, the shear–tension interaction must be considered. From AISC-J3.7, the available stress in tension F'_{nt} per bolt is

$$F'_{nt} = 1.3 F_{nt} - \frac{F_{nt}}{\phi F_{nv}} f_v \leq [F_{nt} = 90 \text{ ksi}]$$

$$= 1.3(90) - \frac{90}{0.75(60)}(19.5) = 78 \text{ ksi}$$

$$\phi R_{nt} = \phi F'_{nt} A_b$$

$$= 0.75(78)0.6013 = 35.2 \text{ kips} > [T_u = 15.6 \text{ kips/bolt}] \qquad \textbf{OK}$$

The connection is very conservatively designed as a bearing-type connection.

(b) Check connection as a slip-critical connection (A325-SC). For strength, a slip-critical connection must satisfy the same strength-related criteria of a bearing-type connection. The shear strength was investigated in (a); bearing strength was given in the problem statement as not controlling.

For serviceability, the adequacy of slip resistance must be checked. Since standard holes are used, design for the serviceability limit state. The reduction factor k_s is calculated as follows:

$$k_s = 1 - \frac{T_u}{D_u N_b T_b}$$

$$= 1 - \frac{93.6}{1.13(6)39} = 0.646$$

and the reduced design slip resistance is

$$\phi R_{nv} = \phi k_s \mu D_u h_{sc} N_s T_b$$

$$= 1.0(0.646)0.35(1.13)(1.0)(1)39$$

$$= 9.96 \text{ kips/bolt} < [V_u = 11.7 \text{ kips/bolt}] \qquad \textbf{NG}$$

Since the reduced ϕR_{nv} is less than the factored shear V_u the bolts are not satisfactory as a slip-critical connection. ∎

EXAMPLE 4.14.2

Determine the maximum value of the load P in Example 4.14.1 assuming (a) a slip-critical connection, and (b) a bearing-type connection (A325-N) with threads possible in the shear planes. Use *AISC Specification* LRFD Method and assume Class A surface condition with standard holes.

Solution:

(a) Slip-critical connection. The factored service load force P_u is

$$P_u = 1.2(0.1)P + 1.6(0.9)P = 1.56P$$

$$P_{ux} = \text{tension component} = 0.8P_u = 1.25P$$

$$P_{uy} = \text{shear component} = 0.6P_u = 0.94P$$

The factored load per bolt in shear is

$$V_u = 0.94P/6 = 0.156P \text{ per bolt}$$

The factored load per bolt in tension is

$$T_u = 1.25P/6 = 0.208P \text{ per bolt}$$

The design strength in shear ϕR_{nv} is

$$\phi R_{nv} = \phi k_s \mu D_u h_{sc} N_s T_b$$

$$= 1.0\left(1 - \frac{1.248P}{1.13(39)6}\right)(0.35)(1.13)(1.0)(1)39$$

$$= 15.43 - 0.0728P$$

Equating the factored shear V_u to the design strength ϕR_{nv} gives

$$0.1560P = 15.43 - 0.0728P$$
$$P = 67.4 \text{ kips}$$

The strength in shear, tension, and bearing must also be checked as if this were a bearing-type connection. For some very slip-resistant surface conditions the strength as a bearing-type connection may control instead of slip resistance; a relatively unusual situation.

(b) Bearing-type connection (A325-N) with threads possible in the shear planes. The factored loads are

$$P_u = 1.2(0.1)P + 1.6(0.9)P = 1.56P$$
$$P_{ux} = \text{tension component} = 0.8P_u = 1.25P$$
$$P_{uy} = \text{shear component} = 0.6P_u = 0.94P$$

The factored load per bolt in shear is

$$V_u = 0.94P/6 = 0.156P \text{ per bolt}$$

The factored load per bolt in tension is

$$T_u = 1.25P/6 = 0.208P \text{ per bolt}$$

From AISC-J3.7, the available stress in tension F'_{nt} per bolt is

$$F'_{nt} = 1.3F_{nt} - \frac{F_{nt}}{\phi F_{nv}}f_v \leq [F_{nt} = 90 \text{ ksi}]$$

$$f_v = \frac{V_u}{A_b} = \frac{0.156P}{0.6013} = 0.26P$$

$$F'_{nt} = 1.3(90) - \frac{90}{0.75(48)}(0.26P)$$

$$= 117 - 0.65P$$

The design strength ϕR_{nt} in tension is

$$\phi R_{nt} = \phi F'_{nt}A_b = 0.75(117 - 0.65P)0.6013 = 52.8 - 0.293P$$

Equate T_u to ϕR_{nt}, and solve for P

$$0.208P = 52.8 - 0.293P$$
$$P = 105 \text{ kips}$$

Check maximum nominal tensile strength

$$F'_{nt} = 117 - 0.65(P)$$

$$= 117 - 0.65(105) = 48.7 \text{ ksi} < 90 \text{ ksi} \qquad \textbf{OK}$$

Thus, the maximum value of the service load P is 67 kips as a slip-critical connection and 105 kips as a bearing-type (A325-N) connection.   ∎

EXAMPLE 4.14.3

Determine the number of $\frac{3}{4}$-in.-diam A325 bolts required to carry a shear force consisting of 14 kips dead load and 56 kips live load, and a tension force of 24 kips dead load and 96 kips live load. The connection is to be designed such that the resultant force acts through the centroid of the connection. Use Load and Resistance Factor Design: (a) design as a bearing-type (A325-X) connection, and (b) design as a slip-critical (A325-SC) connection having Class A surface condition and standard holes.

Solution:

In this design it seems apparent that since both the tension and the shear forces are of comparable magnitude, it is likely that neither the maximum shear strength nor the maximum tension strength of the fasteners may be used, as may be observed from Fig. 4.14.3. The following approach [4.24] may be used when design aids are not available.

(a) Bearing-type connection. According to AISC-J3.7, and using $\phi = 0.75$, the interaction criterion for maximum nominal stress in tension under factored loads (threads excluded from the shear plane) is

$$\phi F'_{nt} = 87.8 - 1.5 f_v \le 67.5 \text{ ksi} \qquad \text{(A325-X)}$$

Similarly, the maximum nominal shear stress is

$$\phi F'_{nv} = \phi \left(1.3 F_{nv} - \frac{F_{nv}}{\phi R_{nt}} f_t \right) \le \phi F_{nv}$$

$$= 0.75 \left[1.3(60) - \frac{60}{0.75(90)} f_t \right]$$

$$= 58.5 - 0.67 f_t \le 45 \text{ ksi} \qquad \text{(A325-X)}$$

The linear interaction equations for $\phi F'_{nt}$ may, in general, be expressed as

$$\phi F'_{nt} = C_1 - C_2 f_v \le \phi F_{nt} \qquad \text{(a)}$$

where C_1 and C_2 are constants.

Convert Eq. (a) into a force equation by multiplying by the bolt area A_b,

$$\phi F'_{nt} A_b = C_1 A_b - C_2 f_v A_b \le \phi F_{nt} A_b \qquad \text{(b)}$$

or, if ΣA_b represents the total area of all bolts, $\phi F'_{nt} \Sigma A_b = \Sigma T_u$ and $f_v \Sigma A_b = \Sigma V_u$, giving

$$\Sigma T_u = C_1 \Sigma A_b - C_2 \Sigma V_u \le \phi R_{nt} \qquad \text{(c)}$$

where ΣT_u and ΣV_u are the total factored tension and shear forces, respectively, applied to the connection. Solving Eq. (c) for ΣA_b gives

$$\Sigma A_b = \frac{\Sigma T_u + C_2 \Sigma V_u}{C_1} \qquad \text{(d)}$$

Alternatively, a similar equation for the required total area based on the maximum nominal stress in shear can be used, as follows:

$$\Sigma A_b = \frac{\Sigma V_u + C_2 \Sigma T_u}{C_1} \qquad \text{(e)}$$

which are the basic design equations for bearing-type connections. The values of C_1 are 87.7 and 58.5 for Eqs. (d) and (e), respectively. The values of C_2 are 1.5 and 0.67 for Eqs. (d) and (e), respectively.

For A325-X connection, Eq. (d) becomes

$$\Sigma A_b = \frac{\Sigma T_u + 1.5 \Sigma V_u}{87.8} \tag{f}$$

and Eq. (e) becomes

$$\Sigma A_b = \frac{\Sigma V_u + 0.67 \Sigma T_u}{58.5} \tag{g}$$

For this example, the factored service loads are

$$\Sigma T_u = 1.2(24) + 1.6(96) = 182 \text{ kips}$$

$$\Sigma V_u = 1.2(14) + 1.6(56) = 106 \text{ kips}$$

and using Eq. (f) which is expected to govern gives

$$\text{Required } \Sigma A_b = \frac{\Sigma T_u + 1.5 \Sigma V_u}{87.8} = \frac{182 + 1.5(106)}{87.8} = 3.9 \text{ sq in.}$$

Alternatively, Eq. (g) can be used to give the same result,

$$\text{Required } \Sigma A_b = \frac{\Sigma V_u + 0.67 \Sigma T_u}{58.5} = \frac{106 + 0.67(182)}{58.5} = 3.9 \text{ sq in.}$$

Check maximum nominal stresses,

$$\phi F'_{nt} = \phi \left[1.3 F_{nt} - \frac{F_{nt}}{\phi F_{nv}} f_v \right] < \phi F_{nt}$$

$$= 0.75 \left[1.3(90) - \frac{90}{0.75(60)} \times \frac{106}{3.9} \right] = 41.0 \text{ ksi} < 0.75(90) \leq 67.5 \text{ ksi} \qquad \textbf{OK}$$

$$\phi F'_{nv} = \phi \left[1.3 F_{nv} - \frac{F_{nv}}{\phi F_{nt}} f_t \right] \leq \phi F_{nv}$$

$$= 0.75 \left[1.3(60) - \frac{60}{0.75(90)} \times \frac{180}{3.9} \right] = 27.7 \text{ ksi} < \phi F_{nv} = 0.75(60) = 45 \text{ ksi } \textbf{OK}$$

As expected, the tension force limitation based on the linear interaction equation controlled. The number n of bolts required is

$$\text{Required } n = \frac{3.9}{0.4418} = 8.8 \text{ bolts}$$

Use 9—$\frac{3}{4}$-in. diam A325-X bolts for a bearing-type connection.

(b) Slip-critical connection. All strength requirements as checked in (a) for a bearing-type connection apply to the slip-critical connection.

Estimate the number of bolts to be 12 and use the reduction factor, Eq. 4.14.7 (AISC-J3.8).

$$k_s = 1 - \frac{T_u}{D_u N_b T_b} = 1 - \frac{182}{1.13(12)28} = 0.52$$

$$\phi R_{nv} = \phi k_s \mu D_u h_{sc} N_s T_b$$
$$= 1.0(0.52)(0.35)(1.13)1.0(1)28 = 5.76 \text{ kips/bolt}$$

$$\text{Required number of bolts } n = \frac{106}{5.76} = 18+$$

Recompute the reduction factor using 19 bolts,

$$k_s = 1 - \frac{T_u}{D_u N_b T_b} = 1 - \frac{182}{1.13(19)28} = 0.7$$

and the revised ϕR_{nv} becomes

$$\phi R_{nv} = \left(\frac{0.7}{0.52}\right)5.76 = 7.75 \text{ kips/bolt}$$

$$\text{Required number of bolts } n = \frac{106}{7.75} = 13+$$

Try 16 bolts,

$$k_s = 1 - \frac{T_u}{D_u N_b T_b} = 1 - \frac{182}{1.13(16)28} = 0.64$$

$$\phi R_{nv} = \left(\frac{0.64}{0.52}\right)5.76 = 7.09 \text{ kips/bolt}$$

$$\text{Required number of bolts } n = \frac{106}{7.09} = 15$$

Use 16—$\frac{3}{4}$-in. diam A325 bolts. ∎

4.15 SHEAR AND TENSION FROM ECCENTRIC LOADING

In a bracket connection such as in Fig. 4.14.1a and d, the eccentric load produces both shear and tension in the upper fasteners. As in most other connections, the manner in which the pieces behave is complex. However, nominal forces carried by the fasteners are usually determined by using one of two approaches: (1) that of neglecting any initial tension in the fasteners or (2) that of considering the initial pretension forces in the fasteners. When fasteners such as A307 bolts are used, the amount of initial tension present is usually small and of an indeterminable amount. Therefore, in this case, the neglect of any initial tension is reasonable and gives conservative results. On the other hand, when high-strength bolts are used the initial pretension forces exist and should be recognized.

If initial tension does not exist to any appreciable degree, the application of moment Pe (Fig. 4.14.1d) will produce a tension that is maximum at the top bolts. Near the bottom of the connection, compression would exist between the pieces being joined with little effect directly on the bolts. The direct shear would be carried nearly entirely by the bolts since little friction would exist from bolt installation. The use of A307 bolts having little initial tension is rare in important connections having shear in combination with moment-induced tension; thus no further treatment is given to the analysis neglecting initial tension.

Veillette and DeWolf [4.24] have conducted tests on tee connections with bolts loaded in shear and tension.

Figure 4.15.1
Stresses on contact surface of
moment-resisting connection,
considering initial tension in
the bolts.

Initial compressive
stress on contact area

(c)

Stress on contact
area due to applied load

(d)

Final stress on
contact area

(e)

Tension from Bending Moment Considering Initial Tension

Consider the service load moment M applied to the bracket of Fig. 4.15.1 to cause tension on
the upper bolts (maximum in the top one of each row of bolts shown as black dots). High-
strength bolts used as such fasteners are required to be installed with a prescribed initial tension in them whether the joint is considered a bearing-type or a slip-critical connection (RCSC
[4.7], Sec. 8). This tension in each bolt will precompress the plates or sections being joined.
For the situation of Fig. 4.15.1, the neutral axis under the action of moment M will occur at the
centroid (CG) of the contact area; that is, at $d/2$ for the rectangular contact area shown.

The initial bearing pressure f_{bi} as shown in Fig. 4.15.1c is assumed to be uniform
over the contact area bd and is equal to

$$f_{bi} = \frac{\Sigma T_b}{bd}$$

(4.15.1)

where ΣT_b = the pretension load times the number of bolts. The tensile stress f_{tb} at the
top due to the applied moment is

$$f_{tb} = \frac{Md/2}{I} = \frac{6M}{bd^2}$$

(4.15.2)

and should not exceed f_{bi} if compression between the pieces is to remain at the top.

The load T on the top bolt is equal to the product of the bolt tributary area (width b
times bolt spacing p) times f_{tb}. Thus,

$$T = f_{tb}bp$$

(4.15.3)

Substituting Eq. 4.15.2 into Eq. 4.15.3 gives the load on the top bolt as

$$T = \frac{6M}{bd^2}bp = \frac{6Mp}{d^2} \tag{4.15.4}$$

Assuming the top bolt is approximately $p/2$ from the top, the value of T can be modified to be

$$T_{\text{modified}} = T\left(\frac{d-p}{d}\right)$$

$$= \frac{6Mp}{d^2}\left(\frac{d-p}{d}\right) \tag{4.15.5}$$

There was logic for using this procedure in Allowable Stress Design where service loads were used directly. The authors consider this method acceptable for use in Load and Resistance Factor Design.

EXAMPLE 4.15.1

Determine the service load capacity P for the connection of Fig. 4.15.2 if the fasteners are $\frac{3}{4}$-in.-diam A325-X bolts subject to shear and tension in a bearing-type connection with no threads in the shear plane. Use AISC LRFD Method, assuming the load is 20% dead load and 80% gravity live load.

Solution:

(a) Compute factored load P_u.

$$P_u = 1.2(0.2P) + 1.6(0.8P) = 1.52P$$

(b) Compute the factored shear and tension on the bolts using the assumption that applied loads do not overcome initial compression between the pieces being joined. Referring to Fig. 4.15.2, the neutral axis for flexure is at mid-depth of the contact area. Equation 4.15.5, using the factored moment M_u and the maximum factored load T_u per bolt for M and T_{modified}, respectively, and noting the moment M_u equals the load P_u times the eccentricity e of 3 in., gives

$$T_u = \frac{6M_up}{d^2}\left[\frac{d-p}{d}\right] = \frac{6(1.52P)(3)3}{2(12)^2}\left[\frac{12-3}{12}\right] = 0.214P$$

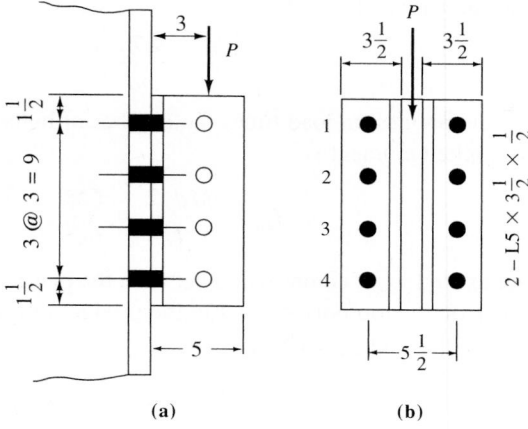

Figure 4.15.2
Shear and moment-induced tension connection for Example 4.15.1.

Note that since there are two vertical lines of fasteners, there is a 2 in the denominator of the above equation.

The shear component taken equally by all bolts is

$$V_u = \frac{P_u}{\Sigma n} = \frac{1.52\,P}{8} = 0.190P$$

(c) Design strengths ϕR_n of bolts in shear and tension. In tension, the design strength ϕR_{nt} is

$$\phi R_{nt} = \phi F'_{nt} A_b$$

where

$$\phi F'_{nt} = \phi\left(1.3F_{nt} - \frac{F_{nt}}{\phi F_{nv}} f_v\right) \le \phi F_{nt}$$

Multiply by A_b and substitute $F_{nt} = 90$ ksi and $F_{nv} = 60$ ksi,

$$\phi F'_{nt} A_b = 0.75\left[1.3(90)A_b - \frac{90}{0.75(60)} f_v A_b\right]$$

$$= 87.8A_b - 1.5V_u \le 67.5A_b$$

$$= 87.8(0.4418) - 1.5V_u \le 67.5(0.4418)$$

$$\text{Max } T_u = \phi F'_{nt} A_b = 38.8 - 1.5V_u \le 29.8 \text{ kips}$$

Equating $T_u = 0.214P$ to Max T_u and then substitute $V_u = 0.190P$ gives

$$0.214P = 38.8 - 1.5(0.190P)$$

$$P = 78 \text{ kips}$$

Then also check limits on $V_u = 19.9$ kips/bolt and $T_u = 29.8$ kips/bolt.

$$[V_u = 0.190P = 0.190(78) = 14.8 \text{ kips/bolt}] \le 19.9 \text{ kips/bolt} \qquad \textbf{OK}$$

$$[T_u = 0.214P = 0.214(78) = 16.6 \text{ kips/bolt}] \le 29.8 \text{ kips/bolt} \qquad \textbf{OK}$$

Therefore, the service load capacity P is 78 kips.   ∎

Considering Initial Tension—Simplified Procedure

As long as the initial compression between plates resulting from initial tension in installed bolts is not totally counteracted by external load, one may compute tensile stress in a bolt by the flexure formula $f = My/I$ as if only the bolt cross-sectional areas comprise the resisting section:

$$f_t = \frac{My}{I} = \frac{My}{\Sigma A_b y^2} \tag{4.15.6}$$

When all fasteners are the same size (as is usual), A_b may be combined with f_t to obtain the tensile force T in a bolt. Thus,

$$T = A_b f_t = \frac{My}{\Sigma y^2} \tag{4.15.7}$$

To show that Eq. 4.15.7 is identical to Eq. 4.15.5, let the depth d of the contact area equal np, where n is the number of fasteners in one line; Eq. 4.15.5 then becomes

$$T = \frac{6Mp}{n^2 p^2}\left(\frac{np - p}{np}\right) = \frac{12M}{n^3 p^2}\left(\frac{p(n - 1)}{2}\right) \tag{4.15.8}$$

Note that $p(n - 1)/2$ is the distance from mid-depth to the outermost fastener and corresponds to y of Eq. 4.15.6. Further, a single line of fasteners spaced at p apart may be treated as a rectangular resisting section of width A/p and depth np. The moment of inertia of such a section would be

$$I = \frac{1}{12}\left(\frac{A}{p}\right)(np)^3 \tag{4.15.9}$$

which corresponds approximately to the moment of inertia of the bolt areas, $\Sigma A_b y^2$. Thus, Eqs. 4.15.7 and 4.15.8 are essentially the same. Based on this reasoning, design Eq. 4.12.28 may also be used to estimate required number of fasteners when fasteners are subject to moment causing tension or shear and tension.

EXAMPLE 4.15.2

For the connection of the bracket of Fig. 4.15.3 to the column, determine the number of $\frac{7}{8}$-in.-diam A325 bolts required to transmit the shear and tension forces. Use 3-in. vertical pitch. (a) Use bearing-type connection (A325-X) connection with threads excluded from the shear plane, and (b) use slip-critical (A325-SC) connection. Use AISC LRFD Method, assuming the load is 52 kips gravity live load and 8 kips dead load.

Solution:

(a) Bearing-type (A325-X) connection. The factored load P_u is

$$P_u = 1.2(8) + 1.6(52) = 93 \text{ kips}$$

The design shear strength ϕR_n for bolts subject to shear alone is

$$\phi R_{nv} = \phi F_{nv} A_b$$
$$= 0.75(60)0.6013 = 27.1 \text{ kips/bolt}$$

Figure 4.15.3
Example 4.15.2 Design for shear and tension.

The design tension strength ϕR_n for bolts subject to tension alone is

$$\phi R_{nt} = \phi F_{nt} A_b$$
$$= 0.75(90)0.6013 = 40.6 \text{ kips/bolt}$$

Noting that M for Eq. 4.12.29 is $92.8(8)/2 = 371$ in.-kips per vertical line of bolts, the number n of bolts per line is approximately

$$n = \sqrt{\frac{6M}{Rp}} = \sqrt{\frac{6(371)}{40.6(3)}} = 4.3 \qquad \text{required for } M \text{ alone}$$

$$n = \frac{P}{2R} = \frac{92.8}{2(27.1)} = 1.7 \qquad \text{required for shear } V \text{ alone}$$

Try 10 bolts (5 per line):

$$\Sigma y^2 = 4(3)^2 + 4(6)^2 = 180 \text{ in.}^2$$

$$R_{ut} = \frac{M_u y}{\Sigma y^2} = \frac{92.8(8)6}{180} = 24.7 \text{ kips/bolt} < 40.6 \text{ kips/bolt} \qquad \textbf{OK}$$

$$R_{uv} = \frac{P_u}{\Sigma n} = \frac{92.8}{10} = 9.3 \text{ kips/bolt} < 27.1 \text{ kips/bolt}$$

Next check the interaction between shear and tension. According to AISC-J3.7, the nominal tensile stress modified in the presence of shear is

$$F'_{nt} = 1.3 F_{nt} - \frac{F_{nt}}{\phi F_{nv}} f_v \le F_{nt} \qquad \text{AISC Formula (J3-3a)}$$

$$F'_{nt} A_b = 1.3(90)0.6013 - \frac{90}{0.75(60)}(9.3) \qquad \textbf{OK}$$

$$= 51.8 \text{ kips} < [F_{nt} A_b = 90(0.6013) = 54.1 \text{ kips}]$$
$$\phi F'_{nt} A_b = 0.75(51.8) = 38.9 \text{ kips} < [R_{nt} = 24.7 \text{ kips}] \qquad \textbf{OK}$$

One could check 4 bolts per line,

$$\Sigma y^2 = 4(1.5)^2 + (4.5)^2 = 90 \text{ in.}^2$$

$$R_{ut} = \frac{M_u y}{\Sigma y^2} = \frac{92.8(8)4.5}{90} = 37.1 \text{ kips} < [\phi R_{nt} = 38.9 \text{ kips}] \qquad \textbf{OK}$$

Clearly, 4 bolts per line is acceptable.
Use 8—$\frac{7}{8}$-in.-diam A325 bolts, 4 per row.

(b) Slip-critical connection. The number of bolts required for slip-critical connections is typically higher than for bearing-type connections. So the 10-bolt arrangement could be satisfactory. When the slip-critical connection is subjected to tension, the available slip-resistance per bolt is reduced. The slip-resistance as determined by AISC-J3.8 is multiplied by the factor k_s given in AISC-J3.9,

$$k_s = 1 - \frac{T_u}{D_u N_b T_b} = 1 - \frac{24.7}{1.13(10)39} = 0.94$$

where N_b is the number of bolts subject to tension.

The reduced design strength then becomes

$$\phi R_{nv} = \phi k_s \mu D_u h_{sc} N_s T_b$$
$$= 1.0(0.94)(0.35)(1.13)(1.00)(1)39$$
$$= 14.5 \text{ kips} > [R_{uv} = 9.3 \text{ kips}] \qquad \textbf{OK}$$

Thus, 5 bolts per line is more than adequate to meet the slip-resistance requirement for combined shear and tension.

Use $10 - \frac{7}{8}$-in.-diam A325 bolts, 5 per row. ∎

SELECTED REFERENCES

4.1. C. Batho and E. H. Bateman. *Investigations on Bolts and Bolted Joints,* Second Report of the Steel Structures Research Committee. London; His Majesty's Stationery Office, 1934.

4.2. W. M. Wilson and F. P. Thomas. "Fatigue Tests on Riveted Joints," *Bulletin 302.* Urbana, IL: University of Illinois, Engineering Experiment Station, 1938.

4.3. A. E. R. De Jonge. Riveted Joints: *A Critical Review of the Literature Covering Their Development.* New York: American Society of Mechanical Engineers, 1945.

4.4. AISC. "Symposium on High-Strength Bolts," *Proceedings of AISC National Engineering Conference.* New York: American Institute of Steel Construction, 1950, 22–43.

4.5. William H. Munse. "Research on Bolted Connections," *Transactions,* ASCE, **121** (1956), 1255–1266.

4.6. "Rivets and High-Strength Bolts, A Symposium," *Transactions,* ASCE, **126,** Part II (1961), 693–820.

4.7. Research Council on Structural Connections. *Load and Resistance Factor Design Specification for Structural Joints Using ASTM A325 or A490 Bolts.* Chicago, IL: American Institute of Steel Construction, 1988.

4.8. Research Council on Structural Connections. *Allowable Stress Design Specification for Structural Joints Using ASTM A325 or A490 Bolts.* Chicago, IL: American Institute of Steel Construction, November 13, 1985.

4.9. Research Council on Structural Connections, *Specification for Structural Joints Using ASTM A325 or A490 Bolts.* Chicago IL: American Institute of Steel Construction, 2004.

4.10. John W. Fisher, Theodore V. Galambos, Geoffrey L. Kulak, and Mayasandra K. Ravindra. "Load and Resistance Factor Design Criteria for Connectors," *Journal of the Structural Division,* ASCE, **104,** ST9 (September 1978), 1427–1441.

4.11. J. W. Fisher, P. O. Ramseier, and L. S. Beedle. "Strength of A440 Steel Joints Fastened with A325 Bolts," *Publications,* IABSE, **23** (1963).

4.12. John L. Rumpf and John W. Fisher, "Calibration of A325 Bolts," *Journal of the Structural Division,* ASCE, **89,** ST6 (December 1963), 215–234.

4.13. John H. A. Struik, Abayomi O. Oyeledun, and John W. Fisher. "Bolt Tension Control with a Direct Tension Indicator," *Engineering Journal,* AISC, **10,** 1 (First Quarter 1973), 1–5.

4.14. Desi D. Vasarhelyi and Kah Ching Chiang. "Coefficient of Friction in Joints of Various Steels," *Journal of the Structural Division,* ASCE, **93,** ST4 (August 1967), 227–243.

4.15. Joseph A. Yura and Karl H. Frank. "Testing Method to Determine the Slip Coefficient for Coatings Used in Bolted Joints," *Engineering Journal,* AISC, **22,** 3 (3rd Quarter 1985), 151–155.

4.16. Theodore V. Galambos, T. A. Reinhold, and Bruce Ellingwood. "Serviceability Limit States: Connection Slip," *Journal of the Structural Division,* ASCE, **108,** ST12 (December 1982), 2668–2680.

4.17. Sherwood F. Crawford and Geoffrey L. Kulak. "Eccentrically Loaded Bolted Connections," *Journal of the Structural Division,* ASCE, **97,** ST3 (March 1971), 765–783.

4.18. Geoffrey L. Kulak. "Eccentrically Loaded Slip-Resistant Connections," *Engineering Journal,* AISC, **12,** 2 (2nd Quarter 1975), 52–55.

4.19. G. Donald Brandt. "Rapid Determination of Ultimate Strength of Eccentrically Loaded Bolt Groups," *Engineering Journal,* AISC, **19,** 2 (2nd Quarter 1982), 94–100. Disc. by Cedric Marsh, *Engineering Journal,* **19,** 4 (4th Quarter 1982), 214–215; Nestor Iwankiw, *Engineering Journal,* **20,** 1 (1st Quarter 1983), 46; 2 (2nd Quarter 1983), 88.

4.20. Avigdor Rutenberg. "Nonlinear Analysis of Eccentric Bolted Connections," *Engineering Journal,* AISC, **21,** 4 (4th Quarter 1984), 227–236.

4.21. J. W. Fisher. "Behavior of Fasteners and Plates with Holes," *Journal of the Structural Division,* ASCE **91,** ST6 (December 1965), 265–286.

4.22. Carl L. Shermer. "Plastic Behavior of Eccentrically-Loaded Connections," *Engineering Journal,* AISC, **8,** 2 (April 1971), 48–51.

4.23. Eugene Chesson, Jr., Norberto L. Faustino, and William H. Munse. "High-Strength Bolts Subjected to Tension and Shear," *Journal of the Structural Division,* ASCE, **91,** ST5 (October 1965), 155–180.

4.24. John R. Veillette and John T. DeWolf. "Eccentrically Loaded High Strength Bolted Connections," *Journal of Structural Engineering,* ASCE, **111,** 5 (May 1985), 1003–1018.

PROBLEMS

All problems are to be done according to the AISC LRFD Method or AISC ASD Method, as indicated by the instructor. All given loads are service loads unless otherwise indicated. All holes are standard holes and surface condition is clean mill scale (Class A) unless otherwise indicated. When an ultimate strength analysis is requested in an Allowable Strength Design problem, use a factor of safety of 2.5 to obtain the allowable value. Where needed, assume distance from center of hole to nearest edge (edge or end distance) is $1\frac{1}{2}$ in. unless otherwise given.

Values of yield stress F_y and tensile strength F_u for member steels are available in Table 2.1.1. For A325 bolts, $F_u^b = 120$ ksi minimum for bolts $\frac{1}{2}$-to 1-in. diameter, and 105 ksi minimum for $1\frac{1}{8}$-to $1\frac{1}{2}$-in. diameter. For A490 bolts, $F_u^b = 150$ ksi minimum for $\frac{1}{2}$-to $1\frac{1}{2}$-in. diameter.

4.1. Determine the service load tension capacity of the connection of the accompanying figure for the case assigned by the instructor. Investigate as bearing-type connection with threads excluded (X) from the shear planes, bearing-type with threads included (N) in the shear planes, or as a slip-critical (SC) connection, as indicated under the heading "Type Connection." Specify the minimum dimensions A and B appropriate for the connection.

Case	% Dead load	% Live load	Plate steel	Bolt diameter (in.)	Type connection
1	10	90	A36	3/4	A325-X
2	15	85	A36	3/4	A325-SC
3	15	85	A572 Gr 50	3/4	A325-N
4	20	80	A572 Gr 50	3/4	A325-SC
5	20	80	A572 Gr 60	7/8	A325-X
6	40	60	A572 Gr 60	7/8	A325-SC
7	15	85	A572 Gr 65	7/8	A490-X
8	15	85	A572 Gr 65	7/8	A490-SC

Problem 4.1

4.2. For any of the cases solved in Prob. 4.1, determine the factor of safety against slip at service load. Do you expect slip to occur at service load? If bolt strength had controlled would slip have been expected at service load?

4.3. Determine the service load capacity T for the butt splice of Prob. 3.15 when $s_1 = s_2 = 2$ in. and A325 bolts are used with no threads in the shear planes. Specify the end distances required and evaluate whether or not the given stagger of 2 in. is sufficient. If the 2 in. is not adequate, specify the value to be used. Use (a) a bearing-type connection (A325-X), and (b) a slip-critical (A325-SC) connection.

4.4. For the case assigned by the instructor, determine the number of bolts required to develop the full capacity of the double angle tension member shown in the accompanying figure. Use a double row of bolts without stagger. Detail the connection.

Case	% Dead load	% Live load	Angle steel	Bolt diameter (in.)	Type connection
1	10	90	A572 Gr 50	3/4	A325-X
2	15	85	A572 Gr 50	3/4	A325-SC
3	30	70	A572 Gr 50	7/8	A325-N
4	40	60	A572 Gr 50	7/8	A325-SC

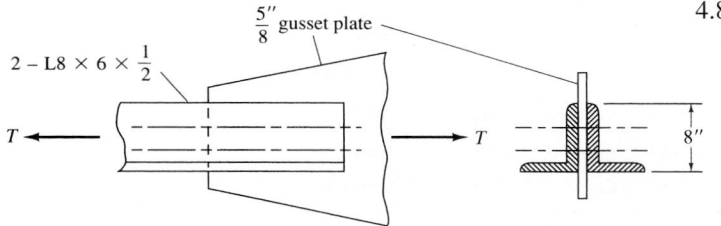

Problem 4.4

4.5. For the single angle tension member of A572 Grade 50 steel in the accompanying figure, how many $\frac{3}{4}$-in.-diam A325 bolts are required for the connection? The load T is 80 kips live load and 20 kips dead load. Assume a slip-critical connection (A325-SC) is to be used. Design the shortest feasible overlap of pieces for the connection, and detail it.

4.6. Solve Prob. 4.5 as a bearing-type (A490-X) connection using $\frac{5}{8}$-in.-diam A490 bolts.

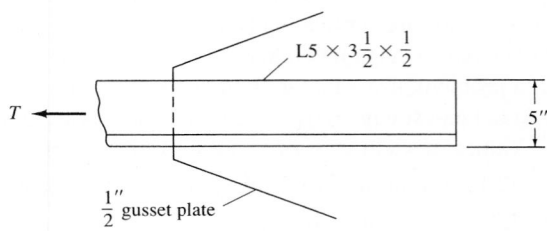

Problems 4.5 and 4.6

4.7. For the single angle tension member of A36 steel in the accompanying figure, determine the number of $\frac{7}{8}$-in.-diam A325 bolts required in a bearing-type connection (A325-N) where threads may exist in the shear plane. The load is 7 kips dead load and 70 kips live load. Assume there are three $\frac{15}{16}$-in.-diam *empty holes* in the outstanding leg that are not a part of the connection. The bolts carrying the load T are to be in a single line with the first bolt located a distance s ahead of the first empty hole. Detail the connection.

4.8. Design and detail the double lap splice shown, to develop maximum tension capacity assuming the load is 20% dead load and 80% live load. The steel is A36 and $\frac{7}{8}$-in.-diam A325 bolts are to be used in a bearing-type connection (A325-X) with no threads in the shear planes. What is the resulting service load capacity of the joint?

4.9. Determine the service load capacity of 2-C10×20 channels as a truss member attached to a $\frac{7}{8}$-in. gusset plate as shown in the accompanying figure. The steel is A572 Grade 65, and the $\frac{7}{8}$-in.-diam A325 bolts are in a bearing-type connection (A325-X) with no threads in the shear planes. The load is 25% dead load and 75% live load.

Problem 4.7

Problem 4.8

Problem 4.9

4.10. Compute the maximum service load P causing eccentric shear on the connection of the accompanying figure. The loading is 25% dead load and 75% live load. The bolts are $\frac{7}{8}$-in.-diam A325 bolts in a bearing-type connection (A325-X) with threads excluded from the shear plane. Assume the bracket plate has adequate strength.

(a) Use the elastic (vector) method.

(b) Use the strength method with rotation about the instantaneous center.

(c) Use *AISC Manual* Tables 7–7 to 7–14, with interpolation.

4.11. For the A36 steel bracket plate of the accompanying figure, calculate the maximum service load P (15% dead load and 85% live load) when $\frac{7}{8}$-in.-diam A325 bolts are used in a bearing-type connection (A325-N) with threads included in the shear planes.

(a) Use the elastic (vector) method.

(b) Use the ultimate strength method.

(c) Use *AISC Manual* Tables 7–7 to 7–14.

4.12. Repeat Prob. 4.11, except use only 6 bolts instead of 8; that is, 2 at 3 in. vertically instead of 3 at 3 in.

Problem 4.10

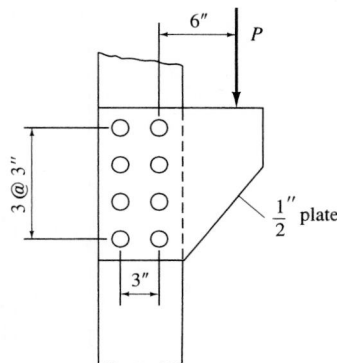

Problem 4.11

4.13. Repeat Prob. 4.11 as a slip-critical (A325-SC) connection instead of a bearing-type connection.

4.14. Select the proper diameter A490 bolts for a bearing-type connection (A490-X) if the loading is 10% dead load and 90% live load.

(a) Use the elastic (vector) method.

(b) Use ultimate strength analysis.

(c) Use *AISC Manual* Tables 7-7 to Table 7-14.

4.15. Assuming the fasteners control the capacity, determine the bolt size required for the connection shown when the load is 10% dead load and 90% live load. The connection is a bearing-type (A325-X) containing A325 bolts.

(a) Use the elastic (vector) method.

(b) Use ultimate strength analysis.

(c) Use *AISC Manual* Tables 7-7 to Table 7-14.

4.16. For the connection of the accompanying figure containing $\frac{3}{4}$-in.-diam A325 bolts in a slip-critical connection, determine the service load capacity P (15% dead load and 85% live load) by the following methods, and compare the results:

(a) Use the elastic (vector) method.

(b) Use the instantaneous center—constant slip resistance method for slip-critical connections (see Example 4.12.4).

(c) Use *AISC Manual* Tables 7–7 to Table 7–14.

Problem 4.14

Problem 4.15

Problem 4.16

4.17. For the eccentric shear loading of the accompanying figure, two vertical lines of $\frac{7}{8}$-in.-diam A325 bolts are used having a 3-in. spacing. Select the proper number of bolts for a bearing-type (A325-X) connection. The load is 40% dead load and 60% live load.

(a) Use the elastic (vector) method.

(b) Use the ultimate strength method.

(c) Use *AISC Manual* Tables 7-7 to Table 7-14.

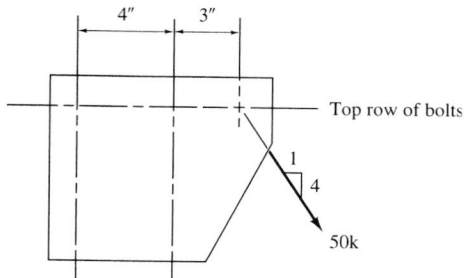

Problem 4.17

4.18. Repeat Prob. 4.17 using $\frac{3}{4}$-in.-diam A490 bolts.

4.19. For the eccentric shear loading of the accompanying figure, $\frac{7}{8}$-in.-diam A325 bolts are used in two vertical lines in a bearing-type (A325-X) connection. The loading is 10 kips dead load and 30 kips live load.

(a) Determine the adequacy of the design using basic principles of the elastic (vector) method.

(b) Compute the service load capacity of the connection using ultimate strength analysis.

(c) Use *AISC Manual* Tables 7-7 to Table 7-14.

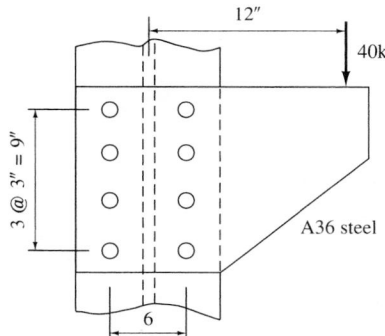

Problem 4.19

4.20. For the eccentric shear loading of the accompanying figure, $\frac{7}{8}$-in.-diam A325 bolts are used in a single vertical line in a bearing-type (A325-X) connection. The loading is 7 kips dead load 33 kips live load. Determine the number of bolts required. What thickness of pieces is required to avoid having bearing control and still use minimum edge distances (see AISC-Table J3.4).

(a) Use the elastic (vector) method.

(b) Use the ultimate strength method.

(c) Use *AISC Manual* Tables 7-7 to Table 7-14.

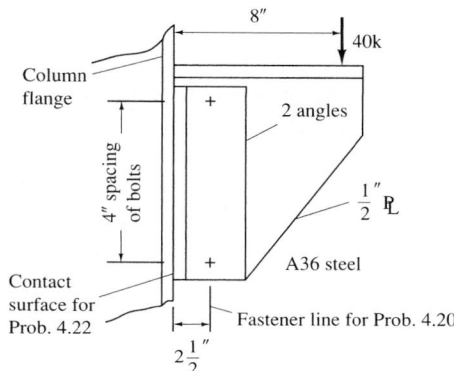

Problems 4.20 and 4.22

4.21. For the connection of the accompanying figure, subject to direct tension and shear, two angles, L4×3$\frac{1}{2}$×$\frac{5}{8}$, are used to carry their maximum capacity as a tension member of A36 steel. Assume the connection of the angles to the structural tee (WT) web will be along a single gage line as shown. Determine the number and positioning of $\frac{7}{8}$-in.-diam A325 bolts to attach the WT to the flange of a W section. The flanges of the WT and the W shape are both $\frac{3}{4}$-in. thick and A36 steel is used.

(a) Use a bearing-type (A325-X) connection.

(b) Use a slip-critical (A325-SC) connection (20% DL; 80% LL).

Problem 4.21

4.22. For the eccentric connection of the accompanying figure, causing shear and tension on the fasteners connecting the angles to the column flange; determine the number and spacing of $\frac{7}{8}$-in.-diam bolts required to make the connection to the $\frac{3}{4}$-in. column flange. The load is 5 kips dead load and 35 kips live load.

(a) Use a bearing-type (A325-X) connection.

(b) Use a slip-critical (A325-SC) connection.

4.23. For the eccentric connection of the accompanying figure, causing shear and tension on the fasteners connecting the structural tee (WT) to the column flange, determine the number of $\frac{7}{8}$-in.-diam bolts spaced vertically at 3 in. required to make the connection to the $\frac{5}{8}$-in. column flange. The load is 14 kips dead load and 80 kips live load.

(a) Use a bearing-type (A325-X) connection.

(b) Use a slip-critical (A325-SC) connection.

Problem 4.23

5

Welding

5.1 INTRODUCTION AND HISTORICAL DEVELOPMENT

The process of welding denotes the joining of metal pieces by heating to a plastic or fluid state, with or without pressure. In its simplest form, "welding" has been known and used for several thousand years. Historians have speculated that the early Egyptians may have first used pressure welding about 5500 B.C. in making copper pipes from sheets by overlapping the edges and hammering. Winterton [5.1] has reported that Egyptian art objects dating about 3000 B.C. have been found on which gold foil has been hammered and fused onto the base copper. This type of welding, called *forge welding,* was man's first process to join pieces of metal together. A well-known early example of forge welding is the Damascus sword which was made by forging layers of iron with different properties. Interestingly, forge welding was sufficiently well developed and important enough to the early Romans that they named one of their gods Vulcan (the god of fire and metalworking) to represent that art. In recent times, the word vulcanizing has been used in reference to treating rubber with sulfur but originally was used to mean "to harden." Today, forge welding is practically a forgotten art in which the village blacksmith was the last major practitioner.

Little progress in welding technology had been made until 1877, prior to which time most of the then known processes such as forge welding and brazing had been used for at least 3000 years. The origin of resistance welding began around 1877 when Professor Elihu Thompson began a set of experiments [5.2] reversing the polarity of transformer coils. He received his first patent [5.3] in 1885 and the first resistance butt welding machine was demonstrated at the American Institute Fair in 1887. In 1889 Charles Coffin [5.2] was issued a patent for flash-butt welding and this became one of the important butt joining processes.

Zerner, in 1885, introduced the carbon arc welding process, making use of two carbon electrodes, and N. G. Slavianoff [5.5] in 1888 in Russia was the first to use the metal arc process using uncoated, bare electrodes. Coffin, working independently also investigated the metal arc process and was issued a U.S. Patent in 1892. In 1889, A. P. Strohmeyer [5.2] introduced the concept of coated metal electrodes to eliminate many of the problems associated with the use of bare electrodes.

161

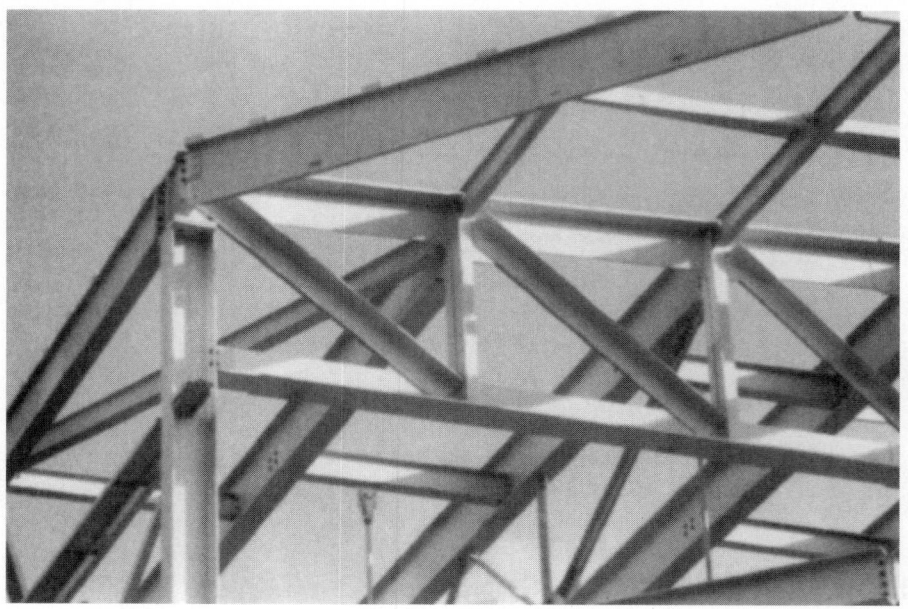

Welded truss as part of roof space truss system. (Photo by C. G. Salmon)

Thomas Fletcher [5.1] in 1887 used a blowpipe, burning hydrogen and oxygen and showed that he could successfully cut or melt metal. In 1901–1903 Fouche and Picard developed torches that could be used with acetylene and, thus, the era of oxyacetylene welding and cutting began.

The period between 1903 and 1918 saw the use of welding primarily as a method of repair, the greatest impetus occurring during World War I (1914–1918). Welding techniques proved to be especially adapted to repairing ships that had been damaged.

After World War I (1918), there was continued experimentation with electrodes and various gases to shield the arc and the weld area, resulting in the development of gas tungsten arc welding and gas metal arc welding (see Sec. 5.2). During the period 1930 to 1950 many improvements occurred, including in 1932 [5.5] the introduction of the use of granular flux to protect the weld, which when coupled to the use of a continuously fed electrode resulted in the development of submerged arc welding (see Sec. 5.2), where the arc is buried under the granular flux. This common method was patented in 1935.

Now (2008) automation has become a significant factor in welding technology and extensive use of welding robotics is occurring.

There are many welding processes available to join various metals and their alloys. Those of particular interest in welding structural steel, and of interest to structural engineers in general, are discussed in Sec. 5.2.

5.2 BASIC PROCESSES

Welding is the process of joining materials (usually metals) by heating them to suitable temperatures such that the materials coalesce into one material. There may or may not be pressure, and there may or may not be filler material applied. *Arc welding* is the general term for the many processes that use electrical energy in the form of an electric arc to generate the heat necessary for welding.

This section treats those processes used in arc welding carbon and low-alloy steel for buildings and bridges. For some situations involving light-gage steel, resistance welding may also be important. More extensive descriptions than those that follow are available in the *Welding Handbook* [2.21, 5.10].

Shielded Metal Arc Welding (SMAW)

Shielded metal arc welding is one of the oldest, simplest, and perhaps most versatile types for welding structural steel. The SMAW process is often referred to as the *manual stick electrode process.* Heating is accomplished by means of an electric arc between a coated electrode and the materials being joined. The welding circuit is shown in Fig. 5.2.1a.

The coated electrode is consumed as the metal is transferred from the electrode to the base material during the welding process. The electrode wire becomes filler material and the coating is converted partly into a shielding gas, partly into slag, and some part is absorbed by the weld metal. The coating is a clay-like mixture of silicate binders and powdered materials, such as fluorides, carbonates, oxides, metal alloys, and cellulose. The mixture is extruded and baked to produce a dry, hard, concentric coating.

The transfer of metal from electrode to the work being welded is induced by molecular attraction and surface tension, without application of pressure. The shielding of the arc prevents atmospheric contamination of the molten metal in the arc stream and in the arc pool. It prevents nitrogen and oxygen from being picked up and forming nitrides and oxides which may cause embrittlement.

The electrode coating may perform the following functions:

1. Produces a gaseous shield to exclude air and stabilize the arc.
2. Introduces other materials, such as deoxidizers, to refine the grain structure of the weld metal.
3. Produces a blanket of slag over the molten pool and the solidified weld to protect it from oxygen and nitrogen in the air, and also retards cooling.

The electrode material is specified under various American Welding Society specifications that are listed in AWS [2.23] Table 3.1, and is summarized in text Table 2.3.1. The designations such as E60XX or E70XX indicate 60 ksi and 70 ksi, respectively, for tensile strength. The X's refer to factors such as the suitable welding positions, recommended power supply, type of coating, and type of arc characteristics. Morgan [5.6] has provided an excellent guide to classification and use of mild steel coated electrodes. Table 5.13.1 indicates which coated electrodes should be used with each particular structural steel.

For welding high-carbon or low-alloy steels, low-hydrogen electrodes are required by AWS [2.23] to be used with SMAW for all steels having yield stresses

Figure 5.2.1
Shielded metal arc welding
(SMAW).

(a) Arc welding circuit

(b) Shielded arc welding

higher than 36 ksi (248 MPa). The low-hydrogen electrode is a rod with a carbonate of soda, or "lime," coating. This electrode requires a different technique from that using the conventional electrode in that a short arc must be made and globular-type, rather than a spray-type, deposition of metal occurs. It is desirable in design because the as-welded mechanical properties have been found to be superior to properties obtained using other types of electrode coatings.

Submerged Arc Welding (SAW)

In the SAW process the arc is not visible because it is covered by a blanket of granular, fusible material, as shown in Fig. 5.2.2. The bare metal electrode is consumable in that it is deposited as filler material. The end of the electrode is kept continuously shielded by the molten flux over which is deposited a layer of unfused flux in its granular condition.

Figure 5.2.2
Submerged arc welding
(SAW).

The granular flux, which is a special feature of this method, is usually laid automatically along the seam ahead of the advancing electrode, and provides a cover that allows the weld to be made without spatter, sparks, or smoke. This flux material protects the weld pool against the atmosphere, serves to clean the weld metal, and modifies the chemical composition of the weld metal [2.21, Vol. 2].

Welds made by the submerged arc process have uniform high quality; exhibiting good ductility, high impact strength, high density, and good corrosion resistance. Mechanical properties of the weld are consistently as good as the base material.

The combinations of bare-rod electrodes and granular flux are classified under AWS A5.17 or A5.23. They are designated FXXX-EXXX where the first X following the F is the first digit of the tensile strength (i.e., 7 for 70 ksi), the second X is a letter indicating the condition of heat treatment (i.e., A for as-welded and P for postweld heat treated), and the third X indicates the lowest temperature at which impact strength of the weld metal meets or exceeds 20 ft-lb (27 J). When the third X is 6, for example, it means the Charpy V-notch impact strength is at least 20 ft-lb (27 J) at $-60°F$ $(-51°C)$. The three Xs following the letter E indicate properties of the electrode.

The submerged arc method is commonly used in shop-fabricated steel operations using automatic or semiautomatic equipment.

Gas Metal Arc Welding (GMAW)

In the GMAW process the electrode is a continuous wire that is fed from a coil through the electrode holder, a gun-shaped device as shown in Fig. 5.2.3. The shielding is entirely from an externally supplied gas or gas mixture. The distinguishing features of this method are the high rates at which filler metal can be transferred, and the gaseous shield that is uniformly provided around the molten weld. Special uses of this method are described by Craig [5.7], Lyttle [5.8], and Dillenbeck and Castagno [5.9].

Figure 5.2.3
Gas metal arc welding
(GMAW).

Originally, this method was used only with inert gas shielding, hence, the name MIG (metal inert gas) has been used. Reactive gases alone are generally not practical; the exception is CO_2 (carbon dioxide). The use of CO_2 has become extensive for welding of steels, either alone or in a mixture with inert gases.

Argon as a shielding gas works for welding virtually all metals; however, it is not recommended for steels because of its expense and the fact that other shielding gases and gas mixtures are acceptable. For welding carbon steel and some low-alloy steels, recent research [5.9] indicates the best overall performance is obtained using 80% CO_2 and 20% helium. Traditionally, it has been recommended [5.10] to use either a mixture of 75% argon and 25% CO_2, or 100% CO_2. For low-alloy steels where toughness is important, the recommendation [5.10] is to use a mixture of 60 to 70% helium, 25 to 30% argon, and 4 to 5% CO_2.

The shielding gas serves the following functions in addition to protecting the molten metal from the atmosphere.

1. Controls the arc and metal-transfer characteristics.
2. Affects penetration, width of fusion, and shape of the weld region.
3. Affects the speed of welding.
4. Controls undercutting.

By mixing an inert gas with a reactive gas the arc may be made more stable and the spatter during metal transfer may be reduced. The use of CO_2 alone for welding steel is the least expensive procedure because of its lower cost for shielding gas, higher welding speed, better joint penetration, and sound deposits with good mechanical properties. The only disadvantage is that it gives harsh and excessive spatter.

The electrode material for welding carbon steels is an uncoated mild steel, deoxidized carbon manganese steel covered under AWS A5.18 and listed in Table 5.13.1 (see also Table 2.3.1). For welding low-alloy steel a deoxidized low-alloy material is necessary.

The GMAW process using CO_2 shielding is good for the lower carbon and low-alloy steels usually used in buildings and bridges.

Flux Cored Arc Welding (FCAW)

The FCAW process, developed in 1958, is similar to GMAW, except that the continuously fed filler metal electrode is tubular and contains the flux material within its core. The core material provides the same functions as does the coating in SMAW or the granular flux in SAW. For a continuously fed wire, an outside coating would not remain bonded to the wire. Gas shielding is provided by the flux core (self-shielded) but additional shielding is frequently provided by CO_2 gas. Flux cored arc welding has become a useful procedure for field welding in severe cold weather conditions [5.11] as well as to speed up high-rise construction [5.12].

The electrodes used for obtaining weld metal having minimum specified yield strengths of 60 ksi (415 MPa) or less are covered under AWS A5.20, and are designated E6XT or E7XT, for tensile strengths of 60 or 70 ksi, respectively. When it is desirable to produce weld metal having yield strength exceeding 60 ksi (415 MPa), the electrodes are covered under AWS A5.29, and are designated E8XT, E10XT, and E11XT, having tensile strengths of 80, 100, and 110 ksi (550, 690, and 760 MPa), respectively.

Electrogas Welding (EGW)

The EGW process is a machine process used primarily for vertical position welding, shown in Fig. 5.2.4. Either flux cored or solid electrodes may be used. This method is used to obtain a single pass weld such as for the splice in a heavy column section. Weld metal is deposited into a cavity created by the separated plate edges on two sides and water-cooled "shoes" or guides to keep the molten metal in its proper location on the other two sides. The gas shielding is provided either by the flux cored electrode, by externally supplied gas, or both.

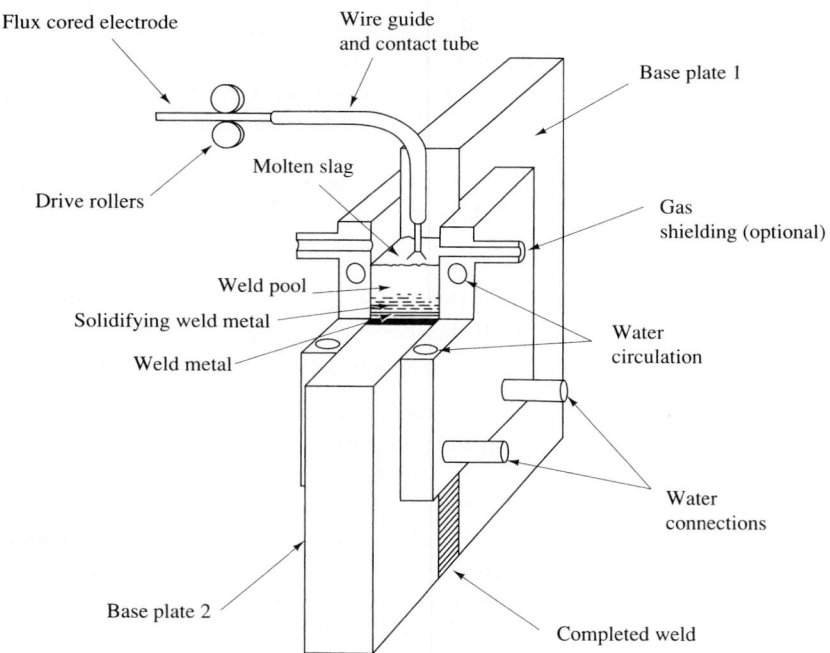

Figure 5.2.4
Electrogas welding with a flux cored electrode. (From *Welding Handbook* [2.23])

Electroslag Welding (ESW)

The ESW process, shown in Fig. 5.2.5, is similar to electrogas welding, except that the welding is actually done by the heat produced through the resistance of the slag to the flow of current. The molten conductive slag protects the weld and melts the filler metal and the plate edges. Since solid slag is not conductive, an arc is required to start the process by melting the slag and heating the plates. Since resistance heating is used for all but this initial stage, the ESW is really not an arc welding process. The side guides, or "shoes", may be nonconsumable as in Fig. 5.2.5, or they may be consumable. The electroslag process

Figure 5.2.5
Nonconsumable guide
method of electroslag welding (three electrodes). (From
Welding Handbook [2.21])

allows welding nearly any thickness of material in one pass; both electrogas and electroslag welding become economical as the thickness of weld required becomes large.
Because of the slow weld travel speed used in this process, a weld with relatively coarse
grain structure and low notch toughness is the result.

An excellent review of electroslag welding has been provided by Raman [5.14].
Schilling and Klippstein [5.15] have reported research on electroslag welding for bridges,
and Pense, Wood, and Fisher [5.16] reported experience with electroslag welding on
welded bridges.

Stud Welding

The most commonly used process of welding a metal stud to a base material is known as
arc stud welding, an essentially automatic process but similar in characteristics to the
SMAW process. The stud serves as the electrode and an electric arc is created from the end
of the stud to the plate. The stud is contained in a gun which controls the timing during the
process. Shielding is accomplished by placing a ceramic ferrule around the end of the stud
in the gun. The gun is placed in position and the arc is created, during which time the
ceramic ferrule contains the molten metal. After a short instant of time, the gun drives the
stud into the molten pool and the weld is completed leaving a small fillet around the stud.
Full penetration across the shank of the stud is obtained and the weld is completed usually
in less than one second.

5.3   WELDABILITY OF STRUCTURAL STEEL

Most of the ASTM-specification construction steels can be welded without special precautions or special procedures. Section 5.13 discusses the need to select the proper electrode
to join a particular grade of steel and a summary of the "matching" electrodes and the base
steel is given in Table 5.13.1.

The *weldability* of a steel is a measure of the ease of producing a crack-free and
sound structural joint. Some of the readily available structural steels are more suited to
welding than others, and are discussed in Chapter 2. Welding procedures should be based
on a steel's chemistry instead of the published maximum alloy content, since most mill
runs are usually below the maximum alloy limits set by its specification. Table 5.3.1 shows

TABLE 5.3.1 Preferred Analysis of Carbon Steel
[5.17] for Good Weldability

Element	Normal range (%)	Percent requiring special care
Carbon	0.06–0.25	0.35
Manganese	0.35–0.80	1.40
Silicon	0.10 max	0.30
Sulfur	0.035 max	0.050
Phosphorus	0.030 max	0.040

the ideal chemical analysis of the carbon steels. Most mild steels fall well within this range, while higher-strength steels may exceed the ideal analysis shown in Table 5.3.1.

When a mill produces a run of steel, it maintains a complete record of its chemical content which follows all shapes made from the particular ingot. If the designer is concerned about the chemistry of a particular grade of steel, a Mill Test Report may be requested. Any variation in chemical content above the ideal values may be evaluated, and special welding procedures be set up to insure a properly welded joint.

5.4   TYPES OF JOINTS

The type of joint depends on factors such as the size and shape of the members coming into the joint, the type of loading, the amount of joint area available for welding, and the relative costs for various types of welds. There are five basic types of welded joints, although many variations and combinations are found in practice. The five basic types are the butt, lap, tee, corner, and edge joints, as shown in Fig. 5.4.1.

Butt Joints

The butt joint is used mainly to join the ends of flat plates of the same or nearly the same thicknesses. The principal advantage of this type of joint is to eliminate the eccentricity developed in single lap joints as shown in Fig. 5.4.1b. When used in conjunction with full penetration groove welds, butt joints minimize the size of a connection and are usually more esthetically pleasing than built-up joints. Their principal disadvantage lies in the fact that the edges to be connected must usually be specially prepared (beveled, or ground flat) and very carefully aligned prior to welding. Little adjustment is possible and the pieces must be carefully detailed and fabricated. As a result, most butt joints are made in the shop where the welding process can be more accurately controlled.

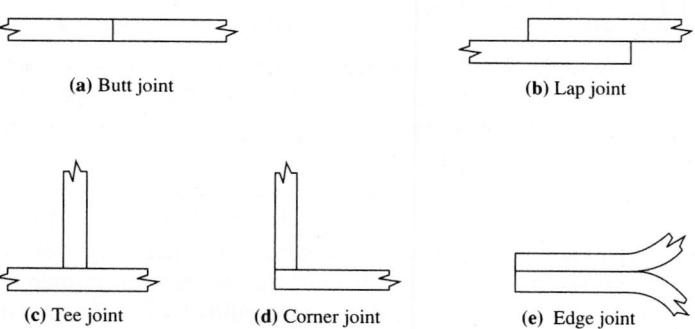

(a) Butt joint   **(b)** Lap joint

Figure 5.4.1
Basic types of welded joints.   **(c)** Tee joint   **(d)** Corner joint   **(e)** Edge joint

Lap Joints

The lap joint, shown in Fig. 5.4.2, is the most common type. It has two principal advantages:

1. *Ease of fitting.* Pieces being joined do not require the preciseness in fabricating as do the other types of joints. The pieces can be slightly shifted to accommodate minor errors in fabrication or to make adjustments in length.

2. *Ease of joining.* The edges of the pieces being joined do not need special preparation and are usually sheared or flame cut. Lap joints utilize fillet welds and are therefore equally well suited to shop or field welding. The pieces being joined are in most cases simply clamped together without the use of special jigs. Occasionally the pieces are positioned by a small number of erection bolts which may be either left in place or removed after the welding is completed.

A further advantage of the lap joint is the ease in which plates of different thickness can be joined, such as in the double lap joint in Fig. 5.4.2e. The reader should especially note the truss connection shown in Fig. 5.4.2c and consider the difficulty in making such a connection by any other type of joint.

Tee Joints

This type of joint is used to fabricate built-up sections such as tees, I-shapes, plate girders, bearing stiffeners, hangers, brackets, and in general, pieces framing in at right angles as shown in Fig. 5.4.1c. This type of joint is especially useful in that it permits sections to be built up of flat plates that can be joined by either fillet or groove welds.

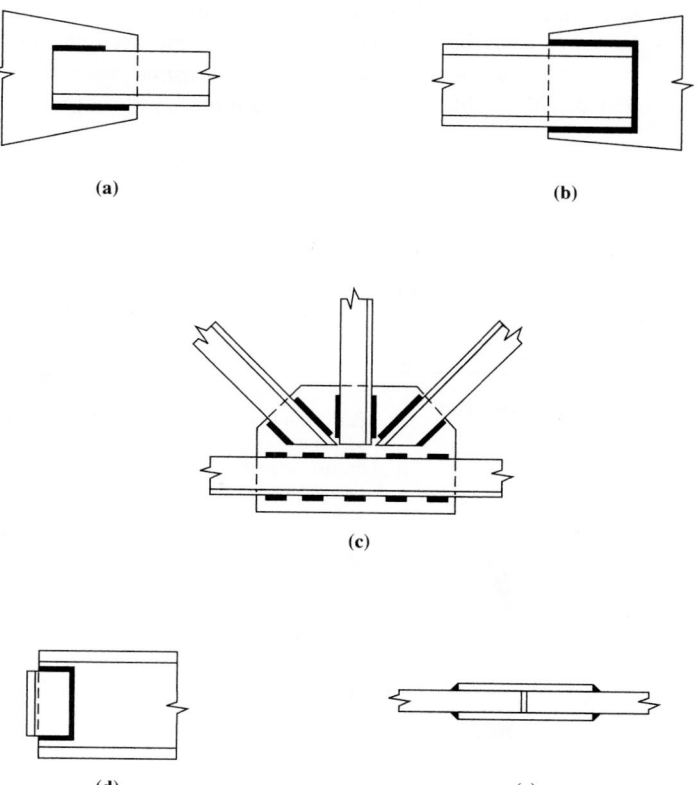

(a)

(b)

(c)

(d)

(e)

Figure 5.4.2
Examples of lap joints.

Corner Joints

Corner joints are used principally to form built-up rectangular box sections such as used for columns and for beams required to resist high torsional forces.

Edge Joints

Edge joints are generally not structural but are most frequently used to keep two or more plates in a given plane or to maintain initial alignment.

As the reader can infer from the previous discussions, the variations and combinations of the five basic types of joints are virtually infinite. Since there is usually more than one way to connect one structural member to another, the designer is left with the decision for selecting the best joint (or combination joints) in each given situation.

5.5   TYPES OF WELDS

The four types of welds are the groove, fillet, slot, and plug welds as shown in Fig. 5.5.1. Each type of weld has specific advantages that determine the extent of its use. Roughly, the four types represent the following percentages of welded construction: groove welds, 15%; fillet welds, 80%; the remaining 5% are made up of the slot, plug, and other special welds.

Groove Welds

The principal use of groove welds is to connect structural members that are aligned in the same plane. Since groove welds are usually intended to transmit the full load of the members they join, the weld should have the same strength as the pieces joined. Such a groove weld is known as a *complete joint penetration groove weld*. When joints are designed so that groove welds do not extend completely through the thickness of the pieces being

Figure 5.5.1
Types of welds.

(a) Groove weld

(b) Fillet welds

Ends shall be semicircular or have corners rounded to a radius not less than thickness of part containing slot

Section A-A

(c) Slot weld

Section A-A

(d) Plug weld

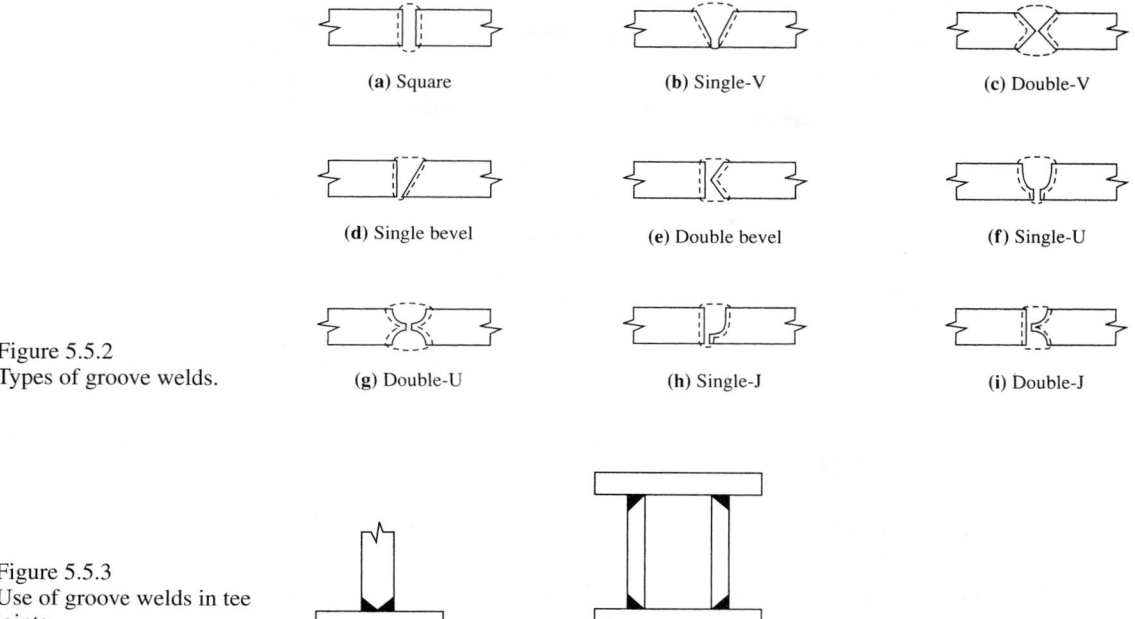

Figure 5.5.2
Types of groove welds.

(a) Square   **(b)** Single-V   **(c)** Double-V

(d) Single bevel   **(e)** Double bevel   **(f)** Single-U

(g) Double-U   **(h)** Single-J   **(i)** Double-J

Figure 5.5.3
Use of groove welds in tee joints.

joined, such welds are referred to as *partial joint penetration groove welds*. For these, special design requirements apply.

There are many variations of groove welds and each is classified according to its particular shape. Most groove welds require a specific edge preparation and are named accordingly. Figure 5.5.2 shows several types of groove welds and indicates the groove preparations required for each. The selection of the proper groove weld is dependent on the welding process used, the cost of edge preparations, and the cost of making the weld. Groove welds may also be used in tee connections as shown in Fig. 5.5.3.

Fillet Welds

Fillet welds owing to their overall economy, ease of fabricating, and adaptability are the most widely used. A few uses of fillet welds are shown in Fig. 5.5.4. They generally require less precision in the "fitting up" because of the overlapping of pieces, whereas the groove weld requires careful alignment with specified gap (root opening) between pieces. The fillet weld is particularly advantageous to welding in the field or in realigning members or connections that were fabricated within accepted tolerances but which may not fit as accurately as desired. In addition, the edges of pieces being joined seldom need special preparation such as beveling or squaring since the edge conditions resulting from flame cutting or from shear cutting procedures are generally adequate.

Slot and Plug Welds

Slot and plug welds may be used exclusively in a connection as shown in Figs. 5.5.1c and d, or they may be used in combination with fillet welds as shown in Fig. 5.5.5. A principal use for plug or slot welds is to transmit shear in a lap joint when the size of the connection limits the length available for fillet or other edge welds. Slot and plug welds are also useful in preventing overlapping parts from buckling.

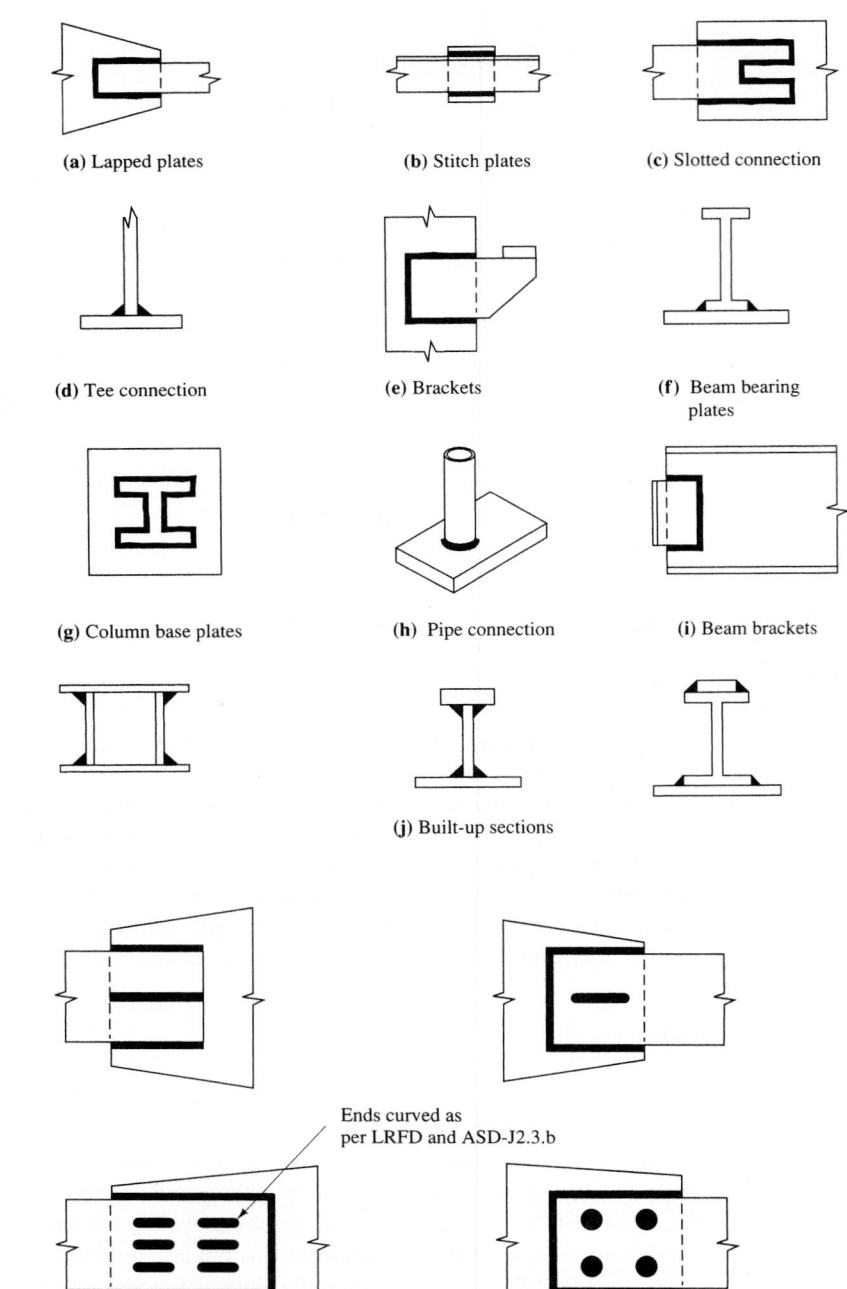

(a) Lapped plates **(b)** Stitch plates **(c)** Slotted connection

(d) Tee connection **(e)** Brackets **(f)** Beam bearing plates

(g) Column base plates **(h)** Pipe connection **(i)** Beam brackets

(j) Built-up sections

Figure 5.5.4
Typical uses of fillet welds.

Ends curved as
per LRFD and ASD-J2.3.b

Figure 5.5.5
Slot and plug welds in
combination with fillet welds.

5.6 WELDING SYMBOLS

Before a connection or joint is welded, the designer must in some way be able to instruct the steel detailer and the fabricator as to the type and size of weld required. The basic types of welds and some of their variations are discussed in Sec. 5.5. If individual and detailed instructions were needed each time a connection was made, the task of providing directions for making the joint would indeed be formidable.

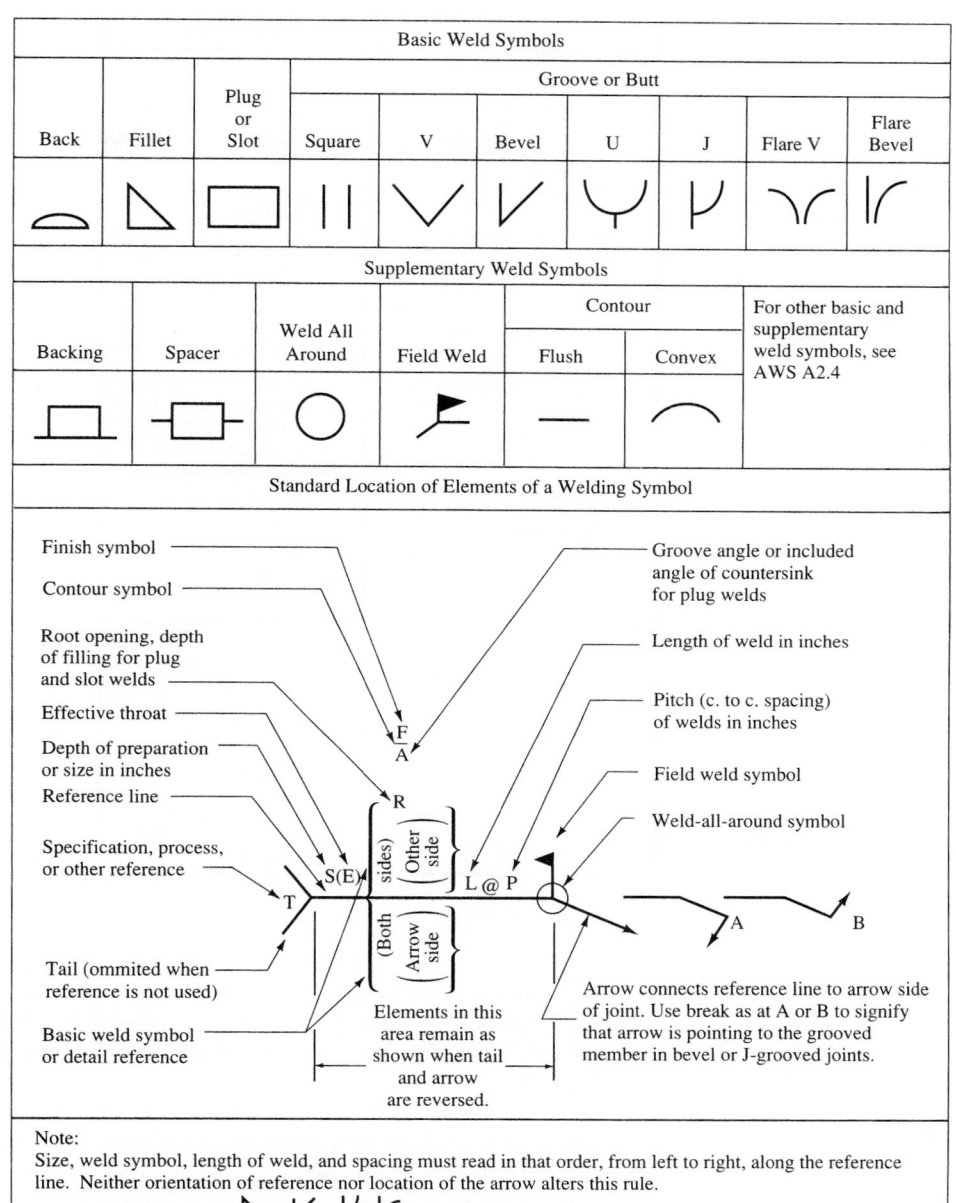

Figure 5.6.1
Standard welding symbols.
(From *AISC Manual* [1.15,
p. 8–35]). Additional welding
symbol information is
available from AWS-A2.4
[5.42].

The need for a simple and yet accurate method for communicating between the designer and fabricator gave rise to the use of shorthand symbols that characterize the type and size of weld. As a result, the American Welding Society standard symbols, shown in Fig. 5.6.1, indicate the type, size, length, and location of weld, as well as any special instructions.

Fillet welds	Groove welds
Fraction indicating leg shown only on one weld when both sides are the same.	Square groove welded from both sides.
Indicates 2-in. long welds are intermittent & staggered on 4-in. centers	Single bevel with $\frac{1}{8}$ in. root opening, 45° bevel on upper piece, and a backing weld used.
Weld all around	Single-V welded from one side having $\frac{3}{8}$ in. root opening and 30° angle within the V.

Figure 5.6.2
Common uses of welding symbols.

Most of the commonly made connections do not require special instructions and are typically specified as shown in Fig. 5.6.2. For a more detailed use of welding symbols the reader is referred to *Standard Symbols for Welding, Brazing and Nondestructive Examination,* A2.4 [5.42].

The reader may feel that the number of symbols is burdensome. However, the system of designating welds is broken down into a few basic types that are built up to give a complete set of instructions. Whenever a particular connection is used in many parts of a structure, it may only be necessary to show a typical detail as shown in Fig. 5.6.3a. Whenever special connections are used, they should be detailed sufficiently to leave no doubt as to the designer's intentions, as shown in Fig. 5.6.3b.

In Fig. 5.6.3b the designer specified that the plug weld be made in the shop and ground flush while the double bevel weld connecting the gusset plate to the column be made in the field. Since the designer did not specify whether the fillet welds attaching the angle to the gusset plate were to be made in the shop or in the field, the steel fabricator would be free to make the decision. However, in this particular detail, it would be better to make the fillet welds in the shop since the plug weld might be overstressed during the field erection process. In general, as many welds as feasible will be made in the shop because of economic considerations. Therefore it is important that the designer specify those welds that are to be *field welded.*

(a)

(b)

Figure 5.6.3
Details showing use of
welding symbols.

5.7   FACTORS AFFECTING THE QUALITY OF WELDED CONNECTIONS

Obtaining a satisfactory welded connection requires the combination of many individual skills, beginning with the actual design of the weld and ending with the welding operation. The structural engineer needs to be aware of the factors that affect the quality of a weld and design the connections accordingly.

Proper Electrodes, Welding Apparatus, and Procedures

After the proper electrode material is specified to match the strength of the steel in the pieces being joined (see Sec. 5.13), the diameter of the welding electrode must be selected. The particular size of the electrode selected is based on the size of the weld to be made and on the electrical current output of the welding apparatus. Since most welding machines have controls for reducing the current output, electrodes smaller than the maximum capability can easily be accommodated and should be used.

Since the weld metal in arc welding is deposited by the electromagnetic field and not by gravity, the welder is not limited to the flat or horizontal welding positions. The four basic welding positions are shown in Fig. 5.7.1. The designer should avoid whenever possible the overhead position, since it is the most difficult one. Joints welded in the shop are usually positioned in the flat or horizontal positions but field welds may require any welding position depending on the orientation of the connection. The welding position for field welds should be carefully considered by the designer.

(a) Flat   (b) Horizontal

(c) Vertical   (d) Overhead

Figure 5.7.1
Welding positions.

Proper Edge Preparation

Typical edge preparations provided for groove welds are shown in Fig. 5.7.2. The root opening R is the separation of the pieces being joined and is provided for electrode accessibility to the base of a joint. The smaller the root opening the greater must be the angle of the bevel. The feathered edge as shown in Fig. 5.7.2a is subject to burn-through unless a backup plate is provided as shown in Fig. 5.7.2b. Backup strips are commonly used when the welding is to be done from one side only. The problem of burn-through is lessened if the bevel is provided a land as shown in Fig. 5.7.2c. The welder should *not* provide a backup plate when a land is provided, since there would be a good possibility that a gas pocket would be formed, preventing a full penetration weld. Occasionally a spacer, as shown in Fig. 5.7.2d, is provided to prevent burn-through but is gouged out before the second side is welded.

(a) Bevel with feathered edge   (b) Bevel with backup plate

Figure 5.7.2
Typical edge preparations for groove welds.

(c) Bevel with a land   (b) Double bevel with a spacer

Control of Distortion

Another factor affecting weld quality is shrinkage. If a single bead is put down in a continuous manner on a plate, it will cause the plate to distort as shown in Fig. 5.7.3. Such distortions will occur unless care is exercised in both the design of the joint and the welding procedure. Figure 5.7.4 shows the result of using unsymmetrical welds as compared to symmetrical welds. Although there are many techniques available for minimizing distortion, the most common one is that of staggering intermittent welds as shown in Fig. 5.7.5a, and then returning to fill in the spaces as shown in Fig. 5.7.5b, a typical sequence being shown. For many structures, such as plate girders, short segments of weld (though not usually regular intermittent welds) may be used at strategic locations to give enough strength to hold all pieces in place; then the continuous lines of weld are placed.

To minimize shrinkage and to insure adequate ductility, the AWS Code (Table 3.2 of Ref. 5.25) has established minimum preheat and interpass temperatures. For welds requiring more than one progression (pass) of a welding operation along a joint, the interpass temperature is the temperature of the deposited weld when the next pass is begun.

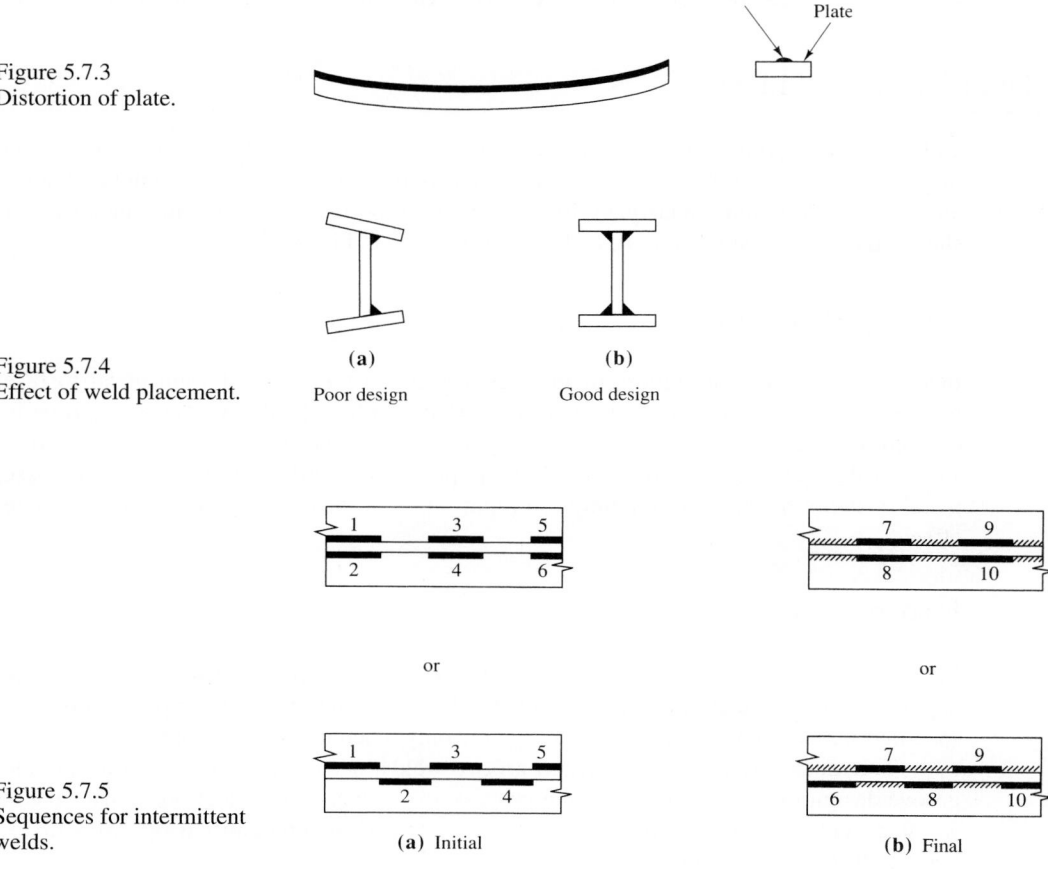

Figure 5.7.3
Distortion of plate.

Figure 5.7.4
Effect of weld placement.

(a)
Poor design

(b)
Good design

Figure 5.7.5
Sequences for intermittent welds.

(a) Initial

(b) Final

The following summarizes ways of minimizing distortion:

1. Reduce the shrinkage forces by
   a. Using minimum weld metal; for grooves use no greater root opening than necessary; do not overweld.
   b. Using as few passes as possible.
   c. Using proper edge preparation and fit-up.
   d. Using intermittent weld, at least for preliminary connection.
   e. Using backstepping; depositing weld segments toward the previously completed weld; i.e., depositing in the direction opposite to the progress of welding the joint.

2. Allow for the shrinkage to occur by
   a. Tipping the plates so after shrinkage occurs they will be correctly aligned.
   b. Using prebending of pieces.

3. Balance shrinkage forces by
   a. Using symmetry in welding; fillets on each side of a piece contribute counteracting effects.
   b. Using scattered weld segments.
   c. Using peening; stretching the metal by series of blows.
   d. Using clamps, jigs, etc.; this forces weld metal to stretch as it cools.

Blodgett [5.18] has provided a more extensive treatment of ways to minimize distortion.

5.8   POSSIBLE DEFECTS IN WELDS

Unless good welding techniques and procedures are used, a number of possible defects may result relating to discontinuities within the weld. Some of the more common defects are: incomplete fusion, inadequate joint penetration, porosity, undercutting, inclusion of slag, and cracks. Examples of these defects are shown in Fig. 5.8.1.

Incomplete Fusion

Incomplete fusion is the failure of the base metal and the adjacent weld metal to fuse together completely. This may occur if the surfaces to be joined have not been properly cleaned and are coated with mill scale, slag, oxides, or other foreign materials. Another cause of this defect is the use of welding equipment of insufficient current, so that base metal does not reach melting point. Too rapid a rate of welding will also have the same effect.

Inadequate Joint Penetration

Inadequate joint penetration means the weld extends a shallower distance through the depth of the groove than specified, as shown in Fig. 5.8.1, where complete penetration was specified. Partial joint penetration is acceptable only when it is so specified.

This defect, relating primarily to groove welds, occurs from use of an unsuitable groove design for the selected welding process, excessively large electrodes, insufficient welding current, or excessive welding rates. Joint designs prequalified by AWS [5.25, Sec. 3.9 through Sec. 3.13] should always be used.

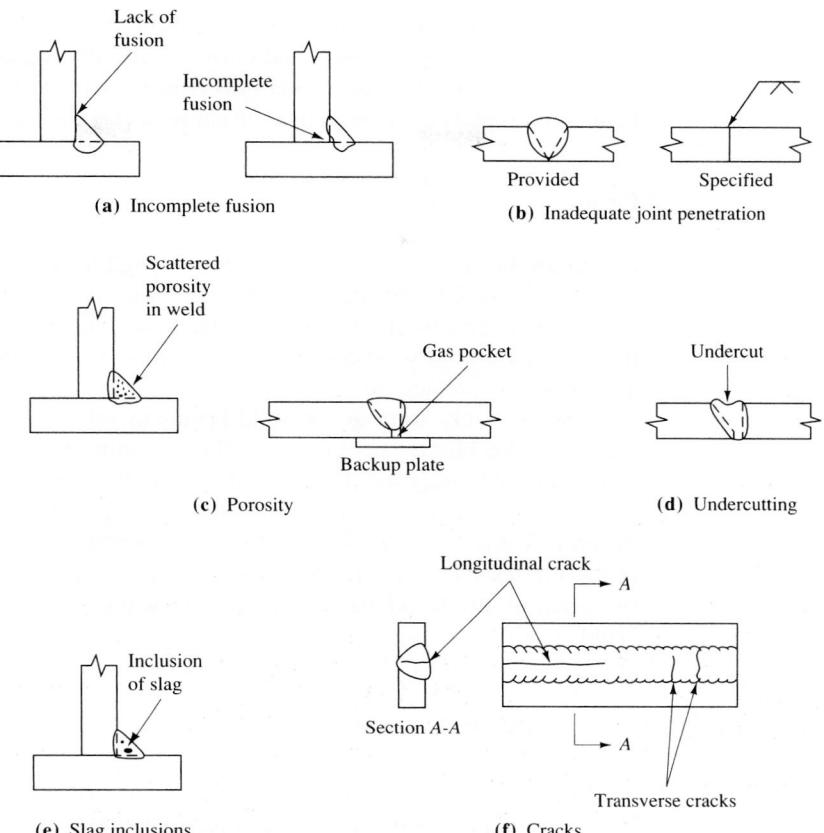

Figure 5.8.1
Possible weld defects.

Porosity

Porosity occurs when voids or a number of small gas pockets are trapped during the cooling process. This defect results from using excessively high current or too long an arc length. Porosity may occur uniformly dispersed through the weld, or it may be a large pocket concentrated at the root of a fillet weld or at the root adjacent to a backup plate in a groove weld. The latter is caused by poor welding procedures and careless use of backup plates.

Undercutting

Undercutting means a groove melted into the base material adjacent to the toe of a weld and left unfilled by weld metal. The use of excessive current or an excessively long arc may burn or dig away a portion of the base metal. This defect is easily detected visually and can be corrected by depositing additional weld material.

Slag Inclusion

Slag is formed during the welding process as a result of chemical reactions of the melted electrode coating and consists of metal oxides and other compounds. Having a lower density than the molten weld metal, the slag normally floats to the surface, where upon cooling, it is easily removed by the welder. However, too rapid a cooling of the joint may trap the slag before it can rise to the surface. Overhead welds as shown in Fig. 5.7.1d are especially subject to slag inclusion and must be carefully inspected. When several passes are necessary to obtain the desired weld size, the welder must remove slag between each pass. Failure to properly do so is a common cause of slag inclusion.

Cracks

Cracks are breaks in the weld metal, either longitudinal or transverse to the line of weld, that result from internal stress. Cracks may also extend from the weld metal into the base metal or may be entirely in the base metal in the vicinity of the weld. Cracks are perhaps the most harmful of weld defects; however, tiny cracks called *microfissures* may not have any detrimental effect.

Some cracks form as the weld begins to solidify, generally caused by brittle constituents, either brittle states of iron or alloying elements, forming along the grain boundaries. More uniform heating and slower cooling will prevent the "hot" cracks from forming.

Cracks may also form at room temperature parallel to but under the weld in the base material. These cracks arise in low-alloy steels from the combined effects of hydrogen, a brittle martensite microstructure, and restraint to shrinkage and distortion. Use of low-hydrogen electrodes along with proper preheating and postheating will minimize such "cold" cracking.

5.9 INSPECTION AND CONTROL

The enormous success and growth in recent years in the area of structural welding of buildings and bridges could not have occurred without some means of inspection and control. The welding industry has led in the development of guidelines which, if followed, virtually insure a sound weld. The inspection and control procedure should begin before the first arc is struck, continue throughout the welding procedure, and if necessary, a pretest of the joint should be made to assure its satisfactory performance. Since such close supervision is not possible on every weld made, the following suggestions will serve as a guideline to achieve good structural welds:

1. Establish good welding procedures.
2. Use only prequalified welders.
3. Use qualified inspectors and have them present.
4. Use special inspection techniques when necessary.

Good welding procedures can be developed from recommendations from the AWS, AISC, and the manufacturers of welding supplies and equipment. The procedure to be followed will depend on the chemical and physical properties of the materials, the types and sizes of weld, and the particular equipment used.

All welders should be required to have passed an American Welding Society Qualification Test before being permitted to make a structural connection. Although this is usually considered adequate, it doesn't prove the ability of the welder to make welds at the actual job site, particularly if the welds are unusual or difficult and were not specified in the Qualification Test. Happily, most welding contractors exercise control over their welders in such situations.

The use of qualified welding inspectors at a job site generally has the effect of causing welders to perform their best work, feeling that the inspector is able to recognize the quality of their welds. The welding inspector should be a competent welder and be able to recognize possible defects. Any poor or suspicious welds should be cut out and replaced.

Figure 5.9.1
Checking size of fillet welds.

(a)   (b)

The simplest and least expensive method of inspection is *visual* but it is dependent on the competence of the observer and weld appearance may be deceiving [5.20]. In some cases a poor weld may be overlooked. A welding gage such as shown in Fig. 5.9.1a offers a rapid means of checking the size of fillet welds.

On more important structures, for welds where fatigue is an important consideration, or welds whose failure could be catastrophic, more rigid inspection techniques [5.20] should be used. Some of the useful ones are the ultrasonic, radiographic, and magnetic particle methods. The *ultrasonic* method [5.21–5.23] passes ultra-high-frequency sound waves through the weldment. Defects in a particular weld will reflect the sound waves while a weld without defects will not impede passage of the waves. *Radiographic* methods include the use of both X-rays and gamma rays. In this method the radiating source is placed on one side of the weld and a photographic plate on the other. This method is expensive and requires special precautions be taken because of the hazards of radiation. However, the method is reliable and furnishes a permanent record. The *magnetic particle* testing method [5.24] uses iron powder which is spread around the welded area and polarized by passing an electric current through the weld. Small local poles will be formed at the edges of any defects, and this may be interpreted by an experienced observer.

AWS D1.1, Chap. 6 [5.25] provides an extensive discussion of the many nondestructive testing methods.

5.10   ECONOMICS OF WELDED BUILT-UP MEMBERS AND CONNECTIONS

Overall economy in welded connections is difficult to evaluate. Some of the factors to be considered such as the amount of electrode material used can easily be computed while other factors such as the value to be placed on esthetics may be intangible. The actual economy of welded connections must be viewed from a broad aspect and include the overall design of the structural system.

Welded connections are usually neater in appearance, providing a less cluttered effect in contrast to bolted connections. Fig. 5.10.1 shows a comparison between a section of a bolted plate girder and a section of a welded plate girder. Besides the neater appearance of the welded joint, welded connections offer the designer more freedom to be innovative in the entire design concept. The designer is not bound to standard sections but may build up any cross-section deemed to be most advantageous. Similarly, the best configuration to transmit the loads from one member to another can be used.

Welded connections generally eliminate the need for holes in members except possibly for erection purposes. Since it is usually the holes at the ends that govern the design of a bolted tension member, a welded connection will generally result in a member with a smaller cross-section.

Welded connections can sometime reduce field construction costs by the fact that members may be shifted slightly to accommodate minor errors in fabrication or erection.

(a) Bolted girder section

(b) Welded girder section

Figure 5.10.1
Comparison between plate
girders.

Also, members may be shortened by cutting and rejoined by suitable welding, as well as lengthened by splicing a piece of the same cross-section.

In addition, several direct factors influence the cost of welding. Generally, welding performed in the shop is less expensive than field welding. Some reasons for this are availability in the shop of automatic welding machines, a more pleasant and less hostile environment (the weather), and the availability of special jigs for holding the pieces to be welded in a more favorable position. Also, work can be scheduled for a continuous operation, whereas field welding must often wait for cranes and special erection equipment. Other operations such as the proper preheating of pieces to be welded can be difficult if not impossible to perform in the field. Other factors that influence welding costs are:

1. Cost of preparing the edges to be welded.
2. The amount of weld material required.
3. The ratio of the actual arc time to overall welding time.
4. The amount of handling required.
5. General overhead costs.

The factors listed above are generally unknown to the designer because the fabricator is usually not selected until after the design has been completed. However, the designer must still make decisions—should short length large size fillet welds or longer length smaller size welds be used? Should large size fillet welds or groove welds be used? If the decision is to use groove welds then the proper and most economical type must be selected.

In most instances the designer is not as concerned with the specific cost of a weld type as with the *relative* cost. Donnelly [5.26] has developed factors relating the cost of fillet and groove welds of common sizes to the cost of a single-pass $\frac{1}{4}$-in. fillet weld.

Currently (2008), welded connections are used for the vast majority of shop connections and a sizable, through not a majority of field connections. More extensive treatment of welding cost is available in the *Welding Handbook,* Vol. 1 [2.21] and from Blodgett [5.27].

5.11 SIZE AND LENGTH LIMITATIONS FOR FILLET WELDS

Since all welding involves the heating of the metal pieces, prevention of too rapid a rate of cooling is of fundamental importance to achieving a good weld. Consider the two extreme thicknesses of plates in Fig. 5.11.1, each of which has received a bead of fillet weld. Most

Figure 5.11.1
Effect of thickness on
cooling rate.

(a) Thick plate   (b) Thin plate

heat energy given off during the welding process is absorbed by plates being joined. The thicker plate dissipates the heat vertically as well as horizontally whereas the thinner plate is essentially limited to a horizontal dissipation. In other words, the thicker the plate, the faster heat is removed from the welding area, thereby lowering the temperature in the region of the weld. Since a minimum temperature is required to cause the base metal to become molten, it is therefore necessary to provide as a minimum, a weld of sufficient size (and heat content) to prevent the plate from removing the heat at a faster rate than it is being supplied. Unless a proper temperature is maintained in the area being welded a lack of fusion will result.

Minimum Weld Size

To help insure fusion and minimize distortion the AWS [5.25] and AISC Specification [1.3] provide for a minimum size weld based on the thicker of the pieces being joined. The requirements for fillet weld based on the leg dimension a of the fillet and for partial-joint-penetration groove weld based on the effective throat (see Sec. 5.12) are given in Tables 5.11.1 and 5.11.2.

Maximum Fillet Weld Size Along Edges

The *maximum* size of fillet weld used *along the edges* of pieces being joined is limited (AISC-J2.2b) in order to prevent the melting of the base material at the location where the

TABLE 5.11.1 Minimum Size* of Fillet Welds (Adapted From
AISC-Table J2.4)

| Material thickness of thicker part joined | | Minimum size** of fillet weld*** | |
(in.)	(mm)	(in.)	(mm)
To 1/4 inclusive	To 6.4 inclusive	1/8	3
Over 1/4 to 1/2	Over 6.4 to 12.7	3/16	5
Over 1/2 to 3/4	Over 12.7 to 19.0	1/4	6
Over 3/4	Over 19.0	5/16	8

* See AISC-J2.2b for maximum size of fillet welds.

** Leg dimension; single pass weld.

*** According to AWS-2.3, weld size need not exceed the thickness of the thinner part joined.

TABLE 5.11.2 Minimum Effective Throat Thickness of Partial-Joint-Penetration Groove Welds (Adapted From AISC Table J2.3)

Material thickness of thicker part joined		Minimum effective throat thickness*	
(in.)	(mm)	(in.)	(mm)
To 1/4 inclusive	To 6.4 inclusive	1/8	3
Over 1/4 to 1/2	Over 6.4 to 12.7	3/16	5
Over 1/2 to 3/4	Over 12.7 to 19.0	1/4	6
Over 3/4 to 1-1/2	Over 19.0 to 38.1	5/16	8
Over 1-1/2 to 2-1/4	Over 38.1 to 57.1	3/8	10
Over 2-1/4 to 6	Over 57.1 to 152	1/2	13
Over 6	Over 152	5/8	16

*See AISC-J2.1 for effective throat thickness. For J- and U-joints, and Bevel- or V-joints having included angle at root of groove $\geq 60°$, effective throat is depth of chamfer. For Bevel- or V-joints having included angle $< 60°$ but $\geq 45°$, effective throat is depth of chamfer minus 1/8 in. (3 mm).

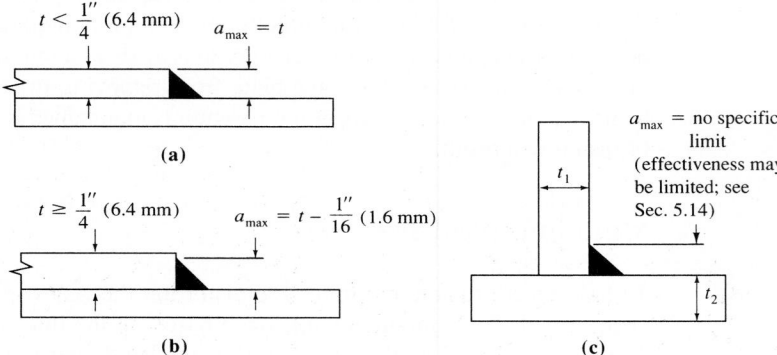

Figure 5.11.2
Maximum weld size.

fillet would meet the corner of the plate if the fillet were made the full plate thickness. The maximum permitted is (see Fig. 5.11.2):

1. Along edges of material *less* than $\frac{1}{4}$-in. (6.4-mm) thick, the maximum size is equal to the thickness of the material.

2. Along edges of material $\frac{1}{4}$ in. (6.4 mm) or more in thickness, the maximum size shall be $\frac{1}{16}$ in. (1.6 mm) less than the thickness of the material, unless the weld is especially designated on the drawings to be built out to obtain full throat thickness.

Minimum Effective Length of Fillet Welds

When placing a fillet weld, the welder builds up the weld to its full dimension as near the beginning of the weld as practicable. However, there is always a slight tapering off in the region where the weld is started and where it ends. Therefore, the minimum effective length of a fillet weld is four times the nominal size (AISC-J2.2.b). If this requirement is not met, the size of the weld shall be considered to be one-fourth of the effective length.

AISC-J2.2.b recommends the use of end returns, whenever practicable, as shown in Fig. 5.11.3. For other limitations the reader is referred to the AISC Specification [1.13].

Figure 5.11.3
Use of end returns.

5.12   EFFECTIVE AREAS OF WELDS

The strengths of the various types of welds discussed in Sec. 5.5 are based on *effective areas*. The effective area of a groove or fillet weld is the product of the *effective throat* dimension t_e times the length of the weld.

The effective throat dimension depends on the nominal size and the shape of the weld, and may be thought of as the minimum width of the expected failure plane.

Groove Welds

The effective throat dimension of a full penetration groove weld is the thickness of the thinner part joined, as shown in Figs. 5.12.1a and b. For a partial joint penetration groove weld, the effective throat may be less than the depth of the chamfer. For example, when bevel- or V-joint grooves have an included angle at the root of the groove less than 60° but not less than 45° when SMAW or SAW processes are used, or when the GMAW or FCAW processes are used in vertical or overhead positions, the effective throat is the depth of the chamfer less $\frac{1}{8}$ in. When the included angle is 60° or more, the effective throat is the full depth of chamfer for all four processes mentioned. Effective throat requirements for the groove situations mentioned, as well as others, are given in AISC-J2.1a. and Table J2.1.

(a)   (b)

Figure 5.12.1
Effective throat dimensions
for groove welds (AISC-
J2.1a) made by SMAW,
SAW, GMAW, and FCAW.

(c) Bevel- or V-joint   (d) Bevel- or V-joint

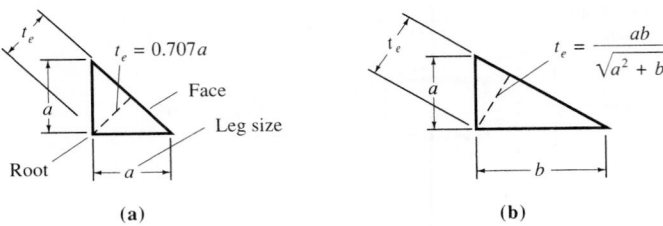

Figure 5.12.2
Effective throat dimensions
for fillet welds.

Fillet Welds

The effective throat dimension of a fillet weld is nominally the shortest distance from the root to the face of the weld, as shown in Fig. 5.12.2. Assuming the fillet weld to have equal legs of nominal size a, the effective throat t_e is $0.707a$. If the fillet weld is designed to be unsymmetrical (a rare situation) with unequal legs, as shown in Fig. 5.12.2b, the value of t_e must be computed from the diagrammatic shape of the weld. The effective throat dimensions for fillet welds made by the submerged arc (SAW) process produces some penetration at the root of the weld, which increases the strength of the weld. Previous AISC provisions allowed for an increased effective throat. The current Specification 2005 requires that the increased penetration must be validated by testing before using an increased effective throat.

Plug and Slot Welds

The effective shearing area of plug or slot welds is their nominal area (sometimes called *faying surface*) in the shearing plane. The resistance of plug or slot welds is the product of the nominal cross-section times the stress on that area.

5.13   NOMINAL STRENGTH OF WELDS

Since welds must transmit the entire load from one member to another, welds must be sized accordingly and be formed from the correct electrode material. For design purposes fillet welds are assumed to transmit loads through *shear stress* on the effective area no matter how the fillets are oriented on the structural connection. Groove welds transmit loads exactly as in the pieces they join.

The electrode material used in welds should have properties of the base material. When properties are comparable, the weld metal is referred to as "matching" weld metal. Table 5.13.1 gives "matching" weld metal for many ASTM structural steels used in buildings and bridges.

Strength of Groove Welds

Complete joint penetration (CJP) groove welds are considered to have the same strength on the *effective area* as the pieces being joined. Therefore, AISC-J2.4 requires no computation of the strength of the CJP groove weld. Welds subject to tension normal to the weld axis must be made with "matching" weld metal, where the properties of the weld metal are comparable to those of the base metal. In compression, where stability of the compression member is usually the major factor, the weld metal strength is permitted to be one classification (10 ksi) lower than the "matching" base metal requirement. The authors recommend, however, that matching electrodes be used when the member may possibly be used in tension in the future.

TABLE 5.13.1 Summary of Matching Filler Metal Requirement as Specified
by AWS D1.1 for the Most Common Types of Structural Steel
(from AWS D1.1 Table 3.1, also *User Note*, AISC. J2.6) [2.23]

Base Metal		Matching Filler Metal
A36 $\leq \frac{3}{4}$ in. thick		60 & 70 ksi Electrodes
A36 $> \frac{3}{4}$ in.	A572 (Gr. 50 & 55)	SMAW: E7015, E7016, E7018, E7028
A588*	A913 (Gr. 50) A992	Other processes: 70 ksi electrodes
A1011	A1018	
A913 (Gr. 60 & 65)		80 ksi electrodes

* For corrosion resistance and color similar to the base see AWS D1.1, Sect. 3.7.3

Notes: 1. Electrodes shall meet the requirements of AWS A5.1, A5.5, A5.17, A5.18, A5.20, A5.23, A5.28 and A5.29.

2. In joints with base metals of different strengths use either a filler metal that matches the higher strength base metal or a filler metal that matches the lower strength and produces a low hydrogen deposit.

When "matching" electrode material is used, the weld material is somewhat stronger than the base material; thus the strength of the welded joint is controlled by the base material properties.

When the groove welded joint is subject to shear, the weld metal is permitted (AWS-Table 2.3 [2.23]) to be of lower strength than the base metal; i.e., less than "matching" weld metal, in which case the strength F_{EXX} of the weld material must be used instead of F_y in computing R_{nw}.

Strength of Fillet Welds

The strength R_{nw} of a fillet weld per inch of length is based on assumption that failure of such a weld is by shear on the effective area whether the shear transfer is parallel to or perpendicular to the axis of the line of fillet weld. In fact, the strength is greater for shear transfer perpendicular to the weld axis; however, for simplicity the situations are treated the same. Thus, fillet weld strength may be controlled by the weld electrode strength *or* by the base material shear strength. The nominal strength R_{nw} per inch of weld length may be expressed,

a. For the weld metal,

$$R_n = F_w A_w = 0.60 F_{EXX} t_e (1 \text{ in.}) = 0.60 t_e F_{EXX} \tag{5.13.1}$$

b. For the base metal,

$$R_n = F_{BM} A_{BM} = 0.6 F_y t (1 \text{ in.}) = 0.6 t F_y \quad \text{yielding} \tag{5.13.2}$$

$$R_n = F_{BM} A_{BM} = 0.6 F_u t (1 \text{ in.}) = 0.6 t F_u \quad \text{rupture} \tag{5.13.3}$$

The stress distribution in welded joints is complex and nonuniform. Figure 5.13.1 shows the typical stress distribution at service load for longitudinal fillet welds in a lap joint. The actual variation of shear stress in the weld from points *A* to *B* depends on the length of the weld as well as the ratio of the widths of the plates being joined. Figure 5.13.2 shows the typical shear variation for fillet welds loaded transverse to the weld axis.

The load-deformation relationship of a fillet weld has been studied by Butler, Pal, and Kulak [5.29], Kulak and Timmler [5.30], Swannell [5.31, 5.32], and Neis [5.33], and is shown in Fig. 5.13.3, where the reader may observe that the strength is related to the angle θ

Figure 5.13.1
Typical stress distribution in a lap joint with longitudinal fillet welds.

Figure 5.13.2
Typical stress distribution in a lap joint with transverse fillet welds.

Note: For a group of tranverse and longitudinal welds AISC-J2.4G gives an alternative method to compute the connection strengrh as follows:

The combined strength, R_n shall be determined as the greater of

$$R_n = R_{wl} + R_{wt}$$

or

$$R_n = 0.85 R_{wl} + 1.5 R_{wt}$$

where:

R_{wl} = the total nominal strength of longitudal welds (kips)

R_{wl} = the total nominal strength of transverse welds (kips)

θ = angle between force and axis of fillet weld

R_i = Eq. 5.17.2

$R_n = 0.6 F_{EXX}(1 + 0.50 \sin^{1.5}\theta) A_w$

Figure 5.13.3
Load-deformation relationship for fillet welds.

to the weld axis at which the weld is loaded. The longitudinal welds in Fig. 5.13.1 are loaded at 0° (parallel to the weld axis); Fig. 5.13.3 shows considerable ductility (large deformation capability) for such welds. Figure 5.13.2 shows transverse welds loaded at 90° (perpendicular to the weld axis); Fig. 5.13.3 shows greater strength but considerably less deformation capability for such welds.

For all AISC Specifications, both ASD and LRFD, prior to the 1993 LRFD Specification, the additional strength of fillet welds loaded at angles θ greater than 0° was not utilized. The shear strength of the weld segment per unit length, taken at $t_e(0.60F_{\text{EXX}})$ (i.e., Eq. 5.13.1), was the upper limit, regardless of the angle θ. The 2005 AISC-J2.4a permits using increased fillet weld strength for weld segments within a group loaded in-plane, through the group center of gravity, at an angle θ measured from the weld axis. This alternate approach is used in the strength method treatment of eccentric shear connections in Sec. 5.17.

Figure 5.13.3 shows that a fillet weld shows maximum deformation Δ_i/a capability before failure from about 0.06 for $\theta = 90°$ to more than 0.15 when $\theta = 0°$. Thus, because of this ductility, lines of weld loaded either parallel or perpendicular to the axis of the weld are assumed for design purposes to resist equally anywhere along their length. Even though the *elastic* distribution of load along the length of weld is not uniform, plastic deformation will permit this simplified designed treatment.

5.14   LOAD AND RESISTANCE FACTOR DESIGN—WELDS

The general philosophy of Load and Resistance Factor Design (LRFD) was described in Secs. 1.8 and 1.9. Equation 1.8.1 gives the structural safety requirement, as follows:

$$\phi R_n \geq \sum \gamma_i Q_i \qquad [1.8.1]$$

where ϕ = resistance factor (strength reduction factor), values of which for welds are given in Table 5.14.1

R_n = nominal resistance (strength) = R_{nw} for welds

γ_i = overload factors (ASCE 7-05)

Q_i = service loads (see Secs. 1.8 and 1.9)

Equation 1.8.1 requires the *design strength* ϕR_n to equal or exceed the summation of *factored loads,* or specifically for welds, Eq. 1.8.1 becomes

$$\phi R_{nw} \geq R_u \qquad (5.14.1)$$

where ϕ = resistance factor (for welds the ϕ factor is the same as for the material it connects; that is, 0.90 for the yielding limit state and 0.75 for the fracture limit state)

R_{nw} = the nominal strength per unit length of weld, but not to exceed the nominal strength per unit length of adjacent base material

R_u = factored load per unit length of weld

Groove Welds

The design strength of complete penetration groove welds, whether loaded in shear, tension or compression, is governed by the strength of the base metal and no computation of the strength of such groove welds is required. For tension normal to the weld, the joints are intended to provide strength equivalent to the base metal. Hence, matching strength filler metal is required.

TABLE 5.14.1 Design Strength of Welds (From *AISC Specification* [1.13], AISC-Table J2.5)

Load Type and Direction Relative to Weld Axis	Pertinent Metal	ϕ and Ω	Nominal Strength (F_{bm} or F_w) kips (N)	Effective Area (A_{BM} or A_W) in.^{2} (mm^2)	Required Filler Metal Strength Level[a, b]
Complete-Joint-Penetration Groove Welds					
Tension normal to weld axis			Strength of the joint is controlled by the base metal		Matching filler metal shall be used. For T and corner joints with backing left in place, notch tough filler metal is required. See Section J2.6.
Compression normal to weld axis			Strength of the joint is controlled by the base metal		Filler metal with a strength level equal to or one strength level less than matching filler metal is permitted.
Tension or compression parallel to weld axis			Tension or compression in parts joined parallel to a weld need not be considered in design of welds joining the parts.		Filler metal with a strength level equal to or less than matching filler metal is permitted.
Shear			Strength of the joint is controlled by the base metal		Matching filler metal shall be used.[c]
Partial-Joint-Penetration Groove Welds Including Flare Vee Groove and Flare Bevel Groove Welds					
Tension normal to weld axis	Base	$\phi = 0.90$ $\Omega = 1.67$	F_y	See J4	
	Weld	$\phi = 0.80$ $\Omega = 1.88$	$0.60F_{EXX}$	See J2.1a	
Compression column to base plate and column splices designed per J1.4(a)			Compressive stress need not be considered in design of welds joining the parts.		
Compression connections of members designed to bear other than columns as described in J1.4(b)	Base	$\phi = 0.90$ $\Omega = 1.67$	F_y	See J4	Filler metal with a strength level equal to or less than matching filler metal is permitted.
	Weld	$\phi = 0.80$ $\Omega = 1.88$	$0.60F_{EXX}$	See J2.1a	
Compression connections not finished-to-bear	Base	$\phi = 0.90$ $\Omega = 1.67$	F_y	See J4	
	Weld	$\phi = 0.80$ $\Omega = 1.88$	$0.90F_{EXX}$	See J2.1a	
Tension or compression parallel to weld axis			Tension or compression in parts joined parallel to a weld need not be considered in design of welds joining the parts.		
Shear	Base		Governed by J4		
	Weld	$\phi = 0.75$ $\Omega = 2.00$	$0.60F_{EXX}$	See J2.1a	

TABLE 5.14.1 (*Continued*)

Fillet Welds Including Fillets in Holes and Slots and Skewed T-Joints					
Shear	Base	Governed by J4			Filler metal with a strength level equal to or less than matching filler metal is permitted.
	Weld	$\phi = 0.75$ $\Omega = 2.00$	$0.60F_{EXX}$[d]	See J2.2a	
Tension or compression parallel to weld axis	Tension or compression in parts joined parallel to a weld need not be considered in design of welds joining the parts.				
Plug and Slot Welds					
Shear parallel to faying surface on the effective area	Base	Governed by J4			Filler metal with a strength level equal to or less than matching filler metal is permitted.
	Weld	$\phi = 0.75$ $\Omega = 2.00$	$0.60F_{EXX}$	J2.3a	

[a] For matching weld metal see AWS D1.1, Section 3.3.

[b] Filler metal with a strength level one strength level greater than matching is permitted.

[c] Filler metals with a strength level less than matching may be used for groove welds between the webs and flanges of built-up sections transferring shear loads, or in applications where high restraint is a concern. In these applications, the weld joint shall be detailed and the weld shall be designed using the thickness of the material as the effective throat, $\phi = 0.80$, $\Omega = 1.88$ and $0.60\ F_{EXX}$ as the nominal strength.

[d] Alternatively, the provisions of J2.4(a) are permitted provided the deformation compatibility of the various weld elements is considered. Alternatively, Sections J2.4(b) and (c) are special applications of J2.4(a) that provide for deformation compatibility.

For partial penetration groove welds, the weld design strength per unit length (Table 5.14.1 and AISC-J4) depends on the type of applied stress:

1. Tension and compression normal to the effective area*:

$$\phi R_{nw} = 0.80(t_e)(0.60F_{EXX}) = 0.48t_e F_{EXX} \qquad \text{weld metal} \qquad (5.14.2)$$

2. Shear on effective area*:

$$\phi R_{nw} = 0.75(t_e)(0.60F_{EXX}) = 0.45t_e F_{EXX} \qquad \text{weld metal} \qquad (5.14.3)$$

The strength of the base materials should also be checked for loads applied parallel to the weld axis, according to AISC-J4.2.

Fillet Welds

The *design strength* (AISC-J2.4.) per unit length of a fillet weld is based on the shear resistance through the throat of the weld, as follows:

$$\phi R_{nw} = 0.75t_e(0.60\ F_{EXX}) \qquad (5.14.4)$$

---

*See Sec. 5.12 for definition of effective area.

but not greater than the shear yield or shear rupture strengths (AISC-J4.2) of the adjacent base materials. The base material design strength is the lower value of:

$$\phi R_n = 1.0(0.6F_y)A_{BM} = 0.6tF_y \qquad \text{yield strength} \qquad (5.14.5)$$

$$\phi R_n = 0.75(0.6F_u)A_{BM} = 0.45tF_u \qquad \text{rupture strength} \qquad (5.14.6)$$

where t_e = effective throat dimension (see Sec. 5.12)

F_{EXX} = tensile strength of electrode material

t = thickness of base material along which weld is placed

F_u = tensile strength of base metal

The governing base material strength will depend on the values of the tensile strength and yield strength for the steel under consideration.

Note that the limit state for fillet welds is fracture through the throat of the fillet; thus, the ϕ factor is 0.75. Values of ϕR_{nw} based on Eq. 5.14.4 for various electrode strengths are given in Table 5.14.2 for shielded metal arc welding (SMAW) and submerged arc welding (SAW).

When a fillet weld is loaded through its center of gravity but at an angle θ measured from the weld axis, or weld segments within a weld group are loaded in-plane, the strength of each weld segment is permitted to be modified from that given by Eq. 5.14.6, in accordance with the provisions of AISC-J2.4. This alternate approach is used in the strength method treatment of eccentric shear connections in Sec. 5.17. Kennedy, Miazga, and Lesik [5.37, 5.38, 5.39] have presented the latest information on fillet weld shear strength.

AISC Table J2.5 containing the design strengths of welds is presented as Table 5.14.1.

TABLE 5.14.2 Design Shear Strength ϕR_{nw} of Fillet Weld, (kips/in.)
(Shielded Metal Arc Welding and Submerged Arc Welding)

Nominal size (in.)	Effective throat (in.)	Minimum tensile strength of weld (ksi)					
		60	70	80	90	100	110
1/8	0.088[a]	2.38[b]	2.77	3.17	3.56	3.96	4.36
3/16	0.133	3.58	4.18	4.77	5.37	5.97	6.56
1/4	0.177	4.77	5.57	6.36	7.16	7.95	8.75
5/16	0.221	5.97	6.96	7.95	8.95	9.94	10.94
3/8	0.265	7.16	8.35	9.54	10.74	11.93	13.12
7/16	0.309	8.35	9.74	11.14	12.53	13.92	15.31
1/2	0.354	9.54	11.14	12.73	14.32	15.91	17.50
9/16	0.398	10.74	12.53	14.32	16.11	17.90	19.69
5/8	0.442	11.93	13.92	15.91	17.90	19.88	21.87
11/16	0.486	13.12	15.31	17.50	19.69	21.87	24.06
3/4	0.530	14.32	16.70	19.09	21.48	23.86	26.25

[a] t_e = 0.707 times leg size a = 0.707(1/8) = 0.088 in.
[b] $\phi t_e(0.60\, F_{EXX})$ = 0.75(0.707a)(0.60 times tensile strength)
   = 0.75(0.707)(1/8)(0.60)(60) = 2.38 kips/in.

TABLE 5.14.3 Design Shear Strength of Fillet Welds, ϕR_{nw} (N/mm)
(Shielded Metal Arc Welding)

Nominal size (mm)	Effective throat (mm)	Minimum tensile strength of weld (MPa)					
		415	485	550	620	690	760
3		396[b]	463	525	592	659	725
4		528	617	700	789	878	967
5		660	772	875	986	1098	1209
6		792	926	1050	1184	1317	1451
8		1056	1234	1400	1578	1756	1934
10		1320	1543	1750	1973	2195	2418
12		1584	1852	2100	2367	2634	2902
14		1848	2160	2450	2762	3073	3385
16		2113	2469	2800	3156	3512	3869
18		2377	2777	3150	3551	3951	4352
20		2641	3086	3500	3945	4390	4836

[a] t_e = 0.707 times leg size a = 0.707(3) = 2.12 mm.
[b] $\phi t_e(0.60 F_{EXX})$ = 0.75(0.707a)(0.60 times tensile strength)
= 0.75(0.707)(2.12)(0.60)415 = 396 N/mm.

EXAMPLE 5.14.1

Determine the effective throat dimension of a $\frac{7}{16}$-in. fillet weld made by (a) shielded metal arc welding (SMAW), and (b) submerged arc welding (SAW), in accordance with the AISC Specification [1.13].

Solution:

(a) $t_e = 0.707a = 0.707(0.4375) = 0.309$ in.

(b) Since consistent penetration beyond the root of the weld has not been demonstrated by test, t_e is equal to 0.309 in. [same as in part (a)]

These values are in agreement with those given in Table 5.14.2 for SMAW. ∎

EXAMPLE 5.14.2

Determine the design shear strength ϕR_{nw} of a $\frac{3}{8}$-in. fillet weld produced by (a) shielded metal arc welding, and (b) submerged arc welding. Assume E70 electrodes having minimum tensile strength F_{EXX} of 70 ksi are used, according to the AISC Specification.

Solution:

(a) SMAW process. $t_e = 0.707a = 0.707(0.375) = 0.265$ in. According to Eq. 5.14.4,

$$\phi R_{nw} = \phi t_e(0.60 F_{EXX}) = 0.75(0.265)(0.60)70 = 8.35 \text{ kips/in.}$$

(b) SAW process. Same as in part (a)

These agree with those in Table 5.14.2. ∎

EXAMPLE 5.14.3

Determine the design shear strength ϕR_n for a $\frac{3}{4}$-in.-diam plug weld using E70 electrode material. Use the AISC Specification.

Solution:
Assuming the weld diameter D satisfies the limitations of AISC-J2.3b relating to the dimension of the piece in which the plug weld is made,

$$\phi R_n = 0.75(\text{area of faying surface, } \pi D^2/4)(0.60F_{EXX})$$
$$= 0.75(0.4418)(0.60)70 = 13.9 \text{ kips} \qquad \blacksquare$$

Maximum Effective Fillet Weld Size

In Sec. 5.11 the limitations on maximum and minimum fillet weld size and length relating to practical design considerations were given. Those requirements relate to the size of weld that is actually placed. Regarding strength, however, no welds of whatever size may be designed using a strength greater than available on the adjacent base material.

Consider the two lines of fillet weld transmitting the shear V across section *a-a* of Fig. 5.14.1a. The design strength ϕR_{nw} of the fillet weld is given by Eq. 5.14.4. The strength of the base metal could be governed by either the rupture or the yield limit state. Equating the capacity per inch of the weld metal to the shear capacity per inch in the base material gives for shielded metal arc welding,

$$\phi R_{nw}(\text{weld}) = \phi R_n(\text{base metal}) \qquad (5.14.7)$$

1. For the base metal strength governed by yielding,

$$0.75(2a)(0.707)0.60F_{EXX} = 1.0(0.6F_y)t_1$$

$$a_{\text{max eff}} = \frac{0.6F_y t_1}{0.75(2)(0.707)0.60F_{EXX}} = 0.943\frac{F_y t_1}{F_{EXX}} \qquad (5.14.8)$$

2. For the base metal strength governed by rupture,

$$0.75(2a)(0.707)0.60F_{EXX} = 0.75(0.6F_u)t_1$$

$$a_{\text{max eff}} = \frac{0.6F_u t_1}{(2)(0.707)0.60F_{EXX}} = 0.707\frac{F_u t_1}{F_{EXX}} \qquad (5.14.9)$$

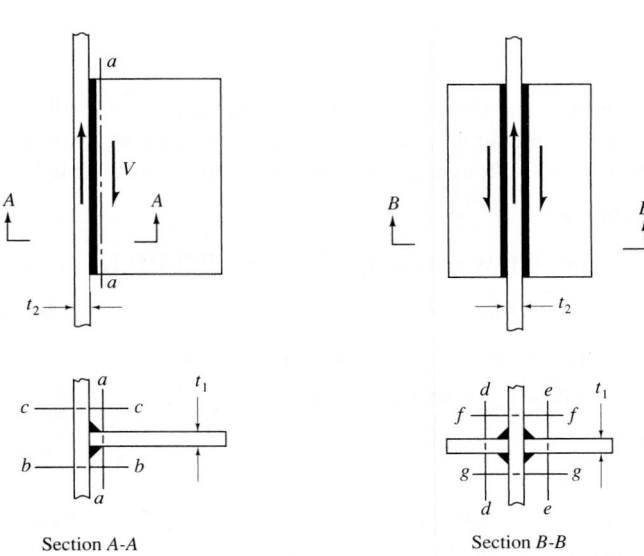

Section A-A

(a)

Section B-B

(b)

Figure 5.14.1
Critical sections for possible overstressing of base material.

where t_1 = thickness of base material

F_y = yield strength of base material

F_u = tensile strength of base material

F_{EXX} = tensile strength of electrode material (70 ksi for E70 electrodes)

Sections *b-b* and *c-c* will not be critical since two lines of weld transfer load across those sections,

1. For yield strength,

$$0.75a(0.707)0.60F_{EXX} = 1.00(0.6F_y)t_2$$

$$a_{max\ eff} = 1.89\frac{F_y t_2}{F_{EXX}} \qquad (5.14.10)$$

2. For rupture strength,

$$0.75a(0.707)0.60F_{EXX} = 0.75(0.6F_u)t_2$$

$$a_{max\ eff} = 1.41\frac{F_u t_2}{F_{EXX}} \qquad (5.14.11)$$

Considering the four fillet welds of Fig. 5.14.1b, sections *d-d* and *e-e* are the same as section *a-a;* therefore, Eq. 5.14.9 applies. On sections *f-f* and *g-g* four fillet welds transfer load across two sections. Thus,

$$0.75(4a)(0.707)(0.60F_{EXX}) = 1.0(2)(0.6F_y)t_1 \qquad (5.14.12)$$

$$0.75(4a)(0.707)(0.60F_{EXX}) = 0.75(2)(0.6F_u)t_1 \qquad (5.14.13)$$

and again Eqs. 5.14.8 and 5.14.9 are the results.

Even when fillet welds connect members that are in tension, the transfer of load by means of the weld is a shear transfer to the base pieces when the fillet welds are parallel to the direction of the load. For such cases, the maximum effective weld size concept still applies.

EXAMPLE 5.14.4

Determine the design shear strength ϕR_{nw} to be used for the flange to web connection in Fig. 5.14.2. The plates are A36 steel and electrodes having F_{EXX} = 70 ksi are to be used with shielded metal arc welding (SMAW).

Solution:
Minimum weld size = a_{min} = $\frac{5}{16}$ in. (AISC-Table J2.4).
Equation 5.14.9 applies,

$$a_{max\ eff} = 0.707\frac{F_u t_w}{F_{EXX}} = 0.707\frac{58t_w}{70} = 0.59t_w$$

$$= 0.59(5/16) = 0.183\ in.$$

$$\phi R_{nw} = 0.75(0.183)(0.707)(0.60)(70) = 4.08\ kips/in.$$

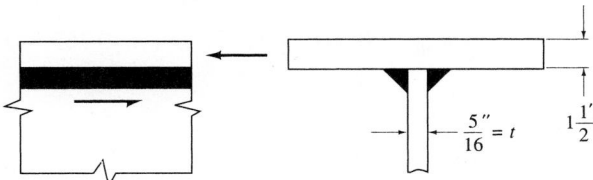

Figure 5.14.2
Example 5.14.4.

or, for two fillets, the design strength is $2(4.08) = 8.16$ kips/in. Thus, even though a $\frac{5}{16}$-in. fillet weld must be placed, its strength in design may not exceed the strength assuming $a = 0.183$ in. ∎

5.15 ALLOWABLE STRENGTH DESIGN—WELDS

The general philosophy of Allowable Strength Design (ASD) was described in Secs. 1.8 and 1.9. Equation 1.8.8 gives the structural safety requirement, as follows:

$$\frac{R_n}{\Omega} \geq R_a \qquad [1.8.8]$$

where Ω is the safety factor.

Table 5.14.1 provides the values of Ω that correspond to both groove and fillet welds under different loading conditions. The welded joint allowable strength is based on the joint nominal strength as defined in Sec. 5.13. The nominal strengths for the base metal and the weld are computed using the values in Table 5.14.1 and AISC-J4. The service load is defined in the AISC ASD Method as the required strength. From Eq. 1.8.8, the allowable strength must equal or exceed the required weld strength R_a. Typically, the design of the weld involves the determination of the required weld length for a predefined weld size. However, in cases of geometric restrictions the required weld size could be determined for a specified weld length. To determine the required weld length to resist a given force, the nominal strength $R_{nw} = F_w A_w$ of the weld is used in Eq. 1.8.8, giving

$$\frac{R_n}{\Omega} = \frac{F_w A_w}{\Omega} = \frac{F_w t_e L_w}{\Omega} \qquad (5.15.1)$$

The required weld length satisfying required strength can then be expressed,

$$L_w \geq \frac{\Omega \Sigma Q_i}{F_w t_e} \qquad (5.15.2)$$

The limitations on maximum and minimum weld size were discussed in Sec. 5.11, and the effective areas were discussed in Sec. 5.12.

The general summary of AISC Allowable Strength Design (ASD) Method provisions relating to welds is in Table 5.14.1.

EXAMPLE 5.15.1

Determine the allowable shear resistance of a $\frac{3}{8}$-in. fillet weld produced by shielded metal arc process. Assume use of E70 electrodes having minimum tensile strength F_{EXX} of 70 ksi, and use AISC ASD Method.

Solution:

SMAW process. For fillet welds, the safety factor given in Table 5.14.1 is 2.0.

$$t_e = 0.707a = 0.707(0.375) = 0.265 \text{ in.}$$

$$R_{nw} = 0.60 F_{EXX} t_e = 0.60(70)0.265 = 11.1 \text{ kips/in.}$$

TABLE 5.15.1 Allowable Resistance R_w of Fillet Welds, (kips/in.)

Nominal size (in.)	Effective throat (in.)	Minimum tensile strength of weld (ksi)					
		60	70	80	90	100	110
1/8	0.088	1.59	1.86	2.12	2.39	2.65	2.92
3/16	0.132	2.38	2.78	3.18	3.58	3.97	4.37
1/4	0.177	3.18	3.71	4.24	4.77	5.30	5.83
5/16	0.221	3.98	4.64	5.30	5.96	6.63	7.30
3/8	0.265	4.77	5.57	6.36	7.16	7.95	8.75
7/16	0.309	5.57	6.49	7.42	8.35	9.28	10.21
1/2	0.353	6.36	7.42	8.48	9.54	10.60	11.66
9/16	0.398	7.16	8.35	9.54	10.74	11.93	13.12
5/8	0.442	7.95	9.28	10.61	11.93	13.26	14.58
11/16	0.486	8.75	10.21	11.67	13.12	14.58	16.04
3/4	0.530	9.54	11.13	12.72	14.31	15.91	17.50

and the allowable strength is

$$\frac{R_{nw}}{\Omega} = \frac{F_w A_w}{2.0} = \frac{11.1}{2.0} = 5.57 \text{ kips/in.}$$

The value may be checked by referring to Table 15.15.1.   ■

Thus, the nominal strength R_{nw} used in LRFD is twice the allowable load R_w used in ASD. Typically, the allowable service load is obtained by dividing the strength by a desired factor; for connections 2.0 has traditionally been used.

5.16   WELDS CONNECTING MEMBERS SUBJECT TO DIRECT AXIAL LOAD

In the design of welds connecting tension or compression members, the welds should be at least as strong as the members they connect and the connection should not introduce significant eccentricity of loading.

Groove Welds

In the case of full joint penetration groove welds as shown in Fig. 5.5.2, the full strength of the cross-section may be developed by selecting the proper electrode material corresponding to the base material as indicated in Table 5.13.1, and specifying an AWS prequalified joint.

EXAMPLE 5.16.1

For the plate tension member (Fig. 5.16.1) carrying axial service loads of 60 kips live load and 12 kips dead load, select the required thickness of the plates (A572 Grade 50 steel), the

Figure 5.16.1
Example 5.16.1.

proper electrode material, and specify a proper AWS prequalified groove joint. Use AISC LRFD Method.

Solution:

(a) Compute the factored load to be carried.

$$T_u = 1.2(12) + 1.6(60) = 110 \text{ kips}$$

(b) Compute the thickness required for the plates.

$$\phi T_n = \phi F_y A_g = 0.90(50)A_g \qquad \text{(yielding limit state)}$$
$$\phi T_n = \phi F_u A_e = 0.75(65)A_e \qquad \text{(fracture limit state)}$$

Since there are no holes and no eccentricity of loading, the effective net area A_e equals the gross area A_g. Thus, from the above two equations it is noted that $0.90(50) < 0.75(65)$; therefore,

$$\text{Required } A_g = \frac{T_u}{0.90(50)} = \frac{110}{45} = 2.45 \text{ sq in.}$$

$$\text{Required } t = \frac{2.45}{6} = 0.41 \text{ in.} \quad \underline{Use \tfrac{7}{16} \times 6 \text{ plates.}}$$

(c) Select electrode and an AWS prequalified joint. From Table 5.13.1, use F7XX-EXXX ($F_{EXX} = 70$ ksi) flux electrode combination. Referring to *AISC Manual* [1.15] section "Prequalified Welded Joints" (Table 8-2) or AWS [2.23] Figs. 2.4 and 2.5, select a prequalified single-V-groove weld designated B-L2a-S. The designation B refers to a butt joint, L refers to limited thickness of material for this weld (in this case 2 in. maximum), and S refers to submerged arc welding. The weld requires that a backup plate be used. The details and welding symbol are shown in Fig. 5.16.2.

Figure 5.16.2
Design sketch for Example
5.16.1.

EXAMPLE 5.16.2

Repeat Example 5.16.1, except use A572 Grade 65 plates, a square-groove weld, and submerged arc welding (SAW).

Solution:

(a) Determine the required plate thickness.

$$\text{Required } A_g = \frac{T_u}{0.90(65)} = \frac{110}{58.5} = 1.89 \text{ sq in.}$$

$$\text{Required } t = \frac{1.89}{6} = 0.31 \text{ in.} \quad \underline{Use \tfrac{5}{16} \times 6 \text{ plates.}}$$

Figure 5.16.3
Design sketch for Example
5.16.2.

(b) Select electrode and specify the proper prequalified joint. From Table 5.13.1, use F8XX-EXXX flux electrode combination. From *AISC Manual* [1.15] or AWS [2.23] Table 2.9.1, select the square-groove weld designated B-L1-S (*AISC Manual* p. 8–37) as indicated in Fig. 5.16.3. This weld has zero root opening and is prequalified for material no thicker than $\frac{3}{8}$ in.   ∎

Note: In Examples 5.16.1 and 5.16.2 it was not necessary to include the weld size or the length of the welds since they are to be made full joint penetration and the full width of the plates unless otherwise specified.

EXAMPLE 5.16.3

Determine the service load capacity, assuming the load is *80%* live load, of the tee connection shown in Figure 5.16.4 and detail the proper double-bevel-groove weld for the SMAW process. Assume the flange of the tee does not control design. Use AISC LRFD Method.

Solution:

(a) Determine the strength of the $\frac{3}{4} \times 8$ plate.

$$\phi T_n = \phi F_y A_g = 0.90(0.75)8 = 194 \text{ kips}$$
$$T_u = 1.2T_D + 1.6T_L = 1.2(0.2T) + 1.6(0.8T) = 1.52T$$
$$\phi T_n = T_u$$
$$T = \frac{194}{1.52} = 128 \text{ kips}$$

The maximum service load tension force permitted by AISC is 128 kips.

(b) Select the electrode material and select a proper prequalified AWS joint. From Table 5.13.1, use E60 electrodes. From the *AISC Manual* [1.15, p. 8–45] or AWS [2.23] Fig. 2.4, select the double-bevel-groove joint designated TC-U5a.

Note: On the basis of strength only, a single $\frac{3}{4}$-in. bevel (TC-U4b) could have been used instead of the double-bevel-groove weld specified. However, welding the stem of the tee from one side only may cause excessive warping and introduces eccentricity into the connection.

Figure 5.16.4
Example 5.16.3.   **(a)** Problem   **(b)** Solution   ∎

Fillet Welds

The design for fillet welds is based on the nominal shear stress on the effective area of the fillet weld as discussed in Sec. 5.12. The selection of the size of the fillet weld is based on the thickness of the pieces being joined and the available length over which the fillet weld can be made. Other factors such as the type of welding equipment used, whether the welds are to be made in the field or in the shop, and the size of other welds being made will also influence the size of fillet specified. Large fillet welds require larger diameter electrodes, which in turn require larger and bulkier welding equipment, not necessarily convenient for field use. The most economical size of fillet weld is usually the one that can be made in one pass; about $\frac{5}{16}$ in. for SMAW and $\frac{1}{2}$ in. for SAW. Also, if a certain size of fillet weld is used in adjacent areas to the particular joint in question, it is advisable to use the same size, since then the same electrodes and welding equipment could be used, and the welder would not have to alter his procedure to accommodate a larger or smaller weld. In addition, inspection of the welds is further simplified.

EXAMPLE 5.16.4

Determine the size and length of the fillet weld for the lap joint shown in Fig. 5.16.5 using the submerged arc (SAW) process if the plates are A36 steel. Use AISC LRFD Method.

Solution:
Referring to Sec. 5.11, AISC-J2.2b gives the following limits:

$$\text{Maximum size} = \tfrac{5}{8} - \tfrac{1}{16} = \tfrac{9}{16} \text{ in.}$$

$$\text{Minimum size} = \tfrac{1}{4} \text{ in.}$$

Use $\frac{1}{2}$-in. fillet weld, since that is about the maximum size that can be made in one pass by the SAW process.

For this SAW weld, since verification of consistent penetration beyond the root of the weld by test has not been confirmed, the effective throat dimension t_e is taken as

$$t_e = 0.707a = 0.707(0.50) = 0.354 \text{ in.}$$

From Table 5.13.1, use F7XX-EXXX flux electrode combination.
The design strength of $\frac{1}{2}$-in. fillet weld per inch of length, according to Eq. 5.14.4, is

$$\phi R_{nw} = \phi t_e(0.60 F_{\text{EXX}}) = 0.75(0.354)42 = 11.2 \text{ kips/in.}$$

Since the weld capacity may not exceed the plate shear rupture strength according to Eq. 5.14.7,

$$\text{Max } \phi R_{nw} = \phi t(0.60 F_u) = 0.75(0.625)(0.60)58 = 16.3 \text{ kips/in.}$$

or

$$\text{Max } \phi R_{nw} = \phi t(0.6 F_y) = 1.0(0.625)(0.6)36 = 13.5 \text{ kips/in.}$$

The weld strength controls.

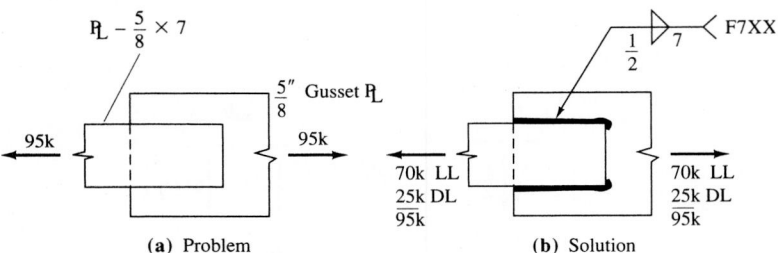

Figure 5.16.5
Example 5.16.4.

(a) Problem   (b) Solution

The factored tensile load to be carried is

$$T_u = 1.2(25) + 1.6(70) = 142 \text{ kips}$$

The total length L_w of fillet weld required is

$$L_w = \frac{142}{11.2} = 12.7 \text{ in.}$$

To satisfy the strength requirement, only 6.5 in. of weld are needed on each side. However, to avoid excessive nonuniformity of loading (i.e., shear lag, Sec. 3.5) in the plates, AISC-J2.2b requires "If longitudinal fillet welds are used alone in end connections of flat-bar tension members, the length of each fillet weld shall be not less than the perpendicular distance between them." In fact, the transverse distance between longitudinal fillet welds may not exceed one-half the length of the weld without reducing the strength of the connection in accordance with AISC-D3.3.

In this example, use $\frac{1}{2}$-in. fillet weld, 7 in. on each side as shown in Fig. 5.16.5b. Note that returns are shown in the figure; however, no dimension is specified and they are *not* included in the strength computation. Returns of this type should be made in order to ensure the full weld strength at the ends of lines of weld.   ∎

EXAMPLE 5.16.5

Rework Example 5.16.4 using $\frac{1}{4}$-in. fillet welds. Consistent fillet weld penetration beyond the root was observed and was found to provide an effective throat equal to the weld size (see AISC-Commentary- J2.2a).

Solution:
The effective weld size is given to be equal to the weld full leg dimension.

$$\phi R_{nw} = \phi t_e(0.60 F_{\text{EXX}}) = 0.75(0.25)42 = 7.87 \text{ kips/in.}$$

which is well below the plate shear yield strength of 13.5 kips/in. The total length L_w of fillet weld required is

$$L_w = \frac{142}{7.87} = 18.0 \text{ in.}$$

Two possible solutions are shown in Fig. 5.16.6, both of which provide 18 in. of $\frac{1}{4}$-in. fillet welds. The solution in Fig. 5.16.6b is preferred since it is more compact and reduces the overall length of the connection, giving better stress distribution.

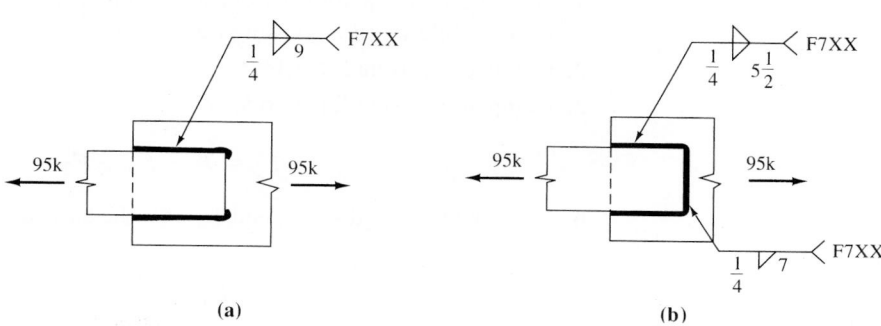

Figure 5.16.6
Solutions to Example 5.16.5.   (a)   (b)   ∎

Figure 5.16.7
Balancing the welds on a
tension member connection.

Balanced Connection

In a number of cases, members subjected to direct axial stress are themselves unsymmetrical and cause eccentricities in welded connections. This discussion relates to largely in-plane forces such as those within a truss. Consider the angle tension member shown in Fig. 5.16.7 welded as indicated. The force T applied at some distance from the connection will act along the centroid of the member as shown. The force T will be resisted by the forces F_1, F_2, and F_3 developed by the weld lines. The forces F_1 and F_3 are assumed to act at the edges of the angle rather than more correctly at the center of the effective throat. The force F_2 will act at the centroid of the weld length which is located at $d/2$. Taking moments about point A located on the bottom edge of the member and considering clockwise moments positive,

$$\sum M_A = -F_1 d - F_2 d/2 + Ty = 0 \tag{5.16.1}$$

or

$$F_1 = \frac{Ty}{d} - \frac{F_2}{2} \tag{5.16.2}$$

The force F_2 is equal to the resistance R_w of the weld per inch times the length L_w of the weld:

$$F_2 = R_w L_{w2} \tag{5.16.3}$$

Horizontal force equilibrium gives

$$\sum F_H = T - F_1 - F_2 - F_3 = 0 \tag{5.16.4}$$

Solving Eqs. 5.16.1 and 5.16.4 simultaneously gives

$$F_3 = T\left(1 - \frac{y}{d}\right) - \frac{F_2}{2} \tag{5.16.5}$$

Designing the connection shown in Fig. 5.16.7 to eliminate eccentricity caused by the unsymmetrical weld is called *balancing the welds*. The procedure for balancing the welds may be summarized as follows:

1. After selecting the proper weld size and electrode, compute the force resisted by the end weld F_2 (if any) using Eq. 5.16.3.
2. Compute F_1 using Eq. 5.16.2.
3. Compute F_3 using Eq. 5.16.5, or

$$F_3 = T - F_1 - F_2 \tag{5.16.6}$$

4. Compute the lengths, L_{w1} and L_{w3}, on the basis of

$$L_{w1} = \frac{F_1}{R_w} \tag{5.16.7a}$$

and

$$L_{w3} = \frac{F_3}{R_w} \tag{5.16.7b}$$

An alternative to the above is to compute the total length L_w of weld required to carry the load, subtract the length on the end, and then allocate the remaining required length to F_1 and F_3 in inverse proportion to the distances from the center of gravity.

Note that approximately balanced welds are "desirable"; however, AISC-J1.7 does *not* require it for "... end connections of statically-loaded single angle, double angle and similar members." Temple and Sakla [5.43] have discussed the effects of balanced connections on angle compression members.

The foregoing discussion of balanced welds is valid for both Load and Resistance Factor Design and Allowable Strength Design. The service load T used in ASD would become factored load T_u used in LRFD; allowable resistance R_w in ASD would become design strength ϕR_{nw} in LRFD. It should be noted that the alternative method of AISC-J2.4C could be used, and can yield more economical design.

EXAMPLE 5.16.6

Design the fillet welds to develop the full strength of the angle shown in Fig. 5.16.8 minimizing the effect of eccentricity. Assume the gusset plate does not govern and the SMAW process is used. Use AISC LRFD Method.

Solution:

(a) Compute the strength of the member. AISC-D3.3 requires all welded and bolted connections to be classified according to Table D3.1. Since the connection is to be made to one leg, case 2 in Table D3.1 is applicable. Accordingly, the U factor is to be determined from $(1 - \bar{x}/\ell)$. The weld length is not known and a U value of 0.90 is assumed. The design strength of the angle member is the smaller of the following:

$$\phi T_n = 0.90 F_y A_g = 0.90(50)3.61 = 162 \text{ kips}$$

$$\phi T_u = 0.75 F_u A_e = 0.75 F_u U A_g$$

$$= 0.75(65)(0.90)3.61 = 158 \text{ kips (controls)}$$

(b) Select weld size and compute strength.

$$\text{Min size fillet weld} = \tfrac{3}{16} \text{ in. (Table 5.11.1)}$$

$$\text{Max size fillet weld} = \tfrac{3}{8} - \tfrac{1}{16} = \tfrac{5}{16} \text{ in. (Fig. 5.11.2)}$$

Use $\tfrac{3}{16}$-in. fillet weld with E70 electrodes. The design strength $f R_{nw}$ is

$$\phi R_{nw} = \phi t_e(0.60 F_{\text{EXX}}) = 0.75\left(\tfrac{3}{16}\right)(0.707)42 = 4.18 \text{ kips/in.}$$

Figure 5.16.8
Example 5.16.6.

Figure 5.16.9
Balancing the welds for
Example 5.16.6.

which cannot exceed the shear rupture strength of the base metal,

$$\text{Max } \phi R_{nw} = \phi t(0.60F_u) = 0.75(0.375)(0.60)65 = 11.0 \text{ kips/in.}$$

The weld controls.

(c) Determine the length of weld to be used for the connection. Referring to Fig. 5.16.9,

$$F_2 = \phi R_{nw}L_w = 4.18(6) = 25.1 \text{ kips}$$

From moment equilibrium about the back of the angle (at F_3),

$$F_1 = \frac{158(1.94) - 25.1(3)}{6} = 38.5 \text{ kips}$$

Summation of forces gives

$$F_3 = T_u - F_1 - F_2 = 158 - 38.5 - 25.1 = 94.4 \text{ kips}$$

$$L_{w1} = \frac{F_1}{\phi R_{nw}} = \frac{38.5}{4.18} = 9.2 \text{ in.} \qquad \underline{\text{Use 10 in.}}$$

$$L_{w3} = \frac{F_3}{\phi R_{nw}} = \frac{94.4}{4.18} = 22.6 \text{ in.} \qquad \underline{\text{Use 23 in.}}$$

Check the eccentricity reduction factor U using an average ℓ value,

$$\ell = \frac{\ell_2}{\ell_3} = \frac{10 + 23}{2} = 16.5 \text{ in.}$$

$$U = 1 - \frac{\overline{x}}{\ell} = 1 - \frac{0.933}{16.5} = 0.94$$

Clearly within design accuracy, the use of $U = 0.85$ or 0.90 would be acceptable. To carry the process further, recompute

$$\phi T_n = \phi F_u A_e = 0.75(65)(0.94)3.61 = 165 \text{ kips}$$

Now yielding controls with $\phi T_n = 162$ kips. The required lengths can be recalculated,

$$F_1 = 39.8 \text{ kips} \Rightarrow L_{w1} = 9.5 \text{ in.} \qquad \underline{\text{Use 10 in.}}$$

$$F_3 = 97.1 \text{ kips} \Rightarrow L_{w3} = 23.2 \text{ in.} \qquad \underline{\text{Use 24 in.}}$$

Could recheck the U value,

$$U = 1 - \frac{\overline{x}}{\ell} = 1 - \frac{0.933}{17} \approx 0.95 \qquad\qquad \textbf{OK}$$

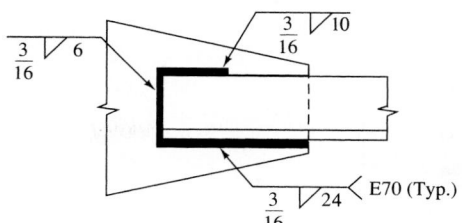

Figure 5.16.10
Solution for Example 5.16.6.

Use welds as shown in Fig. 5.16.10, though for better economy the largest welds that can be placed in one pass are preferred and this would also reduce the connection length. Alternatively, the larger of $(R_{wl} + R_{wt})$ and $(0.85\, R_{wl} + 1.5\, R_{wt})$ could have been used to determine the weld lengths (AISC-J2.4C). ∎

EXAMPLE 5.16.7

Rework Example 5.16.6 if the weld at the end of the angle is omitted, and the SAW process is used instead of SMAW. The fillet weld penetration was observed to be consistent beyond the root of the weld, and was found to provide an effective throat equal to the weld size.

Solution:
This time try $\frac{5}{16}$ in. as more economical and which can still be placed in one pass. Using the forces in Fig. 5.16.11.

Figure 5.16.11
Forces acting for Example 5.16.7.

$$F_1 = \frac{158(1.94)}{6} = 51.1 \text{ kips}$$

$$F_3 = T_u - F_1 = 158 - 51.1 = 106.9 \text{ kips}$$

$$\phi R_{nw} = \phi t_e(0.60 F_{EXX}) = 0.75\left(\tfrac{5}{16}\right)42 = 9.84 \text{ kips/in.}$$

which is less than the shear rupture strength (11.0 kips/in.) of the angle; therefore the weld controls.

$$L_{w1} = \frac{F_1}{\phi R_{nw}} = \frac{51.1}{9.84} = 5.2 \text{ in.}$$

$$L_{w3} = \frac{F_3}{\phi R_{nw}} = \frac{106.9}{9.84} = 10.9 \text{ in.}$$

Check value of U using average ℓ,

$$U = 1 - \frac{\overline{x}}{\ell} = 1 - \frac{0.933}{8.05} = 0.88$$

$$\phi T_n = \phi F_u U A_g = 0.75(65)(0.88)3.61 = 155 \text{ kips} < [T_u = 158 \text{ kips}] \quad \text{Say} \quad \textbf{OK}$$

Use $L_{w1} = 6$ in. and $L_{w3} = 11$ in.

Figure 5.16.12
Solution for Example 5.16.7.

The weld design is summarized in Fig. 5.16.12. Note that small returns at the ends of weld lines are made to ensure full strength of the specified weld lengths. Unless the return length is specified, the returns would not be included in the strength computation. ∎

Slot and Plug Welds

Slot and plug welds have their strength based on the area in the shearing plane between the plates being joined. As indicated in Sec. 5.5, their principal use is in lap joints. Plug welds are also occasionally used to fill up holes in connections, such as beam-to-column angles where temporary erection bolts had been placed to align the members prior to welding. The strength of such welds may or may not be included in the design strength of a joint. As a rule, plug and slot welds are designed to work together with other welds, usually fillet welds, in lap joints as shown in Fig. 5.5.5.

EXAMPLE 5.16.8

Determine the service load T permitted on the connection in Fig. 5.16.13 if the load is 80% live load and 20% dead load. The steel is A36 and the AISC LRFD Method is to be used. See comment regarding returns at end of Ex. 5.16.7.

Solution:
The design strength ϕR_{nw} per inch supplied by the $\frac{1}{2}$-in. fillet welds is

$$\phi R_{nw} = \phi t_e(0.60 F_{\text{EXX}})$$

$$= 0.75[0.5(0.707)]42 = 11.1 \text{ kips/in.}$$

but not to exceed the shear rupture or yield strengths of the plate,

$$\text{Max } \phi R_n = \phi 0.6 F_y A_g = 1.0(0.6)(36)0.625 = 13.5 \text{ kips/in.} \qquad \text{governs}$$

or

$$\text{Max } \phi R_n = \phi 0.6 F_u A_{nv} = 0.75(0.6)(58)0.625 = 16.3 \text{ kips/in.}$$

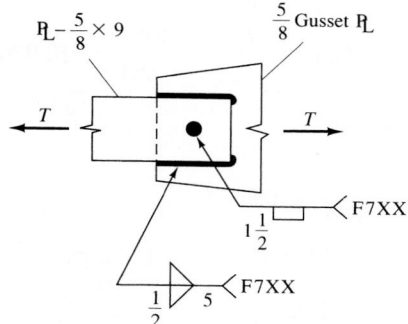

Figure 5.16.13
Example 5.16.8.

The strength provided by the fillet welds is

$$T_1 = L_w(\phi R_{nw}) = 10(11.1) = 111 \text{ kips}$$

The design strength ϕR_n provided by the $1\frac{1}{2}$-in. diam plug weld is

$$T_2 = \phi R_n = 0.75\frac{\pi(1.5)^2}{4}(0.60)70 = 56 \text{ kips}$$

The design strength based on the weld is

$$\phi T_n = T_1 + T_2 = 111 + 56 = 167 \text{ kips (controls)}$$

Check the tensile capacity of the plate:

$$\phi T_n = 0.90 F_y A_g = 0.90(36)(9)0.625 = 182 \text{ kips}$$

$$\phi T_n = 0.75 F_u A_e = 0.75 F_u U A_g = 0.75(58)(1.0)(9)0.625 = 245 \text{ kips}$$

The weld controls, which makes the service load capacity T

$$167 = 1.2(0.2T) + 1.6(0.8T) = 1.52T$$

$$T = 110 \text{ kips}$$ ■

EXAMPLE 5.16.9

Compute the service load capacity of the connection shown in Fig. 5.16.14 when A573 Grade 50 steel and welding by the SMAW process are used. Assume the service load is 83% live load and 17% dead load. Use AISC Load and Resistance Factor Design. See comment regarding weld returns at end of Ex. 5.16.7.

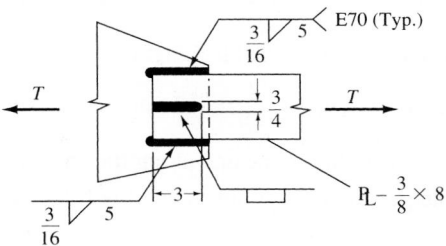

Figure 5.16.14
Example 5.16.9.

Solution:
From Table 5.14.2, a $\frac{3}{16}$-in. E70 fillet weld provides $\phi R_{nw} = 4.18$ kips/in. The resistance T_1 provided by the fillet welds is

$$T_1 = L_w(\phi R_{nw}) = 2(5)(4.18) = 41.8 \text{ kips}$$

The resistance T_2 provided by the $\frac{3}{4}$-in. wide slot weld is

$$\text{Faying area} = \frac{\pi(3/4)^2}{8} + 2.25\left(\tfrac{3}{4}\right) = 1.91 \text{ sq in.}$$

$$T_2 = \phi R_n = 0.75(1.91)(0.60)70 = 60.1 \text{ kips}$$

$$\phi T_n = T_1 + T_2 = 41.8 + 60.1 = 101.9 \text{ kips (controls)}$$

Note that slot has a semicircular end (AISC-J2.3b, par. 4).
Check the tensile capacity of the plate:

$$\phi T_n = 0.90 F_y A_g = 0.90(50)(8)0.375 = 135 \text{ kips}$$

$$\phi T_n = 0.75 F_u A_e = 0.75(65)(8)0.375 = 146 \text{ kips}$$

Thus, the service load capacity T is

$$101.9 = 1.2(0.17T) + 1.6(0.83T) = 1.53T$$

$$T = 67 \text{ kips} \qquad \blacksquare$$

EXAMPLE 5.16.10

Design an end connection to develop the full tensile strength of a C8×13.75 in a lap length of 5 in. as shown in Fig. 5.16.15. The channel of A572 Grade 50 steel is connected to a $\frac{3}{8}$-in. gusset plate, and the fillet welds are to be made by the SMAW process and may not exceed $\frac{3}{8}$-in. Use Load and Resistance Factor Design.

Solution:

(a) Compute the design strength of the channel. The joint length ℓ is taken as 5 in. which is the maximum overlap between the connected elements. The U factor can be computed as follows:

$$U = 1 - \overline{x}/\ell = 1 - 0.554/5 = 0.89$$

$$\phi T_n = 0.90 F_y A_g = 0.90(50)4.04 = 182 \text{ kips}$$

$$\phi T_n = 0.75 F_u A_e = 0.75(65)(0.89)4.04 = 175 \text{ kips}$$

(b) Select fillet weld size a and compute length required.

$$\text{Min } a = \tfrac{3}{16} \text{ in.} \quad \text{(Table 5.11.1)}$$

$$\text{Max } a = 0.303 - \tfrac{1}{16} = 0.24 \text{ in.,} \quad \text{say } \tfrac{1}{4} \text{ in. (Fig. 5.11.2)}$$

While $\frac{1}{4}$-in. weld must be used on one end along the channel web, $\frac{3}{8}$-in. weld could be used along the flanges. It is better not to mix the fillet sizes, so try $\frac{1}{4}$ in. all around.

$$\phi R_{nw} = \phi t_e(0.60 F_{\text{EXX}}) = 0.75(\tfrac{1}{4})(0.707)42 = 5.57 \text{ kips/in.}$$

Figure 5.16.15
Solution for Example
5.16.10.

which cannot exceed the shear yield strength or rupture strength of the base metal,

$$\text{Max } \phi R_n = \phi 0.6 F_y A_g = 1.0(0.6)(50)0.375 = 11.3 \text{ kips/in}$$

or

$$\text{Max } \phi R_n = \phi 0.6 F_u A_e$$

$$= 0.75(0.6)(65)0.375 = 11.0 \text{ kips/in} \qquad \underline{\text{governs}}$$

The weld controls.

$$\text{Required } L_w = \frac{T_u}{\phi R_{nw}} = \frac{175}{5.57} = 31.4 \text{ in.}$$

Since the length all around is only 26 in., additional capacity from fillet weld in a slot, slot welds, or plug welds, is necessary.

(c) Slot weld. Try a slot weld in accordance with AISC-J2.3b.

$$\text{Min width of slot} = \left(t + \tfrac{5}{16}\right) \quad \left(\text{rounded to next odd } \tfrac{1}{16} \text{ in.}\right)$$

$$= 0.303 + 0.3125 = 0.6155, \quad \text{say } \tfrac{11}{16} \text{ in.}$$

$$\text{Max width of slot} = 2\tfrac{1}{4}(\text{weld thickness}) = 2\tfrac{1}{4}(0.303) = 0.68 \text{ in.}$$

Load T_u to be carried by slot weld,

$$\text{Required } T_u = 175 - (26 - 0.68)5.57 = 34 \text{ kips}$$

Try $\tfrac{11}{16}$-in. width of slot and *estimate* the slot area as rectangular even though the end must be rounded:

$$\text{Length required} = \frac{34}{0.75(11/16)42} = 1.57 \text{ in.}$$

$$\text{Max length of slot} = 10 \, (\text{weld thickness}) = 10(0.303) = 3.03 \text{ in.}$$

Use a slot weld $\tfrac{11}{16} \times 2$. The final design is shown in Fig. 5.16.15. Note that the interior end of the slot must be semicircular or have corners rounded to a radius not less than the thickness of the part containing slot (AISC-J2.3b). ■

5.17   ECCENTRIC SHEAR CONNECTIONS—STRENGTH ANALYSIS

There are many situations where the loading of fillet welds is neither parallel to nor transverse to the axis of the fillet welds, as shown in Fig. 5.17.1. Analysis of such eccentric loading cases is complicated by the fact that, as shown in Fig. 5.13.3, the load-deformation behavior is a function of the angle θ between the direction of the resistance and the axis of the fillet weld.

In a manner similar to that used for eccentric loading on bolted connections (see Sec. 4.12), the strength of an eccentrically loaded fillet weld configuration can be determined by locating the instantaneous center of rotation, using the load-deformation relationship of a fillet weld. The resistance R_i of a weld segment at any distance from the instantaneous center is proportional to such distance and acts in a direction perpendicular to the radial distance to the segment. Unlike bolts, however, where the load-deformation relationship is independent of the direction the shear force acts on the circular bolt cross-section, the fillet weld strength depends on the angle between the applied force and the axis of the weld resisting it.

(a) Shear and torsion

(b) Pure torsion

(c) Shear and bending

Figure 5.17.1
Types of eccentric loading.

Until the availability of calculators that could readily obtain the values from complicated formulas, a trial and error procedure to determine the strength of eccentrically loaded connections, either bolted or welded, was not feasible. Thus, traditionally the elastic vector analysis similar to that discussed for bolts was used for eccentrically loaded welds, as is explained in the next section (Sec. 5.18).

Though the AISC Specification [1.13] has in the past not described the method of analysis to be used for fillet welds eccentrically loaded in shear, the most rational procedure is one using strength analysis. The work of Butler, Pal, and Kulak [5.28, 5.29] formed the basis for the 1978 *ASD Manual* tables. Tide [5.34, 5.35] and Brandt [5.36] have presented detailed treatment of the *ASD Manual* approach. Brandt [5.36] has also developed a computer program. The 1986 *LRFD Manual* was based on later work by Kulak and Timmler [5.30]. The equations were revised in 1993 to reflect the work of Miazga, Lesik, and Kennedy [5.37, 5.38, 5.39].

A more complicated situation of eccentric shear parallel to one of the axes of the weld configuration combined with a force acting at 90° to the shear force has been treated by Loomis, Thornton, and Kane [5.40]. Their design aids are primarily for beam connection angle welds, including C-shaped, L-shaped, and lines, where the beam is subject to axial tension or compression.

The general strength analysis procedure is unchanged; however, the strength (kips/in.) of a given segment is not limited to a maximum value of $0.6F_{EXX}t_e$ as was used previously in developing the tables in the AISC Manuals. Instead, the theoretical strength curve can be used.

The design strength of a weld segment per unit length is given by AISC-J2.4 alternative method as

$$\phi R_{nw} = \phi 0.60 F_{EXX}(1.0 + 0.50 \sin^{1.5}\theta)t_e \qquad (5.17.1)$$

where $\phi = 0.75$

θ = angle of loading measured from the weld longitudinal axis, degrees

This strength as a function of θ is used in lieu of the constant strength discussed in Sec. 5.14.

When the weld segment is part of a configuration subject to eccentric shear, loaded in-plane, using an instantaneous center of rotation procedure which satisfies compatibility of deformation along with nonuniform load-deformation behavior, the strength as given by Eq. 5.17.1 is modified by AISC-J2.4 as follows:

$$R_i = 0.60F_{\text{EXX}}t_e(1.0 + 0.50\sin^{1.5}\theta)\left[\frac{\Delta_i}{\Delta_m}\left(1.9 - 0.9\frac{\Delta_i}{\Delta_m}\right)\right]^{0.3} \quad (5.17.2)$$

where

R_i = nominal strength of weld segment i, kips/in.

θ = angle of loading measured from the weld longitudinal axis, degrees

Δ_i = deformation of element i = $r_i\dfrac{\Delta_u}{r_{crit}}$

r_{crit} = distance from instantaneous center of rotation to weld element having minimum ratio Δ_u/r_i

$\Delta_m = 0.209(\theta + 2)^{-0.32}a$

= deformation of element at maximum strength, in.

$\Delta_u = 1.087(\theta + 6)^{-0.65}a \le 0.17a$

= deformation of element when fracture is imminent, usually in element farthest from instantaneous center of rotation

a = leg size of fillet weld, in.

The procedure is as follows:

1. Divide the weld configuration into segments, say 1-in.-long segments.
2. Select a trial location for the instantaneous center of rotation (see Fig. 5.17.2).
3. Assume the resisting force R_i or R_j at any weld segment acts in a direction perpendicular to the radial line from the instantaneous center to the centroid of the weld segment.
4. Compute the angle θ (in degrees) between the direction of the resisting force R_i or R_j and the axis of the weld.
5. Compute the deformations Δ_m and Δ_u which can occur at that particular θ of the weld segment.

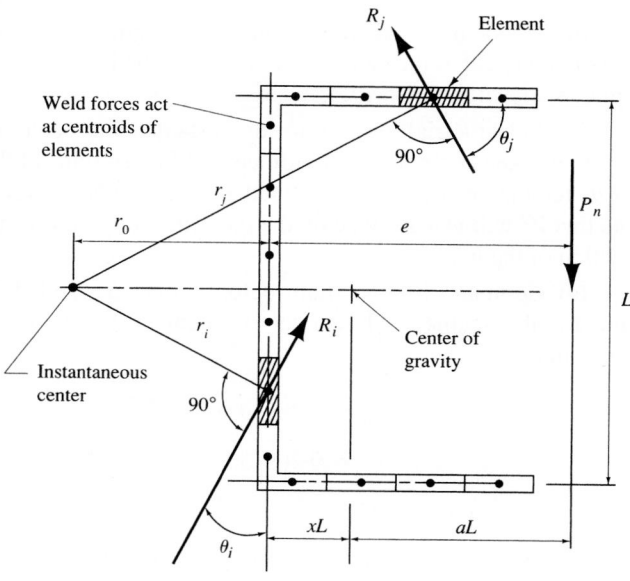

Figure 5.17.2
Resistance R of a fillet weld segment.

6. Deformations on weld segments are assumed to vary linearly with the distance from the instantaneous center to the centroid of the weld segment. Thus, the critical segment is the one where the ratio of its Δ_u to its radial distance r_i is the *smallest*.

7. Compatible deformations Δ_i are then computed at each weld segment.

8. Compute the nominal strength R_i of each weld segment using Eq. 5.17.2.

9. Using statics, compute the load P_n that represents the nominal strength of the connection when the load is applied at the given eccentricity e. For P_n applied in the y-direction as shown in Fig. 5.17.2, the statics equations are

$$\sum M = 0; \qquad P_n(e + r_0) = \sum R_i r_i + \sum R_j r_j \tag{5.17.3}$$

$$P_n = \frac{\sum R_i r_i + \sum R_j r_j}{e + r_0} \tag{5.17.4}$$

$$\sum F_y = 0; \qquad P_n = \sum (R_i)_y + \sum (R_j)_y \tag{5.17.5}$$

$$P_n = \sum R_i \cos \theta_i + \sum R_j \sin \theta_j \tag{5.17.6}$$

10. Compare the values of P_n from Eqs. 5.17.4 and 5.17.6. If they are equal the solution is correct. If they are not equal, revise the trial value of r_0 and repeat the process until Eqs. 5.17.4 and 5.17.6 give the same result.

For cases where P_n is not applied perpendicular to an axis of symmetry of the weld configuration, the three equations of equilibrium must be solved simultaneously for the x and y location of the instantaneous center and the load P_n.

EXAMPLE 5.17.1

Determine the nominal strength P_n of the C-shaped fillet weld configuration shown in Fig. 5.17.3. The horizontal lengths are 3 in. each and the vertical length is 14 in. The eccentric load is applied at 3.5 in. from the 14-in. length. The weld size is $\frac{1}{4}$ in. and E70 electrodes are used in the SMAW process. Assume the base material strength does not govern.

Solution:

(a) Divide the weld configuration into segments 1-in. long. The resisting force will be assumed to act at the center of a segment. The instantaneous center will be used as the origin, as shown in Fig. 5.17.3.

(b) Select a trial location for the instantaneous center (IC) of rotation. Assume say $r_0 = 6$ in. Then compute the coordinates of the centroids of the segments and the angles θ_i and θ_j for the vertical and horizontal segments, respectively, as shown in Table 5.17.1. Note that IC will generally be on the opposite side of the vertical weld line from the point of action of the load.*

(c) Compute the maximum deformations Δ_m and Δ_u which can occur at each θ value for the segments. Illustrate for segment 1, where $\theta = 50.5°$ and $a =$ weld leg $= 0.25$ in.,

$$\Delta_m = 0.209(\theta + 2)^{-0.32}a$$

$$= 0.209(50.5 + 2)^{-0.32}(0.25) = 0.0147$$

---

*The use of a spreadsheet program will allow a rapid solution of this problem by trial and error.

$$\Delta_u = 1.087(\theta + 6)^{-0.65}a \leq 0.17a$$

$$= 1.087(50.5 + 6)^{-0.65}(0.25) = 0.0217 < [0.17(0.25) = 0.0425]$$

OK

(d) Compute the ratio Δ_u/r_i for each segment, determine the minimum value for that ratio. In this case, segment 1 has the minimum value (0.00179). The results are in Table 5.17.2.

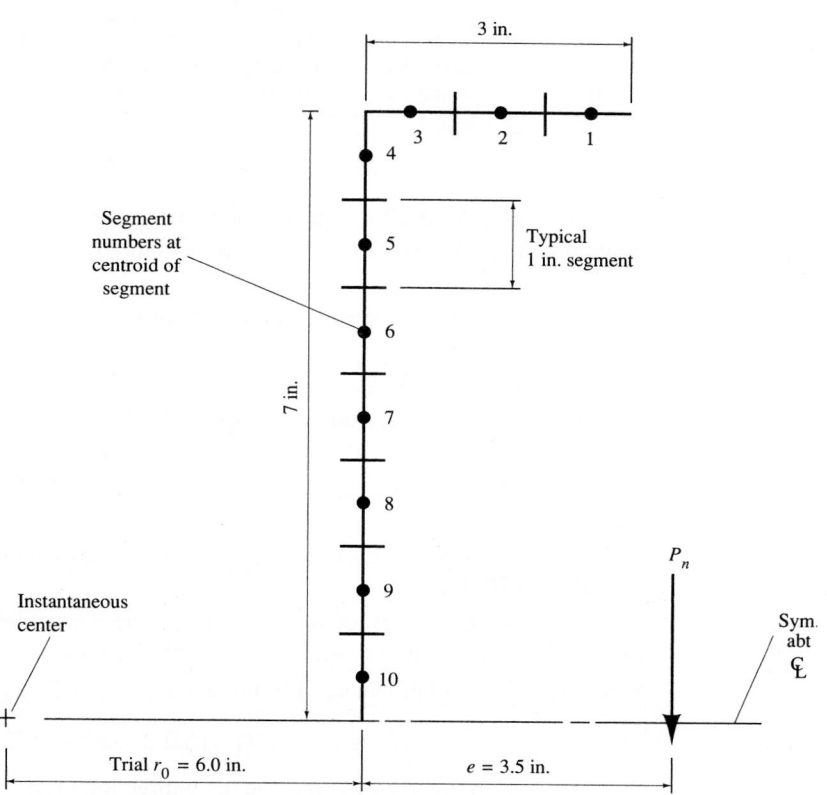

Figure 5.17.3
Weld segments for Example
5.17.1.

TABLE 5.17.1 Geometry For Trial 1 ($r_0 = 6$ in.) of Example 5.17.1

Horizontal segments	Length (in.)	x (in.)	y (in.)	r_i (in.)	θ_j (radians)	θ_j (degrees)
1	1	8.50	7.00	11.011	0.882	50.5
2	1	7.50	7.00	10.259	0.820	47.0
3	1	6.50	7.00	9.552	0.748	42.9

Vertical segments	Length (in.)	x (in.)	y (in.)	r_i (in.)	θ_j (radians)	θ_j (degrees)
4	1	6.00	6.50	8.846	0.825	47.3
5	1	6.00	5.50	8.139	0.742	42.5
6	1	6.00	4.50	7.500	0.644	36.9
7	1	6.00	3.50	6.946	0.528	30.3
8	1	6.00	2.50	6.500	0.395	22.6
9	1	6.00	1.50	6.185	0.245	14.0
10	1	6.00	0.50	6.021	0.083	4.8

TABLE 5.17.2 Resistance at Each Segment-Trial 1 ($r_0 = 6.0$ in.)

Segment	Δ_m	Δ_u	Δ_u/r_i	Δ_i	Δ_i/Δ_m	R_i
1	0.01471	0.01973	0.00179	0.01973	1.34165	9.72
2	0.01504	0.02058	0.00201	0.01839	1.22229	9.68
3	0.01547	0.02169	0.00227	0.01712	1.10673	9.51
4	0.01501	0.02050	0.00232	0.01585	1.05609	9.77
5	0.01551	0.02180	0.00268	0.01459	0.94053	9.46
6	0.01620	0.02362	0.00315	0.01344	0.82987	9.03
7	0.01719	0.02634	0.00379	0.01245	0.72407	8.49
8	0.01874	0.03071	0.00473	0.01165	0.62144	7.87
9	0.02150	0.03872	0.00626	0.01108	0.51549	7.19
10	0.02834	0.05800	0.00963	0.01079	0.38070	6.42

(e) Compute Δ_i. Take the distance r_i to segment 1 as r_{crit} because segment 1 has the minimum Δ_u/r_i of 0.00179. Compute Δ_i as follows:

$$\Delta_i = \text{deformation of element } i = r_i\frac{\Delta_u}{r_{crit}} = r_i(0.00179)$$

Results are in Table 5.17.2.

(f) Compute Δ_i/Δ_m and R_i using Eq. 5.17.2,

$$R_i = 0.60F_{EXX}t_e(1.0 + 0.50\sin^{1.5}\theta)\left[\frac{\Delta_i}{\Delta_m}\left(1.9 - 0.9\frac{\Delta_i}{\Delta_m}\right)\right]^{0.3}$$

In this example, E70 electrodes and 1/4-in. fillet weld are used, making $F_{EXX} = 70$ ksi and $t_e = 0.25(0.707)$ in.

(g) Compute check of statics to determine correctness of trial value of r_0. Using Eqs. 5.17.4 and 5.17.6, the results are in Table 5.17.3.

From Eq. 5.17.4, rotational equilibrium gives

$$P_n = 2(721.94)/(6.0 + 3.5) = 152.0 \text{ kips}$$

From Eq. 5.17.6, force equilibrium in the y-direction gives

$$P_n = 2(69.78) = 139.6 \text{ kips}$$

TABLE 5.17.3 Check of Statics for Trial 1 ($r_0 = 6.0$ in.)

Segment	R_i	r_i	x	y	$(R_i)_x$	$(R_i)_y$	R_ir_i
1	9.72	11.01	8.5	7	6.18	7.51	107.07
2	9.68	10.26	7.5	7	6.60	7.08	99.29
3	9.51	9.55	6.5	7	6.97	6.47	90.83
4	9.77	8.85	6	6.5	7.18	6.63	86.42
5	9.46	8.14	6	5.5	6.39	6.97	76.99
6	9.03	7.50	6	4.5	5.42	7.22	67.71
7	8.49	6.95	6	3.5	4.28	7.33	58.97
8	7.87	6.50	6	2.5	3.03	7.26	51.13
9	7.19	6.18	6	1.5	1.74	6.97	44.46
10	6.42	6.02	6	0.5	0.53	6.40	38.67
					48.32	69.84	721.54

Since the values of P_n from the statics equations are not equal, the assumed location of the instantaneous center is not correct. Try a new value of r_0 and repeat the analysis until the values from Eqs. 5.17.4 and 5.17.6 and identical. Since the applied load P_n is in the y-direction for this example, the summation of forces in the x-direction is automatically satisfied; 48.84 kips acting in one direction from the segments on one side of the axis of symmetry, and the same force acting at 180° from the segments on the other side of the axis of symmetry.

   (h) After several trials the solution for r_0 is determined to be 8.30 in. Table 5.17.4 shows the final solution.

TABLE 5.15.4 Solution For Example 5.17.1 ($r_0 = 8.30$ in.)

Segment	θ_i	Δ_u	Δ_u/r_i	Δ_i	R_i	$(R_i)_y$	$R_i r_i$
1	57.1	0.0184	0.00143	0.0184	10.12	8.49	130.29
2	54.5	0.0189	0.00157	0.0172	10.10	8.22	121.70
3	51.5	0.0195	0.00173	0.0161	10.00	7.83	112.45
4	38.1	0.0232	0.00220	0.0151	9.19	7.24	96.95
5	33.5	0.0249	0.00250	0.0142	8.85	7.38	88.19
6	28.5	0.0272	0.00288	0.0135	8.46	7.43	79.87
7	22.9	0.0306	0.00339	0.0129	8.01	7.38	72.21
8	16.8	0.0357	0.00411	0.0124	7.54	7.22	65.36
9	10.2	0.0444	0.00526	0.0121	7.03	6.92	59.35
10	3.40	0.0631	0.00759	0.0119	6.43	6.42	53.51
					39.10	74.54	879.88

From Eq. 5.17.4, rotational equilibrium gives

$$P_n = 2(879.88)/(8.30 + 3.5) = 149.08 \text{ kips}$$

From Eq. 5.17.6, force equilibrium in the y-direction gives

$$P_n = 2(74.54) = 149.08 \text{ kips}$$

Thus, the nominal strength P_n is 149 kips.   ■

EXAMPLE 5.17.2

Determine the maximum service load P that the eccentric shear connection of Fig. 5.17.4 may be permitted to carry using strength analysis and the AISC LRFD Method. The live load is three times the dead load. The weld is $\frac{1}{4}$ in. and E70 electrodes are used in the SMAW process. Assume the base material strength does not govern.

Solution:

   (a) Strength analysis using basic concepts. The nominal strength of this connection as determined in Example 5.17.1 is

$$P_n = 149 \text{ kips}$$

   (b) Use *AISC Manual* [1.15] tables, p. 8–90. For $\frac{1}{4}$-in. weld using E70 electrodes,

$$a = (e - xL)/L = (3.5 - 0.45)/14 = 0.218$$
$$k = kL/L \qquad = 3.0/14 = 0.214$$

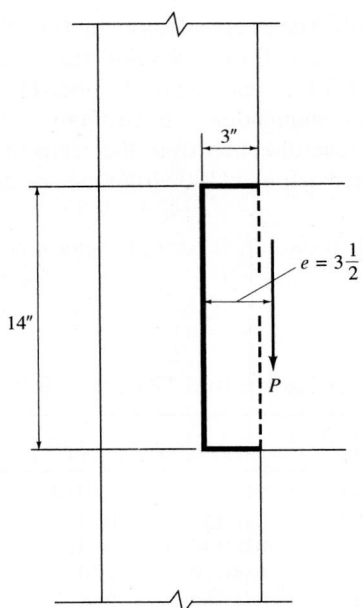

Figure 5.17.4
Loading for Example 5.17.2.

	$k = 0.2$	0.214	0.3	
$a = 0.200$	2.63		3.11	
0.218	2.58	$C = 0.2$	3.05	$C = 2.65$
0.300	2.36		2.79	

Table value $= \phi R_n = \phi C C_1 DL = 0.75(2.65)(1.0)(4)14 = 111$ kips

where $C_1 =$ coefficient for electrode $=$ (Electrode used)$/70$
$D =$ number of $\frac{1}{16}$s of an inch in weld size
$L =$ length of vertical weld (in.)

(c) Summary. Compare the values of the design strength ϕP_n.
1. Strength analysis P_n after completing analysis:

$$\phi P_n = 0.75(149) = 112 \text{ kips}$$

2. *AISC Manual* tables: $\phi P_n = 111$ kips

(d) Compute the safe service load P using the basic factored gravity load combination, For the given 75% live load and 25% dead load,

$$P_u = 1.2D + 1.6L$$
$$\phi P_n = P_u = 1.2(0.25P) + 1.6(0.75P) = 1.5P$$
$$P = \frac{\phi P_n}{1.5} = \frac{112}{1.5} = 75 \text{ kips} \qquad \blacksquare$$

5.18 ECCENTRIC SHEAR CONNECTIONS—ELASTIC (VECTOR) ANALYSIS

The traditional elastic vector analysis is easier than the strength method to carry out when the computer is not available, or when the *AISC Manual* tables are not available. The elastic vector method is conservative, sometimes excessively so.

The elastic method has the following assumptions:

1. Each segment of weld, if of the same size, resists a concentrically applied load with an equal force. This concept was used for welds on tension members in Sec. 5.16.
2. The rotation caused by torsional moment is assumed to occur about the centroid of the weld configuration.
3. The load on a weld segment caused by the torsional moment is assumed to be proportional to the distance from the centroid of the weld configuration.
4. The direction of the force on a weld segment caused by torsion is assumed to be perpendicular to the radial distance from the centroid of the weld configuration.
5. The components of the forces caused by direct load and by torsion are combined vectorially to obtain a resultant force.

It will be convenient to think of this analysis using the principles of mechanics on a homogeneous material, combining direct shear with torsion. Beginning with the stresses on a homogeneous section,

$$f' = \frac{P}{A} = \text{stress due to direct shear} \tag{5.18.1}$$

$$f'' = \frac{Tr}{I_p} = \text{stress due to torsional moment} \tag{5.18.2}$$

where $r = $ radial distance from the centroid to point of stress

$I_p = $ polar moment of inertia

For computing nominal stresses or forces on weld segments the *locations* of the lines of weld are defined by edges along which the fillets are placed, rather than to the center of the effective throat. This makes little difference, since the throat dimension is usually small.

For the general case shown in Fig. 5.18.1, the components of stress caused by direct shear are

$$f'_x = \frac{P_x}{A} \tag{5.18.3a}$$

$$f'_y = \frac{P_y}{A} \tag{5.18.3b}$$

The x- and y-components of f'' resulting from torsion are

$$f''_x = \frac{Ty}{I_p} = \frac{(P_x e_y + P_y e_x)y}{I_p} \tag{5.18.4a}$$

$$f''_y = \frac{Tx}{I_p} = \frac{(P_x e_y + P_y e_x)x}{I_p} \tag{5.18.4b}$$

where

$$I_p = I_x + I_y = \sum I_{xx} + \sum A\bar{y}^2 + \sum I_{yy} + \sum A\bar{x}^2 \tag{5.18.5}$$

In Eq. 5.18.5, $\bar{x}$ and $\bar{y}$ refer to distances from the center of gravity of the weld group to the center of gravity of the individual weld segments. I_{xx} and I_{yy} refer to the moments of inertia of the individual segments with respect to their own centroidal axes.

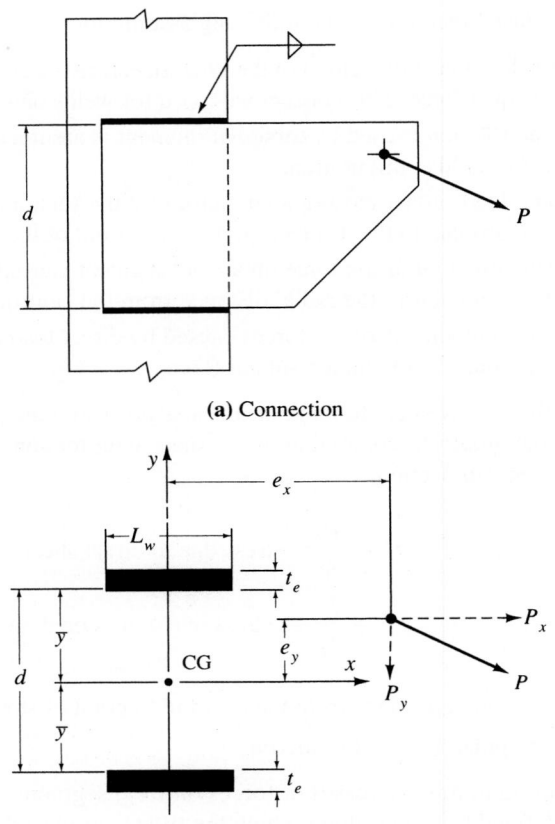

Figure 5.18.1
Eccentric bracket connection.

(a) Connection

(b) Effective cross section

Thus, for the situation of Fig. 5.18.2, Eq. 5.18.5 becomes

$$I_p = 2\left[\frac{L_w(t_e)^3}{12}\right] + 2[L_w(t_e)(\bar{y})^2] + 2\left[\frac{t_e(L_w)^3}{12}\right]$$

$$= \frac{t_e}{6}[L_w(t_e)^2 + 12L_w(\bar{y})^2 + L_w^3] \tag{5.18.6}$$

For practical situations, the first term of Eq. 5.18.6 is neglected because, with t_e small, the term is not significant compared to the other terms. Hence

$$I_p \approx \frac{t_e}{6}[12L_w(\bar{y})^2 + L_w^3] \tag{5.18.7}$$

Note that I_p equals the throat thickness t_e times the property of *lines;* i.e., an element having length but having a width of unity. Actually, the area A in Eqs. 5.18.3 equals the thickness t_e times the total length of the weld configuration; and in Eq. 5.18.5 the polar moment of inertia equals the thickness t_e times the polar moment of inertia of the configuration as *lines.* When the stress f is multiplied by t_e, it becomes a force R per unit length, say, kips/in.

Figure 5.18.2
Treatment of weld
configuration as lines
having unit thickness.

Treating the welds making up the effective cross-section in Fig. 5.18.2 as line welds (i.e., as in deriving Eq. 5.18.7 with $t_e = 1$) and using the general terms b and d, as shown in Fig. 5.18.2, Eq. 5.18.7 becomes

$$I_p \approx \frac{1}{6}\left[12b\left(\frac{d}{2}\right)^2 + b^3\right] = \frac{b}{6}[3d^2 + b^2]$$

(5.18.8)

Table 5.18.1 gives I_p values treated as properties of lines for other common weld configurations.

TABLE 5.18.1 Properties of Welds Treated as Lines

Section b = width; d = depth		Section modulus $I_x/\bar{y}$	Polar moment of inertia, I_p about center of gravity
1.		$S = \dfrac{d^2}{6}$	$I_p = \dfrac{d^3}{12}$
2.		$S = \dfrac{d^2}{3}$	$I_p = \dfrac{d(3b^2 + d^2)}{6}$
3.		$S = bd$	$I_p = \dfrac{b(3d^2 + b^2)}{6}$
4.	$\bar{y} = \dfrac{d^2}{2(b + d)}$ $\bar{x} = \dfrac{b^2}{2(b + d)}$	$S = \dfrac{4bd + d^2}{6}$	$I_p = \dfrac{(b + d)^4 - 6b^2d^2}{12(b + d)}$
5.	$\bar{x} = \dfrac{b^2}{2b + d}$	$S = bd + \dfrac{d^2}{6}$	$I_p = \dfrac{8b^3 + 6bd^2 + d^3}{12} - \dfrac{b^4}{2b + d}$
6.	$\bar{y} = \dfrac{d^2}{b + 2b}$	$S = \dfrac{2bd + d^2}{3}$	$I_p = \dfrac{b^3 + 6b^2d + 8d^3}{12} - \dfrac{d^4}{2d + b}$
7.		$S = bd + \dfrac{d^2}{3}$	$I_p = \dfrac{(b + d)^3}{6}$
8.	$\bar{y} = \dfrac{d^2}{b + 2d}$	$S = \dfrac{2bd + d^2}{3}$	$I_p = \dfrac{b^3 + 8d^3}{12} - \dfrac{d^4}{b + 2d}$
9.		$S = bd + \dfrac{d^2}{3}$	$I_p = \dfrac{b^3 + 3bd^2 + d^3}{6}$
10.		$S = \pi r^2$	$I_p = 2\pi r^3$

EXAMPLE 5.18.1

Compute the maximum load (kips/in.) on the weld configuration shown for the bracket in Fig. 5.18.3 using the elastic (vector) method. Assume the plate thickness does not affect the result.

Solution:

(a) Locate the centroid of the configuration. The maximum force R will occur at points A and B. The properties of lines will be used. Taking moments about the vertical weld,

$$\bar{x} = \frac{2(6)3}{2(6) + 8} = 1.8 \text{ in.}$$

(b) Compute the area (length) and the polar moment of inertia about the centroid of the configuration.

$$L = 2(6) + 8 = 20 \text{ in.}$$

$$I_p = \underbrace{\frac{(8)^3}{12} + 2[6(4)^2]}_{I_x} + \underbrace{2\left[\frac{(6)^3}{12}\right] + 2[6(1.2)^2] + 8(1.8)^2}_{I_y} = 314 \text{ in.}^3$$

(c) Compute the components of force on the weld at points A and B. From the direct shear,

$$R_v = \frac{P}{L} = \frac{15}{20} = 0.75 \text{ kips/in.} \downarrow$$

From the torsion T about the centroid of the configuration,

$$R_x = \frac{Ty}{I_p} = \frac{15(12.2)4}{314} = 2.33 \text{ kips/in.} \rightarrow$$

$$R_y = \frac{Tx}{I_p} = \frac{15(12.2)4.2}{314} = 2.45 \text{ kips/in.} \downarrow$$

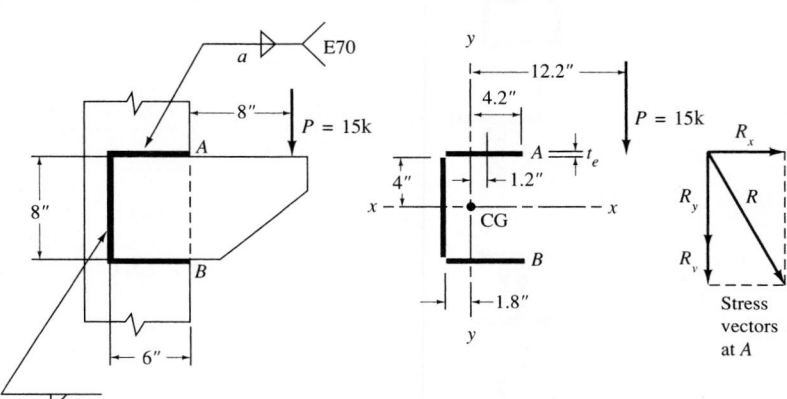

Figure 5.18.3
Example 5.18.1.

The vector sum gives the resultant force R,

$$R = \sqrt{(2.33)^2 + (2.45 + 0.75)^2} = 3.96 \text{ kips/in.}$$ ∎

EXAMPLE 5.18.2

Determine the weld size required for the bracket of Fig. 5.18.3 when the service load P is 15 kips (80% live load and 20% dead load). Compare the results using (a) elastic (vector) analysis from Example 5.18.1 and (b) strength analysis as described in Sec. 5.17, both with AISC LRFD Method. Assume the plate thickness does not affect the result (stiffened seats and bracket plates are treated in Secs. 13.4 and 13.5.)

Solution:

(a) Elastic (vector) method. The basic gravity factored load is computed,

$$P_u = 1.2D + 1.6L = 1.2(0.2)15 + 1.6(0.8)15 = 22.8 \text{ kips}$$

The maximum load (kips/in.) on the weld due to the factored P will be (using the result from Example 5.18.1 for $P = 15$ kips),

$$R_u = 3.96(22.8/15) = 6.02 \text{ kips/in.}$$

$$\text{Weld resistance } \phi R_{nw} = 0.75a(0.707)(0.60\ F_{\text{EXX}})$$

$$= 0.75a(0.707)(0.60)70 = 22.3a$$

The weld size required is obtained by equating the design strength and the factored load.

$$\phi R_{nw} = R_u$$

$$22.3a = 6.02; \quad a = 0.27, \quad \text{say } \tfrac{5}{16} \text{ in.}$$

Use $\tfrac{5}{16}$-in. E70 fillet welds.

(b) Strength analysis. Divide the horizontal 6-in. weld into six parts and the vertical 8-in. length into 16 segments of 1/2-in. each, though it is considered adequate to use 1-in. segments. Using Eq. 5.17.2 and solving by trial in the manner used for Example 5.17.1, the instantaneous center is found to be located -0.1134 in. from the vertical weld line. Negative means the center is on the same side of the vertical weld line as the applied load P. The solution details are in Tables 5.18.2 and 5.18.3.

From Eq. 5.17.4, and multiplying by 2 because of symmetry, gives

$$P_n = \frac{\Sigma R_i r_i}{e + r_0} = \frac{2(354.35)}{14.0 - 0.1134} = 51.03 \text{ kips}$$

$$P_n = \Sigma (R_i)_y = 2(25.52) = 51.04 \text{ kips}$$

The nominal strength $P_n = 51$ kips. This is the strength using $\tfrac{1}{4}$-in. weld with E70 electrodes and the SMAW process. The design strength ϕP_n is $0.75(51) = 38.3$ kips. The weld size required is

$$P_u = 22.8 \text{ kips} = \phi P_n = 38.3 \frac{a}{0.25}$$

$$\text{Required weld size } a = \frac{22.8(0.25)}{38.3} = 0.15 \text{ in.}$$

TABLE 5.18.2 Geometry and Partial Results, Example 5.18.2

Seg. No.	Length (in.)	x (in.)	y (in.)	r_i (in.)	θ (deg)	Δ_u (in.)	Δ_u/r_i
1	1.0	0.387	4.00	4.02	5.5	0.0555	0.01381
2	1.0	1.387	4.00	4.23	19.1	0.0334	0.00790
3	1.0	2.387	4.00	4.66	30.8	0.0261	0.00560
4	1.0	3.387	4.00	5.24	40.3	0.0225	0.00429
5	1.0	4.387	4.00	5.94	47.6	0.0204	0.00344
6	1.0	5.387	4.00	6.71	53.4	0.0191	0.00285
7	0.5	−0.113	3.75	3.75	88.3	0.0142	0.00377
8	0.5	−0.113	3.25	3.25	88.0	0.0142	0.00436
9	0.5	−0.113	2.75	2.75	87.6	0.0142	0.00516
10	0.5	−0.113	2.25	2.25	87.1	0.0143	0.00633
11	0.5	−0.113	1.75	1.75	86.3	0.0143	0.00818
12	0.5	−0.113	1.25	1.26	84.8	0.0145	0.01155
13	0.5	−0.113	0.75	0.76	81.4	0.0149	0.01960
14	0.5	−0.113	0.25	0.27	65.6	0.0169	0.06165

TABLE 5.18.3 Solution for Example 5.18.2 ($r_0 = -0.1134$ in.)

Segment	Δ_m	Δ_i	R_i	$(R_i)_y$	$R_i r_i$
1	0.0274	0.0114	6.58	0.63	26.44
2	0.0197	0.0121	7.67	2.51	32.46
3	0.0171	0.0133	8.60	4.41	40.07
4	0.0158	0.0149	9.33	6.03	48.89
5	0.0150	0.0169	9.78	7.22	58.03
6	0.0145	0.0191	9.91	7.95	66.46
7	0.0124	0.0107	5.52	−0.17	20.69
8	0.0124	0.0093	5.42	−0.19	17.64
9	0.0124	0.0078	5.28	−0.22	14.54
10	0.0124	0.0064	5.09	−0.26	11.46
11	0.0125	0.0050	4.81	−0.31	8.44
12	0.0125	0.0036	4.43	−0.40	5.56
13	0.0127	0.0022	3.85	−0.58	2.92
14	0.0136	0.0008	2.72	−1.12	0.75
				25.52	354.35

(c) Use *AISC Manual* [1.15] tables, p. 8–90. For $\frac{1}{4}$-in. weld using E70 electrodes,

$$a = (e - xL)/L = (14.0 - 1.80)/8 = 1.525$$

$$k = kL/L \qquad = 6.0/8 = 0.75$$

	$k = 0.7$	075	0.8	
$a = 1.4$	1.61		1.82	
1.525	1.50	1.59	1.69	$C = 1.59$
1.6	1.43		1.61	

Table value $= \phi R_n = \phi C C_1 D L = 0.75(1.59)(1.0)(4)8 = 38.2$ kips

As may be noted, the AISC tables give essentially the same result as obtained in part (b).

(d) Summary. The weld size a required using E70 electrodes and the SMAW process is

$$\text{Elastic (vector) method, required } a = \tfrac{5}{16} \text{ in.}$$
$$\text{Strength analysis,} \qquad \text{required } a = \tfrac{3}{16} \text{ in.}$$

As will always be the case, the elastic vector method is conservative. ∎

5.19 LOADS APPLIED ECCENTRIC TO THE PLANE OF WELDS

When an applied load is eccentric to the *plane* of the weld configuration, as in Fig. 5.19.1, the strength method of analysis may still be used as long as the plane of the welds is rigid. The weld plane is rigid in Fig. 5.19.1 because the welds are on each side of a plate; i.e., there is sufficient rigidity between the two lines of weld such that there will be no bending *of the material being welded* in the plane of the welds.

As discussed in Sec. 5.17, the strength of a segment of weld depends on the angle θ_i of the resisting force R_i to the axis of the weld. It makes no difference whether R_i acts at an angle to the plane of the welds (Fig. 5.19.1) or whether it acts in the plane of the welds (eccentric shear as in Secs. 5.17 and 5.18).

The situation in Fig. 5.19.1 is commonly referred to as shear and bending, which is actually the stress condition on the bracket plate supporting the load P. The welds must carry the loads in the same manner that the members being connected carry them. The stresses are shown in Fig. 5.19.2.

For loading of the weld configuration in shear and tension, one must realize that the weld segments subject to compression are not free to rotate; thus, if a strength analysis is made, the compression region should be assumed to have a compressive stress distribution

Figure 5.19.1
Welds in shear and bending.

(a)

(b)

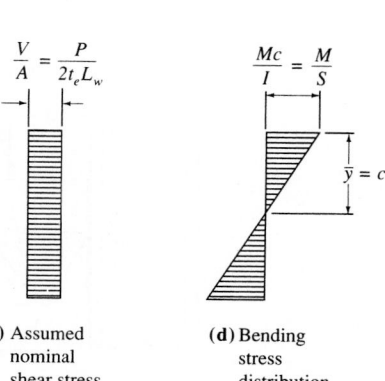

Figure 5.19.2
Stresses on vertical lines of weld acting in shear and bending.

(a) Weld group from Fig. 5.19.1

(b) Actual shear stress distribution

(c) Assumed nominal shear stress distribution

(d) Bending stress distribution

between the pieces being welded. Dawe and Kulak [5.41] reported that relatively good agreement with tests was obtained using any of a triangular, parabolic, or rectangular stress distribution on the compression side of the neutral axis.

The *AISC Manual* [1.15] does not contain tables for the case of shear and bending. The tables for eccentric shear are suggested [1.15] to be used for all cases where "the connection material between the welds is solid and does not bend in the plane of the welds."

Alternatively, the elastic (vector) method is conservative and relatively easy to use for loading in shear and tension.

Thus, there are four methods (Nos. 2 and 3 are really identical) suggested for use:

1. Strength analysis dividing the weld on the tension side of the neutral axis into segments the resistance of each depending on the angle the resistance makes with the weld axis using the formulas in Sec. 5.17. The weld on the compression side of the neutral axis is assumed to have only a resistance parallel to the weld axis. The compression force from bending is assumed to be carried by direct compression of the pieces being welded using a triangular distribution with the yield stress at the extreme compression fiber. The instantaneous center is then located by trial in a manner similar to that illustrated for eccentric shear. Details of the procedure are described by Dawe and Kulak [5.41].

2. Strength analysis exactly as for eccentric shear. In this procedure all weld segments, both on the tension and compression sides of the neutral axis, are assumed to have resistance. This procedure will give the values in the *AISC Manual* tables.

3. *AISC Manual* tables.

4. Elastic (vector) analysis.

EXAMPLE 5.19.1

Compute the size of E70 fillet weld required for the shear and tension connection in Fig. 5.19.3a using the SMAW process. Assume the column and the bracket plate do not control (stiffened seats and bracket plates are treated in Secs. 13.4 and 13.5.). The load is 80% live load and 20% dead load, and the AISC LRFD Method is to be used.

Solution:

(a) Compute the factored load P_u.

$$P_u = 1.2(0.2)10 + 1.6(0.8)10 = 15.2 \text{ kips}$$

(b) Use the traditional elastic (vector) method. The weld segments are treated as lines having a thickness of unity (1.0). The direct shear component is assumed to be carried equally by each segment of weld,

$$(R_n)_v = \frac{P}{A} = \frac{P}{2(1)L_w} = \frac{15.2}{2(1)10} = 0.76 \text{ kips/in.}$$

(a) Bracket

(b) Assumed weld cross section

Figure 5.19.3
Example 5.19.1.

The tension component (horizontal) due to the moment Pe is

$$(R_n)_t = \frac{Mc}{I} = \frac{15.2(6)5}{166.7} = 2.74 \text{ kips/in.}$$

where $I = \frac{1}{12}[2(1)(10)^3] = 166.7 \text{ in.}^4$
The resultant force is

$$\text{Required } \phi R_n = \sqrt{(0.76)^2 + (2.74)^2} = 2.84 \text{ kips/in.}$$

The design strength of E70 electrode fillet weld is

$$\phi R_{nw} = \phi t_e(0.60 F_{\text{EXX}}) = 0.75(0.707)a(0.60)70 = 22.3a$$

and the fillet weld size a required is

$$\text{Required } a = \frac{2.84}{22.3} = 0.13 \text{ in.}, \quad \text{say } \tfrac{3}{16} \text{ in.}$$

(c) Use *AISC Manual* [1.15] tables, p. 8–66. For weld using E70 electrodes.

$$e/L \text{ (Table symbol } a) = 6/10 = 0.60$$

$$k = 0$$

$$\text{Find } C = 2.00$$

$$\text{Table value} = \phi R_n = \phi CC_1 DL = 0.75(2.00)(1.0)(D)10 = 15.0D \text{ kips}$$

$$\text{Required } D = 15.2/15.0 = 1.0$$

$$\text{Required weld size } a = 1.0/16 = 0.06 \text{ in., say } \tfrac{1}{8} \text{ in.}$$

The elastic (vector) method is as expected more conservative than the strength method represented by the *AISC Manual* tables. The minimum desirable size to be used for this situation is probably $\frac{3}{16}$ in.
Use $\frac{3}{16}$-in. E70 fillet welds.   ■

Design for Lines of Weld Subject to Bending Moment

Even when there are moderate returns at the top of lines of fillet weld, an estimate of the length required may be obtained by using the same approach as used to determine the number of bolts in a line in Sec. 4.12. In Fig. 4.12.9, R/p has units kips/in. which becomes ϕR_{nw}, the design strength at the top of the lines of weld.
   For moment alone on one line of weld,

$$R = \frac{M}{S} = \frac{M}{\left(\frac{1}{6}L_w^2\right)} \text{kips/in.} \tag{5.19.1}$$

Since the maximum value of R is ϕR_{nw}

$$\phi R_{nw} = \frac{6M}{L_w^2}$$

$$\text{Required } L_w = \sqrt{\frac{6M}{\phi R_{nw}}} \tag{5.19.2}$$

Equation 5.19.2 for welds corresponds to Eq. 4.12.29 for bolts. Since it is correct only for moment alone, R_{nw} should be entered as a reduced value to account for direct shear.

EXAMPLE 5.19.2

Determine the length L required to carry the load indicated in Fig. 5.19.4, when 75% of the load is live load and 25% is dead load. The weld to be used is $\frac{5}{16}$-in. E70 fillet weld. Use AISC LRFD Method

Figure 5.19.4
Example 5.19.2.

Solution:

(a) Compute factored load P_u:

$$P_u = 1.2(0.25)40 + 1.6(0.75)40 = 60 \text{ kips}$$

(b) Estimate length L of weld required by using Eq. 5.19.2:

$$\phi R_{nw} = 0.75(a)(0.707)(0.60)70 = 22.3a \text{ kips/in.}$$
$$= 22.3(5/16) = 6.96 \text{ kips/in.}$$
$$M_u = 60(4) = 240 \text{ in.-kips per 2 lines of weld}$$
$$\text{Required } L \approx \sqrt{\frac{6M_u}{\phi R_{nw}}} = \sqrt{\frac{6(240/2)}{6(\text{est.})}} = 11 \text{ in.}$$

A reduced value of ϕR_{nw} has been used to account for the direct shear effect. Since the $\frac{1}{2}$-in. returns at the top add something, try $L = 10$ in.

When the returns have a specified dimension as is the case here, the weld is to be placed to provide full strength over the specified dimension. The returns resist the moment $P_n e$ but probably do not do not much to resist P_n as a shear force. For these small returns, it is optional with the designer whether or not to include them in the strength computation.

(c) If the returns are neglected, the *AISC Manual* [1.15] tables, p. 8–66, can be used to obtain an approximate result, as follows:

$$e/L \ (\text{Table symbol } a) = 4/10 = 0.40$$
$$k = 0$$
$$\text{Find } C = 2.66$$
$$\text{Table value} = \phi R_n = \phi C C_1 DL = 0.75(2.66)(1.0)(D)10 = 20.0D \text{ kips}$$
$$\text{Required } D = 60/20 = 3.0$$
$$\text{Required weld size } a = \frac{3.0}{16} = \tfrac{3}{16} \text{ in.}$$

For $L = 8$ in.,

$$\text{Find } C = 2.29, \quad \text{required } a = 0.22 \text{ in.,} \quad \text{say } \tfrac{1}{4} \text{ in.}$$

(d) Elastic (vector) method. The actual weld configuration has the $\tfrac{1}{2}$-in. returns which make the center of gravity of the weld configuration lie closer to the top than the mid-depth assumed in part (c). AISC Tables 8–4, "Coefficients C for Eccentrically Loaded Weld Groups," are indicated to be used also when load is not in the plane of weld group. This is acceptable for two vertical lines. However, for other weld configurations having the load applied out of plane, it is prudent to use a conservative elastic (vector) analysis.

Locate the center of gravity of the configuration,

$$\bar{y} = \frac{2(10)5}{2(10 + 0.5)} = \frac{100}{21} = 4.76 \text{ in.}$$

The direct shear component $(R_n)_v$ is computed assuming that none of the shear is carried by the returns,

$$(R_n)_v = \frac{P}{2L} = \frac{60}{2(10)} = 3.00 \text{ kips/in.}$$

The tension component $(R_n)_t$ due to the moment Pe is

$$I_x = \frac{2L^3}{12} + 2L(5 - 4.67)^2 + 2(0.5)(4.76)^2$$

$$= \frac{(10)^3}{6} + 20(0.24)^2 + (4.76)^2 = 190.5 \text{ in.}^3$$

$$(R_n)_t = \frac{60(4)4.76}{190.5} = 6.00 \text{ kips/in.}$$

The resultant force is

$$\text{Required } \phi R_n = \sqrt{(3.00)^2 + (6.00)^2} = 6.71 \text{ kips/in.}$$

The design strength of E70 electrode $\tfrac{5}{16}$-in. fillet weld is

$$\phi R_{nw} = \phi t_e(0.60 F_{EXX})$$

$$= 0.75(0.707)\left(\tfrac{5}{16}\right)(0.60)70 = 6.97 \text{ kips/in.}$$

The design strength ϕR_{nw} exceeds the 6.71 kips/in. required; thus, $L = 10$ in. is adequate. *Use $L = 10$ in.* ∎

Additional treatment of eccentric load on welds is to be found in Chapter 13 on connections.

SELECTED REFERENCES

5.1. K. Winterton. "A Brief History of Welding Technology," *Welding and Metal Fabrication,* November 1962; December 1962.

5.2. "100 Years of Metalworking—Welding, Brazing and Joining," *The Iron Age,* June 1955.

5.3. H. Carpmael. *Electric Welding and Welding Appliances.* London: D. Van Nostrand Company, 1920.

5.4. Preston M. Hall. "77 Years of Resistance Welding," *The Welding Engineer,* February 1954, 54–55; March 1954, 36–37; April 1954, 62–63.

5.5. W. L. Miskoe. "The Centenary of Modern Welding, 1885–1985—A Commemoration." *Welding Journal,* **65,** April 1986, 19–24.

5.6. D. W. Morgan. "Classification and Use of Mild Steel Covered Electrodes." *Welding Journal,* **55,** December 1976, 1035–1038.

5.7. E. Craig. "A Unique Mode of GMAW Transfer." *Welding Journal,* **66,** September 1987, 51–55.

5.8. K. A. Lyttle. "GMAW—A Versatile Process on the Move," *Welding Journal,* **62,** March 1983, 15–23.

5.9. V. R. Dillenbeck and L. Castagno. "The Effects of Various Shielding Gases and Associated Mixtures in GMA Welding of Mild Steel." *Welding Journal,* **66,** September 1987, 45–49. (See also letter from Nils Larson, Chairman, Committees C50 and A55, American Welding Society, and author's reply, **67,** March 1988, 6.)

5.10. AWS. *Welding Handbook,* Rev 8th ed., Vols. 2 and 3. Miami, FL: American Welding Society, 2004.

5.11. ——. "Office Building Columns Field Spliced with Self-Shielded Welding Wire," *Welding Journal,* **65,** October 1986, 53–54.

5.12. ——. "Self-Shielded FCAW Speeds High-Rise Construction," *Welding Journal,* **63,** April 1984, 47–49.

5.13. ——. "Self-Shielded FCA Welding is a Breeze in the Windy City," *Welding Journal,* **67,** March 1988, 47–48.

5.14. A. Raman. "Electroslag Welds: Problems and Cures," *Welding Journal,* **60,** December 1981, 17–21.

5.15. C. G. Schilling and K. H. Klippstein. "Tests of Electroslag-Welded Bridge Girders," *Welding Journal,* **60,** December 1981, 23–30.

5.16. A. W. Pense, J. D. Wood, and J. W. Fisher. "Recent Experiences with Electroslag Welded Bridges," *Welding Journal,* **60,** December 1981, 33–42.

5.17. Omer W. Blodgett, "Distortion . . . How to Minimize it with Sound Design Practices and Controlled Welding Procedures Plus Proven Methods for Straightening Distorted Members," *Bulletin G261.* Cleveland, OH: The Lincoln Electric Company. (No date)

5.18. Omer W. Blodgett. "Shrinkage Control in Welding," *Civil Engineering,* November 1960, 56–61.

5.19. E. R. Holby. "Weld Appearances May Be Deceiving," *Welding Journal,* **63,** May 1984, 33–36.

5.20. J. E. Jones. "Inspecting for Fatigue," *Welding Journal,* **62,** May 1983, 21–24.

5.21. R. Fenn. "Ultrasonic Monitoring and Control During Arc Welding," *Welding Journal,* **64,** September 1985, 18–22.

5.22. Paul D. Watson. "Design for Welding Examination," *Welding Journal,* **61,** February 1982, 32–35.

5.23. C. M. Fortunko and R. E. Schramm. "Ultrasonic Nondestructive Evaluations of Butt Welds Using Electromagnetic-Acoustic Transducers," *Welding Journal,* **61,** February 1982, 39–46.

5.24. Ronald Selner. "Dye Penetrant and Magnetic Particle Inspection," *Welding Journal,* **61,** February 1982, 28–31.

5.25. AWS. *Structural Welding Code—Steel* (AWS D1.1/D1.1M:2004), 19th edition. Miami, FL: American Welding Society, 2004 [same as 2.23].

5.26. J. A. Donnelly. "Determining the Cost of Welded Joints," *Engineering Journal,* AISC, **5,** 4 (October 1968), 146–147.

5.27. Omer W. Blodgett. "How to Determine the Cost of Welding," *Bulletin G610.* Cleveland, OH: The Lincoln Electric Company. (No date)

5.28. L. J. Butler and G. L. Kulak. "Strength of Fillet Welds as a Function of Direction of Load," *Welding Journal* (Welding Research Supplement), **36,** May 1971, 231s–234s.

5.29. Lorne J. Butler, Shubendu Pal, and Geoffrey L. Kulak. "Eccentrically Loaded Welded Connections," *Journal of the Structural Division,* ASCE, **98,** ST5 (May 1972), 989–1005.

5.30. G. L. Kulak and P. A. Timmler. "Tests on Eccentrically Loaded Fillet Welds," Structural Engineering Report No. 124, Department of Civil Engineering, University of Alberta, Edmonton, Alberta, December, 1984 (23 pp).

5.31. Peter Swannel. "Rational Design of Fillet Weld Groups," *Journal of the Structural Division.* ASCE, **107,** ST5 (May 1981), 789–802. Disc. **108,** ST5 (May 1982), 1197–1198.

5.32. Peter Swannel. "Weld Group Behavior," *Journal of the Structural Division,* ASCE, **107,** ST5 (May 1981), 803–815.

5.33. Vernon V. Neis. "New Constitutive Law for Equal Leg Fillet Welds," *Journal of Structural Engineering,* **111,** 8, August 1985, 1747–1759.

5.34. Raymond H. R. Tide. "Eccentrically Loaded Weld Groups—AISC Design Tables," *Engineering Journal,* AISC, **17,** 4 (4th Quarter 1980), 90–95.

5.35. Raymond H. R. Tide. Disc. of "Rational Design of Fillet Weld Groups." *Journal of the Structural Division,* ASCE, **108,** ST5 (May 1982), 1197–1198.

5.36. G. Donald Brandt. "A General Solution for Eccentric Loads on Weld Groups." *Engineering Journal,* AISC, **19,** 3 (3rd Quarter 1982), 150–159.

5.37. G. S. Miazga and D. J. L. Kennedy. "Behavior of Fillet Welds as a Function of the Angle of Loading," *Canadian Journal of Civil Engineering,* **16,** 1989, 583–599.

5.38. D. F. Lesik and D. J. L. Kennedy. "Ultimate Strength of Fillet Welded Connections Loaded in Plane," *Canadian Journal of Civil Engineering,* **17,** 1990, 55–67.

5.39. D. J. L. Kennedy, G. S. Miazga, and D. F. Lesik. "Discussion of Fillet Weld Shear Strength," *Welding Journal,* **55,** May 1990, 44–46.

5.40. Kenneth M. Loomis, William A. Thornton, and Thomas Kane. "A Design Aid for Connection Angle Welds Subjected to Combined Shear and Axial Loads," *Engineering Journal,* AISC, **22,** 4 (4th Quarter 1985), 178–196.

5.41. John L. Dawe and Geoffrey L. Kulak. "Welded Connections under Combined Shear and Moment," *Journal of the Structural Division,* ASCE, **100,** ST4 (April 1974), 727–741.

5.42. AWS. *Symbols for Welding, Brazing and Nondestructive Examination* (A2.4–98). Miami, FL: American Welding Society, 1998.

5.43. Murray E. Temple and Sherief S. S. Sakla. "Balanced and Unbalanced Welds for Angle Compression Members," *Canadian Journal of Civil Engineering,* **21,** 1994, 396–403.

PROBLEMS

All problems are to be done according to the AISC LRFD Method or ASD Method, as indicated by the instructor. All given loads are service loads unless otherwise indicated. Whenever possible, show all answers on a design sketch (drawn to scale) using appropriate welding symbols.

5.1. Specifically identify the AWS D1.1 Joint Designation for each of the following "prequalified" butt joints made by the submerged arc process (SAW). Specify the proper thickness for each of the plates, as well as the groove angle, root opening, and other requirements for the welds. Draw the cross-section for each weld. The given loads are 85% live load and 15% dead load. Refer to AWS D1.1 [5.25], *AISC Manual* [1.15]

Table 8–32 "Prequalified Welded Joints". The given joints are (a) Square-groove weld, complete penetration; (b) Single-V-groove weld, complete penetration; (c) Single-V-groove weld, complete penetration; and (d) Double-V-groove weld, complete penetration.

5.2. Specifically identify the AWS D1.1 Joint Designation for each of the following "prequalified"

Problem 5.1

(a)

(b)

(c)

(d)

Problem 5.2

butt joints made by the shielded metal arc process (SMAW). Specify the proper thickness for each of the plates, or determine the service load capacity assuming the load is 85% live load and 15% dead load. Indicate any thickness-related limitations and draw the cross-section for each weld. Refer to AWS D1.1 [5.25], *AISC Manual* [1.15] Table 8-32 "Prequalified Welded Joints." The given joints are (a) Square-groove weld, complete penetration; (b) Double-bevel-groove weld, complete penetration; (c) Double-bevel-groove weld, partial penetration; and (d) Single-V-groove weld, partial penetration.

5.3. Determine the service load capacity T of the connection shown when the submerged arc process (SAW) is used. The load is 85% live load and 15% dead load. Use (a) A36 steel, and (b) A572 Grade 65 steel. Assume appropriate electrode material is used.

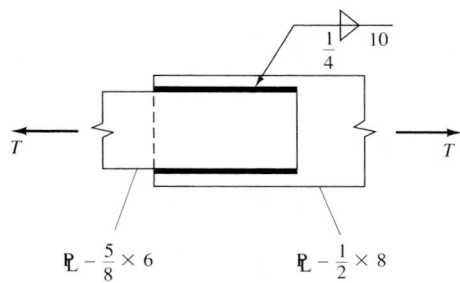

Problem 5.3

5.4. Specify the fillet welds required to develop the strength of the connection shown. State the maximum service load T permitted to be carried. The load is 90% live load and 10% dead

load. Specify the proper flux-electrode material using the submerged arc process.

Problem 5.4

5.5. Specify the plate thickness and weld size to be used for the joints in the accompanying figure. The loads are 70% live load and 30% dead load. State weld material to be used for the shielded metal arc process (SMAW). Compare A36 and A572 Grade 50 steels for each joint. Indicate the preferred design for each joint.

5.6. For the joints in the accompanying figure, satisfy the requirements stated in Prob. 5.5.

5.7. Design the reinforced lap joint shown in the accompanying figure. The plates are 7 in. wide of A36 steel and the SMAW process is used. Refer to AWS Joint Designation BTC-P4 (*AISC Manual* p. 8–35). The given load is 25% dead load and 75% live load.

5.8. Select a pair of channels and design the fillet welds using the SMAW process. The loading is 85% live

Problem 5.5

Problem 5.6

Problem 5.7

load and 15% dead load. Compare for (a) A36 steel and (b) A572 Grade 60 steel.

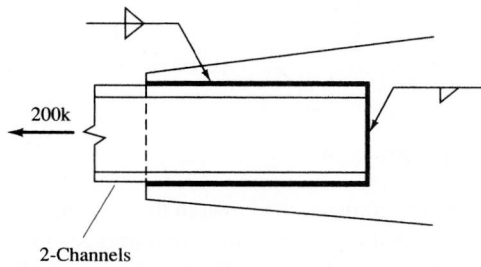

Problem 5.8

5.9. Design the tension plate attached to the wide-flange (W) section as well as the welds, assuming the SMAW process is used. The load is 70% live load and 30% dead load.

(a) Use A572 Grade 42 steel.

(b) Use A572 Grade 65 steel.

(c) Use A572 Grade 42 steel, with fillet welds instead of groove weld.

(d) Use A572 Grade 65 steel, with fillet welds instead of groove weld.

Problem 5.9

5.10. A $5 \times 3\frac{1}{2} \times \frac{3}{8}$ angle of A572 Grade 50 steel is connected by its long leg to a $\frac{5}{16}$-in. gusset plate. Develop the maximum service load capacity (25% dead load; 75% live load) of the angle and use a balanced fillet welded connection with the SMAW process. State the service load capacity. Use the following arrangements:

(a) $\frac{5}{16}$-in. weld on toe and back, with none on end.

(b) $\frac{1}{4}$-in. weld on toe and $\frac{3}{8}$-in. weld on back, and none on end.

Problem 5.10

5.11. Design a balanced connection for two $7 \times 4 \times \frac{1}{2}$ angles connected by their long legs to a $\frac{3}{8}$-in. gusset plate. Develop the maximum service load capacity (20% dead load; 80% live load) and state its value. Use A572 Grade 60 steel and the SMAW process. Detail the joint to balance the loads using the shortest possible overlap.

5.12. Design the welds indicated to develop the full strength of the angles and minimize eccentricity. Assume service load is 20% dead load and 80% live load. Use the SAW welding process.

(a) Use A36 steel.

(b) Use A572 Grade 50 steel.

(c) Use A572 Grade 65 steel.

(d) Use A36 steel, but omit weld on the end of angles.

(e) Use A572 Grade 65 steel, but omit weld on the ends of angles.

Problem 5.12

5.13. Assume a 9-in.-wide plate used in a lap joint must carry 30 kips dead load and 115 kips live load, and a possibility exists of some accidental eccentricity that cannot be computed. To insure a tighter joint, a $1\frac{1}{4}$-in.-diam plug weld is to be used. Determine the thickness of the plate, the amount of lap, and the weld size for the best joint. Assume the gusset plate to which the 9-in. plate is welded does not control any of the design. Use A572 Grade 50 steel and the SMAW process.

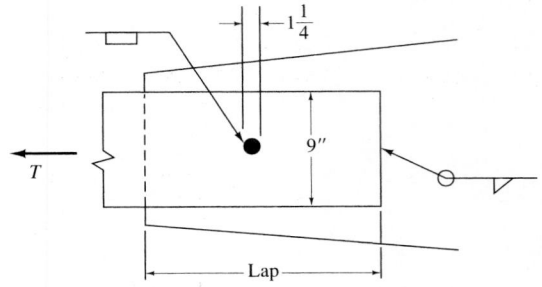

Problem 5.13

5.14. Determine the minimum length of slot in order to develop the full strength of a C12×20.7 welded to a $\frac{3}{8}$-in plate. Use the same size fillet weld over the entire length, and assume it is to be placed by the SMAW process. Assume service load is 35% dead load and 65% live load.

Problem 5.14

5.15. For the connection shown in the accompanying figure, what is the maximum required design strength ϕR_{nw}? The load is 90% live load and 10% dead load. What weld size is indicated if E70 electrodes are used with the SMAW process?

(a) Use strength analysis (i.e., locate the instantaneous center by trial).

(b) Use *AISC Manual* Table 8–4, "Coefficients C for Eccentrically Loaded Weld Groups."

(c) Use elastic (vector) method.

Problem 5.15

Problem 5.17

5.16. For the connection in the accompanying figure, satisfy the requirements of Prob. 5.15.

Problem 5.16

(b) Use *AISC Manual* Table 8–8, "Coefficient C for Eccentrically Loaded Weld Groups."

(c) Use elastic (vector) method.

5.19. Repeat Prob. 5.18, except use $\frac{3}{8}$-in. fillet weld on the side and $\frac{1}{4}$-in. on the end. The steel is A572 Grade 50.

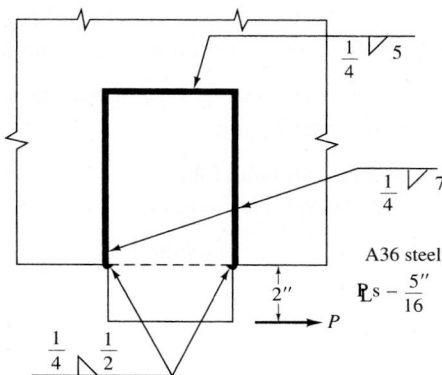

Problems 5.18 and 5.19

5.17. For the bracket shown in the accompanying figure, calculate the service load capacity P (90% live load and 10% dead load) based on the weld. Neglect the returns at ends and assume the SMAW process is to be used.

(a) Use strength analysis (i.e., locate the instantaneous center by trial).

(b) Use *AISC Manual* Table 8–4, "Coefficient C for Eccentrically Loaded Weld Groups."

(c) Use elastic (vector) method.

5.18. Compute the maximum acceptable service load P for the connection of the accompanying figure. The load is 85% live load and 15% dead load. *Ignore the effect of returns at the lower end of the connection.*

(a) Use strength analysis (i.e., locate the instantaneous center by trial).

5.20. Compute the service load capacity P for the welded bracket of the accompanying figure. The load is 70% live load and 30% dead load. Neglect the returns at the outer ends of the C-shaped weld configuration. The weld size is $\frac{3}{8}$ in. and E70 electrodes are used in the shielded metal arc process.

(a) Use strength analysis (i.e., locate the instantaneous center by trial).

(b) Use *AISC Manual* Table 8–8, "Coefficient C for Eccentrically Loaded Weld Groups."

(c) Use elastic (vector) method.

Problem 5.20

5.21. Repeat Prob. 5.20 if the vertical dimension of the weld configuration is 12 in. instead of 8 in.

5.22. Repeat Prob. 5.20 if the vertical dimension of the weld configuration is 4 in. instead of 8 in.

5.23. Compute the theoretical weld size required for the bracket of the accompanying figure when the SMAW process is used. The load is 60% live load and 40% dead load. Neglect the returns at the outer ends of the C-shaped weld configuration.

(a) Use strength analysis (i.e., locate the instantaneous center by trial).

(b) Use AISC Manual Tables 8–8, "Eccentric Loads on Weld Groups."

(c) Use elastic (vector) method.

Problem 5.23

5.24. Repeat Prob. 5.23 if the vertical dimension is 12 in. instead of 16 in.

5.25. Repeat Prob. 5.23 if the vertical dimension is 8 in. instead of 16 in.

5.26. Repeat Prob. 5.23 if the horizontal dimension is 4 in. instead of 16 in.

5.27. Use the elastic (vector) method to derive a general expression for the required weld size on the seat angle of the accompanying figure in terms of the factored load P_u, the leg length L, and the eccentricity e of the applied load. Assume E70 electrodes with the SMAW process are used. Use the following assumptions:

(a) Ignore the returns at the top;

(b) Use an average return of $L/12$; and

(c) Use a return equal to twice the weld size.

If $e = 2\frac{3}{4}$ in. and $L = 6$ in., determine the weld size needed to carry $P = 38$ kips (80% live load and 20% dead load). For the weld size selected, check the capacity P using all three assumptions, and also compare with the result using the AISC Manual Table 10–6, "All-Welded Unstiffened Seated Connections."

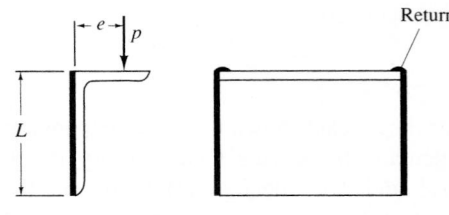

Problem 5.27

5.28. For the bracket of the accompanying figure, determine the length L required when using $\frac{5}{16}$-in. fillet weld with the SMAW process. The load is 70% live load and 30% dead load. Verify your result using the following procedures:

(a) Strength analysis (i.e., locate the instantaneous center by trial).

Problem 5.28

Problem 5.30

(b) *AISC Manual* Table 8–4.

(c) Elastic (vector) method.

5.29. Repeat Prob. 5.28 if the service load is 40 kips.

5.30. For the bracket of the accompanying figure, satisfy the requirements of Prob. 5.28. Note that A572 Grade 50 steel is used instead of A36.

6

Compression Members

PART I: COLUMNS

6.1 GENERAL

In this chapter, members subjected to axial compression forces are to be treated. Referred to by various terms, such as *column, stanchion, post,* and *strut,* these members are rarely if ever actually carrying only axial compression. However, whenever the loading is so arranged that either the end rotational restraint is negligible or the loading is symmetrically applied from members framing in at the column ends, and bending may be considered negligible compared to the direct compression, the member can safely be designed as a concentrically loaded column.

It is well known from basic mechanics of materials that only very short columns can be loaded to their yield stress; the usual situation is that buckling, or sudden bending as a result of instability, occurs prior to developing the full material strength of the member. Thus, a sound knowledge of compression member stability is necessary for those designing in structural steel.

6.2 EULER ELASTIC BUCKLING AND HISTORICAL BACKGROUND

Column buckling theory originated with Leonhard Euler in 1744 [6.1]. An initially straight concentrically loaded member, in which all fibers remain elastic until buckling occurs, is slightly bent as shown in Fig. 6.2.1. Although Euler dealt with a member built-in at one end and simply supported at the other, the same logic is applied here to the pin-end column, which having zero end rotational restraint is the member having least buckling strength.

At any location z, the bending moment M_z on the member bent slightly about the x principal axis is

$$M_z = Py \qquad (6.2.1)$$

and since

$$\frac{d^2y}{dz^2} = -\frac{M_z}{EI} \qquad (6.2.2)$$

Steel framework, showing particularly the exterior columns, Southeast Recreation Facility, University of Wisconsin-Madison Campus. (Photo by C. G. Salmon)

the differential equation becomes

$$\frac{d^2y}{dz^2} + \frac{P}{EI}y = 0 \tag{6.2.3}$$

After letting $k^2 = P/EI$, the solution of this second-order linear differential equation may be expressed

$$y = A\sin kz + B\cos kz \tag{6.2.4}$$

Applying boundary conditions, (a) $y = 0$ at $z = 0$; and (b) $y = 0$ at $z = L$, one obtains for condition (a), $B = 0$; and for condition (b),

$$0 = A\sin kL \tag{6.2.5}$$

Satisfaction of Eq. 6.2.5 may be accomplished in three possible ways; (a) constant $A = 0$, i.e., no deflection; (b) $kL = 0$, i.e., no applied load; and (c) $kL = N\pi$, the requirement for buckling to occur. Thus

$$\left(\frac{N\pi}{L}\right)^2 = \frac{P}{EI}$$

$$P = \frac{N^2\pi^2 EI}{L^2} \tag{6.2.6}$$

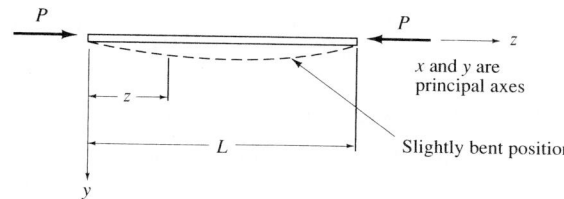

Figure 6.2.1
Euler column.

The fundamental buckling mode, a single-curvature deflection ($y = A \sin \pi z/L$ from Eq. 6.2.4.), will occur when $N = 1$; thus the Euler critical load for a column with both ends pinned is

$$P_{\text{cr}} = \frac{\pi^2 EI}{L^2} \tag{6.2.7}$$

or in terms of average compressive stress, using $I = A_g r^2$

$$F_{\text{cr}} = \frac{P_{\text{cr}}}{A_g} = \frac{\pi^2 E}{(L/r)^2} \tag{6.2.8}$$

Euler's approach was generally ignored for design because test results did not agree with it; columns of ordinary length used in design were not as strong as Eq. 6.2.7 would indicate.

Considère and Engesser [6.2, 6.3] in 1889 independently realized that portions of usual length columns become inelastic prior to buckling and that a value of E should be used that could account for some of the compressed fibers being strained beyond the proportional limit. It was thus consciously recognized that in fact *ordinary length* columns fail by inelastic buckling rather than by elastic buckling.

Complete understanding of the behavior of concentrically loaded columns, however, was not achieved until 1946 when Shanley [6.4, 6.5] offered the explanation that now seems obvious. He reasoned that it was actually possible for a column to bend and still have increasing axial compression, but that it *begins* to bend upon reaching what is commonly referred to as the *buckling load*, which includes inelastic effects on some or all fibers of the cross-section. These inelastic effects are discussed in detail in Sec. 6.4.

An extensive historical review of the development of column theory is given by B. G. Johnston [6.6].

6.3 BASIC COLUMN STRENGTH

To determine a basic column strength, certain conditions may be assumed for the ideal column [6.7]. With regard to material, it may be assumed (1) there are the same compressive stress-strain properties throughout the section; (2) no initial internal stresses exist such as those due to cooling after rolling and those due to welding. Regarding shape and end conditions, it may be assumed (3) the column is perfectly straight and prismatic; (4) the load resultant acts through the centroidal axis of the member until the member begins to bend; (5) the end conditions must be determinate so that a definite equivalent pinned length may be established. Further assumptions regarding buckling may be made, as (6) the small deflection theory of ordinary bending is applicable and shear may be neglected; and (7) twisting or distortion of the cross-section does not occur during bending.

Once the foregoing assumptions have been made, it is now agreed [6.8] that the strength of a column may be expressed by

$$P_{\text{cr}} = \frac{\pi^2 E_t}{(KL/r)^2} A_g = F_{\text{cr}} A_g \tag{6.3.1}$$

where E_t = tangent modulus of elasticity at stress P_{cr}/A_g

A_g = gross cross-sectional area of member

KL/r = effective (or equivalent pinned-end) slenderness ratio

K = effective length factor (treated in Sec. 6.9)

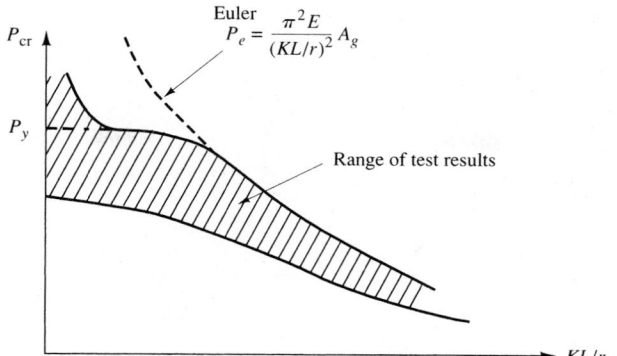

Figure 6.3.1
Typical range of column
strength vs slenderness ratio.

L = length of member

r = radius of gyration = $\sqrt{I/A_g}$

I = moment of inertia

It is well known that long compression members fail by elastic buckling and that short stubby compression members may be loaded until the material yields or perhaps even into the strain-hardening range. However, in the vast majority of usual situations, failure occurs by buckling after a portion of the cross-section has yielded. This is known as *inelastic buckling*.

Actually, buckling under axial load occurs only when the aforementioned assumptions (1) through (7) apply. Columns are usually an integral part of a structure and as such *cannot* behave entirely independently. The practical use of the term *buckling* is that it is the boundary between stable and unstable deflections of a compression member, rather than the instantaneous condition that occurs in the isolated slender elastic rod.

As previously mentioned, for many years theoretical determinations of column strength did not agree with test results. Test results included effects of initial crookedness of the member, accidental eccentricity of load, end restraint, local or lateral buckling, and residual stress. A typical curve of observed strengths was as shown in Fig. 6.3.1. Design formulas, therefore, were based on such empirical results. Various straight-line and parabolic formulas have been used, as well as other more complex expressions, in order to fit the curve of test results in a reasonably accurate, yet practical manner.

In summary, Euler elastic buckling governs the strength for large slenderness ratios, yield strength $P_y = F_y A_g$ controls for short columns, and a transition curve must be used for inelastic buckling.

6.4*   INELASTIC BUCKLING

Since ordinary length columns buckle when some of their fibers are inelastic, having a modulus of elasticity less than their initial elastic value, the logic of Engesser, Considère, and Shanley is explained in this section, generally following Bleich [6.9, pp. 8–20].

Basic Tangent Modulus Theory

Euler's theory pertained only to situations where compressive stress below the elastic limit acts uniformly over the cross-section when unstable equilibrium occurs. Engesser [6.3]

---

*Sections so marked may be omitted without loss of continuity.

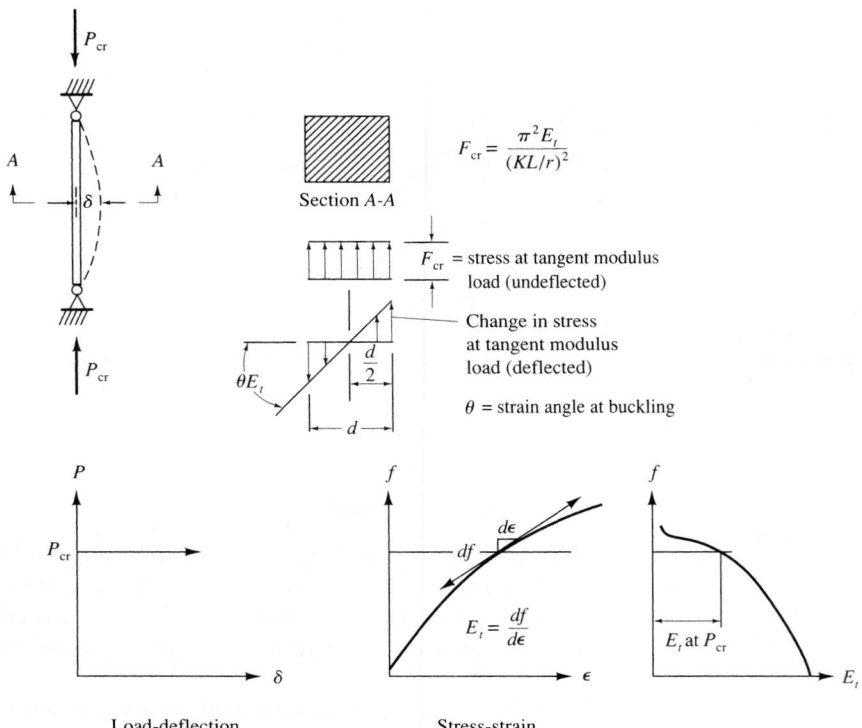

Figure 6.4.1
Engesser original tangent
modulus theory, 1889.

and Considère [6.2] were the first to utilize the possibility of a variable modulus of elasticity. In Engesser's tangent modulus theory the column remains straight up to the instant of failure and the modulus of elasticity at failure is the tangent to the stress-strain curve. The relationships are shown in Fig. 6.4.1. The theory prescribed that at a certain stress, $F_{cr} = P_{cr}/A_g$, the member could acquire an unstable deflected shape and that the deformation at F_{cr} is governed by $E_t = df/d\epsilon$. Thus Engesser modified Euler's equation to become

$$F_{cr} = \frac{P_t}{A_g} = \frac{\pi^2 E_t}{(KL/r)^2} \tag{6.4.1}$$

where P_t is the tangent modulus load, and E_t is the tangent modulus of elasticity at stress F_{cr}.

This theory, however, still did not agree with test results, giving computed loads lower than measured ultimate capacities. The principal assumption that caused this tangent modulus theory to be considered erroneous is that as the member changes from a straight to bent form, no strain reversal takes place. In 1895 Engesser changed his theory, reasoning that during bending some fibers undergo increased strain (lowered tangent modulus) and some fibers are unloaded (higher modulus at the reduced strain); therefore a combined value should be used for the modulus.

Double Modulus Theory

To examine the process of column bending at stresses beyond the elastic limit, consider the section of Fig. 6.4.2 from which Engesser's double modulus, or "reduced" modulus, is developed. This concept had logic to it which was generally accepted but gave computed

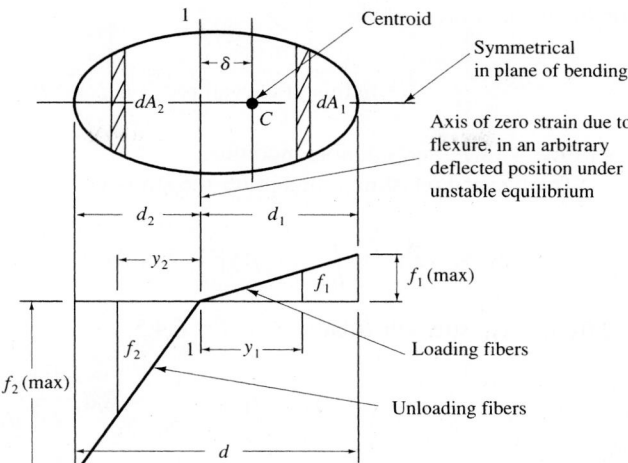

Figure 6.4.2
Stress distribution in
condition of unstable
equilibrium (double
modulus theory).

strengths higher than test values. Not until Shanley's explanation was the inconsistency resolved.

At unstable equilibrium, the stress at the neutral axis (section 1–1 of Fig. 6.4.2) remains as it was prior to the deflection δ occurring. On the loading fibers where strain is increasing, the stress increase is proportional to $E_t = df/d\epsilon$, whereas on the unloading fibers the decrease in strain relieves the elastic part of the strain; thus the stress decrease is proportional to the elastic modulus E.

As shown in Fig. 6.4.3, the strain on the cross-section will be linear. At the extreme unloaded fiber, applying Hooke's Law the stress becomes

$$f_{2(\text{max})} = (\text{unit strain})E = \frac{\Delta\,dz}{dz}E \tag{6.4.2}$$

and at the loaded fiber,

$$f_{1(\text{max})} = \frac{\Delta\,dz\,d_1}{d_2}\frac{E_t}{dz} \tag{6.4.3}$$

$$\frac{\Delta\,dz}{d_2} = d\theta \tag{6.4.4}$$

Thus

$$f_{2(\text{max})} = Ed_2\frac{d\theta}{dz}; \qquad f_{1(\text{max})} = E_td_1\frac{d\theta}{dz} \tag{6.4.5}$$

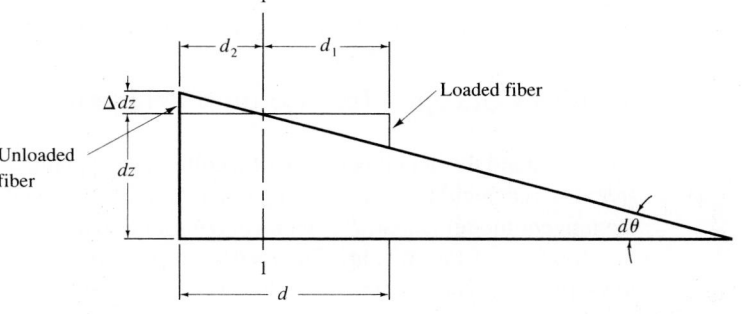

Figure 6.4.3
An element dz along the axis
of the column in the unstable
equilibrium position.

For small curvature,

$$\frac{1}{\text{radius of curvature}} = \frac{M}{E_r I} = \frac{d\theta}{dz} = \frac{d^2 y}{dz^2} \tag{6.4.6}$$

where E_r = Engesser's double modulus.

The internal resisting moment for the stress condition of Fig. 6.4.2 gives

$$M = -Py = \int_0^{d_1} (f_1)(y_1 - \delta)\, dA_1 + \int_0^{d_2} (f_2)(y_2 + \delta)\, dA_2 \tag{6.4.7}$$

and from linear stress distribution and Eq. 6.4.5,

$$f_1 = f_{1(\text{max})}\frac{y_1}{d_1} = E_t d_1 \frac{d^2 y}{dz^2}\frac{y_1}{d_1}$$

$$f_2 = f_{2(\text{max})}\frac{y_2}{d_2} = E d_2 \frac{d^2 y}{dz^2}\frac{y_2}{d_2} \tag{6.4.8}$$

Thus Eq. 6.4.7 becomes

$$-Py = E_t \frac{d^2 y}{dz^2}\int_0^{d_1} y_1(y_1 - \delta)\, dA_1 + E\frac{d^2 y}{dz^2}\int_0^{d_2} y_2(y_2 + \delta)\, dA_2 \tag{6.4.9}$$

Force equilibrium requires

$$\int_0^{d_1} f_1\, dA_1 = \int_0^{d_2} f_2\, dA_2 \tag{6.4.10}$$

which, using Eqs. 6.4.8, gives

$$E_t \frac{d^2 y}{dz^2}\int_0^{d_1} y_1\, dA_1 = E\frac{d^2 y}{dz^2}\int_0^{d_2} y_2\, dA_2 \tag{6.4.11}$$

Using Eq. 6.4.11, it is seen the terms involving δ cancel each other in Eq. 6.4.9, thus giving

$$-Py = E_t \frac{d^2 y}{dz^2}\int_0^{d_1} y_1^2\, dA_1 + E\frac{d^2 y}{dz^2}\int_0^{d_2} y_2^2\, dA_2$$

$$\therefore\ \frac{d^2 y}{dz^2}\left[E_t \int_0^{d_1} y_1^2\, dA_1 + E\int_0^{d_2} y_2^2\, dA_2\right] + Py = 0 \tag{6.4.12}$$

Equation 6.4.12 is obviously of the same form as the elastic buckling equation, Eq. 6.2.3. Thus for the double modulus theory,

$$P_{\text{cr}} = \frac{\pi^2}{L^2}\left[E_t \int_0^{d_1} y_1^2\, dA_1 + E\int_0^{d_2} y_2^2\, dA_2\right] \tag{6.4.13}$$

Shanley Concept—True Column Behavior

To understand the actual behavior of a column as explained by Shanley [6.4] in 1946, consider the rectangular section of Fig. 6.4.4 subjected to axial compression. For loads below the tangent modulus load P_t, the ideal column remains perfectly straight with zero deflection (point A of Fig. 6.4.4a). The load P_t at point A may most correctly be defined as follows [6.10]: "The tangent modulus load is the smallest value of axial load at which

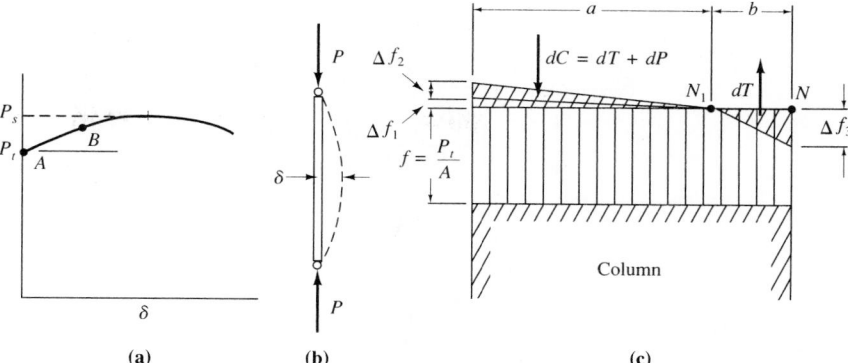

Figure 6.4.4
Shanley concept—true
column behavior.

bifurcation of the equilibrium positions can occur regardless of whether or not the transition to the bent position requires an increase of axial load." Consider that at onset of bending (infinitesimal curvature) there will be an infinitesimal increase in axial strain and stress Δf_1. By the time the curvature becomes finite, i.e., the point N moves to N_1, some strain reversal must occur if the column cross-section is to develop a resisting moment to maintain equilibrium with the moment due to the external load $P\delta$. For small but finite values of curvature, the increment of load represented by stress on the area of increasing strain exceeds the increment of load represented by stress on the area of decreasing strain; thus P is increased by an amount dP (point B of Fig. 6.4.4a). As each increment of curvature takes place P will further increase as long as $dC > dT$. The increased compressive force dC is computed using the tangent modulus E_t while in the region of strain reversal the elastic modulus E is used to compute dT. The double modulus theory, which similarly treated loading and unloading fibers, did not accept $dC > dT$, but rather only considered equilibrium positions near the perfectly straight one.

For practical purposes, the increase of capacity from P_t to P_s (Fig. 6.4.4a) can be neglected for design use. Therefore, the tangent modulus load may be treated as the critical load, i.e., the load at which bending begins.

6.5   RESIDUAL STRESS

Residual stresses are stresses that remain in a member after it has been formed into a finished product. Such stresses result from plastic deformations, which in structural steel may result from several sources: (1) uneven cooling which occurs after hot rolling of structural shapes; (2) cold bending or cambering during fabrication; (3) punching of holes and cutting operations during fabrication; and (4) welding. Under ordinary conditions those residual stresses resulting from uneven cooling and welding are the most important. Actually the important residual stresses due to welding are really the result of uneven cooling.

The mechanism of residual stress due to uneven cooling is treated in the *Welding Handbook* [2.21, Volume 1] and the effect of residual stress on compression structural members appears in the *Guide to Stability Design Criteria for Metal Structures* [6.8, pp. 30–42], prepared by the Structural Stability Research Council. This latter publication will be extensively referred to as the *SSRC Guide*.

In wide-flange or H-shaped sections, after hot rolling, the flanges, being the thicker parts, cool more slowly than the web region. Furthermore, the flange tips having greater exposure to the air cool more rapidly than the region at the junction of flange to web. Consequently, compressive residual stress exists at flange tips and at mid-depth of the web (the regions that cool fastest), while tensile residual stress exists in the flange and the web at the

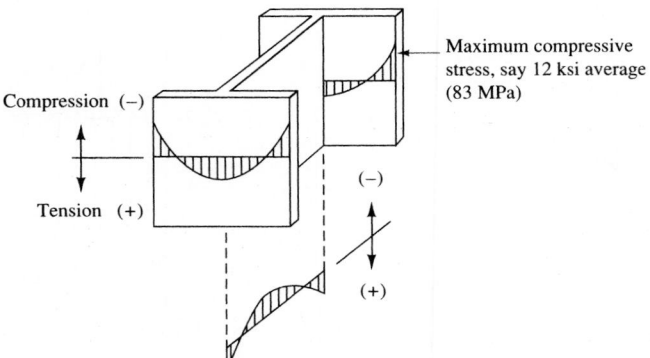

Figure 6.5.1
Typical residual stress pattern
on rolled shapes.

regions where they join. Figure 6.5.1 shows typical residual stress distribution on rolled beams. Considerable variation can be expected as the true pattern will be a function of the dimensions of the section.

At this point one might wonder whether the general column strength equation (Eq. 6.3.1) discussed in the preceding section still is applicable. The theory is applicable, but all fibers in the cross-section cannot be considered as stressed to the same level under the action of the compressive service load. The tangent modulus E_t on one fiber is not the same as that on an adjacent fiber.

In a rolled steel shape the influence of residual stress on the stress-strain curve is shown in Fig. 6.5.2. using average stress on the gross area as the ordinate. It is noted that residual stress in an elastic-plastic material such as steel gives the same effect as that obtained for a material such as aluminum, which is not linearly elastic when it contains no residual stress. Thus, assuming the tangent modulus concept applies, column strength may be said to be based on inelastic buckling because the average stress-strain curve is nonlinear when maximum column strength is reached.

Whereas it was once believed the nonlinear portion of the average stress-strain curve for axially loaded compression members was due entirely to initial curvature and accidental eccentricity, Huber and Beedle [6.11] have verified that residual stress is the primary cause, and the other factors have a relatively minor effect. Residual stresses at flange tips of rolled shapes have been measured as high as 20 ksi (138 MPa), a high percentage of the minimum specified yield stress for steels such as A36. Residual stresses are essentially independent of yield stress, depending instead on cross-sectional dimensions and configuration since those factors govern cooling rates [6.12].

Welding of built-up shapes is an even greater contributor to residual stress than cooling of hot-rolled H-shapes [6.13]. The plates themselves generally have little residual

Figure 6.5.2
Influence of residual stress on
average stress-strain curve.

Figure 6.5.3
Typical residual stress
distribution in welded shapes.

stress initially because of relatively uniform cooling after rolling. However, after the heat is applied to make the welds, the subsequent nonuniform cooling and restraint against distortion cause high residual stresses. Figure 6.5.3 shows typical residual stress patterns on welded H and box built-up shapes.

One should note that compressive residual stresses typically occurring at flange tips are higher in welded than in rolled H-shaped sections. Thus the column strength of such welded shapes will be lower than corresponding rolled shapes. On the other hand, the welded-box shape, having tensile residual stress in the corner regions that contribute most to the stiffness as a column, will be stronger than a rolled shape having the same slenderness ratio. Sherman [6.14, 6.15] has studied residual stresses on rolled tubular members.

Having accepted that residual stresses exist, such information must be used to obtain a column strength curve (average stress vs slenderness ratio) that can form the basis for design. Until the early 1950s, column design was based on many formulas, all of which tried to empirically account for column behavior exhibited by tests. By clearly indicating that the tangent modulus was the proper criterion for strength and by identifying the role of residual stress, the Column Research Council (now Structural Stability Research Council [6.8] has made a significant contribution.

6.6*   DEVELOPMENT OF COLUMN STRENGTH CURVES INCLUDING RESIDUAL STRESS

The following analytical approach, patterned after Huber and Beedle [6.11] and Beedle and Tall [6.16], is intended to show the logic to obtain a column strength equation. Column strength can be obtained by two general methods. One method is to use the residual stress distribution, either the actual variation from measurements or a mathematical model, along with the stress-strain relationship for the material (say a small test specimen of the steel).

The other method is to determine experimentally an average stress-strain relationship by testing short lengths of rolled shapes containing residual stress. Column strength can then be determined from the test results using the tangent modulus of the experimental curve in combination with the appropriate slenderness ratio. Knowledge of the residual stress pattern is not used in this second approach.

Yu and Tall [6.17] have discussed these approaches in detail. Johnston [6.18] and Batterman and Johnston [6.19] have treated the tangent modulus application to inelastic buckling of columns.

The following development is made with the objective of obtaining a relationship between average externally applied stress and the slenderness ratio. Thus, the capacity of

Figure 6.6.1
Comparison of coupon
with H-shaped rolled section
containing residual stress.

a member can be obtained by a simple multiplication of safe stress times gross area, without regard to what the actual stress is at any point in the cross-section or what is the true residual stress pattern.

As a starting point consider steel which as a material is perfectly elastic until a certain strain ϵ_y is achieved and then is plastic (i.e., constant stress with increasing strain). A coupon cut from the web of a rolled shape exhibits such behavior, as shown by the dotted lines in Fig. 6.6.1. The solid lines indicate the behavior of an H-shaped rolled section including residual stress.

To account for the effects of early yielding because of residual stress, consider one fiber at a distance x from the axis of zero strain caused by bending (Fig. 6.6.2). The bending is taken as an infinitesimal amount consistent with equilibrium at the tangent modulus load. The bending moment contribution from stress on the one fiber is

$$dM = (\text{stress})(\text{area})(\text{moment arm}) = (\theta E_t x)(dA)(x) \tag{6.6.1}$$

which for the entire cross-section becomes

$$M = \int_A \theta E_t x^2 \, dA = \theta \int_A E_t x^2 \, dA \tag{6.6.2}$$

From elementary bending theory, the radius of curvature is

$$R = \frac{1}{\theta}$$

$$\theta = \frac{1}{R} = \frac{M}{\text{equivalent } EI} = \frac{M}{E'I} \tag{6.6.3}$$

Thus

$$E'I = \frac{M}{\theta} = \int_A E_t x^2 \, dA$$

$$E' = \frac{1}{I} \int_A E_t x^2 \, dA \tag{6.6.4}$$

Figure 6.6.2
Stress on fiber at distance x
from axis of zero strain due
to bending.

which may be called the *effective modulus* and used in Fg. 6.4.1 as an equivalent value for E_t.

If the idealized elastic-plastic f-ϵ curve of Fig. 6.6.1a (dotted) is used (for $f < F_y$, $E_t = E$ and for $f = F_y$, $E_t = 0$) the bending stiffness of yielded parts becomes zero; however, the buckling strength will be the same as a column whose moment of inertia I_e is the moment of inertia of the portion remaining elastic. Equation 6.6.4 then becomes

$$E' = \frac{E}{I} \int_{A \text{ (elastic part only)}} x^2 \, dA = \qquad E\frac{I_e}{I} \tag{6.6.5}$$

The stress at which the column may begin to bend, from Eq. 6.3.1, is

$$P_{\text{cr}} = \left(\frac{\pi^2 E}{(KL/r)^2} \frac{\int_A x^2 \, dA}{I} \right) A_g = F_{\text{cr}} A_g \tag{6.6.6}$$

$$P_{\text{cr}} = \left(\frac{\pi^2 E(I_e/I)}{(KL/r)^2} \right) A_g = F_{\text{cr}} A_g \tag{6.6.7}$$

In order for Eq. 6.6.7 to be useful, the relationship between F_{cr} and I_e must be established.

Case A. Buckling about Weak Axis

A reasonable assumption will be that the flanges become fully plastic before the web yields (see Fig. 6.6.3).

Let k = proportion of the flange remaining elastic = $2x_0/b = A_e/A_f$. Then Eq. 6.6.5 becomes

$$E\frac{I_e}{I} = E\frac{t_f(2x_0)^3}{12} \left(\frac{12}{t_f b^3} \right) = Ek^3 \tag{6.6.8}$$

if the web is neglected in computing I. Applying the tangent modulus definition,

$$E_t = \frac{\text{nominal incremental stress}}{\text{incremental elastic strain}} = \frac{dP/A}{\dfrac{dP/A_e}{E}} = \frac{A_e E}{A} \tag{6.6.9}$$

$$E_t A = A_e E = (A_w + 2kA_f)E \tag{6.6.10}$$

where A_w = web area
$\quad\quad\quad A_f$ = gross area of one flange
$\quad\quad\quad A$ = total gross area of section

$2x_0$ b

d

Yielded flange tips

Figure 6.6.3
Portion of section that has yielded.

Solving Eq. 6.6.10 for k and using Eq. 6.6.8 in Eq. 6.6.7 gives

$$k = \frac{E_t A}{2 E A_f} - \frac{A_w}{2 A_f} \qquad (6.6.11)$$

$$F_{cr} = \frac{\pi^2 E k^3}{(KL/r)^2} = \frac{\pi^2 E}{(KL/r)^2} \left[\frac{A E_t}{2 A_f E} - \frac{A_w}{2 A_f} \right]^3 \qquad (6.6.12)$$

which includes the elastic web effect, for buckling with respect to the weak axis (y–y).

Case B. Buckling about Strong Axis

Again, assuming the web is elastic, but neglecting its contribution toward the moment of inertia gives approximately

$$E \frac{I_e}{I} \approx E \frac{2 A_e (d/2)^2}{2 A_f (d/2)^2} = E k \qquad (6.6.13)$$

If the elastic web is included,

$$E \frac{I_e}{I} = E \left[\frac{2 k A_f (d^2/4) + t_w d^3/12}{2 A_f (d^2/4) + t_w d^3/12} \right] \qquad (6.6.14)$$

$$= E \left[\frac{2 k A_f + A_w/3}{2 A_f + A_w/3} \right] \qquad (6.6.15)$$

Using tangent modulus definition and Eq. 6.6.10,

$$2 k A_f = \frac{E_t A}{E} - A_w$$

which, upon eliminating the $2 k A_f$ term from Eq. 6.6.15, gives

$$E \frac{I_e}{I} = \left[\frac{E_t A/E - 2 A_w/3}{2 A_f + A_w/3} \right] E \qquad (6.6.16)$$

Thus

$$F_{cr} = \frac{\pi^2 E k}{(KL/r)^2} \qquad (6.6.17)$$

is the approximate equation using Eq. 6.6.11 for k, or more exactly using Eq. 6.6.16 in Eq. 6.6.7 gives

$$F_{cr} = \frac{\pi^2 E}{(KL/r)^2} \left[\frac{E_t A/E - 2 A_w/3}{2 A_f + A_w/3} \right] \qquad (6.6.18)$$

for buckling with respect to the strong axis (x–x).

From the foregoing development it is apparent that two equations are necessary to properly determine column strength of I-shaped sections: one for strong-axis buckling and one for weak-axis buckling. Although the value I_e/I is not itself a function of the residual stress distribution provided that it satisfies the general geometric requirement as shown in Fig. 6.6.3; nevertheless, the critical stress F_{cr}, computed as the buckling load divided by the gross area, has a relationship with KL/r that does depend on residual stress.

Note that if the stress-strain curve for the material is not elastic-plastic, i.e., if E_t is neither E nor zero, then the more general equation, Eq. 6.6.4, must be used instead of Eq. 6.6.5.

EXAMPLE 6.6.1

Establish the column strength curve (F_{cr} vs KL/r) for weak axis buckling of an H-shaped section of steel having a yield stress of 100 ksi (690 MPa) exhibiting perfect elastic-plastic strength in a coupon test (Fig. 6.6.4b), and having the very simplified residual stress pattern shown in Fig. 6.6.4a. Neglect the contribution of the web.

Figure 6.6.4
Data for Example 6.6.1.

(a) Residual-stress pattern

(b) Coupon stress-strain diagram

Solution:

For any external load the strain on every fiber is the same. Until a fiber reaches the strain ϵ_y at first yield, the applied load is

$$P = \int_A f\, dA = f A$$

After a portion has become plastic, the applied load is

$$P = (A - A_e)F_y + \int_{A_e} f\, dA$$

In this problem for $F_{cr} = P/A \le 2F_y/3$ the entire section remains elastic, $E_t = E$, in which case E' is EI_e/I and $I_e = I$; thus,

$$F_{cr} = \frac{2F_y}{3} = \frac{\pi^2 E}{(KL/r)^2}$$

$$\frac{KL}{r} = \sqrt{\frac{\pi^2(29,000)}{(2/3)(100)}} = 65.5 \quad \text{(point 1, Fig. 6.6.5)}$$

When $F_{cr} = P/A > 2F_y/3$, the flange tips have yielded, making I_e less than I; thus

$$\frac{I_e}{I} = \frac{(b/2)^3}{b^3} = \frac{1}{8}$$

$$F_{cr} = \frac{2F_y}{3} = \frac{\pi^2 E(I_e/I)}{(KL/r)^2} = \frac{\pi^2 E}{8(KL/r)^2}$$

$$\frac{KL}{r} = 23.2 \quad \text{(point 2, Fig. 6.6.5)}$$

Figure 6.6.5
Column strength curve for
Example 6.6.1.

for average stress infinitesimally greater than $2F_y/3$. When $F_{cr} = P/A = F_y$,

$$F_{cr} = F_y = \frac{\pi^2 E}{8(KL/r)^2}$$

$$\frac{KL}{r} = 18.9 \quad (\text{point 3, Fig. 6.6.5})$$

when the total load $P = F_y A$. The results are shown in Fig. 6.6.5. If there had been no residual stress at $F_{cr} = F_y$,

$$\frac{KL}{r} = 53.5 \quad (\text{point 4, Fig. 6.6.5})$$ ∎

EXAMPLE 6.6.2

Establish the column strength curve for the more realistic linear distribution of residual stress shown in Fig. 6.6.6. Consider weak-axis buckling of an H-shaped section of steel for both (a) $F_y = 36$ ksi (250 MPa) and (b) $F_y = 100$ ksi (690 MPa). Neglect the effect of the web.

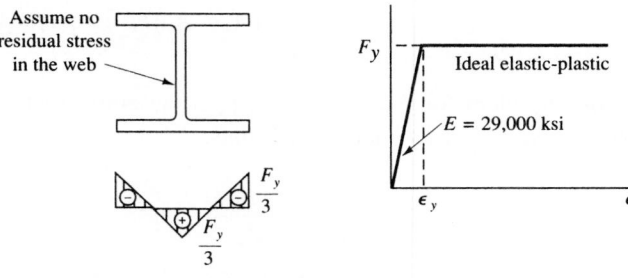

(a) Residual-stress pattern   (b) Coupon stress-strain diagram

Figure 6.6.6
Data for Example 6.6.2.

Solution:
For an average superimposed stress $f = P/A \leq 2F_y/3$, the entire section remains elastic (Fig. 6.6.7a); therefore $E_t = E$, and

$$F_{cr} = \frac{2F_y}{3} = \frac{\pi^2 E}{(KL/r)^2}$$

For an average stress due to applied load greater than $2F_y/3$, part of the cross-section is plastic and part elastic, as in Fig. 6.6.7b. During this stage, the *change in stress is not the same on all fibers,* because the modulus of elasticity is not the same on all fibers.

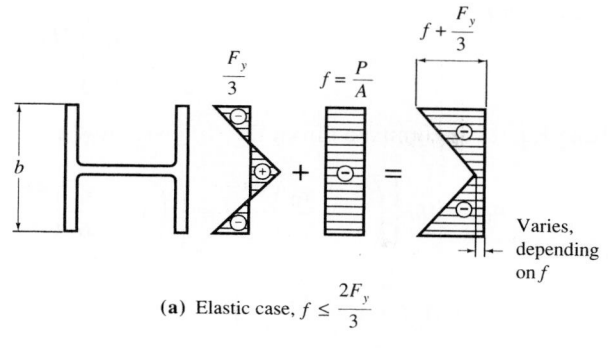

(a) Elastic case, $f \leq \dfrac{2F_y}{3}$

(b) Elasto-plastic case, $f > \dfrac{2F_y}{3}$

f = superimposed stress on elastic fibers

Figure 6.6.7
Stress distribution with linear
residual stress.

$$F_{cr} = \frac{\pi^2 E I_e / I}{(KL/r)^2}$$

$$\frac{I_e}{I} = \frac{2(1/12)(2x_0)^3 t}{2(1/12) b^3 t} = \frac{8 x_0^3}{b^3}$$

neglecting the effect of the web,

$$F_{cr} = \frac{8 \pi^2 E (x_0/b)^3}{(KL/r)^2} \tag{a}$$

which gives F_{cr} as a function of two variables, x_0/b and KL/r. An additional relationship is
required. The total load during the elasto-plastic stage can be expressed

$$P_{cr} = 2 \left[fbt - 2\left(\frac{1}{2}\right)\left(f - \frac{2F_y}{3}\right)\left(\frac{1}{2} - \frac{x_0}{b}\right)bt \right] \tag{b}$$

which is the shaded area of the stress diagram in Fig. 6.6.7b. From similar triangles on the
dotted triangle of Fig. 6.6.7b,

$$\frac{f - 2F_y/3}{\left(\dfrac{1}{2} - \dfrac{x_0}{b}\right)b} = \frac{2F_y/3}{b/2}$$

Solving for f,

$$f = \left[1 - \frac{x_0}{b}\right]\frac{4F_y}{3} \tag{c}$$

Using Eq. (c) to eliminate f from Eq. (b) gives

$$P_{cr} = 2bt\left\{\left(1 - \frac{x_0}{b}\right)\frac{4F_y}{3} - \left[\left(1 - \frac{x_0}{b}\right)\frac{4F_y}{3} - \frac{2F_y}{3}\right]\left(\frac{1}{2} - \frac{x_0}{b}\right)\right\}$$

$$= A_g F_y\left[1 - \frac{4}{3}\left(\frac{x_0}{b}\right)^2\right] \tag{d}$$

Thus

$$F_{cr} = \frac{P}{A_g} = F_y\left[1 - \frac{4}{3}\left(\frac{x_0}{b}\right)^2\right] \tag{e}$$

which is used in combination with Eq. (a). The results are presented in Fig. 6.6.8.

$\dfrac{x_0}{b}$	F_{cr}	F_{cr} for $F_y = 36$ ksi (ksi)	$\dfrac{KL}{r}$	F_{cr} for $F_y = 100$ ksi (ksi)	$\dfrac{KL}{r}$
0.50	$0.667F_y$	24.0	109.2	66.7	65.4
0.45	0.730	26.3	89.0	73.0	53.4
0.40	0.787	28.3	72.0	78.7	43.1
0.35	0.837	30.2	57.0	83.7	34.2
0.30	0.880	31.7	44.1	88.0	26.5
0.25	0.917	33.0	32.9	91.7	19.7
0.20	0.947	34.1	23.2	94.7	13.9
0.10	0.987	35.5	8.0	98.7	4.8

Figure 6.6.8
Column strength curves showing effect of residual stress ($E = 29,000$ ksi). Solution for Example 6.6.2.

If the web of the section were to be included, I_e/I could easily include the web terms. Furthermore, Eq. (b) could also have included the web terms. Such inclusion of the effect of the web brings in the variable A_w/A_f and in most cases the effect is small.

Finally, curves similar to those of Fig. 6.6.8 can be obtained by using an average stress-strain curve for a short length of rolled shape as referred to earlier in this section, in which case Eqs. 6.6.12 and 6.6.18 can be used with the E_t obtained from the "cross-section" stress-strain curve.   ■

6.7   STRUCTURAL STABILITY RESEARCH COUNCIL (SSRC) STRENGTH CURVES

Parabolic Equation—Basis for Allowable Stress Design (1960–2004)

Based upon the methods discussed in Sec. 6.6, column strength curves can be obtained for weak- or strong-axis buckling with various distributions of residual stress. For most practical situations it has been reported that an assumed linear distribution of residual stress in the flanges results in a reasonable average column curve [6.20]. Furthermore, the development in the previous section (Eqs. 6.6.12 and 6.6.17) shows that for the *same* slenderness ratio, H-shaped column sections allowed to bend in the weak direction can carry less load than columns permitted to bend only in the strong direction. Compressive residual stress which is greatest at the flange tips accounts for this strength difference.

Typical column strength curves for parabolic and linear distribution of residual stress across the flange are shown in Fig. 6.7.1. For structural carbon steels, the average value of

Figure 6.7.1
Column strength curves for H-shaped sections having compressive residual stress at flange tips. (Adapted from Ref. 6.20, p. 39)

$$\lambda_c = \frac{KL}{r}\sqrt{\frac{F_y}{\pi^2 E}}$$

the maximum compressive residual stress is 12 to 13 ksi (83 to 90 MPa), corresponding roughly to $0.3F_y$. For the high-strength steels residual stress will generally be a lower fraction of the yield stress.

Since 1960, and until the AISC 2005 Specification structural steel design according to the AISC Allowable Stress Design Specification has used the SSRC parabolic equation based on the one proposed by Bleich [6.9]. The SSRC parabolic curve is

$$F_{cr} = F_y\left[1 - \frac{F_y}{4\pi^2 E}\left(\frac{KL}{r}\right)^2\right] \tag{6.7.1}$$

Equation 6.7.1 is compared in Fig. 6.7.1 with other curves that distinguish between residual stress patterns and axes of bending. The SSRC curve gives fairly good agreement with the weak axis curve for H-shaped sections, particularly when it is noted that parabolic residual stress is more representative of the actual situation than linear residual stress.

From Fig. 6.7.1 one may note that to provide the same degree of safety for all columns, different strength curves would be required depending on the expected residual stress distribution, the shape of the section, and the axis of bending when the column buckles.

Note is made that Fig. 6.7.1 introduces the slenderness function λ_c, which was used as the slenderness parameter (instead of KL/r) in all previous editions of the AISC LRFD Specification. The 2005 *AISC Specification* replaces λ_c by KL/r. The slenderness parameter λ_c is defined as

$$\lambda_c^2 = \frac{F_y}{F_{cr}(\text{Euler})} = \frac{F_y}{\dfrac{\pi^2 E}{(KL/r)^2}} \tag{6.7.2}$$

$$\lambda_c = \frac{KL}{r}\sqrt{\frac{F_y}{\pi^2 E}} \tag{6.7.3}$$

In terms of the slenderness parameter λ_c, the SSRC parabola becomes

$$\frac{F_{cr}}{F_y} = 1 - \frac{\lambda_c^2}{4} \qquad \text{for } \lambda_c \leq \sqrt{2} \tag{6.7.4}$$

Note that $\lambda_c = \sqrt{2}$ when the parabola and the Euler hyperbola become tangent to each other. Thus Eq. 6.7.4 applies for $\lambda_c \leq \sqrt{2}$; for greater values of λ_c, the Euler equation applies,

$$\frac{F_{cr}}{F_y} = \frac{1}{\lambda_c^2} \qquad \text{for } \lambda \geq \sqrt{2} \tag{6.7.5}$$

Strength Equation—Basis for AISC LRFD and ASD Design Methods (2005)

The parabolic equation, Eq. 6.7.4, provides a reasonable approximation for a column strength curve that provides a transition between elastic buckling and yielding, reflecting essentially the effect of residual stress. Traditionally, accidental eccentricity and initial crookedness were accounted for by using as increased safety factor as the slenderness increased. Load and resistance factor design philosophy provided for a constant margin of safety for *all* columns. If the strength truly varies with slenderness, then the nominal strength P_n should account for it.

Bjorhovde, as reported in the *SSRC Guide* [6.8], showed that three column strength curves would be sufficient to approximate the strength for all practical steel shapes. These are referred to as SSRC Curves 1, 2, and 3 and the details are to be found in the *SSRC Guide* [6.8, pp. 52–61].

In the development of the 2005 AISC Specification, the specification committee decided to continue using only one column strength curve for steel design. Thus, the same equation used in the 1999 LRFD Specification was adopted for the 2005 AISC Specification. The equation fits the SSRC Curve 2 modified to reflect an initial out-of-straightness of about 1/1500. However, the slenderness parameter λ_c was replaced by the more familiar slenderness ratio *KL/r*.

The nominal strength P_n of rolled shape compression members (AISC-E3) is given by

$$P_n = F_{cr}A_g \tag{6.7.6}$$

1. For $\dfrac{KL}{r} \leq 4.71\sqrt{\dfrac{E}{F_y}}$ or $F_e \geq 0.44F_y$

$$F_{cr} = \left[0.658^{\frac{F_y}{F_e}} \right] F_y \tag{6.7.7}$$

2. For $\dfrac{KL}{r} > 4.71\sqrt{\dfrac{E}{F_y}}$ or $F_e < 0.44F_y$

$$F_{cr} = 0.877F_e \tag{6.7.8}$$

where F_e is the elastic buckling stress (Euler equation, Eq. 6.2.8), as follows:

$$F_e = F_{cr} = \frac{\pi^2 E}{\left(\dfrac{KL}{r}\right)^2} \tag{6.7.9}$$

The establishment of an acceptable *single* equation of the critical stress F_{cr} has been the subject of some controversy since, as has been shown, the shape of the cross-section and the method of manufacture (i.e., hot-rolling or welding) influence the strength. Furthermore, residual stress and out-of-straightness are significant influencing parameters but are not quantifiable. Another major factor affecting column strength is end restraint, particularly in situations where the joints are *not* rigid; Bjorhovde [6.21] has provided extensive treatment relating to practical design situations. Bjorhovde [6.22, 6.23] also has reviewed the entire subject in the context of load and resistance factor design.

More recently, Hall [6.24] has compared data from physical tests and has presented statistically developed expressions for F_{cr}. Though the actual AISC equations are not referred to by Hall, Fig. 6.7.2 is presented here to compare the experimental data [6.24] with the AISC equation for F_{cr}. Other recent proposals for column strength curves have been given by Rondal and Maquoi [6.26] and Rotter [6.25], as well as the three SSRC Curves 1, 2, and 3 mentioned previously [6.8]. Lui and Chen [6.27] have discussed the design of columns with imperfections using a beam-column (see Chap. 12) approach.

Columns Having Varying Axial Load, Stepped Columns, and Tapered Columns

Special treatment is required when the axial load varies along the length and/or the member is not prismatic over the entire column length. The reader is referred to Anderson and Woodward [6.28] and Castiglioni [6.29] for stepped columns, Shrivastava [6.30] for

Figure 6.7.2
Comparison of AISC equations for F_{cr} for columns with data from physical tests. (Test data from Hall [6.24])

columns having varying axial load, Sandhu [6.31] for columns with an intermediate axial load, and Ermopoulos [6.32] for tapered bars under stepped axial loads.

6.8   LOAD AND RESISTANCE FACTOR DESIGN

The strength requirement in load and resistance factor design according to AISC-E3 may be stated

$$\phi_c P_n \geq P_u \qquad (6.8.1)$$

where ϕ_c = resistance factor = 0.90
   P_n = nominal strength = $F_{cr} A_g$
   F_{cr} = Eqs. 6.7.7 and 6.7.8
   P_u = factored service load (see Sec. 1.9)

   Eqs. 6.7.7 and 6.7.8 for F_{cr} are applicable in design of ordinary rolled H-shaped section columns; however when thin-walled plate elements are used in the cross-section AISC-E3 and AISC-E7 provide for using a reduced efficiency of the section. When a thin element exhibits instability (local buckling) such an element does not carry its proportionate share of the load. This may occur when the element width/thickness ratio exceeds the λ_r values given in AISC Table B4.1.

   AISC-E7 introduces a reduction factor Q into Eqs. 6.7.7 and 6.7.8 when the limits λ_r of AISC Table B4.1 are exceeded. Thus, when local buckling of one or more plate components of the cross-section may occur prior to achieving the overall buckling strength of the member, the F_y in Eqs. 6.7.7 and 6.7.8 becomes QF_y, giving for the critical stress, the following:

   1. For $\dfrac{KL}{r} \leq 4.71\sqrt{\dfrac{E}{QF_y}}$ or $F_e \geq 0.44 QF_y$

$$F_{cr} = \left[0.658^{\frac{QF_y}{F_e}} \right] QF_y \qquad (6.8.2)$$

Figure 6.8.1
Critical column stress F_{cr} vs KL/r according to Load and Resistance Factor Design, for various yield stresses.

2. For $\dfrac{KL}{r} > 4.71\sqrt{\dfrac{E}{QF_y}}$ or $F_e < 0.44QF_y$

$$F_{cr} = 0.877F_e \tag{6.8.3}$$

Since $Q = 1$ for the overwhelming majority of rolled H-shaped sections (standard W, S, and M shapes) listed in the *AISC Manual* Column Load Tables for $F_y = 50$ ksi, the development of the logic behind the use of the Q factor is reserved for Sec. 6.18 in Part II on plate buckling.

The critical stress F_{cr} equations, Eqs. 6.7.7 and 6.7.8, are shown in Fig. 6.8.1 for $F_y = 36, 50, 60, 70,$ and 100 ksi. *AISC Specification* section references for axial compression are given in Table 6.8.1.

Singly Symmetric Double Angles and Tees

Hot-rolled double angles and tee sections usually buckle in the flexural mode based on KL/r with respect to the *x*- or *y*-axis. However, since the shear center (see Sec. 8.4) does not coincide with the centroid of the section, a torsional effect is possible. Though the subject of torsion is not treated until Chapter 8, the reader should note that the *AISC Manual* tables "Available Strength in Axial Compression" giving design axial strength $\phi_c P_n$ for double angles and structural tees include the effect of the flexural-torsional buckling limit state in the calculation of F_{cr}. Since torsional stiffness is significantly related to the cube of the thickness of the elements (flanges and web) this effect will only be significant on sections having very thin components, such as light-gage sections. The double-angle member has

TABLE 6.8.1 Axial Compression—AISC Specification References

Topic	Specification sections AISC [1.13]
Local buckling limits for "noncompact" sections	B4
Local buckling limits for "compact" sections	B4
Slenderness limits	E2
Moment resisting frame, definition	C1.3a
Unbraced frame, definition	C1.3b
Effective length factors	C2
Column formulas, basic	E3
Torsional and flexural torsional buckling	E4
Single angles	E5
Built-up members	E6
Slender compression elements	E7
Alignment chart	Commentary C2

been specifically treated by Astaneh, Goel, and Hanson [6.40], Zahn and Haaijer [6.41], and Galambos [6.87], and the effect of balanced welded connections has been treated by Temple and Sakla [5.43].

Built-up Sections

The compression member strength of built-up sections is affected by the shear transfer strength of the fasteners attaching the elements together, and the slip-resistance of these connectors. A double angle compression chord member of a truss can behave as a single element compression member only when the two angles are adequately attached to each other so that when the flexural buckling limit state is reached there will be no relative axial movement (shear deformation) of one angle relative to another. The strength of built-up sections as affected by fastener strength, spacing, and installation (slip-resistance) of bolts has been studied by Libove [6.39] and Aslani and Goel [6.89, 6.95], and design rules are given by Duan and Chen [6.82].

Column Strength as Affected by Connector Spacing

When the controlling buckling mode "involves relative deformation that produces shear forces in the connectors between individual shapes," a modified slenderness ratio $(KL/r)_m$ is used. For the double angle section, this will be the case when the controlling flexural buckling occurs in the plane parallel to the outstanding legs (y-axis for *AISC Manual* properties). When the controlling flexural buckling limit state is based on slenderness KL/r in the plane parallel to the back-to-back legs (x-axis for *AISC Manual* properties), the angles will move parallel to each other and the connectors will not affect the compression strength.

The modified slenderness ratio $(KL/r)_m$ is also dependent on the slip-resistance of the connections. AISC-E6 provides:

1. For *snug-tight bolted connections:*

$$\left(\frac{KL}{r}\right)_m = \sqrt{\left(\frac{KL}{r}\right)_0^2 + \left(\frac{a}{r_i}\right)^2} \tag{6.8.4}$$

2. For *welded connectors* and for *fully-tightened bolted connections:*

$$\left(\frac{KL}{r}\right)_m = \sqrt{\left(\frac{KL}{r}\right)_0^2 + 0.82\frac{\alpha^2}{(1+\alpha^2)}\left(\frac{a}{r_{ib}}\right)^2} \tag{6.8.5}$$

where
$\left(\dfrac{KL}{r}\right)_0$ = slenderness ratio of the built-up member acting as a unit (for the y-axis in double angle members)

$\dfrac{a}{r_i}$ = largest slenderness ratio of an individual component (for the z-axis of a single angle for the double angle member)

$\left(\dfrac{KL}{r}\right)_m$ = modified (increased) slenderness ratio based on the connectors (for double angle members this replaces KL/r based on y-axis)

a = distance between connectors measured along member length

r_i = minimum radius of gyration of individual component (the z-axis for an angle)

r_{ib} = radius of gyration of individual component relative to its centroidal axis parallel to member axis of buckling (for double angle members this is r_y)

$\alpha = h/(2r_{ib})$ = separation ratio

h = distance between centroids of individual components perpendicular to the member axis of buckling

$\dfrac{a}{r_{ib}}$ = column slenderness ratio of individual component relative to its centroidal axis parallel to member axis of buckling

The application of the modified slenderness ratio to the double angle member applies only when $(KL/r)_y$ exceeds $(KL/r)_x$; however, the modified ratio should also be checked when $(KL/r)_x$ is the larger by only a small amount. In effect, the modified value *always* replaces the $(KL/r)_y$ for the double angle member.

Tubular Sections

The formulas for F_{cr} are applicable for HSS sections and other tubular sections. Additional information on tubular compression members is available in the work of Sherman [6.14, 6.15, 6.33], as well as Snyder and Lee [6.34, 6.35], Chen and Ross [6.36], and Ross, Chen, and Tall [6.37]. For round columns, see Galambos [6.38].

Single Angle Sections

The strength of single-angle struts (i.e. compression members) has received considerable attention in recent years. For many years AISC recommended against using single-angle compression members because of their having the torsional (see Chap. 8) or flexural-torsional (see Chap. 9) limit states for strength.

The nominal strength P_n in compression is determined in accordance with AISC-E3 or AISC-E7, and then the slenderness modifications of AISC-E5 are applied if necessary.

For the design of single-angle compression members, AISC-E5 allows for the effect of eccentricity to be neglected when the members are loaded through one leg (the same one leg), members are attached by a minimum of two-bolt connections or welding, and there are no intermediate transverse loads.

The effective slenderness ratio is calculated as follows:

1. For equal-leg angles and unequal-leg angles connected through the longer leg that are *individual members or web members of a planar truss* with adjacent web members attached to the same side of the gusset plate:

a. When $0 \leq \dfrac{L}{r_x} \leq 80$,

$$\frac{KL}{r} = 72 + 0.75\frac{L}{r_x} \tag{6.8.6}$$

b. When $\dfrac{L}{r_x} \geq 80$,

$$\frac{KL}{r} = 32 + 1.25\frac{L}{r_x} \leq 200 \tag{6.8.7}$$

where L = length of member between work points at truss chord centerlines

r_x = radius of gyration about geometric axis parallel to connected leg

r_y = radius of gyration about the minor principal axis

When the unequal legs length ratio is less than 1.7 and connected through the shorter leg, the KL/r value, as computed in Eqs. 6.8.6 or 6.8.7, shall be increased by adding the following term,

$$4\left[(b_\ell/b_s)^2 - 1\right] \tag{6.8.8}$$

where b_ℓ and b_s are the lengths of the longer and shorter leg, respectively, of the angle.

The minimum KL/r value, after adjustment, is

$$\frac{KL}{r} \geq 0.95\frac{L}{r_x} \tag{6.8.9}$$

2. For equal-leg angles and unequal-leg angles, connected through the longer leg, that are web members of *box or space trusses* with adjacent web members attached to the same side of the gusset plate or chord:

a. When $0 \leq \dfrac{L}{r_x} \leq 75$,

$$\frac{KL}{r} = 60 + 0.8\frac{L}{r_x} \tag{6.8.10}$$

b. When $\dfrac{L}{r_x} \geq 75$,

$$\frac{KL}{r} = 45 + \frac{L}{r_x} \leq 200 \tag{6.8.11}$$

where the definition of symbols is given following Eq. 6.8.7.

When the unequal legs length ratio is less than 1.7 and connected through the shorter leg, the KL/r value, as computed in Eqs. 6.8.10 or 6.8.11, shall be increased by adding the following term,

$$6\left[(b_\ell/b_s)^2 - 1\right] \qquad (6.8.12)$$

where b_ℓ and b_s are the lengths of the longer and shorter leg, respectively, of the angle.

The minimum KL/r value, after adjustment, is

$$\frac{KL}{r} \geq 0.82\frac{L}{r_x} \qquad (6.8.13)$$

3. Single angles having other end conditions, with leg length ratios larger than 1.7, or with transverse loading, must be evaluated using AISC-Chapter H.

Many researchers contributed to development of the single angle research, including Kennedy and Murty [6.42, 6.43], Woolcock and Kitipornchai [6.44], El-Tayem and Goel [6.45, 6.46], Chuenmei [6.47], Galambos [6.87], Elgaaly, Dagher, and Davids [6.90, 6.91], Adluri and Madugula [6.92], Zureick [6.96], and Bathon, Mueller, and Kempner [6.97].

Other Non-Symmetrical Sections

Sections such as cruciform (see Smith [6.48]) and Z-sections require special consideration using the provisions of AISC-E4. The strength of built-up sections connected intermittently by fasteners is affected by such fastener spacing as discussed above.

6.9   EFFECTIVE LENGTH

Discussion of column strength to this point has assumed hinged ends where no moment rotational restraint exists. Zero moment restraint at the ends constitutes the weakest situation for compression members having no transverse movement of one end relative to the other. For such pinned-end columns the equivalent pinned-end length KL is the actual length L; thus $K = 1.0$ as shown in Fig. 6.9.1a. The equivalent pinned-end length is referred to as the *effective length*.

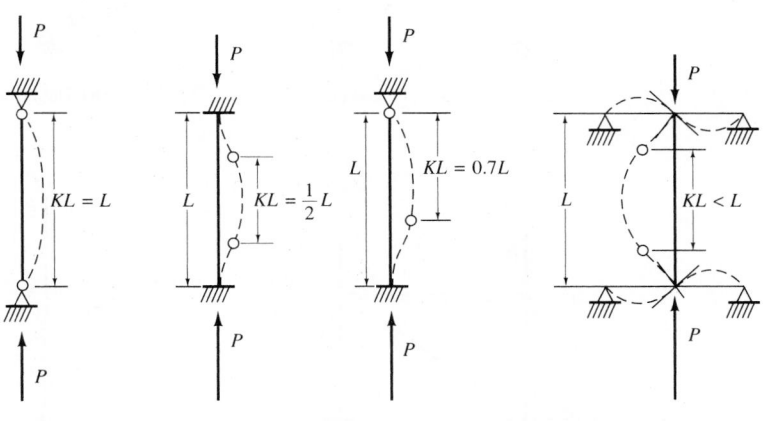

Figure 6.9.1
Effective length KL; no joint translation.

(a) End rotations unrestrained

(b) End rotations fully restrained

(c) One end restrained, other unrestrained

(d) Partially restrained at each end

For most real situations moment restraint at the ends does exist causing the points of zero moment (inflection points) to move away from the restrained ends as shown in Fig. 6.9.1b, c, and d where the effective lengths KL then are reduced.

In many situations it is difficult, or perhaps impossible, to adequately evaluate the degree of moment restraint contributed by adjacent members framing into a column, by a footing and soil under it, and indeed the full interaction of all members of a steel frame.

Whether or not the degree of end restraint can be ascertained accurately, the designer must understand the concepts of *braced frame* and *unbraced frame.* A more extensive treatment of frames is given in Chap. 14.

Braced Frame

A *braced frame,* according to AISC-C1.3a, is one in which "lateral stability is provided by diagonal bracing, shear walls or equivalent means." The vertical bracing system must be "adequate" as determined by structural analysis "... to prevent buckling of the structure and to maintain the lateral stability of the structure, including the overturning effects of drift, under the factored loads...." Note that a vertical column in a braced frame would have no sideways movement of its top relative to its bottom.

Figure 6.9.1 illustrates effective lengths for columns in a braced frame. Once it has been determined that a frame is braced, the bracing is presumed to provide any needed lateral restraint, as in Fig. 6.9.2a and c; therefore, the joints are assumed not to move laterally (at least in first order structural analysis) and an individual column may be designed as if isolated once the effective length factor K has been determined.

From Figs. 6.9.1 and 6.9.2a and c, one may observe that end restraint in braced frames always reduces the distance between inflection points; that is, reduces the effective length KL from the pinned-end condition. The effective length factor K will always be less than unity.

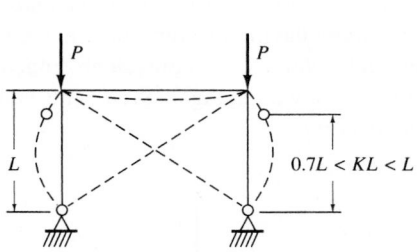

(a) Braced frame, hinged base

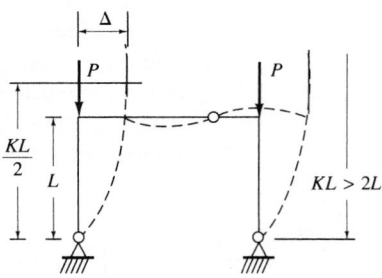

(b) Unbraced frame, hinged base

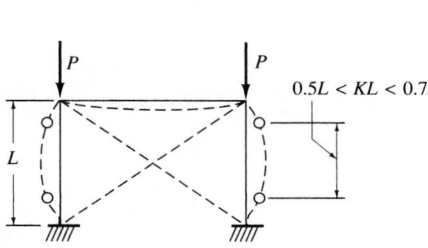

(c) Braced frame, fixed base

(d) Unbraced frame, fixed base

Figure 6.9.2
Effective length KL for frames.

Unbraced Frame

An *unbraced frame,* according to AISC-C1.3b, is one in which "lateral stability depends upon the bending stiffness of rigidly connected beams and columns." The buckling of an unbraced frame is one of sidesway where, for example, the top of a column moves to the side relative to the bottom. In Fig. 6.9.2b and d an unbraced frame is shown having sidesway buckling. The buckled shape and therefore the effective length of the columns will depend on the stiffnesses of the participating members in flexure. The effective length KL may be obtained by matching the buckled shape of a column with a portion of the pinned-end column buckled shape. As shown in Fig. 6.9.2, KL will always exceed L.

To understand why the *minimum* value of K in an unbraced frame is theoretically 1.0, examine the rectangular frame of Fig. 6.9.2d. The stiffest situation would be when the beam is infinitely stiff, that is, it cannot bend. The inflection point would then be at mid-height and the buckled shape would be as in Fig. 6.9.3a.

The practical situation in an unbraced frame is that K is always *greater than unity.* Furthermore, there is no simple way of obtaining a value other than evaluating the end restraint. AISC-C1.3a requires that K "shall be determined according to the methods of analysis suggested in AISC-C2-Commentary."

Alignment Charts for Evaluating Effective Length Factor K

For ordinary design, it is entirely impractical to analyze an entire frame to determine its buckling strength and the effective lengths for the members.

Various investigators have provided charts to permit easy determination of frame buckling loads and effective lengths for commonly encountered situations. Effective length factors K are given by Anderson and Woodward [6.28] for stepped columns, Sandhu [6.31] for columns having an intermediate axial load, Lu [6.49] for gabled frames, Fraser [6.84] for pin-based crane columns, Stoman [6.85] for cross bracings, Rutenberg and Scarlat [6.86] for columns in one-story buildings, and Hassan [6.50] for one-story, one-bay frames, having vertical loads applied to the columns at an intermediate point in addition to the load at the top. Galambos [6.51] has presented them for one- and

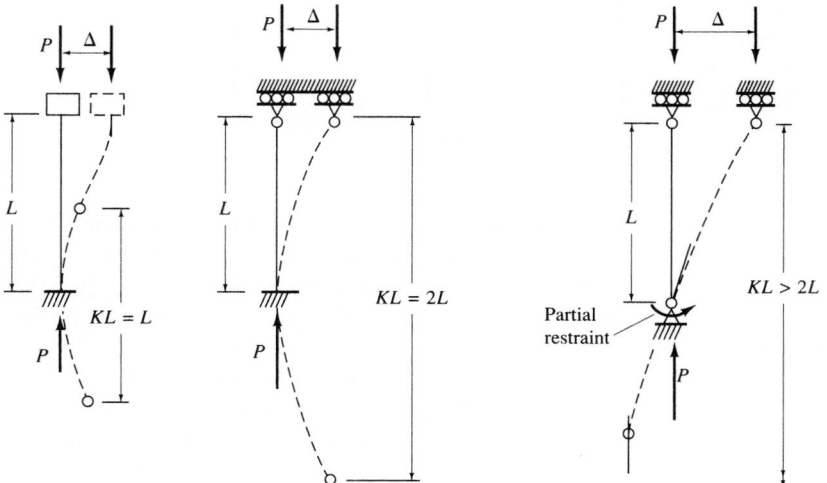

Figure 6.9.3
Effective length *KL*; joint translation possible.

(**a**) End rotation
fully restrained

(**b**) One end rotation fully
restrained, other
unrestrained

(**c**) One end rotation partially restrained,
other end unrestrained

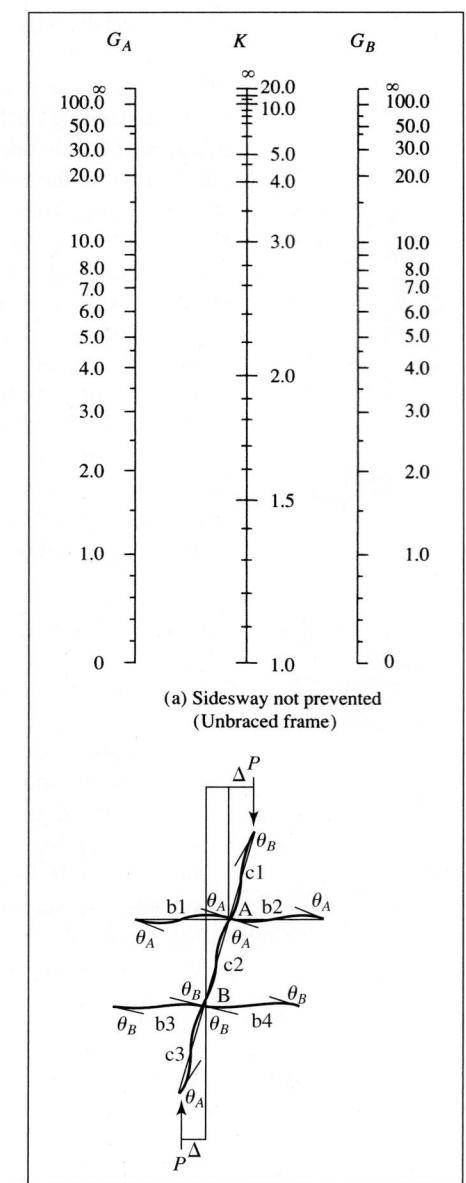

Figure 6.9.4
Alignment Charts for
effective length of columns
in continuous frames
[6.54 or *AISC Manual*
pp. 16.1-241, 242], where

$$G = \frac{\Sigma I/L, \text{ columns}}{\Sigma I/L, \text{ girders}}$$

two-story, one-bay-wide frames, and Gurfinkel and Robinson [6.52] have given *K* values for the general case of an elastic rotationally restrained column (both with and without sidesway elastic restraint). Switzky and Wang [6.53] have summarized buckling load data from which effective lengths can be obtained for frames one-bay wide and up to five stories high.

The most commonly used procedure for obtaining effective length is to use the alignment charts originally developed by O. J. Julian and L. S. Lawrence, and presented in detail by T. C. Kavanagh [6.54].[*] The alignment chart method using Fig. 6.9.4 is also

---

[*]The derivation was presented in detail in the 2nd edition (1980) of this text (pp. 843–851).

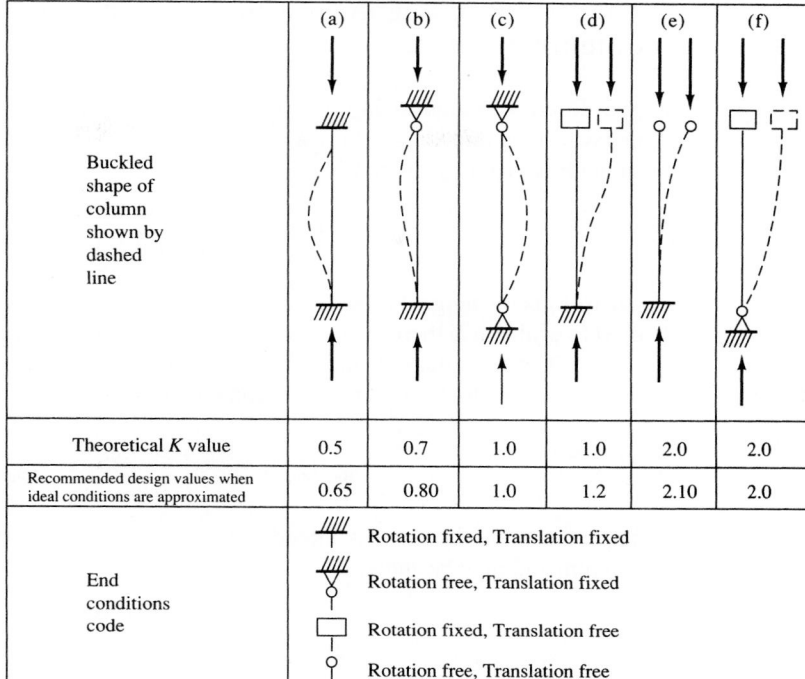

	(a)	(b)	(c)	(d)	(e)	(f)
Buckled shape of column shown by dashed line						
Theoretical *K* value	0.5	0.7	1.0	1.0	2.0	2.0
Recommended design values when ideal conditions are approximated	0.65	0.80	1.0	1.2	2.10	2.0

End conditions code

Rotation fixed, Translation fixed

Rotation free, Translation fixed

Rotation fixed, Translation free

Rotation free, Translation free

Figure 6.9.5
Effective length factors for centrally loaded columns having various idealized end conditions. (Adapted from Ref. 6.8, p. 49)

suggested by the AISC Commentary as satisfying the "analysis" requirement of AISC-C2.2 to get "adequate" *K* values. Dumonteil [6.93] has presented equations equivalent to the alignment charts, and Lui [6.94] has given a simplified formula alternative.

It is emphasized that these alignment charts are based on idealized conditions which seldom exist in real structures. The alignment charts were based on the deflected shapes of the sub assemblages shown in Fig. 6.9.4, in addition to a list of nine conditions, presented in the AISC Commentary-C2.2b. For different end beam conditions, AISC provides a set of girder stiffness multipliers that could be used for improved accuracy in the value of *K*. The development of the alignment charts is primarily based on the assumption that all beam-column connections are fully restrained. With complex framing schemes, the limitations of the alignment charts to model the stability of frames becomes more prevalent. In such situations, AISC Commentary-C2.2b provides alternative approaches for the determination of the effective length. However these methods can be tedious and impractical to apply in routine design. AISC 2005 includes an alternative improved method, called the Direct Analysis Method, that determines the required strength more accurately and allows the use of *K* = 1 in the compressive strength equations. This new method accounts explicitly for imperfections and inelasticity, prior to buckling, by using reduced stiffness to account for the spread of yielding and by using notional lateral loads to account for member imperfections and out-of plumbness. The method is presented in AISC Appendix 7, and is discussed in more detail in Chapters 12 and 14.

For simple situations, one may use Fig. 6.9.5 from the *SSRC Guide* [6.8]. For braced frames it is always conservative to take the *K* factor as unity, and some interpolation is possible from Fig. 6.9.5. For unbraced frames, except perhaps for the flagpole-type situation of Fig. 6.9.5, case (e), an arbitrary selection of *K* is not satisfactory for design.

Adjustment of Alignment Chart K Factors for Inelastic Column Behavior

The design of columns in ordinary construction involves slenderness ratios KL/r in the inelastic buckling range (i.e., $\lambda_c < 1.5$ or $F_{cr} >$ about $0.5F_y$). The inelastic buckling strength is given by Eq. 6.3.1,

$$F_{cr} = \frac{\pi^2 E_t}{(KL/r)^2} \qquad [6.3.1]$$

where E_t is the tangent modulus of elasticity at stress P_{cr}/A_g. When the alignment chart is used to evaluate K there is an implicit assumption that elastic buckling controls.

When the column is inelastic and the beam is elastic, an adjustment may be made in the restraint factor G used in the alignment chart. The G factor would then become

$$G_{inelastic} = \frac{\Sigma(E_t I/L)_{col}}{\Sigma(EI/L)_{beam}} = G_{elastic}\, \tau_a \qquad (6.9.1)$$

The adjustment factor τ_a is intended to account for the reduced stiffness of the inelastic column, and may be stated as

$$\tau_a = \frac{E_t}{E} = \frac{F_{cr,\,inelastic}}{F_{cr,\,elastic}} \qquad (6.9.2)$$

Equation 6.9.2 can be expressed in several ways; in the fourth edition of the book, the authors used a different expression than AISC, using

$$\lambda_c = \frac{F_y}{F_{cr}(\text{Euler})} \qquad (6.9.3)$$

as the parameter to determine the reduction factor τ_a. In the 2005 *AISC Manual* [1.15], the factored service load P_u divided by the gross area A_g is used as the main parameter (along with F_y) in AISC Table 4-21, "Stiffness Reduction Factor" on p. 4-317 (see Fig. 6.9.1).

The factor τ_a is intended to account for "a reduction in the stiffness of columns due to geometric imperfections and spread of plasticity from residual stresses under high compression loading" [1.15, p. 16.1-247].

The procedure of using Eq. 6.9.2 and various simplifications for practical use have been discussed by Yura [6.55]. This provoked considerable discussion by Adams [6.56], Johnston [6.57], Disque [6.58], Smith [6.59], Matz [6.60], and Stockwell [6.61]. The AISC-Commentary C2 endorses Yura's procedure as modified by Disque [6.58].

Values of τ_a (Eq. 6.9.1) have been tabulated for steel of Grades 36, 42, and 50 in Table 6.9.1. The *AISC Manual* [1.15] calls these "Stiffness Reduction Factors". Harichandran [6.88] has summarized the procedure, and noted how easily mistakes can be made.

For truss compression members, end restraint may be present and joint translation is prevented so that K might logically be less than 1.0. Under static loading, stresses in all the members remain in the same proportion to one another for various loads. If all members are designed for minimum weight, they will achieve ultimate capacity simultaneously under live load. Thus restraint offered by members framing at a joint disappears or at least is greatly reduced. The SSRC, therefore, recommends using $K = 1.0$ for members of a truss designed for fixed-position loading. When designing for moving load systems on trusses, K can be reduced to 0.85 because conditions causing maximum stress in the member under consideration will not cause maximum stress in the members framing in to provide restraint [6.8].

TABLE 6.9.1 Adjustment τ_a of Restraint Factor G to Account for Inelastic Buckling (Adapted from Ref. 1.15, Table 4-21, p.4-317)

		τ_a = Ratio of the Tangent Modulus E_t to the Elastic Modulus E					
ASD	LRFD	F_y, ksi					
$\dfrac{P_a}{A_g}$	$\dfrac{P_u}{A_g}$	36		42		50	
		ASD	LRFD	ASD	LRFD	ASD	LRFD
45		—	—	—	—	—	—
44		—	—	—	—	—	0.0599
43		—	—	—	—	—	0.118
42		—	—	—	—	—	0.175
41		—	—	—	—	—	0.231
40		—	—	—	—	—	0.285
39		—	—	—	—	—	0.338
38		—	—	—	—	—	0.389
37		—	—	—	0.0570	—	0.438
36		—	—	—	0.127	—	0.486
35		—	—	—	0.194	—	0.532
34		—	—	—	0.260	—	0.577
33		—	—	—	0.323	—	0.620
32		—	0.0334	—	0.384	—	0.660
31		—	0.115	—	0.443	—	0.699
30		—	0.194	—	0.500	—	0.736
29		—	0.270	—	0.554	0.0842	0.771
28		—	0.344	—	0.606	0.171	0.804
27		—	0.414	—	0.655	0.254	0.835
26		—	0.481	—	0.701	0.334	0.863
25		—	0.545	0.0162	0.745	0.410	0.890
24		—	0.606	0.122	0.786	0.483	0.913
23		—	0.663	0.223	0.823	0.552	0.934
22		—	0.716	0.319	0.858	0.617	0.953
21		0.0695	0.766	0.410	0.890	0.678	0.969
20		0.189	0.811	0.496	0.917	0.734	0.982
19		0.303	0.853	0.577	0.942	0.786	0.992
18		0.410	0.890	0.652	0.962	0.833	0.998
17		0.510	0.922	0.721	0.979	0.875	1.00
16		0.603	0.949	0.784	0.991	0.912	
15		0.687	0.971	0.840	0.999	0.943	
14		0.704	0.988	0.888	1.00	0.968	
13		0.831	0.998	0.929		0.987	
12		0.888	1.00	0.962		0.998	
11		0.935		0.985		1.00	
10				0.999			
9				1.00			
8							
7							
6							
5		↓	↓	↓	↓	↓	↓

— Indicates stiffness reduction factor is not applicable because the required strength exceeds the available strength for $K/r = 0$.

When the adjacent members framing at the ends of a column are heavily loaded compression members, they may have a destabilizing effect instead of stabilizing the member being considered: in effect a *negative G*-factor. Bridge and Fraser [6.81] have presented a procedure to account for such negative *G*-factors.

The application of the *K* factor approach to the unbraced frame containing a "leaner" column is presented in Sec. 14.5.

6.10 LOAD AND RESISTANCE FACTOR DESIGN OF ROLLED SHAPES (W, S, AND M) SUBJECT TO AXIAL COMPRESSION

All examples of this section use W shapes that satisfy the $b_f/2t_f$ and d/t_w limits to preclude local buckling; thus, $Q = 1.0$. Section 6.18 treats the subject when $Q < 1.0$.

General Procedure

Whether Load and Resistance Factor Design or Allowable Strength Design is used, the strength of a compression member is based on its gross area A_g. The strength is always a function of the effective slenderness ratio KL/r, and for short columns the yield stress F_y of the steel. Since the radius of gyration r depends on the section selected, the design of compression members is an indirect process unless column load tables are available. The general procedure to satisfy Eq. 6.8.1 is:

1. Compute the factored service load P_u using all appropriate load combinations, as discussed in Sec. 1.8.
2. Assume a critical stress F_{cr} based on an assumed KL/r.
3. Compute the gross area A_g required from $P_u/(\phi_c F_{cr})$.
4. Select a section. Note that the width/thickness λ_r limitations of AISC-Table B4.1 to prevent local buckling must be satisfied. This is discussed in Part II of this chapter, particularly Sec. 6.16.
5. Based on the larger of $(KL/r)_x$ or $(KL/r)_y$ for the section selected, compute the critical stress F_{cr}.
6. Compute the design strength $\phi_c P_n = \phi_c F_{cr} A_g$ for the section.
7. Compare $\phi_c P_n$ with P_u. When the strength provided exceeds the strength required (and is reasonably close) the design would be acceptable, otherwise repeat Steps 2 through 7.

EXAMPLE 6.10.1

Select the lightest W section of A992 steel to serve as a pinned-end main member column 16 ft long to carry an axial compression load of 115 kips dead load and 125 kips live load in a braced structure, as shown in Fig. 6.10.1. Use Load and Resistance Factor Design and indicate the first three choices.

Solution:

(a) Obtain factored loads.

$$P_u = 1.2D + 1.6L = 1.2(115) + 1.6(125) = 338 \text{ kips}$$

(b) Estimate slenderness ratio and obtain estimated F_{cr}. Since the assumption of hinged ends is made, the effective length KL equals the actual length L, i.e., $K = 1.0$. Considering $KL = 16$ ft as a moderately long length, the slenderness ratio might be estimated

Figure 6.10.1
Example 6.10.1.

at about 70 to 80. For rolled W shapes contained in the *AISC Manual* Column Load Tables for $F_y = 50$ ksi, $Q = 1.0$, thus the special provisions of AISC-E7 are not involved. From Fig. 6.8.1 one might estimate $F_{cr} \approx 34$ ksi $(KL/r \approx 75)$. Alternatively, AISC Formula (E3-2) or AISC "Available Critical Stress" Table 4-22 could be used. The required area may then be computed.

$$\text{Required } A_g = \frac{P_u}{\phi_c F_{cr}} = \frac{338}{0.90(34)} = 11.1 \text{ sq in.}$$

(c) Select a section. Since buckling in the weak direction [based on $(KL/r)_y$] will control the strength for W shapes when KL is the same with respect to both *x*- and *y*-axes, the lightest sections for this design will be those having the least r_x/r_y. A high r_x indicates excessive strength with respect to the strong axis, which may be utilized only by providing additional bracing in the weak direction.

Using the "Available Strength in Axial Compression" Tables in the *AISC Manual* [1.15], one might select a W8×48, the lightest W8 that has at least the required area. Furthermore, the W8 sections have the lowest r_x/r_y for a given area. Try W8×48 section.

(d) Check the W8×48 section. $A = 14.1$ sq in.

$$\frac{KL}{r_y} = 92; \qquad F_{cr} = 26.8 \text{ ksi}$$

$$\phi_c P_n = \phi_c F_{cr} A_g = 0.90(26.8)14.1 = 340 \text{ kips}$$

$$[\phi_c P_n = 340 \text{ kips}] \geq [P_u = 338 \text{ kips}] \qquad\qquad \textbf{OK}$$

Note that *AISC Manual* "Available Critical Stress" Table 4-22 provides a tabulation of $\phi_c F_{cr}$ for values of *KL/r.*

Since the design strength ϕP_n exceeds the factored load P_u, and since no other section having this area has a lower r_x/r_y, the W8×48 is the lightest section available. Deeper sections will be heavier, as follows:

Section	Area (sq in.)	KL/r_y	F_{cr} (ksi)	$\phi_c P_n$ (kips)	
W8×48	14.1	92	26.8	338	1st choice
W10×49	14.4	76	32.9	428	2nd choice
W12×50	14.7	98	24.8	326	NG 1% short
W14×53	15.6	100	24.1	338	3rd choice

EXAMPLE 6.10.2

Select the lightest W section of A992 steel to serve as a main member 28 ft long to carry an axial compression load of 65 kips dead load and 145 kips live load in a braced structure, as shown in Fig. 6.10.2. The member is assumed pinned at top and bottom and in addition has weak direction support at mid-height. The mid-height brace also prevents twisting of the column at the brace location. Use AISC Load and Resistance Factor Design.

Solution:

(a) Obtain factored loads.

$$P_u = 1.2D + 1.6L = 1.2(65) + 1.6(145) = 310 \text{ kips}$$

(b) Select from AISC "Available Strength in Axial Compression" Tables. The effective length factors K for buckling in either the strong or the weak direction equal unity; i.e., $K_x = K_y = 1.0$. Since the "Available Strength in Axial Compression" Tables are computed assuming $(KL/r)_y$ controls, enter these tables with the effective length $(KL)_y$. Thus enter with

$$P_u = 310 \text{ kips}; \qquad (KL)_y = 14 \text{ ft}$$

Starting with the shallow W8 sections and working toward the deeper sections, find

W8×40	$\phi_c P_n = 322$ kips	$r_x/r_y = 1.73$
W10×45	$\phi_c P_n = 358$ kips	$r_x/r_y = 2.15$
W12×45	$\phi_c P_n = 343$ kips	$r_x/r_y = 2.65$

Since the actual support conditions are such that $(KL)_x = 2(KL)_y$, if $r_x/r_y \geq 2$, weak axis controls and tabular loads give the correct answer. Thus, W10×45 and W12×45 are obviously acceptable.

Since r_x/r_y for W8×40 is less than 2, strong axis bending controls. The strength may be obtained from the tables by entering with the equivalent $(KL)_y$ that

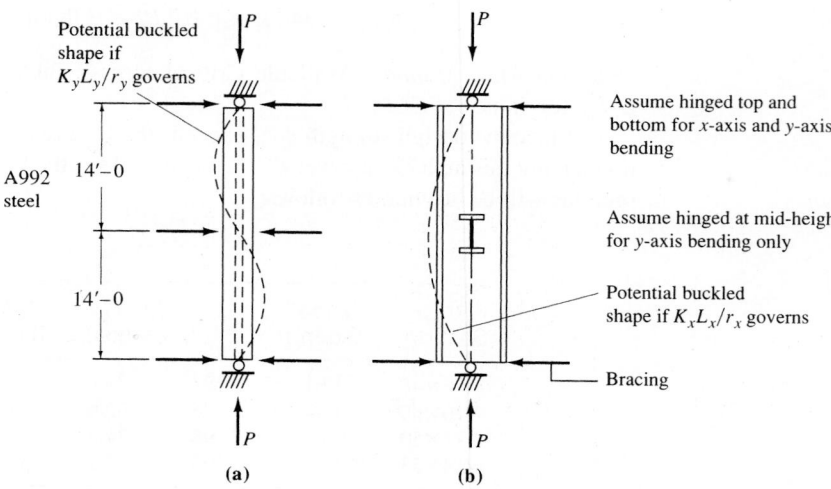

Figure 6.10.2
Example 6.10.2.

corresponds to $(KL)_x = 28$ ft. For equal strength with respect to the x- and y-axes,

$$\frac{(KL)_x}{r_x} = \frac{(KL)_y}{r_y}$$

$$\text{Equivalent } (KL)_y = \frac{(KL)_x}{r_x/r_y} = \frac{28}{1.73} = 16.2 \text{ ft}$$

For $(KL)_y = 16.2$ ft, the W8×40 has a design strength $\phi_c P_n$ of only 272 kips; therefore it is not acceptable.

  (c) Check of section. It is *always* advisable to make a final check of the apparently acceptable section:

$$W10\times45, \quad (KL/r)_y = 14(12)/2.01 = 83.6, \quad \phi_c F_{\text{cr}} = 27.0 \text{ ksi}$$

$$\phi_c P_n = \phi_c F_{\text{cr}} A_g$$

$$= 27(13.3) = 359 \text{ kips} > [P_u = 310 \text{ kips}] \qquad \textbf{OK}$$

Use W10×45. ∎

EXAMPLE 6.10.3

Select the lightest W section of A572 Grade 60 steel to serve as a main member 22 ft long to carry an axial compression load of 100 kips dead load and 200 kips live load in a braced structure, as shown in Fig. 6.10.3. Assume the member hinged at the top and fixed at the bottom for buckling in either principal direction. Use the AISC Load and Resistance Factor Design Method.

Solution:

  (a) Obtain factored loads.

$$P_u = 1.2D + 1.6L = 1.2(100) + 1.6(200) = 440 \text{ kips}$$

  (b) Select the section. For this problem there are no AISC Available Strength Tables available for the direct selection of a section. To get an estimate of the required section, use the Available Strength Tables for $F_y = 50$ ksi. Since the member

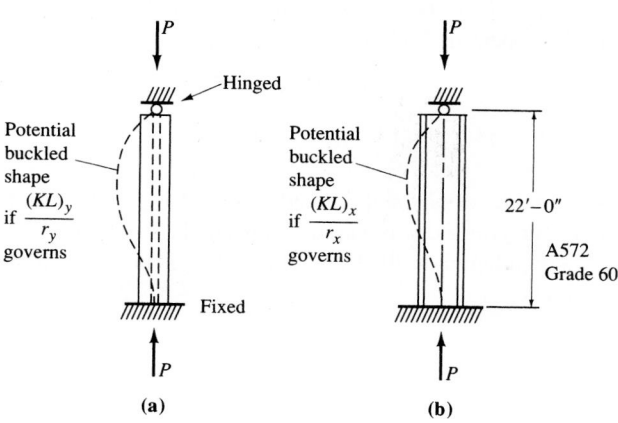

Figure 6.10.3
Example 6.10.3.

is fixed at one end, in accordance with Fig. 6.10.3, assume $K = 0.8$. Select for $(KL)_y = 0.8(22) = 17.6$ ft and

$$\text{Required } \phi_c P_n \approx P_u\left[\frac{F_y = 50}{F_y = 60}\right] = 440\left[\frac{50}{60}\right] = 367 \text{ kips}$$

Try
$$\text{W10}\times49: \quad \phi_c P_n = 392 \text{ kips for } F_y = 50 \text{ ksi}$$
$$\text{W12}\times53: \quad \phi_c P_n = 412 \text{ kips for } F_y = 50 \text{ ksi}$$

There can be no assurance that sections selected using the ratio 50/60 will satisfy design requirements; however, such a procedure will serve as a first trial. Check W10×49 for $P_u = 440$ kips with $F_y = 60$ ksi.

$$\left[\frac{KL}{r_y} = \frac{17.6(12)}{2.54} = 83.2\right] < \left[4.71\sqrt{\frac{E}{F_y}} = 4.71\sqrt{\frac{29,000}{60}} = 103.6\right]$$

$$\therefore \ F_{cr} = \left[0.658^{\frac{F_y}{F_e}}\right]F_y$$

$$F_e = \frac{\pi^2 E}{\left(\frac{KL}{r}\right)^2} = \frac{\pi^2(29,000)}{(83.2)^2} = 41.4 \text{ ksi}$$

$$\phi F_{cr} = 0.90\left[0.658^{\frac{60}{41.4}}\right]60 = 29.5 \text{ ksi}$$

$$\phi P_n = \phi F_{cr} A_g = 29.5(14.4) = 425 \text{ kips}$$

Thus, the W10×49 is not acceptable.
Check W10×54,

$$\frac{KL}{r_y} = \frac{17.6(12)}{2.56} = 82.5; \quad F_e = 42.1 \text{ ksi}; \quad \phi F_{cr} = 29.7 \text{ ksi}$$

$$\phi P_n = \phi F_{cr} A_g = 29.7(15.8) = 469 \text{ kips} > [P_u = 440 \text{ kips}] \qquad \textbf{OK}$$

Use W10×54, A572 Grade 60. Though the calculation is not shown here, $Q = 1.0$ for this section when $F_y = 60$ ksi. The limits for $b_f/2t_f$ and d/t_w are discussed in Sec. 6.16, and the limits appear in Table 6.16.2. ∎

EXAMPLE 6.10.4

Design column A of the unbraced frame of Fig. 6.10.4 as an axially loaded compression member carrying a dead load of 75 kips and a live load of 290 kips using A992 steel. In the plane perpendicular to the frame the system is braced, with supports at top and bottom of a 21-ft height.

Figure 6.10.4
Unbraced frame for
Example 6.10.4.

Solution:

(a) Obtain factored loads.

$$P_u = 1.2D + 1.6L = 1.2(75) + 1.6(290) = 554 \text{ kips}$$

(b) Select a preliminary size as a basis for evaluating the effective length factors K. While it is rare that a frame member would be designed as axially loaded, it may occasionally be proper for some interior members having symmetrical loading. Note is also made that the axially loaded member is one boundary for the more typical beam-column interaction formula discussed in Chapter 12.

It is given that $(KL)_y = 21$ ft for the plane perpendicular to the frame: thus, a preliminary member might be determined from interpolation in the *AISC Manual* "Available Strength in Axial Compression" Tables.

$$(KL)_y = 21 \text{ ft} \qquad \text{Find W12×72,} \qquad \phi_c P_n = 576 \text{ kips}$$

(c) Evaluate the effective length factor K. Using $I = 597$ in.^{4} for W12×72, compute G_{top} in accordance with Fig. 6.9.4 and as discussed in LRFD Commentary C2.

$$G_{\text{top}} = \frac{\Sigma(EI/L)_{\text{columns}}}{\Sigma(EI/L)_{\text{beams}}} = \frac{597/21}{2(881)/40} = 0.65$$

$$G_{\text{bottom}} = 10 \text{ (according to AISC-Fig. C-C 2.3)}$$

Using Fig. 6.9.4,

$$\text{Find } K_x = 1.8$$

$$(KL)_x = 1.8(21) = 37.8 \text{ ft}$$

For equal strength about each axis,

$$\frac{r_x}{r_y} = \frac{37.8}{21} = 1.8 \quad \text{(coincidentally the same as } K_x)$$

(d) Select sections from AISC Available Strength Tables.

W12×72: $r_x/r_y < 1.8$ Equivalent $(KL)_y = 37.8/1.75 = 21.6$ ft

$\phi_c P_n = 559$ kips   **OK**

W14×74: $r_x/r_y > 1.8$ $(KL)_y = 21$ ft $\phi_c P_n = 461$ kips   **NG**

(e) Make final check on W12×72, $I = 597$ in^4. At this point, it may be noted that the stiffness reduction factor could be used to modify the G value to account for inelastic buckling. AISC Table 4-21 is used to determine the reduction value corresponding to the value of P_u/A_g, as follows:

$$\frac{P_u}{A_g} = \frac{559}{21.1} = 26.5 \text{ ksi} \quad \tau_a = 0.849$$

$$G_{\text{inelastic}} = \tau_a G_{\text{elastic}} = 0.849(0.64) = 0.54$$

From the Alignment Chart, Fig. 6.9.4, find $K_x = 1.78$. This will make no significant difference to ϕP_n for this problem. KL/r becomes 84.5,

$\phi_c F_{\text{cr}} = 26.7$ ksi from AISC "Available Critical Stress" Table 4-22

$\phi_c P_n = \phi_c F_{\text{cr}} A_g = 26.7(21.1) = 563$ kips $> [P_u = 554$ kips$]$ **OK**

Use W12×72, A992.   ∎

6.11   ALLOWABLE STRENGTH DESIGN

To satisfy the safety requirement of the Allowable Strength Design (ASD) the allowable strength as defined in AISC-E1 must be equal to or greater than the required strength based on the ASD load combinations and may be stated as

$$\frac{P_n}{\Omega} \geq P_a$$

where P_a = maximum compressive load using ASD load combination
P_n = nominal strength
Ω = safety factor equal to 1.67

Equations 6.7.7 and 6.7.9 are used for typical rolled W sections satisfying the local buckling limitations on width/thickness ratios for plate compression elements given in AISC-E7.

EXAMPLE 6.11.1

Check the adequacy of the W10×45 section for the conditions of Fig. 6.10.2 if the AISC Allowable Strength Method Design was used. This is the section selected by LRFD in Example 6.10.2. A992 steel is used and the service loads are 65 kips dead load and 145 kips live load.

Solution:

(a) Compute slenderness ratios.

$$\frac{(KL)_x}{r_x} = \frac{1.0(28)12}{4.32} = 77.8 \qquad \frac{(KL)_y}{r_y} = \frac{1.0(14)12}{2.01} = 83.6$$

(b) Compute the allowble strength P_n/Ω.

$$\left[\frac{(KL)_y}{r_y} = 83.6\right] < \left[4.71\sqrt{\frac{E}{F_y}} = 4.71\sqrt{\frac{29,000}{50}} = 113.4\right]$$

$$\therefore \ F_{cr} = \left[0.658^{\frac{F_y}{F_e}}\right]F_y$$

$$F_e = \frac{\pi^2 E}{\left(\dfrac{KL}{r}\right)^2} = \frac{\pi^2(29,000)}{(83.6)^2} = 41.0 \text{ ksi}$$

$$F_{cr} = \left[0.658^{\frac{50}{41.0}}\right]50 = 30.0 \text{ ksi}$$

$$P_n/\Omega = F_{cr}A_g/\Omega = 30.0(13.3)/1.67 = 239 \text{ kips}$$

From the AISC "Available Critical Stress" TABLE 4-22

$$\frac{KL}{r_y} = 83.6 \qquad F_{cr}/\Omega = 18.0$$

and

$$P_n/\Omega = (F_{cr}/\Omega)A_g = 18.0(13.3) = 239 \text{ kips} > 220 \qquad \textbf{OK}$$

The allowable strength based on the ASD method exceeds the required strength by 8.6%, while the LRFD method gives a design strength higher than required by 13.7%.

In this case Allowable Strength Design is more conservative than Load and Resistance Factor Design.

Use W10×45. ■

6.12* SHEAR EFFECT

When built-up members are connected together by means of lacing bars, the objective is to make all of the components act as a unit. As a compression member bends, a shearing component of the axial load arises. The magnitude of the shear effect in reducing column strength is proportional to the amount of deformation that can be attributed to shear.

According to the *SSRC Guide* [6.8], shear in columns is caused by:

1. Lateral load, resulting from wind, earthquake gravity, or other cause.
2. Slope, with respect to the line of thrust, due both to unintentional initial curvature and added curvature developed during the buckling process.

3. End eccentricity of load, introduced either by the end connections or fabrication imperfections.

Certainly the shear from lateral load must always be considered in design. Items 2 and 3 should at the least be estimated. *AISC Specification* [1.13] requires providing an arbitrary shear resistance (in addition to any computable shear) of 2% of the compressive strength of the member for lacing bars in latticed columns (see Fig. 6.12.1).

Solid-webbed sections, such as W shapes, have less shear deformation than do latticed columns using lacing bars and/or batten plates.

Furthermore, as shown later, shear has an insignificant effect on reducing column strength for solid-webbed shapes and may safely be neglected. The shear effect should not, however, be neglected for latticed columns.

To include the effect of shear, the curvature resulting from shear should be added to the buckling curvature to obtain the total curvature. It is well known that

$$V = \frac{dM_z}{dz} = P\frac{dy}{dz} \tag{6.12.1}$$

after recognizing that $M_z = Py$ from Eq. 6.2.1 (see Fig. 6.2.1).

The slope θ due to shear deformation is

$$\theta = \frac{\text{shear stress}}{\text{shear modulus}} = \frac{\beta_v V}{AG} \tag{6.12.2}$$

where β_v is a factor to correct for nonuniform stress across various cross-sectional shapes. The shear contribution to curvature becomes

$$\frac{d\theta}{dz} = \frac{\beta_v}{AG}\frac{dV}{dz} = \frac{P\beta_v}{AG}\frac{d^2y}{dz^2} \tag{6.12.3}$$

The total curvature is the sum of Eqs. 6.2.3 and 6.12.3,

$$\frac{d^2y}{dz^2} = -\frac{Py}{EI} + \frac{P\beta_v}{AG}\frac{d^2y}{dz^2}$$

which gives

$$\frac{d^2y}{dz^2} + \frac{P}{EI}\left[\frac{1}{1 - P\beta_v/AG}\right]y = 0 \tag{6.12.4}$$

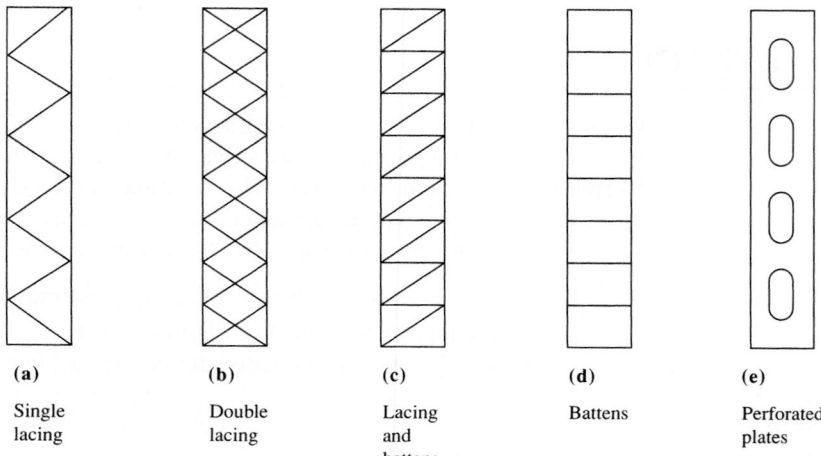

(a)	(b)	(c)	(d)	(e)
Single lacing	Double lacing	Lacing and battens	Battens	Perforated plates

Figure 6.12.1
Types of latticed columns.

which is of the same form as Eq. 6.2.3; therefore the modified form of the Euler critical load is

$$P_{cr} = \frac{\pi^2 EI}{L^2} \underbrace{\left[\frac{1}{1 + \dfrac{\beta_v}{AG} \dfrac{\pi^2 EI}{L^2}} \right]}_{\substack{\text{modification for} \\ \text{shear effect}}} \tag{6.12.5}$$

In accordance with the previous discussion on basic column strength, G and E can be replaced by the tangent modulus values, G_t and E_t, and $E_t/G_t = 2(1 + \mu)$, and L can be replaced by the effective length KL. Further, combining the shear effect with KL gives

$$F_{cr} = \frac{P_{cr}}{A} = \frac{\pi^2 E_t}{(\alpha_v KL/r)^2} \tag{6.12.6}$$

where $\alpha_v = \sqrt{1 + 2(1 + \mu)\pi^2 \beta_v/(KL/r)^2}$. Thus the shear effect may be accounted for by an adjustment to the effective length. For W shapes when bending about the weak axis, β_v averages about 2. Using $\mu = 0.3$ for steel, typical values for α_v are

$$
\begin{aligned}
KL/r = \quad 50 \qquad & \alpha_v = 1.01 \\
= \quad 70 \qquad & = 1.005 \\
= \quad 100 \qquad & = 1.003
\end{aligned}
$$

For slenderness ratios less than about 50, yielding controls, so that the shear effect on solid H-shaped columns is equivalent to an increase in effective length of less than 1%, which can be safely neglected.

Latticed Columns

The lacing or batten plates used to tie together the main longitudinal compression elements are themselves subject to axial deformation. For instance, from Fig. 6.12.2a, the lengthening of the diagonal gives a slope γ_1 over the panel length a and from Fig. 6.12.2b the shortening of the horizontal bars gives a slope γ_2 over panel length a. Because the lacing elements are relatively small in cross-section, the stiffness of such members to resist the transverse shear is considerably less than for solid-webbed members. Detailed treatment of the columns with lacing, battens, or perforated plates is available in the *SSRC Guide* [6.8] and elsewhere [6.62–6.64].

Figure 6.12.2
Shear deformation in laced column.

(a) Effect of lengthening of diagonal bars

(b) Effect of shortening of horizontal bars

The Structural Stability Research Council [6.8] reports the suggestion of Bleich [6.9] that "a conservative estimate of the influence of 60° or 45° lacing, as generally specified in bridge design practice, can be made by modifying the effective length factor" K to a new factor $\alpha_v K$, as follows:

$$\text{For } \frac{KL}{r} > 40, \qquad \alpha_v = \sqrt{1 + 300/(KL/r)^2} \qquad (6.12.7)$$

$$\text{For } \frac{KL}{r} \leq 40, \qquad \alpha_v = 1.1 \qquad (6.12.8)$$

Such effective length modification will rarely affect the design of short columns in braced systems.

6.13* DESIGN OF LATTICED MEMBERS

Under most specifications, latticed members are designed according to detailed empirical rules, most of which are related to local buckling requirements. Two examples follow that illustrate some of the provisions of AISC-E7 as well as general procedures for built-up sections. The reader is referred to Blodgett [6.65] who has summarized the AISC provisions along with other information concerning built-up section design.

EXAMPLE 6.13.1

Design a laced column as shown in Fig. 6.13.1, consisting of four angles to carry 100 kips dead load and 475 kips live load axial compression, with an effective length KL of 30 ft. Assume all connections will be welded and use A572 Grade 50 steel.

Solution:

(a) Establish the depth h of section. The radius of gyration of a four angle column depends only on h and essentially is independent of thickness. Thus selecting h establishes the slenderness ratio, and vice versa. Text Appendix Table A1 shows the

Figure 6.13.1
Details of Example 6.13.1.

(a)   (b)   (d)

relationships between the radius of gyration and the geometry of the cross-section. Thus from text Appendix Table A1,

$$r \approx 0.42h$$

$$\text{Approximate } \frac{KL}{r} = \frac{360}{0.42h} = \frac{857}{h}$$

$$P_u = 1.2(100) + 1.6(475) = 880 \text{ kips}$$

h (in.)	Approx. KL/r	$\phi_c F_{cr}^\dagger$ (ksi)	Required area* A_g (sq in.)	Angles
24	35.7	41.0	21.5	L6×6×$\frac{1}{2}$ $A = 23.0$
				L5×5×$\frac{5}{8}$ $A = 23.44$
22	39.0	40.3	21.8	Same as above
21	40.8	39.9	22.1	Same as above
20	42.9	39.3	22.4	Same as above

†From AISC "Available Critical Stress for Compression Members" Table 4-22.
*Required $A_g = P_u/(\phi_c F_{cr})$.

The 20-in. section is preferred, since the floor area occupied is likely more important than the additional 0.44 sq in. of cross-section. Investigate 4—L5×5×$\frac{5}{8}$, assuming $Q = 1.0$,

$$I_x = I_y = 4[13.6 + 5.86(10.0 - 1.48)^2] = 1756 \text{ in.}^4$$

$$r_x = r_y = \sqrt{\frac{I}{A}} = \sqrt{\frac{1756}{23.44}} = 8.66 \text{ in.} \quad [\text{Approx } 0.42(20) = 8.40 \text{ in.}]$$

$$\frac{KL}{r} = \frac{1.0(30)12}{8.66} = 41.6; \qquad \phi_c F_{cr} = 39.6 \text{ ksi (AISC-TABLE 4-22)}$$

$$\phi_c P_n = \phi_c F_{cr} A_g = 39.6(23.44) = 928 \text{ kips} > (P_u = 880 \text{ kips}) \qquad \textbf{OK}$$

Use 4—L5×5×$\frac{5}{8}$ with $h = 20$ in.

   (b) Local buckling control. For $Q = 1.0$ to apply, angles must satisfy width/thickness limits of AISC Table B4.1, Case 5 to prevent local buckling. In this case,

$$\left[\frac{b}{t} = \frac{5}{0.625} = 8\right] < \left[\lambda_r = 0.45\sqrt{\frac{E}{F_y}} = 0.45\sqrt{\frac{29,000}{50}} = 10.8\right] \qquad \textbf{OK}$$

Development of AISC limits appears in Chapter 6, Part II on plate strength.

   (c) Design single lacing. According to the *User Note* in AISC-E6, "The inclination of lacing bars to the axis of the member shall preferably be not less than 60° for single lacing ...."

   For an inclination of 60° ($\xi = 30°$) (Fig. 6.13.1b), $b = 20 - 2(3) = 14$ in., assuming use of distance between standard gage lines for bolts. This would be approximately center-to-center of welded connection. Thus

$$L_1 \approx 2b \tan 30° = 2(14)0.577 = 16.2 \text{ in.} \quad \text{Use 16 in.}$$

One might prefer to use 15 in. since the *User Note* in AISC-E6.2 states, "When the distance between the lines of welds ... is more than 15 in., the lacing shall preferably be double or be made of angles."

For a single angle,

$$\frac{L_1}{r_z} = \frac{16}{0.978} = 16.4 < 41.6 \left[\begin{array}{c} \text{slenderness ratio for} \\ \text{overall member} \end{array}\right] \quad \textbf{OK}$$

According to AISC-E6.2, "Lacing shall be proportioned to provide a shearing strength normal to the axis of the member equal to 2% of the compressive design strength of the member."

$$V_u = 0.02(880) = 17.6 \text{ kips} \quad (8.8 \text{ kips per side})$$

The force in one bar is

$$P_u = V_u/\cos \xi \approx V_u/\cos 30° = 8.8/0.866 = 10.2 \text{ kips}$$

$$\frac{L}{r} \leq 140 \quad \text{for single lacing}$$

$$r = \sqrt{\frac{I}{A}} = \sqrt{\frac{\frac{1}{12}bt^3}{bt}} = 0.288t \quad (\text{Fig. 6.13.1 d})$$

$$t_{min} = \frac{16}{0.288(140)} = 0.397 \text{ in.} \quad \text{Use } \tfrac{7}{16} \text{ in.}$$

$$\text{For } t = \tfrac{7}{16} \text{ in.,} \quad \frac{L}{r} = \frac{16}{0.288(0.4375)} = 127$$

From AISC "Available Critical Stress for Compression Members" Table 4-22,

$$\phi_c F_{cr} = 13.8 \text{ ksi}$$

$$\text{Required } A = \frac{10.2}{13.8} = 0.74 \text{ sq in.}$$

$$\text{Width } b = \frac{0.74}{0.4375} = 1.69 \text{ in.}$$

Since no holes are required for connectors, tension on net section need not be investigated for this design.
Use bars $\tfrac{7}{16} \times 1\tfrac{3}{4}$.

(d) Design the plates at ends (AISC-E6.2). The tie plates should extend along the length of the member a distance equal to the distance b (Fig. 6.13.1) from the end of the member. Use a length of 14 in.

$$t \geq \frac{b}{50} = \frac{14}{50} = 0.28 \text{ in.}$$

Use tie plates $\tfrac{5}{16} \times 14 \times 1'\text{-}8''$.

(e) Examine the effect of shear on the effective length, Eqs. 6.12.7 or 6.12.8.

$$\frac{KL}{r} = 41.6 > 40, \quad \text{Use Eq. 6.12.7}$$

$$\alpha_v = \sqrt{1 + 300/(KL/r)^2}$$

$$= \sqrt{1 + 300/(41.6)^2} = 1.08$$

Thus the effective length should have been increased 8% due to shear. The neglect of end restraint probably is, in most cases, equal to about the same increase in effective length. ■

EXAMPLE 6.13.2

Redesign the column of Example 6.13.1 using a welded perforated box shape (Fig. 6.13.2). The factored load P_u is 880 kips as computed in Example 6.13.1.

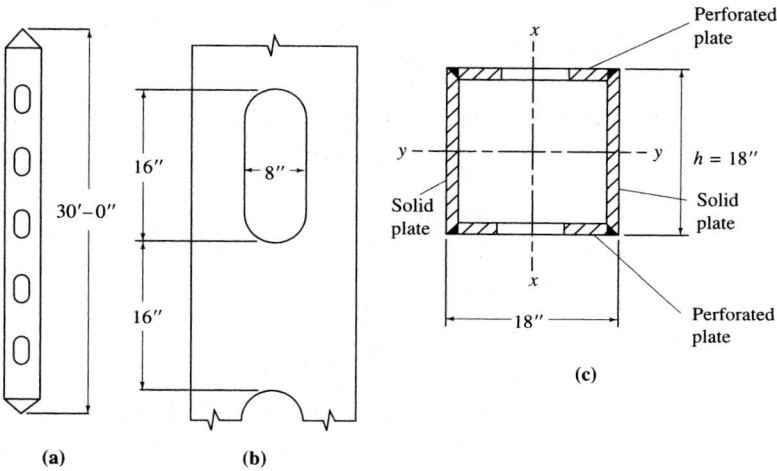

Figure 6.13.2
Details of Example 6.13.2.

(a)   **(b)**

Solution:
One important advantage of a welded shape is the number of individual components in the shape is minimized. Four plates can be used for this welded shape, whereas a bolted or riveted section requires four angles in addition to plates.

(a) Select trial section. Using text Appendix Table A1 for a box shape,

$$r \approx 0.40h$$

$$\frac{KL}{r} = \frac{360}{0.4h} = \frac{900}{h}$$

$$\text{Area} \approx 4ht$$

h (in.)	Approx. KL/r	$\phi_c F_{cr}^{\dagger}$ (ksi)	Required area* A_g (sq in.)	Plate thickness (in.)
21	42.9	39.3	22.4	0.27
$\dfrac{b}{t}$ (solid plate) $= \dfrac{21}{5/16} = 67 > \left(1.49\sqrt{\dfrac{29{,}000}{50}} = 35.9\right)$				NG
18	50	37.5	23.5	0.35
16	56.3	35.7	24.6	0.41
15	60	34.6	23.4	0.45

†From AISC "Available Critical Stress for Compression Members" Table 4-22
*Required $A_g = P_u/(\phi_c F_{cr})$.

Without perforations, 2PLs—$\frac{7}{16}$ × 16 and 2PLs—$\frac{7}{16}$ × 15 are probably acceptable, satisfying b/t ratios in accordance with AISC-B4.2.

Usually, however, such shapes are used on bridges where access is required for maintenance so that perforations may be desirable.

Assume perforations to be 8 in. wide (frequently one-half or less of total width) and $h \approx 18$ in.

$$\text{Net area available} = 2(18)\left(\tfrac{7}{16}\right) + 2(17-8)\left(\tfrac{1}{2}\right) = 24.75 \text{ sq in.}$$

$$I_x = 2(18)\left(\tfrac{7}{16}\right)(8.719)^2 + \tfrac{2}{12}\left[(17)^3 - (8)^3\right]\tfrac{1}{2} = 1564 \text{ in.}^4$$

$$I_y = \tfrac{2}{12}(18)^3\left(\tfrac{7}{16}\right) + 2(9)\left(\tfrac{1}{2}\right)(8.75)^2 = 1114 \text{ in.}^4$$

$$r_y = \sqrt{\frac{1114}{24.75}} = 6.71 \text{ in.}$$

$$\frac{KL}{r} = \frac{1.0(30)12}{6.71} = 53.7; \qquad \phi_c F_{cr} = 36.5 \text{ ksi (AISC TABLE 4-22)}$$

$$\phi_c P_n = \phi_c F_{cr} A_g = 36.5(24.75) = 903 \text{ kips} > [P_u = 880 \text{ kips}] \qquad \textbf{OK}$$

(b) Check proportioning of plates.

1. Check b/t ratio for entire perforated plate (AISC Table B4.1, Case 14):

$$\left[\frac{b}{t} = \frac{17.125}{7/16} = 39.1\right] > \left[\lambda_r = 1.49\sqrt{\frac{E}{F_y}} = 1.49\sqrt{\frac{29,000}{50}} = 35.9\right] \qquad \textbf{NG}$$

Must use a thicker plate: try $t = \frac{1}{2}$ in.

$$\frac{b}{t} = \frac{17}{0.5} = 34 < 35.9 \qquad \textbf{OK}$$

2. Check b/t ratio for unstiffened portion within hole (AISC-Table B4.1):

$$\left[\frac{b}{t} = \frac{4.5}{0.5} = 9\right] < \left[\lambda_r = 0.45\sqrt{\frac{E}{F_y}} = 0.45\sqrt{\frac{29,000}{50}} = 10.8\right] \qquad \textbf{OK}$$

3. Check proportions of access holes (AISC-E6.2):

$$\frac{\text{Length of hole}}{\text{Width of hole}} \le 2 \qquad \text{Length of hole} = 2(8) = 16 \text{ in. max}$$

$$\left(\begin{array}{c}\text{Clear distance}\\ \text{between holes}\end{array}\right) \ge \left(\begin{array}{c}\text{Transverse distance}\\ \text{between nearest lines}\\ \text{of connecting welds}\end{array}\right)$$

$$\text{Clear spacing between holes} = 17 \text{ in. min.}$$

Use 2PLs—$\frac{1}{2}$×18 (solid) and 2PLs—$\frac{1}{2}$×17 (perforated) with holes 16″ by 8″ spaced 2′–8″ center-to-center. ∎

PART II: PLATES

6.14*   INTRODUCTION TO STABILITY OF PLATES

All column sections, whether rolled shapes or built-up sections, are composed of plate elements. Up to this point in the chapter, consideration has been given only to the possibility of buckling of the member based on the slenderness ratio for the entire cross-section. It may be, however, that a local buckling will occur first in one of the plate elements that make up the cross-section. Such local buckling means that the buckled element will no longer take its proportionate share of any *additional* load the column is to carry; in other words, the efficiency of the cross-section is reduced.

The theory of bending of plates and elastic stability of plates are subjects that should be studied in depth by the advanced student in structural engineering. The brief treatment that follows is intended to give the reader the general idea of plate buckling necessary to properly use and understand current steel specifications. The general approach and terminology follow that of Timoshenko [6.66, 6.67].

Before one can treat the stability problem, the differential equation for bending of plates is required, just as the differential equation for the bending of a beam, Eq. 6.2.2, was used in the slender column stability treatment in Sec. 6.2.

Differential Equation for Bending of Homogeneous Plates

First, the strains will be obtained in terms of displacements. Let h = plate thickness and u, v, and w equal the displacements in the x, y, and z directions, respectively. Referring to Fig. 6.14.1, consider an element of a plate $dx\ dy$, and assume no stretching of the neutral plane at $z = 0$. Examining a slice $dx\ dy\ dz$ of the plate element located at a distance z from the neutral plane shows, in Fig. 6.14.2, the coordinate unit strains ϵ_x, ϵ_y, and

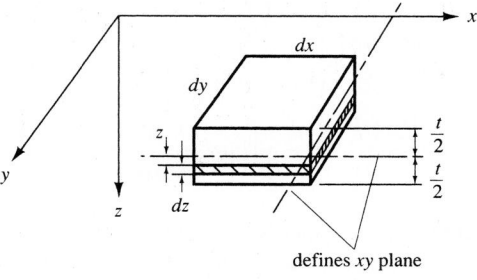

Figure 6.14.1
Plate element, coordinate system definition.

Figure 6.14.2
Deformations of plate element in xy plane.

$$\gamma_1 = \frac{\frac{\partial v}{\partial x}\,dx}{dx} = \frac{\partial v}{\partial x}$$

$$\gamma_2 = \frac{\frac{\partial u}{\partial y}\,dy}{dy} = \frac{\partial u}{\partial y}$$

Figure 6.14.3
Deformation of plate element
perpendicular to the xy plane.

shearing strain γ_{xy}. Thus

$$\epsilon_x = \frac{u + \dfrac{\partial u}{\partial x}dx - u}{dx} = \frac{\partial u}{\partial x} \tag{6.14.1a}$$

$$\epsilon_x = \frac{\partial v}{\partial y} \tag{6.14.1b}$$

$$\gamma_{xy} = \gamma_1 + \gamma_2 = \frac{\partial v}{\partial x} + \frac{\partial u}{\partial y} \tag{6.14.1c}$$

Expressing the displacements in the plane of the plate in terms of the lateral deflection w, as shown in Fig. 6.14.3, and recognizing that positive slope gives a negative displacement u or v, one establishes

$$-u = z\frac{\partial w}{\partial x}; \qquad -v = z\frac{\partial w}{\partial y} \tag{6.14.2}$$

Substitution of Eqs. 6.14.2 into Eqs. 6.14.1 gives strains in terms of curvatures for x-direction bending, y-direction bending, and twisting

$$\epsilon_x = \frac{\partial u}{\partial x} = -z\frac{\partial^2 w}{\partial x^2} \tag{6.14.3a}$$

$$\epsilon_y = \frac{\partial v}{\partial y} = -z\frac{\partial^2 w}{\partial y^2} \tag{6.14.3b}$$

$$\gamma_{xy} = \frac{\partial v}{\partial x} + \frac{\partial u}{\partial y} = -z\left(\frac{\partial^2 w}{\partial x\,\partial y} + \frac{\partial^2 w}{\partial x\,\partial y}\right) = -2z\frac{\partial^2 w}{\partial x\,\partial y} \tag{6.14.3c}$$

Next, making use of Hooke's Law expressing strains in terms of the stress σ_x, σ_y, normal stresses in the x- and y-directions, and τ_{xy}, the shear stress,

$$\epsilon_x = \frac{1}{E}\left[\sigma_x - \mu\sigma_y\right] \tag{6.14.4a}$$

$$\epsilon_y = \frac{1}{E}\left[-\mu\sigma_x + \sigma_y\right] \tag{6.14.4b}$$

$$\gamma_{xy} = \tau_{xy}/G \tag{6.14.4c}$$

where μ = Poisson's ratio (see Sec. 2.6) and G = shear modulus of elasticity.

For any stress condition, such as $\sigma_y = -\sigma_x$ that gives pure shear on an element rotated 45° to the x-axis, the work done by the equivalent systems of Fig. 6.14.4 must be a constant:

$$\tfrac{1}{2}\sigma_x\epsilon_x - \tfrac{1}{2}\sigma_x\epsilon_y = \tfrac{1}{2}\tau\gamma \tag{6.14.5}$$

Figure 6.14.4
Equivalent systems.
(a)   **(b)**

Substituting Eqs. 6.14.4 into Eq. 6.14.5 gives

$$\frac{\sigma_x}{E}(\sigma_x - \mu\sigma_y + \mu\sigma_x - \sigma_y) = \frac{\tau^2}{G}$$

If $\sigma_y = -\sigma_x$, maximum $\tau = \sigma_x$; thus

$$\frac{1}{E}(1 + \mu + \mu + 1) = \frac{1}{G}$$

$$G = \frac{E}{2(1 + \mu)} \qquad (6.14.6)$$

Solving Eqs. 6.14.4 for stresses in terms of strains and substituting Eqs. 6.14.3 for the strains give stresses in terms of curvatures,

$$\sigma_x = \frac{-zE}{1 - \mu^2}\left(\frac{\partial^2 w}{\partial x^2} + \mu\frac{\partial^2 w}{\partial y^2}\right) \qquad (6.14.7a)$$

$$\sigma_y = \frac{-zE}{1 - \mu^2}\left(\mu\frac{\partial^2 w}{\partial x^2} + \frac{\partial^2 w}{\partial y^2}\right) \qquad (6.14.7b)$$

$$\tau_{xy} = -2zG\frac{\partial^2 w}{\partial x\, \partial y} \qquad (6.14.7c)$$

Next it is necessary to relate curvatures and bending moments. Referring to Fig. 6.14.5, and using the right-hand rule for positive twist, it is seen that the moments per unit width are

$$M_x = \int_{-t/2}^{t/2} z\sigma_x\, dz = \frac{-Et^3}{12(1 - \mu^2)}\left(\frac{\partial^2 w}{\partial x^2} + \mu\frac{\partial^2 w}{\partial y^2}\right) \qquad (6.14.8a)$$

$$M_y = \int_{-t/2}^{t/2} z\sigma_y\, dz = \frac{-Et^3}{12(1 - \mu^2)}\left(\mu\frac{\partial^2 w}{\partial x^2} + \frac{\partial^2 w}{\partial y^2}\right) \qquad (6.14.8b)$$

$$M_{xy} = -\int_{-t/2}^{t/2} \tau_{xy}z\, dz = +2G\left(\frac{t^3}{12}\right)\frac{\partial^2 w}{\partial x\, \partial y} \qquad (6.14.8c)$$

Note that plate bending involves double curvature (a dish-shaped deflection surface for the plate). The narrower and longer in plan a plate is, the more the bending causes curvature to be one directional. A beam is a special case of a plate since it has a narrow width and long span. For beams, the Poisson's ratio (μ) effect is neglected. For instance, if the

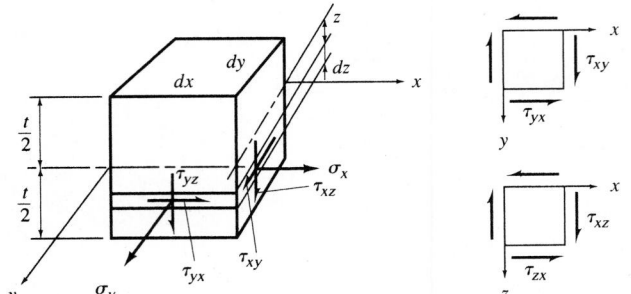

Figure 6.14.5
A plate element in bending.
(Note that forces on faces at
$x = 0$ and $y = 0$ not shown.)

member is narrow in the y-direction and long in the x-direction, Eq. 6.14.8a for the plate would become

$$M_x = \frac{-Et^3}{12} \frac{d^2w}{dx^2} \qquad (6.14.9)$$

where the partial derivatives disappear because w is no longer a function of y. If Eq. 6.14.9 is multiplied by the width b to change from moment per unit width in the y-direction to total moment, Eq. 6.14.9 would be the differential equation for beams,

$$M_x = -EI\frac{d^2w}{dx^2} \qquad (6.14.10)$$

where $I = t^3b/12$.

In theory of plates, the sign convention is that M_x is the bending moment caused by *stresses σ_x* acting in the x-direction. In beam theory, the usual practice is to have the subscript on M refer to the axis of zero stress in bending; that is, the neutral axis. Thus, for the span in the x-direction and the width in the y-direction as assumed for Eq. 6.14.10, M_y (instead of M_x) would have been used in beam theory, since the y-axis is the neutral axis about which bending occurs.

To continue the derivation of the plate bending differential equation, consider the equilibrium of all forces and moments acting on the plate element. Moment summation about the x- and y-axes and force summation in the z-direction gives three equations. Figure 6.14.6 is a free body of the plate element showing only the forces involved with moments about the y-axis.

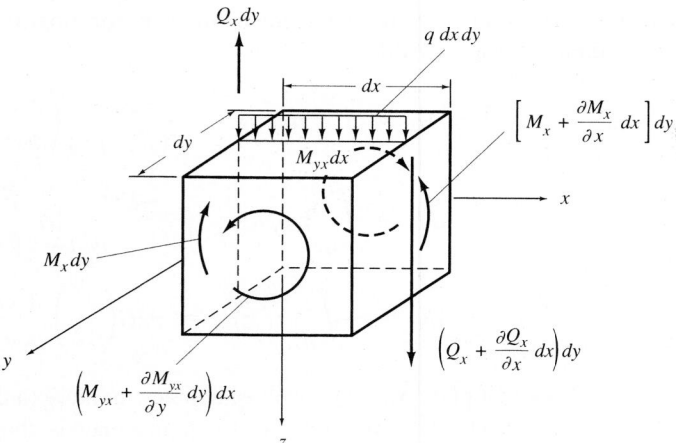

Figure 6.14.6
Free-body forces involved in
rotation about y-axis. (Forces
involved in rotation about the
x-axis not shown.)

Taking moments about the y-axis gives

$$M_x\,dy + \frac{\partial M_x}{\partial x}dx\,dy - M_x\,dy + M_{yx}\,dx + \frac{\partial M_{yx}}{\partial y}dy\,dx - M_{yx}\,dx$$

$$- \left(Q_x\,dy + \frac{\partial Q_x}{\partial x}dx\,dy\right)dx - q\,dx\,dy\frac{dx}{2} = 0$$

$$\nearrow \qquad \uparrow \qquad \uparrow$$
$$\text{neglect} \quad \text{mom.} \quad \text{neglect}$$
$$\text{arm}$$

Neglecting infinitesimals of higher order and dividing by $dx\,dy$ gives

$$\frac{\partial M_x}{\partial x} + \frac{\partial M_{yx}}{\partial y} - Q_x = 0 \qquad\qquad (6.14.11)$$

Similarly for moments about the x-axis,

$$\frac{\partial M_y}{\partial y} + \frac{\partial M_{xy}}{\partial x} - Q_y = 0 \qquad\qquad (6.14.12)$$

Force equilibrium in the z-direction gives

$$\frac{\partial Q_x}{\partial x} + \frac{\partial Q_y}{\partial y} + q = 0 \qquad\qquad (6.14.13)$$

Using Eqs. 6.14.11 and 6.14.12 for Q_x and Q_y, and substitution into Eq. 6.14.13 gives

$$\frac{\partial^2 M_x}{\partial x^2} + \frac{\partial^2 M_y}{\partial y^2} + 2\frac{\partial^2 M_{xy}}{\partial x\,\partial y} = -q \qquad\qquad (6.14.14)$$

Defining $D = Et^3/[12(1 - \mu^2)]$ and substituting Eq. 6.14.8 into Eq. 6.14.14 gives the differential equation for bending of homogeneous plates,

$$D\left(\frac{\partial^4 w}{\partial x^4} + 2\frac{\partial^4 w}{\partial x^2\,\partial y^2} + \frac{\partial^4 w}{\partial y^4}\right) = q \qquad\qquad (6.14.15)$$

Equation 6.14.15, if written for a beam of width b, is the differential equation for load,

$$EI\frac{d^4 w}{dx^4} = qb \qquad\qquad (6.14.16)$$

where qb is the load per unit length along the span of the beam.

Buckling of Uniformly Compressed Plate

The following approach is essentially that of Timoshenko [6.67] as modified by Gerstle [6.68]. Realizing that q is a general term representing the transverse load component causing plate bending, it is desired to find the transverse component of compressive force N_x

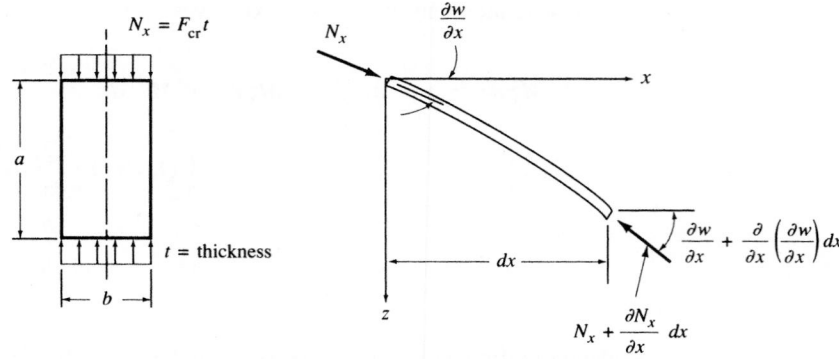

Figure 6.14.7
Uniformly compressed plate.

(a)

(b) Plate element in bent position

when the plate is deflected into a slightly buckled position. Taking summation of forces in the z-direction on the plate element of Fig. 6.14.7b gives

$$N_x\, dy\, \frac{\partial w}{\partial x} - \left(N_x + \frac{\partial N_x}{\partial x}\, dx \right) dy \left(\frac{\partial w}{\partial x} + \frac{\partial^2 w}{\partial x^2}\, dx \right) = q\, dx\, dy$$

$$- \left(N_x \frac{\partial^2 w}{\partial x^2} + \frac{\partial N_x}{\partial x}\, \frac{\partial w}{\partial x} + \frac{\partial N_x}{\partial x}\, dx\, \frac{\partial^2 w}{\partial x^2} \right) dy\, dx = q\, dx\, dy \qquad (16.14.17)$$

which upon neglecting the higher-order infinitesimal terms gives

$$q = -N_x \frac{\partial^2 w}{\partial x^2} \qquad (6.14.18)$$

The differential equation, Eq. 6.14.15, then becomes

$$\frac{\partial^4 w}{\partial x^4} + 2\frac{\partial^4 w}{\partial x^2\, \partial y^2} + \frac{\partial^4 w}{\partial y^4} = -\frac{N_x}{D}\frac{\partial^2 w}{\partial x^2} \qquad (6.14.19)$$

which is a partial differential equation where w is a function of both x and y. The deflection w can be expressed as the product of an x function (X) and a y function (Y). Further, buckling may be assumed to give sinusoidal variation in the x-direction. Thus

$$w = X(x)Y(y) \qquad (6.14.20)$$

letting

$$X(x) = \sin\frac{m\pi x}{a}$$

where the X function satisfies the zero deflection and zero moment conditions of simple support at $x = 0$ and $x = a$. Substitution of Eq. 6.14.20 into Eq. 6.14.19 gives, after cancelling the common term $\sin m\pi x/a$,

$$\left(\frac{m\pi}{a}\right)^4 Y - 2\left(\frac{m\pi}{a}\right)^2 \frac{d^2 Y}{dy^2} + \frac{d^4 Y}{dy^4} = +\frac{N_x}{D}\left(\frac{m\pi}{a}\right)^2 Y$$

$$\frac{d^4 Y}{dy^4} - 2\left(\frac{m\pi}{a}\right)^2 \frac{d^2 Y}{dy^2} + \left[\left(\frac{m\pi}{a}\right)^4 - \frac{N_x}{D}\left(\frac{m\pi}{a}\right)^2 \right] Y = 0 \qquad (6.14.21)$$

an ordinary fourth-order homogeneous differential equation.

The solution may be expressed in the form

$$Y = C_1 \sinh \alpha y + C_2 \cosh \alpha y + C_3 \sin \beta y + C_4 \cos \beta y \qquad (6.14.22)$$

where

$$\alpha = \sqrt{\left(\frac{m\pi}{a}\right)^2 + \sqrt{\frac{N_x}{D}\left(\frac{m\pi}{a}\right)^2}} \quad \text{and} \quad \beta = \sqrt{-\left(\frac{m\pi}{a}\right)^2 + \sqrt{\frac{N_x}{D}\left(\frac{m\pi}{a}\right)^2}}$$

Thus the entire plate deflection equation is

$$w = \left(\sin\frac{m\pi x}{a}\right)(C_1 \sinh \alpha y + C_2 \cosh \alpha y + C_3 \sin \beta y + C_4 \cos \beta y) \quad (6.14.23)$$

which must satisfy boundary conditions. Assuming the x-axis as an axis of symmetry through the plate, i.e., identical support conditions along the two edges parallel to the direction of loading, the odd function coefficients C_1 and C_3 must be zero. Thus

$$w = (C_2 \cosh \alpha y + C_4 \cos \beta y) \sin\frac{m\pi x}{a} \quad (6.14.24)$$

Using simple support conditions at $y = b/2$ and $y = -b/2$, requires that at $y = \pm b/2$,

$$w = 0 = \left(C_2 \cosh \alpha\frac{b}{2} + C_4 \cos \beta\frac{b}{2}\right) \sin\frac{m\pi x}{a}$$

$$\frac{\partial^2 w}{\partial y^2} = 0 = \left(C_2\alpha^2 \cosh \alpha\frac{b}{2} - C_4\beta^2 \cos \beta\frac{b}{2}\right) \sin\frac{m\pi x}{a} \quad (6.14.25)$$

For a solution other than $C_2 = C_4 = 0$ it is necessary for the determinant of the coefficients to be zero. Thus

$$(\alpha^2 + \beta^2) \cosh \alpha\frac{b}{2} \cos \beta\frac{b}{2} = 0 \quad (6.14.26)$$

Since $\alpha^2 \neq -\beta^2$ unless $N_x = 0$ (a trivial solution), and since $\cosh \alpha(b/2) > 1$, the only way Eq. 6.14.26 can be satisfied in the real problem is for

$$\cos \beta\frac{b}{2} = 0$$

Therefore

$$\beta\frac{b}{2} = \frac{\pi}{2}, \ \frac{3\pi}{2}, \ \frac{5\pi}{2}, \ \text{etc.}$$

Using the lowest value of $\beta(b/2)$ and substituting into β as defined below Eq. 6.14.22 gives

$$\frac{b}{2}\sqrt{-\left(\frac{m\pi}{a}\right)^2 + \sqrt{\frac{N_x}{D}\left(\frac{m\pi}{a}\right)^2}} = \frac{\pi}{2}$$

$$\frac{N_x}{D}\left(\frac{m\pi}{a}\right)^2 = \left[\frac{\pi^2}{b^2} + \left(\frac{m\pi}{a}\right)^2\right]^2$$

$$N_x = D\left[\frac{\pi^2 a}{b^2 m\pi} + \frac{m\pi}{a}\right]^2$$

$$N_x = \frac{D\pi^2}{b^2}\left[\frac{1}{m}\frac{a}{b} + m\frac{b}{a}\right]^2 \quad (6.14.27)$$

Since $N_x = F_{cr}t$ and $D = Et^3/[12(1 - \mu^2)]$, the elastic buckling unit stress may be expressed as

$$F_{cr} = k\frac{\pi^2 E}{12(1 - \mu^2)(b/t)^2} \quad (6.14.28)$$

Figure 6.14.8
Buckling coefficient for
uniformly compressed
plate—simple support
longitudinal edges
(Eq. 6.14.29).

where for the specific case treated here, the elastic plate buckling coefficient k is

$$k = \left[\frac{1}{m} \frac{a}{b} + m \frac{b}{a} \right]^2 \qquad (6.14.29)$$

The buckling coefficient k is a function of the type of stress (in this case uniform compression on two opposite edges) and the edge support conditions (in this case simple support on four edges), in addition to the aspect ratio a/b which appears directly in the equation.

The equation for plate buckling, Eq. 6.14.28, is entirely general in terms of k and the development leading up to it for this one case may be considered illustrative of the procedure. The integer m indicates the number of half-waves that occur in the x direction at buckling. Figure 6.14.8 shows that there is a minimum value of k for any given number of half-waves, i.e., the weakest condition. It is noted that this weakest situation occurs when the length is an even multiple of width, and that multiple equals the number of half-waves.

Thus, setting $a/b = m$ gives $k = 4$. Further, as m becomes larger the k equation becomes flatter and approaches a constant value of 4 for large a/b ratio. This gives for the elastic buckling stress equation of plate elements under uniform compression along two edges and simply supported along the two edges parallel to the load,

$$F_{cr} = \frac{4\pi^2 E}{12(1 - \mu^2)(b/t)^2} \qquad (6.14.30)$$

6.15* STRENGTH OF PLATES UNDER UNIFORM EDGE COMPRESSION

Since rolled shapes, as well as built-up shapes, are composed of plate elements, the column strength of the section based on its overall slenderness ratio can only be achieved if the plate elements do not buckle locally. Local buckling of plate elements can cause premature failure of the entire section, or at the least it will cause stresses to become nonuniform and reduce the overall strength.

In Sec. 6.14 the basic approach to elastic stability of plates was developed. The theoretical elastic buckling stress for a plate was shown to be expressible as

$$F_{cr} = k \frac{\pi^2 E}{12(1 - \mu^2)(b/t)^2} \qquad [6.14.28]$$

where k is a constant depending on type of stress, edge support conditions, and length to width ratio (aspect ratio) of the plate, E the modulus of elasticity, μ Poisson's ratio, and b/t the width/thickness ratio.

Figure 6.15.1
Stiffened and unstiffened
compression elements.

(a) Elements supported along
two edges (stiffened elements)

(b) Elements supported along
one edge (unstiffened elements)

In general, plate compression elements can be separated into two categories: (1) stiffened elements; those supported along two edges parallel to the direction of compressive stress; and (2) unstiffened elements; those supported along one edge and free on the other edge parallel to direction of compressive stress. Refer to Fig. 6.15.1 for typical examples of these two situations.

For the elements shown in Fig. 6.15.1 various degrees of edge rotation restraint are present. Fig. 6.15.2 shows the variation in k with aspect ratio a/b for most of the idealized edge conditions, i.e., clamped (fixed), simply supported, and free.

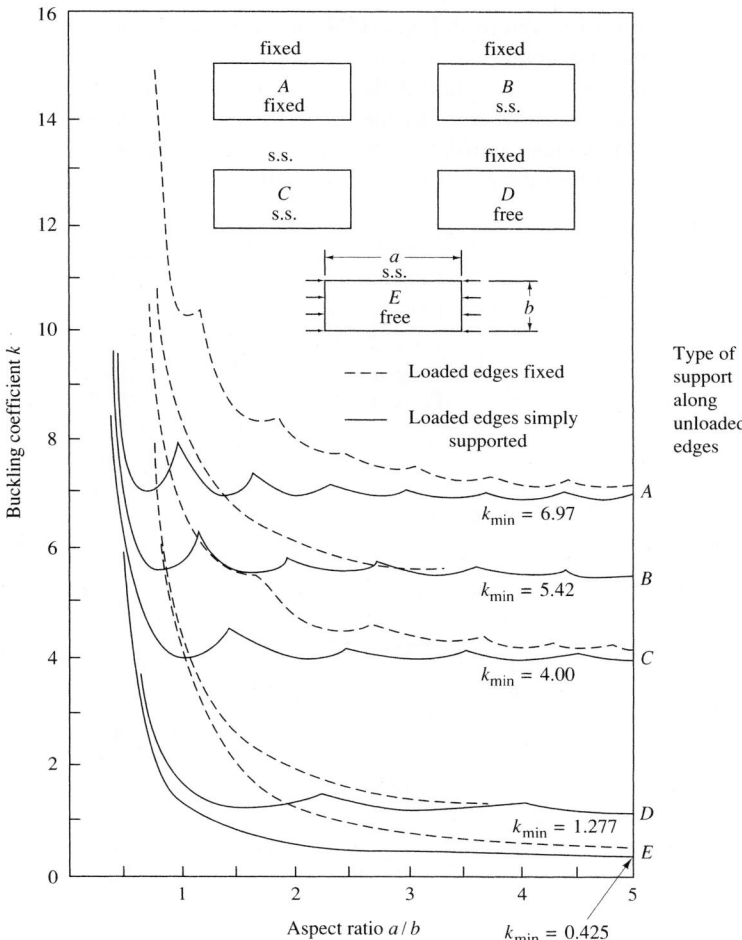

Figure 6.15.2
Elastic buckling coefficients
for compression in flat
rectangular plates.
(Adapted from Gerard
and Becker [6.69])

Actual plate strength in compression is dependent on many of the same factors that affect overall column strength, particularly residual stress. Figure 6.15.3 shows typical behavior of a compressed plate loaded to its ultimate load. Assuming ideal elastic-plastic material containing no residual stress the stress distribution remains uniform until the elastic buckling stress F_{cr} is reached. Further increase in load can be achieved but the portion of the plate farthest from its side supports will deflect out of its original plane. This out-of-plane deflection causes the stress distribution to be nonuniform even though the load is applied through ends which are rigid and perfectly straight.

Figure 6.15.3 shows that plate strength under edge compression consists of the sum of two components; (1) elastic or inelastic buckling stress represented by Eq. 6.14.28, and (2) post-buckling strength. Also one should note the higher post-buckling strength as the width-to-thickness ratio b/t becomes larger. For low values of b/t, not only will post-buckling strength vanish, but the entire plate may have yielded and reached the strain-hardening condition, so that F_{cr}/F_y may become greater than unity. For plates *without* residual stress (referring to Fig. 6.15.4) three regions must be considered for establishing strength; elastic buckling (Euler hyperbola), yielding (segments AB, $A'B$, and $A''B$), and strain hardening.

If F_{cr}/F_y is defined as $1/\lambda_c^2$, Eq. 6.14.28 for plates then becomes

$$\lambda_c = \frac{b}{t}\sqrt{\frac{F_y(12)(1-\mu^2)}{\pi^2 Ek}} \qquad (6.15.1)$$

It is observed from Fig. 6.15.4 that, when compared with columns (curve a), plates (curves b and c) achieve a strain hardening condition at relatively higher values of λ_c. In the earlier discussion on columns the value of λ_c at which strain hardening commences (λ_0) was assumed to be zero because of its relatively small value. The values of λ_0 for columns and plates under uniform edge compression for $F_y = 36$ ksi (248 MPa) are given as follows from Haaijer and Thürlimann [6.70]:

Columns	$\lambda_0 = 0.173$	($KL/r = 15.7$)
Long hinged flanges	$\lambda_0 = 0.455$	($b/t = 8.15$)
Fixed flanges	$\lambda_0 = 0.461$	($b/t = 14.3$)
Hinged webs	$\lambda_0 = 0.588$	($b/t = 32.3$)
Fixed webs	$\lambda_0 = 0.579$	($b/t = 42.0$)

Figure 6.15.3
Behavior of plate under edge compression.

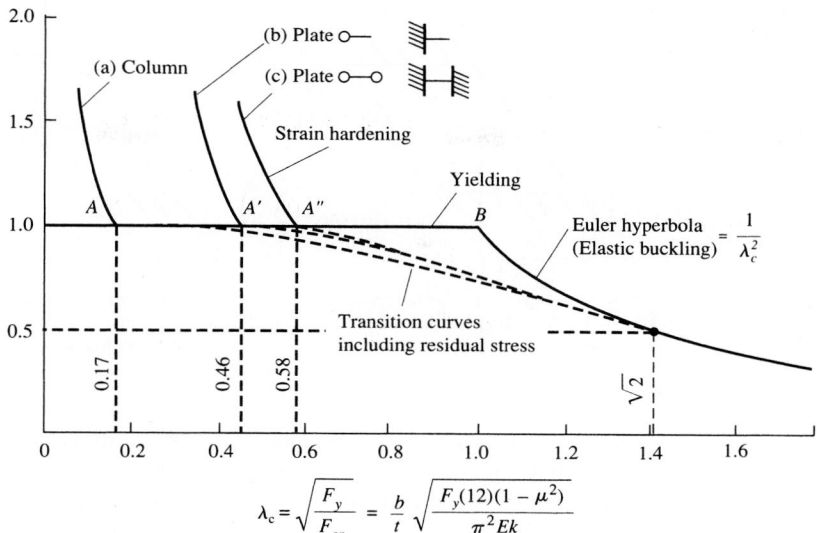

Figure 6.15.4
Plate buckling compared with column buckling. (Adapted from Ref. 6.70)

$$\lambda_c = \sqrt{\frac{F_y}{F_{cr}}} = \frac{b}{t}\sqrt{\frac{F_y(12)(1-\mu^2)}{\pi^2 Ek}}$$

From the table on the previous page, the important factor determining λ_0 is whether the plate element is supported along one or both edges parallel to loading, while the degree of rotational restraint along the loaded edge (simply supported or fixed) has essentially no effect. Thus curves b and c of Fig. 6.15.4 each can represent two cases, where point A' has been taken at $\lambda_c = 0.46$ and point A'' at $\lambda_c = 0.58$.

Since plates as well as rolled shapes contain residual stress the true strength should be represented by a transition curve, Fig. 6.15.4, between the Euler curve and the point at which strain hardening commences.

When considering inelastic behavior, the modulus of elasticity used for calculating strain in the direction of maximum stress σ_x should be the tangent modulus E_t. Examination of Eq. 6.14.4a shows that for inelastic strains in the x-direction but elastic strain in the y-direction, E cannot be factored out. Bleich [6.9] has shown the solution for this case of using different E values, and suggests arbitrarily using $\sqrt{E_t/E}$ as a multiplier for Eq. 6.14.28.

In summary, the strength of plates under edge compression may be governed by (1) strain hardening, low values of λ_c; (2) yielding, at $\lambda_c =$ say 0.5 to 0.6; (3) inelastic buckling, represented by the transition curve (some fibers elastic and some yielded); (4) elastic buckling represented by the Euler hyperbola, at λ_c about 1.4; and (5) post-buckling strength with stress redistribution and large deformation, say for λ_c greater than 1.5.

For design purposes, performance criteria must be established to decide what range of λ_c values may be acceptable in design and how conservative (and simple) or liberal (and relatively complicated) should be the specification expressions for plate strength.

6.16   AISC WIDTH/THICKNESS LIMITS λ_r TO ACHIEVE YIELD STRESS WITHOUT LOCAL PLATE BUCKLING

For a better understanding of the background for these requirements the reader is invited to delve into the subject of plate stability and strength as introduced in Secs. 6.14 and 6.15. However, it may be sufficient for many purposes merely to understand that components such as flanges, webs, angles, and cover plates, which are combined to form a column section may themselves buckle locally prior to the entire section achieving its maximum capacity. Typical elements are shown in Fig. 6.15.1. The buckled deflection of uniformly compressed plates is shown in Fig. 6.16.1 where two categories

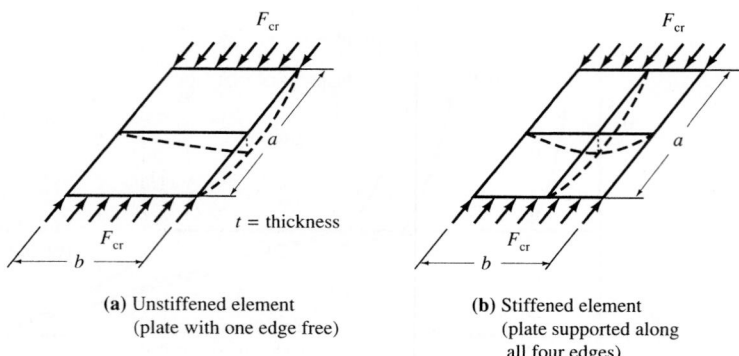

Figure 6.16.1
Buckled deflection of
uniformly compressed
plates.

(a) Unstiffened element
(plate with one edge free)

(b) Stiffened element
(plate supported along
all four edges)

are apparent: (1) "unstiffened" plate elements having one free edge parallel to loading; and (2) "stiffened" plate elements supported along both edges parallel to loading.

Plates in compression behave essentially the same as columns and the basic elastic buckling expression corresponding to the Euler equation for columns has been derived as Eq. 6.14.28,

$$F_{cr} = k \frac{\pi^2 E}{12(1 - \mu^2)(b/t)^2} \qquad [6.14.28]$$

where k is a constant depending on type of stress, edge conditions, and length to width ratio; μ is Poisson's ratio, and b/t is the width/thickness ratio (see Fig. 6.16.1). Typical k values are given in Fig. 6.15.2.

It is known that for low b/t values, strain hardening is achieved without buckling occurring, for medium values of b/t residual stress and imperfections give rise to inelastic buckling represented by a transition curve, and for large b/t buckling occurs in accordance with Eq. 6.14.28. Actual strength for plates with large b/t ratio exceeds buckling strength, i.e., they exhibit post-buckling strength. Thus strength for plates may be shown in a dimensionless fashion as in Fig. 6.16.2.

To establish design requirements, the desired performance must be ascertained. The local buckling of a column component may logically be prevented prior to achieving full strength of the column based on its overall slenderness ratio KL/r. The performance requirement would then be

$$\underset{\substack{\text{component}\\\text{element,}\\\text{i.e., plate}}}{F_{cr}} \geq \underset{\substack{\text{overall}\\\text{column}}}{F_{cr}} \qquad (6.16.1)$$

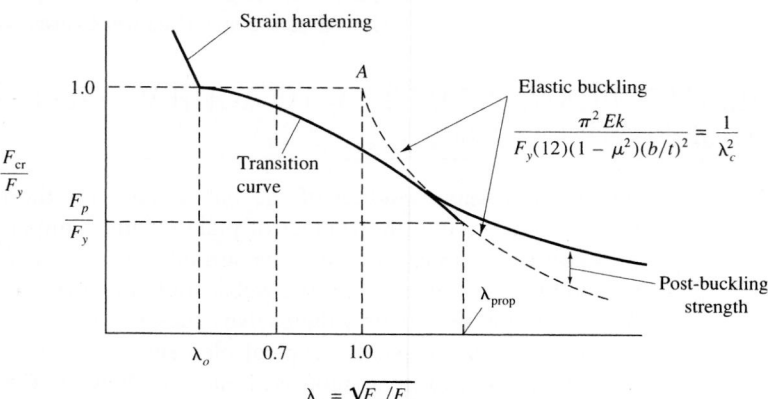

Figure 6.16.2
Dimensionless representation
of plate strength in edge
compression.

meaning that acceptable b/t ratios would vary depending on the overall slenderness ratio of the column. If post-buckling strength were considered, the relationship would be additionally complicated.

For many years the *AISI Specification* [1.8] for cold-formed steel has used the approach of Eq. 6.16.1, and also included treatment of post-buckling strength. AISC-E7 also includes similar provisions to consider post-buckling strength. Once buckling of a compression element has occurred, the efficiency of the element to carry load is reduced.

Design limits are generally simplified to assure the compression element will reach yield stress F_y without local buckling occurring, even though the slenderness ratio of a column may prevent the element from reaching yield stress. The width/thickness (b/t) ratios λ to prevent local buckling until the yield stress is reached are the λ_r values of AISC Table B4.1

The requirement to achieve yield stress without local buckling is

$$F_{cr} = \frac{k\pi^2 E}{12(1 - \mu^2)(b/t)^2} \geq F_y \tag{6.16.2}$$

Using $\mu = 0.3$ for steel, $E = 29{,}000$ ksi, and F_y in ksi,

$$\frac{b}{t} \leq \left[0.951\sqrt{\frac{kE}{F_y}} = 162\sqrt{\frac{k}{F_y,\text{ ksi}}} \right] \tag{6.16.3*}$$

which is represented by point A at $\lambda_c = 1.0$ on Fig. 6.16.2, a point lying above the transition curve. Thus a reduced value of λ_c should be used to minimize the deviation between F_y and the transition curve which accounts for residual stress and imperfections. Thus, $\lambda_c = 0.7$ is taken as a rational value, which gives for b/t

$$\frac{b}{t} \leq \left[0.951\lambda_c\sqrt{\frac{kE}{F_y}} = 113\sqrt{\frac{k}{F_y,\text{ ksi}}} \right] \tag{6.16.4}$$

where F_y is in ksi. Table 6.16.1 shows width/thickness for various situations of uniform compression. The coefficients used by AISC since 1969 tend to imply greater accuracy in criteria than is justified. The original coefficients were established using F_y in psi; after some rounding they formed the basis of the 1963 AISC Specfication. Table 6.16.2 gives evaluated limits for λ_r, the slenderness limit for noncompact elements, from AISC-Table B4-1 (p. 16.1-16)

For rolled and welded box shapes, *tensile* residual stresses are induced at the corners as discussed in Sec. 6.5. When compressive external loads are applied, the tensile residual stress F_r must first be reduced to zero, afterwhich additional compressive load causes compressive stresses at the corners. Thus, when the nominal stress P/A_g is F_y on the cross-section, it is actually $(F_y - F_r)$ at the important regions near the corners of the box shape. Since the superimposed stress level is actually less than F_y, the limit λ_r could be permitted higher than if the residual stress were compressive.

For stiffened elements, such as uniform thickness plates and perforated plates, the 1986 LRFD Specification used $\sqrt{(F_y - F_r)}$ in place of $\sqrt{F_y}$ in the denominator of the width/thickness ratio limits. The 1989 ASD Specification retained the more conservative equations (without the subtracted F_r) used for the past several decades. Since the 1993 LRFD Specification, AISC has reverted to those limit expressions historically used.

For unstiffened elements, the stress at the free edge is predominant and residual stress is generally low or compressive; thus no reduction F_r is (or was) appropriate.

---

*For SI, $\dfrac{b}{t} \leq 425\sqrt{\dfrac{k}{F_y}}$ with F_y in MPa $\hspace{2cm}$ (6.16.3)

TABLE 6.16.1 AISC Width/Thickness Ratio λ_r Limits for Plate Elements Subject to Uniform Compression[a]

Structural elements (1)	Buckling coefficients k (Fig. 6.15.2) (2)	b/t Eq. 6.16.4 (3)	F_y(psi) (4)	AISC-B4 (5)
Unstiffened:				
(a) Single angles	0.425	$2340/\sqrt{F_y}$	$2400/\sqrt{F_y}$	$0.45\sqrt{E/F_y}$
(b) Flanges[d]	0.70^a	$3000/\sqrt{F_y}$	$3000/\sqrt{F_y}$	$0.56\sqrt{E/F_y}$
(c) Stems of tees	1.277	$4050/\sqrt{F_y}$	$4000/\sqrt{F_y}$	$0.75\sqrt{E/F_y}$
Stiffened:				
(a) Uniform thickness flanges, such as tubular sections			$7500/\sqrt{F_y}^{\ c}$	$140\sqrt{E/F_y}$
(b) Webs of W sections and other stiffened elements	5.0^b	$8010/\sqrt{F_y}$	$8000/\sqrt{F_y}$	$1.49\sqrt{E/F_y}$

[a] Arbitrarily selected to be about one-third of the way between simply supported and fixed along the supported edge.

[b] Edge restraint estimated at about $\frac{1}{3}$ fixed ($k = 4.0$ for simple support, and $k = 6.97$ for fixed—see Fig. 6.15.2).

[c] Hollow sections generally receive negligble torsional restraint by the thin supporting edges; thus a coefficient somewhat less than 8000 is used.

[d] AISC-B4.2 explicitly mentions *flanges* of rolled I-shaped sections in uniform compression; included are outstanding legs of pairs of angles in contact, flanges of channels in axial compression, and angles and plates projecting from beams or compression members. All of these are unstiffened elements having substantial rotational restraint along their supported edge.

6.17 AISC WIDTH/THICKNESS LIMIT λ_p TO ACHIEVE SIGNIFICANT PLASTIC DEFORMATION

Sometimes plate elements of the cross-section must not buckle until they have undergone significant compressive strain exceeding the strain ϵ_y at first yield, that is, strain into the plastic region, as shown in Fig. 6.17.1. The lower the width/thickness ratio, the greater the compressive strain ϵ that can be absorbed without buckling. In axially loaded columns there would be no need for the ability to undergo plastic strain because the overall strength of the column based on its KL/r would not require plastic deformation. However in beams, as discussed in Chapter 7, the flanges might be required to undergo significant plastic strain without having local buckling occur.

Referring to Fig. 6.16.2, λ_c must be restricted not to exceed λ_0 if strain hardening is to be reached without plate buckling. From Fig. 6.15.4, λ_c should not exceed about 0.46 for unstiffened compression elements and 0.58 for stiffened compression elements.

For *unstiffened elements*, Eq. 6.16.4 with $\lambda_c = 0.46$ gives

$$\frac{b}{t} \leq \left[0.437\sqrt{\frac{kE}{F_y}} = 74.4\sqrt{\frac{k}{F_y, \text{ksi}}} \right] \qquad (6.17.1)^*$$

TABLE 6.16.2 Width/Thickness Ratio λ_r Limiting Values for Plate Elements to Reach Yield Stress in Axial Compression

Structural elements	AISC-B4 F_y(ksi)					
	36	42	50	60	65	100
Unstiffened:						
(a) Single angles	12.8	11.8	10.8	9.9	9.5	7.7
(b) Flanges	15.9	14.7	13.5	12.3	11.8	9.5
(c) Stems of tees	21.3	19.7	18.1	16.5	15.8	12.8
Stiffened:						
(a) Uniform thickness flanges, as for hollow structural sections	39.7	36.8	33.7	30.8	29.6	23.8
(b) Webs of W sections and other stiffened elements	42.3	39.2	35.9	32.8	31.5	25.4
Unstiffened:						
(a) Single angles	12.6	11.5	10.7	10.0	9.4	7.5
(b) Flanges	15.8	14.4	13.4	12.5	11.8	9.4
(b) Stems of tees	21.1	19.2	17.8	16.7	15.7	12.6
Stiffened:						
(a) Uniform thickness flanges, as for tubular sections	39.5	36.1	33.4	31.3	29.5	23.6
(b) Perforated plates	52.6	48.0	44.5	41.6	39.2	31.4
(c) Others	42.0	38.3	35.5	33.2	31.3	25.1

For unstiffened elements, the stress at the free edge is predominant and residual stress is generally low or compressive; thus no reduction F_r is appropriate.

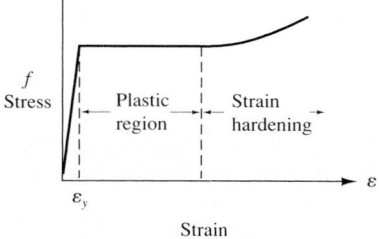

Figure 6.17.1
Plastic and strain-hardening regions of stress-strain relationship for steel.

When $k = 0.425$ (its least value), Eq. 6.17.1 gives

$$\frac{b}{t} \leq \left[0.284\sqrt{\frac{E}{F_y}} = \frac{48.5}{\sqrt{F_y, \text{ksi}}} \right] \qquad (6.17.2)^*$$

$$74.4 \times \sqrt{6.8949} = 195 \text{ etc.}$$

---

*For SI, with F_y in MPa,

$$\frac{b}{t} \leq 195\sqrt{\frac{k}{F_y, \text{MP}_a}} \qquad (6.17.1)$$

$$\frac{b}{t} \leq \frac{127}{F_y, \text{MP}_a} \qquad (6.17.2)$$

Since residual stress effects disappear in the plastic range and material imperfections have less effect, Eq. 6.17.2 is an overly severe limitation. Furthermore, the strain at onset of strain hardening is 15 to 20 times ϵ_y and that amount of plastic strain is not necessary even for achieving the plastic moment strength discussed in Chapter 7. The AISC-B4.1 λ_p compact section limits are intended to achieve compression plastic strain about 7 to 9 times ϵ_y, about one-half the strain necessary to reach strain hardening. Thus, the unstiffened compression element limit λ_p is

$$\frac{b}{t} \leq \left[0.38\sqrt{\frac{E}{F_y}} = \frac{64.7}{\sqrt{F_y, \text{ksi}}} \right] \tag{6.17.3}$$

For stiffened elements, Eq. 6.16.4 with $\lambda_c = 0.58$ gives

$$\frac{b}{t} \leq \left[0.55\sqrt{\frac{kE}{F_y}} = 93.7\sqrt{\frac{k}{F_y, \text{ksi}}} \right] \tag{6.17.4}$$

When $k = 4$, the minimum value assuming edge rotational restraint as the hinged condition (actually the k lies somewhere between values for Cases A and C of Fig. 6.15.2), Eq. 6.17.4 gives

$$\frac{b}{t} \leq \left[1.1\sqrt{\frac{E}{F_y}} = \frac{187}{\sqrt{F_y, \text{ksi}}} \right] \tag{6.17.5}$$

AISC-B4.2 gives the limit for uniformly compressed stiffened flange elements as

$$\frac{b}{t} \leq \left[1.12\sqrt{\frac{E}{F_y}} = \frac{190}{\sqrt{F_y, \text{ksi}}} \right] \tag{6.17.6}$$

A tabulation of the limits, Eq. 6.17.3 for unstiffened elements and Eq. 6.17.6 for stiffened elements, appears in Table 6.17.1. Additional discussions relative to local buckling limitations to develop plastic strength are given by Lay [6.71] and McDermott [6.72].

TABLE 6.17.1 Width/Thickness Ratio λ_p Limits for Plate Elements to Accommodate Plastic Strain in Axial Compression

F_y (ksi)	F_y (MPa)	Unstiffened elements AISC-B4.1 (Eq. 6.17.3)	Stiffened elements AISC-B4.2 (Eq. 6.17.6)
36	250	10.8	31.8
42	290	10.0	29.4
45	310	9.7	28.4
50	340	9.2	27.0
55	380	8.8	25.7
60	410	8.4	24.6
65	450	8.0	23.7

6.18* AISC PROVISIONS TO ACCOUNT FOR THE BUCKLING AND POST-BUCKLING STRENGTHS OF PLATE ELEMENTS

As discussed in Secs. 6.15 and 6.16, plate elements in compression, either "stiffened" or "unstiffened" (see Fig. 6.16.1), have strength after buckling has occurred, i.e., post-buckling strength. Stiffened elements have a large post-buckling strength while unstiffened elements have only a little. However, since the strength of such elements can be evaluated, there is good reason to provide for its use as has been done in the *North American Specification for the Design of Cold-Formed Steel Structural Members* [1.8], first introduced in 1946.

From Fig. 6.18.1a it is apparent that the nominal strength P_n of a stiffened element might be expressed as

$$P_n = t \int_0^b f(x)\, dx \tag{6.18.1}$$

involving an integration of a nonuniform stress distribution; or alternatively, an "effective width" concept (Fig. 6.18.1b) may be used:

$$P_n = tb_E f_{max} = A_{eff} f_{max} \qquad \text{(stiffened element)} \tag{6.18.2}$$

where b_E = effective width over which the maximum stress may be considered uniform and give the correct total capacity.

Figure 6.18.1c shows that Eq. 6.18.1 is equally valid for the unstiffened element except that the stress distribution is not symmetrical about the center of the element. If a reduced stress $f_{avg} < f_{max}$ is used, the unstiffened element capacity could be written (Fig. 6.18.1d),

$$P_n = tbf_{avg} = A_{gross} f_{avg} \ \text{(unstiffened element)} \tag{6.18.3}$$

The AISC and the AISI [1.8] have chosen to treat thin elements according to Eqs. 6.18.2 and 6.18.3, although actually either equation could have been used for either type of element. Because of the large post-buckling strength of the stiffened element, one can imagine that it *has* buckled and part of the element is no longer active. On the other hand, the unstiffened element, with relatively little post-buckling strength may be thought of as not buckling because of the use of a reduced stress.

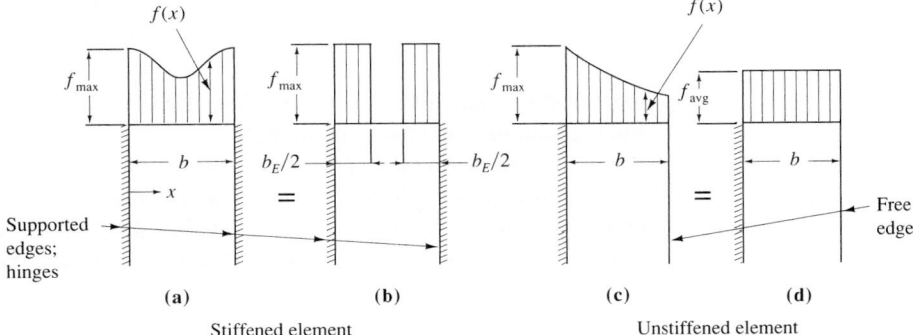

Figure 6.18.1
Plate elements under axial compression, showing actual stress distribution and an equivalent system.

Effect on Overall Column Strength

For design, it is desired to use gross section properties; thus for stiffened elements

$$P_n = \frac{A_{eff}}{A_{gross}} f_{max} A_{gross} = Q_a f_{max} A_{gross} \tag{6.18.4}$$

and for unstiffened elements,

$$P_n = \frac{f_{\text{avg}}}{f_{\text{max}}} f_{\text{max}} A_{\text{gross}} = Q_s f_{\text{max}} A_{\text{gross}} \tag{6.18.5}$$

where Q_a and Q_s may be thought of as shape factors, or form factors.

A compression system composed of both stiffened and unstiffened elements would be treated as unstiffened for establishing the stress f_{avg}; then the effective width for the stiffened elements is determined using $f_{\text{max}} = f_{\text{avg}}$. Thus the total capacity would be

$$P_n = f_{\text{avg}} A_{\text{eff}} \tag{6.18.6}$$

which gives

$$P_n = \frac{f_{\text{avg}}}{f_{\text{max}}} (f_{\text{max}}) \frac{A_{\text{eff}}}{A_{\text{gross}}} (A_{\text{gross}}) = Q_s Q_a f_{\text{max}} A_{\text{gross}} \tag{6.18.7}$$

From Eqs. 6.18.4, 6.18.5, and 6.18.7, it is clear that the effect of premature local buckling before the strength of the overall column has been achieved is to multiply the maximum achievable stress by the form factors Q. Neglecting the possibility of strain hardening, the maximum stress is the yield stress, which is therefore to be multiplied by Q.

The nominal strength of a compression member having slender elements must include the effect of reduced efficiency of the cross-section by including the form, or shape, factor Q in accordance with AISC-E7. The effect of local buckling of slender elements is accounted for by replacing F_y by QF_y. The short column equation, Eq. 6.8.2, is

$$\text{When } \frac{KL}{r} \leq 4.71 \sqrt{\frac{E}{QF_y}} \quad \text{or} \quad F_e \geq 0.44 Q F_y$$

$$F_{\text{cr}} = \left[0.658^{\frac{QF_y}{F_e}} \right] Q F_y \tag{6.8.2}$$

It is only for short columns where the local buckling of cross-sectional elements reduces the efficiency; for long columns where

$$\frac{KL}{r} > 4.71 \sqrt{\frac{E}{QF_y}}$$

local buckling has little effect,; therefore AISC-E7 neglects such effect and considers the Euler equation as the basis for strength.

Since the form factor Q is used to account for the reduction in efficiency of the cross-section instead of modifying the cross-sectional properties, whether or not Q is less than unity, the nominal strength P_n is computed as the product of F_{cr} times A_g, and the radius of gyration r is that of the *gross* section in the computation of KL/r.

Factor Q_s for Unstiffened Elements

AISC-E7.1 gives the equations for Q_s for four different types of shapes; rolled W shapes where the flanges are the unstiffened elements, built-up shapes, single angles and stems of tees. Referring to Fig. 6.16.2, one may note that when the slenderness parameter λ_c is larger than 0.7, Q is less than 1.0. For inelastic buckling, a transition parabola could have been used to compute the reduced stress. For simplification, a straight line has been used, as shown in Fig. 6.18.2. The straight line uses $\lambda_c = 0.7$ as the maximum for which $F_{\text{cr}} = F_y$ and intersects the elastic buckling curve within a small range of possible

Figure 6.18.2
Definition of Q_s for
unstiffened slender
element, [1.15].

maximum slenderness values, dictated by type of shape according to AISC-E7.1. Note that it is not so much the shape *per se* (angle, T-shape, or W shape) that requires a different Q but it is an evaluation of the relative moment restraint along the supported other leg; the legs would tend to buckle simultaneously by rotating about the axis at the junction of the two legs. This case is clearly a weaker one as compared to unstiffened elements (flanges) of W shapes. In addition to elastic buckling strength, there is some post-buckling strength; thus the elastic buckling curve used for design has been raised to account for some post-buckling strength, as shown in Fig. 6.18.2.

Although Q_s has been defined to this point as F_{cr}/F_y, when overall buckling of a column occurs (based on its KL/r) the average stress P_u/A_g is always *less* than F_y. This means local buckling of an unstiffened element will reduce the efficiency of the cross-section *only* when $F_{cr,plate}$ for the plate element is less than $F_{cr,column}$.

Thus, in general, for columns

$$Q_s = \frac{F_{cr,plate}}{F_{cr,column}} \geq \frac{F_{cr,plate}}{F_y} \tag{6.18.8}$$

and for compression flanges of beams,

$$Q_s = \frac{F_{cr,plate}}{F_{cr,beam\ flange}} \geq \frac{F_{cr,plate}}{F_y} \tag{6.18.9}$$

The AISC-E7.1 gives the equations for Q_s:

1. For flanges, angles, and plates projecting from rolled column sections, or other compression members:

a. When $\dfrac{b}{t} \leq \left[0.56\sqrt{\dfrac{E}{F_y}} = \dfrac{95}{\sqrt{F_y, ksi}} \right]$

$$Q_s = 1.0 \tag{6.18.10}$$

b. When $\dfrac{95}{\sqrt{F_y, ksi}} < \dfrac{b}{t} \leq \left[1.03\sqrt{\dfrac{E}{F_y}} = \dfrac{175}{\sqrt{F_y, ksi}} \right]$

$$Q_s = \left[\left(1.415 - 0.74\left(\frac{b}{t}\right)\sqrt{\frac{F_y}{E}} \right) = \left(1.415 - 0.00435\left(\frac{b}{t}\right)\sqrt{F_y, ksi} \right) \right] \tag{6.18.11}$$

c. When $\dfrac{b}{t} > \dfrac{175}{\sqrt{F_y,\text{ksi}}}$

$$Q_s = \left[\frac{0.69E}{F_y\left(\dfrac{b}{t}\right)^2} = \frac{20,000}{(F_y,\text{ksi})\left(\dfrac{b}{t}\right)^2} \right] \tag{6.18.12}$$

where b = width of unstiffened compression element, defined in AISC-B4, and t = thickness of element.

2. For flanges, angles, and plates projecting from built-up columns, or other compression members:

a. When $\dfrac{b}{t} \le \left[0.64\sqrt{\dfrac{Ek_c}{F_y}} = \dfrac{109\sqrt{k_c}}{\sqrt{F_y,\text{ksi}}} \right]$

$$Q_s = 1.0 \tag{6.18.13}$$

b. When $\dfrac{109\sqrt{k_c}}{\sqrt{F_y,\text{ksi}}} < \dfrac{b}{t} \le \left[1.17\sqrt{\dfrac{Ek_c}{F_y}} = \dfrac{199\sqrt{k_c}}{\sqrt{F_y,\text{ksi}}} \right]$

$$Q_s = \left[\left(1.415 - 0.65\left(\frac{b}{t}\right)\sqrt{\frac{F_y}{Ek_c}} \right) = \left(1.415 - 0.00382\left(\frac{b}{t}\right)\sqrt{\frac{F_y,\text{ksi}}{k_c}} \right) \right] \tag{6.18.14}$$

c. When $\dfrac{b}{t} > \dfrac{199\sqrt{k_c}}{\sqrt{F_y,\text{ksi}}}$

$$Q_s = \left[\frac{0.90Ek_c}{F_y\left(\dfrac{b}{t}\right)^2} = \frac{26,100k_c}{(F_y,\text{ksi})\left(\dfrac{b}{t}\right)^2} \right] \tag{6.18.15}$$

where $k_c = \dfrac{4}{\sqrt{h/t_w}}$ (*Note:* for calculation purposes, $0.35 \le k_c \le 0.76$). Note that h = clear distance between flanges less the fillet or corner radius, or clear distance between flanges when welds are used, and t_w = web thickness.

3. For single angles:

a. When $\dfrac{b}{t} \le \left[0.45\sqrt{\dfrac{E}{F_y}} = \dfrac{76.6}{\sqrt{F_y,\text{ksi}}} \right]$

$$Q_s = 1.0 \tag{6.18.16}$$

b. When $\dfrac{76.6}{\sqrt{F_y,\text{ksi}}} < \dfrac{b}{t} \le \left[0.91\sqrt{\dfrac{E}{F_y}} = \dfrac{155}{\sqrt{F_y,\text{ksi}}} \right]$

$$Q_s = \left[\left(1.34 - 0.76\left(\frac{b}{t}\right)\sqrt{\frac{F_y}{E}} \right) = \left(1.34 - 0.00446\left(\frac{b}{t}\right)\sqrt{F_y,\text{ksi}} \right) \right] \tag{6.18.17}$$

c. When $\dfrac{b}{t} > \dfrac{155}{\sqrt{F_y, \text{ksi}}}$

$$Q_s = \left[\frac{0.53E}{F_y \left(\dfrac{b}{t} \right)^2} = \frac{15,400}{(F_y, \text{ksi}) \left(\dfrac{b}{t} \right)^2} \right] \tag{6.18.18}$$

4. For stems of tees:

a. When $\dfrac{d}{t} \leq \left[0.75 \sqrt{\dfrac{E}{F_y}} = \dfrac{128}{\sqrt{F_y, \text{ksi}}} \right]$

$$Q_s = 1.0 \tag{6.18.19}$$

b. When $\dfrac{128}{\sqrt{F_y, \text{ksi}}} < \dfrac{d}{t} \leq \left[1.03 \sqrt{\dfrac{E}{F_y}} = \dfrac{175}{\sqrt{F_y, \text{ksi}}} \right]$

$$Q_s = \left[1.908 - 1.22 \left(\frac{d}{t} \right) \sqrt{\frac{F_y}{E}} = 1.908 - 0.00716 \left(\frac{b}{t} \right) \sqrt{F_y, \text{ksi}} \right] \tag{6.18.20}$$

c. When $\dfrac{d}{t} > \dfrac{175}{\sqrt{F_y, \text{ksi}}}$

$$Q_s = \left[\frac{0.69E}{F_y \left(\dfrac{d}{t} \right)^2} = \frac{20,000}{(F_y, \text{ksi}) \left(\dfrac{b}{t} \right)^2} \right] \tag{6.18.21}$$

where d = the full depth of tee.

Reduction Factor Q_a for Stiffened Elements

The concept of using an effective width over which stress may be considered uniform, even though it is actually nonuniform, was developed by von Kármán [6.73] and later modified by Winter [6.74]. Winter's equation, the format of which has been used by the *AISI Specification* [1.8] since 1946, is

$$\frac{b_E}{t} = 1.9 \sqrt{\frac{E}{f}} \left[1.0 - \frac{0.475}{(b/t)} \sqrt{\frac{E}{f}} \right] \tag{6.18.22}$$

where $f = P_u/A_g$, stress acting on the element ($f_{\max}$ of Fig. 6.18.1)

b/t = actual width/thickness ratio

The form and constants of the equation were essentially determined to agree with experimental results.

Substituting $E = 29,000$ ksi gives

$$\frac{b_E}{t} = \frac{324}{\sqrt{f}} \left[1.0 - \frac{81}{(b/t)\sqrt{f}} \right] \tag{6.18.23}$$

Experience with light gage steel showed that the coefficient of the second term in Eq. 6.18.23 could be reduced. Since 1962 the equation used by both the AISI Specification

[1.8] and the AISC Specification has been essentially unchanged. In AISC-E7.2 are given the following:

1. For uniformly compressed slender elements, except flanges of square and rectangular sections of uniform thickness, when

$$\frac{b}{t} > \left[1.49\sqrt{\frac{E}{f}} = \frac{254}{\sqrt{f,\text{ksi}}} \right]$$

$$\frac{b_E}{t} = \left\{ 1.92\sqrt{\frac{E}{f}}\left[1 - \frac{0.34}{(b/t)}\sqrt{\frac{E}{f}} \right] \quad \text{or} \quad \frac{327}{\sqrt{f,\text{ksi}}}\left[1 - \frac{57.9}{(b/t)\sqrt{f,\text{ksi}}} \right] \right\} \tag{6.18.24}$$

where f is taken as F_{cr} calculated using $Q = 1.0$.

2. For flanges of square and rectangular slender elements of uniform thickness, when

$$\frac{b}{t} > \left[1.40\sqrt{\frac{E}{f}} = \frac{238}{\sqrt{f,\text{ksi}}} \right]$$

$$\frac{b_E}{t} = \left\{ 1.92\sqrt{\frac{E}{f}}\left[1 - \frac{0.38}{(b/t)}\sqrt{\frac{E}{f}} \right] \quad \text{or} \quad \frac{327}{\sqrt{f,\text{ksi}}}\left[1 - \frac{64.7}{(b/t)\sqrt{f,\text{ksi}}} \right] \right\} \tag{6.18.25}$$

where $f = P_n/A_{\text{eff}}$. To avoid iteration in calculating the correct f value, f can conservatively be taken as F_y.

3. For axially-loaded circular sections, when

$$0.11\frac{E}{F_y} < \frac{D}{t} < 0.45\frac{E}{F_y}$$

$$Q = Q_a = \left[\frac{0.038E}{F_y\,(D/t)} + \frac{2}{3} \quad \text{or} \quad \frac{1100}{(F_y,\text{ksi})(D/t)} + \frac{2}{3} \right] \tag{6.18.26}$$

where D is the outside diameter of the section, and t is the wall thickness.

The difference between Eqs. 6.18.25 and 6.18.26 can be explained partly by the difference in the assumed rotational restraint (moment along the supported edges, Fig. 6.18.1b).

Additional discussion of the effective width for stiffened elements in compression is available in the work of Korol and Sherbourne [6.75, 6.76], Dawson and Walker [6.77], and Abdel-Sayed [6.78]. Sharp [6.79] has considered stiffened elements having one edge stiffened by a lip. Kalyanaraman, Pekoz, and Winter [6.80] have proposed an effective width expression for unstiffened elements.

Since a column cross-section may include unstiffened elements, which under present design procedures utilize reduced average stress rather than an effective width, the controlling stress on unstiffened elements is used as the applicable maximum stress acting on the stiffened elements. Thus the stress f is

$$f = \frac{P_u}{A_g} \tag{6.18.27}$$

and the design requirement is that

$$P_u = \phi_c(F_{\text{cr,plate}})A_g \tag{6.18.28}$$

Substituting Eq. 6.18.27 into Eq. 6.18.28 shows that

$$f = \phi_c F_{cr,plate} \tag{6.18.29}$$

and from Eq. 6.18.28,

$$F_{cr,plate} = Q_s F_{cr,column} \tag{6.18.30}$$

Thus,

$$f = \phi_c Q_s F_{cr,column} \tag{6.18.31}$$

Note that $Q_s F_{cr,column}$ is F_{cr} in AISC-E7.2. In other words, when $Q < 1$, F_{cr} is redefined as QF_{cr} in the *AISC Specification*.

Finally, Q_a as defined by Eq. 6.18.4 is

$$Q_a = \frac{A_{eff}}{A_{gross}} = \frac{b_E t}{bt} \tag{6.18.4}$$

where $A_{eff} = A_{gross} - \Sigma(b - b_E)t$.

Design Properties

In computing the nominal strength, the following rules apply in accordance with AISC-E7.
For axial compression:

1. Use gross area A_g for $P_n = F_{cr}A_g$.

2. Use gross area to compute radius of gyration r for *KL/r*.

For flexure:

1. Use reduced section properties for beams with flanges containing stiffened elements.

Since the strengths of beams do not include Q factors relating to thin compression elements, it is appropriate to use section properties based on effective area.
For beam-columns:

1. Use gross area for P_n.

2. Use reduced section properties for flexure involving stiffened compression elements for M_{nx} and M_{ny}.

3. Use Q_a and Q_s for determining P_n.

4. For F_{cr} based on lateral-torsional buckling of beams as discussed in Chapter 9, the maximum value of F_{cr} is $Q_s F_{cr}$ when unstiffened compression elements are involved.

6.19   DESIGN OF COMPRESSION MEMBERS AS AFFECTED BY LOCAL BUCKLING PROVISIONS

Design of single and double angle struts, structural tees, welded built-up I-shapes, and most other built-up sections, including box-type sections, involves the close attention to width/thickness limitations to prevent local buckling. Most rolled W, S, and M shapes have proportions such that local buckling will not occur (that is,) prior to achieving the strength of the section based on the overall slenderness ratio *KL/r*.

The following examples illustrate situations where $Q < 1.0$.

EXAMPLE 6.19.1

A double angle compression chord member for the truss of Fig. 6.19.1 consists of 2—L8×4×$\frac{1}{2}$, having short legs back-to-back. The 28-ft-long member is braced in the plane of the truss every 7 ft, but only at the ends in the transverse direction. Assume the two angles are attached together with fully-tightened bolts and the spacing of bolts is close enough that the double angle member reaches its maximum axial compression strength*. Neglect any contribution to lateral support from the roofing. Compute the maximum axial service compression load this member can be permitted to carry. The service load is 30% dead load and 70% gravity live load. Use A572 Grade 50 steel and the AISC LRFD Method.

Figure 6.19.1
Example 6.19.1.

Solution:

(a) Check local buckling. The 8-in. legs of this double angle member are unstiffened compression elements. Check whether or not the width/thickness ratio λ exceeds λ_r of AISC-E7.1. The authors conservatively use Eqs. 6.18.16, 6.18.17, or 6.18.18 for single angles; one could argue this case should use Eqs. 6.18.10, 6.18.11, 6.18.12. For Eq. 6.18.16,

$$\left(\lambda = \frac{b}{t} = \frac{8}{0.50} = 16.0\right) > \left(\lambda_r = \frac{76.6}{\sqrt{F_y, \text{ksi}}} = \frac{76.6}{\sqrt{50}} = 10.8\right)$$

Since $\lambda > \lambda_r$, local buckling will reduce the section efficiency. Using AISC-E7.1 (or Eq. 6.18.17) check the next limit

$$\left(\frac{b}{t} = 16.0\right) < \left(\frac{155}{\sqrt{F_y, \text{ksi}}} = \frac{155}{\sqrt{50}} = 21.9\right)$$

$$Q_s = \left(1.34 - 0.76\left(\frac{b}{t}\right)\sqrt{\frac{F_y}{E}}\right) \qquad\qquad [6.18.17]$$

$$Q_s = 1.34 - 0.00446\left(\frac{b}{t}\right)\sqrt{F_y, \text{ksi}}$$

$$= 1.34 - 0.00446(16.0)\sqrt{50} = 0.835$$

(b) Compute design strength For axial compression, properties of the gross section are used. From the *AISC Manual* properties of double-angles with short legs back-to-back (p. 1-100),

$$A_g = 11.5 \text{ sq in.}$$

$$r_x = 1.08 \text{ in.} \qquad\qquad r_y = 3.97 \text{ in.} \left(\text{for short legs separated } \tfrac{3}{8} \text{ in.}\right)$$

---

*The possible reduction in strength relating to the connection between the two angles was discussed in Sec. 6.8 and is covered in AISC-E6.

Assuming $K = 1.0$ for truss members as discussed in Sec. 6.9,

$$\frac{(KL)_x}{r_x} = \frac{1.0(7)12}{1.08} = 78 \qquad \frac{(KL)_y}{r_y} = \frac{1.0(28)12}{3.97} = 85$$

Checking the limit of AISC-Formula (E7-2), Eq. 6.8.2,

$$\left(4.71\sqrt{\frac{E}{QF_y}} = 4.71\sqrt{\frac{29{,}000}{0.835(50)}} = 124 \right) > 85$$

Thus,

$$F_{cr} = Q\left[0.658^{\frac{QF_y}{F_c}} \right]F_y$$

where

$$F_e = \frac{\pi^2 E}{(KL/r)^2} = \frac{\pi^2(29{,}000)}{(85)^2} = 39.6 \text{ ksi}$$

$$F_{cr} = 0.835\left[0.658^{\frac{0.835(50)}{39.6}} \right]50 = 26.9 \text{ ksi}$$

$$\phi P_n = \phi F_{cr} A_g = 0.90(26.9)11.5 = 278 \text{ kips}$$

$$P_u = 1.2D + 1.6L = 1.2(0.3P) + 1.6(0.7P) = 1.48P$$

$$\phi P_n = P_u = 1.48P \qquad \therefore\ P = 188 \text{ kips (maximum service load capacity)} \qquad ∎$$

EXAMPLE 6.19.2

Design a double angle compression member for use as a spreader strut for hoisting large loads, as shown in Fig. 6.19.2a. The lifted load is 60 tons, of which 55 tons is live load including impact. The remainder is dead load. Use $F_y = 60$ ksi and Load and Resistance Factor Design.

Solution:

(a) Compute factored load P_u. At 4 to 1 slope of cable the compressive load in the strut is 120 tons (240 kips).

$$P_u = 1.2\left(\frac{5}{60}\right)(240) + 1.6\left(\frac{55}{60}\right)(240) = 376 \text{ kips}$$

(b) Estimate the slenderness ratio. Assume $K = 1.0$ and referring to Fig. 6.19.2b, use text Appendix Table A1 to estimate r,

$$r_x \approx 0.29h$$
$$r_y \approx 0.24b$$

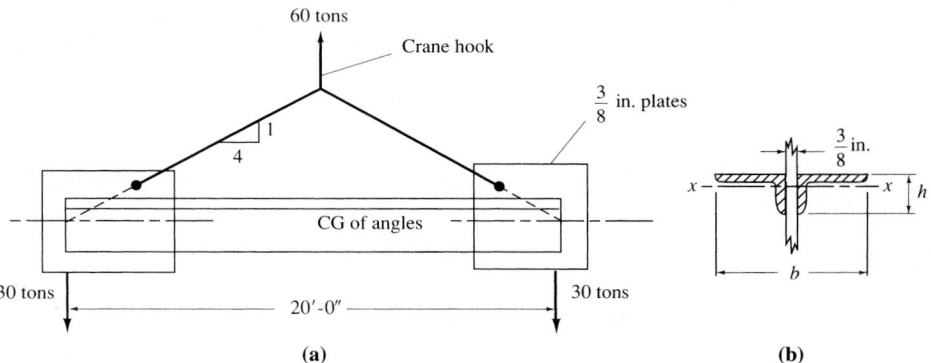

Figure 6.19.2
Example 6.19.2.

If 8-in. legs are used, $r_x \approx 0.29(8) = 2.32$ in.

$$\text{Estimate } \frac{KL}{r} \approx \frac{1.0(20)12}{2.3} = 104$$

(c) Select section. From AISC "Available Critical Stress for Compression Members" Table 4–22 (p. 4–320), estimate $\phi_c F_{cr} \approx 20.4(60/50) = 24.5$.

$$\text{Required } A_g = \frac{P_u}{\phi_c F_{cr}} = \frac{376}{24.5} = 15.4 \text{ sq in.}$$

Try L8×8×$\frac{1}{2}$, $r_{min} = 2.49$ in., $A_g = 15.5$ sq in., $KL/r = 96.4$.
The estimated $\phi_c F_{cr}$ obtained by increasing the $F_y = 50$ ksi value in the ratio of the yield stresses is probably high for KL/r over 100. Applying AISC-E7.1 (the same as Eqs. 6.18.16 and 6.18.17) for single angles,

$$\left(\lambda = \frac{b}{t} = \frac{8}{0.50} = 16.0 \right) > \left(\lambda_r = \frac{76.6}{\sqrt{F_y, \text{ksi}}} = \frac{76.6}{\sqrt{60}} = 9.9 \right)$$

Since $\lambda > \lambda_r$, local buckling will reduce the section efficiency. Using AISC-E7.1 (or Eq. 6.18.17) check the next limit

$$\left(\frac{b}{t} = 16.0 \right) < \left(\frac{155}{\sqrt{F_y, \text{ksi}}} = \frac{155}{\sqrt{60}} = 20.0 \right)$$

$$Q_s = 1.34 - 0.76 \left(\frac{b}{t} \right) \sqrt{\frac{F_y}{E}} \qquad [6.18.17]$$

$$Q_s = 1.34 - 0.00446 \left(\frac{b}{t} \right) \sqrt{F_y, \text{ksi}}$$

$$= 1.34 - 0.00446(16.0) \sqrt{60} = 0.787$$

$$Q = Q_s$$

Check whether or not this is to be treated as a short column,

$$\left(\frac{KL}{r} = 96.4 \right) < \left(4.71 \sqrt{\frac{E}{QF_y}} = 4.71 \sqrt{\frac{29{,}000}{0.787(60)}} = 117 \right)$$

Thus, it is a short column. Evaluating F_{cr},

$$F_e = \frac{\pi^2 E}{\left(\dfrac{KL}{r} \right)^2} = \frac{\pi^2(29{,}000)}{(96.4)^2} = 30.8 \text{ ksi}$$

$$F_{cr} = Q \left[0.658^{\frac{QF_y}{F_e}} \right] F_y = 0.787 \left[0.658^{\frac{0.787(60)}{30.8}} \right] 60 = 24.9 \text{ ksi}$$

$$\phi P_n = \phi_c F_{cr} A_g = 0.90(24.9)15.5 = 347 \text{ kips} < 376 \text{ kips} \qquad \textbf{NG}$$

Increase size of angles to 2-L8×8×$\frac{9}{16}$, $A_g = 17.36$ sq in.,

$$\left(\frac{KL}{r} = 96.4\right) < 117; \frac{b}{t} = 14.2$$

Revised $Q_s = 0.849$; revised $F_{cr} = 25.5$ ksi

$$\phi P_n = \phi_c F_{cr} A_g = 0.90(25.5)17.36 = 398 \text{ kips} > 376 \text{ kips}$$

$\underline{Use\ 2\text{—}L8\times8\times\frac{9}{16}}$ ∎

EXAMPLE 6.19.3

Select the thinnest 12×12 square HSS (hollow structural sections) to carry an axial compression of 60 kips dead load and 250 kips live load. The effective length KL is 18 ft and the member is part of a braced system. Use $F_y = 65$ ksi and the AISC LRFD Method. Note that while using $F_y = 65$ ksi illustrates procedure, Grade 45 is the maximum available for rectangular HSS tubes under ASTM A500.

Solution:

(a) Compute factored load P_u.

$$P_u = 1.2(60) + 1.6(250) = 472 \text{ kips}$$

(b) Obtain estimate of required area. Use text Appendix Table A1 to estimate radius of gyration.

$$r = 0.4h = 0.4(12) = 4.8 \text{ in.}$$

Using Eq. 6.8.2 for F_{cr} with first trial $Q = 1$.

$$\frac{KL}{r} = \frac{18(12)}{4.8} = 45 \qquad\qquad F_{cr} = 53.6 \text{ ksi}$$

$$\text{Required } A_g = \frac{P_u}{\phi_c F_{cr}} = \frac{472}{0.85(53.6)} = 9.78 \text{ sq in.}$$

Try HSS $12\times12\times\frac{1}{4}$ (Fig. 6.19.3):

$$A_g = 10.8 \text{ sq in.}; r_x = r_y = 4.79 \text{ in.}; b/t = 48.5; t = 0.233 \text{ in.}$$

(c) Compute strength of selected section. Check b/t on stiffened elements. Local buckling may reduce the efficiency of this section. Compute the reduction factor Q in accordance with AISC-E7.2b. Because the stress f on the stiffened element of a column section will be less than $\phi_c F_y$, the λ_r limitation will actually be higher than the limitation for Eq. 6.18.25.

Figure 6.19.3
Example 6.19.3.   (a) Gross section   (b) Effective section

$$\text{For } \frac{b}{t} \geq \left[1.40\sqrt{\frac{E}{f}} = \frac{238}{\sqrt{f}, \text{ksi}} \right]$$

$$\frac{b_E}{t} = \left\{ 1.92\sqrt{\frac{E}{f}}\left[1 - \frac{0.38}{(b/t)}\sqrt{\frac{E}{f}} \right] \text{ or } \frac{327}{\sqrt{f}, \text{ksi}}\left[1 - \frac{64.7}{(b/t)\sqrt{f}, \text{ksi}} \right] \right\} \quad [6.18.25]$$

where $f = F_{cr}$ which is a function of Q. For this section having no unstiffened elements $Q_s = 10$; thus $Q = Q_a$. Estimate $Q_a = 0.90$; try $f = Q_a F_{cr} = 0.9(53.6) = 48.2$ ksi using $F_{cr} = 53.6$ ksi as obtained from the first estimate of KL/r,

$$\left(\frac{b}{t} = 48.5 \right) \geq \left[1.40\sqrt{\frac{E}{f}} = \frac{238}{\sqrt{f}, \text{ksi}} = \frac{238}{\sqrt{48.2}} = 34.2 \right]$$

$$\frac{b_E}{t} = \left\{ \frac{327}{\sqrt{f}, \text{ksi}}\left[1 - \frac{64.7}{(b/t)\sqrt{f}, \text{ksi}} \right] = \frac{327}{\sqrt{48.2}}\left[1 - \frac{64.7}{(48.5)\sqrt{48.2}} \right] = 38.0 \right\}$$

Thus local buckling will control the section efficiency for all four sides of the tube. The effective area is

$$A_{eff} = 10.8 - 4\left(\frac{b}{t} - \frac{b_E}{t} \right)t^2$$

$$= 10.8 - 4(48.5 - 38.0)(0.233)^2 = 8.52 \text{ sq in.}$$

$$Q = \frac{A_{eff}}{A_g} = \frac{8.52}{10.8} = 0.789$$

Clearly the section is inadequate because Q is far less than the 1.0 assumed. Increase the section to HSS $12 \times 12 \times \frac{5}{16}$,

$$A_g = 13.4 \text{ sq in.}; \quad r_x = r_y = 4.76 \text{ in.}; \quad b/t = 38.2; \quad t = 0.291 \text{ in.}$$

Recompute the KL/r value for the trial section,

$$\frac{KL}{r} = \frac{18(12)}{4.76} = 45.4$$

Try f less than 48.2 ksi used initially; try 42 ksi,

$$\left(\frac{b}{t} = 38.2 \right) \geq \left[1.40\sqrt{\frac{E}{f}} = \frac{238}{\sqrt{f}, \text{ksi}} = \frac{238}{\sqrt{42}} = 36.7 \right]$$

$$\frac{b_E}{t} = \left\{ \frac{327}{\sqrt{f}, \text{ksi}}\left[1 - \frac{64.7}{(b/t)\sqrt{f}, \text{ksi}} \right] = \frac{327}{\sqrt{42}}\left[1 - \frac{64.7}{(38.2)\sqrt{42}} \right] = 37.3 \right\}$$

$$A_{eff} = 13.4 - 4\left(\frac{b}{t} - \frac{b_E}{t} \right)t^2$$

$$= 13.4 - 4(38.2 - 37.3)(0.291)^2 = 13.1 \text{ sq in.}$$

$$Q_r = \frac{A_{eff}}{A_g} = \frac{13.1}{13.4} = 0.978$$

$$\left(4.71\sqrt{\frac{E}{QF_y}} = 4.71\sqrt{\frac{29,000}{0.978(65)}} = 101\right) > 45.4$$

A short column as expected!

$$F_e = \frac{\pi^2 E}{(KL/r)^2} = \frac{\pi^2(29,000)}{(45.4)^2} = 139 \text{ ksi}$$

$$F_{cr} = Q\left[0.658^{\frac{QF_y}{F_e}}\right]F_y = 0.978\left[0.658^{\frac{0.978(65)}{139}}\right]65 = 52.5 \text{ ksi}$$

$$\phi_c P_n = \phi_c F_{cr} A_g = 0.90(52.5)13.4 = 633 \text{ kips} > 472 \text{ kips}$$

This larger section than the initial selection has a much lower b/t ratio and thus regardless of the assumption on f, Q will be very close to 1.0.

Use HSS $12{\times}12{\times}\frac{5}{16}$.   ■

EXAMPLE 6.19.4

Determine the nominal axial compressive strength P_n for the nonstandard shape of Fig. 6.19.4 for an effective length KL equal to 8 ft. Use $F_y = 100$ ksi and the AISC LRFD Method.

(a) Gross section   (b) Effective section

Figure 6.19.4
Example 6.19.4.

Solution:
In axial compression this section contains unstiffened compression elements (flanges) and a stiffened compression element (the web). Unstiffened elements must be treated first so that the effective stress level may be determined.

   (a) Properties of the gross section.

$$I_y = \text{negligible (web)} + 2\left(\tfrac{1}{12}\right)(0.5)(10)^3 \text{ (flanges)} = 83.4 \text{ in.}^4$$

$$A_g = 11(0.25) + 2(5.0) = 12.75 \text{ sq in.}$$

$$r_y = \sqrt{I/A} = \sqrt{83.4/12.75} = 2.56 \text{ in.}$$

   (b) Unstiffened elements, AISC-E7.lb, Eqs. 6.18.14 and 6.18.15. Check the web slenderness effect k_c on the support restraint of the unstiffened element,

$$\frac{h}{t_w} = \frac{11}{0.25} = 44$$

$$k_c = \frac{4}{\sqrt{h/t_w}} = \frac{4}{\sqrt{44}} = 0.603 < 0.763 \quad \text{Acceptable range!}$$

When k_c lies between 0.35 and 0.763, as it does in this case, the formulation of Q depends on k_c. Using Eq. 6.18.14,

$$\left[\frac{109\sqrt{k_c}}{\sqrt{F_y, \text{ksi}}} = \frac{109\sqrt{0.603}}{\sqrt{100}} = 8.5\right] < \left(\frac{b}{t} = 10.0\right) < \left[\frac{199\sqrt{k_c}}{\sqrt{F_y, \text{ksi}}} = \frac{199\sqrt{0.603}}{\sqrt{100}} = 15.5\right]$$

$$Q_s = 1.415 - 0.00382\left(\frac{b}{t}\right)\sqrt{\frac{F_y, \text{ksi}}{k_c}}$$

$$= 1.415 - 0.00382(10.0)\sqrt{\frac{100}{0.603}} = 0.923$$

(c) Stiffened element, AISC-E7.2a, Eq. 6.18.24,

$$\left(\lambda = \frac{b}{t} = \frac{11}{0.25} = 44\right) > \left(\lambda_r = \frac{254}{\sqrt{100}} = 25.4\right)$$

Thus $Q_a < 1.0$. The stress f that acts on the stiffened element is F_{cr} which includes Q, that is, Eq. 6.8.2. A first trial may be made assuming $Q = Q_s$, that is, $Q_a = 1.0$.

$$\left(\frac{KL}{r} = \frac{8(12)}{2.56} = 37.5\right) < \left(4.71\sqrt{\frac{E}{F_y}} = 4.71\sqrt{\frac{29,000}{100}} = 80.2\right)$$

This is a short column.

$$F_e = \frac{\pi^2 E}{(KL/r)^2} = \frac{\pi^2(29,000)}{(37.5)^2} = 204 \text{ ksi}$$

$$F_{cr} = Q\left[0.658^{\frac{QF_y}{F_e}}\right]F_y = 0.923\left[0.658^{\frac{0.923(100)}{204}}\right]100 = 76.4 \text{ ksi}$$

Trial $\quad f = QF_{cr} = 0.923(76.4) = 70.5 \text{ ksi}$

$$\frac{b_E}{t} = \left\{\frac{327}{\sqrt{f, \text{ksi}}}\left[1 - \frac{57.9}{(b/t)\sqrt{f, \text{ksi}}}\right] = \frac{327}{\sqrt{70.5}}\left[1 - \frac{57.9}{(44)\sqrt{70.5}}\right] = 32.8\right\}$$

$$A_{\text{eff}} = 12.75 - \left(\frac{b}{t} - \frac{b_E}{t}\right)t^2$$

$$= 12.75 - (44 - 32.8)(0.25)^2 = 12.05 \text{ sq in.}$$

$$Q_a = \frac{A_{\text{eff}}}{A_g} = \frac{12.05}{12.75} = 0.945$$

$$Q = Q_a Q_s = 0.945(0.923) = 0.872$$

Another iteration,

$$F_{cr} = Q\left[0.658^{\frac{QF_y}{F_e}}\right]F_y = 0.872\left[0.658^{\frac{0.872(100)}{204}}\right]100 = 72.9 \text{ ksi}$$

$$f = QF_{cr} = 0.872(72.9) = 63.6 \text{ ksi}$$

$$\frac{b_E}{t} = \left\{\frac{327}{\sqrt{f, \text{ksi}}}\left[1 - \frac{57.9}{(b/t)\sqrt{f, \text{ksi}}}\right]\right\} = \frac{327}{\sqrt{63.6}}\left[1 - \frac{57.9}{(44)\sqrt{63.9}}\right] = 34.2$$

$$A_{eff} = 12.75 - \left(\frac{b}{t} - \frac{b_E}{t}\right)t^2$$

$$= 12.75 - (44 - 34.2)(0.25)^2 = 12.14 \text{ sq in.}$$

$$Q_a = \frac{A_{eff}}{A_g} = \frac{12.14}{12.75} = 0.952$$

$$Q = Q_a Q_s = 0.952(0.923) = 0.879 \qquad \text{Close enough!}$$

$$F_{cr} = Q\left[0.658^{\frac{QF_y}{F_e}}\right]F_y = 0.879\left[0.658^{\frac{0.879(100)}{204}}\right]100 = 73.4 \text{ ksi}$$

$$\phi_c P_n = \phi_c F_{cr} A_g = 0.90(73)12.75 = 838 \text{ kips} \qquad \blacksquare$$

SELECTED REFERENCES

6.1. L. Euler. *De Curvis Elasticis, Additamentum I, Methodus Inveniendi Lineas Curvas Maximi Minimive Proprietate Gaudentes.* Lausanne and Geneva, 1744 (pp. 267–268); and "Sur le Forces des Colonnes," *Memoires de l' Academie Royale des Sciences et Belles Lettres,* Vol. 13, Berlin, 1759; English translation of the letter by J. A. Van den Broek, "Euler's Classic Paper 'On the Strength of Columns'," *American Journal of Physics,* **15** (January–February 1947), 309–318.

6.2. A. Considère. "Resistance des pièces comprimées," *Congrès International des Procèdés de Construction,* Paris, 1891, 3, 371.

6.3. F. Engesser. "Ueber die Knickfestigkeit gerader Stabe," *Zeitschrift des Architekten-und Ingenieur—Vereins zu Hannover,* **35** (1889), 455 and 462; also "Die Knickfestigkeit gerader Stabe," *Zentralblatt der Bauverwaltung,* Berlin (December 5, 1891), 483.

6.4. F. R. Shanley. "The Column Paradox," *Journal of the Aeronautical Sciences,* **13,** 12 (December 1946), 678.

6.5. F. R. Shanley. "Inelastic Column Theory," *Journal of the Aeronautical Sciences,* **14,** 5 (May 1947), 261–264.

6.6. Bruce G. Johnston. "Column Buckling Theory: Historic Highlights," *Journal of Structural Engineering,* **109,** 9 (September 1983), 2086–2096; Disc. by T. H. Lin, Zia Razzaq, and B. G. Johnston, **110,** 8 (August 1984), 1930–1933.

6.7. Bruce G. Johnston. "A Survey of Progress, 1944–51," Bulletin No. 1, Column Research Council, January 1952.

6.8. Theodore V. Galambos, ed. *Guide to Stability Design Criteria for Metal Structures,* 5th ed. New York: John Wiley & Sons, 1998.

6.9. Friedrich Bleich. *Buckling Strength of Metal Structures.* New York: McGraw-Hill Book Company, Inc., 1952.

6.10. Bruce G. Johnston. "Buckling Behavior Above the Tangent Modulus Load," *Journal of the Engineering Mechanics Division,* ASCE, **87,** EM6 (December 1961), 79–98.

6.11. A. W. Huber and L. S. Beedle. "Residual Stress and the Compressive Strength of Steel," *Welding Journal,* December 1954, 589s–614s.

6.12. C. H. Yang, L. S. Beedle, and B. G. Johnston. "Residual Stress and the Yield Strength of Steel Beams," *Welding Journal,* April 1952, 205s–229s.

6.13. N. R. Nagaraja Rao, F. R. Estuar, and L. Tall. "Residual Stresses in Welded Shapes," *Welding Journal,* July 1964, 295s–306s.

6.14. Donald R. Sherman. "Residual Stress Measurement in Tubular Members," *Journal of the Structural Division,* ASCE, **95,** ST4 (April 1969), 635–647.

6.15. Donald R. Sherman. "Residual Stresses and Tubular Compression Members," *Journal of the Structural Division,* ASCE, **97,** ST3 (March 1971), 891–904.

6.16. Lynn S. Beedle and Lambert Tall. "Basic Column Strength," *Journal of the Structural Division,* ASCE **86,** ST7 (July 1960), 139–173. Also *Transactions,* ASCE, **127** (1962), part II, 138–179.

6.17. Ching K. Yu and Lambert Tall. "Significance and Application of Stub Column Test Results," *Journal of the Structural Division,* ASCE, **97,** ST7 (July 1971), 1841–1861.

6.18. Bruce G. Johnston. "Inelastic Buckling Gradient," *Journal of the Engineering Mechanics Division,* ASCE, **90,** EM6 (December 1964), 31–47.

6.19. Richard H. Batterman and Bruce G. Johnston. "Behavior and Maximum Strength of Metal Columns," *Journal of the Structural Division,* ASCE, **93,** ST2 (April 1967), 205–230.

6.20. Bruce G. Johnston, ed. *Structural Stability Research Council, Guide to Stability Design Criteria for Metal Structures,* 3rd ed. New York: John Wiley & Sons, Inc., 1976.

6.21. Reidar Bjorhovde. "Effect of End Restraint on Column Strength—Practical Applications," *Engineering Journal,* AISC, **22,** 1 (First Quarter 1984), 1–13.

6.22. Reidar Bjorhovde. "The Safety of Steel Columns," *Journal of the Structural Division,* ASCE, **104,** ST3 (March 1978), 463–477.

6.23. Reidar Bjorhovde. "Columns: From Theory to Practice," *Engineering Journal,* AISC, **26,** 1 (First Quarter 1988), 21–34.

6.24. Dann H. Hall. "Proposed Steel Column Strength Criteria," *Journal of the Structural Division,* ASCE, **107,** ST4 (April 1981), 649–670. Disc. by Bruce G. Johnston, **108,** ST4 (April 1982), 956–957; by Zu-Yan Shen and Le-Wu Lu, **108,** ST7 (July 1982), 1680–1681; by author, **108,** ST12 (December 1982), 2853–2855.

6.25. J. Michael Rotter. "Multiple Column Curves by Modifying Factors," *Journal of the Structural Division,* ASCE, **108,** ST7 (July 1982), 1665–1669.

6.26. Jacque Rondal and René Maquoi. "Single Equation for SSRC Column-Strength Curves," *Journal of the Structural Division,* ASCE, **105,** ST1 (January 1979), 247–250.

6.27. E. M. Lui and W. F. Chen. "Simplified Approach to the Analysis and Design of Columns with Imperfections," *Engineering Journal,* AISC, **22,** 2 (Second Quarter 1984), 99–117.

6.28. John P. Anderson and James H. Woodward. "Calculation of Effective Lengths and Effective Slenderness Ratios of Stepped Columns," *Engineering Journal,* AISC, **9,** 3 (October 1972), 157–166.

6.29. Carlo A. Castiglioni. "Stepped Columns: A Simplified Design Method," *Engineering Journal,* AISC, **24,** 1 (First Quarter 1986), 1–8.

6.30. Suresh C. Shrivastava. "Elastic Buckling of a Column Under Varying Axial Force," *Engineering Journal,* AISC, **18,** 1 (First Quarter 1980), 19–21.

6.31. Balbir S. Sandhu. "Effective Length of Columns with Intermediate Axial Load," *Engineering Journal,* AISC, **9,** 3 (October 1972), 154–156.

6.32. John C. Ermopoulos. "Buckling of Tapered Bars Under Stepped Axial Loads," *Journal of Structural Engineering,* ASCE, **112,** 6 (June 1986), 1346–1354.

6.33. D. R. Sherman. *Tentative Criteria for Structural Applications of Steel Tubing and Pipe,* Washington, D.C.: American Iron and Steel Institute, 1976.

6.34. Julian Snyder and Seng-Lip Lee. "Buckling of Elastic-Plastic Tubular Columns," *Journal of the Structural Division,* ASCE, **94,** ST1 (January 1968), 153–173.

6.35. Seng-Lip Lee and Julian Snyder. "Stability of Strain-Hardening Tubular Columns," *Journal of the Structural Division,* ASCE, **94,** ST3 (March 1968), 683–707.

6.36. Wai F. Chen and David A. Ross. "Tests of Fabricated Tubular Columns," *Journal of the Structural Division,* ASCE, **103,** ST3 (March 1977), 619–634.

6.37. David A. Ross, Wai Fah Chen, and Lambert Tall. "Fabricated Tubular Steel Columns," *Journal of the Structural Division,* ASCE, **106,** ST1 (January 1980), 265–282.

6.38. Theodore V. Galambos. "Strength of Round Steel Columns," *Journal of the Structural Division,* ASCE, **91,** ST1 (February 1965), 121–140.

6.39. Charles Libove. "Sparsely Connected Built-Up Columns," *Journal of Structural Engineering,* ASCE, **111,** 3 (March 1985), 609–627.

6.40. Abolhassan Astaneh-Asl, Subhash C. Goel, and Robert D. Hanson. "Cyclic Out-of-Plane Buckling of Double-Angle Bracing," *Journal of Structural Engineering,* ASCE, **111,** 5 (May 1985), 1135–1153.

6.41. Cynthia J. Zahn and Geerhard Haaijer. "Effect of Connector Spacing on Double Angle Compressive Strength," *Materials and Member Behavior,* Proceedings of Structures Congress '87, Orlando, FL, August 17–20, 1987, pp. 199–212.

6.42. John B. Kennedy and Madugula K. S. Murty. "Buckling of Steel Angle and Tee Struts," *Journal of the Structural Division,* ASCE, **98,** ST11 (November 1972), 2507–2522.

6.43. John B. Kennedy and Madugula K. S. Murty. "Buckling of Angles: State of the Art," *Journal of the Structural Division,* ASCE, **108,** ST9 (September 1982), 1967–1980. Disc., *Journal of Structural Engineering,* ASCE, **109,** 8 (August 1983), 2025–2029.

6.44. Scott T. Woolcock and Sritawat Kitipornchai. "Design of Single Angle Web Struts in Trusses," *Journal of Structural Engineering,* ASCE, **112,** 6 (June 1986), 1327–1345. Disc. **113,** 9 (September 1987), 2102–2107.

6.45. Adel A. El-Tayem and Subhash C. Goel. "Effective Length Factor for Design of X-bracing Systems," *Engineering Journal,* AISC, **24,** 4 (First Quarter 1986), 41–45.

6.46. Subhash C. Goel and Adel A. El-Tayem. "Cyclic Load Behavior of Angle X-Bracing," *Journal of Structural Engineering,* ASCE, **112,** 11 (November 1986), 2528–2539.

6.47. Guo Chuenmei. "Elastoplastic Buckling of Single Angle Columns," *Journal of Structural Engineering,* ASCE, **110,** 6 (June 1984), 1391–1395.

6.48. Erling A. Smith. "Buckling of Four Equal-Leg Angle Cruciform Columns," *Journal of Structural Engineering,* ASCE, **109,** 2 (February 1983), 439–450.

6.49. Le-Wu Lu. "Effective Length of Columns in Gable Frames," *Engineering Journal,* AISC, **2,** 1 (January 1965), 6–7.

6.50. Kamal Hassan. "On the Determination of Buckling Length of Frame Columns," *Publications,* International Association for Bridge and Structural Engineering, **28-**II, 1968, 91–101 (in German).

6.51. Theodore V. Galambos. "Influence of Partial Base Fixity of Frame Stability," *Journal of The Structural Division,* ASCE, **86,** ST5 (May 1960), 85–108.

6.52. German Gurfinkel and Arthur R. Robinson. "Buckling of Elastically Restrained Columns," *Journal of the Structural Division,* ASCE, **91,** ST6 (December 1965), 159–183.

6.53. Harold Switzky and Ping Chun Wang. "Design and Analysis of Frames for Stability," *Journal of the Structural Division,* ASCE, **95,** ST4 (April 1969), 695–713.

6.54. Thomas C. Kavanagh. "Effective Length of Framed Columns," *Transactions,* ASCE, **127** (1962), Part II, 81–101.

6.55. Joseph A. Yura. "The Effective Length of Columns in Unbraced Frames," *Engineering Journal,* AISC, **8,** 2 (April 1971), 37–42; Disc., **9,** 3 (October 1972), 167–168.

6.56. Peter F. Adams. Discussion of "The Effective Length of Columns in Unbraced Frames," by Joseph A. Yura, *Engineering Journal,* AISC, **9,** 1 (January 1972), 40–41.

6.57. Bruce G. Johnston. Discussion of "The Effective Length of Columns in Unbraced Frames," by Joseph A. Yura, *Engineering Journal,* AISC, **9,** 1 (January 1972), 46.

6.58. Robert O. Disque. "Inelastic K-factor for Column Design," *Engineering Journal,* AISC, **10,** 2 (Second Quarter 1973), 33–35.

6.59. C. V. Smith, Jr. "On Inelastic Column Buckling," *Engineering Journal,* AISC, **13,** 3 (Third Quarter 1976), 86–88; Disc., **14,** 1 (First Quarter 1977), 47–48.

6.60. Charles A. Matz. Discussion of "On Inelastic Column Buckling," by C. V. Smith, Jr., *Engineering Journal,* AISC, **14,** 1 (First Quarter 1977), 47–48.

6.61. Frank W. Stockwell, Jr. "Girder Stiffness Distribution for Unbraced Columns," *Engineering Journal,* AISC, **13,** 3 (Third Quarter 1976), 82–85.

6.62. Cyrus Omid'varan. "Discrete Analysis of Latticed Columns," *Journal of the Structural Division,* ASCE, **94,** ST1 (January 1968), 119–132.

6.63. Fung J. Lin, Ernst C. Glauser, and Bruce G. Johnston. "Behavior of Laced and Battened Structural Members," *Journal of the Structural Division,* ASCE, **96,** ST7 (July 1970), 1377–1401.

6.64. Bruce G. Johnston. "Spaced Steel Columns," *Journal of the Structural Division,* ASCE, **97,** ST5 (May 1971), 1465–1479.

6.65. Omer W. Blodgett. *Design of Welded Structures.* Cleveland, Ohio: James F. Lincoln Arc Welding Foundation, 1966.

6.66. S. Timoshenko and S. Woinowsky-Krieger. *Theory of Plates and Shells,* 2nd ed. New York: McGraw-Hill Book Company, Inc., 1959 (pp. 79–82).

6.67. Stephen P. Timoshenko and James M. Gere. *Theory of Elastic Stability,* 2nd ed. New York: McGraw-Hill Book Company, Inc., 1961 (pp. 319–328, 351–356).

6.68. Kurt H. Gerstle. *Basic Structural Design.* New York: McGraw-Hill Book Company, Inc., 1967 (pp. 88–90).

6.69. George Gerard and Herbert Becker. *Handbook of Structural Stability,* Part I—*Buckling of Flat Plates,* Tech. Note 3871, National Advisory Committee for Aeronautics, Washington, D.C., July 1957.

6.70. Geerhard Haaijer and Bruno Thürlimann. "On Inelastic Buckling in Steel," *Transactions,* ASCE, **125** (1960), 308–344.

6.71. Maxwell G. Lay. "Flange Local Buckling in Wide-Flange Shapes," *Journal of the Structural Division,* ASCE, **91,** ST6 (December 1965), 95–116.

6.72. John F. McDermott. "Local Plastic Buckling of A514 Steel Members," *Journal of the Structural Division,* ASCE, **95,** ST9 (September 1969), 1837–1850.

6.73. Theodore von Kármán, E. E. Sechler, and L. H. Donnell. "The Strength of Thin Plates in Compression," *Transactions,* ASME, **54,** APM-54-5 (1932), 53.

6.74. G. Winter. "Strength of Thin Compression Flanges," *Transactions,* ASCE, **112** (1947), 527–576.

6.75. Robert M. Korol and Archibald N. Sherbourne. "Strength Predictions of Plates in Uniaxial Compression." *Journal of the Structural Division,* ASCE, **98,** ST9 (September 1972), 1965–1986.

6.76. Archibald N. Sherbourne and Robert M. Korol. "Post-Buckling of Axially Compressed Plates," *Journal of the Structural Division,* ASCE, **98,** ST10 (October 1972), 2223–2234.

6.77. Ralph G. Dawson and Alastair C. Walker. "Post-Buckling of Geometrically Imperfect Plates," *Journal of the Structural Division,* ASCE, **98,** ST1 (January 1972), 75–94.

6.78. George Abdel-Sayed. "Effective Width of Thin Plates in Compression," *Journal of the Structural Division,* ASCE, **95,** ST10 (October 1969), 2183–2203.

6.79. Maurice L. Sharp. "Longitudinal Stiffeners for Compression Members," *Journal of the Structural Division,* ASCE, **92,** ST5 (October 1966), 187–211.

6.80. V. Kalyanaraman, Teoman Pekoz, and George Winter. "Unstiffened Compression Elements," *Journal of the Structural Division,* ASCE, **103,** ST9 (September 1977), 1833–1848.

6.81. Russell Q. Bridge and Donald J. Fraser. "Improved G-Factor Method for Evaluating Effective Lengths of Columns," *Journal of Structural Engineering,* **113,** 6 (June 1987), 1341–1356.

6.82. Lian Duan and Wai-Fah Chen. "Design Rules of Built-Up Members in Load and Resistance Factor Design," *Journal of Structural Engineering,* **114,** 11 (November 1988), 2544–2554.

6.83. AISC. *Specification for Load and Resistance Factor Design of Single-Angle Members,* effective December 1, 1993. Chicago: American Institute of Steel Construction, 1993.

6.84. Donald J. Fraser. "Uniform Pin-based Crane Columns, Effective Lengths," *Engineering Journal,* AISC, **26,** 2 (2nd Quarter 1989), 61–65.

6.85. Sayed H. Stoman. "Effective Length Spectra for Cross Bracings," *Journal of Structural Engineering,* ASCE, **115,** 12 (December 1989), 3112–3122.

6.86. A. Rutenberg and A. Scarlat. "Roof Bracing and Effective Length of Columns in One-Story Industrial Buildings," *Journal of Structural Engineering,* ASCE, **116,** 10 (October 1990), 2551–2566.

6.87. T. V. Galambos. "Design of Axially Loaded Compressed Angles," *Structural Stability Research Council Annual Technical Session Proceedings,* 1991.

6.88. Ronald S. Harichandran. "Stiffness Reduction Factor for LRFD of Columns," *Engineering Journal,* AISC, **28,** 3 (3rd Quarter 1991), 129–130.

6.89. Farhang Aslani and Subhash C. Goel. "An Analytical Criterion for Buckling Strength of Built-up Compression Members," *Engineering Journal,* AISC, **28,** 4 (4th Quarter 1991), 159–168.

6.90. M. Elgaaly, H. Dagher, and W. Davids. "Behavior of Single-Angle-Compression Members," *Journal of Structural Engineering,* ASCE, **117,** 12 (December 1991), 3720–3741.

6.91. M. Elgaaly, W. Davids, and H. Dagher. "Non-Slender Single Angle Struts," *Engineering Journal,* AISC, **29,** 2 (2nd Quarter 1992), 49–58.

6.92. Seshu Madhava Rao Adluri and Murty K. S. Madugula. "Eccentrically Loaded Steel Single Angle Struts," *Engineering Journal,* AISC, **29,** 2 (2nd Quarter 1992), 59–66.

6.93. Pierre Dumonteil. "Simple Equations for Effective Length Factors," *Engineering Journal,* AISC, **29,** 3 (3rd Quarter 1992), 111–115. Disc. **30,** 1 (1st Quarter 1993), 37; Errata **30,** 1 (1st Quarter 1993), 38.

6.94. Eric M. Lui. "A Novel Approach for K Factor Determination," *Engineering Journal,* AISC, **29,** 3 (3rd Quarter 1992), 150–159.

6.95. Farhang Aslani and Subhash C. Goel. "Analytical Criteria for Stitch Strength of Built-up Compression Members," *Engineering Journal,* AISC, **29,** 4 (4th Quarter 1992), 102–110.

6.96. A. Zureick. "Design Strength of Concentrically Loaded Single Angle Struts," *Engineering Journal,* AISC, **30,** 1 (1st Quarter 1993), 17–30.

6.97. Leander Bathon, Wendelin H. Mueller III, and Leon Kempner, Jr. "Ultimate Load Capacity of Single Steel Angles," *Journal of Structural Engineering,* ASCE, **119,** 1 (January 1993), 279–300.

PROBLEMS

All problems are to be done according to the AISC Load and Resistance Factor Design Method or Allowable Strength Design Method as indicated by the instructor. All given loads are service loads unless otherwise indicated. For each problem, draw the potential buckled shape on a figure showing the column and its restraints for both *x* and *y* principal directions. A final check of strength must be shown in all design problems.

6.1. For the case assigned by the instructor, select the lightest W section to carry an axial compression load as indicated. The member is part of a braced frame. Assume the member as pinned at the top and bottom.

Case	P_D Dead load (kips)	P_L Live load (kips)	Member length (ft)	Steel grade
1	20	80	22	A992 Grade 50
2	20	80	14	A572 Grade 50
3	60	40	30	A992 Grade 50
4	60	40	20	A992 Grade 50
5	20	80	14	A572 Grade 60
6	20	80	14	A572 Grade 65

6.2. Select the lightest W section to carry a compressive load of 200 kips dead load and 625 kips live load. The effective length KL is 25 ft. Use A992 steel.

6.3. Compute the maximum service axial compression load permitted on the built-up cross-section of the accompanying figure. The load is 30% dead load and 70% live load. The steel used is A992, and the effective lengths are $(KL)_y = 14$ ft and $(KL)_x = 42$ ft.

$\frac{5}{8} \times 16$

$\frac{3}{4} \times 26$

$\frac{5}{8} \times 16$

Problem 6.3

6.4. For the data of Prob. 6.1, consider the member to be fixed at the bottom and hinged at the top and part of a braced system. Select the lightest W section for the case assigned.

6.5. Select the lightest W section to serve as an axially loaded column 28 ft long, in a braced frame, with additional lateral support in the weak direction at mid-height. The load to be carried is 65 kips dead load and 150 kips live load. Assume the top and bottom of the column are hinged.

(a) Use A992 and indicate first and second choices.

(b) Use A572 Grade 60, indicating first and second choices.

6.6. Select the most economical W section to carry an axial compression load of 50 kips dead load and 100 kips live load. The member has $(KL)_x = (KL)_y = 18$ ft and is part of a braced system. Assume that relative costs of various steels are as follows: A992, 1.14; A572 Grade 60, 1.20.

6.7. Redesign the column of Prob. 6.6 assuming additional weak direction support at mid-height.

6.8. Select the lightest W section to carry an axial compression load of 60 kips dead load and 250 kips live load. The member is part of a braced frame and is assumed to be pinned at the top and bottom of its 30 ft length, and in addition has lateral support in the weak direction at 14 ft from the bottom. Use (a) A992 steel; (b) $F_y = 65$ ksi.

6.9. Select the lightest W section to carry an axial compression of 90 kips dead load and 320 kips live load. The member is part of a braced frame. The idealized support conditions are that the member is hinged in both principal directions at the top of a 30 ft height; supported in the weak direction at 14 and 22 ft from the bottom; and fixed in both directions at the bottom. Use (a) A992 steel; (b) $F_y = 60$ ksi

6.10. Select the lightest W section for the column shown in the accompanying figure. The loading is 30 kips dead load and 120 kips gravity live load. The member is built into a wall so that it may be considered as continuously braced in the weak direction. Note: Not all of the available W sections are included in the AISC Manual "Available Strength in Axial Compression" tables. Use (a) A992 steel; (b) $F_y = 60$ ksi

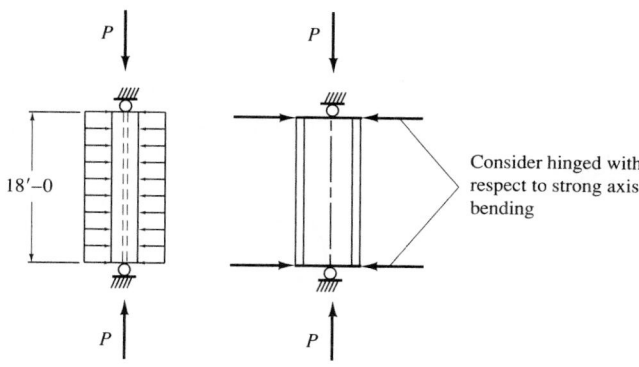

Problem 6.10

6.11. Redesign the column of Prob. 6.2 assuming there is no residual stress or accidental eccentricity such that the column buckling strength may be represented by Euler's equation, Eq. 6.2.8, using KL for L. If designing using the LRFD Method, use $\phi_c = 0.90$. If designing using the ASD Method, use $\Omega_c = 1.67$. Use $F_y = 50$ ksi and take note that F_{cr} cannot exceed F_y.

6.12. Use the tangent modulus theory to compute the column strength relationship (average unit stress F_{cr} on gross area vs slenderness ratio KL/r). Draw the diagram to scale and obtain any F_{cr} values by scaling from your diagram. The steel has $F_y = 50$ ksi but the stress-strain curve for the material is as shown in the accompanying figure. Assume no residual stress. Using your F_{cr} curve, select the lightest W section for the loading and support conditions of Prob. 6.2. If designing using the LRFD Method, use $\phi_c = 0.90$. If designing using the ASD Method, use $\Omega_c = 1.67$.

6.13. Using the tangent modulus theory: (a) Construct a column strength ($F_{cr} = P_n/A_g$ vs KL/r) for an H-shaped section. Assume weak axis bending $(KL/r)_y$ controls and neglect the effect of the web. Assume the idealized stress-strain relationship shown in the accompanying figure is to be used for each fiber of the cross-section, and the residual stress distribution in the flange is as shown. (b) Select the lightest W section to carry a dead load of 100 kips and a live load of 200 kips with an effective length KL of 30 ft. Use your constructed curve as the relationship between F_{cr} and KL/r. If designing using the LRFD Method, use $\phi_c = 0.90$. If designing using the ASD Method, use $\Omega_c = 1.67$. Use $F_y = 50$ ksi. (c) Solve using the AISC Specification and compare with tangent modulus theory result.

6.14. Follow the same requirements as for Prob. 6.13 to construct a column strength curve. This time the residual stress distribution is linearly varying as shown in the accompanying figure. For (b) and (c) use the loading and support conditions of Prob. 6.2.

6.15. Repeat Prob. 6.14, except for (b) and (c) use the loading and support conditions of Prob. 6.13.

6.16. Design an interior column (use W shape) for a multistory rigid frame. No bracing is provided in the plane of the frame. In the plane perpendicular to the frame, bracing is provided at top, bottom, and mid-height of columns and simple flexible beam-to-column connections are used. The axial compressive load is 400 kips dead load and 1100 kips live load, and bending moments are neglected. Use A572 Grade 50 Steel.

Problem 6.12

Problem 6.13

Problem 6.14

Problem 6.16

6.17. An axial compression load consisting of 100 kips dead load and 400 kips live load is to be carried by a column having an effective length $KL = 28$ ft. Use A992 steel. Satisfy the width/thickness limits λ_r of AISC-Table B4.1.

(a) Design a rolled W section.

(b) Design the lightest welded I-shaped section using three plates.

(c) Design a welded laced (single lacing) column consisting of four angles.

(d) Design a welded box with solid plates.

(e) Design a welded box having two perforated plates.

6.18. Design as in Prob. 6.17, except the axial load is 190 kips dead load and 210 kips live load and $KL = 32$ ft.

6.19. Compute the maximum service load (25% dead load and 75% live load) acceptable for a structural tee WT12×38 when used in a truss location where it is braced in the plane of the truss at 20-ft intervals and braced transverse to the plane of the truss at 10-ft intervals. Apply the provisions of AISC-E7 if necessary. Use (a) A992 steel; (b) $F_y = 65$ ksi

6.20. Select the lightest double angle compression member to carry 110 kips dead load and 130 kips live load.

The effective length $KL = 20$ ft. Assume the backs of the angles are separated by a $\frac{3}{8}$-in. gusset plate. Indicate the orientation of the angle legs (i.e., short or long legs back-to-back). If angles are selected for all three steels, indicate the economical choice if relative costs are A36 (1.0), $F_y = 50$ (1.07), and $F_y = 60$ (1.10). Use (a) A36 steel; (b) $F_y = 50$ ksi; and (c) $F_y = 60$ ksi

6.21. Design as in Prob. 6.20, except the member must carry 40 kips dead load and 140 kips live load, and $KL = 16$ ft.

6.22. Design as in Prob. 6.20, except the member must carry 30 kips dead load and 50 kips live load, and $KL = 16$ ft.

6.23. Design a top chord member for a roof truss to carry 40 kips dead load, 80 kips live load, and 40 kips wind load acting simultaneously. Assume this loading combination governs the design. The member is braced in the plane of the truss by adjoining web members connecting in at 5-ft intervals. The chord is braced transverse to the plane of the truss at 10-ft intervals. Neglect bending due to roof loads.

(a) Design a double angle member connected to gusset plates.

(b) Design a structural tee.

6.24. Select the lightest structural tee (WT) for use as a top chord compression member to carry 35 kips dead load and 100 kips live load. Neglect bending. The member has 9-ft effective length for buckling in either the x-x or y-y plane. Use (a) A992 steel; (b) $F_y = 65$ ksi.

6.25. Select the lightest structural tee (WT) to serve as the compression chord of a truss to carry 40 kips dead load and 45 kips live load. In the plane of the truss the chord is braced by adjoining web members that frame in at 5-ft intervals. Perpendicular to the plane of the truss, the chord is braced at 10 ft by a system of lateral purlin supports. Use the most economical of A992 or A572 Grade 65 steels if Grade 65 costs 12% more than A992.

6.26. Compute the service axial compressive load permitted on a $10 \times 10 \times \frac{1}{4}$ structural tube having an effective length $(KL)_y = 8$ ft. The load is 60% live load. Use (a) A992 steel; (b) $F_y = 60$ ksi; and (c) $F_y = 100$ ksi.

6.27. Compute the service axial compressive load permitted on a $12 \times 8 \times \frac{1}{4}$ structural tube having an effective length $(KL)_y = 7$ ft for weak axis bending, and $(KL)_x = 10$ ft for strong axis bending. The load is 35% dead load and 65% live load. Use (a) A992 steel; (b) $F_y = 65$ ksi; and (c) $F_y = 90$ ksi.

6.28. Redesign the column of Prob. 6.1, selecting an HSS instead of a W section.

6.29. Redesign the column of Prob. 6.6, selecting an HSS instead of a W section.

6.30. Redesign the column of Prob. 6.7, selecting an HSS instead of a W section.

6.31. Compute the service axial compressive load permitted on the nonstandard I-shaped section shown in the accompanying figure if the load is 30% dead load and 70% live load. The effective length $(KL)_y = 12$ ft and $(KL)_x = 6$ ft. Use (a) A992 steel; (b) $F_y = 60$ ksi; and (c) $F_y = 100$ ksi.

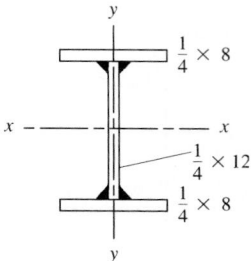

Problem 6.31

6.32. Repeat Prob. 6.31 except use the nonstandard tee section of the accompanying figure.

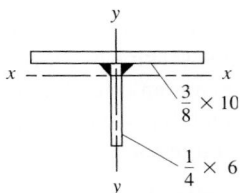

Problem 6.32

7

Beams: Laterally
Supported

7.1 INTRODUCTION

A beam is defined as a member acted upon primarily by transverse loading, often gravity dead and live load effects. The term transverse loading is taken to include end moments. Thus, beams in a structure may also be referred to as *girders* (usually the most important beams which are frequently at wide spacing); *joists* (usually less important beams which are closely spaced, frequently with truss-type webs); *purlins* (roof beams spanning between trusses); *stringers* (longitudinal bridge beams spanning between floor beams); *girts* (horizontal wall beams serving principally to resist bending due to wind on the side of an industrial building; frequently supporting corrugated siding); and *lintels* (members supporting a wall over window or door openings). Other terms, such as header, trimmer, and rafter, are sometimes used.

A beam is a combination of a tension element and a compression element. The concepts of tension members and compression members are now combined in the treatment as a beam. In this chapter, the compression element, (one flange) that is integrally braced perpendicular to its plane through its attachment to the stable tension flange by means of the web, is assumed also to be braced laterally in the direction perpendicular to the plane of the web. Thus, overall buckling of the compression flange as a column cannot occur prior to its full participation to develop the moment strength of the section. While it is likely true that most beams used in practical situations are adequately braced laterally so that such stability need not be considered, the percentage of stable situations is probably not as high as assumed. The important treatment of lateral stability is found in Chapter 9. Galambos [7.1] has provided an interesting history of beam design according to various editions of the AISC Specification.

7.2 SIMPLE BENDING OF SYMMETRICAL SHAPES

The most common design situations involve selection of rolled wide-flange shapes from the AISC tables, which often becomes routine and may lead the designer into overconfidence in treatment of beams. It is well known that the flexure formula ($f = Mc/I$) is

321

Beams, including open-web joists, channels, and W (wide-flange) shapes, along with tubular columns. (Photo by C.G. Salmon)

Figure 7.2.1
Elastic section modulus expressions for symmetrical shapes.

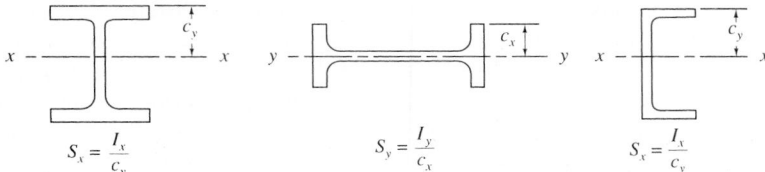

applicable to ordinary situations. The stresses on the common sections of Fig. 7.2.1 may be computed by the simple flexure formula when loads are acting in one of the principal directions. When any section with at least one axis of symmetry and loaded through the shear center is subjected to a bending moment in an arbitrary direction, the components M_{xx} and M_{yy}, in the principal directions, can be obtained and the stress computed as

$$f = \frac{M_{xx}}{S_x} + \frac{M_{yy}}{S_y}$$

(7.2.1)

where S is the *section modulus*, defined as the moment of inertia I divided by the distance c from the center of gravity to the extreme fiber. The subscripts x and y indicate the axis about which the moment of inertia is computed and from which the distance c is measured (see Fig. 7.2.1). For members without at least one axis of symmetry the reader is referred to Sec. 7.10.

7.3 BEHAVIOR OF LATERALLY STABLE BEAMS

When beams have adequate lateral stability of the compression flange, the only stability limit state that might prevent achieving maximum moment strength is local buckling in compression of the flange and/or web plate elements comprising the cross-section.

The stress distribution on a typical wide-flange shape subjected to increasing bending moment is shown in Fig. 7.3.1. In the service load range the section is elastic as in Fig. 7.3.1a, and the elastic condition exists until the stress at the extreme fiber reaches the yield stress F_y

Figure 7.3.1
Stress distribution at different
stages of loading.

(Fig. 7.3.1b). Once the strain ϵ reaches ϵ_y (Fig. 7.3.2), increasing strain induces no increase in stress. This elastic-plastic stress-strain behavior is the accepted idealization for structural steels having yield stresses up to about $F_y = 65$ ksi (448 MPa).

When the yield stress is reached at the extreme fiber (Fig. 7.3.1b), the nominal moment strength M_n is referred to as the *yield moment* M_y and is computed as

$$M_n = M_y = S_x F_y \tag{7.3.1}$$

When the condition of Fig. 7.3.1d is reached, every fiber has a strain equal to or greater than $\epsilon_y = F_y/E_s$, i.e., it is in the plastic range. The nominal moment strength M_n is therefore referred to as the *plastic moment* M_p, and is computed

$$M_p = F_y \int_A y\, dA = F_y Z \tag{7.3.2}$$

where $Z = \int y\, dA$ is called the *plastic modulus*.

Note the ratio M_p/M_y is a property of the cross-sectional shape and is independent of the material properties. This ratio is known as the *shape factor* ξ,

$$\xi = \frac{M_p}{M_y} = \frac{Z}{S} \tag{7.3.3}$$

For wide-flange (W) shapes in flexure about the strong axis (x–x) the shape factor ranges from about 1.09 to about 1.18, with the usual value being about 1.12. One may conservatively say the plastic moment strength M_p of W sections bent about their strong axis is at least 10% greater than the strength M_y when the extreme fiber just reaches the yield stress F_y.

Design procedures have long recognized that beams do exhibit the behavior discussed above. Extensive testing has adequately verified that plastification of the entire cross-section does occur [7.2] (assuming lateral-torsional buckling as treated in Chapter 9 and local buckling as treated in Chapter 6, Part II, do not occur).

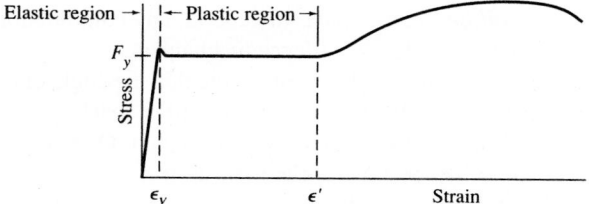

Figure 7.3.2
Stress-strain diagram for
most structural steels.

EXAMPLE 7.3.1

Determine the shape factor for a rectangular beam of width b and depth d.

Solution:

Referring to Fig. 7.3.3a, the moment M_y at first yield is

$$M_y = \int_A f y \, dA$$

and

$$f = F_y \frac{y}{d/2} = F_y \frac{2y}{d}$$

$$M_y = 2 \int_0^{d/2} \frac{2F_y}{d} y^2 b \, dy = F_y \frac{bd^2}{6} = F_y S$$

From Fig. 7.3.3b,

$$M_p = \int_A f y \, dA = 2 \int_0^{d/2} F_y b \, y \, dy = F_y \frac{bd^2}{4} = F_y Z$$

The shape factor is then

$$\xi = \frac{M_p}{M_y} = \frac{Z}{S} = 1.5$$

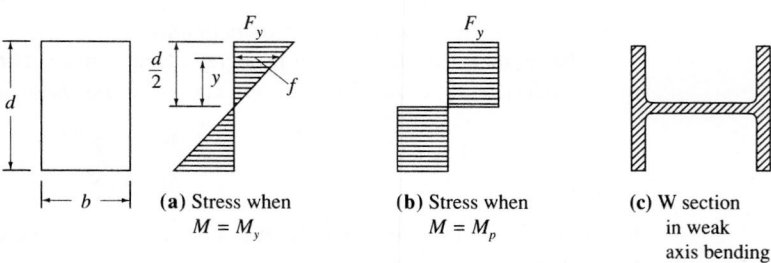

Figure 7.3.3
A rectangular section and a
W section in weak-axis
bending.

(a) Stress when $M = M_y$ (b) Stress when $M = M_p$ (c) W section in weak axis bending

which illustrates that there is a greater reserve beyond first yield in the bending of a rectangular section than in an I-shaped section bending about its strong axis. The reader is alerted to the fact that the W shape bent about its *weak* axis (y–y) is essentially a rectangular section (two rectangles separated by a distance) (Fig. 7.3.3c). ∎

Once the plastic moment strength M_p has been reached, the section can offer no additional resistance to rotation, behaving as a hinge but with constant resistance M_p, a condition known as a *plastic hinge*. In a statically determinate beam, such as a simply supported one, having one plastic hinge form will make the structure unstable; one real hinge at each end and a plastic hinge in the midspan region will create an unstable situation, known as a *collapse mechanism*. In general, any combination of three hinges, real or plastic, in a span will result in a collapse mechanism.

Referring to Fig. 7.3.4, one may note that the angle of rotation θ (radians/inch) is elastic from service load M until the extreme fiber reaches F_y at M_y, then becomes partially inelastic until the plastic moment M_p is reached. Once the plastic hinge has occurred and the $M - \theta$ curve has become horizontal, deflection of the beam (rotation of the plastified cross-section) increases without restraint. At the collapse condition the elastic deformation

Figure 7.3.4
Plastic behavior.

(a)

(b) Moment–rotation characteristics

due to bending on the segments between the ends and midspan is negligible compared to the rotation θ_u occurring at the plastic hinge. Thus, the analysis may treat the collapse situation as two rigid bodies having an angular discontinuity θ_u at midspan. As will be shown later in Sec. 10.2, it is only for statically determinate situations that one can expect every point along the factored moment diagram to be proportional to the elastic moment diagram.

Redistribution of the moments occurs during loading beyond the elastic range in usual statically indeterminate situations; that is, the bending moment diagram after a plastic hinge has occurred will no longer be proportional to the elastic bending moment diagram.

As discussed by Yura, Galambos, and Ravindra [7.3], even if the beam has adequate resistance with regard to lateral-torsional buckling (Chapter 9) and to local buckling (Chapter 6, Part II) to achieve the plastic moment strength, the actual limit state will still be failure by lateral-torsional buckling, compression flange local buckling, or web local buckling, but in the plastic range instead of the elastic range. Thus, prevention of failure by any of these instability modes until adequate rotation θ_u has occurred requires limits on the distance between points of lateral support, on the width/thickness ratio of the compression flange, and on the depth/thickness ratio of the web.

7.4   LATERALLY SUPPORTED BEAMS—LOAD AND RESISTANCE FACTOR DESIGN

The strength requirement for beams in load and resistance factor design according to AISC-F1 may be stated

$$\phi_b M_n \geq M_u \qquad (7.4.1)$$

where ϕ_b = resistance (i.e., strength reduction) factor for flexure = 0.90
   M_n = nominal moment strength
   M_u = factored service load moment (see Sec. 1.9)

Compact Sections

The nominal strength M_n for laterally stable "compact sections" according to AISC-F2.1 may be stated

$$M_n = M_p \qquad (7.4.2)$$

where M_p = plastic moment strength = $F_y Z_x$
Z_x = plastic modulus, Eq. 7.3.2
F_y = specified minimum yield stress

Design must account for the fact that local buckling of the compression flange or local buckling of the web may occur prior to achieving the high compressive strain necessary to develop M_p. When the width/thickness limitations λ_r of AISC-B4 are satisfied, achievement of only M_y is assured (that is, local buckling at a stress below F_y is prevented as discussed in Sec. 6.16). The limits λ_r for preventing local buckling in beams are given in Table 7.4.1. The

TABLE 7.4.1 Width/Thickness Limits λ_r for "Noncompact Section" Beams to Achieve F_y at Extreme Fiber (AISC-Table B4.1)

Yield stress	Flange (Unstiffened) Rolled I-shape			Flange (Unstiffened) Welded I-shape	Flange (Stiffened) HSS box	Web
F_y (ksi)	$\dfrac{b_f}{2t_f} = 1.0\sqrt{\dfrac{E}{F_y}}$ $= \dfrac{170}{\sqrt{F_y}}$	$\dfrac{h}{t_w}$	k_c*	$\dfrac{b_f}{2t_f} = 0.95\sqrt{\dfrac{k_cE}{F_L}}$** $= \dfrac{162k_c}{\sqrt{F_L}}$	$\dfrac{b}{t} = 1.4\sqrt{\dfrac{E}{F_y}}$ $= \dfrac{238}{\sqrt{F_y}}$	$\dfrac{h}{t_w} = 5.7\sqrt{\dfrac{E}{F_y}}$ $= \dfrac{970}{\sqrt{F_y}}$
36	28.4	161.8	0.31	19.1	39.7	161.8
		100	0.40	20.4		
		40	0.63	25.6		
42	26.3	149.8	0.33	17.7	36.8	149.8
		100	0.40	18.9		
		40	0.63	23.7		
45	25.4	144.7	0.33	17.8	35.5	144.7
		100	0.40	18.2		
		40	0.63	22.9		
50	24.1	137.3	0.34	16.2	33.7	137.3
		100	0.40	17.3		
		40	0.63	21.7		
55	23.0	130.9	0.35	15.4	32.1	130.9
		100	0.40	16.5		
		40	0.63	20.7		
60	22.0	125.3	0.36	15.0	30.8	125.3
		100	0.40	15.8		
		40	0.63	19.8		
65	21.1	120.4	0.36	14.6	29.6	120.4
		100	0.40	15.2		
		40	0.63	19.0		
90	18.0	102.3	0.40	12.9	25.1	102.3
		100	0.40	12.9		
		40	0.63	12.9		
100	17.0	97.1	0.41	7.9	23.8	97.1
		100	0.40	7.8		
		40	0.63	12.2		

*$k_c = \dfrac{4}{\sqrt{h/t_w}}$, where $0.35 \le k_c \le 0.763$

**$F_L = 0.7F_y$ for $S_{xt}/S_{xc} \ge 0.7$ and $F_L = F_y(S_{xt}/S_{xc}) \ge 0.5F_y$ for $S_{xt}/S_{xc} < 0.7$. First case was used here.

TABLE 7.4.2 Width/Thickness Limits λ_p for "Compact Section"* Beams to Achieve Significant Plastic Strain (AISC-Table B4.1, with $E = 29{,}000$ ksi and F_y in ksi)

F_y (ksi)	Unstiffened elements (uniform compression) $\dfrac{b_t}{2t_r} \le \dfrac{65}{\sqrt{F_y}}$	Stiffened elements (uniform compression) $\dfrac{b}{t_f} \le \dfrac{190}{\sqrt{F_y}}$	Stiffened elements (bending) $\dfrac{h}{t_w} \le \dfrac{640}{\sqrt{F_y}}$
36	10.8	31.7	107
42	10.0	29.3	98.8
45	9.7	28.3	95.4
50	9.2	26.9	90.5
55	8.8	25.6	86.3
60	8.4	24.5	82.6
65	8.1	23.6	79.4

*Plastic analysis is restricted by AISC-Appendix 1.2 (p. 16.1-151) to steels having $F_y \le 65$ ksi (450 MPa).

extreme fiber strain is assured only of reaching $\epsilon_y = F_y/E_s$. To achieve greater strain, the values of b/t (referred to generally as λ by LRFD) must be further restricted. To undergo large plastic strain the more severe width/thickness limitations λ_p discussed in Sec. 6.17 and prescribed for "compact sections" must be satisfied, as given in Table 7.4.2.

User Note in AISC-F2 states that all current ASTM A6 W, S, C and MC standard sections have compact webs and flanges except for ten sections identified to have flanges with $b_f/2t_f$ ratios larger than λ_p.

For the welded (flange and web continuously attached) I-shape, the $b_f/2t_f$ limit in 2005 is a function of h/t_w. Rolled I-shapes typically have h/t_w less than 40 and nearly all have that ratio 55 or less; welded I-shapes of similar proportions can be expected to have similar h/t_w ratios. The 1993 limits for welded I-shapes are higher for proportions similar to rolled I-shapes, and lower for thin web plate girders. The explanation is that the thinner the web the less rotational restraint it offers to prevent flange local buckling. The reduced limits for $b_f/2t_f$ at the web λ_r limit compared with 1986 LRFD are, for example, 21.7 instead of 24.0 for A36 steel, and 16.6 instead of 18.3 for Grade 50 steel. Since h/t_w is low for rolled I-shapes, the limit independent of h/t_w was retained for simplicity.

Noncompact Sections

Local buckling in hot-rolled I-shaped sections is, for practical purposes, only possible in the flanges. The nominal strength M_n for laterally stable "noncompact sections" whose flange width/thickness λ *exactly equals* the limits λ_r of AISC-Table B4.1 is the moment strength available when the extreme fiber is at the yield stress F_y. Because of residual stress, the strength is expressed as

$$M_n = M_r = (F_y - F_r)S_x = 0.7F_y S_x \qquad (7.4.3)$$

This is the "residual moment" that will cause the extreme fiber stress to rise from its residual stress $0.3F_y$ value when there is no applied load acting to the yield stress F_y. The elastic section modulus S_x equals the moment of inertia I_x divided by the distance from the neutral axis to the extreme fiber.

Partially Compact Sections

The nominal strength M_n for laterally stable "noncompact sections" whose flange width/thickness ratios λ are less than λ_r but not as low as λ_p must be linearly interpolated between M_p and $M_r = 0.7F_yS_x$, as follows according to AISC-F3.

$$M_n = M_p - (M_p - 0.7F_yS_x)\left(\frac{\lambda - \lambda_{pf}}{\lambda_{rf} - \lambda_{pf}}\right) \qquad (7.4.4)$$

where $\lambda = b_f/2t_f$ for I-shaped member flanges
  b_f = flange width
  t_f = flange thickness
  λ_{pf} = compact limit for reaching M_p (AISC-Table B4.1)
  λ_{rf} = noncompact limit for reaching M_r (AISC-Table B4.1)

The reader may note that the λ_r expressions for unstiffened flanges given in Table 7.4.1 differ from the derived

$$\left[0.56\sqrt{\frac{E}{F_y}} = \frac{95}{\sqrt{F_y, \text{ksi}}}\right]$$

given for uniform compression [Table 6.16.1, unstiffened (b) flanges] in Sec. 6.16. In addition to the introduction of h/t_w, as discussed above, the difference arises because the flange is not actually uniformly compressed and in the rolled and welded I-shaped sections residual stress (usually assumed as 10 ksi in rolled sections and 16.5 ksi in welded sections) is present. The limit on h/t_w for the web arises from the buckling of the stiffened plate (the web) under linearly varying stress (bending moment stress), a case which was not included in Sec. 6.16.

Slender Sections

When the width/thickness ratio λ exceeds the limit λ_r of AISC-B4, the section is referred to as "slender" and must be treated in accordance with AISC-F3.2(b). The nominal

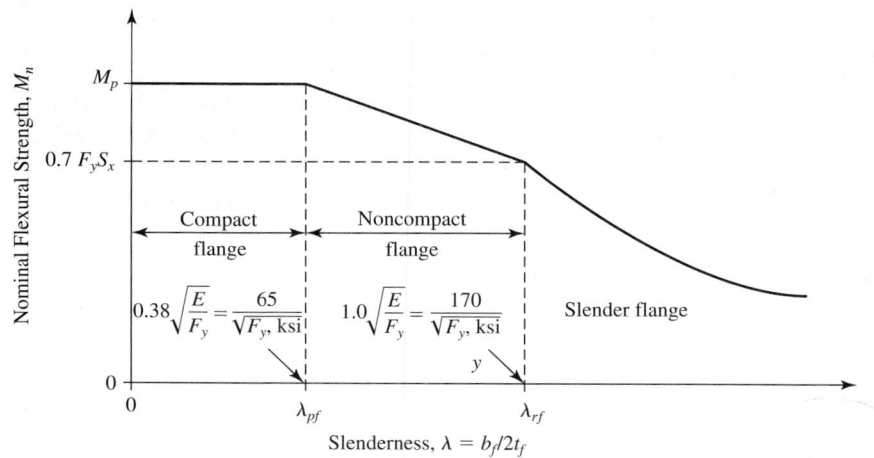

Figure 7.4.1
Nominal flexural strength as a function of the flange slendernes ratio of rolled I-shapes. Adapted from *AISC Manual* [1.15, Table B4.1].

strength of such a section is

$$M_n = \frac{0.9Ek_cS_x}{\sqrt{h/t_w}} = \frac{26,100k_cS_x}{\sqrt{h/t_w}} \qquad (7.4.5)$$

taking $E = 29,000$ ksi.

    Figure 7.4.1 illustrates the three ranges for the limit state of flange local buckling, showing the relationship between nominal flexural strength M_n and the flange width/thickness ratio $b_f/2t_f$.

EXAMPLE 7.4.1

Select the lightest W or M section to carry a uniformly distributed dead load of 0.2 kip/ft superimposed (i.e., in addition to the beam weight) and 0.8 kip/ft live load. The simply supported span (Fig. 7.4.2) is 20 ft. The compression flange of the beam is fully supported against lateral movement. Use Load and Resistance Factor Design, and select for the following steels; A36; A992; and A572 Grade 65.

Figure 7.4.2
Example 7.4.1.

Solution:

    (a) Compute the factored load M_u.

$$w_u = 1.2w_D + 1.6w_L = 1.2(0.2) + 1.6(0.8) = 1.52 \text{ kips/ft}$$

$$M_u = w_u L^2/8 = 1.52(20)^2/8 = 76 \text{ ft-kips (not including beam weight)}$$

    (b) A36 steel. Assume "compact section" since the vast majority of rolled sections satisfy $\lambda \leq \lambda_p$ for both the flange and the web. The design strength $\phi_b M_n$ is

$$\phi_b M_n = \phi_b M_p = \phi_b Z_x F_y$$

The design requirement is

$$\text{Required } \phi_b M_n = M_u$$

$$\text{Required } Z_x = \frac{M_u}{\phi_b F_y} = \frac{76(12)}{0.90(36)} = 28.1 \text{ in.}^3$$

Select from *AISC Manual*-Table 3-2 (pp 3–11 to 3–19) "Selection by Z_x":

$$\text{Try W12×22} \qquad Z_x = 29.3 \text{ in.}^3$$

Check "compact section" flange limit λ_{pf} (Table 7.4.2):

$$\left(\lambda = \frac{b_f}{2t_f} = \frac{4.03}{2(0.425)} = 4.7\right) < \left(\lambda_{pf} = \frac{65}{\sqrt{F_y, \text{ksi}}} = 10.8\right) \qquad \textbf{OK}$$

It is assumed that λ_p for the web is not exceeded; this is automatic for rolled W shapes.

Check the strength. Correct the factored moment M_u to include the beam weight.

$$w_u = 1.2(0.222) + 1.6(0.8) = 1.55 \text{ kips/ft}$$

$$M_u = 76(1.55/1.52) = 77.3 \text{ ft-kips}$$

$$M_n = M_p = Z_x F_y = 29.3(36)/12 = 87.9 \text{ ft-kips}$$

$$\phi_b M_n = 0.90(87.9) = 79.1 \text{ ft-kips} > [M_u = 77.3 \text{ ft-kips}] \qquad \textbf{OK}$$

Use W12×22, F_y = 36 ksi.

   (c) A992, Grade 50 steel.

$$\text{Required } Z_x = \frac{M_u}{\phi_b F_y} = \frac{76(12)}{0.90(50)} = 20.3 \text{ in.}^3$$

Select from *AISC Manual*-Table 3-2 (pp 3–11 to 3–19) "Selection by Z_x";

$$\text{Try W10×19} \qquad Z_x = 21.6 \text{ in.}^3$$

Check "compact section" flange limit λ_{pf} (Table 7.4.2):

$$\left(\lambda = \frac{b_f}{2t_f} = \frac{4.020}{2(0.395)} = 5.1 \right) < \left(\lambda_{pf} = \frac{65}{\sqrt{F_y, \text{ksi}}} = 9.2 \right) \qquad \textbf{OK}$$

Check the strength. Correct the factored moment M_u to include the beam weight.

$$M_u = 76(1.51/1.52) = 75.5 \text{ ft-kips}$$

$$M_n = M_p = Z_x F_y = 21.6(50)/12 = 90 \text{ ft-kips}$$

$$\phi_b M_n = 0.90(90) = 81 \text{ ft-kips} > [M_u = 75.5 \text{ ft-kips}] \qquad \textbf{OK}$$

Use W10×19, F_y = 50 ksi.

   (d) A572 Grade 65 steel.

$$\text{Required } Z_x = \frac{M_u}{\phi_b F_y} = \frac{76(12)}{0.90(65)} = 15.6 \text{ in.}^3$$

Select from *AISC Manual*-Table 3-2 (pp 3–11 to 3–19) "Selection by Z_x":

$$\text{Try W12×14} \qquad Z_x = 17.4 \text{ in.}^3 \qquad S_x = 14.9 \text{ in.}^3$$

Check "compact section" flange limit λ_{pf} (Table 7.4.2):

$$\left(\lambda = \frac{b_f}{2t_f} = \frac{3.97}{2(0.225)} = 8.8 \right) > \left(\lambda_{pf} = \frac{65}{\sqrt{F_y, \text{ksi}}} = 8.1 \right) \qquad \textbf{NG}$$

$$\left[\lambda = \frac{h}{t_w} = 54.3 \text{ (AISC p. 1-25)} \right] < \left[\lambda_{pf} = 79.4 \text{ (Table 7.4.2)} \right] \qquad \textbf{OK}$$

In this case the controlling limit state is local buckling of the flange $\lambda_r > \lambda > \lambda_p$, as above, the section is classified as "noncompact."

Check the strength. The strength is obtained by interpolation between M_p and $M_r = 0.7F_yS_x$ using Eq. 7.4.4. First λ_r for the flange must be obtained (from Table 7.4.1) as 21.1, which exceeds $b_f/2t_f$ of 8.8 and the section is "noncompact":

$$\lambda_{rf} = 1.0\sqrt{\frac{E}{F_y}} = \frac{170}{\sqrt{F_y, \text{ksi}}} = \frac{170}{\sqrt{65}} = 21.1$$

$$M_n = M_p - (M_p - 0.7F_yS_x)\left(\frac{\lambda - \lambda_{pf}}{\lambda_{rf} - \lambda_{pf}}\right)$$

$$M_p = Z_xF_y = 17.4(65)/12 = 94.2 \text{ ft-kips}$$

$$M_n = 94.2 - [94.2 - 0.7(65)14.9/12]\left(\frac{8.8 - 8.1}{21.1 - 8.1}\right) = 92 \text{ ft-kips}$$

$$M_u = 76(1.54/1.52) = 77 \text{ ft-kips}$$

$$\phi_bM_n = 0.90(92) = 83 \text{ ft-kips} > [M_u = 77 \text{ ft-kips}] \qquad \textbf{OK}$$

Use W12×14, $F_y = 65$ ksi.   ∎

7.5   LATERALLY SUPPORTED BEAMS—AISC-ASD METHOD

In accordance with the philosophy of Allowable Strength Design described in Secs. 1.8 and 1.9, Eq. 1.8.8(a) gives the flexural structural safety requirement, as follows:

$$\frac{M_n}{\Omega_b} \geq M_a \qquad (7.5.1)$$

where M_a = required strength, which equals the service load moment
M_n/Ω_b = allowable flexural strength
M_n = nominal flexural strength, computed as in Section 7.4
Ω_b = safety factor equal to 1.67 according to AISC-F1

EXAMPLE 7.5.1

Redesign the beam of Example 7.4.1 using the AISC Allowable Strength Design Method with A36 steel.

(a) A36 steel. Assume "compact section" since nearly all sections satisfy the width/thickness limits λ_p thus, the allowable strength M_n/Ω relationship can be expressed

$$\left[\frac{M_n}{\Omega} = \frac{F_yZ_x}{\Omega}\right] \geq M_a$$

That is, the allowable strength must equal or exceed the factored load M_a.

The superimposed service load (1 kip/ft) bending moment is

$$M_a = wL^2/8 = 1.0(20)^2/8 = 50 \text{ ft-kips}$$

$$\text{Required } Z_x = \frac{\Omega M_a}{F_y} = \frac{1.67(50)12}{36} = 27.8 \text{ in.}^3$$

Select from *AISC Manual* Tables 3-2 "Selection by Z_x" the lightest section having at least $Z_x = 27.8$ in.^{3}

$$\text{Try W12}\times 22 \qquad Z_x = 29.3 \text{ in.}^3$$

Check AISC-Table B4.1 "compact" limit λ_p for flanges (web for A36 automatically satisfies λ_p limit),

$$\left(\lambda = \frac{b_f}{2t_f} = \frac{4.03}{2(0.425)} = 4.7\right) < \left(\lambda_{pf} = \frac{65}{\sqrt{F_y, \text{ksi}}} = 10.8\right) \qquad \textbf{OK}$$

Use W12×22, $F_y = 36$ ksi. ∎

7.6   SERVICEABILITY OF BEAMS

Serviceability, instead of *strength,* may and often does control the design of beams. Excessive deflection may cause damage to supported nonstructural elements such as partitions, may impair the usefulness of the structure by, for instance, distorting door jambs so that doors will not open or close, or may cause "bouncy" floors. These are serviceability problems, often unrelated to the strength of the floor system.

An excellent treatment of design for serviceability of steel structures is provided by Fisher and West [7.55]. An overall appraisal of structural serviceability is provided by an ASCE Ad Hoc Committee [7.4]. Galambos and Ellingwood [7.5] have discussed general serviceability limit states, and Ellingwood [7.6] has provided guidance for steel structures. Excessive deflection is often indicative of excessive vibration and noise transmission, both serviceability issues.

Overall treatment of building floor vibrations, including recommended criteria, is provided by Murray [7.7]. Murray, Allen, and Ungar have an extensive practical treatment of vibrations in the AISC publication, *Floor Vibrations Due to Human Activity* [7.59]. Wright and Walker [7.8], Murray [7.9], Ellingwood and Tallin [7.11], Tolaymat [7.12], and Allen [7.13] also have treated floor vibrations and the related human response. Hatfield [7.14] has provided a design chart for floor vibration. Allen and Murray [7.54] have given design criteria for vibration resulting from walking. Griffis [7.57] has discussed serviceability limit states under wind.

On roofs a major deflection-related concern is ponding of water; this is specifically treated later in this section.

Deflection

Numerous structural analysis methods are available for computing deflections on uniform and variable moment of inertia sections. In general, the maximum deflection in an elastic member may be expressed as

$$\Delta_{\max} = \beta_1 \frac{WL^3}{EI} \qquad (7.6.1)$$

where W = total *service* load on the span

L = span length

E = modulus of elasticity (29,000 ksi or 200,000 MPa for steel)

I = moment of inertia

β_1 = coefficient which depends upon the degree of fixity at supports, the variation in moment of inertia along the span, and the distribution of loading. (For a simply supported beam, $\beta_1 = 5/384$; other values are available in *AISC Manual* Tables 3–23, pp. 3–211 to 3–226)

Figure 7.6.1
Typical bending moment
diagram for interior span
of continuous uniformly
loaded beam.

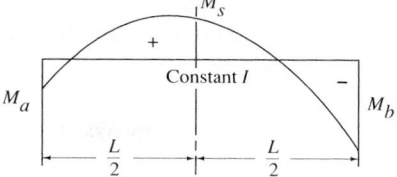

For continuous beams, the midspan deflection in the common situation of a uniform loading on a prismatic beam with unequal end moments (see Fig. 7.6.1) may be expressed* as

$$\Delta_{\text{midspan}} = \frac{5L^2}{48EI}[M_s - 0.1(M_a + M_b)] \tag{7.6.2}$$

Equation 7.6.2 will give satisfactory results when considered to be the maximum deflection for nearly all practical loadings for beams having uniform moment of inertia. Equation 7.6.2 may be verified by the use of a method such as conjugate beam.

For uniformly loaded simply supported beams, Eq. 7.6.1 becomes

$$\Delta_{\text{max}} = \frac{5wL^4}{384EI} \tag{7.6.3}$$

which upon substitution of $M = wL^2/8$, $f = Mc/I$, and $c = d/2$, gives

$$\Delta_{\text{max}} = \frac{10fL^2}{48Ed} \tag{7.6.4}$$

Equation 7.6.4 can be used as a good approximation for any simply supported beam as long as the maximum stress occurs near midspan. Refer to Table 7.6.1 for typical values.

AISC-L3 states, "Deflections in structural members and structural systems under *appropriate service load combinations* shall not impair the *serviceability* of the structure." No specific limits on deflection are given in the 2005 AISC Specification.

Traditionally, a live load deflection limit of $L/360$ was invoked when plastered ceilings were used. This generally was adequately restrictive; in recent years, however, ceiling structures have been more tolerant of larger deflection.

For the $L/360$ limitation, Eq. 7.6.4, using $E = 29,000$ ksi becomes

$$\frac{L}{d} \leq \frac{48(29,000)}{10(360)f} = \frac{387}{f} \tag{7.6.5}^{\dagger}$$

where f is the service load stress in ksi.

TABLE 7.6.1 Deflection Relationships According to Eq. 7.6.4

Δ_{max}	L/d	L/d (f = 22 ksi)	L/d (f = 30 ksi)
$L/360$	$387/f$	17.6	12.9
$L/300$	$464/f$	21.1	15.5
$L/240$	$580/f$	26.3	19.3
$L/200$	$696/f$	31.6	23.2

---

*See Chu-Kia Wang, Charles G. Salmon and José A. Pincheira, *Reinforced Concrete Design,* 7th ed. (John Wiley & Sons, New York, 2007), p. 516.

†For SI; $\dfrac{L}{d} = \dfrac{2666}{f}$ for f in MPa.   (7.6.5)

On the other hand, the AISC-L3 statement above is more general. Each structure must be evaluated based on its own requirements and limitations. In spite of this, the many useful traditional approaches are still useful. Typically, $L/240$ has been acceptable for common roof members. According to AISC Commentary-L3, "Deflections of about $1/300$ of the span (for cantilevers, $1/150$ of the length) are visible and may lead to general architectural damage or cladding leakage. Deflections greater than $1/200$ of the span may impair operation of movable components such as doors, windows, and sliding partitions."

Past guides may still be useful. Fully stressed (say, to a service load stress of about $0.6F_y$) floor beams and girders used the limit,

$$\frac{L}{d} \leq \frac{800}{F_y, \text{ ksi}} \qquad (7.6.6)*$$

For floor beams and girders, subject to shock or vibratory loads, supporting open areas free of partitions or other sources of damping,

$$\frac{L}{d} \leq 20 \qquad (7.6.7)*$$

For roof purlins, fully stressed, except flat roofs,

$$\frac{L}{d} \leq \frac{1000}{F_y, \text{ ksi}} \qquad (7.6.8)$$

Assuming the service load stress f equals the typical maximum value of $0.66F_y$ for "compact sections," the coefficients of 800 and 1000 correspond to L/d of $528/f$ and $660/f$, respectively. Using these L/d values in Eq. 7.6.4 would give simply supported beam limits of about $L/260$ and $L/210$, probably not severe enough limits for most situations.

On continuous spans, it is the actual deflection that is of importance, not the L/d ratio. For continuous beam deflection, a comparison of Eqs. 7.6.2 and 7.6.4 shows that Eq. 7.6.4 can also be used for continuous beams if the stress f is computed using the equivalent bending moment M_e,

$$M_e = M_s - 0.1(M_a + M_b) \qquad (7.6.9)$$

EXAMPLE 7.6.1

Select the lightest W section to carry a uniform dead load of 0.5 kip/ft and a live load of 1.0 kip/ft on a simply supported span of 42 ft. Adequate lateral support is provided. The live load deflection is limited to $L/360$. Use A572 Grade 50 and Load and Resistance Factor Design.

Solution:

(a) Compute the factored moment M_u. Estimating the beam weight at 70 lb/ft:

$$w_u = 1.2w_D + 1.6w_L$$

$$= 1.2(0.5 + 0.07 \text{ est}) + 1.6(1.0) = 2.28 \text{ kips/ft}$$

$$M_u = w_u L^2/8 = 2.28(42)^2/8 = 504 \text{ ft-kips}$$

---

*For SI, with F_y in MPa, $\dfrac{L}{d} \leq \dfrac{5500}{F_y}$ (7.6.6)

$\dfrac{L}{d} \leq \dfrac{6900}{F_y}$ (7.6.7)

(b) Compute required plastic modulus Z_x to satisfy strength requirement. Assuming compact section, the design strength $\phi_b M_n$ is

$$\phi_b M_n = \phi_b M_p = \phi_b F_y Z_x$$

$$\text{Required } Z_x = \frac{M_u}{\phi_b F_y} = \frac{504(12)}{0.90(50)} = 134 \text{ in.}^3$$

Select from *AISC Manual* Table 3-2 "Selection by Z_x"

$$\text{Try W24}\times55: \quad Z_x = 134 \text{ in.}^3$$

The section is compact for $F_y = 50$ ksi.

(c) Compute required moment of inertia I_x to satisfy the deflection limit. The service load moment instead of the factored moment must be used since deflection is of concern when the structure is being used, not when failure is imminent. The service live load moment is

$$M = wL^2/8 = 1.0(42)^2/8 = 221 \text{ ft-kips}$$

$$\Delta = \frac{5wL^4}{384EI} = \frac{5ML^2}{48EI}$$

$$\text{Required } I = \frac{5ML^2}{48E\Delta} = \frac{5(221)(42)^2(144)}{48(29,000)(42/360)} = 1724 \text{ in.}^4$$

Select from *AISC Manual* Table 3-3 "Selection by I_x"

$$\text{Try W24}\times68: \quad I_x = 1830 \text{ in.}^4$$

Note that the section required to control deflection is larger than the section required for strength; i.e., deflection controls.

(d) Check the W24×68 section:

$$w_u = 1.2(0.5 + 0.068) + 1.6(1.0) = 2.28 \text{ kips/ft}$$

$$M_u = 503 \text{ ft-kips}$$

$$M_n = M_p = Z_x F_y = 177(50)/12 = 738 \text{ ft-kips}$$

$$\phi_b M_n = \phi_b(738) = 0.90(738) = 664 \text{ ft-kips} > M_u \qquad \textbf{OK}$$

As expected, the strength considerably exceeds the required strength. Check deflection:

$$M(\text{service live load}) = 221 \text{ ft-kips}$$

$$\Delta = \frac{5ML^2}{48EI} = \frac{5(221)(42)^2 1728}{48(29,000)1830} = 1.32 \text{ in.} < \left[\frac{L}{360} = 1.40 \text{ in.}\right] \qquad \textbf{OK}$$

Note that the live load deflection and the limit are quite close (within 6%) but the strength considerably exceeds the requirement (32% greater); that is, *deflection controls*.

Use W24×68, $F_y = 50$ ksi. ■

Ponding of Water on Flat Roofs

When members of a flat roof system deflect, a bowl-shaped volume is created which is capable of retaining water. As water begins to accumulate, deflection increases to provide an increased volumetric capacity. This cyclical process continues until either (1) the succeeding deflection increments become smaller and equilibrium is reached; or (2) succeeding deflection increments are increasing, the system becomes unstable, and collapse occurs. This retention of water which results solely from the deflection of flat roof framing is what is referred to as *ponding*. From a serviceability standpoint, this ponding of water is a major reason for splitting of roof membranes, resulting in costly replacement of both the membrane and the insulation.

To prevent ponding of water accumulated on flat roofs, the 1963 AISC Specification required supporting members to satisfy the limitation

$$\frac{L}{d} \leq \frac{600}{f_b} \tag{7.6.10}$$

where f_b is the computed service load bending stress in ksi. Using Eq. 7.6.4, this would correspond roughly to a deflection limitation $L/240$ on a simply supported span.

Avoidance of ponding is much more complex than indicated by the above limitation. Marino [7.17] has provided an extensive treatment that forms the basis for the AISC Appendix 2 provisions. The flat roof is treated as a two-way system of secondary members (say, purlins) elastically supported by primary members (say, girders) which are rigidly supported by walls or columns, as shown in Fig. 7.6.2.

AISC-Appendix 2, Sec. 2.1 gives a simple but conservative criterion

$$C_p + 0.9C_s \leq 0.25 \tag{7.6.11}*$$

and

$$I_d \geq 25S^4(10^{-6}) \tag{7.6.12}*$$

where
$$C_p = \frac{32L_s L_p^4}{10^7 I_p} \tag{7.6.13}*$$

$$C_s = \frac{32SL_s^4}{10^7 I_s} \tag{7.6.14}*$$

L_p = length of primary member, ft
L_s = length of secondary member, ft
S = spacing of secondary member, ft
I_p = moment of inertia of primary member, in.^{4}
I_s = moment of inertia of secondary member, in.^{4}
I_d = moment of inertia of steel deck supported on secondary members, in.^{4} per ft

The criterion $C_p + 0.9C_s \leq 0.25$ assumes the supporting members to be loaded to full strength before onset of ponding. The terms C_p and C_s indicate the relative stiffnesses of the primary and secondary support systems, respectively. The right hand side of Eq. 7.6.11

---

*For SI, with L and S in metres, and I in mm^4, and I_d in mm^4/m,
$$C_p + 0.9C_s \leq 0.25 \tag{7.6.11}$$
$$I_d \geq 400S^4 \tag{7.6.12}$$
$$C_p = \frac{5L_s L_p^4}{10^{13} I_p} \tag{7.6.13}$$
$$C_s = \frac{5SL_s^4}{10^{13} I_s} \tag{7.6.14}$$

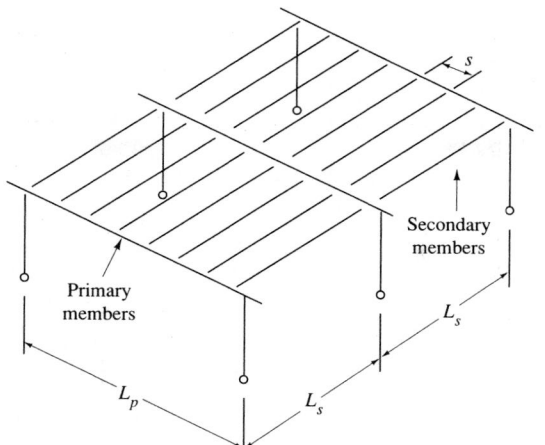

Figure 7.6.2
Flat roof arrangement for
ponding analysis. (From
Marino [7.17])

is a Stress Index U (See AISC-Appendix 2) representing the ratio of the increase in stress as a result of ponded water to the service load allowable stress. For instance, assuming that the service load stress in a member supporting ponded water could be permitted to increase to $0.8F_y$ from $0.6F_y$ or $0.66F_y$, the Stress Indexes U would be

$$U = \frac{0.8F_y - 0.66F_y}{0.66F_y} = 0.212$$

$$U = \frac{0.8F_y - 0.6F_y}{0.6F_y} = 0.33$$

A Stress Index U of 0.25 is a reasonable lower limit for that quantity. Design aids appear in AISC-Appendix 2 as Figs. A-2-1 and A-2-2. These diagrams show Eq. 7.6.11 to be conservative, the more so when stresses at onset of ponding are low. Burgett [7.18] has provided graphs for a fast check using Eqs. 7.6.11 and 7.6.12. More recently, Ruddy [7.19] has illustrated the procedure for a concrete floor over metal decking and supported by steel beams and girders.

Equation 7.6.12 pertains to roof decking that is supported on secondary members. Since this contributes little to ponding, it may be treated as a one-way system in the manner presented by Chinn [7.20]. When roof decking *is* the secondary member, then it should be treated according to Eq. 7.6.11.

More elaborate mathematical treatment of ponding has been given by Salama and Moody [7.21], Sawyer [7.22, 7.23], Chinn, Mansouri, and Adams [7.24], Avent and Stewart [7.25], and Avent [7.26].

7.7   SHEAR ON ROLLED BEAMS

Whereas long beams may be governed by deflection and medium length beams are usually controlled by flexural strength, short-span beams may be governed by shear.

To review the development of the shear stress equation for symmetrical sections, consider the slice dz of the beam of Fig. 7.7.1a, shown as a free body in Fig. 7.71b. If the unit shear stress v at a section y_1 from the neutral axis is desired, it is observed from Fig. 7.7.1c that

$$dC' = vt\, dz \qquad (7.7.1)$$

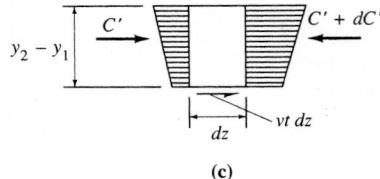

Figure 7.7.1
Flexural stresses involved in
derivation of shear stress
equation.

The horizontal forces arising from bending moment are

$$C' = \int_{y_1}^{y_2} f \, dA$$

$$C' + dC' = \int_{y_1}^{y_2} (f + df) \, dA$$

Subtracting,

$$dC' = \int_{y_1}^{y_2} df \, dA \tag{7.7.2}$$

$$df = \frac{dM y}{I} \tag{7.7.3}$$

$$dC' = \int_{y_1}^{y_2} \frac{dM y}{I} \, dA = \frac{dM}{I} \int_{y_1}^{y_2} y \, dA \tag{7.7.4}$$

Substituting Eq. 7.7.4 into Eq. 7.7.1 and solving for the shear stress v gives

$$v = \frac{dM}{dz} \left(\frac{1}{tI} \right) \int_{y_1}^{y_2} y \, dA \tag{7.7.5}$$

and upon recognizing that $V = dM/dz$, and letting

$$Q = \int_{y_1}^{y_2} y \, dA$$

the familiar equation

$$v = \frac{VQ}{It} \tag{7.7.6}$$

is obtained where Q is the first moment of area about the x-axis of the cross-sectional area
between the extreme fiber at y_2 (Fig. 7.7.1b) and the particular location at y_1 at which the
shear stress is to be determined.

Under usual procedures of steel design, the shear stress is computed as the average value over the gross area of the web neglecting the effect of any fastener holes; thus

$$f_v = \frac{V}{A_w} = \frac{V}{dt_w} \tag{7.7.7}$$

Note that large holes cut in a beam web to permit passage of pipes and ducts require special consideration and their effect may *not* be neglected.

The following example illustrates that in an I-shaped beam most of the shear is carried by the web.

EXAMPLE 7.7.1

Determine the elastic shear stress distribution on a W24×94 beam subjected to a service load shear force of 200 kips. Also compute the portion of the shear carried by the flange and that carried by the web. (See Fig. 7.7.2).

(a) **(b) Shear stress**

Figure 7.7.2
Example 7.7.1.

Solution:

(a) Stress at junction of flange and web.

$$V = 200 \text{ kips}$$

$$Q = 9.065(0.875)(12.155 - 0.4375) = 92.9 \text{ in.}^3$$

$$v = \frac{200(92.9)}{2700(0.515)} = 13.4 \text{ ksi (web)}, \quad v = 0.76 \text{ ksi (flange)}$$

(b) Stress at neutral axis.

$$Q = 92.9 + (12.155 - 0.875)^2(0.515)(0.5) = 92.9 + 32.8 = 125.7 \text{ in.}^3$$

$$v = \frac{200(125.7)}{2700(0.515)} = 18.1 \text{ ksi}$$

(c) Shear carried by flanges and web. Using an approximate linear variation,

$$V(\text{flanges}) = 2\left(\tfrac{1}{2}\right)(0.76)(0.875)(9.065) = 6 \text{ kips}$$

$$V(\text{web}) = 200 - 6 = 194 \text{ kips}$$

In this case, 97% of the shear is carried by the web.

(d) Average shear stress f_v on web.

$$f_v = \frac{V}{dt_w} = \frac{200}{24.31(0.515)} = 16.0 \text{ ksi}$$

which is 11.6% below the maximum value.   ∎

Nominal Shear Strength V_n in Rolled Beams

As shown by Example 7.7.1, the web is the element that primarily carries the shear in I-shaped sections. This is true also for the web (or webs) of "singly or doubly symmetric members and channels subject to shear in the plane of the web." (AISC-G2.1.)

As long as the web is stable, that is, instability resulting from shear stress or a combination of shear and bending stress cannot occur, the shear strength V_n of the section is based on overall shear yielding of the web. Thus,

$$V_n = \tau_y A_w \tag{7.7.8}$$

where τ_y = shear yield stress of the web steel

A_w = area of the web = dt_w for rolled beams

d = overall depth for rolled beams

t_w = web thickness

According to the "energy of distortion" theory (see Sec. 2.6), the shear yield stress τ_y equals the tension-compression yield stress F_y divided by $\sqrt{3}$ when shear stress acts alone, giving from Eq. 2.6.5,

$$\tau_y = 0.58 F_y \tag{2.6.5}$$

Thus, it is logical for AISC-G2.1 to use $\tau_y = 0.6 F_y$. Equation 7.7.8 then becomes

$$V_n = 0.6 F_{yw} A_w \tag{7.7.9}$$

where F_{yw} = yield stress of the web (= F_y for the section in rolled sections).

Equation 7.7.9 implies that h/t_w ratios do not exceed (AISC-G2.1)

$$\frac{h}{t_w} = 2.24 \sqrt{\frac{E}{F_y}} = \frac{380}{\sqrt{F_y, \text{ksi}}} \tag{7.7.10}$$

The development of Eq. 7.7.10 appears in Chapter 11 (Sec. 11.8) on plate girders. However, typical of plates as developed in Chapter 6 the buckling strength depends on a slenderness ratio; in this case h/t_w as shown in Fig. 7.7.3. From that figure one may note that maximum shear strength is available when h/t_w does not exceed 69.7 (for A36 steel). The h/t_w limits below which Eq. 7.7.9 is applicable for rolled beams without stiffeners are given in Table 7.7.1.

Load and Resistance Factor Design for Shear in Rolled Beams

The shear strength requirement in the AISC LRFD Method according to AISC-G1 may be stated

$$\phi_v V_n \geq V_u \tag{7.7.11}$$

where ϕ_v = 1.0 for h/t_w *not* exceeding the limit of Eq. 7.7.10 (basically rolled beams)

ϕ_v = 0.90 for h/t_w exceeding the limit of Eq. 7.7.10 (see plate girders, Chap. 11)

V_n = $0.6 F_y A_w$ (i.e., Eq. 7.7.9) for beams without transverse stiffeners not exceeding h/t_w limits given in Table 7.7.1

V_u = factored service load shear (see Sec. 1.9)

TABLE 7.7.1 Maximum h/t_w Limits When Stiffeners Are Not Used (Based on AISC-G2.1)

F_y (ksi)	F_y (MPa)	h/t_w[†]
36	248	63.6
42	290	58.9
45	310	56.9
50	345	53.9
60	414	49.3
65	448	47.3
100	689	38.2

[†] h/t_w limit = $380/\sqrt{F_y}$, ksi, Eq. 7.7.10; see Chap. 11 (Sec. 11.8) for derivation.

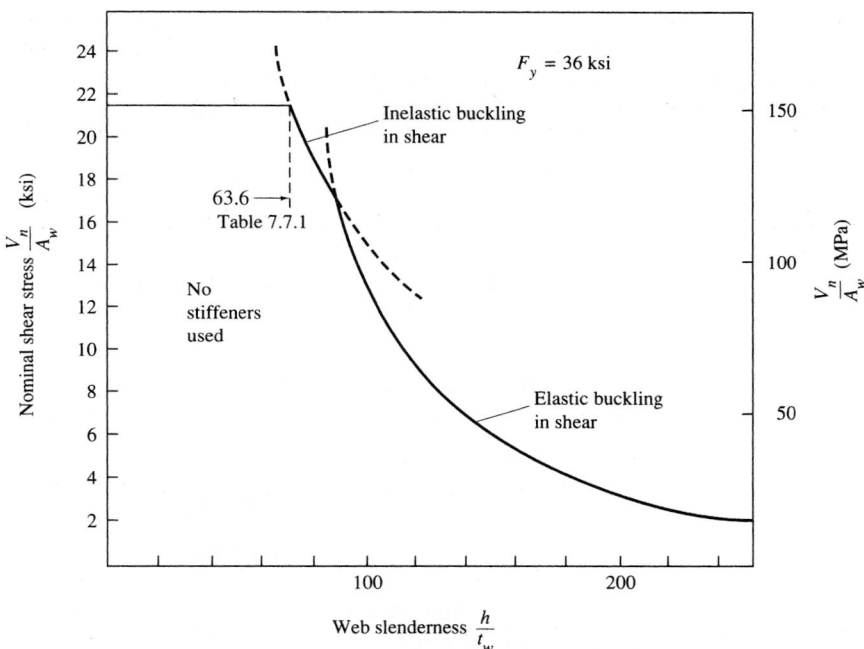

Figure 7.7.3
Nominal shear stress V_n/A_w vs web slenderness h/t_w for A36 steel beams *without* transverse stiffeners.

Allowable Strength Design for Shear in Rolled Beams

The safety requirement for shear in the AISC ASD Method according to AISC-G1 may be stated

$$\frac{V_n}{\Omega} \geq V_a \tag{7.7.12}$$

where $V_n = 0.6F_y A_w$ (i.e., Eq. 7.7.9) for beams without transverse stiffeners not exceeding h/t_w limits given in Table 7.7.1 (same as for the AISC LRFD Method)

V_a = factored service load shear

Ω_v = 1.50 for h/t_w *not* exceeding the limit of Eq. 7.7.10 (basically rolled beams)

Ω_v = 1.67 for h/t_w exceeding the limit of Eq. 7.7.10 (see plate girders, Chap. 11)

EXAMPLE 7.7.2

Select the lightest W section of A992 steel to carry a live load of 26 kips/ft and dead load of 1 kip/ft (in addition to the weight of the beam). The simply supported span is 5 ft. Lateral bracing is adequate for lateral stability. Use Load and Resistance Factor Design.

Solution:
Since the loading is heavy and the span is short, the designer should investigate shear as well as flexure.

(a) Compute factored loads M_u and V_u.

$$w_u = 1.2w_D + 1.6w_L$$
$$= 1.2(1) + 1.6(26) = 42.8 \text{ kips/ft}$$
$$M_u = wL^2/8 = 42.8(5)^2/8 = 133.8 \text{ ft-kips}$$
$$V_u = wL/2 = 42.8(5)/2 = 107 \text{ kips}$$

With these heavy superimposed loads the beam weight will have little effect.

(b) Select a section for flexure. Assume "compact section"; then,

$$\phi_b M_n = \phi_b M_p = \phi_b Z_x F_y$$

The design requirement is that $\phi_b M_n = M_u$; thus,

$$\text{Required } Z_x = \frac{M_u}{\phi_b F_y} = \frac{133.8(12)}{0.90(50)} = 35.7 \text{ in.}^3$$

Try W12×26 from *AISC Manual* Table 3-2 "Selection by Z_x" as the lightest beam having $Z_x \geq 35.7$ in.[3] All sections included in that table are compact.

(c) Check shear.

$$\phi_v V_n = \phi_v(0.6F_y)A_w$$
$$= 1.0(0.6)(50)(12.2)(0.23) = 84.2 \text{ kips}$$

Since $V_u = 107$ kips exceeds the shear strength provided by W12×26, the section is not adequate. The required web area A_w for shear is

$$\text{Required } A_w = \frac{V_u}{\phi_v(0.6F_y)} = \frac{107}{1.0(0.6)50} = 3.57 \text{ sq in.}$$

From the bending moment requirement, the next heavier sections are deeper, such as W14 or W16. If a W16 is selected, its web thickness required will be $3.57/16 = 0.22$ in. Try W14×30, $Z_x = 54.0$ in.[3] For shear,

$$\phi_v V_n = 1.0(0.6)(50)(13.8)(0.27) = 112 \text{ kips}$$

which exceeds the factored shear $V_u = 107.1$ kips (including beam weight) and is acceptable.

For using $\phi = 1$, the condition of $h/t_w \leq 2.24\sqrt{E/F_y}$ must be satisfied:

$$\frac{h}{t_w} = \frac{\text{unsupported height, } T}{\text{web thickness, } t_w} = \frac{11.625}{0.27} = 43.1 < \left[2.24\sqrt{E/F_y} = 54.0\right] \quad \textbf{OK}$$

This is less than the limit value of 54.0 and confirms the use of Eq. 7.7.9 for the nominal strength V_n in shear. The detailed discussion of the use of intermediate stiffeners when h/t_w exceeds the value from Table 7.7.1 appears in Chapter 11 on plate girders. Note that the value for h/t_w tabulated in the *AISC Manual* for the W14×30 is 45.4 based on an underestimate of the radius at the junction of flange to web; when the larger tabulated value is available it should preferably be used.

Use W14×30, $F_y = 50$ ksi.   ∎

7.8   CONCENTRATED LOADS APPLIED TO ROLLED BEAMS

When concentrated loads are applied to beams, beam bearing at supports, and reactions of beam flanges at connections to columns, a localized yielding from high compressive stress followed by inelastic buckling in the web region adjacent to the toe of a fillet occurs in the vicinity of concentrated loads. This entire behavior was formerly combined under the category "web crippling."

Typical of compression-related situations, there are two possible behaviors; yielding and instability. Recent AISC Specifications consider three categories: (a) local web yielding, (b) web crippling, and (c) sidesway web buckling.

The transmission of concentrated loads in beam-to-column connections is treated in Chapter 13 and concentrated loads on plate girders and related design of bearing stiffeners is treated in Chapter 11.

Load and Resistance Factor Design

The requirement of AISC-J10 may be stated

$$\phi R_n \geq R_u \tag{7.8.1}$$

where ϕ = resistance (strength reduction) factor
R_n = nominal reaction strength
R_u = factored reaction

Load and Resistance Factor Design—Local Web Yielding

Referring to Fig. 7.8.1, the concentrated reaction R acting on a beam is assumed critical at the toe of the fillet (a distance k from the face of the beam). The load is assumed to distribute along the web at a slope of 2.5 to 1. Prior to 1985 LRFD Specification, the distribution

Figure 7.8.1
Local web yielding considerations for establishing bearing length.

N = bearing length
k = distance from outer face of flange to web
toe of fillet (property given in *AISC Manual*
with dimensions of rolled sections)
R = concentrated load to be transmitted to girder

was conservatively taken as 45°. The 1978 ASD and earlier Specifications were based on the work of Lyse and Godfrey [7.27]. Investigators [7.28, 7.29] have shown the 45° slope to be overly conservative; the load actually spreads over a distance $(N + 5k)$ to $(N + 7k)$ for an interior load rather than the distance $(N + 2k)$ formerly used.

The nominal reaction strength R_n based on the yield strength at the toe of the fillet on a rolled I-shaped section is as follows:

1. For *interior loads* where the concentrated load is applied at a distance from the end of the member that is greater than the depth of the member,

$$R_n = (5k + N)F_{yw}t_w \qquad (7.8.2)$$

2. For *end reactions*,

$$R_n = (2.5k + N)F_{yw}t_w \qquad (7.8.3)$$

where k = distance from outer face of flange to web toe of fillet

       N = length of bearing $\geq k$ for end beam reaction

    F_{yw} = specified minimum yield stress of the web

     t_w = web thickness

The resistance (strength reduction) factor ϕ for local web yielding is 1.0; this reflects the traditional lower total safety factor used to control local web yielding than used for overall strength of a member.

Allowable Strength Design—Local Web Yielding

The safety requirement in the AISC ASD Method, Eq. 1.8.8 is

$$\frac{R_n}{\Omega} \geq R_a \qquad [1.8.8]$$

where R_n = nominal reaction strength

     R_a = factored service concentrated load

     Ω = 1.50

The required nominal reaction strength R_n based on the yield strength at the toe of the fillet on a rolled I-shaped section is given by Eqs. 7.8.2 and 7.8.3.

Recent research has shown that Eqs. 7.8.2 and 7.8.3 are conservative, with yielding confined to a length defined by a 2.5 to 1 slope.

Load and Resistance Factor Design—Web Crippling

To control the stability of the web at concentrated loads, AISC-J10.3 gives the nominal strength, based on the work of Roberts [7.30]. The nominal reaction strength R_n based on this stability criterion is

1. For *interior loads*, (i.e., concentrated load acts at $d/2$ or more from member end)

$$R_n = 0.80t_w^2\left[1 + 3\left(\frac{N}{d}\right)\left(\frac{t_w}{t_f}\right)^{1.5}\right]\sqrt{\frac{EF_{yw}t_f}{t_w}}$$

$$= 136t_w^2\left[1 + 3\left(\frac{N}{d}\right)\left(\frac{t_w}{t_f}\right)^{1.5}\right]\sqrt{\frac{F_{yw}t_f}{t_w}} \qquad (7.8.4)$$

2. For *end reactions* (i.e., concentrated load acts at less than $d/2$ from member end):

    2a. For $N/d \leq 0.2$

$$R_n = 0.40t_w^2 \left[1 + 3\left(\frac{N}{d}\right)\left(\frac{t_w}{t_f}\right)^{1.5} \right] \sqrt{\frac{EF_{yw}t_f}{t_w}}$$

$$= 68t_w^2 \left[1 + 3\left(\frac{N}{d}\right)\left(\frac{t_w}{t_f}\right)^{1.5} \right] \sqrt{\frac{F_{yw}t_f}{t_w}} \tag{7.8.5}$$

    2b. For $N/d > 0.2$

$$R_n = 0.40t_w^2 \left[1 + \left(\frac{4N}{d} - 0.2\right)\left(\frac{t_w}{t_f}\right)^{1.5} \right] \sqrt{\frac{EF_{yw}t_f}{t_w}}$$

$$= 68t_w^2 \left[1 + \left(\frac{4N}{d} - 0.2\right)\left(\frac{t_w}{t_f}\right)^{1.5} \right] \sqrt{\frac{F_{yw}t_f}{t_w}} \tag{7.8.6}$$

where E = modulus of elasticity of the steel (taken as 29,000 ksi)
          t_f = flange thickness through which the concentrated load is transmitted
       t_w = beam web thickness
        d = overall depth of the beam
    F_{yw} = specified yield stress of the beam web, (taken as ksi when E is inserted)

Allowable Strength Design—Web Crippling

Again, Eq. 1.8.8 expresses the safety requirement in the AISC ASD Method, as follows

$$\frac{R_n}{\Omega} \geq R_a \tag{1.8.8}$$

where R_n = nominal reaction strength
       R_a = factored service concentrated load
      Ω = 2.0

For the AISC Allowable Strength Method, AISC-J10.3 specifies $\Omega = 2.0$. The required nominal strength R_n is given by Eqs. 7.8.4 through 7.8.6.

Load and Resistance Factor Design—Sidesway Web Buckling

The nominal reaction strength R_n based on sidesway web buckling as given by AISC-J10.4 controls for compressive individual concentrated forces applied to members when relative lateral movement between the loaded compression flange and the tension flange is possible. This phenomenon has been studied by Roberts [7.30], Elgaaly [7.31], and Roberts and Chong [7.32]. Rarely will sidesway web buckling control on a rolled I-shaped section; however, it can influence design of a plate girder.

    The nominal reaction strength R_n based on sidesway web buckling in accordance with AISC-J10.4 is as follows:

1. When the compression flange *is restrained* against rotation:
    for $(h/t_w)/(L_b/b_f) \leq 2.3$,

$$R_n = \frac{C_r t_w^3 t_f}{h^2} \left[1 + 0.4\left(\frac{h/t_w}{L_b/b_f}\right)^3 \right] \tag{7.8.7}$$

for $(h/t_w)/(L_b/b_f) > 2.3,$

$$R_n = \text{no limit}$$

2. When the compression flange is *not restrained* against rotation:
for $(h/t_w)/(L_b/b_f) \leq 1.7,$

$$R_n = \frac{C_r t_w^3 t_f}{h^2}\left[0.4\left(\frac{h/t_w}{L_b/b_f}\right)^3\right] \tag{7.8.8}$$

for $(h/t_w)/(L_b/b_f) > 1.7,$

$$R_n = \text{no limit}$$

where L_b = largest laterally unbraced length along either flange at the point of load
b_f = flange width
t_w = web thickness
h = clear distance between flanges less the fillet or corner radius for rolled shapes; distance between adjacent lines of fasteners or the clear distance between flanges when welds are used for built-up shapes
C_r = 960,000 ksi when $M_u < M_y$ at the location of the force
= 480,000 ksi when $M_u \geq M_y$ at the location of the force

For the situation where the flange is restrained against rotation, when the factored force exceeds the design strength ϕR_n, (where $\phi = 0.85$), either more local lateral bracing must be used or stiffeners provided. When the flange *is not* restrained against rotation, and the strength is not adequate, local lateral bracing must be provided at both flanges at the point of application of the concentrated forces.

Allowable Stress Design—Sidesway Web Buckling

For allowable strength design, AISC-J10.4 gives Ω equal to 1.76. The safety requirement is given by Eq. 1.8.8.

EXAMPLE 7.8.1

Determine the size of bearing plate required for an end reaction of 15 kips dead load and 25 kips live load on a W10×26 beam of A992 steel. The beam rests on a concrete wall having a 28-day compressive strength $f'_c = 4000$ psi. Use the AISC LRFD Method.

Solution:

(a) Compute factored reaction R_u and required R_n.

$$R_u = 1.2(15) + 1.6(25) = 58 \text{ kips}$$

$$\text{Required } R_n = R_u/\phi = 58/1.0 = 58 \text{ kips}$$

(b) Determine plan dimensions for bearing plate. The bearing length must satisfy the more severe requirement of Eq. 7.8.3 (local web yielding), and Eqs. 7.8.5 or 7.8.6 (web crippling). Solving the simpler Eq. 7.8.3 for the required bearing length N gives

$$N = \frac{R_n}{F_{yw}t_w} - 2.5k = \frac{58}{50(0.260)} - 2.5(0.875) = 2.3 \text{ in.}$$

Try a 3-in. bearing plate. As a practical matter, 3 in. should be considered as minimum bearing length unless clearances require a lesser length.

Before checking Eqs. 7.8.5 or 7.8.6, whichever applies, for web crippling, investigate the requirement for bearing on the concrete. The nominal bearing strength P_p of concrete is given by AISC-J8 as

$$P_p = 0.85 f_c' A_1$$

where A_1 is the area of steel concentrically bearing on a concrete support. The requirement is that $\phi P_p \geq R_u$ (i.e., the factored bearing load). The strength reduction factor ϕ for bearing is 0.60 (AISC-J8). Thus, solving for the required bearing area A_1 gives

$$\text{Required } A_1 = \frac{\text{Required } P_p}{0.85 f_c'} = \frac{R_u/\phi}{0.85 f_c'} = \frac{58/0.60}{0.85(4)} = 28.4 \text{ sq in.}$$

For a 3-in. bearing length, the width B (see Fig. 7.8.2) would have to be 9.5 in. Since the beam flange width is only 5.77 in., a width B smaller than 9.5 in. is desirable. A 4-in. bearing length requires a width of 7.1 in. Try a plate 4 in. $\times$ 7.5 in. ($A_1 = 30$ sq. in.).

Now check web crippling on the W10×26:

$$N/d = 4/10.33 > 0.2; \text{ Eq. 7.8.6 applies.}$$

$$R_n = 68(0.260)^2 \left[1 + \left(\frac{4(4)}{10.33} - 0.2 \right) \left(\frac{0.260}{0.440} \right)^{1.5} \right] \sqrt{\frac{50(0.440)}{0.260}} = 68.2 \text{ kips}$$

Then the design strength $\phi R_n = 0.75(68.2) = 51.1$ kips. Since this is less than the factored load $R_u = 58$ kips, the bearing length N must be increased. A $5\frac{1}{2}$-in. bearing length N gives $\phi R_n = 59.5$ kips which exceeds R_u and is acceptable. *Use* bearing plate, $N = 5\frac{1}{2}$ in. $\times B = 6$ in.

(c) Determine the bearing plate thickness. The uniform (assumed) bearing pressure under factored load is

$$\text{Uniform bearing pressure } p = \frac{58}{5.5(6)} = 1.76 \text{ ksi}$$

The critical section for bending is taken at the toe of the flange-to-web fillet, a distance k from the mid-thickness of the web, and the beam flange is assumed not to participate. The bending moment is treated as that acting on a cantilever beam. The factored moment M_u is

$$M_u = \frac{p(B/2 - k)^2 N}{2} = \frac{1.76(3 - 0.875)^2 N}{2} = 3.97N \qquad \text{(a)}$$

For a rectangular section, the bending strength requirement is

$$\phi M_n \geq M_u \qquad \text{(b)}$$

Figure 7.8.2
Example 7.8.1.

where $\phi = 0.90$ and $M_n = M_p = ZF_y$. For a rectangular section of width N and thickness t (see Example 7.3.1),

$$Z = Nt^2/4 \tag{c}$$

To satisfy Eq. (b) above, $\phi ZF_y = 3.97N$; thus,

$$\frac{0.90Nt^2 F_y}{4} = 3.97N$$

and for $F_y = 36$ ksi, the required thickness becomes

$$\text{Required } t = \sqrt{\frac{3.97(4)}{0.90(36)}} = 0.70 \text{ in.} \quad \text{Use } t = \tfrac{3}{4} \text{ in.}$$

Use bearing plate, $\tfrac{3}{4} \times 5\tfrac{1}{2} \times 0'-6''$. ∎

Solving for the bearing plate thickness in general, equating ϕM_n to M_u,

$$\phi \frac{Nt^2}{4} F_y = \frac{p(B/2 - k)^2 N}{2}$$

$$\text{Required } t = \sqrt{\frac{2p(B/2 - k)^2}{\phi F_y}} \tag{7.8.9}$$

7.9 HOLES IN BEAMS

Flange Holes

For tension members the effect of fastener holes has been discussed in Chapter 3, where holes are deducted and net section is used. For compression members, since the fasteners occupy most of the space in the hole, the fasteners are assumed in design to completely fill the holes and a deduction for holes is not made.

When the nominal strength M_n reaches the plastic moment M_p, certainly tension flange holes reduce that strength; however, there will be a shift in neutral axis associated with a loss of strength at one flange, an effect that somewhat counteracts the effect of holes. Traditional design has neglected the effect of holes when holes do not represent a significant proportion of the flange area. When the nominal moment strength M_n is limited to less than M_p by some type of instability the effect of holes is reduced.

Traditionally, the AISC Specification has required deduction for flange holes only when the area of the holes exceeds 15% of the gross flange, and then only the area in excess of 15% was deducted.

The 2005 AISC Specification approach represents a departure even from the 1999 AISC Specification. AISC-F13.1. indicates that gross section is to be used except when the _tensile rupture limit state_ is a failure possibility.

1. The tensile rupture limit state does NOT apply when

$$F_u A_{fn} \geq F_y A_{fg} Y_t \tag{7.9.1}$$

2. The tensile rupture limit state MAY apply when

$$F_u A_{fn} < F_y A_{fg} Y_t \tag{7.9.2}$$

When Eq. 7.9.2 applies, the nominal flexural strength M_n at the location of the holes shall not be taken greater than

$$M_n = \frac{F_u A_{fn}}{A_{fg}} S_x \qquad (7.9.3)$$

where A_{fn} = net tension flange area

A_{fg} = gross tension flange area

$Y_t = 1.0$ for $F_y/F_n \leq 0.8$; 1.1 otherwise

Recent tests [*AISC Manual*, Ref. 1.15, p. 16.1-284] "... indicate that the flexural strength on the net section is better predicted by comparison of the quantities $F_y A_{fg}$ and $F_u A_{fn}$, with slight adjustment when the ratio of F_y to F_u exceeds 0.8."

Examples of procedure for considering bolt holes appear in the Chapter 13 section on beam and girder splices.

Web Holes

The *AISC Specification* [1.13] permits neglect of fastener holes located in the web, largely for the same reasons fastener holes in the flange may be neglected. Large holes cut into beam webs are entirely another matter. These holes require special analysis and usually will have to be reinforced by attaching extra plate material, often including stiffeners, around the sides of the hole. Design of large web holes in beams is outside the scope of this text. The recent work of the ASCE Task Committee [7.33, 7.34] should serve as a basic guide for design of beams with web openings. The definitive work related specifically to the AISC LRFD approach is that of Darwin [7.60]. The reader may also refer to the work of Bower [7.35-7.38], Frost and Leffler [7.39], Mandel, Brennan, Wasil, and Antoni [7.40], Cooper and Snell [7.41], Chan and Redwood [7.42, 7.43], Wang, Snell, and Cooper [7.44], Larson and Shah [7.45], Cooper, Snell, and Knostman [7.46], Redwood, Baranda, and Daly [7.47], Redwood and Uenoya [7.48], Daugherty [7.49], and Narayanan and Der-Avanessian [7.50]. Design tables for rectangular holes have been given by Redwood [7.51], and a design example is presented by Kussman and Cooper [7.52].

7.10 GENERAL FLEXURAL THEORY

Thus far, consideration has been given only to symmetrical shapes loaded symmetrically for which $f = Mc/I$ is correct for computing elastic flexural stress. The following development treats the general bending of arbitrary prismatic beams, i.e., beams having any cross-sectional shape with no variation along the length. They are also assumed to be free from twisting.

Consider the straight uniform cross-section beam of Fig. 7.10.1 acted upon by moment applied in the plane *ABCD* which makes the angle γ with the xz plane. The moments are represented by vectors normal to the plane of action (positive moment defined by using right-hand rule for rotation).

Examine next a portion of the beam of length z as shown in Fig. 7.10.2a. To satisfy equilibrium on the free body of Fig. 7.10.2a requires

$$\sum F_z = 0, \qquad \int_A \sigma \, dA = 0 \qquad (7.10.1)$$

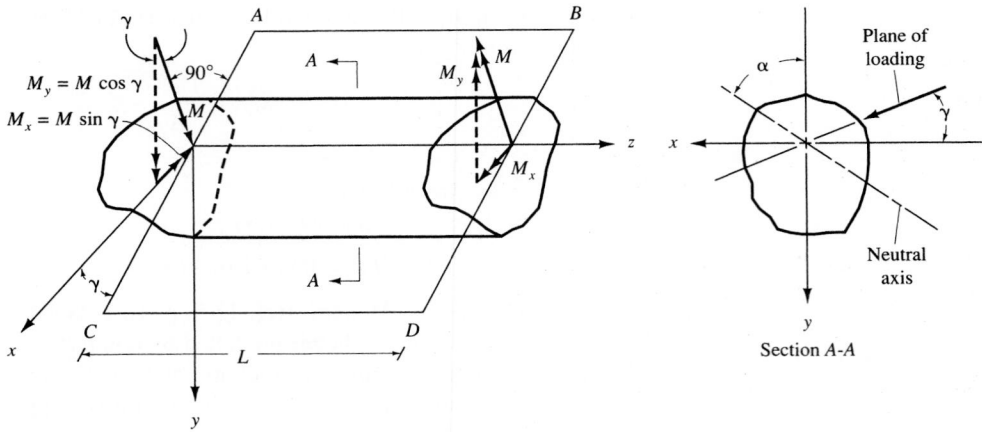

Figure 7.10.1
Prismatic beam under pure bending.

$$\sum M_x = 0, \qquad M_x = \int_A y\sigma \, dA \qquad\qquad (7.10.2)$$

$$\sum M_y = 0, \qquad M_y = \int_A x\sigma \, dA \qquad\qquad (7.10.3)$$

It is to be noted from Fig. 7.10.2b that the moments M_x and M_x are both positive in accordance with the customary convention of calling positive bending that which causes compression in the top portion of the beam. Also, the subscripts for M designate the axis *about which* bending occurs; i.e., the direction of the moment vector.

Bending in the *yz* Plane Only

If bending occurs in the *yz* plane, the stress σ is then proportional to y. Thus

$$\sigma = k_1 y \qquad\qquad (7.10.4)$$

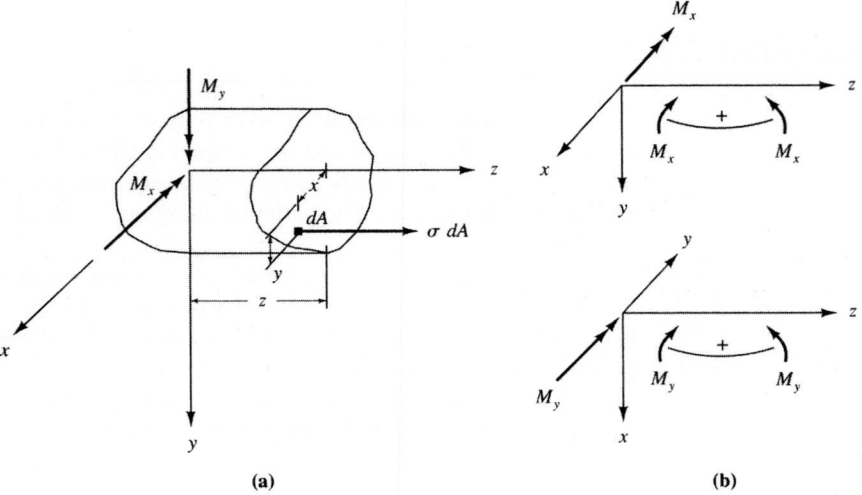

Figure 7.10.2
Free body of a portion of a
beam having length *z*.

(a)

(b)

Using Eqs. 7.10.1 through 7.10.3 gives

$$k_1 \int_A y \, dA = 0 \tag{7.10.5}$$

$$M_x = k_1 \int_A y^2 \, dA = k_1 I_x \tag{7.10.6}$$

$$M_y = k_1 \int_A xy \, dA = k_1 I_{xy} \tag{7.10.7}$$

From the first of the above expressions, $\int_A y \, dA = 0$, meaning x must be a centroidal axis. The stress may then be computed as

$$\sigma = \frac{M_x y}{I_x} \quad \text{or} \quad \frac{M_y y}{I_{xy}} \tag{7.10.8}$$

and the angle γ must be such that

$$\tan \gamma = \frac{M_x}{M_y} = \frac{I_x}{I_{xy}} \tag{7.10.9}$$

As a practical matter it would be unlikely that an unsymmetrical beam section would be located in a plane making the angle γ with the xz plane and have bending occur in the yz plane. Equation 7.10.9 also shows that if a section is used with at least one axis of symmetry (for which $\int_A xy \, dA = I_{xy} = 0$), $\tan \gamma = \infty$, $\gamma = 90°$, meaning the loading and the bending both occur in the yz plane.

　　An important conclusion here is that only if $I_{xy} = 0$ does the bending occur in the plane of loading. For example, on unsymmetrical shapes such as angles or zees loaded in the xz or yz plane, the plane of loading and the plane of bending will be different.

Bending in the xz Plane Only

If bending occurs in the xz plane the stress σ is then proportional to x. Thus

$$\sigma = k_2 x \tag{7.10.10}$$

and from Eqs. 7.10.1 through 7.10.3 are obtained:

$$k_2 \int_A x \, dA = 0 \tag{7.10.11}$$

which means y must be a centroidal axis. Also,

$$M_x = k_2 \int_A xy \, dA = k_2 I_{xy} \tag{7.10.12}$$

$$M_y = k_2 \int_A x^2 \, dA = k_2 I_y \tag{7.10.13}$$

and

$$\tan \gamma = \frac{M_x}{M_y} = \frac{I_{xy}}{I_y} \tag{7.10.14}$$

In the case where $I_{xy} = 0$, $\tan \gamma = 0$, i.e., the loading and bending both occur in the xz plane.

Bending in Neither xz Nor yz Planes

This is the realistic case when considering unsymmetrical sections in flexure. Since it is assumed all stresses are within the elastic limit, the total stress σ is the sum of the stresses due to bending in each of the xz and yz planes. Thus

$$\sigma = k_1 y + k_2 x \qquad (7.10.15)$$

and

$$M_x = k_1 I_x + k_2 I_{xy} \qquad (7.10.16)$$

$$M_y = k_1 I_{xy} + k_2 I_y \qquad (7.10.17)$$

Solving Eqs. 7.10.16 and 7.10.17 for k_1 and k_2 and substituting into Eq. 7.10.15 gives

$$\sigma = \frac{M_x I_y - M_y I_{xy}}{I_x I_y - I_{xy}^2} y + \frac{M_y I_x - M_x I_{xy}}{I_x I_y - I_{xy}^2} x \qquad (7.10.18)$$

which is the general flexure equation. The assumptions inherent in Eq. 7.10.18 are (a) a straight beam; (b) constant cross-section; (c) x- and y-axes are mutually perpendicular centroidal axes; and (d) that stress is proportional to strain and the maximum value is within the proportional limit.

Principal Axes

The principal axes are mutually perpendicular centroidal axes for which the moment of inertia is either a maximum or minimum. Furthermore, these axes are the only mutually perpendicular axes for which the product of inertia I_{xy} is zero. When a section has an axis of symmetry, that axis is a principal axis and Eq. 7.10.18 becomes

$$\sigma = \frac{M_x}{I_x} y + \frac{M_y}{I_y} x \qquad (7.10.19)$$

When there is no axis of symmetry, Eq. 7.10.19 can still be used if the principal axes are located and the quantities M_x, M_y, I_x, I_y, x, and y are all corrected so as to refer to the principal axes. Usually such transformations offer no advantage over direct use of Eq. 7.10.18.

Inclination of the Neutral Axis

When the loads acting on a flexural member pass through the centroid of the cross-section but are inclined with respect to either of the principal axes, the stresses may be determined by using Eq. 7.10.18 or Eq. 7.10.19. However, note should be made that the neutral axis is not necessarily perpendicular to the plane of loading. As shown from Eqs. 7.10.9 and 7.10.14 and Fig. 7.10.1,

$$\tan \gamma = \frac{M_x}{M_y} \qquad (7.10.20)$$

Since the stress along the neutral axis is equal to zero, σ may be set equal to zero in Eq. 7.10.18. With $\sigma = 0$, solving for $-x/y$ gives

$$-\frac{x}{y} = \left(\frac{M_x I_y - M_y I_{xy}}{I_x I_y - I_{xy}^2}\right)\left(\frac{I_x I_y - I_{xy}^2}{M_y I_x - M_x I_{xy}}\right) \qquad (7.10.21)$$

From Fig. 7.10.1, Section *A–A*, it is seen that at any point on the neutral axis, $\tan \alpha = -x/y$. Dividing both numerator and denominator of the right side of Eq. 7.10.21 by M_y gives

$$\tan \alpha = \frac{\dfrac{M_x}{M_y} I_y - I_{xy}}{I_x - \dfrac{M_x}{M_y} I_{xy}} \qquad (7.10.22)$$

Substitution of Eq. 7.10.20 into Eq. 7.10.22 gives

$$\tan \alpha = \frac{I_y \tan \gamma - I_{xy}}{I_x - I_{xy} \tan \gamma} \qquad (7.10.23)$$

When investigating a section with one axis of symmetry, $I_{xy} = 0$; Eq. 7.10.23 then becomes

$$\tan \alpha = \frac{I_y}{I_x} \tan \gamma \qquad (7.10.24)$$

EXAMPLE 7.10.1

A W18×50 used as a beam is subjected to loads inclined at 5° from the vertical axis as shown in Fig. 7.10.3. Locate the inclination of the neutral axis.

Solution:

$$I_x = 800 \text{ in.}^4; \qquad I_y = 40.1 \text{ in.}^4$$

$$\tan 85° = \tan \gamma$$

Using Eq. 7.10.24,

$$\tan \alpha = \frac{I_y}{I_x} \tan \gamma = \frac{40.1}{800} \tan 85° = 0.573$$

$$\alpha = 29.8° \ (29°50')$$

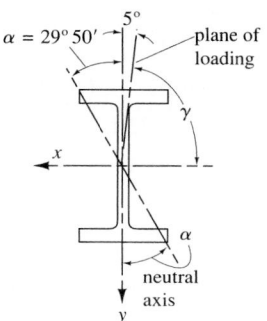

Figure 7.10.3
Biaxially loaded beam of
Example 7.10.1.

EXAMPLE 7.10.2

Compute the maximum flexural stress in a $6 \times 4 \times \frac{1}{2}$ angle with the long leg vertically downward when service loaded with 0.5 kip/ft on a simply supported span of 10 ft (see Fig. 7.10.4). Compare the value assuming the angle is completely free to bend in any direction with that obtained assuming bending in only the vertical plane.

Figure 7.10.4
Data for Example 7.10.2.

Solution:

(a) Angle free to bend in any direction. Using Eq. 7.10.18 with AISC properties

$$I_x = 17.4 \text{ in.}^4, \qquad I_y = 6.27 \text{ in.}^4$$

$$I_{xy} = [6(3.00 - 1.99)(-0.987 + 0.25) + 3.5(2.25 - 0.987)(-1.99 + 0.25)]0.50$$

$$= [(-4.47) + (-7.69)]0.50 = -6.08 \text{ in.}^4$$

$$M_x = \tfrac{1}{8}(0.5)(10)^2 = 6.25 \text{ ft-kips} = 75 \text{ in.-kips}$$

$$M_y = 0$$

Stress at point A:

$$f_A = \frac{M_x(I_y y - I_{xy} x)}{I_x I_y - I_{xy}^2} = \frac{6.25(12)[6.27(+4.01) - (-6.08)(-0.987)]}{17.4(6.27) - (6.08)^2}$$

$$= \frac{75(19.14)}{72.1} = 75(0.265) = +19.9 \text{ ksi (tension)}$$

Stress at point B:

$$f_B = \frac{75[6.27(+4.01) - (-6.08)(-0.487)]}{72.1}$$

$$= 75(0.308) = +23.1 \text{ ksi (tension)}$$

Stress at point C:

$$f_C = \frac{75[6.27(-1.99) - (-6.08)(-0.987)]}{72.1}$$

$$= 75(-0.256) = -19.2 \text{ ksi (compression)}$$

(b) Angle free to bend in any direction. Use alternate method suggested by Gaylord and Gaylord [7.53] (pp. 143–147). Compute the stresses assuming first that the beam bends only in the yz plane. Conveniently let $M_x = 100$ in.-kips. Then, according to Eq. 7.10.9, $M_y = M_x I_{xy}/I_x$ must also be acting (Fig. 7.10.5a) if bending is to occur only in the yz plane.

Since the real loading has only M_x, M_y must be removed by application of an equal and opposite moment, further considering that bending occurs only in the xz plane. This means, according to Eq. 7.10.14, the simultaneous application of $M_x = M_y I_{xy}/I_y$ (see Fig. 7.10.5b).

Bending in vertical plane:

$$M_x = 75.0 \left(\frac{100}{66.16} \right) = 113.4 \text{ in.-kips}$$

$$f_{A1} = f_{B1} = \frac{113.4(4.01)}{17.4} = 26.1 \text{ ksi (tension)}$$

$$f_{C1} = \frac{113.4(1.99)}{17.4} = 13.0 \text{ ksi (compression)}$$

Bending in horizontal plane:

$$M_y = -113.4(-6.08)/17.4 = 39.6 \text{ in.-kips}$$

$$f_{A2} = f_{C2} = \frac{39.6(0.987)}{6.27} = 6.2 \text{ ksi (compression)}$$

$$f_{B2} = \frac{39.6(0.487)}{6.27} = 3.1 \text{ ksi (compression)}$$

Total stresses in general bending:

$$f_A = +26.1 - 6.2 = +19.9 \text{ ksi (tension)}$$

$$f_B = +26.1 - 3.1 = +23.0 \text{ ksi (tension)}$$

$$f_C = -13.0 - 6.2 = -19.2 \text{ ksi (compression)}$$

which agree with the values, 19.9, 23.1, and –19.2 ksi, respectively, as computed by the general formula.

The general equation for stress at any point is seen to be

$$f = \frac{113.4y}{17.4} + \frac{39.6x}{6.27}$$

where if $f = 0$, the neutral axis is $y = -0.969x$.

Figure 7.10.5
Solution by superposition of bending in the vertical and horizontal planes.

(a) Bending in vertical plane   (b) Bending in horizontal plane   (c) General bending

This superimposition method permits the designer to visualize what is taking place. If attached construction constrains an unsymmetrical section to bend in the vertical plane, the restraining moment capacity can be computed by using Eq. 7.10.9.

   (c) Angle restrained to bend in the vertical plane:

$$f_A = f_B = \frac{75(4.01)}{17.4} = +17.3 \text{ ksi (tension)}$$

$$f_C = \frac{75(-1.99)}{17.4} = -8.6 \text{ ksi (compression)}$$

Unless the horizontal constraints actually act, the tensile stress at point B is underestimated by 25% and the compressive stress at C is underestimated by 55%.

Frequently designers assume $f = My/I$ is applicable without considering whether or not adequate horizontal restraints are present. Although usually some degree of restraint is present, care should be exercised when investigating unsymmetrical beams. Neglect of the lateral (horizontal) component is always on the unsafe side. ■

7.11   BIAXIAL BENDING OF SYMMETRIC SECTIONS

Flexural stresses on sections with at least one axis of symmetry and loaded through the centroid may be computed using Eq. 7.10.19, which when modified to give maximum stress σ becomes

$$\sigma = \frac{M_x}{S_x} + \frac{M_y}{S_y} \tag{7.11.1}$$

where $S_x = I_x/(d/2)$ and $S_y = I_y/(b/2)$ are the section modulus values.

Nominal Strength

The nominal strength of a section subject to biaxial bending is not readily determined. Such strength will certainly depend on the proportions of the section and the relative magnitudes of the applied moments M_x and M_y. The use of an interaction equation such as used for beam-columns is not believed adequately conservative for cases where there is no axial compression or tension. In the past, the AISC ASD Specification has traditionally limited the combined stress according to Eq. 7.11.1 to a maximum of $0.60 F_y$. This effectively implies that the nominal strength of the section under combined M_x and M_y is reached when the extreme fiber stress reaches the yield stress F_y. Thus, the entire cross-section will be elastic; no credit is then given to any ability of the cross-section to undergo plastic deformation. Certainly this approach is conservative.

Following the above approach, the elastic stress equation, Eq. 7.11.1, could be used with nominal maximum strength moments, M_{nx} and M_{ny}, and a maximum stress F_y. Thus, the nominal strength could be assumed to be reached when

$$\frac{M_{nx}}{S_x} + \frac{M_{ny}}{S_y} = F_y \tag{7.11.2}$$

Load and Resistance Factor Design

Method 1. Many designers will use the beam-column safety criteria of AISC-H1 for the member loaded in biaxial bending without axial load. In the authors' view that is likely an unconservative approach. The criteria of AISC-H1 were developed as strength relationships for *compression* members, which simultaneously are acted on by bending moments; not very close to treating the strength in biaxial bending. The *AISC Specification* certainly permits biaxial bending to be covered by AISC-H1 (see *User Note* at the beginning of AISC-Chapter F). When one does that, no new principles are encountered in this section of the text.

Method 2. AISC-H2 uses the combined stress approach of Eq. 7.11.2 recognizing that the safety provisions include an overload effect and an understrength (resistance factor) effect. This method is intended to apply for other than doubly and singly symmetric sections. However, AISC-H2 permits use of this method in lieu of AISC-H1. A stress-based simple linear interaction equation (without the axial load term) is used,

$$\frac{f_{bx}}{F_{bx}} + \frac{f_{by}}{F_{by}} \leq 1 \tag{7.11.3}$$

where f_{bx} = flexural stress due to factored moment M_{ux}
   $= M_{ux}/S_x$
   f_{by} = flexural stress due to factored moment M_{uy}
   $= M_{uy}/S_y$
   F_{bx} = flexural stress due to the design moment $\phi_b M_{nx}$
   $= \phi_b M_{nx}/S_x$
   F_{by} = flexural stress due to the design moment $\phi_b M_{ny}$
   $= \phi_b M_{ny}/S_y$

Equation 7.11.3 can also be written in terms of forces (i.e., moments in this case)

$$\frac{M_{ux}}{\phi_b M_{nx}} + \frac{M_{uy}}{\phi_b M_{ny}} \leq 1 \tag{7.11.4}$$

To compensate for the use of the beam-column strength equation, which can be unconservative in the absence of axial load, the nominal strengths about the *x*- and *y*-axes may be taken equal to the nominal strengths based on yielding, which puts a conservatively high (or at least as high) value for F_{bx} and F_{by} in the denominator of the terms of Eq. 7.11.3. Taking this conservative approach,

$$M_{nx} = F_y S_x$$
$$M_{ny} = F_y S_y \tag{7.11.5}$$

Substituting Eqs. 7.11.5 into Eq. 7.11.4, and multiplying by S_x gives an equation for the required S_x

$$S_x \geq \frac{M_{nx}}{\phi_b F_y} + \frac{M_{ny}}{\phi_b F_y}\left(\frac{S_x}{S_y}\right) \tag{7.11.6}$$

Allowable Strength Design

In the AISC ASD Method, Eq. 7.11.4 replaces the factored service load moments M_u (divided by ϕ_b) with the service load moment M_u (here called M_x and M_y) multiplied by the safety factor Ω, which then gives

$$\frac{\Omega M_x}{M_{nx}} + \frac{\Omega M_y}{M_{ny}} \leq 1 \tag{7.11.7}$$

Using the conservative yield moments in the denominator, Eq. 7.11.7 becomes

$$\frac{\Omega M_x}{S_x F_y} + \frac{\Omega M_y}{S_y F_y} \leq 1 \tag{7.11.8}$$

and solving for the required S_x gives

$$S_x \geq \frac{\Omega M_x}{F_y} + \frac{\Omega M_y}{F_y}\left(\frac{S_x}{S_y}\right) \tag{7.11.9}$$

TABLE 7.11.1 Typical S_x/S_y Values

Shape	Depth d (in.)	S_x/S_y
M	6, 8	5–7
M	10, 12	8–11
Light W and M	4–8	3
W	8, 10	3–4
W	12	3–6
W	14 (up to 84 lb/ft)	4–8
W	14 (over 84 lb/ft)	$2\frac{1}{2}$–3
W	16, 18, 21	5–9
W	24, 27	6–10
W	30, 33, 36	7–12
S	6–8	d
S	10–18	$0.75d$
C	up to 7	$1.5d$
C	8–10	$1.25d$
C	12, 15	d

EXAMPLE 7.11.1

Select the lightest W or M section to carry service dead load moments $M_x = 15$ ft-kips and $M_y = 5$ ft-kips, and live load moments $M_x = 45$ ft-kips and $M_y = 20$ ft-kips. Consider that adequate lateral bracing is provided to preclude instability. Use steel having $F_y = 50$ ksi, and the AISC LRFD Method.

Solution:

(a) Compute factored loads M_{ux} and M_{uy}.

$$M_{ux} = 1.2(15) + 1.6(45) = 90 \text{ ft-kips}$$

$$M_{uy} = 1.2(5) + 1.6(20) = 38 \text{ ft-kips}$$

(b) Using Method 2, determine required section modulus S_x and select section. Equation 7.11.6 gives

$$\text{Required } S_x \geq \frac{90(12)}{0.90(50)} + \frac{38(12)}{0.90(50)}\left(\frac{S_x}{S_y}\right)$$

$$\geq 24 + 10 S_x/S_y$$

From Table 7.11.1 the ratio S_x/S_y can be expected to be on the order of 3 to 4; thus $S_x \approx 54$ to 64 in.^{3}

Using the *AISC Manual* [15], one must use the plastic modulus Z_x in AISC Table 3-2 "Selection by Z_x". The section modulus S_x may be approximately converted to Z_x by multiplying by the average shape factor 1.12 (see Sec. 7.3). For this case the initial trial S_x values of 54 to 64 would have become $Z_x \approx 61$ to 72 in.^{3} and the same starting section of W16×40 would have been obtained.

Try W10×60:

$$S_x/S_y = 66.7/23.0 = 2.9$$

Required $S_x = 24 + 10(2.9) = 53$ in.$^3 < 66.7$ in.^{3} **OK**

For W10×49,

$$S_x/S_y = 54.6/18.7 = 2.92$$

Required $S_x = 24 + 10(2.92) = 53.2$ in.$^3 < 54.6$ in.^{3} **OK**

Use W10×49, F_y 50 ksi.

If one had assumed a 10-in. depth as desirable and used Eq. 7.11.6 with a coefficient of 3.5 instead of 3,

$$\text{Required } S_x = 24 + 10(3.5d/b_f)$$

For W10 sections, d/b_f is either 1 or 1.25.

$$\text{Required } S_x \approx 24 + 35 = 59 \text{ in.}^3$$

which would also have required a check of two sections. The more general approach using typical S_x/S_y values seems most useful unless a specific depth is desired. ■

SELECTED REFERENCES

7.1. T. V. Galambos. "History of Steel Beam Design," *Engineering Journal,* AISC, **14,** 4 (Fourth Quarter 1977), 141–147.

7.2. Joint Committee of Welding Research Council and the American Society of Civil Engineers. *Commentary on Plastic Design in Steel,* 2nd ed., ASCE Manual and Reports on Practice No. 41, New York. 1971.

7.3. Joseph A. Yura, Theodore V. Galambos, and Mayasandra K. Ravindra. "The Bending Resistance of Steel Beams," *Journal of the Structural Division,* ASCE, **104,** ST9 (September 1978), 1355–1370.

7.4. Ad Hoc Committee on Serviceability Research, Committee on Research of the Structural Division. "Structural Serviceability: A Critical Appraisal and Research Needs," *Journal of Structural Engineering,* ASCE, **112,** 12 (December 1986), 2646–2664.

7.5. Theodore V. Galambos and Bruce Ellingwood. "Serviceability Limit States: Deflection," *Journal of Structural Engineering,* ASCE, **112,** 1 (January 1986), 67–84.

7.6. Bruce Ellingwood. "Serviceability Guidelines for Steel Structures," *Engineering Journal,* AISC, **26,** 1 (1st Quarter 1989), 1–8.

7.7. Thomas M. Murray. "Building Floor Vibrations," *Engineering Journal,* AISC, **28,** 3 (3rd Quarter 1991), 102–109. Errata, **28,** 4 (4th Quarter 1991), 176.

7.8. Richard N. Wright and William H. Walker. "Vibration and Deflection of Steel Bridges," *Engineering Journal,* AISC, **9,** 1 (January 1972), 20–31.

7.9. Thomas M. Murray. "Design to Prevent Floor Vibrations," *Engineering Journal,* AISC, **12,** 3 (Third Quarter 1975), 82–87.

7.10. Thomas M. Murray. "Acceptability Criterion for Occupant-Induced Floor Vibrations," *Engineering Journal,* AISC, **18,** 2 (Second Quarter 1981), 62–70.

7.11. Bruce Ellingwood and Andrew Tallin. "Structural Serviceability: Floor Vibrations," *Journal of Structural Engineering,* ASCE, **110,** 2 (February 1984), 401–418. Disc. **111,** 5 (May 1985), 1158–1161.

7.12. Raed A. Tolaymat. "A New Approach to Floor Vibration Analysis," *Engineering Journal.* AISC, **25,** 4 (4th Quarter 1988), 137–143.

7.13. D. E. Allen. "Building Vibrations from Human Activities," *Concrete International,* **12,** 6 (June 1990), 66–73.

7.14. Frank J. Hatfield. "Design Chart for Vibration of Office and Residential Floors," *Engineering Journal,* AISC, **29,** 4 (4th Quarter 1992), 141–144.

7.15. J. W. Larson and R. K. Huzzard. "Economical Use of Cambered Steel Beams," presented at AISC National Steel Construction Conference, March 1990. Available as Bethlehem Steel Corp. *Technical Bulletin TB-309,* May 1990.

7.16. David T. Ricker. "Cambering Steel Beams," *Engineering Journal,* AISC, **26,** 4 (4th Quarter 1989), 136–142.

7.17. Frank J. Marino. "Ponding of Two-Way Roof Systems," *Engineering Journal,* AISC, **3,** 3 (July 1966), 93–100.

7.18. Lewis B. Burgett. "Fast Check for Ponding," *Engineering Journal,* AISC, **10,** 1 (First Quarter 1973), 26–28.

7.19. John L. Ruddy. "Ponding of Concrete Deck Floors," *Engineering Journal,* AISC, **23,** 3 (Third Quarter 1986), 107–115.

7.20. James Chinn. "Failure of Simply-Supported Flat Roofs by Ponding of Rain," *Engineering Journal,* AISC, **2,** 2 (April 1965), 38–41.

7.21. A. E. Salama and M. L. Moody. "Analysis of Beams and Plates for Ponding Loads," *Journal of the Structural Division,* ASCE, **93,** ST1 (February 1967), 109–126.

7.22. D. A. Sawyer. "Ponding of Rainwater on Flexible Roof Systems," *Journal of the Structural Division,* ASCE, **93,** ST1 (February 1967), 122–147.

7.23. D. A. Sawyer. "Roof-Structure Roof-Drainage Interaction," *Journal of the Structural Division,* ASCE, **94,** ST1 (January 1968), 175–198.

7.24. James Chinn, Abdulwahab H. Mansouri, and Staley F. Adams. "Ponding of Liquids on Flat Roofs," *Journal of the Structural Division,* ASCE, **95,** ST5 (May 1969), 797–807.

7.25. R. Richard Avent and William G. Stewart. "Rainwater Ponding on Beam-Girder Roof Systems," *Journal of the Structural Division,* ASCE, **101,** ST9 (September 1975), 1913–1927.

7.26. R. Richard Avent. "Deflection and Ponding of Steel Joists," *Journal of the Structural Division,* ASCE, **102,** ST7 (July 1976), 1399–1410.

7.27. I. Lyse and H. J. Godfrey. "Investigation of Web Buckling in Steel Beams," *Transaction,* ASCE, **100** (1935), 675–706.

7.28. J. D. Graham, A. N. Sherbourne, R.N. Khabbaz, and C. D. Jensen. *Welded Interior Beam-to-Column Connections.* New York: American Institute of Steel Construction, Inc., 1959.

7.29. B. G. Johnston and G. G. Kubo. "Web Crippling at Seat Angle Supports," Fritz Laboratory Report No. 192A2, Lehigh University, Bethlehem, Pa., 1941.

7.30. T. M. Roberts. "Slender Plate Girders Subjected to Edges Loading," *Proceedings of the Institution of Civil Engineers,* Part 2, September 1981, 71.

7.31. M. Elgaaly. "Web Design Under Compressive Edge Loads," *Engineering Journal,* AISC, **20,** 4 (Fourth Quarter 1983), 153–171.

7.32. Terence M. Roberts and Chooi K. Chong. "Collapse of Plate Girders Under Edge Loading," *Journal of the Structural Division,* ASCE, **107,** ST8 (August 1981), 1503–1509.

7.33. ASCE Task Committee on Design Criteria for Composite Structures in Steel and Concrete. "Proposed Specification for Structural Steel Beams with Web Openings," *Journal of Structural Engineering,* ASCE, **118,** 12 (December 1992), 3315–3324.

7.34. ASCE Task Committee on Design Criteria for Composite Structures in Steel and Concrete. "Commentary on Proposed Specification for Structural Steel Beams with Web Openings (with Design Example)," *Journal of Structural Engineering,* ASCE, **118,** 12 (December 1992), 3325–3349.

7.35. John E. Bower. "Elastic Stresses Around Holes in Wide-Flange Beams," *Journal of the Structural Division,* ASCE, **92,** ST2 (April 1966), 85–101.

7.36. John E. Bower. "Experimental Stresses in Wide-Flange Beams with Holes," *Journal of the Structural Division,* ASCE, **92,** ST5 (October 1966), 167–186.

7.37. John E. Bower. "Ultimate Strength of Beams with Rectangular Holes," *Journal of the Structural Division,* ASCE, **94,** ST6 (June 1968), 1315–1337.

7.38. John E. Bower, Chairman, Subcommittee on Beams with Web Openings of the Task Committee on Flexural Members. "Suggested Design Guides for Beams with Web Holes," *Journal of the Structural Division,* ASCE, **97,** ST11 (November 1971), 2707–2728. Disc., **99,** ST6 (June 1973), 1312–1315.

7.39. Ronald W. Frost and Robert E. Leffler. "Fatigue Tests of Beams with Rectangular Web Holes," *Journal of the Structural Division,* ASCE, **97,** ST2 (February 1971), 509–527.

7.40. James A. Mandel, Paul J. Brennan, Benjamin A. Wasil, and Charles M. Antoni. "Stress Distribution in Castellated Beams," *Journal of the Structural Division,* ASCE, **97,** ST7 (July 1971), 1947–1967.

7.41. Peter B. Cooper and Robert R. Snell. "Tests on Beams with Reinforced Web Openings," *Journal of the Structural Division,* ASCE, **98,** ST3 (March 1972), 611–632.

7.42. Peter W. Chan and Richard G. Redwood. "Stresses in Beams with Circular Eccentric Web Holes," *Journal of the Structural Division,* ASCE, **100,** ST1 (January 1974), 231–248.

7.43. Richard G. Redwood and Peter W. Chan. "Design Aids for Beams with Circular Eccentric Web Holes," *Journal of the Structural Division,* ASCE, **100,** ST2 (February 1974), 297–303.

7.44. Tsong-Miin Wang, Robert R. Snell, and Peter B. Cooper. "Strength of Beams with Eccentric Reinforced Holes," *Journal of the Structural Division,* ASCE, **101,** ST9 (September 1975), 1783–1800.

7.45. Marvin A. Larson and Kirit N. Shah. "Plastic Design of Web Openings in Steel Beams," *Journal of the Structural Division,* ASCE, **102,** ST5 (May 1976). 1031–1041.

7.46. Peter B. Cooper, Robert R. Snell, and Harry D. Knostman. "Failure Tests on Beams with Eccentric Web Holes," *Journal of the Structural Division,* ASCE, **103,** ST9 (September 1977), 1731–1738.

7.47. Richard G. Redwood, Hernan Baranda, and Michael J. Daly. "Tests of Thin-Webbed Beams with Unreinforced Holes," *Journal of the Structural Division,* ASCE, **104,** ST3 (March 1978), 577–595.

7.48. Richard G. Redwood and Minoru Uenoya. "Critical Loads for Webs with Holes," *Journal of the Structural Division,* ASCE, **105,** ST10 (October 1979), 2053–2067.

7.49. Brian K. Daugherty. "Elastic Deformation of Beams with Web Openings," *Journal of the Structural Division,* ASCE, **106,** ST1 (January 1980), 301–312.

7.50. Rangachari Narayanan and Norire Gara-Verni Der-Avanessian. "Design of Slender Webs Having Rectangular Holes." *Journal of Structural Engineering,* ASCE, **111,** 4 (April 1985), 777–787.

7.51. R. G. Redwood. "Tables for Plastic Design of Beams with Rectangular Holes." *Engineering Journal,* AISC, **9,** 1 (January 1972), 2–19.

7.52. Richard L. Kussman and Peter B. Cooper. "Design Example for Beams with Web Openings," *Engineering Journal,* AISC, **13,** 2 (Second Quarter 1976), 48–56.

7.53. E. H. Gaylord, Jr., and C. N. Gaylord. *Design of Steel Structures.* New York: McGraw-Hill Book Company, Inc., 1957, Chap. 5.

7.54. D. E. Allen and T. M. Murray. "Design Criterion for Vibrations Due to Walking," *Engineering Journal,* AISC, **30,** 4 (4th Quarter 1993), 117–129.

7.55. James M. Fisher and Michael A. West. *Serviceability Design Considerations for Steel Buildings,* Steel Design Guide Series No. 3, 2nd ed. Chicago, IL: American Institute of Steel Construction, 2003

7.56. Lawrence G. Griffis. *Load and Resistance Factor Design of W-Shapes Encased in Concrete,* Steel Design Guide Series No. 6. Chicago, IL: American Institute of Steel Construction, 1992.

7.57. Lawrence G. Griffis. "Serviceability Limit States Under Wind Load," *Engineering Journal,* AISC, **30,** 1(1st Quarter 1993), 1–16.

7.58. C. J. Earls and T. V. Galambos. "Design Recommendations for Equal Leg Single Angle Flexural Members," *Journal of Constructional Steel Research,* **43** (1997), 1–3, 65–85.

7.59. Thomas M. Murray, David E. Allen, and Eric E. Ungar. *Floor Vibrations Due to Human Activity,* Steel Design Guide Series No. 11. Chicago, IL: American Institute of Steel Construction, 1997, 69 pp.

7.60. David Darwin. *Steel and Composite Beams with Web Openings,* Design Guide Series No. 2. Chicago, IL: American Institute of Steel Construction, 1990, 63 pp.

PROBLEMS

All problems are to be done according to the AISC LRFD Method or the AISC ASD Method, as indicated by the instructor. All given loads are service loads unless otherwise indicated. For all problems *assume adequate lateral support of the compression flange such that lateral stability does not control.* Assume all standard sections are equally readily available in the indicated grade of steel (even though actually they are not). A figure showing span and loading is required, and a final check of strength is required.

7.1. For the case (or cases) assigned by the instructor, select the lightest W section to carry a uniformly distributed dead load w_D in addition to the beam weight, and a uniformly distributed live load w_L as indicated. The member is simply supported and deflection is *not* of concern.

Case	w_D Dead load (kip/ft)	w_L Live load (kip/ft)	Span length (ft)	Steel grade
1	0.2	0.8	35	A992
2	0.2	0.8	35	A572 Grade 65
3	0.2	0.8	35	A514 Grade 100
4	0.2	0.8	55	A992
5	0.2	0.8	55	A572 Grade 65
6	0.2	0.8	55	A514 Grade 100
7	0.4	1.6	30	A572 Grade 65
	0.8	3.2	30	
8	0.8	3.2	30	A572 Grade 65
9	0.4	1.4	60	A992
10	0.4	1.4	60	A572 Grade 65

7.2. A simply supported welded I-section beam carries a concentrated load W at midspan (see accompanying figure). The load is 20% dead load and 80% live load. For the case (or cases) assigned by the instructor, determine the maximum service load W that can be permitted to be carried.

Problem 7.2
Case 1 (for other cases see table)

Case	w_D Flange plates (in.)	w_L Web plate (in.)	Span length (ft)	Steel grade
1	1/2 × 14	3/8 × 30	28	A992
2	3/4 × 12	3/8 × 24	28	A607 Grade 70
3	3/4 × 18	1/2 × 30	28	A572 Grade 65

7.3. Repeat Prob. 7.1 (cases 1 to 4 as assigned) additionally assuming the live load deflection may not exceed $L/360$ (*not* an *AISC Specification* requirement).

7.4. Repeat Prob. 7.1 (cases 4 to 6 as assigned) additionally assuming the live load deflection may not exceed $L/300$ (*not* an *AISC Specification* requirement).

7.5. For the case (or cases) assigned by the instructor, select the lightest W section to carry a uniformly distributed dead load w_D in addition to the beam weight, and a uniformly distributed live load w_L as indicated. The member is simply supported and deflection is *not* of concern.

Case	w_D Dead load (kip/ft)	w_L Live load (kip/ft)	Span length (ft)	Steel grade
1	2.0	18.0	7	A992
2	2.0	18.0	7	A572 Grade 60
3	2.0	18.0	7	A514 Grade 100
4	8.0	12.0	5	A992
5	8.0	12.0	5	A572 Grade 65

7.6. A W24×94 beam on a 6-ft span (see accompanying figure) underpins a column that brings 110 kips dead load and 280 kips live load to its top flange at a location 2.5 ft from the left support. The column bearing plate is 12 in. measured along the beam, and the bearing plates at the end supports are each 8 in. Investigate this beam of A992 steel for (a) flexure, (b) shear, and (c) satisfactory transmission of the reactions and concentrated load (i.e., local web yielding and web crippling). Specify changes (if any) required to satisfy the *AISC Specification*.

Problem 7.6

7.7. A W16×77 section of A992 steel is to serve on a 10-ft simply supported span. The wall bearing length is 10 in. What maximum *slowly* moving concentrated service load (25% dead load; 75% live load) may be carried?

7.8. Determine the size bearing plate required for an end reaction of 11 kips dead load and 44 kips live load on a W14×53 beam of A572 Grade 60 steel. The beam rests on a 4-in.-thick concrete wall ($f'_c = 3500$ psi). Specify thickness in multiples of $\frac{1}{4}$ in. and the length and width to whole inches.

7.9. (LRFD Method only) For a W12×87 of A992 steel, calculate the design shear strength ϕV_n, and the bearing length N required when the design end reaction ϕR_n equals the design shear strength.

7.10. through 7.13.

For the section shown in the accompanying figure, assume uniform loading in the plane of the web (*yz* plane) for a simply supported span of 12 ft and neglect any torsional effects. Assume the service load acting is 20% dead load and 80% live load, and that the limit state occurs when the maximum stress reaches the yield stress F_y at one point.

(a) Determine the maximum uniform service load assuming bending occurs in the plane of loading (*yz* plane).

(b) Use the loading determined in (a) to compute the flexural stress at points designated by letters, assuming the beam is free to bend and not restrained to bend in the *yz* plane. Use the flexure formula, Eq. 7.10.18.

(c) Repeat (b) but use method describe in Example 7.10.2(b).

(d) If your instructor specifically assigns and discusses this part, locate principal axes, transform the moment into components M'_x and M'_y about the principal axes, compute moments of inertia I'_x and I'_y with respect to these axes, and use $f = M'_x/S'_x + M'_y/S'_y$.

(e) State conclusions.

7.14. The given 8×6×$\frac{1}{2}$ angle is positioned with its long leg pointing downward and used as a simply supported beam of 12-ft span. The uniform dead load is 0.1 kip/ft (including angle weight) and gravity live load is 0.5 kip/ft. The horizontal leg is to be restrained by attachments to make the angle bend vertically. Assuming the attachment to the horizontal leg is simply supported, for what service load lateral bending moment must the connection be designed? Consider only the unsymmetrical section effect and neglect any torsion. Assume the

$Z6 \times 3\frac{1}{2} \times 21.1$
$I_x = 34.4$ in.^{4}
$I_y = 12.9$ in.^{4}

Problem 7.10

Problem 7.11

Problem 7.12

Problem 7.13

0.6 kip/ft

|←————12'–0————→|

Problem 7.14

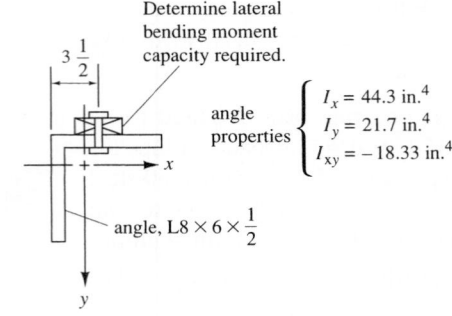

Determine lateral bending moment capacity required.

$3\frac{1}{2}$

angle properties $\begin{cases} I_x = 44.3 \text{ in.}^4 \\ I_y = 21.7 \text{ in.}^4 \\ I_{xy} = -18.33 \text{ in.}^4 \end{cases}$

x

angle, L8 × 6 × $\frac{1}{2}$

y

controlling limit state is the achievement of yield stress F_y at the extreme fiber.

7.15. Select the lightest W8 section of A992 steel to use as a purlin on a roof sloped 30° to the horizontal. The span is 21 ft, the load is uniform 0.18 kip/ft dead load plus the purlin weight and 0.34 kip/ft snow load. Lateral stability is assured by attachment of the roofing to the compression flange. Assume the load acts through the beam centroid, there are no sag rods, and biaxial bending must be assumed. Any torsional effect can be resisted by the roofing and therefore it can be neglected.

7.16. Select the lightest W section to carry moments, $M_x = 145$ ft-kips (15% dead load and 85% live load) and the lateral moment M_f resisted by one flange is 30 ft-kips (20% dead load and 80% live load). To select the beam assume $M_y \approx 2\,M_f$ and that torsion is neglected. Use A992 steel and assume lateral stability does not govern.

7.17. Select the lightest W section to carry $M_x = 275$ ft-kips (30% dead load and 70% live load) and lateral moment M_f resisted by one flange is 100 ft-kips (20% dead load and 80% live load). To select the beam assume $M_y \approx 2M_f$. Use A572 Grade 50 steel and assume lateral stability does not control.

7.18. Repeat Prob. 7.17 but select a combination wide flange section and channel as found in the *AISC Manual* [1.15] under "W Shapes with Cap Channels," pp. 1–112 or 1–113.

8
Torsion

8.1   INTRODUCTION

In structural design, torsional moment may, on occasion, be a significant force for which provision must be made. The most efficient shape for carrying a torque is a hollow circular shaft; extensive treatment of torsion and torsion combined with bending and axial force is to be found in most texts on mechanics of materials [8.1].

Frequently torsion is a secondary, though not necessarily a minor effect that must be considered in combination with the action of other forces. The shapes that make good columns and beams, i.e., those that have their material distributed as far from their centroids as practicable, are not equally efficient in resisting torsion. Thin-wall circular and box sections are stronger torsionally than sections with the same area arranged as channel, I, tee, angle, or zee shapes.

When a simple circular solid shaft is twisted, the shearing stress at any point on a transverse cross-section varies directly as the distance from the center of the shaft. Thus, during twisting, the cross-section which is initially planar remains a plane and rotates only about the axis of the shaft.

In 1853 the French engineer Adhémar Jean Barré de Saint-Venant presented to the French Academy of Sciences the classical torsion theory that forms the basis for present-day analysis.* Saint-Venant showed that when a noncircular bar is twisted, a transverse section that was planar prior to twisting does *not* remain plane after twisting. The original cross-section plane surface becomes a warped surface. In torsion situations the out-of-plane, or warping effect, must be considered in addition to the rotation, or pure twisting, effect.[†]

---

*For a summary of Saint-Venant's work, see Isaac Todhunter and Karl Pearson, *A History of the Theory of Elasticity and of the Strength of Materials,* Vol. II, 1893 (reprinted by Dover Publications, Inc., New York, 1960, pp. 17–51).
[†]Throughout Chapters 8 and 9, the symbol ϕ is used for the angle of twist, and should not be confused with the resistance factor ϕ used for Load and Resistance Factor Design. The resistance factor ϕ in these chapters is used subscripted; ϕ_b to indicate bending.

In this chapter primary emphasis is given to the recognition of torsion on the usual structural members, such as I-shaped, channel, angle, and zee sections; how the torsional stresses may be approximated and how such members may be selected to resist torsional effects.

Also included is a brief treatment of torsional stiffness and the computation of torsional stresses on closed thin-wall sections as well as torsional buckling.

8.2   PURE TORSION OF HOMOGENEOUS SECTIONS

A review of shear stress under torsion alone and of torsional stiffness seems a desirable beginning prior to considering structural shapes in locations where the warping of the cross-section is restrained.

Consider a torsional moment T acting on a solid shaft of homogeneous material and uniform cross-section, as shown in Fig. 8.2.1. Assume no out-of-plane warping, or at least that out-of-plane warping has negligible effect on the angle of twist ϕ. This assumption will be nearly correct so long as the cross-section is small compared to the length of the shaft and also that no significant reentrant corners exist. It is further assumed that no distortion of the cross-section occurs during twisting. The rate of twist (twist per unit length) may therefore be expressed as

$$\theta = \text{rate of twist} = \frac{d\phi}{dz} \tag{8.2.1}$$

which can be thought of as torsional curvature (rate of change of angle). Since it is the relative rotation of the cross-sections at z and $z + dz$ that causes strain, the magnitude of displacement at a given point is proportional to the distance r from the center of twist. The strain angle γ, or unit shear strain, at any element r from the center is

$$\gamma \, dz = r \, d\phi$$
$$\gamma = r(d\phi/dz) = r\theta \tag{8.2.2}$$

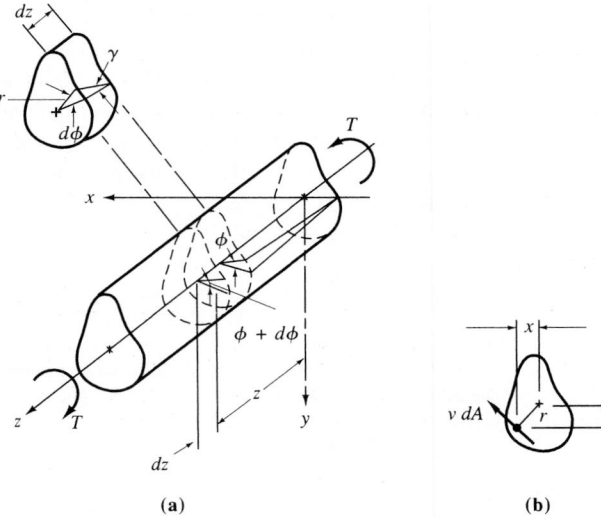

Figure 8.2.1
Torsion of a prismatic shaft.

(a)

(b)

Using the shear modulus G, Hooke's law gives the unit shear stress v as

$$v = \gamma G \tag{8.2.3}$$

Thus, as shown in Fig. 8.2.1b, the elemental torque is

$$dT = rv\, dA = r\gamma G\, dA = r^2(d\phi/dz)G\, dA \tag{8.2.4}$$

The total resisting moment for equilibrium is

$$T = \int_A r^2 \frac{d\phi}{dz} G\, dA$$

and since $d\phi/dz$ and G are constants at any section,

$$T = \frac{d\phi}{dz} G \int_A r^2\, dA = GJ\frac{d\phi}{dz} \tag{8.2.5}$$

where $J = \int_A r^2 dA$. Equation 8.2.5 may be thought of as analogous to flexure, i.e., bending moment M equals rigidity EI times curvature d^2y/dz^2. Here torsional moment T equals torsional rigidity GJ times torsional curvature (rate of change of angle).

Shear stress may then be computed using Eqs. 8.2.2 and 8.2.3,

$$v = \gamma G = r\frac{d\phi}{dz} G \tag{8.2.6}$$

and

$$\frac{d\phi}{dz} = \frac{T}{GJ}$$

which gives

$$v = \frac{Tr}{J} \tag{8.2.7}$$

Thus as long as the assumptions of this development reasonably apply, torsional shear stress is proportional to the radial distance from the center of twist.

Circular Sections

For the specific case of the circular section of diameter t, no warping of the sections occurs (i.e., no assumption is required) and $J =$ polar moment of inertia $= \pi t^4/32$. Thus, for maximum shear stress at $r = t/2$,

$$v_{\max} = \frac{16T}{\pi t^3} \tag{8.2.8}$$

Rectangular Sections

The analysis as applied to rectangles becomes complex since the shear stress is affected by warping, though essentially the angle of twist is unaffected.

As an approximation, consider the element of Fig. 8.2.2 subjected to shear, in which

$$\gamma = t\frac{d\phi}{dz} \tag{8.2.9}$$

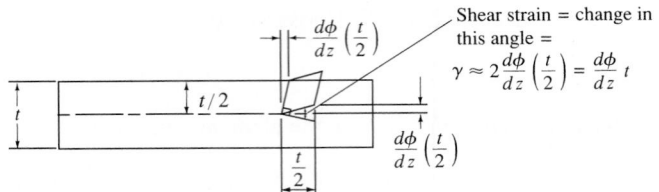

Figure 8.2.2
Torsion of a rectangular
section.

For a thin rectangle, neglecting end effects, the shear stress may be expressed as

$$v = \gamma G = tG\frac{d\phi}{dz} \tag{8.2.10}$$

or using Eq. 8.2.5,

$$v = \frac{Tt}{J} \tag{8.2.11}$$

From the theory of elasticity [8.1–8.3], the maximum shear stress v_{max} occurs at the midpoint of the long side of a rectangle and acts parallel to it. The magnitude is a function of the ratio b/t (length/width) and may be expressed as

$$v_{max} = \frac{k_1 T}{bt^2} \tag{8.2.12}$$

and the torsional constant J may be expressed as

$$J = k_2 bt^3 \tag{8.2.13}$$

where the values of k_1 and k_2 may be found in Table 8.2.1.

TABLE 8.2.1 Values of k_1 and k_2 for Eqs. 8.2.12 and 8.2.13

b/t	1.0	1.2	1.5	2.0	2.5	3.0	4.0	5.0	∞
k_1	4.81	4.57	4.33	4.07	3.88	3.75	3.55	3.44	3.00
k_2	0.141	0.166	0.196	0.229	0.249	0.263	0.281	0.291	0.333

I-shaped, Channel, and Tee Sections

As will be observed from a study of Table 8.2.1 the values of k_1 and k_2 become nearly constant for large ratios b/t. Thus the torsional constants for sections composed of thin rectangles may be computed as the sum of the values for the individual components. Such an approach will give an approximation which neglects the contribution in the fillet region where the components are joined. For most common structural shapes this approximation causes little error, thus

$$J \approx \sum \tfrac{1}{3}bt^3 \tag{8.2.14}$$

where b is the long dimension and t the thin dimension of the rectangular elements.

More accurate expressions for various structural shapes have been developed by Lyse and Johnston [8.4], Chang and Johnston [8.5], Kubo, Johnston, and Eney [8.6], and El Darwish and Johnston [8.7].

In addition to the torsional properties in the *AISC Manual*, torsional design aid publications are available by AISC [8.8], Hotchkiss [8.9], and Heins and Kuo [8.10].

8.3   SHEAR STRESSES DUE TO BENDING OF THIN-WALL OPEN CROSS-SECTIONS

Before treating the computation of stresses due to torsion of thin-wall open sections restrained from warping, a review of shear stress resulting from general flexure will be developed. Recognition of a torsion situation precedes concern about calculation of resulting stresses. Extensive treatment of thin-wall members of open cross-section is given by Timoshenko [8.11].

Referring to the general thin-wall section of Fig. 8.3.1, where x and y are centroidal axes, consider equilibrium of the element $t\,ds\,dz$ acted upon by flexural stress σ_z and shear stress τ, both of which result from bending moment. The shear stress τ multiplied by the thickness t may be termed the *shear flow* τt. Force equilibrium in the z direction requires

$$\frac{\partial(\tau t)}{\partial s}\,ds\,dz + t\frac{\partial \sigma_z}{\partial z}\,dz\,ds = 0 \tag{8.3.1}$$

or

$$\frac{\partial(\tau t)}{\partial s} = -t\frac{\partial \sigma_z}{\partial z} \tag{8.3.2}$$

1. *Assume moment is applied in the yz plane only,* i.e., $M_y = 0$. The flexural stress due to bending, as given by Eq. 7.10.18, is

$$\sigma_z = \frac{M_x}{I_x I_y - I_{xy}^2}(I_y y - I_{xy} x) \tag{7.10.18}$$

$$\frac{\partial \sigma_z}{\partial z} = \frac{\partial M_x/\partial z}{I_x I_y - I_{xy}^2}(I_y y - I_{xy} x) \tag{8.3.3}$$

Recognizing that $V_y = \partial M_x/\partial z$, and substituting Eq. 8.3.3 into Eq. 8.3.2 gives

$$\frac{\partial(\tau t)}{\partial s} = \frac{-t V_y}{I_x I_y - I_{xy}^2}(I_y y - I_{xy} x) \tag{8.3.4}$$

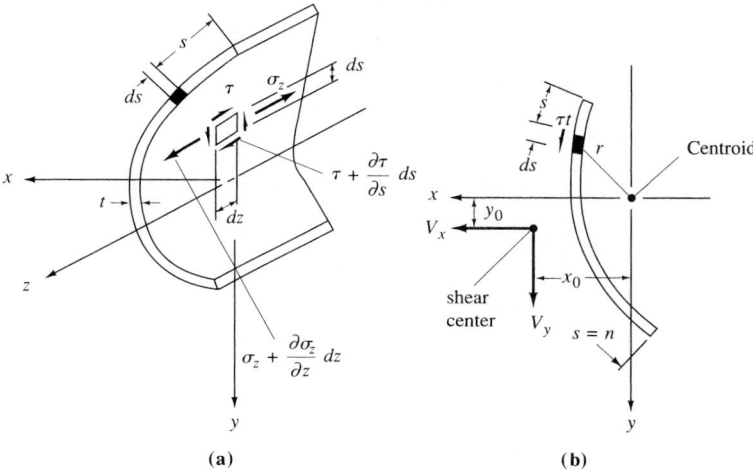

Figure 8.3.1
Stresses on thin-wall open sections in bending.

Integrating to find τt at a distance s from a free edge gives the shear flow τt as

$$\tau t = \frac{-V_y}{I_x I_y - I_{xy}^2} \left[I_y \int_0^s yt \, ds - I_{xy} \int_0^s xt \, ds \right] \qquad (8.3.5)$$

2. *Assume moment is applied in the xz plane only, i.e., $M_x = 0$.* The flexural stress due to bending as given by Eq. 7.10.18 is

$$\sigma_z = \frac{M_y}{I_x I_y - I_{xy}^2} (-I_{xy}y + I_x x) \qquad [7.10.18]$$

Taking $\partial \sigma_z / \partial z$, recognizing that $V_x = \partial M_y / \partial z$, and integrating to get the shear flow τt, gives in a manner similar to Eq. 8.3.5,

$$\tau t = \frac{+V_x}{I_x I_y - I_{xy}^2} \left[I_{xy} \int_0^s yt \, ds - I_x \int_0^s xt \, ds \right] \qquad (8.3.6)$$

3. *Moments applied in both yz and xz planes.* If shear stresses are desired they can be computed by superimposing the results from Eqs. 8.3.5 and 8.3.6.

It is to be observed from Fig. 8.3.1b that equilibrium requires that the shear V_y in the y direction equal the components of τt in the y direction summed over the entire section. Similarly V_x equals the summation of τt components in the x direction. Rotational equilibrium must also be satisfied; the moment about the centroid of the section is (see Fig. 8.3.1b)

$$\int_0^n (\tau t) r \, ds$$

which will be zero in some cases (such as I-shaped and Z-shaped sections). If such rotational equilibrium is automatically satisfied when the flexural shears act through the centroid, then no torsion will occur simultaneously with bending.

8.4 SHEAR CENTER

The shear center is the location in a cross-section where no torsion occurs when flexural shears act in planes passing through that location. In other words, forces acting through the shear center will cause no torsional stresses to develop, i.e.,

$$\int_0^n (\tau t) r \, ds = 0 \qquad (8.4.1)$$

Since the shear center does not necessarily coincide with the centroid of the section, the shear center must be located in order to evaluate the torsional stress. For I-shaped and Z-shaped sections, the shear center coincides with the centroid, but for channels and angles it does not.

Referring to Fig. 8.3.1b, consider the shears V_x and V_y acting at distances from the centroid y_0 and x_0, respectively, such that the torsional moment with respect to the centroid is the same as $\int (\tau t) r \, ds$ integrated from zero to n; thus

$$V_y x_0 - V_x y_0 = \int_0^n (\tau t) r \, ds \qquad (8.4.2)$$

In other words, the torsional moment is $(V_y x_0 - V_x y_0)$ when the loads are applied in planes passing through the centroid but is zero if the loads are in planes passing through the shear center, i.e., the point whose coordinates are (x_0, y_0).

It is observed that the location of shear center is independent of the magnitude or type of loading, but is dependent only on the cross-sectional configuration.

To determine the shear center location, first let one of the shears be zero, say $V_y = 0$; then from Eq. 8.4.2,

$$y_0 = -\frac{1}{V_x} \int_0^n (\tau t) r \, ds \tag{8.4.3}$$

where according to Eq. 8.3.6,

$$\tau t = \frac{V_x}{I_x I_y - I_{xy}^2} \left[I_{xy} \int_0^s yt \, ds - I_x \int_0^s xt \, ds \right]$$

Alternately, letting $V_x = 0$ gives from Eq. 8.4.2,

$$x_0 = \frac{1}{V_y} \int_0^n (\tau t) r \, ds \tag{8.4.4}$$

where according to Eq. 8.3.5,

$$\tau t = \frac{-V_y}{I_x I_y - I_{xy}^2} \left[I_y \int_0^s yt \, ds - I_{xy} \int_0^s xt \, ds \right]$$

EXAMPLE 8.4.1

Locate the shear center for the channel section of Fig. 8.4.1.

Solution:
Many practical cases can be solved without using the general formulas, Eqs. 8.4.3 and 8.4.4. Since the shear center location is a problem in equilibrium, moments may most conveniently be taken through a point that eliminates the greatest number of forces. Thus, letting $V_x = 0$ and taking moments about point A of Fig. 8.4.1a, changes the equilibrium equation, Eq. 8.4.2, to

$$V_y q = V_f h = \int_0^b (\tau t) h \, ds \tag{a}$$

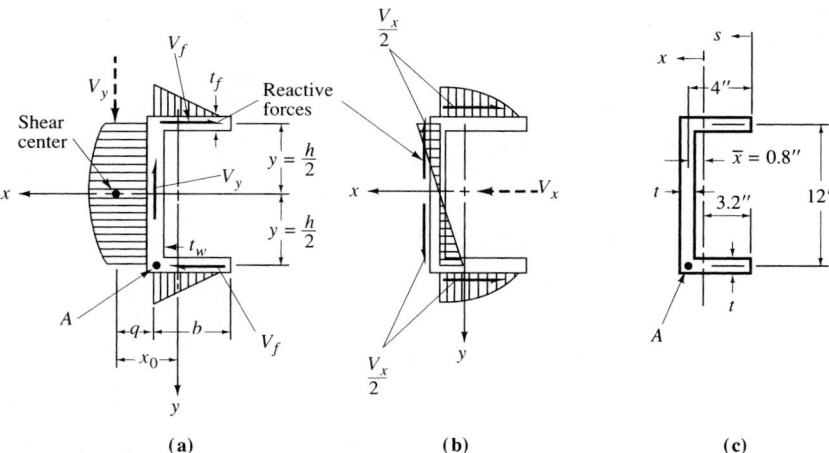

Figure 8.4.1
Channel of Example 8.4.1.

(a)   (b)   (c)

where according to Eq. 8.3.5,

$$\tau t = \frac{-V_y}{I_x} \int_0^s yt \, ds = \frac{-V_y}{I_x} yts \qquad \text{(b)}$$

For these thin-wall sections, the length s along which integration is performed is measured at mid-thickness.

Substituting Eq. (b) into Eq. (a), and using $y = -h/2$ where $t = t_f$ gives

$$V_y q = \int_0^b \frac{-V_y}{I_x}\left(\frac{-h}{2}\right) t_f h s \, ds$$

$$= \frac{V_y t_f h^2}{2I_x} \int_0^b s \, ds = \frac{V_y t_f h^2 b^2}{4I_x} \qquad \text{(c)}$$

Thus the shear center location along the x-axis is

$$q = \frac{t_f h^2 b^2}{4I_x} \qquad \text{(d)}$$

measured in the positive x direction to the left of the channel web.

For the shear center coordinate measured along the y-axis, apply V_x and let $V_y = 0$, and because of symmetry V_x must act at $y = 0$ for equilibrium. To demonstrate, let V_x be applied at the distance y_0 below the x-axis and take moments about point A. Satisfying equilibrium,

$$V_x\left(\frac{h}{2} - y_0\right) = \int_0^b (\tau t) y \, ds \qquad \text{(e)}$$

where according to Eq. 8.3.6,

$$\tau t = \frac{-V_s}{I_y} \int_0^s xt \, ds \qquad \text{(f)}$$

To illustrate numerically, use $b = 4$ in., $h = 12$ in., and $t = t_w$ in which case the centroid of the channel (refer to Fig. 8.4.1c) is located

$$\overline{x} = \frac{\Sigma Ax}{\Sigma A} = \frac{2b(b/2)t_w}{(h + 2b)t_w} = \frac{2(4)(2)}{12 + 2(4)} = 0.8 \text{ in.}$$

Then, $s = x + 3.2$ in.

$$I_y = \left[\tfrac{1}{3}(4)^3 2 - 20(0.8)^2\right]t_w = 29.87t_w$$

$$\tau t = \frac{-V_x t_w}{29.87 t_w} \int_0^s (s - 3.2) \, ds = \frac{-V_x}{29.87}\left(\frac{s^2}{2} - 3.2s\right)$$

Substitution of τt into Eq. (e) gives

$$V_x\left(\frac{h}{2} - y_0\right) = \int_0^4 \frac{-V_x}{29.87}\left(\frac{s^2}{2} - 3.2s\right) h \, ds$$

$$= \frac{-V_x h}{29.87}\left(\frac{s^3}{6} - 3.2\frac{s^2}{2}\right)\Big|_0^4 = \frac{+V_x h}{2}$$

Thus it is shown that y_0 is zero. The shear center may also be located as follows. First compute, by integrating over each stress distribution of Fig. 8.4.1, the shear forces acting in each of the component elements of the section. Then the shear center is located such that V_x or V_y counteract all of the shear forces acting on the components to produce equilibrium. In solving for the shear center location, the solution may be made as illustrated, and then checked by verifying that the forces are in equilibrium.   ■

8.5   TORSIONAL STRESSES IN I-SHAPED STEEL SECTIONS

The structural engineer must recognize a torsion situation and be able to apply approximate design methods and perform a stress analysis when necessary, even though only occasionally will torsion be severe enough to control the design of a section. Rolled steel sections under uniform and nonuniform torsion have been studied analytically and experimentally by many investigators.

The development in this section is similar to that of Timoshenko [8.11], Lyse and Johnston [8.4], Kubo, Johnston, and Eney [8.6], Goldberg [8.12], and Chu and Johnson [8.13]. Discussions of some of the practical aspects, along with solutions for various loading and support cases, are given by Hotchkiss [8.9] and Johnston [8.14]; charts for design are available in the handbook from AISC [8.8] and in the paper by Johnston [8.14]; and design tables using the β modified flexure analogy method developed by the authors are presented in Sec. 8.6. Lin [8.15, 8.16] has given additional and expanded β value tables.

Application of load in a plane other than the one through the shear center (see Fig. 8.5.1) will cause the member to twist unless external restraints prevent such twisting. The torsional stress due to twisting consists of both shear and flexural stresses. These stresses must be superimposed on the shear and flexural stresses that exist in the absence of torsion.

Torsion may be categorized into two types: pure torsion, or as it is often called, *Saint-Venant's torsion,* and warping torsion. Pure torsion assumes that a cross-sectional plane prior to application of torsion remains a plane and only element rotation occurs during torsion. A circular shaft subjected to torsion is a situation where pure torsion exists as the only type. Warping torsion is the out-of-plane effect that arises when the flanges are laterally displaced during twisting, analogous to bending from laterally applied loads.

1. *Pure torsion (Saint-Venant's Torsion).* Just as flexural curvature (change in slope per unit length) can be expressed as $M/EI = d^2y/dz^2$, i.e., moment divided by flexural rigidity equals flexural curvature, in pure torsion the torsional moment M divided by the torsional rigidity GJ equals the torsional curvature (change in angle of twist ϕ per unit

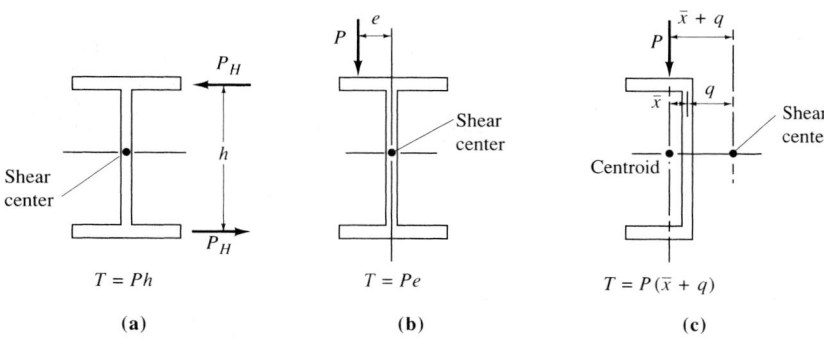

Figure 8.5.1
Common torsional loadings.

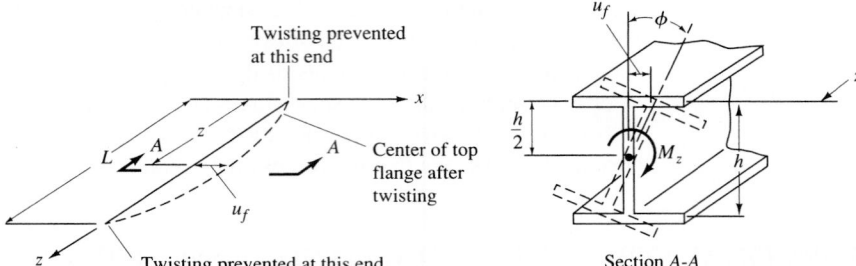

Figure 8.5.2
Torsion of an I-shaped
section.

length). Recalling previously derived Eq. 8.2.5 for T, which now becomes the component M_s due to pure torsion,

$$M_s = GJ\frac{d\phi}{dz} \tag{8.5.1}$$

where M_s = Pure torsional moment (Saint-Venant torsion)

   G = shear modulus of elasticity = $E/[2(1 + \mu)]$, in terms of the tension-compression modulus of elasticity E and Poisson's ratio μ

   J = torsional constant (see Sec. 8.2)

In accordance with Eq. 8.2.7, stress due to M_s is proportional to the distance from the center of twist.

2. *Warping torsion.* A beam subjected to torsion M_z, as in Fig. 8.5.2, will have its compression flange bent in one direction laterally while its tension flange is bent in the other. Whenever the cross-section is such that it would warp (become a nonplanar section) if not restrained, the restrained system has stresses induced. The torsional situation of Fig. 8.5.2 illustrates a beam that is prevented from twisting at each end but the top flange deflects laterally by an amount u_f. This lateral flange bending causes flexural normal stresses (tension and compression) as well as shear stresses across the flange width.

Thus, torsion may be thought of as being composed of two parts: (1) rotation of elements, the pure torsion part, and (2) translation producing lateral bending, the warping part.

3. *Differential equation for torsion on I- and channel-shaped sections.* Consider the deflected position of a flange centerline, as in Fig. 8.5.2, where u_f is the lateral deflection of one of the flanges at a section a distance z from the end of the member; ϕ is the twist angle at the same section, and V_f (Fig. 8.5.3) is the horizontal shear force developed in the flange at the section due to lateral bending. An important assumption is that the web remains a plane during rotation, so that the flanges deflect laterally an equal amount. Thus the web is assumed thick enough compared to the flanges so that is does not bend during twisting as a result of high torsional resistance of the flanges. Except for thin-web plate girders, it has been shown [8.6, 8.17] that assuming no lateral bending in the web, i.e., no

Figure 8.5.3
Warping shear force
on I-shaped section.

effect on the warping torsion component, is sufficiently correct for practical purposes. Since rarely are thin-web plate girders used without stiffeners, and certainly not when torsional stress exists, such cases are not of practical importance.

From geometry,

$$u_f = \phi \frac{h}{2} \tag{8.5.2}$$

for small values of ϕ. For understanding of torsion on I- and channel-shaped sections, Eq. 8.5.2 is the single most important relationship. The twist angle is directly proportional to the lateral deflection. Torsion boundary conditions are analogous to lateral bending boundary conditions.

Differentiating three times with respect to z in Eq. 8.5.2 gives

$$\frac{d^3 u_f}{dz^3} = \left(\frac{h}{2}\right)\frac{d^3 \phi}{dz^3} \tag{8.5.3}$$

For one flange the curvature relationship is

$$\frac{d^2 u_f}{dz^2} = \frac{-M_f}{EI_f} \tag{8.5.4}$$

where M_f is the lateral bending moment on one flange, I_f is the moment of inertia for one flange about the y-axis of the beam, and the minus sign arises from positive bending as shown in Fig. 8.5.2. Also, since $V = dM/dz$,

$$\frac{d^3 u_f}{dz^3} = \frac{-V_f}{EI_f} \tag{8.5.5}$$

Using Eqs. 8.5.3 and 8.5.5 gives

$$V_f = -EI_f\left(\frac{h}{2}\right)\frac{d^3 \phi}{dz^3} \tag{8.5.6}$$

Referring to Fig. 8.5.3, the torsional moment component M_w, causing lateral bending of the flanges, equals the flange shear force V_f times the moment arm h. This assumes no shear resistance to warping is contributed by the web,

$$M_w = V_f h = -EI_f \frac{h^2}{2}\frac{d^3 \phi}{dz^3} \tag{8.5.7}$$

$$= -EC_w \frac{d^3 \phi}{dz^3} \tag{8.5.8}$$

where $C_w = I_f h^2/2$, often referred to as the *warping torsional constant*.

The total torsional moment is composed of the sum of the rotational part M_s and the lateral bending part M_w which from Eqs. 8.5.1 and 8.5.8 give

$$M_z = M_s + M_w = GJ\frac{d\phi}{dz} - EC_w\frac{d^3 \phi}{dz^3} \tag{8.5.9}$$

the differential equation for torsion. The torsional moment M_z depends on the loading and in usual situations will be a polynomial in z. Expressions for the torsion constant J and warping constant C_w for various shapes are to be found in text, Appendix Table A2.

Rewrite Eq. 8.5.9, dividing by EC_w,

$$\frac{d^3\phi}{dz^3} - \frac{GJ}{EC_w}\frac{d\phi}{dz} = \frac{-M_z}{EC_w} \tag{8.5.10}$$

Letting $\lambda^2 = GJ/EC_w$ ($\lambda = 1/a$ of *Torsion Analysis of Steel Members* [8.8]), and for the homogeneous solution of Eq. 8.5.10, let $\phi_h = Ae^{mz}$,

$$\frac{d^3\phi}{dz^3} - \lambda^2\frac{d\phi}{dz} = 0 \tag{8.5.11}$$

which upon substitution of the homogeneous solution gives

$$Ae^{mz}(m^3 - \lambda^2 m) = 0 \tag{8.5.12}$$

which requires

$$m(m^2 - \lambda^2) = 0 ; \quad \therefore\ m = 0, m = \pm\lambda$$

Thus

$$\phi_h = A_1 e^{\lambda z} + A_2 e^{-\lambda z} + A_3 \tag{8.5.13}$$

which upon using the hyperbolic function identities and regrouping the constants may be expressed as

$$\phi_h = A \sinh \lambda z + B \cosh \lambda z + C \tag{8.5.14}$$

where

$$\lambda = \frac{1}{a} = \sqrt{\frac{GJ}{EC_w}}$$

For the particular solution, since M_z is in general some function of z,

$$M_z = f(z)$$

Let $\phi_p = f_1(z)$, and substitute into Eq. 8.5.10, giving

$$\frac{d^3 f_1(z)}{dz^3} - \lambda^2\frac{df_1(z)}{dz} = -\frac{1}{EC_w}f(z) \tag{8.5.15}$$

where terms on the left-hand side must be paired with terms on the right side. Rarely will $f_1(z)$ be required to contain higher than second-degree terms.

EXAMPLE 8.5.1

Develop, using the differential equation, the expressions for the twist angle ϕ, as well as the first, second, and third derivatives, for the case of concentrated torsional moment applied at midspan when the ends are torsionally simply supported.

Solution:
Referring to Fig. 8.5.4, it is apparent that M_z is constant and equal to $T/2$. Thus let

$$\phi_p = C_1 + C_2 z \quad \text{(any polynomial)} \tag{a}$$

(a) Distribution of total torsional moment $M_z = M_s + M_w$ causing shear in flange

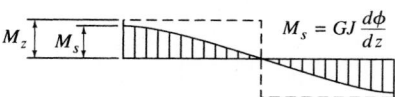

(b) Distribution of portion of torsional moment M_s due to Saint-Venant torsion (pure torsion)

(c) Distribution of portion of torsional moment M_w due to warping torsion

Figure 8.5.4
Case of Example 8.5.1.
Concentrated torsional
moment at midspan;
torsionally simply supported.
(Adapted from Hotchkiss
[8.9], Fig. 3)

Using Eq. 8.5.10 gives

$$-\lambda^2 C_2 = -\frac{1}{EC_w}\left(\frac{T}{2}\right); \qquad \therefore\; C_2 = \frac{T}{2GJ}$$

The other constant C_1 may be combined with C of Eq. 8.5.14. The complete solution for this loading is therefore

$$\phi = A \sinh \lambda z + B \cosh \lambda z + C + \frac{T}{2GJ} z \qquad \text{(b)}$$

Consider the boundary conditions for torsional simple support. Thinking of the lateral bending of the flange (since ϕ is proportional to u_f), simple support conditions mean zero moment and deflection at each end, or for torsion.

$$\phi = 0 \qquad \text{at}\quad z = 0 \qquad \text{and}\quad z = L$$

$$\frac{d^2\phi}{dz^2} = \phi'' = 0 \qquad \text{at}\quad z = 0 \qquad \text{and}\quad z = L$$

In this case the differential equation is discontinuous at $L/2$; thus, using zero slope of the flange at $L/2$, i.e., $\phi' = 0$, along with $\phi = 0$ and $\phi'' = 0$ at $z = 0$ will permit solution for the three constants of Eq. (b).

From $\phi = 0$ at $z = 0$,

$$0 = B + C \qquad \text{(c)}$$

Using $\phi'' = 0$ at $z = 0$,

$$\phi'' = A\lambda^2 \sinh \lambda z + B\lambda^2 \cosh \lambda z$$

$$0 = B \qquad \text{(d)}$$

Thus from Eq. (c),

$$C = 0$$

Using $\phi' = 0$ at $z = L/2$,

$$0 = A\lambda \cosh \lambda L/2 + \frac{T}{2GJ} \tag{e}$$

$$A = -\frac{T}{2GJ\lambda}\left[\frac{1}{\cosh \lambda L/2}\right]$$

Finally, Eq. (b) becomes

$$\phi = \frac{T}{2GJ\lambda}\left[\lambda z - \frac{\sinh \lambda z}{\cosh \lambda L/2}\right] \tag{f}$$

Also

$$\phi' = \frac{T}{2GJ}\left[1 - \frac{\cosh \lambda z}{\cosh \lambda L/2}\right] \tag{g}$$

$$\phi'' = \frac{T\lambda}{2GJ}\left[\frac{-\sinh \lambda z}{\cosh \lambda L/2}\right] \tag{h}$$

$$\phi''' = \frac{T\lambda^2}{2GJ}\left[\frac{-\cosh \lambda z}{\cosh \lambda L/2}\right] \tag{i}$$

Thus the solution of the differential equation is illustrated. The stress equations making use of the derivatives are developed in the next section. ∎

4. *Torsional stresses.* The shear stress v_s resulting from the Saint-Venant torsion M_s is computed in accordance with the form of Eq. 8.2.11,

$$v_s = \frac{M_s t}{J} \tag{8.2.11}$$

and using Eq. 8.5.1 gives

$$v_s = Gt\frac{d\phi}{dz} \tag{8.5.16}$$

whose distribution is shown in Fig. 8.5.5a. Though shown uniform fully across the flange, the stress drops sharply to zero at the flange tips.

The shear stress v_w that results from warping varies parabolically across the width of the rectangular flange as shown in Fig. 8.5.5b and may be computed as

$$v_w = \frac{V_f Q_f}{I_f t_f} \tag{8.5.17}$$

where Q_f = statical moment of area about the y-axis.

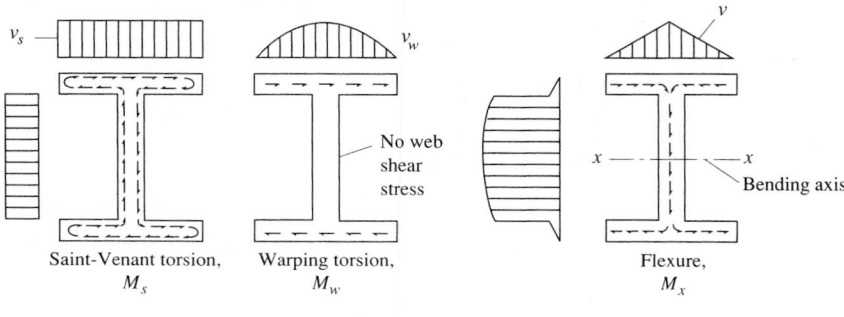

Figure 8.5.5
Direction and distribution of shear stress in I-shaped sections.

Figure 8.5.6
Dimensions for computation of statical moment of area, Q_f.

The negligible shear carried by the web is not considered. For maximum shear stress v_w, which actually acts at the face of web but may be approximated as acting at the mid-width of the flange, take Q_f (see Fig. 8.5.6) as

$$Q_f = A\bar{x} = \frac{bt_f}{2}\left(\frac{b}{4}\right)$$

Substituting Q_f and V_f from Eq. 8.5.6 into Eq. 8.5.17 gives

$$v_w = E\frac{b^2 h}{16}\frac{d^3\phi}{dz^3} \tag{8.5.18}$$

taking the absolute value. The direction of the shear flow has no effect on the combining of shear stresses.

The tension or compression stress due to lateral bending of flanges (i.e., warping of the cross-section as shown in Fig. 8.5.7) may be expressed as

$$f_{bw} = \frac{M_f x}{I_f} \tag{8.5.19}$$

which varies linearly across the flange width as shown in Fig. 8.5.7. The bending moment M_f, the lateral moment acting on one flange, may be obtained by substituting Eq. 8.5.2 into Eq. 8.5.4 and noting that $I_f h^2/2$ is the warping torsional constant C_w,

$$M_f = EI_f\left(\frac{h}{2}\right)\frac{d^2\phi}{dz^2} = \frac{EC_w}{h}\frac{d^2\phi}{dz^2} \tag{8.5.20}$$

The minus sign is dropped since tension occurs on one side while compression occurs on the other.

Normal stress distribution when warping is restrained

Counterclockwise rotation

Rotation axis

Clockwise rotation

Figure 8.5.7
Warping of cross-section.

The maximum stress occurs at $x = b/2$, which when used with Eq. 8.5.20 gives for Eq. 8.5.19,

$$f_{bw} = EI_f \left(\frac{h}{2}\right) \frac{d^2\phi}{dz^2} \left(\frac{b}{2I_f}\right)$$

$$f_{bw} = \frac{Ebh}{4} \frac{d^2\phi}{dz^2} \qquad (8.5.21)$$

In a summary, three kinds of stresses arise in any I-shaped or channel section due to torsional loading: (a) shear stresses v_s in web and flanges due to rotation of the elements of the cross-section (Saint-Venant torsional moment, M_s); (b) shear stresses v_w in the flanges due to lateral bending (warping torsional moment, M_w); and (c) normal stresses (tension and compression) f_{bw} due to lateral bending of the flanges (lateral bending moment on flange, M_f).

EXAMPLE 8.5.2

A W18×71 beam on a 24-ft simply supported span is loaded with a concentrated load of 20 kips at midspan. The ends of the member are simply supported with respect to torsional restraint (i.e., $\phi = 0$) and the concentrated load acts with a 2-in. eccentricity from the plane of the web (see Fig. 8.5.8). Compute combined bending and torsional stresses.

Figure 8.5.8
Data for Example 8.5.2.

Solution:
The differential equation solution for this type of loading and end restraint was obtained in Example 8.5.1. The solution as obtained is

$$\phi = \frac{T}{2GJ\lambda} \left[\lambda z - \frac{\sinh \lambda z}{\cosh \lambda L/2}\right]$$

In accordance with the derivation (see Fig. 8.5.4), T is the applied torsional moment,

$$T = 20(2) = 40 \text{ in.-kips}$$

Recalling from Eq. 8.5.10,

$$\lambda = \sqrt{\frac{GJ}{EC_w}} = \sqrt{\frac{3.39}{2.6(4685)}} = \frac{1}{59.9} = 0.01668$$

where $\dfrac{E}{G} = \dfrac{2E(1+\mu)}{E} = 2.6 \quad \text{for } \mu = 0.3$

$$J \approx \sum \frac{bt^3}{3}, \text{ Eq. 8.2.14}$$

$$= \tfrac{1}{3}\left[2(7.635)(0.810)^3 + (18.47 - 1.620)(0.495)^3\right] = 3.39 \text{ in.}^4$$

$$C_w = \frac{I_f h^2}{2} = \frac{(7.635)^3(0.810)}{12}\frac{(18.47 - 0.810)^2}{2} = 4685 \text{ in.}^6$$

The above values of J and C_w compare with $J = 3.49$ and $C_w = 4700$ given in the *AISC Manual* [1.15] computed for rectangular flanges using a more exact expression for J including the effect of the fillets at the junction of flange to web. Though different sources give slightly different values for these torsional constants, any of the values are satisfactory for design purposes.

The function values required are

$$\lambda L = 24(12)/59.9 = 4.80$$

z	λz	$\sinh \lambda z$	$\cosh \lambda z$
$0.1L$	0.480	0.499	1.118
$0.2L$	0.960	1.116	1.498
$0.3L$	1.441	1.994	2.231
$0.4L$	1.922	3.343	3.489
$0.5L$	2.402	5.477	5.567

(a) Pure torsion (Saint-Venant torsion). Using Eq. 8.5.16,

$$v_s = Gt\, d\phi/dz$$

$$\frac{d\phi}{dz} = \frac{T}{2GJ}\left(1 - \frac{\cosh \lambda z}{\cosh \lambda L/2}\right)$$

$$v_s = \frac{Tt}{2J}\left(1 - \frac{\cosh \lambda z}{5.567}\right) = \frac{40t}{2(3.39)}\left(1 - \frac{\cosh \lambda z}{5.567}\right)$$

The shear stress v_s is a maximum at $z = 0$ and zero at $z = L/2$:

$$v_s \text{ (flange at } z = 0) = \frac{40(0.810)}{2(3.39)}\left(1 - \frac{1}{5.567}\right) = 3.92 \text{ ksi}$$

$$v_s \text{ (web at } z = 0) = 3.92\frac{0.495}{0.810} = 2.40 \text{ ksi}$$

(b) Lateral bending of flanges (warping torsion). Use Eq. 8.5.18 for shear stress in flanges,

$$v_w = E\frac{b^2 h}{16}\frac{d^3\phi}{dz^3}$$

$$\frac{d^3\phi}{dz^3} = \frac{T\lambda^2}{2GJ}\left(\frac{-\cosh \lambda z}{\cosh \lambda L/2}\right)$$

$$v_w = \frac{T}{2C_w}\frac{b^2 h}{16}\left(\frac{-\cosh \lambda z}{\cosh \lambda L/2}\right)$$

This shear stress acts at mid-width of the flange, and the maximum value occurs at $z = L/2$ while the minimum value is at $z = 0$,

$$v_w \text{ (flange at } z = L/2) = \frac{40}{2(4700)}\left(\frac{(7.635)^2 17.660}{16}\right) = 0.27 \text{ ksi}$$

$$v_w \text{ (flange at } z = 0) = 0.27\frac{1.0}{5.567} = 0.05 \text{ ksi}$$

For normal stress in flanges due to warping, use Eq. 8.5.21:

$$f_{bw} = \frac{Ebh}{4}\frac{d^2\phi}{dz^2}$$

$$\frac{d^2\phi}{dz^2} = \frac{T\lambda}{2GJ}\left[\frac{-\sinh\lambda z}{\cosh\lambda L/2}\right]$$

$$f_{bw} = \frac{T(2.6)\lambda bh}{8J}\left[\frac{\sinh\lambda z}{\cosh\lambda L/2}\right]$$

which is a maximum at $z = L/2$ and zero at $z = 0$. Thus

$$f_{bw} \text{ (flanges at } z = L/2) = \frac{40(2.6)(7.635)(17.660)}{8(3.39)(59.9)}\left[\frac{5.477}{5.567}\right] = 8.49 \text{ ksi}$$

(c) Ordinary flexure. Maximum normal stress is

$$f_b \text{ (at } z = L/2) = \frac{PL}{4S_x} = \frac{20(24)(12)}{4(127)} = 11.34 \text{ ksi}$$

The shear stresses due to flexure are constant from $z = 0$ to $L/2$ and are computed by

$$v = \frac{VQ}{It} = \frac{10Q}{1170t} = \frac{Q}{117t}$$

For maximum flange shear stress, taking the more correct value at the face of the web rather than the value at mid-width of the flange,

$$Q = \left(\frac{7.635 - 0.495}{2}\right)(0.810)\left(\frac{17.660}{2}\right) = 25.53 \text{ in.}^3$$

$$v \text{ (flange at } z = 0) = \frac{25.53}{117(0.810)} = 0.27 \text{ ksi}$$

For maximum web shear stress,

$$Q = 7.635(0.810)\left(\frac{17.660}{2}\right) + \frac{16.850}{2}(0.495)\left(\frac{16.850}{4}\right) = 72.18 \text{ in.}^3$$

$$v \text{ (web at } z = 0) = \frac{72.18}{117(0.495)} = 1.25 \text{ ksi}$$

A summary of stresses showing combinations is given in Table 8.5.1.

TABLE 8.5.1 Summary of Stresses for Example 8.5.2

Type of Stress	Support $(z = 0)$	Midspan $(z = L/2)$
Compression and tension maximum stresses:		
Vertical bending, f_b	0	11.34
Torsional bending, f_{bw}	0	8.49
		19.83 ksi
Shear stress, web:		
Saint-Venant torsion, v_s	2.40	0
Vertical bending, v	1.25	1.25
	3.65 ksi	
Shear stress, flange:		
Saint-Venant torsion, v_s	3.92	0
Warping torsion, v_w	0.05	0.27
Vertical bending, v	0.27	0.27
	4.24 ksi	0.54 ksi

8.6   ANALOGY BETWEEN TORSION AND PLANE BENDING

Because the differential equation solution is time consuming, and really suited only for analysis, design of a beam to include torsion is most conveniently done by making the analogy between torsion and ordinary bending.

Consider that the applied torsional moment T of Fig. 8.6.1 can be converted into a couple P_H times h. The force P_H can then be treated as a lateral load acting on the flange of a beam.

The substitute system will have constant shear over one-half the span, a diagram as given in Fig. 8.5.4a. The true distribution of lateral shear which contributes to lateral deflection is only that part due to warping as shown in Fig. 8.5.4c. Thus the substitute system overestimates the lateral shear force and consequently overestimates the lateral bending moment M_f which causes tension and compression stresses.

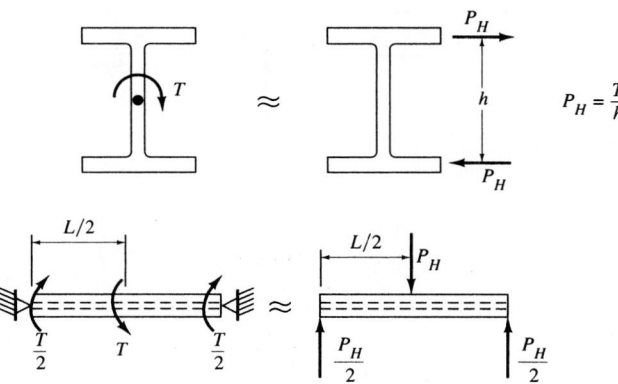

Figure 8.6.1
Analogy between flexure and torsion.

In most practical design situations when it is desirable to include the effect of torsion, the compressive stress due to the warping component is the quantity of most importance. The shear stress contributions are normally not of significance.

EXAMPLE 8.6.1

Compute the stresses on the W18×71 beam of Example 8.5.2 and Fig. 8.5.8 using the flexural analogy rather than the differential equation solution.

Solution:

The substitute system is as shown in Fig. 8.6.2a. The lateral bending moment is then

$$M_f = V_f(L/2) = 1.13(12) = 13.6 \text{ ft-kips}$$

acting on one flange. Twice the moment acting on the entire section gives

$$f_{bw} = \frac{2M_f}{S_y} = \frac{2(13.6)(12)}{15.8} = 20.6 \text{ ksi}$$

For torsional shear stress, since $M_z = T/2 = 20$ in.-kips,

$$v_s = \frac{M_z t}{J} = \frac{20(0.810)}{3.39} = 4.78 \text{ ksi (flange)}$$

$$v_s = 4.78\left(\frac{0.495}{0.810}\right) = 2.92 \text{ ksi (web)}$$

For lateral bending flange shear stress,

$$v_w = \frac{V_f Q_f}{I_f t_f} = \frac{1.13(5.90)}{(30.0)0.810} = 0.27 \text{ ksi}$$

where $Q_f = (7.635/2)(0.810)(7.635/4) = 5.90 \text{ in.}^3$

Figure 8.6.2
Comparison of lateral shear on flange due to warping torsion with that from simple lateral flexure analogy.

(a) Flexure analogy shear

(b) Warping torsion shear

The results of the two methods are compared as follows:

Type of stress	Flexural analogy	Differential equation
Compression/tension stress = $f_b + f_{bw}$ = 11.3 + 20.6 =	31.9 ksi	19.83 ksi
Web shear stress = $v + v_s$ = 1.25 + 2.92 =	4.17 ksi	3.65 ksi
Flange shear stress = $v + v_s + v_w$ = 0.27 + 4.78 + 0.27 =	5.32 ksi	4.24 ksi

The flexure analogy without modification is obviously a very conservative approach. In some situations it is so excessively conservative as to be practically useless. Furthermore, the most important design item, the lateral bending stress f_{bw} is overestimated by the greatest amount.

The relationship between the flexural analogy and the true torsion problem is best illustrated by referring to Fig. 8.5.4a. Note that the full torsional shear resulting from M_s and M_w is analogous to the lateral flexure problem. Figure 8.5.4b shows the portion of the shear that goes into rotation of elements, while Fig. 8.5.4c shows the portion contributing to lateral flange bending. If one could correctly assess how the shear due to warping torsion compares with the lateral flexure situation, design for torsion could be greatly simplified without being grossly conservative.

Figure 8.6.2b shows the accurate variation of V_f for the problem of Example 8.6.1, computed according to Eq. 8.5.7, whereupon

$$V_f = \frac{T}{2h}\left(\frac{\cosh \lambda z}{\cosh \lambda L/2}\right) \tag{8.6.1}$$

in which T/h can be thought of as the lateral load, which means the shear from the lateral bending analogy is $T/2h$, which is then modified by the hyperbolic function.

The lateral bending moment can thus be expressed for this problem as

$$M_f = \beta\frac{T}{2h}\left(\frac{L}{2}\right) \tag{8.6.2}$$

or, in general, the change in lateral moment between the support and location of zero shear is

$$\Delta M_f = \beta \times (\text{area under flexure analogy shear diagram}) \tag{8.6.3}$$

where β is a reduction factor that depends on λL.

It is to be noted that if Eq. 8.6.2 is multiplied by h, and the concentrated moment T is thought of as a concentrated load, the analogous moment M_fh (sometimes referred to as *bimoment*) equals β times the simple beam moment.

Thus the modified flexure analogy gives

$$M_fh = \beta\left(\frac{TL}{4}\right) \tag{8.6.4}$$

for the case of Fig. 8.6.1. ∎

Tables 8.6.1 through 8.6.5 give "exact" values for β for several common loading and restraint conditions. For other cases Table I of Ref. 8.9 (where M_w equals $M_f h$ above) or the curves of *Torsional Analysis of Steel Members* [8.8] may be used. In Tables 8.6.3 and 8.6.4, m is the applied torsional loading per unit length (say, in.-kips/ft).

TABLE 8.6.1 β Values, Concentrated Load, Torsional Simple Support

$$M_f h = \beta(TabL)$$
$$\text{at } z = aL$$

| λL | \multicolumn{5}{c}{β values} |
	$a = 0.5$	$a = 0.4$	$a = 0.3$	$a = 0.2$	$a = 0.1$
0.5	0.98	0.98	0.98	0.99	0.99
1.0	0.92	0.93	0.94	0.95	0.97
2.0	0.76	0.77	0.80	0.84	0.91
3.0	0.60	0.62	0.65	0.72	0.83
4.0	0.48	0.50	0.54	0.62	0.76
5.0	0.39	0.41	0.45	0.54	0.70
6.0	0.33	0.34	0.39	0.47	0.65
8.0	0.25	0.26	0.30	0.37	0.55
10.0	0.20	0.21	0.24	0.31	0.48

TABLE 8.6.2 β Values, Concentrated Load, Torsionally Fixed Supports

$$M_f h = \beta_1(Tab^2 L)$$
$$\text{at } z = 0$$
$$M_f h = \beta_2(Ta^2 bL)$$
$$\text{at } z = L$$

| λL | $a = 0.5$ | $a = 0.4$ | | $a = 0.3$ | | $a = 0.2$ | |
	$\beta_1 = \beta_2$	β_1	β_2	β_1	β_2	β_1	β_2
0.5	0.99	1.00	0.99	1.00	0.99	1.00	0.99
1.0	0.98	0.98	0.98	0.98	0.98	0.99	0.98
2.0	0.92	0.93	0.92	0.94	0.92	0.96	0.92
3.0	0.85	0.86	0.84	0.88	0.84	0.91	0.85
4.0	0.76	0.78	0.75	0.81	0.75	0.86	0.77
5.0	0.68	0.70	0.67	0.74	0.67	0.80	0.69
6.0	0.60	0.63	0.59	0.67	0.60	0.75	0.62
8.0	0.48	0.51	0.47	0.56	0.49	0.65	0.52
10.0	0.39	0.42	0.39	0.47	0.41	0.56	0.44

TABLE 8.6.3 β Values, Uniform Load, Torsional Simple Support

$$M_f h = \beta\left(\frac{m}{2}\, ab L^2\right)$$

at $z = aL$

λL	β values				
	$a = 0.5$	$a = 0.4$	$a = 0.3$	$a = 0.2$	$a = 0.1$
0.5	0.97	0.97	0.98	0.98	0.98
1.0	0.91	0.91	0.91	0.91	0.92
2.0	0.70	0.71	0.71	0.72	0.74
3.0	0.51	0.51	0.52	0.54	0.57
4.0	0.37	0.37	0.38	0.41	0.44
5.0	0.27	0.27	0.29	0.31	0.34
6.0	0.20	0.20	0.22	0.24	0.28
8.0	0.12	0.12	0.13	0.16	0.19
10.0	0.08	0.08	0.09	0.11	0.14

TABLE 8.6.4 β Values, Uniform Load, Torsionally Fixed Supports

$$M_f h = \beta\left(\frac{m}{12}\, L^2\right)$$

at $z = 0$ and $z = L$

λL	0.5	1.0	2.0	3.0	4.0	5.0	6.0	8.0
β	0.99	0.98	0.94	0.88	0.81	0.74	0.67	0.56

TABLE 8.6.5 β Values, Concentrated Load, Torsionally Fixed Supports

$M_f h = \beta$(positive moment by flexure theory)
$= \beta[2Ta^2b^2L]$
at $z = aL$

λL	$a = 0.5$	$a = 0.3$	$a = 0.1$
0.5	0.99	1.00	1.00
1.0	0.98	0.99	1.01
2.0	0.92	0.95	1.05
3.0	0.85	0.91	1.10
4.0	0.76	0.85	1.16
5.0	0.68	0.79	1.21
6.0	0.60	0.73	1.25

EXAMPLE 8.6.2

Recompute the stresses due to torsion on the beam of Example 8.6.1, using the modified flexural analogy method utilizing the β values from Table 8.6.1.

Solution:
The flexure analogy gives

$$M_f = 13.6 \text{ ft-kips}$$

as previously computed.

$$\lambda L = 4.80 \text{ (as computed in Example 8.5.2)}$$

From Table 8.6.1 at $a = 0.5$, $\beta \approx 0.41$, i.e., use about 41 percent of the flexure analogy value. Thus the modified flexure analogy gives

$$M_f = 13.6(0.41) = 5.58 \text{ ft-kips}$$

$$f_{bw} = \frac{2M_f}{S_y} = \frac{2(5.58)12}{15.8} = 8.48 \text{ ksi}$$

which compares favorably with $f_{bw} = 8.49$ ksi as computed by the differential equation solution using $\lambda L = 4.80$. For this case that exactly fits a table case, the β modified flexure analogy is the "exact" value obtained from the differential equation solution value. ∎

8.7   PRACTICAL SITUATIONS OF TORSIONAL LOADING

There are relatively few occasions in actual practice where the torsional load can cause significant twisting, and frequently these situations arise during construction. In most building construction the members are laterally restrained by attachments along the length of the member and therefore they are not free to twist. Even though torsional loading exists, it may be self-limiting because the rotation cannot exceed the end slope of the transverse attached members.

Torsion exists on spandrel beams, where such loading may be uniformly distributed; torsion also exists where a beam frames into a girder on one side only, or where unequal reactions come to opposite sides of a girder. The design of crane runway girders involves the combination of biaxial bending and torsion, and is illustrated in Sec. 8.8 for laterally stable beams. Any situation where the loading or reaction acts eccentrically to the shear center gives rise to torsion.

Analysis for Torsional Moment

The determination of the torsional moment in a framing system involves an elastic analysis where the joints may be rigid or semi-rigid. While the details of such an analysis are outside the scope of this text, some discussion is necessary so that at least the problem is understood. Goldberg [8.12] has discussed this subject and presented an approximate method suitable for design. Spandrel girders have been treated by Lothers [8.18]. Chen and Jolissaint [8.19] have provided a simple analysis technique for rigid frames.

Consider an example of a floor framing system (similar to Goldberg's [8.12]) as shown in Fig. 8.7.1. Spandrel beam *AB* is subjected to torsion because of the floor beams framing on only one side. Contrary to some common belief, however, the torsional moment is *not* equal to the beam reaction times its eccentricity from the centerline of the girder web. Moment is transmitted across the joint, and the end moment on the beam must

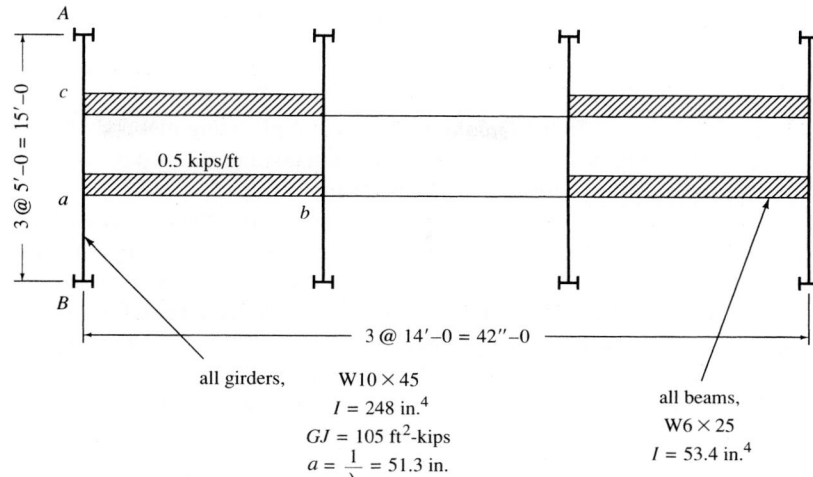

Figure 8.7.1
Plan view of floor framing.

equal the torsional moment on the girder. To attack such a problem one must first determine the relationship between the angle of twist ϕ and the applied torsional moment T.

For example, in Fig. 8.7.1 the loading system causes equal torsional moments at the
$\frac{1}{3}$ points on member AB. Assuming the girder torsionally simply supported at ends A and B,
using either the differential equation solution formulas or the curves of *Torsional Analysis
of Steel Members* [8.8] (Case 4, p. 64), for $\lambda L = 15(12)/51.3 = 3.51$, one finds the angle
of twist ϕ at point a,

$$\phi \frac{GJ}{TL} \approx 0.09$$

or

$$\phi_{aa} = 16.2 \frac{T_a}{GJ}$$

for T applied at a. In addition the value of ϕ at point a for T applied at c is

$$\phi_{ac} = 0.07(180) \frac{T_c}{GJ} = 12.6 \frac{T_c}{GJ}$$

Finally, for $T_a = T_c = T$,

$$\phi_a = (16.2 + 12.6) \frac{T}{GJ} = 28.8 \frac{T}{GJ}$$

The twist angle ϕ_a must be compatible with the end slope of the beam; using slope deflection,

$$M_{ab} = M_{Fab} + \frac{2EI}{L_{ab}} \left[-2\phi_a - \phi_b + \frac{3\Delta}{L_{ab}} \right]$$

where
M_{Fab} = the fixed-end moment for beam ab at a
ϕ_a = beam slope at a
ϕ_b = beam slope at b
Δ = relative deflection between a and b

After having established the necessary slope deflection equations for moments, joint
equilibrium and shear conditions are necessary, after which simultaneous equations must
be solved. One such joint equation is that at joint a,

$$T + M_{ab} = 0$$

After solving for the slopes, then the torsional moments can be found; for the torsional moment at *a*,

$$T = \phi_a \, GJ/28.8$$

In the Goldberg [8.12] example using members having properties similar to those of Fig. 8.7.1, the value of *T* obtained was 1.55 in.-kps using an approximate method of satisfying deformation compatibility.

Suppose one had taken the simple beam reaction for member *ab*, 0.5(7) = 3.5 kips, and assume use of *AISC Manual* Table 10–1, "All-Bolted Double-Angle Connections" (see Fig. 13.21). If the eccentricity had been taken to the bolt line on the outstanding leg $\left(2\frac{1}{4} \text{ in.}\right)$, the torsional moment would have been far too great, while if the eccentricity had been taken as one-half the W10×45 (Fig. 8.7.1) web thickness (0.350/2), the torsional moment would have been far too small.

The proper torsional moment can only be obtained (even approximately) by considering deformation compatibility.

Torsional End Restraint

If a torsional situation is deemed to require analysis, the torsional end restraint must be evaluated. Under AISC-B3.6 two basic types of construction are permitted: Type FR (fully restrained) which is the traditional "rigid frame" construction; and Type PR (partially restrained) which includes "simple" or "conventional" framing where there is assumed to be negligible flexural restraint at the joint, as well as "semi-rigid" framing where a defined flexural restraint exists that is less than Type FR.

The correlation of "simple" and "rigid" framing with torsional restraint is shown in Fig. 8.7.2. Again, the lateral bending analogy will help in visualizing the torsional restraint conditions. Figure 8.7.2a shows the analogy situation of zero deflection and zero moment

(a) AISC simple framing connection
$$M_z = M_s + M_w; \quad M_f = 0$$

Figure 8.7.2
Torsional restraint conditions.
(Adapted from Hotchkiss
[8.9], Fig. 8)

(b) AISC rigid framing connection (AISC Type FR) with additional stiffening plates.

$$M_z = M_w; \quad M_s = 0$$

which correspond torsionally to $\phi = 0$ and $d^2\phi/dz^2 = 0$. It is noted that $\phi = 0$ only if the simple connection extends over a significant portion of the beam depth.

Figure 8.7.2b shows the analogy situation of zero deflection and zero slope which corresponds torsionally to $\phi = 0$ and $d\phi/dz = 0$. Hotchkiss [8.9] states the ends of the beam must be boxed in (plates A, Fig. 8.7.2b) so as to assure $d\phi/dz = 0$. Ojalvo [8.20, 8.22] has discussed torsional restraint and indicates that "boxing" is not sufficient to obtain $d\phi/dz = 0$. When "boxing", that is, welding stiffener plates between the toes of flanges and extending them along the beam for a length equal to the beam depth, is used and the beam is then welded to a thick column flange the authors believe it is essentially "torsionally fixed". Furthermore, if the column has flexible flanges, column stiffeners (plates B, Fig. 8.7.2b) should be provided.

The Ojalvo [8.20] suggestion of welding a length of channel, angle, or bent plate between the flanges on one side of the web seems to be a more efficient and economical solution. The ends of the member must be welded against the insides of the flanges, and there must be a weld vertically along the edges that bear against the web. Tests of several torsional restraints, including that proposed by Ojalvo [8.20], have been reported by Heins and Potocko [8.21]. Vacharajittiphan and Trahair [8.23] also discuss torsional restraint at I-section joints.

The structural engineer should remember that in practical situations where no special design is made at the ends, the torsional restraint is neither simple ($d^2\phi/dz^2 = 0$) nor fixed ($d\phi/dz = 0$) but is, however, usually such that the end twist is nearly zero ($\phi = 0$).

8.8   LOAD AND RESISTANCE FACTOR DESIGN FOR TORSION—LATERALLY STABLE BEAMS

Nominal Strength

The nominal strength of a section subject to torsion or torsion combined with flexure is not readily determined. Such strength will certainly depend on the proportions of the section and the relative magnitudes of the forces applied. The AISC ASD Specification (such as the 1978 Specification, Sec. 1.5.1.4.4) has traditionally limited the combined stress to a maximum of $0.60F_y$. This implies that the nominal strength of the section under combined bending and torsion is reached when the extreme fiber stress reaches the yield stress F_y. Thus, the entire cross-section will be elastic; no credit is given to ability of the cross-section to undergo plastic deformation. Certainly this approach is conservative. An interesting review of the design of I-shaped beams for combined flexure and torsion is given by Driver and Kennedy [8.35].

AISC-H2 uses an interaction equation, where the nominal strength about each of the two major axes can conservatively be taken equal to its yield moments. (Note that in this chapter the strength reduction factor ϕ_b is given the subscript to clearly distinguish it from the angle ϕ of twist.) Thus, the elastic biaxial bending stress equation, Eq. 7.11.1, can be used after converting the torsional moment into a pair of lateral bending moments acting in opposite directions on each flange.

In the examples that follow, the beams are assumed to be stable such that the lateral-torsional buckling limit state does not control (see Chapter 9).

EXAMPLE 8.8.1

Select the lightest W section of A992 steel to carry 0.55 kips/ft dead load, in addition to the weight of the beam, and live load of 2.0 kips/ft. The superimposed load is applied eccentrically 7 in. from the center of the web on the simply supported span of 28 ft as shown in Fig. 8.8.1. Assume the ends of the beam have torsional simple support.

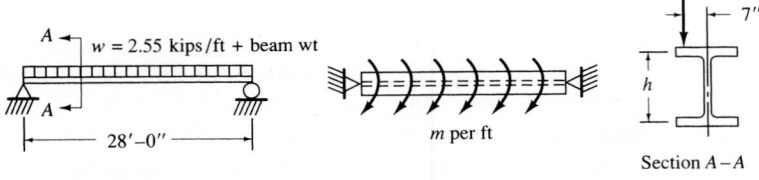

Figure 8.8.1
Conditions for Example 8.8.1.   (a) Vertical support   (b) Torsional support

Solution:

(a) Compute factored loads eccentrically applied.

$$w_u = 1.2(0.55) + 1.6(2.0) = 3.86 \text{ kips/ft}$$

(b) Compute factored moment M_{ux}. Estimating the beam weight as 0.13 kips/ft, the moment M_{ux} is

$$M_{ux} = \tfrac{1}{8}wL^2 = \tfrac{1}{8}[3.86 + 1.2(0.13)](28)^2 = 394 \text{ ft-kips}$$

(c) Consider the torsion effect. The factored uniformly distributed torsional moment is

$$m_u = 3.86(7) = 27.0 \text{ in.-kips/ft}$$

Consider m_u/h as the uniformly distributed lateral load acting on *one* flange of the beam. Then using the flexure analogy, the lateral bending moment M_f acting on one flange is

$$M_f = \frac{1}{8}\frac{m_u}{h}L^2 = \frac{1}{8}\frac{27.0}{h}(28)^2 = \frac{2646}{h} \text{ ft-kips}$$

without regard to the modification factor β.

As a first approximation, assume $h = 14$ in. and $\beta = 0.5$ (approximation from Table 8.6.3, for $\lambda L \approx 3$). Thus the modified flexure analogy gives

$$M_f = \beta\frac{2646}{h} = 0.5\left(\frac{2646}{14}\right) = 94.5 \text{ ft-kips}$$

The design acceptability criterion is

$$\frac{M_{ux}}{\phi_b M_{nx}} + \frac{M_{uy}}{\phi_b M_{ny}} \le 1$$

and using the procedure discussed in Sec. 7.11 gives

$$\text{Required } S_x \ge \frac{M_{ux}}{\phi_b F_y} + \frac{M_{uy}}{\phi_b F_y}\left(\frac{S_x}{S_y}\right)$$

$$= \frac{394(12)}{0.90(50)} + \frac{2(94.5)12}{0.90(50)}(3) = 256 \text{ in.}^3$$

in which the ratio S_x/S_y is estimated at 3 (Table 7.11.1) for medium weight W14 sections, and M_f is doubled to give an equivalent moment acting on two flanges.

This would indicate a W14×176 having an $S_x = 281$ in.³ Since the actual S_x/S_y ratio for W14 sections in this weight range is 2.6, the required S_x is then reduced to 236, indicating W14×145.

Using torsional properties in the *AISC Manual* [1.15, p. 1–21] for the W14×145,

$$\sqrt{GJ/EC_w} = \sqrt{15.2/[2.6(31,700)]} = 1/73.6$$
$$\lambda L = 28(12)/73.6 = 4.6$$

More complete torsional properties are available from Ref. 8.8, pp. 33–53.
Using Table 8.6.3, β is reduced to about 0.33, which further reduces required S_x to about 191 in.[3] Try W14×132.

$$\lambda L = 28(12)/73.4 = 4.58$$

$$\beta \approx 0.31 \quad \text{(Table 8.6.3)}$$

$$M_f = \beta \frac{m_u L^2}{8h} = 0.31 \frac{27.0(28)^2}{8(14.66 - 1.030)} = 60.2 \text{ ft-kips}$$

Check design strength criterion under AISC-H1. Compute the factored bending stress f_{un},

$$\frac{M_{ux}}{\phi_b M_{nx}} + \frac{M_{uy}}{\phi_b M_{ny}} = \frac{M_{ux}}{S_x} + \frac{M_{uy}}{S_y} \le \phi_b F_y$$

$$f_{un} = \frac{M_{ux}}{S_x} + \frac{M_{uy}}{S_y} = \frac{394(12)}{209} + \frac{60.2(12)}{74.5/2}$$

$$= 22.6 + 19.4 = 42.0 \text{ ksi} < (\phi_b F_y = 45.0 \text{ ksi}) \qquad \textbf{OK}$$

Use W14×132.

Where high torsional strength is required, the wide W14 sections are most suitable. For the *same weight per foot,* deeper sections give a reduced stress from in-plane (of web) flexure but an increased stress from restraint of torsional warping. The W21×132 (f_{un} = 39.5 ksi) and the W24×131 (f_{un} = 40.8 ksi) give about the same maximum total flexural stress as the above selected beam.

The differential equation solution gives for the factored lateral bending stress f_{un} due to warping torsion 19.1 ksi as compared with 19.4 ksi computed above. The maximum factored flange shear stress f_{uv} is 22.3 ksi, while that in the web is 18.1 ksi, both computed from the differential equation solution. These are acceptable under AISC-H1,

$$f_{uv} = 22.3 \text{ ksi} < [\phi_b \tau_y = \phi_b(0.6F_y) = 27.0 \text{ ksi}] \qquad \textbf{OK} \quad \blacksquare$$

EXAMPLE 8.8.2

Design a beam having torsionally fixed ends to carry two concentrated loads of 28 kips (8 kips dead load and 20 kips live load) acting eccentric to the plane of the web by 6 in. as shown in Fig. 8.8.2. Assume for conservatism that for in-plane (of web) flexure the beam is simply supported. Use A992 steel and the AISC LRFD Method.

Figure 8.8.2
Loading and factored
moments for Example 8.8.2.

Solution:

(a) Compute factored loads eccentrically applied.

$$W_u = 1.2(8) + 1.6(20) = 41.6 \text{ kips}$$

(b) Compute factored moment M_{ux}. Estimating the beam weight as 0.15 kips/ft, the moment M_{ux} is

$$M_{ux} = W_u(12) + \tfrac{1}{8}wL^2 = 41.6(12) + \tfrac{1}{8}[1.2(0.15)](36)^2 = 528 \text{ ft-kips}$$

(c) Consider the torsion effect. The factored concentrated torsional moment is

$$T_u = 41.6(6/12) = 20.8 \text{ ft-kips}$$

Considering T_u/h as the analogous *lateral* concentrated loads acting at *one* flange, the fixed-end moments are computed; thus

$$M_f h \text{ (at ends)} = \frac{T_u ab^2}{L^2} + \frac{T_u a^2 b}{L^2} = \frac{20.8(12)(24)}{(36)^2}(24 + 12)$$

$$= 110.9 + 55.5 = 166 \text{ ft}^2\text{-kips}$$

and in the positive moment zone (midspan region),

$$M_f h \begin{pmatrix} \text{at concentrated} \\ \text{loads} \end{pmatrix} = \frac{T_u L}{3} - 166$$

$$= 20.8(12) - 166 = 83 \text{ ft}^2\text{-kips}$$

The above moments are computed without regard for the β reduction factor; the flexure analogy gives $M_f h$ values as shown in Fig. 8.8.2b. These values are more appropriate than using the expression in Table 8.6.5 because that expression is for *one* concentrated load. The β values from Table 8.6.5 are reasonable, however, since the effect of one load on the torsional stress at the other load is small.

Estimating average λL at about 3, and using $aL = 0.3L$ in Table 8.6.2 for end moments, the modified analogous fixed-end moments become

$$M_f h \text{ (at ends)} = 0.88(110.9) + 0.84(55.5) = 98 + 47 = 145 \text{ ft}^2\text{-kips}$$

For positive moment at 12 ft from the support, refer to Table 8.6.5 and estimate β as 0.9, though the exact case being treated is not covered in any of β tables. Thus

$$M_f h \text{ (at } z = 0.3 L) \approx 0.9(83) = 75 \text{ ft}^2\text{-kips}$$

which is known to be conservatively high (see Fig. 8.8.3) because the value 75 includes the effects of both concentrated torsional moments. (*Torsional Analysis of Steel Members* [8.8], Case 6, indicates that T_u applied at $z = 0.3L$ has negligible effect at 0.7L.)

(d) Select the member. Assume $h \approx 14$ in., which makes

$$M_f = 75(12)/14 = 64 \text{ ft-kips}$$

$$\text{Required } S_x \geq \frac{M_{ux}}{\phi_b F_y} + \frac{M_{uy}}{\phi_b F_y}\left(\frac{S_x}{S_y}\right)$$

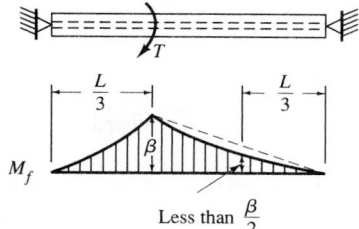

Figure 8.8.3
M_f variation for concentrated
torsional moment T.

$$= \frac{528(12)}{0.90(50)} + \frac{2(64)(12)}{0.90(50)}(2.5)$$

$$= 141 + 86 = 227 \text{ in.}^3$$

A preliminary check of Table 8.6.5 gives about 0.8 for a β estimate. This would reduce the required S_x to 217 in.^{3} and indicates a W14×132. Try W14×132: $S_x = 209$ in.^{3}:

$$\lambda L = 36(12)/73.2 = 5.90$$

Using Table 8.6.5, find $\beta \approx 0.74$ which gives

$$M_f \text{ (at } z = 0.3\, L) = \beta \frac{M_f h}{h} = 0.74 \frac{83(12)}{14.66 - 1.03} = 53.9 \text{ ft-kips}$$

$$\text{Required } S_x = 141 + \frac{2(53.9)12}{0.90(50)}(2.8) = 222 \text{ in.}^3$$

Thus the next heavier section is indicated; *Use* W14×145.
   For a more accurate check of the W14×132 using *Torsional Analysis of Steel Sections* [8.8], Case 6 for $\lambda L = 5.90$ at $z = 0.3L$,

$$M_f h = \left(\begin{array}{c}\text{Ref. 8.8}\\\text{coeff}\end{array}\right)\frac{T_u}{\lambda} = \frac{0.37(20.8)}{(1/73.2)12} = 47.0 \text{ ft}^2\text{-kips}$$

$$M_f = 47.0(12)/13.63 = 41.3 \text{ ft-kips}$$

$$f_{un} = \frac{528(12)}{209} + \frac{41.3(12)}{74.5/2} = 43.7 \text{ ksi}$$

Since f_{un} does not exceed $\phi_b F_y = 0.90(50) = 45.0$ ksi, the W14×132 would be acceptable by the more exact check.
   Also, the stress under factored moment at the supports must be checked. Using Table 8.6.2, find $\beta_1 \approx 0.68$ and $\beta_2 \approx 0.61$ for $\lambda L = 5.87$ (W14×145) and $aL = 0.3L$. Then

$$M_f h = 0.68(110.9) + 0.61(55.5) = 75.4 + 33.9 = 109.3 \text{ ft}^2\text{-kips}$$

$$M_f = 109.3(12)/13.69 = 95.8 \text{ ft-kips}$$

Thus the factored moment M_f about the y-axis resisted by one flange gives the factored stress f_{un}

$$f_{un} = \frac{M_f}{S_y/2} = \frac{95.8(12)}{87.3/2} = 26.3 \text{ ksi} < (\phi_b F_y = 45.0 \text{ ksi}) \qquad \textbf{OK} \quad \blacksquare$$

These two examples illustrate that using approximate β values, along with the flexure analogy for lateral bending due to warping torsion, gives sufficiently quick and accurate results for ordinary design. Furthermore, the designer can better visualize what is happening using the flexure analogy rather than working with the hyperbolic functions for ϕ.

The β modified flexure analogy has been expanded by Lin [8.15] where additional β tables are provided. Johnston [8.14] has provided more detailed design aids to compute torsional functions other than the compressive or tensile stress due to restraint of warping; i.e., particularly the shear stress. Johnston also has several excellent detailed design examples. Salmon [8.24] and Lin [8.16] have provided additional insight in their discussions of Johnston's paper. Additional approximate formulas for design are provided by Johnston, Lin, and Galambos [8.25, pp. 330–331].

For additional treatment of combined torsion and flexure, particularly on channel and zee sections, the reader is referred to the work of Lansing [8.26]. For nonprismatic open section members, Evick and Heins [8.27] present solution techniques and give some design information.

Another topic, outside the scope of this text, is the secondary lateral bending moment that arises from the torsional deflection of the compression flange laterally. In the deflected position the compressive force resulting from ordinary flexural moment M_x times the lateral flange deflection gives rise to the secondary lateral moment which in turn causes greater lateral deflection. Discussion of this topic appears elsewhere [8.28, 8.14] and is similar to the secondary bending moment that occurs in beam-columns, a subject treated in Chapter 12.

8.9 ALLOWABLE STRENGTH DESIGN FOR TORSION— LATERALLY STABLE BEAMS

In Allowable Strength Design, the safety criterion is Eq. 7.11.7,

$$\frac{\Omega M_x}{M_{nx}} + \frac{\Omega M_y}{M_{ny}} \leq 1 \qquad [7.11.7]$$

Using the yield moments for M_{nx} and M_{ny}, the equation becomes

$$\frac{\Omega M_x}{S_x F_y} + \frac{\Omega M_y}{S_y F_y} \leq 1 \qquad [7.11.8]$$

$$S_x \geq \frac{\Omega M_x}{F_y} + \frac{\Omega M_y}{F_y}\left(\frac{S_x}{S_y}\right) \qquad [7.11.9]$$

The torsion is converted into equivalent M_y by using the flexure analogy as discussed in Sec. 8.6. The procedure is the same as illustrated for LRFD in Sec. 8.8 except in ASD the service loads are used instead of factored loads.

EXAMPLE 8.9.1

Investigate the W14×132 selected in Example 8.8.1 for the loading and conditions of Fig. 8.8.1. Use the AISC ASD Method.

Solution:

(a) Compute the service load moment M_x. Including the 0.13 kip/ft beam weight,
$$M_x = \tfrac{1}{8}wL^2 = \tfrac{1}{8}(2.55 + 0.13)(28)^2 = 263 \text{ ft-kips}$$

(b) Consider the torsion effect. The unformly distributed service load torsional moment m is
$$m = 2.55(7) = 17.9 \text{ in.-kips/ft}$$

Consider m/h as the uniformly distributed lateral load acting on *one* flange of the beam. Then using the β modified flexure analogy, the lateral bending moment M_f acting on one flange is

$$M_f = \beta \frac{1}{8}\frac{m}{h}L^2 = 0.31\frac{1}{8}\frac{17.9}{13.63}(28)^2 = 39.9 \text{ ft-kips}$$

(c) Check AISC-H1 interaction criterion, as follows

$$\frac{\Omega M_x}{M_{nx}} + \frac{\Omega M_y}{M_{ny}} = \frac{1.67(263)(12)}{209(50)} + \frac{2(1.67)(39.9)(12)}{74.5(50)} = 0.934 < 1 \qquad \textbf{OK}$$

Thus the W14×132 is acceptable! ∎

8.10 TORSION IN CLOSED THIN-WALL SECTIONS

In general, where a high torsional stiffness is required, a closed section is preferred over the ordinary open section, such as the I-shape or channel. An excellent general discussion of the torsion phenomena with comparative behavior of open and closed sections is given by Tamberg and Mikluchin [8.29]. Some practical comments relating to closed sections in torsion are given by Siev [8.30]. The high torsional stiffness exhibited by closed sections makes them ideal for aircraft structural components and curved girders in bridges and buildings. This subject is treated in a number of textbooks [8.1–8.3] so that only a brief treatment follows.

In the closed section of Fig. 8.10.1, the walls are assumed thin so that the shearing stress may be assumed uniformly distributed across the thickness t. If the shear stress is τ, then τt is the shear force per unit distance along the wall, usually referred to as *shear flow*. Since only torsional stress is presently being considered, the normal stresses (σ_z of Fig. 8.10.1b) are zero. Since $\sigma_z = 0$, the shear flow τt cannot vary along the wall; i.e., τt is constant.

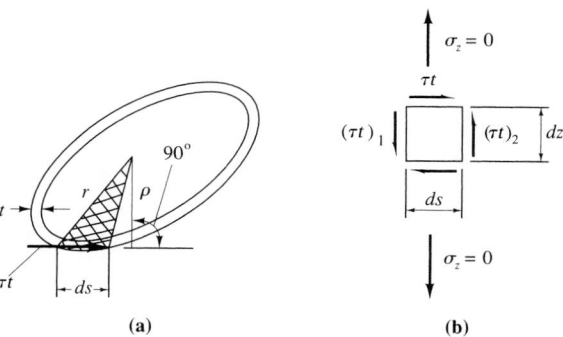

Figure 8.10.1
Shear flow in a closed thin-wall section.

Referring to Fig. 8.10.1a, the increment of torsional moment contributed by each element is

$$dT = \tau t \rho \, ds \qquad (8.10.1)$$

Integrating gives the full torsional moment, which is in effect the same as Eq. 8.2.5,

$$T = \tau t \int_s \rho \, ds \qquad (8.10.2)$$

Note that $\frac{1}{2}\rho \, ds$ is the cross-hatched area of the triangular segment in Fig. 8.10.1. Thus the integral

$$\int_s \rho \, ds = 2A \qquad (8.10.3)$$

where A = area enclosed by the walls. Finally,

$$T = 2\tau t A \qquad (8.10.4)$$

If a cut is made in the wall of a closed thin-wall section (Fig. 8.10.2), a relative movement (as in Fig. 8.10.2b) will be produced between the two sides in the axial direction of the member. The unit shear strain along the perimeter is

$$\gamma = \tau/G \qquad (8.10.5)$$

The internal strain energy for any elemental length ds along the perimeter is

$$dW_i = \tfrac{1}{2}\tau t \gamma \, ds \qquad (8.10.6)$$

$$= \frac{1}{2}\left(\frac{T}{2A}\right)\frac{\tau}{G}\, ds \qquad (8.10.7)$$

The twisting moment T about point 0 can now be replaced by a couple, T/r. The external work done by the couple is

$$dW_e = \frac{1}{2}\left(\frac{T}{r}\right)n = \frac{T\theta}{2} \qquad (8.10.8)$$

Equating internal and external work per unit length gives

$$\frac{T\theta}{2} = \frac{T}{4AG}\int_s \tau \, ds \qquad (8.10.9)$$

$$\theta = \frac{\int_s \tau \, ds}{2AG} = \frac{\tau t \int_s ds/t}{2AG} \qquad (8.10.10)$$

since τt is a constant.

Figure 8.10.2
Forces on a cut thin-wall section.

(a)　　　　(b)

In order to obtain more useful forms of the equations, recall from Sec. 8.2 that

$$T = GJ\theta \qquad\qquad [8.2.5]$$

and using Eq. 8.10.10 gives

$$T = GJ\frac{\tau t \int ds/t}{2AG} \qquad\qquad (8.10.11)$$

and eliminating T between Eqs. 8.10.4 and 8.10.11 gives, when solving for the torsional constant J,

$$J = \frac{4A^2}{\int_s ds/t} \qquad\qquad (8.10.12)$$

The design of closed sections for bending and torsion is outside the scope of this text, and the reader is referred to the work of Felton and Dobbs [8.31]. Shermer [8.32] has shown applications of the closed thin-wall tube concept of torsion applied to trussed structures and structures containing plates and truss forms.

The development of the equations for torsion stiffness and stress in closed thin-wall sections may be developed alternatively using the "membrane analogy" developed by Prandtl [8.1–8.3].

The objective here is to illustrate the high degree of torsional stiffness that closed sections exhibit as compared to open ones.

EXAMPLE 8.10.1

Compare the torsional resisting moment T and the torsional constant J for the sections of Fig. 8.10.3 all having about the same cross-sectional area. The maximum shear stress τ is 14 ksi.

Solution:

(a) Circular thin-wall section. Using Eq. 8.10.4,

$$T = 2\tau t A = 2(14)(0.5)[\pi(10)^2/4]\tfrac{1}{12} = 91.6 \text{ ft-kips}$$

$$J = \frac{4A^2}{\int ds/t} = \frac{4(25\pi)^2}{20\pi} = 393 \text{ in.}^4$$

where $\int ds/t = 2\pi(5)/0.5 = 20\pi$.

Figure 8.10.3
Sections for Example 8.10.1.

(b) Rectangular box section.

$$T = 2\tau tA = 2(14)(0.5)(72)\tfrac{1}{12} = 84.0 \text{ ft-kips}$$

$$J = \frac{4A^2}{\displaystyle\int ds/t} = \frac{4(72)^2}{(36/0.5)} = 288 \text{ in.}^4$$

(c) Channel section. Since for this open section,

$$\tau = \frac{Tt}{J}$$

the maximum shear stress will be in the flange. Also,

$$J = \sum \tfrac{1}{3}bt^3$$

$$J = \tfrac{1}{3}[10(0.5)^3 + 2(5.5)(1)^3] = 4.1 \text{ in.}^4$$

$$T = \frac{J\tau}{t_f} = \frac{4.1(14)}{(1)(12)} = 4.8 \text{ ft-kips}$$

The circular section is best for torsional capacity, the rectangular box is next; these closed sections have the torsional constant J equal to 96 and 71 times that of the channel, respectively. The resisting moments are 19 and 18, respectively, times that of the channel. ∎

8.11   TORSION IN SECTIONS WITH OPEN AND CLOSED PARTS

Generally this problem is treated by combining the principles discussed separately for open and closed parts. The procedure to be used for determining resisting moment, stiffness, and shear center location for such sections is presented with examples by Chu and Longinow [8.33]. The following is a summary of pertinent equations:

Total resisting moment is

$$T = \sum_{i=1}^{n} 2\tau_i t_i A_i + GJ\theta \tag{8.11.1}$$

where $J = \Sigma \tfrac{1}{3}bt^3$ for open parts only.

In addition, each of the closed cells must satisfy Eq. 8.10.10:

$$\int_s \tau_i t_i \frac{ds}{t_i} = 2GA_i\theta \tag{8.11.2}$$

8.12   TORSIONAL BUCKLING

Since the differential equation for torsion was developed earlier in this chapter, and buckling of axially loaded columns has previously been treated, torsional buckling may now be examined. The strength of most centrally loaded columns is reached at the tangent-modulus Euler load with a reduced efficiency if local buckling occurs before overall column buckling occurs, as discussed in Chapter 6. However, some thin-wall sections such as angles, tees,

Figure 8.12.1
Torsional buckling.

(a)   (b)

zees, and channels, having relatively low torsional stiffness may, under axial compression, buckle torsionally while the longitudinal axis remains straight.

The subject of torsional buckling is treated extensively by Timoshenko and Gere [6.67, pp. 225–250] and Bleich [6.9].

Using concepts previously developed, it is the objective here to show mathematically that such buckling can occur and identify situations where the designer should be cautious.

Consider the doubly symmetrical section in the shape of a cross given in Fig. 8.12.1, whose shear center and centroid coincide. Recalling the Euler equation,

$$EI\frac{d^2y}{dz^2} + Py = 0 \qquad\qquad [6.2.3]$$

which differentiated twice becomes

$$EI\frac{d^4y}{dz^4} = -P\frac{d^2y}{dz^2} \qquad\qquad (8.12.1)$$

Since $EI\,d^4y/dz^4$ is the loading, the Euler column can be thought of as a beam laterally loaded with the fictitious loading $-P(d^2y/dz^2)$. Thus with the section put in the slightly buckled position (i.e., rotated the angle ϕ at distance z from the end) the compressive force $\sigma_z t\,dr$ on the element $dr\,dz$ is statically equivalent to a lateral load whose intensity per unit length is

$$-(\sigma_z t\,dr)\frac{d^2(r\phi)}{dz^2}$$

The increment of torsional moment about the z-axis tributary to the length dz equals the load times the moment arm r; thus

$$dm_z = -(\sigma_z tr\,dr)\frac{rd^2\phi}{dz^2}dz \qquad\qquad (8.12.2)$$

The total torsional moment for the slice dz of the column is

$$m_z = -\sigma_z \frac{d^2\phi}{dz^2} dz \int_A r^2 t\, dr \qquad (8.12.3)$$

Equation 8.12.3 represents the contribution to the torsional moment M_z tributary to the element dz:

$$dM_z = m_z \qquad (8.12.4)$$

The differential equation for torsion on I-shaped sections, Eq. 8.5.10, is

$$M_z = GJ \frac{d\phi}{dz} - EC_w \frac{d^3\phi}{dz^3} \qquad [8.5.10]$$

which when differentiated once becomes

$$\frac{dM_z}{dz} = GJ \frac{d^2\phi}{dz^2} - EC_w \frac{d^4\phi}{dz^4} \qquad (8.12.5)$$

Referring to Fig. 8.5.2, positive M_z at the section z gives a clockwise rotation; whereas in Fig. 8.12.1 at the section z there is counterclockwise rotation. Thus the Eq. 8.12.4 relationship requires a minus sign for use in Eq. 8.12.5. The differential equation for torsional buckling is then

$$\sigma_z \frac{d^2\phi}{dz^2} \int_A r^2 t\, dr = GJ \frac{d^2\phi}{dz^2} - EC_w \frac{d^4\phi}{dz^4}$$

or

$$EC_w \frac{d^4\phi}{dz^4} - \left(GJ - \sigma_z \int_A r^2 t\, dr \right) \frac{d^2\phi}{dz^2} = 0 \qquad (8.12.6)$$

in which $\int_A r^2 t\, dr = I_p$, the polar moment of inertia *about the shear center.* When the centroid coincides with the shear center, Eq. 8.12.6 alone determines the buckling condition.

Note is made that the warping rigidity EC_w is zero (text Appendix Table A2) for shapes consisting of thin rectangular elements intersecting at a common point.

For other cases, Eq. 8.12.6 may be written as

$$\frac{d^4\phi}{dz^4} + p^2 \frac{d^2\phi}{dz^2} = 0 \qquad (8.12.7)$$

where

$$p^2 = \frac{\sigma_z I_p - GJ}{EC_w}$$

for which the general solution is

$$\phi = A_1 \sin pz + A_2 \cos pz + A_3 z + A_4 \qquad (8.12.8)$$

Considering the pin-end column, with rotation about z prevented at each end, but with warping not restricted at the ends gives in a manner similar to the Euler column derivation in Chapter 6, that

$$A_2 = A_3 = A_4 = 0$$

and since A_1 cannot also be zero,

$$\sin pL = 0, \quad pL = n\pi$$

The elastic buckling stress $\sigma_{z\,\text{critical}}$ at which torsional buckling occurs is

$$\frac{\pi^2}{L^2} = \frac{\sigma_z I_p - GJ}{EC_w}$$

$$\sigma_{z\,\text{critical}} = \left[\frac{\pi^2 EC_w}{L^2} + GJ\right]\frac{1}{I_p} = F_{ez} \tag{8.12.9}$$

which is accurate for doubly symmetrical sections whose shear center and centroid coincide, such as I-shaped sections. The symbol F_{ez} is used instead of $\sigma_{z\,\text{critical}}$ in AISC-E4. Equation 8.12.9 is used with $I_p = I_x + I_y$ and L is the effective length $K_2 L$. For the common single-angle strut, since the distance from centroid to shear center is small, Eq. 8.12.9 will provide a reasonable approximation for the torsional buckling stress. Expressions for the warping constant C_w and the torsion constant J for various shapes are to be found in text Appendix Table A2. Though not in that table, values of the warping constant C_w for a combination W-section with a channel cap (see *AISC Manual* Tables 1-19 and 1-20 pp. 1–112–115) have been given by Lue and Ellifritt [8.34].

The reader should not lose sight of the fact that the most probable buckling mode is still that occurring at the tangent-modulus Euler load because of lateral bending about the x- or y-axis. Thus the problem involves three critical values of axial load; bending about either principal axis and twisting about the longitudinal axis. On wide-flange sections, torsional buckling may be important on sections with extra wide flanges and short lengths [6.67, pp. 225–250].

In the general case where the shear center does not coincide with the centroid, the buckling failure is actually a combination of torsion and flexure. For this case, the three differential equations, (1) buckling by lateral bending about the x-axis; (2) buckling by lateral bending about the y-axis; and (3) twisting about the shear center, are interdependent. Thus three simultaneous differential equations must be solved to get the buckling loads. The development and solution of the these equations is outside the scope of this text and is adequately treated elsewhere [6.9; 6.67, pp. 225–250].

LRFD Design for Torsional and Flexural-Torsional Buckling

For design, AISC-E3 implies that the nominal strength of compression members with compact and noncompact sections (as defined in AISC-B4) can be computed *without considering flexural-torsional buckling.*

Compression members without slender elements that are singly symmetric or unsymmetrical, and certain doubly symmetric sections, such as cruciform and built up sections, must be designed to include the flexural-torsional buckling limit state according to AISC-E4. Single angle compression members are excluded from these provisions and are to be designed according to AISC-E3, E5 and E7.

For doubly symmetric sections (such as built-up I-sections), when the flexural-torsional limit state is evaluated, an equivalent radius of gyration r_E can be compared with r_x and r_y to reduce the computations. The alternative is to compute the elastic flexural-torsional buckling stress F_e from Eq. 8.12.9 [identical to AISC Formula (E4-4)] and compare with F_{cr} computed using the larger of $K_x L_x / r_x$ and $K_y L_y / r_y$ in the column formulas of AISC-E3 (also text Chapter 6).

To develop the r_E equation, set Eq. 8.12.9 equal to the Euler equation,

$$\frac{\pi^2 E}{(L/r_E)^2} = \frac{EC_w \pi^2}{I_p L^2} + \frac{GJ}{I_p}$$

$$r_E = \sqrt{\frac{C_w}{I_p} + \frac{GJL^2}{EI_p \pi^2}}$$

which for steel with $E/G = 2.6$ gives

$$r_E = \sqrt{\frac{C_w}{I_p} + 0.04 \frac{JL^2}{I_p}}$$

(8.12.10)

for doubly symmetric sections. It has been demonstrated that only for short lengths will r_E be lower than r_x and r_y for W shapes [6.9].

Singly symmetric sections may buckle in a combination flexural-torsional mode, which will depend on the Euler column buckling stress F_{ey} for axis of symmetry, and the torsional buckling stress F_{ez} (Eq. 8.12.9). AISC-E4 gives the elastic buckling stress F_e for the combined mode as

$$F_e = \frac{F_{ey} + F_{ez}}{2H}\left(1 - \sqrt{1 - \frac{4F_{ey}F_{ez}H}{(F_{ey} + F_{ez})^2}}\right)$$

(8.12.11)

where

$$F_{ey} = \frac{\pi^2 E}{(K_y L/r_y)^2}$$

(8.12.12)

$$F_{ez} = \left(\frac{\pi^2 E C_w}{(K_z L)^2} + GJ\right)\frac{1}{I_p}$$

(8.12.13)

$$H = 1 - \left(\frac{x_0^2 + y_0^2}{\bar{r}_0^2}\right) = \frac{I_x + I_y}{I_p}$$

(8.12.14)

$$I_p = A\bar{r}_0^2 = I_x + I_y + A(x_0^2 + y_0^2)$$

(8.12.15)

E = tension-compression modulus of elasticity, ksi
G = shear modulus of elasticity, ksi
C_w = torsional warping constant, in.^{6}
J = torsion constant, in.^{4}
I_x, I_y = moment of inertia about principal axes, x and y
I_p = polar moment of inertia, Eq. 8.12.15, in.^{4}
K_y, K_z = effective length factors in the y-direction, and for torsional buckling (z-axis)
x_0, y_0 = coordinates of shear center with respect to centroid of section, in.
r_y = radius of gyration about the *axis of symmetry*
r_0 = polar radius of gyration about the shear center

EXAMPLE 8.12.1

For the sections given in Fig. 8.12.2, determine under what conditions, if any, torsional or flexural-torsional buckling is likely to occur under axial compression loading. Assume the members pinned at the ends of the unbraced lengths, and free to warp at the ends, fully recognizing that these two assumptions minimize buckling strength.

Solution:

(a) W8×31. Since the centroid and shear center coincide, use Eq. 8.12.10 to get the equivalent r_E:

$$J = \tfrac{1}{3}\left[2(7.995)(0.435)^3 + 7.13(0.285)^3\right] = 0.494 \text{ in.}^4$$

$$(AISC \ Manual \ \text{Table 1-1}, J = 0.536 \text{ in.}^4)$$

Figure 8.12.2
Sections for Example 8.12.1.

(a) W8 × 31
$A = 9.13$ sq in.
$I_x = 110$ in.^{4}
$I_y = 37.1$ in.^{4}
$r_y = 2.02$ in.

(b) WT8 × 28.5
$A = 8.38$ sq in.
$I_x = 48.7$ in.^{4}
$I_y = 21.6$ in.^{4}
$r_x = 2.41$ in.
$r_y = 1.60$ in.

(c) L8 × 8 × $\frac{1}{2}$
$A = 7.75$ sq in.
$I_x = I_y = 48.6$ in.^{4}
$r_x = r_y = 2.50$ in.
$r_z = 1.59$ in.

$$C_w = h^2 I_y/4 = (8.00 - 0.435)^2(37.1)/4 = 531 \text{ in.}^6$$

$$(AISC\ Manual, C_w = 530 \text{ in.}^6)$$

$$I_p = I_x + I_y = 110 + 37.1 = 147.1 \text{ in.}^4$$

$$r_E = \sqrt{\frac{C_w}{I_p} + 0.04\frac{JL^2}{I_p}} = \sqrt{\frac{531}{147.1} + 0.04\frac{0.494L^2(144)}{147.1}}$$

$$= \sqrt{3.61 + 0.0193L^2}$$

where L is the unsupported length, in feet. Only when L is less than 4.9 ft does r_y exceed r_E; only for a very short column is torsional buckling a possibility, and even when r_E approaches its minimum value (when $L = 0$), which is $r_E \text{(min)} = 1.90$ in., it is only 6% less than r_y.

The result using this section is typical of standard W and S shapes and indicates torsional buckling may generally be neglected for them.

(b) WT8×28.5. The centroid and shear center do not coincide, but the section has one axis of symmetry; use Eq. 8.12.11:

$$J = \tfrac{1}{3}\left[(7.12)(0.715)^3 + 7.50(0.430)^3\right] = 1.066 \text{ in.}^3$$

Using the formula from text Appendix Table A2,

$$C_w = \frac{1}{36}\left[\frac{(7.120)^3(0.715)^3}{4} + (7.50)^3(0.430)^3\right] = 1.85 \text{ in.}^6$$

The *AISC Manual* [1.15, Table 1-1] gives $J = 1.10$ in.^{3} and $C_w = 1.99$ in.^{6}

$$I_p = I_x + I_y + Ay_0^2 = 48.7 + 21.6 + 8.38(1.58)^2 = 91.2 \text{ in.}^4$$

For use in Eq. 8.12.11, compute the critical stresses F_{ey} and F_{ez} from Eqs. 8.12.12 and 8.12.13 using for an effective length KL (for y- and z-axes) the common length of 10 ft:

$$F_{ey} = \frac{\pi^2(29,000)}{[1.0(10)(12)/1.60]^2} = 50.9 \text{ ksi}$$

$$F_{ez} = \left(\frac{\pi^2 E C_w}{(K_z L)^2} + GJ \right) \frac{1}{I_p} \qquad [8.12.13]$$

$$F_{ez} = \left(\frac{\pi^2 (29{,}000) 1.85}{[1.0(10)12]^2} + \frac{29{,}000(1.066)}{2.6} \right) \frac{1}{91.2} = 131 \text{ ksi}$$

The distance y_0 from the centroid of the section to the shear center at the junction of the mid-thicknesses of the flange and the web is

$$y_0 = y - t_f/2 = 1.94 - 0.715/2 = 1.58 \text{ in.}$$
$$x_0 = 0$$
$$I_p = 91.2 \text{ in.}^4 \text{ (computed above)}$$
$$H = \frac{I_x + I_y}{I_p} = \frac{48.7 + 21.6}{91.2} = 0.77$$

Then using Eq. 8.12.11,

$$F_e = \frac{50.9 + 131}{2(0.77)} \left(1 - \sqrt{1 - \frac{4(50.9)(131)0.77}{(50.9 + 131)^2}} \right) = 45.3 \text{ ksi}$$

Equating F_e to Euler's formula, an equivalent radius r_E of gyration for flexural-torsional buckling can be obtained to compare with r_x and r_y,

$$45.3 = \frac{\pi^2 E}{(KL/r_E)^2} = \frac{\pi^2 29{,}000}{(120/r_E)^2}$$
$$r_E = 1.51 \text{ in.}$$

In this case r_E is less than $r_y = 1.60$ in. and flexural-torsional buckling is critical for ordinary lengths. The same situation will occur with double angle compression members. An equal leg angle (such as in Fig. 8.12.2c) will have z_2 as its axis of symmetry, and the radius of gyration with respect to that axis must be used for r_y in Eq. 8.12.12. ■

As a conclusion to this treatment, the designer is cautioned about using open sections (torsionally weak sections) in compression having less than two axes of symmetry, particularly when high width/thickness ratios exist for the elements. The width/thickness ratio limits (λ_r in AISC-B4) for control of local buckling, if not exceeded, provide some control since local buckling of sections such as angles, flanges, and tees is closely related to torsional buckling.

SELECTED REFERENCES

8.1. Arthur P. Boresi, Richard J. Schmidt, and Omar M. Sidebottom. *Advanced Mechanics of Materials,* 5th ed. New York: John Wiley and Sons, Inc., 1993, Chap. 6.

8.2. S. Timoshenko. *Strength of Materials,* Part II, 2nd ed. New York: D. Van Nostrand Company, Inc., 1941 Chap. 6.

8.3. William McGuire. *Steel Structures.* Englewood Cliffs, NJ: Prentice-Hall, Inc., 1968, pp. 346–400.

8.4. Inge Lyse and Bruce G. Johnston. "Structural Beams in Torsion," *Transactions,* ASCE, **101** (1936), 878–926 (includes Discussions).

8.5. F. K. Chang and Bruce G. Johnston. "Torsion of Plate Girders," *Transactions,* ASCE, **118** (1953), 337–396.

8.6. Gerald G. Kubo, Bruce G. Johnston, and William J. Eney. "Nonuniform Torsion of Plate Girders," *Transactions,* ASCE, **121** (1956), 759–785. (Good summary of torsion theory.)

8.7. I. A. El Darwish and Bruce G. Johnston. "Torsion of Structural Shapes," *Journal of the Structural Division,* ASCE, **91,** ST1 (February 1965), 203–227. Errata: **92,** ST1 (February 1966), 471. See also *Transactions,* ASCE, **131** (1966), 428–429 for summary of equations.

8.8. Paul A. Seaburg and Charles J. Carter. *Torsional Analysis of Steel Members,* Steel Design Guide Series No. 9, Chicago IL: American Institute of Steel Construction, 116 pp., 1997.

8.9. John G. Hotchkiss. "Torsion of Rolled Steel Sections in Building Structures," *Engineering Journal,* AISC, **3,** 1 (January 1966), 19–45.

8.10. Conrad P. Heins, Jr. and John T. C. Kuo. "Torsional Properties of Composite-Girders," *Engineering Journal,* AISC, **9,** 2 (April 1972), 79–85.

8.11. S. Timoshenko. "Theory of Bending, Torsion, and Buckling of Thin-Walled Members of Open Cross-Section," *J. Franklin Inst.,* **239,** 3, 4, and 5 (1945), 201–219, 249–268, and 343–361.

8.12. John E. Goldberg. "Torsion of I-Type and H-Type Beams," *Transactions,* ASCE, **118** (1953), 771–793.

8.13. Kuang-Han Chu and Robert B. Johnson. "Torsion in Beams with Open Sections," *Journal of the Structural Division,* ASCE, **100,** ST7 (July 1974), 1397–1419.

8.14. Bruce G. Johnston. "Design of W-Shapes for Combined Bending and Torsion," *Engineering Journal,* AISC, **19,** 2 (Second Quarter 1982), 65–85.

8.15. Philip H. Lin. "Simplified Design for Torsional Loading of Rolled Steel Members," *Engineering Journal,* AISC, **14,** 3 (Third Quarter 1977), 98–107.

8.16. Phil H. Lin. Discussion of "Design of W-Shapes for Combined Bending and Torsion," by Bruce G. Johnston, *Engineering Journal,* AISC, **20,** 2 (Second Quarter 1983), 82–87.

8.17. J. N. Goodier and M. V. Barton. "The Effects of Web Deformation on the Torsion of I-Beams," *J. Appl. Mech.,* March 1944, p. A-35.

8.18. J. E. Lothers. "Torsion in Steel Spandrel Girders," *Transactions,* ASCE, **112** (1947), 345–376.

8.19. Min-Tse Chen and Donald E. Jolissaint, Jr. "Pure and Warping Torsion Analysis of Rigid Frames," *Journal of Structural Engineering,* ASCE, **109,** 8 (August 1983), 1999–2003.

8.20. Morris Ojalvo. Discussion of "Warping and Distortion at I-Section Joints," by P. Vacharajittiphan and N. S. Trahair, *Journal of the Structural Division,* ASCE, **101,** ST1 (January 1975), 343–345.

8.21. Conrad P. Heins and Robert A. Potocko. "Torsional Stiffening of I-Girder Webs," *Journal of the Structural Division,* ASCE, **105,** ST8 (August 1979), 1689–1698.

8.22. Morris Ojalvo. Discussion of "Torsional Stiffening of I-Girder Webs," by C. P. Heins and R. A. Potocko, *Journal of the Structural Division,* ASCE, **106,** ST4 (April 1980), 939.

8.23. Porpan Vacharajittiphan and Nicholas S. Trahair. "Warping and Distortion at I-Section Joints," *Journal of the Structural Division,* ASCE, **100,** ST3 (March 1974), 547–564.

8.24. Charles G. Salmon. Discussion of "Design of W-Shapes for Combined Bending and Torsion," by Bruce G. Johnston, *Engineering Journal,* AISC, **19,** 4 (Fourth Quarter 1982), 215–216.

8.25. Bruce G. Johnston, F. J. Lin, and T. V. Galambos. *Basic Steel Design,* 3rd ed. Englewood Cliffs, NJ: Prentice-Hall, Inc., 1986.

8.26. Warner Lansing. "Thin-Walled Members in Combined Torsion and Flexure," *Transactions,* ASCE, **118** (1953), 128–146. (Particular emphasis on channel and zee sections.)

8.27. Donald R. Evick and Conrad P. Heins, Jr. "Torsion of Nonprismatic Beams of Open Section," *Journal of the Structural Division,* ASCE, **98,** ST12 (December 1972), 2769–2784.

8.28. Basil Sourochnikoff. "Strength of I-Beams in Combined Bending and Torsion," *Transactions,* ASCE, **116** (1951), 1319–1342.

8.29. K. G. Tamberg and P. T. Mikluchin, "Torsional Phenomena Analysis and Concrete Structure Design," *Analysis of Structural Systems for Torsion,* SP-35, American Concrete Institute, 1973, 1–102.

8.30. Avinadav Siev. "Torsion in Closed Sections," *Engineering Journal,* AISC, **3,** 1 (January 1966), 46–54.

8.31. Lewis P. Felton and M. W. Dobbs. "Optimum Design of Tubes for Bending and Torsion," *Journal of the Structural Division,* ASCE, **93,** ST4 (August 1967), 185–200.

8.32. Carl L. Shermer. "Torsional Strength and Stiffness of Steel Structures," *Engineering Journal,* AISC, **17,** 2 (Second Quarter 1980), 33–37.

8.33. Kuang-Han Chu and Anatole Longinow. "Torsion in Sections with Open and Closed Parts," *Journal of the Structural Division,* ASCE, **93,** ST6 (December 1967), 213–227.

8.34. Tony Lue and Duane S. Ellifritt. "The Warping Constant for the W-Section with a Channel Cap," *Engineering Journal,* AISC, **30,** 1 (First Quarter 1993), 31–33.

8.35. Robert G. Driver and D. J. Laurie Kennedy. "Combined Flexure and Torsion of I-shaped Steel Beams," *Canadian Journal of Civil Engineering,* **16** (1989), 124–139.

PROBLEMS

All design problems are to be done according to the AISC LRFD Method or the AISC ASD Method, as indicated by the instructor. All given loads are service loads unless othewise indicated. For all problems *assume adequate lateral support of the compression flange such that lateral stability does not control.* Assume all standard sections are equally readily available in the indicated grade of steel (even though actually they are not). A figure showing span and loading is required.

8.1. For the channel shown in the accompanying figure, separately apply V_x and V_y through the centroid of the section. For each shear compute and draw to scale the shear flow τt distribution along each of the elements of the cross-section. On the two separate diagrams (one for V_x and one for V_y) of shear flow distribution compute the total shear force in each element of the cross-section in terms of the applied shear V_x or V_y. Using these computed shear forces calculate the two coordinates of the shear center.

$I_x = 192.7$ in.⁴
$I_y = 19.14$ in.⁴
$\bar{x} = 1.087$ in.

Problem 8.1

8.2. Repeat the requirements of Prob. 8.1 for the channel with sloping flanges. Comment on the effect of using average thickness instead of the actual sloping flanges for determining shear center on standard rolled channels.

Problem 8.2

8.3. Repeat the requirements of Prob. 8.1 for the section of the accompanying figure.

$I_x = 17.25$ in.⁴
$I_y = 3.91$ in.⁴
$I_{xy} = +3.25$ in.⁴

Dimensions refer to centerlines of elements

Problem 8.3

8.4. Repeat the requirements of Prob. 8.1 for the section of the accompanying figure.

Standard angle

Problem 8.4 $L8 \times 6 \times \frac{1}{2}$

8.5. Repeat the requirements of Prob. 8.1 for the section of the accompanying figure.

Z section
$I_x = 19.19$ in.⁴
$I_y = 9.05$ in.⁴

Problem 8.5

8.6. Locate the shear center for the combined W and channel crane girder section. Is there significant error in assuming the shear center lies at the centroid? Use average thickness and constant depth for the channel.

C12 × 20.7

W27 × 94

Problem 8.6

8.7. An MC18×58 is to be used on a 24-ft simply supported span to carry a total load of 0.8 kips/ft, with the load applied in the plane of the web. Suppose the flanges are to have attachments so that the channel will bend vertically about the x-axis. What lateral moment capacity M_y should the attachments be capable of resisting? What percent of M_x does this represent?

8.8. (a) Develop the torsion differential equation solution for the W section having torsionally fixed ends and an eccentrically applied concentrated load at midspan.

(b) Compute the torsion constant J, the warping constant C_w, and λ (i.e., $1/a$); then use them in parts (c) through (f).

(c) Compute the combined bending stress, including warping torsion and ordinary flexure components, at $z = 0, 0.3L$, and $0.5L$ unless otherwise instructed.

(d) Compute the maximum shear stress in the web, including Saint-Venant torsional shear and flexural shear, at the same locations indicated in (c).

(e) Compute the flange shear force V_f due to warping restraint at the same locations indicated in (c).

(f) Compute the maximum shear stress in the flange, including Saint-Venant torsion, warping shear, and vertical flexure flange shear, at the same locations indicated in (c).

(g) Give a tabular summary of all stresses.

8.9. Develop the torsion differential equation solution for the cantilever beam with an eccentric concentrated load at its end, and compute constants and stresses as given in items (b) through (e) of Prob. 8.8. Is there any relationship between this problem and Prob. 8.8?

W14 × 90

Problem 8.8

W12 × 96

Problem 8.9

8.10. Develop the torsion differential equation solution for the uniformly loaded beam with loading applied eccentrically to the web. Consider the ends torsionally simply supported. Compute constants and stresses as given in items (b) through (e) of Prob. 8.8.

Problem 8.10

8.11. Repeat Prob. 8.10, considering the ends torsionally fixed.

8.12. Repeat Prob. 8.8, with the load applied in the plane of the y-axis of a channel, C15×50.

8.13. For the case (or cases) assigned by the instructor, select the lightest W14 section, using the b modified flexure analogy approach, to carry a concentrated load W at midspan, in addition to the weight of the beam. The ends of the simply supported span are assumed to have torsional simple support. Check the stresses in the section selected using the "exact" solution for assumed conditions as developed in Example 8.5.2.

Case	W_D Dead load (kips)	W_L Live load (kips)	Span length (ft)	F_y Steel yield stress (ksi)	e Eccentricity of loading (in.)
1	7	22	24	50	2
2	7	22	24	50	3
3	7	22	24	50	4
4	10	15	26	50	2
5	10	15	26	50	3
6	10	15	26	50	4
7	10	15	26	50	5
8	10	15	26	50	6
9	10	15	26	50	7

8.14. For the case (or cases) assigned by the instructor, select the lightest W section, using the b modified flexure analogy approach, to carry uniform loading w, in addition to the weight of the beam.

The ends of the span are assumed to be *fixed* for both flexure and torsion. If "exact" solution was obtained in Prob. 8.11, check maximum stress using that solution.

Case	w_D Dead load (kips/ft)	w_L Live load (kips/ft)	Span length (ft)	Steel yield stress (ksi)	Eccentricity of loading (in.)
1	0.35	1.4	26	50	7
2	0.35	1.4	26	50	5
3	0.35	1.4	26	50	3
4	0.35	1.4	26	50	2

8.15. For the case (or cases) listed in Prob. 8.14 and assigned by the instructor, select the lightest W section, using the b modified flexure analogy approach, to carry uniform loading w, in addition to the weight of the beam. The ends of the span are *simply supported* for both flexure and torsion. If the "exact" solution is available (solution to Prob 8.11), check stresses using the solution.

8.16. For the case (or cases) listed in Prob. 8.14 and assigned by the instructor, compare the solution using A992 and A572 Grade 65 steels. Assume the Grade 65 costs 7% more per pound than A36. Indicate the economical solution to the problem. Assume ends simply supported for flexure but fixed for torsion.

8.17. Given the 40-ft simply supported span carrying two symmetrically placed concentrated loads, 20% dead load and 80% live load, as shown in the accompanying figure. If the loads are eccentric to the web by 5 in., and the member is torsionally simply supported, select the lightest W14 section suitable using the b modified flexure analogy method.

Problem 8.17

8.18. The 30-ft simply supported (for M_x) span is to carry two symmetrically placed concentrated loads of 8 kips dead load and 14 kips live load located 10 ft from the supports. The loads are 6-in. eccentric to the web, and full fixity is assumed for torsional restraint. Select the lightest W section using the β modified flexure analogy method.

Problem 8.18

8.19. Estimate, using the modified flexure analogy design procedure used for W shapes and crane girders, the uniformly distributed load capacity, 20% dead load and 80% live load, for a C15×50 of A36 steel. The ends of the beam are restrained against rotation ($\phi = 0$), but the ends are free to warp. Check stresses using the differential equation solution.

Problem 8.19

8.20. A simply supported beam is to carry 0.2 kip/ft dead load and 0.6 kip/ft live load on a span of 24 ft. A channel section is to be used, and since it is not laterally restrained, torsion is to be considered in the design. Assume the member is torsionally simply supported. After computing the torsional moment, use the generally accepted approximate flexure analogy to make a selection of a channel section from the *AISC Manual*. Investigate the stress on the selected section using the exact solution based on the stated loading and support condition. Use A36 steel.

Problem 8.20

8.21. Assume a single W section is to serve as a crane runway girder which carries a vertical loading, as shown. In addition, design must include an axial compressive force of 14 kips and a horizontal force of 4 kips on each wheel applied at $4\frac{1}{4}$ in. above the top of the compression flange. Assume torsional simple support at the ends of the beam. Select the lightest W14 section of A992 steel using the β modified flexure analogy approach. *Note:* All loads except weight of the crane runway girder are live loads.

8.22. Compare the torsional constants J for each of the sections of the accompanying figure. If the maximum shear stress is 14 ksi, compute the torsional moment capacity of each section. Can this be done for the W30×99? Explain.

Problem 8.21

(a)   (b)

W 30 × 99

Problem 8.22

8.23. Investigate the possibility of torsional buckling occurring on the following beams: (a) W16×31; (b) W6×16; (c) M8×6.5. For each, plot equivalent slenderness ratio r_E against length of column. Give conclusions.

8.24. Estimate the buckling load, assuming zero residual stress and elastic buckling so that Euler's equation applies, on the following sections for a pin-end length of 10 ft. At what length would torsional buckling be likely to control?

(a) MC10×6.5

(b) L4×4×$\frac{1}{4}$

(c) WT7×15

(d) Zee section shown

r_x = 1.62 in.
r_y = 1.33 in.
r_z = 0.67 in.
I_p = 10.51 in.^{4}

Problem 8.24

9

Lateral-Torsional Buckling of Beams

9.1 RATIONAL ANALOGY TO PURE COLUMNS

Emphasis in this chapter is on the lateral stability considerations associated with bending about the strong axis of the section. In beams, as in axially loaded columns, it is not possible to achieve perfect loading, i.e., beams are never perfectly straight, not perfectly homogeneous, and are usually not loaded in exactly the plane that is assumed for design and analysis.

Consider the compression zone of the laterally unsupported beam of Fig. 9.1.1. With the loading in the plane of the web, according to ordinary beam theory, points *A* and *B* are equally stressed. Imperfections in the beam and accidental eccentricity in loading actually result in different stresses at *A* and *B*. Furthermore, residual stresses as discussed in Chapter 6 contribute to unequal stresses across the flange width at any distance from the neutral axis.

Figure 9.1.1
Beam laterally supported only at its ends.

In a qualitative way one may look upon the compression flange of a beam as a column, with all the considerations treated in chapter 6. The rectangular flange as a column would ordinarily buckle in its weak direction, by bending about an axis such as 1–1 of Fig. 9.1.1b, but the web provides continuous support to prevent such buckling. At higher compressive loads the rectangular flange will tend to buckle by bending about axis 2–2 of Fig. 9.1.1b. It is this sudden buckling of the flange about its strong axis in a lateral direction that is commonly referred to as lateral buckling. The analogy between the compression flange of a beam and a column is intended to present only the general behavior for lateral buckling.

In order to evaluate this behavior more precisely, one must realize that the compression flange is not only braced in its weak direction by its attachment via the web to the stable tension flange, but the web also provides continuous restraint (rotational and transverse) along the junction of the flange and web (assuming the web and flange plates are continuously attached). Thus the bending stiffness of the web brings the entire section into action when lateral motion commences.

9.2 LATERAL SUPPORT

Rarely does a beam exist with its compression flange entirely free of all restraint. Even when it does not have a positive connection to a floor or roof system, there is still friction between the beam flange and whatever it supports. There are two categories of lateral support that are definite and adequate; these are:

1. Continuous lateral support by embedment of the compression flange in a concrete floor slab (Fig. 9.2.1a and b).

2. Lateral support at intervals (Fig. 9.2.1c through g) provided by cross beams, cross frames, ties, or struts, framing in laterally, where the lateral system is itself adequately stiff and braced.

It is necessary to examine not only the individual beam for adequate bracing, but also the entire system. Figure 9.2.2a shows beam *AB* with a cross beam framing in at midlength, but buckling of the entire system is still possible unless the system is braced, such as shown in Fig. 9.2.2b.

All too frequently in design, the engineer encounters situations that are none of these well-defined cases. A common unknown situation occurs when heavy beams have light-gage steel decking resistance welded to them; certainly providing a degree of restraint along the member. However the stiffness and lateral strength may be questioned. Other questionable cases are (a) where bracing frames into the beam in question, but it is at or near the tension flange; (b) timber or light-gage decking floor systems that rest on but are not solidly attached to the beams; (c) rigid frames that are enclosed in light-gage metal sheathing.

However, it is better to assume no lateral support in doubtful situations. Alternatively, it may be possible in some cases to evaluate it as an elastic restraint. Such an analytical approach is discussed in Sec. 9.13.

Lateral support must not be ignored; probably most failures in steel structures are the result of inadequate bracing against lateral instability of some type. The engineer is also reminded to consider carefully the construction stage when all of the restraints which may eventually act are not yet in place.

Large girder with stud shear connectors on top flange to be embedded in concrete to provide continuous lateral support (see Fig. 9.2.1b). (Photo by C. G. Salmon)

Figure 9.2.1
Types of definite lateral support.

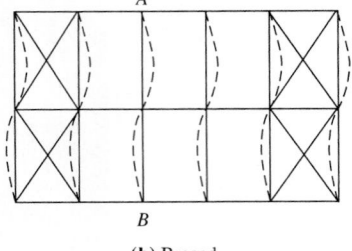

Figure 9.2.2
Lateral buckling of a roof
or floor system.

(a) Unbraced   (b) Braced

9.3 STRENGTH OF I-SHAPED BEAMS UNDER UNIFORM MOMENT

In the development of design equations. the case of constant (uniform) moment along a laterally unbraced length is usually used as the basic case for lateral-torsional buckling. Using the analogy of the compression flange as a column, the uniform moment causes constant compression in one flange over the entire unbraced length. When there is a moment gradient (variation in moment) the compression force in the flange varies along the unbraced length, resulting in a lower average compression force over that length. The lower average compression force means less likelihood of lateral-torsional buckling.

Lateral-torsional buckling is a limit state that may control the strength of a beam. The general behavior of a beam may be represented by Fig. 9.3.1 from Yura, Galambos, and Ravindra [9.1]. As discussed in Chapter 6, Part II, local buckling of the plate elements (such as the flange or web) in compression may limit the strength of a section. The maximum strength of a beam will be its plastic moment strength M_p as discussed in Chapter 7.

Whether or not the plastic moment strength is reached, failure will be one of the following modes:

1. Local buckling of the flange in compression.
2. Local buckling of the web in flexural compression.
3. Lateral-torsional buckling.

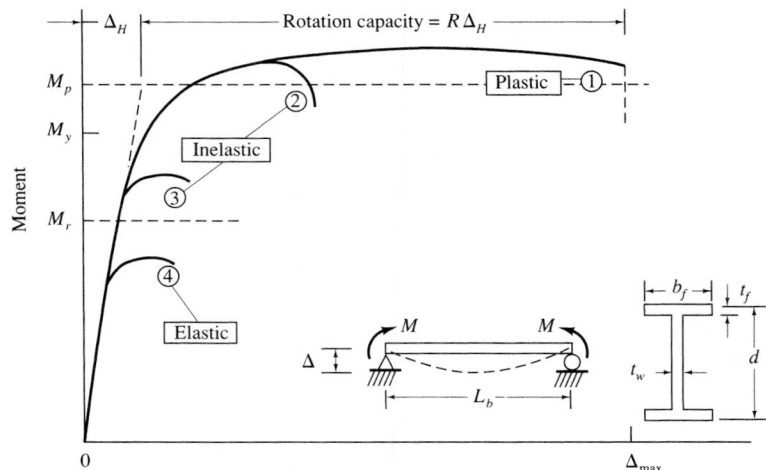

Figure 9.3.1
Beam behavior (From
Yura, Galambos, and
Ravindra [9.1])

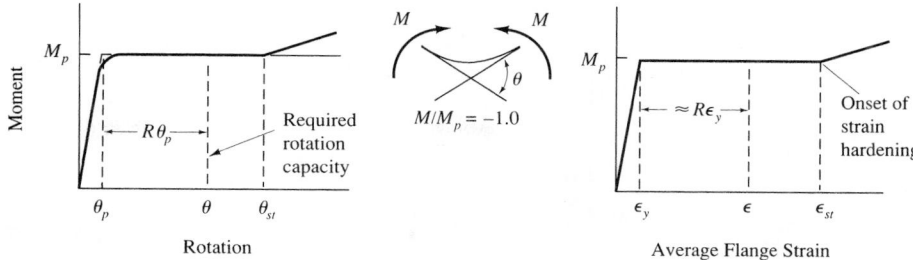

Figure 9.3.2
Deformation requirements for
developing plastic strength.

Four categories of behavior are represented in Fig. 9.3.1:

1. Plastic moment strength M_p is achieved *along with* large deformation. Deformation ability, called *rotation capacity* as shown in Fig. 9.3.2, is essentially the ability to undergo large flange strain without instability.

2. Inelastic behavior where plastic moment strength M_p is achieved but little rotation capacity is exhibited, because of inadequate stiffness of the flange and/or web to resist local buckling, or inadequate lateral support to resist lateral-torsional buckling, while the flange is inelastic.

3. Inelastic behavior where the moment strength M_r, the moment above which residual stresses cause inelastic behavior to begin, is reached or exceeded; however, local buckling of the flange or web, or lateral-torsional buckling prevent achieving the plastic moment strength M_p.

4. Elastic behavior where moment strength M_{cr} is controlled by elastic buckling; any or all of local flange buckling, local web buckling, or lateral-torsional buckling.

Most rolled W shapes have low enough slenderness ratios ($b_f/2t_f$ for flange and h/t_w or d/t_w for the web) such that they are categorized as "compact" as discussed in Chapter 7. For such cases, achievement of the plastic moment M_p depends on the laterally unbraced length L_b. A long laterally unbraced length will indicate moment strength M_{cr} controlled by *elastic* lateral-torsional buckling.

9.4   ELASTIC LATERAL-TORSIONAL BUCKLING

Because structural design requires a thorough understanding of stability, the following development is presented of the basic equation for the elastic lateral-torsional buckling strength M_{cr} of an I-shaped beam under the action of constant (uniform) moment acting in the plane of the web.

More detailed treatment of this uniform moment case as well as other common loading cases is to be found in Timoshenko and Gere [6.67], Bleich [6.9], de Vries [9.2], Hechtman et al. [9.3], Austin et al. [9.4], Clark and Jombock [9.5], Salvadori [9.6], and Galambos [9.7].

Differential Equation for Elastic Lateral-Torsional Buckling

Referring to Fig. 9.4.1, which shows the beam in a buckled position, it is observed that the applied moment M_0 in the yz plane will give rise to moment components $M_{x'}$, $M_{y'}$, and $M_{z'}$, about the x'-, y'-, and z'- axes, respectively. This means there will be bending curvature in both the $x'z'$ and $y'z'$ planes as well as torsional curvature about the z'-axis.

Figure 9.4.1
I-shaped beam in slightly
buckled position.

(a) Top view

(c)

Section A-A

(b) Side view

Direction Cosines			
	x	y	z
x'	1	ϕ	$-\dfrac{du}{dz}$
y'	$-\phi$	1	$-\dfrac{dv}{dz}$
z'	$\dfrac{du}{dz}$	$\dfrac{dv}{dz}$	1

Assuming small deformation, the bending in the $y'z'$ plane (considering the direction cosine is 1 between y'- and y-, and z'- and z-axes) may be written

$$EI_x \frac{d^2v}{dz^2} = M_{x'} = M_0 \tag{9.4.1}$$

where v is the displacement of the centroid in the y direction (see Fig. 9.4.1b).

Also, the curvature in the $x'z'$ plane is

$$EI_y \frac{d^2u}{dz^2} = M_{y'} = M_0\phi \tag{9.4.2}$$

as is seen from Fig. 9.4.1c, where u is the displacement of the centroid in the x direction.

The differential equation for torsion of I-shaped beams was developed in Chapter 8 as Eq. 8.5.9, as follows:

$$M_{z'} = GJ \frac{d\phi}{dz} - EC_w \frac{d^3\phi}{dz^3} \tag{8.5.10}$$

From Fig. 9.4.1 and the direction cosines, the torsional component of M_0 when the beam is slightly buckled is proportional to the slope of the beam in the xz plane:

$$M_{z'} = -\frac{du}{dz} M_0 \tag{9.4.3}$$

which gives for the torsional differential equation

$$-\frac{du}{dz}M_0 = GJ\frac{d\phi}{dz} - EC_w\frac{d^3\phi}{dz^3} \tag{9.4.4}$$

Two assumptions are inherent in Eqs. 9.4.1 and 9.4.2, both of which relate to the assumption of small deformation. It is assumed that properties $I_{x'}$ and $I_{y'}$ equal I_x and I_y, respectively; and also the I_x is large compared to I_y, so that Eq. 9.4.1 is not linked to Eqs. 9.4.2 and 9.4.4. Thus displacement v in the plane of bending does not affect the torsional function ϕ.

Differentiating Eq. 9.4.4 with respect to z gives

$$-\frac{d^2u}{dz^2}M_0 = GJ\frac{d^2\phi}{dz^2} - EC_w\frac{d^4\phi}{dz^4} \tag{9.4.5}$$

From Eq. 9.4.2,

$$\frac{d^2u}{dz^2} = \frac{M_0\phi}{EI_y}$$

which when substituted into Eq. 9.4.5 gives

$$EC_w\frac{d^4\phi}{dz^4} - GJ\frac{d^2\phi}{dz^2} - \frac{M_0^2}{EI_y}\phi = 0 \tag{9.4.6}$$

which is the differential equation for the angle of twist.

To obtain a solution for Eq. 9.4.6, divide by EC_w and let

$$2\alpha = \frac{GJ}{EC_w} \quad \text{and} \quad \beta = \frac{M_0^2}{E^2C_wI_y} \tag{9.4.7}$$

Equation 9.4.6 then becomes

$$\frac{d^4\phi}{dz^4} - 2\alpha\frac{d^2\phi}{dz^2} - \beta\phi = 0 \tag{9.4.8}$$

Let

$$\left.\begin{array}{l} \phi = Ae^{mz} \\[4pt] \dfrac{d^2\phi}{dz^2} = Am^2e^{mz} \\[4pt] \dfrac{d^4\phi}{dz^4} = Am^4e^{mz} \end{array}\right\} \tag{9.4.9}$$

Substitution of Eqs. 9.4.9 into Eq. 9.4.8 gives

$$Ae^{mz}(m^4 - 2\alpha m^2 - \beta) = 0 \tag{9.4.10}$$

Since e^{mz} cannot be zero and A can be zero only if no buckling has occurred (a trivial solution) the bracket expression of Eq. 9.4.10 must be zero:

$$m^4 - 2\alpha m^2 - \beta = 0$$

which gives for the solution

$$m^2 = \alpha \pm \sqrt{\beta + \alpha^2}$$

or

$$m = \pm\sqrt{\alpha \pm \sqrt{\beta + \alpha^2}} \tag{9.4.11}$$

It is apparent from Eq. 9.4.11 that m will consist of two real and two complex roots because

$$\sqrt{\beta + \alpha^2} > \alpha$$

Let

$$n^2 = \alpha + \sqrt{\beta + \alpha^2} \quad \text{(both real roots)} \tag{9.4.12}$$

$$q^2 = -\alpha + \sqrt{\beta + \alpha^2} \quad \text{(real part of complex roots)} \tag{9.4.13}$$

Using the four values for m, the expression for ϕ from Eq. 9.4.9 becomes

$$\phi = A_1 e^{nz} + A_2 e^{-nz} + A_3 e^{iqz} + A_4 e^{-iqz} \tag{9.4.14}$$

The complex exponential functions may be expressed in terms of circular functions,

$$\left.\begin{array}{l} e^{iqz} = \cos qz + i \sin qz \\ e^{-iqz} = \cos qz - i \sin qz \end{array}\right\} \tag{9.4.15}$$

By using Eqs. 9.4.15 and defining new constants A_3 and A_4 which equal $(A_3 + A_4)$ and $(A_3 i - A_4 i)$, respectively, one obtains

$$\phi = A_1 e^{nz} + A_2 e^{-nz} + A_3 \cos qz + A_4 \sin qz \tag{9.4.16}$$

The constants A_1 and A_4 are determined by the end support conditions. For the case of torsional simple support, i.e., beam ends may not twist, but are free to warp, the conditions are

$$\phi = 0, \quad \frac{d^2\phi}{dz^2} = 0 \quad \text{at } z = 0 \quad \text{and} \quad z = L$$

For $\phi = 0$ at $z = 0$, Eq. 9.4.16 gives

$$0 = A_1 + A_2 + A_3 \tag{9.4.17}$$

For $d^2\phi/dz^2 = 0$ at $z = 0$,

$$0 = A_1 n^2 + A_2 n^2 - A_3 q^2 \tag{9.4.18}$$

Multiplying Eq. 9.4.17 by n^2 and subtracting Eq. 9.4.18 gives

$$0 = A_3(q^2 + n^2), \qquad \therefore \underline{A_3 = 0}$$

Then, from Eq. 9.4.17,

$$A_1 = -A_2 \tag{9.4.19}$$

Thus Eq. 9.4.16 becomes

$$\phi = A_1(e^{nz} - e^{-nz}) + A_4 \sin qz \tag{9.4.20}$$

which may be written

$$\phi = 2A_1 \sinh nz + A_4 \sin qz \tag{9.4.21}$$

At $z = L$, $\phi = 0$; therefore, from Eq. 9.4.21,

$$0 = 2 A_1 \sinh nL + A_4 \sin qL \tag{9.4.22}$$

Also, at $z = L$, $d^2\phi/dz^2 = 0$, which gives

$$0 = 2A_1 n^2 \sinh nL - A_4 q^2 \sin qL \tag{9.4.23}$$

Multiplying Eq. 9.4.22 by q^2 and adding to Eq. 9.4.23 gives

$$2A_1(n^2 + q^2)\sinh nL = 0 \tag{9.4.24}$$

Since $(n^2 + q^2)$ cannot be zero, and sinh nL can be zero only if $n = 0$, therefore A_1 must be zero:

$$A_1 = -A_2 = 0$$

Finally, from Eq. 9.4.21,

$$\phi = A_4 \sin qL = 0 \tag{9.4.25}$$

If lateral-torsional buckling occurs, A_4 cannot be zero, so that

$$\sin qL = 0$$
$$qL = N\pi \tag{9.4.26}$$

where N is any integer.

The elastic buckling condition is defined by

$$q = \frac{N\pi}{L} \tag{9.4.27}$$

which for the fundamental buckling mode $N = 1$.

The value of M_0 which satisfies Eq. 9.4.27 is said to be the critical moment:

$$q = \sqrt{-\alpha + \sqrt{\beta + \alpha^2}} = \frac{\pi}{L} \tag{9.4.28}$$

Squaring both sides, and substituting the definitions of α and β, from Eqs. 9.4.7,

$$-\frac{GJ}{2EC_w} + \sqrt{\frac{M_0^2}{E^2 C_w I_y} + \left(\frac{GJ}{2EC_w}\right)^2} = \frac{\pi^2}{L^2} \tag{9.4.29}$$

Solving for $M_0 = M_{cr}$ gives

$$M_{cr}^2 = E^2 C_w I_y \left[\left(\frac{\pi^2}{L^2} + \frac{GJ}{2EC_w}\right)^2 - \left(\frac{GJ}{2EC_w}\right)^2\right] \tag{9.4.30}$$

$$M_{cr} = \sqrt{\frac{\pi^2 EI_y GJ}{L^2} + \frac{\pi^4 E^2}{L^4} I_y C_w} \tag{9.4.31}$$

Factoring π/L from inside the root sign gives

$$M_{cr} = \frac{\pi}{L}\sqrt{EI_y GJ + \left(\frac{\pi E}{L}\right)^2 I_y C_w} \tag{9.4.32}$$

Equation 9.4.32 is the elastic lateral-torsional buckling strength for an I-shaped section under the action of constant moment in the plane of the web over the laterally unbraced length L. To adjust for moment gradient, Eq. 9.4.32 may be multiplied by a factor C_b. Thus, in general

$$M_{cr} = C_b \frac{\pi}{L}\sqrt{EI_y GJ + \left(\frac{\pi E}{L}\right)^2 I_y C_w} \tag{9.4.33}$$

and the elastic lateral torsional buckling stress can be expressed as

$$F_{cr} = \frac{M_{cr}}{S_x} = \frac{C_b \pi}{LS_x}\sqrt{EI_y GJ + \left(\frac{\pi E}{L}\right)^2 I_y C_w} \tag{9.4.34}$$

9.5 INELASTIC LATERAL-TORSIONAL BUCKLING

When the moment strength is based on some of the fibers of the cross-section reaching a strain ϵ (see Fig. 9.3.2) that is greater than ϵ_y (that is $\epsilon > F_y/E$), buckling is more likely to occur than when the strain $\epsilon < \epsilon_y$. When elements are inelastic the stiffness as related to the modulus of elasticity decreases; therefore, buckling strength decreases. The larger the strain requirement the lower must be the slenderness ratios related to the various types of buckling.

Studies of inelastic lateral-torsional buckling have been made by Galambos [9.8], Lay and Galambos [9.9, 9.10], Massey and Pitman [9.11], Trahair and Kitipornchai [9.12], Hartmann [9.13], and Ojalvo and Weaver [9.14]. Lateral-torsional buckling as used in Load and Resistance Factor Design is summarized by Yura et al. [9.1].

The strength of I-shaped beams as related to laterally unbraced length L_b as studied by Nethercot and Trahair [9.15] and reported by Yura et al. [9.1] is shown in Fig. 9.5.1. Though the torsional stiffness is not greatly affected by residual stress, the column-related strength is very much affected, as discussed in Chapter 6. In the presence of residual stress, the maximum *elastic* moment strength M_r is given by

$$M_r = S_x(F_y - F_r) \qquad (9.5.1)^*$$

For the same reasons (such as unknown residual stress magnitude and variation, accidental eccentricity, and initial crookedness) as for inelastic column buckling, the range between M_r and M_p is not readily analyzed. Note from Fig. 9.5.1 that when there is uniform moment the decrease in strength due to residual stress is greatest.

When the bending moment acts with a gradient having maximum moment at one end to one-half that amount at the other end the residual stress effect is relatively insignificant. In this case only a small portion of the span near maximum moment will be inelastic; when the applied load stress is lower, the addition of residual stress will be less likely to cause a fiber to reach yield stress.

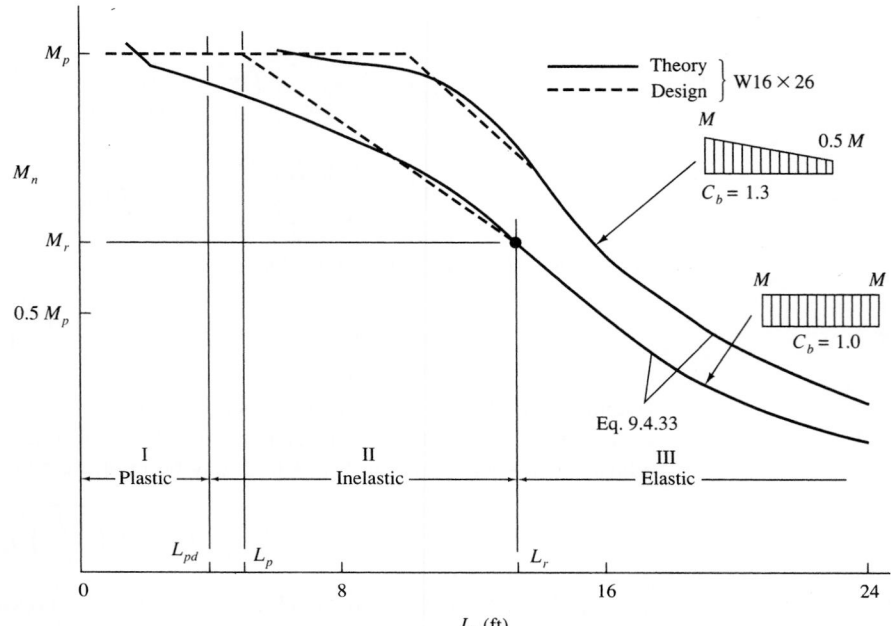

Figure 9.5.1
Beam behavior as related to lateral support. (From Yura, Galambos, and Ravindra [9.1])

---

*Note that AISC-F2.2 uses $0.7F_y$ in place of $(F_y - F_r)$ to compute M_r.

Lateral Bracing Requirements In the Inelastic Range

To gain an idea of the type of lateral bracing requirements that are needed to achieve the strength and rotation capacity, one may use the elastic lateral-torsional buckling equation, Eq. 9.4.31, but redefine the rigidities EI_y and GJ to include values in the inelastic range. Since generally lateral support will be provided at locations where the plastic moment M_p is expected to occur, and the distances between lateral support points will be relatively short, it has been determined [9.10] that the term involving torsional rigidity GJ may be neglected. Thus, Eq. 9.4.31, neglecting the first term, becomes

$$M_{cr} = \frac{\pi^2 E}{L^2}\sqrt{C_w I_y} \tag{9.5.2}$$

Since M_{cr} must reach M_p, substitute $M_p = Z_x F_y$ for M_{cr}. Also, $C_w = I_y h^2/4$, $I_y = A r_y^2$, and replace L by L_b representing laterally unbraced length. Equation 9.5.2 then gives the maximum slenderness ratio L_b/r_y to achieve M_p,

$$\frac{L_b}{r_y} \leq \sqrt{\frac{\pi^2 E}{2F_y}\left(\frac{hA}{Z_x}\right)} \tag{9.5.3}$$

which applies for uniform bending moment. An upper limit to Eq. 9.5.3 is obtained if one assumes perfect elasto-plastic steel without residual stress. Taking in Eq. 9.5.3 a conservative (low) value of 1.5 for $\sqrt{hA/Z_x}$ gives (the numerical 570 arises from $E = 29{,}000$ ksi)

$$\frac{L_b}{r_y} \leq \left[\sqrt{\frac{\pi^2 E(1.5)}{2F_y}} = 2.7\sqrt{\frac{E}{F_y}} = \frac{570}{\sqrt{F_y, \text{ksi}}}\right] \tag{9.5.4}$$

Experimental data [9.16, 9.17] show that a lower limit than Eq. 9.5.4 is necessary to achieve adequate rotation capacity R (Fig. 9.3.2). Thus AISC-F2.2 has set the limit L_p/r_y as

$$\frac{L_p}{r_y} = 1.76\sqrt{\frac{E}{F_y}} = \frac{300}{\sqrt{F_y, \text{ksi}}} \tag{9.5.5}*$$

The foregoing is for I-shaped members bent about the x-axis (i.e., strong axis) where the limit state is achievement of plastic moment strength M_p.

When the ability to absorb additional plastic strain is desired, the limit must be lower. In terms of Fig. 9.3.2, L_b/r_y at the Eq. 9.5.5 limit will give rotation capacity somewhat greater than θ_p. When more rotation is required as when plastic analysis, as discussed in Chapter 10, is used the limit must be reduced. The *AISC Specification* [1.13] is based on a rotation capacity factor R (Fig. 9.3.2) approximately 3 when plastic analysis is used.

The elastic modulus of elasticity E in Eq. 9.5.3 should be replaced by the strain-hardening modulus of elasticity E_{st} when the actual strain ϵ approaches the strain hardening strain. Lay and Galambos [9.9] suggested that E_{st} can be approximated as E/F_y. When that is done, evaluating Eq. 9.5.3 using $\sqrt{hA/Z_x} = 1.5$ and L_b for L gives,

$$\frac{L_b}{r_y} \leq \frac{5600}{F_y, \text{ksi}} \tag{9.5.6}$$

---

*For SI, with F_y in MPa and $E = 200{,}000$ MPa,

$$\frac{L_p}{r_y} = 1.76\sqrt{\frac{E}{F_y}} = \frac{790}{\sqrt{F_y, \text{MPa}}} \tag{9.5.5}$$

which makes the limit vary inversely in a linear manner with F_y instead of inversely as the square root of F_y. Again, there are neglected strength factors such as torsional stiffness and end restraint. Tests by Bansal [9.16] at the University of Texas have established the limit using the same form of Eq. 9.5.6, additionally including provision for moment gradient, as

$$\frac{L_b}{r_y} \le \left[0.12 + 0.076\left(\frac{M_1}{M_p}\right)\right]\left(\frac{E}{F_y}\right) = \frac{3500 + 2200M_1/M_p}{F_y, \text{ksi}} \qquad (9.5.7)*$$

where M_1 = smaller moment at the ends of the laterally unbraced segment (taken positive when moments cause *reverse* curvature).

9.6 LOAD AND RESISTANCE FACTOR DESIGN—I-SHAPED BEAMS SUBJECTED TO STRONG-AXIS BENDING

This section considers the full range of strength from laterally stable beams to situations where lateral-torsional buckling causes considerable strength reduction. Loading in the plane of the web is assumed.

The strength requirement according to AISC-B3.3 may be stated

$$\phi_b M_n \ge M_u \qquad (9.6.1)$$

where ϕ_b = strength reduction factor for flexure = 0.90
M_n = nominal moment strength
M_u = factored service load moment (see Sec. 1.8)

Figure 9.6.1 shows the effect of laterally unbraced length L_b on the lateral-torsional buckling strength. Of course, local buckling may result in lower moment strength M_n if the plate element (flange or web) width/thickness ratio is too high.

Case 1: Plastic Moment is Reached ($M_n = M_p$) Along With Large Plastic Rotation Capacity ($R \ge 3$ in Fig. 9.3.2)

The section must be "compact" to prevent local buckling; that is, λ for the flange ($b_f/2t_f$) and for the web (h/t_w) must not exceed λ_p as discussed in Sec. 6.17 (values in Table 9.6.1) and lateral bracing must be provided such that the laterally unbraced length L_b does not exceed L_{pd}, where from Eq. 9.5.7.

$$\frac{L_{pd}}{r_y} \le \left[0.12 + 0.076\left(\frac{M_1}{M_p}\right)\right]\left(\frac{E}{F_y}\right) = \frac{3500 + 2200M_1/M_p}{F_y, \text{ksi}} \qquad (9.6.2)$$

AISC-Appendix 1.7 uses M_2 instead of M_p; however, M_2 will always be M_p under Case 1. In this category, plastic analysis involving redistribution of moments as discussed in Chapter 10 is an option.

---

*For SI, with F_y in MPa and $E = 200,000$ MPa,

$$\frac{L_b}{r_y} \le \left[0.12 + 0.076\left(\frac{M_1}{M_p}\right)\right]\left(\frac{E}{F_y}\right) = \frac{24,000 + 15,200M_1/M_p}{F_y, \text{MPa}} \qquad (9.5.7)$$

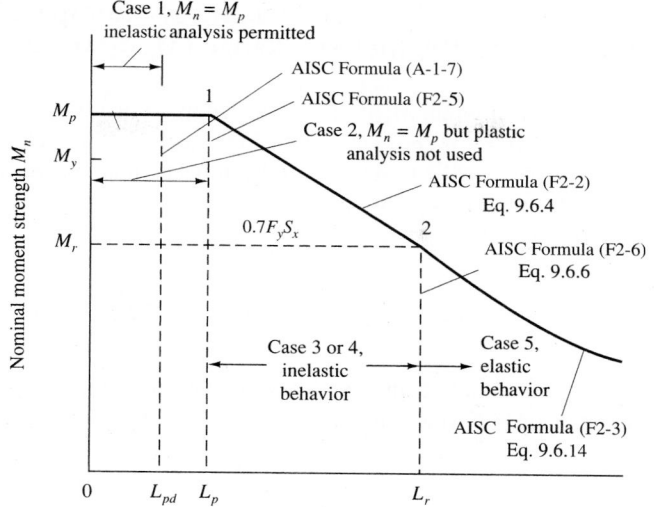

Figure 9.6.1
Nominal strength M_n of "compact" sections as affected by lateral-torsional buckling.

TABLE 9.6.1 Slenderness Ratio Limits λ_p for "Compact" I-Shaped Beams to Achieve Plastic Moment Strength M_p (AISC-Table B4.1)[†]

Yield stress F_y	Flange local buckling $\dfrac{b_f}{2t_f} = 0.38\sqrt{\dfrac{E}{F_y}}$ $= \dfrac{65}{\sqrt{F_y}}$	Web local buckling $\dfrac{h}{t_w} = 3.76\sqrt{\dfrac{E}{F_y}}$ $= \dfrac{640}{\sqrt{F_y}}$	Lateral-torisional buckling $\dfrac{L_b}{r_y} = 1.76\sqrt{\dfrac{E}{F_y}}$ $= \dfrac{300}{\sqrt{F_y}}$
36	10.8	107	50.0
42	10.0	98.8	46.3
45	9.7	95.4	44.7
50	9.2	90.5	42.4
55	8.8	86.3	40.5
60	8.4	82.6	38.7
65	8.1	79.4	37.2

[†]Taking $E = 29{,}000$ ksi and F_y in ksi

Case 2: Plastic Moment is Reached ($M_n = M_p$) But With Relatively Little Rotation Capacity ($R < 3$ in Fig. 9.3.2)

The section must be "compact" to prevent local buckling; that is, λ for the flange $(b_f/2t_f)$ and for web (h/t_w) must not exceed λ_p as discussed in Sec. 6.17 and lateral bracing must be provided such that λ (i.e., L_b/r_y) does not exceed λ_p. The three slenderness limits for various yield stresses are given in Table 9.6.1. This latter limit λ_p, from Eq. 9.5.5, means L_b may not exceed L_p when $C_b = 1.0$.

$$\frac{L_p}{r_y} = 1.76\sqrt{\frac{E}{F_y}} = \frac{300}{\sqrt{F_y, \text{ksi}}} \qquad (9.6.3)$$

Case 3: Lateral-Torsional Buckling of "Compact" Sections May Occur in the Inelastic Range ($M_p > M_n \geq M_r$)

This moment strength M_n is approximated as a linear relationship between points 1 (M_p at L_p) and 2 ($M_r = 0.7F_yS_x$ at L_r) on Fig. 9.6.1. Local buckling must be precluded. Since nearly all of the rolled sections are "compact," that is, they have $\lambda \leq \lambda_y$ as discussed in Sec. 6.16, the nominal strength M_x is a linear function of the lateral-torsional buckling strength. Thus, taking $M_r = 0.7F_yS_x$,

$$M_n = C_b\left[M_p - (M_p - 0.7F_yS_x)\left(\frac{L_b - L_p}{L_r - L_p}\right)\right] \leq M_p \qquad (9.6.4)$$

where M_r is the moment strength available for service loads when extreme fiber reaches the yield stress F_y (including the residual stress), and may be expressed

$$M_r = 0.7F_yS_x \qquad (9.6.5)$$

where F_y = minimum specified yield stress for the steel
  S_x = elastic section modulus = $I_x/(d/2)$
  I_x = moment of inertia about the x-axis
  d = overall depth of the section

The length L_r is obtained by equating the maximum elastic moment strength M_r (Eq. 9.6.5) to the elastic lateral-torsional buckling strength M_{cr} (Eq. 9.4.33) and solving for L. Upon performing that operation, one obtains L_r (L in Eq. 9.4.33) as given by AISC-F2.2b,

$$L_r = 1.95r_{ts}\frac{E}{0.7F_y}\sqrt{\frac{Jc}{S_xh_o}}\sqrt{1 + \sqrt{1 + 6.76\left(\frac{0.7F_y}{E}\frac{S_xh_o}{Jc}\right)^2}} \qquad (9.6.6)$$

where $r_{ts}^2 = \dfrac{\sqrt{I_yC_w}}{S_x}$ $\qquad\qquad\qquad\qquad\qquad\qquad (9.6.7)$

$c = 1$ $\qquad\qquad\qquad$ for a doubly symmetric I-shape

$c = \dfrac{h_o}{2}\sqrt{\dfrac{I_y}{C_w}}$ $\qquad$ for a channel

h_o = distance between the flange centroids, in.

It is noted that r_{ts} may be approximated as the radius of gyration of the compression flange plus one-sixth of the web, as stated in the *User Note* of AISC-F2.2:

$$r_{ts} = \frac{b_f}{\sqrt{12\left(1 + \dfrac{1}{6}\dfrac{ht_w}{b_ft_f}\right)}} \qquad (9.6.8)$$

Case 4: General Limit State Where Nominal Moment Strength M_u Occurs in the Inelastic Range ($M_p > M_n \geq M_r$)

This condition is relatively uncommon for rolled shapes having $F_y \leq 50$ ksi, because only eleven sections are in that category, as per AISC-F3 *User Note*.

With noncompact or slender flanges, the nominal flexural strength M_u is the lower value obtained according to the limit states of lateral-torsional buckling (as discussed for the previous two Cases) and compression flange local buckling. When $\lambda_p < (\lambda = b_f/2t_f) < \lambda_r$ the strength will be in this category. These slenderness limits are given for various yield stresses in Table 9.6.2. AISC has used the symbol λ to represent the general stability slenderness: $\lambda = L_b/r_y$ when considering lateral-torsional buckling; $\lambda = b_f/2t_f$ when considering flange local buckling; and $\lambda = h/t_w$ when considering web local buckling.

The variation in strength with the generalized slenderness ratio λ is shown in Fig. 9.6.2. When λ lies between λ_p and λ_r for either of the two limit states considered

TABLE 9.6.2 Slenderness Ratio Limits λ_r for "Noncompact Section" I-Shaped Beams to Achieve F_y at Extreme Fiber (AISC-B4.1)

Yield stress F_y (ksi)	Flange local buckling Rolled I-shape $\dfrac{b_f}{2t_f} = 1.0\sqrt{\dfrac{E}{F_y}}$ $= \dfrac{170}{\sqrt{F_y}}$	$\dfrac{h}{t_w}$	k_c^*	Flange local buckling Welded I-shape $\dfrac{b_f}{2t_f} = 0.95\sqrt{\dfrac{k_cE}{F_L^{**}}}$ $= \dfrac{162}{\sqrt{0.7F_y^{**}/k_c}}$	Web local buckling $\dfrac{h}{t_w} = 5.70\sqrt{\dfrac{E}{F_y}}$ $= \dfrac{970}{\sqrt{F_y}}$
36	27.7	161.7	0.35	19.1	161.7
		100	0.40	20.4	
		40	0.63	25.6	
42	24.9	149.7	0.35	17.7	149.7
		100	0.40	18.9	
		40	0.63	23.7	
45	23.8	144.6	0.35	17.1	144.6
		100	0.40	18.2	
		40	0.63	22.9	
50	22.3	137.2	0.35	16.2	137.2
		100	0.40	17.3	
		40	0.63	21.7	
55	21.0	130.8	0.35	15.4	130.8
		100	0.40	16.5	
		40	0.63	20.7	
60	19.9	125.2	0.36	15.0	125.2
		100	0.40	15.8	
		40	0.63	19.8	
65	19.0	120.3	0.36	14.4	120.3
		100	0.40	15.2	
		40	0.63	19.0	
90	15.8	102.2	0.40	12.9	102.2
		100	0.40	12.9	
		40	0.63	16.2	
100	14.9	97.0	0.41	12.4	97.0
		100	0.40	12.2	
		40	0.63	15.3	

$^*k_c = \dfrac{4}{\sqrt{h/t_w}}$, where $0.35 \leq k_c \leq 0.763$

$^{**}F_L = 0.7F_y$

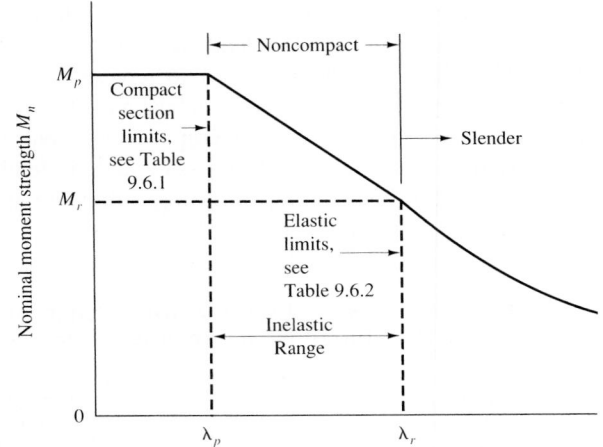

Figure 9.6.2
Nominal strength M_n vs
generalized slenderness ratio λ
for limit states of flange local
buckling, web local buckling,
and lateral-torsional buckling.

by AISC (web local buckling is excluded here) (1) flange local buckling, or (2) lateral-torsional buckling, the relationship for nominal moment strength is taken linear. For the limit state of flange local buckling, AISC-F3 prescribes

$$M_n = M_p - (M_p - 0.7F_yS_x)\left(\frac{\lambda - \lambda_{pf}}{\lambda_{rf} - \lambda_{pf}}\right) \qquad (9.6.9)$$

where $\lambda = b_f/2t_f$

         $\lambda_{pf} =$ limiting slenderness λ_p for a compact flange, AISC-Table B4.1

         $\lambda_{rf} =$ limiting slenderness λ_r for a noncompact flange, AISC-Table B4.1

For the limit state of lateral-torsional buckling, the linear relationship of Eq. 9.6.9 is used; however, the result is increased by C_b when there is moment gradient. Of course, the maximum M_u is M_p regardless of how steep is the moment gradient. The expression according to AISC-F3 is identical to Eq. 9.6.4, except expressed in the form of the ratios λ,

$$M_n = C_b\left[M_p - (M_p - 0.7F_yS_x)\left(\frac{\lambda - \lambda_{pf}}{\lambda_{rf} - \lambda_{pf}}\right)\right] \le M_p \qquad (9.6.10)$$

where

$$C_b = \frac{12.5M_{\max}}{2.5M_{\max} + 3M_A + 4M_B + 3M_C}R_m \le 3.0 \qquad (9.6.11)$$

where $C_b =$ a modification factor for non-uniform bending moment variation for a beam segment laterally unbraced except at the segment ends, based on *absolute values* of bending moments (C_b can always conservatively be taken as 1.0)

    $M_{\max} =$ maximum moment in the unbraced segment

      $M_A =$ moment at 1/4 point of unbraced segment

      $M_B =$ moment at midpoint of unbraced segment

      $M_C =$ moment at 3/4 point of unbraced segment

      $R_m =$ a new term in the 2005 AISC Specification to account for possible mono symmetry in the cross-section. The value is 1.0 for doubly symmetric sections and for singly symmetric sections in single curvature bending; reverse curvature of singly symmetric sections uses a reduction expression (given in AISC-F1)

Equation 9.6.11 (AISC-Formula F1-1) was introduced in 1993 (without the maximum value of 3.0 and without the factor R_m), replacing the following equation which had been used to adjust for non-uniform moment since 1961,

$$C_b = 1.75 + 1.05\left(\frac{M_1}{M_2}\right) + 0.3\left(\frac{M_1}{M_2}\right)^2 \leq 2.3 \qquad (9.6.12)$$

The newer equation, Eq. 9.6.11, based on the work of Kirby and Nethercot [9.18], allows a more accurate determination of the modification factor C_b when the unbraced segment has a nonlinear moment variation; however, when the precision with which the loading, the structural model, and the material properties are known, the more detailed equation, Eq. 9.6.11, seems questionable.

The AISC Commentary-F1 says, "It is still satisfactory to use C_b from Equation C-F1-1 for straight-line moment diagrams within the unbraced length." (*Author note;* that is, text Eq. 9.6.12.) In Eq. 9.6.12, the ratio M_1/M_2 is negative when the moments cause single curvature; that is, the most severe loading with constant bending moment gives $C_b = 1.0$. A comparison of Eqs. 9.6.11 and 9.6.12 is shown in Table 9.6.3 and Fig. 9.6.3.

According to AISC-F1, C_b is permitted to be taken as 1.0 for all cases; consistent with assuming the most severe loading (constant bending moment). For cantilevers or overhangs where the free end is unbraced, C_b is prescribed to be taken as 1.0.

For laterally unbraced segments having a parabolic moment variation, Eq. 9.6.11 is superior (and more conservative) to using Eq. 9.6.12. Table 9.6.4 gives the 2005 design values from Eq. 9.6.11 for several situations of parabolic unbraced segments.

Case 5: General Limit State Where Nominal Moment Strength M_n Equals the Elastic Buckling Strength M_{cr} (i.e., $M_n < M_r$, where $M_r = 0.7F_yS_x$)

There are two subcategories for this case; (1) slenderness ratios λ for flange or web local buckling *do not exceed* λ_r; and (2) slenderness ratios exceed λ_r. When λ exceeds λ_r, local buckling will occur prior to the extreme fiber reaching the yield stress and the efficiency of the cross-section is reduced. Such elements are known as "slender" compression elements; a subject treated in Sec. 6.18.

TABLE 9.6.3 Comparison of C_b for Linear Moment Variation

$\dfrac{M_1}{M_2}$	C_b 2005 Eq. 9.6.11	C_b 1986 Eq. 9.6.12
−1.00	1.00	1.00
−0.75	1.11	1.13
−0.50	1.25	1.30
−0.25	1.43	1.51
0.00	1.67	1.75
0.25	2.00	2.03
0.50	2.17	2.30
0.75	2.22	2.30
1.00	2.27	2.30

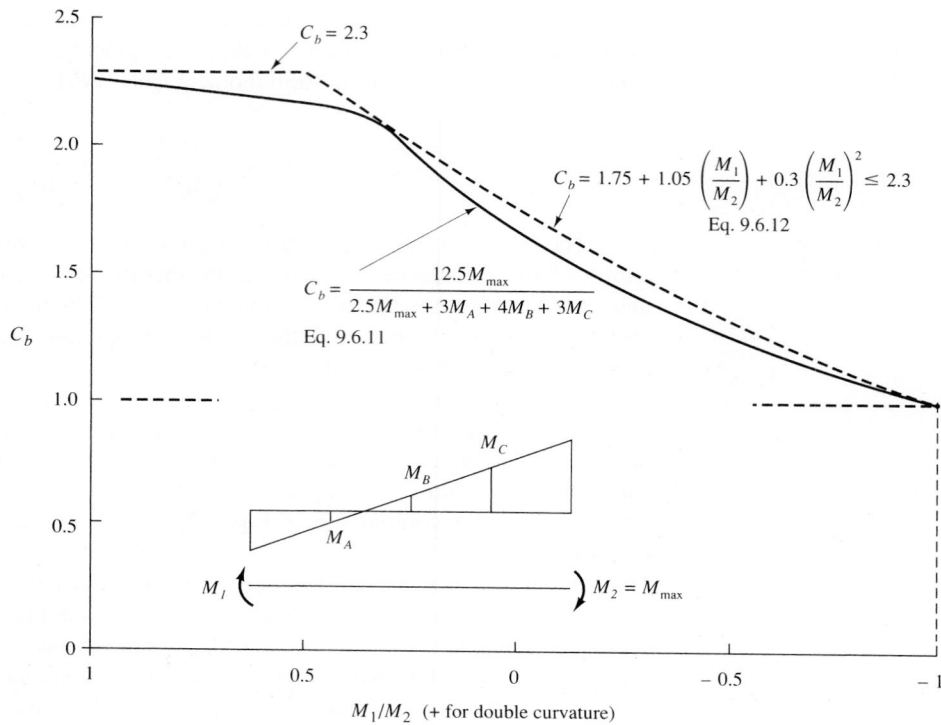

Figure 9.6.3
Comparison of C_b equations for linear variation of moment over laterally unbraced length.

TABLE 9.6.4 C_b for Parabolic Segments Using AISC-Commentary-F1, Formula (C-F1-2), Eq. 9.6.11*

Case 1	Laterally braced at ends; points 1 and 5 only; M_{max} at 3	$C_b = 1.14$	
Case 2	Laterally braced at ends and midspan; points 1, 3, and 5 only; M_{max} at 3	$C_b = 1.30$	
Case 3	Laterally braced at end and 1st quarter point; bracing at points 1 and 2; M_{max} at 2	$C_b = 1.52$	
Case 4	Laterally braced at 1st and 2nd quarter points; bracing at points 2 and 3; M_{max} at 3	$C_b = 1.06$	
Case 5	Laterally braced at 1st and 3rd quarter points; bracing at points 2 and 4; M_{max} at 3	$C_b = 1.03$	

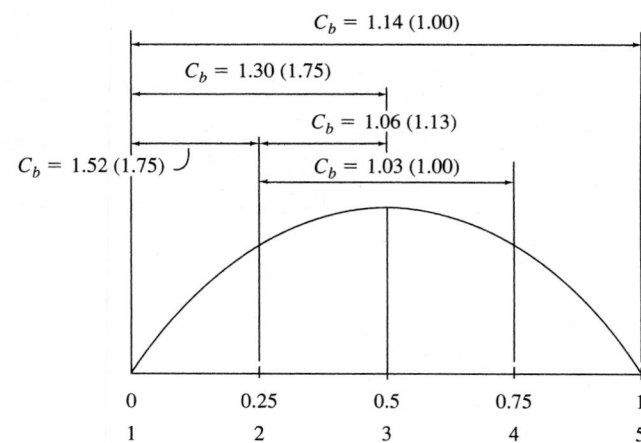

*Values from 1986 LRFD, Eq. 9.6.12 shown in parenthesis.

All rolled I-shaped sections have $\lambda < \lambda_r$ for local buckling; therefore *beams* (as opposed to plate girders) having $\lambda > \lambda_r$ are not treated in this text. When $\lambda < \lambda_r$ for local buckling (see Table 9.6.2), the nominal critical stress F_{cr} is given by Eq. 9.4.34 (Also AISC-F2.2b),

$$F_{cr} = \frac{C_b \pi}{LS_x} \sqrt{EI_y GJ + \left(\frac{\pi E}{L}\right)^2 I_y C_w} \qquad [9.4.34]$$

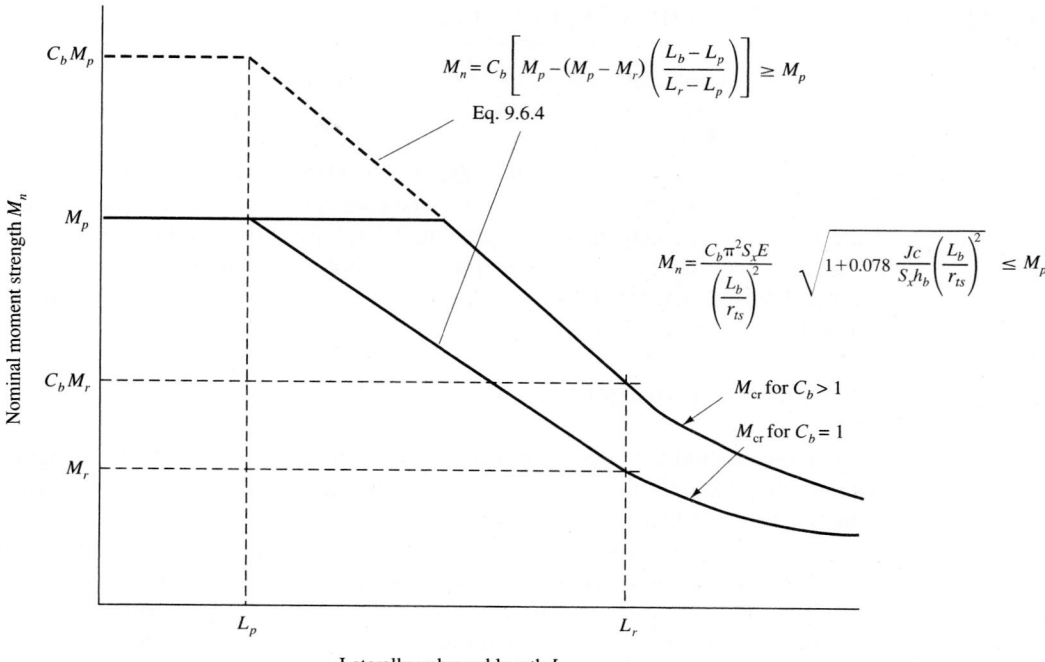

Figure 9.6.4
Nominal moment strength as affected by C_b.

Using the properties r_{ts} and h_o, defined in Eqs. 9.6.7 and 9.6.8, Eq 9.4.34 becomes

$$F_{cr} = \frac{C_b \pi^2 E}{\left(\dfrac{L_b}{r_{ts}}\right)^2} \sqrt{1 + 0.078 \frac{Jc}{S_x h_o}\left(\frac{L_b}{r_{ts}}\right)^2} \qquad (9.6.13)$$

and the nominal moment strength is given as:

$$M_n = F_{cr} S_x \le M_p \qquad (9.6.14)$$

Note that M_{cr} cannot logically be greater than $C_b M_r$, nor greater than M_p; therefore, Eq. 9.6.13 has those as the meaningful upper limits. Fig. 9.6.4 shows the nominal strength M_n variation with L_b, emphasizing the effect of C_b.

9.7 ALLOWABLE STRENGTH DESIGN METHOD—I-SHAPED BEAMS SUBJECTED TO STRONG-AXIS BENDING

The general philosophy of Allowable Strength Design (ASD) for beams was presented in Sec. 7.5. The nominal strength expressions for M_n may be divided by the flexural safety factor Ω_b, to determine the available strength. The same limit states defined in the previous section for the LRFD Method are also used for the ASD Method. The beam design tables and charts in the new *AISC Manual* (2005) allow for using the required moment M_a based on ASD load combination in the factor of safety.

9.8   EFFECTIVE LATERALLY UNBRACED LENGTH

Design equations, such as those of AISC-F2, are based on the assumption of torsional simple support (see Figs. 8.5.4 and 8.7.2) at the ends of the unbraced segment. This means that *for torsional behavior only* the effective laterally unbraced length KL_b equals the actual laterally unbraced length L_b; that is, the effective length factor $K = 1.0$.

Nearly any type of lateral brace or end connection will prevent rotation (that is, keep the angle ϕ of twist equal to zero; see Fig. 9.4.1c) about the longitudinal axis of the beam. However, only rarely will there be much restraint against rotation of the flanges about a vertical axis as discussed in regard to Fig. 8.5.7 (i.e., warping restraint); thus $K = 1.0$ is reasonable for most cases.

Nonuniform Moment

The moment gradient does not directly affect laterally unbraced effective length; however, it does affect lateral-torsional buckling strength and is accounted for by the factor C_b when there is non-constant bending moment.

When a beam has transverse loading (not just end moments) the strength in lateral-torsional buckling will depend upon whether the load is applied at the top flange, the centroid, or the bottom flange. The weakest situation will be when the load is applied at the top flange, and the strongest when the load is applied at the bottom flange. The *SSRC Guide* [6.8, pp. 192–208] provides for adjustment to the moment gradient factor C_b to provide for these effects.

Continuous Beams

A continuous beam has lateral end restraint moments develop as a result of continuity over several spans. When the adjacent spans are shorter than the span being considered, or at least braced laterally at closer intervals; or the adjacent spans are less severely loaded, some lateral moment restraint may develop.

Typically, however, such end restraint about the y–y axis should not be assumed present since alternate unbraced spans could buckle in opposite directions. For treatment of lateral buckling on continuous beams, the reader is referred to Salvadori [9.6, 9.19], Hartmann [9.20], Trahair [9.21, 9.23], Powell and Klingner [9.22], Roeder and Assadi [9.24], and Fukumoto, Itoh, and Hattori [9.25]. The *SSRC Guide* [6.8, pp. 192–208] presents a method for evaluating the strength of a continuous beam span having intermediate points of lateral support.

In continuous beams, the point of inflection has often been treated as a braced point when design equations did not provide for the effect of moment gradient. In current design using the *AISC Specification* [1.13], the effect of moment gradient appears in all equations except in the requirements to establish the ability to qualify as a "compact section". The requirements of AISC-F2.2 where L_p is prescribed and do not include the moment gradient. When applying these provisions one may wish to *consider* the inflection point as a possible braced point.

The authors' present opinion is that whenever moment gradient is explicitly provided for in a design equation (no matter how approximately), the inflection point *should not* be considered a braced point. However, when provision is not made for the effect of moment gradient, generally the inflection may be considered as a braced point. The combination of torsional restraint provided by the floor system attachments and the continuity at the support point of maximum negative moment will normally be adequate to justify treating the inflection point as if it were actually braced *for the purpose of evaluating whether or not L_b exceeds L_p*. The important factor is the amount of torsional restraint provided by the floor system at the inflection point.

Regarding endorsement of this suggestion, AISC-Symbols defines L_p as used in AISC-F2.2 as "Limiting laterally unbraced length for limit state of yielding." Thus, there is no implication in AISC for the use of the inflection point as a braced point.

Cantilever Beams Without Lateral Bracing

Unlike the flagpole column where the effective pinned-end length is twice the actual length, the lateral-torsional buckling of a cantilever beam is not even as severe as the unbraced segment under uniform moment. If one considers the analogy to a column as discussed in Sec. 9.1, such a result is logical. Since the moment at the free end of the cantilever is zero, the compression force in the flange decreases from a maximum at one end to zero at the free end; thus, the loading is less severe than if the compression force were constant over the entire length.

Clark and Hill [9.26] and *SSRC Guide,* 3rd ed. [6.20] reported that it is conservative to use the full length as the effective laterally unbraced length for lateral-torsional buckling of cantilever beams. However, the conservativeness of using the actual length for a cantilever depends in large part on having *fixed torsional restraint* at the supported end, as well as having the loading applied at the shear center or at the bottom flange. Since such torsional fixity rarely exists, the authors recommend using the actual cantilever length as the effective laterally unbraced length.

Furthermore, the moment gradient factor C_b, Eqs. 9.6.11 and 9.6.12, is *not* applicable for a cantilever; thus C_b should be taken as unity. Obviously, moment gradient has some effect; however, a cantilever inherently has moment that varies from maximum at the support to zero at the free end. The use of the actual length as the effective length already recognizes the moment gradient.

When a more detailed design treatment of lateral-torsional buckling of cantilevers is desired, the *SSRC Guide,* 5th ed. [6.8, pp. 192–208] provides (from a recommendation by Nethercot) effective length factors K in order to use KL instead L in Eq. 9.4.32. Kitipornchai, Dux, and Richter [9.27] have an excellent overall review of the subject. Again in that method, $C_b = 1$ would be used. For nonuniform cantilevers, refer to the work of Massey and McGuire [9.28].

9.9   EXAMPLES: LOAD AND RESISTANCE FACTOR DESIGN

Several examples of I-shaped beams are presented to illustrate the Load and Resistance Factor Design Method to include the effect of lateral-torsional buckling. Other considerations, such as deflection, shear, and web local yielding and crippling were treated and illustrated in Chapter 7.

EXAMPLE 9.9.1

A simply supported beam is loaded as shown in Fig. 9.9.1. The uniform load is 15% dead load and 85% live load, and the concentrated load is 40% dead load and 60% live load. The beam has transverse lateral support at the ends and every 7'-6" along the span. Select the lightest W section of A992 steel, using AISC Load and Resistance Factor Design.

Solution:

(a) Determine the factored moment M_u at midspan and the required nominal strength M_n. Using ASCE 7-Section 2.3 [1.2],

$$w_u = 1.2(0.15)(1.4) + 1.6(0.85)(1.4) = 2.16 \text{ kips/ft}$$

Figure 9.9.1
Examples 9.9.1 and 9.10.1.

$$W_u = 1.2(0.40)(48) + 1.6(0.60)(48) = 69 \text{ kips}$$

$$M_u = \frac{1}{8}(2.16)(30)^2 + \frac{69(30)}{4} = 761 \text{ ft-kips}$$

Since the LRFD requirement is

$$\phi_b M_n \geq M_u \qquad\qquad [9.6.1]$$

where $\phi_b = 0.90$, the strength reduction factor for flexure, the required nominal strength is

$$\text{Required } M_n = M_u/\phi_b = 761/0.90 = 846 \text{ ft-kips}$$

(b) Estimate whether or not lateral supports are close enough to design the beam using plastic analysis (AISC-Appendix 1.7) or to use the full plastic moment strength M_p (AISC-F2.1) without plastic analysis. To use either of these provisions the section must be "compact" for local buckling in accordance with AISC-B4. Since the beam is simply supported, plastic analysis (as described in Chapter 10) does not apply. Assume the beam will be in Case 2 (see Sec. 9.6) where $M_n = M_p$. The maximum laterally unbraced length L_b is given by Eq. 9.6.3,

$$L_p = 1.76 r_y \sqrt{\frac{E}{F_y}} = \frac{300}{\sqrt{F_y}} r_y = \frac{300}{\sqrt{50}} r_y = 42 r_y$$

If one assumes that $r_y \approx 0.22 b_f$ (text Appendix Table A1),

$$\text{Min } b_f = \frac{L_p}{42(0.22)} = \frac{7.5(12)}{9.24} = 9.7 \text{ in.} \qquad \text{if } L_p = 7.5 \text{ ft}$$

Assuming the flange width is at least 9.7 in., the strength can reach the plastic moment M_p.

(c) Select from *AISC Manual* Table 3-2, "Selection by Z_x". This is an efficient procedure when the designer is certain that the section is "compact" and that $L_b < L_p$. The required plastic modulus Z_x is

$$\text{Required } M_u = \text{Required } M_p = 846 \text{ ft-kips}$$

$$\text{Required } Z_x = \frac{\text{Required } M_p}{F_y} = \frac{846(12)}{50} = 203 \text{ in.}^3$$

The lightest section that has $Z_x \geq 203$ in.^{3} is W24×84 having $Z_x = 224$ in.^{3}; in addition b_f is less than the required minimum of 9.7 in. The next lightest section found in *AISC Manual* Table 3-2 (p. 3-16) is W18×97 with $Z_x = 211$ in.^{3} and $b_f > 9.7$ in.,

(d) Check the section. The dead weight of the beam must be included; it could have been estimated at the beginning of the design.

$$M_u(\text{dead load}) = 1.2(0.084)(30)^2/8 = 11 \text{ ft-kips}$$

$$M_u = 761 + 11 = 772 \text{ ft-kips}$$

$$M_n = M_p = Z_x F_y = 211(50)/12 = 879 \text{ ft-kips}$$

$$L_p = 42r_y = 42(2.65)/12 = 9.3 \text{ ft} > (L_b = 7.5 \text{ ft})$$

The other requirements (see Table 9.6.1) for "compact section" are satisfied.

$$[\phi_b M_n = 0.90(879) = 791 \text{ ft-kips}] > [M_u = 772 \text{ ft-kips}] \qquad \textbf{OK}$$

In this case the strength provided is nearly 3% high; however, the W18×97 is the lightest section satisfying the requirements. The flange and web of this section satisfy the "compact section" requirements to prevent local buckling; that is, $\lambda \le \lambda_p$ as given in Table 9.6.1.

Use W18×97 ∎

EXAMPLE 9.9.2

Select the lightest W section for the simply supported beam of Fig. 9.9.2. The superimposed load is 0.4 kip/ft dead load and 1.0 kip/ft live load. Lateral support is provided at the ends and at midspan. Assume deflection limitations need not be considered. Use A992 steel and AISC Load and Resistance Factor Design.

Solution:

(a) Determine the factored moment M_u at midspan and the required design strength $\phi_b M_n$. Estimate the beam weight about 90 lb/ft. Using ASCE 7-Section 2.3 [1.2],

$$w_u = 1.2(0.4 + 0.09) + 1.6(1.0) = 2.19 \text{ kips/ft}$$

$$M_u = 2.19(50)^2/8 = 684 \text{ ft-kips}$$

$$\text{Required } \phi_b M_n = M_u = 684 \text{ ft-kips}$$

(b) Use the *AISC Manual* beam curves, Table 3-10, "Available Moment vs. Unbraced Length." These curves are plots of the design moment strength $\phi_b M_n$ vs laterally unbraced length L_b for rolled W and M shapes for $F_y = 50$ ksi. The curves are the nominal strength M_n relationship shown in Fig. 9.6.1 multiplied by the strength reduction factor $\phi_b (= 0.90$ for flexure). The moment gradient factor $C_b = 1.0$, the most severe loading case (except for loads applied at the top flange in laterally unbraced locations), is used for the curves.

Figure 9.9.2
Data for Examples 9.9.2,
9.9.4, 9.10.2, and 9.10.4.

For this example, C_b will be greater than 1.0 computed in accordance with Eq. 9.6.11 for the unbraced half-span segment,

$$C_b = \frac{12.5M_{max}}{2.5M_{max} + 3M_A + 4M_B + 3M_C} \qquad [9.6.11]$$

where
$M_{max} = 684$ ft-kips, maximum at midspan
$M_A = 300$ ft-kips at 1/4 point of unbraced segment
$M_B = 513$ ft-kips at midpoint of unbraced segment
$M_C = 641$ ft-kips at 3/4 point of unbraced segment

$$C_b = \frac{12.5(684)}{2.5(684) + 3(300) + 4(513) + 3(614)} = 1.30$$

The same result can be obtained from Case 3 in Table 9.6.4. This is significantly more conservative (i.e., lower) than using Eq. 9.6.12 from the 1986 LRFD Specification, which gives $C_b = 1.75$.

Referring to Fig. 9.6.4, for situations where $L_b > L_p$, the nominal strength M_n is proportional to C_b up to the maximum strength M_p. To use the *AISC Manual* beam curves based on $C_b = 1.0$ for C_b greater than 1.0, divide the M_n equations by C_b. The moment ordinate in the *AISC Manual* beam curves is the design strength $\phi_b M_n$, which must equal or exceed the factored moment M_u.

The significance and application of C_b in design is discussed in detail by Zuraski [9.29].

For the example,

$$\frac{\text{Required } \phi_b M_n}{C_b} = \frac{M_u}{C_b} = \frac{684}{1.30} = 526 \text{ ft-kips}$$

Entering the curves with factored moment $M_u/C_b = 526$ ft-kips and $L_b = 25$ ft, find

$$W18\times97, \quad \phi_b M_n \approx 569 \text{ ft-kips}$$

This may well be too conservative; M_n increases linearly with C_b. When $L_b \leq L_r$ and $M_r C_b = M_p$ no further beneficial effect of increasing C_b is possible.

Determine the value of L_r (see Fig. 9.6.1) for $W18\times97$ using Eq. 9.6.6.

$$L_r = 1.95 r_{ts} \frac{E}{0.7F_y} \sqrt{\frac{Jc}{S_x h_o}} \sqrt{1 + \sqrt{1 + 6.76\left(\frac{0.7F_y}{E}\frac{S_x h_o}{Jc}\right)^2}}$$

where $r_{ts} = 3.08$ in. and $h_o = 17.7$ in. as properties given in the *AISC Manual*.
Also, $c = 1$, for a doubly symmetric I-shape. Thus,

$$L_r = 1.95(3.08)\frac{29,000}{0.7(50)}\sqrt{\frac{5.86(1)}{188(17.7)}}\sqrt{1 + \sqrt{+ 16.76\left(\frac{0.7(50)}{29,000}\frac{188(17.7)}{5.86(1)}\right)^2}}\frac{1}{12}$$

$$= 30.3 \text{ ft}$$

The value of L_r for a rolled section is also available for $F_y = 50$ ksi in *AISC Manual* Table 3-6, "Maximum Total Uniform Loads".

Since L_b in this example (25 ft) is close to L_r, determine the maximum effective C_b. Setting $M_r C_b = 0.7 F_y S_x C_b = M_p = Z_x F_y$ gives

$$0.7 S_x C_b = Z_x$$

$$\text{Max effective } C_b = \frac{Z_x}{0.7 S_x} = 1.43 \frac{Z_x}{S_x}$$

Since $Z_x / S_x \approx 1.12$ as an average, the maximum effective C_b for A992 steel will be about 1.60. Thus, in this example where $C_b = 1.30$, the nominal strength M_n will not likely reach M_p. If it seemed that M_p would be reached, *AISC Manual* Table 3-6, "Maximum Total Uniform Loads" could be used to select the section.

(c) Check the W18×97 section. Including the new dead weight of the beam,

$$M_u = 2.20(50)^2/8 = 688 \text{ ft-kips}$$

$$L_r = 29 \text{ ft} > [L_b = 25 \text{ ft}]$$

$$M_p = Z_x F_y = 211(50)/12 = 879 \text{ ft-kips}$$

From the previous example, L_p was computed to be 9.3 ft. Thus, $L_p < L_b < L_r$, which means M_n should be determined using Eq. 9.6.4 which includes C_b as follows:

$$M_n = C_b \left[M_p - (M_p - 0.7 F_y S_x) \left(\frac{L_b - L_p}{L_r - L_p} \right) \right] \leq M_p \qquad [9.6.4]$$

$$M_n = 1.30 \left[10{,}550 - [10{,}550 - 0.7 \, (50)188] \left(\frac{25 - 9.36}{30.3 - 9.36} \right) \right] \frac{1}{12} = 822 \text{ ft-kips}$$

Since computed M_n of 822 ft-kips does not exceed $M_p = 879$ ft-kips; then

$$M_n = 822 \text{ ft-kips}$$

$$[\phi_b M_n = 0.90(822) = 740 \text{ ft-kips}] > [M_u = 688 \text{ ft-kips}] \qquad \textbf{OK}$$

The W18×97 is the lightest section satisfying the requirements. In addition to what has been checked, the section must be "compact" to assure that the computed strength is correct. As are nearly all A992 sections, this section is "compact" for local buckling; that is, it satisfies the limits for web and flange given in Table 9.6.1.

Use W18×97.   ■

EXAMPLE 9.9.3

Select an economical W section for the beam of Fig. 9.9.3. Lateral support is provided at the vertical supports, concentrated load points, and at the end of the cantilever. The 76-kip load contains 16 kips dead load and the 40-kip load includes 12 kips dead load; the remainder is live load. Use A992 steel and the AISC Load and Resistance Factor Design Method.

Solution:

(a) Obtain the bending moment envelope resulting from factored loads. The moment envelope will include two loading cases: (1) dead load plus live load on the 52-ft

Figure 9.9.3
Data for Example 9.9.3.

span and no live load on the cantilever; and (2) dead load plus live load on cantilever and no live load on the 52-ft span.

$$W_{u1} = 1.2(16) + 1.6(60) = 115 \text{ kips}$$

$$W_{u2} = 1.2(12) + 1.6(28) = 59.2 \text{ kips}$$

For loading case 1, the maximum factored moment under W_1 is

$$M_u \text{ on cantilever} = 1.2(12)21 = 302 \text{ ft-kips}$$

$$M_u \text{ at } W_1 = 115(24)(28)/52 - 302(24/52) = 1350 \text{ ft-kips}$$

For loading case 2, the maximum factored moment at the cantilever support is

$$M_u = 59.2(21) = 1240 \text{ ft-kips}$$

The factored moment envelope is shown in Fig. 9.9.3.

   (b) Segment A. Three laterally unbraced segments must be examined since each length is different and is subject to a different maximum factored end moment M_u. By inspection, segment A controls over segment C; both the moment and the unbraced length are larger on segment A, and the moment gradient is the same; thus C_b is the same. From the approximate Eq. 9.6.12 of the 1986 *LRFD Specification*, $C_b = 1.75$ by inspection when the moment at one end of the segment is zero. Using the more accurate 2005 Eq. 9.6.11, or from Table 9.6.3, $C_b = 1.67$.

$$M_u = 1350 \text{ ft-kips} \qquad C_b = 1.67 \qquad L_b = 24 \text{ ft}$$
$$\text{Required } \phi_b M_n = 1350 \text{ ft-kips}$$

Use the *AISC Manual* Table 3-10 charts, "Allowable Moment vs. Unbraced Length." As in Example 9.9.2, the C_b factor may be combined with the factored moment M_u,

$$\frac{M_u}{C_b} = \frac{1350}{1.67} = 808 \text{ ft-kips}$$

Enter the charts with required $\phi_b M_n = M_u/C_b = 808$ ft-kips and $L_b = 24$ ft; find

W24×117 $\phi_b M_n = 912$ ft-kips   for $C_b = 1.0$

For $C_b = 1.67$,

$$\phi_b M_n = C_b (\phi_b M_u \text{ for } C_b = 1.0) = 1.67(912)$$

$$= 1520 \text{ ft-kips} > (\phi_b M_p = 1230 \text{ ft-kips})$$

$$\phi_b M_n = \phi_b M_p = 1230 \text{ ft-kips} < (\text{Required } \phi_b M_n = 1350 \text{ ft-kips}) \qquad \textbf{NG}$$

Determine $L_r = 30.4$ ft for this section from Eq. 9.6.6 or more easily from *AISC Manual* Table 3-6, "Maximum Total Uniform Load." For segment A where $L_b = 24$ ft the C_b must be considerably less than the maximum effective value of 1.6 obtained in Example 9.9.2 in order for the nominal strength, modified with C_b, not to exceed the plastic moment M_p. For the given unbraced length and a C_b value larger than 1.60, it is more likely that the section will have its available nominal strength capped at M_p. Therefore, use of *AISC Manual* Table 3-2 "Selection by Z_x" seems appropriate.

$$\text{Required } Z_x = \frac{\text{Required } M_p}{F_y} = \frac{(1350/0.90)12}{50} = 360 \text{ in.}^3$$

The lightest section having $Z_x \geq 360$ in.^{3} is W30×116 which has $Z_x = 378$ in.^{3} Because this section has $L_r = 22.6$ ft, lower than the L_r for the W24×117, the strength with $L_b = 24$ ft may not be adequate.
Check the adequacy of the selected W30×116 section. For $C_b = 1.67$,

$$\phi_b M_n = C_b (\phi_b M_n \text{ for } C_b = 1.0) = 1.67(786)$$

$$= 1310 \text{ ft-kips} < (\phi_b M_p = 1420 \text{ ft-kips})$$

$$\phi_b M_n = 1310 \text{ ft-kips} < (\text{Required } \phi_b M_n = 1350 \text{ ft-kips}) \qquad \textbf{NG}$$

The design strength is less than the required strength (though marginally). Hence try the next heavier section from W30 or W33 series. Examine W33×118 with $\phi_b M_p = 1560$ ft-kips and $\phi_b M_u$ (at $L_b = 24$ ft) = 873 ft-kips.
Again for $C_b = 1.67$,

$$\phi_b M_u = C_b (\phi_b M_n \text{ for } C_b = 1.0) = 1.67(873)$$

$$= 1460 \text{ ft-kips} < (\phi_b M_p = 1560 \text{ ft-kips})$$

$$\phi_b M_n = 1460 \text{ ft-kips} > (\text{Required } \phi_b M_n = 1350 \text{ ft-kips}) \qquad \textbf{OK}$$

When L_b is near or larger than L_r, the best way of selecting the lightest section will be using the charts (*AISC Manual* Table 3-10), "Available Moment vs. Unbraced

Figure 9.9.4
Comparison of design
strengths $\phi_b M_n$ for
W33×118 and W30×116 for
$F_y = 50$ ksi—Example 9.9.3.

Length". To help understand the strength relationships, examine Fig. 9.9.4, where the
W30×116 and W24×117 design strengths $\phi_b M_n$ are shown for $C_b = 1.0$ and 1.67.

(c) Segment B. This segment has a steeper moment gradient than segment A, i.e.,
$C_b > 1.67$; however, the maximum C_b usable is 1.60 for A992 steel. The laterally
unbraced length is the longer 28 ft, indicating that segment B will likely control. Using
Eq. 9.6.11,

$$C_b = \frac{12.5 M_{\max}}{2.5 M_{\max} + 3 M_A + 4 M_B + 3 M_C} \qquad [9.6.11]$$

or for the ratio $M_1/M_2 = +302/1350 = +0.22$ the C_b can be obtained from Fig. 9.6.3 as
$C_b = 2.0$. Or linearly interpolating the values in Fig. 9.6.3, gives $C_b = 1.96$. Such precision
is unjustified. Use $C_b = 2.0$.

An entry to the *AISC Manual* Table 3-10, "Available Moment vs. Unbraced Length"
charts would use $M_u/C_b = 1350/2.0 = 675$ ft-kips at $L_b = 28$ ft and find W24×104 for
segment B, which is not satisfactory for segment A. Try a W33×118.

(d) Check the W33×118. From the *AISC Manual*, p. 3–43, find

$$L_r = 23.5 \text{ ft} \qquad \phi_b M_p = 1560 \text{ ft-kips} \qquad \phi_b M_r = 942 \text{ ft-kips}$$

For segments A and B, L_b exceeds L_r indicating elastic lateral-torsional buckling is the controlling limit state. Thus, M_{cr} must be computed for both the 24- and 28-ft unbraced lengths
(segments A and B). Using Eq. 9.6.13 multiplied by S_x, to give M_{cr},

$$M_{cr} = F_{cr} S_x = \frac{C_b S_x \pi^2 E}{\left(\dfrac{L_b}{r_{ts}}\right)^2} \sqrt{1 + 0.078 \frac{Jc}{S_x h_o}\left(\frac{L_b}{r_{ts}}\right)^2}$$

The properties needed are $S_x = 359$ in.^{3}; $r_{tx} = 2.89$ in.; $J = 5.31$ in.^{4}; and $h_o = 32.1$ in.
For the two possible controlling segments A and B $\phi_b M_n = \phi_b M_{cr}$, as follows:

Segment A, $C_b = 1.67$ and $L_b = 24$ ft,

$$\phi_b M_{cr} = \frac{0.90(1.67)(359)\pi^2 29{,}000}{\left(\dfrac{24(12)}{2.89}\right)^2}\sqrt{1 + 0.078\frac{5.31(1.0)}{359(32.1)}\left(\frac{24(12)}{2.89}\right)^2\frac{1}{12}}$$

$$= 1510 \text{ ft-kips}$$

which is less than $\phi_b M_p$.

Segment B, $C_b = 2.0$ and $L_b = 28$ ft,

$$\phi_b M_{cr} = \frac{0.90(2.0)(359)\pi^2 29{,}000}{\left(\dfrac{28(12)}{2.89}\right)^2}\sqrt{1 + 0.078\frac{5.31(10)}{359(32.1)}\left(\frac{28(12)}{2.89}\right)^2\frac{1}{12}}$$

$$= 1390 \text{ ft-kips}$$

which is less than $\phi_b M_p$.

Segment B controls;

$$(\phi_b M_n = 1390 \text{ ft-kips}) > (M_u = 1380 \text{ ft-kips}) \qquad \textbf{OK}$$

Thus, the W33×118 is acceptable. The section satisfies the limits for flange and web given in Table 9.6.1 to preclude local buckling, and therefore is "compact"; i.e., $\lambda \le \lambda_p$. Therefore, the lateral-torsional buckling limit state controls, as assumed. The W30×116 has $\phi_b M_n$ only 1310 ft-kips for segment A; thus, it is not acceptable.

Use W33×118. ■

EXAMPLE 9.9.4

Repeat Example 9.9.2 (See Fig. 9.9.2) except use A572 Grade 60 steel.

Solution:

(a) Obtain the factored moment M_u at midspan and the required design strength $\phi_b M_n$. From Example 9.9.2, including the beam weight now estimated as 90 lb/ft (For A992 steel, the beam is W18×97).

$$w_u = 1.2(0.4 + 0.09) + 1.6(1.0) = 2.19 \text{ kips/ft}$$

$$M_u = 2.19(50)^2/8 = 684 \text{ ft-kips}$$

$$\text{Required } \phi_b M_n = M_u = 684 \text{ ft-kips}$$

(b) Select trial section using beam charts (*AISC Manual* Table 3-10, "Available Moment vs. Unbraced Length".) Since there are no LRFD tables or charts for $F_y = 60$ ksi, a more general approach must be used. When reaching the plastic moment strength M_p is expected to be the controlling limit state, that strength is proportional to the yield stress. Thus, when $L_b < L_p$, use *AISC Manual* Table 3-2, "Selection by Z_x." When L_b is expected to exceed L_p, adjust the required design moment in the ratio of the yield stress used to $F_y = 50$ ksi for which there are charts. Thus, for this example,

$$\text{Adjusted } M_u \text{ for } F_y = 50 \text{ ksi} = 684(50/60) = 570 \text{ ft-kips}$$

The C_b factor may be combined with the factored moment M_u,

$$\frac{\text{Adjusted } M_u \text{ for } F_y = 50 \text{ ksi}}{C_b} = \frac{570}{1.3} = 439 \text{ ft-kips}$$

Entering the charts with required $\phi_b M_n = M_u/C_b = 439$ ft-kips and $L_b = 25$ ft, find

W18×86	$\phi_b M_n = 484$ ft-kips
W16×89	$\phi_b M_n = 467$ ft-kips
W24×94	$\phi_b M_n = 454$ ft-kips

The sections are all located in the charts above and to the right of the entry point. From *AISC Manual* Table 3-2, "Selection by Z_x" find for $F_y = 50$ ksi:

$$W18\times86, \quad L_r = 28.5 \text{ ft}, \quad \phi_b M_p = 698 \text{ ft-kips}$$
$$\phi_b M_r = 436 \text{ ft-kips}$$
$$W16\times89, \quad L_r = 30.2 \text{ ft}, \quad \phi_b M_p = 656 \text{ ft-kips}$$
$$\phi_b M_r = 407 \text{ ft-kips}$$
$$W24\times94, \quad L_r = 21.2 \text{ ft}, \quad \phi_b M_p = 953 \text{ ft-kips}$$
$$\phi_b M_r = 583 \text{ ft-kips}$$

Try the moderately deep W18. Since L_b is moderately lower than L_r of 28.5 ft for $F_y = 50$ ksi, and realizing that L_r will be lower for $F_y = 60$ ksi, the controlling lateral-torsional buckling strength M_{cr} could be either elastic or inelastic. The variation of L_r with F_y for W18×86 is shown in Fig. 9.9.5; the two *AISC Manual* tabulated values are marked as points A and B. Typically, as the yield stress is higher, the L_r is lower, but there is less decrease than if the relationship were linear. For this example, because L_r is higher but close to $L_b = 25$ ft, the section could either fall in the transition domain or in the elastic buckling domain.

(c) Check W18×86. Compute M_p,

$$M_p = Z_x F_y = 186(60)/12 = 930 \text{ ft-kips}$$

$$\phi_b M_p = 0.90(930) = 837 \text{ ft-kips}$$

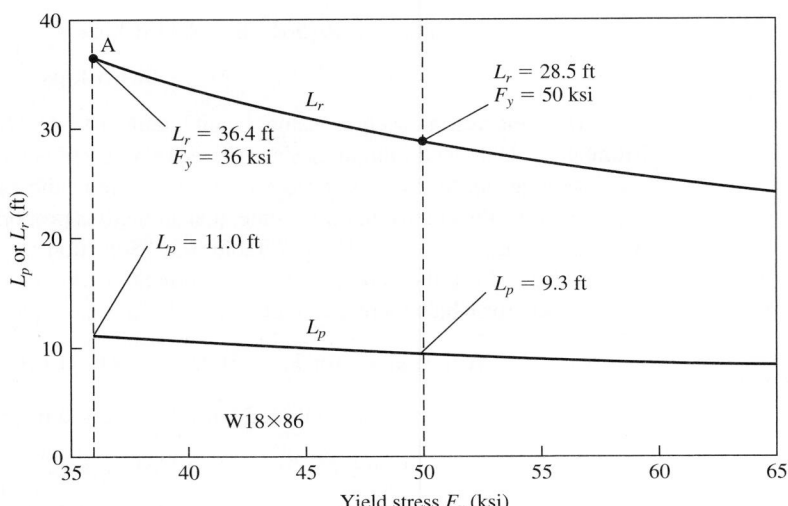

Figure 9.9.5
Variation in L_r and L_p for various yield stresses- W18×86.

Compute L_r for F_y = 60 ksi from Eq. 9.6.6 below using r_{ts} = 3.05 in., J = 4.1 in.^{4}, S_x = 166 in.^{3}, c = 1, and h_o = 17.6 in. for W18×86

$$L_r = 1.95 r_{ts} \frac{E}{0.7 F_y} \sqrt{\frac{Jc}{S_x h_o}} \sqrt{1 + \sqrt{1 + 6.76 \left(\frac{0.7 F_y}{E} \frac{S_x h_o}{Jc} \right)^2}}$$

$$= 25.2 \text{ ft} > L_b = 25 \text{ ft}$$

Since L_b = 25 ft for this example does not exceed L_r, the inelastic lateral-torsional buckling strength must be computed to make the check of this section. Using Eq. 9.6.4 with C_b = 1.30 gives

$$M_n = C_b \left[M_p - (M_p - 0.7 F_y S_x) \left(\frac{L_b - L_p}{L_r - L_p} \right) \right]$$

$$= 1.30 \left[930 - (930 - 0.7(60)(166/12)) \left(\frac{25 - 8.5}{25.2 - 8.5} \right) \right]$$

$$= 761 \text{ ft-kips} < 930 \text{ ft-kips} \qquad \textbf{OK}$$

$$[\phi M_n = 0.90(761) = 685 \text{ ft-kips}] > [M_u = 684 \text{ ft-kips}] \qquad \textbf{OK}$$

If minimum deflection is desired a deeper section could be selected. Consider the alternative section W24×94. Compute L_r for F_y = 60 ksi from Eq. 9.6.6 below using r_{ts} = 2.4 in., J = 5.26 in.^{4}, S_x = 222 in^3, c = 1, h_o = 23.4 in. for W24×94

$$L_r = 1.95 r_{ts} \frac{E}{0.7 F_y} \sqrt{\frac{Jc}{S_x h_o}} \sqrt{1 + \sqrt{1 + 6.76 \left(\frac{0.7 F_y}{E} \frac{S_x h_o}{Jc} \right)^2}}$$

$$= 18.9 \text{ ft} < [L_b = 25 \text{ ft}]$$

$$F_{cr} = \frac{C_b \pi^2 E}{\left(\frac{L_b}{r_{ts}} \right)^2} \sqrt{1 + 0.078 \frac{Jc}{S_x h_o} \left(\frac{L_b}{r_{ts}} \right)^2} = 35.6 \text{ ksi}$$

and the available moment strength is

$$\phi M_n = \phi F_{cr} S_x = 0.90(35.6)(222)/12 = 593 \text{ ft-kips} < [\phi M_p = 837 \text{ ft-kips}]$$

$$[\phi M_n = 593 \text{ ft-kips}] < [M_u = 684 \text{ ft-kips}] \qquad \textbf{NG}$$

The next heavier section is W24×103, giving L_r = 9.4 ft and

$$[\phi_b M_n = 691 \text{ ft-kips}] > [M_u = 688 \text{ ft-kips, adjusted for beam weight}] \qquad \textbf{OK}$$

Investigating the W21 series does not give a lighter section.

The W18×86 is satisfactory with less than 1% overstrength. The section also satisfies the "compact section" requirements $\lambda \leq \lambda_p$ for flange and web as given in Table 9.6.1.

For the lightest section, *Use* W18×86 if deflection can be tolerated; or *use* W24×103 if minimum deflection is desired.   ∎

EXAMPLE 9.9.5

What W section can be used for the beam of Example 9.9.4 if lateral support is provided every 5 ft?

Solution:

In this case, since the lateral-torsional stability has been improved by reducing the unbraced length, the deeper sections can carry greater loads. The deeper sections will also be the lighter ones.

(a) Factored moment M_u. From Example 9.9.4 and assuming a lighter section than the 103 or 86-lb/ft sections used in Example 9.9.4,

$$w_u = 1.2(0.4 + 0.07) + 1.6(1.0) = 2.16 \text{ kips/ft}$$

$$M_u = 2.16(50)^2/8 = 675 \text{ ft-kips}$$

$$\text{Required } \phi_b M_n = M_u = 675 \text{ ft-kips}$$

(b) Select a trial section. Assume the section will be adequately braced to achieve its plastic moment strength M_p,

$$\text{Required } Z_x = \frac{\text{Required } M_p}{F_y} = \frac{(675/0.90)12}{60} = 150 \text{ in.}^3$$

From *AISC Manual* Table 3-2, "Selection by Z_x", the lightest section that has $Z_x \geq 150$ in.^{3} is W24×62 having $Z_x = 153$ in.^{3} Note that $L_p = 4.87$ ft for $F_y = 50$ ksi, and will be lower for $F_y = 60$ ksi. Thus, a section having somewhat larger Z_x must be used. The next lightest section is W21×68 having $Z_x = 160$ in.^{3}.

(c) Check the section. Note that the 5-ft laterally unbraced segment adjacent to midspan will have C_b very close to 1.0 (actually it is 1.02 for $M_1/M_2 = -0.96$); if the check shows the strength to be slightly low, the correct C_b can be obtained from Table 9.6.3 (or it can be computed). Assume $C_b = 1.0$ for the check. The dead weight assumed is almost equal to the actual weight of the beam

For the W21×68, using $S_x = 140$ in.^{3}, $r_{ts} = 2.17$ in., $h_o = 20.4$ in., and $J = 2.45$ in.^{4}

$$L_p = 1.76 r_y \sqrt{\frac{E}{F_y}} = \frac{300}{\sqrt{F_{y,\text{ksi}}}} r_y = \frac{300}{7.75}(1.8)\frac{1}{12} = 5.8 \text{ ft}$$

$$L_r = 1.95 r_{ts} \frac{E}{0.7 F_y} \sqrt{\frac{Jc}{S_x h_o}} \sqrt{1 + \sqrt{1 + 6.76\left(\frac{0.7 F_y}{E} \frac{S_x h_o}{Jc}\right)^2}} = 16.8 \text{ ft}$$

Since L_b is less than L_p

$$\phi M_n = \phi M_p = \phi Z_x F_y = 0.9(160)(60)/12$$

$$= 720 \text{ ft-kips} > [M_u = 675 \text{ ft-kips}] \qquad \textbf{OK}$$

Use W21×68 ($F_y = 60$ ksi) with about 7% overstrength. ∎

EXAMPLE 9.9.6

Given the welded I-shaped section of Fig. 9.9.6 used as a 45-ft simply supported beam laterally supported at the one-third points. Determine the service live load the beam may be

Figure 9.9.6
Data for Example 9.9.6.

Welded section

permitted to carry if the dead load is 0.15 kip/ft including the beam weight. Use Load and Resistance Factor Design. The steel has $F_y = 65$ ksi.

Solution:

(a) Compute cross-sectional properties.

$$\text{Area,} \quad A = 28.1 \text{ sq. in.}$$

$$I_x = \frac{1}{12}\left[\left(26 + \frac{5}{8} + \frac{5}{8}\right)^3 (16) - (26)^3\left(16 - \frac{5}{16}\right)\right] = 4003 \text{ in.}^4$$

$$S_x = \frac{I_x}{d/2} = \frac{4003}{13.625} = 294 \text{ in.}^3$$

$$I_y = 2\left(\frac{1}{12}\right)(16)^3(0.625) = 427 \text{ in.}^4$$

$$r_y = \sqrt{\frac{I_y}{A}} = \sqrt{\frac{427}{28.1}} = 3.90 \text{ in.}$$

$$J = \frac{1}{3}\left[2(16)(0.625)^3 + 26(0.3125)^3\right] = 2.87 \text{ in.}^4$$

$$C_w = \frac{I_y h^2}{4} = \frac{427(26 + 0.625)^2}{4} = 75,670 \text{ in.}^6$$

$$r_{ts} = \sqrt{\frac{\sqrt{I_y C_w}}{S_x}} = \sqrt{\frac{\sqrt{427(75,670)}}{294}} = 4.40 \text{ in.}$$

$$h_o = 26.6 \text{ in.}$$

(b) Investigate the local flange buckling and local web buckling limit states. Check $\lambda \le \lambda_p$ according to AISC-B4; the limits are given in Table 9.6.1.

For flange local buckling,

$$\left(\lambda = \frac{b_f/2}{t_f} = \frac{16/2}{0.625} = 12.8\right) > \left(\lambda_p = 0.38\sqrt{\frac{E}{F_y}} = \frac{65}{\sqrt{F_y}} = 8.0\right) \qquad \textbf{NG}$$

For web local buckling,

$$\left(\lambda = \frac{h}{t_w} = \frac{26}{0.3125} = 83.2\right) > \left(\lambda_p = 3.76\sqrt{\frac{E}{F_y}} = \frac{640}{\sqrt{F_y}} = 79.4\right) \qquad \textbf{NG}$$

Thus, the section is not "compact" with regard to either the flange or the web. Next, check whether the flange and/or the web must be classified as "slender" elements; that is, determine whether λ exceeds λ_r (AISC-B4.1 and Table 9.6.2).

For flange local buckling,

$$\frac{h}{t_w} = 83.2; \qquad k_c = \frac{4}{\sqrt{h/t_w}} = \frac{4}{\sqrt{83.2}} = 0.44 \quad \begin{matrix} >0.35 \\ <0.763 \end{matrix}$$

$$F_L = 0.7F_y = 0.7\,(65) = 45.5 \text{ ksi} \qquad \text{(Note b of AISC Table B4.1)}$$

$$\lambda_r = 0.95\sqrt{\frac{k_c E}{F_L}} = \frac{162}{\sqrt{F_L/k_c}} = \frac{162}{\sqrt{45.5/0.44}} = 15.9$$

$$\left(\lambda = \frac{b_f/2}{t_f} = 12.8\right) < (\lambda_r = 15.9) \qquad\qquad\qquad \textbf{OK}$$

For web local buckling,

$$\left(\lambda = \frac{h}{t_w} = 83.2\right) < \left(\lambda_p = 5.70\sqrt{\frac{E}{F_y}} = \frac{970}{\sqrt{F_y}} = 120\right) \qquad \textbf{OK}$$

Thus, the slenderness ratios λ for both the flange and the web lie between λ_p and λ_r; the section is "noncompact" with regard to both the flange and the web, and two of the possible controlling limit states for nominal strength will be local buckling of the flange or the web in the inelastic range as shown in Fig. 9.6.2.

(c) Compute the plastic moment strength M_p and the moment strength M_r at the elastic limit (see Fig. 9.6.2).

$$M_p = Z_x F_y = 319(65)/12 = 1728 \text{ ft-kips}$$

$$M_r = (0.7F_y)S_x = (45.5)294/12 = 1115 \text{ ft-kips}$$

(d) Compute the nominal strength M_n based on the limit state of *local buckling of the flange*. Using Eq. 9.6.9,

$$M_n = M_p - (M_p - 0.7F_y S_x)\left(\frac{\lambda - \lambda_{pf}}{\lambda_{rf} - \lambda_{pf}}\right) \qquad [9.6.9]$$

$$M_n = 1728 - \left[1728 - 0.7(65)294\left(\frac{1}{12}\right)\right]\left(\frac{12.8 - 8.0}{15.9 - 8.0}\right) = 1355 \text{ ft-kips}$$

(e) Compute the nominal strength M_n based on the limit state of *local buckling of the web*. Using Eq. 9.6.9,

$$M_n = 1728 - \left[1728 - 0.7(65)294\left(\frac{1}{12}\right)\right]\left(\frac{83.2 - 79.4}{120.4 - 79.4}\right) = 1671 \text{ ft-kips}$$

(f) Examine the lateral-torsional buckling limit state. Compute L_p using Eq. 9.6.3,

$$L_p = 1.76r_y\sqrt{\frac{E}{F_y}} = 1.76(3.9)\sqrt{\frac{29{,}000}{65}}\frac{1}{12} = 12.1 \text{ ft}$$

Since $L_b = 15$ ft exceeds L_p, the lateral-torsional buckling limit state must be examined further. Compute L_r using Eq. 9.6.6, with, $S_x = 294$ in.^{3}, $r_{ts} = 4.37$ in., $h_o = 26.6$ in., $J = 2.87$ in.^{4}, and $F_y = 65$ ksi

$$L_r = 1.95 r_{ts} \frac{E}{0.7F_y} \sqrt{\frac{Jc}{S_x h_o}} \sqrt{1 + \sqrt{1 + 6.76\left(\frac{0.7F_y}{E}\frac{S_x h_o}{Jc}\right)^2}} = 30.4 \text{ ft}$$

Thus, the nominal strength M_n based on the limit state of *lateral-torsional buckling* is linearly interpolated between M_p and $M_r = 0.7F_y S_x$ according to Eq. 9.6.4 using $C_b = 1.0$ for the center 15-ft laterally unbraced segment of the 45-ft span of this example (computed $C_b = 1.01$),

$$M_n = C_b\left[M_p - (M_p - 0.7F_y S_x)\left(\frac{L_b - L_p}{L_r - L_p}\right)\right] \le M_p \qquad [9.6.4]$$

$$M_n = 1.0\left[1728 - \left[1728 - 0.7(65)294\left(\frac{1}{12}\right)\right]\left(\frac{15 - 12.1}{30.4 - 12.1}\right)\right]$$

$$= 1630 \text{ ft-kips}$$

(g) Final evaluation. The nominal strength M_n is the lowest value among the possible controlling limit states: 1355 ft-kips for flange local buckling; 1671 ft-kips for web local buckling: and 1630 ft-kips for lateral-torsional buckling. Thus,

$$M_n = 1355 \text{ ft-kips}$$

$$\phi_b M_n = 1220 \text{ ft-kips}$$

$$\text{Maximum } M_u = \phi_b M_n = 1220 \text{ ft-kips}$$

$$M_u = 1.2 M_D + 1.6 M_L$$

$$M_D = 0.15(45)^2/8 = 38 \text{ ft-kips}$$

$$M_L = \frac{M_u - 1.2 M_D}{1.6} = \frac{1220 - 1.2(38)}{1.6} = 734 \text{ ft-kips}$$

$$w_L = \frac{8 M_L}{L^2} = \frac{8(734)}{(45)^2} = \underline{2.90 \text{ kips/ft}} \qquad \blacksquare$$

9.10   EXAMPLES: ALLOWABLE STRENGTH DESIGN

Several examples are presented to illustrate the Allowable Strength Design Method to include lateral-torsional buckling as a factor determining the allowable stress. Other considerations, such as deflection, shear, and web crippling were treated and illustrated in Chapter 7.

EXAMPLE 9.10.1

A simply supported beam is loaded as shown in Fig. 9.9.1. The beam has transverse lateral support at the ends and every $7'-6''$ along the span. Select the lightest W section of A992 steel, using Allowable Strength Design.

Solution:

  (a) Calculate the service load moment.

$$M_a = \tfrac{1}{8}(1.4)(30)^2 + \tfrac{1}{4}48(30) = 518 \text{ ft-kips}$$

The ASD requirement is

$$\frac{M_n}{\Omega} \geq M_a$$

where $\Omega = 1.67$, the safety factor for flexure. Therefore, the required nominal strength M_n is

$$\text{Required } M_n = \Omega M_a = 1.67(518) = 865 \text{ ft-kips}$$

  (b) Assume the selected section will be in Case 2 where $M_n = M_p$. As in Example 9.9.1, for $L_b \leq L_p$, and the minimum b_f must at least equal 9.7 in. Select from *AISC Manual* Table 3-2, "W Shapes Selection by Z_x"

$$\text{Required } Z_x = \frac{\text{Required } M_n}{F_y} = \frac{865(12)}{50} = 208 \text{ in.}^3$$

As in Example 9.9.1, the lightest section would be W24×84, with $b_f = 9.02$ in., which is below the minimum required to attain the plastic moment strength M_p. From *AISC Manual* Table 3-2 "W-Shapes Selection by Z_x", one can find W18×97, as in the LRFD example, to be the lightest suitable section, with $Z_x = 211$ in.^{3} and $L_p = 9.36$ ft.
  Include the beam weight and check the allowable strength,

$$M_a = \tfrac{1}{8}(1.4 + 0.097)(30)^2 + \tfrac{1}{4}48(30) = 528 \text{ ft-kips}$$

From *AISC Manual* Table 3-2,

$$(M_p/\Omega = 526 \text{ ft-kips}) \approx (M_a = 528 \text{ ft-kips}) \qquad\qquad \text{say } \textbf{OK}$$

Local buckling limits for compact section are also satisfied.

Use W18×97   ■

EXAMPLE 9.10.2

Repeat the selection of the lightest W section for the simply supported beam of Fig. 9.9.2 designed using AISC Load and Resistance Factor Design, except here use Allowable Strength Design. Lateral support is provided at the ends and at midspan. Assume deflection limitations need not be considered. Use A992 steel.

Solution:

  (a) Determine the required nominal moment strength M_n at midspan. Assume the beam weight to be about 90 lb/ft.

$$M_a = \tfrac{1}{8}(1.49)(50)^2 = 466 \text{ ft-kips}$$

$$L_b = \text{Laterally unbraced length} = 25 \text{ ft}$$

$$C_b = 1.30 \text{ (as in Example 9.9.2)}$$

The ASD requirement (with $\Omega = 1.67$) is

$$\frac{M_n}{\Omega} \geq M_a$$

The required allowable strength is,

$$\text{Required } \frac{M_n}{\Omega} = M_a = 466 \text{ ft-kips}$$

(b) Use *AISC Manual* Table 3-10, "Available Moment vs. Unbraced Length". These curves are plots of the allowable flexural strength vs laterally unbraced length L_b for W shapes having $F_y = 50$ ksi, the same curves used with the LRFD Method examples. With the ASD Method, the green columns in the tables are used for the allowable strength values.

Entering the curves with

$$\frac{M_a}{C_b} = \frac{466}{1.30} = 359 \text{ ft-kips and } L_b = 25 \text{ ft}$$

Find W18×97,

$$(M_n/\Omega = 379 \text{ ft-kips}) > (\text{Required } M_n/\Omega = 359 \text{ ft-kips})$$

Note that the W18×97 is within the linear transition portion of the strength curve.

Check W18×97 having $Z_x = 211$ in.^{3}, $S_x = 188$ in.^{3}, $L_p = 9.36$ ft, and $L_r = 30.3$ ft.

$$M_n = C_b \left[M_p - (M_p - 0.7F_y S_x)\left(\frac{L_b - L_p}{L_r - L_p}\right) \right] \leq M_p$$

$$= 1.3 \left[879 - \left(879 - 0.7(50)188(\tfrac{1}{12})\right)\left(\frac{25 - 9.36}{30.3 - 9.36}\right) \right] = 822 \text{ ft-kips}$$

The nominal moment strength M_n computed above is less than the upper limit $M_p = 879$ ft-kips. Thus,

$$\left(\frac{M_n}{\Omega} = \frac{822}{1.67} = 492 \text{ ft-kips}\right) > (M_a = 466 \text{ ft-kips}) \qquad \textbf{OK}$$

Use W18×97 with only 6% overstrength.   ∎

EXAMPLE 9.10.3

Select an economical W section for the beam of Fig. 9.9.3. Lateral support is provided at the vertical supports, concentrated load points, and at the end of the cantilever. Use A992 steel and the AISC Allowable Strength Design Method.

Solution:

Three cases must be considered since each of the three laterally unbraced lengths is different and is subject to a different maximum bending moment. Assume the segment containing the largest moment governs.

As discussed in Example 9.9.3, start by considering segment A, which has the largest maximum service load moment of 866 ft-kips, as shown in Fig. 9.10.1. Enter *AISC*

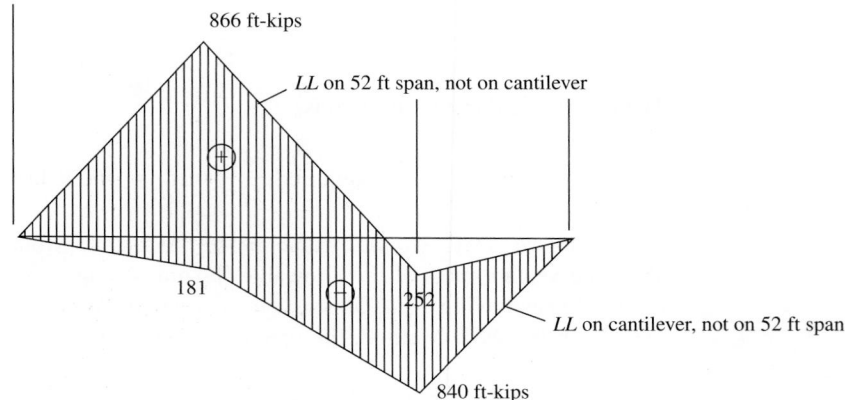

866 ft-kips

LL on 52 ft span, not on cantilever

181

252

LL on cantilever, not on 52 ft span

840 ft-kips

Figure 9.10.1
Bending Moment M_a
envelope due to superimposed
ASD load combinations.

Manual Table 3-10, "Available Moment vs. Unbraced Length" with the modified required allowable strength, the length of segment A (unbraced length), and $C_b = 1.67$.

$$\frac{M_a}{C_b} = \frac{866}{1.67} = 519 \text{ ft-kips} \quad L_b = 24 \text{ ft}$$

Two solid lines close to each other are above and to the right of this point, namely, W24×104 and W21×111. Examine the lighter W24×104. For an unbraced length of 24 ft, this section is within inelastic lateral-torsional buckling domain with a $M_n/\Omega = 526$ ft-kips. The allowable strength with $C_b = 1.67$ is

$$\frac{M_n}{\Omega} = 1.67\left(\frac{M_n}{\Omega} \text{ for } C_b = 1.0\right) = 1.67(526) = 878 \text{ ft-kips}$$

$$\frac{M_p}{\Omega} = 721 \text{ ft-kips} < 878 \text{ ft-kips}; \frac{M_p}{\Omega} \text{ controls!}$$

$$\therefore \quad \left(\frac{M_n}{\Omega} = 721 \text{ ft-kips}\right) < (M_a = 866 \text{ ft-kips}) \qquad \textbf{NG}$$

Must choose a deeper section with comparable weight or may use a heavier section with the same depth. Try W33×118 with $M_n/\Omega = 602$ ft-kips at $L_b = 24$ ft.

$$\frac{M_n}{\Omega} = 1.67\left(\frac{M_n}{\Omega} \text{ for } C_b = 1.0\right) = 1.67(602) = 1005 \text{ ft-kips}$$

$$\frac{M_p}{\Omega} = 1040 \text{ ft-kips} > 1005 \text{ ft-kips}; \quad \text{computed } \frac{M_n}{\Omega} \text{ controls!}$$

$$\left(\frac{M_n}{\Omega} = 1005 \text{ ft-kips}\right) > (M_a = 866 \text{ ft-kips}) \qquad \textbf{OK}$$

Use W33×118.   ∎

EXAMPLE 9.10.4

Investigate using Allowable Strength Design the W18×86 section of $F_y = 60$ ksi steel that was acceptable by Load and Resistance Factor Design in Example 9.9.4 (see Fig. 9.9.2). The simply supported span of 50 ft has lateral support at the ends and midspan.

Solution:

(a) Determine the service load moment M at midspan.

$$M_a = wL^2/8 = 1.486(50)^2/8 = 464 \text{ ft-kips}$$

(b) Check the allowable strength, M_n/Ω against the required strength M_a. All the values associated with the nominal strength of the beam, as determined in Example 9.9.4 with the LRFD Method are valid for the ASD Method. So, the following values are adopted for ASD verification

$$L_r = 25.2 \text{ ft.} > L_b = 25 \text{ ft.}$$

The inelastic lateral buckling is the controlling limit state (as in Example 9.9.4) with $M_n = 761$ ft-kips

The allowable moment strength is given as:

$$\left[\frac{M_n}{\Omega} = \frac{761}{1.67} = 456 \text{ ft-kips} \right] < M_a = 464 \text{ ft-kips}$$

The beam is understrength by only 1.7%; thus it is acceptable.   ∎

9.11   WEAK-AXIS BENDING OF I-SHAPED SECTIONS

The treatment thus far in this chapter has dealt with lateral-torsional buckling where instability might occur in a direction perpendicular to the plane of strong-axis bending (that is, the buckling occurs in the weaker direction). When an I-shaped beam is bent about its weak axis (y-axis), that is, bending in a plane *perpendicular* to the plane of the web, making the y-axis the neutral axis, lateral-torsional instability is no longer of concern. The beam will tend to deflect only in the direction of the loading since that is the principal axis orientation offering least resistance. Since lateral instability will not occur on doubly symmetric I-shaped sections bent about their weak axis, the only factor that might prevent them from achieving the plastic moment condition would be local buckling of the compression portion (unstiffened element) of the flanges.

The lateral-torsional buckling limit state, defined in AISC-F2, applies only to members subjected to bending about the major axis. In addition, the web will not be a compression element when the loading is in a plane parallel to the flanges. Thus, the limit states for I-shaped sections in weak axis bending are (1) development of the plastic moment strength and (2) flange local buckling. AISC-F6 defines the nominal flexural strength as follows

1. Yielding

$$M_n = M_p = F_y Z_y \leq 1.6 F_y S_y \tag{9.11.1}$$

For sections with compact flanges this limit state applies.

2. Local flange buckling

(a) For sections with noncompact flanges

$$M_n\left[M_p - (M_p - 0.7F_yS_y)\left(\frac{\lambda - \lambda_{pf}}{\lambda_{rf} - \lambda_{pf}}\right)\right]\tag{9.11.2}$$

(b) For sections with slender flanges

$$F_{cr} = \frac{0.69E}{\left(\dfrac{b_f}{2t_f}\right)^2}\tag{9.11.3}$$

9.12* LATERAL BUCKLING OF CHANNELS, ZEES, MONOSYMMETRIC I-SHAPED SECTIONS, AND TEES

The basic development of lateral buckling strength-related criteria has assumed that loads are applied vertically through the shear center. Furthermore, the resistance to lateral buckling assumed that the shear forces which developed in the flanges were equal and the center of twist was located at mid-height.

Channels

Unless loaded through the shear center, a channel is subjected to combined bending and torsion. Since the shear center is not in the plane of the web (see Fig. 8.5.1), usual loadings through the centroid or in the plane of the web give rise to such combined stress. For loads in a plane parallel to the web, lateral buckling must be considered, even if the torsional moment may properly be neglected. The 1976 *SSRC Guide* [6.20] states "if an otherwise laterally unsupported channel has concentrated loads brought in by other members that frame into it, such loads can be considered as being applied at the shear center, *provided that the span of the framing member is measured from the channel shear center and the framing connections are designed for the moment and shear at the connection.*"

For design purposes, Hill [9.32] indicates that lateral-torsional buckling equations for symmetric I-shaped sections may be applied for channels. Such a procedure is stated to err on the unsafe side by about 6% in extreme cases. AISC 2005 adopts this approach. Equation 9.4.33 can be used to represent the elastic lateral-torsional buckling strength of a channel loaded essentially in the plane of its web. The torsion warping constant C_w for a channel has a different expression than the expression for an I-shaped section; the expression is available in text Appendix Table A2. The torsion properties for rolled channels are available in the *AISC Manual* [1.15].

EXAMPLE 9.12.1

Determine the nominal strength M_n for a channel, C12×20.7 of steel having $F_y = 50$ ksi, used on a span of 24 ft with concentrated loads at the one-third points as shown in Fig. 9.12.1. It is assumed that the loads act at the shear center of the channel.

Solution:

(a) Determine whether or not the section satisfies the "compact section" requirements. For local buckling limit states the flange and web must satisfy the same requirements as for I-shaped sections; i.e., as given in Table 9.6.1. For the flange,

Figure 9.12.1
Example 9.12.1.

$$\left(\lambda = \frac{b_f}{t_f} = \frac{2.942}{0.501} = 5.9\right) < \left(\lambda_p = \frac{65}{\sqrt{F_y}} = 9.2\right) \qquad \textbf{OK}$$

For the web, λ must not exceed λ_p as given in Table 9.6.1, as follows:

$$\left(\lambda = \frac{h}{t_w} \approx \frac{T}{t_w} = \frac{9.75}{0.282} = 34.6\right) < \left(\lambda_p = \frac{640}{\sqrt{F_y}} = 90.5\right) \qquad \textbf{OK}$$

Note that neither b_f/t_f nor h/t_w is given in the *AISC Manual* as a property for channel sections. Thus, for h the authors recommend using the dimensional property T to compute the ratio. The webs of channels usually have low ratios h/t_w so that one might almost *assume* $\lambda < \lambda_p$. This channel is a "compact section" for the local buckling limit states.

   (b) Examine the lateral-torsional buckling limit state. Plastic moment strength can develop when $L_b \leq L_p$, where according to Eq. 9.6.3,

$$L_p = \frac{300}{\sqrt{F_{y, ksi}}}r_y = \frac{300(0.80)/12}{\sqrt{50}} = 2.8 \text{ ft}$$

$$M_p = Z_x F_y = 25.4(50)/12 = 106 \text{ ft-kips}$$

Using the properties $I_y = 3.86$ in.^{4}, $C_w = 112$ in.^{6}, $S_x = 21.5$ in.^{3}, $r_{ts} = 0.983$ in., $h_o = 11.5$ in., $J = 0.369$ in.^{4}, one obtains

$$M_r = 0.7F_yS_x = 0.7(50)(21.5)/12 = 62.7 \text{ ft-kips}$$

$$c = \frac{h_o}{2}\sqrt{\frac{I_y}{C_w}} = 1.07$$

$$L_r = 1.95r_{ts}\frac{E}{0.7F_y}\sqrt{\frac{Jc}{S_xh_o}}\sqrt{1 + \sqrt{1 + 6.76\left(\frac{0.7F_y}{E}\frac{S_xh_o}{Jc}\right)^2}} = 9.5 \text{ ft} \qquad [9.6.6]$$

Thus,

$$(L_p = 2.8 \text{ ft}) < (L_b = 8 \text{ ft}) < (L_r = 9.5 \text{ ft})$$

The nominal strength M_n is a linear interpolation between M_p and M_r according to Eq. 9.6.4. The governing laterally unbraced segment in Fig. 9.12.1 is the center one where the maximum moment occurs with a constant moment; therefore, $C_b = 1.0$ for that segment. Evaluating the nominal strength gives

$$M_n = C_b\left[M_p - (M_p - 0.7F_yS_x)\left(\frac{L_b - L_p}{L_r - L_p}\right)\right] \le M_p \qquad [9.6.4]$$

$$M_n = 1.0\left[107 - (107 - 62.7)\left(\frac{8.0 - 2.8}{9.1 - 2.8}\right)\right] = 70.4 \text{ ft-kips} \qquad ■$$

Zees

The zee-section lateral buckling strength is complicated by the fact that loading in the plane of the web causes unsymmetrical bending, resulting because a principal axis does not lie in that plane. The general treatment of buckling under biaxial bending is found in Sec. 9.14. The effect of biaxial bending on zee sections was found by Hill [9.32] to reduce the critical moment M_x to 90–95 percent of the value given by Eq. 9.4.33. In addition, the torsion-bending constant C_w, is different than for channels or I-shaped sections.

For design purposes, in view of the fact that unbraced zees are relatively rare, AISC does not provide for them. The authors have long suggested using one-half the values obtained using F_{cr} from Eq. 9.6.13, which is AISC Formula (F2-4).

Monosymmetric I-shapes

I-shaped sections symmetrical about the y-axis, but unsymmetrical about the x-axis, are summarized in the *SSRC Guide* [6.8] and by Clark and Hill [9.26]. The additional variable involved is y_0, the distance from the centroid of the girder cross-section to the shear center (positive if the shear center lies between the centroid and compression flange, otherwise negative).

The LRFD 1999 Specification for flexure of single symmetric I-shapes, as was given in LRFD-A-F1.1, is replaced in the 2005 Specification by a new method based on the work of White [9.75]. These new provisions are in AISC-F4 [see AISC-Table in *User Note* F1.1]. Four possible failure limit states are defined for singly symmetric I-shapes with flanges that could be (a) compact, (b) noncompact, or (c) slender, and webs that could be either compact or noncompact. The categories are (1) compression flange yielding (CFY), (2) lateral-torsional buckling (LTB), (3) compression flange local buckling (FLB), and (4) tension flange yielding (TFY). The possible local buckling of the web is taken into account by using the factors R_{pc} and R_{pt}. These two factors can vary from 1.0 to 1.6. Conservatively these factors can be assumed to be unity (1.0). The nominal flexural strength for each limit state is as follows:

1. *Compression flange yielding (CFY).* The nominal strength from AISC-F4.1 is

$$M_n = R_{pc}M_{yc} = R_{pc}F_yS_{xc} \qquad (9.12.1)$$

where S_{xc} is the section modulus I_x/y_c referred to the compression flange. This limit state occurs only when $S_{xt} > S_{xc}$.

2. *Lateral-torsional buckling (LTB).* The nominal strength from AISC-F4.2 is

(a) *Inelastic LTB with $L_p < L_b < L_r$.*

$$M_n = C_b \left[R_{pc}M_{yc} - \left(R_{pc}M_{yc} - F_L S_{xc} \right) \left(\frac{L_b - L_p}{L_r - L_p} \right) \right] \leq -R_{pc}M_{yt} \quad (9.12.2)$$

(b) *Elastic LTB with $L_b > L_r$,*

$$M_n = F_{cr}S_{nt} \leq R_{pc}M_{yc} \quad (9.12.3)$$

where $M_{yc} = F_y S_{xc}$

$$F_{cr} = \frac{C_b \pi^2 E}{\left(\dfrac{L_b}{r_t} \right)^2} \sqrt{1 + 0.078 \frac{J}{S_{xc}h_o} \left(\frac{L_b}{r_t} \right)^2} \qquad J = 0 \text{ if } \frac{I_{yc}}{I_y} \leq 0.23$$

$$F_L = 0.7F_y \text{ when } S_{xt}/S_{xc} \geq 0.7$$

$$= F_y S_{xt}/S_{xc} \text{ but not less than } 0.5F_y \text{ when } S_{xt}/S_{xc} < 0.7$$

$$L_p = 1.1r_t \sqrt{\frac{E}{F_L}} = \frac{187}{\sqrt{F_L}} \quad (9.12.4)$$

$$L_r = 1.95r_t \frac{E}{F_L} \sqrt{\frac{J}{S_{xc}h_o}} \sqrt{1 + \sqrt{1 + 6.76 \left(\frac{F_L}{E} \frac{S_{xc}h_o}{J} \right)^2}} \quad (9.12.5)$$

(c) *Web plastification factor* (as defined by AISC) R_{pc}, which accounts for premature failure due to web buckling, is

(aa) When $\dfrac{h_c}{t_w} \leq \left(\lambda_{pw} = \dfrac{640}{\sqrt{F, \text{ksi}}} \right)$,

$$R_{pc} = \frac{M_p}{M_{yc}} \quad (9.12.6)$$

(bb) When $\dfrac{h_c}{t_w} > \left(\lambda_{pw} = \dfrac{640}{\sqrt{F, \text{ksi}}} \right)$,

$$\left. \begin{aligned} R_{pc} &= \left[\frac{M_p}{M_{yc}} - \left(\frac{M_p}{M_{yc}} - 1 \right) \left(\frac{\lambda - \lambda_{pw}}{\lambda_{rw} - \lambda_{pw}} \right) \right] \leq \frac{M_p}{M_{yc}} \\ R_{pc} &= \left[\frac{M_p}{M_{yc}} - \left(\frac{M_p}{M_{yc}} - 1 \right) \left(\frac{\lambda - \dfrac{640}{\sqrt{F, \text{ksi}}}}{\dfrac{970}{\sqrt{F, \text{ksi}}} - \dfrac{640}{\sqrt{F, \text{ksi}}}} \right) \right] \leq \frac{M_F}{M_{yc}} \end{aligned} \right\} \quad (9.12.7)$$

where $M_p = F_y Z_x \leq 1.6 F_y S_{xc}$

The values inserted for λ_{pw} and λ_{rw} apply to doubly symmetric I-shaped sections bent about their major axis.

(d) *I-shapes having rectangular compression flange.* The effective radius of gyration r_t for lateral-torsional buckling is

$$r_t = \frac{b_{fc}}{\sqrt{12\left(\dfrac{h_o}{d} + \dfrac{1}{6}a_w\dfrac{h^2}{h_o d}\right)}} \tag{9.12.8}$$

where $\quad a_w = \dfrac{h_c t_w}{b_{fc} t_{fc}}$

$\qquad b_{fc} =$ compression flange width

$\qquad t_{fc} =$ compression flange thickness

For cases having channel caps and cover plates, properties are as defined in AISC-F4.2 (p. 16.1-51).

3. *Compression Flange Local Buckling (FLB).* The nominal strength from AISC-F4.3 is

(a) *Noncompact flanges* $\lambda_{pf} < \lambda < \lambda_{rf}$.

$$\left.\begin{array}{l} M_n = \left[R_{pc}M_{yc} - \left(R_{pc}M_{yc} - F_L S_{xc}\right)\left(\dfrac{\lambda - \lambda_{pf}}{\lambda_{rf} - \lambda_{pf}}\right)\right] \\[3em] M_n = \left[R_{pc}M_{yc} - \left(R_{pc}M_{yc} - F_L S_{xc}\right)\left(\dfrac{\dfrac{b_f}{2t_f} - \dfrac{65}{\sqrt{F_y}}}{\dfrac{170}{\sqrt{F_y}} - \dfrac{65}{\sqrt{F_y}}}\right)\right] \end{array}\right\} \tag{9.12.9}$$

The values inserted for λ_{pf} and λ_{rf} apply to flanges of rolled I-shaped sections and channels, and flanges of tees.

(b) *Slender flanges* $\lambda > \lambda_{rf}$,

$$M_n = \frac{0.9Ek_c S_{xc}}{\left(\lambda = \dfrac{b_{fc}}{2t_{fc}}\right)^2} \tag{9.12.10}$$

where $\quad k_c = \dfrac{4}{\sqrt{h/t_w}}$.

4. *Tension Flange Local Yielding (TFY).* The nominal strength from AISC-F4.4 is

$$M_n = R_{pt}M_{yt} = R_{pt}F_y S_{xt} \tag{9.12.11}$$

(a) When $\dfrac{h_c}{t_w} \leq \left(\lambda_{pw} = \dfrac{640}{\sqrt{F, \text{ksi}}}\right)$,

$$R_{pt} = \frac{M_p}{M_{yt}} \tag{9.12.12}$$

(b) When $\dfrac{h_c}{t_w} > \left(\lambda_{pw} = \dfrac{640}{\sqrt{F, \text{ksi}}}\right)$,

$$R_{pt} = \left[\frac{M_p}{M_{yt}} - \left(\frac{M_p}{M_{yt}} - 1 \right)\left(\frac{\lambda - \lambda_{pw}}{\lambda_{rw} - \lambda_{pw}} \right) \right] \leq \frac{M_p}{M_{yt}}$$

$$R_{pt} = \left[\frac{M_p}{M_{yt}} - \left(\frac{M_p}{M_{yt}} - 1 \right)\left(\frac{\dfrac{h_c}{t_w} - \dfrac{640}{\sqrt{F, \text{ksi}}}}{\dfrac{970}{\sqrt{F, \text{ksi}}} - \dfrac{640}{\sqrt{F, \text{ksi}}}} \right) \right] \leq \frac{M_p}{M_{yt}}$$

(9.12.13)

The values inserted for λ_{pw} and λ_{rw} apply to doubly symmetric I-shaped sections bent about their major axis.

Tee Sections

A T-section may be thought of as a monosymmetric I-shaped section that has the moment of inertia I_y of one flange equal to zero. Traditionally, both the ASD and LRFD Methods have been vague about how a T-section is to be treated. Rolled structural tees will rarely have strength controlled by the lateral-torsional buckling limit state. Whenever a tee section is loaded in the plane of its web (moment about the x-axis and r_x is less than r_y, there should be no limit on laterally unbraced length. Previously 1999 LRFD-F1 stated, "The lateral-torsional buckling limit state is not applicable to members subject to bending about the minor axis ..." This does not seem to appear in the 2005 AISC Specification or Commentary. However, the authors believe the statement is still valid. A significant number of rolled tees are in this category. Ellifritt, Wine, Sputo, and Samuel [9.74] have discussed the flexural strength of WT sections.

The lateral-torsional buckling (LTB) strength of a tee with the stem in compression is another matter. According to the AISC Commentary-F9, "... the LTB strength of a tee with the stem in compression may be only about one-fourth of the strength for the stem in tension." The portion of the stem in compression typically controls the strength for T-sections having reverse curvature. The moment gradient factor C_b is always to be taken as 1.0 when the stem is in compression. AISC-F9 identifies three possible limit states:

1. *Yielding (plastic moment strength is reached).* The nominal strength is

$$M_n = M_p \tag{9.12.14}$$

where $M_p = F_y Z_x \leq 1.6 M_y$ for stems in tension

$M_p \leq M_y$ for stems in compression

2. *Elastic lateral-torsional buckling.* The nominal strength is

$$M_n = M_{cr} = \frac{\pi \sqrt{E I_y G J}}{L_b} \left[B + \sqrt{1 + B^2} \right] \tag{9.12.15}$$

where $B = \pm 2.3 \left(\dfrac{d}{L_b} \right)\sqrt{\dfrac{I_y}{J}}$ \hspace{2cm} (9.12.16)

The plus sign is used when the stem is in tension, and the minus sign applies when the stem is in compression. AISC-F9 states, "If the tip of the stem is in compression anywhere along the *unbraced length*, the negative value of B shall be used."

3. *Flange local buckling.* The nominal strength is

$$M_n = F_{cr} S_{sc} \tag{9.12.17}$$

where S_{xt} = the elastic section modulus referred to the compression flange

$$F_{cr} = F_y \left(1.19 - 0.50 \frac{b_f}{2t_f} \sqrt{\frac{F_y}{E}} \right) \qquad \text{for noncompact sections} \quad (9.12.18)$$

$$F_{cr} = \frac{0.69E}{\left(\dfrac{b_f}{2t_f} \right)^2} \qquad \text{for slender sections} \quad (9.12.19)$$

EXAMPLE 9.12.2

Investigate the moment strength of a structural tee section WT7×19 both when used with its flange in compression and with the flange in tension. Use AISC Specification and A992 steel. Show how the strength is affected by lateral bracing.

Solution:

Case 1: Flange in Compression.

  (a) Maximum strength. The flange must satisfy that $\lambda \leq \lambda_p$ to preclude the flange local buckling limit state from reducing the strength.

$$\left(\lambda = \frac{b_f}{2t_f} = \frac{6.770}{2(0.515)} = 6.6 \right) < \left(\lambda_p = \frac{65}{\sqrt{F_y}} = 9.2 \right) \qquad \textbf{OK}$$

Since the stem (web) is primarily in tension, λ for the web does not have to satisfy any limit. Thus, AISC-F9.1 indicates

$$\text{Max } M_n = 1.6M_y = 1.6 S_x F_y = 1.6(4.22)(50)\frac{1}{12} = 28.1 \text{ ft-kips}$$

Also, the plastic moment strength cannot be exceeded,

$$M_p = F_y Z_x = 50(7.45)\frac{1}{12} = 31.0 \text{ ft-kips}$$

In this case, the shape factor is $31(1.6)/28.1 = 1.77$. Because of concern about inelastic deformation at service load, AISC limits the usable increase above first yield to $1.6M_y$ rather than M_p, which is $1.77M_y$.

  (b) The lateral torsional buckling strength controls the nominal strength, if a limiting unbraced length, similar to L_p, is exceeded. The limiting length can be determined by equating the nominal strengths defined by AISC Formula (F9-2) and Formula (F9-4) and solve for $L_{b,\text{limit}}$, as follows

$$F_y Z_x \text{ (or } 1.6M_y) = \frac{\pi \sqrt{EI_y GJ}}{L_{b,\text{limit}}} \left[B + \sqrt{1 + B^2} \right]$$

  (c) Determine the nominal strength expression for $L_b > L_{b,\text{ limit}}$.

$$M_{cr} = \frac{\pi \sqrt{EI_y GJ}}{L_b} \left[B + \sqrt{1 + B^2} \right] \leq 1.6M_y \qquad \text{AISC Formula (F9-4)}$$

where $B = \pm 2.3 \left(\frac{d}{L_b} \right) \sqrt{\frac{I_y}{J}}$ $\qquad$ AISC Formula (F9-5)

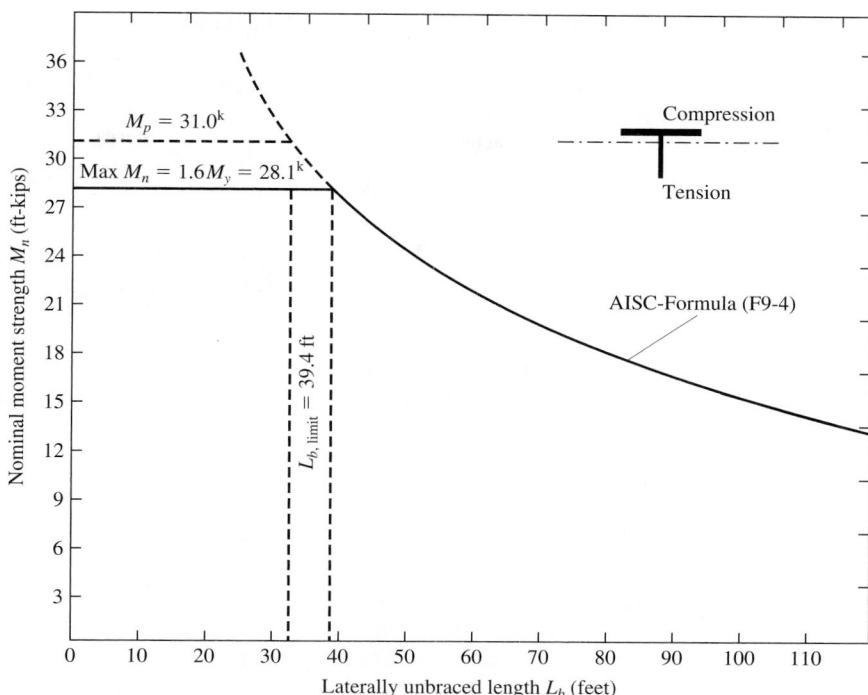

Figure 9.12.2
Nominal moment strength M_n vs laterally unbraced length L_b for WT7×19 having flange in compression.

The plus sign applies when the stem is in tension as for this case. Evaluating for the WT7×19, taking $E/G = 2.6$ gives,

$$M_{cr} = \frac{\pi E \sqrt{13.3(0.398/2.6)}}{L_b}[B + \sqrt{1 + B^2}]$$

$$M_{cr} = \frac{130,000}{L_b}[B + \sqrt{1 + B^2}]$$

where

$$B = +2.3\left(\frac{7.05}{L_b}\right)\sqrt{\frac{13.3}{0.398}} = +\frac{93.7}{L_b}$$

The nominal strength M_n variation with L_b is shown in Figure 9.12.2. The value of L_b at which the nominal strength is governed by both yielding and lateral torsional buckling limit states is dictated by the maximum nominal strength value of 28.1 ft-kip,

$$28.1(12) = \frac{130,000}{L_{b,\text{limit}}}\left[\frac{93.7}{L_{b,\text{limit}}} + \sqrt{1 + \left(\frac{93.7}{L_{b,\text{limit}}}\right)^2}\right]$$

By trial one gets

$$L_{b,\text{limit}} = 39.4 \text{ ft}$$

Case 2: Stem in compression

  (d) The maximum nominal strength when stem is in compression for the yielding limit state is capped by AISC-F9.1 at the yielding moment, M_y. Local buckling of the stem, when it is in compression, is not considered by AISC. The reason for not considering the stem slenderness is attributed to the significant reduction of the nominal strength when the stem is in compression.

  The next step would be to compute L_r, the value of L_b at which $M_{cr} = M_r$.

(e) Elastic lateral-torsional buckling. AISC Formula (F9-4) for tee sections as used in part (d) above is still valid when the stem is in compression. However, the constant B must be taken negative $(-)$; thus, the buckling stress will be lower for the tee in this orientation.

$$M_{cr} = \frac{130,000}{L_b}[B + \sqrt{1 + B^2}]$$

$$B = -2.3\left(\frac{7.05}{L_b}\right)\sqrt{\frac{13.3}{0.398}} = -\frac{93.7}{L_b}$$

The nominal strength relationship for the tee having its stem in compression is not presented in graphical form; however, M_n remains constant at $F_y S_{xc}$ to $L_{b,\,limit}$, then decreases according to M_{cr}. ∎

9.13* LATERAL BRACING DESIGN

The questions of what constitutes bracing and how to design bracing continue to be major concerns of practicing engineers. The subject is included in this chapter because a major item of concern in lateral bracing design is the restraint required to prevent lateral-torsional buckling in beams. The development of this section, however, is applicable to the bracing of columns as well as beams. In Sec. 6.8 the concept of *braced* and *unbraced* systems was briefly discussed in regard to the effective length factor K. In the following discussion, the emphasis in on *braced* systems; that is, the overall structural system is braced by cross bracing or attachment to an adjoining system that is braced. The bracing requirements for frames are treated in Chapter 14. Bracing for individual beams or columns may consist of cross bracing where the axial stiffness of the bracing elements is utilized; it may be provided at discrete locations by flexural members framing in transverse to the member being braced, wherein both axial and flexural stiffnesses of the bracing member are utilized; or such bracing may be provided continously by material such as light gage roof decking or wall panels.

AISC 2005 Appendix 6 contains provisions for the design of beam and column point bracing. Also point bracing has been treated by Zuk [9.39], Winter [9.40], Massey [9.41], Pincus [9.42], Galambos [9.43], Urdal [9.44], Lay and Galambos [9.45], Taylor and Ojalvo [9.46], Hartmann [9.47], Mutton and Trahair [9.48], Medland and Segedin [9.49], and Plaut [9.73]. Recent practical treatment has been provided by Yura [9.50], Lutz and Fisher [9.51], Ales and Yura [9.52], Clarke and Bridge [9.53], and Yura [9.72]. What follows is largely a combination of the work of Winter [9.40], Galambos [9.43], and Yura [9.50, 9.72].

Point Bracing for Elastic Columns and Beams

Consider the axially loaded column of Fig. 9.13.1a where the top and bottom of the member are assumed to be supported in such a way that no side movement occurs at one end relative to the other. Such restraint would constitute a *braced* system. The bracing to create such restraint may be considered as a spring at the top that is capable of developing a horizontal reaction equal to the spring constant β times the deflection Δ. When the brace has a large spring constant (that is, the brace is very stiff) the deflection Δ could be close to zero and yet the spring may provide a large enough horizontal force to prevent any side motion (sideway) at the top. This would be the situation in Fig. 9.13.1b. The equilibrium

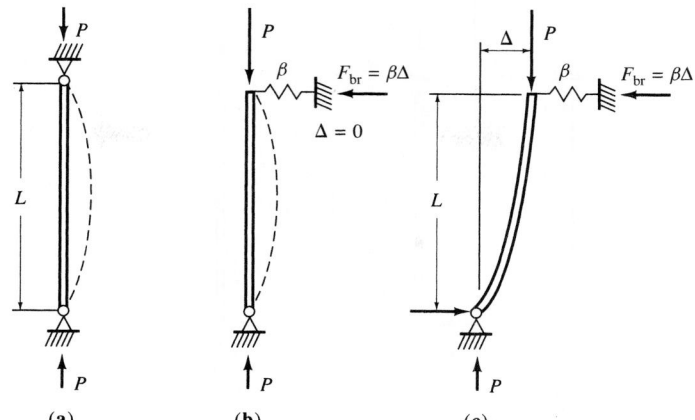

Figure 9.13.1
Bracing for a single-story
column.

(a)   (b)   (c)

requirement is shown in Fig. 9.13.1c, wherein a sidesway is shown. If one imagines this as
a slightly deflected position, then in order to have equilibrium, it is required that

$$P\Delta = F_{br}L = (\beta\Delta)L \tag{9.13.1}$$

If $(\beta\Delta)L$ is less than $P\Delta$, sidesway occurs. If $(\beta\Delta)L$ is greater than $P\Delta$, no sidesway occurs, and the column would be considered braced. The ideal brace, then, would be
one that has just enough stiffness β to prevent movement (at the top in this example); that is,

$$\beta = \frac{P}{L} \tag{9.13.2}$$

The maximum load for which bracing would be required is the elastic bucking load P_{cr}, or
the load causing yielding or inelastic buckling if that is lower than the elastic P_{cr}. Thus the
largest required stiffness β_{ideal} is

$$\beta_{ideal} = \frac{P_{cr}}{L} \tag{9.13.3}$$

The concept is shown by the plot of P vs βL in Fig. 9.13.2, wherein when β exceeds
β_{ideal}, P_{cr} is reached and the column buckles without end translation (sidesway); in other
words, it is a *braced* system. When β is less than β_{ideal}, a sidesway deflection will occur
such that $P = \beta L$; in other words, a so-called *unbraced* system. The major treatment of
unbraced systems is in Chapter 14, devoted to rigid frames.

Figure 9.13.2
Brace stiffness relative to
concept of "braced"
$(K = 1.0)$ and "unbraced"
$(K > 1.0)$ systems for
column hinged at top and
bottom.

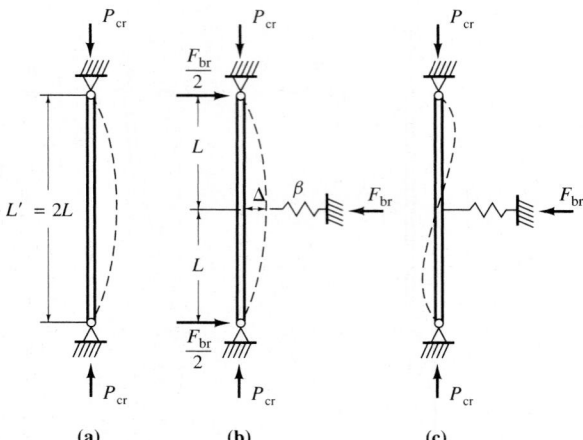

Figure 9.13.3
Mid-height brace for a
two-story column.

(a)   (b)   (c)

Next, extend the concept to a two-story column within a braced system, as shown in Fig. 9.13.3. When no displacement occurs at mid-height, i.e., full bracing is provided, the column will buckle at a load nearly equal to

$$P_{cr} = \frac{\pi^2 EA}{(L/r)^2}$$ (9.13.4)

In other words, one may imagine that a hinge exists at mid-height. Buckling occurs when the column snaps into the two half-wave mode of Fig. 9.13.3c.

Taking moments about the imaginary hinge location with the column deflected by an amount Δ, as in Fig. 9.13.3b, gives

$$P_{cr}\Delta = \frac{F_{br}}{2}L$$ (9.13.5)

Since $F_{br} = \beta\Delta$,

$$P_{cr}\Delta = \frac{(\beta\Delta)L}{2}$$ (9.13.6)

As in the one-story column, if β_{ideal} is the necessary stiffness to create a nodal point (zero deflection) at mid-height of the two-story column, then

$$\beta_{ideal} = \frac{2P_{cr}}{L}$$ (9.13.7)

For situations with more than two equal spans, the same procedure may be used to obtain β_{ideal}. Examination of Fig. 9.13.4 for three equal spans will show that the spring forces F_{br} can act either in the same or in opposite directions. Assuming they act in the *same* direction (Fig. 9.13.4a), using imaginary hinges at one-third span points, and taking moments when slightly deflected at the brace points, gives

$$F_{br}L = (\beta\Delta)L = P_{cr}\Delta; \qquad \beta_{ideal} = \frac{P_{cr}}{L}$$ (9.13.8)

Assuming the forces F_{br} acting in *opposite* directions (Fig. 9.13.4b) gives

$$F_{br}L/3 = (\beta\Delta)L = P_{cr}\Delta; \qquad \beta_{ideal} = \frac{3P_{cr}}{L}$$ (9.13.9)

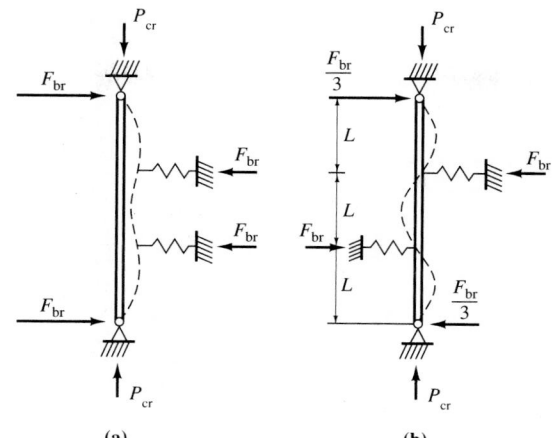

Figure 9.13.4
Column braced to make three
equal spans.

The configuration requiring the highest spring constant is the correct one, that which will permit the highest critical load. If a lesser stiffness is used, an alternate buckling mode will occur at a lower load, accompanied by displacement at the springs.

By the same process, β_{ideal} may be determined for any number of equal spans. In general,

$$\beta_{\text{ideal}} = \frac{\alpha P_{\text{cr}}}{L} \qquad (9.13.10)$$

where α varies from 1 for one span to 4 for infinite equal spans. The variation is given in Fig. 9.13.5.

Thus Eqs. 9.13.3, 9.13.7, 9.13.9, and 9.13.10 give the ideal brace *stiffness* to prevent translation at the points where the braces act.

In addition to stiffness, a brace must provide adequate *strength*. The strength F_{br} required of an ideal brace is

$$F_{\text{br}} = \beta_{\text{ideal}}\Delta \qquad (9.13.11)$$

but until buckling occurs, Δ is zero (see Fig. 9.13.6); therefore there will be no brace force in the ideal system until buckling occurs.

Compression members in real structures are not perfectly straight, perfectly aligned vertically, nor perfectly loaded as assumed in calculations; there is always an initial crookedness. In other words, Δ is not zero even when there is no compressive load P acting. Reexamine the single-story column of Fig. 9.13.1 assuming there is an initial deflection Δ_0 that exists even when P is zero. Then, as shown in Fig. 9.13.7, equilibrium requires

$$(\beta\Delta)L = P(\Delta + \Delta_0) \qquad (9.13.12)$$

Figure 9.13.5
Variation of required spring
constant for column with
number of equal unbraced
spans.

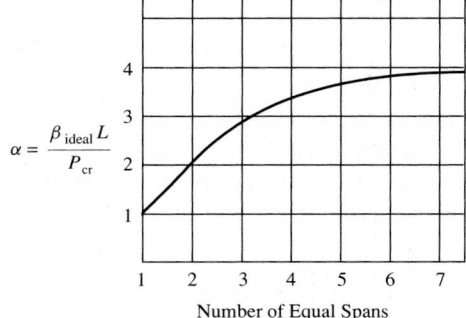

$$\alpha = \frac{\beta_{\text{ideal}}\,L}{P_{\text{cr}}}$$

Number of Equal Spans

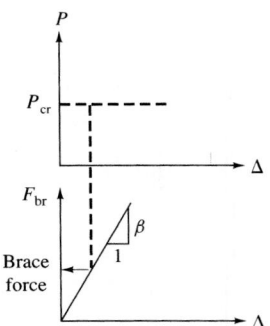

Figure 9.13.6
Brace force relative to
column load for ideal system.

Figure 9.13.7
Column with initial
crookedness Δ_0.

(a) No load applied   (b) When load is applied

for $P = P_{cr}$,

$$\beta_{reqd} = \frac{P_{cr}}{L}\left(1 + \frac{\Delta_0}{\Delta}\right) \tag{9.13.13}$$

Since $\beta_{ideal} = P_{cr}/L$, Eq. 9.13.13 then becomes

$$\beta_{reqd} = \beta_{ideal}\left(1 + \frac{\Delta_0}{\Delta}\right) \tag{9.13.14}$$

which is the *stiffness* requirement for compression members having initial crookedness Δ_0.

The *strength* requirement is then

$$F_{br} = \beta_{reqd}\Delta = \beta_{ideal}\left(1 + \frac{\Delta_0}{\Delta}\right)\Delta$$

$$F_{br} = \beta_{ideal}(\Delta + \Delta_0) \tag{9.13.15}$$

Normal tolerances on crookedness of compression members would vary from 1/500 to 1/1000 of the length [9.41]. The AISC *Code of Standard Practice** indicates acceptable out-of-plumbness to be $L/1000$. Considering accidental eccentricity of loading, Winter [9.40] suggests taking Δ_0 from 1/250 to 1/500 of the length.

**Code of Standard Practice for Steel Buildings and Bridges,* American Institute of Steel Construction, March 18, 2005 (AISC-Section 16.3).

AISC Provisions—Stability Bracing Design for Beams

The following applies to the AISC Load and Resistance Factor Design Method. (Similar equations available in AISC-Appendix 6.3.) To obtain design equations, Winter [9.40] suggested $\Delta = \Delta_0 = L/500$. Substitution of this in Eqs. (9.13.14) and (9.13.15) gives the design equations.

1. For *stiffness* β_{reqd},

$$\beta_{reqd} = 2\beta_{ideal} \tag{9.13.16}$$

2. For *nominal strength* F_{br},

$$F_{br} = \beta_{ideal}(2\Delta_0)$$
$$F_{br} = \beta_{ideal}(0.004L_b) \tag{9.13.17}$$

where $\beta_{ideal} = P_{cr}/L_b$.

AISC-Appendix 6 (p. 16.1-191) identifies two types of bracing to achieve point restraint against (a) lateral movement, (b) torsional movement, and (c) lateral-torsional buckling. In this section, only lateral bracing is discussed. Lateral bracing is classified as either relative bracing or nodal bracing. In floor systems, relative bracing uses diagonal struts, whereas nodal bracing uses transverse bracing supported by rigid supports, as shown in Fig. 9.13.8. For lateral bracing, the brace must be attached near the compression flange, except for a cantilever, where an end brace must be attached to the top flange. For continuous beams, braces must be attached to both flanges at the brace point nearest the inflection point (AISC-Appendix 6.3.1)

Using Winter's approach, AISC developed the following design equations to obtain the required nominal strength and stiffness for lateral bracing:

1. Relative bracing.   The required brace *strength* is

$$P_{br} = \frac{0.008M_rC_d}{h_o} \tag{9.13.18}$$

where
M_r = required flexural strength (factored service loads) using either the AISC LRFD Method or the AISC ASD Method

C_d = 1.0 for single curvature = 2.0 for double curvature (This value is to be applied to the brace point nearest the point of inflection.)

h_o = distance between flange centroids

Assuming nominal strength is controlled by elastic buckling,

$$P_u = \frac{\pi^2 EI}{L_b^2}$$

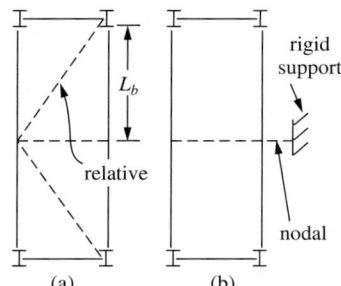

Figure 9.13.8
Types of lateral bracing
(a) relative bracing,
(b) nodal bracing

Equate required strength to available nominal strength and solve for the required moment of inertia I

$$\frac{\pi^2 EI}{L_b^2} = \frac{0.008 M_r C_d}{h_o}$$

$$\text{Required } I = \frac{0.008 M_r C_d L_b^2}{\pi^2 E h_o} \tag{9.13.19}$$

The required brace *stiffness* is

$$\beta_{\text{br}} = \frac{1}{\phi}\left(\frac{4 M_r C_d}{L_b h_o}\right) \tag{9.13.20}$$

Recognizing that axial stiffness β is EA/L which can be equated to the required brace stiffness,

$$\frac{EA}{L_b} = \frac{1}{\phi}\left(\frac{4 M_r C_d}{L_b h_o}\right)$$

Solving for the required area of the brace,

$$\text{Required } A = \frac{1}{\phi}\frac{L_b}{E}\left(\frac{4 M_r C_d}{L_b h_o}\right) \tag{9.13.21}$$

2. Nodal bracing. The required brace *strength* is

$$P_{\text{br}} = \frac{0.02 M_r C_d}{h_o} \tag{9.13.22}$$

Assuming nominal strength is controlled by elastic buckling,

$$P_u = \frac{\pi^2 EI}{L_b^2}$$

Equate required strength to available nominal strength and solve for the required moment of inertia I

$$\frac{\pi^2 EI}{L_b^2} = \frac{0.02 M_r C_d}{h_o}$$

$$\text{Required } I = \frac{0.02 M_r C_d L_b^2}{\pi^2 E h_o} \tag{9.13.23}$$

The required brace *stiffness* is

$$\beta_{\text{br}} = \frac{1}{\phi}\left(\frac{10 M_r C_d}{L_b h_o}\right) \tag{9.13.24}$$

Recognizing that axial stiffness β is EA/L which can be equated to the required brace stiffness,

$$\frac{EA}{L} = \frac{1}{\phi}\left(\frac{10 M_r C_d}{L_b h_o}\right)$$

Solving for the required area of the brace,

$$\text{Required } A = \frac{1}{\phi}\frac{L}{E}\left(\frac{10 M_r C_d}{L_b h_o}\right) \tag{9.13.25}$$

where $\phi = 0.75$

L = length of brace

L_b = laterally unbraced length

The other terms are defined following Eq. 9.13.18.

EXAMPLE 9.13.1

Design a brace (brace A) to provide lateral support for a W27×84 beam (beam A) positioned as shown in Fig. 9.13.9. Assume the braces are 7.5 ft long, are attached near the compression flange, and are located along the supported beam at the one-third point of a 48-ft span. Use the AISC Load and Resistance Factor Method and A992 steel.

Solution:

Since the bracing locations are given, the first step is to estimate the force in the compression zone of the beam or beams when the nominal strength M_n of the beams is reached. Since the braces are 16 ft apart, the strength of the beam could be based on F_{cr} less than F_y. For this W27×84 section, $L_p = 6.89$ ft and $L_r = 20.3$ ft. Thus, when braced every 16 ft this beam will have M_n less than M_p.

This is a case of nodal bracing. Hence, the required brace strength and stiffness are determined using Eqs. 9.13.10 and 9.13.11. The required factored load strength $M_r = M_u$ (M_r is used as a symbol in AISC-Appendix 6 for *required flexural strength*) is assumed to be equal to the design strength $\phi_b M_n$. Therefore,

$$M_n = C_b\left[M_p - (M_p - 0.7 F_y S_x)\left(\frac{L_b - L_p}{L_r - L_p}\right)\right]$$

Figure 9.13.9
Data for Example 9.13.1.

(a) Plan view

(b)

where C_b is taken as 1.0, and

$$M_p = F_y Z_x = 50(244)\tfrac{1}{12} = 1017 \text{ ft-kips}$$

Thus,

$$M_n = 1.0 \left\{ 1017 - \left[1017 - 0.7(50)(213)\tfrac{1}{12} \right] \left(\frac{16 - 6.89}{20.3 - 6.89} \right) \right\} = 748 \text{ ft-kips}$$

Required design strength $M_r = \phi_b M_n = 0.90\,(748) = 673$ ft-kips

For the W27×84 with $h_o = 26.1$ in., required moment of inertia I is

$$\text{Required } I = \frac{0.02 M_r C_d L_b^2}{\pi^2 E h_o} = \frac{0.02(673)(1.0)(16)^2}{(3.14)^2 29{,}000(26.1)}(12)^3 = 0.8 \text{ in.}^4$$

and required area A is

$$\text{Required } A = \frac{1}{\phi}\frac{L}{E}\left(\frac{10 M_r C_d}{L_b h_o} \right) = \frac{1}{0.75}\frac{7.5(12)}{29{,}000}\left(\frac{10(673)(12)(1.0)}{16(12)(26.1)} \right) = 0.07 \text{ sq. in.}$$

Use C6×10.5 as the lightest section having $I_y = 0.86$ in.$^4 > 0.8$in.^{4} and $A = 3.08$ sq in. $>$ 0.07 sq in. ∎

EXAMPLE 9.13.2

For the beam laterally braced by joists as shown in Fig. 9.13.10, determine the amount of weld required so that the joist will adequately brace the beam. The service bending moment is 25 ft-kips dead load and 100 ft-kips live load. The steel is A36. Use the Load and Resistance Factor Design Method.

Solution:

(a) Determine stiffness required. Whereas in Example 9.13.1 the compression strength P_{cr} of the flange was used, in this example the strength to be accommodated is computed from the factored load force in the flange. The factored moment M_u is

$$M_u = 1.2(25) + 1.6(100) = 190 \text{ ft-kips}$$

(b) The joist bracing system will be assumed to provide relative bracing in this case, therefore the required brace stiffness using the LRFD Method is:

$$\beta_{br} = \frac{1}{\phi}\left(\frac{4 M_r C_d}{L_b h_o} \right)$$

4'–0" Joist spacing

W14×34

Figure 9.13.10
Beam laterally braced by joists.

Figure 9.13.11
Diaphragm action of metal
deck attached to joists.

$L_b = 4$ ft, $C_d = 1.0$, $M_r = 190$ ft-kips, and $h_o = 13.5$ in. for W14×34

$$\beta_{br} = \frac{1}{0.75}\left(\frac{4(190)(12)(1.0)}{(4)(13.5)}\right) = 225 \text{ kips/ft}$$

Assuming the joists have metal deck adequately attached to them, there will develop in the decking an in-plane shear, known as *diaphragm action*, that will restrain the relative axial movement of two adjacent joists, thereby restraining relative lateral movement between two adjacent laterally braced points on the beam being braced, such as points A and B of Fig. 9.13.11. The adequacy of the attachment of the decking to the joists will determine the degree to which diaphragm action prevents the relative motion of points, such as A and B. If there is zero relative motion, then the requirement for lateral support may be based on $\alpha = 1$, i.e., the same as the bracing of a single-story column. In which case, the required stiffness would be

$$\beta_{\text{reqd}} = 2\beta_{\text{ideal}} = \frac{2\alpha P_{\text{cr}}}{L} = \frac{2(1)(181)}{4} = 90 \text{ kips/ft}$$

Using diaphragm action can greatly reduce the stiffness requirement for point bracing. In this case, if diaphragm action can be developed, the stiffness requirement is one-fourth as much as required without any diaphragm action. Metal deck with $2\frac{1}{2}$-in. concrete fill would provide many times the required stiffness. Even the diaphragm action of the metal deck without the concrete slab is likely to provide more than the 90 kips/ft required. Treatment of diaphragm bracing is generally outside the scope of this text; however, a number of references [9.54–9.65] on diaphragm action are included at the end of this chapter.

(c) Determine weld required to attach joists to top flange of W14×34. The force required is, according to Eq. 9.13.22,

$$\text{Required } F_{\text{br}} = 0.004\alpha P_{\text{cr}} \qquad [9.13.17]$$

Conservatively take $\alpha = 4$ as for a series of point braces,

$$\text{Required } F_{\text{br}} = 0.004(4)181 = 2.9 \text{ kips}$$

Using a $\frac{3}{16}$-in. fillet weld (satisfies AISC-Table J2.4 for minimum weld size on $\frac{7}{16}$-in. flange thickness) with E70 electrodes and the SMAW process, the nominal strength R_{nw} per inch is

$$R_{nw} = \tfrac{3}{16}(0.707)0.60(70) = 5.57 \text{ kips/in.}$$

$$\text{Required } L_w = \frac{F_{\text{br}}}{R_{nw}} = \frac{2.9}{5.57} = 0.5 \text{ in.}$$

Use $\frac{3}{16}$-in. weld, E70, $L = 0' - 0\frac{3}{4}''$ on each side of joist bearing seat. This is needed to provide minimum length of 4 times weld size, satistying AISC-J2.2b. ∎

Bracing Requirements for Inelastic Steel Beams

When ability to accommodate large plastic strain is desired at bracing points, such as when plastic strength or plastic analysis is used in design, the procedure in the previous section may not be adequate. Lay and Galambos [9.45] have developed a set of rules for design in cases where such high plastic strain (rotation capacity) is required to be accommodated.

In effect, bracing requirements are based on a rotation capacity R consistent with the beam unbraced length slenderness ratio given by Eq. 9.5.7. It has been found that within the laterally unbraced length "local buckling causes a curtailment of the load capacity of the member and therefore defines the rotation capacity of the beam" [9.45].

The derivation of Lay and Galambos [9.45] has determined the maximum lateral moment that can develop in the compression flange under a _uniform_ moment, $M_x = M_p$, by using the strain distributions on the compression flange due to (a) compression due to $M_x = M_p$ and (b) the lateral bending strains when local buckling occurs on the "compression" side.

The design recommendations are:

1. For axial _strength,_ the required cross-sectional area is

$$\text{Required } A_{\text{brace}} = \left[\frac{\alpha_{st} - 1}{\alpha_e - \sqrt{\alpha_e}}\right]\left[\frac{2}{3}\right]\frac{A_{\text{comp}}}{(L_{\text{av}}/b)} \qquad (9.13.26)$$

where $\quad L_{\text{av}} = \dfrac{2L_L L_R}{L_L + L_R}$

L_L = unbraced length to left of braced point

L_R = unbraced length to right of braced point

b = width of compression flange

α_{st} = strain at strain hardening divided by yield strain, ϵ_{st}/ϵ_y (A value of 12 may be reasonable for steels up to $F_y = 60$ ksi [9.46].)

α_e = elastic modulus divided by strain hardening modulus of elasticity, E_s/E_{st}

2. The axial _stiffness_ requirement is satisfied when

$$\frac{L_{\text{brace}}}{L_{\text{av}}} \leq 0.57\left[\frac{\alpha_{st} - 1}{\alpha_e - \sqrt{\alpha_e}}\right]\left[\frac{\text{actual } A_{\text{brace}}}{\text{required } A_{\text{brace}}}\right]\left[\frac{L_a}{b}\right] \qquad (9.13.27)$$

where $\quad L_a$ = longer of the two adjacent unbraced lengths.

In addition to the axial strength and stiffness requirements, Lay and Galambos [9.45] indicated that when only the compression flange is braced there are additional flexural strength and stiffness requirements that must be satisfied. These flexural requirements (not given here) give overly large and deep bracing members. Recent studies [9.47] indicate that flexural requirements are unnecessary for lateral bracing locations away from beam vertical reactions. When the compression flange is braced, point restraint giving the necessary axial strength and stiffness is sufficient. Lateral bracing at vertical supports undoubtedly does need some flexural strength and stiffness to prevent a beam from tipping, but ordinary framing at such locations generally provides adequate flexural strength and stiffness.

EXAMPLE 9.13.3

For the beam of Example 9.13.1 (Fig. 9.13.9) determine the brace size required if plastic hinge rotation is required at the bracing points.

Solution:

(a) Determine the section required for axial *strength.* Use Eq. 9.13.26,

$$\text{Required } A_{\text{brace}} = \left(\frac{\alpha_{st} - 1}{\alpha_e - \sqrt{\alpha_e}} \right) \left(\frac{2}{3} \right) \frac{A_{\text{comp}}}{(L_{\text{av}}/b)}$$

where $\alpha_{st} = \epsilon_{st}/\epsilon_y$ which may be taken as 12 for $F_y = 36$ ksi and can probably also be used for steels to about $F_y = 60$ ksi. For $\alpha_e = E/E_{st}$, use $E_{st} = 450$ ksi, giving $\alpha_e = 29,000/450 = 64$. Certainly, the use of such values is accurate enough for design purposes. For A36 steel, $\alpha_{st} = 12$, $\alpha_e = 64$.

$$\frac{\alpha_{st} - 1}{\alpha_e - \sqrt{\alpha_e}} = \frac{12 - 1}{64 - \sqrt{64}} = 0.2$$

$$L_{\text{av}} = \frac{2L_L L_R}{L_L + L_R} = \frac{2(16)(16)(12)}{16 + 16} = 192 \text{ in.}$$

$$\text{Required } A_{\text{brace}} = 0.2 \left(\frac{2}{3} \right) \frac{A_{\text{comp}}}{192/9.96} = 0.007 A_{\text{comp}}$$

$$= 0.007(24.8/2) = 0.09 \text{ sq in.}$$

Practically any rolled shape will satisfy this small requirement.

(b) Examine the requirement for axial *stiffness.* Use Eq. 9.13.27, assuming conservatively the area A_{brace} provided equals exactly the required,

$$\frac{L_{\text{brace}}}{L_{\text{av}}} \leq 0.57 \left[\frac{\alpha_{st} - 1}{\alpha_e - \sqrt{\alpha_e}} \right] \left[\frac{\text{actual } A_{\text{brace}}}{\text{required } A_{\text{brace}}} \right] \frac{L_a}{b}$$

$$\leq 0.57(0.2)(1.0)(192/9.96) = 2.2$$

$$L_{\text{brace}} \leq 2.2 L_{\text{av}} = 2.2(16) = 35 > 7.5 \text{ ft for brace} \qquad \textbf{OK}$$

As in Example 9.13.1, the slenderness ratio recommended limit of 200 for compression members will give a member adequate for bracing purposes.

Use C4×5.4. ■

One conclusion regarding lateral bracing for beams and columns is that the requirements for such bracing are easily met. It is more important to provide a brace of *some size* than to be overly concerned about what the size should be.

Empirical procedures have long been used in lieu of a rational investigation of the strength and stiffness requirements for braces. A summary of some of these rules is given by Lay and Galambos [9.45]. The typical rule-of-thumb has been to use a brace having a strength equal to or greater than 2 percent of the compressive strength of the compression element being braced. This seems to be a conservative alternative to an analytical study.

9.14* BIAXIAL BENDING OF DOUBLY SYMMETRIC I-SHAPED SECTIONS

When an I-shaped section is loaded in the plane of its major axis, that is, a moment M_y causes weak axis bending, in combination with strong axis bending M_x from loading in the plane of its web, the strength limit state may be controlled by yielding under combined stress or by lateral-torsional buckling. Two common situations are roof purlins and crane support girders. How does the moment M_y affect lateral stability? When M_y acts alone lateral-torsional buckling is *not* a possible limit state, and most rolled shapes will develop their plastic moment. When M_x acts alone, lateral-torsional buckling may well be the controlling limit state.

This subject of biaxial bending, including torsion, has been treated by Dohrenwend [9.66], Gaylord and Gaylord [9.67], Springfield [9.68], Pastor and DeWolf [9.69], and Razzaq and Galambos [9.70, 9.71]. Many others, including the *SSRC Guide* [6.8, pp. 344–351; 6.20], have treated the subject in the context of beam-columns, which are discussed in Chapter 12.

From Eq. 9.4.32 for pure bending M_x with respect to the strong axis,

$$M_{cr}^2 = M_x^2 = \frac{\pi^2}{L^2}\left[\frac{\pi^2 E^2 C_w I_y}{L^2} + EI_y GJ\right] \tag{9.14.1}$$

or

$$\frac{M_x^2}{EI_y GJ} = \frac{\pi^2}{L^2}\left[1 + \left(\frac{\pi}{\lambda L}\right)^2\right] \tag{9.14.2}$$

where $\lambda = 1/a = \sqrt{GJ/EC_w}$.

When simultaneously a moment M_y is applied, Eq. 9.14.2 becomes [9.65]

$$\frac{M_x^2}{EI_y GJ} + \frac{M_y^2}{EI_x GJ} = \frac{\pi^2}{L^2}\left[1 + \left(\frac{\pi}{\lambda L}\right)^2\right] \tag{9.14.3}$$

which is applicable to I-shaped sections with two axes of symmerty. Furthermore, it is applicable for sections with point symmetry (such as the zee), and is approximately valid for channels when M_y does not exceed $0.25M_x$ [9.67].

Combinations of M_x and M_y which satisfy Eq. 9.14.3 will plot as an ellipse, as shown in Fig. 9.14.1 for a W14×74.

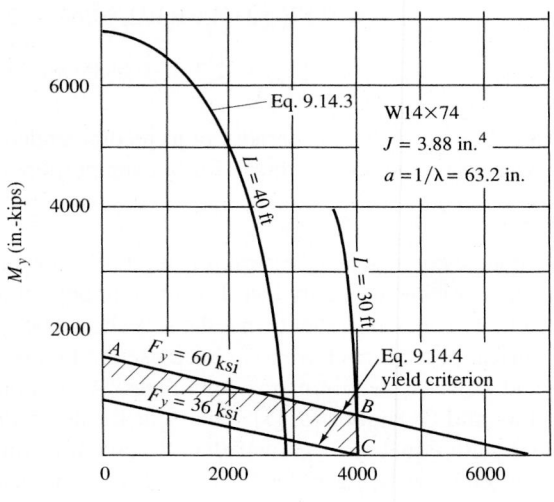

Figure 9.14.1
Lateral buckling strength for biaxial bending of doubly symmetric I-shaped sections.

The usual limit state accepted for biaxial bending is the achievement of the yield stress F_y at the extreme fiber under combined stress. Therefore, no matter how stable a beam may be, the combination of moments must satisfy

$$\frac{M_x}{S_x} + \frac{M_y}{S_y} \leq F_y \qquad (9.14.4)$$

Since the relationships shown in Fig. 9.14.1 for W14×74 are typical, certain conclusions may be drawn. Consider a laterally unbraced length of 30 ft and steel with $F_y = 60$ ksi. Ideally, the ultimate condition is defined by the lines AB (yielding controls) and BC (buckling controls).

The most important observation is that for ordinary laterally unbraced lengths (say 25 ft or less) the line BC is nearly vertical; therefore *simultaneous application of M_y does not appreciably affect the critical moment M_x.*

AISC LRFD Design Method Provisions

For biaxial bending, the first *User Note* in AISC-F (p. 16.1-44) refers to AISC-H1 to H3, which apply for doubly symmetric members subject to bending and axial force. The interaction equations for combined bending and axial load are treated in Chap. 12. When torsion is included in the loading, the *User Note* (p. 16.1-70) indicates to use AISC-H3, which applies to "Members under Torsion and Combined Torsion, Flexure, Shear, and/or Axial Force."

Without axial load P_u, AISC-Formula (H1-1b) would be applicable, as follows:

$$\left(\frac{M_{ux}}{\phi_b M_{nx}} + \frac{M_{uy}}{\phi_b M_{ny}} \right) \leq 1.0 \qquad (9.14.5)$$

where M_{ux} = factored service load moment about the x-axis (AISC uses M_{rx})
   M_{uy} = factored service load moment about the y-axis (AISC uses M_{ry})
   M_{nx} = nominal strength for bending about the x-axis (AISC uses M_{cx}/ϕ_b)
   M_{ny} = nominal strength for bending about the y-axis (AISC uses M_{cy}/ϕ_b)
   ϕ_b = resistance factor in bending, 0.90

It is logical to use such an interaction equation as Eq. 9.14.5; however, from discussion at the beginning of this section, such a procedure is conservative. Note that M_{nx} and M_{ny} can be as high as M_{px} and M_{py}, respectively.

Earlier ASD Specifications did not specifically direct the designer to the combined bending and axial load interaction equations for biaxial bending *without* axial load.

The authors, and others [9.67] have long recommended the following procedure to be used for biaxial bending of symmetric I-shaped sections subject to biaxial bending without axial compression.

1. For the *yielding limit state* controlling (line AB of Fig. 9.14.1):

$$\left(f_{un} = \frac{M_{ux}}{S_x} + \frac{M_{uy}}{S_y} \right) \leq \phi_b F_y \qquad (9.14.6)$$

2. For the *lateral-torsional buckling limit state* controlling (line BC of Fig. 9.14.1):

$$\phi M_{nx} \geq M_{ux} \qquad (9.14.7)$$

where f_{un} = factored service load normal (compressive or tensile) stress

M_{ux} = factored service load moment about the x-axis

M_{uy} = factored service load moment about the y-axis

M_{nx} = nominal moment strength for a member loaded only in the plane of the web (i.e., strong axis bending according to AISC-F1).

AISC-H1 uses the assumption that the *yielding limit state* controls; thus, the provision requires the combined normal (compressive or tensile) stress f_{un} under factored loads not to exceed $\phi_b F_y$.

A common application of the biaxial bending analysis is when the applied loading includes torsional loading, which contributes to y-axis bending, as discussed in Chap. 8. Equations 9.14.6 and 9.14.7 seem particularly applicable for that situation; though AISC-H2 indicates to use Eq. 9.14.6 alone. Note that AISC-H1 (Eq. 9.14.5) is *not* applicable when torsion is included.

It is explicitly stated in AISC Commentary-H3.3 that torsional strength for "odd-shaped" built-up cross-sections can be computed using AISC-H3.3. This implies that hot-rolled sections subject to combined stress and torsion must be investigated using AISC-H1 and H2.

EXAMPLE 9.14.1

Design a W section to serve as a crane support girder to carry a live load moment M_x of 301 ft-kips (without impact) from the crane wheels. In addition, a moment M_f of 30 ft-kips acts in the plane of the top flange as a result of movement of the crane back and forth between the support girders. This moment M_f is based on a lateral force acting on the top flange equal to 10% (the total is 20% with one-half at each end) of the lifted load and crane trolley weight in accordance with ASCE 7-05, Section 4.10. The moment M_f about the y-axis is assumed to be resisted by one flange; in effect, this is to account for the torsional effect by using the flexure analogy (see Sec. 8.6). The approximation of equivalent systems is shown in Fig. 9.14.2. Assume the simply supported span of 24 ft is laterally braced only at the ends. Use Load and Resistance Factor Design and $F_y = 50$ ksi.

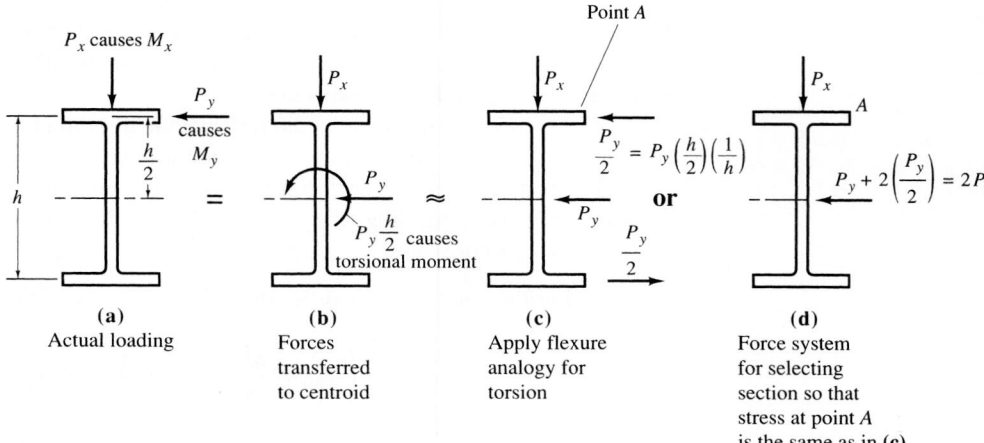

Figure 9.14.2
Approximate equivalent system for biaxial bending and torsion.

Solution:

  (a) Compute factored moments. Using ASCE 7 LRFD load combinations,

$$\text{Estimated dead load moment} = 10 \text{ ft-kips}$$

$$M_{ux} = 1.2(10) + 1.6(301)1.25 \quad (\text{using 25\% impact increase})$$

$$= 614 \text{ ft-kips}$$

$$M_{uf} = 1.6(30) = 48 \text{ ft-kips}$$

$$M_{uy} = 2M_{uf} = 96 \text{ ft-kips} \quad (\text{see Fig. 9.14.2b})$$

  (b) Select a section. Use the approach of Sec. 7.11 with the criterion of Eq. 9.14.6,

$$\text{Required } S_x \geq \frac{M_{ux}}{\phi_b F_y} + \frac{M_{uy}}{\phi_b F_y}\left(\frac{S_x}{S_y}\right)$$

$$\text{Required } S_x \geq \frac{614(12)}{0.90(50)} + \frac{96(12)}{0.90(50)}\left(\frac{S_x}{S_y}\right)$$

$$\geq 164 + 26(\approx 6) = 320 \text{ in.}^3$$

Try W24×131: $S_x = 329$ in.^{3}

$$\text{Required } S_x \geq \frac{614(12)}{0.90(50)} + \frac{2(48)(12)6.2}{0.90(50)} = 322 \text{ in.}^3$$

  (c) Check the strength. The yield limit state appears to be satisfied; however, the stability limit state must be checked and a final check of all criteria should always be made.

    Check the *yield limit state* criterion, Eq. 9.14.6, adjusting M_{ux} slightly to reflect the correct beam weight,

$$f_{un} = \frac{M_{ux}}{S_x} + \frac{M_{uy}}{S_y} = \frac{613(12)}{329} + \frac{96(12)}{53.0} = 22.4 + 21.7 = 44.1 \text{ ksi}$$

$$(f_{un} = 44.1 \text{ ksi}) < [\phi_b F_y = 0.90(50) = 45 \text{ ksi}] \qquad\qquad \textbf{OK}$$

    Check the *lateral-torsional buckling limit state,* Eq. 9.14.7

$$\phi_b M_{nx} \geq M_{ux} \qquad\qquad [9.14.7]$$

For the W24×131,

$$M_p = F_y Z_x = 50(370)\tfrac{1}{12} = 1542 \text{ ft-kips}$$

$$L_p = 1.76 r_y \sqrt{\frac{E}{F_y}} = \frac{300 r_y}{\sqrt{F_y, \text{ksi}}} = \frac{300(3.01)}{\sqrt{50}}\frac{1}{12} = 10.5 \text{ ft}$$

$$L_r = 31.9 \text{ ft (from AISC Table 3-2 "Selection by } Z_x\text{" or Eq. 9.6.6)}$$

Then using Eq. 9.6.4 with $C_b = 1$,

$$M_n = C_b\left[M_p - (M_p - 0.7F_yS_x)\left(\frac{L_b - L_p}{L_r - L_p}\right)\right] \le M_p \qquad\qquad [9.6.4]$$

$$M_n = 1.0\left\{ 1542 - \left[1542 - 0.7(50)(329)\tfrac{1}{12}\right]\left(\frac{24 - 10.5}{31.9 - 10.5}\right)\right\} = 605 \text{ ft-kips}$$

$$\phi M_n = 0.90(605) = 545 \text{ ft-kips} > M_{ux} = 614 \qquad\qquad \textbf{OK}$$

Use W24×131, $F_y = 50$ ksi. ∎

SELECTED REFERENCES

9.1. Joseph A. Yura, Theodore V. Galambos, and Mayasandra K. Ravindra. "The Bending Resistance of Steel Beams," *Journal of the Structural Division,* ASCE, **104,** ST9 (September 1978), 1355–1370.

9.2. Karl de Vries. "Strength of Beams as Determined by Lateral Buckling," *Transactions,* ASCE, **112** (1947), 1245–1320.

9.3. R. A. Hechtmann, J. S. Hattrup, E. F. Styer, and J. L. Tiedemann. "Lateral Buckling of Rolled Steel Beams," *Transactions,* ASCE, **122** (1957), 823–843.

9.4. W. J. Austin, S. Yegian, and T. P. Tung, "Lateral Buckling of Elastically End-Restrained I Beams," *Transactions,* ASCE, **122** (1957), 374–388.

9.5. J. W. Clark and J. R. Jombock. "Lateral Buckling of I-beams Subjected to Unequal End Moments," *Journal of the Engineering Mechanics Division,* ASCE, **83,** EM3 (July 1957).

9.6. Mario G. Salvadori. "Lateral Buckling of I Beams Under Thrust and Unequal End Moments," *Transactions,* ASCE, **120** (1955), 1165–1182.

9.7. Theodore V. Galambos. *Structural Members and Frames.* Englewood Cliffs, NJ: Prentice-Hall, Inc., 1968.

9.8. Theodore V. Galambos. "Inelastic Lateral Buckling of Beams," *Journal of the Structural Division,* ASCE, **89,** ST5 (October 1963), 217–242.

9.9. Maxwell G. Lay and Theodore V. Galambos. "Inelastic Steel Beams Under Uniform Moment," *Journal of the Structural Division,* ASCE, **91,** ST6 (December 1965), 67–93.

9.10. Maxwell G. Lay and Theodore V. Galambos. "Inelastic Beams Under Moment Gradient," *Journal of the Structural Division,* ASCE, **93,** ST1 (February 1967), 381–399.

9.11. Campbell Massey and F. S. Pitman. "Inelastic Lateral Stability Under a Moment Gradient," *Journal of the Engineering Machanics Division,* ASCE, **92,** EM2 (April 1966), 101–111.

9.12. Nicholas S. Trahair and Sritawat Kitipornchai. "Buckling of Inelastic I-Beams Under Uniform Moment," *Journal of the Structural Division,* ASCE, **98,** ST11 (November 1972), 2551–2566.

9.13. A. J. Hartmann. "Inelastic Flexural-Torsional Buckling," *Journal of the Engineering Mechanics Division,* ASCE, **97,** EM4 (August 1971), 1103–1119.

9.14. Morris Ojalvo and Ronald R. Weaver. "Unbraced Length Requirements for Steel I-Beams," *Journal of the Structural Division,* ASCE, **104,** ST3 (March 1978), 479–490.

9.15. David A. Nethercot and Nicholas S. Trahair. "Inelastic Lateral Buckling of Determinate Beams," *Journal of the Structural Division,* ASCE, **102,** ST4 (April 1976), 701–717.

9.16. J. Bansal. "The Lateral Instability of Continuous Beams," *AISI Report No.* 3, American Iron and Steel Institute, New York, August 1971.

9.17. Peter W. Hoadley. "Practical Significance of LRFD Beam Buckling Factors," *Journal of Structural Engineering,* ASCE, **117,** 3 (March 1991), 988–996.

9.18. P. A. Kirby and D. A. Nethercot. *Design for Structural Stability.* New York: Wiley, 1979.

9.19. Mario G. Salvadori. "Lateral Buckling of Eccentrically Loaded I-Columns," *Transactions,* ASCE, **121** (1956), 1163–1178.

9.20. A. J. Hartmann. "Elastic Lateral Buckling of Continuous Beams," *Journal of the Structural Division,* ASCE, **93,** ST4 (August 1967), 11–26.

9.21. Nicholas S. Trahair, "Elastic Stability of Continuous Beams," *Journal of the Structural Division,* ASCE, **95,** ST6 (June 1969), 1295–1312.

9.22. Graham Powell and Richard Klingner. "Elastic Lateral Buckling of Steel Beams," *Journal of the Structural Division,* ASCE, **96,** ST9 (September 1970), 1919–1932.

9.23. N. S. Trahair and M. A. Bradford. *The Behavior and Design of Steel Structures*, 2nd ed. London: Chapman and Hall, 1988.

9.24. Charles W. Roeder and Mahyar Assadi. "Lateral Stability of I-beams with Partial Support," *Journal of the Structural Division,* ASCE, **108,** ST8 (August 1982), 1768–1779.

9.25. Yushi Fukumoto, Yoshito Itoh, and Ryoji Hattori. "Lateral Buckling on Welded Continuous Beams," *Journal of the Structural Division,* ASCE, **108,** ST10 (October 1982), 2245–2262.

9.26. J. W. Clark and H. N. Hill. "Lateral Buckling of Beams," *Journal of the Structural Division,* Proceedings ASCE, **86,** ST7 (July 1960), 175–196. Also, *Transactions,* ASCE, **127** (1962), Part II, 180–201.

9.27. Sritawat Kitipornchai, Peter F. Dux, and Nevelle J. Richter. "Buckling and Bracing of Cantilevers," *Journal of Structural Engineering,* ASCE, **110,** 9 (September 1984), 2250–2262.

9.28. Campbell Massey and Peter J. McGuire. "Lateral Stability of Nonuniform Cantilevers," *Journal of the Engineering Mechanics Division,* ASCE, **97,** EM3 (June 1971), 673–686.

9.29. Patrick D. Zuraski. "The Significance and Application of C_b in Beam Design," *Engineering Journal,* AISC, **29,** 1 (First Quarter, 1992), 20–25.

9.30. Joseph A. Yura. "Bracing for Stability—State-of-the-Art," *Proceedings of Structures Congress XIII,* American Society of Civil Engineers, New York, (April 2-5, 1995), 88–103.

9.31. James M. Fisher and Michael A. West. *Erection Bracing of Low-Rise Structural Steel Buildings,* Steel Design Guide Series No. 10. Chicago: American Institute of Steel Construction, 1997.

9.32. H. N. Hill. "Lateral Buckling of Channels and Z-Beams," *Transactions,* ASCE, **119** (1954), 829–841.

9.33. John M. Anderson and Nicholas S. Trahair. "Stability of Monosymmetric Beams and Cantilevers," *Journal of the Structural Division,* ASCE, **98,** ST1 (January 1972), 269–286.

9.34. Sritawat Kitipornchai and Nicholas S. Trahair. "Buckling Properties of Monosymmetric I-Beams," *Journal of the Structural Division,* ASCE, **106,** ST5 (May 1980), 941–957.

9.35. D. A. Nethercot. "Elastic Lateral Buckling of Beams," *Beams and Beam Columns—Stability in Strength* (ed. R. Narayanan). Barking, Essex, England: Applied Science Publishers, 1983.

9.36. Sritawat Kitipornchai, Chien Ming Wang, and Nicholas S. Trahair. "Buckling of Monosymmetric I-Breams Under Moment Gradient," *Journal of Structural Engineering,* ASCE, **112,** 4 (April 1986), 781–799.

9.37. Chien Ming Wang and Sritiwat Kitipornchai. "Buckling Capacities of Monosymmetric I-Beams," *Journal of Structural Engineering.* ASCE, **112** (November 1986), 2373–2391.

9.38. Sritawat Kitipornchai and Alain D. Wong-Chung. "Inelastic Buckling of Welded Monosymmetric I-Beams," *Journal of Structural Engineering,* **113,** 4 (April 1987), 740–756.

9.39. William Zuk. "Lateral Bracing Forces on Beams and Columns," *Journal of the Engineering Mechanics Division,* ASCE, **82,** EM3 (July 1956), Proc. Paper No. 1032, 16 pp.

9.40. George Winter. "Lateral Bracing of Columns and Beams," *Transactions,* ASCE, **125** (1960), 807–845.

9.41. Campbell Massey. "Lateral Bracing Force of Steel I-Beams," *Journal of the Engineering Mechanics Division,* ASCE, **88,** EM6 (December 1962), 89–113.

9.42. George Pincus. "On the Lateral Support of Inelastic Columns," *Engineering Journal,* AISC, **1,** 4 (October 1964), 113–115.

9.43. Theodore V. Galambos. "Lateral Support for Tier Building Frames," *Engineering Journal,* AISC, **1,** 1 (January 1964) 16–19; Disc, **1,** 4 (October 1964), 141.

9.44. Tor B. Urdal. "Bracing of Continuous Columns," *Engineering Journal,* AISC, **6,** 3 (July 1969), 80–83.

9.45. Maxwell G. Lay and T. V. Galambos. "Bracing Requirements for Inelastic Steel Beams," *Journal of the Structural Division,* ASCE, **92,** ST2 (April 1966), 207–228.

9.46. Arthur C. Taylor, Jr. and Morris Ojalvo. "Torsional Restraint of Lateral Buckling," *Journal of the Structural Division,* ASCE, **92,** ST2 (April 1966), 115–129.

9.47. A. J. Hartmann. "Experimental Study of Flexural-Torsional Buckling," *Journal of the Structural Division,* ASCE, **96,** ST7 (July 1970), 1481–1493.

9.48. Bruce R. Mutton and Nicholas S. Trahair. "Stiffness Requirements for Lateral Bracing," *Journal of the Structural Division,* ASCE, **99,** ST10 (October 1973), 2167–2182.

9.49. Ian C. Medland and Cecil M. Segedin. "Brace Forces in Interbred Column Structures," *Journal of the Structural Division,* ASCE, **105,** ST7 (July 1979), 1543–1556.

9.50. Joseph A. Yura. "Fundamentals of Beam Bracing," *Is Your Structure Suitably Braced?,* Structural Stability Research Council, 1993 Conference, Milwaukee, Wisconsin, April 6–7, 1993, 20 pp.

9.51. LeRoy A. Lutz and James M. Fisher. "A Unified Approach for Stability Bracing Requirements," *Engineering Journal,* AISC, **22,** 4 (4th Quarter 1985), 163–167.

9.52. Joseph M. Ales, Jr. and Joseph A. Yura. "Bracing Design for Inelastic Structures," *Is Your Structure Suitably Braced?,* SSRC 1993 Conference, Milwaukee, Wisconsin, April 6–7, 1993, 29–37.

9.53. M. J. Clarke and R. Q. Bridge. "Bracing Force and Stiffness Requirements to Develop the Design Strength of Columns," *Is Your Structure Suitably Braced?,* SSRC 1993 Conference, Milwaukee, Wisconsin, April 6–7, 1993, 75–86.

9.54. George Pincus and Gordon P. Fisher, "Behavior of Diaphragm-Braced Columns and Beams," *Journal of the Structural Division,* ASCE, **92,** ST2 (April 1966), 323–350.

9.55. Samuel J. Errera, George Pincus, and Gordon P. Fisher. "Columns and Beams Braced by Diaphragms," *Journal of the Structural Division,* ASCE, **93,** ST1 (February 1967), 295–318.

9.56. Larry D. Luttrell. "Strength and Behavior of Light-Gage Steel Shear Diaphragms," *Cornell Engineering Research Bulletin* No. 67–1, July 1967.

9.57. T. V. S. R. Apparao, Samuel J. Errera, and Gordon P. Fisher. "Columns Braced by Girts and a Diaphragm," *Journal of the Structural Division,* ASCE, **95,** ST5 (May 1969), 965–990.

9.58. Arthur H. Nilson and Albert R. Ammar. "Finite Element Analysis of Metal Deck Shear Diaphragms," *Journal of the Structural Division,* ASCE, **100,** ST4 (April 1974), 711–726.

9.59. David A. Nethercot and Nicholas S. Trahair. "Design of Diaphragm-Braced I-Beams," *Journal of the Structural Division,* ASCE, **101,** ST10 (October 1975), 2045–2061.

9.60. Amir Simaan and Teoman B. Pekoz. "Diaphragm Braced Members and Design of Wall Studs," *Journal of the Structural Division,* ASCE, **102,** ST1 (January 1976), 77–92.

9.61. Samuel J. Errera and Tamirisa V. S. R. Apparao. "Design of I-Shaped Beams with Diaphragm Bracing," *Journal of the Structural Division,* ASCE, **102,** ST4 (April 1976), 769–781.

9.62. J. Michael Davies. "Calculation of Steel Diaphragm Behavior," *Journal of the Structural Division,* ASCE, **102,** ST7 (July 1976), 1411–1430.

9.63. Samuel J. Errera and Tamirisa V. S. R. Apparao. "Design of I-Shaped Columns with Diaphragm Bracing," *Journal of the Structural Division,* ASCE, **102,** ST9 (September 1976), 1685–1701.

9.64. John T. Easley. "Strength and Stiffness of Corrugated Metal Shear Diaphragms," *Journal of the Structural Division,* ASCE, **103,** ST1 (January 1977), 169–180.

9.65. SDI. *Tentative Recommendations for the Design of Steel Deck Diaphragms.* Westchester, IL: Steel Deck Institute, October 1972.

9.66. C. O. Dohrenwend. "Action of Deep Beams Under Combined Vertical, Lateral and Torsional Loads," *Journal of Applied Mechanics,* **8** (1941), A-130.

9.67. E. H. Gaylord, Jr. and C. N. Gaylord. *Design of Steel Structures.* New York: McGraw-Hill Book Company, Inc., 1957 (pp. 169–170).

9.68. John Springfield. "Design of Columns Subject to Biaxial Bending," *Engineering Journal,* AISC, **12,** 3 (Third Quarter 1975), 73–81.

9.69. Thomas P. Pastor and John T. DeWolf. "Beams with Torsional and Flexural Loads," *Journal of the Structural Division,* ASCE, **105,** ST3 (March 1979), 527–538.

9.70. Zia Razzaq and Theodore V. Galambos. "Biaxial Bending of Beams with or without Torsion," *Journal of the Structural Division,* ASCE, **105,** ST11 (November 1979), 2145–2162.

9.71. Zia Razzaq and Theodore V. Galambos. "Biaxial Bending Tests with or without Torsion," *Journal of the Structural Division,* ASCE, **105,** ST11 (November 1979), 2163–2185.

9.72. Joseph A. Yura. "Winter's Bracing Approach Revisited," *Proceedings, 50th Anniversary Conference,* Structural Stability Research Council, Lehigh University, 21–22 June 1994, pp. 375–382.

9.73. R. H. Plaut. "Requirements for Lateral Bracing of Columns with Two Spans," *Journal of Structural Engineering,* ASCE, **119,** 10 (October 1993), 2913–2931.

9.74. Duane S. Ellifritt, Gregory Wine, Thomas Sputo, and Santosh Samuel. "Flexural Strength of WT Sections," *Engineering Journal,* AISC, **29,** 2 (Second Quarter 1992), 67–74.

9.75. Donald W. White. "Unified Flexural Resistance Equations for Stability Design of Steel I-Section Members Overview," *Structural Engineering, Mechanics and Materials Report No. 24a,* School of Civil & Environmental Engineering, Georgia Institute of Technology, Atlanta, GA, 2005.

PROBLEMS

All problems are to be done according to the AISC LRFD Method or the ASD Method, as indicated by the instructor. All given loads are service loads unless otherwise indicated. Assume lateral support consists of translational restraint but not moment (rotational) restraint, unless otherwise indicated. Assume all standard sections are equally readily available in the indicated grade of steel (even though actually they are not). A figure showing span and loading is required, and after making a design selection, a final check of strength (ϕM_n compared with M_u for LRFD) or (M_n/Ω compared with M_a for ASD) is required. *Note:* Live load must always be applied (or not) such that it causes maximum (or minimum) effect.

9.1. For the case (or cases) assigned by the instructor, plot design strength $\phi_b M_n$ (ft-kips) for the AISC LRFD Method vs laterally unbraced length L_b (ft). Show the portions controlling $\phi_b M_n$ with solid lines and the noncontrolling parts with dashed lines. For any case assigned, show relationships for both $C_b = 1.0$ and 2.3.

Case	Section	F_y (ksi)
1	Plate girder: 5/8×12 flanges; 5/16×30 web	36
2	Plate girder: 7/8×16 flanges; 3/4×26 web	36
3	W16×26	50
4	W14×145	50
5	W21×62	50
6	W24×84	50
7	W16×26	65
8	W14×145	65
9	W24×84	65

9.2. For the case (or cases) assigned by the instructor, determine the maximum concentrated service load P that can act at midspan on a simply supported span. Lateral supports exist only at the ends of the span. The service load is 65% live load and 35% dead load.

Case	Section	Span (ft)	F_y (ksi)
1	W21×62	20	50
2	W24×84	24	50
3	W30×99	30	50

9.3. For the case (or cases) assigned by the instructor, select the lightest W section as a beam. Assume only flexure must be considered; i.e., omit treating shear and deflection. The dead load given is *in addition* to the weight of the beam.

Case	w_D Dead load (kip/ft)	w_L Live load (kip/ft)	Span length (ft)	F_y (ksi)	Lateral support
1	0.9	2.0	20	50	Continuous
2	0.9	2.0	20	50	Ends and midspan
3	0.9	2.0	20	50	Ends only
4	0.7	1.4	28	50	Ends and midspan
5	0.7	1.4	28	60	Ends and midspan
6	0.3	0.9	35	50	Continuous
7	0.3	0.9	35	50	Every 7 feet
8	0.3	0.9	35	50	Ends and midspan
9	0.3	0.9	35	65	Continuous
10	0.3	0.9	35	65	Every 7 feet
11	0.3	0.9	35	65	Ends and midspan
12	0.3	0.9	35	100	Continuous
13	0.3	0.9	35	100	Every 7 feet
14	0.3	0.9	35	100	Ends and midspan
15	0	1.0	35	50	Every 5 ft
16	0	1.0	35	50	Ends only
17	0.7	2.8	48	50	Every 16 feet
18	0.7	2.8	48	60	Every 16 feet

9.4. Select the lightest W sections for the situation shown in the accompanying figure, under the following conditions:

(a) A992 steel; continuous lateral support

(b) A992 steel; lateral support at ends only

(c) A992 steel; lateral support at ends and at point A

(d) A572 Grade 60 steel; lateral support at ends and point A

28 kips dead load
7 kips live load

0.7 kips/ft dead load
3.5 kips/ft live load
(not including beam weight)

Problem 9.4

9.5. Select the lightest W section for the situation shown in the accompanying figure using (a) A572 Grade 50 steel and (b) steel with $F_y = 65$ ksi, assuming lateral support at the ends and at point A only.

W = 15 k dead load
= 11 k live load
w = 0.2 k/ft dead load
(not including beam weight)
= 0.8 k/ft live load

Problem 9.5

9.6. Select the lightest W section for the conditions shown in the accompanying figure. Assume there is no deflection limitation. Use (a) A992 steel and (b) A572 Grade 60 steel.

W = 15 k dead load
15 k live load
w = 0.4 k/ft dead load
(including beam)
= 1.1 k/ft live load
Lateral support
Vertical support

Problem 9.6

9.7. A floor beam, laterally supported at the ends only and supporting vibration inducing heavy machinery, is subject to the loads shown in the accompanying figure. Select the lightest W section of A992 steel. Compare the result when there is no deflection limit with that when L/d is limited to a maximum of 20 under full load, a traditional limit to minimize perceptible vibration due to pedestrian traffic.

14 k   14 k ── Gravity live loads

Problem 9.7

9.8. For the case (or cases) assigned by the instructor, select the lightest W section to serve as a uniformly loaded library floor beam on a simply supported beam. Lateral support occurs at the ends and at $L/4$, $L/2$, and $3L/4$. Live load deflection may not exceed $L/300$ (not an AISC LRFD Method or an ASD Method limit but a design limit for this design). Given dead load moment does *not* include beam weight. Assume $C_b = 1.0$.

Case	M_D Dead load moment (ft-kips)	M_L Live load moment (ft-kips)	Span length (ft)	F_y (ksi)	Deflection limit
1	49	98	28	50	$L/360$
2	49	98	28	60	$L/360$
3	0	240	48	50	$L/300$
4	0	240	48	65	$L/300$
5	50	190	48	50	$L/360$
6	50	190	48	65	$L/360$
7	80	750	60	50	$L/400$
8	80	750	60	60	$L/400$

9.9. A beam is to serve as a floor beam on a simply supported span of 20 ft. The live load consists of a movable concentrated load (no impact) of 50 kips. Live load deflection may not exceed $L/360$.

(a) Select the lightest W section of A992 steel when continuous lateral support is provided.

(b) Repeat (a) if lateral support is provided only at the ends.

9.10. A W10×33 is to be used as a simply supported beam on a span of 25 ft with lateral support at the ends only. The beam is required to support a plastered ceiling. If the dead load is 0.15 kips/ft (including beam weight), what is the maximum uniform service live load permitted on the beam, using A992 steel? What percentage increase in live load can be gained if the beam is A572 Grade 65 steel? Comment.

9.11. Redesign for loading Cases 1 and/or 2, as assigned, of Prob. 9.8 when lateral support is provided only at the ends and at midspan. In addition to the L/360 traditional deflection limit for a plastered ceiling, the architect requires the beam to be no deeper than nominal W12.

9.12. Investigate the beam of the accompanying figure for bending and shear if the section is A992 steel. External lateral support for the beam is provided only at the vertical supports and at the tip of the cantilever. If one additional lateral support were provided at the 12-kip load, how much lighter, if any, could the W14 section be made?

9.13. Select the lightest W section for each of the situations shown in the accompanying figure. The concentrated load W is 5 kips dead load and 15 kips live load. Assume lateral support is provided at the reactions and at the concentrated loads. Use A992 steel.

9.14. Select the lightest W section for the beam of the accompanying figure. Lateral support is provided at concentrated load points, reactions, and at end of cantilever. Use (a) A992 steel and (b) A572 Grade 65 steel.

9.15. Determine the nominal strength M_n for the AISC LRFD Method for the channel section and loading in Prob. 8.19 if lateral support exists only at the ends. *Neglecting torsion,* how much larger, if any, section would be required? Use A992 steel.

9.16. Select for the LRFD Method a channel for the conditions of Prob. 8.20, assuming lateral support at the ends only. Consider torsion in accordance with Chapter 8. Use the safety criterion that the factored load stress f_{un} may not exceed $\phi_b F_y$.

W = 2 k dead load
10 k live load
w = 0.3 k/ft dead load
(including beam weight)
= 1.1 k/ft live load

Problem 9.12

(a)   (b)

Problem 9.13

W = 10 k dead load
20 k live load
w = 3.0 k/ft dead load
(not including beam weight)

Problem 9.14

9.17. Design a built-up I-shaped beam with different sized flanges for the conditions of Prob. 9.4, part (b). What percent weight can be saved, if any, by using different sized flanges? Use beam depth and web thickness approximately the same as for the lightest rolled W shape that satisfies loading conditions. A36 steel.

9.18. For the beam selected for a case assigned from Prob. 9.3, estimate the size of bracing (i.e., select a section) required. The bracing frames into both sides and is attached to the compression flange. Length of bracing is 6 ft.

9.19. For the beam designed in the case assigned in Prob. 9.8, estimate the size bracing required. Assume bracing is 12 ft long, frames into both sides of the beam, and is attached to the compression flange. Preferably select channels.

9.20. Determine the adequacy of a W24×84 (with rail, 20 lb/ft) serving as a crane support girder of A992 steel. The simple span is 20 ft with lateral support at the ends only. Use accepted good practice in accounting for the torsional effect of lateral loading.

Maximum moments occurring near midspan are
M_x (live load plus impact) = 125 ft-kips
M_x (dead load) = 14 ft-kips
M_f = 14 ft-kips (assumed live load resisted by one flange using top flange lateral loading in accordance with the *AISC Specification.*

9.21. Select the lightest W8 section to be used in an inclined position such that the plane of the web makes an angle of 30 degrees with the plane of loading. The beam is to be of A992 steel, and it has lateral support only at the ends of the 22-ft simple span. The uniform gravity load is 0.3 kips/ft snow load and 0.1 kip/ft dead load, in addition to the beam weight.

9.22. For the case assigned by instructor, design the lightest W section to serve as a crane support girder as shown in the accompanying figure. Assume lateral support at the ends only and that deflection need not be restricted. Assume one-half the lateral force on the crane rail. Assume a cab-operated crane acts at each runway girder. Use A992 steel.

Problem 9.22

For Problem 9.22

Case	Crane capacity W (tons)	Maximum gravity end-truck wheel load R_1 (kips)	Trolley weight (kips)	Span L_{crane} (ft)	Crane end-truck wheel spacing L_{truck} (ft)
1	15	18	6	40	9.5
2	20	27	6	36	8.0
3	30	36	6	32	6.0
4	20	25	5	30	6.0

9.23. Redesign the crane support girder of Prob. 9.22, using a combination channel and W section.

9.24. Design the crane runway girder indicated on the accompanying figure. The two cranes are each 40-ton capacity, with the end-truck wheel spacing as given in the figure. The rails are ASCE 60 lb with clamps (see *AISC Manual*). The dead load of each crane is 20 kips, equally distributed to its four wheels. Each crane trolley weighs 4 kips. Use A36 steel for channels and A992 for W sections and (a) select a single W section and (b) select a combination W section and channel, where the channel would have its web flat against the top flange of the W section.

Assume crane runway girders are simply supported

Problem 9.24

10

Continuous Beams

10.1 INTRODUCTION

This chapter brings together theoretical concepts relating to the unified design specification which is the basis for the AISC LRFD Method and the ASD Method that have been presented in Chapters 7 and 9. In addition, plastic analysis is introduced for application to continuous beams; this procedure may be applicable with "compact sections" having adequate lateral bracing. Plastic analysis is permitted under Appendix 1 of the *AISC Specification*. The reader is assumed to be familiar with elastic methods of statically indeterminate analysis.

Before proceeding into the chapter, the reader should review Secs. 7.3 through 7.5 where the strengths of the cross-section in flexure and in shear are treated, and where LRFD and ASD as applied to laterally stable beams are presented.

The reader is also expected to be familiar with the lateral stability criteria treated in Chapter 9, particularly Secs. 9.1, 9.2, and 9.6 through 9.10.

For the reader who wishes a package study of plastic analysis and the related beam design, Secs. 7.3 and 7.4 provide the basic plastic moment and shear strengths, including the concepts of plastic moment, plastic hinge, and shape factor. Section 7.4 contains the basic approach of using factored loads to obtain the required nominal strength, as well as design examples for a simply supported beam. The plastic analysis and LRFD design sections of this chapter can then form the body of the study.

10.2 PLASTIC STRENGTH OF A STATICALLY INDETERMINATE BEAM

In Sec. 7.3 it was shown that a simply supported beam supporting a concentrated load reached its limit state (i.e., collapse condition) when the concentrated load is large enough to cause the plastic moment M_p to develop; that is, cause a plastic hinge to develop under the concentrated load.

Referring to Fig. 10.2.1, as the concentrated load increases toward the value W_n that causes the collapse condition, the portion of the beam where the moment does not exceed M_y is elastic (meaning straight line θ). In the region near the plastic moment M_p there is an

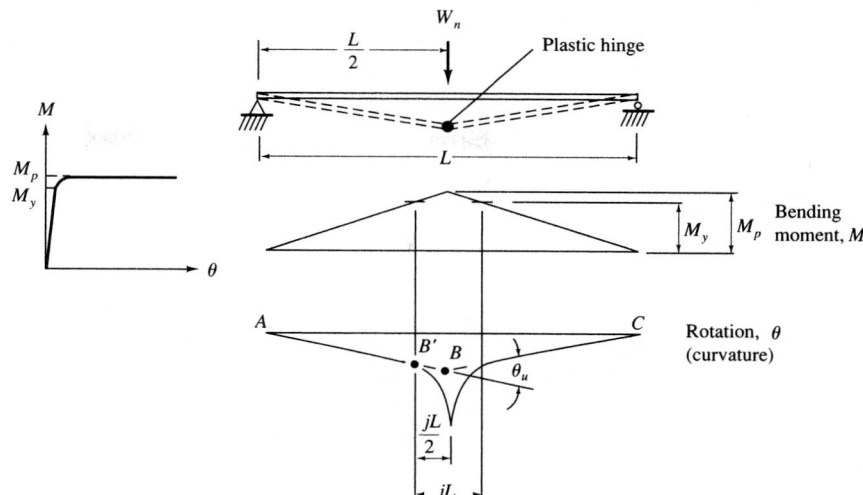

Figure 10.2.1
Moment—curvature
relationships for a plastic
hinge.

inelastic length jL, and the midspan curvature is large. For practical purposes the beam may be treated as two rigid parts (straight lines) connected by a hinge at B, known as a *plastic hinge*, and having a concentrated angle change, or hinge rotation, θ_u.

The actual length of a plastic hinge is dependent on the shape of the cross-section and can vary from about one-tenth to as much as one-third of the span. As an example, consider the relationship between the bending moments and the beam segments AB and AB' in Fig. 10.2.1:

$$\frac{AB}{M_p} = \frac{AB'}{M_y} \quad \text{or} \quad \frac{L/2}{M_p} = \frac{L/2 - jL/2}{M_y} \tag{10.2.1}$$

Solving for j,

$$j = 1 - \frac{M_y}{M_p} = 1 - \frac{1}{\xi} \tag{10.2.2}$$

where ξ is the shape factor as discussed in Sec. 7.3.

EXAMPLE 10.2.1

Compute the length of the plastic hinge for the beam shown in Fig. 10.2.1 for (a) a W16×40 and (b) a rectangular beam having a width b and a depth d.

Solution:

(a) For W16×40,

$$\xi = \frac{M_p}{M_y} = \frac{Z}{S} = \frac{72.9}{64.7} = 1.13$$

From Eq. 10.2.2,

$$jL = L\left(1 - \frac{1}{\xi}\right) = L\left(1 - \frac{1}{1.13}\right) = 0.115L$$

(b) For rectangular section,

$$\xi = \frac{Z}{S} = \frac{\dfrac{bd^2}{4}}{\dfrac{bd^2}{6}} = 1.5$$

$$jL = L\left(1 - \frac{1}{1.5}\right) = 0.333L \qquad \blacksquare$$

Even though the distance jL may be as much as one-third of the span, as shown in Example 10.2.1, the simple assumption of a plastic hinge at a point has been amply demonstrated by tests. Beedle [10.1] and Massonnet and Save [10.2] have extensive discussions of theoretical and experimental verification of plastic analysis procedures so individual research references are not included here.

Plastic Limit Load—Equilibrium Method

At the collapse condition with the plastic limit load W_n acting, the requirements of equilibrium are still applicable. Consider first a statically determinate simply supported beam, as shown in Fig. 10.2.2. A collapse condition is achieved when the load W_n is large enough to cause the plastic moment M_p to occur at one location (in this case under the load). When the sufficient number of plastic hinges have been developed to allow instantaneous hinge rotations without developing increased resistance, a mechanism is said to have occurred.

EXAMPLE 10.2.2

Determine the collapse mechanism load W_n for the W21×62 beam of Fig. 10.2.2. Assume $F_y = 50$ ksi.

Solution:
Using equilibrium,

$$M_p = \frac{W_n L}{4}$$

$$W_n = \frac{4M_p}{L} = \frac{4(F_y Z)}{L} = \frac{4(50)(144)}{24(12)} = 100 \text{ kips}$$

Using the simplified procedure of considering the behavior as ideally elastic-plastic, the deflection occurring when M_p is reached is based on the elastic equation, which is

Figure 10.2.2
Example 10.2.2.

strictly valid only until M_y (neglecting residual stress) is reached. At a maximum moment of M_y,

$$\Delta_y = \frac{WL^3}{48EI} = \frac{M_yL^2}{12EI} = \frac{F_yIL^2}{c(12)EI} = \frac{F_yL^2}{12cE}$$

$$= \frac{50(24)^2(144)}{12(10.5)(29,000)} = 1.14 \text{ in.}$$

Assuming a linear extension until $M = M_p$,

$$\Delta_p = 1.14\frac{Z}{S} = 1.14\frac{144}{127} = 1.14(1.14) = 1.30 \text{ in.}$$

The deflection when M_p is achieved will be higher than this. However, it is the service load deflection that is normally of concern, and this is correctly computed by the usual elastic procedures.

Note that the collapse mechanism moment diagram for this statically determinate beam is the same shape as that which occurs if the maximum moment is down in the service load range. At whatever load level, from an infinitesimal load to collapse mechanism load, the bending moment at every point remains in a constant proportion to the load.   ∎

EXAMPLE 10.2.3

Determine the collapse mechanism load W_n for the fixed end W16×40 beam of Fig. 10.2.3. Assume $F_y = 50$ ksi.

Solution:
Since the beam is statically indeterminate, three plastic hinges are required to form a mechanism. Using equilibrium,

$$2M_p = \frac{W_nL}{4}$$

$$M_p = F_yZ = 50(72.9)\tfrac{1}{12} = 304 \text{ ft-kips}$$

$$W_n = 8M_p/L = 8(304)/24 = 101 \text{ kips}$$

In order to determine the load–deflection diagram, it would be necessary to know the loading history up to the collapse condition. In this case the elastic bending moments, as shown in Fig. 10.2.3, give equal positive and negative bending moment; therefore the three plastic hinges form simultaneously. Moments will increase simultaneously in direct proportion until the collapse condition is reached. Again, as for the statically determinate case, as load is applied the bending moment at every point remains in a constant proportion to the load. For this special statically indeterminate case, plastic analysis has no advantage over elastic analysis.

$$\Delta_y = \frac{WL^3}{192EI} = \frac{M_yL^2}{24EI} = \frac{F_yL^2}{24Ec}$$

$$= \frac{50(24)^2(144)}{24(29,000)8} = 0.75 \text{ in.}$$

Assuming a linear extension until $M = M_p$,

$$\Delta_p = 0.75\frac{Z}{S} = 0.75\frac{73.0}{64.7} = 0.85 \text{ in.}$$

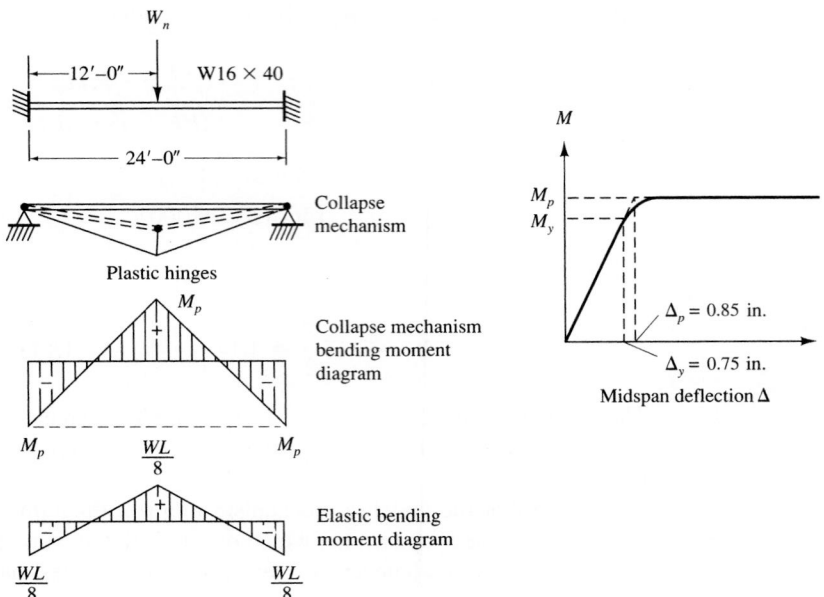

Figure 10.2.3
Example 10.2.3.

EXAMPLE 10.2.4

Determine the load–deflection diagram for the W16×40 beam of Fig. 10.2.4 for loading up to the collapse condition. Assume $F_y = 50$ ksi.

Solution:

The collapse mechanism load W_n can be found directly without knowing the sequence of plastic hinge formation. In this case, since the elastic moments are different at the three locations where plastic hinges will form, the plastic hinges will not form simultaneously.

Figure 10.2.4
Example 10.2.4.

(a)

(b)

The equilibrium requirement at the collapse mechanism condition gives

$$M_p = \frac{W_n(10)(20)}{30} - M_p$$

$$W_n = 2M_p \frac{3}{20} = 0.3M_p$$

$$M_p = 304 \text{ ft kips} \quad \text{(from Example 10.2.3)}$$

$$W_n = 0.3(304) = 91.2 \text{ kips}$$

The load–deflection diagram, however, requires examination of the loading stages.

Stage 1. From the elastic bending moments, the first plastic hinge will form at point A,

$$\frac{40W_1}{9} = M_p = 304 \text{ ft-kips}$$

$$W_1 = 9(304)/40 = 68.4 \text{ kips}$$

$$\Delta_1 = \frac{W_1(10)^3(20)^3}{3(30)^3 \, EI} = \frac{68.4(8000)(1728)}{3(27)(29,000)(518)} = 0.78 \text{ in.}$$

Stage 2. With $W_1 = 68.4$ kips one can consider that part of the available moment strength at points B and C has been used up. The strength remaining is

$$M_B = M_p - 80(68.4)/27 = 304 - 203 = 101 \text{ ft-kips}$$

$$M_C = M_p - 20(68.4)/9 = 304 - 152 = 152 \text{ ft-kips}$$

As the load is increased above $W_1 = 68.4$ kips, the added load acts on a different elastic system. The moments caused by the additional load are not distributed over the span in the same manner as moments caused by the first 68.4 kips; thus the term "redistribution of moments" is applied. Figure 10.2.5a shows the elastic system and moments for stage 2. It is apparent that the next plastic hinge will form under the load.

$$\frac{140W_2}{27} = 101 \text{ ft-kips}$$

$$W_2 = 27(101)/140 = 19.5 \text{ kips}$$

$$\Delta_2 = \frac{W_2a^2b^3(3L + a)}{12L^3EI} = \frac{19.5(10)^2(20)^3(90 + 10)(1728)}{12(30)^3(29,000)(518)} = 0.55 \text{ in.}$$

Stage 3. With a total of $W_1 + W_2 = 68.4 + 19.5 = 87.9$ kips applied, the strength available at C is

$$M_C = M_p - 152 \text{ (Stage 1)} - 40(19.5)/9 \text{ (Stage 2)} = 65.3 \text{ ft-kips}$$

$$20W_3 = 65.3 \text{ ft-kips}; \quad W_3 = 3.3 \text{ kips}$$

$$\Delta_3 = \frac{W_3b^3}{3EI} = \frac{3.3(20)^3(1728)}{3(29,000)(518)} = 1.01 \text{ in.}$$

The complete load–deflection diagram appears in Fig. 10.2.4b. The total load to cause the collapse mechanism to form equals

$$W_n = W_1 + W_2 + W_3 = 68.4 + 19.5 + 3.3 = 91.2 \text{ kips}$$

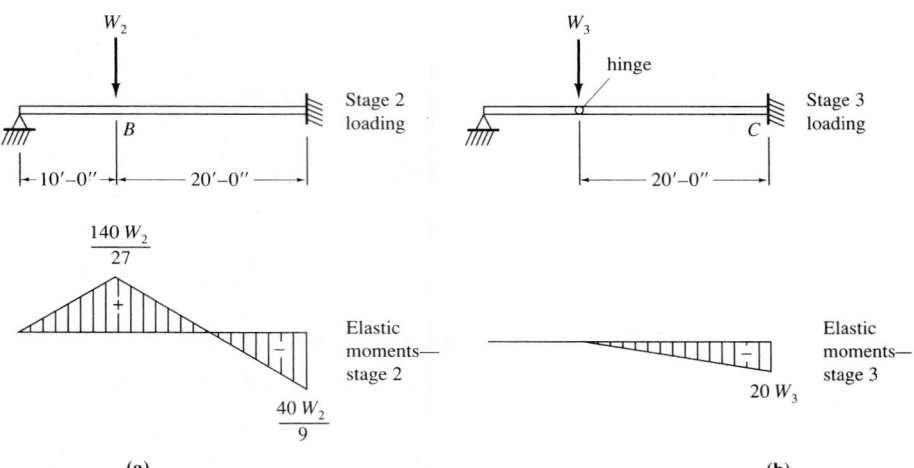

Figure 10.2.5
Stages 2 and 3 for
Example 10.2.4.

(a)   (b)

Note that this statically indeterminate case, unlike the previous special case, exhibits the multistage load–deflection diagram typical of statically indeterminate systems. As load is increased moments vary in a different ratio for each stage of loading. Thus the collapse mechanism moment diagram is *not* equal to a constant times the elastic moments. Only for statically determinate cases, and a few special statically indeterminate ones can the collapse mechanism moment diagram be obtained by multiplying the elastic moments by a constant. ∎

EXAMPLE 10.2.5

Using the equilibrium method for the continuous beam of Fig. 10.2.6, determine the collapse mechanism load W_n. Assume $F_y = 50$ ksi and that the controlling limit state is the achievement of plastic moment strength.

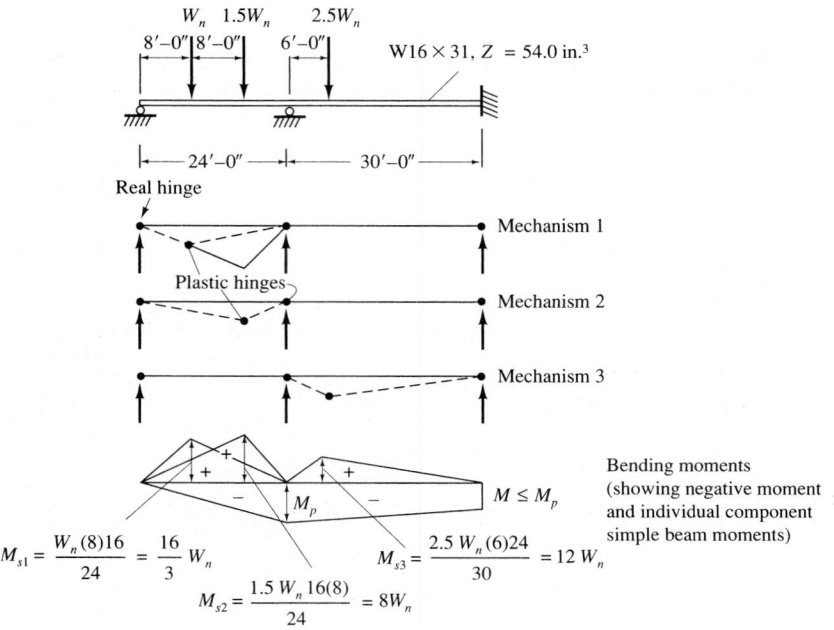

Figure 10.2.6
Example 10.2.5.

Solution:

Whereas in the previous examples only one mechanism was possible, and therefore obviously, there are many situations where the collapse mechanism is not obvious. Each of the three possible mechanisms will be investigated for this example.

Mechanism 1: Positive moment plastic hinge under the load W_n with a negative moment plastic hinge at the center support,

$$M_p = M_{s1} + \tfrac{1}{2}M_{s2} - \tfrac{1}{3}M_p$$
$$\tfrac{4}{3}M_p = \tfrac{16}{3}W_n + \tfrac{1}{2}(8W_n)$$
$$M_p = 7W_n$$

Mechanism 2: Positive moment plastic hinge at the load $1.5W_n$, with a negative moment plastic hinge at the center support,

$$M_p = \tfrac{1}{2}M_{s1} + M_{s2} - \tfrac{2}{3}M_p$$
$$\tfrac{5}{3}M_p = \tfrac{8}{3}W_n + 8W_n$$
$$M_p = \tfrac{32}{5}W_n = 6.4W_n$$

Mechanism 3: Positive moment plastic hinge at the load $2.5W_n$ with negative moment plastic hinges at the ends of the 30-ft span,

$$M_p = M_{s3} - M_p$$
$$2M_p = 12W_n$$
$$M_p = 6W_n$$

Since the plastic moment capacity required is largest for Mechanism 1, it controls:

$$W_n = \frac{M_p}{7} = \frac{50(54.0)}{7(12)} = 32 \text{ kips}$$

In this case the collapse mechanism occurs in the 24-ft span while the 30-ft span remains stable and elastic. A different section in each span may be a more economical solution. ∎

Plastic Limit Load—Energy Method

The principle of virtual work may also be applied to obtain the plastic limit load W_n in an analysis of a given structure, or to find the required plastic moment M_p in a design problem.

Consider that as the collapse mechanism (i.e., plastic limit) load is reached the beam moves through a virtual displacement δ. For equilibrium, the external work done by the load moving through the virtual displacement must equal the internal strain energy due to the plastic moment rotating through small angles (hinge rotations).

EXAMPLE 10.2.6

Determine the plastic limit load W_n in Example 10.2.2 by the virtual work principle. Referring to Fig. 10.2.7,

Figure 10.2.7
Example 10.2.6.

$$\text{External work} = \text{Internal work}$$

$$W_n \delta = M_p \, 2\theta$$

$$W_n \left(\frac{\theta L}{2} \right) = 2 M_p \theta$$

$$W_n = 4 M_p / L$$

the same as previously obtained. ∎

EXAMPLE 10.2.7

Determine the plastic limit load W_n in Example 10.2.3 by the virtual work principle. Referring to Fig. 10.2.8,

$$\text{External work} = \text{Internal work}$$

$$W_n \frac{\theta L}{2} = 2 M_p \theta + M_p (2\theta)$$

$$W_n = 8 M_p / L$$

which agrees with the previous solution.

Figure 10.2.8
Example 10.2.7.

∎

EXAMPLE 10.2.8

Determine the collapse mechanism (i.e., plastic limit) load W_n for Example 10.2.4 by the virtual work method. Referring to Fig. 10.2.9,

$$\text{External work} = \text{Internal work}$$

$$W_n \frac{2\theta L}{3} = M_p (2\theta + \theta + 3\theta)$$

$$W_n = \frac{3}{2L} (6) M_p = 9 M_p / L$$

For $L = 30$ ft, $W_n = 0.3 M_p$ as before.

Figure 10.2.9
Example 10.2.8.

∎

EXAMPLE 10.2.9

Determine the plastic limit load W_n for the continuous beam of Example 10.2.5. Referring to Fig. 10.2.10,

Figure 10.2.10
Example 10.2.9.

Mechanism 1: $\qquad W_n(2\theta)(8) + 1.5W_n\theta(8) = M_p(3\theta + \theta)$

$$28W_n = 4M_p$$

$$M_p = 7W_n$$

Mechanism 2: $\qquad W_n(\theta)(8) + 1.5W_n2\theta(8) = M_p(3\theta + 2\theta)$

$$32W_n = 5M_p$$

$$M_p = 6.4W_n$$

Mechanism 3: $\qquad 2.5W_n(4\theta')(6) = M_p(4\theta' + 5\theta' + \theta')$

$$60W_n = 10M_p$$

$$M_p = 6W_n \qquad\qquad ∎$$

The equlibrium and energy procedures are equally applicable for frames where degrees of freedom in sidesway must be considered. In multistory braced frames, the design procedure is to cause the plastic hinges to form in the girders rather than in the columns; thus the girders are treated as continuous beams. Where bracing contributes an axial force to the girder, the beam-column principles of Chapter 12 apply.

10.3   PLASTIC ANALYSIS—LOAD AND RESISTANCE FACTOR DESIGN EXAMPLES

An elementary example of the design of a simply supported beam that achieves its collapse mechanism was presented in Secs. 7.4 and 7.5. Prior to the adoption of AISC Load and Resistance Design Specification in 1986, plastic analysis was used in *plastic design* as discussed in Sec. 1.8. For additional elementary material on the philosophy, procedure, and experimental verification of plastic design, Refs. 10.3 through 10.10 are suggested.

The factored load plastic analysis option within Load and Resistance Factor Design is the most rational approach and is the authors' recommended procedure for continuous beams. Alternatively, factored load *elastic* analysis may be used even when the conditions for plastic analysis are satisfied.

When elastic analysis is used under either LRFD or ASD, adjustments to the elastic analysis bending moment diagram may be made under conditions specified by AISC-Appendix-1.3. These adjustments attempt to bring the result close to the AISC inelastic design and the LRFD plastic analysis result.

The use of plastic analysis to evaluate the moment strength limit state does not preclude other limit states, such as shear, flange and/or web local buckling, lateral-torsional buckling, and particularly the serviceability limit state of deflection, from controlling.

EXAMPLE 10.3.1

Design the lightest W section of A992 steel for use as the two-span continuous beam of Fig. 10.3.1. Lateral support is provided at the ends and midpoint of each span. Use Load and Resistance Factor Design utilizing plastic analysis requirements under AISC-Appendix-1.7, even though the unbraced lengths L_b are relatively long.

Solution:

(a) Compute the factored loads, w_u and W_u. Temporarily neglecting the weight of the beam,

$$w_u = 1.2(1.5) + 1.6(5) = 9.80 \text{ kips/ft}$$

$$W_u = 1.2(30) + 1.6(60) = 132 \text{ kips}$$

The required nominal strength is the factored load divided by ϕ_b. In this case,

$$\text{Required } w_n = w_u/\phi_b = 9.80/0.90 = 10.89 \text{ kips/ft}$$

$$\text{Required } W_n = W_u/\phi_b = 132/0.90 = 147 \text{ kips}$$

As indicated in the problem statement, plastic analysis is to be used. Evaluate the possible collapse mechanisms; in this case there are two, as shown in Fig. 10.3.1b.

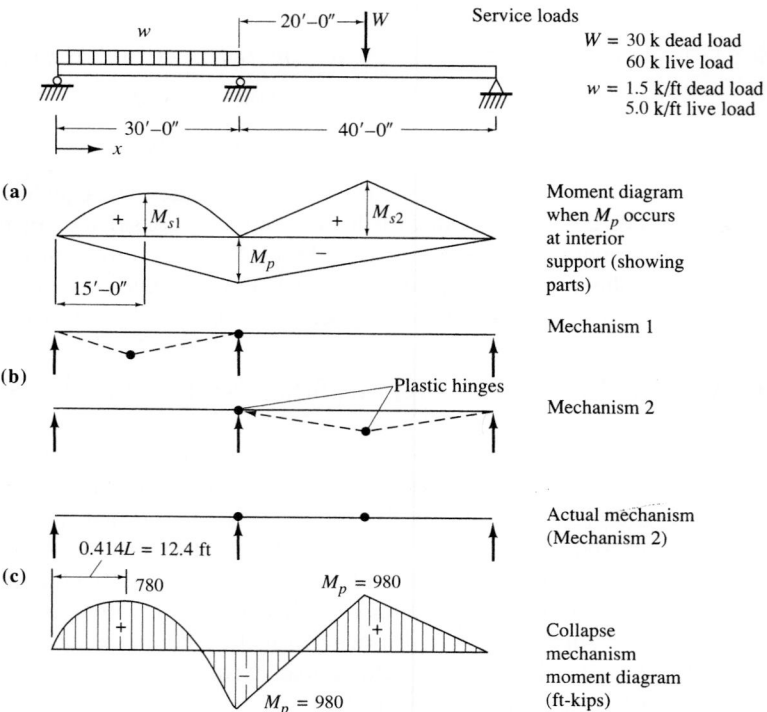

Figure 10.3.1
Example 10.3.1.

(b) *Mechanism 1:* For a span with uniform load, simply supported at one end, and expected to develop a plastic hinge at the other, the location of maximum positive moment is not self-evident. At an unknown distance x from the discontinuous end, the net positive moment is

$$M_x = \frac{w_n}{2}x(L - x) - \frac{x}{L}M_p \qquad \text{(a)}$$

Taking $dM_x/dx = 0$ and solving for x, the location of maximum moment is

$$x = \frac{L}{2} - \frac{M_p}{w_n L} \qquad \text{(b)}$$

Substituting Eq. (b) into Eq. (a) gives

$$M_x(\text{max}) = \frac{w_n}{2}\left(\frac{L}{2} - \frac{M_p}{w_n L}\right)\left(L - \frac{L}{2} + \frac{M_p}{w_n L}\right) - \frac{M_p}{L}\left(\frac{L}{2} - \frac{M_p}{w_n L}\right) \qquad \text{(c)}$$

When a plastic hinge develops, $M_x(\text{max}) = M_p$; thus Eq. (c) becomes

$$M_p = \frac{w_n L^2}{8} - \frac{M_p^2}{2w_n L^2} - \frac{M_p}{2} + \frac{M_p^2}{w_n L^2} \qquad \text{(d)}$$

Substitute $M_s = w_n L^2/8$ into Eq. (d), giving

$$M_p^2 - 24M_p M_s + 16M_s^2 = 0 \qquad \text{(e)}$$

Solving the quadratic for M_p gives

$$M_p = 0.686 M_s \qquad \text{(10.3.1)}$$

and the location of the plastic hinge is at

$$x = \frac{L}{2} - \frac{M_p}{w_n L}\left(\frac{L}{L}\right)\frac{8}{8} = \frac{L}{2} - \frac{M_p}{M_s}\left(\frac{L}{8}\right)$$

$$= L\left(\frac{1}{2} - \frac{0.686}{8}\right) = 0.414L$$

The results of this development are shown in Fig. 10.3.2.

Figure 10.3.2
Collapse condition for uniformly loaded span continuous at one end and simply supported at the other.

For the specific example, Mechanism 1 (Fig. 10.3.1b) requires

$$M_p = 0.686 M_{s1}; \qquad M_{s1} = \frac{w_n L^2}{8} = \frac{10.9(30)^2}{8} = 1225 \text{ ft-kips}$$

$$M_p = 0.686(1225) = 840 \text{ ft-kips}$$

(c) *Mechanism 2:* For the 40-ft span,

$$M_{s2} = \frac{W_n L}{4} = \frac{147(40)}{4} = 1470 \text{ ft-kips}$$

Equilibrium requires

$$M_p = M_{s2} - \frac{M_p}{2} = 1470 - \frac{M_p}{2}$$

$$M_p = 1470/1.5 = 980 \text{ ft-kips}$$

The largest M_p requirement obtained from the possible mechanisms determines the controlling condition. Thus Mechanism 2 governs. A check by determining the complete collapse mechanism moment diagram should be made to insure that M_p is the maximum moment. If a location is found where the computed nominal moment exceeds M_p, the mechanism used is not correct. In this case, Fig. 10.3.1c shows $M_p = 980$ ft-kips is the maximum, so that Mechanism 2 is the correct one.

The maximum positive moment on the 30-ft span using Eq. (a) is

$$M_u/\phi = \frac{10.9}{2}(12.4)(17.6) - 0.414 M_p$$

$$= 1190 - 0.414(980) = 780 \text{ ft-kips} < 980 \qquad \textbf{OK}$$

(d) Select section:

$$\text{Required } Z_x = \frac{M_p}{F_y} = \frac{980(12)}{50} = 235 \text{ in.}^3$$

Try W27×84 with $Z_x = 244$ in.^{3} as indicated by the *AISC Manual,* Table 3-4, "Selection by Z_x". Because of the relatively long laterally unbraced lengths, a section having larger r_y may be required.

(e) Check lateral support. When plastic analysis is used, more rotation capacity at the plastic hinges is relied upon to develop the mechanism. AISC-Appendix 1.7 requires that L_b not exceed L_{pd} given by Eq. 9.5.7 [AISC Formula (A-1-7)],

$$L_{pd} = \left[0.12 + 0.076 \left(\frac{M_1}{M_p} \right) \right]\left(\frac{E}{F_y} \right) r_y = \frac{3500 + 2200(M_1/M_p)}{F_y} r_y \qquad [9.5.7]$$

M_p is used instead of M_2 as in AISC-Appendix 1.7 because when the formula applies, M_2 will always be M_p. In the region of the last hinge to form, and in regions not adjacent to a plastic hinge, Eq. 9.5.7 does not apply.

At unbraced segments adjacent to the last plastic hinge to form, the laterally unbraced length L_b may not exceed L_p if C_b is 1.0; however, when the moment gradient is favorable (i.e., $C_b > 1$), M_p may still be achieved even though $L_b > L_p$. L_p is given by Eq. 9.5.5 [AISC Formula (F2-5)],

$$L_p = 1.76 r_y \sqrt{\frac{E}{F_y}} = \frac{300 r_y}{\sqrt{F_y}, \text{ksi}} \qquad [9.5.5]$$

For this problem, four segments must be considered, as follows:

1. *15 ft at left end:* L_{pd} does not apply for segments containing no plastic hinges. Since the maximum moment of 780 ft-kips is less than M_p, L_b can be greater than L_p. This segment will be checked last since it is not expected to control the design.

2. *15 ft adjacent to interior support:*

$$M_1 = 780 \text{ ft-kips}; \qquad M_p = 980 \text{ ft-kips}$$

$$L_{pd} = \frac{3500 + 2200(780/980)}{50} r_y = 105\, r_y$$

$$\text{Minimum } r_y = \frac{L_b}{105} = \frac{15(12)}{105} = 1.71 \text{ in.}$$

3. *20 ft adjacent to interior support:*

$$M_1 = 980 \text{ ft-kips}; \qquad M_p = 980 \text{ ft-kips}$$

$$L_{pd} = \frac{3500 + 2200(980/980)}{50} r_y = 114\, r_y$$

$$\text{Minimum } r_y = \frac{L_b}{114} = \frac{20(12)}{114} = 2.11 \text{ in.}$$

4. *20 ft at right end:* Since the maximum positive moment plastic hinge is the last one to form, L_{pd} does not apply. However when L_b exceeds L_p, either the strength decreases due to inelastic lateral-torsional buckling limit state, or the strength still may reach M_p because of a favorable moment gradient. When L_{pd} does not apply, the section must satisfy AISC-F2. In this design, the moment gradient gives $C_b = 1.67$ from Table 9.6.3.

Assume the minimum r_y required is 2.11 in. for segment 3; in which case the W27×84 ($r_y = 2.07$ in.) is *not* acceptable for the segments adjacent to the interior support. Increase size to W27×94 ($r_y = 2.12$ in.).

Investigate further for the W27×94,

$$M_p = Z_x F_y = 278(50)/12 = 1160 \text{ ft-kips}$$

$$L_p = 1.76(2.11)\sqrt{29,000/50}\,\tfrac{1}{12} = 7.5 \text{ ft}$$

$$M_r = 0.7 F_y S_x = 0.70(50)243/12 = 710 \text{ ft-kips}$$

$$L_r = \text{Eq. } 9.6.6 = 21.6 \text{ ft}$$

(f) Check strength for $L_b = 20$ ft at the right end of the 40-ft span. M_n is controlled by inelastic lateral-torsional buckling since $L_b < L_r$. The strength lies between M_p and M_r when $C_b = 1.0$; however, in this case with $C_b = 1.67$, AISC Formula (F2-2) or Eq. 9.6.10 gives

$$M_n = 1.67\left[1160 - (1160 - 710)\left(\frac{20 - 7.5}{21.6 - 7.5}\right)\right]$$

$$= 1.67(760) = 1270 \text{ ft-kips}$$

Since computed M_n exceeds M_p, $M_n = M_p$. Thus, the 20-ft segment at right end is satisfactory for lateral bracing.

$$(M_n = M_p = 1160 \text{ ft-kips}) > (\text{Required } M_p = 980 \text{ ft-kips}) \qquad \textbf{OK}$$

(g) Check strength for $L_b = 15$ ft segment at left end of span 1. Since L_b lies between L_p (7.5 ft) and L_r (21.6 ft), M_n is controlled by inelastic lateral-torsional buckling. For this case, C_b is

$$C_b = \frac{12.5M_{max}}{2.5M_{max} + 3M_A + 4M_B + 3M_C} \qquad [9.6.11]$$

The moment gradient is parabolic; thus, from Table 9.6.4, C_b is 1.30. To illustrate the calculation,

$$M_{max} = \text{maximum moment in the unbraced segment}$$
$$M_A = \text{moment at } 1/4 \text{ point} = 0.4375M_{max}$$
$$M_B = \text{moment at midpoint} = 0.75M_{max}$$
$$M_C = \text{moment at } 3/4 \text{ point} = 0.9375M_{max}$$
$$C_b = \frac{12.5M_{max}}{2.5M_{max} + 3(0.4375M_{max}) + 4(0.75M_{max}) + 3(0.9375M_{max})}$$
$$= \frac{12.5M_{max}}{9.625M_{max}} = 1.2987 \qquad \therefore C_b = 1.30.$$

Using Eq. 9.6.10,

$$M_n = C_b\left[M_p - (M_p - 0.7F_yS_x)\left(\frac{L_b - L_p}{L_r - L_p}\right)\right] \le M_p$$

$$M_n = 1.30\left[1160 - (1160 - 710)\left(\frac{15 - 7.5}{21.6 - 7.5}\right)\right]$$

$$= 1.30(920) = 1200 \text{ ft-kips} > (M_p = 1160 \text{ ft-kips})$$

The segment is satisfactory with $M_n = M_p = 1160$ ft-kips.

(h) Check moment strength. Note that the factored loads were divided by ϕ_b to obtain the required nominal loads that can be carried when the collapse mechanism (required nominal moment strengths M_n) is reached. Alternatively, the factored moment M_u diagram could have been obtained; that would be appropriate when elastic analysis is used. However, when plastic analysis is used it seems more appropriate to use the M_n diagram involving the collapse mechanism, since only at the collapse mechanism do the plastic hinge and M_p have meaning. In this case, after the beam weight is added, a recalculation of required M_p for Mechanism 2 gives

$$M_{s2} = \frac{W_nL}{4} + \frac{(w_u/\phi)L^2}{8} = 1470 + \frac{0.094(1.2/0.90)(40)^2}{8} = 1495 \text{ ft-kips}$$

$$M_p = M_{s2} - 0.5M_p = M_{s2}/1.5 = 1000 \text{ ft-kips}$$

$$(\text{Provided } M_p = 1160 \text{ ft-kips}) > (\text{Required } M_p = 1000 \text{ ft-kips}) \qquad \textbf{OK}$$

It is clear that the beam weight has little effect.

(i) Check shear. At the interior end of the 40-ft span the factored load shear V_u is

$$V_u = 73.4 + \frac{980}{40} = 97.9 \text{ kips}$$

$$\frac{h}{t_w} = 49.5 < \left[2.24\sqrt{\frac{E}{F_y}} = \frac{382}{\sqrt{F_y, \text{ksi}}} = 54.0 \right] \Rightarrow \phi = 1$$

$$V_n = 0.6F_yA_w = 0.6(50)(26.9)0.490 = 395 \text{ kips}$$
$$[\phi V_n = 1.0(395) = 395 \text{ kips}] > 97.8 \text{ kips} \qquad \textbf{OK}$$

As expected, except on short spans with heavy loading, shear does not control.

(j) Deflection. If excessive deflection appears to be a problem, *service load* deflection must be computed. Deflection when collapse is imminent is not easily determined, nor is it generally of interest. In this case,

$$\frac{L}{d} = \frac{40(12)}{27} = 17.8$$

Though *L/d* is not any guarantee regarding deflection, it does provide some indication of difficulty as discussed in Sec. 7.6. In this case deflection is not likely a problem.
Use W27×94, $F_y = 50$ ksi ■

EXAMPLE 10.3.2

Design the lightest constant moment of inertia section required for the continuous beam of Fig. 10.3.3. Lateral supports are provided at the vertical supports and at each concentrated load. The steel has $F_y = 60$ ksi. Use Load and Resistance Factor Design utilizing plastic analysis under AISC-Appendix 1.

Solution:

(a) Determine factored loads W_u and associated nominal loads W_n that must be carried when a collapse mechanism is imminent.

Figure 10.3.3
Example 10.3.2.

$$W_{u1} = 1.2(3) + 1.6(12) = 22.8 \text{ kips}$$
$$W_{u2} = 1.2(7) + 1.6(18) = 37.2 \text{ kips}$$
$$W_{u3} = 1.2(10) + 1.6(40) = 76.0 \text{ kips}$$

Using $\phi_b = 0.90$, the required nominal ultimate loads W_n to be carried are

$$\text{Required } W_{n1} = W_{u1}/\phi_b = 22.8/0.90 = 25.3 \text{ kips}$$
$$\text{Required } W_{n2} = W_{u2}/\phi_b = 37.2/0.90 = 41.3 \text{ kips}$$
$$\text{Required } W_{n3} = W_{u3}/\phi_b = 76.0/0.90 = 84.4 \text{ kips}$$

As indicated in the problem statement, plastic analysis is to be used. Evaluate the possible collapse mechanisms; in this case there are three, as shown in Fig. 10.3.3.

$$M_{s1} = \frac{25.3(10)15}{25} = 152 \text{ ft-kips}$$

$$M_{s2} = \frac{41.3(20)5}{25} = 165 \text{ ft-kips}$$

$$M_{s3} = \frac{84.4(10)(20)}{30} = 563 \text{ ft-kips}$$

(b) *Mechanism 1:* Assume moment under W_1 load equals M_p,

$$M_p = M_{s1} + \tfrac{1}{2}M_{s2} - \tfrac{10}{25}M_p$$
$$\tfrac{35}{25}M_p = 152 + \tfrac{1}{2}(165) = 235$$
$$M_p = \tfrac{25}{35}(235) = 168 \text{ ft-kips}$$

(c) *Mechanism 2:* Assume moment under W_2 load equals M_p,

$$M_p = \tfrac{1}{3}M_{s1} + M_{s2} - \tfrac{20}{25}M_p$$
$$\tfrac{45}{25}M_p = \tfrac{1}{3}(152) + 165 = 216$$
$$M_p = \tfrac{25}{45}(216) = 120 \text{ ft-kips}$$

(d) *Mechanism 3:* Assume moment under W_3 load equals M_p,

$$M_p = M_{s3} - M_p$$
$$2M_p = 563; \quad M_p = 282 \text{ ft-kips (Controls)}$$

If a section is selected that has $M_p = 282$ ft-kips, the resulting moments under the W_1 and W_2 loads are

$$M_1 = M_{s1} + \tfrac{1}{2}M_{s2} - \tfrac{10}{25}M_p$$
$$= 235 - \tfrac{10}{25}(282) = 122 \text{ ft-kips} < 282 \qquad \textbf{OK}$$
$$M_2 = \tfrac{1}{3}M_{s1} + M_{s2} - \tfrac{20}{25}M_p$$
$$= 216 - \tfrac{20}{25}(282) = -10 \text{ ft-kips} < 282 \qquad \textbf{OK}$$

(e) Select beam.

$$\text{Required } Z_x = \frac{282(12)}{60} = 56.4 \text{ in.}^3$$

Try W18×35 having $Z_x = 66.5$ in.^{3} This section *is compact* for $F_y = 60$ ksi; for flange local buckling,

$$\left(\lambda = \frac{b_f}{2t_f} = 7.06\right) < \left(\lambda_p = 0.38\sqrt{\frac{E}{F_y}} = \frac{65}{\sqrt{F_y,\text{ksi}}} = 8.4\right) \qquad \textbf{OK}$$

For web local buckling,

$$\left(\lambda = \frac{h}{t_w} = 53.5\right) < \left(\lambda_p = 3.76\sqrt{\frac{E}{F_y}} = \frac{640}{\sqrt{F_y,\text{ksi}}} = 82.7\right) \qquad \textbf{OK}$$

(f) Check lateral support. Although it is likely the plastic hinge at W_3 is the last to form, the designer may not be sure. *The authors suggest that it is prudent to apply the L_{pd} criterion of AISC-Appendix 1.7 to all segments having M_p at either end, eliminating the need for knowing the sequence of plastic hinge formation.* The 20-ft segment, being considerably longer than the other unbraced segments, will undoubtedly control lateral-torsional buckling.

$$L_{pd} = \frac{3500 + 2200(M_1/M_p)}{F_y} r_y \qquad [9.5.7]$$

Since the moment gradient in this segment is one of reverse curvature, with $M_1/M_p = +1$,

$$L_{pd} = 5700r_y/60 = 95r_y$$

which means for $L_b = 20$ ft, the r_y required would be

$$\text{Required } r_y = 20(12)/95 = 2.5 \text{ in.}$$

The W18×35 having r_y only 1.33 in. is not acceptable. None of the W shapes in the range 35 to 45 lb/ft have r_y any higher than 2.0. Since a realistic design having a section with r_y greater than 2.5 in. cannot be achieved, additional bracing must be provided.

If one additional lateral support is provided at 10 ft from the right end of the structure, determine the W shape that will be satisfactory. For the resulting two 10 ft segments that replace the 20-ft segment, $M_1/M_p = 0$; consequently,

$$L_{pd} = 3500r_y/60 = 58r_y$$
$$\text{Required } r_y = 10(12)/58 = 2.1 \text{ in.}$$

This would require W16×67 or W18×76, the same sections needed for maximum strength when L_b was 20 ft. Assume lateral bracing at 5-ft intervals in the 30-ft span. Then at the most critical location near the right support,

$$M_1/M_p = -141/282 = -0.5$$
$$L_{pd} = 2400r_y/60 = 40r_y$$
$$\text{Required } r_y = 5(12)/40 = 1.5 \text{ in.}$$

The lightest section, W18×35, for the M_p requirement has $r_y = 1.22$ in. and still will not qualify. However, W16×36, the next lightest section has $r_y = 1.52$ in. and is also compact with $F_y = 60$ ksi.

Use 5-ft spacing of lateral bracing on the 30-ft span.

On the 25-ft span, check the laterally unbraced 10-ft segment at the left end. This segment does not have a plastic hinge associated with it. The strength of the W16×36 having $L_b = 10$ ft must be determined.

$$M_p = Z_x F_y = 64(60)/12 = 320 \text{ ft-kips}$$

$$L_p = [300(1.52)/\sqrt{60}]/12 = 4.9 \text{ ft}$$

$$M_r = 0.7 F_y S_x = 0.70(60)56.5/12 = 198 \text{ ft-kips}$$

$$L_r = \text{Eq. 9.6.6} = 13.7 \text{ ft}$$

$$M_n = 1.67\left[320 - (320 - 198)\left(\frac{10 - 4.9}{13.7 - 4.9}\right)\right] = 416 \text{ ft-kips}$$

Since computed M_n exceeds M_p, $M_n = M_p$. Since $M_n = 320$ ft-kips exceeds 122 ft-kips under the W_1 load the 10-ft segment is satisfactory.

Use W16×36. $F_y = 60$ ksi. ∎

EXAMPLE 10.3.3

Redesign the beam of Example 10.3.2, using either cover plates or two different sections butt-spliced together. Use plastic analysis in Load and Resistance Factor Design under AISC-Appendix 1. Use $F_y = 60$ ksi.

Solution:

Assume that the change of section occurs near the inflection point to the right of the central support, giving rise to different plastic moments M_{p1} and M_{p2} at the two restrained supports (see Fig. 10.3.4.).

(a) Consider first the 25-ft span supporting the W_1 and W_2 loads. Assume the positive moment under the W_1 load reaches M_{p1},

$$M_{p1} = M_{s1} + \tfrac{1}{2}M_{s2} - \tfrac{10}{25}M_{p1}$$
$$= 152 + 82.5 - 0.4M_{p1}$$
$$M_{p1} = 168 \text{ ft-kips}$$

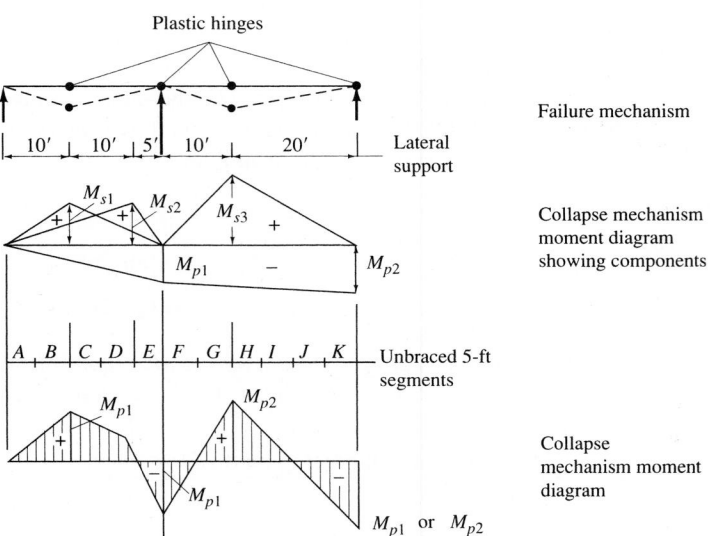

Figure 10.3.4
Example 10.3.3.

Under the W_2 load the moment is

$$M = \tfrac{1}{3}M_{s1} + M_{s2} - \tfrac{20}{25}M_{p1}$$

$$= 50.7 + 168 - 0.8(168) = 84 \text{ ft kips} < 168 \qquad \textbf{OK}$$

(b) Lateral support requirements. Examine the 10-ft laterally unbraced length between the W_1 and W_2 loads, where the ratio $M_1/M_p = -84/168 = -0.5$. Using Eq. 9.5.7,

$$L_{pd} = \frac{3500 + 2200(-0.5)}{60}r_y = 40r_y$$

For $L_b = 10$ ft, the required r_y is $120/40 = 3.0$ in. which cannot be achieved by any realistic design. Increase the lateral support to be every 5 ft. This will be necessary in the 30-ft span also, as shown in Example 10.3.2.

Assuming the plastic hinge $(+M_p)$ between segments B and C (see Fig. 10.3.4) is the last hinge to form, Eq. 9.5.7 is not required by AISC-Appendix 1.7 to be satisfied for those segments. However, in order for the nominal moment strength M_n to reach M_p, L_b may not exceed L_p when $C_b = 1.0$. In this case segment C has the flatter moment gradient, having $M_1/M_2 = -126/168 = -0.75$. Thus, from Table 9.6.3, $C_b = 1.11$; this would mean that L_b could exceed L_p by a small amount such that $1.11M_n$ could still equal M_p, even though M_n might be less than M_p. It will be conservative to require L_b to not exceed L_p for this segment,

$$L_p = 1.76r_y\sqrt{\frac{E}{F_y}} = \frac{300}{\sqrt{F_y, \text{ksi}}} = \frac{300}{\sqrt{60}}r_y = 38.7r_y$$

$$\text{Minimum } r_y = \frac{L_b}{38.7} = \frac{5(12)}{38.7} = 1.55 \text{ in.}$$

Referring to text Appendix Table A1, $r_y \approx 0.22b_f$ to $0.25b_f$. Thus,

$$\text{Minimum } b_f \text{ required} \approx \frac{1.55}{0.25} \text{ to } \frac{1.55}{0.22} = 6.2 \text{ to } 7 \text{ in.}$$

This requirement applies for segments B and C adjacent to the last plastic hinge to form.

For segment E and F, rotation capacity must be assured by satisfying AISC-Appendix 1.7; i.e., L_b may not exceed L_{pd}. Segment F has the flatter moment gradient and controls; $M_1/M_p = 0$. Thus,

$$L_{pd} = 3500r_y/60 = 58r_y$$

$$\text{Required } r_y = 5(12)/58 = 1.03 \text{ in.}$$

(c) Select section for the 25-ft span.

$$\text{Required } Z_x = \frac{\text{Required } M_p}{F_y} = \frac{168(12)}{60} = 33.6 \text{ in.}^3$$

$$\text{Required } b_f \approx 6.2 \text{ to } 7 \text{ in.}$$

$$\text{Required } r_y = 1.55 \text{ in.}$$

Using the *AISC Manual* Table 3-2, "Selection by Z_x" find several sections that satisfy the required Z_x:

$$\text{W12×26,} \qquad Z_x = 37.2 \text{ in.}^3$$
$$\text{W14×26,} \qquad Z_x = 40.2 \text{ in.}^3$$
$$\text{W16×26,} \qquad Z_x = 44.2 \text{ in.}^3$$

These sections all have $r_y < 1.55$ in., though W12×26 is close with $r_y = 1.51$ in.

Try W12×26, $r_y = 1.51$ in., and investigate segment C:

$$L_{pd} = \frac{3500 + 2200(-126/168)}{60}r_y = 30.5r_y = 3.8 \text{ ft}$$

$$L_p = \frac{300}{\sqrt{F_y, \text{ksi}}}r_y = \frac{300(1.51)}{\sqrt{60(12)}} = 4.9 \text{ ft}$$

$C_b = 1.11$ from Table 9.6.3

$M_p = Z_xF_y = 37.2(60)/12 = 186$ ft-kips

$M_r = 0.7F_yS_x = 0.70(60)33.4/12 = 117$ ft-kips

$L_r = $ Eq. 9.6.6 $= 13.3$ ft

Since L_b is *between* L_p and L_r, inelastic lateral-torsional buckling controls.

AISC-Formula (F2-2) linear relationship applies:

$$M_n = 1.11\left[186 - (186 - 117)\left(\frac{5 - 4.9}{13.3 - 4.9}\right)\right]$$

$$= 1.11(185) = 205 \text{ ft-kips} > M_p$$

Since computed M_n exceeds M_p, $M_n = M_p$. Thus segments B and C are satisfactory for lateral bracing.

Use W12×26, $F_y = 60$ ksi, for the 25-ft span. For the 30-ft span, top and bottom cover plates will be added. An alternative would be to use a heavier section having the same proportions and butt join them by welding or by splice plates.

Since the section used has a greater plastic moment strength than required, the moment diagram under loads W_u/ϕ_b is as shown in Fig. 10.3.5b. The diagram assumes the first plastic

Figure 10.3.5
Comparison of design condition and actual condition for 25-ft span of Example 10.3.3.

hinge is at the support, and that it does form under the application of the required ultimate loads W_u/ϕ_b. The $M_{p1} = 186$ ft-kips as provided is then used in the analysis of the 30-ft span.

In the actual condition, segment E is more severely loaded than in the design condition. However it is still not as severely loaded as segment F.

(d) Determine plastic moment requirement on the 30-ft span. When M_{p2} forms under the W_3 load, equilibrium requires

$$M_{p2} = M_{s3} - \tfrac{1}{3}(M_{p2} - M_{p1}) - M_{p1} \quad \text{(long cover plates)}$$

if M_{p2} is developed at the fixed end. This would require cover plates to extend to the support. If the cover plates are not extended to the support, equilibrium requires

$$M_{p2} = M_{s3} - M_{p1} \quad \text{(short cover plates)}$$

If long cover plates are used,

$$M_{p2} = 563 - \tfrac{1}{3}M_{p2} + \tfrac{1}{3}(186) - 186; \quad \tfrac{4}{3}M_{p2} = 439, \quad M_{p2} = 329 \text{ ft-kips}$$

If shorter cover plates are used,

$$M_{p2} = 563 - 186 = 377 \text{ ft-kips}$$

Since there is little difference in requirements, the cover plates will be made as short as possible. The cover plates, top and bottom, must provide

$$M_p \text{ (cover)} = 377 - 186 = 191 \text{ ft-kips}$$

(e) Select plate size. For an area A_p representing the cover plate at one flange, and a distance d center-to-center of cover plates, the plastic modulus Z becomes

$$Z = 2A_p\left(\frac{d}{2}\right) = A_p d$$

$$\text{Required } A_p = \frac{Z}{d} = \frac{191(12)}{60d} = \frac{38.2}{d} = \frac{38.2}{13.0} = 2.9 \text{ sq in.}$$

assuming $d \approx 12.5 + 0.5$ ($\tfrac{1}{2}$-in. plates). If the plate width is somewhat less than the flange width of W12×26, say 6 in., the area A_p provided will be adequate.

Check local buckling on plate as a stiffened element welded along two sides, according to AISC-Table B4.1-Case 7:

$$\left(\frac{b}{t} = \frac{6}{0.5} = 12\right) < \left(\lambda_p = 1.12\sqrt{\frac{E}{F_y}} = \frac{190}{\sqrt{F_y, \text{ksi}}} = \frac{190}{\sqrt{60}} = 24.5\right) \qquad \textbf{OK}$$

$\underline{Use -\tfrac{1}{2} \times 6 \text{ plates}}$, top and bottom. These would probably be welded continuously along the sides in the length direction. Discussion of weld sizes and other requirements is contained in Chapter 5.

(f) Determine plate length. Referring to Fig. 10.3.6 showing the collapse condition for the 30-ft span, the distance L_1 is the theoretical length required for cover plates; this distance may easily be scaled from the diagram. For this straight line moment diagram, the distance L_1 is easily computed,

$$L_1 = \left(1 - \frac{186 + 186}{186 + 377}\right)(10 + 20) = 0.34(30) = 10.2 \text{ ft}$$

Figure 10.3.6
Collapse condition for 30-ft span of Example 10.3.3.

Extension of the plates in each direction must develop the cover plate's proportion of the strength required and satisfy the requirements of AISC-F13.3. The tension or compression strength required from one plate is

$$\text{Required } Z_x = \frac{377(12)}{60} = 75.4 \text{ in.}^3$$

$$\%Z \text{ from plates} = \frac{A_p}{A_p + A_f}(100) = \frac{3.00}{3.00 + 2.47}(100) = 55\%$$

neglecting the web effect. The nominal tensile force T_n required to be developed is

$$T_n \approx \frac{0.55ZF_y}{d} = \frac{0.55(75.4)60}{12.22 + 0.5} = 196 \text{ kips}$$

Assuming no weld across the end of the cover plate and continuous fillet weld along the sides, determine the number of inches of weld necessary to develop the desired force in the plates. As treated in Sec. 5.14, the strength per inch of $\frac{3}{16}$-in. fillet weld made by the submerged arc process using E70 electrodes is (using subscript v for the shear resistance factor),

$$\phi_v R_{nw} = 0.75(3/16)(0.60)(70) = 5.91 \text{ kips/in.}$$

Equating the design strength of L_w inches of fillet weld to the design tensile strength $\phi_b T_n$ required to be developed,

$$L_w(5.91) = \phi_b T_n = 0.90(196) = 176 \text{ kips}$$

$$L_w = 176/5.91 = 30 \text{ in.} \quad (2 \text{ lines of } 15 \text{ in.})$$

The length of plates required is

$$\text{Length} = L_1 + 2(15 \text{ in. on each end}) = 10.2 + 2.5 = 12.7 \text{ ft}$$

Use $-\frac{1}{2} \times 6$ plates, 13 ft long, beginning 5'-4" from support as shown in Fig. 10.3.6.
In this example, the member was not spliced; however, butt splicing the member might be a more economical choice to the use of cover plates. A compromise solution would be to use a section having a Z_x larger than the minimum for the 25-ft span, but not as large as would be required to omit plates entirely. The cover plates could then be thinner and shorter. Splices are discussed in Sec. 10.6. ∎

10.4   ELASTIC ANALYSIS—AISC LRFD METHOD

Within Load and Resistance Factor Design, either inelastic analysis under AISC-Appendix 1 or elastic analysis under AISC-F2 may be the appropriate procedure. When sections are "compact" with respect to local buckling and the laterally unbraced length L_b does not exceed L_{pd}, plastic analysis *is permitted* to be used. Plastic analysis is *not required* to be used. Elastic analysis is *always permitted*.

EXAMPLE 10.4.1

Redesign the beam of Example 10.3.1 using Load and Resistance Factor Design but *not utilizing* plastic analysis. Use A992 steel.

Solution:

(a) The factored loads W_u and w_u were computed in Example 10.3.1, as follows:

$$w_u = 1.2(1.5) + 1.6(5) = 9.8 \text{ kips/ft}$$

$$W_u = 1.2(30) + 1.6(60) = 132 \text{ kips}$$

(b) The elastic analysis bending moment diagram under factored service loads must be computed using a method of statically indeterminate structural analysis. This appears in Fig. 10.4.1.

The four separate laterally unbraced segments and their corresponding C_b values and maximum factored moments M_u are shown in Fig. 10.4.1.

(c) Select a section based on the maximum negative moment. Assuming "compact" section for local buckling, the required plastic modulus Z_x would be

$$M_u = \phi_b M_p = \phi_b Z_x F_y$$

$$\text{Required } Z_x = \frac{M_u}{\phi_b F_y} = \frac{1038(12)}{0.90(50)} = 277 \text{ in.}^3$$

Figure 10.4.1
Example 10.4.1.

Using *AISC Manual* Table 3-2, "Selection by Z_x" find

$$W30\times99, \qquad Z_x = 312 \text{ in.}^3, \qquad r_y = 2.1 \text{ in.}$$

(d) Check lateral support. In order to obtain the plastic moment strength M_p, the laterally unbraced length L_b cannot exceed L_p when $C_b = 1.0$.

$$L_p = 1.76r_y\sqrt{\frac{E}{F_y}} = \frac{300}{\sqrt{F_y, \text{ksi}}}r_y = \frac{300(2.1)}{\sqrt{50}(12)} = 7.4 \text{ ft}$$

At first it may appear that L_p is less than 15 ft and therefore, M_n must be less than M_p. Traditionally since moment gradient does not appear in the L_p equation, the designer would evaluate the rotational restraint at the point of inflection (which in this case is at 7.1 ft from the interior support). If the designer decided that twist of the cross-section was adequately prevented the inflection point could be treated as a braced point. Under AISC-F2.2c, the full moment gradient across the inflection point to the next actual lateral support is accounted for by C_b. Thus, even when L_b exceeds L_p, the C_b greater than 1.0 may bring the available strength to M_p anyway.

(e) Evaluate the available strength for the W30×90 near the interior support. Using Eq. 9.6.11 for C_b with $M_1/M_2 = +801/1038 = +0.77$,

$$C_b \approx 2.23$$
$$M_p = Z_xF_y = 283(50)/12 = 1180 \text{ ft-kips}$$
$$L_p = 7.4 \text{ ft} \quad [\text{computed in part (d)}]$$
$$M_r = 0.7F_yS_x = 0.7(50)(245)/12 = 715 \text{ ft-kips}$$
$$L_r = \text{Eq. 9.6.6} = 20.9 \text{ ft}$$

Since L_b is *between* L_p and L_r, inelastic lateral-torsional buckling controls. The AISC-Formula (F2-2) linear relationship applies

$$M_n = 2.23\left[1180 - (1180 - 715)\left(\frac{20 - 7.4}{20.9 - 7.4}\right)\right]$$
$$= 2.23(746) = 1660 \text{ ft-kips} > M_p$$

Since computed M_n exceeds M_p, $M_n = M_p$. Thus, segments B and C are satisfactory for lateral bracing.

$$[\phi_bM_p = 0.90(1180) = 1060 \text{ kips}] > (M_u = 1038 \text{ kips}) \qquad \textbf{OK}$$

(f) Check segment D. As in segments B and C, L_b is between L_p and L_r; thus, M_n is obtained for this inelastic buckling case by linear interpolation, this time using $C_b = 1.67$ with AISC-Formula (F1-1),

$$M_n = 1.67(746) = 1246 \text{ kips} > M_p$$

Since computed M_n exceeds M_p, $M_n = M_p = 1180$ ft-kips. Thus, segment D is satisfactory for lateral bracing.

(g) Check segment A. In this segment, the maximum moment occurs *between* the braced points; thus, C_b could be conservatively taken as 1.0. When computed using Eq. 9.6.11, $C_b = 1.17$. The linear interpolated value of M_n using AISC-Formula (F2-2), with $L_b = 15$ ft is

$$M_n = 1.17(918) = 1074 \text{ kips} < M_p$$
$$[\phi_bM_n = 0.90(1047) = 967 \text{ kips}] > (M_u = 645 \text{ ft-kips}) \qquad \textbf{OK}$$

When using plastic analysis with these long laterally unbraced lengths, the design was impossible. ∎

10.5   ELASTIC ANALYSIS—ALLOWABLE STRENGTH DESIGN

The design of usual continuous beams having relatively close spacing of lateral supports should be done using plastic analysis under the AISC LRFD Method. AISC-Appendix 1.1 does not permit the use of the inelastic analysis with the ASD Method. However, for continuous beams with the unbraced length requirements as defined in AISC-Appendix 1.7, the negative moment under the support, computed using elastic analysis, can be reduced by 10%, according to AISC-Appendix 1.3. This moment redistribution is permitted for both LRFD and ASD Methods. So, using the ASD Method for continuous beams is limited to elastic analysis. The ASD procedure for design is similar to the LRFD Method. Allowable Strength Design uses adjustments of the elastic moment diagram to indirectly account for plastic behavior and redistribution of moments. Such adjustments cannot reflect true behavior. Occasionally, for designs involving long laterally unbraced lengths where lateral-torsional buckling is the limit state instead of plastic moment strength, Allowable Strength Design may be a practical alternative to elastic analysis under Load and Resistance Factor Design.

EXAMPLE 10.5.1

Redesign the beam of Example 10.3.1 (Fig. 10.3.1) using the Allowable Strength Design Method with A992 steel.

Solution:
The elastic bending moment diagram (see Fig. 10.5.1) under service loads must first be computed using a method of statically indeterminate analysis.

The four separate unbraced segments with their C_b values and design moments are shown in Fig. 10.5.1. If lateral bracing is sufficient to permit a moment redistribution the negative moment will multiplied by 0.9 and the positive moment will be multiplied by 1.1. After the selection of a section, AISC-Appendix 1.7 should be checked to verify if the section qualifies for moment redistribution.

(a) Determine required plastic section modulus assuming $L_b < L_p$ which allows a reduction in $(-M)$. Assuming $M_n = M_p$,

$$M_a \leq \frac{M_n}{\Omega} = \frac{F_y Z_x}{\Omega}$$

and

$$Z_{x,req} = \frac{M_a}{F_y/\Omega} = \frac{0.9(692)(12)}{50/1.67} = 250 \text{ in.}^3$$

which is the minimum plastic section modulus required assuming the most favorable conditions. The laterally unbraced length L_b is 15 ft for segment B which is sufficiently large that it is likely to exceed both L_p for any practical choice of beam.

(b) Whenever L_b exceeds L_p the suitable approach is to use AISC Table 3-10, "Available Moment vs. Unbraced Length". For segment C, enter with,

$$L_b = 20 \text{ ft} \qquad \text{and modified } M_a = \text{actual } M_a/C_b = 692/2.23 = 310 \text{ ft-kips}$$

Section W18×76 is the lightest that satisfies the above requirements.

From the charts, the allowable strength is

$$\frac{M_n}{\Omega} = 316 \text{ ft-kips} \qquad \text{at } L_b = 20 \text{ ft}$$

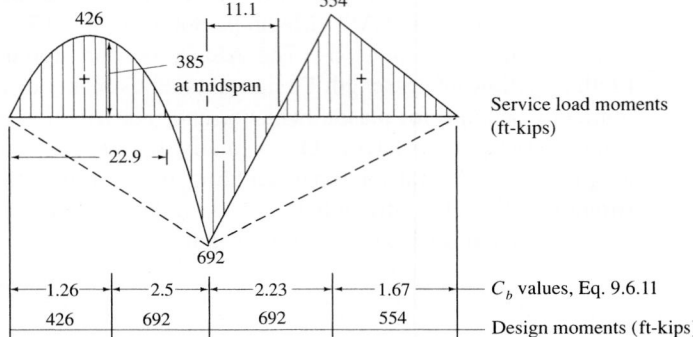

Figure 10.5.1
Example 10.5.1.

and
$$\left[\frac{C_b M_n}{\Omega} = 2.23(316) = 705 \text{ ft-kips}\right] > \left[\frac{M_p}{\Omega} = 407 \text{ ft-kips}\right]$$
$$< 0.9(692) = 623 \text{ ft-kips} \qquad\qquad \textbf{NG}$$

Since C_b is very high, a W section can be determined using the Z_x required, as shown in part (a). From AISC Table 3-2, "Selection by Z_x," find

W30×90 with $L_p = 7.4$ ft, $L_r = 20.9$ ft, and $M_p/\Omega_b = 706$ ft-kips

Since L_b is between L_p and L_r inelastic lateral-torsional buckling controls.

The AISC- Formula (F2-2) linear relationship applies

$$M_n = 2.23\left[1180 - (1180 - 715)\left(\frac{20 - 7.4}{20.9 - 7.4}\right)\right]$$
$$= 2.23(746) = 1660 \text{ ft-kips} > M_p = 1180 \text{ ft-kips}$$

Since Computed M_n exceeds M_p, The allowable strength for segment C is:
$$[M_n/\Omega = M_p/\Omega = 1180/1.67 = 708 \text{ ft-kips}] > M_{a,adjusted}$$
$$= 0.9(692) = 623 \text{ ft-kips} \qquad\qquad \textbf{OK}$$

(c) Check segment D
$$M_n = 1.67(746) = 1246 \text{ ft-kips} > M_p = 1180 \text{ ft-kips, So}$$
$$[M_n/\Omega = M_p/\Omega = 1180/1.67 = 708 \text{ ft-kips}] > M_{a,adjusted}$$
$$= 1.1(554) = 609 \text{ ft-kips} \qquad\qquad \textbf{OK}$$

(d) Check segment A, It should be noted that with an $L_b = 15$ ft., M_n must be recalculated using AISC Formula (F2-2), which gives $M_n = 1.26(918) = 1157$ ft-kips $< M_p = 1180$ ft-kips, so

$$[M_n/\Omega = 1157/1.67 = 693 \text{ ft-kips}] > M_{a,adjusted} = 1.1(426) = 469 \text{ ft-kips} \qquad \textbf{OK}$$
Use W30×90, $F_y = 50$ ksi. ∎

A comparison with Example 10.3.1 where LRFD using plastic analysis was used shows that elastic analysis obtained the lighter section. W30×90 compared to W27×94. Primarily the difference results from the minimum requirement on r_y of 2.11 in. that is needed to satisfy L_{pd}. However, the results are extremely close and in other situations the final design could be the same.

10.6 SPLICES

While the design of connections is outside the scope of this chapter, the location of and strength requirements for beam splices are appropriately discussed here. It is obvious that if a splice is designed for the moment and shear capacity for the member spliced, full continuity is maintained and no special precautions are necessary. Some designers prefer to use shear splices at points of contraflexure and thus introduce a real hinge at a point of zero moment. There are two reasons why this should be avoided: (1) the point of contraflexure under service load is not at the same location that it occurs under factored loads (i.e., at its mechanism condition); (2) moments obtained assuming continuity are invalid if real hinges are inserted. Hart and Milek [10.11] have provided a good discussion of this problem.

According to AISC-J6, connections must be designed for the moments, shears, and axial loads to which they are to be subjected.

If full continuity is assumed when determining moments, either under service loads or under factored loads, then splices should provide that continuity. *A reduced stiffness at a splice may prevent or reduce transmission of moment across that section and significantly change the resulting moment diagram.*

EXAMPLE 10.6.1

Examine the effect of a shear splice at the point of contraflexure in the 40-ft span of the two-span continuous beam of Example 10.3.1. Use A992 steel.

Solution:

(a) Full dead load plus live load. If full continuity is maintained, partial loading in some spans to account for live load in various locations is unnecessary. Note that the moment $M_p = 978$ ft-kips (see Fig. 10.6.1a) can develop even when the adjacent span has a reduced load. In other words, each span may be treated separately, as long as continuity exists so that the negative moment may be assumed to reach M_p.

(b) Effect of hinge at inflection point. Use of a shear splice in effect creates a hinge for all stages of loading (i.e., transforms the system into a statically determinate one). The maximum negative moment that can develop is limited to the shear at the hinge times the distance to the support. When load on the span with the hinge is reduced, the negative moment that can develop is reduced. As shown in Fig. 10.6.1b, when only the dead load is on the 40-ft span, the nominal ultimate load acting is

$$W_n = W_u/\phi_b = 1.2(30)/0.90 = 40.0 \text{ kips}$$
$$\text{Shear at splice} = 40(20/30) = 26.7 \text{ kips}$$
$$\text{Moment at support} = 26.7(10) = 267 \text{ ft-kips}$$

With the full factored dead plus live load ($w_u = 9.8$ kips/ft from Example 10.3.1) acting in the adjacent span, the maximum required nominal positive moment strength in the 30-ft span is

$$V_{nAB} = \frac{(9.8/0.90)30}{2} - \frac{267}{30} = 163 - 8.9 = 154 \text{ kips}$$

Figure 10.6.1
Example 10.6.1.

$$\text{Required } M_n = \frac{(154)^2}{2(9.8/0.90)} = 1095 \text{ ft-kips} > M_p \qquad \textbf{NG}$$

Since the original design required $M_p = 978$ ft-kips, the 30-ft span is now inadequate as a result of the inability of the negative moment to exceed 267 ft-kips with the reduced load on the 40-ft span.

(c) Minimum strength required for splice. If the splice design is to provide less than full continuity, Fig. 10.6.1c illustrates the minimum capacity needed. If the section being spliced provides $M_p = 978$ ft-kips, the positive moment in the 30-ft span may not exceed that value. Thus, referring to Fig. 10.6.1c, and assuming the maximum occurs at approximately $0.45L_1$,

$$M_1 = \frac{(9.8/0.90)(0.45)0.55}{2}(30)^2 - 0.45 M_B = 978 \text{ ft-kips}$$

$$M_B = \frac{1213 - 978}{0.45} = \frac{235}{0.45} = 522 \text{ ft-kips}$$

In order to have this moment develop, a nominal moment strength must be provided at the splice point:

$$M_2 = \frac{(36/0.90)30}{4} - \frac{M_B}{2} = 300 - 261 = 39 \text{ ft-kips}$$

$$\text{Required } M_n \text{ at splice} = \frac{522 + 39}{2} - 39 = 242 \text{ ft-kips}$$

If the splice in this LRFD inelastic analysis problem is designed for the shear as determined from Fig. 10.6.1a, plus the 242 ft-kip moment, the reduced loading of span *BC* will not cause unanticipated overload in span *AB*.

Finally, it is noted that reduced load in span *AB* has no detrimental effect on span *BC*. The authors prefer that continuity be maintained by the design of splices for 100 percent of the moment capacity of members spliced; however, reduced capacities providing more economical designs may be used as long as the resulting designs are checked under possible partial loadings. Other examples and a more detailed procedure are given in Ref. 10.11. ∎

SELECTED REFERENCES

10.1. Lynn S. Beedle. *Plastic Design of Steel Frames*. New York: John Wiley & Sons, Inc., 1958.

10.2. C. E. Massonnet and M. A. Save. *Plastic Analysis and Design*, Vol. I, Beams and Frames. Waltham, Mass: Blaisdell Publishing Company, 1965.

10.3. Bruce G. Johnston. "Strength as a Basis for Structural Design," *Proc. AISC National Engineering Conf.*, 1956, pp. 7–13.

10.4. Bruno Thürlimann. "Simple Plastic Theory," *Proc. AISC National Engineering Conf.*, 1956, pp 13–18.

10.5. Robert L. Ketter. "Analysis and Design Examples," *Proc. AISC National Engineering Conf.*, 1956, pp. 19–35.

10.6. Lynn S. Beedle. "Experimental Verification of Plastic Theory," *Proc. AISC National Engineering Conf.*, 1956, pp 35–49.

10.7. Bruno Thürlimann. "Modifications to 'Simple Plastic Theory'," *Proc. AISC National Engineering Conf.*, 1956, pp. 50–57.

10.8. Frederick S. Merritt. "How to Design Steel by the Plastic Theory." *Engineering News-Record* (Apr. 4, 1957), 38–43.

10.9. Bruce G. Johnston, C. H. Yang, and Lynn S. Beedle. "An Evaluation of Plastic Analysis as Applied to Structural Design," *Welding Journal Research Suppl.* (May 1953), 1–16.

10.10. Joint Committee of Welding Research Council and the American Society of Civil Engineers. *Commentary on Plastic Design in Steel*, 2nd ed., ASCE Manual and Reports on Practice No. 41, New York, 1971.

10.11. Willard H. Hart and William A. Milek. "Splices in Plastically Designed Continuous Structures," *Engineering Journal*, AISC **2**, 2 (April 1965), 33–37.

PROBLEMS

All problems are to be done according to the AISC LRFD Method. Use *plastic analysis* if possible unless otherwise indicated. All given loads are service loads unless otherwise indicated, and assume the loads are always in the position shown even though uniform live load would rarely exist that way. Assume all standard sections are equally readily available in the indicated grade of steel (even though actually they are not). A figure showing span and loading is required, and after making a design selection a final check of strength, (ϕM_n compared with M_u) is required.

10.1. Determine the maximum value for the service load *P* of the beam in the accompanying figure. Assume adequate lateral support such that $L_b < L_p$. The load *P* is 20% dead load and 80% live load.

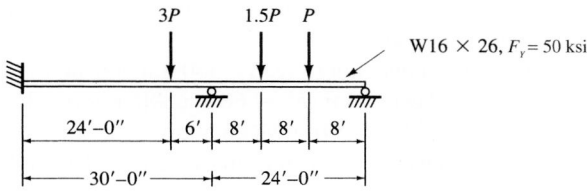

Problem 10.1

10.2. Determine the uniformly distributed service load (40% dead load, 60% live load) a W24×104, $F_y = 50$ ksi, may be permitted to carry as a two-span continuous beam having equal spans of 40 ft. Assume deflection does not control, and that lateral support is provided at the vertical supports and at 15 and 30 ft from the simply supported ends.

10.3. Select the lightest W section for the two-span continuous beam of the accompanying figure. Use $F_y = 50$ ksi. The concentrated loads are 16 kips dead load and 24 kips live load; the uniform load is 0.5 kips/ft dead load and 2.0 kips/ft live load. Specify alternative lateral support if necessary in

Problem 10.3

order to use maximum plastic strength in the design.

10.4. Select the lightest W section for the two-span continuous beam of the accompanying figure. Use $F_y = 60$ ksi. The concentrated loads are 12 kips dead load and 60 kips live load. Lateral support is provided at the vertical supports and under the concentrated load. Specify alternative lateral support if necessary in order to use maximum plastic strength in the design.

Problem 10.4

10.5. Select the lightest W section for the two-span continuous beam of the accompanying figure. Use $F_y = 50$ ksi. The uniform load is 1 kip/ft dead load and 2 kip/ft live load. Lateral support is provided at the vertical supports and at 10-ft intervals. Specify alternative lateral support if necessary in order to use maximum plastic strength in the design.

Problem 10.5

10.6. Select the lightest W section for the two-span continuous beam of the accompanying figure. Use $F_y = 50$ ksi. The concentrated load is 20 kips dead load and 30 kips live load; the uniform load is 0.8 kip/ft dead load and 2.7 kips/ft live load. Specify the uniform laterally unbraced lengths L_b to give the optimum design.

Problem 10.6

10.7. Select the lightest W section for the three-span continuous beam of the acompanying figure. Use $F_y = 50$ ksi. The 50-kip concentrated loads are 30% dead load and 70% live load, the 20-kip concentrated load is all dead load, and the uniform load is 20% dead load and 80% live load.

10.8. Select the lightest W section for a three-span (50 ft-65 ft-50 ft) continuous beam to carry a uniform dead load of 2 kips/ft in addition to the beam weight, and a uniform live load of 1.5 kips/ft. In this problem, live load is to be treated in its usual manner; that is, of indefinite length and positioned to give maximum effects. Live load deflection (maximum) may not exceed $L/360$. Lateral support is provided every 5 ft. $F_y = 60$ ksi. After doing the design with $L_b = 5$ ft, give an alternate more economical design prescribing the location of lateral supports.

Problem 10.7

10.9. Repeat Prob. 10.8 using steel having $F_y = 50$ ksi and specify lateral bracing for an economical design, instead of using the 5-ft spacing of Prob. 10.8.

10.10. Assume one splice is required for the beam selected for Prob. 10.3. Specify its location and the shear and moment for which it should be designed. Assume any load or loads may be reduced to their dead load value while other loads remain at their maximum values.

10.11. Same as Prob. 10.10. but splice the beam of Prob. 10.7.

10.12. For the beam shown in the accompanying figure, (a) select the lightest W section assuming the same section extending over all three span and (b) redesign, using a smaller base section with welded cover plates where needed. For the cover plate, specify the size, length, and location of the plate. The concentrated loads are 10 kips dead load and 25 kips live load, and the uniform loading is 0.5 kip/ft dead load and 1.5 kip/ft fixed position live load. $F_y = 50$ ksi. Specify the necessary lateral support.

Problem 10.12

11

Plate Girders

11.1 INTRODUCTION AND HISTORICAL DEVELOPMENT

A plate girder is a beam built up from plate elements to achieve a more efficient arrangement of material than is possible with rolled beams. Plate girders are economical where spans are long enough to permit saving in cost by proportioning for the particular requirements. Plate girders may be of riveted, bolted, or welded construction. Beginning with early railroad bridges during the period 1870–1900, riveted plate girders (Fig. 11.1.1) composed of angles connected to a web plate, with or without cover plates, were extensively used in the United States on spans from about 50 to 150 ft.

Beginning in the 1950s when welding became more widely used (because of improved quality of welding and shop-fabricating economies resulting from increased use of automatic equipment) shop-welded plate girders composed of three plates (Fig. 11.1.2) gradually replaced riveted girders. During this period also, high-tensile-strength bolts were

Figure 11.1.1
Typical components of riveted plate girder.

(a) Cross-section   (b) Elevation at end of span

516

Plate girders, showing welded stiffeners in place, rocker bearings for vertical supports at the pier, transverse cross bracing between girders, and hinges to provide a simple support for the spans to the right of the hinges. (Photo by C. G. Salmon)

displacing rivets in field construction. Since the 1960s nearly all plate girders are shop welded using two flange plates and one web plate to make an I-shaped cross section.

Where practically all *riveted* girders were composed of plate and angle components having the same material yield strength, the tendency now with *welded* girders is to combine materials of different strength. By changing materials at various locations *along the span* so that higher strength materials are available at locations of high moment and/or shear, or by using different strength material for flanges than for web (hybrid girders), more efficient and economical girders can be obtained.

Because few railroad bridges are being built today, discussion of economical spans and other dimensioning comments in this chapter is limited to highway bridges, where

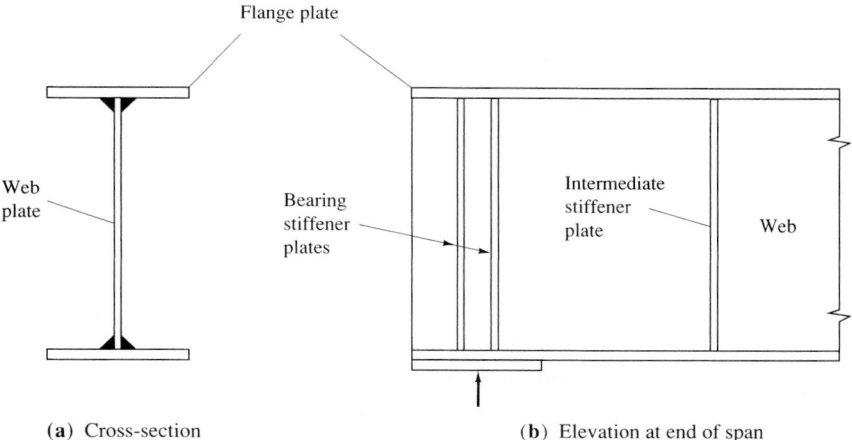

Figure 11.1.2
Typical components of a welded plate girder.

(a) Cross-section

(b) Elevation at end of span

Figure 11.1.3
Other types of welded plate girders.

(a) Box girder   **(b)** Hybrid girder   **(c)** Delta girder

(labels within figure: Compression flange; Higher-strength flanges (say, structural tee); Lower-strength web)

most are continuous over two or more spans; or to buildings where some spans may be assumed as simply supported but, more frequently, are part of a rigid frame system.

Better understanding of plate girder behavior, higher strength steels, and improved welding techniques have combined to make plate girders economical in many situations formerly thought to be ideal for the truss. Generally, simple spans of 70 to 150 ft (20 to 50 m) have traditionally been the domain for the plate girder. For bridges, continuous spans frequently using haunches (variable depth sections) are now the rule for spans 90 ft or more. There are several three-span continuous plate girders in the United States with center spans exceeding 400 ft, and longer spans are likely to be feasible in the future. The longest plate girder in the world is a three-span continuous structure over the Save River at Belgrade, Yugoslavia, with spans 246–856–246 ft (75–260–75 m). It is a double box girder in cross-section varying in depth from 14 ft 9 in. (4.5 m) at midspan to 31 ft 6 in. (9.6 m) at the pier. The structure replaced a suspension bridge destroyed in World War II.

Three types of plate girders whose design is outside the scope of this chapter are shown in Fig. 11.1.3: (a) the box girder, providing improved torsional stiffness for long-span bridges; (b) the hybrid girder, providing variable material strength in accordance with stresses; and (c) the delta girder, providing improved lateral rigidity for long lengths of lateral unsupport.

Prior to studying the theoretical development in this chapter the reader is advised to review Chapter 6, Part II, where the basic elastic stability of plates is treated.

Since the design of riveted girders has been extensively treated in older texts [11.1, 11.2] and such girders are rarely used at present, emphasis is placed on welded girders. No example of bolted or riveted girder design is given; however, high-strength bolted splices, commonly found in field connections, are treated in Chapter 13.

11.2 DIFFERENCE BETWEEN BEAM AND PLATE GIRDER

A plate girder is actually a deep beam. The limit states treated in Chapter 9 with regard to beams are still applicable for plate girders. Figure 11.2.1 shows the nominal flexural strength M_n vs the slenderness ratio λ for the basic flexure limit states, lateral-torsional buckling, flange local buckling, and web local buckling. The relationship for lateral-torsional buckling in Fig. 11.2.1a is valid when a section is "compact" (i.e., $\lambda \leq \lambda_p$ in AISC-Table B4.1) with regard to the flange and web local buckling limit states. When the section is "noncompact" (i.e., $\lambda_p > \lambda \leq \lambda_r$ in AISC-Table B4.1), the nominal strength M_n must be determined for all three limit states in Fig. 11.2.1; the lowest value controls.

The web local buckling limit state is treated in AISC-F4 and F5. The effect of inelastic buckling of a noncompact web is accounted for indirectly by multiplying the moment causing yielding in the compression or tension flange by a "web plastification factor" R_{pc} or R_{tc}. Elastic buckling of slender webs is accounted for by the bend-buckling strength

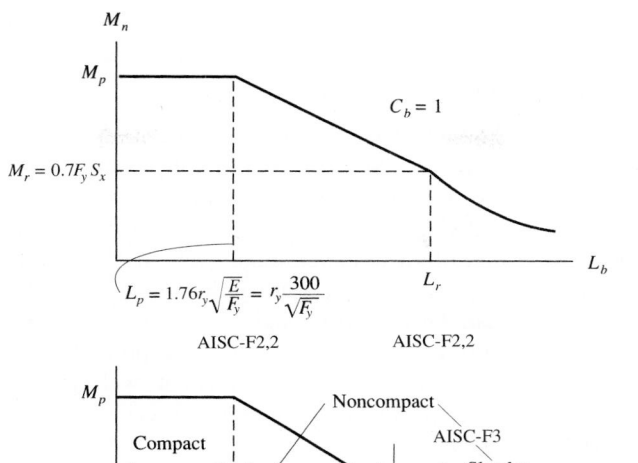

(a) Lateral-torsional buckling limit state (for shapes "compact" for flange or web local buckling)

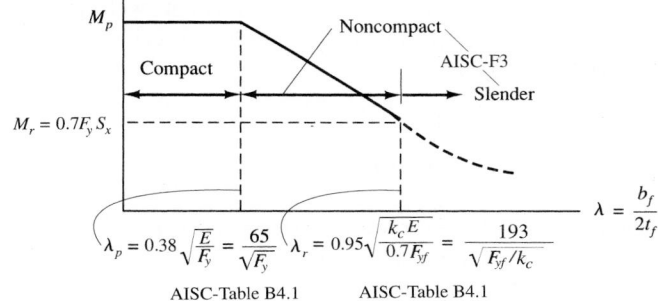

(b) Flange local buckling limit state

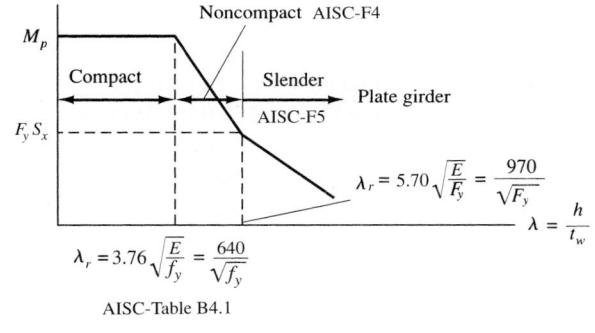

(c) Web local buckling limit state

Figure 11.2.1
Limit states in flexure. (for doubly symmetric I-shaped sections)

reduction factor R_{pg}. This adjusted moment value is then used as the maximum capacity of the section instead of the plastic moment M_p. These R factors are discussed in detail in the next sections.

All rolled W (wide-flange shapes) (with F_y of 36 or 50 ksi) have "compact" webs and only a few sections have "noncompact" flanges; this has previously been discussed in Chapters 7 and 9. If the flange is "noncompact" or "slender" (and the web is "compact") the nominal strength is the lower of the values from the lateral-torsional limit state and the flange local buckling limit state (AISC-F2 and F3).

When the web is "noncompact" or "slender," AISC-F4 and F5 give provisions to account for local and bend-buckling of the web. AISC-F4 for "noncompact" webs allows such sections to be instead conservatively designed according to AISC-F5, which is specifically intended for "slender" webs. In general, plate girder webs are typically "slender."

The flexural and shear strengths of a plate girder are largely related to the web. The "slender" web may cause several problems:

1. Buckling due to bending in the plane of the web will reduce the efficiency of the web to carry its elastic share of the bending moment.

2. Buckling of the compression flange in the vertical direction due to inadequate stiffness of the web to prevent such buckling.

3. Buckling due to shear.

This chapter is largely devoted to treatment of these problems.

Since the 1961 AISC Specification, the design of plate girders has used the same strength approach developed by Basler and others at Lehigh University.

The most distinguishing feature of a plate girder is the use of regularly spaced transverse stiffeners. Stiffeners increase the strength of the web to carry shear. The elastic or inelastic buckling strength of a plate girder web in shear does not represent the maximum strength in shear. There will be significant *post-buckling strength* after buckling (slight out-of-plane deformation) has occurred when properly designed transverse stiffeners are used. The girder will behave like a truss with its web carrying diagonally the tension forces and the stiffeners carrying the compression forces. This truss-like behavior is referred to as *tension-field action.*

11.3 VERTICAL FLANGE BUCKLING LIMIT STATE

The maximum limit on the web slenderness h/t_w is based on the stiffness needed in the plane of the web to prevent the compression flange from buckling vertically (Fig. 11.3.1c). Note that h, the clear unsupported height of the web in a rolled section, is the depth h of the web plate in a welded I-shaped section. Furthermore, some flexural stiffness is needed from the web along the flange-to-web connection to preclude torsional buckling of the flange (Fig. 11.3.1b).

For the purpose of this development, one may imagine that the flange is a compression member independent of the rest of the girder (see Fig. 11.3.2). When the girder is bent, as exaggerated in Fig. 11.3.3, the curvature gives rise to flange force components that cause compression on the edges of the web adjacent to the flanges. When the web remains stable when subject to those compressive flange force components, the flange cannot buckle vertically. In the following derivation the flange itself is assumed to have *zero* stiffness to resist vertical buckling; a conservative procedure.

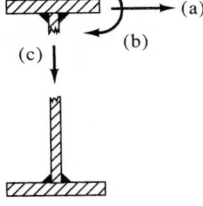

Figure 11.3.1
(a) Lateral buckling;
(b) torsional buckling;
(c) vertical buckling.

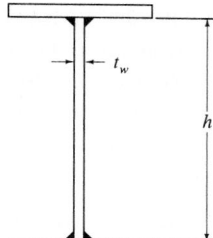

Figure 11.3.2
Vertical buckling of the compression flange.

Flange acts independently when h/t_w is large

Strength from web required for bracing is small

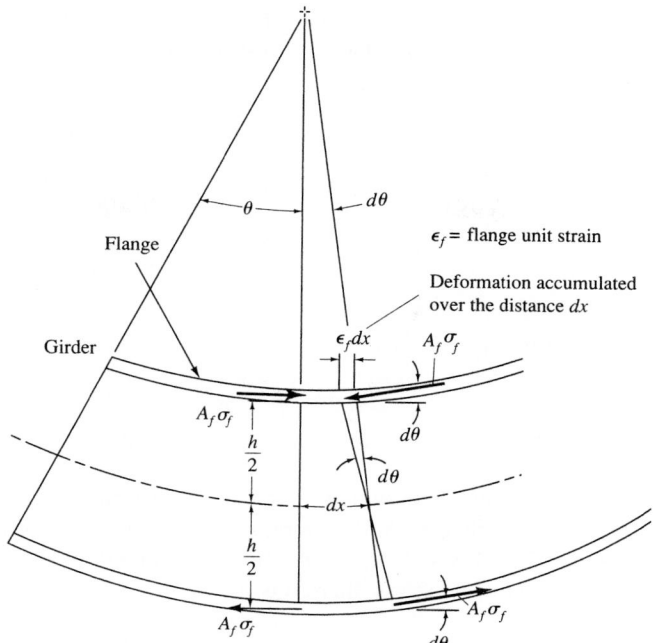

Figure 11.3.3
Flange forces arising from
girder curvature.

Referring to Fig. 11.3.3, the deformation $\epsilon_f\, dx$ accumulated over the distance dx is

$$\epsilon_f\, dx = d\theta\frac{h}{2} \tag{11.3.1}$$

$$d\theta = \frac{2\epsilon_f}{h}dx \tag{11.3.2}$$

As shown in Fig. 11.3.4a, the vertical component causing compression is $\sigma_f A_f\, d\theta$. After dividing by the area $t_w\, dx$ to obtain the compressive stress f_c as shown in Fig. 11.3.4b, one may substitute Eq. 11.3.2 for $d\theta$,

$$f_c = \frac{\sigma_f A_f\, d\theta}{t_w\, dx} = \frac{2\sigma_f A_f \epsilon_f}{t_w h} \tag{11.3.3}$$

Referring again to Eq. 6.14.28, the elastic buckling stress for a plate,

$$F_{\text{cr}} = \frac{k\pi^2 E}{12(1 - \mu^2)(b/t)^2} \tag{6.14.28}$$

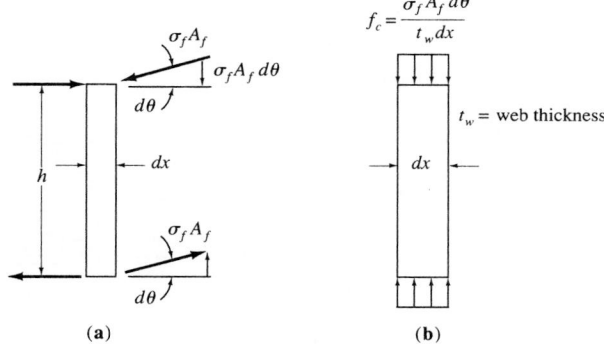

Figure 11.3.4
Effect of flange force
component normal to
flange plate.

where $b = h$, $t = t_w$, and $k = 1$ for the case of the Euler plate assumed free along edges parallel to loading, and pinned top and bottom. Thus

$$F_{cr} = \frac{\pi^2 E}{12(1 - \mu^2)(h/t_w)^2} \qquad (11.3.4)$$

Equating applied stress, Eq. 11.3.3, to the critical stress, Eq. 11.3.4, gives

$$\frac{2\sigma_f A_f \epsilon_f}{t_w h} = \frac{\pi^2 E}{12(1 - \mu^2)(h/t_w)^2} \qquad (11.3.5)$$

which, letting $t_w h = A_w$, gives

$$\frac{h}{t_w} = \sqrt{\frac{\pi^2 E}{24(1 - \mu^2)}\left(\frac{A_w}{A_f}\right)\left(\frac{1}{\sigma_f \epsilon_f}\right)} \qquad (11.3.6)$$

Conservatively assume that σ_f must reach the flange yield stress F_y to achieve the strength of the flange. Furthermore, if residual stress F_r exists in the flange distributed as shown in Fig. 11.3.5, then the total flange strain will be that due to the sum of the residual stress plus the yield stress; therefore

$$\epsilon_f = (F_r + F_y)/E \qquad (11.3.7)$$

It is the strain adjacent to the web that is of concern; in Fig. 11.3.5c the change is from F_r in tension (point A) to F_y in compression (point B).

Substitution of $\sigma_f = F_y$, $\epsilon_f =$ Eq. 11.3.7, $\mu = 0.3$, into Eq. 11.3.6 gives

$$\frac{h}{t_w} = \frac{0.672 E \sqrt{A_w/A_f}}{\sqrt{F_y(F_y + F_r)}} \qquad (11.3.8)$$

a conservative estimate of the maximum h/t_w to prevent vertical buckling. Basler [11.5] suggests A_w/A_f will rarely be below 0.5 and that $F_r = 0.3F_y$ is realistic. If these substitutions are made,

$$\frac{h}{t_w} = \frac{0.475E}{\sqrt{F_y(F_y + 0.3F_y)}} \qquad (11.3.9)$$

Equation 11.3.9 has been developed without regard to placement of stiffeners. The effect of stiffeners would certainly be to increase the strength above the elastic buckling strength based on F_{cr} of Eq. 11.3.4. Tests reported by Frost and Schilling [11.6] on hybrid girders having A514 ($F_y = 100$ ksi) flanges indicate that h/t_w can conservatively be

Figure 11.3.5
Effect of residual stress.

accepted in design as 250 if $a/h \leq 1.0$, and 200 for a/h between 1.0 and 1.5. The limitation of 200 may be given for other flange yield stresses as $2000/\sqrt{F_y}$ for $a/h \leq 1.5$.

Maximum h/t_w—AISC Design

Equation 11.3.9 when simplified becomes

$$\frac{h}{t_w} = \frac{0.42E}{F_y} \qquad (11.3.10)$$

which is the AISC-F13.2 maximum slenderness limitation (F_y in ksi).

In the presence of transverse stiffeners, higher ratios are permitted. Based on the recommendations of the ASCE-AASHO Joint Committee, Subcommittee 1 on Hybrid Girder Design [11.10], AISC-F13.2 gives the maximum h/t_w as

$$\frac{h}{t_w} \leq \left[11.7\sqrt{\frac{E}{F_y}} = \frac{2000}{\sqrt{F_y, \text{ksi}}} \right] \qquad (11.3.11)$$

when stiffener spacing a to web depth h does not exceed 1.5. Values for Eqs. 11.3.10 and 11.3.11 are given in Table 11.3.1.

TABLE 11.3.1 Maximum h/t_w Limitations—AISC-F13.2

F_y (ksi)	h/t_w for Eq. 11.3.10 for $a/h > 1.5$	h/t_w for Eq. 11.3.11 for $a/h \leq 1.5$	F_y (MPa)
36	338	333	248
42	290	309	290
45	271	298	310
50	244	283	345
55	222	270	379
60	203	258	414
65	187	248	448
100	122	200	689

11.4 NOMINAL MOMENT STRENGTH—AISC DESIGN

The concepts relating to the nominal moment strength M_n have been presented in Chapters 7 and 9, and reviewed in Sec. 11.2. Complexity in design arises from using M_r and L_r, expressed by Eqs. 9.6.5 and 9.6.6, respectively, for plate girders. For rolled beams, not only are the values of M_r (actually $\phi_b M_r$) and L_r available in the *AISC Manual* Tables 3-6, but also all of the properties, including the torsion properties, are available. For plate girders all of the properties must be computed for each girder.

Furthermore, webs of plate girders are typically "slender," though in a few cases they may be "noncompact," but they are never "compact." The limit states of AISC-F4 for shapes having "noncompact" webs and AISC-F5 for shapes having "slender" webs are to be used for plate girders. According to an AISC-F4 "*User Note,*" shapes having "noncompact" webs can be conservatively designed using AISC-F5.

Since plate girders will usually have "slender" webs; that is, λ will exceed λ_r for web local buckling, the strength cannot exceed that based on reaching yield stress F_y at the extreme fiber. No inelastic behavior is considered possible for design purposes.

The nominal moment strength M_n of slender web plate girders is controlled either by the limit state of yielding at the tension flange or that of buckling at the compression flange, according to AISC-F5:

For *yielding of the tension flange* (AISC-F5.4),

$$M_n = F_y S_{xt} R \tag{11.4.1}*$$

For *vertical buckling of the compression flange* (AISC-F5.2),

$$M_n = F_y S_{xc} R_{pg} \tag{11.4.2}$$

where F_y = yield stress of the tension flange
F_{cr} = buckling stress at the compression flange, controlled by lateral-torsional buckling, flange local buckling, or yielding.
S_{xt} = section modulus referred to the tension flange, I_x/y_t
S_{xc} = section modulus referred to the compression flange, I_x/y_c
I_x = moment of inertia with respect to the x-axis
y_t = distance from the CG of the section to the tension extreme fiber
y_c = distance from the CG of the section to the compression extreme fiber

also

$$R_{pg} = 1 - \frac{a_w}{1200 + 300a_w}\left(\frac{h_c}{t_w} - 5.7\sqrt{\frac{E}{F_y}}\right)$$

$$= 1 - \frac{a_w}{1200 + 300a_w}\left(\frac{h_c}{t_w} - \frac{970}{\sqrt{F_y, \text{ksi}}}\right) \le 1.0 \tag{11.4.3}$$

where $a_w = A_w/A_{fc} = (h_c t_w)/(b_{fc} t_{fc})$
A_w = web area
A_{fc} = compression flange area
h_c = twice the distance from the section centroid to the inside faces of the flanges less the fillet or corner radius, for rolled sections; or the inside faces of the flanges when welds are used for built-up sections
= h, height of web plate for a symmetrical I-shaped plate girder

The factor R_{pg} is to account for the "bend-buckling effect" on the "slender" web and its reduced ability to carry its elastic share of the bending moment. The development is presented in Section 11.6. Equation. 11.4.3 from the original work of Basler [11.5] is revised from that used in the 1986 LRFD Specification in order to give appropriate R_{pg} values for a_w between 0 and 10; a_w cannot exceed 10 according to AISC-F5.2. The previous equation was specifically applicable for $a_w \le 2.0$.

The determination of the "critical stress" F_{cr} to be used is obtained by dividing the nominal strength M_n by the section modulus S_x. For the "slender" web, the maximum F_{cr} is the yield stress F_y in the flange.

---

*Note that the 2005 AISC Specification does not use R_{pg} in Eq. 11.4.1 relating to the tension flange. This omission seems inappropriate because the reduction in moment strength R_{pg} resulting from bend-buckling of the web, as discussed in Sec. 11.6, is not related to a specific flange, but rather to the effective moment of inertia, which is reduced making both S_{xt} and S_{xc} reduced.

For the limit state of *lateral-torsional buckling*, AISC-F5.2 gives

1. For $L_b \leq L_p$,

$$L_b \leq \left(L_p = 1.1r_t\sqrt{\frac{E}{F_y}} = r_t\frac{187}{\sqrt{F_y,\text{ksi}}} \right) \tag{11.4.4}$$

$$F_{\text{cr}} = F_y \tag{11.4.5}$$

2. For $L_p < L_b \leq L_r$,

$$\left(L_p = r_t\frac{187}{\sqrt{F_y,\text{ksi}}} \right) < L_b \leq \left(L_r = \pi r_t\sqrt{\frac{E}{0.7F_y}} = r_t\frac{640}{\sqrt{F_y,\text{ksi}}} \right) \tag{11.4.6}$$

$$F_{\text{cr}} = C_b F_y \left[1 - 0.3\left(\frac{L_b - L_p}{L_r - L_p}\right) \right] \leq F_y \tag{11.4.7}$$

3. For $L_b > L_r$,

$$L_b > \left(L_r = r_t\frac{640}{\sqrt{F_y,\text{ksi}}} \right) \tag{11.4.8}$$

$$F_{\text{cr}} = \frac{C_b \pi^2 E}{\left(\dfrac{L_b}{r_t}\right)^2} \leq F_y \tag{11.4.9}$$

For the limit state of *flange local buckling*, AISC-F5.3 gives

1. For $\lambda \leq \lambda_p$ (i.e., "compact" flange),

$$\left(\lambda = \frac{b_f}{2t_f} \right) \leq \left(\lambda_p = 0.38\sqrt{\frac{E}{F_y}} = \frac{65}{\sqrt{F_y,\text{ksi}}} \right) \tag{11.4.10}$$

$$F_{\text{cr}} = F_y \tag{11.4.11}$$

2. For $\lambda_p < \lambda \leq \lambda_r$ (i.e., "noncompact" flange),

$$\left(\lambda_p = \frac{65}{\sqrt{F_y,\text{ksi}}} \right) < \left(\lambda = \frac{b_f}{2t_f} \right) \leq \left(\lambda_r = 0.95\sqrt{\frac{E}{0.7F_y/k_c}} = \frac{193}{\sqrt{F_y,\text{ksi}/k_c}} \right)$$

$$\tag{11.4.12}$$

$$F_{\text{cr}} = C_b F_y \left[1 - 0.3\left(\frac{\lambda - \lambda_p}{\lambda_r - \lambda_p}\right) \right] \leq F_y \tag{11.4.13}$$

3. For $\lambda > \lambda_r$ (i.e., "slender" flange),

$$\left(\lambda = \frac{b_f}{2t_f} \right) > \left(\lambda_r = \frac{193}{\sqrt{F_y,\text{ksi}/k_c}} \right) \tag{11.4.14}$$

$$F_{\text{cr}} = \frac{0.9Ek_c}{\left(\dfrac{b_f}{2t_f}\right)^2} \tag{11.4.15}$$

Note that $0.35 \leq \left(k_c = 4/\sqrt{h/t_w} \right) \leq 0.76$.

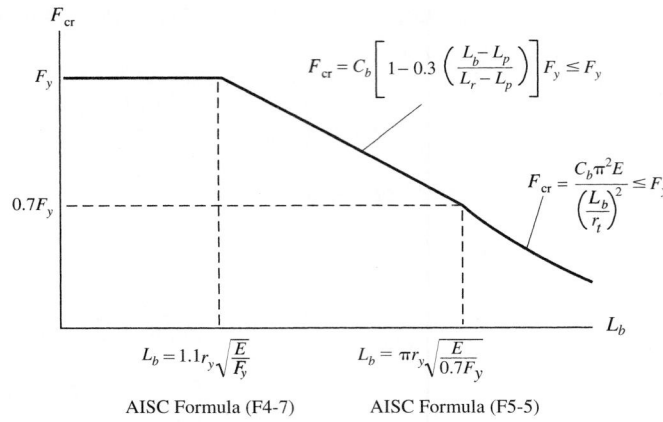

(a) Lateral-torsional buckling limit state

$$M_n = R_{pg} F_{cr} S_x$$

$$F_{cr} = C_b \left[1 - 0.3 \left(\frac{L_b - L_p}{L_r - L_p} \right) \right] F_y \leq F_y$$

$$F_{cr} = \frac{C_b \pi^2 E}{\left(\frac{L_b}{r_t} \right)^2} \leq F_y$$

$$L_b = 1.1 r_y \sqrt{\frac{E}{F_y}}$$

$$L_b = \pi r_y \sqrt{\frac{E}{0.7 F_y}}$$

AISC Formula (F4-7)   AISC Formula (F5-5)

(b) Flange local buckling limit state

$$M_n = R_{pg} F_{cr} S_x$$

$$F_{cr} = \left[1 - 0.3 \left(\frac{\lambda - \lambda_{pf}}{\lambda_{rf} - \lambda_{pf}} \right) \right] F_y$$

$$F_{cr} = \frac{0.9 E k_c}{\lambda^2}$$

$$\lambda_p = 0.38 \sqrt{\frac{E}{F_y}}$$

$$\lambda_r = 0.95 \sqrt{\frac{k_c E}{F_L}}$$

$$\lambda = \frac{b_f}{2 t_f}$$

Figure 11.4.1
Limit states in flexure for plate girder with slender web (AISC-F5) where $k_c = 4/\sqrt{h/t_w}$ and $0.35 \leq k_c \leq 0.76$.

The general relationships of Eqs. 11.4.4 through 11.4.15 are shown in Fig. 11.4.1.

The provisions for beams having "slender" webs (AISC-F5) approximate the M_n vs L_b relationship for rolled beams as shown in Fig. 9.6.1. Note that L_p of Eq. 11.4.4, 11.4.6, and 11.4.8 uses r_t instead of the r_y used in Eq. 9.6.3. r_t is defined by AISC Formula (F4-10) but can be approximated conservatively (i.e., low) as the radius of gyration of the compression flange plus one-third of the compression portion of the web taken about the y-axis. Equation 11.4.6 for L_r avoids the more complicated Eq. 9.6.6 for rolled beams; and also avoids the necessity of computing torsion properties for plate girders. These new provisions for beams having "slender" webs depart from the simplified relationships given in earlier specifications and adopt expressions analogous to those used for rolled beams.

The slender web plate would restrict the maximum moment strength to $F_y S_x$ if the beam provisions of AISC-F5 were used, rather than M_p when the web plate is "compact" or "noncompact" (that is, when $\lambda \leq \lambda_r$ for the web). A comparison in Fig. 11.4.2 of the lateral-torsional buckling provisions of AISC-F5 with AISC-F2 for a thin web $\left(\frac{3}{8} \times 100 \right)$ plate girder shows good agreement.

For the *flange local buckling limit state,* the I-shaped built up girders limit λ_p is identical to that used for rolled beams (see Table 9.6.1). The limit λ_r for I-shaped built up girders the expressions given in Table 9.6.2 for beams.

TABLE 11.4.1 Slenderness Ratio Limits λ_p and λ_r for Flange Local Buckling and Lateral-Torsional Buckling Limit States Under AISC-F5 For Doubly Summetric I-Girders With Slender Web

Yield stress	Flange local buckling				Lateral-torsional buckling	
	λ_p			λ_r	L_p	L_r
F_y (ksi)	$\dfrac{b_f}{2t_f} = 0.38\sqrt{\dfrac{E}{F_y}}$ $= \dfrac{65}{\sqrt{F_y}}$, ksi	$\dfrac{h}{t_w}$	k_c^*	$\dfrac{b_f}{2t_f} = 0.95\sqrt{\dfrac{k_c E}{0.7 F_y}}$ $= \dfrac{193}{\sqrt{F_y/k_c}}$	$L_b = 1.1 r_t \sqrt{\dfrac{E}{F_y}}$ $= r_t \dfrac{187}{\sqrt{F_y}}$, ksi	$L_b = \pi r_t \sqrt{\dfrac{E}{0.7 F_y}}$ $= r_t \dfrac{639}{\sqrt{F_y}}$, ksi
36	10.8	161.7[†]	0.35	19.0	$31.2 r_t$	$106.5 r_t$
		100	0.40	20.3		
		40	0.63	25.5		
42	10.0	149.7	0.35	17.6	$28.9 r_t$	$98.6 r_t$
		100	0.40	18.8		
		40	0.63	23.6		
45	9.7	144.6	0.35	17.0	$27.9 r_t$	$95.3 r_t$
		100	0.40	18.2		
		40	0.63	22.8		
50	9.2	137.2	0.35	16.1	$26.5 r_t$	$90.4 r_t$
		100	0.40	17.3		
		40	0.63	21.7		
55	8.8	130.8	0.35	15.4	$25.2 r_t$	$86.2 r_t$
		100	0.40	16.5		
		40	0.63	20.7		
60	8.4	125.2	0.36	15.0	$24.1 r_t$	$82.5 r_t$
		100	0.40	15.8		
		40	0.63	19.8		
65	8.1	120.3	0.36	14.4	$23.2 r_t$	$79.3 r_t$
		100	0.40	15.1		
		40	0.63	19.0		
90	6.9	102.2	0.40	12.9	$19.7 r_t$	$67.4 r_t$
		100	0.40	12.9		
		40	0.63	16.1		
100	6.5	97.0	0.41	12.4	$18.7 r_t$	$63.9 r_t$
		100	0.40	12.2		
		40	0.63	15.3		

$^*k_c = \dfrac{4}{\sqrt{h/t_w}}$, where $0.35 \le k_c \le 0.76$

[†]Maximum h/t_w given for each yield stress $= 970/\sqrt{F_y} = \lambda_r$ (AISC Table-B4.1 for "webs in flexural compression").

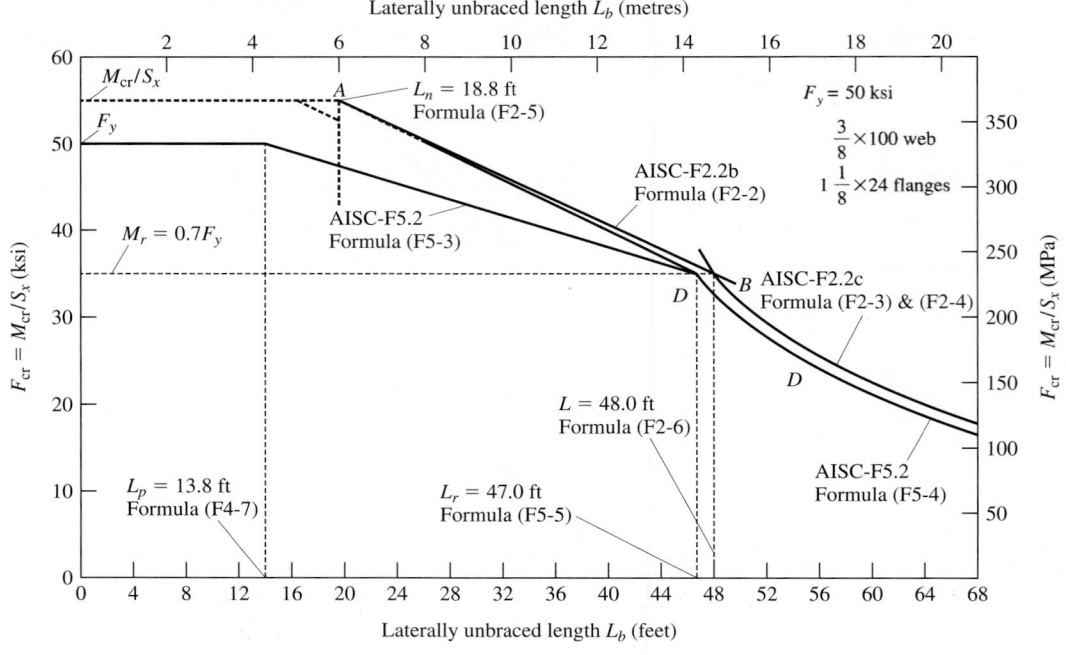

Figure 11.4.2
Comparison of lateral-torsional buckling limit state using AISC-F5 for beams with slender web with AISC-F2 for compact beams. Properties of girder: $r_y = 5.32$ in., $r_t = 6.24$ in., $M_p = 15,280$ ft-kips, and $M_r = 0.7F_yS_x = 9660$ ft-kips

11.5 MOMENT STRENGTH REDUCTION DUE TO BEND-BUCKLING OF THE WEB

Since the plate girder web usually has a high h/t_w ratio, buckling may occur as a result of bending in the plane of the web (see Fig. 11.5.1). The slenderness ratio λ_r above which such buckling may occur is developed in what follows. Furthermore, after this elastic buckling occurs there is post-buckling strength. When the plate girder is proportioned to most efficiently carry load, the web will buckle before the nominal moment strength of the girder is reached.

Typical of any plate stability situation, the elastic buckling stress is represented by Eq. 6.14.28,

$$F_{cr} = k \frac{\pi^2 E}{12(1 - \mu^2)(b/t)^2} \qquad [6.14.28]$$

where for this case $b = h$.

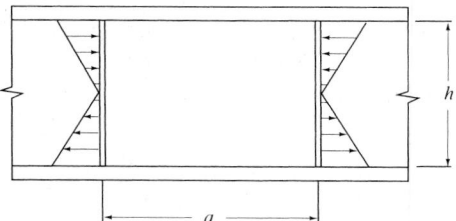

Figure 11.5.1
Web plate under pure
moment.

The theoretical development of the k values for bending in the plane of the plate is given by Timoshenko and Woinowski-Krieger.* For any given type of loading, k varies with the aspect ratio, a/h (see Fig. 11.5.2), and with the support conditions along the edges. If the plate can be considered to have full fixity (full moment resistance against edge rotation) along the edges parallel to the direction of loading (i.e., at the edges joined to the flanges), the minimum value of k is 39.6 for any a/h ratio. If the flanges are assumed to offer no resistance to edge rotation, the minimum k value is 23.9. The variation of k with a/h ratio is given in Fig. 11.5.2.

Thus the critical stress (using $E = 29{,}000$ ksi and $\mu = 0.3$) may be said to lie between

$$F_{cr} = \frac{627{,}000}{(h/t)^2} \text{ ksi}\quad \text{for } k = 23.9 \text{ (simple support at flanges)}$$

and

$$F_{cr} = \frac{1{,}038{,}000}{(h/t)^2} \text{ ksi}\quad \text{for } k = 39.6 \text{ (full fixity at flanges)}$$

While each particular girder will have a different degree of flange restraint, fully welded flange to web connections will surely approach the full fixity case. It will be reasonable then to arbitrarily select a k value closer to 39.6, say 80 percent of the difference toward the higher value. One might say that

$$F_{cr} = \frac{950{,}000}{(h/t)^2} \text{ ksi} \tag{11.5.1}$$

is representative of the stress when elastic buckling is imminent due to bending in the plane of the web. Such "bend buckling" cannot occur if

$$\frac{h}{t_w} \leq \sqrt{\frac{950{,}000}{F_{cr}, \text{ksi}}} = \frac{970}{\sqrt{F_{cr}}} \text{ or } 5.7\sqrt{\frac{E}{F_{cr}}} \tag{11.5.2}$$

where t_w = web plate thickness.

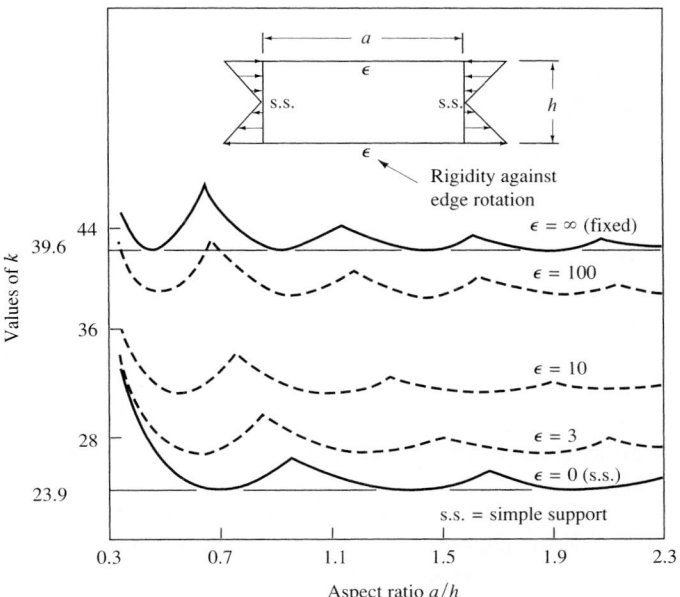

Figure 11.5.2
Buckling coefficients for plates subject to bending in the plane of the plate. (From *Handbook of Structural Stability,* Vol. 1 [6.69], p. 92)

*Reference 6.66, pp. 373–379.

Figure 11.5.3
Buckling of plates subject to bending in the plane of the web.

Figure 11.5.4
Nominal moment strength M_n of girders as affected by strength of the web plate resisting bending moment in the plane of the web: A36 steel.

Since the web carries only a small part of the total bending moment to which the girder is subjected, neglect of the transition zone arising from inelastic buckling will not be significant.

Buckling of the web does not end the usefulness of the girder. Fig. 11.5.4 shows the relationship between nominal moment strength M_n vs h/t_w. This figure assumes the lateral-torsional buckling and local flange buckling limit states do not control.

When the post-buckling strength of the girder is considered, the strength is raised from line *BC* of Fig. 11.5.4 to line *BD*. The actual position of line *BD* varies with A_w/A_f, the ratio of the web area to the compression flange area.

EXAMPLE 11.5.1

Using $h/t_w = 320$, determine the expression for M_n/M_y (point D of Fig. 11.5.4) in terms of A_w/A_f.

Solution:

With $h/t_w = 320$, "bend-buckling" occurs at a low level of flexural stress. Such buckling does not signify the maximum bending moment that can be carried; however, under additional load the flexural stress on the compression side of the neutral axis becomes nonlinear. In order to retain the use of the flexure formula, Mc/I, a reduced effective section must be used. The reduced section entirely neglects much of the web in the region where the buckling (out-of-planeness) has occurred. The effective section shown in Fig. 11.5.5 was proposed by Basler [11.5].

(a) Determine location of neutral axis. Equating static moments about the neutral axis gives

$$A_f(kh) + \frac{t_w(kh)^2}{2} = A_f(1-k)h + \frac{3}{32}h\left(\frac{61}{64}h - kh\right)t_w$$

Divide by $A_f h$:

$$k + k^2\frac{t_w h}{2A_f} = (1-k) + \frac{3}{32}\left(\frac{61}{64} - k\right)\frac{t_w h}{A_f}$$

Noting that $t_w h = A_w$ and letting $a_r = A_w/A_f$ gives

$$k^2 + k\left(\frac{4}{a_r} + \frac{3}{16}\right) = \frac{2}{a_r} + \frac{183}{1024}$$

$$k = \sqrt{\frac{192}{1024} + \frac{38}{16a_r} + \frac{4}{a_r^2}} - \left(\frac{3}{32} + \frac{2}{a_r}\right) \tag{a}$$

(b) Determine the effective moment of inertia.

$$I_e = A_f(kh)^2 + \frac{1}{3}t_w(kh)^3 + A_f(1-k)^2h^2 + \frac{3t_w h}{32}\left(\frac{61}{64}h - kh\right)^2$$

Using $t_w h = A_w$ and $a_r = A_w/A_f$,

$$I_e = A_f h^2\left[\frac{a_r}{3}k^3 + k^2 + (1-k)^2 + \frac{3a_r}{32}\left(\frac{61}{64} - k\right)^2\right] \tag{b}$$

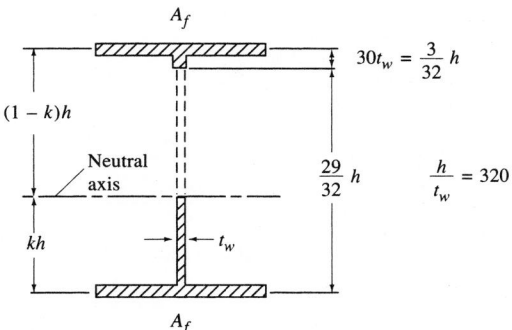

Figure 11.5.5
Effective section in bending when vertical flange buckling is imminent.

(c) Determine the nominal moment strength M_n. Assuming the extreme fiber in compression stressed to the yield stress F_y,

$$M_n = \frac{F_y I_e}{(1-k)h} \tag{c}$$

(d) Determine the moment strength M_y assuming the entire section elastic (and therefore effective) with the extreme fiber stress equal to F_y. In developing the expression the flange-area concept is used as shown in Fig. 11.5.6. The moment strength of the web is approximately (Fig. 11.5.6a)

$$M_{\text{web}} = fS_x = f\left(\tfrac{1}{6}t_w h^2\right) \tag{d}$$

which assumes web depth, distance between flange centroids, and overall depth are the same. The moment strength of the equivalent flange area system (Fig. 11.5.6b) is

$$M_{\text{equiv}} = fA_f'h \tag{e}$$

Equating Eqs. (d) and (e) gives for the equivalent flange area, A_f',

$$A_f' = \tfrac{1}{6}t_w h = \tfrac{1}{6}A_w \tag{f}$$

The total moment strength of a girder where the stress $f = F_y$ then becomes

$$M_y = F_y\left[A_f + \frac{A_w}{6}\right]h \tag{g}$$

$$= F_y A_f h\left(1 + \frac{a_r}{6}\right) \tag{h}$$

The vertical ordinate of point D in Fig. 11.5.4 is obtained by dividing Eq. (c) by Eq. (h):

$$\frac{M_n}{M_y} = \frac{\dfrac{a_r}{3}k^3 + k^2 + (1-k)^2 + \dfrac{3a_r}{32}\left(\dfrac{61}{64} - k\right)^2}{(1-k)(1 + a_r/6)} \tag{i}$$

which is plotted in Fig. 11.5.7.

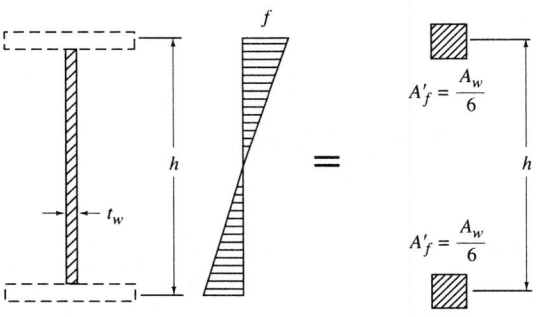

Figure 11.5.6
Equivalent flange area to replace web.

(a) Actual condition   **(b)** Equivalent condition

Figure 11.5.7
Reduction in nominal
moment strength considering
post-buckling strength at
maximum h/t_w for A36 steel.

From Fig. 11.5.7, the variation in M_n/M_y might be approximated by a straight line for A_w/A_f from zero to three with a slope of $-(1.00 - 0.73)/3.0 = -0.09$. Thus, at $h/t_w = 320$,

$$\frac{M_n}{M_y} = 1.0 - 0.09\frac{A_w}{A_f} \tag{j}$$

It may be observed that the straight line agrees better for $h/t_w = 360$, the situation for which this linear equation was originally developed [11.5], than it does for $h/t_w = 320$. For higher strength steels, for which the maximum h/t_w to prevent vertical buckling of the flange is less than 360, more of such a stiffer web participates with the compression flange, causing a greater reduction in M_n/M_y.

The linear reduction based on Eq. (j) does not seem conservative, but is within several percent of the more accurate curve using $30t_w$ as the effective depth of web participating with the compression flange.

Tests [11.5] have verified the correctness of this linear reduction method using $h/t_w = 360$ as its basis.   ∎

Reduced Nominal Strength M_n when $h/t_w > \left[5.7\sqrt{E/F_y}, = 970/\sqrt{F_y}\right]$

By reference of Fig. 11.5.4, it may be reasonably assumed that M_n/M_y varies linearly from point B to D. Thus the reduction in M_n/M_y per A_w/A_f per h/t greater than that at point B is

$$\frac{\text{Slope per } A_w/A_f}{320 - 162} = \frac{0.09}{158} = 0.00057 \quad (\text{say } 0.0005)$$

Thus M_n/M_y for the region from point B to D (Fig. 11.5.4) assuming linear variation, is

$$\frac{M_n}{M_y} = 1.0 - 0.0005\frac{A_w}{A_f}\left(\frac{h}{t_w} - \frac{970}{\sqrt{F_y}}\right) \tag{11.5.3}$$

Since $M_y = F_yS_x$, Eq. 11.5.3 may be written

$$M_n = S_xF_y\left[1.0 - 0.0005\frac{A_w}{A_f}\left(\frac{h}{t_w} - \frac{970}{\sqrt{F_y}}\right)\right] \tag{11.5.4}$$

As shown in the development, Eq. 11.5.4 is based on the assumption that the ratio A_w/A_f does not exceed about 3. In the 1993 LRFD Specification, the relationship was revised to accommodate A_w/A_f up to 10. Thus, in place of the coefficient 0.0005, the coefficient becomes

$$\frac{a_r}{1200 + 300a_r}$$

in accordance with the original development by Basler [11.5]. Thus the more general form of Eq. 11.5.4 is

$$M_n = S_x F_y \left[1 - \frac{a_r}{1200 + 300a_r} \left(\frac{h}{t_w} - \frac{970}{\sqrt{F_y}} \right) \right] \tag{11.5.5}$$

Equation 11.5.5 assumes no influence of the lateral-torsional buckling and flange local buckling limit states. When the controlling limit state prevents the flange stress from reaching F_y, then *that* controlling limit state critical stress F_{cr} should replace F_y in Eq. 11.5.5. Thus, Eq. 11.5.5 becomes

$$M_n = S_x F_{cr} \left[1 - \frac{a_r}{1200 + 300a_r} \left(\frac{h}{t_w} - \frac{970}{\sqrt{F_{cr}}} \right) \right] \tag{11.5.6}$$

$$M_n = S_x F_{cr} R_{pg} \tag{11.5.7}$$

where $R_{pg} = 1 - \dfrac{a_r}{1200 + 300a_r} \left(\dfrac{h}{t_w} - \dfrac{970}{\sqrt{F_{cr}}} \right) \leq 1.0$ [11.4.3]

$a_r = A_w/A_f \leq 10$

$A_f =$ compression flange area

$A_w =$ web area

This reduction factor R_{pg} is given in AISC-F5.2.

    In summary, when h/t_w exceeds $970/\sqrt{F_{cr}}$, one may view Eq. 11.5.7 either as (1) the full section modulus S_x multiplied by a reduced stress $F_{cr} R_{pg}$, or (2) as a reduced section modulus $S_x R_{pg}$ multiplied by the full stress F_{cr}. Philosophically it should be considered the latter. The idea of using a reduced section when buckling has occurred to cause a non-linear stress distribution, is the same concept used for the stiffened plate element in Chapter 6, Part II.

11.6 NOMINAL MOMENT STRENGTH—HYBRID GIRDERS

As discussed in the last section, a girder having large h/t_w may have its web buckle due to flexural stress, thereby increasing the load-carrying requirement of the compression flange. This extra load-carrying requirement for the flange also may occur when a hybrid girder is used. A hybrid girder is one in which the flanges are of a higher strength steel than the web. The use of a hybrid girder has particular economic advantages in composite construction, as described in Sec. 16.9.

    The special behavioral feature of the hybrid girder is the yielding of the lower strength web before the maximum flange strength has been reached. When the moment strength of the hybrid girder is achieved, the web will have participated to a lesser extent than in a girder using only one grade of steel.

    Frost and Schilling [11.6], Schilling [11.7], Carskaddan [11.8], and Toprac and Natarajan [11.9] have studied the hybrid girder. The state-of-the-art and design recommendations for hybrid girders have been given by a Joint ASCE-AASHO Joint Committee [11.10].

The principal effect of using a lower yield stress web than is used in the flange is that the onset of yielding in the web will occur prior to yielding in the flanges. For example, when the web is A36 steel and the flanges have $F_y = 100$ ksi, the web may yield at about 40% of the nominal strength M_n based on yielding of the flanges.* This means that the web will yield *even at service load.*

The design of hybrid girders, according to Subcommittee 1 of ASCE-AASHO Joint Committee [11.10], should be based on the moment causing the initiation of flange yielding. This may be accomplished by either of two procedures:

1. The nominal strength M_n may be computed by setting the extreme fiber strain ϵ_{yf} equal to F_{yf}/E_s, where F_{yf} is the yield stress of the flange steel (see Fig. 11.6.1b). At this stage much of the lower yield stress web will have exceeded its yield strain $\epsilon_{yw} = F_{yw}/E_s$, in which case the stress distribution over the depth of the section is nonlinear, as shown in Fig. 11.6.1c.

2. The nominal strength M_n may be computed as a homogeneous elastic section entirely of the flange steel, that is $F_{yf}S_x$, reduced by multiplying by a reduction factor.

A reduction factor to account for yielding of web prior to reaching the flange yield stress has been removed from the AISC 2005 Specification. No further advice is given for hybrid girders in the Specification. It is the conclusion of the authors that the *AISC Specification* does not recognize designing hybrid girders as a viable alternative. However, if designing a hybrid girder with significant difference in the yield strengths of the flange and the web, an additional reduction factor given in the LRFD 1999 Specification can be used. This procedure is described next.

Thus, the moment strength M_n of an hybrid girder may be expressed as Eq. 11.4.2 (or Eq. 11.5.6) multiplied by a reduction factor R_e to account for yielding in the web prior to reaching the the extreme fiber,

$$M_u = F_{cr}\, S_{xc} R_{pg} R_e \qquad (11.6.1)$$

where,

$$R_e = \frac{12 + a_r(3m - m^3)}{12 + 2a_r} \le 1.0 \qquad (11.6.2)$$

where $a_r = A_w/A_f$ = ratio of the cross-sectional area of the web to the cross-sectional area of one flange

$m = F_{yw}/F_{yf}$ = ratio of the yield stress of the web steel to the yield stress of the flange steel

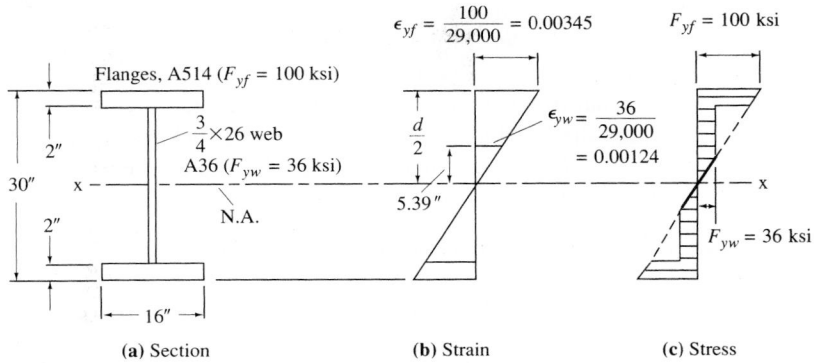

Figure 11.6.1
Hybrid section showing stress and strain when F_{yf} is reached at extreme fiber of section.

(a) Section   (b) Strain   (c) Stress

*See 2nd edition, pp. 590–592.

R_{pg} = reduction for web instability when $h/t_w > 970/\sqrt{F_{cr}}$ (see Eq. 11.4.3)

F_{cr} = critical compression flange stress based on lowest value obtained from the lateral-torsional buckling or the flange local buckling limit states.

The reduction factor R_{pg} was recommended by the ASCE-AASHO Joint Committee [11.10], and is somewhat different than the 1986 LRFD Specification expression recommended by Cooper, Galambos, and Ravindra [11.11].

In applying the reduction factor to R_{pg} account for bend-buckling of the web, the derivation of the web slenderness limit of $970/\sqrt{F_y}$ involved the plate yield stress F_y, which means the web. On the other hand, examination of Fig. 11.6.1 shows that when the flange reaches F_{yf} the strain on the portions of the web adjacent to the flanges will have strain exceeding the yield strain $\epsilon_{yw} = F_{yw}/E_s$ in the web. Logically the slenderness limit to prevent buckling must then be lower than $970/\sqrt{F_{yw}}$. Dawe and Kulak [11.12] have indicated that because of the restraining effect of sturdy flanges the web may be expected to undergo plastic strain without buckling even when the h/t_w ratio is $800/\sqrt{F_{yw}}$. Zahn [11.13] has noted that when $F_{yf} = 50$ ksi (flanges) and $F_{yw} = 36$ ksi (web) the same limiting slenderness ratio is obtained,

$$\left(\frac{800}{\sqrt{F_{yw}}} = 133\right) \approx \left(\frac{970}{\sqrt{F_{yf}}} = 137\right)$$

Thus, rather than have a separate limit equation relating to bend-buckling of the web of a hybrid girder using F_{yw}, AISC concluded (presumably) it would be satisfactory and simpler to use the same limit equation for both homogeneous and hybrid girders. The proper result would be obtained by always using the *flange yield stress* F_{yf} in the h/t_w slenderness limit equation,

$$\left(\frac{h}{t_w}\right)_{\substack{\text{bend-buckling}\\\text{limit}}} = \frac{970}{\sqrt{F_{yf}}}$$

The special features of hybrid girders relating to composite construction are contained in Chapter 16.

11.7 NOMINAL SHEAR STRENGTH—ELASTIC AND INELASTIC BUCKLING

Typical of I-shaped sections, the web carries most of the shear. Since the plate girder inherently has a thin web $\left(h/t_w > 970/\sqrt{F_y}, \text{ksi}\right)$, stability is of primary concern.

Consider a web plate panel of length a between transverse stiffeners and having a clear height h between longitudinal plate supports (i.e., between flanges, flange and longitudinal stiffener, or between longitudinal stiffeners), as shown in Fig. 11.7.1. In regions of high shear and low bending moment, the buckling strength of the panel may be investigated assuming it is acted upon by shear alone (pure shear), as in Fig. 11.7.1.

Elastic Buckling Under Pure Shear

The elastic buckling stress for any plate is given by Eq. 6.14.28 as

$$F_{cr} = k\frac{\pi^2 E}{12(1-\mu^2)(b/t)^2}$$ [6.14.28]

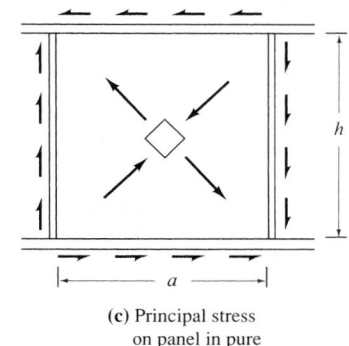

Figure 11.7.1
Classical shear theory applied
to plate girder web panel.

(a) Element in
pure shear

(b) Principal stress
on element in
pure shear

(c) Principal stress
on panel in pure
shear

where for the case of pure shear (see Fig. 11.7.1), Eq. 6.14.28 may be written (using τ in place of F for shear stress and k_v for k)

$$\tau_{cr} = k_v \frac{\pi^2 E}{12(1 - \mu^2)\left(\dfrac{\text{short dimension}}{t}\right)^2} \tag{11.7.1}$$

where for the case of edges simply supported (i.e., displacement prevented but rotation about edges unrestrained)

$$k_v = 5.34 + 4.0\left(\frac{\text{short dimension}}{\text{long dimension}}\right)^2 \tag{11.7.2}$$

the development of which has been given by Timoshenko and Woinowski-Krieger.*

For design purposes it may be desirable to put Eqs. 11.7.1 and 11.7.2 in terms of h, the unsupported web height, and a, the stiffener spacing. When this is done two cases must be considered.

1. If $a/h \le 1$ (see Fig. 11.7.2a), Eq. 11.7.1 becomes

$$\tau_{cr} = \frac{\pi^2 E[5.34 + 4.0(a/h)^2]}{12(1 - \mu^2)(a/t)^2} \frac{(h/a)^2}{(h/a)^2} \tag{11.7.3}$$

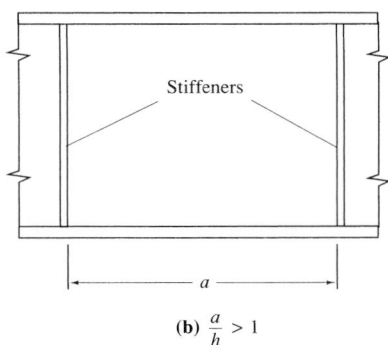

Figure 11.7.2
Two cases of intermediate
stiffener spacing.

Stiffeners

Stiffeners

(a) $\dfrac{a}{h} < 1$

(b) $\dfrac{a}{h} > 1$

*Reference 6.66, pp. 379–385.

2. If $a/h \geq 1$ (see Fig. 11.7.2b), Eq. 11.7.1 becomes

$$\tau_{cr} = \frac{\pi^2 E[5.34 + 4.0(h/a)^2]}{12(1 - \mu^2)(h/t)^2} \qquad (11.7.4)$$

It is apparent from Eqs. 11.7.3 and 11.7.4 that if one desires to use h/t as the stability ratio in the denominator, then two expressions for k_v are necessary. For all ranges of a/h, Eqs. 11.7.3 and 11.7.4 may be written

$$\tau_{cr} = \frac{\pi^2 E k_v}{12(1 - \mu^2)(h/t)^2} \qquad (11.7.5)$$

where

$$k_v = 4.0 + 5.34/(a/h)^2 \qquad \text{for } a/h \leq 1 \qquad (11.7.6)$$

$$k_v = 4.0/(a/h)^2 + 5.34 \qquad \text{for } a/h \geq 1 \qquad (11.7.7)$$

These theoretical expressions for k_v are replaced in AISC-G2.1b by the following:

$$\left.\begin{array}{l} k_v = 5 \text{ for } \textit{unstiffened} \text{ webs having } h/t_w < 260 \\[2mm] k_v = 5 + \dfrac{5}{(a/h)^2} \text{ for } \textit{stiffened} \text{ webs} \\[4mm] k_v = 5 \text{ for } \textit{stiffened} \text{ webs when } \dfrac{a}{h} > 3.0 \text{ or } \dfrac{a}{h} > \left(\dfrac{260}{h/t_w}\right)^2 \end{array}\right\} \qquad (11.7.8)$$

For use in design, Eq. 11.7.5 has been put into nondimensional form, defining C_v as the ratio of shear stress τ_{cr} at buckling to shear yield stress τ_y,

$$C_v = \frac{\tau_{cr}}{\tau_y} = \frac{\pi^2 E k_v}{\tau_y(12)(1 - \mu^2)(h/t)^2} \qquad (11.7.9)$$

which is C_v for *elastic* stability. Substitution of $E = 29,000$ ksi, $\mu = 0.3$, $\tau_y = 0.6F_{yw}$ (see Eq. 7.7.9), and using the subscript w for both the yield stress F_{yw} and the plate

TABLE 11.7.1 Elastic Buckling Coefficient k_v for Shear Strength

Ratio a/h	AISC Eq. 11.7.8
0.2	130.0
0.4	36.3
0.6	18.9
0.8	12.8
1.0	10.0
1.2	8.5
1.4	7.6
1.6	7.0
1.8	6.5
2.0	6.3
2.2	6.0
2.4	5.9
2.6	5.7
2.8	5.6
3.0	5.6

thickness t_w to identify this behavior as occurring in the *web*, gives

$$C_v = \frac{\pi^2(29,000)k_v}{0.6F_{yw}(12)(1-0.09)(h/t_w)^2}$$

The following simplification of the equation gives C_v as given by AISC-G2.1b, valid for τ_{cr} below the proportional limit as shown in Fig. 11.7.3,

$$C_v = \frac{1.51Ek_v}{(h/t_w)^2 F_y} = \frac{43,800k_v}{(h/t_w)^2(F_y,\text{ksi})} \tag{11.7.10}$$

Inelastic Buckling under Pure Shear

As in all stability situations, residual stresses and imperfections cause inelastic buckling as critical stresses approach yield stress. A transition curve for inelastic buckling was given by Basler [11.3] based on curve fitting and using test results from Lyse and Godfrey [11.4]. In the transition zone between elastic buckling and yielding,

$$\tau_{cr} = \sqrt{\tau_{\substack{\text{prop.}\\\text{limit}}} \cdot \tau_{\substack{\text{cr(ideal)}\\\text{elastic}}}} \tag{11.7.11}$$

The proportional limit is taken as $0.8\tau_y$, higher than for compression in flanges, because the effect of residual stress is less. Dividing Eq. 11.7.11 by τ_y to obtain C_v and using Eq. 11.7.10 gives

$$C_v = \frac{\tau_{cr}}{\tau_y} = \sqrt{(0.8)\frac{1.51Ek_v}{(h/t_w)^2 F_y}}$$

$$= \frac{1.1\sqrt{k_v E/F_y}}{h/t_w} = \frac{187}{h/t_w}\sqrt{\frac{k_v}{F_y,\text{ksi}}} \tag{11.7.12}$$

which is given by AISC-G2.1b and is shown schematically in Fig. 11.7.3.

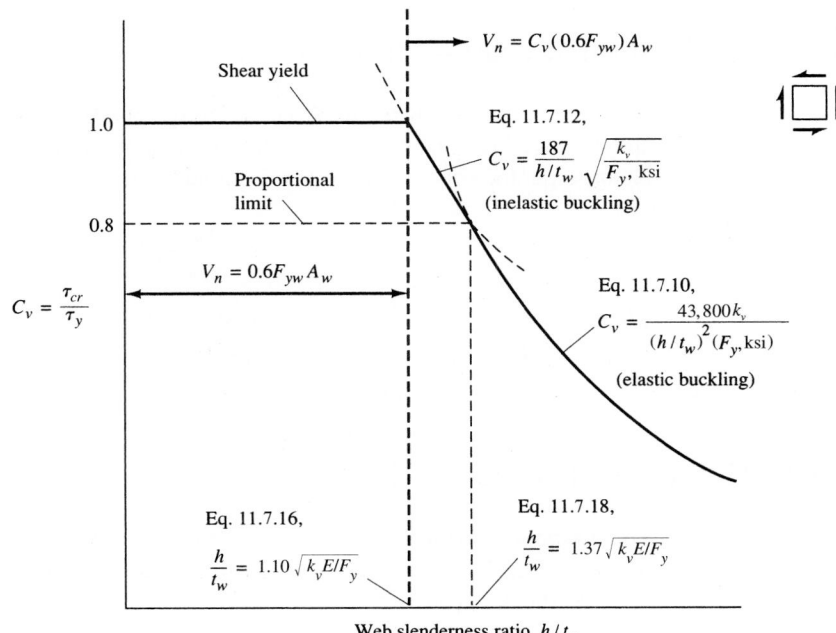

Figure 11.7.3
Buckling of plate girder web resulting from shear alone—AISC-G2

Nominal Shear Strength

The nominal shear strength V_n of a girder based on inelastic or elastic buckling of the web may be expressed

$$V_n = \tau_{cr}A_w \tag{11.7.13}$$

or using $C_v = \tau_{cr}/\tau_y$,

$$V_n = C_v\tau_y A_w \tag{11.7.14}$$

Approximating τ_y as $0.6F_y$ gives

$$V_n = C_v(0.6F_y)\,A_w \tag{11.7.15}$$

which is AISC Formula (G2-1). In Eq. 11.7.15, C_v is Eq. 11.7.10 for elastic buckling when $C_v \le 0.8$, and is Eq. 11.7.12 when $C_v > 0.8$.

Equation 11.7.15 will apply also to rolled beams since rarely would transverse stiffeners be used. If one wishes to have an explicit expression for h/t_w corresponding to $C_v = 1$ (i.e., the web yields in shear and no buckling occurs), Eq. 11.7.12 may be solved for h/t_w when $C_v = 1$,

$$\frac{h}{t_w} = 1.10\sqrt{\frac{k_v E}{F_y}} = 187\sqrt{\frac{k_v}{F_y}}, \text{ksi} \tag{11.7.16}$$

When h/t_w does not exceed the value from Eq. 11.7.16, the nominal shear strength is

$$V_n = 0.6F_y A_w \tag{11.7.17}$$

which is AISC Formula (G2-1) with $C_v = 1$. This was first discussed in Chapter 7 (Sec. 7.7).

The h/t_w relationship that divides elastic and inelastic buckling may be obtained by setting C_v equal to 0.8 in Eq. 11.7.10, giving

$$\frac{h}{t_w} = 1.37\sqrt{\frac{k_v E}{F_y}} = 233\sqrt{\frac{k_v}{F_y}}, \text{ksi} \tag{11.7.18}$$

The AISC relationship between buckling strength in shear and web slenderness ratio h/t_w is shown in Fig. 11.7.3.

Shear Strength—Allowable Strength Design

In accordance with Allowable Strength Design philosophy presented in Secs. 1.8 and 1.9, and discussed for shear in Sec. 7.7, the strength relationship may be divided by the safety factor to put the equation in the service load range,

$$\frac{V_n}{\Omega_v} \ge V_r \tag{11.7.19}$$

where $\Omega_v = 1.50$ according to AISC-G2.1a
$V_n/\Omega_v =$ allowable shear strength
$V_r =$ required shear strength

11.8 NOMINAL SHEAR STRENGTH—INCLUDING TENSION-FIELD ACTION

The elastic and inelastic buckling strength of the web subject to shear is represented by *ABCD* in Fig. 11.8.1. A plate stiffened by flanges and transverse stiffeners has considerable post-buckling strength.

Figure 11.8.1
Shear capacity available,
considering post-buckling
strength.

According to Basler [11.3], the ability of a plate girder to behave in a manner similar to a truss was recognized as early as 1898. As shown in Fig. 11.8.2, the tension forces are carried by membrane action of the web (referred to as *tension-field action*) while the compression forces are carried by the transverse stiffeners. The work of Basler [11.3] led to a theory that agreed with tests and provides criteria to ensure that truss action can develop. The inclusion of truss action raises the shear strength from that based on buckling (*ABCD* on Fig. 11.8.1) to approach a condition corresponding to shear yield in classical beam theory (*ABE* of Fig. 11.8.1).

The nominal shear strength V_n may be expressed as the sum of the buckling strength V_{cr} and the post-buckled strength V_{tf} from tension-field action,

$$V_n = V_{cr} + V_{tf} \tag{11.8.1}$$

The nominal buckling strength is given by Eq. 11.7.14 with $V_n = V_{cr}$,

$$V_{cr} = C_v \tau_y A_w \tag{11.8.2}$$

where $C_v = \tau_{cr}/\tau_y$ and is given by Eqs. 11.7.10 and 11.7.12 for elastic and inelastic buckling, respectively.

The shear strength V_{tf} arising from the tension-field action in the web develops a band of tensile forces that occur after the web has buckled under diagonal compression (principal stresses in ordinary beam theory). Equilibrium is maintained by the transfer of force to the vertical stiffeners. As the girder load increases, the angle of the tension-field changes to accommodate the greatest carrying capacity. Figure 11.8.3 shows a 50 × 50-in. (approx. 1.3 × 1.3 m) panel with a $\frac{1}{4}$-in. (6.4 mm) web which has buckled under diagonal compression when subjected to pure shear. It also illustrates the anchorage requirement wherein the longitudinal component of the tension-field must be transmitted to the flange in the adjacent panel, as shown by the vertical breaks in the whitewash at the flange in the corner of the adjacent panel where the tension-field intersects the stiffener and flange.

Figure 11.8.2
Tension-field action.

Figure 11.8.3
Tension-field in test plate
girder. (From Ref. 11.3,
Courtesy of Lehigh
University)

Tension-Field Action: Optimum Direction

Consider the tensile membrane stress σ_t which develops in the web at the angle γ, as shown in Fig. 11.8.4. If such tensile stresses can develop over the full height of the web, then the total diagonal tensile force T would be

$$T = \sigma_t t_w h \cos \gamma \tag{11.8.3}$$

the vertical component of which is the shear force V, given by

$$V = T \sin \gamma = \sigma_t t_w h \cos \gamma \sin \gamma \tag{11.8.4}$$

If such diagonal tensile stresses could develop *along* the flanges, vertical stiffness of the flanges would be required. Since the flanges have little vertical stiffness and are acting to their capacity in resisting flexure on the girder, the tension-field actually can develop only over a band width such that the vertical component can be transferred at the vertical stiffeners. The stiffeners can be designed to carry the necessary compressive force. It will be assumed that the tension-field (or partial tension-field as some may prefer to call it) may develop over the band width s, shown in Fig. 11.8.5a.

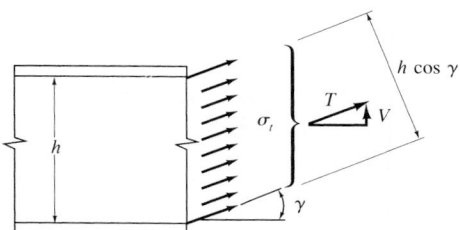

Figure 11.8.4
Membrane stresses in
tension-field action.

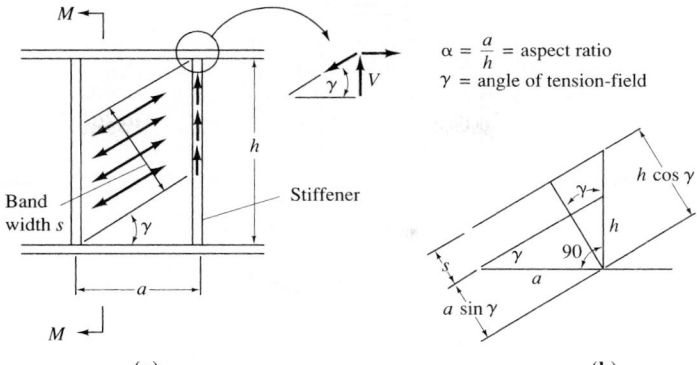

Figure 11.8.5
Forces arising from
tension-field.

The membrane tensile force tributary to one stiffener is $\sigma_t s t_w$, and the partial shear force ΔV_{tf} developed by compression in the stiffener is

$$\Delta V_{tf} = \sigma_t s t_w \sin \gamma \tag{11.8.5}$$

and the angle γ is the angle providing the maximum shear component from the partial tension-field.

From the geometry shown in Fig. 11.8.5b,

$$s = h \cos \gamma - a \sin \gamma \tag{11.8.6}$$

where a = stiffener spacing. Substitution of Eq. 11.8.6 into Eq. 11.8.5 gives

$$\Delta V_{tf} = \sigma_t t_w (h \cos \gamma - a \sin \gamma) \sin \gamma$$

$$= \sigma_t t_w \left(\frac{h}{2} \sin 2\gamma - a \sin^2 \gamma \right) \tag{11.8.7}$$

For maximum ΔV_{tf}, it is required that $d(\Delta V_{tf})/d\gamma = 0$. Thus

$$\frac{d(\Delta V_{tf})}{d\gamma} = \sigma_t t_w \left(\frac{h}{2} (2) \cos 2\gamma - 2a \sin \gamma \cos \gamma \right) = 0 \tag{11.8.8}$$

$$0 = h \cos 2\gamma - a \sin 2\gamma$$

or

$$\tan 2\gamma = \frac{h}{a} = \frac{1}{a/h} \tag{11.8.9}$$

From the trigonometry of Eq. 11.8.9,

$$\sin 2\gamma = \frac{1}{\sqrt{1 + (a/h)^2}} \tag{11.8.10}$$

also

$$\sin^2 \gamma = \frac{1 - \cos 2\gamma}{2} = \frac{1}{2} \left[1 - \frac{a/h}{\sqrt{1 + (a/h)^2}} \right] \tag{11.8.11}$$

The maximum contribution ΔV_{tf} from tension-field action is then obtained by substituting Eqs. 11.8.10 and 11.8.11 into Eq. 11.8.7 giving

$$\Delta V_{tf} = \sigma_t \frac{h t_w}{2} \left[\sqrt{1 + (a/h)^2} - a/h \right] \tag{11.8.12}$$

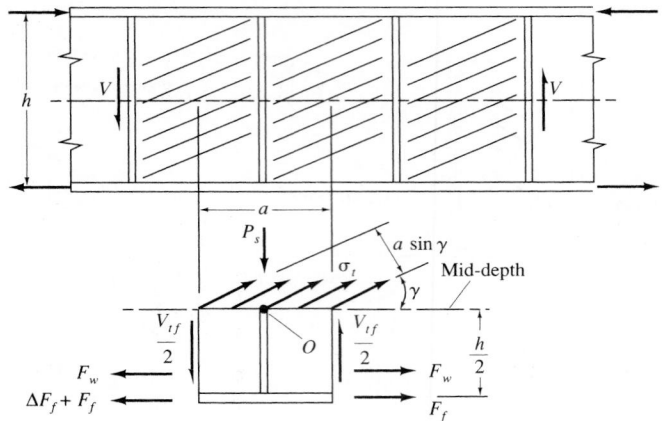

Figure 11.8.6
Force in stiffener resulting
from tension-field action.

It is not practical to use Eq. 11.8.12 directly, since the shear contribution from the part of the section (such as M–M of Fig. 11.8.5) that cuts through the triangles outside the band s must be added. The state of stress in these triangles is unknown, requiring an alternate approach to finding the total shear V_{tf} when the optimum angle γ is reached.

An alternate way, as used by Basler [11.3], is to cut a free body as in Fig. 11.8.6. The section is taken vertically midway between two adjacent stiffeners and horizontally at mid-depth. The mid-depth cut provides access to the tension-field where the state of stress is known, and the shear resultant on each vertical face equals $V_{tf}/2$ from symmetry.

Shear Strength from Tension-Field Action

Using the free body of Fig. 11.8.6, horizontal force equilibrium requires

$$\Delta F_f = (\sigma_t t_w \, a \sin \gamma)\cos \gamma$$
$$= \sigma_t \frac{t_w a}{2} \sin 2\gamma \tag{11.8.13}$$

The incremental web force ΔF_w is not used because the web contributes little to the flexural strength of the girder. Rotational equilibrium, taken about point O, requires

$$\Delta F_f \frac{h}{2} - \frac{V_{tf}a}{2} = 0 \tag{11.8.14}$$

Solving Eq. 11.8.14 for ΔF_f and substituting into Eq. 11.8.13 gives

$$\frac{V_{tf}a}{h} = \sigma_t \frac{t_w a}{2} \sin 2\gamma \tag{11.8.15}$$

Solving for V_{tf} and using Eq. 11.8.10 for $\sin 2\gamma$ gives

$$V_{tf} = \sigma_t \frac{ht_w}{2}\left[\frac{1}{\sqrt{1 + (a/h)^2}}\right] \tag{11.8.16}$$

Failure Condition

The actual state of stress in the web involves both shear stress τ and normal stress σ_t; thus the failure of an element subjected to shear in combination with an inclined tension must be considered, as shown in Fig. 11.8.7. Two basic assumptions are involved: first, τ_{cr} remains

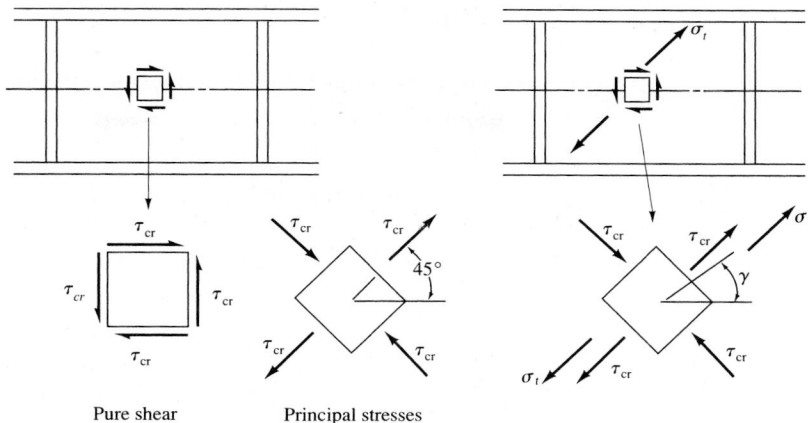

Figure 11.8.7
State of stress.

(a) At imminent buckling   (b) At ultimate shear

at constant value from buckling load to ultimate load and therefore the tension-field stress σ_t acts in addition to the principal stress τ_{cr}; second, the angle γ in Fig. 11.8.7b will be conservatively taken as 45° even though it will always be less than that value.

The generally accepted relationship for failure in plane stress is the "energy of distortion" theory (discussed in Sec. 2.7) shown as the ellipse in Fig. 11.8.8, which may be written

$$\sigma_1^2 + \sigma_2^2 - \sigma_1\sigma_2 = F_y^2 \qquad (11.8.17)$$

where σ_1 and σ_2 are principal stresses. Point A represents the case of shear alone and point B represents tension alone. The actual states of stress in plate girder webs fall on the ellipse between points A and B, and a straight line is a reasonable approximation of the segment AB,

$$\sigma_1 = F_y + \sigma_2\left(\sqrt{3} - 1\right) \qquad (11.8.18)$$

and for the stress condition that $\sigma_1 = \tau_{cr} + \sigma_t$ and $\sigma_2 = -\tau_{cr}$, Eq. 11.8.18 becomes

$$\frac{\sigma_t}{F_y} = 1 - \frac{\tau_{cr}}{F_y/\sqrt{3}} = 1 - C_v \qquad (11.8.19)$$

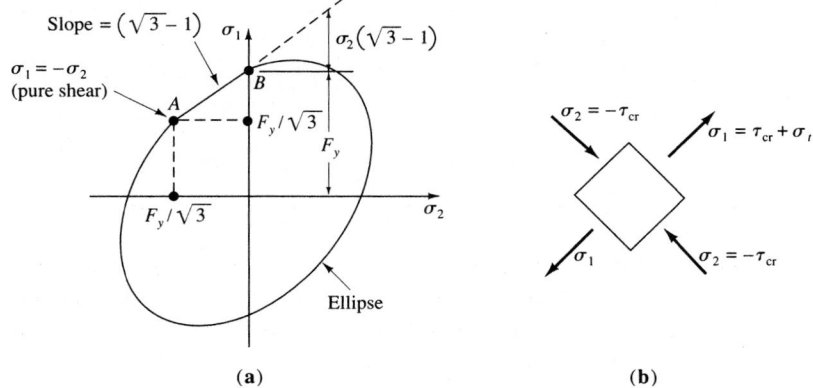

Figure 11.8.8
Energy-of-distortion failure
criterion.

(a)   (b)

Force in Stiffener

Using Fig. 11.8.6, vertical force equilibrium requires the force P_s in the stiffener to be

$$P_s = (\sigma_t t_w a \sin \gamma) \sin \gamma \tag{11.8.20}$$

Then, substituting the trigonometric identity,

$$\sin^2 \gamma = \frac{1 - \cos 2\gamma}{2}$$

gives

$$P_s = \sigma_t \left(\frac{a t_w}{2}\right) \left[1 - \frac{a/h}{\sqrt{1 + (a/h)^2}}\right] \tag{11.8.21}$$

Substituting Eq. 11.8.19 into Eq. 11.8.21 gives

$$P_s = \frac{F_y(1 - C_v)a t_w}{2} \left[1 - \frac{a/h}{\sqrt{1 + (a/h)^2}}\right] \tag{11.8.22}$$

which is the force in the stiffener when nominal shear strength V_n is reached, including tension-field action.

More recent work as discussed in the *SSRC Guide* [6.8] has shown that Eq. 11.8.22 may be simplified by using $a/h = 1$; in which case,

$$P_s = 0.5 F_y(1 - C_v) a t_w \left(1 - \frac{1}{\sqrt{2}}\right) \tag{11.8.23}$$

$$P_s = 0.15 F_y(1 - C_v) a t_w \tag{11.8.24}$$

Nominal Shear Strength, Including Both Buckling and Post-Buckling Strengths

Since thin-web plate girders exhibit some strength in shear before diagonal buckling occurs (V_n from Sec. 11.7) and additional strength in the post-buckling range (V_{tf} from Eq. 11.8.16), their actual strength is the sum of both components. Substituting Eqs. 11.8.2 and 11.8.16 into 11.8.1 gives

$$V_n = h t_w \left[\tau_y C_v + \frac{\sigma_t}{2\sqrt{1 + (a/h)^2}}\right] \tag{11.8.25}$$

Substituting Eq. 11.8.19 for σ_t and using $\tau_y = F_y/\sqrt{3}$ gives

$$V_n = F_y h t_w \left[\frac{C_v}{\sqrt{3}} + \frac{1 - C_v}{2\sqrt{1 + (a/h)^2}}\right] \tag{11.8.26}$$

A summary of studies of tension-field mechanisms is presented in the *SSRC Guide* [6.8, p. 221].

AISC Design

When tension-field action is developed by using appropriately spaced and sized transverse stiffeners, the nominal strength in shear may be expressed by Eq. 11.8.26. Factoring $\sqrt{3}$

from the denominator, and then approximating $F_y/\sqrt{3}$ as $0.6F_y$, gives

$$V_n = 0.6F_y A_w \left(C_v + \frac{1 - C_v}{1.15\sqrt{1 + (a/h)^2}} \right) \qquad (11.8.27)$$

which is AISC Formula (G3-2). Curves for Eq. 11.8.27 are presented in Fig. 11.8.9 for $F_y = 50$ ksi steel.

The force P_s in the intermediate stiffeners when tension-field action is utilized is given by Eq. 11.8.24. The stiffener area A_{st} required is

$$\text{Required } A_{st} = \frac{P_s}{F_{yst}} = \frac{0.15F_y(1 - C_v)at_w}{F_{yst}} \qquad (11.8.28)$$

When the panel has more shear strength $\phi_v V_n$ than is needed to carry the factored shear V_u, the stiffener area A_{st} may be reduced by multiplying by $(V_u/\phi_v V_n)$. Also, the derivation assumed the stiffener was aligned with the center of the web; when stiffeners are used only on one side or if angle stiffeners are used, there is an eccentric effect and the stiffener area must be increased. In addition, the area $(18t_w \times t_w)$ of the web tributary to the stiffener may be subtracted from the required area A_{st}. Thus, AISC-G3.3 gives the requirement (using the LRFD ratio $V_u/\phi_v V_n$) as

$$\text{Required } A_{st} = \frac{F_y}{F_{yst}} \left[0.15 D_s h t_w (1 - C_v) \frac{V_u}{\phi_v V_n} - 18t_w^2 \right] \geq 0 \qquad (11.8.29)$$

where D_s = factor to account for eccentric loading on stiffeners
  = 1.0 for stiffeners in pairs on each side of web
  = 1.8 for single angle stiffeners
  = 2.4 for single plate stiffeners

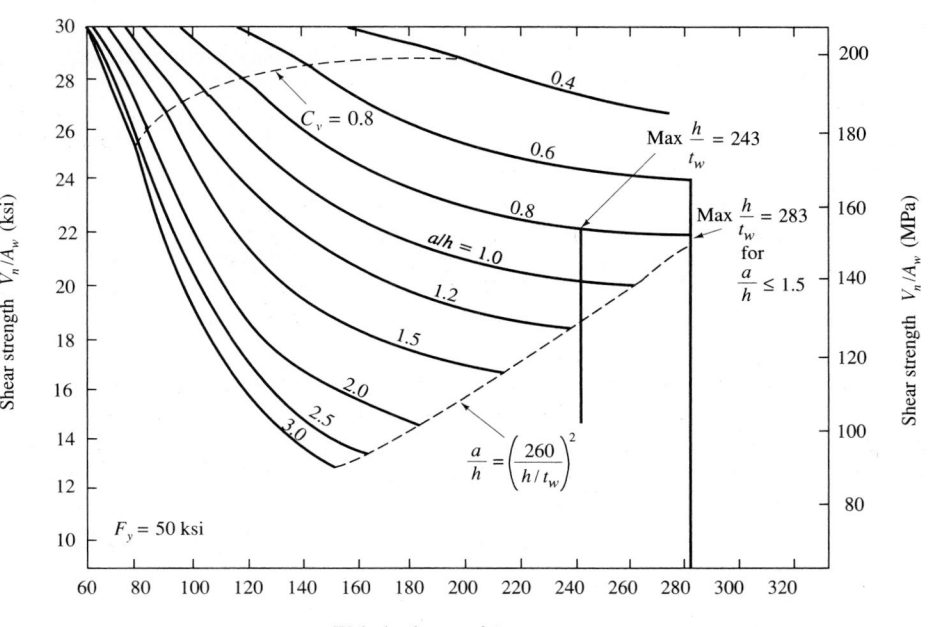

Figure 11.8.9
Nominal shear strength V_n, including tension-field action, divided by web area A_w for plate girders having $F_y = 50$ ksi (345 MPa), according to AISC-G3.

$$F_{yst} = \text{specified yield stress for stiffener steel}$$

V_u = factored service load shear at the location of the stiffener (AISC uses V_r as the *required* strength in shear)

V_n = nominal strength in shear, as defined in AISC-G3.2

ϕ_v = resistance factor for shear

The term $\phi_v V_n$ has long been called the *design* shear strength in LRFD and is called V_c the *available* shear strength in AISC-G3.3. To the authors, the additional symbols only add confusion, but apparently were deemed necessary for the Allowable Strength Design Method.

Note that Eq. 11.8.29 uses h instead of a, a reasonable approximation and a simplification that makes the stiffener area A_{st} required proportional to the web area A_w.

Sometimes stiffeners are alternated on each side of the web to gain better economy or they are placed all on one side to improve esthetics. Referring to Fig. 11.8.10a, the symmetrical pair of stiffeners reaches its yield condition with the force P_s,

$$P_s = 2wt F_{yst} = A_{st} F_{yst} \quad \text{(for concentric load)} \qquad (11.8.30)$$

On the other hand, an eccentrically loaded stiffener becomes plastic with a stress distribution as shown in Fig. 11.8.10b. For this case, force equilibrium requires

$$P_s = (w - x)t F_{yst} - xt F_{yst} \qquad (a)$$

and moment equilibrium requires

$$xt F_{yst}\left(w - \frac{x}{2}\right) = F_{yst}(w - x)t\left(\frac{w - x}{2}\right) \qquad (b)$$

Solving the quadratic gives $x = 0.293w$. Substitution for x in Eq. (a) gives

$$P_s = [(w - 0.293w)t - 0.293wt] F_{yst}$$

$$= 0.414wt F_{yst} = 0.414 A'_{st} F_{yst} \quad \text{(for eccentric load)} \qquad (11.8.31)$$

If single plate stiffeners are used on one side only, equating Eqs. 11.8.30 and 11.8.31 shows

$$A'_{st} = \frac{A_{st}}{0.414} = 2.4 A_{st} \qquad (11.8.32)$$

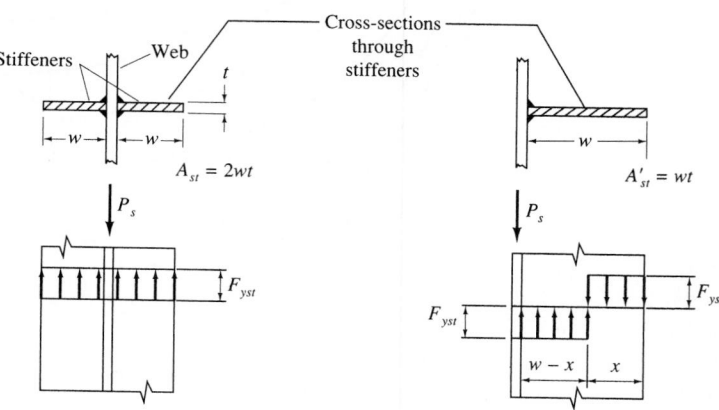

Figure 11.8.10
Intermediate stiffeners at
nominal shear strength V_n,
including tension-field
action.

(a) Axially loaded stiffener   **(b)** Eccentrically loaded stiffener

To correct for eccentric loading of stiffeners, the factor D is used in Eq. 11.8.29; 2.4 for single plate stiffeners. For a single angle whose center of gravity is closer to the web, the multiplier D reduces to 1.8.

11.9   STRENGTH IN COMBINED BENDING AND SHEAR

In the vast majority of cases the nominal strength M_n in bending is not influenced by shear, nor is the nominal shear strength V_n influenced by moment. Particularly, in very slender webs where "bend-buckling" may occur, the bending stress is redistributed as discussed in Sec. 11.5, so that the flanges carry an increased share. The shear strength of the web, however, is not reduced as a result of "bend-buckling" because most of the shear strength is from tension-field action with only a small contribution from the portion of the web adjacent to the flange. Since plate girders are usually designed using slender webs, AISC Commentary-G2 explicitly states that the effect of bending on shear is deemed negligible and need not be considered. In stockier webs no "bend-buckling" may occur, but high web shear in combination with bending may cause yielding of the web adjacent to the flange; again resulting in a transfer of part of the web's share of the bending moment to the flange. The strength of girders subject to combined bending and shear is the subject of the third major paper by Basler [11.17].

Since instability is precluded, a plastic analysis may be used. When subjected to high bending moment, the web yields adjacent to the flange and is, therefore, unable to carry shear. In the mid-depth region of the web the shear causes yielding; thus this part of the web is unable to carry bending moment.

Referring to Fig. 11.9.1, the nominal shear strength V'_n in the presence of bending moment may be expressed as

$$V'_n = \tau_y y_0 t_w \tag{11.9.1}$$

When *no* bending moment is present, that is $y_0 = h$, the nominal shear strength V_n would be

$$V_n = \tau_y t_w h \tag{11.9.2}$$

Eliminating τ_y from Eqs. 11.9.1 and 11.9.2 gives

$$y_0 = \left(\frac{V'_n}{V_n}\right)h \tag{11.9.3}$$

The nominal moment strength M'_n in the presence of shear from Fig. 11.9.1a is

$$M'_n = A_f F_y h + F_y t_w \left(\frac{h}{2}\right)\left(\frac{h}{2}\right) - F_y t_w \left(\frac{y_0}{2}\right)\left(\frac{y_0}{2}\right) \tag{11.9.4}$$

Figure 11.9.1
Shear and moment strengths under combined bending and shear.

Normal stress   Shear stress

(a) Simple procedure

Normal stress   Shear stress

(b) More accurate procedure

which upon substitution of Eq. 11.9.3 into Eq. 11.9.4 and letting $A_w = ht_w$ gives

$$M'_n = F_y A_f h \left\{ 1 + \frac{1}{4}\frac{A_w}{A_f}\left[1 - \left(\frac{V'_n}{V_n}\right)^2\right]\right\} \tag{11.9.5}$$

The nominal strength M_n equals M_y when the extreme fiber reaches the yield stress F_y, and with the web fully participating, is, according to Eq. (g), Example 11.5.1,

$$M_n = M_y = F_y h\, A_f\left(1 + \frac{1}{6}\frac{A_w}{A_f}\right) \tag{11.9.6}$$

As the percentage of the shear strength V_n that is utilized increases, the nominal moment strength M'_n decreases. In the absence of instability, but in the presence of high shear, the nominal bending moment strength may be expressed as

$$M'_n = M_n \left[\frac{1 + \frac{1}{4}a_r[1 - (V'_n/V_n)^2]}{1 + \frac{1}{6}a_r}\right] \tag{11.9.7}$$

where $a_r = A_w/A_f$.

When $M'_n = M_n$, $V'_n/V_n = 0.577$, or approximately 0.6. When more than 60 percent of the maximum shear strength V_n is used, the available nominal strength M'_n becomes less than M_n. Table 11.9.1 gives some values for M'_n/M_n for various values of a_r in the practical range, with graphical illustration in Fig. 11.9.2.

The relationship in Fig. 11.9.2 may be expressed:

For $V'_n/V_n \le 0.60$,

$$M'_n = M_n \tag{11.9.8}$$

For $M'_n/M_n \le 0.75$,

$$V'_n = V_n \tag{11.9.9}$$

When Eqs. 11.9.8 or 11.9.9 are not applicable, the interaction relationship must be used. If one uses a conservative value of $a_r = A_w/A_f = 2.0$ and considers the strength reduction from points A to B of Fig. 11.9.2 as a straight line, the slope of AB would be

$$\text{Slope of } AB = \frac{-0.25}{0.40} = -\frac{5}{8}$$

TABLE 11.9.1 Values of M'_n/M_n in Accordance with Eq. 11.9.7 for $V'_n/V_n \ge 0.6$

$a_r = \dfrac{A_w}{A_f}$	For $\dfrac{V'_n}{V_n} = 0.8$	For $\dfrac{V'_n}{V_n} = 1.0$
0	1.0	1.0
0.5	0.965	0.923
1.0	0.934	0.857
1.5	0.908	0.800
2.0	0.885	0.750

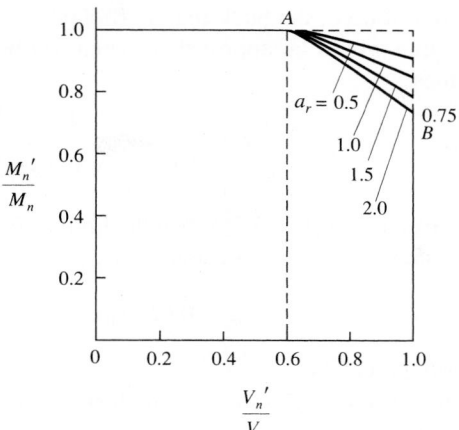

Figure 11.9.2
Moment–shear strength
interaction relationship.

The reduction equation then becomes

$$\frac{M'_n}{M_n} = 1 - \frac{5}{8}\left(\frac{V'_n}{V_n} - 0.6\right) \le 1.0 \tag{11.9.10}$$

or

$$\frac{M'_n}{M_n} + \frac{5}{8}\left(\frac{V'_n}{V_n}\right) \le \left(1.0 + \frac{5}{8}(0.6) = 1.375\right) \tag{11.9.11}$$

where M'_n = nominal flexural strength *in the presence of shear*

V'_n = nominal shear strength *in the presence of flexure*

M_n = maximum nominal flexural strength

V_n = maximum nominal shear strength

11.10 INTERMEDIATE TRANSVERSE STIFFENERS

Plate girders will usually be designed to have intermediate stiffeners as shown in Fig. 11.1.2 and the photo on p. 652. The two stability parameters for the web are h/t_w and a/h as discussed in Secs. 11.7 and 11.8. Buckling resulting from shear can be avoided when these stability parameters are kept low enough; alternatively, the shear stress can be kept below the critical buckling stress τ_{cr}. Since rolled beams have low h/t_w ratios, buckling resulting from shear will not occur. When the spacing a of stiffeners makes a/t_w low enough, and their size is adequate to allow them to act as compression verticals in a truss as discussed in Sec. 11.8, post-buckling strength (tension-field action) is available and may be utilized in design.

Requirements to Omit Intermediate Stiffeners—AISC

Stiffeners need not be used when the flexural strength of the section can be achieved without diagonal buckling resulting from shear. Stiffeners would not be required, according to AISC-G2.2, and shown in Fig. 11.7.3, when

$$\frac{h}{t_w} \le \left(1.10\sqrt{\frac{k_v E}{F_y}} = 187\sqrt{\frac{k_v}{F_y, \text{ksi}}}\right) \tag{[11.7.16]}$$

When stiffeners are not used, the buckling coefficient k_v, given by Eq. 11.7.8, is to be taken as 5. This value would be approached when a/h becomes large. When $k_v = 5$, Eq. 11.7.16 becomes

$$\frac{h}{t_w} \le \left(2.46\sqrt{\frac{E}{F_y}} = \frac{419}{\sqrt{F_y, \text{ksi}}} \right) \tag{11.10.1}$$

which is the limit given by AISC-G2.2. When the limit of Eq. 11.10.1 is not exceeded, the maximum nominal shear strength V_u is achieveable,

$$V_n = 0.6F_y A_w C_v \tag{[11.7.17]}$$

which is AISC Formula (G2-1).

When less than the maximum shear strength $\phi_v V_n$ is required, intermediate stiffeners are not required when

$$V_u \le (\phi_v V_n = 1.0(0.6F_y)A_w C_v) \tag{11.10.2}$$

unless $h/t_w > 260$. Intermediate stiffeners *are required* when h/t_w exceeds 260.

The expressions for C_v to be used for unstiffened girders are Eqs. 11.7.12 for inelastic buckling and 11.7.10 for elastic buckling with $k_v = 5$, as follows (with $E = 29{,}000$ ksi inserted when using F_y in ksi):

1. When $\dfrac{419}{\sqrt{F_y, \text{ksi}}} \le \dfrac{h}{t_w} \le \dfrac{521}{\sqrt{F_y, \text{ksi}}}$ (i.e., inelastic buckling)

$$C_v = \frac{419}{(h/t_w)\sqrt{F_y, \text{ksi}}} \tag{11.10.3}$$

where the limit containing the 521 arises from AISC-G2.1 when $k_v = 5$, as follows:

$$\frac{h}{t_w} \text{ limit} = 1.37\sqrt{\frac{k_v E}{F_y}} = 233\sqrt{\frac{k_v}{F_y, \text{ksi}}} = \frac{233\sqrt{5}}{\sqrt{F_y, \text{ksi}}} = \frac{521}{\sqrt{F_y, \text{ksi}}}$$

2. When $\dfrac{h}{t_w} > \dfrac{521}{\sqrt{F_y, \text{ksi}}}$ (i.e., elastic buckling)

$$C_v = \frac{7.55E}{(h/t_w)^2 F_y} = \frac{219{,}000}{(h/t_w)^2 F_y, \text{ksi}} \tag{11.10.4}$$

In summary, intermediate stiffeners are not required when both of the following requirements are satisfied:

1. $$\frac{h}{t_w} > 260 \tag{11.10.5}$$

2. $$\frac{V_u}{\phi_v} \le 0.6F_y A_w C_v \tag{11.10.6}$$

where C_v is given by Eqs. 11.10.3 and 11.10.4. Equation 11.10.5 was recommended by Basler [11.14] as a practical limit. He recommended that fabrication, handling, and erection are facilitated when the smaller panel dimension, a or h, does not exceed $260t_w$. When stiffeners are used, h is less than a.

Placement Criteria Including Tension-Field Action—AISC LRFD Method

When the factored shear V_u (called by AISC the required strength V_r) exceeds $\phi_r V_n$ with $\phi_v = 1.0$ and V_n given by Eq. 11.7.17 with $k_v = 5$, stiffeners are required. When $h/t_w > 260$, stiffeners are always required. The use of intermediate stiffeners reduces the a/h ratio and increases the nominal strength V_n. Equation 11.10.6 logically applies for situations *with* and *without* intermediate stiffeners when the objective is to prevent buckling resulting from shear.

Under AISC-G3 both buckling strength and post-buckling strength are recognized. The post-buckling behavior, known as tension-field action, is similar to truss action as shown in Figs. 11.8.2 and 11.8.3, and the total nominal strength V_n is given by Eq. 11.8.27,

$$V_n = 0.6F_y A_w \left(C_v + \frac{1 - C_v}{1.15\sqrt{1 + (a/h)^2}} \right) \qquad [11.8.27]$$

which is AISC Formula (G3-2). Since C_v is a function of h/t_w, V_u is a function of both h/t_w and a/h, making evaluation difficult without a design aid. In "stress" format, the design strength $\phi_v V_n / A_w$ for plate girders is given in AISC Tables 3-16 for $F_y = 36$ ksi and 50 ksi.

While theoretically the only upper limits on h/t_w are those of AISC-F13.2 to prevent vertical buckling of the flange, practical considerations relating to fabrication, handling and erection [11.14] give rise to the traditional ASD restriction

$$\frac{a}{h} \le \left(\frac{260}{h/t_w} \right)^2 \le 3.0 \qquad (11.10.7)$$

AISC-G3.1 does not allow the use of tension-field when Eq. 11.10.7 is exceeded; it also requires using $k_v = 5$ when the above a/h limit is exceeded. The *Proposed Criteria for Load and Resistance Factor Design of Steel Building Structures* [11.15], the source for the AISC LRFD plate girder provisions, indicates Eq. 11.10.7 as a limit. The AISC Tables 3-16 also use Eq. 11.10.7 as a limit. The authors recommend not exceeding the a/h limit of Eq. 11.10.7 even if the Specification may permit it.

End Panels and Interior Panels Having an Opening in an Adjacent Panel

Figure 11.8.5 shows that at the junction of intermediate stiffener and flange, equilibrium requires an axial tension to develop in the flange of the *adjacent* panel. When no such flange is available, as in an end panel, the tension-field cannot adequately develop. AISC, therefore, considers that only buckling strength (*no* tension-field action) is available in that end panel. Thus, for end panels (that is, panels having *no adjacent panel* and interior panels having a significant hole in an adjacent panel:

$$V_n = 0.6F_y A_w C_v \qquad [11.7.17]$$

where C_v is given by Eqs. 11.10.3 and 11.10.4.

Stiffness Requirement

Intermediate stiffeners must be sufficiently rigid to keep the web *at the stiffener* from deflecting out-of-plane when buckling of the web occurs. The stiffener must have a rigidity EI_{st} that is related to the web plate rigidity $Et^3a/[12(1-\mu^2)]$. AISC-G2.2 gives

$$I_{st} \geq jat_w^3 \tag{11.10.8}$$

where I_{st} = moment of inertia of the cross-sectional area of a transverse stiffener taken about the center of the web thickness when the stiffener consists of a pair of plates, or about the face of the stiffener in contact with the web when single plate stiffeners are used

$$j = \frac{2.5}{(a/h)^2} - 2 \leq 0.5 \text{ which is AISC Formula (G2-6)} \tag{11.10.9}$$

Various theoretical relationships have been developed for the ratio γ_0 of the stiffener rigidity EI_{st} to one panel of web plate rigidity, which may be expressed

$$\gamma_0 = \frac{EI_{st}}{Da} = \frac{EI_{st}[12(1-\mu^2)]}{Et_w^3 a} \tag{11.10.10}$$

where I_{st} = optimum stiffener moment of inertia
$D = Et^3/[12(1-\mu^2)]$ = flexural rigidity per unit length of web plate

Equation 11.10.8 is essentially the following proposed by Bleich [6.9, p. 417]

$$\gamma_0 = 4\left(\frac{7}{(a/h)^2} - 5\right) \tag{11.10.11}$$

According to Eq. 11.10.10 with $\mu = 0.3$,

$$\text{Required } I_{st} = \frac{\gamma_0 t_w^3 a}{12(1-\mu^2)} = \frac{\gamma_0 t_w^3 a}{10.92} \tag{11.10.12}$$

Substitution of Eq. 11.10.11 into Eq. 11.10.12 gives

$$I_{st} \geq \left(\frac{2.56}{(a/h)^2} - 1.83\right)at_w^3 \tag{11.10.13}$$

which was simplified by Vincent [11.16] for use in AASHTO Load Factor Design for bridges to become Eq. 11.10.8 used in AISC-G2.2.

Strength Requirement

Intermediate stiffeners carry a compression load only after buckling of the web has occurred. As the post-buckling truss-like "tension-field action" increases, the stiffener force increases. The maximum force P_s in the stiffener, reached simultaneously with reaching the nominal shear strength V_n, is given by Eq. 11.8.22. The stiffener area required will be the force P_s divided by the yield stress F_{yst} of the stiffener steel, as follows (using LRFD terminology with V_u and $\phi_v V_n$ instead of AISC V_r and V_c).

$$\text{Required } A_{st} = \frac{F_y}{F_{yst}}\left[0.15 D_s h t_w(1-C_v)\frac{V_u}{\phi_v V_n} - 18t_w^2\right] \geq 0 \tag{11.8.29}$$

Connection to Web

When tension-field action is used, the force P_s which equals the required area A_{st} times the stiffener yield stress F_{yst}, must be transferred between the web and the stiffener. Basler [11.3] recommended the force P_s be considered transferred over one-third the girder height. On that basis, Basler [11.3] recommended that intermediate stiffeners be designed to provide a shear *flow* (kips/in.) strength f_{nv} given by

$$f_{nv} = 0.045h\sqrt{\frac{F_y}{E}} \qquad\qquad (11.10.14)$$

There is no 2005 AISC Specification shear flow requirement, though one could use Eq. 11.10.14, considering f_{nv} as ϕR_{nw} for fillet welds (see Chap. 5). In general, when intermittent fillet welds are used, the minimum segment length and minimum clear between segments will likely be sufficient (AISC-J2.2).

For calculation, h is in inches and f_{nv} is in kips/in. As with the area requirement, when panels adjacent to the stiffener are not loaded to their full strength, the shear flow used as the connection requirement may be reduced in proportion that the strength required is less than that provided.

Connection to Flanges

Intermediate stiffeners are provided to assist the web; to stiffen and create nodal lines during buckling of the web and to accept compression forces transmitted directly from the web. At the compression flange, welding of the stiffener across the flange as shown in Fig. 11.10.1 provides stability to the stiffener and holds it perpendicular to the web; in addition, such welding provides restraint against torsional buckling (Fig. 11.3.1b) of the compression flange.

On the tension flange, the effects of stress concentration increase the fatigue or brittle fracture possibilities, i.e., welding in no way helps the tension flange. Since the work of Basler [11.17] has shown that welding of stiffeners to the tension flange is unnecessary for proper functioning of stiffeners, AISC-G2.2 permits stopping stiffeners "short of the tension flange provided bearing is not needed to transmit a concentrated load or reaction." The weld by which the stiffener is attached to the web "shall be terminated not closer than 4 times the web thickness nor more than 6 times the web thickness from the near toe of the web-to-flange weld."

For situations where the stiffener serves as the attachment for lateral bracing, the welding to the compression flange should be designed to transmit 1 percent (rule-of-thumb) of the compressive force in the flange. For important lateral bracing design in situations involving long unsupported lengths, the strength of lateral bracing connections should be designed using the principles of Sec. 9.13.

Figure 11.10.1
Intermediate stiffener
connection to flange.

11.11 BEARING STIFFENER DESIGN

Concentrated loads, such as at unframed end reactions, must be carried by stiffeners placed in pairs. Whenever concentrated loads, such as end reactions or columns supported by plate girders, exceed the local web yielding, web crippling, or sidesway web buckling strengths, bearing stiffeners must be provided. Local web yielding and web crippling were discussed in Sec. 7.8 since they are also of concern on rolled beams. Local web yielding is provided for in AISC-J10.2; web crippling is in AISC-J10.3. Sidesway web buckling generally is of concern only on narrow flange plate girders and is in AISC-J10.4. These three phenomena all are related to the strength of a thin web in the vicinity of concentrated loads.

Bearing stiffeners, unlike intermediate stiffeners, should be close fitting at the bearing end, and when the concentrated load is compression against the flange, the stiffener may either be connected to, or bear against, the flange transmitting the concentrated load. When the concentrated load is a tension pull on the flange, the stiffener must be attached to the flange being pulled. For plate girders the usual situation is compression against the flange. In general, compression load transmitting bearing stiffeners should extend "approximately to the edge of the flange plates ..." according to the 1978 AISC Specification (ASD-1.10.5). There is no such requirement in the current Specification, but the authors believe it to be good practice.

Column Stability Criterion

Bearing stiffeners transmitting compression loads are designed as columns under the provisions of AISC-J10.8. The column consists of the stiffeners, plus a portion of the web tributary to them, as defined in AISC-J10.8, and shown in Fig. 11.11.1.

The effective length KL of the "column" is less than the depth h of the web plate because of the restraint provided by the flanges. The effective length KL, according to AISC-J10.8, is to be taken "equal to $0.75h$".

The slenderness ratio is computed,

$$\frac{KL}{r} = 0.75\frac{h}{r} \qquad (11.11.1)$$

where h = web plate depth

r = radius of gyration of the shaded portion shown in Fig. 11.11.1 taken about the mid-thickness of the web

The effective area A_e required is then computed using the column strength P_n in accordance with AISC-E3.

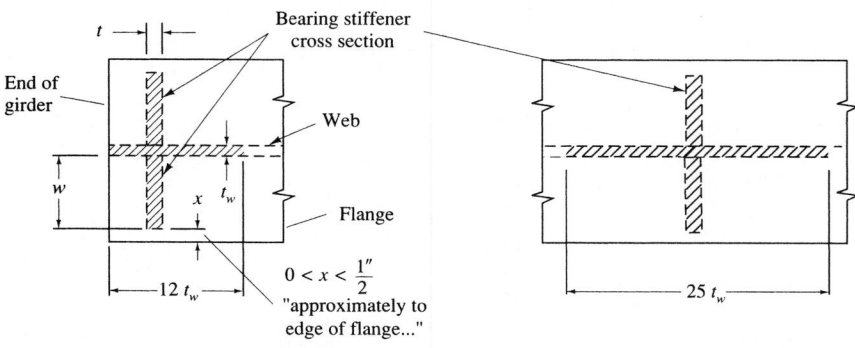

Figure 11.11.1
Bearing stiffener effective cross-sections.

(a) End stiffener   (b) Interior stiffener

The strength requirement for the AISC LRFD Method is

$$\phi_c P_n \geq P_u \tag{11.11.2}$$

where ϕ_c = resistance factor = 0.90

$P_n = F_{cr} A_e$

P_u = factored concentrated compression load

F_{cr} = column buckling stress according to AISC-E3

A_e = column area; i.e., the shaded area of Fig. 11.11.1 which includes the stiffener plates plus the tributary web area.

Thus, the required effective area A_e is

$$\text{Required } A_e = \frac{P_u}{\phi_c F_{cr}} \tag{11.11.3}$$

Local Buckling Criterion

Since the width w of the stiffener plates is governed by the plate girder flange width (see Fig. 11.11.1), the minimum thickness to prevent local buckling is

$$\text{Min } t = \frac{w}{0.56\sqrt{E/F_y}} = \frac{w}{95/\sqrt{F_y}, \text{ksi}} \tag{11.11.4}$$

as governed by $\lambda \leq \lambda_r$ for a uniformly stressed unstiffened compression element according to AISC-B4. When the thickness exceeds the limit of Eq. 11.11.6, the "slender" compression element will have reduced efficiency (i.e., $Q < 1$ as discussed in Sec. 6.18) and must be treated in accordance with AISC-E7. Because a bearing stiffener is an important element of a plate girder, the authors recommend satisfying Eq. 11.11.4 so that $Q = 1$.

Bearing Criterion

In order to bring bearing stiffener plates tight against the flanges, one corner of each stiffener plate must be cut off so as to clear the flange-to-web fillet weld. The remaining area of direct bearing is less than the gross area of the stiffener plates. The strength in bearing under AISC-J7 must be satisfactory.

The AISC LRFD Method requirement is

$$\phi R_n \geq P_u \tag{11.11.5}$$

where $\phi = 0.75$

R_n = nominal bearing strength = $1.8 F_y A_{pb}$

P_u = factored concentrated load (which AISC calls the "required strength")

A_{pb} = contact area of stiffener bearing against the flange

11.12 LONGITUDINAL WEB STIFFENERS

Longitudinal stiffeners, as shown in Fig. 11.12.1, can increase the bending and shear strengths of a plate girder. In general, they are not as effective as transverse stiffeners; however, they are frequently desired on highway bridge girders for esthetic reasons. Studies of

Figure 11.12.1
Effect of longitudinal stiffener
on plate girder web stability.

(a) Side view   **(b)** Cross-section   **(c)** Buckled shape

longitudinal stiffener effectiveness, as related to stiffener size and location, have been made by Cooper [11.18, 11.19] and others at Lehigh University. These studies and others are summarized in the *SSRC Guide* [6.8, pp. 211–223] and by Bleich [6.9, pp. 418–423]. The ASCE-AASHTO Task Committee [11.20] provides a full review of the theory and design of longitudinally stiffened plate girders.

The principal use of longitudinal stiffeners is in highway bridge design, where transverse stiffeners are used on both sides of a steel girder except on the exterior side of the exterior girder for more pleasing appearance. Rarely are longitudinal stiffeners used on *both* sides of a web as shown in Fig. 11.12.1.

The primary function of longitudinal stiffeners is to control lateral web deflections [11.20], and hence improve the bend-buckling strength as discussed in Sec. 11.5.

As discussed in Sec. 11.5, the elastic buckling strength of the web plate in bending (Fig. 11.5.1) may be written

$$F_{cr} = \frac{\pi^2 E k}{12(1 - \mu^2)(h/t)^2} \tag{11.12.1}$$

If the plate is stiffened by a longitudinal stiffener, as shown in Fig. 11.12.1, the value of k will be significantly greater than for the unstiffened case. The stiffener used should be stiff enough so that when buckling occurs, a nodal line will be formed along the line of the stiffener.

Under bending alone, the value of the buckling coefficient k has been found to be as high as 142.6 for the case where the flanges are assumed to provide full restraint to rotation at points A and B of Fig. 11.12.1c and $m = h/5$. For the case where the flanges provide no moment restraint at A and B (simply supported) the stiffener located at $m = h/5$ is also the optimum location. Such stiffener placement in the compression zone serves the purpose of maintaining the full effectiveness of the web in resisting bending stress, which is really the stiffener's principal function.

For webs subjected to shear alone, the longitudinal stiffener should be located at mid-height. For combined shear and bending the stiffener should be located so that $h/5 < m < h/2$; because of its principal function, however, it should preferably be closer to $h/5$.

For design there are two requirements: (1) a moment of inertia to insure adequate stiffness to create a nodal line along the stiffener, and (2) an area adequate to carry axial compression force while acting integrally with the web.

The design requirement for stiffness can be expressed as a function of the rigidity of the web, using the same approach as discussed for transverse stiffeners. Substituting the web height h for the transverse stiffener spacing a in Eq. 11.10.12 gives

$$\text{Required } I_{st} = \frac{\gamma_0 t_w^3 h}{10.92} \tag{11.12.2}$$

The AISC Specification [1.13] gives no information regarding longitudinal stiffeners. For highway bridges, AASHTO has given the following expression for Load Factor Design:

$$\text{Required } I_{st} = t_w^3 h \left[2.4 \left(\frac{a}{h} \right)^2 - 0.13 \right] \tag{11.12.3}$$

The moment of inertia I_{st} is to be that of the stiffener plate(s) combined with a centrally located web strip not more than $18t_w$ in width.

The location of the longitudinal stiffener shall be at $m = h/5$, and the local buckling requirement as from AASHTO,

$$\text{Min } t = \frac{w}{82/\sqrt{F_y, \text{ksi}}} \tag{11.12.4}$$

In addition, the radius of gyration r of the stiffener combined with a centrally located web strip not more than $18t_w$ in width shall be at least

$$\text{Min } r \geq \frac{a}{727/\sqrt{F_y, \text{ksi}}} \tag{11.12.5}$$

Note the AASHTO coefficients have been converted from their stated values to accommodate using F_y in ksi as used throughout this text instead of psi as in AASHTO.

11.13 PROPORTIONING THE SECTION

The cross-section of a girder must be selected such that it adequately performs its functions and requires minimum cost. The function requirements may be summarized as:

1. Strength to carry bending moment (adequate section modulus S_x).
2. Vertical stiffness to satisfy any deflection limitations (adequate moment of inertia I_x).
3. Lateral stiffness to prevent lateral-torsional buckling of compression flange (adequate lateral bracing or low L_b/r_t).
4. Strength to carry shear (adequate web area).
5. Stiffness to improve buckling or post-buckling strength of the web (related to h/t and a/h ratios).

To satisfy these function requirements at minimum cost, it will be assumed in what follows that minimum cost is equivalent to minimum weight.

Flange-Area Formula

For simplicity in design it is convenient to replace the real system of Fig. 11.13.1a with a substitute system, Fig. 11.13.1b, which allows the moment to be replaced by a couple with the forces of the couple acting at the flange centroids. The forces can then be treated as axial load situations. If the distance between flange forces is approximately $(h + d)/2$, the forces of the couple are

$$C = T = \frac{M}{(h + d)/2} \tag{11.13.1}$$

The effective area on which these forces act is equal to the flange plate area A_f plus additional area A_f' to represent the effectiveness of the web in resisting moment.

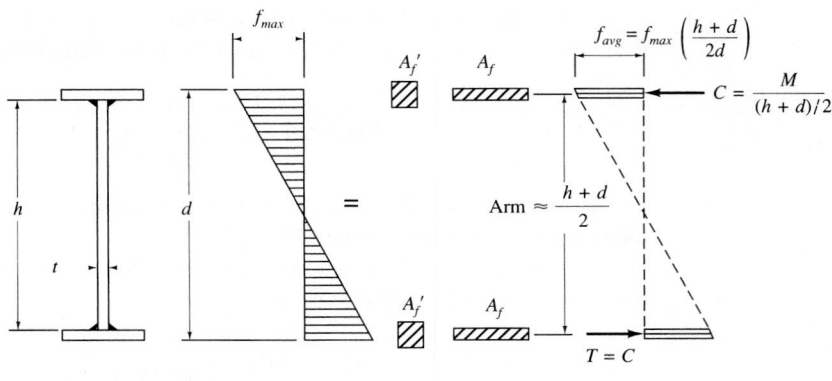

Figure 11.13.1
Flange-area formula
development.

(a) Actual system   (b) Flange-area system

The average stress on the total effective area is

$$f_{avg} = \frac{\text{Force}}{\text{Area}} = \frac{M}{(h + d)/2}\left(\frac{1}{A_f + A_f'}\right) \qquad (11.13.2)$$

The area A_f' must be taken such that the bending moment carried by the web is the same for both the real and substitute systems:

$$M_{\text{real system}} = f_{max}\left(\frac{h}{d}\right)\frac{th^2}{6} \qquad (11.13.3)$$

$$M_{\text{substitute system}} = f_{max}\left(\frac{h + d}{2d}\right)A_f'\left(\frac{h + d}{2}\right) \qquad (11.13.4)$$

Equating Eqs. 11.13.3 and 11.13.4 gives

$$A_f' = \frac{h}{d}\left(\frac{th^2}{6}\right)\left(\frac{2d}{h + d}\right)\left(\frac{2}{h + d}\right) = \frac{th}{6}\left(\frac{2h}{h + d}\right)^2 \qquad (11.13.5)$$

which if $A_w = th$, and the squared factor is neglected, becomes

$$A_f' = \frac{A_w}{6} \qquad (11.13.6)$$

Next, solving Eq. 11.13.2 for A_f gives

$$A_f = \frac{M}{[(h + d)/2]f_{avg}} - A_f' \qquad (11.13.7)$$

which, using Eq. 11.13.5 and $f_{avg} = f_{max}(h + d)/2d$, gives

$$A_f = \left[\frac{M}{f_{max}h}\left(\frac{d}{h}\right) - \frac{A_w}{6}\right]\left(\frac{2h}{h + d}\right)^2 \qquad (11.13.8)$$

Letting the squared factor equal unity overestimates slightly the value of A_f, while letting $d/h = 1$ underestimates the value. For preliminary design purposes these simplifications are justified to give a simple expression for the required area of one flange plate,

$$A_f = \frac{M}{fh} - \frac{A_w}{6} \qquad (11.13.9)$$

In the use of Eq. 11.13.9, if f is taken as the average stress on the flange, the d/h term will be nearly accounted for. When checking a section, of course, the correct moment of inertia must be obtained and the maximum strength computed.

Optimum Girder Depth

The variation in girder cross-sectional area is to be examined as a function of web depth to determine the depth which will give minimum area. Extended treatment of this subject has been given by Shedd [11.1], Bresler, Lin, and Scalzi [11.21] and Blodgett [11.22]. Schilling [11.23], Azad [11.24], Fleischer [11.25], and Anderson and Chong [11.26] have provided extended treatment of optimum proportioning, including the hybrid girder having a web of lower yield strength than the flanges.

The average gross area A_g of the girder for the entire span may be expressed

$$A_g = 2C_1 A_f + C_2 ht \qquad (11.13.10)$$

where C_1 = factor to account for reducing flange size at regions of lower than maximum moment

C_2 = factor to account for reducing web thickness at regions of reduced shear

Substituting Eq. 11.13.9 into Eq. 11.13.10 gives

$$A_g = 2C_1\left(\frac{M}{fh} - \frac{ht}{6}\right) + C_2 ht \qquad (11.13.11)$$

To find the minimum average gross area,

$$\frac{\partial A_g}{\partial h} = 0 \qquad (11.13.12)$$

(a) *Case 1.* No depth restriction; desire large h/t. Assume β_w = constant = h/t; $t = h/\beta_w$. Equation 11.13.11 becomes

$$A_g = 2C_1\left(\frac{M}{fh} - \frac{h^2}{6\beta_w}\right) + C_2\frac{h^2}{\beta_w} \qquad (11.13.13)$$

$$\frac{\partial A_g}{\partial h} = 0 = \frac{-2C_1 Mf}{f^2 h^2} - \frac{4C_1 h}{6\beta_w} + \frac{2C_2 h}{\beta_w} \qquad (11.13.14)$$

$$0 = -6C_1 M\beta_w - 2C_1 h^3 f + 6C_2 h^3 f \qquad (11.13.15)$$

from which

$$h = \sqrt[3]{\frac{3MC_1\beta_w}{f(3C_2 - C_1)}} \qquad (11.13.16)$$

and if one neglects the section reduction in regions of lower stress, $C_1 = C_2 = 1$, Eq. 11.13.16 becomes

$$h = \sqrt[3]{\frac{3M\beta_w}{2f}} \qquad (11.13.17)$$

where $M = M_u/\phi$ = factored service load moment divided by $\phi = 0.90$ for the AISC LRFD Method

f = average stress on flange using F_{cr} as extreme fiber value when nominal moment strength is achieved according to AISC-F5 or in some rare cases AISC-F4

Using Eq. 11.13.13 with $C_1 = C_2 = 1$ and substituting for M/f from Eq. 11.13.17 gives

$$A_g = \frac{4h^2}{3\beta_w} - \frac{h^2}{3\beta_w} + \frac{h^2}{\beta_w} = \frac{2h^2}{\beta_w} \qquad (11.13.18)$$

from which the girder weight per foot can be estimated using the fact that steel weight is 3.4 lb/sq in./linear ft (0.00784 kg/mm^2/linear metre).

$$\text{lb/ft} = 3.4A_g = \frac{6.8h^2}{\beta_w} = 8.9 \sqrt[3]{\frac{M^2}{f^2\beta_w}} \qquad (11.13.19)*$$

using inch units for the variables. Stiffeners will generally increase this value by 5 to 10 percent. M and f are defined for the AISC LRFD Method following Eq. 11.13.17.

(b) *Case 2.* Minimum web thickness; $t = $ const. Differentiating Eq. 11.13.11, $\partial A_g/\partial h = 0$, gives

$$\frac{-2C_1Mf}{f^2h^2} - \frac{C_1t}{3} + C_2t = 0 \qquad (11.13.20)$$

$$h = \sqrt{\frac{6C_1M}{ft(3C_2 - C_1)}} \qquad (11.13.21)$$

If $C_1 = C_2 = 1$,

$$h = \sqrt{\frac{3M}{ft}} \qquad (11.13.22)$$

where M and f are defined for LRFD following Eq. 11.13.17. The weight per foot can be obtained using A_g from Eq. 11.13.11,

$$\text{lb/ft} = 3.4A_g = 4.53ht = 7.85\sqrt{\frac{Mt}{f}} \qquad (11.13.23)*$$

using inch units for the variables. Again, an estimate for the weight of stiffeners should be added to the equation value.

(c) *Case 3.* Heavy shear which governs web area; $A_w = $ constant = web area, ht. Equation 11.13.11 becomes

$$A_g = 2C_1\left(\frac{M}{fh} - \frac{A_w}{6}\right) + C_2A_w \qquad (11.13.24)$$

from which it is apparent that minimum A_g results from maximum depth h. This case usually does not govern.

If the same kind of steel is used throughout, the value of C_1 may vary from 0.7 to 0.9 when used with the maximum positive moment: 0.85 to 0.90 is the usual range. The value of C_2 is not as likely to vary except on continuous structures where it might be 1.05 when used with maximum positive moment, or 0.95 when used with maximum negative moment.

---

*For SI units, the mass per metre is

$$\text{kg/m} = 1.72 \sqrt[3]{\frac{M^2}{f^2\beta_w}} \qquad (11.13.19)$$

$$\text{kg/m} = 0.0181\sqrt{\frac{Mt}{f}} \qquad (11.13.23)$$

using mm units for the variables.

Because of the complexity of evaluating C_1 and C_2 for continuous structures, it might be well to take them as unity.

Flange Plate Changes in Size

It is usually economical to reduce the size of flange plates in the region of low moment. While no specific rules can be made to help the designer to determine when it is desirable to change flange plate size, certain simple relationships are possible if only one change in flange size is desired.

  (a) *Case 1.* Linear variation in moment—two flange plate sizes. Consider the situation of Fig. 11.13.2a, and assuming both plates are fully utilized for bending moment, the flange-area formula can be used for each plate.

$$A_f = \frac{M}{hf} - \frac{A_w}{6} \tag{11.13.25}$$

$$A_{f1} = \frac{M(x/L)}{hf} - \frac{A_w}{6} \tag{11.13.26}$$

The total flange volume in the length L is

$$\text{Vol} = A_f(L - x) + A_{f1}x$$

$$= \frac{M(L - x)}{hf} - \frac{A_w}{6}(L - x) + \frac{M}{hf}\left(\frac{x^2}{L}\right) - \frac{A_w}{6}x$$

$$= \frac{M}{hf}\left(\frac{L^2 - xL + x^2}{L}\right) - \frac{A_wL}{6} \tag{11.13.27}$$

For minimum volume,

$$\frac{\partial(\text{Vol})}{\partial x} = 0 = 2x - L; \qquad x = \frac{L}{2}$$

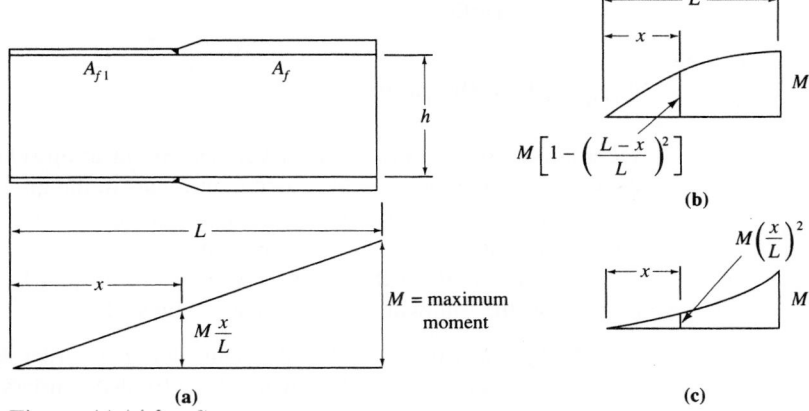

Figure 11.13.2
Common moment variations for determining changes in flange plate size.

which means

$$\frac{A_{f1}}{A_f} = \frac{\dfrac{M}{2hf} - \dfrac{A_w}{6}}{\dfrac{M}{hf} - \dfrac{A_w}{6}} \approx \frac{1}{2} \qquad (11.13.28)$$

(b) *Case 2.* Parabolic variation as for uniformly loaded simple beam (see Fig. 11.13.2b). The total volume in the length L is

$$\text{Vol} = \frac{M}{hf}\left(\frac{L^3 - L^2x + 2Lx^2 - x^3}{L^2}\right) - \frac{A_w L}{6}$$

$$\frac{\partial(\text{Vol})}{\partial x} = 0 = x^2 - \frac{4}{3}Lx + \frac{L^2}{3}; \qquad x = \frac{L}{3}$$

and

$$\frac{A_{f1}}{A_f} \approx \frac{5}{9} \qquad (11.13.29)$$

(c) *Case 3.* Parabolic variation as for uniformly loaded cantilever (see Fig. 11.13.2c). The total volume in the length L is

$$\text{Vol} = \frac{M}{hf}\left(\frac{L^3 - L^2x + x^3}{L^2}\right) - \frac{A_w L}{6}$$

$$\frac{\partial(\text{Vol})}{\partial x} = 0 = 3x^2 - L^2; \qquad x = \frac{L}{\sqrt{3}}$$

and

$$\frac{A_{f1}}{A_f} \approx \frac{1}{3} \qquad (11.13.30)$$

The foregoing developments can provide a guide for change of plate sizes. Since making a change involves a groove welded butt joint of the flange plates, enough material must be saved to more than offset the welding cost.

As a rule of thumb, unless 200 to 300 lb of material are saved in a flange plate per added splice the added cost of the butt splice (considering a plate about 2 ft wide and 2 in. thick) is not justified.

Flange Plate Proportions

According to most of the theory developed earlier in this chapter, the flange plate can be any width and thickness as long as it contributes to the girder the properties necessary to satisfy the functional requirements. However, nearly all testing has used flange plate dimensions which were considered reasonable.

In order to assist the engineer who is unsure of what constitutes such reasonable dimensions, the following guidelines are suggested.

1. Typically the ratio of girder flange width to girder depth, b_f/d, varies from about 0.3 for shallow girders to about 0.2 for deep girders.
2. Plate widths should be in 2-in. increments.

3. Plate thickness increments should be as follows:

$$\frac{1}{16}\text{ in.} \qquad t \leq \frac{9}{16}\text{ in.}$$

$$\frac{1}{8}\text{ in.} \qquad \frac{5}{8} \leq t \leq 1\frac{1}{2}\text{ in.}$$

$$\frac{1}{4}\text{ in.} \qquad t > 1\frac{1}{2}\text{ in.}$$

4. Where lateral stability is of concern, the flange plate width-to-thickness ratios $b_f/2t_f$ should be kept at about the λ_p value of AISC-B4, *in the maximum moment regions.* This will permit reducing the thickness of flange plates in regions of low moment. For such cases, flange plate area reduction should be made by reducing the thickness rather than the width.

5. For laterally stable girders, the flange plate area reduction in regions of lower moment may be accomplished by reducing the thickness, reducing the width, or reducing both thickness and width. A slight advantage in fatigue strength accrues by reducing the width rather than the thickness [11.22]. The transition slope should not exceed 1 in $2\frac{1}{2}$ for either width or thickness, and is usually 1 in 4 to 1 in 12 for the transition in width [11.22].

11.14 PLATE GIRDER DESIGN EXAMPLE—LRFD

Partially design a two-span continuous welded plate girder to support uniform load w of 0.8 kips/ft dead load and 3.2 kips/ft live load, plus two fixed position concentrated loads W of 15 kips dead load and 60 kips live load in each span as shown in Fig. 11.14.1. Lateral support is provided at each support and every 25 ft between supports. The girder is to have constant depth web plate for the two spans. Use A36 steel in the positive moment zone and A572 Grade 50 for the negative moment zone. Use Load and Resistance Factor Design.

Additional specifications and general comments:

1. The live load is to be applied in its correct manner; that is, applied as necessary to obtain the maximum and minimum moments and shears at every location along the girder. This will mean two live load cases for the factored bending moment envelope as shown in Fig. 11.14.2. For the shear envelope, the maximum values at the exterior and interior supports are obtained from the two bending moment loading cases. For all other shear envelope values, the live load was placed using partial loading in the manner dictated by the shear influence lines.*

2. Assume $\frac{5}{16}$ in. is minimum practical web plate thickness.

3. Assume no depth restriction; also, that any deep thin-web girder that is selected can be feasibly fabricated, transported to the construction site, and erected without excessive difficulty or cost.

Figure 11.14.1
Girder loading and support
for design example.

---

*See Chu-Kia Wang and Charles G. Salmon, *Introductory Structural Analysis.* Englewood Cliffs, NJ: Prentice-Hall, Inc., 1984, pp. 258–275.

Figure 11.14.2
Factored moment and
factored shear envelopes for
two-span continuous beam of
illustrative example.

Solution:

(a) *Factored Loads.* Using the basic factored gravity load combination of ASCE 7,

$$\text{Uniform live load } w_u = 1.6(3.2) = 5.12 \text{ kips/ft}$$
$$\text{Uniform dead load } w_u = 1.2(0.8) = 0.96 \text{ kips/ft}$$
$$\text{Concentrated live load } W_u = 1.6(60) = 96 \text{ kips}$$
$$\text{Concentrated dead load } W_u = 1.2(15) = 18 \text{ kips}$$

(b) *Structural Analysis.* Any method of statically indeterminate structural analysis can be used to obtain the elastic moment and shear envelopes under factored loads. The results are presented in Fig. 11.14.2.

(c) *Estimate of Weight.* The maximum factored moments due to the super-imposed loads (that is, without girder weight) of Fig. 11.14.1 for the positive and negative moment regions are

$$+M_u = +7640 \text{ ft-kips}$$
$$-M_u = -10{,}800 \text{ ft-kips}$$

Since there is no depth limitation, the depth based on maximum h/t_w may be desired. The AISC-F13.2 limits are obtained, referring to Table 11.3.1,

$$\text{Max } h/t_w = 322 \quad (332 \text{ when } a/h \leq 1.5) \qquad \text{A36 steel}$$
$$\text{Max } h/t_w = 243 \quad (282 \text{ when } a/h \leq 1.5) \qquad \text{A572 Grade 50}$$

There will be strength reduction from the bend-buckling limit state when h/t_w exceeds $970/\sqrt{F_{cr}}$, ksi according to AISC-F5. For this design those limits are

$$h/t_w > 162 \qquad \text{A36 when } F_{cr} = F_y$$
$$h/t_w > 137 \qquad \text{A572 Grade 50 when } F_{cr} = F_y$$

For a weight estimate, try $\beta_w = h/t_w = 300$ and use Eq. 11.13.19, $M = $ Required $M_n = M_u/\phi_b = 7640/0.90 = 8490$ ft-kips, and $f = R_{pg}F_{cr} \approx 34$ ksi (somewhat less than F_y):

$$\text{Wt/ft} = 8.9 \sqrt[3]{\frac{(\text{Required } M_n)^2}{(R_{pg}F_{cr})^2 \beta_w}} = 8.9 \sqrt[3]{\frac{[8490(12)]^2}{(34)^2 300}} = 276 \text{ lb/ft}$$

Assuming the girder weight is 300 lb/ft gives the maximum factored positive moment $M_u = 7900$ ft-kips. Recomputing the above formula gives

$$\text{Wt/ft} \approx 280 \text{ lb/ft} \quad (\text{from formula})$$

The negative moment will require a slightly heavier section even though the ratio of maxi-mum positive to maximum negative moment is approximately the same as the ratio (i.e., 50/36) of the yield stresses for the materials used. The web of the Grade 50 region of the girder must be thicker than the A36 region.

Allowing something extra (10% is a reasonable estimate) for the stiffeners would give a value slightly above 300 lb/ft. Use $w = 300$ lb/ft as the estimated girder weight. The factored moment M_u and factored shear V_u envelopes are given in Fig. 11.14.2.

(d) *Determine Web Plate Sizes.* For $+M_u$ with A36 steel assume $C_1 = C_2 = 1$ and use Eq. 11.13.17. Evaluate the optimum value for h using various h/t_w values, for example, using $h/t_w = 320$ gives

$$h = \sqrt[3]{\frac{3(\text{Required } M_n)\beta_w}{2R_{pg}F_{cr}}} = \sqrt[3]{\frac{3(7900/0.90)(12)320}{2(34)}} = 114 \text{ in.}$$

Based on the positive moment requirement, the economical depth appears at first to be 114 in. from the formula. However, if the lightest weight girder is desired, the web area must be used efficiently. Once the $\frac{3}{8}$ in. thickness is obtained from the 114 in. depth, the max-mum depth for a $\frac{3}{8}$-in. web should be determined, which gives an h of 120 to 124 in. This is considerably deeper than the formula value. Also, there is considerable reserve shear strength;

$\dfrac{h}{t_w} = \beta_w$	Formula h (in.)	h (in.)	t (in.)	A_w (sq in.)	$\dfrac{V_u}{A_w}$ (ksi)	Actual $\dfrac{h}{t}$
320	114.1	114	3/8	42.8	8.6	304
333	115.7	114	3/8	42.8	8.6	304
Max for 3/8 in.		125.0 using $h/t_w = 332$				
		124	3/8	46.5	7.9	331
Max for 3/8 in.		120.8 using $h/t_w = 322$				
		120	3/8	45.0	8.2	320
Max for 5/16 in.		104.2 using $h/t_w = 332$				
		104	5/16	32.5	11.4	333
Max for 5/16 in.		100.6 using $h/t_w = 322$				
		100	5/16	31.3	11.8	320

V_u/A_w will typically be satisfactory as high as 12 to 16 ksi. Note that $\phi V_n/A_w$ can be about 18 ksi for $h/t_w = 320$ according to AISC-Table 3-16. With the design factored shear stress only about 8 to 9 ksi with the $\frac{3}{8}$-in. web, it isprobable that a lighter girder will result using a thinner (say $\frac{5}{16}$ in.) web. The $\frac{5}{16}$-in. web would indicate a girder shallower than 114 in., say around 100 in.

Evaluate optimum h for the negative moment region initially using Eq. 11.13.17 with $M = M_u/\phi = 11,260/0.90$ and $h/t_w = 240$.

On the basis of these preliminary computations, a depth of 100 in. is chosen. The formula indicates somewhat deeper; however, the low shear indicates that the web can be thinner. Since intermediate stiffeners are to be used, the reader may note that when h/t_w ratios are larger the stiffener spacing requirement over most of the girder length will be controlled by Eq. 11.10.10 and will not be dependent on the magnitude of the shear force.

$\dfrac{h}{t_w} = \beta_w$	Formula h (in.)	h (in.)	t (in.)	A_w (sq in.)	$\dfrac{V_u}{A_w}$ (ksi)	Actual $\dfrac{h}{t}$
240	104.0	104	7/16	45.5	12.1	238
283	109.9	108	7/16	47.3	11.6	247
Max for 7/16 in.		123.7 using $h/t_w = 282$				
		122	7/16	53.4	10.3	279
Max for 7/16 in.		106.2 using $h/t_w = 243$				
		106	7/16	46.4	11.8	242
Max for 3/8 in.		106.1 using $h/t_w = 282$				
		106	3/8	39.8	13.8	283
Max for 3/8 in.		91.0 using $h/t_w = 243$				
		90	3/8	33.8	16.3	240

Thus the deeper the girder, the longer must be the stiffener plates, with little advantage to offset the extra weight. Generally, multiples of 2 in. should be used for the web plate depth.

Allowing for some flexibility in the design, use $h = 100$ in. which will mean that the web slenderness h/t_w is near but not at the upper limit. Try

$$\tfrac{5}{16} \times 100 \qquad (h/t_w = 320) \text{ for } +M, \qquad F_y = 36 \text{ ksi}$$

$$\tfrac{3}{8} \times 100 \qquad (h/t_w = 267) \text{ for } -M, \qquad F_y = 50 \text{ ksi}$$

(e) *Select Flange Plates for Negative Moment.* $M_u = 11{,}260$ ft-kips, web plate $= \tfrac{3}{8} \times 100$ ($A_w = 37.5$ sq in.). The lateral-torsional buckling limit state must be examined with regard to the 25-ft distance between lateral supports. The slenderness parameter λ involves the radius of gyration r_t, as discussed following Eq. 11.4.15. If a typical ratio $b_f/d \approx 0.25$ is assumed, then $b_f = 24$ in. Using the radius of gyration of a rectangle about its mid-depth,

$$r = b_f/\sqrt{12} = 0.289 b_f$$

which gives $r_t \approx 6.9$ in., say 7 in.

Check L_p for lateral-torsional buckling using AISC-F4.2,

$$\text{Estimated } L_p = 1.1 r_t \sqrt{\frac{E}{F_y}} = r_t \frac{187}{\sqrt{F_y, \text{ksi}}}$$

$$= \frac{7(187)}{\sqrt{50}} \frac{1}{12} = 15.5 \text{ ft} < (L_b = 25 \text{ ft})$$

$$\text{Estimated } L_r = \pi r_t \sqrt{\frac{E}{0.7 F_y}} = r_t \frac{640}{\sqrt{F_y, \text{ksi}}} = \frac{7(640)}{\sqrt{50}} \frac{1}{12} = 53 \text{ ft}$$

Thus, $L_p < L_b < L_r$. L_b is significantly lower than L_r; thus, it is anticipated that the plate girder will be governed by inelastic lateral-torsional buckling obtaining F_{cr} according to AISC-F5.2. However, note is made that the moment gradient is favorable giving a C_b greater than 1.0. Referring to the 25-ft laterally unbraced segment adjacent to the interior support in Fig. 11.14.2, where maximum M_u occurs, the moment gradient is closely a linear one varying from 11,260 ft-kips to zero. Thus, $C_b = 1.67$ for that segment when maximum moment is acting. It is highly probable that the lateral-torsional limit state does not affect the strength of this girder.

The high web slenderness h_c/t_w, will, however, reduce the moment strength. To obtain an estimated reduction from the bend-buckling limit state, use Eq. 11.4.3 (AISC-F5) with estimated h_c/t_w of 267 and $a_w = A_w/A_f \approx 1.5$

$$R_{pg} = 1 - \frac{a_w}{1200 + 300 a_w}\left(\frac{h_c}{t_w} - 5.7\sqrt{\frac{E}{F_y}}\right)$$

$$= 1 - \frac{a_w}{1200 + 300 a_w}\left(\frac{h_c}{t_w} - \frac{970}{\sqrt{F_y, \text{ksi}}}\right)$$

$$\approx 1 - \frac{1.5}{1200 + 300(1.5)}\left(267 - \frac{970}{\sqrt{\approx 48 \text{ ksi}}}\right) = 0.90$$

which makes $R_{pg} F_{cr} \approx 42$ ksi. (based on an estimated F_{cr} value of 46)

Using the flange-area formula, Eq. 11.13.9, gives the requirement for one flange as

$$A_f = \frac{M_u/\phi_b}{R_{pg}F_{cr}h} - \frac{A_w}{6} = \frac{11{,}260(12)/0.90}{42(101)} - \frac{37.5}{6} = 29.1 \text{ sq in.}$$

Some possible choices considering 42 ksi to be a low estimate:

$$1\tfrac{1}{4} \times 22, \quad A_f = 27.5 \text{ sq in.,} \quad b_f/2t_f = 8.8$$

$$1\tfrac{1}{8} \times 24, \quad A_f = 27.0 \text{ sq in.,} \quad b_f/2t_f = 10.7$$

$$1\tfrac{1}{8} \times 26, \quad A_f = 29.25 \text{ sq in.,} \quad b_f/2t_f = 11.6$$

Try plates—$1\tfrac{1}{4} \times 22$: $\quad A_f = 27.5$ sq in.

In this selection, the width-to-thickness ratio $\lambda = b_f/2t_f$ should be kept near the λ_p value (9.2 for $F_y = 50$ ksi from Table 11.4.1). Some strength reduction may be tolerated from the flange local buckling limit state, but preferably not.

Check F_{cr} for the lateral-torsional buckling limit state (AISC-F5), Eqs. 11.4.4 through 11.4.9:

$$a_w = \frac{h_c t_w}{b_f t_f} = \frac{100(0.375)}{22(1.25)} = 1.36$$

$$r_t = \sqrt{\frac{\text{Flange } I_y}{A_f + A_w/6}} = \sqrt{\frac{\tfrac{1}{12}(22)^3(1.25)}{27.5 + 37.5/6}} = 5.73 \text{ in.}$$

$$L_p = r_t \frac{187}{\sqrt{F_y, \text{ksi}}} = \frac{5.73(187)}{\sqrt{50}}\frac{1}{12} = 12.6 \text{ ft} < (L_b = 25 \text{ ft})$$

$$L_r = r_t \frac{640}{\sqrt{F_y, \text{ksi}}} = \frac{5.73(640)}{\sqrt{50}}\frac{1}{12} = 43 \text{ ft}$$

Since $L_p < L_b < L_r$, F_{cr} may be below F_y. However, in this case at the negative moment zone, $C_b = 1.67$ from Table 9.6.3 assuming moment varies linearly over the 25-ft segment; thus, $C_b F_{cr}$ is quite likely above F_y (the upper limit). Check Eq. 11.4.7 for the lateral-torsional buckling limit state.

$$F_{cr} = C_b F_y \left[1 - 0.3 \left(\frac{L_b - L_p}{L_r - L_p} \right) \right]$$

$$= 1.67(50) \left[1 - 0.3 \left(\frac{25 - 12.7}{43 - 12.7} \right) \right] = 73 \text{ ksi} > (F_y = 50 \text{ ksi})$$

Thus, F_{cr} based on the lateral-torsional buckling limit state is $F_y = 50$ ksi.

Compute F_{cr} for the flange local buckling limit state (AISC-F53) when, according to Eq. 11.4.10,

$$\left(\lambda = \frac{b_f}{2t_f} \right) > \left(\lambda_p = \frac{65}{\sqrt{F_y, \text{ksi}}} \right).$$

This was satisfied in this design when the flange plates were selected; otherwise this limit state must be treated according to Eqs. 11.4.10 through 11.4.13.

Next, evaluate the strength reduction resulting from the bend-buckling limit state when $h_c/t_w > 970/\sqrt{F_y}$, ksi, using Eq. 11.4.3,

$$R_{pg} = 1 - \frac{a_w}{1200 + 300a_w}\left(\frac{h_c}{t_w} - \frac{970}{\sqrt{F_y,\text{ksi}}}\right) \leq 1.0 \qquad [11.4.3]$$

$$a_w = \frac{A_w}{A_f} = \frac{37.5}{27.5} = 1.36$$

In evaluating Eq. 11.4.3, the F_{cr} is the lesser value from the lateral-torsional buckling (LTB) and the flange local buckling (FLB) limit states. In this design, $F_{cr} = F_y = 50$ ksi. Thus, Eq. 11.4.3 is evaluated using $a_w = 1.36$, $h_c/t_w = 267$, and $F_{cr} = 50$ ksi, to obtain $R_{pg} = 0.89$.

The moment of inertia of the cross-section must be evaluated,

$$1\tfrac{1}{4} \times 22: \quad 27.5(2)(101.25/2)^2 = 140,960$$
$$\tfrac{3}{8} \times 100: \quad (0.375)(100)^3/12 = \underline{31,250}$$
$$I = 172,210 \text{ in.}^4$$

$$S_x = \frac{I}{(h/2 + t_f)} = \frac{172,210}{50 + 1.25} = 3360 \text{ in.}^3$$

Then, using Eq. 11.4.2, the nominal strength M_n can be evaluated,

$$M_n = F_{cr}S_{xc}R_{pg} = 50(3360)(0.89)/12 = 12,460 \text{ ft-kips}$$
$$\phi M_n = 0.90(12,460) = 11,200 \text{ ft-kips} \approx (M_u = 11,260 \text{ ft-kips})$$

Accept plates $1\tfrac{1}{4} \times 22$.

(f) *Select Flange Plates for Positive Moment.* $M_u = 7900$ ft-kips, web plate $= \tfrac{5}{16}$ $\times 100$ ($A_w = 31.25$ sq in.). Estimate the flange width to be about 22 in., the same as for negative moment region, which makes $r_t \approx 5.7$ in as computed in part (e) above. In this positive moment region $C_b \approx 1.0$ conservatively, which means F_{cr} for the lateral-torsional buckling limit state might be a little below F_y.

Again estimate $R_{pg} \approx 0.90$ for the effect of the bend-buckling limit state,

$$R_{pg}F_{cr} \approx 0.90(36) = 32 \text{ ksi}$$

Using the flange-area formula, Eq. 11.13.9, gives the requirement for one flange as

$$A_f = \frac{M_u/\phi_b}{R_{pg}F_{cr}h} - \frac{A_w}{6} = \frac{7900(12)/0.90}{32(101)} - \frac{31.25}{6} = 27.4 \text{ sq in.}$$

Some possible choices:

$$1\tfrac{1}{4} \times 22, \quad A_f = 27.5 \text{ sq in.}, \quad b_f/2t_f = 8.8$$
$$1\tfrac{1}{8} \times 24, \quad A_f = 27.0 \text{ sq in.}, \quad b_f/2t_f = 10.7$$
$$1\tfrac{1}{8} \times 26, \quad A_f = 29.25 \text{ sq in.}, \quad b_f/2t_f = 11.6 > 10.8 \qquad \textbf{NG}$$

Try plates—$1\tfrac{1}{8} \times 24$: $\quad A_f = 27.0$ sq in.

The choices here are identical to those listed for the negative moment zone. Here, however, the $1\frac{1}{8} \times 24$ may provide enough strength because $\lambda = b_f/2t_f = 10.7$ does not exceed $\lambda_p = 10.8$ for A36 steel. That lighter plate will be investigated in this example; however, the practical choice would be the same $1\frac{1}{4} \times 22$ used in the negative moment zone.

Check F_{cr} for the lateral-torsional buckling limit state (AISC-F5), Eqs. 11.4.4 through 11.4.9:

$$a_w = \frac{h_c t_w}{b_f t_f} = \frac{100(0.3125)}{24(1.125)} = 1.16$$

$$r_t = \sqrt{\frac{\text{Flange } I_y}{A_f + A_w/6}} = \sqrt{\frac{\frac{1}{12}(24)^3(1.125)}{27 + 31.25/6}} = 6.34 \text{ in.}$$

$$L_p = r_t \frac{187}{\sqrt{F_y, \text{ksi}}} = \frac{6.34(187)}{\sqrt{36}}\frac{1}{12} = 16.5 \text{ ft} < (L_b = 25 \text{ ft})$$

$$L_r = r_t \frac{640}{\sqrt{F_y, \text{ksi}}} = \frac{6.34(640)}{\sqrt{36}}\frac{1}{12} = 56 \text{ ft}$$

Since $L_p < L_b < L_r$, F_{cr} must be computed in accordance with AISC-F5, Eqs. 11.4.4 through 11.4.9. In this case, $C_b = 1.0$,

$$F_{cr} = C_b F_y \left[1 - 0.3\left(\frac{L_b - L_p}{L_r - L_p}\right) \right]$$

$$= 1.0(36)\left[1 - 0.3\left(\frac{25 - 16.5}{56 - 16.5}\right) \right] = 33.7 \text{ ksi} < (F_y = 36 \text{ ksi})$$

Then $F_{cr} = 33.7$ ksi.

Next, evaluate the strength reduction resulting from the bend-buckling limit state when $h/t_w > 970/\sqrt{F_{cr}}$, ksi, using Eqs. 11.4.2 and 11.4.3 with $h_c/t_w = 320$.

$$a_w = \frac{A_w}{A_f} = \frac{31.25}{27.0} = 1.16$$

$$R_{pg} = 1 - \frac{1.16}{1200 + 300(1.16)}\left(320 - \frac{970}{\sqrt{36}} \right) = 0.88$$

The moment of inertia of the cross-section must be evaluated,

$$1\frac{1}{8} \times 24: \quad 27.0(2)(101.125/2)^2 = 138,050$$

$$\tfrac{5}{16} \times 100: \quad (0.3125)(100)^3/12 = \underline{26,040}$$

$$I = 164,090 \text{ in.}^4$$

$$S_x = \frac{I}{(h/2 + t_f)} = \frac{164,090}{50 + 1.125} = 3210 \text{ in.}^3$$

Then, using Eq. 11.4.2, the nominal strength M_n can be evaluated,

$$M_n = F_{cr}S_{xc}R_{pg} = 33.7(3210)(0.88)/12 = 7930 \text{ ft-kips}$$

$$\phi_b M_n = 0.90(7930) = 7140 \text{ ft-kips} < [M_u = 7900 \text{ ft-kips}] \qquad \textbf{NG}$$

This understrength is not acceptable. Increase flange plates to $1\frac{1}{4} \times 24$, with $A_f = 30.0$ sq in., $I_x = 179,800$ in.[4], and $S_x = 3509$ in.[3] The $b_f/2t_f = 9.6$ does not exceed $\lambda_p = 10.8$; thus, local flange buckling does not control. For this section,

$$a_w = \frac{100(0.3125)}{24(1.25)} = 1.04$$

$$r_t = \frac{24}{\sqrt{12\left(1 + \frac{1}{6}(1.04)\right)}} = 6.40 \text{ in.}$$

$$L_p = 1.1(6.40)\sqrt{\frac{29,000}{36}} = 200 \text{ in.} = 16.7 \text{ ft} < L_b = 25 \text{ ft}$$

$$L_r = \pi(6.40)\sqrt{\frac{29,000}{0.7(36)}} = 682 \text{ in.} = 57 \text{ ft} > L_b$$

$$F_{cr} = 1.0\left[36 - 0.3(36)\left(\frac{25 - 16.7}{57 - 16.7}\right)\right] = 33.8 \text{ ksi}$$

$$R_{pg} = 1 - \frac{a_w}{1200 + 300a_w}\left(\frac{h_c}{t_w} - \frac{970}{\sqrt{F_{cr}}}\right) \leq 1.0 \qquad [11.4.3]$$

$$= 1 - \frac{1.04}{1200 + 300(1.04)}\left(320 - \frac{970}{\sqrt{36}}\right) = 0.89$$

The nominal strength M_n is

$$M_n = F_{cr}S_{xc}R_{pg} = 33.8(3509)(0.89)/12 = 8800 \text{ ft-kips}$$

$$\phi_b M_n = 0.90(8800) = 7920 \text{ ft-kips} > [M_u = 7900 \text{ ft-kips}] \qquad \textbf{OK}$$

Use flange plates $1\frac{1}{4} \times 24$.

(g) *Intermediate Stiffeners—Placement in Positive Moment Zone.* Web plate $= \frac{5}{16} \times 100$, $A_w = 31.25$ sq in., $F_y = 36$ ksi.

Exterior end, $V_u = 369$ kips (Fig. 11.14.2). In end panels post-buckling strength (i.e., tension-field action) is *not permitted to be used.* Equation 11.10.5 represents the nominal shear strength V_n of such an end panel,

$$V_n = C_v(0.6F_y)A_w$$

$$\text{Required } C_v = \frac{V_u/\phi_v}{0.6F_yA_w} = \frac{369/1.0}{0.6(36)31.25} = 0.547 \qquad [11.10.5]$$

The coefficient C_v is given by Eq. 11.7.10 for $C_v < 0.8$,

$$C_v = \frac{43,800k_v}{(h/t_w)^2F_y} = \frac{43,800k_v}{(320)^236} = \frac{k_v}{84.2}$$

$$\text{Required } k_v = 0.547(84.2) = 46.1$$

$$k_v = 46.1 = 5 + \frac{5}{(a/h)^2}$$

$$\text{Max} \frac{a}{h} = 0.35$$

$$\text{Max } a = 0.35(100) = 35 \text{ in.}$$

Use 2'-9".

Frequently, this maximum *a/h* can be obtained by using *AISC Manual* Tables 3-16a, which contains $\phi_v V_n/A_w$ values as a function of h/t_w and *a/h*. In this case,

$$\frac{\text{Required } \phi_v V_n}{A_w} = \frac{V_u}{A_w} = \frac{369}{31.25} = 11.8 \text{ ksi}$$

Then enter *AISC Manual* Tables 3-16 with $h/t_w = 320$, look for 11.8 ksi, and determine *a/h*. In this case, follow the curve corresponding to 12 ksi and follow the curve down to the *x*-axis, and the corresponding *a/h* values that can be read is 0.35, which agrees with the above calculation.

Panel 2: Assume the shear envelope is linear (a conservative assumption) over the 25 ft to the bearing stiffener under the concentrated load. Thus, at 2.75 ft from the support,

$$V_{u2} = 369 - \frac{369 - 228}{25}(2.75) = 354 \text{ kips}$$

Equation 11.8.27 which includes tension-field action is applicable in all panels of this girder, except the exterior panel treated above. AISC Tables 3-16b, entering with $h/t_w = 320$ and

$$\frac{\text{Required } \phi_v V_n}{A_w} = \frac{V_u}{A_w} = \frac{354}{31.25} = 11.3 \text{ ksi}$$

The curve corresponding to 12 ksi, is governed by a maximum limit on *a/h* for h/t_w values below 228. This is indicated by the 12 ksi- curve dotted portion. Following the solid line leads to a value of 0.66 for *a/h*. This means that the upper limit according to Eq. 11.10.10 controls,

$$\frac{a}{h} \leq \left(\frac{260}{h/t_w}\right)^2 = \left(\frac{260}{320}\right)^2 = 0.66 \leq 3.0$$

$$a = 0.66(100) = 66 \text{ in. } (5.5 \text{ ft})$$

Since lateral support occurs every 25 ft and a bearing stiffener is required at concentrated loads, the stiffener spacing is usually arranged to fit these limitations. Considering the first stiffener at 2'-9", 22'-3" remains. Four spaces at the maximum of 5.5 ft will be less than 22'-3". Thus, five spaces must be used. It is preferable to use 3-in. multiples for stiffener spacing.

Use stiffener spaces as follows starting from the exterior support: 2@2'-6", 4@5'-0".

For the region from 25 to 50 ft from the end, the *a/h* limitation of 0.66 still governs. *Use* 5 spaces @5'-0".

For the region from 50 to 75 ft, the maximum shear stress is still below the value in panel 2, so that *a/h* maximum still governs. *Use* 5 spaces @5'-0".

(h)   *Intermediate Stiffeners—Placement in Negative Moment Zone.*   Web plate = $\frac{3}{8}$ × 100, A_w = 37.5 sq in., F_y = 50 ksi.

Interior end:

$$V_u = 549 \text{ kips}$$

Use Eqs. 11.8.27 and 11.10.7 (AISC-G3) for this interior panel (there is an adjacent panel in the other span):

$$\text{Required } \frac{\phi_v V_n}{A_w} = \frac{V_u}{A_w} = \frac{549}{37.5} = 14.6 \text{ ksi} \qquad \frac{h}{t_w} = \frac{100}{0.375} = 267$$

Using *AISC Manual* Tables 3-16a, again find the maximum *a/h* is controlled by the arbitrary limit (see also Fig. 11.8.9).

$$\text{Max} \frac{a}{h} = \left(\frac{260}{h/t_w}\right)^2 = \left(\frac{260}{267}\right)^2 = 0.95$$

This panel adjacent to the interior support on a continuous span has high shear and bending moment at the same location. In previous AISC Specifications, moment-shear interaction was considered in determining the stiffener spacings. The 2005 AISC Specification considers the effect negligible. According to the AISC Commentary-G2, "Consideration of the effect of bending on the shear strength is not required because the effect is deemed negligible." It is the authors' opinion that the selected spacing of stiffeners in zones of high shear and moment should be conservatively closer than permitted by the *AISC Specification. Use* 4 spaces @ 5 ft.

(i)   *Location of Flange and Web Splices.*   The location of splices is partially dependent on the type of splices used; for a bolted field splice joining the A36 with the A572 sections, both flanges and web usually would be spliced at the same location; for a field welded splice it might be preferable to splice the flanges at a location offset from the web splice by as much as 10 ft. Such an offset reduces stress concentration and probably assists in getting proper alignment at the splice. For this example assume splices for flanges and web are to occur at the same location.

From previous calculations for stiffener spacing it is apparent that splicing could be done at, say, 26 ft from the interior support, just beyond the bearing stiffener. Perhaps, however, it can be made closer to the interior support.

Try 5'-6" from the concentrated load (19'-6" from interior support) which would be the maximum spacing based on the maximum a/h = 0.66 for the $\frac{5}{16}$-in. web. See part (g).

Examine a panel of $\frac{5}{16}$-in. web immediately to the right of the concentrated load that is 25 ft from the interior support. The maximum moment of about 3500 ft-kips occurs only when no live load is on span 1; and it is lower than $\phi_b M_n$ = 8430 ft-kips [from part(f)], making splicing at this location acceptable.

A summary showing stiffener spacing appears in Fig. 11.14.6.

(j)   *Size of Intermediate Stiffeners.*   Frequently, A36 stiffeners are suitable, with higher yield stress material offering little if any saving. Try A36 steel for all stiffeners.

Panel 2 from exterior support: This is the first panel in which tension-field action is presumed to occur.

Having reduced the spacing of the first stiffener to 2′-6″ from 2′-9″, the maximum shear force in the panel will be somewhat higher than used for the panel 2 computation in part (g). Thus,

$$V_u = 355 \text{ kips}$$
$$V_u/A_w = \text{Required } \phi_v V_n/A_w = 355/31.25 = 11.4 \text{ ksi}$$
$$a/h = 30/100 = 0.30$$
$$h/t_w = 100/0.3125 = 320$$

Entering *AISC Manual* Table 3-16 with the above values, and using AISC Formula (G3-3), one gets

$$\phi_v V_n/A_w \approx 15.0 \text{ ksi}$$

$$A_{st} > \frac{F_y}{F_{yst}}\left[0.15 D_s h t_w (1 - C_v)\frac{V_u}{\phi V_n} - 18t_w^2\right]$$

$$A_{st} = \frac{36}{36}\left[0.15(1)(100)(0.3125)(1-0.544)\frac{355}{469} - 18(0.3125)^2\right] = \text{negative number}$$

If the web contribution is disregarded,

$$A_{st} = 1.62 \text{ sq in.}$$

The requirement may be reduced in proportion that the shear strength is underutilized,

$$\frac{V_u/A_w}{\phi_v V_n/A_w} = \frac{11.4}{15.0} = 0.76$$

The stiffener area A_{st} required is

$$\text{Required } A_{st} = 0.76(1.62) = 1.23 \text{ sq in.}$$

This area requirement assumes stiffeners are to be used in pairs. Furthermore, local buckling of this unstiffened element must be precluded; i.e., $\lambda \leq \lambda_r$, according to AISC-B4. The width b and thickness t, as shown in Fig. 11.14.3, must satisfy

$$\left(\lambda = \frac{b}{t}\right) \leq \left(\lambda_r = 0.56\sqrt{\frac{E}{F_y}} = \frac{95}{\sqrt{F_y, \text{ksi}}} = \frac{95}{\sqrt{36}} = 15.8\right)$$

The requirement for stiffness, Eq. 11.10.8 gives

$$I_{st} \geq ja t_w^3 \qquad [11.10.8]$$

where

$$j = \frac{2.5}{(a/h)^2} - 2 = \frac{2.5}{(30/100)^2} - 2 = 25.8$$

$$I_{st} \geq 25.8(30)(0.3125)^3 = 23.6 \text{ in.}^4$$

Figure 11.14.3
Cross-section of intermediate stiffener plates.

To find minimum acceptable stiffener width (Fig. 11.14.3),

$$\text{Required } r^2 = \frac{I_{st}}{A_{st}} = \frac{23.6}{1.23} = 19.2 \text{ in.}^2$$

$$\text{Provided } r^2 = \frac{tW^3}{12tW} = \frac{W^2}{12}$$

$$\text{Required } W = \sqrt{12(19.2)} = 15.2 \text{ in.}$$

This would indicate 8-in. wide plates, and to satisfy $b/t \leq 15.8$ the thickness would be about 0.5 in. These would be inordinately large and A_{st} would be about 8 sq in. Try pairs of plates $\frac{3}{8} \times 5$, giving $b/t = 13.3$ which is less than λ_r.

Check the moment of inertia I_{st},

$$I_{st} = \frac{tW^3}{12} = \frac{0.375(10.3125)^3}{12} = 34.3 \text{ in.}^4$$

Thus, the provided I_{st} exceeds the required 23.5 in.[4] If the b/t limit of 15.8 can be accepted as 16 (i.e., 1% high), plate $\frac{5}{16} \times 5$ will be acceptable; the moment of inertia I_{st} for those plates is 28.6 in.[4] The authors prefer the $\frac{5}{16} \times 5$ plates.

Since the stiffeners for the panel are about the minimum size based on local buckling provisions, the reduced area required for the more interior panels on the $\frac{5}{16}$-in. web is of little consequence.

Use 2PLs—$\frac{5}{16} \times 5$ for all intermediate stiffeners attached to the $\frac{5}{16}$-in. web.

Examine the end panel adjacent to the interior support. Because of the shear—flexure interaction requirement the design shear strength $\phi_v V_n$ required is:

$$\text{Required } \phi_v V_n = 549 \text{ kips } [\text{from part (h)}]$$

$$\text{Required } \phi_v V_n / A_w = 549/37.5 = 14.6 \text{ ksi}$$

$$a/h = 60/100 = 0.6$$

$$h/t_w = 100/0.375 = 267$$

As in the calculation to determine the stiffener spacing, at $a/h = 0.60$ the panel uses less than 100% of the shear strength. Entering *AISC Manual* Table 3-17 with the above values, find Required $\phi V_n / A_w = 24.2$ ksi. Check by calculating Required A_{st} from Eq. 11.8.29,

$$k_v = 5 + \frac{5}{(0.6)^2} = 18.9$$

$$\left(\frac{h}{t_w} = 267 \right) > \left(1.37 \sqrt{\frac{k_v E}{F_y}} = \frac{233}{\sqrt{F_{y,\,ksi}/k_v}} = \frac{233}{\sqrt{50/18.9}} = 143 \right)$$

AISC-Formula (G2-5) applies,

$$C_v = \frac{1.51 E k_v}{(h/t_w)^2 F_y} = \frac{43,800 k_y}{(h/t_w)^2 F_{y,\,ksi}} = \frac{43,800(18.9)}{(267)^2 50} = 0.232$$

$$V_n = 0.6F_y A_w \left(C_v + \frac{1 - C_v}{1.15\sqrt{1 + (a/h)^2}} \right)$$

$$= 0.6(50)37.5 \left(0.232 + \frac{1 - 0.231}{1.15\sqrt{1 + (0.60)^2}} \right) = 905 \text{ kips}$$

$$\phi_v V_n = 1.0(905) = 905 \text{ kips}$$

$$\text{Required } A_{st} = \frac{F_y}{F_{yst}} \left[0.15DA_w(1 - C_v)\frac{V_u}{\phi_v V_n} - 18t_w^2 \right] \qquad [11.8.29]$$

$$= \frac{50}{36} \left[0.15(1.0)37.5(1 - 0.232)\frac{549}{905} - 18(0.375)^2 \right] = 0.12 \text{ sq in.}$$

This value for the required stiffener area is small and will not control. Thus stiffness controls rather than strength. Use A36 stiffeners with limit $\lambda_r = 15.8$. Try $\frac{5}{16} \times 5$ plates in pairs.

Check required I_{st} using stiffeners spaced at 5 ft (i.e., $a = 60$ in.),

$$I_{st} \geq jat_w^3 = 4.94(60)(0.375)^3 = 15.7 \text{ in.}^4$$

where

$$j = \frac{2.5}{(a/h)^2} - 2 = \frac{2.5}{(60/100)^2} - 2 = 4.94$$

$$I_{st} = \frac{tW^3}{12} = \frac{0.3125(10.375)^3}{12} = 29.0 \text{ in.}^4 > 15.7 \text{ in.}^4 \qquad \textbf{OK}$$

The $\frac{5}{16} \times 5$ plates are satisfactory for use on the $\frac{3}{8}$-in. web of Grade 50 in the negative moment zone.

Use 2 PLs—$\frac{5}{16} \times 5$ of A36 steel for all intermediate stiffeners connected to the $\frac{3}{8}$ in. web.

(k) *Connections of Intermediate Stiffeners to Web.* For the $\frac{5}{16}$-in. A36 web, using Eq. 11.10.14,

$$f_{nv} = 0.045h\sqrt{\frac{F_y^3}{E}} \qquad [11.10.14]$$

Even though there is no specific force requirement for connecting intermediate stiffeners except the tension-field force P_s used for the A_{st} requirement, the use of Eq. 11.10.14 as recommended by Basler [11.3] seems appropriate.

$$f_{nv} = 0.045(100)\sqrt{\frac{(36)^3}{29,000}} = 5.7 \text{ kips/in.}$$

The stiffener spacing at the first panel at the exterior support was finally made 2'-6" instead of the 2'-9" permitted. Thus the f_{nv} value could logically be reduced in proportion that the shear is underutilized. Using Eq. 11.7.10 for $h/t_w > \left[1.1\sqrt{k_v E /F_y} = 243 \right]$ with $a/h = 30/100 = 0.30$, $h/t_w = 320$, and $k_v = 60.6$ from Eq. 11.7.8 gives

$$C_v = \frac{43,800k_v}{(h/t_w)^2 F_y, \text{ksi}} = \frac{43,800(60.6)}{(320)^2(36)} = 0.72$$

$$V_n = C_v(0.6F_{yw})A_w = 0.72(0.6)(36)(31.25) = 486 \text{ kips}$$

$$\frac{V_u}{\phi V_n} = \frac{369}{1.0(486)} = 0.76$$

Required $f_{nv} = 5.7(0.76) = 4.3$ kips/in.

Min weld size $a = \frac{3}{16}$ in. (AISC-Table J2.4)

Determine maximum effective weld size (AISC-J2.4 using shear strength on throat of fillet with E70 electrodes) (see also text, Sec. 5.14), Eq. 5.14.9,

$$a_{\text{max eff}} = 0.707\frac{F_u t_1}{F_{\text{EXX}}} = 0.707\frac{58(5/16)}{70} = 0.183 \text{ in.}$$

Try $\frac{3}{16}$-in. weld of E70 electrodes. The nominal weld strength R_{nw} is

$$R_{nw} = a(0.707)(0.6F_{\text{EXX}})$$
$$= 0.183(0.707)(0.6)70 = 5.44 \text{ kips/in.}$$

Since four lines of fillet welds (Fig. 11.14.3) are to provide 4.8 kips/in., then 4.8/4 = 1.2 kips/in. are required along each line.

$$\% \text{ of continuous } \tfrac{3}{16}\text{-in. weld required} = \frac{1.2}{5.44}(100) = 22\%$$

For intermittent welding, minimum segment is 1.5 in. (or 4 times the weld size, if longer), according to AISC-J2.2b; the nominal strength of this segment is

$$L_w R_{nw} = 1.5(5.44) = 8.2 \text{ kips}$$

$$\text{Max pitch} = \frac{8.2}{1.2} = 6.8 \text{ in.}$$

Use $\frac{3}{16}$ in.—$1\frac{1}{2}$ in. segments @$6\frac{1}{2}''$ pitch, E70 electrodes, for connecting $\frac{5}{16} \times 5$ plates to $\frac{5}{16}$ in. web.

For this $\frac{3}{8}$-in. A572 Grade 50 web:

$$\text{Required } f_{nv} = 0.045(100)\sqrt{\frac{(50)^3}{29,000}} = 9.3 \text{ kips/in.}$$

where the yield stress of the web is used in f_{nv}.

At the interior support, the actual shear strength equals approximately the required strength; thus no reduction in f_{nv} is appropriate.

$$\text{Min weld size } a = \tfrac{3}{16} \text{ in.}$$

$$a_{\text{max eff}} = 0.707\frac{F_u t_1}{F_{\text{EXX}}} = 0.707\frac{58(3/8)}{70} = 0.183 \text{ in.}$$

based on the A36 stiffeners. Use $\frac{3}{16}$-in. fillet weld with E70 electrodes,

$$R_n = 0.183(0.707)(0.6)70 = 5.44 \text{ kips/in.}$$

For $1\frac{1}{2}$-in. segments, the maximum pitch is

$$\text{Max pitch} = \frac{1.5(5.44)}{9.3/4} = 3.5 \text{ in.}$$

which means nearly continuous $\frac{3}{16}$-in. weld.

Use $\frac{3}{16}$ *in. continuous weld, E70 electrodes, for connecting* $\frac{5}{16} \times 5$ *plates* (A36) *to* $\frac{3}{8}$ *in. web* ($F_y = 50$ *ksi*).

(l) *Flange to Web Connection—A36 steel region.* The welding of the flanges to the web must provide for the factored horizontal shear flow $V_u Q / I_x$ at the joint. Though this requirement does not explicitly occur in the *AISC Specification*, it is certainly implied in a provision such as AISC-E6.2, 1st Par. for built-up members where connection is required to be "... adequate to provide for the transfer of the required forces." The factored load shear flow is

$$\text{Shear flow} = \frac{V_u Q}{I_x} \text{ kips/in.}$$

where V_u = factored shear at section

Q = statical moment of flange area about *x*-axis

I_x = moment of inertia of section about *x*-axis

Welding along both sides of the web provides a shear flow strength, which, if it exceeds the $V_u Q / I_x$ requirement, may be reduced by the use of intermittent welding. Normally flange to web welding is made continuous, primarily because automatic fabricating procedures usually make it more economical to do so. However, for design purposes the minimum percent of continuous weld acceptable in each panel between stiffeners will be prescribed. If it is more economical for the fabricator to use more weld, it is permissible to do so. The following calculations conservatively assume shielded metal arc welding (SMAW) is to be used.

$$\text{Min weld size } a = \tfrac{5}{16} \text{ in.} \quad (\text{AISC-Table J2.4})$$

$$a_{\text{max eff}} = 0.707 \frac{F_u t_1}{F_{\text{EXX}}} = 0.707 \frac{58(5/16)}{70} = 0.183 \text{ in.}$$

for E70 electrodes. The shear flow $V_u Q / I_x$ is

$$Q = A_f \left(\frac{h}{2} + \frac{t_f}{2} \right) = 24(1.125)(50.65) = 1370 \text{ in.}^3$$

$$\frac{V_u Q}{I_x} = \frac{369(1370)}{164,000} = 3.1 \text{ kips/in.}$$

Equating the strength of two fillets to $V_u Q / I_x$ gives

$$2a(0.707)(0.6)70 = 3.1$$

$$\text{Required } a = 0.05 \text{ in.} < 0.183 \text{ in.}$$

Use $\frac{5}{16}$ *in. weld, E70 electrodes* (effective size = 0.183 in.). The design strength ϕR_{nw} for continuous weld on both sides of the web is

$$\phi R_{nw} = 0.75(2)(0.707)(0.183)(0.6)70 = 8.2 \text{ kips/in.}$$

$$\text{Min \% continuous weld} = \frac{3.1}{8.2}(100) = 38\%$$

The maximum spacing of weld segments, according to AISC-E6.2, 3rd par., is

$$\text{Spacing} \le \left(0.75t \sqrt{\frac{E}{F_y}} = 0.75 \left(\frac{5}{16} \right) \sqrt{\frac{29,000}{36}} = 6.6 \text{ in.} \right) \le 12 \text{ in.}$$

(m)   *Flange to Web Connection—A572 steel region.*

$$\text{Min weld size } a = \tfrac{5}{16} \text{ in.}\quad \text{(AISC-Table J2.4)}$$

$$a_{\max \text{ eff}} = \frac{F_u t_1}{2(0.707)F_{\text{EXX}}} = 0.707\frac{65(3/8)}{70} = 0.246 \text{ in.}$$

The maximum shear flow strength requirement is

$$\frac{V_u Q}{I_x} = \frac{549(1390)}{172{,}000} = 4.4 \text{ kips/in.}$$

where

$$Q = A_f(h/2 + t_f/2) = 22(1.25)(50.625) = 1390 \text{ in.}^3$$

The design strength ϕR_{nw} for continuous $\tfrac{5}{16}$-in. weld on both sides of the web is

$$\phi R_{nw} = 0.75(2)(0.707)(0.246)(0.6)70 = 11.0 \text{ kips/in.}$$

$$\text{Min \% continuous weld} = \frac{4.4}{11.0}(100) = 40\%$$

The maximum spacing of weld segments, according to AISC-6.2 is

$$\text{Spacing} \le \left(0.75t\sqrt{\frac{E}{F_y}} = 0.75\left(\frac{3}{8}\right)\sqrt{\frac{29{,}000}{50}} = 6.7 \text{ in.}\right) \le 12 \text{ in.}$$

Use $\tfrac{5}{16}$ in. weld, E70 electrodes. The minimum percentage of continuous weld required for each panel along the $\tfrac{3}{8}$-in. A572 Grade 50 web is summarized in Fig. 11.14.6. AISC-J2b requires at least 22% of continuous weld (that is, $1\tfrac{1}{2}$-in. segments @ 6.7 in. maximum pitch)

(n)   *Design of Bearing Stiffeners—Interior Support.*   As discussed in Sec. 11.11, bearing stiffeners generally are required on plate girders at all concentrated loads and reactions. Specifically, AISC-J10.7 requires a pair of transverse stiffeners extending the full depth of the web "at unframed ends of beams and girders not otherwise restrained against rotation about their longitudinal axes." On rolled beams it may be possible to satisfy the requirements of AISC-J10.2 to prevent yielding, AISC-J10.3 to prevent web crippling, and AISC-J10.4 to prevent sidesway web buckling; however, satisfying AISC-J10.2, J10.3, and J10.4 on a plate girder would be rare. AISC-J10 provides the specific requirements for bearing stiffeners.

The interior factored reaction is

$$\text{Reaction} = 2(549) = 1098 \text{ kips}$$

Bearing stiffeners should extend approximately to the edge of the flange plate even though not required to do so by AISC. The width b should then be

$$b = \frac{22 - 0.375}{2} = 10.8 \text{ in.}\quad \text{(say 10 in.)}$$

Column strength criterion according to AISC-J4.4. (Fig. 11.14.4 for cross-section):

$$r \approx 0.25(20.375) = 5.1 \text{ in.}$$

$$\frac{KL}{r} \approx \frac{0.75(100)}{5.1} = 14.7$$

Figure 11.14.4
Cross-section (trial) of
bearing stiffener at interior
support.

Using AISC "Available Critical Stress for Compression Members" Table 4-22, find

$$\phi_c\, F_{cr} = 32.0 \text{ ksi} \quad \text{(for A36 steel)}$$

$$\text{Required } A_{st} = \frac{1098}{32.0} = 34.3 \text{ sq in.}$$

$$\text{Required } t = \frac{34.3 - 9.37(0.375)}{2(10)} = 1.54 \text{ in.}$$

Local buckling criterion:

$$\left(\lambda = \frac{b}{t}\right) \le \left(\lambda_t = 0.56\sqrt{\frac{E}{F_y}} = \frac{95}{\sqrt{F_y,\text{ksi}}} = \frac{95}{\sqrt{36}} = 15.9\right)$$

$$\text{Required } t = \frac{10}{15.9} = 0.63 \text{ in.}$$

Note that the proportioning requirements of AISC-J10.8 that refer to the "flange" are primarily relating to a rigid beam-to-column connection of the type treated in Sec. 13.6; the term "flange" in this context does *not* mean the plate girder flange.

Bearing criterion (AISC-J7) using the contact area A_{pb}:

$$\text{Reqd } A_{pb} = \frac{1098}{\phi(1.8)F_y} = \frac{1098}{0.75(1.8)36} = 22.6 \text{ sq in.}$$

$$A_{pb} = 2(10 - 0.5)t$$
$$\nearrow$$
$$\text{est. to allow for fillet weld}$$

$$\text{Required } t = \frac{22.6}{2(9.5)} = 1.19 \text{ in.}$$

Column strength criterion governs. So as to properly distribute this reaction, try 4PLs—$\frac{7}{8}$ × 10 to serve as bearing stiffeners. (Alternative, 2PLs—$1\frac{3}{4}$ × 10.)

Recheck r for columns strength:

$$r = \sqrt{\frac{\frac{1}{12}(1.75)(20.375)^3}{20(1.75) + 9.375(0.375)}} = 5.7 \text{ in.} > \text{estimated 5.1 in.}$$

By inspection, column strength is adequate.

Use 4PLs—$\frac{7}{8}$ × 10 for $R = 1098$ kips (A36 steel), as shown in Fig. 11.14.5.

(o) *Connection for Bearing Stiffeners to Web.*

$$\text{Required } f_{nv} = \frac{1098}{8(100)} = 1.37 \text{ kips/in.}$$

Figure 11.14.6 Design sketch.

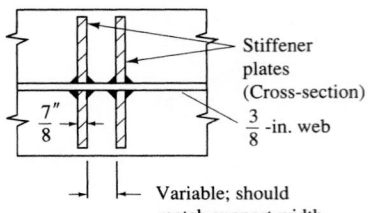

Figure 11.14.5
Cross-section (final) of
bearing stiffener at
interior support.

for eight lines of fillet weld (Fig. 11.14.5). Use $\frac{5}{16}$ in., with maximum effective size = 0.246 and E70 electrodes, as determined in part (m). The design strength ϕR_{nw} for one line of continuous weld is

$$\phi R_{nw} = 0.75(0.707)(0.246)(0.6)70 = 5.5 \text{ kips/in.}$$

$$\text{Min \% continuous weld} = \frac{1.37}{5.5}(100) = 25\%$$

(p) *Bearing Stiffeners at Concentrated Loads 25 ft from Supports.* At these locations, check whether or not the intermediate stiffeners will be adequate to serve as bearing stiffeners. The factored loads W_u are

$$W_u = 1.2(15) + 1.6(60) = 114 \text{ kips}$$

$$r = \sqrt{\frac{\frac{1}{12}(5/16)(10.3125)^3}{10(0.3125) + 25(0.3125)(0.3125)}} = 2.27 \text{ in.}$$

$$\frac{KL}{r} = \frac{0.75(100)}{2.27} = 33$$

Using AISC "Available Critical Stress for Compression Members" Table 4-22 find

$$\phi_c F_{cr} = 30.6 \text{ ksi (for A36 steel)}$$

$$A_{st} = 10(5/16) + 25(5/16)(5/16) = 5.57 \text{ sq in.}$$

$$\phi_c P_n = 30.6(5.57) = 170 \text{ kips} > (W_u = 114 \text{ kips}) \qquad \textbf{OK}$$

The $\frac{5}{16} \times 5$ intermediate stiffeners are satisfactory for bearing stiffeners under the W_u concentrated loads.

(q) *Design Sketch.* Every design must have all decisions summarized on a design sketch, such as Fig. 11.14.6. The design of the bearing stiffeners for the exterior support has been omitted because the procedure has been adequately illustrated in parts (o) and (p). The girder weight, multiplying a plate cross-sectional area by 3.4 to get lb/ft, is (for one span):

A572 plates:	$1\frac{1}{4} \times 22 \times 20'$	= 93.5(40)	= 3,740
	$\frac{3}{8} \times 100 \times 20'$	= 127.5(20)	= 2,550
A36 plates:	$1\frac{1}{4} \times 24 \times 80'$	= 102.1(160)	= 16,328
	$\frac{5}{16} \times 100 \times 80'$	= 106.2(80)	= 8,500
Stiffeners:	$\frac{5}{16} \times 5 \times 8.33 \times 38$	= 5.31(8.33)38	= 1,682
	$\frac{7}{8} \times 10 \times 8.33 \times 4$	= 29.75(8.33)4	= 991
		Total	= 33,791 lb

Average weight = 338 lb/ft.

SELECTED REFERENCES

11.1. Thomas C. Shedd, *Structural Design in Steel.* New York: John Wiley & Sons, Inc., 1934, Chap. 3.

11.2. Edwin H. Gaylord, Jr. and Charles N. Gaylord, *Design of Steel Structures.* New York: McGraw-Hill Book Company, Inc., 1957, Chap. 8.

11.3. Konrad Basler, "Strength of Plate Girders in Shear," *Transactions*, ASCE, **128,** Part II (1963), 683–719. (Also as Paper No. 2967, *Journal of the Structural Division*, ASCE, October 1961.)

11.4. I. Lyse and H. J. Godfrey. "Investigation of Web Buckling in Steel Beams," *Transactions*, ASCE, **100** (1935), 675–706.

11.5. Konrad Basler. "Strength of Plate Girders in Bending." *Transactions*, ASCE, **128,** Part II (1963), 655–686. (Also as Paper No. 2913, *Journal of the Structural Division*, ASCE, August 1961.)

11.6. Ronald W. Frost and Charles G. Schilling. "Behavior of Hybrid Beams Subjected to Static loads," *Journal of the Structural Division*, ASCE, **90,** ST3 (June 1964), 55–88.

11.7. Charles G. Schilling. "Web Crippling Tests on Hybrid Girders," *Journal of the Structural Division*, ASCE, **93,** ST1 (February 1967), 59–70.

11.8. Phillip S. Carskaddan. "Shear Buckling of Unstiffened Hybrid Beams," *Journal of the Structural Division*, ASCE, **94,** ST8 (August 1968), 1965–1990.

11.9. A. Anthony Toprac and Murugesam Natarajan, "Fatigue Strength of Hybrid Plate Girders," *Journal of the Structural Division*, ASCE, **97,** ST4 (April 1971), 1203–1225.

11.10. C. G. Schilling, Chairman. "Design of Hybrid Steel beams," Report of Sub-committee 1 on Hybrid Beams and Girders, Joint ASCE-AASHO Committee on Flexural Members, *Journal of the Structural Division*, ASCE, **94,** ST6 (June 1968), 1397–1426.

11.11. Peter B. Cooper, Theodore V. Galambos, and Mayasandra K. Ravindra. "LRFD Criteria for Plate Girders," *Journal of the Structural Division*, ASCE, **104,** ST9 (September 1978), 1389–1407.

11.12. John L. Dawe and Geoffery L. Kulak. "Local Buckling of W Shape Columns and Beams," *Journal of Structural Engineering*, **110,** 6 (June 1984), 1292–1304.

11.13. Cynthia J. Zahn. "Plate Girder Design Using LRFD," *Engineering Journal*, AISC, **24,** 1 (First Quarter 1987), 11–20.

11.14. Konrad Basler. "New Provisions for Plate Girder Design," *Proceedings, AISC National Engineering Conference.* Chicago, IL: American Institute of Steel Construction, 1961, 65–74.

11.15. AISI. "Proposed Criteria for Load and Resistance Factor Design of Steel Building Structures," *Bulletin No. 27.* Washington, DC: American Iron and Steel Institute, January 1978.

11.16. George S. Vincent. "Tentative Criteria for Load Factor Design of Steel Bridges," *Bulletin No. 15.* Washington, DC: American Iron and Steel Institute, March 1969.

11.17. Konrad Basler. "Strength of Plate Girders Under Combined Bending and Shear," *Journal of the Structural Division*, ASCE, **87,** ST7 (October 1961), 181–197. See also *Transactions*, ASCE, **128** (1963), Part II, 720–735.

11.18. Peter B. Cooper. "Strength of Longitudinally Stiffened Plate Girders," *Journal of the Structural Division*, ASCE, **93,** ST2 (April 1967), 419–451.

11.19. M. A. D'Apice, D. J. Fielding, and P. B. Cooper. "Static Tests on Longitudinally Stiffened Plate Girders," *Welding Research Council Bulletin No. 117*, October 1966. Also, *Strength of Plate Girders with Longitudinal Stiffeners, Bulletin No. 16*, American Iron and Steel Institute, April 1969, Paper No. III. (Includes historical survey and bibliography on longitudinally stiffened plates.)

11.20. Task Committee on Longitudinally Stiffened Plate Girders of the ASCE-AASHTO Committee on Flexural Members of the Committee on Metals of the Structural Division. "Theory and Design of Longitudinally Stiffened Plate Girders," *Journal of the Structural Division*, ASCE, **104,** ST4 (April 1978), 697–716.

11.21. Boris Bresler, T. Y. Lin, and John Scalzi. *Design of Steel Structures*, 2nd ed. New York: John Wiley & Sons, Inc., 1968, pp. 497–554.

11.22. Omer W. Blodgett. *Design of Welded Structures.* Cleveland, Ohio: James F. Lincoln Arc Welding Foundation, 1966.

11.23. Charles G. Schilling. "Optimum Proportions for I-shaped Beams," *Journal of the Structural Division*, ASCE, **100,** ST12 (December 1974), 2385–2401.

11.24. Abul K. Azad. "Continuous Steel I-Girders: Optimum Proportioning," *Journal of the Structural Division*, ASCE, **106,** ST7 (July 1980), 1543–1555.

11.25. Walter H. Fleischer. "Design and Optimization of Plate Girders and Weld-fabricated Beams for Building Construction," *Engineering Journal*, AISC, **22,** 1 (First Quarter 1985), 1–10.

11.26. Katherine E. Anderson and Ken P. Chong. "Least Cost Computer-aided Design of Steel Girders," *Engineering Journal*, AISC, **23,** 4 (Fourth Quarter 1986), 151–156.

PROBLEMS

All problems are to be done according to the AISC Load and Resistance Factor Design Method. All given loads are service loads unless otherwise indicated. A design sketch showing all decisions is required in all design problems.

Analysis

11.1. (This problem solution is used for Probs. 11.2 to 11.6.) A plate girder supported as shown must carry a dead load of 2 kips/ft and live load of 8 kips/ft, not including the girder weight. In addition, a concentrated dead load of 100 kips must be carried at the end of the cantilever. Compute and draw to scale for later use the factored moment and shear envelopes for this girder.

11.2. (Use results of Prob. 11.1.) For the given plate girder designed for the conditions of Prob. 11.1, compute and draw to scale the moment capacity diagram, $\phi_b M_n$ vs location along span. Compare capacity with moment envelope requirements from Prob. 11.1.

11.3. (Use results of Prob. 11.1.) For the given plate girder, compute and draw to scale the shear capacity diagram $\phi_v V_n$ vs location along span *based on location of intermediate stiffeners*. Compare shear strength with shear envelope requirements from Prob. 11.1.

11.4. (Use results of Prob. 11.1.) Compute and draw to scale the shear strength diagram $\phi_v V_n$ vs location along span, for the girder *based on the flange-to-web connection*.

11.5. (Use results of Prob. 11.1.) Check the adequacy of each of the bearing stiffeners, including connection to the web.

11.6. Given the data for the 50-ft simply supported span, having lateral support at the ends and at the concentrated loads located 18 ft from each end;

   (a) Investigate the acceptability of the 84-in. spacing for stiffener panel 4.

   (b) Investigate combined shear and bending at its most critical location in the girder.

   (c) Investigate the adequacy of the size of intermediate stiffeners.

   (d) Investigate the adequacy of the size of the bearing stiffener $\left(2\text{PLs}-\frac{1}{2} \times 8\right)$ at the support.

   (e) Specify the flange-to-web connection.

   (f) Specify the connection for intermediate stiffeners.

   (g) Specify the connection for the support bearing stiffener.

11.7. Using the given information concerning the portion of the plate girder, determine how close stiffener B must be to stiffener A in order for the design to be acceptable according to Load and Resistance Factor Design.

11.8. Using the given information concerning the portion of the plate girder, determine how close stiffener B must be to stiffener A in order for the design to be acceptable.

Problem 11.1

Bearing stiffeners

$2 \text{ Ls} - \frac{7}{8} \times 10$

A 36 steel

$\text{L} - 1\frac{1}{8} \times 22$

12'-6"

Flange plate - $1\frac{7}{8} \times 22$ (both flanges)

E70 $\frac{5}{16}$ ▷ 2 – 5

1	2	3 Web $\frac{7}{16} \times 78$	4	5	6	7	8
60	60	50	50	40	40	40	50

Percent of continuous weld-flange to web, 3/8 in. fillet (E70)

4'-0"

10 @ 6'-4"

75'-0"

c-c bearing

Bearing stiffener 6 Ls $- \frac{5}{8} \times 10$

weld $\frac{5}{16}$ (E70), 2" @ $6\frac{1}{2}$, each side, each plate

A

Web spice

$\text{L} - 1\frac{7}{8} \times 22$

10'-0"

19'-6"

$\text{L} - 1\frac{1}{8} \times 22$

Bearing stiffener

$2 \text{ Ls} \frac{5}{8} \times 10$

9	10 Web $\frac{7}{16} \times 78$	11	12	13	14	15 Web $\frac{1}{2} \times 78$	16	17
50	60	100	100	100	50	50	40	30

Panel number

$\frac{1}{4}$ ▷ 2 – 7 E70

4'-8" 3'-0" 6'-0" 3 @ 6'-4" = 19'-0"

Intermediate stiffeners $- \frac{5}{16} \times 5$ plates, in pairs.

Problems 11.2 through 11.5

Bearing stiffeners $2 \text{ Ls} - \frac{1}{2} \times 10$

0.8 k/ft DL, 1.9 k/ft LL

20 k DL
60 k DL

Sym. about ₵

$2 \text{ Ls} - \frac{5}{16} \times 4$

$2 \text{ Ls} - \frac{5}{16} \times 4$

Bearing stiffeners
$2 \text{ Ls} - \frac{1}{2} \times 8$

1	2	3	4

3'-0" 7'-6" 7'-6" 7'-0"

50'-0" simple span

Lateral support occurs at ends and at concentrated loads.

$I_x = 46,500$ in.[4]

A36 steel

$\text{L} - \frac{3}{4} \times 20$

$\frac{5}{16} \times 70$

$\text{L} - \frac{3}{4} \times 20$

Problem 11.6

$L - 2\frac{3}{8} \times 20$

$L - \frac{5}{16} \times 76$

$L - 2\frac{3}{8} \times 20$

Determine max value

Given: Steel, $F_y = 50$ ksi
$h/t = 243$
$I = 157{,}400$ in.^{4}
$R_{pg}F_{cr} = 48$ ksi according to AISC-F5

12,800 ft-kips

Factored moment M_u

7000 ft-kips

10'-0"

Bending moment

620 k

550 k

Factored shear V_u

Shear

Problem 11.7

$L - 2\frac{1}{4} \times 28$

$L - \frac{3}{8} \times 92$

$L - 2\frac{1}{4} \times 28$

Determine max. spacing

Given: Steel, $F_y = 50$ ksi
$h/t = 245$
$I = 303{,}300$ in.^{4}
$R_{pg}F_{cr} = 48$ ksi according to AISC-F5

20,400 ft.-kips

13,800 ft-kips

Factored moment M_u

10'-0"

Bending moment

Slope 6.5 k/ft

655 k

720 k

Factored shear V_u

Shear

Problem 11.8

11.9. Given the plate girder interior panel as shown, of A36 material.

(a) Determine the shear strength V_n (kips) of the given panel. What percent of the capacity represents elastic buckling strength and what percent comes from "tension-field action"?

(b) If $M_u/\phi M_n = 0.92$ at the extreme fiber of the web, what is the design shear strength ϕV_n (kips)?

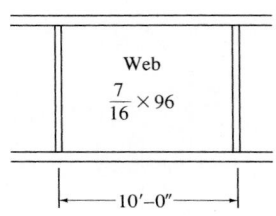

Web

$\frac{7}{16} \times 96$

10'-0"

Problem 11.9

11.10. The cross-section near the end of a simply supported girder consists of flange plates, $1\frac{1}{2} \times 18$, and a web plate, $\frac{3}{8} \times 90$ ($I = 135,780$ in.^{4}). The girder is of A36 steel and has lateral support of the compression flange every 14 ft.

(a) Determine the service moment capacity of this section if the loading is 70% live load and 30% dead load.

(b) Determine the maximum distance from the reaction to the first intermediate stiffener if the end reaction is 120 kips dead load and 200 kips live load.

(c) Using E70 electrodes, determine the necessary flange to web welding in the end panel. If intermittent welding can be used, indicate the length of weld segment and the pitch distance.

Design

The following requirements apply to all design problems:

(a) Consider live load to be applied as necessary to give maximum range of internal forces (moments and shears).

(b) Moment and shear envelopes are to be drawn to be scale.

(c) For *continuous girders*, use $F_y = 50$ ksi in the negative moment zone and A36 in the positive moment zone. For statically determinate girders use the steel specified in the problems.

(d) Consider $\frac{5}{16}$ in. as the minimum web thickness available.

(e) Follow the proportioning guidelines of Chapter 11, unless you clearly indicate that you are doing otherwise for a specific stated reason.

(f) Use A36 steel for stiffeners if possible.

(g) Specify intermittent welding for connections if any material saving results, even though a cost analysis may later dictate continuous welding.

(h) Submit design sketch to approximate scale on $8\frac{1}{2} \times 11$ paper showing all final decisions.

(i) Compute the total average weight per foot of the girder, including all stiffeners.

(j) For the *two-span continuous girder* designs, assume one splice is required in each span. Any extra butt-welded flange splice (two flanges)

should be treated as adding 10 lb/ft to the average weight. (This may approximate the added cost effect of such splices.) Web splices in excess of one required per span should be considered as adding 6 lb/ft to the average girder weight.

11.11. For the case indicated by the instructor, design a simply supported I-shaped plate girder cross-section and specify the location and size of intermediate stiffeners. Omit design of connections and bearing stiffeners. The cross-section is to be constant over the entire length. Use a depth as dictated by the optimum depth equations discussed in Sec. 11.13. The girder must carry a uniform dead load of 0.5 kip/ft in addition to the girder weight.

Case	Span (ft)	Live loading (kips/ft)	Yield stress (ksi)	Lateral support
1	60	3	36	Continuous
2	70	3	50	Every 17.5 ft
3	80	3	60	Every 20 ft
4	60	5	36	Every 20 ft
5	70	5	50	Every 17.5 ft
6	80	5	60	Every 20 ft
7	90	3	36	Every 15 ft

11.12. For the case assigned by the instructor, completely design a *two-span continuous* girder, including bearing stiffeners and all connections. In addition to the uniform loading, two concentrated loads W (60% dead load, 40% live load) are located at the distance a from each end in each span. Lateral support is provided at all vertical supports, at concentrated loads, and at the intervals indicated over the interior of the spans.

Case	w_D Dead load (kip/ft)	w_L Live load (kip/ft)	Span length (ft)	W (kips)	a (ft)	Lateral support
1	1.25	2.25	140	60	35	Every 35 ft
2	1.25	2.25	125	60	25	Every 25 ft
3	1.25	2.25	130	70	30	Every 25 ft
(for Case 3, only *one* concentrated load at 30 ft from exterior supports)						
4	1.00	2.50	100	100	20	Every 30 ft
5	0.80	3.00	100	80	20	Every 30 ft
6	0.80	3.00	125	0	—	Every 25 ft
7	0.80	4.00	125	0	—	Every 25 ft
8	1.25	2.25	200	0	—	Continuous
9	1.25	2.25	200	0	—	Every 40 ft

Review of Theory

11.13. Explain the physical significance of the following C_v values for shear in plate girders. Describe the limit state based on shear strength in each of the following categories: (a) $C_v \leq 0.8$; (b) $0.8 < C_v \leq 1.0$; (c) $C_v > 1.0$

11.14. What is the specific behavior that is prevented when h/t_w is kept below $970/\sqrt{F_y}$, ksi?

11.15. Show by a diagram of forces in equilibrium why stiffener spacing requirements are different for end panels and panels containing large holes than they are for interior panels of a plate girder.

11.16. Consider the AISC (AISC-F13.2) limitation on the clear distance between flanges.

   (a) Show the stress condition on the web plate that gives rise to the limitation equation.

   (b) If h/t_w equals the limit value, show the effective girder cross-section that might be used to compute moment capacity.

11.17. On a plate girder with $F_y = 50$ ksi, if the web $h/t_w = 185$, what will happen to the girder before the nominal moment strength is reached? Particularly describe what happens in a panel between intermediate stiffeners where the moment is high and the shear is low. Be specific and use a sketch.

12

Combined Bending and Axial Load

12.1 INTRODUCTION

Nearly all members in a structure are subjected to both bending moment and axial load—either tension or compression. When the magnitude of one or the other is relatively small, its effect is usually neglected and the member is designed as a beam, as an axially loaded column, or as a tension member. For many situations neither effect can properly be neglected and the behavior under the combined loading must be considered in design. The member subjected to axial compression and bending is referred to as a *beam-column*, and is the major element treated in this chapter. The general subject of strength and stability considerations and design procedures for beam-columns has been extensively treated by Massonnet [12.1] and Austin [12.2], and a recent summary is given in the *SSRC Guide* [6.8]. The history of steel beam-column design has been reviewed by Sputo [12.56].

Since bending is involved, all of the factors considered in Chapters 7 and 9 apply here also; particulary, the stability related factors, such as lateral-torsional buckling and local buckling of compression elements. When bending is combined with axial tension, the chance of instability is reduced and yielding usually governs the design. For bending in combination with axial compression, the possibility of instability is increased with all of the considerations of Chapter 6 applying. Furthermore, when axial compression is present, a secondary bending moment arises equal to the axial compression force times the deflection.

A number of categories of combined bending and axial load along with the likely mode of failure may be summarized as follows:

1. Axial tension and bending; failure usually by yielding.
2. Axial compression and bending about one axis; failure by instability in the plane of bending, without twisting. (Transversely loaded beam-columns that are stable with regard to lateral-torsional buckling are an example of this category.)
3. Axial compression and bending about the strong axis: failure by lateral-torsional buckling.
4. Axial compression and biaxial bending—torsionally stiff sections; failure by instability in one of the principal directions. (W shapes are usually in this category.)

5. Axial compression and biaxial bending—thin-walled open sections; failure by combined twisting and bending on these torsionally weak sections.

6. Axial compression, biaxial bending, and torsion; failure by combined twisting and bending when plane of bending does not contain the shear center.

Because of the number of failure modes, no simple design procedure is likely to account for such varied behavior. Design procedures generally are in one of three categories: (1) limitation on combined stress; (2) semi-empirical interaction formulas, based on working stress procedures, and (3) semi-empirical interaction procedures based on strength. Limitations on combined stress ordinarily cannot provide a proper criterion unless instability is prevented, or large safety factors are used. Interaction equations come closer to describing the true behavior since they account for the stability situations commonly encountered. The *AISC Specification* formulas for beam-columns are of the interaction type.

12.2 DIFFERENTIAL EQUATION FOR AXIAL COMPRESSION AND BENDING

In order to understand the behavior, the basic situation of case 2, Sec. 12.1, will be treated. Failure by instability in the plane of bending is assumed. Consider the general case of Fig. 12.2.1, where the lateral loading $w(z)$ in combination with any end moments, M_1 and M_2, constitute the primary bending moment M_i which is a function of z. The primary moment causes the member to deflect y giving rise to a secondary moment Py. Stating the moment M_z at the location z of Fig. 12.2.1, gives

$$M_z = M_i + Py = -EI\frac{d^2y}{dz^2} \tag{12.2.1}$$

for sections with constant EI, and dividing by EI gives

$$\frac{d^2y}{dz^2} + \frac{P}{EI}y = -\frac{M_i}{EI} \tag{12.2.2}$$

For design purposes, the general expression for moment M_z is of greater importance than the deflection y. Differentiating Eq. 12.2.2 twice gives

$$\frac{d^4y}{dz^4} + \frac{P}{EI}\frac{d^2y}{dz^2} = -\frac{1}{EI}\frac{d^2M_i}{dz^2} \tag{12.2.3}$$

From Eq. 12.2.1,

$$\frac{d^2y}{dz^2} = -\frac{M_z}{EI} \quad \text{and} \quad \frac{d^4y}{dz^4} = -\frac{1}{EI}\frac{d^2M_z}{dz^2}$$

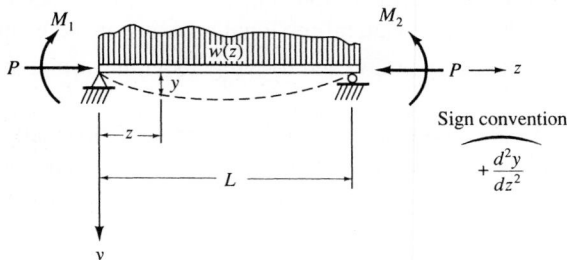

Figure 12.2.1
General loading of beam-column.

Substitution in Eq. 12.2.3 gives

$$-\frac{1}{EI}\frac{d^2M_z}{dz^2} + \frac{P}{EI}\left(\frac{-M_z}{EI}\right) = -\frac{1}{EI}\frac{d^2M_i}{dz^2}$$

or, simplifying and letting $k^2 = P/EI$,

$$\frac{d^2M_z}{dz^2} + k^2M_z = \frac{d^2M_i}{dz^2} \qquad (12.2.4)$$

which is of the same form as the deflection differential equation, Eq. 12.2.2.

The homogeneous solution for Eq. 12.2.4 is

$$M_z = A\sin kz + B\cos kz$$

as first discussed in Sec. 6.2. To this must be added the particular solution that will satisfy the right-hand side of the differential equation. Since $M_i = f(z)$, where $f(z)$ is usually a polynomial in z, the particular solution will be of the same form; thus the complete solution may be written

$$M_z = A\sin kz + B\cos kz + f_1(z) \qquad (12.2.5)$$

where $f_1(z)$ = value of M_z satisfying Eq. 12.2.4. When M_z is a continuous function, the maximum value of M_z may be found by differentiation:

$$\frac{dM_z}{dz} = 0 = Ak\cos kz - Bk\sin kz + \frac{df_1(z)}{dz} \qquad (12.2.6)$$

For most ordinary loading cases, such as concentrated loads, uniform loads, end moments, or combinations thereof, it may be shown that

$$\frac{df_1(z)}{dz} = 0$$

in which case a general expression for maximum M_z can be established; from Eq. 12.2.6,

$$Ak\cos kz = Bk\sin kz$$

$$\tan kz = \frac{A}{B}$$

$$(12.2.7)$$

At maximum M_z,

$$\sin kz = \frac{A}{\sqrt{A^2 + B^2}}, \qquad \cos kz = \frac{B}{\sqrt{A^2 + B^2}} \qquad (12.2.8)$$

and substitution of Eqs. 12.2.8 into Eq. 12.2.5 gives

$$M_{z\,max} = \frac{A^2}{\sqrt{A^2 + B^2}} + \frac{B^2}{\sqrt{A^2 + B^2}} + f_1(z)$$

$$= \sqrt{A^2 + B^2} + f_1(z) \qquad (12.2.9)$$

It is noted that whenever $df_1(z)/dz \neq 0$, Eq. 12.2.6 must be solved for kz and the result substituted into Eq. 12.2.5.

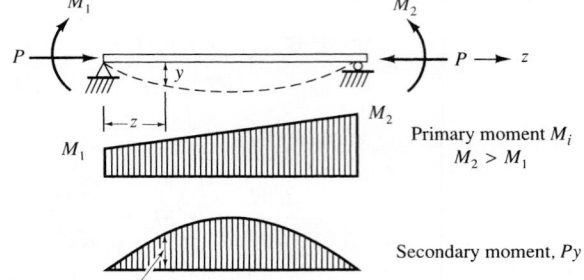

Figure 12.2.2
Case 1—end moments
without transverse loading.

Case 1—Unequal End Moments Without Transverse Loading

Referring to Fig. 12.2.2, the primary moment M_i may be expressed

$$M_i = M_1 + \frac{M_2 - M_1}{L} z \tag{12.2.10}$$

Since

$$\frac{d^2 M_i}{dz^2} = 0$$

Eq. 12.2.4 becomes a homogeneous equation, in which case $f_1(z) = 0$ for Eq. 12.2.5. The maximum moment, Eq. 12.2.9, is

$$M_{z\,max} = \sqrt{A^2 + B^2} \tag{12.2.11}$$

The constants A and B are evaluated by applying the boundary conditions to Eq. 12.2.5. The general equation is

$$M_z = A \sin kz + B \cos kz$$

and the conditions are

1. at $z = 0$, $M_z = M_1$

                     $\therefore B = M_1$

2. at $z = L$, $M_z = M_2$

                     $M_2 = A \sin kL + M_1 \cos kL$

$$\therefore A = \frac{M_2 - M_1 \cos kL}{\sin kL}$$

so that

$$M_z = \left(\frac{M_2 - M_1 \cos kL}{\sin kL} \right) \sin kz + M_1 \cos kz \tag{12.2.12}$$

and

$$M_{z\,max} = \sqrt{\left(\frac{M_2 - M_1 \cos kL}{\sin kL} \right)^2 + M_1^2}$$

$$= M_2 \sqrt{\frac{1 - 2(M_1/M_2)\cos kL + (M_1/M_2)^2}{\sin^2 kL}} \tag{12.2.13}$$

Case 2—Transverse Uniform Loading

Referring to Fig. 12.2.3, the primary moment M_i may be expressed as

$$M_i = \frac{w}{2} z(L - z) \tag{12.2.14}$$

Figure 12.2.3
Case 2—transverse uniform loading.

Since

$$\frac{d^2 M_i}{dz^2} = -w$$

$f_1(z) \neq 0$; the particular solution for the differential equation is required. Let $f_1(z) = C_1 + C_2 z$; i.e., any polynomial. Substitute the particular solution into Eq. 12.2.4,

$$\frac{d^2[f_1(z)]}{dz^2} = 0$$

$$0 + k^2(C_1 + C_2 z) = -w$$

Thus

$$C_1 = -w/k^2$$
$$C_2 = 0$$

Equation 12.2.5 then becomes

$$M_z = A \sin kz + B \cos kz - w/k^2 \tag{12.2.15}$$

Applying the boundary conditions,

1. at $z = 0$,

$$M_z = 0$$
$$0 = B - w/k^2$$
$$\therefore B = w/k^2$$

2. at $z = L$,

$$M_z = 0$$
$$0 = A \sin kL + \frac{w}{k^2} \cos kL - \frac{w}{k^2}$$
$$\therefore A = \frac{w}{k^2} \left(\frac{1 - \cos kL}{\sin kL} \right)$$

Since $df_1(z)/dz = 0$, Eq. 12.2.9 gives maximum moment,

$$M_{z\,max} = \frac{w}{k^2} \sqrt{\left(\frac{1 - \cos kL}{\sin kL} \right)^2 + 1} - \frac{w}{k^2}$$

$$M_{z\,max} = \frac{w}{k^2}\left(\sec\frac{kL}{2} - 1\right)$$

$$= \frac{wL^2}{8}\left(\frac{8}{(kL)^2}\right)\left(\sec\frac{kL}{2} - 1\right) \tag{12.2.16}$$

$$\underbrace{\phantom{\frac{wL^2}{8}\left(\frac{8}{(kL)^2}\right)\left(\sec\frac{kL}{2} - 1\right)}}_{\substack{\text{magnification factor due} \\ \text{to axial compression}}}$$

Case 3—Equal End Moments Without Transverse Loading (Secant Formula)

Consider that $M_1 = M_2 = M$ in Fig. 12.2.2, in which case Eq. 12.2.13 becomes

$$M_{z\,max} = M\sqrt{\frac{2(1 - \cos kL)}{\sin^2 kL}} \tag{12.2.17}$$

$$= M\sqrt{\frac{2(1 - \cos kL)}{1 - \cos^2 kL}} = M\left(\frac{1}{\cos kL/2}\right)$$

$$= M \sec\frac{kL}{2} \tag{12.2.18}$$

Loading a beam-column with constant moment over its length causes a constant compression along one entire flange and constitutes the most severe loading on such a member. In view of this it would always be conservative to multiply the maximum primary moment due to *any* loading by sec $kL/2$, which is excessively conservative in most cases.

12.3 MOMENT MAGNIFICATION—SIMPLIFIED TREATMENT FOR MEMBERS IN SINGLE CURVATURE WITHOUT END TRANSLATION

As an alternate to the differential equation approach, a simple approximate procedure is satisfactory for many common situations.

Assume a beam-column is subject to lateral loading $w(z)$ that causes a deflection δ_0 at midspan, as shown in Fig. 12.3.1. The secondary bending moment may be assumed to

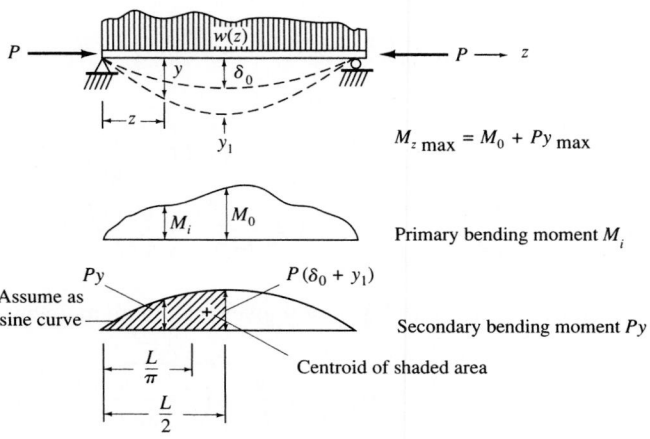

Figure 12.3.1
Primary and secondary
bending moment.

vary as a sine curve, which is nearly correct for members with no end restraint whose primary bending moment and deflection are maximum at midspan.

The portion of the midspan deflection y_1, due to the secondary bending moment, equals the moment of the M/EI diagram between the support and midspan (shaded portion of Fig. 12.3.1) taken about the support, according to the moment-area principle:

$$y_1 = \frac{P}{EI}(y_1 + \delta_0)\left(\frac{L}{2}\right)\frac{2}{\pi}\left(\frac{L}{\pi}\right) = (y_1 + \delta_0)\frac{PL^2}{\pi^2 EI} \tag{12.3.1}$$

or

$$y_1 = (y_1 + \delta_0)\frac{P}{P_e} \tag{12.3.2}$$

where $P_e = \pi^2 EI/L^2$. Solving for y_1,

$$y_1 = \delta_0\left[\frac{P/P_e}{1 - P/P_e}\right] = \delta_0\left(\frac{\alpha}{1 - \alpha}\right) \tag{12.3.3}$$

where $\alpha = P/P_e$. Since y_{max} is the sum of δ_0 and y_1,

$$y_{max} = \delta_0 + y_1 = \delta_0 + \delta_0\left(\frac{\alpha}{1 - \alpha}\right) = \frac{\delta_0}{1 - \alpha} \tag{12.3.4}$$

The maximum bending moment including the axial effect becomes

$$M_{z\,max} = M_0 + Py_{max} \tag{12.3.5}$$

Substituting the expression for y_{max} in Eq. 12.3.5 and setting $P = \alpha P_e = \alpha \pi^2 EI/L^2$, Eq. 12.3.5 may be written as the primary moment M_0 multiplied by a magnification factor B_1; thus

$$M_{z\,max} = M_0 B_1 \tag{12.3.6}$$

where

$$B_1 = \frac{C_m}{1 - \alpha} \tag{12.3.7}$$

and

$$C_m = 1 + \left(\frac{\pi^2 EI\,\delta_0}{M_0\,L^2} - 1\right)\alpha \tag{12.3.8}$$

which can be expressed as

$$C_m = 1 + \psi\alpha \tag{12.3.9}$$

For usual cases of single curvature, the magnification factor B_1 to be applied to the primary bending moment is equal to $C_m/(1 - \alpha)$. Rigorous differential equation solutions for Cases 1 to 7 shown in Table 12.3.1 are available in Timoshenko and Gere [6.67, Chap. 1]. Calculation of approximations for ψ of Eq. 12.3.9 have been made by Yura, whose calculations are presented by Iwankiw [12.3].

The approximate values for C_m for positive moment shown in Table 12.3.1 are computed using Eq. 12.3.8 and they are in general agreement with theoretical results even though Eq. 12.3.8 is derived using a sine curve deflection. The negative moment values have been included by AISC beginning with the 1980 AISC Manual. It is noted that the magnification of negative moment is necessary in order to maintain zero slope at the fixed supports when the beam deflects in the positive moment region. From a practical point of view, it is doubtful that zero slope is maintained; thus, the negative moment magnification is probably overestimated.

The correct value for C_m will be close to 1.0 for all cases because α rarely will exceed about 0.3. For this reason AISC-C2.1b provides that C_m can be determined by "rational analysis" or 1.0 can be used for all members. The authors consider that C_m from Table 12.3.1 will satisfy the requirement of "rational analysis."

TABLE 12.3.1 Suggested Values for C_m for Situations with No Joint Translation[a]

	Case	C_m (positive) moment)	C_m (negative moment)	Primary Bending Moment
1		$1 + 0.2\alpha$[†]	—	M_m
2		1.0	—	M_m
3		$1 - 0.2\alpha$	—	M_m
4		$1 - 0.3\alpha$	$1 - 0.4\alpha$	M_m
5		$1 - 0.4\alpha$	$1 - 0.4\alpha$	M_m
6		$1 - 0.4\alpha$	$1 - 0.3\alpha$	M_m
7		$1 - 0.6\alpha$	$1 - 0.2\alpha$	M_m
8		Eq. (12.3.8)	not available	M_m

[a]Adapted from AISC Commentary Table C-C2.1

[†]$\alpha = \dfrac{P_u}{P_e} = \dfrac{P_u}{\pi^2 E/(KL/r)^2}$

EXAMPLE 12.3.1

Compare the differential equation magnification factor for the loading of Fig. 12.2.3, Eq. 12.2.16, with the approximate value, Eq. 12.3.7.

Solution:

For the differential equation,

$$B_1 = \text{magnification factor} = \frac{2}{(kL/2)^2}\left(\sec\frac{kL}{2} - 1\right) \tag{a}$$

where

$$\frac{kL}{2} = \frac{L}{2}\sqrt{\frac{P}{EI}} = \frac{\pi}{2}\sqrt{\alpha}$$

For the approximate solution,

$$B_1 = \frac{C_m}{1 - \alpha}$$

$$\delta_0 = \frac{5wL^4}{384EI}; \qquad M_0 = \frac{wL^2}{8}$$

$$\frac{\delta_0}{M_0} = \frac{5L^2}{48EI}$$

$$C_m = 1 + \left(\frac{\pi^2 EI}{L^2}\frac{5L^2}{48EI} - 1\right)\alpha = 1 + 0.028\alpha$$

$$B_1 = \frac{1 + 0.028\alpha}{1 - \alpha} \tag{b}$$

α	sec $kL/2$	Eq. (a)	Eq. (b)
0.1	1.137	1.114	1.114
0.2	1.310	1.257	1.257
0.3	1.533	1.441	1.441
0.4	1.832	1.686	1.685
0.5	2.252	2.030	2.028
0.6	2.884	2.546	2.542
0.7	3.941	3.405	3.399
0.8	6.058	5.125	5.112
0.9	12.419	10.284	10.253

Obviously, there is no significant difference between the differential equation solution and the approximate solution for this case.   ■

12.4   MOMENT MAGNIFICATION—MEMBERS SUBJECT TO END MOMENTS ONLY; NO JOINT TRANSLATION

For the situation shown in Fig. 12.2.2, which has no transverse loading, the theoretical maximum moment is given by Eq. 12.2.13,

$$M_{z\,\text{max}} = M_2\sqrt{\frac{(M_1/M_2)^2 - 2(M_1/M_2)\cos kL + 1}{\sin^2 kL}} \tag{12.2.13}$$

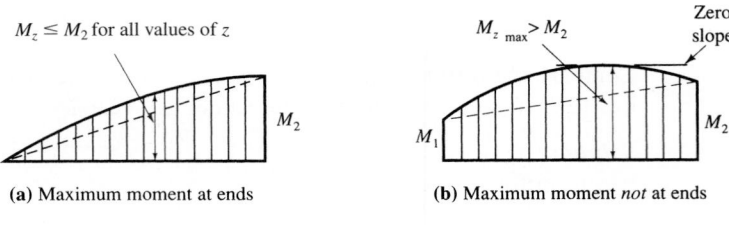

(a) Maximum moment at ends   (b) Maximum moment *not* at ends

(c) Equivalent uniform moment with
maximum magnified moment at midspan

Figure 12.4.1
Primary plus secondary
bending moment for
members acted upon by
end moments only.

For this situation the maximum moment may be either (1) the larger end moment M_2 at the braced location (Fig. 12.4.1a), or (2) the magnified moment given by Eq. 12.2.13 which occurs at various locations out along the span (Fig. 12.4.1b), depending on the ratio M_1/M_2 and the value of α, since $kL = \pi\sqrt{\alpha}$. In order to make an analysis, one needs to know whether the maximum moment occurs at a location away from the support, and if so, the correct *distance*. To eliminate the need for such information, the concept of equivalent uniform moment (Fig. 12.4.1c) is used. Thus when investigating a member at a location away from the supported point, use of the equivalent moment assumes $M_{z\,max}$ to be at midspan.

To establish the equivalent moment M_E, let the solution for uniform moment, Eq. 12.2.17 with $M = M_E$, be equated with Eq. 12.2.13. One obtains

$$M_E = M_2\sqrt{\frac{(M_1/M_2)^2 - 2(M_1/M_2)\cos kL + 1}{2(1 - \cos kL)}} \qquad (12.4.1)$$

By the procedure used in Example 12.3.1 it may be shown that for uniform moment, the magnification factor is obtained from Eq. 12.2.18:

$$B_1 = \sec\frac{kL}{2} = \frac{1}{1 - \alpha} \qquad (12.4.2)$$

and using the equivalent uniform moment M_E to replace M_1 and M_2, the full maximum moment may be expressed as

$$M_{z\,max} = M_E\left(\frac{1}{1 - \alpha}\right) \qquad (12.4.3)$$

which when compared with Eq. 12.3.6 may be written

$$M_{z\,max} = C_m M_2\left(\frac{1}{1 - \alpha}\right) \qquad (12.4.4)$$

where $\qquad C_m = M_E/M_2$

$$= \sqrt{\frac{(M_1/M_2)^2 - 2(M_1/M_2)\cos kL + 1}{2(1 - \cos kL)}} \qquad (12.4.5)$$

Equation 12.4.5 does not consider lateral-torsional buckling, or fully cover the double-curvature cases where M_1/M_2 lies between -0.5 and -1.0. The actual failure of members bent in double curvature with bending moment ratios -0.5 to -1.0 is generally one of "unwinding" through from double to single curvature in a sudden type of buckling, as discussed by Ketter [12.4], among others.

Austin [12.2] has discussed proposals of Massonnet [12.1] and Horne [12.5] to approximate C_m. For many years the AISC Specifications have used the following simple approximation,

$$C_m = 0.6 + 0.4\frac{M_1}{M_2} \qquad (12.4.6)$$

where M_1 = smaller bending moment at one end of a member

M_2 = larger bending moment at one end of a member

Note that M_1 and M_2 use bending moment signs rather than rotational moment signs. Prior to the 1986 LRFD Specification, Eq. 12.4.6 was not permitted to be less than 0.4; a very conservative procedure. However, since C_m as used in current design interaction relationships by LRFD and ASD is *not* directly related to lateral-torsional buckling in the manner of the Massonnet and Horne equations, there is no reason for the lower limit. A study of C_m has been presented by Chen and Zhou [12.6], and the inelastic amplification factor has been discussed by Sohal and Syed [12.55].

The earlier 0.4 lower limit on C_m has been eliminated [1.13] since the 1993 LRFD Specification.

A comparison of Eqs. 12.4.5 and 12.4.6 is shown in Fig. 12.4.2. Note that for a given value of $\alpha = P/P_e$, the curve terminates when the moment M_2 at the end of the member exceeds the magnified moment within the span. The most important situations are those where the magnified moment within the span exceeds the moment at the end. The straight line used by AISC lies near the upper limit for C_m at any given bending moment ratio, and thus seems to be a reasonable approximation.

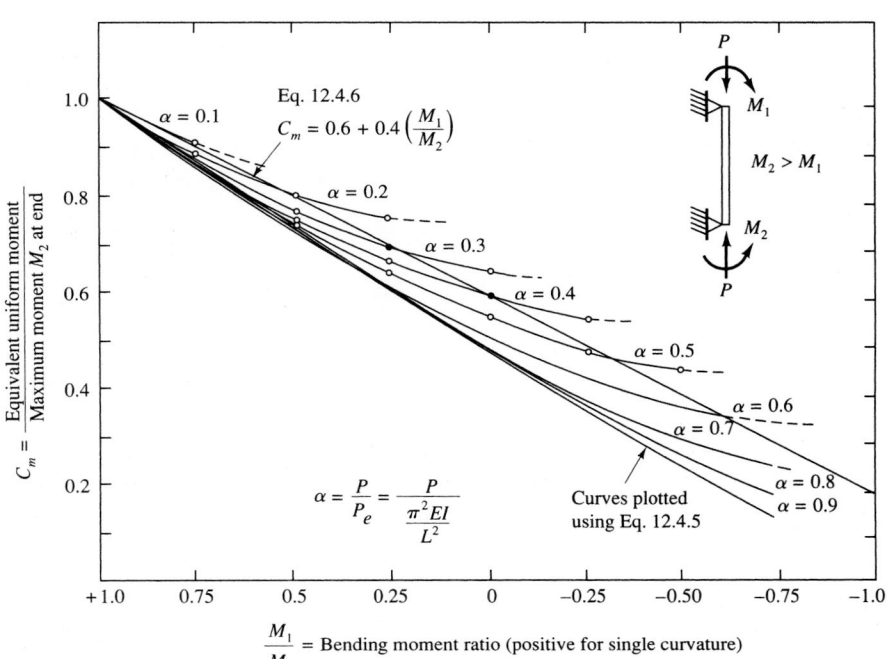

Figure 12.4.2
Comparison of theoretical C_m with AISC straight line approximation for members acted upon by end moments only, without joint translation.

12.5 MOMENT MAGNIFICATION—MEMBERS WITH SIDESWAY POSSIBLE

The unbraced frame, i.e., the frame where joint translation may occur when instability arises due to slenderness of the compression elements, does not lend itself to the simple but relatively accurate treatment presented in the last two sections. More complete treatment of braced and unbraced elastic frames is in Chapter 14 and in the *SSRC Guide* [6.8].

A simple approximation of C_m for this case may be obtained by starting with Eq. 12.3.6 which applies for the single curvature case,

$$M_{\max} = M_0 B = M_0\left(\frac{C_m}{1-\alpha}\right) \tag{12.5.1}$$

where B = magnification factor.

Next consider the situation of Fig. 12.5.1. Whatever the degree of restraint at the top and bottom of the two-story member, the deflection curve, and therefore the secondary bending moment (P times deflection), may reasonably be assumed to be a sine curve, in which case, the development used when no sidesway occurs (Fig. 12.3.1) is also valid here. Since $2L$ from Fig. 12.5.1 equals L for Fig. 12.3.1, Eq. 12.3.8 for C_m becomes

$$C_m = 1 + \left(\frac{\pi^2 EI\delta_0}{4L^2 M_0} - 1\right)\alpha \tag{12.5.2}$$

The larger effective length ($2L$ instead of L) is also used in the computation of α. Next, referring to Fig. 12.5.1,

$$\delta_0 = \frac{(H/4)L^3}{3EI} \tag{12.5.3}$$

$$M_0 = \frac{HL}{4} \tag{12.5.4}$$

Substitution of Eqs. 12.5.3 and 12.5.4 into Eq. 12.5.2 gives

$$C_m = 1 + \left[\frac{\pi^2 EI}{4L^2}\left(\frac{HL^3}{6EI}\right)\left(\frac{2}{HL}\right) - 1\right]\alpha$$

or

$$C_m = 1 + \left(\frac{\pi^2}{12} - 1\right)\alpha = 1 - 0.18\alpha \tag{12.5.5}$$

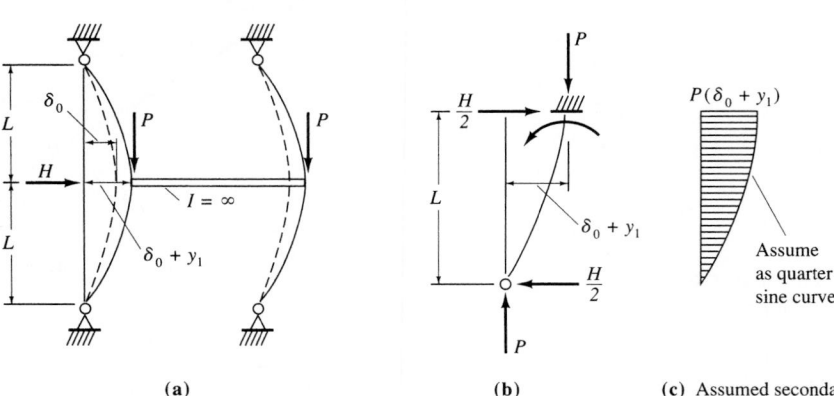

Figure 12.5.1
Beam-column with sidesway.

(a)   (b)   (c) Assumed secondary bending moment

However, the 2005 *AISC Specification* neglects the effect of α on C_m and uses a value of $C_m = 1.0$ for frames subject to joint translation.

12.6   NOMINAL STRENGTH—INSTABILITY IN THE PLANE OF BENDING

The basic strength of a beam-column where lateral-torsional buckling and local buckling are adequately prevented, and bending is about one axis, will be achieved when instability occurs *in the plane of bending* (*without twisting*) (case 2 of Sec. 12.1). The differential equation solution, which includes the second-order P times deflection y term, shows that the axial compression effect and the bending moment effect cannot be determined separately and combined by superposition. It is a nonlinear relationship.

Furthermore, residual stresses cause some fibers to reach the yield stress before others, even when the elastically computed stresses due to applied load are the same at those fibers. This is similar to the effect on axially loaded compression members discussed in Secs. 6.5 and 6.6.

An analysis to determine the strength interaction between axial compression P and bending moment M for a beam-column is complicated. First, the $M-\theta-P$ (moment–curvature–axial compression) relationship must be developed. This can be done by assuming the yield penetration depth γh (Fig. 12.6.1) at various values. For each γh there will be a complete range of $M-\theta-P$, or $M/M_p-\theta/\theta_y-P/P_y$ in nondimensional form, where M_p is the plastic moment, θ_y is the curvature when the extreme fiber reaches stress F_y, and $P_y = A_g F_y$. From these $M/M_p-\theta/\theta_y-P/P_y$ curves, a series of $M/M_p-\theta/\theta_y$ curves can be obtained; one for each value of P/P_y.

Once $M/M_p-\theta/\theta_y$ curves have been obtained, a specific combination of P_u/P_y and slenderness ratio KL/r is selected; the moment M/M_p is then applied incrementally to as high a value (M_u/M_p) such that the deflection is still stable. This combination of P_u/P_y, M_u/M_p, and KL/r represents one point on a strength interaction relationship, such as shown in Fig. 12.6.2. Various aspects of this procedure have been explained by Ketter, Kaminsky, and Beedle [12.7], Galambos and Ketter [12.8], and Ketter [12.4].

Similar procedures to obtain interaction relationships have been presented by Hauck and Lee [12.9, 12.10], Lee and Anand [12.11], and Rossow, Barney, and Lee [12.12].

An interaction equation representing this behavior is the following,

$$\frac{P_u}{P_n} + \frac{M_u}{M_n} \leq 1 \qquad (12.6.1)$$

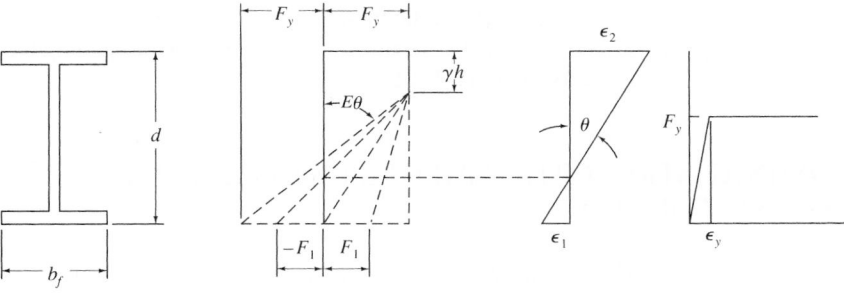

Figure 12.6.1
Member under axial compression and bending, for $\epsilon_1 \leq \epsilon_y$.

(a) Section

(b) Assumed yield penetration and series of stress conditions

(c) Strain

(d) Stress-strain diagram for material

Figure 12.6.2
Strength interaction curves
(W8×31, F_y = 33 ksi, linear
residual stress = $F_r = 0.3F_y$,
strong axis bending) for braced
frame members. *Note:* For
F_y > 33 ksi, use adjusted
$L/r = $ (actual L/r) $\sqrt{F_y/33}$.
(Adapted from Ketter [12.4])

(a) Moment ratio = −1.0

(b) Moment ratio = +0.5

where P_u = maximum axial compression load when nominal strength is reached based
on the interaction relationship

P_n = nominal strength of an axially loaded compression member based on
slenderness ratio *KL/r*

M_u = maximum moment when the nominal strength is reached based on the
interaction relationship, *including the second-order effect* (*the so-called*
P−Δ effect)

= Eq. 12.3.6 = $M_0C_m/(1 - \alpha)$

M_0 = primary bending moment

$\alpha = P_uL^2/\pi^2EI$

C_m = Eq. 12.3.9

M_n = maximum moment strength = M_p for the laterally stable situation
discussed in this section

Thus, Eq. 12.6.1 may be written

$$\frac{P_u}{P_n} + \frac{M_o}{M_n}\left(\frac{C_m}{1 - \alpha}\right) \leq 1 \qquad (12.6.2)$$

In studying the interaction curves of Fig. 12.6.2, the reader is reminded that the
strength P_n when M_u = 0 is based on the slenderness ratio KL/r_x. As stated at the beginning of the section, the member was assumed to fail by instability in the plane of bending.
Other studies of instability in the plane of bending have included the effects of transverse loading [12.4, 12.13–12.17].

12.7 NOMINAL STRENGTH—FAILURE BY COMBINED BENDING AND TORSION

The ordinary beam-column unbraced over a finite length involves consideration of
instability transverse or oblique to the plane of bending, involving torsional effects.
This subject is an extension of lateral-torsional buckling in beams (Chapter 9) and
involves both elastic and inelastic considerations. Interaction curves for a number of
elastic buckling situations have been developed, including (1) I-section columns with

eccentric end loads in the plane of the web [12.18]; (2) I-columns with unequal end moments but without restraint to rotation about principal axes at the ends of the member [12.19]; and (3) I-columns with unequal end moments—hinged at ends for rotation about strong axis, but elastically restrained for rotation about the weak axis at the ends [12.20]. An excellent summary of the topic is presented by Massonnet [12.1] which includes discussion of plastic effects.

A number of studies are available containing both analytical and experimental treatments of inelastic lateral-torsional buckling [12.21–12.26].

Tests by Massonnet [12.1] and computer studies [12.21, 12.22] including inelastic effects indicate that interaction diagrams similar to those in Fig. 12.6.2 will result when including lateral-torsional buckling. The main difference is that P_u/P_y will be lower when based on KL/r_y (weak axis) rather than on KL/r_x (strong axis). Of course, M_n may also be less than M_p because of lateral-torsional buckling.

Torsional-Flexural Buckling of Thin-Walled Open Sections

A singly symmetric section subject to flexure and axial compression will have a torsional moment acting even without being in the slightly buckled position. This is because the shear center and center of gravity do not coincide (see Chap. 8 on torsion). The differential equation development has been given by Pekoz and Winter [12.27], Pekoz and Celebi [12.28], and the topic has been well summarized by Yu [12.29]. The AISI Specification [1.8] provides detailed rules for design to include the possibility of torsional-flexural buckling.

12.8   NOMINAL STRENGTH—INTERACTION EQUATIONS

Strength interaction equations relating axial compression P_u to bending moment M_u have long been recognized as the practical procedure for design.

Case 1—No Instability

For the braced location where instability cannot occur (i.e., $Kl/r = 0$), the uppermost curves of Fig. 12.6.2 apply and may be approximated by

$$\frac{P_u}{P_y} + \frac{M_u}{1.18M_p} = 1.0 \tag{12.8.1}$$

and $M_u/M_p \le 1.0$. In the above equation, $P_y = A_g F_y$ and M_p = maximum moment strength of the member in the absence of axial load (equals the plastic moment for all cases where premature local buckling is prevented). The comparison with the theoretical result in Fig. 12.8.1 shows Eq. 12.8.1 to be a good approximation.

Case 2—Instability in the Plane of Bending

The curves of Fig. 12.6.2 for various combinations of moments and values of L/r_x may be approximated by

$$\frac{P_u}{P_n} + \frac{M_E}{M_p(1 - P_u/P_e)} = 1.0 \tag{12.8.2}$$

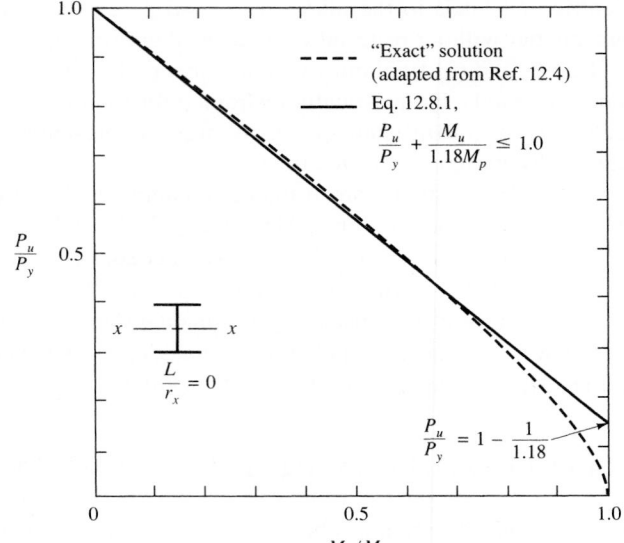

Figure 12.8.1
"Exact" nominal strength interaction relationship for typical wide-flange sections (including residual stress) compared with interaction equation—Case 1, no instability.

where P_n = compression nominal strength under axial load based on slenderness ratio for the *axis of bending*

M_E = Eq. 12.4.1 (or its alternate, $C_m M_2$, where C_m = Eq. 12.4.6)

$P_e = \pi^2 EI/L^2$

Massonnet [12.1] has shown that Eq. 12.8.2 is a good approximation by comparing it with the curves of Galambos and Ketter [12.8]. A comparison in Fig. 12.8.2 of Eq. 12.8.2 with some curves from Fig. 12.6.2a shows the correlation.

For primary bending moment from transverse loading, Lu and Kamalvand [12.13] have shown that when M_E is replaced by $C_m M_{ui}$, using C_m as given by Eq. 12.3.9, Eq. 12.8.2

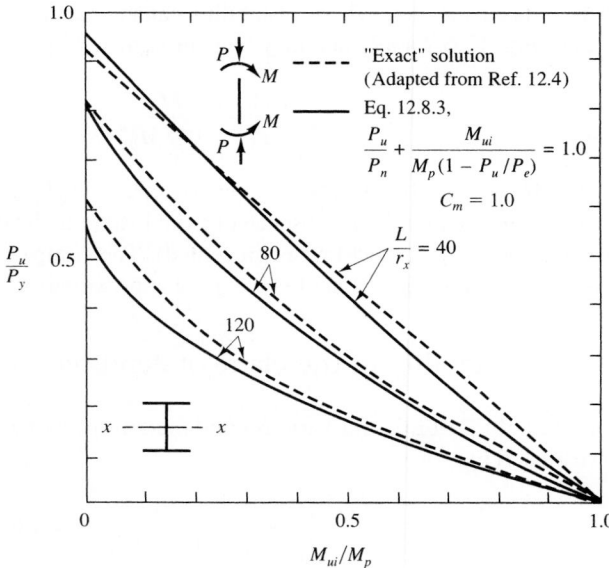

Figure 12.8.2
"Exact" nominal strength interaction curves for typical wide-flange sections (including residual stress) compared with interaction equation—Case 2, instability in the plane of bending ("exact" solution from Fig. 7 of Ref. 12.4). *Note:* For actual use, use adjusted $L/r = $ (actual $L/r)\sqrt{F_y/33}$.

(actually Eq. 12.8.3 with $C_m = 1.0$) is also a good representation for the "exact" solutions. Thus, in general, the interaction equation may be written

$$\frac{P_u}{P_n} + \frac{C_m M_{ui}}{M_p(1 - P_u/P_e)} = 1.0 \tag{12.8.3}$$

for all cases of instability *in the plane of bending.*

Case 3—Instability by Lateral-Torsional Buckling

Massonnet [12.1] has shown that, with only slight error, the form of Eq. 12.8.2 may also be used for this case. P_n for this case may be governed by the slenderness ratio for the axis normal to the axis of bending. Further, since lateral-torsional buckling as a beam may occur for a moment less than M_p, use M_n instead of M_p in Eq. 12.8.3. Using the various definitions of C_m from Secs. 12.3 through 12.5, the general notation $C_m M_{ui}$ should be used. Thus for instability under Cases 2 or 3, the following interaction equation may be considered to apply:

$$\frac{P_u}{P_n} + \frac{M_{ui} C_m}{M_n(1 - P_u/P_e)} = 1.0 \tag{12.8.4}$$

where P_u = applied axial compression load

M_{ui} = applied primary bending moment

$P_n = A_g F_{cr}$ = nominal strength for axially loaded compression member (Eqs. 6.7.7 or 6.7.8 for F_{cr} would be used in LRFD)

M_n = nominal moment strength in the absence of axial load computed by methods of Chapter 9. For adequately braced members of low slenderness ratio where local buckling is precluded, $M_n = M_p$.

C_m = factors discussed in Secs. 12.3–12.5

$P_e = \dfrac{\pi^2 EI}{L^2}$

Other proposals for interaction equations have been given by Cheong-Siat-Moy and Downs [12.30], Duan and Chen [12.31], Sohal, Duan, and Chen [12.32], and Cal, Liu, and Chen [12.33].

12.9   BIAXIAL BENDING

The nominal strength of members under axial compression and biaxial bending has been studied by Birnstiel and Michalos [12.34], Culver [12.35, 12.36], Harstead, Birnstiel, and Leu [12.37], Syal and Sharma [12.38], Santathadaporn and Chen [12.39], and Chen and Atsuta [12.40]. Even with a number of simplifying assumptions, the analysis is complex. Some tests have been performed [12.41] which, though limited, have shown agreement with computer studies. Tebedge and Chen [12.43] have given interaction surfaces in the form of tables for design. The status of work on biaxial bending of compression members is summarized by Chen and Santathadaporn [12.42] and more recently by the *SSRC Guide* [6.8].

Simple plastic theory becomes inadequate when moments exist about two principal axes. When only one moment exists, plastic behavior (constant moment with increasing rotation θ) is exhibited no matter what the value of axial compression. The effective plastic moment reduces as axial compression increases, but plastic behavior does occur.

Figure 12.9.1
Contours on strength
interaction surface for short
members where instability
does not occur. (Adapted from
Chen and Santathadaporn
[12.42])

Upon application of an additional moment about the other principal axis, one might consider an interaction surface relating P, M_x, and M_y. Even for ideal elastic-plastic material, however, present plastic analysis theorems neglect the influence of deformation on equilibrium. For zero length compression members, the concept of an interaction surface (see Fig. 12.9.1) may be thought of as a first step to obtaining the strength under biaxial bending.

While few designers concern themselves greatly about the sequence of load application, nevertheless loading sequence affects strength. This is also true for uniaxial bending and compression, but it has less effect on that case than for the biaxial loading.

Figure 12.9.2 illustrates several loading sequences to reach point A, a particular value of P_u, M_{ux}, and M_{uy}. Point A may be reached by the following paths: (1) Apply P_u first, then M_{uy}, then M_{ux} (path 0–1–2–A); (2); apply P_u first, then apply M_{ux} and M_{uy} proportionally (path 0–1–A); (3) apply P_u, M_{ux} and M_{uy} by increasing magnitude in constant proportion (path 0–A); (4) apply M_{ux} and M_{uy} in constant proportion, then apply P_u (path 0–3–A).

Other combinations are possible and in general the loading may become applied via any path through space to get from 0 to A on Fig. 12.9.2. A given section will exhibit a different strength for each path of loading. Nearly all investigators to date (2008) have used proportional loading (path 0–A).

The strength of compression members under biaxial bending is not sufficiently well known to make use of it for plastic analysis of rigid space frames; therefore, plastic analysis

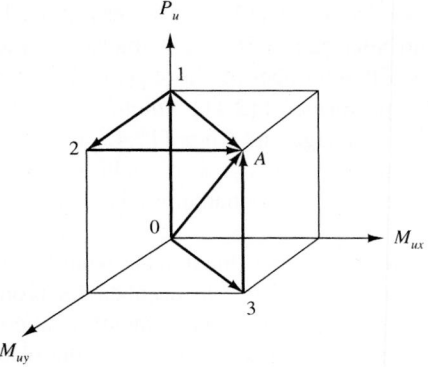

Figure 12.9.2
Paths of loading for biaxial
bending combined with axial
force. (Adapted from Chen
and Santathadaporn [12.42])

should be restricted to planar structures, or ones for which planar behavior represents a reasonable approximation.

For lack of any contrary information, an interaction formula, such as Eq. 12.8.4, is usually assumed to apply for biaxial bending. Computer studies and some tests indicate that such a procedure is realistic for those cases investigated. Thus the full interaction equation would be

$$\frac{P_u}{P_n} + \frac{M_{ux}C_{mx}}{M_{nx}(1 - P_u/P_{ex})} + \frac{M_{uy}C_{my}}{M_{ny}(1 - P_u/P_{ey})} \leq 1 \qquad (12.9.1)$$

where all terms are as defined following Eq. 12.8.4, except that now the quantities subscripted x and y must be evaluated for bending about the axis indicated by subscript.

12.10 AISC DESIGN CRITERIA

Long tradition in Allowable Stress Design of beam-columns has used two equations for design; one for yielding or maximum strength and one for stability. That would correspond to using Eq. 12.8.1 for yielding and Eq. 12.8.4 for stability. The early LRFD recommendations [12.44] used those equations.

According to Yura [12.45], Task Group 10 developing LRFD design rules for beam-columns had the following objectives:

1. The rules should apply to a wide range of variables, such as strong and weak axis bending, effect of residual stress, braced (nonsway) and unbraced (sway possible) structures, inelastic behavior, all column slenderness ratios, leaned column systems, and second-order effects.

2. The second-order effect should be kept separate and identifiable in the criterion so that the designer could perform a second-order analysis if desired.

3. Elastic analysis, including elastic second-order analysis if desired, should be the basis for determining internal forces, since practical inelastic analysis techniques are not available for office use.

4. The strength interaction equation using elastic second-order analysis should not be greater than 5% unconservative when compared with "exact" inelastic second-order analysis theoretical solutions.

5. Mathematically identical problems should give the same results using the strength interaction criterion.

These design rules formed the basis for what is now referred to as the traditional "Effective Length Method." This method has been adopted by LRFD since 1993. The 2005 *AISC Specification* provides three alternatives for stability design including an improved version of the traditional Effective Length Method. Attempts over the past few years to eliminate the use of the effective length factor K in determining the nominal compression strength P_n, have lead to the development of a new method called the "Direct Analysis Method." Since 1993, a single criterion that combines yielding and stability equations has been adopted. A thorough review of the LRFD provisions for beam-columns has been provided by Chen and Lui [12.46].

For design, AISC-H1.1 prescribes the following for compression and bending moment. Since the emphasis in this text is on the LRFD Method, the AISC-H1.1 equations are presented in the useful LRFD format, using factored service loads P_u instead of the general P_r, and the design strength $\phi_c P_n$ instead of the general P_c in the compression-related terms; and factored moments M_{ux} and M_{uy} instead of the general M_{rx} and M_{ry}, and the design strengths $\phi_b M_{nx}$ and $\phi_b M_{ny}$ instead of the general M_{cx} and M_{cy}, in the moment-related terms. Thus,

1. For $\dfrac{P_u}{\phi_c P_n} \geq 0.2$

$$\frac{P_u}{\phi_c P_n} + \frac{8}{9}\left(\frac{M_{ux}}{\phi_b M_{nx}} + \frac{M_{uy}}{\phi_b M_{ny}}\right) \leq 1.0 \qquad (12.10.1)$$

2. For $\dfrac{P_u}{\phi_c P_n} < 0.2$

$$\frac{P_u}{2\phi_c P_n} + \left(\frac{M_{ux}}{\phi_b M_{nx}} + \frac{M_{uy}}{\phi_b M_{ny}}\right) \leq 1.0 \qquad (12.10.2)$$

where P_u = factored axial compressive load

P_n = nominal compressive strength considering the member as loaded by axial compression only in accordance with AISC-E3 (see Chapter 6)

ϕ_c = resistance factor for compression = 0.90

M_{ux} = factored bending moment about the x-axis, *including second-order effects*

= $B_{1x} M_{ntx} + B_{2x} M_{\ell tx}$, if moment magnification is used in lieu of computing the elastic second-order moments. (The term $B_{2x} M_{\ell tx}$ applies only for the unbraced frame member; i.e., where sway is possible. This case is treated in Sec. 12.11.) Provisions for computing B_1 and B_2 are in AISC-C2.1b.

M_{nx} = nominal moment strength for bending about the x-axis, in accordance with AISC-F (see Chapter 9)

ϕ_b = resistance factor for flexure = 0.90

M_{uy} = same as M_{ux} except referred to the y-axis

M_{ny} = same as M_{nx} except referred to the y-axis

Moment Magnifier—Braced Frame

The moment magnifier B_1 for members having *no* joint translation was treated in Secs. 12.3 and 12.4. For this nonsway case, AISC Formula (C2-2) gives the magnifier as (subscript 1 refers to the braced case)

$$B_1 = \frac{C_m}{1 - \alpha P_u/P_{e1}} \geq 1.0 \qquad (12.10.3)$$

Note that α in Eq. 12.10.3 is unrelated to the use of α in the general development of equations in the earlier parts of this chapter. Equation 12.10.3 leads to confusion when using the AISC LRFD Method because α there is a device necessary when using the AISC ASD Method, where it has a safety-related purpose. For the LRFD Method, α in Eq. 12.10.3 should not be there, as the development of Eq. 12.3.7 shows (AISC says take α as 1.0). Thus, for AISC LRFD Eq. 12.10.3 can be written, and used in the remainder of this text, as

$$B_1 = \frac{C_m}{1 - P_u/P_{e1}} \geq 1.0 \qquad (12.10.4)$$

where C_m = factors discussed in Secs. 12.3 and 12.4, to be taken as follows:

1. For *braced frame* members having transverse loading between supports, C_m is an integral part of the moment magnifier B_1, whose value may be conservatively taken as

$C_m = 1.0$, or according to the AISC-C2.lb definition of C_m "shall be determined by analysis." An "analysis" could include the development in Secs. 12.3 and 12.4 to give the following:

$$C_m = 1 + \psi\alpha = 1 + \psi\frac{P_u}{P_{e1}} \qquad (12.10.5)$$

where P_u = factored axial compression load

P_{e1} = Euler load using KL/r for the axis of bending with $K \leq 1.0$ (for *braced frame*)

$$P_{e1} = \frac{\pi^2 E A_g}{(KL/r)^2} \qquad (12.10.6)$$

Values of ψ are available from AISC-Commentary Table C-C2.1 in an abbreviated form, or from Table 12.3.1

2. For *braced frame* members without transverse loading between supports but having end moments M_1 (smaller one) and M_2 (larger one), C_m converts the linearly varying primary bending moment into an equivalent uniform moment $M_E = C_m M_2$; thus,

$$C_m = 0.6 - 0.4(M_1/M_2) \qquad (12.10.7)$$

The moments M_1 and M_2 are rotational moments, rather than bending moments as used in Sec. 12.4. Therefore, the ratio is negative $(-)$ for single curvature and positive $(+)$ for double curvature. Since M_2 is the primary moment, C_m is really not part of the magnification factor.

The braced frame (no translation) beam-column total factored moment is

$$M_u = B_1 M_{nt} \qquad (12.10.8)$$

where M_{nt} = the primary factored moment (for the *no translation* case; hence the subscript *nt*)

EXAMPLE 12.10.1

Investigate the acceptability of a W16×67 used as a beam-column in a braced frame under the loading shown in Fig. 12.10.1. The steel is A572 Grade 60.

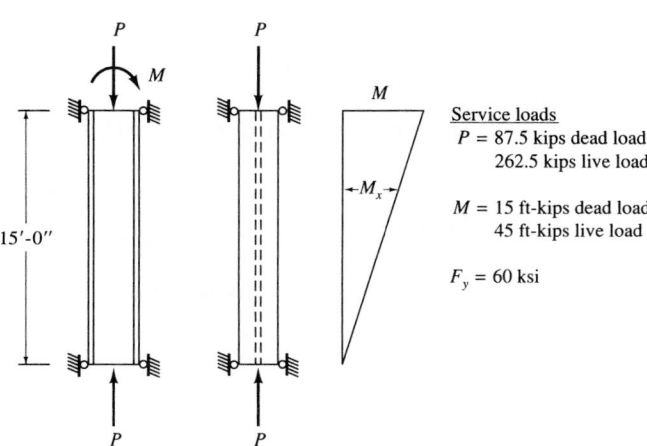

Service loads
P = 87.5 kips dead load
      262.5 kips live load

M = 15 ft-kips dead load
      45 ft-kips live load

F_y = 60 ksi

Figure 12.10.1
Beam-column for Examples
12.10.1 and 12.14.1.

Solution:

(a) Factored loads.

$$P_u = 1.2P_D + 1.6P_L = 1.2(87.5) + 1.6(262.5) = 525 \text{ kips}$$

$$M_{nt} = 1.2M_D + 1.6M_L = 1.2(15) + 1.6(45) = 90 \text{ ft-kips}$$

$$M_u = B_1 M_{nt}$$

Note that the factored primary moment is referred to using the subscript *nt* (*no translation*).

(b) Column effect. Calculate the slenderness ratio KL/r,

$$\text{Largest } \frac{KL}{r} = \frac{KL}{r_y} = \frac{1(15)12}{2.46} = 73$$

Since the *AISC Manual* Tables 4-22 are not available for $F_y = 60$ ksi, AISC-E3 must be used to compute the critical stress F_{cr}. Thus,

$$\left(\frac{KL}{r} = 73\right) < \left(4.71\sqrt{\frac{E}{F_y}} = \frac{800}{\sqrt{F_y, \text{ksi}}} = 104\right)$$

which makes it a case of inelastic buckling,

$$F_e = \frac{\pi^2 E}{(KL/r)^2} = \frac{\pi^2(29,000)}{(73)^2} = 53.4 \text{ ksi}$$

$$\phi_c F_{cr} = \phi_c\left(0.658^{\frac{F_y}{F_e}}\right)F_y = 0.90\left(0.658^{\frac{60}{53.4}}\right)60 = 33.7 \text{ ksi}$$

$$\phi_c P_n = \phi_c F_{cr} A_g = 33.7(19.7) = 664 \text{ kips}$$

Check $P_u/\phi_c P_n \geq 0.2$,

$$\frac{P_u}{\phi_c P_n} = \frac{525}{664} = 0.79 > 0.2 \qquad\qquad \underline{\textit{Use Eq.12.10.1}}$$

Note that for the web $\lambda = h/t_w = 35.9$, which exceeds the λ_r limit of $1.49\sqrt{E/F_y}$ $= 254/\sqrt{F_y}, \text{ksi} = 32.7$ given in AISC-Table B4.1. However, Q will beless than 1.0 only when h/t_w exceeds $254/\sqrt{F_{cr}}, \text{ksi} = 41.6$. (See Sec. 6.18 andAISC-E7.2).

(c) Beam effect. The laterally unbraced length L_b is 15 ft.

$$M_p = F_y Z_x = 60(130)/12 = 650 \text{ ft-kips}$$

$$M_r = 0.7F_y S_x = 0.7(60)(117)/12 = 410 \text{ ft-kips}$$

$$L_p = 1.76r_y\sqrt{\frac{E}{F_y}} = r_y\frac{300}{\sqrt{F_y, \text{ksi}}} = \frac{300(2.46)}{\sqrt{60}}\frac{1}{12} = 7.9 \text{ ft}$$

Since L_b exceeds L_p, L_r must be evaluated. Using $r_{ts} = 2.82$ in., $h_o = 15.7$ in., $J = 2.39$ in^4, and $c = 1$;

$$L_r = 1.95r_{ts}\frac{E}{0.7F_y}\sqrt{\frac{Jc}{S_x h_o}}\sqrt{1 + \sqrt{1 + 6.76\left(\frac{0.7F_y}{E}\frac{S_x h_o}{Jc}\right)^2}}$$

$$L_r = 1.95(2.82)\frac{29{,}000}{0.7(60)}\sqrt{\frac{2.39(1.0)}{117(15.7)}}\sqrt{1+\sqrt{1+6.76\left(\frac{0.7(60)}{29{,}000}\frac{117(15.7)}{2.39(1.0)}\right)^2}}\;\frac{1}{12} = 23 \text{ ft}$$

Thus, $L_p < L_b < L_r$; therefore M_n lies between M_p and M_r for $C_b = 1.0$. In this case, however, $C_b = 1.67$ using Eq. 9.6.11 or Table 9.6.3, thus, it is probable that M_n will equal M_p. Check whether or not W16×67 is "compact" for $F_y = 60$ ksi according to AISC-Table B4.1. For the flange,

$$\left(\frac{b_f}{2t_f} = \frac{10.235}{2(0.665)} = 7.7\right) < (\lambda_p = 8.4 \text{ from Table 9.6.1}) \qquad \textbf{OK}$$

For the web,

$$\lambda_p = 3.76\sqrt{\frac{E}{F_y}} = \frac{640}{\sqrt{F_y, \text{ksi}}} = \frac{640}{\sqrt{60}} = 83$$

$$\left[\lambda = \frac{h}{t_w} = 35.9 \text{ from } AISC\ Manual\right] < [\lambda_p = 83] \qquad \textbf{OK}$$

The section is "compact". Next use Eq. 9.6.4 to compute M_n,

$$M_n = C_b\left[M_p - (M_p - 0.7F_yS_x)\left(\frac{L_b - L_p}{L_r - L_p}\right)\right] \le M_p \qquad [9.6.4]$$

$$= 1.67\left[650 - (650 - 410)\left(\frac{15 - 7.9}{23 - 7.9}\right)\right] = 897 \text{ ft-kips}$$

Since M_n cannot exceed M_p, $M_n = M_p$ for this analysis.

   (d) Moment magnification. The slenderness ratio KL/r involved in moment magnification must relate to the *axis of bending*, in this case the x-axis,

$$\text{Axis of bending } \frac{KL}{r} = \frac{KL}{r_x} = \frac{1.0(15)12}{6.96} = 25.9$$

$$C_m = 0.6 - 0.4(M_1/M_2) = 0.6 - 0.4(0/90) = 0.6$$

$$P_{e1} = \frac{\pi^2 E A_g}{(KL/r)^2} = \frac{\pi^2(29{,}000)19.7}{(25.9)^2} = 8430 \text{ kips}$$

$$B_1 = \frac{C_m}{1 - P_u/P_{e1}} = \frac{0.6}{1 - 525/8430} = 0.6(1.07) = 0.64$$

   Since B_1 is computed to be less than 1.0, use 1.0. In this case, the moment varying from 90 ft-kips to zero over the 15 ft length is the same as if $C_mM_2 = 0.6(90) = 54$ ft-kips were constant over that length. The 54 ft-kips is then magnified to $54(1.07) = 58$ ft-kips, *but it is still less than the actual value 90 ft-kips at the end of the member.* The use of $B_1 = 1.0$ is the same as saying the magnified value out in the span is less than the value M_2 at the end of the member.

(e) Check AISC Formula (H1-1a), Eq. 12.10.1, omitting the bending term for the y-axis,

$$\frac{P_u}{\phi_c P_n} + \frac{8}{9}\left(\frac{M_{ux}}{\phi_b M_{nx}}\right) \leq 1.0 \qquad [12.10.1]$$

$$0.791 + \frac{8}{9}\left(\frac{90}{0.90(650)}\right) = 0.791 + 0.137 = 0.928 < 1 \qquad \textbf{OK}$$

The W16×67 section *is* acceptable. ∎

12.11 UNBRACED FRAME—AISC DESIGN

As shown in Figs. 6.9.2b and c and 6.9.3, an unbraced frame must rely on the flexural interaction of its beams and columns to limit horizontal displacement. Under lateral loads, a "braced" frame will resist the lateral force by such components as diagonal bracing or shear walls so that lateral distortion will be of small magnitude. Thus secondary bending moments $P\Delta$ from sidesway (the $P-\Delta$ effect) may ordinarily be neglected. However, for "unbraced" frames, the relatively larger sidesway deflection Δ due to lateral load will give rise to secondary moments $P\Delta$ (P is the gravity load) that must be provided for in design. Thus, an unbraced frame requires an analysis to accomplish the following tasks:

1. Provide strength under factored loads to resist gravity load, neglecting any side-sway effect except in cases of unbalanced loading or unsymmetric structural configuration where a "significant restraining force" (AISC-Commentary C2, par. 6) would be necessary to prevent sway. Out-of-plumbness can typically be expected to have greater effect than sway under gravity load.

2. Provide strength under factored loads to resist lateral load (i.e., factored wind or earthquake load). The moments under lateral load include the primary moments from first-order elastic analysis plus secondary moments due to $P-\Delta$ effect.

3. Provide stiffness such that the relative horizontal deflection (sway) between adjacent floors, and for the entire frame, is within specified limits (usually, say, equal to the height L_s divided by 400 or 500 under service loads).

Referring to Fig. 12.11.1, first-order equilibrium requires

$$M_{\ell t1} + M_{\ell t2} = H_u L_s \qquad (12.11.1)$$

The first-order sway deflection Δ_{1u} causes the total gravity load ΣP_u to be acting at the eccentricity Δ_{1u}. The lateral load moment $H_u L_s$ is thus increased by the amount $\Sigma P_u \Delta_{1u}$. Since the total moment now acting is $H_u L_s + \Sigma P_u \Delta_{1u}$, the relative lateral deflection (sway) will increase to Δ_{2u} when the structure reaches equilibrium in the final displaced position, as shown in Fig. 12.11.1b.

The final (including $P-\Delta$ effect) moment equilibrium may be expressed

$$B_2(M_{\ell t1} + M_{\ell t2}) = H_u L_s + \Sigma P_u \Delta_{2u} \qquad (12.11.2)$$

in which B_2 is the magnification factor, and $M_{\ell t1}$ and $M_{\ell t2}$ are primary moments (called $M_{\ell t}$ in AISC-C2.1b, with the subscript ℓt referring to lateral translation) resulting from a sway analysis under factored loads.

Comparing Eq. 12.11.2 with Eq. 12.11.1 gives

$$B_2 = \frac{H_u L_s + \Sigma P_u \Delta_{2u}}{H_u L_s} \qquad (12.11.3)$$

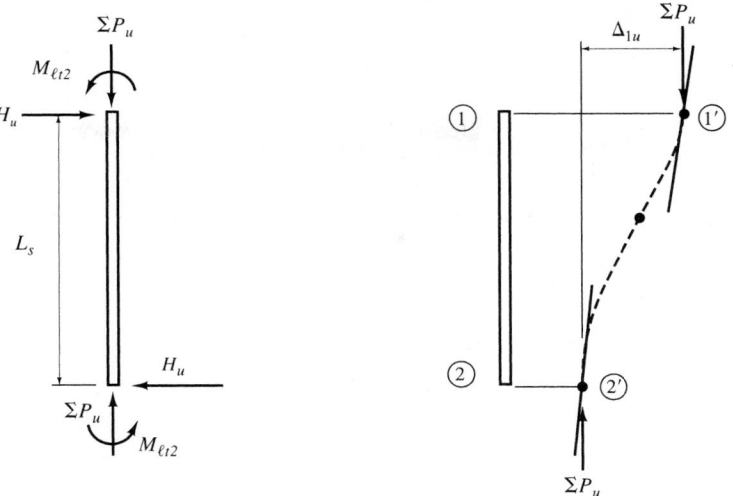

(a) First-order analysis: For equilibrium, $M_{\ell t1} + M_{\ell t2} = H_u L_s$

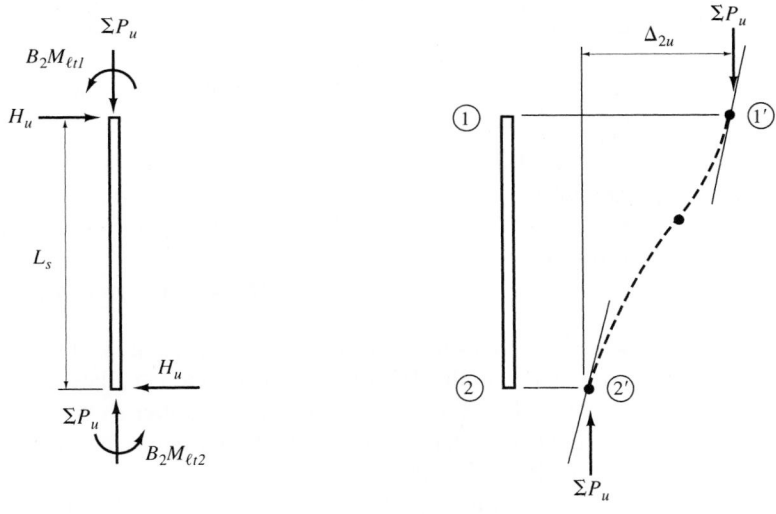

Figure 12.11.1
Summation of forces acting
on all columns in one story of
a multistory building frame.

(b) Second-order analysis: For equilibrium, $B_2(M_{\ell t1} + M_{\ell t2}) = H_u L_s + \Sigma P_u \Delta_{2u}$

Referring to Fig. 12.11.1, and using a proportionality factor η, let

$$\Delta_{u1} = \eta H_u \qquad (12.11.4)$$

which is valid for first-order analysis. The equivalent magnified lateral load in Fig. 12.11.1b, that is, the total moment divided by L_s, may be taken as

$$\text{Equivalent lateral load} = H_u + \frac{\Sigma P_u \, \Delta_{2u}}{L_s} \qquad (12.11.5)$$

Thus,

$$\Delta_{2u} = \eta \, (\text{Equivalent lateral load})$$

$$= \eta \left(H_u + \frac{\Sigma P_u \, \Delta_{2u}}{L_s} \right) \qquad (12.11.6)$$

Substituting Δ_{1u} for ηH_u gives

$$\Delta_{2u} = \Delta_{1u} + \frac{\Delta_{1u}\,\Sigma P_u\,\Delta_{2u}}{H_u L_s} \tag{12.11.7}$$

from which solving for Δ_{2u} gives

$$\Delta_{2u} = \frac{\Delta_{1u}}{1 - \Sigma P_u\left(\dfrac{\Delta_{1u}}{H_u L_s}\right)} \tag{12.11.8}$$

Substituting Eq. 12.11.8 into Eq. 12.11.3 gives

$$B_2 = \frac{1}{1 - \Sigma P_u\left(\dfrac{\Delta_{1u}}{H_u L_s}\right)} \tag{12.11.9}$$

Note that H_u in Eq. 12.11.9 is the *total* lateral force (shear) acting in the story, called ΣH in AISC-C2.1b. Even though the deflection Δ_{1u} and force H_u are for factored loads in the derivation, the ratio Δ_{1u}/H_u for factored loads or Δ_1/H for service loads will be the same, since the first-order analysis is to be *elastic*.

There are two ways in which the magnified end moments $B_2 M_{\ell t1}$ and $B_2 M_{\ell t2}$ may be obtained:

1. *Amplified First-Order Elastic Analysis Method (Moment Magnifier Method).* AISC-C2.1b indicates the required second-order flexural strength, M_r, for use in AISC Formulas (H1-1a) or (H1-1b), Eqs. 12.10.1 or 12.10.2, is obtained as follows:

$$M_u = B_1 M_{nt} + B_2 M_{\ell t} \tag{12.11.10}$$

where *two* first-order elastic analyses are required; (1) a gravity-only analysis assuming no sway to obtain the M_{nt} values and B_1; and (2) a lateral-load-only sway analysis to obtain the $M_{\ell t}$ values and B_2. The nonsway magnifier B_1 has been treated in Secs. 12.3, 12.4, and 12.10. Duan, Sohal, and Chen [12.47] have presented a discussion of the B_1 factor and suggested alternative formulation.

The sway magnifier B_2 is given by AISC-C2.1b as

$$B_2 = \frac{1}{1 - \dfrac{\alpha \Sigma P_{nt}}{\Sigma P_{e2}}} \geq 1 \tag{12.11.11}$$

The authors believe Eq. 12.11.11 is confusing for those particularly interested in Load and Resistance Factor Design. The term α is necessary in the AISC Allowable Strength Design Method where a value of $\alpha = 1.6$ must be used. For the LRFD Method, $\alpha = 1$ and could be omitted from the equation. The α used here (a one-time use) should not be confused with the α used in the earlier sections of this chapter. The term P_{nt} uses the subscript nt to signify the axial force is the factored axial force P_u, assuming *no* lateral *t*ranslation of the frame. Eq. 12.11.11 could be written,

$$B_2 = \frac{1}{1 - \dfrac{\Sigma P_u}{\Sigma P_{e2}}} \geq 1 \tag{12.11.12}$$

when focusing on the LRFD Method. The term P_{e2} can be defined (AISC-C2.1b) as follows:

P_{e2} = elastic critical buckling resistance for the story determined by sidesway buckling analysis

$$\Sigma P_{e2} = \Sigma \frac{\pi^2 EI}{(K_2 L)^2} \qquad (12.11.13)$$

(Eq. 12.11.13 is used where sidesway buckling effective length factors K_2 are determined for columns in moment frames.)

K_2 = effective length factor in the plane of bending, based on sidesway buckling (i.e., for the unbraced frame)

$$\Sigma P_{e2} = R_M \frac{\Sigma HL}{\Delta_H} \qquad (12.11.14)$$

(Eq. 12.11.14 can be used for all types of lateral load resisting systems.)

R_M = 1.0 for a braced frame system

= 0.85 for moment-frame and combined systems

ΣH = story shears produced by the lateral forces used to compute Δ_H

Δ_H = first-order interstory drift resulting from lateral forces

2. *Second-Order Analysis.* An alternate to computing the magnified primary moments using Eq. 12.11.10 is to directly use a second-order analysis under factored loads. A second-order analysis is one in which the equations of equilibrium are based on the deformed structure instead of the original undeformed geometry as in first-order analysis. The reader is referred to the methods of MacGregor and Hage [12.48], LeMessurier, McNamara, and Scrivener [12.49], LeMessurier [12.50, 12.51], Wood, Beaulieu, and Adams [12.52, 12.53], and Scholz [12.54] for suggestions on second-order analysis.

Equation 12.11.11 is of the form derived in Secs. 12.3 and 12.5, and Eq. 12.11.12 was derived as Eq. 12.11.9, replacing Δ_{1u} by the symbol Δ_H recognizing that H_u is the total horizontal force acting in the story by using the Σ sign. Further, noting that the quantity $\Delta_H / \Sigma H$ may be either for factored loads or for service loads, the subscript u is not used.

The story stiffness concept of assuming each story behaves independently of the other stories and that the $P-\Delta$ effect is equivalent to the effect of a lateral force $\Sigma P_u \, \Delta / L$ has been discussed by Chen and Lui [12.46] and MacGregor and Hage [12.48], among others.

AISC Provisions for Stability Analysis and Design

The 2005 *AISC Specification* includes significant changes in the provisions for stability analysis and design of building structures and their components. The traditional approach for overall stability design, adopted by previous AISC Specifications, is still accepted by the new specification but has been subjected to modifications and restrictions. The updated traditional approach is now referred to as the Effective Length Method. An alternative, more rational new method of analysis and design has been incorporated in the Specification, referred to as the Direct Analysis Method.

The old traditional method accounted for the effects of geometric imperfections and distributed yielding (including residual stress effects) only within interaction strength equations. The structural analysis is carried out using the nominal structure geometry and elastic stiffness. A new approach has also been included (referred to as Direct Analysis Method) in Appendix 7 of the 2005 *AISC Specification*. The Direct Analysis Method accounts for nominal geometric imperfection and stiffness reduction effects directly in the structural analysis.

The Direct Analysis Method is a rigorous analysis method that allows formulations of more realistic limit state models. The new Direct Analysis Method is presented as an alternative method in AISC, with the intention of improving and simplifying design for stability. This improvement is accomplished through the inclusion of nominal geometric imperfections and member stiffness reduction effects directly in the analysis. The more accurate required strength values account for all possible second order effects and hence lead to the use of $K = 1.0$ in calculating the in-plane column strength P_n for use in the interaction equations of AISC-Chapter H. On the other hand, the Effective Length Method includes the above effects indirectly in the interaction equations. As a result, lengthy and error-prone calculations are required to compute nomograph-based effective length factors to assess the in-plane strength of unbraced steel frames.

It is important to recognize that initial imperfections in the structure (i.e. out-of-plumbness, fabrication tolerances, member residual stresses and general softening of the structure at the strength limit state) tend to increase the magnitude of load effects in the structure above those predicted by the Effective Length Method traditional analysis. This becomes important for stability-sensitive structures containing large vertical loads with small lateral load requirements, leading to relatively low lateral load resistance. This results in underestimating the actual internal moments. The reduced required moments do not impact the design of beam-columns, since the column curves are calibrated to account for this reduction through the interaction equations. However, the reduced moment can affect the design of beams and connections, which provide rotational restraint to the column. To limit the potential for gross underestimation of the load effects in stability-sensitive structures, a maximum limit of 1.5 on B_2 has been imposed by AISC-C2 on the use of the Effective Length Method.

The Effective Length Method

Traditional Effective Length Method procedure is based on first-order analysis of the ideal, geometrically perfect, nominally elastic structure. The method accounts for residual stress and geometric imperfection effects on the sidesway stability of unbraced moment frames in an implicit fashion in the calculation of the effective length factor K, and the use of the AISC column strength equations. For most structures, the limitations placed on this method by the 2005 *AISC Specification* are normally satisfied. Nevertheless, for some cases, this method could underestimate the internal forces, as explained earlier. This limitation has been addressed in the *AISC Specification* by including a minimum notional lateral load (equivalent to the effect of a nominal out-of-plumbness) in gravity-only load combinations, and by restricting the use of the method to frames in which the amplification of the story drifts is less than or equal to 1.5.

The following steps describe the new Effective Length Method:

1. Use the nominal geometry and member properties for the second-order analysis.
2. Perform a second-order anaylis for each of the load cases can be first-order analysis with B_1 and B_2.
3. Apply N_i as a minimum lateral load in all gravity-only load combinations ($N_i = 0.002Y_i$- Notional load at level i and Y_i = gravity load at level i from the LRFD load combination being considered).
4. All load combinations must be subjected to the second-order analysis (using any method that properly considers both $P-\Delta$ and $P-\delta$ effects).
5. Design the beams and girders according to the requirements of AISC-Chapter F, AISC-Chapter 1 and other relevant sections of the Specification.

6. Check for each level of the frame that second-order effects as measured by the ratio of the average second- to first-order story drifts (B_2) are less than or equal to 1.5.

7. For all beam-columns in moment frames apply the interaction Formulas (H1-1) (or where applicable, Formula (H1-2))

   (a) For unbraced moment frames, determine the Effective Length Factor K_2 for each column.

   (b) When second-order effects are small such that $\Delta_{2nd}/\Delta_{1st}$ or $B_2 \leq 1.1$, K may be taken equal to 1.0 in the calculation of P_n (but not in the calculation of B_2) for all types of frames.

   (c) Adjustments to K_2 for column inelasticity are permissible according to the procedure outlined in the AISC Chapter C Commentary.

   (d) When using the sidesway-uninhibited alignment chart for calculation of K_2, adjustments are typically necessary to account for non-ideal framing conditions other than those assumed in the development of the alignment chart.

8. For determining member axial nominal compression strength, use the relevant equations of AISC-Chapter E, Sections E3 through E7.

EXAMPLE 12.11.1

Investigate the acceptability of W14×145 columns in a single-bay multistory unbraced frame, part of which is shown in Fig. 12.11.2. The axial compression P is the total load acting on the bottom story columns and the wind load H is the total horizontal service wind load acting on the entire multi-story frame. The steel is A992. Use AISC LRFD Method, and the Effective Length Method.

Solution:

(a) Compute factored loads. In accordance with ASCE 7-05, at least two factored load combinations are necessary when wind or earthquake act on the structure,

1. Gravity Load (LRFD Load Combination 2-ASCE 7-05)

$$P_u = 1.2(230) + 1.6(92) = 423 \text{ kips}$$

$$w_u = 1.2(0.5) + 1.6(1.5) = 3.0 \text{ kips/ft on each floor}$$

Minimum notional load is to be applied only with the gravity-only load combination (AISC-C2.2a(3)).

$$H_u = 0.002(2 \times 423 + 3 \times 28) = 1.9 \text{ kips}$$

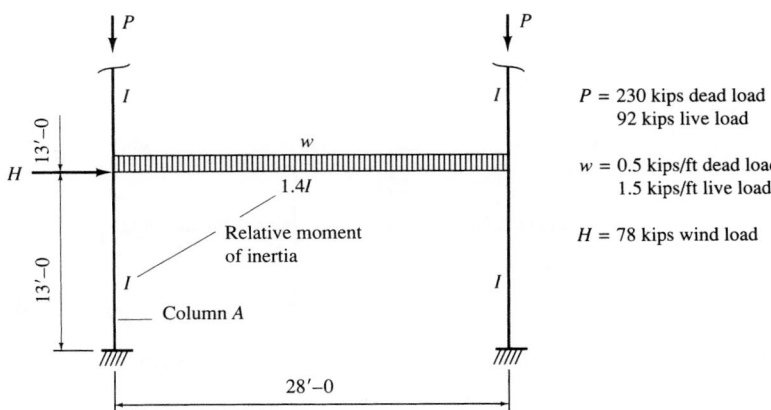

Figure 12.11.2
Lowest portion of single-bay multistory unbraced frame for Example 12.11.1.

P = 230 kips dead load
      92 kips live load

w = 0.5 kips/ft dead load
      1.5 kips/ft live load

H = 78 kips wind load

2. Gravity Load + Wind (LRFD Load Combination 4-ASCE 7-05)

$$P_u = 1.2(230) + 1.0(92) = 368 \text{ kips}$$

$$H_u = 1.6(78) = 124.8 \text{ kips}$$

$$w_u = 1.2(0.5) + 1.0(1.5) = 2.1 \text{ kips/ft on all floors except roof}$$

(b) First-order elastic structural analysis. Unless a second-order analysis is to be performed, the factored moments used in the investigation must be obtained using moment magnification, Eq. 12.11.10. The factored gravity load plus the resulting notional load (loading 1) in part (a) is divided into two separate first-order elastic analyses under factored loads; a non-sway analysis under gravity loads, and a sway analysis under lateral notional load only. The sway analysis causes an additional factored compressive load of 3 kips. The factored moments of part (a) are shown in Figs. 12.11.3b and c.

The gravity plus wind (loading 2) in part (a) is also divided into two separate first-order elastic analyses under factored loads; a nonsway analysis under gravity loads only, and a sway analysis under lateral load only. The results on the column to be investigated are as shown in Fig. 12.11.3e and f; the sway analysis also causes an additional factored compressive load of 18.5 kips on the leeward column. Loading 2 will be initially assumed to govern the beam-column strength.

(c) Column strength. Verify that local buckling will not occur; i.e., that $Q = 1.0$:

$$\left[\lambda = \frac{b_f}{2t_f} = 7.8\right] < \left[\lambda_r = \frac{95}{\sqrt{F_y, \text{ksi}}} = \frac{95}{\sqrt{50}} = 13.5\right] \qquad \textbf{OK}$$

$$\left[\lambda = \frac{h}{t_w} = 19.3\right] < \left[\lambda_r = \frac{253}{\sqrt{F_y, \text{ksi}}} = \frac{253}{\sqrt{50}} = 35.6\right] \qquad \textbf{OK}$$

Figure 12.11.3
Forces from first-order elastic analyses under factored loads—Example 12.11.1.

(a) Gravity-only + notional   (b) Nonsway part   (c) Sway part

(d) Gravity + wind   (e) Nonsway part   (f) Sway part

The effective length factor K_x in the plane of the frame is determined using restraint factors G with Fig. 6.9.4.

$$G_{\text{top}} = \frac{\Sigma I/L, \text{columns}}{\Sigma I/L, \text{beams}} = \frac{2(I/13)}{1.4I/28} = 3.08$$

$G_{\text{bottom}} = 1.0$ (The practical recommendation of AISC Commentary-Fig. C-C2.3 for nominally fixed base.)

Find $K_x = 1.57$ (From Fig. 6.9.4)

Since in the y-direction the column is considered pinned at the top and bottom, $K_y = 1.0$; thus,

$$\frac{K_xL_x}{r_x} = \frac{1.57(13)12}{6.33} = 38.7 \qquad \frac{K_yL_y}{r_y} = \frac{1.0(13)12}{3.98} = 39.2$$

$$\phi_c F_{\text{cr}} = 40.2 \text{ ksi (AISC Table 4-22)}$$

$$\phi_c P_n = \phi_c F_{\text{cr}} A_g = 40.2(42.7) = 1720 \text{ kips}$$

Check $P_u/(\phi_c P_n) \geq 0.2$ for the gravity load case

$$\left(\frac{P_u}{\phi_c P_n} = \frac{423 + 0.3 + 3.0(14)}{1720} = 0.27 \right) > 0.2; \quad \text{Use Eq. 12.10.1}$$

(d) Beam effect. The laterally unbraced length $L_b = 13$ ft.

$$M_p = F_y Z_x = 50(260)/12 = 1080 \text{ ft-kips}$$

$$L_p = 1.76 r_y \sqrt{\frac{E}{F_y}} = \frac{300}{\sqrt{F_y, \text{ksi}}} r_y = \frac{300(3.98)}{\sqrt{50(12)}} = 14.1 \text{ ft}$$

Since $L_b < L_p$, $M_u = M_p$ assuming the section is "compact" for flange and web local buckling. Check whether or not the W14×145 is "compact" for $F_y = 50$ ksi according to AISC-Table B4.1. Note that when $h/t_w < \lambda_r$ for axial compression alone, as in part (c), then $\lambda < \lambda_p$ for beam action is automatically satisfied. For most rolled I-shaped sections, this is the case. Also, AISC-F2 "*User Note*" indicates that the W14×145 section is "compact"; therefore,

$$M_n = M_p = 1080 \text{ ft-kips} \qquad \phi_b M_n = 0.90(1083) = 975 \text{ ft-kips}$$

(e) Moment magnification—nonsway magnifier B_1 for the structure of Fig. 12.11.3b. The slenderness ratio KL/r involved in moment magnification must relate to the *axis of bending*. In this case the x-axis. Thus $K_x \leq 1.0$ for the nonsway part. Since the structure is actually an *unbraced* one, the authors recommend using 1.0 rather than any value less than 1.0 (say, from Fig. 6.9.4a).

$$\text{Axis of bending } \frac{KL}{r} = \frac{KL}{r_x} = \frac{1.0(13)12}{6.33} = 24.6$$

$$C_m = 0.6 - 0.4(M_1/M_2) = 0.6 - 0.4(29/57) = 0.4$$

$$P_{e1} = \frac{\pi^2 EI}{(KL)^2} = \frac{\pi^2 (29,000)(1380)}{[13(12)]^2} = 16,210 \text{ kips}$$

$$P_u = 423 + 14(3.0) = 465 \text{ kips}$$

Using Eq. 12.11.12,

$$B_1 = \frac{C_m}{1 - P_u/P_{e1}} = \frac{0.4}{1 - 465/16,210} = 0.4(1.02) < 1.0$$

When B_1 is computed to be less than 1.0, the magnified moment between the ends of the column in Fig. 12.11.3c is less than the moment at the end of the member (57 ft-kips). Use $B_1 M_{nt} = 57$ ft-kips

(f) Moment magnification—sway magnifier B_2 for the structure of Fig. 12.11.3d. The total factored compression load ΣP_u to be carried by all columns of the frame within one story (in this case two columns) is

$$\sum P_u = 2(423) + 28(3.0) = 930 \text{ kips}$$

The Euler load P_{e2} for the column being investigated must be computed using KL/r for the axis of bending, and the K value must be for the unbraced frame, that is, $K \geq 1.0$. In this case $K = 1.57$ and $KL = 1.57(13)12 = 245$ in as determined in part (c). For one column,

$$P_{e2} = \frac{\pi^2 EI}{(KL)^2} = \frac{\pi^2 (29,000)1710}{(245)^2} = 8160 \text{ kips}$$

Since two identical columns resist sway, $\Sigma P_{e2} = 2(8160) = 16,320$ kips. Thus, the sway magnifier is

$$B_2 = \frac{1}{1 - \dfrac{\Sigma P_u}{\Sigma P_{e2}}} = \frac{1}{1 - \dfrac{930}{16,320}} = 1.06$$

Since B_2 is less than 1.5, use of this method is permitted.

The maximum magnified moment M_u, Eq. 12.11.10, for column A is

$$M_u = B_1 M_{nt} + B_2 M_{\ell t} = 1.0(57) + 1.06(8.5) = 66 \text{ ft-kips}$$

Note that AISC-C2.1b requires the maximum M_{nt} and maximum $M_{\ell t}$ to be used in computing M_u. Logically, the values of M_{nt} and $M_{\ell t}$ simultaneously should be at the same end of the member. Even that procedure is not entirely correct because the maximum magnified moment in the nonsway case may occur out in the span (and will when B_1 exceeds 1.0) while the magnified sway moment occurs at the end of the member.

(g) Check AISC Formula (H1-1a), omitting the bending term for the y-axis and adjusting the P_u to account for second order effect.

$$P_u = B_1 P_{nt} + B_2 P_{lt} = 1.0(465) + 1.06(0.3) = 465.0 \text{ kips}$$

$$\frac{P_u}{\phi P_n} + \frac{8}{9}\left(\frac{M_{ux}}{\phi_b M_{nx}}\right) \leq 1.0$$

$$0.27 + \frac{8}{9}\left(\frac{66}{975}\right) = 0.33 < 1.0 \qquad\qquad \textbf{OK}$$

For the gravity + wind loading case, steps (c) through (f) are repeated as steps (h) through (k) as follows

(h) Check $P_u/(\phi_c P_n) \geq 0.2$,

$$\frac{P_u}{\phi_c P_n} = \frac{368 + 18.5 + 14(2.1)}{1720} = 0.24 > 0.2; \quad \text{Use Eq. 12.10.1}$$

(i) Moment magnification–nonsway magnifier B_1.

$$C_m = 0.6 - 0.4(M_1/M_2) = 0.6 - 0.4(21/40) = 0.4$$

$$B_1 = \frac{C_m}{1 - P_u/P_{e1}} = \frac{1.0}{1 - 416/20,111} = 0.4 < 1.0$$

(j) Moment magnification–sway magnifier B_2.

$$\sum P_u = 2(368) + 2.1(28) = 795 \text{ kips}$$

$$B_2 = \frac{1}{1 - \dfrac{\sum P_u}{\sum P_{e2}}} = \frac{1}{1 - \dfrac{795}{16318}} = 1.05 < 1.50$$

$\therefore$ Effective Length Method is permitted

$$M_{ux} = B_1 M_{nt} + B_2 M_{\ell t}$$
$$= 1.0(40) + 1.05(554) = 622 \text{ ft-kips}$$

(k) Check AISC Formula (H1-1a).

$$P_u = B_1 P_{nt} + B_2 P_{\ell t}$$
$$= 1.0(398) + 1.05(18.5) = 417 \text{ kips}$$

$$\frac{P_u}{\phi_c P_n} + \frac{8}{9}\left(\frac{M_{ux}}{\phi_b M_n}\right) \leq 1.0$$

$$\frac{417}{1720} + \frac{8}{9}\left(\frac{622}{975}\right) = 0.81 < 1.0 \qquad\qquad\qquad \textbf{OK}$$

The W14×145 section is acceptable according to AISC LRFD Method. ∎

The Direct Analysis Method

The new Direct Analysis Method, presented in AISC-Appendix 7, accounts for nominal geometric imperfection and stiffness reduction effects directly within the structural analysis. When the required strengths of members have been determined from an analysis conforming to these new requirements, individual members can be designed using an effective length factor of unity in calculating the nominal strengths of members subject to compression.

1. Use reduced member properties $EI^* = 0.8\tau_b EI$ and $EA^* = 0.8EA$ for all members participating in the lateral load resistance of the frame.
2. Apply the yielding reduction factor τ_b. In lieu of applying $\tau_b < 1.0$, it is permissible to increase the notional loads N_i by $0.001Y_i$.

3. Determine the notional lateral load to be applied to each level and perform a second-order analysis including the effects of all other lateral loads and all gravity loads stabilized by the structure.

4. Notional loads are applied as follows

$$N_i = 0.002Y_i \text{ - Notional load at level } i$$

where Y_i = gravity load (including dead, live, snow, and vertical wind if any) at level i from each LRFD load combination being considered.

5. Apply N_i as a minimum load in all *gravity-only load combinations* when second-order effects as measured by the ratio of the average second-order to first-order story drifts ($\Delta_{2nd}/\Delta_{1st}$ or B_2 *based on the nominal member properties*) is ≤ 1.5.

The second-order analysis must be carried out for all applicable load combinations (ASCE 7-05 Section 2.3.2). The design of the beams and girders is based on the requirements of AISC-Chapter F, AISC-Chapter I, and other sections of the Specification as applicable. For all beam-columns in moment frames apply the interaction AISC Formulas (H1-1) [or where applicable, AISC Formula (H1-2)] using an effective length factor $K = 1.0$. For determining member resistances in axial compression, use the equations of AISC-Chapter E, Sections E3 through E7 as applicable.

Note that the Specification allows the frame model to be set up to incorporate out-of-plumbness explicitly in the analysis without the need for applying notional lateral loads. However, it is usually easier to apply notional loads corresponding to a specified nominal out-of-plumbness along with other applied loads on the structure rather than modify the structure geometry.

EXAMPLE 12.11.2

Repeat Example 12.11.1 and investigate the acceptability of the W12×145 beam-column using the Direct Analysis Method of AISC-Appendix 7 with reduced flexural stiffness: $EI^* = 0.8\tau_b EI$.

Solution:
Calculate reduced stiffness for all members participating in lateral resistance according to AISC-A7.3.3. In this example all beams and columns are assumed to participate. It is noted that one can argue that only the columns are the structural elements resisting lateral forces and accordingly reduce the column stiffness only.

Check if $\dfrac{P_u}{P_y} \leq 0.50$

$$P_y = F_y A_g = 50(42.7) = 2135 \text{ kips}$$

$$P_u = 423 + 3(14) + 0.3 = 465 \text{ kips}$$

$$\frac{P_u}{P_y} = \frac{465}{2135} = 0.22 < 0.5 \quad \therefore \tau_b = 1.0$$

Therefore use $E^* = 0.8E = 0.8(29,000) = 23,200$ ksi for all beams and columns in the analysis of the frame. This reduction will not impact the moment distribution at the joint but will increase the lateral deflection due to the lateral forces and will reduce P_{e1} and P_{e2}. The end moments are the same as previously determined in Example 12.11.1. Calculate B_2 using the reduced stiffness:

$$\sum P_{nt} = 2(368) + 28(2.1) = 795 \text{ kips}$$
$$K = 1.57 \text{ (as calculated previously)}$$

$$\sum P_{e2} = \frac{\pi^2 E^*}{(KL)^2} = 2\frac{\pi^2(23,200)1710}{(1.57(13)12)^2} = 13,050 \text{ kips}$$

$$B_2 = \frac{1}{1 - \dfrac{\sum P_{nt}}{\sum P_{e2}}} = \frac{1}{1 - \dfrac{795}{13,050}} = 1.07$$

The magnifier B_2 can also be determined using the following expression for P_{e2}:

$$P_{e2} = R_M \frac{\sum HL}{\Delta_H}$$

where $R_M = 1.0$ for braced frames and 0.85 for moment frames

$\sum H = story$ shear produced by lateral forces $= 125$ kips

$L =$ story height $= 13(12) = 156$ in.

$\Delta_H = first$ order interstory drift due to lateral forces $= 1.13$ in. (results of elastic frame analysis)

$$P_{e2} = 0.85\frac{125(156)}{1.13} = 14,670 \text{ kips}$$

$$B_2 = \frac{1}{1 - \dfrac{1(795)}{14,670}} = 1.06$$

$$B_1 = \frac{C_m}{1 - \dfrac{P_u}{P_{e1}}}$$

$C_m = 0.4$ as previously calculated

$P_u = P_{ut} + B_2 P_u = 397 + 1.07(18.5) = 418$ kips

$$P_{e1} = \frac{\pi^2 EI^*}{(KL)^2} = \frac{\pi^2(23,200)1710}{[1.0(13)12]^2} = 16,090 \text{ kips}$$

$$B_1 = \frac{0.4}{1 - \dfrac{418}{16,090}} = 0.41 \quad \text{Use } B_1 = 1.0$$

$$M_u = B_1 M_{nt} + B_2 M_{\ell t} = 1.0(40) + 1.07(554) = 633 \text{ ft-kips}$$

Using AISC-Formula (H1-1a):

$$\frac{P_u}{\phi P_n} + \frac{8}{9}\frac{M_u}{\phi M_{ny}} = \frac{418}{1720} + \frac{8}{9}\frac{633}{975} = 0.82 < 1.0$$

W12×145 is adequate. It is worth noting that both methods produced the same interaction formula value. The advantage however is associated with appropriate bending moment values for the design of frame joints and beams.   ∎

The First-Order Method

According to AISC-C2.2b, a simplified first-order method can be used, when the required compressive strength is less than half the yield strength in all members whose flexural stiffnesses contribute to the lateral stability of the structure. In this method, the structure is analyzed using the nominal geometry and nominal elastic stiffness of all members; required member strengths are determined from a first-order analysis; all load combinations include an additional lateral load at each frame level of a magnitude based on the gravity load applied at that level and the lateral stiffness of the structure. The nominal strengths of compression members may be determined assuming $K = 1$ and the beam-column moments must be adjusted to account for non-sway amplification, using B_1.

This section provides a method for designing frames using a first-order elastic analysis with $K = 1.0$, provided the sidesway amplification (2nd order / 1st order) ≤ 1.5 (or $B_2 \leq 1.5$), and the required compressive strength of all members that are part of the lateral load resisting frame (other than truss members whose flexural stiffness is neglected in the analysis) have $P_u < 0.5P_y$. All load combinations must include an additional lateral load N_i applied in combination with other loads at each level of the structure specified by AISC Formula (C2-8). This equation is based on the clause within AISC-Appendix 7 that permits the notional load to be applied as a minimum lateral load in the gravity load only combinations and not in combination with other lateral loads when (2nd order/1st order) (or B_2) ≤ 1.5. The minimum value of N_i of $0.0042Y_i$ is based on the assumption of a minimum first-order drift ratio due to any effects of $\Delta/L = 0.002$. Note that a target maximum drift ratio, corresponding to drifts under the LRFD strength load combinations, can be assumed at the start of design to determine the additional lateral load N_i. As long as that drift ratio is not exceeded at any strength load level, the design will be conservative. The nonsway amplification of beam-column moments is addressed within the procedure specified in this section by applying the B_1 amplifier of AISC-C2.1 conservatively to the total member moments.

12.12 DESIGN PROCEDURE—AISC LRFD METHOD

Various design procedures for selecting an economical beam-column section have been proposed. The *AISC Specification* requirement is to satisfy Eq. 12.10.1 or 12.10.2, as appropriate (AISC-H1). In the past, the authors believed it was advantageous to convert, at least in an approximate way, the factored bending moment M_u into an equivalent axial compression load P_{uEQ}. Occasionally, conversion of the factored axial compression P_u into equivalent moment M_{uEQ} will speed the selection process. That approach is still appropriate, and was described in detail in the 4th edition (Sec. 12.12) of this text.

The more recent method of Aminmansour [12.57] is preferred by AISC in the authoring of the 2005 *AISC Manual* [1.15] (see Part 6). The end result must be the same; that is, to satisfy the requirements of AISC-H1.

The authors present Aminmansour's [12.57] development of design aids using the AISC LRFD Method formulation, Eq. 12.10.1, [which is AISC Formula (H1-1a)],

1. For $\dfrac{P_u}{\phi_c P_n} \geq 0.2$

$$\frac{P_u}{\phi_c P_n} + \frac{8}{9}\left(\frac{M_{ux}}{\phi_b M_{nx}} + \frac{M_{uy}}{\phi_b M_{ny}}\right) \leq 1.0 \qquad [12.10.1]$$

Rearrange Eq. 12.10.1 as follows,

$$\left(\frac{1}{\phi_c P_n}\right)P_u + \left(\frac{8}{9\phi_b M_{nx}}\right)M_{ux} + \left(\frac{8}{9\phi_b M_{ny}}\right)M_{uy} \leq 1 \qquad (12.12.1)$$

Define:

$$p = \frac{1}{\phi_c P_n} \qquad b_x = \frac{8}{9\phi_b M_{nx}} \qquad b_y = \frac{8}{9\phi_b M_{ny}} \qquad (12.12.2)$$

Note that p, b_x, and b_y are unique properties for each section, which vary with axial compression load, effective unbraced length KL, and for flexure with the laterally unbraced length L_b. In other words, these properties p, b_x, and b_y are functions of the design strength in axial compression, and design strengths in moment about the x- and y-axes. The *AISC Manual* Table 6-1 provides these properties for all W sections having $F_y = 50$ ksi. Equation 12.12.1 becomes

$$pP_u + b_xM_{ux} + b_yM_{uy} \le 1 \tag{12.12.3}$$

Similarly, using Eq. 12.10.2,

2. For $\dfrac{P_u}{\phi_c P_n} < 0.2$

$$\frac{P_u}{2\phi_c P_n} + \left(\frac{M_{ux}}{\phi_b M_{nx}} + \frac{M_{uy}}{\phi_b M_{ny}} \right) \le 1.0 \tag{12.10.2}$$

Substituting Eqs. 12.12.2 into Eq. 12.10.2 gives

$$\frac{1}{2}pP_u + \frac{9}{8}b_xM_{ux} + \frac{9}{8}b_yM_{uy} \le 1 \tag{12.12.4}$$

For the initial trial section, Aminmansour [12.57] compiled a set of average values for p, b_x, and b_y; these appear in Table 12.12.1. For every depth group of W sections, average values are given for a range of unbraced column lengths KL and a range of laterally unbraced lengths L_b. Table 12.12.1 gives values for lengths from 8 ft to 20 ft.

TABLE 12.12.1 Average Values of the p, b_x and b_y Coefficients $\times$ 1000*

| | Effective length KL (ft) with least radius of gyration or unbraced length L_b for x-axis bending | | | | | | | | | | | | | |
| | 8 | | 10 | | 12 | | 14 | | 16 | | 18 | | 20 | | |
	p	b_x	p	b_x	p	b_x	p	b_x	p	b_x	p	b_x	p	b_x	b_y
W40	0.19	0.20	0.19	0.20	0.20	0.20	0.21	0.21	0.22	0.21	0.23	0.22	0.24	0.23	1.10
W36	0.20	0.21	0.20	0.21	0.21	0.22	0.22	0.23	0.23	0.23	0.24	0.24	0.26	0.25	1.10
W33	0.22	0.26	0.22	0.26	0.23	0.27	0.24	0.28	0.25	0.29	0.26	0.30	0.28	0.31	1.20
W30	0.25	0.37	0.26	0.38	0.27	0.39	0.28	0.41	0.29	0.43	0.31	0.45	0.33	0.40	1.70
W27	0.29	0.41	0.30	0.42	0.31	0.43	0.33	0.44	0.35	0.46	0.37	0.48	0.40	0.41	1.90
W24	0.37	0.67	0.38	0.70	0.40	0.73	0.42	0.62	0.45	0.65	0.49	0.62	0.53	0.49	2.80
W21	0.48	0.96	0.51	1.01	0.54	1.08	0.59	0.82	0.64	0.86	0.72	0.67	0.81	0.56	3.80
W18	0.69	1.40	0.74	1.48	0.81	1.42	0.91	1.22	1.03	1.18	1.20	0.83	1.40	0.86	4.70
W16	1.22	3.07	1.34	3.32	1.51	2.74	1.74	2.93	2.06	1.98	2.46	1.76	2.92	1.83	10.80
W14	0.61	1.74	0.64	1.60	0.68	1.47	0.74	1.37	0.81	1.13	0.90	1.17	1.02	0.98	3.50
W12	0.94	2.54	1.00	2.27	1.08	2.37	1.19	2.22	1.32	1.85	1.50	1.92	1.73	1.80	7.0
W10	2.57	5.99	3.20	4.72	4.03	4.99	5.10	4.13	6.38	3.82	7.88	3.95	9.60	3.66	13.50
W8	5.20	11.18	7.04	8.68	9.44	9.20	12.43	7.47	15.99	7.80	20.09	7.25	24.76	6.72	21.80
W6	10.27	27.28	14.69	24.60	20.46	22.17	27.53	21.38	35.94	18.34	45.49	16.43	56.16	17.28	47.80
W5	7.02	24.34	8.87	25.48	11.79	26.74	16.05	28.12	20.96	29.66	26.53	27.51	32.75	29.0	48.10
W4	12.05	42.04	17.62	44.29	25.38	46.80	34.54	49.62	45.11	52.79	57.09		70.49		83.20

*Units; p in kips^{-1}; b_x and b_y in ft-kips^{-1}

12.13 EXAMPLES—AISC LRFD METHOD

Several examples are included to illustrate application of the interaction formulas using principles and procedures for columns from Chapter 6, and beams from Chapters 7 and 9.

EXAMPLE 12.13.1

Select the lightest W14 section to carry an axial compression load of 170 kips dead load and 40 kips live load, in combination with a bending moment of 140 ft-kips dead load, 140 ft-kips live load, and 420 ft-kips wind load. The member is part of a *braced* system, with transverse support provided in each principal direction at the top and bottom of a 14-ft length. Conservatively assume the moment causes single curvature and varies as shown in Fig. 12.13.1. Use A992 steel and Load and Resistance Factor Design.

Solution:

(a) Compute factored loads. Assume a first-order elastic analysis was performed to obtain the given forces. Using ASCE 7 LRFD Load Combinations 2 and 4,

1. Gravity Load

$$P_u = 1.2(170) + 1.6(40) = 268 \text{ kips}$$

$$M_{nt} = 1.2(140) + 1.6(140) = 392 \text{ ft-kips}$$

2. Gravity Load + Wind

$$P_u = 1.2(170) + 0.5(40) = 224 \text{ kips}$$

$$M_{nt} = 1.2(140) + 0.5(140) + 1.6(420) = 910 \text{ ft-kips}$$

It appears that loading Case 2 will be the more severe loading; thus, use that case to select the section. Then any other loading cases can be checked.

(b) Since the bending moment is dominant, estimate an intermediate p value and solve for the required b_x. Using the unbraced length $KL = 14$ ft, enter the *AISC Manual* Table 6-1, and find a p value for a W14 section. Estimate $p = 0.8$. (Note that *AISC Manual* Table 6-1 gives $p \times 1000$.)

Check

$$pP_u = \frac{P_u}{\phi_c P_n} = \frac{0.8}{1000}(224) = 0.179 < 0.2$$

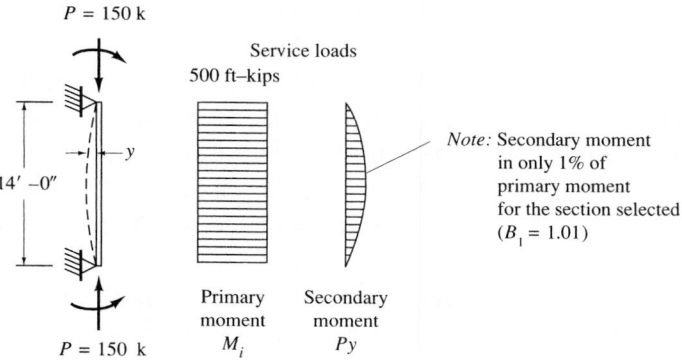

Figure 12.13.1
Examples 12.13.1
and 12.15.1.

AISC Formula (H1-1b) [Eq. 12.10.2] applies

$$\frac{1}{2}pP_u + \frac{9}{8}b_xM_{ux} = \frac{1}{2}\left(\frac{0.8}{1000}\right)224 + \frac{9}{8}\frac{b_x}{1000}(910) \leq 1.0$$

$$\text{Required } b_x = 0.89$$

Enter *AISC Manual* "Combined Axial and Bending—W shapes" Table 6-1, with $b_x = 0.89$ and $KL = 14$ ft to select a W14 section. Find W14×159 ($p = 0.541$, and $b_x = 0.826$ for $KL = 14$ ft. Check interaction equation,

$$pP_u = \frac{P_u}{\phi_c P_n} = \frac{0.541}{1000}(224) = 0.121 < 0.2$$

$$\frac{1}{2}pP_u + \frac{9}{8}b_xM_{ux} = \frac{1}{2}\left(\frac{0.541}{1000}\right)224 + \frac{9}{8}\left(\frac{0.826}{1000}\right)910 = 0.91 \qquad \textbf{OK}$$

Try the next lighter section; W14×145 ($p = 0.593$ and $b_x = 0.912$)

$$\frac{1}{2}pP_u + \frac{9}{8}b_xM_{ux} = \frac{1}{2}\left(\frac{0.593}{1000}\right)224 + \frac{9}{8}\left(\frac{0.912}{1000}\right)910 = 1.0 \qquad \textbf{OK}$$

Check W14×145 in detail using behavioral concepts.

(c) Column effect. Both the flange and the web satisfy $\lambda < \lambda_r$ to preclude local buckling prior to achieving column strength based on KL/r. Using $KL = 14$ ft with the minimum r_y for W14×145 in *AISC Manual* Table 4-1, find

$$\phi_c P_n = 1690 \text{ kips}$$

Check $P_u/(\phi_c P_n) \geq 0.2$,

$$\left(\frac{P_u}{\phi_c P_n} = \frac{224}{1690} = 0.132\right) < 0.2; \qquad \text{Use Eq. 12.10.2}$$

(d) Beam effect. The laterally unbraced length L_b is 14 ft.

$$M_p = F_y Z_x = 50(287)/12 = 1196 \text{ ft-kips}$$

$$L_p = \frac{300}{\sqrt{F_y}, \text{ksi}}r_y = \frac{300(3.98)}{\sqrt{50}(12)} = 14.1$$

Since $L_b < L_p$, $M_n = M_p$ regardless of C_b. In this case, $C_b = 1.0$. Check the "compact" section requirement; i.e., is $\lambda \leq \lambda_p$ for both flange and web? For $F_y = 50$ ksi steel, *AISC Specification* "User Note in F3" indicates nearly all current sections have "compact" flanges, and the exceptions are indicated. Regarding the "compact" web, *AISC Specification* "User Note in G2.1" gives the few exceptions. This W14×145 section is "compact". The detailed check is shown in Example 12.11.1, part (d). Thus,

$$\phi_b M_n = \phi_b M_p = 0.90(1196) = 1076 \text{ ft-kips}$$

$$\frac{M_u}{\phi_b M_n} = \frac{910}{1076} = 0.845$$

Quite likely this section will work.

(e) Moment magnification for W14×145. The slenderness ratio KL/r involved in moment magnification must relate to the *axis of bending*, in this case the *x*-axis.

$$\text{Axis of bending } \frac{KL}{r} = \frac{KL}{r_x} = \frac{1.0(14)12}{6.33} = 26.5$$

$$C_b = 1.0 \quad \text{(constant moment)}$$

$$P_{e1} = \frac{\pi^2 E A_g}{(KL/r)^2} = \frac{\pi^2(29,000)42.7}{(26.5)^2} = 17,400 \text{ kips}$$

$$B_1 = \frac{C_m}{1 - P_u/P_{e1}} = \frac{1.0}{1 - 224/17,400} = 1.01$$

(f) Check AISC-Formula (H1-1b), Eq. 12.10.2, omitting the bending term for the *y*-axis,

$$M_{ux} = M_{nt}B_1 = 910(1.01) = 919 \text{ ft-kips}$$

$$\frac{P_u}{2\phi_c P_n} + \left(\frac{M_{ux}}{\phi_b M_{nx}}\right) = \frac{0.132}{2} + 0.845 = 0.91 < 1.0 \qquad \textbf{OK}$$

The other loading case from part (a) clearly does not control.

Use W14×145, A992 steel. ∎

EXAMPLE 12.13.2

Select the lightest W section to carry axial compression of 30 kips dead load and 130 kips live load applied at an eccentricity $e = 5$ in. as shown in Fig. 12.13.2. The member is part of a braced frame, and is conservatively assumed loaded in single curvature with constant *e*. Use A992 steel and the AISC LRFD Method.

Solution:

(a) Compute factored loads. Assume a first-order elastic analysis was performed to obtain the given forces. Using ASCE 7 basic gravity load factors,

$$P_u = 1.2(30) + 1.6(130) = 244 \text{ kips}$$

$$M_u = P_u(e) = 244(5/12) = 102 \text{ ft-kips}$$

(b) The axial force in this case is dominant. Select an average b_x value and solve for the required *p*, using the appropriate interaction equation.

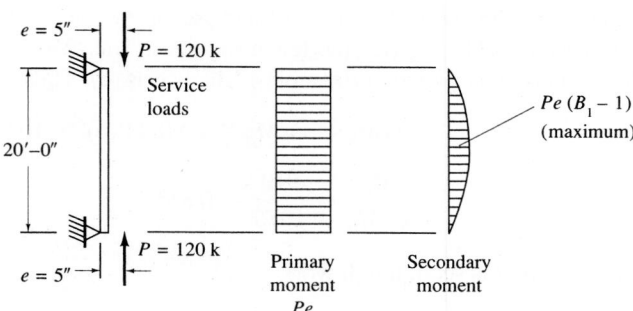

Figure 12.13.2
Example 12.13.2.

For W10 sections using $KL = 20$ ft, select from Table 12.12.1, an average value $b_x = 4.13$. (Note that *AISC Manual* Table 6-1 gives p and $b_x \times 1000$.) Assume AISC Formula (H1-1a), Eq. 12.10.1, controls.

$$pP_u + b_x M_{ux} = \left(\frac{p}{1000}\right)244 + \frac{4.13}{1000}(102) \le 1.0$$

$$\text{Required } p = 2.37$$

Find W10×60, $p = 2.38$ and $b_x = 3.73$. Check the section,

$$pP_u = \frac{P_u}{\phi_c P_n} = \frac{2.38}{1000}(244) = 0.58 > 0.2 \quad \text{Use Eq. 12.10.1,}$$

$$pP_u + b_x M_{ux} = 0.58 + \frac{3.73}{1000}(102) = 0.96 < 1.0 \qquad \qquad \textbf{OK}$$

Check other similar weight/ft deeper sections. Try W12×58 having $p = 2.54$ and $b_x = 3.41$.

$$pP_u = \frac{P_u}{\phi_c P_n} = \frac{2.54}{1000}(244) = 0.62 > 0.2 \quad \text{Use Eq. 12.10.1,}$$

$$pP_u + b_x M_{ux} = 0.62 + \frac{3.41}{1000}(102) = 0.97 < 1.0 \qquad \qquad \textbf{OK}$$

Both the W10×60 and the W12×58 are acceptable. Check the W12×58 in detail using behavioral concepts.

(c) Column effect. Both the flange and the web satisfy $\lambda < \lambda_r$ to preclude local buckling prior to achieving column strength based on KL/r. Using $KL = 20$ ft with the minimum r_y for W12×58 in *AISC Manual* Table 4-1, find

$$\phi_c P_n = 393 \text{ kips}$$

Check $P_u/(\phi_c P_n) \ge 0.2$,

$$\left(\frac{P_u}{\phi_c P_n} = \frac{244}{393} = 0.62\right) > 0.2; \qquad \text{Use Eq. 12.10.1}$$

(d) Beam action.

$$(L_p = 8.87 \text{ ft}) < (L_b = 20 \text{ ft}) < (L_r = 29.9 \text{ ft})$$

Therefore, $\phi_b M_n$ must be linearly interpolated between $\phi_b M_p$ (324 ft-kips from *AISC Manual* p. 3-70) and $\phi_b M_r$ (205 ft-kips from *AISC Manual* p. 3-70) according to Eq. 9.6.4. (Note that $\phi_b M_r$ is $\phi_b(0.7F_y S_x)$.)

$$M_n = C_b\left[M_p - (M_p - 0.7F_y S_x)\left(\frac{L_b - L_p}{L_r - L_p}\right)\right] \le M_p \qquad [9.6.4]$$

$$\phi_b M_n = C_b\left[\phi_b M_p - (\phi_b M_p - \phi_b M_r)\left(\frac{L_b - L_p}{L_r - L_p}\right)\right] \le \phi_b M_p$$

$$= 1.0\left[324 - (324 - 205)\left(\frac{20 - 8.87}{29.9 - 8.87}\right)\right] = 261 \text{ ft-kips}$$

(e) Moment magnification for W12×58. The slenderness ratio KL/r involved in moment magnification must relate to the *axis of bending*, in this case the x-axis.

$$\text{Axis of bending } \frac{KL}{r} = \frac{KL}{r_x} = \frac{1.0(20)12}{5.28} = 45.5$$

$$C_b = 1.0 \text{ (constant moment)}$$

$$P_{e1} = \frac{\pi^2 E A_g}{(KL/r)^2} = \frac{\pi^2 (29,000)17.0}{(45.5)^2} = 2350 \text{ kips}$$

$$B_1 = \frac{C_m}{1 - P_u/P_{e1}} = \frac{1.0}{1 - 244/2350} = 1.12$$

(f) Check AISC-Formula (H1-1a), Eq. 12.10.1, omitting the bending term for the y-axis,

$$M_{ux} = M_{nt}B_1 = 102(1.12) = 114 \text{ ft-kips}$$

$$\frac{P_u}{\phi_c P_n} + \frac{8}{9}\left(\frac{M_{ux}}{\phi_b M_{nx}}\right) = 0.62 + \frac{8}{9}\left(\frac{114}{261}\right) = 1.01 \approx 1.0 \qquad \textbf{OK}$$

The other loading case from part (a) clearly does not control.

<u>*Use*</u> W12×58, A992 steel. The W10×60 was also shown above to be acceptable. ■

EXAMPLE 12.13.3

Design a beam-column W section for the service load conditions shown in Fig. 12.13.3. The compression load P is 3 kips dead load and 25 kips live load, as might be caused by a crane, and the horizontal load H is 7 kips live load, as might be the horizontal effect of a crane. The member is part of a *braced* system, has support in the weak direction at mid-height, but only at the top and bottom for the strong direction. Use A992 steel and the AISC LRFD Method.

Solution:

The particular features of this example are (a) the bracing is not at the same locations for both principal directions; and (b) the lateral transverse loading causes the primary bending moment.

Assume hinged supports in both directions, top and bottom.

(a) Loading   **(b)** Idealized primary moment   **(c)** Assumption for C_m

Figure 12.13.3
Examples 12.13.3 and
12.15.2.

(a) Compute factored loads. Assume a first-order elastic analysis was performed to obtain the given forces. Using ASCE 7-LRFD load combination 2,

$$P_u = 1.2(40) + 1.6(100) = 208 \text{ kips at top}$$

$$W_u = 1.2(3) + 1.6(25) = 44 \text{ kips}$$

$$P_u = 208 + 44 = 252 \text{ kips at bottom } (controls)$$

$$H_u = 1.6(7) = 11.2 \text{ kips}$$

The horizontal reaction at the bottom of the column is $(H/2 + W/16)$, making the maximum moment,

$$\text{Maximum moment} = (H/2 + W/16)8$$

$$M_{nt} = (11.2/2 + 44/16)8 = 66.8 \text{ ft-kips}$$

(b) Establish effective lengths. The member must be viewed as a column without bending moment, then as a beam without column load, as in Fig. 12.13.4.
   For column action:

$$K_x L_x = 16 \text{ ft} \qquad K_y L_y = 8 \text{ ft}$$

It is required that $r_x/r_y \geq 2$ if $K_y L_y = 8$ ft is valid for entering *AISC Manual* Table 4-1.
   For beam action:

$$\text{Laterally unbraced length } L_b = 8 \text{ ft}$$

$$KL \text{ for moment magnification} = K_x L_x = 16 \text{ ft}$$

(c) Select a suitable W section using *AISC Manual* Table 6-1. The axial load is dominant for this design. Therefore, select an average b_x value for $L_b = 8$ ft and determine the maximum p permitted to satisfy the interaction equations. For W10 sections using $L_b = 8$ ft, select from Table 12.12.1, an average value $b_x = 5.99$.

$$pP_u + b_x M_{ux} = \left(\frac{p}{1000}\right)252 + \frac{5.99}{1000}(66.8) \leq 1.0$$

$$\text{Required } p \leq 2.38$$

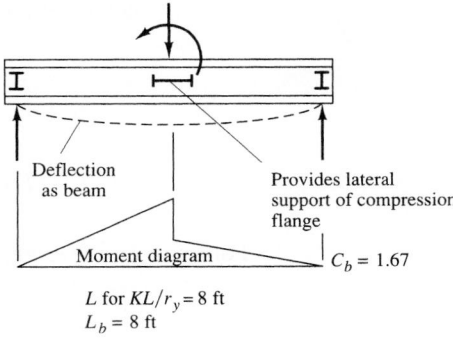

Figure 12.13.4
Separate beam action and column action from Examples 12.13.3 and 12.15.2.

8'–0"

8'–0"

$KL_x = 16$ ft $KL_y = 8$ ft

(a) Column action

Deflection as beam

Provides lateral support of compression flange

Moment diagram

$C_b = 1.67$

L for $KL/r_y = 8$ ft
$L_b = 8$ ft

(b) Beam action

From *AISC Manual* Table 6-1, select W10×49, $p = 2.34$ and $b_x = 3.92$.

$$pP_u = \frac{P_u}{\phi_c P_n} = \frac{2.34}{1000}(252) = 0.59 > 0.2 \text{ Use Eq. 12.10.1}$$

$$pP_u + b_x M_{ux} = 0.59 + \frac{3.92}{1000}(66.8) = 0.85 < 1.0$$

Try the next lighter section, W10×45, $p = 3.27$, and $b_x = 4.39$,

$$pP_u = \frac{P_u}{\phi_c P_n} = \frac{3.27}{1000}(252) = 0.82 > 0.2$$

$$pP_u + b_x M_{ux} = 0.82 + \frac{4.39}{1000}(66.8) = 1.11 > 1.0 \qquad \textbf{NG}$$

The W10×49 seems acceptable; also the W12×50 and the W14×53.

    (d) Check W10×49 for column action.

$$\frac{K_y L_y}{r_y} = \frac{1(8)12}{2.54} = 37.8 \qquad \frac{K_x L_x}{r_x} = \frac{1(16)12}{4.35} = 44.1$$

Since x-axis controls, can either evaluate the formula of AISC-E3, or modify the KL to enter the *AISC Manual* Table 4-1;

    Equivalent KL = Actual KL for x-axis divided by $r_x/r_y = 16/(1.71) = 9.4$ ft.

Find $\qquad\qquad\qquad\qquad \phi_c P_n = 562$ kips

Check $P_u/(\phi_c P_n) \geq 0.2$,

$$\left(\frac{P_u}{\phi_c P_n} = \frac{252}{562} = 0.45\right) > 0.2; \qquad \text{Use Eq. 12.10.1}$$

    (e) Beam action.

$$(L_p = 8.97 \text{ ft}) > (L_b = 8 \text{ ft})$$

Therefore,

$$\phi_b M_n = \phi_b M_p = \phi_b F_y Z_x = 0.90(50)60.4/12 = 227 \text{ ft-kips}$$

    (f) Moment magnification for W10×49. The slenderness ratio KL/r involved in moment magnification must related to the axis of bending, in this case the x-axis,

$$\text{Axis of bending } \frac{KL}{r} = \frac{KL}{r_x} = \frac{1.0(16)12}{4.35} = 44.1$$

$$P_{e1} = \frac{\pi^2 E A_g}{(KL/r)^2} = \frac{\pi^2(29{,}000)14.4}{(44.1)^2} = 2120 \text{ kips}$$

For transverse loading, C_m may be evaluated using Table 12.3.1, Case 3

$$C_m = 1 - 0.2\frac{P_u}{P_{e1}} = 1 - 0.2\left(\frac{252}{2120}\right) = 0.976$$

$$B_1 = \frac{C_m}{1 - P_u/P_{e1}} = \frac{0.976}{1 - 252/2120} = 1.11$$

(g) Check AISC Formula (H1-1a), Eq. 12.10.1, omitting the bending term for the y-axis,

$$M_{ux} = M_{nt}B_1 = 66.8(1.11) = 74.1 \text{ ft-kips}$$

$$\frac{P_u}{\phi_c P_n} + \frac{8}{9}\left(\frac{M_{ux}}{\phi_b M_{nx}}\right) = 0.45 + \frac{8}{9}\left(\frac{74.1}{227}\right) = 0.74 < 1.0 \qquad \text{Too Low}$$

(h) Try W10×45. Since $(r_x/r_y = 2.15) > 2$, KL for column action = 8 ft (rather than 9.4 ft when r_x/r_y was less than 2 for the W10×49).

$\phi_c P_n = 505$ kips from *AISC Manual* Table 4-1 for $K_y L_y = 8$ ft

$$\left(\frac{P_u}{\phi_c P_n} = \frac{252}{505} = 0.50\right) > 0.2$$

$$(L_p = 7.10 \text{ ft}) < (L_b = 8 \text{ ft}) < (L_r = 26.9 \text{ ft})$$

$\phi_b M_p = 206$ ft-kips $\phi_b M_r = 129$ ft-kips from *AISC Manual* Table 3-6

$$\phi_b M_n = C_b\left[\phi_b M_p - (\phi_b M_p - \phi_b M_r)\left(\frac{L_b - L_p}{L_r - L_p}\right)\right] \leq \phi_b M_p$$

$$= 1.0\left[206 - (206 - 129)\left(\frac{8 - 7.10}{26.9 - 7.10}\right)\right] = 203 \text{ ft-kips}$$

$$\text{Axis of bending } \frac{KL}{r} = \frac{KL}{r_x} = \frac{1.0(16)12}{4.32} = 44.4$$

$$P_{e1} = \frac{\pi^2 E A_g}{(KL/r)^2} = \frac{\pi^2(29,000)13.3}{(44.4)^2} = 1930 \text{ kips}$$

$$B_1 = \frac{C_m}{1 - P_u/P_{e1}} = \frac{0.976}{1 - 252/1930} = 1.12$$

$$M_{ux} = M_{nt}B_1 = 66.8(1.12) = 75 \text{ ft-kips}$$

$$\frac{P_u}{\phi_c P_n} + \frac{8}{9}\left(\frac{M_{ux}}{\phi_b M_{nx}}\right) = 0.50 + \frac{8}{9}\left(\frac{75}{203}\right) = 0.83 < 1.0 \qquad \text{Still Low}$$

(i) Try W10×39. Since $(r_x/r_y = 2.16) > 2$, KL for column action = 8 ft (rather than 9.4 ft when r_x/r_y was less than 2 for the W10×49).

$\phi_c P_n = 435$ kips from *AISC Manual* Table 4-1 for $K_y L_y = 8$ ft

$$\left(\frac{P_u}{\phi_c P_n} = \frac{252}{435} = 0.58\right) > 0.2$$

$$(L_p = 6.99 \text{ ft}) < (L_b = 8 \text{ ft}) < (L_r = 24.2 \text{ ft})$$

$\phi_b M_p = 176$ ft-kips $\phi_b M_r = 111$ ft-kips

$$\phi_b M_n = C_b\left[\phi_b M_p - (\phi_b M_p - \phi_b M_r)\left(\frac{L_b - L_p}{L_r - L_p}\right)\right] \leq \phi_b M_p$$

$$\phi_b M_n = 1.0\left[176 - (176 - 111)\left(\frac{8 - 6.99}{24.2 - 6.99}\right)\right] = 172 \text{ ft-kips}$$

$$\text{Axis of bending } \frac{KL}{r} = \frac{KL}{r_x} = \frac{1.0(16)12}{4.27} = 45.0$$

$$P_{e1} = \frac{\pi^2 EA_g}{(KL/r)^2} = \frac{\pi^2(29{,}000)11.5}{(45.0)^2} = 1630 \text{ kips}$$

$$B_1 = \frac{C_m}{1 - P_u/P_{e1}} = \frac{0.976}{1 - 252/1630} = 1.15$$

$$M_{ux} = M_{nt} B_1 = 66.8(1.15) = 72.1 \text{ ft-kips}$$

$$\frac{P_u}{\phi_c P_n} + \frac{8}{9}\left(\frac{M_{ux}}{\phi_b M_{nx}}\right) = 0.58 + \frac{8}{9}\left(\frac{72.1}{172}\right) = 0.98 < 1.0 \qquad \textbf{OK}$$

Use W10×39, A992 steel. ∎

EXAMPLE 12.13.4

Investigate the adequacy of the W8×24 section in Fig. 12.13.5, of A572 Grade 50 steel loaded as a beam-column. The axial compression is 15 kips dead load and 60 kips live load, and the uniformly distributed superimposed lateral load of 0.1 kips/ft dead load and 0.4 kips/ft live load causes bending about the weak axis. Use the AISC LRFD Method.

Solution:

(a) Compute factored loads. Assume a first-order elastic analysis was performed to obtain the given forces. Using ASCE 7-LRFD load combination 2,

$$w_u = 1.2(0.124) + 1.6(0.4) = 0.79 \text{ kips/ft}$$

(adding in 0.024 kip/ft beam weight)

$$M_{nt} = w_u L^2/8 = 0.79(10)^2/8 = 9.86 \text{ ft-kips}$$

$$P_u = 1.2(15) + 1.6(60) = 114 \text{ kips}$$

(b) Column action. In this example, buckling as a column occurs in the plane of bending, whereas in the three previous examples column buckling occurred about the axis which was not the axis of bending. Using *AISC Manual* Table 4-22 "Available Critical Stress for Compression Members,"

$$\text{Largest } \frac{KL}{r} = \frac{KL}{r_y} = \frac{1.0(10)12}{1.61} = 74.5; \qquad \phi_c F_{cr} = 30.0 \text{ ksi}$$

$$\phi_c P_n = \phi_c F_{cr} A_g = 30.0(7.08) = 212 \text{ kips}$$

Figure 12.13.5
Examples 12.13.4 and
12.15.3.

The ratio $P_u/\phi_c P_n = 114/212 = 0.54$, which exceeds 0.2; therefore AISC Formula (H1-1a), Eq. 12.10.1, applies.

(c) Beam action. Because bending accurs in the weak direction, lateral-torsional buckling is not a possible controlling limit state; nor is web local buckling. The section is "compact" according to AISC-B4 if flange local buckling is precluded; that is, when $\lambda \leq \lambda_p$ for the flange,

$$\left(\lambda = \frac{b_f}{2t_f} = \frac{6.50}{2(0.400)} = 8.1\right) \leq \left(\lambda_p = \frac{65}{\sqrt{F_y,\ \text{ksi}}} = \frac{65}{\sqrt{50}} = 9.2\right) \quad \textbf{OK}$$

Therefore, according to AISC-F6,

$$M_n = M_p = F_y Z_y = 50(8.57)/12 = 35.7 \text{ ft-kips}$$

There is a maximum value for M_n in weak-axis bending,

$$\text{Maximum } M_n = 1.6 F_y S_x = 1.6(50)5.63/12 = 37.5 \text{ ft-kips} > 35.7 \text{ ft-kips} \quad \textbf{OK}$$

$$\phi_b M_n = 0.90(35.7) = 32.1 \text{ ft-kips}$$

This value cannot be obtained from the *AISC Manual* Table 3-2 "Selection by Z_x" because those tables are only for *x*-axis bending.

(d) Moment magnification. The slenderness ratio *KL/r* involved in moment magnification must relate to the *axis of bending*, in this case the *y*-axis.

$$\text{Axis of bending } \frac{KL}{r} = \frac{KL}{r_y} = \frac{1.0(10)12}{1.61} = 74.5 \quad [\text{computed in part (b)}]$$

$$P_{e1} = \frac{\pi^2 E A_g}{(KL/r)^2} = \frac{\pi^2(29{,}000)7.08}{(74.5)^2} = 365 \text{ kips}$$

For transverse loading, C_m may be evaluated using Table 12.3.1, Case 1, which indicates

$$C_m = 1.0$$

$$B_1 = \frac{C_m}{1 - P_u/P_{e1}} = \frac{1.0}{1 - 114/365} = 1.45$$

(e) Check AISC Formula (H1-1a), Eq. 12.10.1, omitting the bending term for the *x*-axis,

$$M_{uy} = M_{nt} B_1 = 9.86(1.45) = 14.3 \text{ ft-kips}$$

$$\frac{P_u}{\phi_c P_n} + \frac{8}{9}\left(\frac{M_{uy}}{\phi_b M_{ny}}\right) = 0.54 + \frac{8}{9}\left(\frac{14.3}{32.1}\right) = 0.94 < 1.0 \quad \textbf{OK}$$

The W8×24, A572 Grade 50, *is acceptable.* ■

EXAMPLE 12.13.5

Select a suitable W section for the column member of the frame shown in Fig. 12.13.6. The joints are rigid to give frame action in the plane of bending; however, in the transverse direction sway bracing is provided and the attachments are assumed to be hinged. The frame is attached to other braced construction such that this frame is considered part of a *braced* system. Assume a structural analysis has been performed using factored loads and the results are as follows: factored axial compression $P_u = 128$ kips; and factored moment

Figure 12.13.6
Examples 12.13.5, 12.13.6,
and 12.15.4.

(a) Frame   **(b)** End view   **(c)** Loading and moment
diagram

at the top of column M_{nt2} = 234 ft-kips and at the bottom M_{nt1} = 117 ft-kips. Use A992 steel and the AISC LRFD Method.

Solution:

(a) Establish effective lengths. The member must be viewed as a column without bending moment, then as a beam without column load. For the plane of bending, the effective length factor K may be estimated as ≈ 0.7 from Fig. 6.9.5, or may be obtained from the Alignment Chart, Fig. 6.9.4a, as

$$G_A = 1.0 \text{ (fixed)} \qquad G_B = \frac{I/20}{3I/30} = 0.5 \qquad K_x = 0.73$$

For the transverse direction, the member is assumed hinged at the top and bottom; thus, $K_y = 1.0$. For beam action, $L_b = 10$ ft and for moment magnification,

$$K_x L_x = 0.73(20) = 14.6 \text{ ft}$$

(b) Select a trial section using *AISC Manual* Table 6-1. The bending moment in this case is dominant compared to the axial compression. From Table 12.12.1, select an average p value for $KL = 10$ ft (assuming y-axis buckling controls) and determine the maximum b_x permitted to satisfy the interaction equations.

For W10 sections, select $p = 3.2$ for $KL = 10$ ft

Assume AISC Formula (H1-1a), Eq. 12.10.1, applies

$$pP_u + b_x M_{ux} = \left(\frac{3.2}{1000}\right)128 + \frac{b_x}{1000}(234) \le 1.0$$

Required $b_x \le 2.52$

Select W10×77; $p = 1.15$ for $KL = 10$ ft; $b_x = 2.45$ for $L_b = 10$ ft
Check interaction equation:

$$pP_u = \frac{P_u}{\phi_c P_n} = \frac{1.15}{1000}(128) = 0.147 < 0.2$$

AISC Formula (H1-1b), Eq. 12.10.2, applies.

$$\frac{1}{2}pP_u + \frac{9}{8}b_x M_{ux} = \frac{1}{2}\left(\frac{1.15}{1000}\right)128 + \frac{9}{8}\left(\frac{2.45}{1000}\right)234 = 0.72 \le 1.0 \quad \text{Too Low}$$

Try W10×60; $p = 1.48$ for $KL = 10$ ft; $b_x = 3.22$ for $L_b = 10$ ft

Check interaction equation:

$$\frac{1}{2}pP_u + \frac{9}{8}b_xM_{ux} = \frac{1}{2}\left(\frac{1.48}{1000}\right)128 + \frac{9}{8}\left(\frac{3.22}{1000}\right)234 = 0.94 \leq 1.0 \qquad \textbf{OK}$$

The W10×60 is acceptable; the W12×53 and W14×53 may also be acceptable.

(c) Check the W12×53 in detail using behavioral concepts; column effect. Both the flange and the web satisfy $\lambda < \lambda_r$ to preclude local buckling prior to achieving column strength based on KL/r. Using $KL = 10$ ft with the minimum r_y for W12×53 in *AISC Manual* Table 4-1, find

$$\phi_c P_n = 590 \text{ kips}$$

Check $P_u/(\phi_c P_n) \geq 0.2$,

$$\left(\frac{P_u}{\phi_c P_n} = \frac{128}{590} = 0.217\right) > 0.2; \qquad \text{Use Eq. 12.10.1}$$

(d) Beam action. The laterally unbraced length L_b is 10 ft.

$$(L_p = 8.76 \text{ ft}) < (L_b = 10 \text{ ft}) < (L_r = 28.2 \text{ ft})$$

Therefore, $\phi_b M_n$ must be linearly interpolated between $\phi_b M_p$ (292 ft-kips from *AISC Manual* p. 3-71) and $\phi_b M_r$ (185 ft-kips from *AISC Manual* p. 3-71) according to Eq. 9.6.4. Note that $\phi_b M_r$ is $\phi_b(0.7F_y S_x)$.

$$M_n = C_b\left[M_p - (M_p - 0.7F_y S_x)\left(\frac{L_b - L_p}{L_r - L_p}\right)\right] \leq M_p \qquad [9.6.4]$$

$$\phi_b M_n = C_b\left[\phi_b M_p - (\phi_b M_p - \phi_b M_r)\left(\frac{L_b - L_p}{L_r - L_p}\right)\right] \leq \phi_b M_p$$

$$= 1.0\left[292 - (292 - 185)\left(\frac{10 - 8.76}{28.2 - 8.76}\right)\right] = 285 \text{ ft-kips}$$

(e) Moment magnification for W12×53. The slenderness ratio KL/r involved in moment magnification must relate to the *axis of bending*, in this case the *x*-axis.

$$\text{Axis of bending } \frac{KL}{r} = \frac{KL}{r_x} = \frac{0.7(20)12}{5.23} = 32.1$$

$$P_{e1} = \frac{\pi^2 E A_g}{(KL/r)^2} = \frac{\pi^2(29{,}000)15.6}{(32.1)^2} = 4330 \text{ kips}$$

For end moment loading, C_m is AISC Formula (C2-4), Eq. 12.4.6,

$$C_m = 0.6 - 0.4\frac{M_1}{M_2} = 0.6 - 0.4\left(\frac{117}{234}\right) = 0.4$$

$$B_1 = \frac{C_m}{1 - P_u/P_{e1}} = \frac{0.4}{1 - 128/4330} = 0.4(1.03) = 0.41 < 1.0$$

Thus, the magnified moment out away from the support does not exceed the primary moment at the support; thus, $B_1 = 1.0$.

(f) Check AISC Formula (H1-1a), Eq. 12.10.1, omitting the bending term for the y-axis.

$$M_{ux} = M_{nt}B_1 = 234(1.0) = 234 \text{ ft-kips}$$

$$\frac{P_u}{\phi_c P_n} + \frac{8}{9}\left(\frac{M_{ux}}{\phi_b M_{nx}}\right) = 0.216 + \frac{8}{9}\left(\frac{234}{285}\right) = 0.95 \le 1.0 \qquad \textbf{OK}$$

Use W12×53, A992 steel; could use W10×60 (checked above). The check of W14×53 gives 0.93 for Eq. 12.10.1. ∎

EXAMPLE 12.13.6

Repeat Example 12.13.5 except treat the frame as unbraced in the plane of bending. Though the total factored loads acting in this example are the same as in Example 12.13.5, they would have been determined by an entirely different procedure. Two elastic analyses under factored loads are required; a nonsway analysis for gravity loads, and a sway analysis for lateral loads. Assume the factored moments M_{nt} for the nonsway analysis are 42 ft-kips and 21 ft-kips at the top and bottom of the column, respectively. The factored moments $M_{\ell t}$ from the sway analysis are 192 ft-kips both at the top and the bottom. All moments cause double curvature in the column. Continue to use A992 steel and the AISC LRFD Method.

Solution:

(a) Effective lengths for column and beam action. A significant difference between this case and the preceding one is the effective length factor K_x for the plane of bending exceeds 1.0. Using the Alignment Chart, Fig. 6.9.4b for the *unbraced* frame (sidesway not prevented),

$$G_A = 1.0 \text{ (fixed)} \qquad G_B = \frac{I/20}{3I/30} = 0.5 \qquad K_x \approx 1.24$$

This value of K may be adjusted for inelastic buckling according to the discussion in Sec. 6.9.

(b) Select initial trial sections in W10, W12, and W14 using the interaction equations and the *AISC Manual* Table 6-1.

This time the M_u equation includes both the sway and nonsway terms. It is relatively easier to estimate B_1 for the braced frame (and for the nonsway part of the unbraced frame analysis) than it is to estimate B_2 for the sway analysis. Often B_1 is 1.0 or not much larger; however, B_2 is commonly in the range 1.2 to 1.5 and it may be larger. Practical designs should have B_2 not exceeding about 1.5.

For this design, estimate $B_1 = 1.0$ and $B_2 = 1.3$,

$$M_{ux} = M_{nt}B_1 + M_{\ell t}B_2$$

Estimated $M_{ux} = 42(1.0) + 192(1.3) = 292 \text{ ft-kips}$

Next, note that the *AISC Manual* Table 6-1 can be used directly for $KL = 10$ ft only when r_x/r_y exceeds

$$K_x L_x / K_y L_y = 1.24(20)/10 = 2.48$$

Selection of W10 or W12 sections may likely have their column strength controlled by column buckling in the plane of bending, while deeper sections will be controlled by weak-axis buckling.

In order to initiate the selection process, it is assumed that weak-axis buckling controls. The factored moment M_u is dominant compared to the factored axial compression P_u. Thus, assume p value for $KL = 10$ ft and solve for the corresponding maximum b_x. From Table 12.12.1, select for W10, $p = 3.2$ for $KL = 10$ ft. Solving for maximum b_x gives

$$pP_u + b_x M_{ux} = \left(\frac{3.2}{1000}\right)128 + \frac{b_x}{1000}(292) \le 1.0$$

$$\text{Required } b_x \le 2.02$$

Select from *AISC Manual* Table 6-1, W10×88; $p = 0.999$ for $KL = 10$ ft; $b_x = 2.11$ for $L_b = 10$ ft. It is noted that r_x/r_y for this section is 1.73, which means that column buckling strength is governed by strong axis buckling and the initial assumption is not correct. Thus, the unbraced length to be used is

$$\text{Equivalent } KL = 1.24(20)/(r_x/r_y) = 1.24(20)/1.73 = 14.3 \text{ ft}$$

For the W10×88, $p = 1.18$ for $KL = 14.3$,

$$pP_u = \frac{P_u}{\phi_c P_n} = \frac{1.18}{1000}(128) = 0.151 < 0.2$$

AISC Formula (H1-1b), Eq. 12.10.2, applies.

$$\frac{1}{2}pP_u + \frac{9}{8}b_x M_{ux} = \frac{1}{2}\left(\frac{1.18}{1000}\right)128 + \frac{9}{8}\left(\frac{2.11}{1000}\right)292 = 0.77 \le 1.0 \quad \text{Too Low}$$

Check the W10×68, $p = 1.53$ for $KL = 14.3$; $b_x = 2.81$ for $L_b = 10$ ft:

$$pP_u = \frac{P_u}{\phi_c P_n} = \frac{1.53}{1000}(128) = 0.196 < 0.2$$

AISC Formula (H1-1b), Eq. 12.10.2, applies.

$$\frac{1}{2}pP_u + \frac{9}{8}b_x M_{ux} = \frac{1}{2}\left(\frac{1.53}{1000}\right)128 + \frac{9}{8}\left(\frac{2.81}{1000}\right)292 = 1.02 \approx 1.0 \qquad \textbf{OK}$$

The W10×68 seems acceptable. It is always wise to check against the actual AISC formula after using tables to arrive at a tentative choice.

(c) Check the column effect.

Both the flange and the web satisfy $\lambda < \lambda_r$ to preclude local buckling prior to achieving column strength based on KL/r. Using $KL = 10$ ft with the minimum r_y for W10×68 in *AISC Manual* Table 4-1, find

$$\text{Equivalent } KL = 1.24(20)/(r_x/r_y) = 1.24(20)/1.71 = 14.5 \text{ ft}$$

$$\phi_c P_n = 617 \text{ kips}$$

Check $P_u/(\phi_c P_n) \ge 0.2$,

$$\left(\frac{P_u}{\phi_c P_n} = \frac{128}{617} = 0.207\right) > 0.2; \qquad \text{Use Eq. 12.10.1}$$

(d) Beam action. The laterally unbraced length L_b is 10 ft.

$$(L_p = 9.15 \text{ ft}) < (L_b = 10 \text{ ft}) < (L_r = 40.6 \text{ ft})$$

Therefore, $\phi_b M_n$ must be linearly interpolated between $\phi_b M_p$ (320 ft-kips from *AISC Manual* p. 3-73) and $\phi_b M_r$ (199 ft-kips from *AISC Manual* p. 3-73) according to Eq. 9.6.4. Note that $\phi_b M_r$ is $\phi_b(0.7 F_y S_x)$.

$$M_n = C_b \left[M_p - (M_p - 0.7 F_y S_x) \left(\frac{L_b - L_p}{L_r - L_p} \right) \right] \le M_p \qquad [9.6.4]$$

$$\phi_b M_n = C_b \left[\phi_b M_p - (\phi_b M_p - \phi_b M_r) \left(\frac{L_b - L_p}{L_r - L_p} \right) \right] \le \phi_b M_p$$

$$= 1.0 \left[320 - (320 - 199) \left(\frac{10 - 9.15}{40.6 - 9.15} \right) \right] = 317 \text{ ft-kips}$$

(e) Moment magnification for W10×68—nonsway part B_1. For end moment loading in the nonsway part, C_m will be 0.4, the same as in Example 12.13.5, part (e). In this example it is prefereable to use $K_x = 1.0$ for P_{e1}. However, $B_1 = 1.0$ whether $K_x = 0.7$ or $K_x = 1.0$.

(f) Moment magnification—sway part B_2. The slenderness ratio in the plane of the frame increases from the nonsway part because here $K_x = 1.24$; thus,

$$\text{Axis of bending} \quad \frac{KL}{r} = \frac{KL}{r_x} = \frac{1.24(20)12}{4.44} = 67.0$$

$$P_{e1} = \frac{\pi^2 E A_g}{(KL/r)^2} = \frac{\pi^2 (29{,}000) 20.0}{(67.0)^2} = 1275 \text{ kips}$$

Since in this example, two equal sized columns having equal loading are participating in the sway resistance,

$$\Sigma P_{e2} = 2(1275) = 2550 \text{ kips}$$

$$\Sigma P_u = 2(128) = 256 \text{ kips}$$

Then Eq. 12.11.12 for the sway magnifier B_2 gives

$$B_2 = \frac{1}{1 - \dfrac{\Sigma P_u}{\Sigma P_{e2}}} = \frac{1}{1 - \dfrac{256}{2550}} = 1.11$$

(g) Check AISC Formula (H1-1a), Eq. 12.10.1, omitting the bending term for the y-axis

$$M_{ux} = M_{nt} B_1 + M_{\ell t} B_2 = 42(1.0) + 192(1.11) = 255 \text{ ft-kips}$$

$$\frac{P_u}{\phi_c P_n} + \frac{8}{9} \left(\frac{M_{ux}}{\phi_b M_{nx}} \right) = 0.207 + \frac{8}{9} \left(\frac{255}{317} \right) = 0.92 \le 1.0 \qquad \textbf{OK}$$

Use W10×68, A992 steel. ∎

EXAMPLE 12.13.7

Select the lightest W12 section to carry an axial compression in addition to biaxial bending, loaded as shown in Fig. 12.13.7. Assume a first-order structural analysis has been performed using factored loads. The results give $P_u = 375$ kips, $M_{ntx} = 38$ ft-kips about the

Figure 12.13.7
Example 12.13.7.

x-axis at the top of the column, and M_{nty} = 14 ft-kips about the y-axis at the top of the column. Use A572 Grade 50 steel and the AISC LRFD Method.

Solution:

(a) Effective lengths for column and beam action.

Column action:
$$K_x L_x = 1.0(15) = 15 \text{ ft}$$
$$K_y L_y = 1.0(7.5) = 7.5 \text{ ft}$$

Beam action:
$$L_b = 7.5 \text{ ft for } x\text{-axis bending}$$
$$C_{bx} = 1.25 \text{ (Eq. 9.6.11)(AISC-F1)}$$
(from Table 9.6.3 for
$$M_1/M_2 = -19/38 = -0.5)$$
$$L_b = \text{not applicable for } y\text{-axis bending}$$
$$C_{by} = \text{not applicable for } y\text{-axis bending}$$

Moment magnification:
$$KL = 15 \text{ ft for } P_{ex}$$
$$C_{mx} = 0.6$$
$$KL = 7.5 \text{ ft for } P_{ey}$$
$$C_{my} = 0.6 - (3/14) = 0.4$$

Note that lateral-torsional buckling is not a possible limit state for weak-axis bending of W sections; therefore, L_b and C_b are not applicable for the y-axis.

(b) Select a trial section. Since the axial compression force is large, select b_x and b_y values and solve for the maximum p value. From Table 12.12.1, select trial b_x = 2.54 for L_b = 8 ft and trial b_y = 7.0 (values are times 1000).

$$pP_u + b_x M_{ux} + b_y M_{uy} = \left(\frac{p}{1000}\right)375 + \frac{2.54}{1000}(38) + \frac{7.0}{1000}(14) \le 1.0$$

$$\text{Required } p \le 2.15$$

From *AISC Manual* Table 6-1, assume weak-axis buckling controls (i.e., use $K_y L_y = 1.0(7.5)$ = 7.5 ft); try W12×45; p = 1.98, b_x = 3.75, b_y = 12.5.

$$pP_u = \frac{P_u}{\phi_c P_n} = \frac{1.98}{1000}(375) = 0.743 > 0.2; \quad \text{Use Eq. 12.10.1}$$

$$pP_u + b_x M_{ux} + b_y M_{uy} = \left(\frac{1.98}{1000}\right)375 + \frac{3.75}{1000}(38) + \frac{12.5}{1000}(14) = 1.06 > 1.0$$

Try W12×50, $p = 1.78$, $b_x = 3.34$, $b_y = 11.1$. Using Eq. 12.10.1,

$$pP_u + b_x M_{ux} + b_y M_{uy} = \left(\frac{1.78}{1000}\right)375 + \frac{3.34}{1000}(38) + \frac{11.1}{1000}(14) = 0.95 < 1.0 \quad \textbf{OK}$$

(c) Column action.

$$\frac{KL}{r_x} = \frac{15(12)}{5.18} = 34.7 \qquad \frac{KL}{r_y} = \frac{7.5(12)}{1.96} = 45.9$$

Find from *AISC Manual* Table 4-22 (p. 4-319), $\phi_c F_{cr} = 38.5$ ksi.

$$\phi_c P_n = \phi_c F_{cr} A_g = 38.5(14.7) = 566 \text{ kips}$$

The ratio $P_u/\phi_c P_n = 375/566 = 0.66$ which exceeds 0.2; therefore, AISC Formula (H1-1a), Eq. 12.10.1, applies.

(d) Beam action—*x*-axis. From *AISC Manual* Table 3-2 (p. 3-17) "Selection by Z_x," for W12×50 find

$$L_p = 6.92 \text{ ft}$$
$$L_r = 23.9 \text{ ft}$$
$$\phi_b M_{px} = 270 \text{ ft-kips}$$
$$\phi_b M_{rx} = 169 \text{ ft-kips}$$
$$(L_p = 6.92 \text{ ft}) < (L_b = 7.5 \text{ ft}) < (L_r = 23.9 \text{ ft}),$$

Therefore, $\phi_b M_n$ must be linearly interpolated between $\phi_b M_p$ (270 ft-kips from *AISC Manual* p. 3-17) and $\phi_b M_r$ (169 ft-kips from *AISC Manual* p. 3-17) according to Eq. 9.6.4. Note that $\phi_b M_r$ is $\phi_b(0.7 F_y S_x)$.

$$\phi_b M_n = C_b \left[\phi_b M_p - (\phi_b M_p - \phi_b M_r)\left(\frac{L_b - L_p}{L_r - L_p}\right) \right] \leq \phi_b M_p$$

$$= 1.25 \left[270 - (270 - 169)\left(\frac{7.5 - 6.92}{23.9 - 6.92}\right) \right] = 333 \text{ ft-kips}$$

The result for $\phi_b M_n$ cannot exceed $\phi_b M_p = 270$ ft-kips. Thus, $\phi_b M_n = 270$ ft-kips.

(e) Beam action—*y*-axis. Since lateral-torsional buckling and web local buckling cannot be applicable limit states, and $\lambda < \lambda_p$ for the flange precludes local flange buckling,

$$M_{ny} = M_{py} = F_y Z_y = 50(21.3)/12 = 88.8 \text{ ft-kips}$$

$$\phi_b M_{ny} = 0.90(88.8) = 80 \text{ ft-kips}$$

Checking AISC-F6.1,

$$M_{py} = 88.8 \text{ ft-kips} < [1.6 F_y S_y = 1.6(50)13.9/12 = 92.7 \text{ ft-kips}] \quad \textbf{OK}$$

All standard I-shaped sections have M_p slightly greater than the previous $1.5 F_y S_y$ limit with respect to *y*-axis bending. Since the 1.5 number is a judgment value, raising it to 1.6 seems reasonable; it will not often control using 1.6.

(f) Moment magnification—*x*-axis. The slenderness ratio *KL/r* involved in moment magnification must relate to the *axis of bending*, in this case the *x*-axis,

$$\text{Axis of bending } \frac{KL}{r} = \frac{KL}{r_x} = \frac{15(12)}{5.18} = 34.7$$

$$P_{e1} = \frac{\pi^2 E A_g}{(KL/r)^2} = \frac{\pi^2(29,000)14.7}{(34.7)^2} = 3500 \text{ kips}$$

For end moment loading, C_m is AISC Formula (C2-4), Eq. 12.4.6,

$$C_m = 0.6 - 0.4\frac{M_1}{M_2} = 0.6 - 0.4\left(\frac{0}{38}\right) = 0.6$$

$$B_1 = \frac{C_m}{1 - P_u/P_{e1}} = \frac{0.6}{1 - 375/3500} = 0.67 < 1.0$$

Thus, the magnified moment out away from the support does not exceed the primary moment at the support; thus, $B_{1x} = 1.0$.

(g) Moment magnification—*y*-axis. The slenderness ratio *KL/r* involved in moment magnification must relate to the *axis of bending*, in this case the *y*-axis,

$$\text{Axis of bending } \frac{KL}{r} = \frac{KL}{r_y} = \frac{7.5(12)}{1.96} = 45.9$$

$$P_{e1} = \frac{\pi^2 E A_g}{(KL/r)^2} = \frac{\pi^2(29,000)14.7}{(45.9)^2} = 2000 \text{ kips}$$

For end moment loading, C_m is AISC Formula (C2-4), Eq. 12.4.6,

$$C_m = 0.6 - 0.4\frac{M_1}{M_2} = 0.6 - 0.4\left(\frac{3}{14}\right) = 0.51$$

$$B_1 = \frac{C_m}{1 - P_u/P_{e1}} = \frac{0.51}{1 - 375/2000} = 0.63 < 1.0$$

Thus, the magnified moment out away from the support does not exceed the primary moment at the support; thus, $B_{1y} = 1.0$.

(h) Check AISC Formula (H1-1a), Eq. 12.10.1,

$$M_{ux} = M_{ntx}B_{1x} = 38(1.0) = 38 \text{ ft-kips}$$

$$M_{uy} = M_{nty}B_{1y} = 14(1.0) = 14 \text{ ft-kips}$$

$$\frac{P_u}{\phi_c P_n} + \frac{8}{9}\left(\frac{M_{ux}}{\phi_b M_{nx}}\right) + \frac{8}{9}\left(\frac{M_{uy}}{\phi_b M_{ny}}\right) \le 1.0$$

$$\frac{375}{566} + \frac{8}{9}\left(\frac{38}{270}\right) + \frac{8}{9}\left(\frac{14}{80}\right) = 0.94 \le 1.0 \qquad \textbf{OK}$$

Use W12×50, A992 steel.   ∎

SELECTED REFERENCES

12.1. Charles Massonnet. "Stability Considerations in the Design of Steel Columns," *Journal of the Structural Division*, ASCE, **85**, ST7 (September 1959), 75–111.

12.2. Walter J. Austin. "Strength and Design of Metal Beam-Columns," *Journal of the Structural Division*, ASCE, **87**, ST4(April 1961), 1–32.

12.3. Nestor R. Iwankiw. "Note on Beam-Column Moment Amplification Factor," *Engineering Journal*, AISC, **21**, 1 (First Quarter 1984), 21–23; Disc. by Le-Wu Lu, **22**, 1 (First Quarter 1985), 47–48; Disc. by Joseph A. Yura, **22**, 1 (First Quarter 1985), 48.

12.4. Robert L. Ketter. "Further Studies of the Strength of Beam-Columns," *Journal of the Structural Division*, ASCE, **87**, ST6 (August 1961), 135–152. Also *Transactions*, ASCE, **127** (1962), Part II, 244–266.

12.5. M. R. Horne. "The Stanchion Problem in Frame Structures Designed According to Ultimate Carrying Capacity," *Proc. Inst. Civil Engrs.*, **5**, 1, Part III (April 1956), 105–146.

12.6. Wai-Fah Chen and Suiping Zhou. "C_m Factor in Load and Resistance Factor Design," *Journal of Structural Engineering*, ASCE, **113**, 8 (August 1987), 1738–1754.

12.7. Robert L. Ketter, Edmund L. Kaminsky, and Lynn S. Beedle. "Plastic Deformation of Wide-Flange Beam-Columns," *Transactions*, ASCE, **120** (1955), 1028–1069.

12.8. Theodore V. Galambos and Robert L. Ketter. "Columns Under Combined Bending and Thrust," *Journal of the Engineering Mechanics Division*, ASCE, **85**, EM2 (April 1959), 1–30. Also *Transactions*, ASCE, **126** (1961), Part 1, 1–25.

12.9. George F. Hauck and Seng-Lip Lee. "Stability of Elasto-Plastic Wide-Flange Columns," *Journal of the Structural Division*, ASCE, **89**, ST6 (December 1963), 297–324.

12.10. S. L. Lee and G. F. Hauck. "Buckling of Steel Columns Under Arbitrary End Loads," *Journal of the Structural Division*, ASCE, **90**, ST2 (April 1964), 179–200.

12.11. S. L. Lee and S. C. Anand. "Buckling of Eccentrically Loaded Steel Columns," *Journal of the Structural Division*, ASCE, **92**, ST2 (April 1966), 351–370.

12.12. Edwin C. Rossow, George B. Barney, and Seng-Lip Lee. "Eccentrically Loaded Steel Columns with Initial Curvature," *Journal of the Structural Division*, ASCE, **93**, ST2 (April 1967), 339–358.

12.13. Le-Wu Lu and Hassan Kamalvand. "Ultimate Strength of Laterally Loaded Columns," *Journal of the Structural Division*, ASCE, **94**, ST6 (June 1968), 1505–1524.

12.14. W. F. Chen. "Further Studies of Inelastic Beam-Column Problem," *Journal of the Structural Division*, ASCE, **97**, ST2 (February 1971), 529–544.

12.15. Gordon W. English and Peter F. Adams. "Experiments on Laterally Loaded Steel Beam-Columns," *Journal of the Structural Division*, ASCE, **99**, ST7 (July 1973), 1457–1470.

12.16. Francois Cheong-Siat-Moy. "Methods of Analysis of Laterally Loaded Columns," *Journal of the Structural Division*, ASCE, **100**, ST5 (May 1974), 953–970.

12.17. Francois Cheong-Siat-Moy. "General Analysis of Laterally Loaded Beam-Columns," *Journal of the Structural Division*, ASCE, **100**, ST6 (June 1974), 1263–1278.

12.18. B. G. Johnston. "Lateral Buckling of I-Section Columns with Eccentric End Loads in the Plane of the Web," *Transactions*, ASME, **62** (1941), A–176.

12.19. Mario G. Salvadori. "Lateral Buckling of I-Beams," *Transactions*, ASCE, **120** (1955), 1165–1182.

12.20. M. Salvadori. "Lateral Buckling of Eccentrically Loaded I-Columns," *Transactions*, ASCE, **121** (1956), 1163–1178.

12.21. Constancio Miranda and Morris Ojalvo. "Inelastic Lateral-Torsional Buckling of Beam Columns," *Journal of the Engineering Mechanics Division*, ASCE, **91**, EM6 (December 1965), 21–37.

12.22. Yushi Fukumoto and T. V. Galambos. "Inelastic Lateral-Torsional Buckling of Beam Columns," *Journal of the Structural Division*, ASCE, **92**, ST2 (April 1966), 41–61.

12.23. T. V. Galambos, P. F. Adams, and Y. Fukumoto. "Further Studies on the Lateral-Torsional Buckling of Steel Beam-Columns," *Bulletin No. 115*, Welding Research Council, July 1966.

12.24. Ralph C. Van Kuren and T. V. Galambos. "Beam Column Experiments," *Journal of the Structural Division*, ASCE, **90**, ST2 (April 1964), 223–256.

12.25. Mark A. Bradford and Nicholas S. Trahair. "Inelastic Buckling Tests on Beam-Columns," *Journal of Structural Engineering*, ASCE, **112**, 3 (March 1986), 538–549.

12.26. Mark A. Bradford, Peter E. Cuk, Marian A. Gizejowski, and Nicholas Trahair. "Inelastic Lateral Buckling of Beam-Columns," *Journal of Structural Engineering*, ASCE, **113**, 11 (November 1987), 2259–2277.

12.27. T. B. Pekoz and G. Winter. "Torsional-Flexural Buckling of Thin-Walled Sections Under Eccentric Load," *Journal of the Structural Division*, ASCE, **95**, ST5 (May 1969), 941–963.

12.28. Teoman B. Pekoz and N. Celebi. "Torsional Flexural Buckling of Thin-Walled Sections Under Eccentric Load," *Cornell Engineering Research Bulletin 69-1.* Ithaca, NY: Cornell University, 1969.

12.29. Wei-Wen Yu. *Cold-Formed Steel Design*, 2nd ed. New York: John Wiley & Sons, Inc., 1991, Chap. 6.

12.30. Francois Cheong-Siat-Moy and Tom Downs. "New Interaction Equation for Steel Beam-Columns," *Journal of the Structural Division*, ASCE, **106,** ST5 (May 1980), 1047–1061.

12.31. Lian Duan and Wai-Fah Chen. "Design Interaction Equation for Steel Beam-Columns," *Journal of Structural Engineering*, ASCE, **115,** 5 (May 1989), 1225–1243. Disc. **117,** 8 (August 1991), 2553–2561.

12.32. Iqbal S. Sohal, Lian Duan, and Wai-Fah Chen. "Design Interaction Equations for Steel Members," *Journal of Structural Engineering*, ASCE, **115,** 7 (July 1989), 1650–1665. Disc., **117,** 7 (July 1991), 2189–2196.

12.33. C. S. Cal, X. L. Liu, and W. F. Chen. "Further Verifications of Beam-Column Strength Equations," *Journal of Structural Engineering*, ASCE, **117,** 2 (February 1991), 501–513.

12.34. Charles Birnstiel and James Michalos. "Ultimate Load of H-Columns Under Biaxial Bending," *Journal of the Structural Division*, ASCE, **89,** ST2 (April 1963), 161–197.

12.35. Charles G. Culver. "Exact Solution of Biaxial Bending Equations," *Journal of the Structural Division*, ASCE, **92,** ST2 (April 1966), 63–83.

12.36. Charles G. Culver, "Initial Imperfections in Biaxial Bending," *Journal of the Structural Division*, ASCE, **92,** ST3 (June 1966), 119–135.

12.37. Gunnar A. Harstead, Charles Birnstiel, and Keh-Chun Leu. "Inelastic Behavior of H-Columns Under Biaxial Bending," *Journal of the Structural Division*, ASCE, **94,** ST10 (October 1968), 2371–2398.

12.38. Ishwar C. Syal and Satya S. Sharma. "Biaxially Loaded Beam-Column Analysis," *Journal of the Structural Division*, ASCE, **97,** ST9 (September 1971), 2245–2259.

12.39. Sakda Santathadaporn and Wai F. Chen. "Analysis of Biaxially Loaded Steel H-Columns," *Journal of the Structural Division*, ASCE, **99,** ST3 (March 1973), 491–509.

12.40. Wai F. Chen and Toshio Atsuta. "Ultimate Strength of Biaxially Loaded Steel H-Columns," *Journal of the Structural Division*, ASCE, **99,** ST3 (March 1973), 469–489.

12.41. Charles Birnstiel. "Experiments on H-Columns Under Biaxial Bending," *Journal of the Structural Division*, ASCE, **94,** ST10 (October 1968), 2429–2449.

12.42. Wai F. Chen and Sakda Santathadaporn. "Review of Column Behavior Under Biaxial Loading," *Journal of the Structural Division*, ASCE, **94,** ST12 (December 1968), 2999–3021.

12.43. Negussie Tebedge and Wai F. Chen. "Design Criteria for H-Columns Under Biaxial Loading," *Journal of the Structural Division*, ASCE, **100,** ST3 (March 1974), 579–598.

12.44. Reidar Bjorhovde, Theodore V. Galambos, and Mayasandra K. Ravindra. "LRFD Criteria for Steel Beam-Columns," *Journal of the Structural Division*, ASCE, **104,** ST9 (September 1978), 1371–1387.

12.45. Joseph A. Yura. "Combined Bending and Axial Load," Notes distributed by AISC at 1988 National Steel Construction Conference, Miami Beach, FL to assist classroom teaching of LRFD. Chicago, IL: American Institute of Steel Construction, 1988.

12.46. W. F. Chen and E. M. Lui. "Columns with End Restraint and Bending in Load and Resistance Factor Design," *Engineering Journal*, AISC, **22,** 3 (Third Quarter 1985), 105–132. Errata: **25,** 2 (Second Quarter 1988), 59.

12.47. Lian Duan, Iqbal S. Sohal, and Wai-Fah Chen. "On Beam-Column Moment Amplification Factor," *Engineering Journal*, AISC, **26,** 4 (Fourth Quarter 1989), 130–135. Disc. **27,** 4 (Fourth Quarter 1990), 168–172.

12.48. James G. MacGregor and Sven E. Hage. "Stability Analysis and Design of Concrete Frames," *Journal of the Structural Division*, ASCE, **103,** ST10 (October 1977), 1953–1970.

12.49. William J. LeMessurier, Robert J. McNamara, and J. C. Scrivener. "Approximate Analytical Model for Multistory Frames," *Engineering Journal*, AISC, **11,** 4 (Fourth Quarter 1974), 92–98.

12.50. Wm. J. LeMessurier. "A Practical Method of Second Order Analysis: Part 1—Pin Jointed Systems," *Engineering Journal*, AISC, **13,** 4 (Fourth Quarter 1976), 89–96.

12.51. Wm. J. LeMessurier. "A Practical Method of Second Order Analysis: Part 2—Rigid Frames," *Engineering Journal*, AISC, **14,** 2 (Second Quarter 1977), 49–67.

12.52. Brian R. Wood, Denis Beaulieu, and Peter F. Adams. "Column Design by *P* Delta Method," *Journal of the Structural Division*, ASCE, **102,** ST2 (February 1976), 411–427.

12.53. Brian R. Wood, Denis Beaulieu, and Peter F. Adams. "Further Aspects of Design by *P*-Delta Method," *Journal of the Structural Division*, ASCE, **102,** ST3 (March 1976), 487–500.

12.54. H. Scholz. "*P*-Delta Effect in Elastic Analysis of Sway Frames," *Journal of Structural Engineering*, ASCE, **113,** 3 (March 1987), 534–545.

12.55. I. S. Sohal and N. A. Syed. "Inelastic Amplification Factor for Design of Steel Beam-Columns," *Journal of Structural Engineering*, ASCE, **118,** 7 (July 1992), 1822–1839.

12.56. Thomas Sputo. "History of Steel Beam-Column Design," *Journal of Structural Engineering*, ASCE, **119,** 2 (February 1993), 547–557.

12.57. Abbas Aminmansour, "A New Approach for Design of Steel Beam-Columns," *Engineering Journal*, AISC, **37,** 2 (Second Quarter 2000), 41–72.

PROBLEMS

All problems (except Probs. 12.24–12.28) are to be done according to AISC Load and Resistance Factor Design Method. The requirement of W section is intended to include W, S, and M sections. All given loads are service loads unless otherwise indicated. Assume lateral support consists of translational restraint but not moment (rotational) restraint, unless otherwise indicated. Assume all standard sections are equally readily available in the indicated grade of steel (even though actually they are not). A figure showing span and loading is required, and after making a design selection a final check of the strength interaction equation (involving factored loads P_u and M_u, and design strengths ϕP_n and ϕM_n).

Problems 12.1 through 12.23 relate to design considerations. Problems 12.24 through 12.28 relate to other theoretical considerations.

12.1. Investigate the adequacy of the section in the accompanying figure assuming the loading is 30% dead load and 70% live load. No translation of joints can occur, and external lateral support is provided at the ends only.

12.2. Investigate the given section in the accompanying figure with regard to safety if primary bending is

in the weak direction and the loading w is 0.02 kip/ft dead load and 0.08 kip/ft live load and P is 2 kips dead load and 18 kips live load. A572 Grade 50 steel.

12.3. Determine the maximum service load W (kips) at the mid-height of the beam-column shown in the accompanying figure. Assume the member is hinged with

Problem 12.1

Problem 12.2

Problem 12.3

respect to bending in both x and y directions at the top and bottom. Additionally, lateral support occurs in the weak direction at mid-height.

12.4. Investigate the adequacy of the given section of the accompanying figure. No joint translation can occur and external lateral support is provided at the ends only.

12.5. Determine the safe service load W permitted for this braced frame beam-column of the accompanying figure.

12.6. Determine the service axial load P which the W12×45 of the accompanying figure may be permitted to carry. Lateral support is provided at ends and at midspan. Compare for A572 Grade 50 and Grade 65 steels.

12.7. Select the lightest W14 section to carry a service load P as shown in the accompanying figure, with an eccentricity $e = 12$ in. with respect to the strong axis. Assume the member is part of a braced system, and conservatively assume the effective

W8 × 28
A992

Bending axis

Service loads

M_b = 2.00'k dead load
         2.50'k live load

M_a = 2.90'k dead load
         3.80'k live load

P = 14.0'k dead load
       50.0'k live load

Problem 12.4

Assume hinged for both principal directions

W12 × 96
A992

Fixed for both principal directions

Service loads

P = 56 kips dead load
       80 kips live load

W = 20% dead load
        80% live load

Problem 12.5

W12 × 45

Lateral support

Vertical support

Service loads

W = 0.2 kips/ft dead load
        0.5 kips/ft live load

P = 40% dead load
       60% live load

Problem 12.6

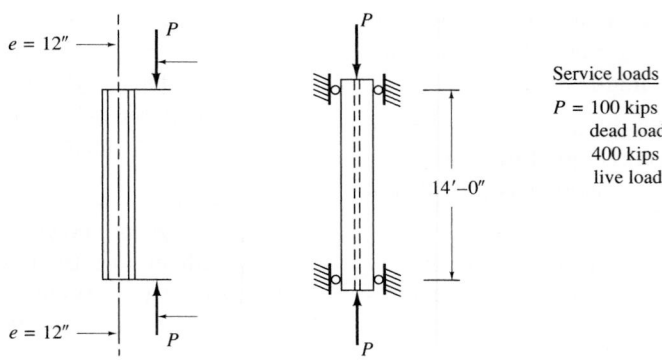

Problem 12.7

length equals the unbraced height. Use A572 Grade 60 steel.

12.8. Select the lightest W14 section to carry an axial compression P of 100 kips dead load and 140 kips live load along with a bending moment M of 125 ft-kips dead load and 325 ft-kips live load, which, for conservative simplicity, is assumed to be constant along the 15-ft equivalent pin-end length of the member in the braced structure. Use A992.

12.9. If the service load P is 30 kips dead load and 95 kips live load and w is 0.8 kips/ft dead load and 1.2 kips/ft live load for the beam-column span and support conditions of Prob. 12.6, select the lightest W section. Use A572 Grade 60 steel.

12.10. Select the lightest W14 section for the beam-column of the accompanying figure. Assume lateral buckling is adequately prevented such that $L_b < L_p$. Use A992 steel.

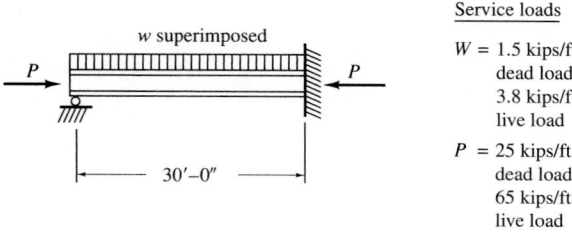

Service loads

$W = 1.5$ kips/ft dead load
3.8 kips/ft live load

$P = 25$ kips/ft dead load
65 kips/ft live load

Problem 12.10

12.11. For the member of a braced system in the accompanying figure, select the lightest W section. Use A572 Grade 60 steel.

12.12. Redesign the section for Example 12.13.5 for A572 Grade 60 steel.

Service loads

$P = 85$ kips dead load
15 kips live load

$M = 18$ ft-kips dead load
50 ft-kips live load

Problem 12.11

12.13. A frame braced against sidesway has a beam-column loaded as in the accompanying figure resulting from an elastic analysis. The horizontal member has lateral support at its ends and every 9 ft. Select the lightest W section acceptable using A992 steel.

12.14. Redesign the column member of Prob. 12.13 as part of an unbraced frame, but disregard the service loads of that problem. Assume two factored load first-order analyses have been performed: (1) the gravity load nonsway analysis giving $P_u = 85$ kips and $M_{nt} = 130$ ft-kips; and (2) a sway analysis giving $M_{\ell t} = 375$ ft-kips and $P_u = 14$ kips. Assume both columns must carry the same load.

12.15. Design the column A as a W section for the unbraced frame member of the accompanying figure. The frame is unbraced in the plane of the frame and analysis has determined $K_x = 1.4$; in the perpendicular direction the structure is braced such that $K_y = 1.0$. First-order factored load

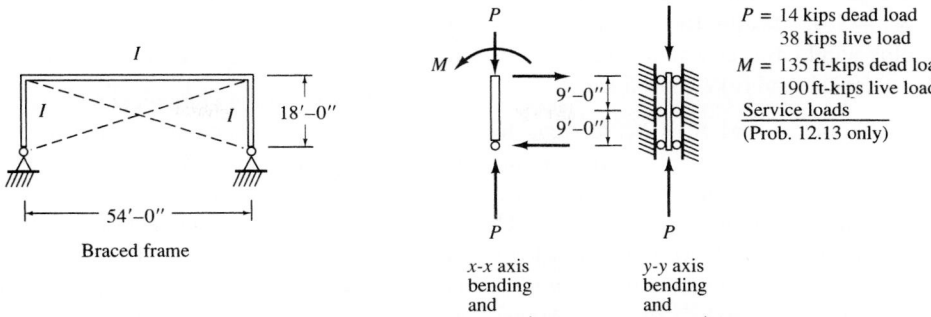

Problems 12.13 and 12.14

analyses under dead load, live load, and wind $(1.2D + 1.6W + 1.0L)$ have been performed; the *nonsway* gravity analysis $(1.2D + 1.0L)$, and the *sway* wind $(1.6W)$ analysis giving the values as shown. Assume all columns in the story are the same

and the load on each is identical. Use steels: (a) $F_y = 50$ ksi; (b) $F_y = 60$ ksi; and (c) $F_y = 65$ ksi.

12.16. Design columns A and B as W sections for the unbraced frame of the accompanying figure. Perpendicular to the frame, assume the system braced

Problem 12.15

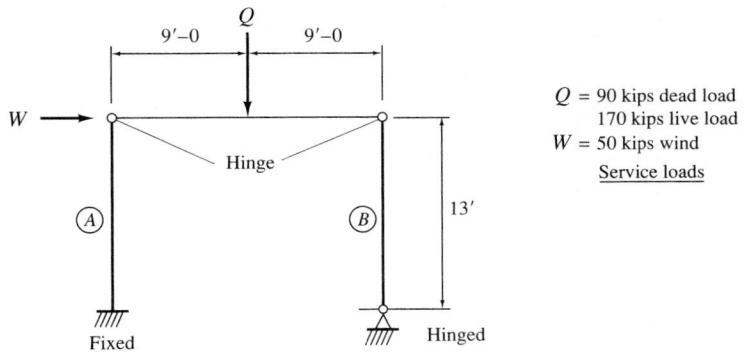

Problem 12.16

and the columns additionally braced at mid-height with $K_y = 1.0$. Design for the dead load, live load, plus wind case. Use steels: (a) $F_y = 50$ ksi; (b) $F_y = 60$ ksi; and (c) $F_y = 65$ ksi.

12.17. Design columns A and B as W sections for the unbraced frame shown in the accompanying figure. The system is braced in the direction perpendicular to the frame such that $K_y = 1.0$. Design for the dead load, live load, plus wind load case using the uniform live load acting on *both* spans. Use steels: (a) $F_y = 50$ ksi; (b) $F_y = 60$ ksi; and (c) $F_y = 65$ ksi.

12.18. Design the lightest W12 section for column A of the unbraced frame in the accompanying figure. Use the dead load plus live load plus wind loading case. Preliminary design has selected W27×94 for all adjacent beams, and it has been decided column A will be approximately the same size as those above and below it. Assume the forces given are the result of two first-order analyses; a nonsway analysis giving factored moments M_{nt}, and a sway analysis giving factored moments $M_{\ell t}$. In the

weak direction, assume the system braced such that $K_y = 1.0$ and $L_y = 13$ ft. Use steels: (a) $F_y = 50$ ksi; (b) $F_y = 60$ ksi; and (c) $F_y = 65$ ksi.

12.19. For the vierendeel truss (rigid frame) shown in the accompanying figure, investigate the adequacy of members A and B. The uniform loading is 90% live load and 10% dead load including the weight of the steel section. Assume all other forces are 10% dead load and 90% live load. The steel is A992. Assume simple cross-bracing between given frame and an adjacent parallel one.

12.20. Redesign member A for A572 Grade 60 steel.

12.21. Design the lightest W section to carry two eccentric 26-kip loads causing primary bending plus an axial compression of 350 kips. Assume all loads are 20% dead load and 80% live load. Assume torsional fixity at the vertical supports; assume lateral bracing at midspan is adequate to prevent lateral-torsional buckling, but does permit enough rotation for torsional moments to develop. Use A36 steel.

Problem 12.17

Problem 12.18

Member A   Member B

Problems 12.19 and 12.20

Problem 12.21

12.22. A column in a building has the factored load reactions at the top from beams framing into it, as shown in the accompanying figure. Assume the framing beams contribute moments at the top of the column, but the bottom of the column is hinged (no moments). The beams framing into the web are assumed to rest on seats, where reactions are assumed to be 2 in. from the center of the web. The reaction from the other beam is assumed to be acting at the face of the flange. Use A992 steel and select lightest W section.

12.23. For the factored loading shown, select lightest sections for the following conditions:

(a) W14 using A992 steel

(b) W in any depth using $F_y = 50$ ksi

(c) W14 using $F_y = 60$ ksi

(d) W14 using $F_y = 70$ ksi

Problem 12.22

Problem 12.23

Problems Relating to Theoretical Considerations

12.24. through 12.28. For the given loading and support conditions, develop the differential equation for the moment M_z in the plane of bending, and determine the maximum value for M_z.

Problem 12.26

Problem 12.24

Problem 12.27

Problem 12.25

Problem 12.28

13
Connections

13.1 TYPES OF CONNECTIONS

Steel connections are categorized by AISC-B3.6 depending on the amount of restraint developed by the connections. These classifications are:

Fully-Restrained (FR) Moment Connections. This situation occurs when full continuity is provided at the connection so that original angles between intersecting members are maintained essentially constant during loading of the structure; i.e., with rotational restraint on the order of 90% or more of that necessary to prevent any angle change.

Simple Connections (also called *Simple Framing*). This situation occurs where rotational restraint at the ends of members is as little as practicable. For beams, a "Simple Connection" is intended to provide only shear transfer at the ends. A "Simple Connection" is usually assumed to exist when the original angle between intersecting members may change approximately 80% or more of the amount it theoretically would change if frictionless hinged connections could be used. When a simply supported beam is designed, "Simple Connections" must be used. When plastic analysis is used, continuity is inherently assumed; therefore in that case it is inappropriate to use "Simple Connections." Two or more planar systems designed using plastic analysis may, however, be linked together by "Simple Connections" combined with a bracing system (such as cross-bracing).

AISC design requires that when a "Simple Connection" is used, three specific requirements (AISC-B3.6a) apply, paraphrased as follows:

a. The simply supported beam reactions under factored loads must be adequately carried by such connections.

b. The structure and its connections must be adequate to resist factored lateral loads.

c. Connections must have sufficient inelastic rotation capacity so that angle changes inherent in the "Simple Connection" assumption can occur under the combination of factored gravity and lateral loading *without overloading the end fastening system.*

Welded connections for rigid frame construction, showing beam-to column connections with column web stiffeners. Rural Mutual Insurance Building, Madison, Wis. (Photo by C. G. Salmon)

Partially-Restrained (PR) Moment Connections. Formerly called *semi-rigid fram-ing*, this situation occurs when the connection transfers moment accompanied with rota-tion between connected members. The moment-rotation response characteristics of the connection must be "... documented in the technical literature or established by analyti-cal or experimental means." [AISC-B6b.(b)]. Partially-Restrained connection occurs when rotational restraint is approximately between 20% and 90% of that necessary to prevent relative angle change. This means that with partially-restrained connection the moment transmitted across the joint is neither zero (or a small amount) as in a "simple connection," nor is it the full continuity moment as assumed in elastic rigid-frame analysis.

Partially-Restrained connections are not used in structures when plastic analysis is used in design, and have not been commonly used in traditional Allowable Stress Design because of the difficulty in obtaining the moment-rotation relationship for a given connection. However, with greater availability of higher strength steels (such as A992) required in designing this type of connection, the authors believe use of Partially-Restrained connections will increase.

Beam Line

In order to better understand the practical distinction between the AISC framing types, the beam line developed by Batho and Rowan [13.1] and used by Sourochnikoff [13.5] is a useful graphical device.

As shown in Fig. 13.1.1, consider a beam AB loaded in any manner and subject to end moments M_a and M_b, and with end slopes θ_a and θ_b. The moments necessary to have $\theta_a = \theta_b = 0$ are designated M_{Fa} and M_{Fb}, the fixed-end moments. Writing the slope deflection equations,*

$$\left. \begin{array}{l} M_a = M_{Fa} + \dfrac{4EI}{L}\theta_a + \dfrac{2EI}{L}\theta_b \\[3mm] M_b = M_{Fb} + \dfrac{2EI}{L}\theta_a + \dfrac{4EI}{L}\theta_b \end{array} \right\} \tag{13.1.1}$$

Solving Eqs. 13.1.1 for θ_a and θ_b gives

$$\left. \begin{array}{l} \dfrac{6EI}{L}\theta_a = 2(M_a - M_{Fa}) - (M_b - M_{Fb}) \\[3mm] \dfrac{6EI}{L}\theta_b = -(M_a - M_{Fa}) + 2(M_b - M_{Fb}) \end{array} \right\} \tag{13.1.2}$$

Subtracting the second equation from the first gives

$$\frac{6EI}{L}(\theta_a - \theta_b) = 3(M_a - M_b) - 3(M_{Fa} - M_{Fb}) \tag{13.1.3}$$

If symmetrical loading is considered, then

$$M_b = -M_a, \qquad \theta_b = -\theta_a, \qquad M_{Fb} = -M_{Fa} \tag{13.1.4}$$

M_a

A

θ_b

θ_a

B

M_b

(a) Final end moments and slopes

M_{Fa}

M_{Fb}

Fixed-end moments

Figure 13.1.1
Moments and rotations for slope-deflection equations (shown with positive signs).

---

*For instance, see Chu-Kia Wang and Charles G. Salmon, *Introductory Structural Analysis*. Englewood Cliffs, NJ: Prenctice-Hall, Inc., 1984 (Chap. 9).

Figure 13.1.2
Moment—rotation
characteristics of AISC
connection types.

in which case Eq. 13.1.3 becomes

$$\frac{2EI}{L}\theta_a = M_a - M_{Fa}$$

or

$$M_a = M_{Fa} + \frac{2EI}{L}\theta_a \qquad (13.1.5)$$

which may be called the *beam-line equation*. When $\theta_a = 0$ (a full fixity condition), $M_a = M_{Fa}$; and for a hinged end where $M_a = 0$, the slope becomes $\theta_a = -M_{Fa}/(2EI/L)$.

Figure 13.1.2 shows a diagram of the beam-line equation and also the moment-rotation behavior of typical connections of AISC Types FR, PR, and Simple. The typical FR (rigid) connection would have to carry an end moment M_1 about 90% or more of M_{Fa}, hence its degree of restraint may be said to be 90%. The Simple Connection may have to resist only 20% or less of the moment M_{Fa}, as indicated by the moment M_2, while the PR (semi-rigid) connection would be expected to resist some intermediate value M_3 at perhaps 50% of the fixed-end moment M_{Fa}. AISC Commentary-B3.6 identifies the three types of connections by the level of their secant rotational stiffness, $K_S = M_S/\theta_S$, under service load, where M_S and θ_S are the moment and rotation, respectively, at service loads.

When $K_S > 20EI/L$, the connection is considered to be fully restrained; when $K_S < 2EI/L$, the connection is considered to be simple, and connections having stiffness between the two limits are considered partially restrained.

When the moment–rotation characteristics of a particular connection are available, the strength can be designed so that the resulting end rotation θ is compatible with that caused by the loads. Discussions of semi-rigid connections and moment–rotation characteristics of various connection arrangements are given by Hechtman and Johnston [13.6], Schenker, Salmon, and Johnston [13.7], Lindsay, Ioannides, and Goverdhan [13.8], Brown [13.9], Ammerman and Leon [13.10], Leon, Ammerman, Lin, and McCauley [13.11], Nethercot, Davison, and Kirby [13.12], Chen and Kishi [13.13], Azizinamini and Radziminski [13.14], Leon and Ammerman [13.15], Bjorhovde, Colson, Brozzetti [13.16], Zandonini and Zanon [13.17], and Leon [13.18].

13.2 SIMPLE SHEAR CONNECTIONS

Simple shear connections (commonly known as *framed beam* connections), and now (2008) referred to as "Simple Connections" (AISC-B3.6), are used to connect beams to other beams or to column flanges or webs when simple support of the beam has been assumed. Design of such connections has become somewhat standardized using Part 10 of the *AISC Manual*, which provides the design methods and tables for Simple Shear Connections: Table 10-1 "All Bolted Double-Angle Connections," Table 10-2 "Bolted/Welded Double-Angle Connections," Table 10-3 "All-Welded Double-Angle Connections," Table 10-4 "Bolted/Welded Shear End-Plate Connections," Table 10-5 "All-Bolted Unstiffened Seated Connections," Table 10-6 "All-Welded Unstiffened Seated Connections," Table 10-7 "All-Bolted Stiffened Seated Connections," and Table 10-8 "Bolted/Welded Stiffened Seated Connections."

Typical bolted and welded framed connections are shown in Fig. 13.2.1. It is intended in such connections that the angles be as flexible as possible. The connection to the column (2 rows of 5 fasteners shown in Fig. 13.2.1a) is usually made in the field, while the connection to the beam web (one row of 5 fasteners shown in Fig. 13.2.1b) is usually made in the shop. Generally on plans, shop fastener holes are shown as in Fig. 13.2.1b, while field fastener holes are shown as solid black dots. There have been many studies [13.19–13.28] of simple shear double-angle connections.

In today's fabrication practice, the shop connection is usually welded, while the field connection may be either bolted or welded; thus, any combination in Fig. 13.2.1 of (a) with (b) or (c); or (d) with (b) or (c) may be used.

(a)   (b)   (c)

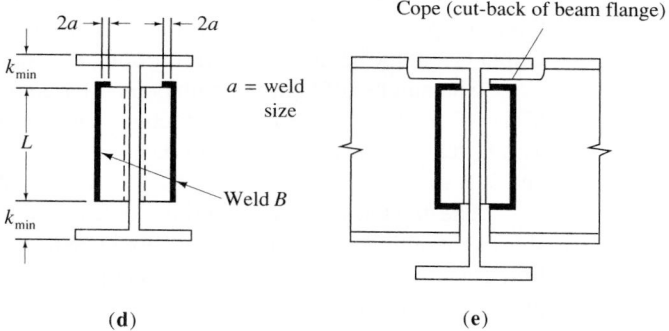

(d)   (e)

Figure 13.2.1
Simple shear double-angle
connections.

The *single-plate framing connection* is a modification where a single plate (instead of the pair of angles) is bolted flat against the beam web and then is welded perpendicular to the beam web or column flange or web to which it is attached. The design of single plate framing connections has been studied by Richard, Gillett, Kriegh, and Lewis [13.29], Young and Disque [13.30], Richard, Kriegh, and Hormby [13.31], Hormby, Richard, and Kriegh [13.32], and Astaneh, Call, and McMullin [13.33]. *AISC Manual* Table 10-9 "Single Plate Connections," provides a standardized design method.

Another type of simple shear connection is the *tee framing connection* as studied by Astaneh and Nader [13.35, 13.36], where the tee flange attaches to the supporting column (or beam) and the tee web laps against the loaded beam to transmit its shear.

Another single-plate framing connection, studied by Kennedy [13.34], uses the plate in the vertical position welded flat against the end of the beam with the connection to the beam or column made with bolts.

When angles, sometimes known as *clip angles*, are used to attach a beam to a column, there is a clearance setback of about $\frac{1}{2}$ in. so that if the beam is too long, within acceptable tolerances, the angles may be relocated without cutting off a piece of the beam. When beams intersect and are attached to other beams so that the flanges of both are at the same elevation, as in Fig. 13.2.1e, the beams framing in have their flanges coped, or cut away. The loss of section is primarily loss of flange that carries little shear anyway, so that normally a cope results in little loss of shear strength. Birkemoe and Gilmor [13.20] have shown that a coped web subject to high bearing stress in a high-strength bolted beam end connection may fail in a tearing mode (known as "block shear") along a line through the holes, as shown in Fig. 13.2.2. Additional study of block shear in such situations has been made by Yura, Birkemoe, and Ricles [13.21] and Ricles and Yura [13.23].

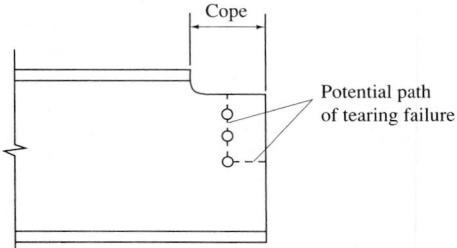

Figure 13.2.2
Tearing failure at coped ends of framed beam connection.

"Block shear" was first discussed in regard to tension members in Sec. 3.6. Block shear may be critical in simple shear connections when there is a short connection using few bolts that do not extend uniformly over the entire depth of the web. AISC-J4.3 requires consideration of the block shear limit state "where the top flange is coped and in similar situations, such as tension members and gusset plates."

In addition to block shear, copes of the flanges of beams may affect local web buckling, as reported by Cheng and Yura [13.24], and lateral-torsional buckling, as reported by Gupta [13.25], Cheng, Yura, and Johnson [13.26], and Cheng and Yura [13.27].

The number of high-strength bolts is based on the direct shear, neglecting any eccentricity of loading, while the weld length and size include the effect of eccentric loading. The fasteners, bolts or welds, are designed in accordance with procedures of Chapters 4 and 5, respectively.

The thickness of the framing angles or plate is usually controlled by the "block shear" strength. In addition, angles should be thick enough such that bearing does not control. The angles are *expected to bend* so that the assumed rotation of the supported beam at its ends can occur.

Flexural Deformation and Strength of Connection Angles

Referring to Fig. 13.2.3, the tensile force T per inch acts at the top of framing angles of length L as shown in Fig. 13.2.1 when an end moment acts. This end moment arises from the reaction P acting at an eccentricity e measured, as in Fig. 13.2.1b or c, from the point of action of P to the centroid of the fastener line A or to the centroid of weld A.

The concentrated load T acting on the connection angles may be considered as acting on a fixed-end beam for the bolted connection and on a simply supported beam for a welded connection. The true situation for each case is partially end restrained. The higher restraint for the bolted connection arises from the clamping action between the pieces caused by initial tension in the bolts, while the welding causes little clamping action.

A simple approximation of the deflection Δ at the tension end of the framing angles will therefore be that of a fixed-end beam (high-strength bolted connection) having concentrated load T at mid-length of a span g,

$$\Delta = \frac{Tg^3}{192EI} \tag{13.2.1}$$

and for the concentrated load T acting at mid-length of a simply supported span g (welded connection),

$$\Delta = \frac{Tg^3}{48EI} \tag{13.2.2}$$

Noting that the maximum force T will occur when the top of the angle yields; i.e., Max $T = 2F_y t$ per unit length at the top, and that $I = t^3/12$ per unit length, Eqs. 13.2.1 and 13.2.2 become

$$\Delta = \frac{2F_y t g^3}{192E(t^3/12)} = \frac{F_y g^3}{8Et^2} \tag{13.2.3}$$

(a) High-strength bolted

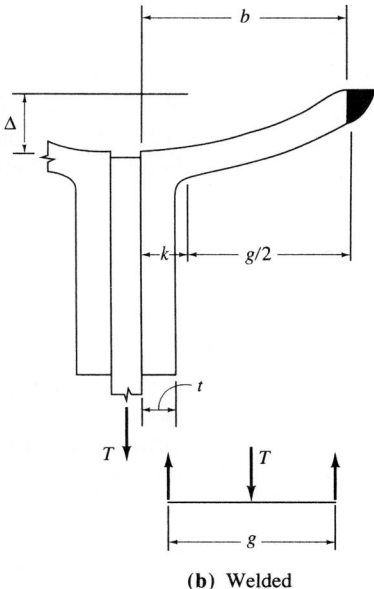

(b) Welded

Figure 13.2.3
Behavior at tension edge of framing angles.

and

$$\Delta = \frac{2F_y t g^3}{48E(t^3/12)} = \frac{F_y g^3}{2Et^2} \tag{13.2.4}$$

Note that the more rotation required at the end of the beam the greater must be the deformation Δ and the thinner (low t) must be the angles. In general, this deformation is self-limiting because the mid-length portion of the angles will remain elastic even though the top end may be inelastic. Furthermore, the use of thick angles would mean that the stress at the top of the angles due to T might be less than F_y, thus reducing Δ (and the corresponding end rotation).

EXAMPLE 13.2.1

Compute the factored load P_u capacity for the reaction on the 5 row Simple Connection of Fig. 13.2.1 for connecting a W30×99 beam to a column having a $\frac{3}{4}$-in. flange. Use $\frac{3}{4}$-in.-diam A325 bolts as a *bearing-type connection* (A325-X) having no threads in the shear planes. The connection uses standard holes with clean mill scale (Class A) surface condition, and assume that the bolt hole deformation is a design consideration at service load. Use the AISC LRFD Method with A36 angles and A992 beam and column.

Solution:

(a) Design strengths ϕR_n per bolt for the connection to the web of the W30×99, $t_w = 0.520$ in. Using Eq. 4.7.3, the design strength in double shear ($m = 2$) is

$$\phi R_n = \phi F_{nv} m A_b \tag{4.7.3}$$
$$= 0.75(60)(2)0.4418 = 39.8 \text{ kips/bolt}$$

Since the bolt spacing exceeds $3d_b = 2.25$ in., the nominal strength in bearing on the 0.520-in. web, from Eq. 4.7.9, is

$$R_n = 1.2 L_c t F_u \le 2.4 dt F_u \tag{4.7.9}$$
$$\phi R_n = \phi 2.4 dt F_u = 0.75(2.4)(0.75)(0.520)65 = 45.6 \text{ kips/bolt}$$

(b) Total connection factored load P_u capacity based on the web connection to the W30×99 beam. Common practice has been to neglect the eccentricity e with respect to the fastener line. Thus, the maximum factored reaction P_u would equal the design strength per bolt times the number n of bolts,

$$\text{Maximum } P_u = n\phi R_n = 5(39.8) = 199 \text{ kips}$$

If the eccentricity with respect to the fastener line were considered,

$$e = 2\frac{1}{4} \text{ in.}$$

assuming the reaction to be along fastener line A. For this eccentric shear loading, ultimate strength analysis or elastic vector analysis, as discussed in Sec. 4.12, may be used. Using ultimate strength analysis as represented by the *AISC Manual* Table 7-7, "Coefficients C for Eccentrically Loaded Bolt Groups," p. 7-32, for vertical spacing of fasteners $s = 3$ in., the number of fasteners $n = 5$, and eccentricity $e_x = 2.25$ in.,

$$\text{Find coefficient } C = 4.27$$
$$P_u = C\phi R_n = 4.27(39.8) = 170 \text{ kips}$$

This is 15% less than when eccentricity is neglected.

(c) Design strengths ϕR_n per bolt for the connection to the $\frac{3}{4}$-in. flange of the column. The design strength in single shear $(m = 1)$ is

$$\phi R_n \text{ (single shear)} = \phi F_{nv} m A_b$$

$$\phi R_n = 0.75(60)(1)0.4418 = 19.9 \text{ kips/bolt}$$

The design strength in bearing on the angles will be less than bearing on the $\frac{3}{4}$-in. flange which has $F_y = 50$ ksi. The minimum thickness for the angles so that bearing will not govern is obtained by setting the strength in bearing equal to the strength (19.9 kips/bolt) in shear.

$$\phi R_n \text{ (bearing)} = \phi 2.4 dt F_u = 19.9 \text{ kips/bolt}$$

$$\text{Min } t = \frac{19.9}{\phi 2.4 d F_u} = \frac{19.9}{0.75(2.4)(0.75)65} = 0.23 \text{ in.}$$

The design strength in tension will be dependent on the applied shear load; however, the upper limit on that strength is

$$\phi R_n \text{ (tension)} = \phi F_{nt} A_b$$

$$= 0.75(90)0.4418 = 29.8 \text{ kips/bolt}$$

(d) Total connection factored load P_u capacity based on the connection to the $\frac{3}{4}$-in. flange. Neglecting eccentricity, the ten fasteners in single shear give

$$P_u = 10(19.9) = 199 \text{ kips}$$

If the combined shear and tension effect is considered, and the elastic vector method is used, the factored tension T_u at the most heavily loaded bolt is

$$T_u = \frac{Mc}{I} A = \frac{P_u ec}{I} A = \frac{P_u(2.25)6}{4(3)^2 + 4(6)^2} = 0.075 P_u$$

and the direct shear per bolt is

$$V_u = \frac{P}{A} A = \frac{P_u}{10} = 0.10 P_u$$

Using AISC-Formulas (J3-2) and (J3-3a), the nominal load F'_{nt} in tension permitted in the presence of shear is, for bearing-type A325-X connections using $\frac{3}{4}$-in.-diam bolts $(A_b = 0.4418$ sq in.$)$,

$$\phi F'_{nt} A_b = \phi \left(1.3 F_{nt} - \frac{F_{nt}}{\phi F_{nv}} f_v \right) A_b \leq \phi F_{nt} A_b$$

$$= 0.75 \left[1.3(90) - \frac{90}{0.75(60)} f_v \right] A_b \leq 29.8 \text{ kips/bolt}$$

Let $\phi F'_{nt} A_b = 0.075 P_u$, and $f_v A_b = 0.10 P_u$; then

$$0.075 P_u = 38.8 - 0.150 P_u$$

$$P_u = 172 \text{ kips}$$

Check that $0.075 P_u$ does not exceed 29.8 kips; $0.075 P_u = 0.075 (172) = 12.9$ kips, which means the upper limit on $\phi F'_{nt} A_b$ is not exceeded. In addition to that check, the strength based on combined tension and shear cannot exceed the strength based on shear alone. Referring to Fig. 4.14.3, f_{uv} cannot exceed 45 ksi; however, $f_{uv} = 0.10P_u/A_b$ $= 0.10(172)/0.4418 = 38.9$ ksi from the above computation. Thus, P_u cannot exceed 172 kips based on combined tension and shear.

(e) Summary of factored load capacity results.

Connection to beam web:	$P_u = 199$ kips (neglect e)
	$P_u = 170$ kips (consider e)
Connection to column flan	$P_u = 199$ kips (neglect e)
	$P_u = 172$ kips (consider e)

For the illustration of edge and end distance requirements, the value of 199 kips neglecting eccentricity will be used.

(f) End and edge distances. When bearing strength based on $2.4F_u$ is used according to AISC-J3.10, as shown in parts (a) and (c), a minimum end distance of $1.5d_b$ is acceptable. In this case,

$$\text{Minimum end distance } L_e \geq [1.5d_b = 1.5(0.75) = 1.13 \text{ in.}]$$

When a lesser distance is desired, Eq. 4.7.16 can be used, where P_u (i.e., ϕR_n) is the factored load per bolt. Equation 4.7.16 should also be used to check the end distance on the beam web (using $K = 1.2$).

$$\text{End distance } L_e \geq \frac{R_n}{KF_u t} + \frac{d}{2} \qquad [4.7.16]$$

$$L_e \geq \left[\frac{P}{\phi 1.2 F_u t} + \frac{d_b}{2} = \frac{199/5}{0.75(1.2)(58)0.52} + \frac{0.75}{2} = 1.84 \text{ in.} \right]$$

The end distance requirement is conservative (i.e., large) because the full load on the uppermost bolt is assumed to be directed toward the nearest edge; actually only the horizontal component need be used. The 1.75 in. provided is certainly adequate.

(g) Shear on the net section through the angles. For shear as well as tension, the net area A_{nv} according to AISC-B3.13 is based on the nominal diameter of the hole plus $\frac{1}{16}$-in. Thus, for standard holes,

$$A_{nv} = t\left[14.5 - 5\left(\tfrac{13}{16} + \tfrac{1}{16}\right) \right] = 10.13t \text{ per angle}$$

and
$$A_{gv} = 14.5\, t \text{ per angle}$$

Shear yield design strength (AISC-J4.2a) is

$$\text{Required } t = \frac{\text{Reaction}}{\phi(0.6F_y)A_{gv}} = \frac{199/2}{1.0(0.6)(36)14.5} = 0.318 \text{ in}$$

Shear rupture design strength (AISC-J4.2b) is

$$\phi R_n = \phi(0.6F_u)A_{nv}$$

$$\text{Required } t = \frac{\text{Reaction}}{\phi(0.6F_u)A_{nv}} = \frac{199/2}{0.75(0.6)(58)10.13} = 0.376 \text{ in. governs}$$

Select $\frac{3}{8}$-in. angles.

(h) Examine the degree of simple support provided by $\frac{3}{8}$-in. angles. Using Eq. 13.2.3, the elastic deflection is

$$\Delta = \frac{F_y g^3}{8Et^2} = \frac{36[2(2.5 - 13/16)]^3}{8(29,000)(0.375)^2} = 0.04 \text{ in.}$$

The longest beam span uniformly loaded for a W30×99 having a factored reaction of 199 kips is

$$\phi_b M_p = 0.90 Z_x F_y = 0.90(312)36/12 = 842 \text{ ft-kips}$$

$$M_u = \frac{W_u L}{8} = \frac{199(2)L}{8}$$

$$L = \frac{8(842)}{199(2)} = 16.9 \text{ ft}$$

Assuming the service moment is about 0.5 of M_p, the elastically computed end slope is

$$\theta = \frac{WL^2}{24EI} = \frac{ML}{3EI} = \frac{(842/2)(12)(16.9)12}{3(29,000)3990} = 0.0030 \text{ radian}$$

Assuming rotation about the bottom of the angle, the deformation required at the top of the angles to accommodate the service load rotations at the ends of the beam would be

$$\Delta = (d/2)\theta = 14.5(0.0030) = 0.04 \text{ in.} \approx 0.04$$

Based on these approximate calculations, the angles must yield at service load to accommodate rotation at the ends of the beam. Even if such rotation occurs, the full simply supported beam moment is not likely to occur; there will be some end moment.   ■

The conclusion is that simple shear connections should use the thinnest angles consistent with the bearing and shear fracture limit states, as well as practical limitations. Bertwell [13.37] has provided some additional discussion on the behavior of simple shear connections.

EXAMPLE 13.2.2

Investigate the 5 row simple shear connection of Example 13.2.1 (Fig. 13.2.1) as a *slip-critical connection* (A325-SC) assuming 75% live load and 25% dead load and using standard holes.

Solution:

(a) Strength in shear and bearing. A slip-critical connection has the same strength requirements for shear and bearing as a bearing-type connection. From Example 13.2.1,

$$\phi R_n = 39.8 \text{ kips/bolt} \quad \text{(double shear)}$$

$$\phi P_n = 45.6 \text{ kips/bolt} \quad \text{(bearing on 0.52-in. web)}$$

Using $\frac{3}{8}$-in.-thick angles means bearing on the angles will not control, nor will the shear fracture limit state.

(b) Capacity based on strength limit states. The maximum factored load P_u that can be carried was determined in Example 13.2.1 to be 199 kips (neglecting eccentricity). Based on this the service load capacity is

$$1.2(0.25P) + 1.6(0.75P) = 199 \text{ kips}$$

$$P = 133 \text{ kips}$$

(c) Check the slip serviceability limit state. The service load capacity per bolt is, using AISC Formula (J3-4), Eq. 4.9.1, with Class A surface condition and standard holes,

$$R_n = \mu D_u h_{sc} T_b N_s = 0.35(1.13)(1.0)(28)2 = 22.2 \text{ kips/bolt}$$

$$\phi = 1.0 \text{ for prevention of slip as a serviceability limit state}$$

$$P = \phi R_n(\text{number of bolts}) = 1.0(22.2)5 = 111 \text{ kips}$$

Thus, comparing the results from parts (b) and (c), the serviceability (slip resistance) limit state controls. This will usually be the case for slip-critical connections. When a surface condition is used having a high slip-resistance coefficient and/or threads may exist in the shear planes, it will be possible for the strength limit state to control.

(d) Consider eccentricity for the slip-critical connection. The combination of shear and tension on the attachment to the column flange will be more critical than eccentric shear at bolt line A (Fig. 13.2.1). The service load components at the most heavily loaded bolts are

$$T = \frac{Mc}{I}A = \frac{P(2.25)6}{4(3)^2 + 4(6)^2} = 0.075P \qquad (\text{tension})$$

and the direct shear per bolt is

$$V = \frac{P}{A}A = \frac{P}{10} = 0.10P \qquad (\text{direct shear})$$

The service load capacity in shear for a bolt simultaneously subject to tension is, according to AISC-J3.9,

$$R_n (\text{with tension}) = k_s R_n(\text{without tension})$$

where, with one slip plane ($N_s = 1$),

$$k_s = 1 - \frac{T_u}{D_u T_b N_s} = 1 - \frac{0.075(133)}{1.13(28)1} = 0.69$$

Thus,

$$R_n = 0.69(11.2) = 7.66 \text{ kips} \left[\text{using one-half the part (a) value of 22.2}\right]$$

$$P = \phi R_n(\text{number of bolts}) = 1.0(7.66)10 = 77 \text{ kips}$$

Note that when designing to prevent slip at service load, the unfactored (i.e., the service load) tensile force must be used to calculate the k_s factor. Also, it is noted that this reduced capacity is conservative because it applies to all bolts, the majority of which are subject to a lower level of tensile force.

(e) Summary.

$P = 133$ kips	(based on strength limit states)
$P = 111$ kips	(based on serviceability limit state of slip neglecting accentricity)
$P = 77$ kips	(based on serviceability limit state of slip considering eccentricity)

The *AISC Manual* Tables 10-1 implicitly accept the neglect of eccentricity for double angle simple shear connections. To verify this, one must use the tables to determine the design strength. For 5 rows with $\frac{3}{4}$-in. A325 bolts, enter *AISC Manual* Table 10-1 (p. 10–20) for SC Class A, to find the strength equal to 111 kips. This value matches the calculated value of part (c) where the eccentricity was neglected.   ∎

EXAMPLE 13.2.3

Design a double-angle simple shear connection for a W10×68 beam having a factored load reaction P_{u1} of 70 kips and a W24×104 beam having a factored load reaction P_{u2} of 210 kips. These two beams are to frame into opposite sides of a plate girder having a $\frac{3}{8}$-in. web as shown in Fig. 13.2.4. The connection is to be of $\frac{3}{4}$-in.-diam A325 bolts in a bearing-type connection with threads excluded from the shear planes. Use AISC LRFD Method and A992 steel.

Solution:

(a) Compute design bolt values relating to connections to webs of W10×68 and W24×104. Initially assume that $s \geq 3d_b$, $L_c \geq 2.0d_b$, and two or more bolts in a line of force with standard holes and hole deformation being a design consideration, so that $R_n = 2.4F_u d_b t$ for bearing,

$$\phi R_n \ (\text{bearing}) = \phi(2.4F_u)d_b t$$

$$\phi R_n = 0.75(2.4)(65)(0.75)0.470 = 41.2 \ \text{kips/bolt} \quad (\text{W10})$$

$$\phi R_n = 0.75(2.4)(65)(0.75)0.500 = 43.9 \ \text{kips/bolt} \quad (\text{W24})$$

$$\phi R_n \ (\text{double shear}) = \phi F_{nv} m A_b$$

$$\phi R_n = 0.75(60.0)(2)0.4418 = 39.8 \ \text{kips/bolt}$$

Figure 13.2.4
Framed beam connection having unequal reactions.

$$\text{Number of bolts} = \frac{70}{39.8} = 1.8, \quad \text{say 2 (W10)}$$

$$\text{Number of bolts} = \frac{210}{39.8} = 5.3, \quad \text{say 6 (W24)}$$

Use 3-in. bolt spacing with 1.5-in. end distance at cope (top of angles).

(b) Check block shear on W10 according to AISC-J4.3. This is discussed in Sec. 3.6. Combining Eqs. 3.6.1 and 3.6.2, or use AISC Formula (J4-5), to compute the nominal strength for the limit state of block shear rupture,

$$R_n = (0.6F_u A_{nv} + F_u U_{bs} A_{nt}) \leq (0.6F_y A_{gv} + F_u U_{bs} A_{nt})$$

Obtain the areas to be used in the computation,

$$A_{gv} = \text{shear yield area} = 4.5(0.470) = 2.11 \text{ sq in.}$$

$$A_{nv} = \text{shear fracture area} = \left[4.5 - 1.5\left(\tfrac{13}{16} + \tfrac{1}{16}\right)\right]0.470 = 1.50 \text{ sq in.}$$

$$A_{nt} = \text{tension fracture area} = \left[2.0 - 0.5\left(\tfrac{13}{16} + \tfrac{1}{16}\right)\right]0.470 = 0.73 \text{ sq in.}$$

With a single row of bolts the tensile stress is assumed uniform and $U_{bs} = 1.0$.

$$R_n = 0.6F_u A_{nv} + F_u U_{bs} A_{nt}$$
$$= 0.6(65)1.50 + 65(1.0)0.73 = 106 \text{ kips}$$
$$R_n = 0.6F_y A_{gv} + F_u U_{bs} A_{nt}$$
$$= 0.6(50)2.11 + 65(1.0)0.73 = 111 \text{ kips (upper limit)}$$

The factored reaction capacity $P_u = \phi R_n$, where $\phi = 0.75$,

$$P_u = \phi R_n = 0.75(106) = 79 \text{ kips} > 70 \text{ kips required} \qquad \textbf{OK}$$

Use 2 bolts to connect the W10 section.

(c) Check block shear on W24 according to AISC-J4.3. From Eqs. 4.6.1 and 4.6.2,

$$R_n = (0.6F_u A_{nv} + F_u U_{bs} A_{nt}) \leq (0.6F_y A_{gv} + F_u U_{bs} A_{nt})$$

Obtain the areas to be used in the computation,

$$A_{gv} = \text{shear yield area} = 16.5(0.50) = 8.25 \text{ sq in.}$$

$$A_{nv} = \text{shear fracture area} = \left[16.5 - 5.5\left(\tfrac{13}{16} + \tfrac{1}{16}\right)\right]0.50 = 5.84 \text{ sq in.}$$

$$A_{nv} = \text{tension fracture area} = \left[2.0 - 0.5\left(\tfrac{13}{16} + \tfrac{1}{16}\right)\right]0.50 = 0.78 \text{ sq in.}$$

With a single row of bolts the tensile stress is assumed uniform and $U_{bs} = 1.0$.

$$R_n = 0.6F_u A_{nv} + F_u U_{bs} A_{nt}$$
$$= 0.6(65)5.84 + 65(1.0)0.78 = 279 \text{ kips}$$
$$R_n = 0.6F_y A_{gv} + F_u U_{bs} A_{nt}$$
$$= 0.6(50)8.25 + 65(1.0)0.78 = 298 \text{ kips (upper limit)}$$

The factored reaction capacity $P_u = \phi R_n$, where $\phi = 0.75$,

$$P_u = \phi R_n = 0.75(279) = 209 \text{ kips} \approx 210 \text{ kips (required)} \qquad \textbf{OK}$$

Use 6 bolts to connect the W24 section.

(d) Connections to plate girder web. For this connection, the two bolts common to both sides will be governed by double shear or bearing on the $\frac{3}{8}$-in. plate, while the remainder are governed by single shear or bearing on the $\frac{3}{8}$-in. plate. Again, the connection is detailed using standard holes with bolt hole deformation being a design consideration.

$$\phi R_n \text{ (bearing)} = \phi(2.4F_u)dt$$

$$\phi R_n = 0.75(2.4)(65)(0.75)0.375 = 32.9 \text{ kips/bolt}$$

$$\phi R_n \text{ (double shear)} = 39.8 \text{ kips/bolt} \quad [\text{from (a) above}]$$

$$\phi R_n \text{ (single shear)} = 39.8/2 = 19.9 \text{ kips/bolt}$$

For the bolts common to both sides, bearing governs. The four bolts in common carry $70/4 = 17.5$ kips from the W10×68. The remainder is available for the W24×104 reaction; i.e., $32.9 - 17.5 = 15.4$ kips.

If all bolts were to carry equal load,

$$\text{Number of bolts} = \frac{210}{19.9} = 10.6, \quad \text{say } 12$$

When 12 bolts are used, the average load per bolt will be $210/12 = 17.5$ kips. Even though this exceeds the 15.4 kips available on the top 4 bolts, the lower 8 bolts are not fully loaded. Accept this arrangement.

(e) Angles thickness. Bearing will not control angles thickness unless end distance $L_e < 1.5d_b$ or bolt spacing $s < 3d_b$, according to AISC-J4.10 because $2.4F_u$ was used to compute bearing strength. For the $\frac{3}{4}$-in. diam bolts, $1.5d_b = 1.13$ in. and $3d_b = 2.25$ in. In this design, the end distance $L_e = 1.25$ in. and the minimum spacing is 2.5 in.

The thickness of angles might be controlled by either their shear rupture strength or shear yield strength in accordance with AISC-J4.2

$$[\phi(0.6F_y)A_{gv} \text{ and } \phi(0.6F_u)A_{nv}] \geq P_u$$

For the angles connected to the W10,

$$1.0(0.6)(50)(6)2t \geq 70$$

$$t \geq \frac{70}{1.0(0.6)(50)(2)6.0} = 0.195 \text{ in.}$$

or

$$0.75(0.6)(65)\left[6.0 - 2\left(\tfrac{13}{16} + \tfrac{1}{16}\right)\right]2t \geq 70$$

$$t \geq \frac{70}{0.75(0.6)65[6.0 - 1.75]2} = 0.28 \text{ in.} \qquad \text{Controls}$$

For the angles connected to the W24,

$$1.0(0.6)(50)(18)2t \geq 210$$

$$t \geq \frac{210}{1.0(0.6)(50)(2)18} = 0.194 \text{ in.}$$

or

$$0.75(0.6)(65)\left[18.0 - 6\left(\tfrac{13}{16} + \tfrac{1}{16}\right)\right]2t \geq 210$$

$$t \geq \frac{210}{0.75(0.6)65[18.0 - 5.25]2} = 0.28 \text{ in.} \qquad \text{Controls}$$

The bearing value ϕR_n for $\tfrac{5}{16}$-in. angles is 27.4 kips/bolt, which is more than adequate to carry the factored load of 17.5 kips/bolt from the beams (a coincidence that the same factored load per bolt is contributed by both the W10 and the W24); therefore, $\tfrac{5}{16}$-in. angles are satisfactory as connecting angles.

Use 2—L4×3$\tfrac{1}{2}$×$\tfrac{5}{16}$×0′–6″ for W10×68.
Use 2—L4×3$\tfrac{1}{2}$×$\tfrac{5}{16}$×1′–6″ for W24×104.

The length of angle should not exceed the dimension T, which is $7\tfrac{5}{8}$ in. for the W10×68. The girder flange thickness is such that a cope is required on the beams that encroaches on the T dimension. ∎

Weld Capacity in Eccentric Shear on Angle Connections

Since no initial tension is involved with welded connections, the eccentricity of loading, even though small, is considered. The principles of Chapter 5 (See. 5.18) are used with the welds treated as lines.

EXAMPLE 13.2.4

Compute the factored load P_u capacity for weld A on the angle connection shown in Fig. 13.2.1. The beam is a W30×99 and the weld is $\tfrac{1}{4}$ in. with E70 electrodes. The angles are 4×3$\tfrac{1}{2}$×$\tfrac{5}{16}$×1′–2$\tfrac{1}{2}$″ in length. Use A36 steel for the angles and A992 steel for the W section and the AISC LRFD Method.

Solution:

Analysis of this eccentric shear situation may be done using strength analysis as presented in Sec. 5.17 or the elastic (vector) method presented in Sec. 5.18.

(a) Elastic (vector) method. Using I_p from Table 5.18.1 and referring to Fig. 13.2.1c,

$$I_p = \frac{8(3)^3 + 6(3)(14.5)^2 + (14.5)^3}{12} - \frac{(3)^4}{2(3) + 14.5} = 583.5 \text{ in.}^3$$

Using the moment of inertia computed with a 1-in. effective throat, the force per unit length at critical locations can be computed.

$$R_v = \frac{P_u}{2(20.5)} = 0.0244P_u \quad \text{(direct shear component)} \downarrow$$

$$\overline{x} = \frac{(3)^2}{2(3) + 14.5} = 0.44 \text{ in.}$$

The x and y components of force due to torsional moment are

$$R_y = \frac{P_u(3.50 - 0.44)(3.50 - 0.44 - 0.50)}{2(583.5)} = 0.00671P_u \downarrow$$

$$R_x = \frac{P_u(3.50 - 0.44)(7.25)}{2(583.5)} = 0.0190P_u \rightarrow$$

$$R_u = P_u\sqrt{(0.0244 + 0.0067)^2 + (0.0190)^2} = 0.0364P_u$$

The design strength ϕR_{nw} per inch of weld is

$$\phi R_{nw} = \phi(0.707a)(0.60F_{EXX})$$
$$= 0.75(0.707)\left(\tfrac{1}{4}\right)(0.60)70 = 5.57 \text{ kips/in.}$$

Check base metal shear strength for the beam web and the angles,
For shear rupture

$$(\phi R_{nw})_{\text{base metal}} = 0.75(0.60F_u)t = 0.45F_u t$$
$$(\phi R_{nw})_{\text{angle}} = 0.45(58)0.3125 = 8.16 \text{ kips/in.}$$
$$(\phi R_{nw})_{\text{web}} = 0.45(65)(0.520/2) = 7.61 \text{ kips/in.}$$

For shear yield

$$(\phi R_{nw})_{BM} = 1.0(0.6F_y)A_{BM} = 0.60F_u t$$
$$(\phi R_{nw})_{\text{angle}} = 0.60(36)0.3125 = 6.75 \text{ kips/in.}$$
$$(\phi R_{nw})_{\text{web}} = 0.6(50)(0.520/2) = 7.8 \text{ kips/in.}$$

Weld strength controls; $\phi R_{nw} = 5.57$ kips/in.

$$P_u = \frac{5.57}{0.0364} = 153 \text{ kips}$$

(b) Strength analysis. Use *AISC Manual* [1.15], Table 8–8, "Coefficients C for Eccentrically Loaded Weld Groups" with $\theta = 0°$. For $\tfrac{1}{4}$-in. weld using E70 electrodes,

$$a = (e - xL)/L = (3.5 - 0.44)/14.5 = 0.211$$
$$k = kL/L \qquad = 3.0/14.5 = 0.207$$

	$k = 0.2$	0.207	0.3	
$a = 0.2$	2.63		3.11	
0.211	2.60	2.63	3.08	$C = 2.63$
0.25	2.51		2.96	

Table value $= P_n$

$$P_n = CC_1DL = 2.63(1.0)(4)14.5 = 153 \text{ kips}$$
$$\phi P_n = 0.75(153) = 115 \text{ kips}$$

where $C_1 =$ coefficient for electrode $=$ (Electrode used)/70
   $D =$ number of $\tfrac{1}{16}$s of an inch in weld size
   $L =$ length of vertical weld, in.

Since there are two angles, the factored load reaction capacity is

$$P_u = 2(115) = 230 \text{ kips}$$

As expected, the strength analysis gives the higher value.   ∎

Tests of welded angle connections by Johnston and Green [13.19] and Johnston and Diets [13.38] have demonstrated that performance of web angles agrees generally with assumptions.

Weld Capacity in Tension and Shear on Angle Connections

This is the field-welded connection shown in Fig. 13.2.1d. There is no agreement regarding the strength analysis for this situation. Blodgett [13.2] considers the strength as an eccentric shear situation in the plane of the welds. With the eccentric load as in Fig. 13.2.5b, the angles bear against themselves for a distance of $L/6$ from the top, and the torsional stress over the remaining $\frac{5}{6}$ of the length L is resisted by the weld. Neglecting the effects of the returns at the top, the horizontal component R_x can be obtained from moment equilibrium. Equilibrium in the plane of the load P and weld leg B requires

$$\underbrace{\frac{1}{2}R_x\left(\frac{5}{6}L\right)}_{\text{force}}\underbrace{\frac{2}{3}L}_{\text{arm}} = \frac{P}{2}e_2 \tag{13.2.5}$$

$$R_x = \frac{9Pe_2}{5L^2}\text{force/unit length} \tag{13.2.6}$$

The direct shear component is

$$R_v = \frac{P}{2L}\text{force/unit length} \tag{13.2.7}$$

$$\text{Actual } R = \sqrt{\left(\frac{P}{2L}\right)^2 + \left(\frac{9}{5}\frac{Pe_2}{L^2}\right)^2}$$

$$R = \frac{P}{2L^2}\sqrt{L^2 + 12.9e_2^2}\text{ force/unit length} \tag{13.2.8}$$

Equation 13.2.8 neglects eccentricity e_1, which tends to cause tension at the top of the weld lines. The authors believe it is more appropriate to consider the flexural stress distribution of Fig. 13.2.5c to be a more appropriate approach. The flexural tension component R_x at the top of the weld B is

$$R_x = \frac{Mc}{I} = \frac{Pe_1(L/2)}{2L^3/12} = \frac{3Pe_1}{L^2} \tag{13.2.9}$$

when the returns at the tops of the angles are neglected. The direct shear component R_v is

$$R_v = \frac{P}{2L}\text{force/unit length} \tag{13.2.10}$$

Figure 13.2.5
Field-welded connection for
web framing angles.

(a)   (b)   (c)

Figure 13.2.6
Weld configuration for web
angles and beam seats.

$$\text{Actual } R = \sqrt{\left(\frac{P}{2L}\right)^2 + \left(\frac{3Pe_1}{L^2}\right)^2}$$

$$R = \frac{P}{2L^2}\sqrt{L^2 + 36e_1^2} \text{ force/unit length} \tag{13.2.11}$$

Or, if returns are considered (distance b of Fig. 13.2.6) the expression becomes complicated. The *AISC Manual* Table 10-3, p. 10-47 indicates the returns to be twice the weld size. The returns have the greatest effect when the angle length L is short. It may be reasonable to consider the returns to be $L/12$ (2 times $\frac{1}{4}$ in. weld for $L = 6$ in.).

Using, from Table 5.18.1 (Case 4), $S = I/\bar{y}$ referred to the tension fiber at the top of the configuration,

$$S = 2\left(\frac{4bd + d^2}{6}\right) \tag{13.2.12}$$

which for $d = L$ and $b = L/12$ becomes

$$S = \frac{4L^2}{9} \tag{13.2.13}$$

The flexural component, as shown in Fig. 13.2.6, is

$$R_x = \frac{M}{S} = \frac{Pe_1}{S} = \frac{Pe_1}{4L^2/9} = \frac{9Pe_1}{4L^2} \tag{13.2.14}$$

Since little of the shear is carried by the returns, they are neglected for the direct shear component, giving

$$R_v = \frac{P}{2L} \tag{13.2.15}$$

$$\text{Actual } R = \sqrt{\left(\frac{P}{2L}\right)^2 + \left(\frac{9Pe_1}{4L^2}\right)^2}$$

$$R = \frac{P}{2L^2}\sqrt{L^2 + 20.25e_1^2} \text{ kips/in.} \tag{13.2.16}$$

EXAMPLE 13.2.5

Determine the factored load capacity P_u of weld B on Fig. 13.2.5 if $\frac{5}{16}$-in. weld is used and $L = 20$ in. E70 electrodes are used in shielded metal arc welding (SMAW). $4 \times 3 \times \frac{3}{8}$ angles are used. Assume base material is thick enough to preclude shear fracture as the controlling limit state; i.e., the fillet weld strength controls. Use the AISC LRFD Method.

Solution:

(a) Best procedure, Eq. 13.2.16

$$\phi R_{nw} = 0.75(0.707)\left(\tfrac{5}{16}\right)(0.60)70 = 6.96 \text{ kips/in.}$$

$$\text{Actual } R_u = \frac{P_u}{2L^2}\sqrt{L^2 + 20.25 e_1^2}$$

$$e_1 = 3.00 - \bar{x} = 3.00 - 0.25 = 2.75 \text{ in.}$$

$$\bar{x} = \frac{2(2.5)(1.25)}{2(2.5) + 20} = 0.25 \text{ in.}$$

$$\text{Actual } R_u = \frac{P_u}{2(20)^2}\sqrt{(20)^2 + 20.25(2.75)^2} = 0.0294 P_u$$

$$P_u = \frac{6.96}{0.0294} = 237 \text{ kips}$$

(b) Neglecting returns entirely, Eq. 13.2.11,

$$\text{Actual } R_u = \frac{P_u}{2(20)^2}\sqrt{(20)^2 + 36(2.75)^2} = 0.0324 P_u$$

$$P_u = \frac{6.96}{0.0324} = 215 \text{ kips}$$

(c) Using Ref. 13.2 equation, Eq. 13.2.8,

$$\text{Actual } R_u = \frac{P_u}{2(20)^2}\sqrt{(20)^2 + 12.9 e_2^2}$$

$$e_2 = 4\text{-in. leg}$$

$$\text{Actual } R_u = 0.0308 P_u$$

$$P_u = \frac{6.96}{0.0308} = 226 \text{ kips}$$

The authors believe method (a) to be appropriate, $P_u = 237$ kips. ∎

Note that when the elastic (vector) method is used, the same formulas apply for Allowable Strength Design and for Load and Resistance Factor Design. In ASD, P is the service load connection capacity, R is the service load resultant force per unit length at the most highly stressed weld segment, and the allowable strength is R_w. In LRFD, P_u and R_u represent factored loads, and the design strength ϕR_{nw} is used.

13.3 SEATED BEAM CONNECTIONS—UNSTIFFENED

As an alternative to simple shear connections using web angles, or other attachments to the beam web, a beam may be supported on a seat, either unstiffened or stiffened. In this section the unstiffened seat is treated, as shown in Fig. 13.3.1. The unstiffened seat (an angle) is shown in Fig. 13.3.1 and is designed to carry the entire reaction. This type of connection must always, however, be used with a top clip angle, whose intended function is to provide lateral support of the compression flange.

Figure 13.3.1
Seated beam connections—
unstiffened.

As with the case of the simple shear angle connection, the seated connection is intended to transfer only the vertical reaction and should not give significant restraining moment on the end of the beam; thus the seat and the top angle should be relatively flexible. The behavior of welded seat angle connections has been studied by Lyse and Schreiner [13.39], Roeder [13.40] Roeder and Dailey [13.41], and Carter et al [13.101].

The thickness of seat angle is determined by the flexural strength at a critical section of the angle, as shown in Fig. 13.3.2. If a bolted connection is used without attachment to the beam (Fig. 13.3.2a), the critical section should probably be taken as the net section through the upper bolt line. When the beam is attached to the seat (as it should be) as in Fig. 13.3.2b, the rotation of the beam at the end creates a force that tends to restrain the pull away from the column. The critical section for flexure will then be at or near the base of the fillet on the outstanding leg. Similarly for the welded seat, the weld completely along the end holds the angle tight against the column, in which case the critical section is as shown in Fig. 13.3.2c, whether or not the beam is attached to the seat. As a practical matter, rarely will the beam be left unattached from the seat, so the design procedures of this section use a critical section as in Figs. 13.3.2b and c, taken at $\frac{3}{8}$ in. from the face of the angle.

The bending moments on the critical section of the angle and on the connection to the column flange are determined by taking the beam reaction times the distances to the critical sections. The beam reaction occurs at the centroid of the bearing stress distribution, as

Figure 13.3.2
Critical section for flexure
on seats.

(a) Bolted—beam *not* connected to seat (b) Bolted—beam connected to seat (c) Welded

Figure 13.3.3
Bearing stress assumptions
for seated connections.

(a) Conservative assumption   (b) AISC & Ref. 13.23 assumption   (c) Flexible angle   (d) Thick stiff angle

shown in Fig. 13.3.3. While the *AISC Specification* does not state how the computation of this bending moment is to be made, a conservative approach is to assume the reaction at the center of the full contact width (Fig. 13.3.3a). This will lead to excessively thick angles in most cases. The less conservative approach of assuming the reaction at the center of the *required* bearing length N measured from the end of the beam (Fig. 13.3.3b) has been used by Blodgett [13.2] and has been the approach used for *AISC Manual* tables. Another rational distribution for a flexible seat angle is the triangular distribution of Fig. 13.3.3c, and if the angle is very stiff the reaction may become heavier on the outer edge, as in Fig. 13.3.3d.

The design of unstiffened seats involves the following steps:

1. Determine the seat width.
2. Determine the moment arms e and e_f.
3. Determine the length and thickness of the angle.
4. Determine the supporting angle leg dimension, and the weld size; or the number and placement of bolts.

The design of seated beam connections and the background for the AISC load tables have been reviewed by Garrett and Brockenbrough [13.42] and by Brockenbrough [13.43] and Carter et al [13.100].

The seat width is determined from the bearing length N required based on (a) the local web yielding limit state, as given by AISC-J10.2, or (b) the web crippling limit state, as given by AISC-J10.3. Local web yielding usually controls the bearing length.

The *AISC Manual* Part 9 presents a simplified form of the nominal strength equations for local web yielding and web crippling at beam ends. The simplified equations express the nominal strength in terms of design parameters. These parameters are given for all W sections in *AISC Manual* Table 9-4, "Beam Design Constants."

Local Web Yielding—LRFD Method [AISC Formula (J10-3)].

The required bearing length

$$N = \frac{P_u}{\phi F_{yw} t_w} - 2.5k \qquad (13.3.1)$$

where P_u = factored load reaction

       ϕ = resistance factor = 1.0

       t_w = web thickness of supported beam

F_{yw} = yield stress of web of supported beam

k = distance from outer face of flange to web toe of fillet

The required bearing length N may not be taken less than k for beam end reactions. Generally the seat width should not be less than 3 in. with *AISC Manual* Table 10–5 "All-Bolted Unstiffened Seated Connections," indicating a standard 4-in. seat width.

Local Web Crippling—LRFD Method [AISC Formulas (J10-5a) and (J10-5b)].

For $N/d \leq 0.2$,

$$P_u = \phi R_n = \phi 0.40 t_w^2 \left[1 + 3 \left(\frac{N}{d} \right) \left(\frac{t_w}{t_f} \right)^{1.5} \right] \sqrt{\frac{E F_{yw} t_f}{t_w}} \tag{13.3.2}$$

Solving for N, using the parameters available (and defined therein) in the *AISC Manual* Table 9-4,

$$N \geq \frac{P_u - \phi R_3}{\phi R_4} \tag{13.3.3}$$

Note that $\phi = 0.75$ for the local web crippling limit state in the AISC LRFD Method. For $N/d > 0.2$,

$$P_u = \phi R_n = \phi 0.40 t_w^2 \left[1 + 3 \left(\frac{4N}{d} - 0.2 \right) \left(\frac{t_w}{t_f} \right)^{1.5} \right] \sqrt{\frac{E F_{yw} t_f}{t_w}} \tag{13.3.4}$$

Solving for N, using the parameters available in the *AISC Manual* Table 9–4,

$$N \geq \frac{P_u - \phi R_5}{\phi R_6} \tag{13.3.5}$$

where

P_n = factored load reaction

ϕ = resistance factor = 0.75

t_w = web thickness of supported beam

t_f = flange thickness of supported beam

F_{yw} = yield stress of web of supported beam

E = steel modulus of elasticity (29,000 ksi)

The moment arms e and e_f are obtained as follows, referring to Fig. 13.3.3a,

$$e_f = \text{erection clearance} + \frac{N}{2} \tag{13.3.6}$$

$$e = e_f - t - \frac{3}{8} \tag{13.3.7}$$

The bending moment on the critical section of the angle is

$$M_u = P_u e \tag{13.3.8}$$

The thickness t of the angle is obtained by letting $M_u = \phi_b M_n$.

Angle Thickness—LRFD Method (AISC-F11.1),

$$\phi_b M_n = \phi_b M_p = \phi_b Z F_y = \phi_b \frac{L t^2}{4} F_y \tag{13.3.9}$$

$$t = \sqrt{\frac{4P_u e}{\phi_b L F_y}}$$ (13.3.10)

where P_u = factored load reaction to be carried

ϕ_b = resistance factor for flexure = 0.90

e = eccentricity of load to the critical section on angle, such as Eq. 13.3.7

L = length of seat angle (i.e., width of rectangular section being bent)

F_y = yield stress of seat angle steel

This length of the seat angle is generally taken as either 6 in. or 8 in. for a beam gage g of $3\frac{1}{2}$-in. and $5\frac{1}{2}$-in., respectively.

The number of bolts, which are in combined shear and tension, is determined in accordance with the principles of Sec. 4.15.

The weld size and length are obtained using the principles of Sec. 5.19 with Eq. 13.2.16 applicable to this case; direct shear and bending about the x–x axis for the configuration of Fig. 13.2.5 with the returns $b \approx L/12$.

EXAMPLE 13.3.1

Design the seat angle to support a W12×40 beam on a 25-ft span, assuming the beam has adequate lateral support. Use A36 steel and Load and Resistance Factor Design.

Solution:

In many cases it will be wise practice to design the seat for the maximum reaction when the beam is fully loaded in flexure.

(a) Determine the seat width, length, and thickness. The flexural strength is

$$\phi_b M_n = \phi_b M_p = \phi_b Z_x F_y = 0.90(57.5)36/12 = 155 \text{ ft-kips}$$

$$P_u = \frac{w_u L}{2} = \frac{8\phi_b M_n}{2L} = \frac{8(155)}{2(25)} = 24.8 \text{ kips}$$

Bearing length N required based on local web yielding, Eq. 13.3.1, is

$$N \geq \frac{P_u - \phi R_1}{\phi R_2}$$

$$R_1 = 2.5 k_b F_{yw} t_w = 2.5(1.25)(50)0.295 = 46.1 \text{ kips}$$

$$R_2 = f_{yw} t_w = 50(0.295) = 14.75 \text{ kips/in}$$

$$N \geq \frac{24.8 - 0.90(46.1)}{0.90(14.75)} = \text{negative}$$

A minimum bearing length must be used. Following *AISC Manual*, Table 10–5, "All-Bolted Unstiffened Seated Connections," suggestion, use 4-in. seat (i.e., angle leg). According to AISC-J10.2, $N \geq k$. With a 4-in. angle leg, N will exceed k (i.e., 1.25 in.). Check web crippling N_{req} against actual N using a clearance of $\frac{3}{4}$-in. to allow possible mill underrun.

$$N = 4 - 0.75 = 3.25 \text{ in.}$$

$$\text{Compute } \frac{N}{d} = \frac{3.25}{11.94} = 0.27 > 0.2, \text{ use Eq. 13.3.3}$$

$$N \geq \frac{P_u - \phi R_5}{\phi R_6}$$

$$R_3 = 0.40t_w^2 \left[\left(1 - 0.2\left(\frac{t_w}{t_f}\right)^{1.5} \right) \right] \sqrt{\frac{EF_{yw}t_f}{t_w}}$$

$$= 0.40(0.295)^2 \left[\left(1 - 0.2\left(\frac{0.295}{0.515}\right)^{1.5} \right) \right] \sqrt{\frac{29,000(50)0.515}{0.295}} = 50.6 \text{ kips}$$

$$R_4 = 0.40t_w^2 \left(\frac{4}{d}\right)\left(\frac{t_w}{t_f}\right)^{1.5} \sqrt{\frac{EF_{yw}t_f}{t_w}}$$

$$= 0.40(0.295)^2 \left(\frac{4}{11.94}\right)\left(\frac{0.295}{0.515}\right)^{1.5} \sqrt{\frac{29,000(50)0.515}{0.295}} = 8.04 \text{ kips/in}$$

$$N \geq \frac{24.8 - 0.75(50.6)}{0.75(8.04)} = \text{negative}$$

∴ 4 in outstanding leg is OK.

As expected, web crippling does not control. Following the usual practice $N = k$ is used for determining angle thickness,

$$e_f = \frac{1.25}{2} + \frac{3}{4} = 1.375 \text{ in.}$$

Trying $t = \frac{1}{2}$ in.,

$$e = e_f - t - \tfrac{3}{8} = 1.375 - 0.50 - 0.375 = 0.50 \text{ in.}$$

Since the usual gage $g = 5\frac{1}{2}$ in. (*AISC Manual* p. 1-24) for W12×40, use angle length of 8 in. The angle thickness required is then, by Eq. 13.3.9,

$$t^2 = \frac{4P_u e}{\phi_b F_y L} = \frac{4(24.8)0.50}{0.90(36)8} = 0.19 \text{ in.;} \qquad t = 0.44 \text{ in.}$$

Use seat angle, $\frac{1}{2}$ in. thick and 8 in. long.

(b) Determine bolted connection to column, using $\frac{3}{4}$-in.-diam A325 bolts in a bearing-type connection with no threads in the shear plane.

$\phi R_n = 19.9$ kips   (single shear) (*AISC Manual*, Table 7-1)

$\phi R_n = 39.1$ kips   (bearing on $\frac{1}{2}$-in.-thick angle) (*AISC Manual*, Table 7-5)

$\phi R_n = 29.8$ kips   (tension) (*AISC Manual*, Table 7-2)

Using Eq. 4.12.29, obtain a rough estimate of the number of bolts per vertical line for two vertical rows of fasteners at a 3-in. pitch.

$$n \approx \sqrt{\frac{6M}{Rp}} = \sqrt{\frac{6(24.8)1.38}{19.9(3)2}} = 1.3$$

Try 2 bolts (i.e., $n = 1$) as shown in Fig. 13.3.4. The direct shear component is

$$R_v = \frac{P_u}{n} = \frac{24.8}{2} = 12.4 \text{ kips} < 19.9 \text{ kips} \qquad\qquad \textbf{OK}$$

(a) Bolted seat   **(b)** Welded seat

Figure 13.3.4
Designs for Example 13.3.1.

Since the bolts lie on the center of gravity, no moment of inertia can be computed using Σy^2. However, since initial tension exists, the initial compression, according to Eq. 4.15.1, is

$$f_{bi} = \frac{\Sigma T_b}{bd} = \frac{2(28)}{8(3)} = 2.33 \text{ ksi}$$

The change in stress due to moment, Eq. 4.15.2, is

$$f_{tb} = \frac{6M}{bd^2} = \frac{6Pe}{bd^2} = \frac{6(24.8/1.5)1.38}{8(3)^2} = 1.9 \text{ ksi}$$

Since $1.9 < 2.33$, the initial precompression is not eliminated and the connection can be considered safe. The factored load reaction $P_u = 24.8$ kips was divided by an average overload factor of 1.5 to approximate service load. Under full factored load the initial tension would be overcome at the top of the angle; however, compression would exist at the toes of the angles and the tension would be carried in the bolts. The initial tension approach seems justified for this case.
Use 2 bolts, with seat angle, L4×3×$\frac{1}{2}$×0′-8″.

(c) Determine welded connection to column, using Eq. 13.2.16, with E70 electrodes with shielded metal arc welding:

Max weld size $= \frac{1}{2} - \frac{1}{16} = \frac{7}{16}$ in.

Min weld size = AISC-Table J2.4 based on thickest material being joined.

Try $L = 4$-in. supported leg:

$$R_u = \frac{P_u}{2L^2}\sqrt{L^2 + 20.25e_1^2} \qquad [13.2.16]$$

where $e_1 = e_f$

$$R_u = \frac{24.8}{2(4)^2} \sqrt{(4)^2 + 20.25(1.38)^2} = 5.72 \text{ kips/in.}$$

$$\phi R_{nw} = \phi(0.707a)(0.60F_{\text{EXX}}) = 0.75(0.707a)42.0 = 22.3a$$

Base metal shear fracture does not control on $\frac{1}{2}$-in. angles.

$$\text{Weld size } a = \frac{5.72}{22.3} = 0.26 \text{ in.,} \quad \text{say } \tfrac{5}{16} \text{ in.}$$

A more conservative approach is to measure e_f to the center of the contact bearing width of the seat (Fig. 13.3.3a). This traditional AISC method for tables giving weld capacity for seats gives

$$e_f = \frac{N}{2} + \frac{3}{4} = \frac{3.5 - 0.75}{2} + 0.75 = 2.13 \text{ in.}$$

which upon substitution into Eq. 13.2.16 with $\phi R_{nw} = 22.3a = 22.3(0.3125) = 6.97$ kips/in. (for $\frac{5}{16}$ in. weld), gives a factored load capacity of

$$P_u = \frac{\phi R_{nw}(2L^2)}{\sqrt{L^2 + 20.25e_f^2}} = \frac{6.97(2)(4)^2}{\sqrt{(4)^2 + 20.25(2.13)^2}} = 21.5 \text{ kips}$$

The weld size required would then be

$$\text{Required } a = \tfrac{5}{16}\left(\frac{24.8}{21.5}\right) = 0.36 \text{ in., say } \tfrac{3}{8} \text{ in.}$$

Use L4×$3\frac{1}{2}$×$\frac{1}{2}$×0'-8″ with $\frac{3}{8}$ in. weld. Since the W12×40 flange width is 8 in., the beam must be "blocked" or "cut" to have a reduced flange width over the seat so that the necessary welding can be done, or if the column flange permits, the seat may be longer than 8 in. The final designs are shown in Fig. 13.3.4.   ∎

13.4   STIFFENED SEAT CONNECTIONS

When reactions become heavier than desirable for unstiffened seats, stiffeners may be used with the seat angle in bolted construction, or a T-shaped stiffened seat may be used in welded construction. The unstiffened seat may become excessively thick when the factored beam reaction exceeds about 60 kips. There are no AISC restrictions, however, to the maximum load that may be carried by unstiffened seats.

The stiffened seat as discussed herein is not intended to be part of a moment resisting connection, but rather it is only to support vertical loads. Here the stiffened seat is treated as "simple framing". Behavior of welded brackets has been studied by Jensen [13.44].

There are two basic types of loading used on stiffened seats; the common one where the reaction is carried with the beam web directly in line with the stiffener, as shown in Fig. 13.4.1; the other is with a beam oriented so the plane of the web is at 90 degrees to the plane of the stiffener, as in Fig. 13.4.2. Furthermore, a difference in behavior arises depending on the angle at which the stiffener is cut, as shown in Fig. 13.4.3. If the angle θ is approximately 90 degrees, the stiffener behaves similarly to an unstiffened element under uniform compression, and local buckling may be prevented by satisfying AISC-B4. When the supporting plate is cut to create a triangular bracket plate a different behavior results, and this case is discussed in Sec. 13.5.

Figure 13.4.1
Stiffened seat—beam web in line with stiffener.

Top angles must be used, as in Fig. 13.3.4.

(a) Bolted

(b) Welded

Figure 13.4.2
Bracket supporting concentrated load.

1" min. clearance

P

Figure 13.4.3
Two cases of inclination angle at free edge of stiffener.

e

P

$\theta = 90°$

W

(a)

e

P

$\theta \neq 90°$

W

(b)

The steps in the design of stiffened seats are as follows:

1. Determine the seat width.
2. Determine the eccentricity e_s of load.
3. Determine the stiffener thickness t_s.
4. Determine the angle sizes and arrangement of bolts; or the weld size and length.

The seat width is based on the required bearing length N (1) to prevent local web yielding, according to AISC-J10.2, Eqs. 13.3.1 or 13.3.2; and (2) to prevent web crippling, according to AISC-J10.3, Eqs. 13.3.3 through 13.3.5. Because of the rigidity of the stiffener, the most highly stressed portion is at the edge of the seat rather than at the interior side as it was for the unstiffened seat (see Fig. 13.4.4).

For stiffened seats, the *AISC Manual*, Table 10–7, "All-Bolted Stiffened Seated Connections," and Table 10–8, "Bolted/Welded Stiffened Seated Connections," contain maximum factored load reactions (i.e., design strength) for bolted connection stiffener angle outstanding legs of $3\frac{1}{2}$, 4, and 5 in. and welded seat widths of 4 to 9 in.

Assuming the beam reaction P is located at $N/2$ from the edge of the seat, the stiffener thickness t_s should satisfy several criteria:

1. Stiffener thickness t_s should be equal to or greater than the thickness t_w of the supported beam web,

$$t_s \geq t_w \qquad (13.4.1)$$

2. Local buckling of the stiffener must be prevented in accordance with AISC-B4 and AISC-J10.8,

$$t_s \geq \left[\frac{W}{0.56\sqrt{E/F_y}} = \frac{W}{95/F_y,\text{ksi}} \right] \qquad (13.4.2)$$

where W = width of stiffener.

3. Bearing on the contact area of stiffener must satisfy AISC-J7. For *angle* stiffeners, it is assumed that $\frac{1}{2}$-in. of the angle is cut off in order to get close bearing under the seat angle. Under the AISC LRFD Method ($\phi = 0.75$),

$$t_s \geq \frac{P_u}{\phi(1.8F_y)2(W - 0.5)} \qquad \text{(for angle stiffeners)} \qquad (13.4.3)$$

A structural tee might be used instead of two angles; in which case the 2 in the denominator of Eq. 13.4.3 would *not* be used. Eq. 13.4.3 assumes no eccentricity of load with respect to center of bearing contact length assumed.

4. For eccentric loading on stiffener, bearing strength according to AISC-J7 must be satisfied. In this situation, a single welded plate stiffener is generally used. The highest bearing stress at the outer edge of the stiffener may not exceed $\phi(1.8F_y)$ under factored load. Using combined stress (see Fig. 13.4.4),

$$f_b = \frac{P}{A} + \frac{M}{S} = \frac{P_u(e_s - W/2)}{t_sW^2/6} = \frac{P_u}{t_sW^2}(6e_s - 2W) \qquad (13.4.4)$$

$$t_s \geq \frac{P_u(6e_s - 2W)}{\phi(1.8F_y)W^2} \qquad \text{(for welded stiffener)} \qquad (13.4.5)$$

Figure 13.4.4
Bearing load distribution on stiffened seats.

5. Plate thickness must be adequate to develop the fillet welds used to attach it, according to AISC-J2.4. The concept of maximum effective weld size was developed in Sec. 5.14, giving Eqs. 5.14.8 and 5.14.9 relating weld size to base material thickness.

Since the net shear area for the base material is equal to the gross shear area, the nominal strength is governed by shear yielding for both A36 and A992. Thus, from Eq. 5.14.8,

$$a_{\text{max eff}} = \frac{0.60 F_y t_1}{2(0.75)(0.707)0.60 F_{\text{EXX}}} = 0.943 \frac{F_y t_1}{F_{\text{EXX}}} \qquad [5.14.8]$$

where
t_1 = thickness of base material ($= t_s$ here)
F_u = tensile strength of base material
F_{EXX} = tensile strength of electrode material (70 ksi for E70 electrodes)

Assuming two lines of fillet weld of size a using E70 electrodes, the stiffener thickness t_s required such that the stiffener plate will not be overloaded in shear is, from Eq. 5.14.8,

$$t_s \geq 2.06a \quad \text{(for A36 steel)} \qquad (13.4.6)$$

$$t_s \geq 1.48a \quad \text{(for A992 steel)} \qquad (13.4.7)$$

Once the stiffener dimensions have been established, the connection must be designed to transmit the reaction at the moment arm e_s. For the bolted connection, *AISC Manual*, Table 10–7, "All-Bolted Stiffened Seated Connections," uses only direct shear in determining fastener group capacities. One may reason, as in Example 13.3.1, that as long as initial compression between the pieces in contact is not reduced to zero due to flexure, the moment component need not be considered.

For the welded connection suggested by the *AISC Manual*, p. 10–92, as shown in Fig. 13.4.1b, the weld configuration is subject to direct shear and flexure using the combined stress at the top of the weld as the critical one. Thus, the configuration is identical to that used for web framing angles (see Fig. 13.2.1d) except the return is longer. Using $d = L$ and $b = 0.2L$ in the S values for Case 4 from Table 5.18.1 gives

$$\bar{y} = \frac{L^2}{2(L+b)} = \frac{L^2}{2(1.2L)} = \frac{L}{2.4}$$

$$S_x = \frac{2(4bL + L^2)}{6} = \frac{4(0.2L)L + L^2}{3} = 0.6L^2$$

Then,

$$R_x = \frac{M}{S_x} = \frac{P e_s}{0.6L^2} \quad \text{force/unit length} \rightarrow$$

$$R_y = \frac{P}{2(L + 0.2L)} = \frac{P}{2.4L} \quad \text{force/unit length} \downarrow$$

$$R = \sqrt{\left(\frac{P e_s}{0.6L^2}\right)^2 + \left(\frac{P}{2.4L}\right)^2}$$

$$R = \frac{P}{2.4L^2}\sqrt{16e_s^2 + L^2} \quad \text{force/unit length} \qquad (13.4.13)$$

Equation 13.4.13 is used for obtaining loads for welded stiffened beam seats in *AISC Manual*, Table 10–8, "Bolted/Welded Stiffened Seated Connections," when e_s is taken as $0.8W$. For the AISC LRFD Method, R is the design strength ϕR_{nw} and P is the factored load reaction P_u.

EXAMPLE 13.4.1

Design a welded stiffened seat to support a W30×99 beam having a factored load reaction $P_u = 160$ kips. Use A992 steel and Load and Resistance Factor Design.

Solution:

The bearing length N required is obtained from (a) local web yielding and (b) web crippling criteria. From local web yielding (AISC-J10.2),

$$\phi = 1.0$$

$$R_1 = 2.5kF_{yw}t_w = 2.5\left(1\frac{7}{16}\right)(50)(0.520) = 93.4 \text{ kips}$$

$$R_2 = F_{yw}t_w = 50(0.520) = 26.0 \text{ kips/in}$$

$$N \geq \frac{P_u - \phi R_1}{\phi R_2} = \frac{160 - 1.0(93.4)}{1.0(26.0)} = 2.6 \text{ in.}$$

From the web crippling criterion (AISC-J10.3), assuming $N/d \leq 0.2$ use Eq. 13.3.3,

$$\phi = 0.75$$

$$R_3 = 0.40t_w^2\sqrt{\frac{EF_{yw}t_f}{t_w}} = 0.4(0.52)^2\sqrt{\frac{29,000(50)0.670}{0.520}} = 147.8 \text{ kips}$$

$$R_4 = 0.40t_w^2\left(\frac{3}{d}\right)\left(\frac{t_w}{t_f}\right)^{1.5}\sqrt{\frac{EF_{yw}t_f}{t_w}}$$

$$= 0.40(0.52)^2\left(\frac{3}{29.65}\right)\left(\frac{0.520}{0.67}\right)^{1.5}\sqrt{\frac{29,000(50)0.670}{0.520}} = 10.2 \text{ kips/in.}$$

$$N \geq \frac{P_u - \phi R_3}{\phi R_4} = \frac{160 - 0.75(147.8)}{0.75(10.2)} = 6.42 \text{ in.} \qquad \text{Controls}$$

With $N = 6.5$ in., $N/d = 6.5/29.65 = 0.22 > 0.2$. This means Eq. 13.3.4a must be used to determine the N_{reqd}. Thus,

$$R_5 = 0.40t_w^2\left[1 - 0.2\left(\frac{t_w}{t_f}\right)^{1.5}\right]\sqrt{\frac{EF_{yw}t_f}{t_w}}$$

$$= 0.40(0.52)^2\left[1 - 0.2\left(\frac{0.520}{0.67}\right)^{1.5}\right]\sqrt{\frac{29,000(50)0.670}{0.520}} = 127.6 \text{ kips}$$

$$R_6 = 0.40t_w^2\left(\frac{4}{d}\right)\left(\frac{t_w}{t_f}\right)^{1.5}\sqrt{\frac{EF_{yw}t_f}{t_w}}$$

$$= 0.40(0.52)^2\left(\frac{4}{29.65}\right)\left(\frac{0.520}{0.67}\right)^{1.5}\sqrt{\frac{29,000(50)0.670}{0.520}} = 13.6 \text{ kips/in.}$$

$$N \geq \frac{P_u - \phi R_5}{\phi R_6} = \frac{160 - 0.75(127.6)}{0.75(13.6)} = 6.30 \text{ in.} \qquad \therefore N = 6.5 \text{ in. is adequate}$$

Required $W = 6.5 + 0.5(\text{setback}) = 7.0$ in.   *Use* 7 in.

For the seat plate thickness, use a thickness comparable to the flange of the W30×99 supported beam; *use* $\frac{5}{8}$ in. Minimum weld size for welding on $\frac{5}{8}$-in. seat and 0.67-in. flange is $\frac{1}{4}$ in.

The stiffener thickness is next to be established:

$$t_s \geq t_w = 0.520 \text{ in.} \qquad [13.4.1]$$

$$t_s \geq \frac{W}{95/\sqrt{F_y},\text{ksi}} = \frac{7}{13.4} = 0.52 \text{ in.} \qquad [13.4.2]$$

$$e_s = W - \frac{N}{2} = 7.0 - \frac{6.5}{2} = 3.8 \text{ in.}$$

$$t_s \geq \frac{P_u(6e_s - 2W)}{\phi(1.8F_y)W^2} = \frac{160(22.5 - 14)}{0.75(90)(7)^2} = 0.41 \text{ in.} \qquad [13.4.6]$$

The use of a $\frac{5}{8}$-in. stiffener plate would mean a maximum effective weld size of

$$t_s \geq 1.49a$$

$$a_{\text{max eff}} = \frac{t_s}{1.49} = \frac{0.625}{1.49} = 0.42 \text{ in.} \qquad [13.4.9]$$

Thus, weld size is not of concern since a weld smaller than 0.42 in. would be preferred. Generally, the maximum weld that can be placed in one pass would be used; in this case, $\frac{5}{16}$ in.

For estimating the length L of weld required, assume that e_s in Eq. 13.4.13 is approximately $L/4$; i.e., that $e_s = 0.8W = 0.8(7) = 5.6$ in. is roughly $L/4$.

$$R = \frac{P}{2.4L^2}\sqrt{16\left(\frac{L^2}{16}\right) + L^2} = 0.59\frac{P}{L}$$

For LRFD, R becomes ϕR_{nw} and using $\frac{5}{16}$-in. E70 weld,

$$\phi R_{nw} = \phi(0.707a)(0.60F_{\text{EXX}})$$

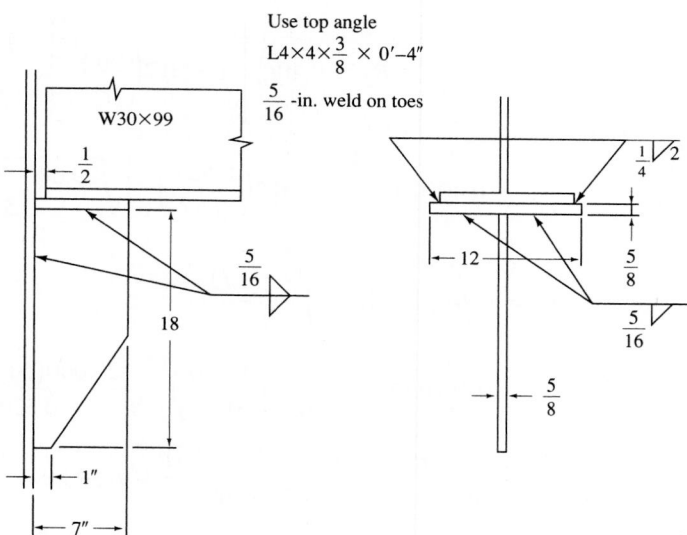

Figure 13.4.5
Design for Example 13.4.1.

$$\phi R_{nw} = 0.75(0.707)\left(\tfrac{5}{16}\right)(0.60)70 = 6.96 \text{ kips/in.}$$

$$\text{Required } L \approx \frac{0.59(160)}{6.96} = 13.6 \text{ in.}$$

For $L = 14$ in., 5.6 in. is $0.4L$; which when used as e_s/L in Eq. 13.4.13 gives required $L \approx 18$ in. The answer lies between 14 and 18 in. Try $L = 16$ in. with $\tfrac{5}{16}$-in. weld,

$$R_u = \frac{160}{2.4(16)^2}\sqrt{16(5.6)^2 + (16)^2} = 7.2 \text{ kips/in.} > 6.96 \text{ kips/in.} \qquad \textbf{NG}$$

The length $L = 16$ in. is not adequate; one could use 17 in. but 18 in. may be preferred. *Use* $\tfrac{5}{16}$-in. weld with $L = 18$ in. Use stiffener plate, $\tfrac{5}{8}\times 7\times 1'\text{-}6''$; and seat plate $\tfrac{5}{8}\times 7\times 1'\text{-}0''$. The seat plate width equals the flange width (10.45 in.) plus enough to easily make the welds (approx. 4 times the weld size is often used). The final design is shown in Fig. 13.4.5.   ∎

13.5   TRIANGULAR BRACKET PLATES

When the stiffener for a bracket is cut into a triangular shape, as in Fig. 13.4.3b, the plate behaves in a different manner than when the free edge is parallel to the direction of applied load in the region where the greatest stress occurs, as in Fig. 13.4.5. The triangular bracket plate arrangement and notation are shown in Fig. 13.5.1.

The behavior of triangular bracket plates has been studied analytically by Salmon [13.45] and experimentally by Salmon, Buettner, and O'Sheridan [13.46] and design suggestions have been proposed by Beedle et al. [13.47]. For small stiffened plates to support beam reactions there is little danger of buckling or failure of the stiffener if cut into a triangular shape.

Most Exact Analysis and Design Recommendations

For many years, design of such brackets was either empirical without benefit of theory or tests, or when in doubt, angle or plate stiffeners were used along the diagonal edge. The recommendations presented here are based on certain assumptions: (1) the top plate is solidly attached to the support; (2) the load P is distributed (though not necessarily uniformly) and has its centroid at approximately $0.6b$ from the support; and (3) the ratio b/a, loaded edge to supported edge, lies between 0.50 and 2.0.

The original theoretical analysis was concerned with elastic buckling; however, the experimental work showed that triangular bracket plates have considerable post-buckling

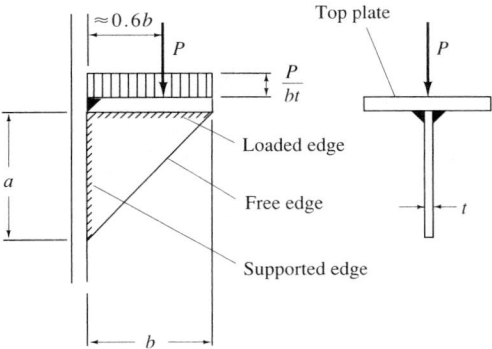

Figure 13.5.1
Triangular bracket plate.

strength. Yielding along the free edge frequently occurs prior to buckling, at which point redistribution of stresses occurs. A considerable margin of safety against collapse was observed indicating the ultimate capacity may be expected to be at least 1.6 times the buckling load.

The maximum stress was found to occur at the free edge; however because of the complex nature of the stress distribution, the stress on the free edge is not obtainable by any simple process. Because of this difficulty, a ratio z was established between the average stress, P/bt, on the loaded edge to the maximum stress f_{max} on the free edge. The original theoretical expression [13.45] for z was revised as a result of the tests [13.46], which conformed closely to what one could realistically expect in practice. The relationship is given [13.46] as

$$z = \frac{P/bt}{f_{max}} = 1.39 - 2.2\left(\frac{b}{a}\right) + 1.27\left(\frac{b}{a}\right)^2 - 0.25\left(\frac{b}{a}\right)^3 \tag{13.5.1}$$

which for practical purposes may be obtained from Fig. 13.5.2.

The nominal strength P_n when the free edge reaches the yield stress is

$$P_n = F_y z bt \tag{13.5.2}$$

For the plate buckling limit state, the width/thickness ratio b/t must be restricted in accordance with a relationship of the type of Eq. 6.16.4,

$$\frac{b}{t} \leq \frac{\text{constant}}{\sqrt{F_y}} \tag{13.5.3}$$

Figure 13.5.3 gives the variation in $(b/t)\sqrt{F_y}$ with b/a for the theoretical studies [13.45] (fixed and simply supported), the welded bracket tests result [13.46], and the authors' suggested design curve. The design requirement may be expressed (with F_y in ksi) as

$$\text{For } 0.5 \geq \frac{b}{a} \leq 1.0; \qquad \frac{b}{t} \leq 1.47\sqrt{\frac{E}{F_y}} = \frac{250}{\sqrt{F_y,\text{ksi}}} \tag{13.5.4a}$$

$$\text{For } 1.0 \leq \frac{b}{a} \leq 2.0; \qquad \frac{b}{t} \leq 1.47\left(\frac{b}{a}\right)\sqrt{\frac{E}{F_y}} = \frac{250(b/a)}{\sqrt{F_y,\text{ksi}}} \tag{13.5.4b}$$

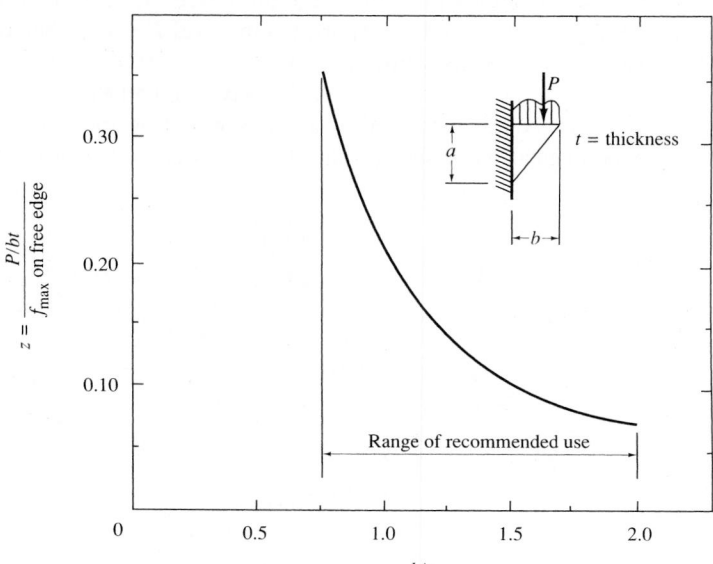

Figure 13.5.2
Coefficient used to obtain
maximum stress on free edge.

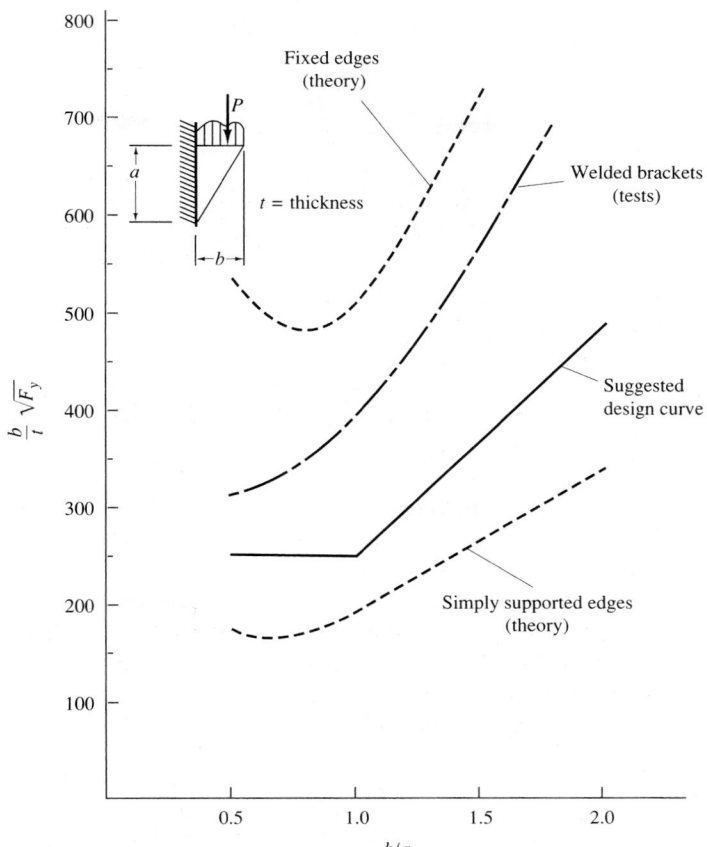

Figure 13.5.3
Critical *b/t* values so that yield stress is reached along diagonal free edge without buckling.

Satisfying the above limits means that yielding along the *diagonal free edge* will occur prior to buckling, and with conservatism compared to the welded tests, as shown by Fig. 13.5.3. The authors note that the conversion of formulas into a non-dimensional format (not just here but throughout the *AISC Specification*) has resulted in implied accuracy far in excess of the original derivations or recommendations. Eqs. 13.5.4 have the ridiculous coefficient 1.47; this is a backward derivation from the original approximate 250 coefficient. Design recommendations strived to be simple and within the scope of the accuracy of the background work.

It is noted that the *b/t* limits suggested here are higher than those of Ref. 13.47 (p. 552), which were based solely on the theoretical studies [13.45]. Reference 13.47 suggests a coefficient of 180 instead of 250 in Eq. 13.5.4 and reaches a maximum of 300 instead of 500 as indicated by Eq. 13.5.4b for $b/a = 2.0$. The reason for the higher values is found in the test results which showed the principal stress along the diagonal free edge to be lower relative to the stress on the loaded edge than had been established by the theoretical study. In other words, the *z* value, Eq. 13.5.1, as determined by tests is substantially smaller than assumed for the design suggestion of Ref 13.46.

Plastic Strength of Bracket Plates

Reference 13.47 suggests that to develop the full plastic strength of brackets the *b/t* ratios should be restricted to about $\frac{1}{3}$ of those limitations for achieving first yield on the free edge. The test results [13.46] indicated that ultimate strengths of at least 1.6 times

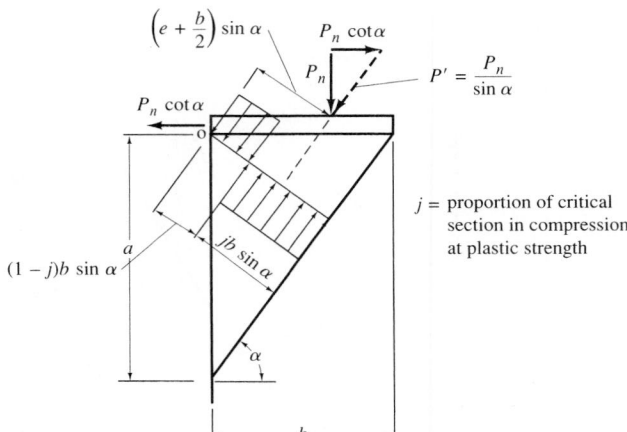

Figure 13.5.4
Plastic strength analysis.

buckling strengths could be achieved due to post-buckling strength. To be certain of developing the *plastic capacity* of the bracket, it may be realistic to use half of the limitations of Eqs. 13.5.4a and b.

To establish the plastic strength of a bracket plate used in rigid-frame structures, one may follow the approach of Beedle et al. [13.47] as shown in Fig. 13.5.4. This method assumes that plastic strength develops on the critical section. Taking force equilibrium parallel to the free edge and moment equilibrium about point O gives the Beedle et al. [13.47] equation for the ultimate load,

$$P_n = F_y t \sin^2\alpha \left(\sqrt{4e^2 + b^2} - 2e \right) \qquad (13.5.5)$$

In addition to the triangular plate being adequate, the top plate must carry the nominal load $P_n \cot \alpha$.

EXAMPLE 13.5.1

Determine the thickness required for a triangular bracket plate 25 in. by 20 in. to carry a factored load of 60 kips. Assume the load is located 15 in. from the face of support as shown in Fig. 13.5.5, and that A36 material is used. Use Load and Resistance Factor Design.

Solution:

(a) Use the more exact method. Since the load is approximately at the 0.6 point along the loaded edge, the bracket fits the assumption of this method. Using Eq. 13.5.2,

$$P_n = F_{yz} bt \qquad [13.5.2]$$

Figure 13.5.5
Bracket for Example 13.5.1.

Since this is a compression situation, the design strength ϕP_n should be equated to P_u. It may be reasonable to use the ϕ for compression; $\phi_c = 0.85$. From Fig. 13.5.2, $b/a = 25/20 = 1.25$, find $z = 0.135$. Then the strength requirement gives

$$P_u = \phi_c P_n = 0.85 F_y \, zbt = 60 \text{ kips}$$

$$t \geq \frac{P_u}{\phi_c F_y \, zb} = \frac{60}{0.85(36)(0.135)25} = 0.58 \text{ in.}$$

The stability requirement, Eq. 13.5.4b, gives

$$t \geq \frac{b\sqrt{F_y}}{250(b/a)} = \frac{25\sqrt{36}}{250(1.25)} = 0.48 \text{ in.}$$

Use $\frac{5}{8}$-in. plate.

(b) Plastic strength method. Using Eq. 13.5.5 for the strength requirement,

$$t \geq \frac{P_u}{\phi F_y \sin^2 \alpha \left[\sqrt{4e^2 + b^2} - 2e\right]}$$

Using $e = 15 - 25/2 = 2.5$ in.,

$$t \geq \frac{60}{0.85(36)(0.39)\left[\sqrt{4(2.5)^2 + (25)^2} - 2(2.5)\right]} = 0.25 \text{ in.}$$

For stability, using *one-half* of Eq. 13.5.4b

$$t \geq \frac{b\sqrt{F_y}}{125(b/a)} = \frac{25\sqrt{36}}{125(1.25)} = 0.96 \text{ in.}$$

Use 1-in. plate as a conservative practice to assure deformation well beyond first yield along the free edge.

The authors note that if a 1-in. plate is just stable enough to inhibit buckling until the plastic strength is obtained, that strength would be 4 times (1.0/0.25) the factored load P_u. ∎

13.6   CONTINUOUS BEAM-TO-COLUMN CONNECTIONS

In continuous beam-to-column connections, it is the design intent to have full transfer of moment and little or no relative rotation of members within the joint [i.e., AISC-Fully Restrained Moment Connections (FR)—rigid-frame connections]. Since the flanges of a beam carry most of the bending moment via tension and compression flange forces acting at a moment arm approximately equal to the beam depth, it is the transfer of these essentially axial forces for which provision must be made. Since the shear is carried primarily by the web of a beam, full continuity requires that it be transferred directly from the web.

Columns being rigidly framed by beams may have attachments to both flanges, as in Fig. 13.6.1a, b, and c, or to only one flange, as in Fig. 13.6.1d, and in Fig. 13.6.2. Alternatively, the rigid attachment of beams may be to the web, from either or both sides, as in Fig. 13.6.3. When the rigid system has rigid attachments *either* to the flanges or the web (but not to both) the system is said to be a two-way, or planar, rigid frame. When the rigid frame

Back up plates

(a) No column stiffeners; direct flange to flange groove welds

(b) Horizontal plate stiffeners

(c) Vertical tee stiffeners

(d) Seat and top plate

Angle or bracket

Figure 13.6.1
Continuous beam-to-column connections: welded attachment to column flange.

(a) Top and bottom plates

(b) Top plate and seat

(c) Split tees, with or without web angles

(d) Welded end plates

Figure 13.6.2
Continuous beam-to-column connections: bolted attachments.

(a) Top plates and seat bracket

(b) Top and bottom plates, web plate; and seat plate

T-section

(c) Tee section with flange back-up plates and small web plates

Back-up plate

Figure 13.6.3
Continuous beam-to-column connections: welded attachment to column web.

system consists of continuous connections to both flange (or flanges) and web (either or both sides), the system becomes a three-dimensional system called a space frame.

The variety of arrangements for a continuous beam-to-column connection is so great as to preclude any complete listing or illustration; however, those shown in Figs. 13.6.1, 13.6.2, and 13.6.3 are believed to be common in current (2008) design use. Most connections are partly shop welded and then completed in the field by either welding or fastening with high-strength bolts.

The principal design concern is with transmission of concentrated loads caused by flange forces on beams to the adjacent columns. The web of a column may be unable to accept the compression force from a beam flange without additional stiffening; the flange of a column may exhibit excessive deformation caused by a tension force from a beam flange.

Rigid framing is used to greatest advantage in (a) structures designed using plastic analysis under the AISC Load and Resistance Factor Method (AISC-B3.1), or (b) structures designed using elastic analysis where the sections are "compact" and the 10% reduction is permitted under AISC-Appendix 1. In these cases, the objective of the connection will be to develop the full plastic moment strength at the joint, and in addition, be able to undergo plastic hinge rotation.

Many researchers [7.28, 13.48–13.59] have adequately demonstrated the ability of beam-to-column rigid frame connections to develop the plastic moment, as well as exhibit adequate rotation capacity (ductility). A summary of bolted beam-to-column connections is given by Kulak, Fisher, and Struik [3.1].

Horizontal Stiffener in Compression Region of Connection

When the forces in beam flanges are transmitted to column flanges as compression or tension forces, horizontal stiffeners, as in Figs. 13.6.1b and d and Fig. 13.6.2a, may be required. Such stiffeners prevent local flange bending (AISC-J10.1) from the tension force, local web yielding (AISC-J10.2), web crippling (AISC-J10.3), and compression buckling (AISC-J10.5) caused by the compression force. Local web yielding and web crippling were treated in Sec. 7.8.

Local Web Yielding.

AISC treatment of this phenomenon is based on work of Graham, Sherbourne, Khabbaz, and Jensen [7.28].

Consider a beam compression flange bearing against a column as in Fig. 13.6.4a. When the maximum local yielding strength of the column web is reached, the load will have been distributed along the base of the fillet (k from the face of the flange) on a 1:2.5 slope. Using Eq. 7.8.2 for an interior concentrated load according to AISC and taking the bearing length N as the beam flange thickness t_{fb} and the reaction R or ϕR_n as the flange force P_{bf} gives (for the AISC LRFD Method)

$$P_{bf} = \phi(5k + t_{fb})F_{yw}t_{wc} \qquad (13.6.1)$$

where
P_{bf} = design strength of the column web to resist a beam flange transmitted force under factored loads
ϕ = resistance factor = 1.0
F_{yw} = yield stress of the column web
t_{wc} = column web thickness
t_{fb} = beam compression flange thickness

The required P_{bf} could be as high as $F_{yb}A_f$ when the beam is expected to develop its plastic moment strength.

Web Crippling.

In accordance with Eq. 7.8.4, using t_{fb} for the bearing length N,

$$P_{bf} = \phi 0.80 t_{wc}^2 \left[1 + 3\left(\frac{t_{fb}}{d}\right)\left(\frac{t_{wc}}{t_{fc}}\right)^{1.5} \right] \sqrt{\frac{EF_{yw}t_{fc}}{t_{wc}}} \qquad (13.6.2)$$

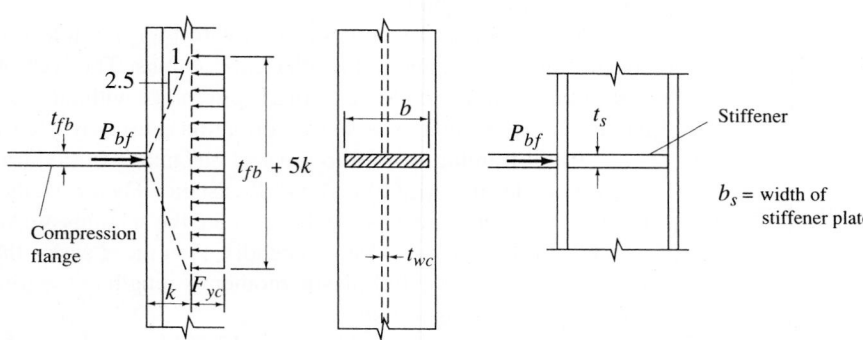

Figure 13.6.4
Strength of column web in compression region of connection—AISC approach.

(a) *Without* column web stiffener   (b) *With* column web stiffener

$$\phi = \text{resistance factor} = 0.75$$
$$t_{fc} = \text{column flange thickness}$$
$$d = \text{column overall depth}$$

The other terms are defined above following Eq. 13.6.1.

Compression Buckling of the Web.

Overall buckling of the column web must also be prevented as a controlling limit state. This limit state is of concern when concentrated loads from beam flanges are applied to both column flanges. When only one column flange is subjected to a concentrated load, the overall web buckling limit state need not be checked.

Equation 6.14.28 applies to elastic buckling of a plate.

$$F_{cr} = \frac{k\pi^2 E}{12(1 - \mu^2)(b/t)^2} \qquad [6.14.28]$$

Chen and Newlin [13.60] and Chen and Oppenheim [13.61] have treated web buckling in a column as analogous to a plate subject to equal and opposite concentrated loads as shown in Fig. 13.6.5. The elastic buckling load P_{cr} for that situation using a simply supported plate having a large ratio a/h, is given by Timoshenko and Gere [6.67, pp. 387–389] as

$$P_{cr} = \frac{4\pi^2 E t^3}{12(1 - \mu^2)h} = \frac{1.152 E t^3}{h} = \frac{33,400 t^3}{h} \qquad (13.6.3)$$

If the rotational restraint provided by the column flanges were the fully fixed condition, the buckling strength would be theoretically twice as great as that given by Eq. 13.6.2. Experimental work [13.52, 13.60, 13.61] has shown that when lower yield stress steel (such as A36) is involved, yielding of the web along the junction with the flange occurs at a load level corresponding closely to the simple support case, i.e., Eq. 13.6.3. When 100 ksi (690 MPa) yield stress steel was used in tests, the higher yield stress steel provided a high degree of rotational restraint along the junction of flange to web, giving a buckling strength about twice that obtained when A36 steel was used. Thus, Chen and Newlin [13.60] suggested that the increase in effective degree of fixity as the loaded edge may be accounted for in practical design by making the strength proportional to the square root of the yield stress. Their stability criterion is a modification of Eq. 13.6.3.

$$P_{cr} = \frac{33,400 t^3}{h} \sqrt{\frac{F_{yw}, \text{ksi}}{36}} = \frac{5570 t^3 \sqrt{F_{yw}, \text{ksi}}}{h} \qquad (13.6.4)$$

The coefficient 5570 of this semirational expression was adjusted downward to 4100, representing a lower bound for all test results. Then for the beam-to-column connection t is

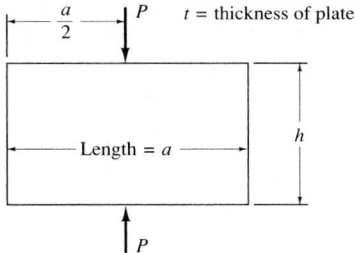

Figure 13.6.5
Plate subject to equal and opposite concentric concentrated loads.

the column web thickness t_{wc} and the concentrated load P_{cr} is the nominal reaction strength P_n, giving the AISC-J10.5 expression

$$P_n = \frac{24t_w^3\sqrt{EF_{yw}}}{h} = \frac{4100t_w^3\sqrt{F_{yw}}, \text{ksi}}{h} \quad (13.6.5)$$

where $\quad P_n$ = nominal strength provided by column web

$\quad\quad h$ = web depth clear of fillets = $d - 2k$

$\quad\quad t_w$ = column web thickness

$\quad\quad F_{yw}$ = column web yield stress

Substitution of the required nominal strength P_{bf}/ϕ for P_n, and using $\phi = 0.90$, according to AISC-J10.5, gives the factored load P_{bf} that can be carried without overall column web buckling when no stiffeners are used,

$$P_{bf} = \frac{\phi 24t_w^3\sqrt{EF_{yw}}}{h} = \frac{\phi 4100t_w^3\sqrt{F_{yw}}, \text{ksi}}{h} \quad (13.6.6)$$

where $\quad \phi = 0.90$ (for the AISC LRFD Method)

$\quad\quad P_{bf}$ = maximum factored service concentrated load from beam flanges

When any or all of the three concentrated load related limit states (i.e., local web yielding, web crippling, or compression buckling of the web) indicate inadequacy of the column web to transmit the beam flange compression force P_{bf}, stiffeners in pairs (total area A_{st}) must be provided on the column to resist the force P_{bf}.

Compression Stiffener Rules-AISC.

The following are the AISC requirements for stiffener design:

1. When the web buckling (AISC-J10.3) or compression buckling of the web (AISC-J10.5) limit states indicate the need for stiffeners, they are to be designed as axially loaded compression members (AISC-J10.8). Alternatively, doubler plates may be used in accordance with AISC-J10.3 or AISC-J10.5.

2. When compression buckling of the web (AISC-J10.8) controls, the stiffeners (if used) must extend the entire depth of the column.

3. When local web yielding (AISC-J10.2) controls and the concentrated load P_{bf} is applied at only one column flange, the stiffeners need not extend more than one-half the depth of the column web.

4. When local web yielding (AISC-J10.2) controls, the area A_{st} of stiffeners (in pairs) required is the excess of the factored force P_{bf} over the design resistance ϕR_u, divided by the stiffener design yield stress ϕF_{yst}; thus,

$$A_{st} \geq \frac{P_{bf} - \phi F_{yst}(t_{fb} + 5k)t_{wc}}{\phi F_{yst}} \quad (13.6.7)$$

5. Proportioning of stiffeners, The following proportioning requirement appears in AISC-J10.8. The stiffener width b_{st} plus $\frac{1}{2}$ the column web thickness t_{wc} may not be less than $\frac{1}{3}$ of the beam flange width b_{fb} or moment plate connection width delivering the force P_{bf}.

$$\left(b_{st} + \frac{t_{wc}}{2}\right) \geq \frac{b_{fb}}{3} \quad (13.6.8)$$

which makes

$$b_{st} \geq \left(\frac{b_{fb}}{3} - \frac{t_{wc}}{2}\right) \quad (13.6.9)$$

6. The local buckling limits of AISC-B4, restated in AISC-J10.8(2), for unstiffened compression elements must be satisfied. Since these limit states relate to $b_f/2t_f$ for the beam transmitting the force to the column, the limit will be satisfied when the stiffener thickness t_s is not less than one-half the beam flange thickness t_{fb}. Thus,

$$t_s \geq \frac{t_{fb}}{2} \tag{13.6.10}$$

7. The weld joining stiffeners to the column web should be sized to carry the force in the stiffener caused by unbalanced moments on opposite sides of the column.

Horizontal Stiffener in Tension Region of Connection

At the beam tension flange attachment to a column, the pull on the column flange, as shown in Fig. 13.6.6, may cause sufficient deformation as to impair the strength of the column. A yield line analysis was performed by Graham et al. [7.28] on the portion of the column flange of width q and length p, as in Fig. 13.6.6. Placing a line load on the system, the nominal strength P_n was approximated as

$$P_n = 7t_{fc}^2 F_{yf} + t_{fb}mF_{yw} \tag{13.6.11}$$

where the first term represents the bending resistance of the column flange as two plate elements (one on each side of the web) and the second term is the portion of the load that goes directly into the column web. Using conservatism comparable to the local web yielding in compression criterion, the right side of Eq. 13.6.11 was multiplied by 0.8, then solving for t_{fc} gives

$$t_{fc} = \sqrt{\frac{P_n}{7F_{yf}}\left(1.25 - \frac{t_{fb}mF_{yw}}{P_n}\right)} \tag{13.6.12}$$

From tests [7.28] the minimum value of $t_{fb}mF_{yw}/P_n$ was determined to be 0.15. Thus, using 0.15 for the second term in the bracket of Eq. 13.6.12 gives the conservative expression used by AISC for the minimum column flange thickness t_{fc} to avoid the need of a column stiffener to assist in carrying the tension force from a beam flange,

$$t_{fc} \geq 0.4\sqrt{\frac{P_n}{F_{yf}}} \tag{13.6.13}$$

Figure 13.6.6
Strength of column flange in tension region of connection.

Section A-A

Local Flange Bending from Tension Force.

In order for more general applicability to any situation, Eq. 13.6.13 is solved for the nominal strength P_n in AISC-J10.1.

$$P_n = 6.25t_{fc}^2 F_{yf} \tag{13.6.14}$$

Since the factored flange force P_{bf} cannot exceed the design strength ϕP_n, either Eq. 13.6.14 can be used to determine the available strength or Eq. 13.6.13 can be used to compute the minimum flange thickness needed to avoid stiffeners. Thus, Eq. 13.6.13 for minimum column flange thickness t_{fc} in the AISC LRFD Method would be

$$t_{fc} \geq 0.4\sqrt{\frac{P_{bf}}{\phi F_{yf}}} \tag{13.6.15}$$

where P_{bf} = factored concentrated tension load
ϕ = 0.90 (for the AISC LRFD Method)
F_{yf} = yield stress of the column flange being bent

Since, generally, the compression-related stiffener requirements are more likely to control, the reader should particularly note those requirements. The same proportioning requirements for compression stiffeners should be used for tension stiffeners. Usually the same size stiffeners would be used for both compression and tension (if any tension stiffener is needed). Examples of the design of beam-to-column moment connections have been presented by Miller [13.62].

EXAMPLE 13.6.1

Design the connection for the rigid framing of two W16×40 beams to the flanges of a W12×65 column using A992 steel, as shown in Fig. 13.6.7. Use A36 steel for stiffeners if needed. Use the AISC LRFD Method.

Solution:

(a) Compression region. Design for the maximum transmitted by the beam. Thus the maximum factored force will be assumed to be the design strength of the beam flange.

$$P_{bf} = \phi A_f F_y = 0.90(6.995)(0.505)50 = 159 \text{ kips}$$

since the beam is "compact" with If the beam were "noncompact," this approach might still be reasonable unless the actual factored moment is used to establish the connection force.

Figure 13.6.7
Connection with horizontal stiffeners for Example 13.6.1.

To avoid stiffeners, the column web-related strength must satisfy AISC-J10.2 (local web yielding), AISC-J10.3 (web crippling), and AISC J10.5 (compression buckling of the web). Solving Eq. 13.6.1, 13.6.2, and 13.6.6 for the maximum factored load P_{bf} that can be carried without stiffeners,

$$P_{bf} = \phi \, (5k + t_{fb})F_{yw}t_{wc} \qquad\qquad [13.6.1]$$

$$= 1.0[5(1.3125) + 0.505](50)0.390 = 138 \text{ kips}$$

$$P_{bf} = \phi \, 0.80t_{wc}^2 \left[1 + 3\left(\frac{t_{fb}}{d}\right)\left(\frac{t_{wc}}{t_{fc}}\right)^{1.5} \right] \sqrt{\frac{EF_{yw}t_{fc}}{t_{wc}}} \qquad [13.6.2]$$

$$= 0.75(0.80)(0.390)^2 \left[1 + 3\left(\frac{0.505}{12.12}\right)\left(\frac{0.390}{0.605}\right)^{1.5} \right] \sqrt{\frac{29{,}000(50)0.605}{0.390}}$$

$$= 146 \text{ kips}$$

$$P_{bf} = \frac{\phi 24t_w^3 \sqrt{EF_{yw}}}{h} = \frac{\phi 4100t_w^3 \sqrt{F_{yw}},\text{ksi}}{h} \qquad\qquad [13.6.6]$$

Note that in Eq. 13.6.6, h is the "clear distance between flanges less the fillet or corner radius for rolled shapes: distance between adjacent lines of fasteners or the clear distance between flanges when welds are used for built-up shapes." For rolled shapes h/t_w is a property given in the *AISC Manual*. Using $h/t_w = 24.9$, from the *AISC Manual* p. 1–23 in, Eq. 13.6.6 gives

$$P_{bf} = \frac{\phi 4100t_w^3 \sqrt{F_{yw}},\text{ksi}}{h} = \frac{0.90(4100)(0.390)^2 \sqrt{50}}{24.9} = 159 \text{ kips}$$

Since the factored load capacities P_{bf} without stiffeners for local web yielding ($P_{bf} = 138$ kips) and for web crippling ($P_{bf} = 146$ kips) are less than the 159 kips required, compression stiffeners are required under AISC-J10.2 and AISC-J10.3. The extra strength must be obtained from stiffeners.

Local web yielding (Eq. 13.6.1) indicates the need for compression stiffeners. Taking the lowest ($P_{bf} = 138$ kips) of the three strengths in Eq. 13.6.7,

$$\text{Required } A_{st} = \frac{\text{Actual } P_{bf} - 138}{\phi F_{yst}} = \frac{159 - 138}{0.85(36)} = 0.7 \text{ sq in.}$$

Stiffeners to prevent local web yielding need not extend the full depth of the column web unless the force is applied at both column flanges; in this case, the stiffeners must extend full depth.

Stiffeners are also needed to prevent web crippling, and such stiffeners must extend full depth and be designed as columns the same as bearing stiffeners in a plate girder, as discussed in Sec. 11.12. The factored column load that must be carried is the full $P_{bf} = 159$ kips.

Size Limitations. Using Eq. 13.6.9, the minimum stiffener width b_{st} is

$$b_{st} \geq \left(\frac{b_{fb}}{3} - \frac{t_{wc}}{2}\right) = \frac{6.995}{3} - \frac{0.390}{2} = 2.1 \text{ in., say 3 in.}$$

The minimum stiffener thickness t_s according to Eq. 13.6.10 is

$$\text{Min } t_s = \frac{t_{fb}}{2} = \frac{0.505}{2} = 0.252 \text{ in., say } \tfrac{5}{16} \text{ in.}$$

Two plates—$\tfrac{5}{16} \times 3$ give more than the minimum area $A_{st} = 0.7$ sq in. required.

Local Buckling Limits. When elastic analysis is used,

$$\left(\lambda = \frac{b_{st}}{t_s}\right) \le \left(\lambda_r = \frac{95}{\sqrt{F_y}} = \frac{95}{\sqrt{36}} = 15.8\right)$$

$$\text{Min } t_s = \frac{3}{15.8} = 0.19 \text{ in.}$$

or, when plastic analysis is used,

$$\left(\lambda = \frac{b_{st}}{t_s}\right) \le \left(\lambda_p = \frac{65}{\sqrt{F_y}} = \frac{65}{\sqrt{36}} = 10.8\right)$$

$$\text{Min } t_s = \frac{3}{10.8} = 0.28 \text{ in.}$$

Column Strength of Stiffeners. The column consists of the stiffeners acting in combination with a length of web equal to $25t_{wc} = 25(0.390) = 9.75$ in. (see AISC-J10.8 and text Sec. 11.8).

$$r = \sqrt{\frac{\frac{1}{12}(0.3125)(3 + 3 + 0.39)^3}{2(3)0.3125 + 9.75(0.390)}} = 1.1 \text{ in.}$$

Using an effective length KL equal to 0.75 of the column depth: i.e., approximately 9 in.,

$$\frac{KL}{r} = \frac{9}{1.1} \approx 8; \qquad \text{Thus, } \phi_c F_{cr} = 32.3 \text{ ksi}$$

$$A_g = 2(3)0.3125 + 9.75(0.390) = 5.68 \text{ sq in.}$$

$$\phi_c P_n = A_g(\phi_c F_{cr}) = 5.68(32.3) = 184 \text{ kips} > 159 \text{ kips} \qquad \textbf{OK}$$

(b) Tension region. To prevent excessive bending of a column flange from the action of a beam flange tension force, the minimum column flange thickness required by AISC-J10.1 is given by Eq. 13.6.15,

$$t_{fc} \ge 0.4\sqrt{\frac{P_{bf}}{\phi F_{yf}}} = 0.4\sqrt{\frac{159}{0.90(50)}} = 0.75 \text{ in.}$$

Since the flange thickness of the W12×65 ($t_f = 0.605$ in.) is less than the required $t_{fc} = 0.75$ in., tension stiffeners are required. The local buckling limits do not apply to tension elements; however, for practical purposes the same stiffeners used as compression stiffeners should be used here.

Use 2 PLs—$\frac{5}{16}$×3×10$\frac{7}{8}$, A36, for both compression and tension sides.

(c) Connection of plates to column. The forces to be used for the design of welding are shown in Fig. 13.6.8. When beams frame from both sides and contribute equal flange forces P_{bf}, the welds on the ends of the plates must carry the portion of the force P_{bf} which is not taken directly by the web.

For this example, Fig. 13.6.8a applies. The weld should be designed to carry the maximum stiffener force $A_{st}F_{yst}$: actually, when the stiffeners act in combination with the web, each part will try to carry load in proportion to its stiffness. Thus, the stiffeners will likely carry more than just the excess that the web alone could not carry. Maximum

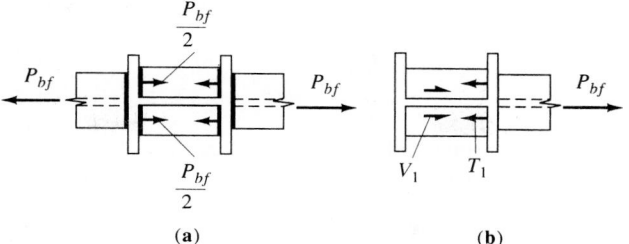

Figure 13.6.8
Weld requirements for
horizontal plates.

(a)

(b)

effective weld size along ends of stiffener plates in tension, using SMAW with E70 electrodes, is

$$a_{\text{max eff}} = \frac{0.90F_{yst}t_s}{2(0.707)0.75(0.60F_{EXX})} = 1.414\frac{F_{yst}t_s}{F_{EXX}} = 1.414\frac{36t_s}{70} = 0.73t_s$$

Using $\frac{5}{16}$-in. stiffener plates.

$$a_{\text{max eff}} = 0.73t_s = 0.73(0.3125) = 0.23 \text{ in.}$$

$$\text{Min weld size } a = \tfrac{1}{4} \text{ in.}$$

The maximum factored force to be carried by the stiffeners is

$$\text{Force} = \phi_c A_{st}F_{yst} = 0.90(2)(3)(0.3125)36 = 61 \text{ kips}$$

For fillet welds top and bottom of plates, the strength required per inch is

$$\text{Required } \phi R_{nw} = \frac{61}{2(6)} = 5.1 \text{ kips/in.}$$

From Table 5.14.2, $\frac{1}{4}$-in. weld provides $\phi R_{nw} = 5.57(0.23/0.25) = 5.1$ kips/in. *Use* $\frac{1}{4}$-in. fillet weld, top and bottom, on both tension and compression plates where they bear against the column flanges. Along the column web, fillet weld is required only on one side of the plate.

When a beam frames in from one side only, as in Fig. 13.6.8b, the weld forces T_1 are designed as in the symmetrical case. Additionally, however, shear forces V_1 must be developed in accordance with AISC-J10.6 to take the proportion of the unbalanced flange force which comes to the stiffener plates, in this case $V_1 = T_1$. ∎

Vertical Plate and Tee Stiffeners

Sometimes it may be desirable to use vertical plates, or structural tee sections as shown in Fig. 13.6.1c. Particularly, they may be useful in a four-way system where beams are attached to the tee sections. Research [7.28] indicates that a vertical stiffener at the toe of the flange is only one-half as effective as is the web of the column.

Thus, assuming two vertical stiffeners (one at each flange toe) each having one-half the web strength indicated by Eq. 13.6.1, gives the following for the maximum factored load P_{bf} that can be transmitted based on local web yielding:

$$P_{bf} = \phi F_{yw}(t_{fb} + 5k)t_{wc} + 2\phi\left(\frac{F_{yst}}{2}\right)(t_{fb} + 5k)t_s \qquad (13.6.16)$$

which upon solving for the required stiffener thickness t_s gives

$$t_s \geq \frac{P_{bf}}{\phi(t_{fb} + 5\,k)F_{yw}} - t_{wc}\left(\frac{F_{yw}}{F_{yst}}\right) \qquad (13.6.17)$$

where P_{bf} = factored concentrated load from beam flange
ϕ = 0.90 (for the AISC LRFD Method)
t_{fb} = flange thickness of beam transmitting force
t_{wc} = column web thickness
k = distance from outer face of column flange to web toe of fillet
F_{yst} = yield stress of stiffener
F_{yw} = yield stress of the column web

If the vertical stiffener is a plate, then compression buckling of the column web must be prevented by satisfying Eq. 13.6.8 under AISC-J10.5. When structural tees are used, the web attachment precludes compression buckling of the web. Regarding web crippling (AISC-J10.3), it is unlikely this limit state will control when vertical stiffeners are used at the column flange tips.

For the design of the tee stiffener and connection when a beam is attached to it, as in Fig. 13.6.3c, some special considerations are necessary. The beam flange, if equal in width to the tee, acts on the tee as shown in Fig. 13.6.9a. For analysis, one might consider uniform loading on a two-span beam as shown in Fig. 13.6.9b, wherein the load would be transmitted $\frac{5}{8}$ into the tee web and $\frac{3}{16}$ each to the column flanges. Blodgett [13.2] suggests when the beam flange extends full width of the tee, one may assume the effective flange width b_E (Fig. 13.6.9a) tributary to the tee web to be $\frac{3}{4}$ of the beam flange width.

One may summarize the LRFD design requirements when the *beam flange width is approximately the same as the tee flange width:*

1. The tee web thickness t_w must satisfy Eq. 13.6.1:

$$t_w \geq \frac{0.75\,P_{bf}}{(t_{fb} + 5\,k)F_{yst}} \qquad (13.6.18)$$

where $0.75\,P_{bf}$ = the portion of the factored load transmitted by the beam flange tributary to the tee stiffener
k = distance to the root of fillet: tee section
t_{fb} = flange thickness of beam transmitting force

Figure 13.6.9
Structural tee stiffener.

2. The structural tee flange thickness t_s must be able to carry the beam tensile flange force without excessive deformation; hence Eq. 13.6.15 should be satisfied. This will be conservative since the equation was derived for the free-edge condition at the flange toes, whereas here there is a welded connection. Using $0.75\,P_{bf}$ as in **1.**

$$t_{fc} \geq 0.4\sqrt{\frac{0.75\,P_{bf}}{\phi F_{yst}}} = 0.35\sqrt{\frac{P_{bf}}{\phi F_{yst}}} \qquad (13.6.19)$$

3. The structural tee flange width b_s must extend fully between the column flanges:

$$b_s = d_{col} - 2t_{fc} \qquad (13.6.20)$$

where d_{col} = column overall depth
t_{fc} = flange thickness of column

4. The structural tee depth d_s must be adequate to be nearly flush with the outer edges of the column flanges:

$$d_s = \frac{b_{fc} - t_{wc}}{2} \qquad (13.6.21)$$

where b_{fc} = column flange width
t_{wc} = column web width

When the beam flange width is significantly less than the tee flange width (say, more than an inch or two), P_{bf} instead of $0.75\,P_{bf}$ should be used in Eqs. 13.6.18 and 13.6.19.

In making the welded connection when beam flange and tee flange are nearly equal in width, the weld on the tee web (two segments of two fillets) is to resist the moment assuming $0.75\,P_{bf}$ must be carried at the flange tips of the tee, Blodgett suggests [13.2, p. 5.7–13] to design for $\frac{1}{3}$ of the beam flange force (somewhat greater than the $\frac{3}{16}$ of Fig. 13.6.9b) to be carried.

EXAMPLE 13.6.2

Design a vertical tee stiffener connection to frame a W14×61 beam into the web of a W12×65 column. Use A992 steel. Use the type of connection shown in Fig. 13.6.9, and the AISC LRFD Method.

Solution:
Since the beam flange width (9.995 in.) is approximately the same as the clear distance between column flanges ($12.12 - 1.21 = 10.91$ in.) Eqs. 13.6.18 and 13.6.19 may be applied.

(a) Determine the stiffener web thickness t_w required to prevent local web yielding (AISC-J10.1). The maximum *factored* beam flange force P_{bf} that can be carried is

$$P_{bf} = \phi A_f F_y = 0.90(9.995)(0.645)50 = 290 \text{ kips}$$

Using Eq. 13.6.18, and estimating $k = 1$ in. for the tee section.

$$\text{Required } t_w \geq \frac{0.75\,P_{bf}}{(t_{fb} + 5k)F_{yst}} = \frac{0.75(290)}{[0.645 + 5(1.0)]50} = 0.77 \text{ in.}$$

(b) Determine stiffener flange thickness to prevent distortion under tension. Using Eq. 13.6.19.

$$\text{Required } t_s = 0.35 \sqrt{\frac{P_{bf}}{\phi F_{yst}}} = 0.35 \sqrt{\frac{290}{0.90(50)}} = 0.89 \text{ in.}$$

(c) Select a structural tee section. The maximum flange width permitted is

$$\text{Max } b_s = d_{col} - 2t_{fc} = 12.12 - 2(0.605) = 10.91 \text{ in.}$$

$$\text{Max depth} = 0.5(12.00 - 0.390) = 5.81 \text{ in.}$$

Try W12×96 cut into tees:

$$t_s = 0.900 \text{ in.} \qquad (\text{flange thickness})$$

$$t_w = 0.550 \text{ in. with } k = 1\tfrac{5}{8} \text{ in.}$$

Rechecking Eq. 13.6.18,

$$\text{Required } t_w \geq \frac{0.75\,(290)}{[0.645 + 5(1.625)]50} = 0.50 \text{ in.} < \text{provided } t_w \qquad \textbf{OK}$$

Try W12×96 cut as shown in Fig. 13.6.10a.

(d) Welding on stiffener web. Referring to Fig. 13.6.9c, the length of tee stiffener required is

$$\text{Length} = d_{bm} + 5\,k = 13.89 + 5(1.625) = 22.02 \text{ in.}$$

Try length of tee = 2′–0″.

Length of weld, upper and lower ends:

$$t_{fb} + 5\,k = 0.645 + 5(1.625) = 8.77 \text{ in.}$$

Try 9-in. weld at each end, connecting web of tee to web of column.

Figure 13.6.10
Example 13.6.2. Tee
selection and welding to web.

Assume the moment contributed by the center $\frac{1}{4}$ of the flange is tributary to the tee web.

$$\phi M_n = 0.75\phi F_y Z_x = \left[0.75(0.90)(50)102\right]\tfrac{1}{12} = 287 \text{ ft-kips}$$

where Z_x is the plastic modulus for the W14×61 beam, which is "compact" for $F_y = 50$ ksi. The strength of the welding is conservatively treated with the elastic "vector" analysis using section modulus S of the two 9-in. weld segments treated as lines,

$$S = 2\left(\frac{1}{12}\right)\left[\frac{(24)^3 - (24 - 18)^3}{12}\right] = 189 \text{ in.}^2$$

The factored load R_u at the top of the weld is

$$R_u = \frac{\phi M_n}{S} = \frac{287(12)}{189} = 18.2 \text{ kips/in.}$$

The weld size a required is

$$a = \frac{R_u}{\phi(0.707)(0.60\ F_{\text{EXX}})} = \frac{18.2}{0.75(0.707)(0.60)70} = 0.82 \text{ in.}$$

Check flexural strength of stiffener web as a rectangular section of width $t_w = 0.55$ in. and depth 24 in.

$$\phi M_p = \phi F_y Z = 0.90(50)\left(\frac{0.55(24)^2}{4}\right)\frac{1}{12} = 297 \text{ ft-kips} > 287 \text{ ft-kips} \qquad \textbf{OK}$$

Though the elastic "vector" analysis is a conservative treatment for welds, nevertheless the weld force to be transmitted to the base material would not exceed the base material strength. At the top of the weld lines, the maximum effective weld size is

$$\frac{\text{Tension strength}}{\text{of stiffener web}} = \text{Strength of two fillets}$$

$$t_w \phi_t F_y = 2[\phi(0.707\ a)0.60\ F_{\text{EXX}}]$$

$$0.55(0.90)50 = 2(0.75)(0.707\ a)(0.60)70$$

$$24.75 = 44.54a: \qquad a_{\text{max eff}} = 0.56 \text{ in.}$$

The use of $\frac{13}{16}$-in. welds on the 0.55-in. web is not acceptable; therefore, increase the section to W12×120 from which to cut the tee. The web is still not quite thick enough; however, this seems acceptable because of the uncertainty in the proportion of the beam moment transmitted through the tee web and that transmitted between flanges of the tee and the column, as well as the conservatism in the weld resistance calculation.

Use W12×120 as shown in Fig. 13.6.10b.

(e) Welding on stiffener flanges. Assume conservatively that the tee connection to the column flange (Figs. 13.6.9a and b) may have to carry as much as $\frac{1}{3}$ of the beam flange force. The concentrated forces from the beam flange may be considered as distributed along the column flange over the distance $t_{fb} + 5t_s$, as shown in Fig. 13.6.11.

Assuming the factored force T_u attributable to the groove weld along one flange is

$$T_u = \phi F_y t_{fb}\left(\frac{b}{3}\right) = 0.90(50)(0.645)\left(\frac{9.995}{3}\right) = 97 \text{ kips}$$

Figure 13.6.11
Forces carried by weld along
tee stiffener flanges.

$$t_{fb} + 5t_s = 0.645 + 5(1.105) = 6.2 \text{ in.}$$

$$\text{Factored weld force } R_u = \frac{97}{6.2} = 15.6 \text{ kips/in.}$$

Using a partial joint penetration U-groove weld with E70 electrode material, the design strength ϕR_{nw} is

$$\phi R_{nw} = 0.80(0.60F_{\text{EXX}}) = 0.80(0.60)70 = 33.6 \text{ kips/in.}$$

for tension normal to effective area (AISC-Table J2.5). The required effective throat dimension is

$$\text{Required effective throat} = \frac{15.6}{33.6} = 0.46 \text{ in.}$$

According to AISC-Table J2.1, the effective throat for a partial joint penetration U-groove is the "depth of groove." The minimum effective throat (AISC-Table J2.3) is $\frac{5}{16}$ in. for the stiffener flange of 1.105 in.
Use $\frac{11}{16}$-in. single-U-groove (AWS-1.1, BC-P6 prequalified joint, also *AISC Manual*, p. 8-57) weld along the edges of the flange of the tee.

   (f) Effect of beam shear force. Ordinarily the length of the weld is so large that the additional capacity required to carry end shear is negligible. A W14×61 might be expected to carry something on the order of 70 kips factored load shear, in which case it is resisted by

$$\text{Total length of weld} = 4(10)[\text{for web}] + 2(24)[\text{for flanges}] = 88 \text{ in.}$$

$$\text{Required extra weld strength} = \frac{70}{88} = 0.80 \text{ kips/in.}$$

$$\text{Added size for web fillets} = \frac{0.80}{0.75(0.707)42} = 0.04 \text{ in.}$$

$$\text{Added size for flange groove welds} = \frac{0.80}{33.6} = 0.03 \text{ in.}$$

   The shear component and the flexure component actually act at 90 degrees to one another so that adding the requirements algebraically would be overly safe. The authors consider the weld design adequate without an increase for direct shear.   ∎

Top Tension Plates

When the beam is connected to the column flange and the column is stiffened by vertical or horizontal plate stiffeners, or when the beam is connected to the column web through a vertical tee stiffener, a simple means of transmitting the moment from the beam is with a tension plate at the top of the beam combined with (1) a bottom compression plate and web plates for shear; (2) a bottom seat angle; or (3) a bottom bracket (stiffened seat). See Figs. 13.6.1d and 13.6.2a and b. The behavior of such top plate connections has been studied by Pray and Jensen [13.63] and Brandes and Mains [13.64].

The design of seat angles and brackets has been treated in Secs. 13.3 and 13.4. Transmission of tension and compression forces into the column has been treated earlier in this section. Emphasis here is on the top plate, a tension member whose design is illustrated in the following example.

EXAMPLE 13.6.3

Design the top tension plate and its connection by welding or by A325 high-strength bolts for transferring the full end moment from a W14×61 to a column. Use A992 steel, and the AISC LRFD Design Method. Assume the connection to be of the type shown in Figs. 13.6.1d or 13.6.2a.

Solution:

(a) Design plate as a tension member.

Since the W14×61 is a "compact section" for $F_y = 50$ ksi, the factored flange force T_u is approximately

$$T_u = \frac{\phi M_n}{d_{bm}} = \frac{\phi F_y Z_x}{d_{bm}} = \frac{0.90(50)102}{\approx 14} = 328 \text{ kips}$$

When there are no holes, the yielding limit state for tension members will control over the fracture on net section limit state. Thus, using Eq. 3.9.2, the required gross plate area A_g is

$$A_g = \frac{T_u}{\phi_t F_y} = \frac{328}{0.90(50)} = 7.3 \text{ sq in.}$$

The plate width must be less than the W14×61 flange width of 10.0 in.
Use PL—$\frac{7}{8}$×9, $A_g = 7.9$ sq in., for welded connection.

(b) Determine welding for plate to beam flange. Try $\frac{3}{8}$-in. fillet weld with E70 electrodes using the shielded metal are process.

$$\phi R_{nw} = \phi(0.707\, a)(0.60\, F_{EXX}) = 0.75(0.707\, a)(0.60)70 = 22.3\, a$$

$$= 22.3(0.375) = 8.35 \text{ kips/in.}$$

The required length L_w of weld is

$$L_w = \frac{T_u}{\phi R_{nw}} = \frac{328}{8.35} = 39 \text{ in.}$$

To reduce the length of weld, use $\frac{1}{2}$-in. weld. This will require $L_w = 29.4$ in.
Use $\frac{1}{2}$-in. weld. 9 in. on end and 11 in. on each side for a total of 31 in. The design is summarized in Fig. 13.6.12a.

Figure 13.6.12
Example 13.6.3. Moment connection using top plate.

The plate is connected to the column flange by a full penetration single-bevel groove weld, using a backup bar. When a seat is used it will serve as backup to make the flange groove weld. Otherwise, a backup plate is needed below the compression flange. Frequently it may be satisfactory to use fillet welds along the top and bottom of the plate to make the connection to the column flange.

(c) Design bolted connection for plate to beam flange. Use $\frac{7}{8}$-in.-diam A325 bolts in a bearing-type connection with threads excluded from the shear plane (A325-X).

$$\phi R_n = \phi F_{nv} m A_b = 0.75(60)1(0.6013) = 27.1 \text{ kips/bolt}$$
(single shear)

$$\phi R_n = \phi(2.4\, F_u d_b t) = 0.75(2.4)(65)(0.875)0.645 = 66.0 \text{ kips/bolt}$$
(bearing on 0.645-in. beam flange)

The use of 2.4 F_u assumes $s \geq 3d_b$ and $L_e \geq 1.5d_b$.

When bolts are used, the holes in the W14×61 may indicate a reduction in strength according to AISC-F13, as discussed in Sec. 7.9. No reduction is to be made when the following condition is satisfied.

$$F_u A_{fn} \geq Y_t F_y A_{fg}$$

where A_{fn} = net flange area
A_{fg} = gross flange area
Y_t = 1.0 for $F_y/F_u \leq 0.80$

$$65[9.995(0.645) - 2(1)0.645] \geq 50(9.995)0.645$$
$$335 > 322$$

Thus, the limit state of tensile rupture does not apply and the design strength of the flange in tension is

$$\phi T_n = \phi F_y A_{fg} = 0.90(50)(9.995)0.645 = 290 \text{ kips}$$

For the plate as a tension member, the effect of holes must be deducted in computing the effective net area A_e,

$$A_g = \frac{T_u}{\phi_t F_y} = \frac{252}{0.90(50)} = 5.6 \text{ sq in.}$$

$$A_e = \frac{T_u}{\phi_t F_u} = \frac{252}{0.75(65)} = 5.2 \text{ sq in.}$$

The effective net area A_e equals the actual net area A_n in this concentric loading situation, according to AISC-D3.3.

Try PL—1×9. Check A_n and AISC-J4.4,

$$A_n = [9.00 - 2(1)]0.875 = 6.1 \text{ sq in.} > [A_e = 5.2 \text{ sq in.}] \qquad \textbf{OK}$$

$$0.85 A_g = 0.85(9.0)0.875 = 6.7 \text{ sq in.} > A_n \qquad \textbf{OK}$$

$$\text{No. of bolts required} = \frac{252}{27.1} = 9.3: \quad \text{Say } 10$$

Use 10—$\frac{7}{8}$-in.-diam A325 bolts, as shown in Fig. 13.6.12b.   ■

Split-Beam Tee Connections—Prying Action

For bolted moment connections, the split-beam tee as shown in Fig. 13.6.2c is not often any longer used. However, the design of a connection involving the transfer of a tensile force through a thick-plate bolted connection, such as the tee connected to the flange, will illustrate treatment of "prying action." Prying action was first mentioned in Sec. 4.13.

Consider the deformation of a tee section, as in Fig. 13.6.13. where as the pull on the web deforms the flange and deflects it outward, the edges of the flange tips bear against the connected piece giving rise to the force Q, known as the "prying force." Inclusion of this force is required by AISC-J3.6 wherein it states, "The required tensile strength shall include any tension resulting from prying action produced by deformation of the connected parts."

When bolted connections are subject to distortion such as in Fig. 13.6.13, the treatment of Sec. 4.13 for bolts in tension is not valid. When a thick-flanged tee distorts, the flange tips tend to dig in. giving rise to the force Q.

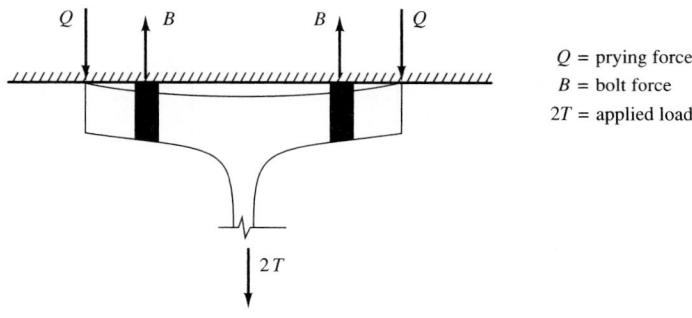

Q = prying force
B = bolt force
$2T$ = applied load

Figure 13.6.13
Prying action.

Kulak, Fisher, and Struik [3.1, pp. 274–282] have reviewed the various theories relating to prying action. They have recommended the procedure used in the *AISC Manual* [1.15]. The model, formulation, and design procedure is well explained by Thornton [13.65, 13.67] and Astaneh [13.66].

The analytical model used in the *AISC Manual* [1.15] procedure is shown in Fig. 13.6.14. Moment equilibrium requires

$$M_1 + M_2 - Tb = 0 \qquad (13.6.22)$$

The cantilever moment is

$$M_2 = Qa \qquad (13.6.23)$$

and force equilibrium requires

$$T + Q - B = 0 \qquad (13.6.24)$$

where T = externally applied factored load on one bolt

Q = prying force tributary to one bolt

B = tensile load in the bolt (to be compared with the design strength ϕR_n of the bolt)

Eqs. 13.6.22 through 13.6.24 are the independent equations of equilibrium that must be satisfied.

Next, referring to "width" as the dimension along the length of the tee section tributary to one bolt, let

$$\delta = \frac{\text{net width at bolt line}}{\text{gross width at critical section near web face}} \qquad (13.6.25)$$

M_2 may be multiplied and divided by δM_1 without any effect, and the ratio $M_2/\delta M_1$ may be called α: thus,

$$M_2 = M_2\left(\frac{\delta M_1}{\delta M_1}\right) = \left(\frac{M_2}{\delta M_1}\right)\delta M_1 = \alpha\delta M_1 \qquad (13.6.26)$$

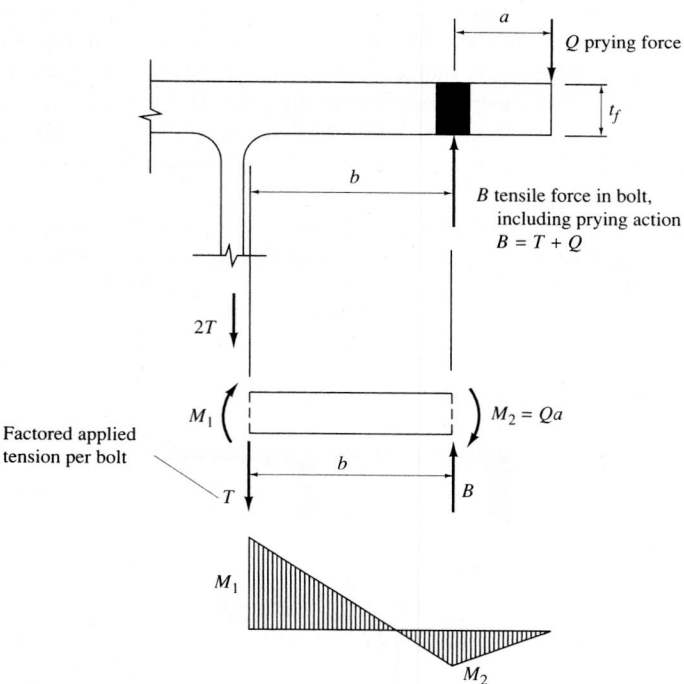

Figure 13.6.14
Analytical model for prying action.

Substituting Eq. 13.6.26 for M_2 into Eqs. 13.6.22 and 13.6.23 gives the following two equations for M_1,

$$M_1 = \frac{Tb}{1 + \alpha\delta} \tag{13.6.27}$$

$$M_1 = \frac{Qa}{\alpha\delta} \tag{13.6.28}$$

Eliminating M_1 from Eqs. 13.6.27 and 13.6.28 gives the prying force Q as,

$$Q = T\left(\frac{\alpha\delta}{1 + \alpha\delta}\right)\left(\frac{b}{a}\right) \tag{13.6.29}$$

Substitution of Eq. 13.6.29 into the force equilibrium equation, Eq. 13.6.24, gives

$$T + Q - B = 0 \tag{13.6.24}$$

$$T + T\left(\frac{\alpha\delta}{1 + \alpha\delta}\right)\left(\frac{b}{a}\right) - B = 0$$

and solving for B gives

$$B = T\left[1 + \left(\frac{\alpha\delta}{1 + \alpha\delta}\right)\left(\frac{b}{a}\right)\right] \tag{13.6.30}$$

There are two design requirements that must be satisfied: (1) the moment strength of the flange must be adequate, and (2) the bolt strength in tension must be adequate.

Moment Strength of Tee Flanges.

The AISC LRFD requirement would be

$$\phi M_n \geq M_1 \tag{13.6.31}$$

$$\phi M_n = M_p = \phi Z F_y = \phi \frac{wt_f^2}{4} F_y \tag{13.6.32}$$

where w is the length of flange, parallel to stem, tributary to one bolt, and ϕ is the resistance factor $\phi_b = 0.90$ for flexure. The *AISC Manual* p. 9-10 on "prying action" modified the moment strength equation previously preferred by AISC, by replacing the yield stress F_y with the ultimate tensile strength F_u to reflect the fact that experimental results have shown consistently higher strength values than predicted by the previous preferred prying model. Prying action has been studied by Thornton [13.65], Astaneh [13.66], and Swanson [13.102]. Putting Eqs. 13.6.32 and 13.6.27 into Eq. 13.6.31, and replacing F_y with F_u gives the design requirement for the flange thickness t_f.

$$t_f \geq \sqrt{\frac{4Tb}{\phi_b w F_u(1 + \alpha'\delta)}} \tag{13.6.33}$$

where α' is defined following Eq. 13.6.40.

Tension Strength of Bolts.

The AISC LRFD Method requirement would be

$$\phi R_n \geq B \tag{13.6.34}$$

where B is the factored load force on one bolt, given by Eq. 13.6.30.

The design strength of bolts in tension is given by Eq. 4.7.8.

$$\phi R_n = \phi F_{nt} A_b \tag{4.7.4}$$

where $\phi = 0.75$, or for A325 bolts having $F_{nt} = 90$ ksi, and A490 bolts having $F_{nt} = 113$ ksi,

$$\phi R_n = 0.75(90)A_b = 67.5A_b \qquad (\text{A325}) \qquad (13.6.35)$$

$$\phi R_n = 0.75(113)A_b = 84.8A_b \qquad (\text{A490}) \qquad (13.6.36)$$

Thornton [13.65] has shown there is no unique solution for Eqs. 13.6.33 and 13.6.34. Thus, a trial procedure to obtain one satisfactory solution is necessary. One may note that when $\alpha = 0$ there will be no prying action and single curvature bending: and when $\alpha = 1.0$ there will be maximum prying action and double curvature bending.

The aforementioned model and development has been calibrated by tests [3.1] indicating the b dimension in Fig. 13.6.14 should be larger and the a dimension smaller by an amount equal to one-half the bolt diameter d_b. Thus, a' and b' are used in place of a and b in *all* of the foregoing equations,

$$a' = a + \frac{d_b}{2} \qquad (13.6.37)$$

$$b' = b - \frac{d_b}{2} \qquad (13.6.38)$$

Flange Thickness.

The flange thickness t_f required from modified (using b' for b) Eq. 13.6.33 is

$$t_f \geq \sqrt{\frac{4Tb'}{\phi_b w F_u(1 + \alpha'\delta)}} \qquad (13.6.39)$$

and using $\phi_b = 0.90$ for flexure, Eq. 13.6.39 becomes

$$t_f \geq \sqrt{\frac{4.44Tb'}{w F_u(1 + \alpha'\delta)}} \qquad (13.6.40)$$

where T = factored externally applied tension per bolt
   $b' = b - d_b/2$
   d_b = bolt diameter
   w = width of resisting section (i.e., length of tee section) tributary to one bolt
   $\alpha' = 1.0$ if $\beta \geq 1$

   = the lesser of 1 and $\dfrac{1}{\delta}\left(\dfrac{\beta}{1 - \beta}\right)$ if $\beta < 1$

$\beta = \dfrac{1}{\rho}\left(\dfrac{B}{T} - 1\right)$

$\rho = b'/a'$

$a' = \left(a + \dfrac{d_b}{2}\right) \leq \left(1.25b + \dfrac{d_b}{2}\right)$

   δ = ratio of net area at bolt line (where M_2 acts) to gross area where
     M_1 acts $= (w - d_b)/w$

The bolt design strength required according Eq. 13.6.34 is

$$\phi R_n \geq \left(\frac{1}{Q}\right)T \qquad (13.6.41)$$

where

$$\frac{1}{Q} = \left[1 + \frac{\alpha'\delta}{1 + \alpha'\delta}\left(\frac{b'}{a'}\right)\right] \qquad (13.6.42)$$

In the 2005 *AISC Manual* (p. 9-12) section on "prying action", Q is redefined for use in design, as follows:

$$1. \; Q = 1 \qquad \qquad \text{if } \alpha' > 0 \qquad \qquad (13.6.43)$$

Category 1. implies sufficient stiffness and strength of the fitting to develop full bolt nominal strength.

$$2. \; Q = \left(\frac{t}{t_c}\right)^2 (1 + \alpha'\delta) \quad \text{if } 0 \le \alpha' \le 1 \qquad \qquad (13.6.44)$$

Category 2. implies the fitting is sufficient to develop full bolt nominal strength but not sufficient stiffness.

$$3. \; Q = \left(\frac{t}{t_c}\right)^2 (1 + \delta) \qquad \text{if } \alpha' > 1 \qquad \qquad (13.6.45)$$

Category 3. implies insufficient strength to develop the bolt nominal strength.

Note that t_c is the required thickness of fitting to develop the bolt nominal strength without prying action (a modification of Eq. 13.6.40); thus,

$$t_c = \sqrt{\frac{4.44 B b'}{w F_u}} \qquad \qquad (13.6.46)$$

and

$$\alpha' = \frac{1}{\delta (1 + \rho)}\left[\left(\frac{t_c}{t}\right)^2 - 1\right] \qquad \qquad (13.6.47)$$

where $\rho = b'/a'$.

EXAMPLE 13.6.4

Design a split-beam tee connection, such as in Fig. 13.6.2c, to enable a plastic hinge to develop in a W14×61 beam framing to the flange of a W14×159 column. Use A992 steel with $\frac{7}{8}$-in.-diam A325 bolts in a bearing-type connection (A325-N). Use the AISC LRFD Design Method.

Solution:

(a) Compute the factored tensile force to be carried. The W14×61 is compact (i.e., $\lambda \le \lambda_p$ for local flange buckling and local web buckling): in addition, assume that lateral-torsional buckling is precluded (i.e., $L_b \le L_p$).

Since there are flange holes in the W14×61 for this bolted connection, the nominal moment strength M_n of the beam may be less than M_p. Check in accordance with AISC-F13.1. As discussed in Sec. 7.9, in order to use full gross properties, Eq. 7.9.1 must be satisfied,

$$F_u A_{fn} \ge F_y A_{fg} Y_t \qquad \qquad [7.9.1]$$

$$F_y/F_u = 50/65 = 0.83 > 0.8; \text{ therefore, } Y_t = 1.1$$

$$F_u A_{fn} = 65(5.32) = 346 \text{ kips}$$

$$F_y A_{fg} Y_t = 50(6.45)1.1 = 355 \text{ kips}$$

$$(F_u A_{fn} = 346 \text{ kips}) \text{ is not } > (F_y A_{fg} Y_t = 355 \text{ kips})$$

This means gross properties cannot be used to determine M_n. Instead, AISC Formula (F13-1), Eq. 7.9.2, must be used to determine the nominal moment strength M_n at the location of the holes, as follows:

$$M_n = \frac{F_u A_{fn}}{A_{fg}} S_x \qquad [7.9.2]$$

$$M_n = \frac{65(5.32)}{6.45} 92.1\left(\frac{1}{12}\right) = 412 \text{ ft-kips}$$

$$\phi M_n = 0.90(412) = 371 \text{ ft-kips}$$

If all the bending moment is carried by the tees, the force of the internal couple is

$$\text{Force} = \text{Factored load to be carried} = \frac{M_u}{d_{bm}} = \frac{371(12)}{13.89} = 321 \text{ kips}$$

(b) Check whether or not the tensile force can be accommodated by the bolts in tension.

$$\phi R_n = 67.5 A_b \qquad [13.6.35]$$

$$\phi R_n = 67.5 A_b = 67.5(0.4418) = 29.8 \text{ kips}$$

Only 8 bolts will fit, as shown in Fig. 13.6.15; therefore the maximum factored tensile force that may be carried is

$$T_u = 8(29.8) = 238 \text{ kips} < 321 \text{ kips} \qquad \textbf{NG}$$

Figure 13.6.15
Example 13.6.4. Split-beam
tee connection with tee stub.

When this difficulty arises, one may use a stub beam or a tee stub attached to the bottom of the main beam (Fig. 13.6.15) to increase the moment arm of the couple. Actually, when designing for the support moment, one might have used a beam size required for the midspan moment and then used the stub beam to gain the increased capacity required at the support. The necessary moment arm is

$$\text{Required arm} = \frac{371(12)}{238} = 18.7 \text{ in.}$$

$$\text{Extra depth required} = 18.7 - 13.89 = 4.8 \text{ in.}$$

Try as stub beam a WT5×24.5, $t_w = 0.340$ in., $t_f = 0.560$ in., $b_f = 10.0$ in., $d = 4.990$ in., whose dimensions are comparable to the main W14×61 beam.

$$\text{Force of couple} = \frac{371(12)}{13.89 + 4.990} = 236 \text{ kips}$$

Using 8 bolts in tension,

$$\left(R_u = \frac{236}{8} = 29.5 \text{ kips} \right) < (\phi R_n = 29.8 \text{ kips}) \qquad \textbf{OK}$$

(c) Check shear strength on web (section c-c of Fig. 13.6.15) of WT5×24.5. Applying AISC-G2.1 the length L of tee required is

$$\text{Required } L = \frac{\text{Force}}{\phi(0.6 \ F_y)t_w} = \frac{236}{1.0(0.6)(50)0.340} = 23.1 \text{ in.}$$

(d) Determine bolts required to transmit tension and compression forces at the top and bottom of the beam.

$$\phi R_n = 0.4418(0.75)48 = 15.9 \text{ kips} \qquad \text{(controls)}$$

(single shear, *AISC Manual*, Table 7-1)

$$\phi R_n = 0.75(2.4 \ F_u d_b t) = 0.75(2.4)(65)(0.75)t = 87.8t$$

(bearing assuming $s \geq 3d_b$, $L_e \geq 2.0d_b$)

(deformation at hole is a design consideration)

$$\text{Number of bolts} = \frac{236}{15.9} = 14.8 \qquad \text{Use 16}$$

The minimum length of WT5×24.5 required using 8 bolts per line at 3-in. spacing is 24 in.

Use WT5×24.5 stub tee, 2'-3" long, welded to the bottom of the W14×61, as shown in Fig. 13.6.15.

(e) Determine thickness required to transmit tension on section *a-a* (Fig. 13.6.15):

$$\text{Required } A_g = \frac{T_u}{0.90 \ F_y} = \frac{236}{0.90(50)} = 5.2 \text{ sq in.}$$

and

$$\text{Required } A_n = \frac{T_u}{0.75 \ F_y} = \frac{236}{0.75(65)} = 4.8 \text{ sq in.}$$

Using the length of section *a-a* as 13 in. (column flange width = 15.565 in.), and deducting two holes, gives

$$t \geq \frac{4.8}{13 - 2(0.875)} = 0.43 \text{ in.}$$

$$t \geq \frac{15.9}{87.8} = 0.18 \text{ in.} \quad \text{(bearing does not control)}$$

(**f**) Determine the flange thickness for the tee section attached to the column. Equation 13.6.40 has been developed to provide flexural strength on section *b-b* of Fig. 13.6.15. Estimate the adjusted distance b'. Estimate the usual gage g as about 4 in.; thus,

$$b' = b - \frac{d_b}{2} = \frac{g}{2} - \frac{t_w}{2} - \frac{d_b}{2} = \frac{4}{2} - \frac{t_w}{2} - \frac{3}{8} \approx \text{ say } 1.25 \text{ in.}$$

Using *AISC Manual* procedure, p. 9-10, determine the minimum thickness required to eliminate prying action: thus $Q = 0$ in which case $\alpha = 0$. Then, from Eq. 13.6.40, assuming a length w of 14 in. at the critical section (section *b-b*) of the tee,

$$t_f \geq \sqrt{\frac{4.44 \, Tb'}{wF_u(1 + \alpha\delta)}} = \sqrt{\frac{4.44(236/2)1.25}{14(65)(1 + 0)}} = 0.85 \text{ in.}$$

Alternatively, if the required thickness to eliminate prying action is not practical, or if a thinner flange thickness is desired, the *AISC Manual* provides a method which is based on the recommendations of Thornton [13.65] to compute β, which from Eq. 13.6.30 (again using a' and b' instead of a and b) is a function of $\alpha\delta$.

$$\beta = \left(\frac{B}{T} - 1\right)\frac{a'}{b'} = \left(\frac{29.8}{29.5} - 1\right)(\approx 1.25) = 0.013$$

Then use α as follows:

$$\text{if } \beta \geq 1 \quad \text{use } \alpha = 1 \, (\text{meaning large prying force})$$

$$\text{if } \beta < 1, \quad \text{use } \alpha = \text{lesser of } \frac{1}{\delta}\left(\frac{\beta}{1 - \beta}\right) \text{ and } 1.0$$

In this example, δ may be estimated using Eq. 13.6.25 as

$$\delta = \frac{3 - (0.75 + 1/16)}{3} = 0.73$$

$$\frac{1}{\delta}\left(\frac{\beta}{1 - \beta}\right) = \frac{1}{0.73}\left(\frac{0.013}{1 - 0.013}\right) = 0.018 < 1: \quad \text{use } \alpha = 0.018$$

In which case, $\alpha\delta = 0.018(0.73) = 0.013$ for use in Eq. 13.6.40 for t_f.

$$t_f \geq \sqrt{\frac{4.44Tb'}{wF_u(1 + \alpha\delta)}} = \sqrt{\frac{4.44(236/2)1.25}{14(65)(1 + 0.013)}} = 0.84 \text{ in.}$$

Try a tee cut from S24×80, $t_f = 0.870$ in., $t_w = 0.500$ in.

(g) Check the prying force using Eq. 13.6.29 with a' and b' instead of a and b,

$$a' = a + \frac{d_b}{2} = \frac{b_f - g}{2} + \frac{d_b}{2} = \frac{7.000 - 4}{2} + \frac{3}{8} = 1.875$$

$$b' = b - \frac{d_b}{2} = \frac{g}{2} - \frac{t_w}{2} - \frac{d_b}{2} = \frac{4}{2} - \frac{0.500}{2} - \frac{3}{8} = 1.375$$

Kulak, Fisher, and Struik [3.1] recommended that $a \leq 1.25b$. In this example, $a = 1.5$ in. and $b = 1.75$ in.; thus, $a \leq 1.25b$. Taking the length of the tee section as 14 in. (see top view of Fig. 13.6.15) with four holes deducted.

$$\delta = \frac{14 - 4(0.75 + 1/16)}{14} = 0.77$$

Using the same trial value $\alpha = 0.018$ as previously gives $\alpha\delta = 0.013$. Then find Q.

$$Q = T\left(\frac{\alpha\delta}{1 + \alpha\delta}\right)\left(\frac{b'}{a'}\right) = T\left(\frac{0.013}{1 + 0.013}\right)\frac{1.375}{1.875} = 0.01T$$

Then compare $(T + Q)$ per bolt to ϕR_n.

$$T + Q = 1.01T$$

$$1.01 R_u = 1.01(29.5) = 29.8 \text{ kips} = \phi R_n \qquad \textbf{OK}$$

Since the factored load R_u per bolt, increased by the minor prying force, exactly equals the design tension strength ϕR_n of a bolt, the bolts are satisfactory. Thornton [13.65] has shown that when the actual t_f exceeds the required t_f, the actual Q will be less than the Q computed above. More detail on this procedure is provided in Ref. 13.65 and in the *AISC Manual*.

(h) Recheck the thickness t_f required.

$$t_f \geq \sqrt{\frac{4.44\,Tb'}{wF_u(1 + \alpha'\delta)}} = \sqrt{\frac{4.44(236/2)1.375}{14(65)(1 + 0.013)}} = 0.88 \text{ in.}$$

The flange thickness t_f provided is 0.870 in. The design can be considered satisfactory because the provided thickness is within 1% of the computed requirement.

Use tees cut from S24×80 to carry tensile and compressive forces.

   Not discussed in this example is the development of the shear strength of the W14×61. A pair of angles may be attached to the beam web for the purpose of providing whatever shear is required. The final design is shown in Fig. 13.6.15. ∎

End-Plate Connections

The practical alternative to the split-beam tee connection is the end-plate moment connection, as shown in Fig. 13.6.2d. Having much simpler fabrication details, a single plate welded on the end of a beam has become relatively common.

   An excellent summary of the history of the use of end-plate connections is given by Griffiths [13.68]. The behavior of end-plate moment connections has been studied by Beedle and Christopher [13.49], Onderdonk, Lathrop, and Coel [13.48], Douty and McGuire [13.50], Agerskov [13.69, 13.70], Krishnamurthy [13.71], Krishnamurthy, Huang, Jeffrey, and Avery [13.72], Mann and Morris [13.73], Murray [13.74], Hendrick and Murray [13.75], Yee and Melchers [13.76], Murray and Kukreti [13.77], and Curtis and Murray [13.78], among others.

A conservative approach to end-plate connection design is to use the prying action concept discussed previously in this section. The region near the tension flange of the beam is designed similarly to the split-beam tee connection. This fastener group is designed for shear and tension, including any effect of prying action.

Krishnamurthy [13.71] has recommended a modified split-beam tee method based on analytical study and correlation with available tests. This method involves equations containing statistical coefficients and is suitable only with adequate design aids. Murray et al [13.74, 13.77] has proposed procedures for 4-bolt and 8-bolt stiffened end plates.

The 2005 *AISC Manual* refers to the procedure of AISC *Design Guide Series 4*, (2nd Ed) "Extended End-Plate Moment Connections, Seismic and Wind Applications" by Thomas M. Murray and Emmett A. Sumner [13.79], as well as AISC *Design Guide Series 16*, "Flush and Extended Multiple-Row Moment End-Plate Connections" by Thomas M. Murray and W. Lee Shoemaker [13.103]. These design guides give procedures which must be considered recommended practice. The *AISC Specification* does not prescribe procedures and requirements for end-plate connections. As research continues, design guides are modified and additional ones are developed. The reader must keep in mind the difference between the *AISC Specification* which is the "law" (so to speak) and the *AISC Manual* which is a "handbook" giving tables and recommended practice to aid designers.

Regarding end-plate connections, procedures are presented that were originally proposed by Borgsmiller and Murray [13.104] which are based on yield-line analysis of the plate for determining its thickness, and a modified tee-hanger analysis to determine required bolt strength. Detailed treatment is given in the *Guides*. According to the *Guides*, the design can be based on maximum prying forces and minimum plate thickness, or can be based on no prying forces with a corresponding thicker plate. These alternatives allow the designer to optimize either the bolt diameter or the end-plate thickness, as desired.

It is often desirable to extend the plate outside the beam flange so that all of the welding can be done in the shop. See *AISC Manual*, p. 12-8 for diagrams and design assumptions for the extended end-plate FR moment connections.

EXAMPLE 13.6.5

Design an end-plate connection for a W14×53 beam attachment to a W14×176 column, both of A992 steel. Use A572 Grade 50 for the end plate. Design for maximum factored beam moment and 85 kips factored shear. Use A490 bolts in a bearing-type connection (A490-X). Use the AISC LRFD Design Method.

Solution:

(a) Determine the diameter of the bolt assuming a 4-bolt pattern required to resist the factored bending moment. The design moment strength $\phi_b M_n$ of the beam is

$$\phi_b M_n = \phi_b M_p = \phi_b Z_x F_y = 0.90(87.1)(50)\tfrac{1}{12} = 327 \text{ ft-kips}$$

$$\text{Max } T_u = \frac{\phi_b M_n}{d - t_{fb}} = \frac{327(12)}{13.92 - 0.660} = 296 \text{ kips}$$

For A490 bolts,

$$\phi R_n = \phi(0.75 A_b)F_u^b = 0.75(113)A_b = 84.8 A_b$$

$$\text{Required bolt area} = \frac{\text{Max } T_n}{\phi R_n} = \frac{296}{84.8} = 3.5 \text{ sq in.}$$

Figure 13.6.16
End-plate connection for
Example 13.6.5.

The required bolt diameter when a 4-bolt pattern is used.

$$d_b = \sqrt{\frac{\text{Total Area}}{4}\left(\frac{4}{\pi}\right)} = \sqrt{\frac{3.5}{\pi}} = 1.06 \text{ in.} \qquad \text{Use } 1\tfrac{1}{8} \text{ in.}$$

(b) Establish the plan dimensions of the end plate. For determining the distance s (Fig. 13.6.16) the fillet weld size (for E70 electrodes) and the bolt installation clearance are needed.

$$\text{Required } \phi R_{nw} = \frac{T_u}{L_w} = \frac{T_u}{2b_f - t_w} = \frac{296}{2(8.060) - 0.370} = 18.8 \text{ kips/in.}$$

Because the maximum effective fillet weld size along the 0.660 in. flange is 0.48 in., 18.8 kips/in. cannot be developed (see Table 5.14.2, which indicates $\tfrac{3}{4}$-in. weld using E80 electrodes is needed to obtain 18.8 kips/in.). Thus, use full penetration groove weld. The minimum assembling clearance for $1\tfrac{1}{8}$-in.-diam bolts is given by *AISCManual* Table 7–16, p. 7-81 "Entering and Tightening Clearance," as $1\tfrac{9}{16}$-in. (typically about $\tfrac{1}{2}$ in. more than bolt diameter).

$$\text{Distance } s \text{ (Fig. 13.6.16)} = 1\tfrac{9}{16} + \tfrac{5}{8} = 2.19 \text{ in.}$$

Try a plate 9 in. wide and about 18 in. long.

(c) Estimate plate thickness t_p required. A conservative procedure is to use Eq. 13.6.40 with the prying force Q set equal to zero. The distance b' will be

$$b' = s - \frac{d_b}{2} = 2.19 - \frac{1.125}{2} = 1.63 \text{ in.}$$

$$t_p \geq \sqrt{\frac{4.44 \, Tb'}{wF_u(1 + \alpha\delta)}} = \sqrt{\frac{4.44(296/2)1.63}{9(65)(1 + 0)}} = 1.35 \text{ in.}$$

The designer could probably use the procedure in Example 13.6.4 relating to the prying action Q to justify a thinner plate.

(d) Check combined shear and tension on bolts. Referring to Sec. 4.14, compute stress f_{uv} resulting from factored shear.

$$f_{uv} = \frac{V_u}{A_b} = \frac{85}{6(0.994)} = 14.3 \text{ ksi}$$

The factored load tensile stress limit F'_{nt} from AISC Formula (J3-3a) for A490-X is

$$F'_{nt} = 1.3F_{nt} - \frac{F_{nt}}{\phi F_{nv}} f_{uv} = 1.3(113) - \left(\frac{113}{0.75(75)}\right)14.3 = 118 \text{ ksi}$$

The upper limit is $F_{nt} = 113$ ksi. Combined shear and tension is not a problem.

(e) Length of end plate. At the end near the compression flange of the beam, it is often desirable to extend the plate outside the beam flange an amount equal to the end plate thickness t_p (or even farther). This will increase the length of the critical section used in computing the strength based on local web yielding. The strength in local web yielding under a concentrated load is given by AISC-J10.2; however, a concentrated load passing through a thick end-plate distributes over a longer critical length. Hendrick and Murray [13.75] have recommended the following expression for the *factored* compression reaction strength P_{bf}.

$$P_{bf} = F_{yc}t_{wc}(t_{fb} + 6k + 2t_p + 2a)$$

where P_{bf} = factored compression force from beam flange

$\quad t_{wc}$ = column web thickness

$\quad t_{fb}$ = beam flange thickness

$\quad k$ = distance from face of column flange to root of fillet

$\quad t_p$ = thickness of end-plate

$\quad a$ = fillet weld leg dimension for beam flange to end-plate weld

In this example,

$$P_{bf} = 50(0.830)[0.660 + 6(2.0) + 2(1.375) + 2(0.75)] = 700 \text{ kips}$$

This far exceeds the factored applied compression force of 296 kips and is therefore satisfactory without stiffeners.

The overall web buckling of the column should be checked according to AISC-J10.5, as well as web crippling according to AISC-J10.3. These checks and the design of any stiffeners required have been previously shown at the beginning of this section in Example 13.6.1.

Use PL$-1\frac{3}{8}\times9\times1'-6\frac{1}{2}''$, with $6-1\frac{1}{8}$-in.-diam bolts in a bearing-type connection (A490-X), as shown in Fig. 13.6.16.

Murray [13.79] has provided an overall treatment of extended end-plate moment connections that is endorsed by AISC. ∎

Beam to Column-Web Direct Connection

Instead of using vertical tee stiffeners (Fig. 13.6.1c) or complicated details such as shown in Fig. 13.6.3, occasionally the designer would prefer to attach a beam directly to the column web, using either a welded or bolted connection. The column web must then resist the moment effect by plate action.

Abolitz and Warner [13.80], Stockwell [13.81], and Kapp [13.82] have presented yield line analyses to determine the strength of a column web when directly attached by a moment connection or a direct tension connection. Rentschler, Chen, and Driscoll [13.58, 13.83], Hoptay and Ainso [13.84], and Hopper, Batson, and Ainso [13.85] have conducted tests of such connections. Attachment of beams to box columns presents a similar situation. Practical design data is presented in Ref. 13.81; other design recommendations appear in Ref. 13.85. A detailed discussion of yield line analysis and design of these connections is outside the scope of this text.

Figure 13.7.1
Intersecting beam
connections: tension
flanges *not attached* to
each other.

(a)   (b)

13.7   CONTINUOUS BEAM-TO-BEAM CONNECTIONS

When beams frame transversely to other beams or girders, they may be attached to either or both sides of the girder web using simple shear (framed beam) connections (Sec. 13.2) or using simple seats in combination with a framed beam connection (Sec. 13.3). When full continuity of the beam is desired, the connection must develop a higher degree of rigidity than provided by simple shear connections. For beam-to-beam connections, the principal objective is to provide a means of allowing the tensile force developed in one beam flange to be carried across to the adjacent beam framing opposite the girder web. These connections may be divided into two categories; (1) those with intersecting tension flanges not rigidly attached to one another, as in Fig. 13.7.1; and (2) those with rigidly attached intersecting tension flanges, as in Fig. 13.7.2.

When the intersecting tension flanges are not rigidly attached (Fig. 13.7.1) the connection design is essentially a tension member design at the tension flange along with a shear connection.

On the other hand, the designer should be cautious when designing intersecting beams with tension flanges attached, since this case becomes one of biaxial instead of uniaxial stress. With biaxial stress the possibility of brittle fracture increases (see Sec. 2.9). In addition, a biaxial stress yield criterion must be used, such as Eq. 2.6.2, to establish safety:

$$\sigma_y^2 = \sigma_1^2 + \sigma_2^2 - \sigma_1\sigma_2 \qquad [2.6.2]$$

where σ_1, σ_2 are the principal stresses acting. When the beams frame at 90 degrees to one another the axial stresses in the flanges are principal stresses.

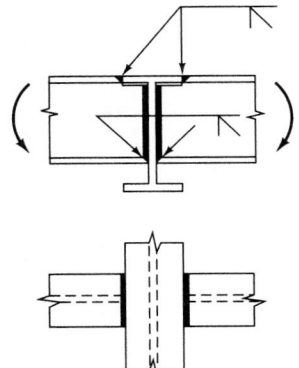

Figure 13.7.2
Intersecting beam
connection: tension flanges
attached to each other.

The designer should make certain the σ_y of Eq. 2.6.2 does not exceed the tension–compression yield stress F_y under factored loads for the AISC LRFD Method.

13.8   RIGID-FRAME KNEES

In the design of rigid frames, the safe transmission of load at the junction of beam and column is of great importance. When members join with their webs lying in the plane of the frame, the junction is frequently referred to as a knee joint. Typical knee joints are (1) the square knee, with or without a diagonal or other stiffener (Fig. 13.8.1a and b); (2) the square knee with a bracket (Fig. 13.8.1c); (3) the straight haunched knee (Fig. 13.8.1d); and (4) the curved haunched knee (Fig. 13.8.1e).

In the analysis of a structure, the designer commonly assumes the span of a member to extend from center-to-center of joints; in this case the knees. The moment of inertia would be assumed to vary in accordance with the moment of inertia of a cross-section taken at right angles to lines extending center-to-center of knees. Internal forces are then determined using statically indeterminate analysis, including the variable moment of inertia effect when the knees are haunched or curved.

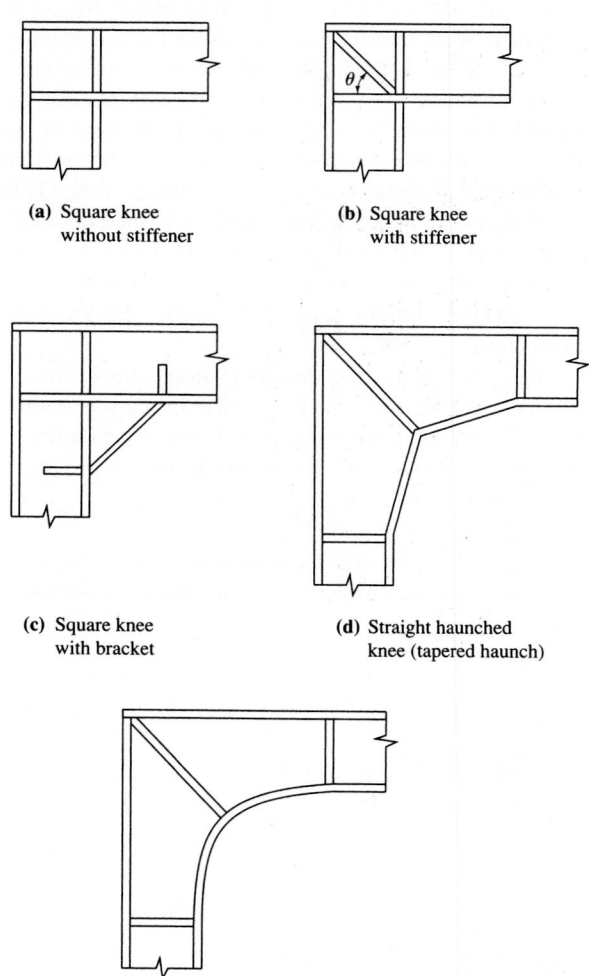

(a) Square knee without stiffener

(b) Square knee with stiffener

(c) Square knee with bracket

(d) Straight haunched knee (tapered haunch)

(e) Curved haunched knee

Figure 13.8.1
Rigid-frame knees.

The general design concepts applicable for rigid frame knees are summarized in *ASCE Manual* 41 [7.2, pp. 167–186] which forms the basis for much of what follows.

To be adequately designed, a knee connection must (1) transfer the end moment between the beam and the column, (2) transfer the beam end shear into the column, and (3) transfer the shear at the top of the column into the beam. Furthermore, in performing the three functions relating to strength, the knee must deform in a manner consistent with the analysis by which moments and shears were determined.

If a plastic hinge associated with the failure mechanism is expected to form at or near the knee, adequate rotation capacity must be built into the connection. Square knees have the greatest rotation capacity but are also the most flexible (i.e., deform elastically the most under service load conditions.) Curved knees are the stiffest but have the least rotation capacity. Since straight tapered knees provide reasonable stiffness along with adequate rotation capacity, in addition to the fact that they are cheaper than curved haunches to fabricate, the straight haunched knees are commonly used.

Shear Transfer in Square Knees

In the design of a rigid frame having square knees, two rolled sections may come together at right angles as shown in Fig. 13.8.1a. A frame analysis, either elastic or plastic, will have established what moments and shears act on the boundaries of the square knee region, as shown in Fig. 13.8.2a. The forces carried by the flanges must be transmitted by shear into the web, as shown in Fig. 13.8.2b.

(a) Unreinforced square knee

(b) Forces on corner region

(c) Web deformed due to shear

(d) Equivalent web deformation due to shear

Figure 13.8.2
Shear transfer in a square knee.

Assuming all bending moment to be carried by the flanges, and approximating the distance between flange centroids as $0.95d_b$, the flange force is

$$T = \frac{M}{0.95d_b} \qquad (13.8.1)$$

where d_b is the overall depth of the beam. The nominal strength in shear of the web across AB is

$$V_n = V_{ab} = \tau_y t_w d_c \qquad (13.8.2)$$

where d_c is the overall depth of the column.

The factored moment M_u gives rise to the factored tension force T_u. The design strength ϕV_n may then be equated to T_u to give the required web thickness t_w.

$$\phi V_n = T_u \qquad (13.8.3)$$

$$\phi \tau_y t_w d_c = \frac{M_u}{0.95d_b} \qquad (13.8.4)$$

In accordance with AISC-J10.6, $\tau_y = 0.6F_y$, and $\phi = 0.90$ for the yielding limit state. Equation 13.8.4 when solved for the required web thickness t_w gives

$$\text{Required } t_w = \frac{1.95M_u}{F_y d_b d_c} = \frac{1.95M_u}{F_y A_{bc}} \qquad (13.8.5)$$

where $A_{bc} = d_b d_c = $ the planar area within the knee. Equations 13.8.4 and 13.8.5 assume that interaction between axial load P_u and column shear V_u does not control, i.e., that $P_u \le 0.4P_y$ for the column.

This same approach must be used when beams frame on opposite faces of a column producing a high shear on the column web, as shown in Fig. 13.8.3. The total shear to be transferred across AB is

$$\text{Total factored shear} = C_{u1} + T_{u2} - V_u \qquad (13.8.6)$$

where $C_{u1} = $ compression force produced by M_{u1}
$\quad\quad\quad T_{u2} = $ tensile force produced by M_{u2}
$\quad\quad\quad V_u = $ factored shear in the column

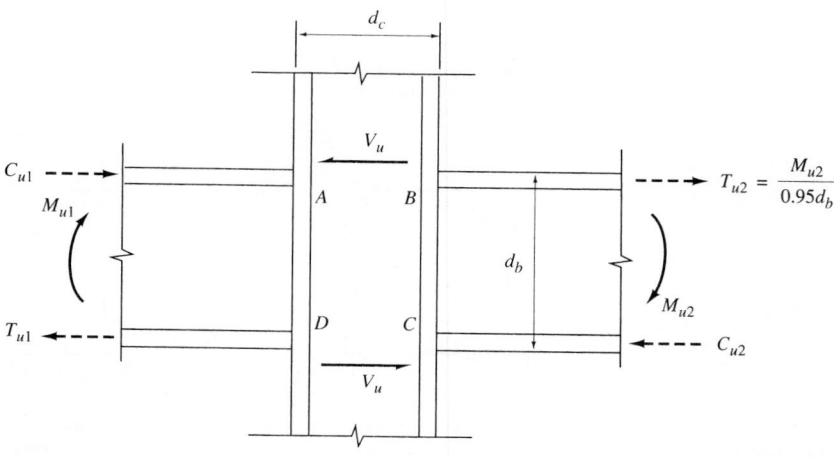

Figure 13.8.3
General case of shear transfer within a rigid joint.

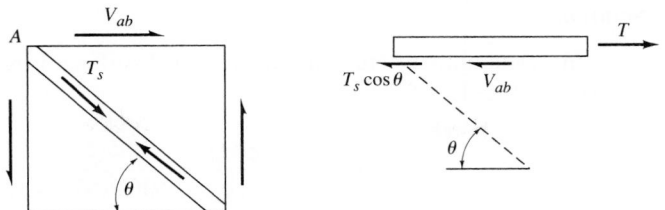

Figure 13.8.4
Diagonal Stiffener Effect

The total factored shear given by Eq. 13.8.6 would replace T_u in the development of Eq. 13.8.5. Thus, for two beams of equal depth framing on opposite faces of a column,

$$\text{Required } t_w = \frac{1.95(M_{u1} + M_{u2})}{F_y A_{bc}} - \frac{1.85 V_u}{F_y d_c} \qquad (13.8.7)$$

In a rigid frame knee, the required web thickness will typically exceed that provided by a W section; thus, reinforcement will be required. A doubler plate is sometimes used to thicken the web region; a generally impractical solution because of the difficulty making the attachment to the column web. Usually a pair of diagonal stiffeners is the best solution.

When diagonal stiffeners are used, the horizontal component $C_s \cos \theta$ of the stiffener force C_s participates with the web. Equilibrium (Fig. 13.8.4) then requires

$$T = V_{ab} + C_s \cos \theta \qquad (13.8.8)$$

or

$$\frac{M_u}{0.95 d_b} = \phi_v(0.60 F_y) t_w d_c + A_{st} \phi_c F_{cr} \cos \theta \qquad (13.8.9)$$

Solving for required A_{st} gives

$$\text{Required } A_{st} = \left(\frac{1}{\phi_c F_{cr} \cos \theta} \right) \left[\frac{M_u}{0.95 d_b} - \phi_v(0.60 F_y) t_w d_c \right] \qquad (13.8.10)$$

where ϕ_v = 0.90 for the yield limit for shear

ϕ_c = 0.90 for compression elements

F_{cr} = compression limit state stress computed using AISC-E3 in the manner similar to the stabilty limit state for bearing stiffeners, as discussed in Sec. 11.11

EXAMPLE 13.8.1

Using the LRFD Design Method, design the squre knee connection to join a W27×94 girder to a W14×74 column. The factored moment M_u to be carried through the joint is 522 ft-kips. Use A992 for the W sections, A572 Grade 50 for the plates, and E70 electrodes with shielded metal arc welding (SMAW).

Solution:

(a) Determine whether or not a diagonal stiffener is required. Using Eq. 13.8.4, and referring to Fig. 13.8.5,

$$\text{Required } t_w = \frac{1.95M_u}{F_y A_{bc}} = \frac{1.95(522)12}{50(26.92)14.17} = 0.64 \text{ in.}$$

$$\text{Actual } t_w = 0.490 \text{ in. for W27×94} < 0.64 \text{ in.}$$

A diagonal stiffener is required.

(b) Determine stiffener size.

$$\tan \theta = \frac{26.92}{14.17} = 1.900; \qquad \cos \theta = 0.466$$

Using Eq. 13.8.10 with the assumption $F_{cr} \approx F_y$,

$$\text{Reqd } A_{st} = \left(\frac{1}{0.85(50)0.466}\right)\left(\frac{522(12)}{0.95(26.92)} - 0.90(0.6)(50)(0.49)14.17\right)$$

$$\text{Required } A_{st} = 2.9 \text{ sq in.}$$

(c) Evaluate the strength of the stiffeners acting as a column. The slenderness ratio *KL/r* for a compression member consisting of two plates 3 in. wide attached to the web is

$$\text{Overall depth} = 2(3.0) + t_w = 6.49 \text{ in.}$$

$$r = \sqrt{\frac{t_{st}(\text{depth})^3}{12 t_{st}(\text{depth})}} = \text{depth}\sqrt{\frac{1}{12}} = 6.49\sqrt{\frac{1}{12}} = 1.9 \text{ in.}$$

$$\frac{KL}{r} = \frac{d_c/\cos \theta}{r} = \frac{14.17/0.466}{1.9} = 16; \qquad \phi F_{cr} = 44.2 \text{ ksi}$$

Note that the upper bound for $\phi_c F_{cr} = 0.90(50) = 45.0$ ksi. As a practical matter, except for unusually long stiffeners, F_{cr} may be approximated as F_y for design of diagonal stiffeners. Thus, the required A_{st} as computed in part (b) is essentially correct.

Use 2 PLs—$\frac{1}{2}$×3. Note that $\lambda = 3.0/0.5 = 6.0$ does not exceed λ_r of 15.8 from AISC-B4; thus, local buckling is precluded.

Figure 13.8.5
Square knee of
Example 13.8.1.

(d) Determine the fillet weld size along length *AB* in Fig. 13.8.5. The weld must transmit the factored flange force into the beam web. The maximum design flange force that can develop is $\phi_t F_y A_f$,

$$\text{Flange force } \phi_t F_y A_f = 0.90(50)(10.070)0.785 = 356 \text{ kips}$$

The design strength of fillet welds along both sides of web is

$$\phi R_{nw} = 0.75t_e(0.60F_{EXX}) = 0.75(0.707)2a(42) = 44.5a$$

$$\text{Available length for weld} = 26.92 - 2(0.745) = 25.4 \text{ in.}$$

$$\text{Required } a = \frac{356}{25.4(44.5)} = 0.31 \text{ in.}$$

$$\text{Check: } a_{\text{max eff}} = 0.707\frac{F_u t_{wb}}{F_{EXX}} = 0.707\frac{65(0.490)}{70} = 0.32 \text{ in.} > 0.31 \text{ in.} \quad \textbf{OK}$$

$$\text{Min } a = \tfrac{5}{16}\text{-in. from Table 5.11.1} \quad (\text{for } t_{fc} = 0.785 \text{ in.})$$

$\underline{Use\ \tfrac{5}{16}\text{-in. E70 fillet weld along length } AB \text{ (both sides of girder web).}}$

(e) Determine fillet weld size along length *BC*. The connection of the column web to the beam flange must carry the force resulting from flexure and axial load combined with the shear. At the most highly stressed location there will be tension and shear acting simultaneously on the weld. A conservative approach will be to compute the resultant of the shear and tension components.

$$\text{Tensile component} = \phi_t F_y t_w = 0.90(50)0.450 = 20.3 \text{ kips/in.}$$

$$\text{Shear component} = \frac{V_u}{d_c - 2t_f} = \frac{21}{14.17 - 2(0.785)} = 1.7 \text{ kips/in.}$$

$$\text{Resultant loading} = \sqrt{(20.3)^2 + (1.7)^2} = 20.4 \text{ kips/in.}$$

$$\text{Required } a = \frac{20.4}{2(0.707)(0.75)42} = 0.46 \text{ in.}$$

Since the tensile component is dominant, maximum effective weld size is

$$t_{wc}\phi_t F_y = 2\phi(0.707)a(0.60F_{EXX})$$

$$a_{\text{max eff}} = \frac{0.450(0.90)50}{2(0.75)(0.707)(0.60)70} = 0.46 \text{ in.} \quad \textbf{OK}$$

$\underline{Use\ \tfrac{1}{2}\text{-in. E70 fillet weld along length } BC \text{ (both sides of column web).}}$

(f) Determine the weld required along stiffeners. The weld must develop the stiffener strength,

$$\phi C_s = \phi F_{yst} A_{st} = 0.90(50)(2)(3.0)0.50 = 135 \text{ kips}$$

$$\text{Required } a = \frac{135/(14.17/\cos\theta)}{4(0.707)(0.75)42} = 0.05 \text{ in.}$$

$$\text{Min } a = \tfrac{3}{16}\text{-in. (Table 5.11.1 for } \tfrac{1}{2}\text{-in. stiffener)}$$

$\underline{Use\ \tfrac{3}{16}\text{-in. E70 fillet weld along diagonal } AC \text{ (both sides of girder web).}}$

(g) Determine the extent of stiffener required from point C vertically into girder. The design strength based on local web yielding AISC-J10.2 of the W27×94 web to carry the design compression force P_{bf} from the inside column flange at C is, using Eq. 13.6.1,

$$P_{bf} = \phi(5k + t_{fb})F_{yw}t_{wc} \qquad [13.6.1]$$

$$= 1.0[5(1.4375) + 0.785](50)0.490 = 195 \text{ kips}$$

where t_{fb} is the thickness of the W14×74 flange transmitting the force, and t_{wc} is the web thickness of the W27×94 resisting the force.

Since the flange concentrated load from the column is 356 kips, as computed in part (d), and the diagonal stiffeners are already performing another task, vertical stiffeners must be used along CD. When stiffeners are needed to prevent local flange yielding (AISC-J10.2) or local flange bending (AISC-J10.1), they must extend "at least one-half the depth of the web". Perhaps the designer should also investigate the strength based on web crippling (AISC-J10.3); however, with the stiffeners along CD and the diagonal stiffeners along AC (which were investigated as a column) web crippling is adequately prevented. When the local web yielding limit state does not require stiffeners, the web crippling limit state should be investigated.

(h) Establish plate size for stiffeners along CD. The required area of stiffeners,

$$\text{Required area} = \frac{(356 - 195)/2}{\phi_c F_y} = \frac{81}{0.90(50)} = 1.8 \text{ sq in./plate}$$

$$\text{Width available} = \frac{b_{fb} - t_{wb}}{2} = \frac{9.990 - 0.490}{2} = 4.8 \text{ in.}$$

$$\text{Minimum } t_s = \frac{1.8}{4.0} = 0.45 \text{ in.} \quad \text{assuming 4-in. plates}$$

Use $\frac{1}{2}$-in. plates. To preclude the local buckling limit state, λ must not exceed λ_r according to AISC-B4. Here $\lambda = 8.0$, which does not exceed 15.8 (i.e., λ_r) and is acceptable. When the strength M_n required within the knee is the plastic momentstrength M_p, it will be preferable to keep $\lambda \leq \lambda_p$. Since $\lambda_p = 65/\sqrt{F_y,\text{ksi}} = 9.2$, in this case $\lambda < \lambda_p$. *Use* 2PLs—$\frac{1}{2}$×4×1'-0", tapered from full width at C to zero width at D (Fig. 13.8.5). ∎

Straight Haunch and Curved Haunch Knees

Straight haunch knees (also called *tapered haunches*), as shown in Fig. 13.8.1d, and curved haunch knees, as shown in Fig. 13.8.1e, may extend over a significant portion of a span; in which case they are not really connections but rather an integral part of a variable moment of inertia frame. Detailed treatment of tapered and curved haunches is available in Blodgett [13.2] and *ASCE Manual 41* [7.2]. For analysis of rigid frames, the reader is referred to *Single Span Rigid Frames in Steel* [13.86].

13.9 COLUMN BASE PLATES

Column base plates distribute the concentrated loads of columns to the supporting medium, commonly a concrete pedestal or footing. These heavy loads must be distributed to prevent crusing of the concrete support. Another concern is the connection, or anchorage of the base plate and column to the concrete foundation. In frame analysis, evaluation of degree of fixity may be of interest. The presentation that follows is according to the AISC LRFD Method.

Base Plates Under Axial Load

The design of base plates involves several considerations:

1. The area of the base plate will depend on the bearing strength of the concrete or grout under the steel plate.

2. The thickness of the plate will be controlled by the bending strength of the plate. When the plate dimensions B and N, shown in Fig. 13.9.1a, are significantly larger than the profile dimensions b and d of the steel section, the traditional approach is to design the plate as having cantilever spans m and n uniformly loaded.

3. For plates not extending much beyond the profile limits of the steel section, an alternate approach is required. This situation arises when the column load is relatively small. In this case the lightly loaded plate may be treated as uniformly loaded over an H-shaped area A_H adjacent to the inside perimeter of the column, as shown in Fig. 13.9.1b.

The *AISC Specification* does not give a specific design procedure for base plates. The *AISC Manual* section "COLUMN BASE PLATES FOR AXIAL COMPRESSION," pp. 14–4 to 14–7, contains the procedure presented in 1990 by Thornton [13.87], which includes design of both heavily and lightly loaded plates. Lightly loaded base plates have traditionally given designers difficulty because the cantilever distances n and m of Fig. 13.9.1a were small, giving little or no calculated thickness to resist bending. Stockwell [13.88] suggested a yield-line solution, and Murray [13.89] significantly extended the yield-line approach. Ahmed and Kreps [13.90] demonstrated the inconsistencies in the early design suggestions. DeWolf [13.91] and Fling [13.92] have also treated the subject. Thornton [13.87, 13.93] summarized and synthesized these studies to obtain the currently accepted procedure.

Figure 13.9.1
Column base plates.

(a) Cantilever approach for large plates

(b) Yield line approach for small plates

Establish the Plan Dimensions B × N.

This will be controlled by bearing on the concrete in accordance with AISC-J8. The design bearing strength $\phi_c P_p$ must at least equal the factored column load P_u,

$$\phi_c P_p \geq P_u \tag{13.9.1}$$

There are two categories for nominal strength P_p in bearing:

1. Bearing on the full area A_1 of a concrete support,

$$P_p = 0.85 f'_c A_1 \tag{13.9.2}$$

2. Bearing on area A_1 which is less than full area A_2 of a concrete support,

$$P_p = 0.85 f'_c A_1 \sqrt{A_2/A_1} \leq 0.85 f'_c (2A_1) \tag{13.9.3}$$

where $\phi_c = 0.60$ for bearing on concrete

         A_1 = area ($B \times N$ in Fig. 13.9.1) of steel plate concentrically bearing on a concrete support

         A_2 = maximum area of the portion of the supporting surface that is geometrically similar to and concentric with the loaded area

         f'_c = specified 28-day compressive strength for conete

         $f_p = P_u/BN$

Plate Thickness According to the Cantilever Method.

Assuming the bearing pressure is uniform over the entire bearing plate, the factored cantilever moment is

$$M_u = f_p \left(\frac{Nn^2}{2} \right) \qquad \text{(on section parallel to column web)} \tag{13.9.4a}$$

$$M_u = f_p \left(\frac{Bm^2}{2} \right) \qquad \text{(on section parallel to column flanges)} \tag{13.9.4b}$$

The yielding limit state for bending on the plate requires

$$\phi_b M_n \geq M_u \tag{13.9.5}$$

where

$$\phi_b M_n = \phi_b M_p = \phi_b Z F_y = \phi_b (N t_p^2/4) F_y \tag{13.9.6a}$$

$$\phi_b M_n = \phi_b M_p = \phi_b Z F_y = \phi_b (B t_p^2/4) F_y \tag{13.9.6b}$$

Equating $\phi_b M_p$ with M_u, Eq. 13.9.4a with Eq. 13.9.6a, and Eq. 13.9.4b with Eq. 13.9.6b, gives the following two equations

$$\phi_b \left(\frac{N t_p^2}{4} \right) F_y = f_p \left(\frac{Nn^2}{2} \right) \tag{13.9.7a}$$

$$\phi_b \left(\frac{B t_p^2}{4} \right) F_y = f_p \left(\frac{Bm^2}{2} \right) \tag{13.9.7b}$$

Solving for t_p in Eqs. 13.9.7 gives

$$t_p = n\sqrt{\frac{2f_p}{\phi_b F_y}} \qquad \text{or} \qquad t_p = m\sqrt{\frac{2f_p}{\phi_b F_y}} \qquad (13.9.8)$$

or, with $\phi_b = 0.90$, Eqs. 13.9.8 become

$$t_p = 1.5n\sqrt{\frac{f_p}{F_y}} \qquad \text{or} \qquad t_p = 1.5m\sqrt{\frac{f_p}{F_y}} \qquad (13.9.9)$$

Plate Thickness According to Yield Line Approach.

In this semi-empirical approach, the effective cantilever length $\lambda n'$ replaces m or n in Eq. 13.9.9. Thornton's formulation [13.93] of the yield-line analysis indicates,

$$n' = \tfrac{1}{4}\sqrt{db_f} \qquad (13.9.10)$$

where d = overall depth of the steel W section

b_f = flange width of the steel W section

The factor λ relates to the load P_u carried by the area db_f actually loaded. Thornton [13.87] gives

$$\lambda = \frac{2\sqrt{X}}{1 + \sqrt{1 - X}} \qquad (13.9.11)$$

where

$$X = \frac{4P_0}{(d + b_f)^2 F_p} \leq 1.0 \qquad (13.9.12)$$

where F_p = limiting bearing pressure $0.85f'_c$ from AISC-J8

$P_0 = P_u(db_f/BN)$

Substituting for P_0 in Eq. 13.9.12 and noting that $P_p = F_p(BN)$ gives

$$X = \frac{4}{(d + b_f)^2 F_p}\left(P_u\frac{db_f}{BN}\right) = \frac{4db_f}{(d + b_f)^2}\frac{P_u}{P_p} \leq 1.0 \qquad (13.9.13)$$

Thornton [13.87] developed λ as the link between the Murray [13.89]-Stockwell [13.88] method and his method [13.93]. Essentially, $\lambda = 1$ is the divide between lightly loaded ($\lambda < 1$) and heavily loaded situations. From Eq. 13.9.11, $\lambda = 1$ when $X = 0.64$.

For the AISC LRFD Method design purposes, the *AISC Manual* (p. 14–5) placed the ϕ_c with P_p in the denominator of Eq. 13.9.13. Thus,

$$X = \frac{4db_f}{(d + b_f)^2}\frac{P_u}{\phi_c P_p} \leq 1.0 \qquad (13.9.14)$$

The introduction of ϕ_c is unrelated to the derivation of X, and within the accuracy of the method makes no difference; however, in the use of Eq. 13.9.14, the link to Eq. 13.9.1 is a practical one.

Design Equation.

The required thickness for an axially loaded base plate is, referring to Eqs. 13.9.9,

$$t_p = 1.5 \,(\text{largest of } n, m, \text{ or } \lambda n')\sqrt{\frac{f_p}{F_y}} \qquad (13.9.15)$$

where $n = (B - 0.8b_f)/2$

$m = (N - 0.95d)/2$

$n' = \frac{1}{4}\sqrt{db_f}$

$\lambda = \dfrac{2\sqrt{X}}{1 + \sqrt{1 - X}} \le 1$; conservatively $\lambda = 1$

$X = \dfrac{4db_f}{(d + b_f)^2} \cdot \dfrac{P_u}{\phi_c P_p} \le 1.0$

Inherently, λ can exceed 1.0; however, when it does, the plate is heavily loaded and the simple cantilever method is appropriate.

Some practical aspects of column base selection are given by Ricker [13.94]. De Wolf and Ricker [13.95] have presented an overall treatment of column base plate design.

EXAMPLE 13.9.1

Design a base plate for a W14×145 column of A992 steel to carry factored axial loads of 400 kips dead load, 275 kips live load, and 100 kips wind. Assume a concrete pedestal will be under the base plate and the pedestal will have a dimension 6 in. larger than the base plate in each direction. The steel is A36 and the concrete has $f'_c = 3$ ksi. Use the AISC LRFD Method.

Solution:

(a) Compute the factored load P_u.

$$P_u = 1.2P_D + 1.6P_L = 1.2(400) + 1.6(275) = 920 \text{ kips}$$

$$P_u = 1.2P_D + 0.5P_L + 1.6P_W$$

$$= 1.2(400) + 0.5(275) + 1.6(100) = 778 \text{ kips}$$

The factored load P_u to be carried is 920 kips.

(b) Determine the maximum base plate area A_1 required. Assuming the pedestal is exactly the dimensions of the base plate, the required area using Eq. 13.9.2 would be

$$\text{Required } A_1 = \frac{P_u}{\phi(0.85)f'_c} = \frac{920}{0.60(0.85)3} = 601 \text{ sq in.}$$

The plate should be approximately square, with slight differences in dimensions B and N to give nearly equal values for m and n.

$$0.80b_f = 0.80(15.50) = 12.40 \text{ in.}$$

$$0.95d = 0.95(14.78) = 14.04 \text{ in.}$$

Try $B = 24$ in. and $N = 25$ in. This would make the concrete pedestal 30 in. by 31 in. with an area $A_2 = 930$ sq in. Using Eq. 13.9.3,

$$\sqrt{A_2/A_1} = \sqrt{930/601} = 1.24 \le 2.0$$

$$\text{Required } A_1 = \frac{\text{Preliminary required } A_1}{1.24} = \frac{601}{1.24} = 485 \text{ sq in.}$$

Use $B = 21$ in. and $N = 23$ in. giving $A_1 = 483$ sq. in. Adding the 6 in. to each dimension will make the pedestal 27 in. by 29 in. and $A_2 = 783$ sq in. Using Eq. 13.9.3 again gives required $A_1 = 472$ sq in. Keeping bearing plate dimensions in whole inches, this calculation converges quickly.

(c) Determine the plate thickness. Using Eq. 13.9.15,

$$t_p = 1.5(\text{largest of } n, m, \text{ or } \lambda n')\sqrt{\frac{f_p}{F_y}} \qquad\qquad [13.9.15]$$

$$n = 0.5(B - 0.8b_f) = 0.5(21 - 12.40) = 4.30 \text{ in.}$$

$$m = 0.5(N - 0.95d) = 0.5(23 - 14.04) = 4.48 \text{ in.}$$

Using the selected $B = 21$ in. and $N = 23$ in.,

$$\phi_c P_p = 0.60(0.85\, f_c')A_1\sqrt{A_2/A_1}$$

$$= 0.60(0.85)(3)(483)\sqrt{783/483}$$

$$= 1.53(483)1.27 = 941 \text{ kips}$$

$$X = \frac{4db_f}{(d + b_f)^2}\frac{P_u}{\phi_c P_p} = \frac{4(14.78)15.50}{(14.78 + 15.50)^2}\frac{920}{941} = 0.97 > 0.64$$

When $X > 0.64$, $\lambda = 1$. Then

$$n' = \tfrac{1}{4}\sqrt{db_f} = \tfrac{1}{4}\sqrt{14.78(15.50)} = 3.78 \text{ in.}$$

The largest of n, m, and n' is 4.48 in.

$$t_p = 1.5(\text{largest of } n, m, \text{ or } \lambda n')\sqrt{\frac{f_p}{F_y}}$$

$$= 1.5(4.48)\sqrt{\frac{P_u}{BNF_y}} = 1.5(4.48)\sqrt{\frac{920}{21(23)36}} = 1.55 \text{ in.}$$

Thus, the cantilever method requires the greater thickness (1.55 in.). *Use* plate $1\tfrac{1}{2} \times 21 \times 1'\text{-}11''$. Only rarely will the yield line approach formulas give the greater thickness. ∎

Column Bases to Resist Moment

Column bases frequently must resist moment in addition to axial compression. The situation has some similarities to the behavior of bolted connections discussed in Chapter 4, and in many respects is analogous to the situation of reinforcing bars in concrete construction. The axial force causes a precompression between the base plate and the contact surface (frequently a concrete wall or footing). When the moment is applied, the precompression on the tension side in flexure is reduced, often to zero, leaving only the anchor bolt to provide the tensile force resistance. On the compression side, the contact area remains in compression. The anchorage will have an ability to undergo rotational deformation, depending primarily on the length of anchor bolt available to deform elastically. Also, the behavior is influenced by whether or not the anchor bolts are given an initial pretension (similar to the installation of high-strength bolts as discussed in Chapter 4). The moment-rotation characteristics of column anchorages are treated in detail by Salmon, Schenker, and Johnston [13.96].

A number of elaborate methods are available for designing moment-resisting bases, with variations depending on the magnitude of the eccentricity of loading and the specific details of the anchorage. Some simple details are shown in Fig. 13.9.2.

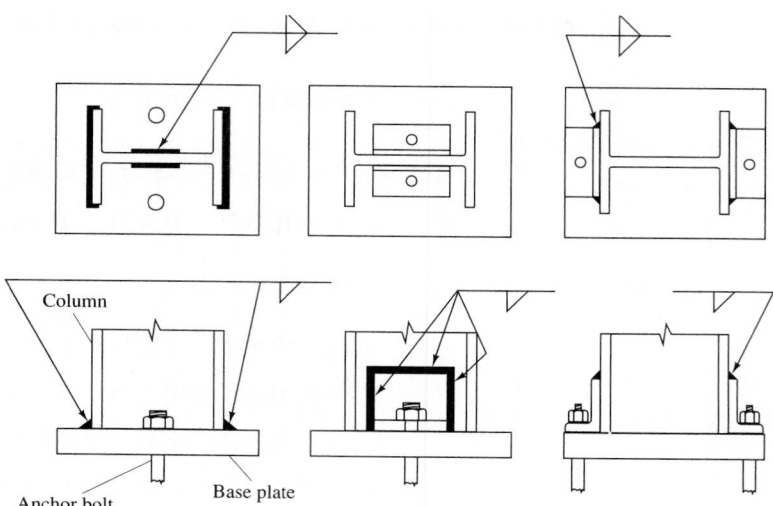

Figure 13.9.2
Column bases.

When the eccentricity of loading, $e = M/P$, is small such that it does not exceed $\frac{1}{6}$ of the plate dimension N in the direction of bending (i.e., within the middle third of the plate dimension), the ordinary combined stress formula applies. Thus for small e,

$$f_p = \frac{P}{A} \pm \frac{M}{S} \qquad (13.9.16)$$

where $M = Pe$.

$$S = Ar^2/(N/2) = AN/6$$
$$r^2 = N^2/12$$
$$f_p = \frac{P}{A} \pm \frac{6Pe}{AN} = \frac{P}{A}\left[1 \pm \frac{6e}{N}\right] \qquad (13.9.17)$$

Equation 13.9.17 is correct for $e \leq N/6$ when there is no bolt pretension, and is considered satisfactory for practical purposes at least up to $e = N/2$ without serious error.

Figure 13.9.3
Moment-resisting column
anchorage.

When the eccentricity e exceeds $N/6$, part of the base plate at the tension face becomes inactive and the stress distribution becomes as shown in Fig. 13.9.3. A simple practical assumption to use for such a situation is that the resultant of the triangular distribution is located directly under the compression flange of the column. When large moment is designed for, generally the attachment arrangement to the base of the column becomes more complex than those of Fig. 13.9.2. For a more detailed presentation on designing column bases for moment, the reader is referred to Blodgett [13.2, Sec. 3.3]. Experimental studies of base plates subject to axial load and moment have been reported by DeWolf and Sarisley [13.97] and Thambiratnam and Paramasivan [13.98].

13.10 BEAM SPLICES

There are several reasons why a rolled beam or plate girder may be spliced, such as: (1) the full length may not be available from the mill; (2) the fabricator may find it economical to splice even when the full length could be obtained; (3) the designer may desire to use splice points as an aid to cambering; and (4) the designer may desire a change in section to fit the variation in strength required along the span.

On each side of a joint, one must provide for the transverse shear and bending moment. For welded plate girders, and frequently for rolled beams, the splice may be accomplished by a full-penetration groove weld. Splices of material made entirely in the shop are nearly always groove welded. Field splices are becoming more frequently all welded, though usually using lapping of pieces and fillet-welded connections, instead of groove welding where dimensional control is a critical factor. Since full-penetration groove-welded splices are as strong as the base material, they require no further comment here.

Many field splices for beams and welded plate girders use laping splice plates and high-strength bolts as connectors. In this section the four-plate beam splice is treated.

Splices are designed for the moment M and the shear V occurring at the spliced point, or they are designed for some arbitrary or specification-prescribed higher values. AISC-J6 requires groove-welded splices to "develop the full strength of the smaller spliced section." "Other types of splices in cross sections of plate girders and beams shall develop the strength required by the forces at the point of splice." Earlier AISC Specifications required splices to develop not less than 50% of the effective strength of the material spliced.

For beam splices, each element of the splice is designed to do the work the material underlying the splice plates could do if uncut. Plates on the flange do the work of the flange, while plates over the web do the work of the web. One typical arrangement of a four-plate splice is shown in Fig. 13.10.1.

When determining the forces for which to design, one should recognize that the splice actually covers a finite distance along the span (perhaps 1 or 2 ft). Along this distance the moment and shear vary. In accordance with the principles used for designing bolted connections, the forces designed for should be those acting on the center of gravity of the bolt group. From Fig. 13.10.2 it may be noted that theoretically the moment at the centroid of the connectors on one side of the splice is different from that at the centroid on the other. Some designers, therefore, proportion the connection on each side of the splice for a moment, $M_1 = M + Ve$. Such a procedure rarely seems justified. Design of such splices has been treated by Kulak and Green [13.99]. In those unusual situations when a

Figure 13.10.1
Four-plate beam splice.

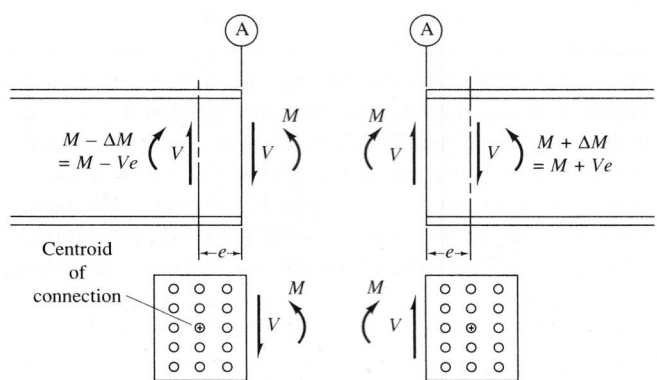

Figure 13.10.2
Forces acting on web splice
plates.

splice is located where both shear and bending moment are high, such a procedure might seem desirable.

Most splices are located in regions where either the shear or the bending moment are low so that often the design is for a specification-required minimum strength. As discussed in Sec. 10.6 for plastic analysis, one must be wary of designing a splice for a low moment just because the splice is near an inflection point. *If moments for the structure were computed using theory of statically indeterminate structures without a hinge within the span, one should not later design a splice which has low stiffness to act as a hinge.*

For most situations the authors recommend designing for the actual or specification-required forces *at the splice* and neglecting any eccentricity effect.

EXAMPLE 13.10.1

Design a rolled beam splice (four-plate type of Fig. 13.10.1) for a W24×84 beam to be located where the factored moment M_u is 510 ft-kips and the factored shear V_u is 195 kips. Use A992 steel for the beam and A572 Grade 50 for the plates and $\frac{3}{4}$-in.-diam A490 bolts in a bearing-type connection (A490-X). Use AISC LRFD Method.

Solution:

(a) Check the maximum design strengths ϕM_n and ϕV_n of the W24×84 beam.

$$\phi_b M_n = \phi_b M_p = 0.90 Z_x F_y = [0.90(224)(50)]\tfrac{1}{12} = 840 \text{ ft-kips}$$
$$\phi_v V_n = \phi_v(0.60 F_y) A_w = 1.0(0.60)(50)(24.1)0.47 = 340 \text{ kips}$$

The factored loads are 61% and 57% of the design moment and design shear strengths, respectively. Even though AISC does not require any minimum proportion of the strength to be developed by a splice, it is prudent to design splices for a significant (say, at least 50% of strength; the authors prefer at least 2/3 of the strength) proportion of the member strength.

(b) Web plates. The web plates must carry all the shear. The design strength ϕR_n for bolts in double shear is

$$\phi R_n = \phi F_{nv} m A_b = 0.75(75)(2)0.4418 = 49.7 \text{ kips}$$
$$\text{Number of bolts} = \frac{195}{49.7} = 3.9, \text{ say } 4$$

Since 4 bolts are needed for shear alone, and the web plates will also carry of portion of the bending moment the beam web would carry, try 2 lines of 4 bolts each as shown in

Figure 13.10.3
Design for Example 13.10.1.

Fig. 13.10.3. There will be 3 spaces vertically of approximately 6 in. when the plate depth is made 21 in. (which is the available flat dimension T of the beam web).

The web plate thickness to prevent shear rupture along a net section is, according to AISC-J4.2,

$$\phi(0.6F_u)A_{nv} = V_u$$

$$\text{Required } A_{nv} = \frac{V_u}{\phi(0.6F_u)} = \frac{195}{0.75(0.6)65} = 6.7 \text{ sq in.}$$

Using 4 bolts, the thickness required for each of two plates based on the shear rupture limit state will be

$$\text{Required } t = \frac{A_{nv}}{2[21 - 4(13/16)]} = \frac{6.7}{2(17.8)} = 0.19 \text{ in.}$$

A thickness of $\frac{1}{4}$ in. should be considered the practical minimum.
Use 2 PLs—$\frac{1}{4}$×21.

(c) Flange plates. The splice plates are designed as tension members. The plate width is made 9 in.; this is slightly less than the beam flange width of 9.02 in. Both the fracture and yielding limit states must be investigated in accordance with AISC-J4.1,

$$\text{Flange force} = \frac{M_u}{\text{Arm}} = \frac{M_u}{d + \text{est. } t} = \frac{510(12)}{24.10 + 0.5} = 249 \text{ kips}$$

$$\phi T_n = \phi A_g F_y \quad \text{yielding of the splice plate } (\phi = 0.90)$$

$$\phi T_n = \phi A_n F_u \quad \text{fracture of the splice plate } (\phi = 0.75)$$

$$A_n \leq 0.85 A_g$$

In addition, "block shear" must be considered, as shown in Example 4.8.3.

$$\text{Required } A_g = \frac{T_u}{\phi F_y} = \frac{249}{0.90(50)} = 5.53 \text{ sq in.}$$

$$\text{Required } A_n = \frac{T_u}{\phi F_u} = \frac{249}{0.75(65)} = 5.11 \text{ sq in.}$$

$$\text{Minimum } A_g = \frac{\text{Required } A_n}{0.85} = \frac{5.11}{0.85} = 6.01 \text{ sq in.}$$

$$\text{Required } t = \frac{5.11}{9 - 2(3/4 + 1/8)} = 0.71 \text{ in.}$$

Block shear will not be of concern with plates as thick as $\frac{3}{4}$ in.; particularly here where 5 bolts per line are used.

Use 2PLs—$\frac{3}{4}$×9.

(d) Flange bolts. The bolts are in single shear,

$$\phi R_n = 24.9 \text{ kips} \qquad (\text{single shear, } AISC \ Manual, \text{Table 7-1})$$

$$\phi R_n = \phi(2.4 d_b t F_u) \quad (\text{bearing: AISC-J4.10-hole deformation is not a consideration})$$

$$= 0.75(3.0)(3/4)(0.625)65 = 68.6 \text{ kips}$$

$$\text{Number of bolts} = \frac{T_u}{\phi R_n} = \frac{249}{24.9} = 10 \text{ bolts}$$

Use 2 rows of 5 bolts, each side of splice.

(e) Web bolts. Compute the moment carried by the web plates when F_y is reached at the center of the tension flange plates,

$$\phi_b M_n = \phi_b \left(\frac{td^2}{6}\right) F_y \left(\frac{10.5}{12.3}\right) = 0.90\left(\frac{2(0.25)(21)^2}{6}\right)\frac{50(0.85)}{12} = 117 \text{ ft-kips}$$

A conservative approach is to determine the force on the web bolts nearest the flange using the elastic vector method,

$$\Sigma x^2 + \Sigma y^2 = 8(1.5)^2 + 4(9)^2 + 4(3)^2 = 378 \text{ in.}^2$$

$$R_{ux} = \frac{My}{\Sigma x^2 + \Sigma y^2} = \frac{117(12)9}{378} = 33.4 \text{ kips} \rightarrow$$

$$R_{uy} = \frac{Mx}{\Sigma x^2 + \Sigma y^2} = \frac{117(12)1.5}{378} = 5.6 \text{ kips} \downarrow$$

$$R_{uv} = \frac{P}{\Sigma N} = \frac{195}{8} = 24.4 \text{ kips} \downarrow$$

$$R_u = \sqrt{[R_{uy} + R_{uv}]^2 + R_{ux}^2}$$

$$R_u = \sqrt{(5.6 + 24.4)^2 + (33.4)^2} = 44.9 \text{ kips} < 51.6 \text{ kips} \qquad \textbf{OK}$$

where, for the W24×84 web (0.470 in.), assuming deformation at bolt hole is not a design consideration (AISC-J4.10).

$$\phi R_n = 0.75(3.0)(0.75)(0.470)65 = 51.6 \text{ kips.}$$

Figure 13.10.4
Check for adequacy of bolts
at top of web plates,
Example 13.10.1.

One further check should be made to see whether the design bolt strength used for the flange splice plate bolts is compatible with the design bolt strength used at the outer web bolts. If a linear deformation relationship as would be caused by bending moment is assumed (Fig. 13.10.4),

$$\text{Flange horizontal force} = 2(24.9) = 49.8 \text{ kips} \quad \text{(two bolts)}$$

$$\text{Max } R_{nH} = 49.8\left(\frac{9}{12.05}\right) = 37.2 \text{ kips} > (R_{ux} = 33.4 \text{ kips}) \qquad \textbf{OK}$$

Use 2 vertical rows of 4 bolts each, each side of joint.

The final design is shown in Fig. 13.10.3. End and edge distances should also be checked in accordance with procedures discussed in Chapter 4. ∎

SELECTED REFERENCES

13.1. C. Batho and H. C. Rowan. "Investigations of Beam and Stanchion Connections," 2nd Report, Steel Structures Research Committee, Dept. of Scientific and Industrial Research of Great Britain, His Majesty's Stationery Office, London, 1934.

13.2. Omer W. Blodgett. *Design of Welded Structures.* Cleveland, Ohio: James F. Lincoln Arc Welding Foundation, 1966.

13.3. AISC. *Manual of Steel Construction, Volume II Connections,* ASD 9th edition/LRFD 1st edition. Chicago: American Institute of Steel Construction, 1992.

13.4. William A. Thornton. "Load and Resistance Factor Design of Connections," *Proceedings, Solutions in Steel,* National Engineering Conference. Chicago, IL: American Institute of Steel Construction, June 12–14, 1986 (pp. 33–1 to 33–22).

Semi-Rigid Connections

13.5. Basil Sourochnikoff. "Wind Stresses in Semi-Rigid Connections of Steel Framework," *Transactions,* ASCE, **115** (1950), 382–402.

13.6. Robert A. Hechtman and Bruce G. Johnston. *Riveted Semi–Rigid Beam-to-Column Building Connections.* Progress Report Number 1. Chicago, IL: American Institute of Steel Construction, November 1947 (118 pp.).

13.7. Leo Schenker, Charles G. Salmon, and Bruce G. Johnston. *Structural Steel Conections,* Armed Forces Special Weapons Project, Report No. 352, Engineering Research Institute, University of Michigan, June 1954.

13.8. Stanley D. Lindsey, Socrates A. Ioannides, and Arvind Goverdhan. "LRFD Analysis and Design of Beams with Partially Restrained Connections," *Engineering Journal,* AISC, **19,** 4 (Fourth Quarter 1985), 157–162.

13.9. Jack H. Brown. "Moments on Beam-Columns with Flexible Connections in Braced Frames," *Engineering Journal,* AISC, **20,** 4 (Fourth Quarter 1986), 157–165.

13.10. Douglas J. Ammerman and Roberto T. Leon. "Behavior of Semi-rigid Composite Connections," *Engineering Journal,* AISC, **21,** 2 (Second Quarter 1987), 53–62.

13.11. Roberto T. Leon, Douglas J. Ammerman, Jihshya Lin, and Robert D. McCauley. "Semi-rigid Composite Steel Frames," *Engineering Journal,* AISC, **21,** 4 (Fourth Quarter 1987), 147–155.

13.12. David A. Nethercot, J. Buick Davison, and Patrick A. Kirby. "Connection Flexibility and Beam Design in Non-sway Frames," *Engineering Journal,* AISC, **22,** 3 (Third Quarter 1988), 99–108.

13.13. Wai-Fah Chen and N. Kishi. "Semirigid Steel Beam-to-Column Connections: Data Base and Modeling," *Journal of Structural Engineering,* ASCE, **115,** 1 (January 1989), 105–119.

13.14. Atorod Azizinamini and James B. Radziminski. "Static and Cyclic Performance of Semirigid Steel Beam-to-Column Connections," *Journal of Structural Engineering*, ASCE, **115,** 12 (December 1989), 2979–2999.

13.15. Roberto T. Leon and Douglas J. Ammerman. "Semi-Rigid Composite Connections for Gravity Loads," *Engineering Journal*, AISC, **27,** 1 (First Quarter 1990), 1–11.

13.16. Reidar Bjorhovde, Andre Colson, and Jacques Brozzetti. "Classification System for Beam-to-Column Connections," *Journal of Structural Engineering*, ASCE, **116,** 11 (November 1990), 3059–3076.

13.17. Riccardo Zandonini and Paolo Zanon. "Beam Design in PR Braced Steel Frames," *Engineering Journal*, AISC, **25,** 3 (Third Quarter 1991), 85–97.

13.18. Roberto Leon. "Composite Semi-Rigid Connections," *Modern Steel Construction*, AISC, October 1992, 18–23.

Simple Shear Connections—Double Angle

13.19. Bruce Johnston and Lloyd F. Green. "Flexible Welded Angle Connections," *Welding Journal*, **19,** 10 (October 1940), 402s–408s.

13.20. Peter C. Birkemoe and Michael I. Gilmor. "Behavior of Bearing Critical Double-Angle Beam Connections," *Engineering Journal*, AISC, **15,** 4 (4th Quarter 1978), 109–115.

13.21. Joseph A. Yura, Peter C. Birkemoe, and James M. Ricles. "Beam Web Shear Connections: An Experimental Study," *Journal of the Structural Division*, ASCE, **108,** ST2 (February 1982), 311–325.

13.22. AISC. "Predesigned Bolted Framing Angle Connections," *Engineering Journal*, AISC, **19,** 1 (First Quarter 1982), 1–11.

13.23. James M. Ricles and Joseph A. Yura. Strength of Double-Row Bolted-Web Connections, *Journal of Structural Engineering*, ASCE, **109,** 1 (January 1983), 126–142. (Also Ref. 3.12).

13.24. Jung-June R. Cheng and Joseph A. Yura. "Local Web Buckling of Coped Beams," *Journal of Structural Engineering*, ASCE, **112,** 10 (October 1986), 2314–2331.

13.25. Ajaya K. Gupta. "Buckling of Coped Steel Beams," *Journal of Structural Engineering*, ASCE, **110,** 9 (September 1984), 1977–1987; Disc. by Jung-June Cheng and Joseph A. Yura, **112,** 1 (January 1986), 201–204.

13.26. Jung-June R. Cheng, Joseph A. Yura, and C. Philip Johnson. "Lateral Buckling Coped Steel Beams," *Journal of Structural Engineering*, ASCE, **114,** 1 (January 1988), 1–15.

13.27. Jung-June R. Cheng and Joseph A. Yura. "Lateral Buckling Tests on Coped Steel Beams," *Journal of Structural Engineering*, ASCE, **114,** 1 (January 1988), 16–30.

13.28. Abolhassan Astaneh, Marwan N. Nader, and Lincoln Malik. "Cyclic Behavior of Double Angle Connections," *Journal of Structural Engineering*, ASCE, **115,** 5 (May 1989), 1101–1118.

Simple Shear Connections—Single Plate

13.29. Ralph M. Richard, Paul E. Gillett, James D. Kriegh, and Brett A. Lewis. "The Analysis and Design of Single Plate Framing Connections," *Engineering Journal*, AISC, **17,** 2 (Second Quarter 1980), 38–51.

13.30. Ned W. Young and Robert O. Disque. "Design Aids for Single Plate Framing Connections," *Engineering Journal*, AISC, **18,** 4 (Fourth Quarter 1981), 129–148.; Disc. by John D. Griffiths, **19,** 3 (Third Quarter 1982), 179.

13.31. Ralph M. Richard, James D. Kriegh, and David E. Hormby. "Design of Single Plate Framing Connections with A307 Bolts," *Engineering Journal*, AISC, **19,** 4 (Fourth Quarter 1982), 209–213; Disc. by Edward P. Becker, **22,** 1 (First Quarter 1985), 50–51.

13.32. David E. Hormby, Ralph M. Richard, and James D. Kriegh. "Single-Plate Framing Connections with Grade-50 Steel and Composite Construction," *Engineering Journal*, AISC, **21,** 3 (Third Quarter 1984), 125–138.

13.33. Abolhassan Astaneh, Steven M. Call, and Kurt M. McMullin. "Design of Single Plate Shear Connections," *Engineering Journal*, AISC, **26,** 1 (1st Quarter 1989), 21–32. Disc. by Ralph M. Richard, **27,** 3 (Third Quarter 1990), 121–126.

13.34. D. J. L. Kennedy. "Moment–Rotation Characteristics of Shear Connections," *Engineering Journal*, AISC, **6,** 4, (October 1969), 105–115.

Simple Shear Connections—Tee Framing

13.35. Abolhassan Astaneh and Marwan N. Nader. "Design of Tee Framing Shear Connections," *Engineering Journal*, AISC, **26,** 1 (First Quarter 1989), 9–20.

13.36. Abolhassan Astaneh and Marwan N. Nader. "Experimental Studies and Design of Steel Tee Shear Connections," *Journal of Structural Engineering*, ASCE, **116,** 10 (October 1990), 2882–2902.

Seated Beam Connections

13.37. W. Bertwell. Discussion of "Design of Bolts or Rivets Subject to Combined Shear and Tension," by Alfred Zweig, *Engineering Journal*, AISC, **3,** 4 (October 1966), 165–167.

13.38. Bruce G. Johnston and Gordon R. Diets. "Tests of Miscellaneous Welded Building Connections," *Welding Journal*, **21,** 1 (January 1942), 5s–27s.

13.39. Inge Lyse and Norman G. Schreiner. "An Investigation of Welded Seat Angle Connections," *Welding Journal*, **14,** 2 (February 1935), Research Supplement, 1–15.

13.40. Charles W. Roeder. "Results of Experiments on Seated-beam Connections," *Proceedings*, National Engineering Conference & Conference of Operating Personnel. Chicago, IL: American Institute of Steel Construction, April 29–May 2, 1987 (pp. 43-1 to 43-12).

13.41. Charles W. Roeder and Ronald H. Dailey. "The Results of Experiments on Seated Beam Connections," *Engineering Journal*, AISC, **26,** 3 (Third Quarter 1989), 90–95.

13.42. J. H. Garrett, Jr. and R. L. Brockenbrough. "Design Loads for Seated-beam Connections in LRFD," *Engineering Journal*, AISC, **23,** 2 (Second Quarter 1986) 84–88.

13.43. Roger L. Brockenbrough. "Design Loads for Seated-beam Connections," *Proceedings*, National Engineering Conference & Conference of Operating Personnel. Chicago, IL: American Institute of Steel Construction, April 29–May 2, 1987 (pp. 12-1 to 12-16).

13.44. Cyril D. Jensen. "Welded Structural Brackets," *Welding Journal*, **15,** 10 (October 1936), Research Supplement, 9–15.

Triangular Bracket Plates

13.45. Charles G. Salmon. "Analysis of Triangular Bracket-Type Plates," *Journal of the Engineering Mechanics Division*, ASCE, **88,** EM6 (December 1962), 41–87.

13.46. Charles G. Salmon, Donald R. Buettner, and Thomas C. O'Sheridan. "Laboratory Investigation of Unstiffened Triangular Bracket Plates," *Journal of the Structural Division*, ASCE, **90,** ST2 (April 1964) 257–278.

13.47. Lynn S. Beedle et al. *Structural Steel Design.* New York: The Ronald Press Company, 1964 (pp. 550–555).

Continuous Beam-to-Column Connections

13.48. A. B. Onderdonk, R. P. Lathrop, and Joseph Coel. "End Plate Connections in Plastically Designed Structures," *Engineering Journal*, AISC, **1,** 1 (January 1964), 24–27.

13.49. Lynn S. Beedle and Richard Christopher. "Tests of Steel Moment Connections," *Engineering Journal*, AISC, **1,** 4 (October 1964), 116–125.

13.50. Richard T. Douty and William McGuire. "High Strength Bolted Moment Connections," *Journal of the Structural Division*, ASCE, **91,** ST2 (April 1965). 101–128.

13.51. J. S. Huang, W. F. Chen, and L. S. Beedle. "Behavior and Design of Steel Beam-to-Column Moment Connections," *WRC Bulletin 188*, Welding Research Council, New York, October 1973, 1–23.

13.52. J. E. Regec, J. S. Huang, and W. F. Chen. "Test of a Fully Welded Beam-to-Column Connection," *WRC Bulletin 188*, Welding Research Council, New York, October 1973, 24–35.

13.53. John Parfitt, Jr. and Wai F. Chen. "Test of Welded Steel Beam-to-Column Connections," *Journal of the Structural Division*, ASCE, **102,** ST1 (January 1976), 189–202.

13.54. Bruce D. Macdonald. "Moment Connections Weakened by Laminations," *Journal the Structural Division*, ASCE, **105,** ST8 (August 1979), 1605–1619.

13.55. Paul Grundy, Ian R. Thomas, and Ian D. Bennetts. "Beam-to-Column Moment Connections," *Journal of the Structural Division*, ASCE, **106,** ST1 (January 1980), 313–330. Disc. by N. Krishnamurthy, **106,** ST12 (December 1980), 2573–2574.

13.56. W. F. Chen and K. V. Patel. "Static Behavior of Beam-to-Column Moment Connections," *Journal of the Structural Division*, ASCE, **107,** ST9 (September 1981), 1815–1838.

13.57. Jelle Witteveen, Jan W. B. Stark, Frans S. K. Bijlaard, and Piet Zoetemeijer. "Welded and Bolted Beam-to-Column Connections," *Journal of the Structural Division*, ASCE, **108,** ST2 (February 1982), 433–455.

13.58. G. P. Rentschler, W. F. Chen, and G. C. Driscoll. "Beam-to-Column Web Connection Details," *Journal of the Structural Division*, ASCE, **108,** ST2 (February 1982), 393–409.

13.59. Kirit V. Patel and Wai F. Chen. "Nonlinear Analysis of Steel Moment Connections," *Journal of Structural Engineering*, ASCE, **110,** 8 (August 1984), 1861–1874.

13.60. Wai F. Chen and David E. Newlin. "Column Web Strength in Beam-to-Column Connections," *Journal of the Structural Division*, ASCE, **99,** ST9 (September 1973), 1978–1984.

13.61. Wai F. Chen and Irving J. Oppenheim. "Web Buckling Strength of Beam-to-Column Connections," *Journal of the Structural Division*, ASCE, **100,** ST1 (January 1974), 279–285.

13.62. Eugene W. Miller. "Load and Resistance Factor Design of Moment Connections," *Proceedings*, *Solutions in Steel*, National Engineering Conference. Chicago, IL: American Institute of Steel Construction, June 12–14, 1986 (pp. 29–1 to 29–22).

13.63. R. Ford Pray and Cyril Jensen. "Welded Top Plate Beam-Column Connections," *Welding Journal*, **35,** 7 (July 1956), 338s–347s.

13.64. J. L. Brandes and R. M. Mains. "Report of Tests of Welded Top-Plate and Seat Building Connections," *Welding Journal*, **24,** 3 (March 1944), 146s–165s.

13.65. William A. Thornton. "Prying Action—A General Treatment," *Engineering Journal*, AISC, **22,** 2 (Second Quarter 1985), 67–75.

13.66. Abolhassan Astaneh. "Procedure for Design and Analysis of Hanger-type Connections," *Engineering Journal*, AISC, **22,** 2 (Second Quarter 1985), 63–66.

13.67. W. A. Thornton. "Strength and Serviceability of Hanger Connections," *Engineering Journal*, AISC, **29,** 4 (Fourth Quarter 1992), 145–149.

End-Plate Moment Connections

13.68. John D. Griffiths. "End Plate Moment Connections–Their Use and Misuse," *Engineering Journal*, AISC, **21,** 1 (First Quarter 1984), 32–34.

13.69. Henning Agerskov. "High-Strength Bolted Connections Subject to Prying," *Journal of the Structural Division*, ASCE, **102,** ST1 (January 1976), 161–175.

13.70. Henning Agerskov. "Analysis of Bolted Connections Subject to Prying," *Journal of the Structural Division*, ASCE, **103,** ST11 (November 1977), 2145–2163.

13.71. N. Krishnamurthy. "A Fresh Look at Bolted End-Plate Behavior and Design," *Engineering Journal*, AISC, **15,** 2 (2nd Quarter 1978), 39–49. Disc. by H. Agerskov, W. McGuire, J. D. Griffiths and J. M. Wooten, N. W. Rimmer, and author, **16,** 2 (2nd Quarter 1979), 54–64.

13.72. Natarajau Krishnamurthy, Horng-Te Huang, Paul K. Jeffrey, and Louie K. Avery. "Analytical $M-\theta$ Curves for End-Plate Connections," *Journal of the Structural Division*, ASCE, **105,** ST1 (January 1979), 133–145.

13.73. Allan P. Mann and Linden J. Morris. "Limit Design of Extended End-Plate Connections," *Journal of the Structural Division*, ASCE, **105,** ST3 (March 1979), 511–526.

13.74. Thomas M. Murray. "Beam-to-Column End-Plate Connections—Column Design Considerations," presented at AISC National Engineering Conference, March 28–30, 1984, Tampa, FL (10 pp.).

13.75. Alan Hendrick and Thomas M. Murray. "Column Web Compression Strength at End-Plate Connections," *Engineering Journal*, AISC, **21,** 3 (Third Quarter 1984), 161–169.

13.76. Yoke Leong Yee and Robert E. Melchers. "Moment–Rotation Curves for Bolted Connections," *Journal of Structural Engineering*, ASCE, **112,** 3 (March 1986), 615–635.

13.77. Thomas M. Murray and Anant R. Kukreti. "Design of 8-bolt Stiffened Moment End Plates," *Engineering Journal*, AISC, **25,** 2 (Second Quarter 1988) 45–53.

13.78. Larry E. Curtis and Thomas M. Murray. "Column Flange Strength at Moment End-plate Connections," *Engineering Journal*, AISC, **26,** 2 (Second Quarter 1989), 41–50.

13.79. Thomas M. Murray and Emmett A. Sumner. *Extended End-Plate Moment Connections, Seismic and Wind Applications,* Steel Design Guide Series 4 (2nd Ed.). Chicago: American Institute of Steel Construction, Inc., 2003.

Beam-to-Column Web Moment Connections

13.80. A. Leon Abolitz and Marvin E. Warner. "Bending Under Seated Connections," *Engineering Journal*, AISC, **2,** 1 (January 1965), 1–5.

13.81. Frank W. Stockwell, Jr. "Yield Line Analysis of Column Webs with Welded Beam Connections," *Engineering Journal*, AISC, **11,** 1 (First Quarter 1974), 12–17.

13.82. Richard H. Kapp. "Yield Line Analysis of a Web Connection in Direct Tension," *Engineering Journal*, AISC, **11,** 1 (Second Quarter 1974), 38–41.

13.83. Glenn P. Rentschler, Wai F. Chen, and George C. Driscoll. "Tests of Beam-to-Column Web Moment Connections," *Journal of the Structural Division*, ASCE, **106,** ST5 (May 1980), 1005–1022.

13.84. Joseph M. Hoptay and Heino Ainso. "An Experimantal Look at Bracket-Loaded Webs," *Engineering Journal*, AISC, **18,** 1 (First Quarter 1981), 1–7.

13.85. Bruce E. Hopper, Gordon B. Batson, and Heino Ainso. "Bracket Loaded Webs with Low Slenderness Ratios," *Engineering Journal*, AISC, **22,** 1 (First Quarter 1985), 11–18.

Rigid Frame Knees

13.86. John D. Griffiths. *Single Span Rigid Frames in Steel.* New York: American Institute of Steel Construction, Inc., 1948.

Column Base Plates

13.87. W. A. Thornton. "Design of Base Plates for Wide Flange Columns—A Concatenation of Methods," *Engineering Journal*, AISC, **27,** 4 (Fourth Quarter 1990), 173–174.

13.88. Frank W. Stockwell, Jr. "Preliminary Base Plate Selection," *Engineering Journal*, AISC, **12,** 3 (Third Quarter 1975), 92–99.

13.89. Thomas M. Murray. "Design of Lightly Loaded Steel Column Base Plates," *Engineering Journal*, AISC, **20,** 4 (Fourth Quarter 1983), 143–152.

13.90. Salahuddin Ahmed and Robert R. Kreps. "Inconsistencies in Column Base Plate Design in the New AISC ASD Manual," *Engineering Journal*, AISC, **27**, 3 (Third Quarter 1990). 106–107.

13.91. John T. DeWolf. "Axially Loaded Column Base Plates," *Journal of the Structural Division*, ASCE, **104**, ST5 (May 1978), 781–794.

13.92. Russell S. Fling. "Design of Steel Bearing Plates," *Engineering Journal*, AISC, **7**, 2 (April 1970), 37–40.

13.93. W. A. Thornton. "Design of Small Base Plates for Wide Flange Columns," *Engineering Journal*, AISC, **27**, 3 (Third Quarter 1990), 108–110.

13.94. David T. Ricker. "Some Practical Aspects of Column Base Selection," *Engineering Journal*, AISC, **26**, 3 (Third Quarter 1989), 81–89.

13.95. John T. DeWolf and David T. Ricker. *Column Base Plates*, Steel Design Guide Series 1 (Publ. No. D801). Chicago: American Institute of Steel Construction, Inc., 1990.

13.96. Charles G. Salmon, Leo Schenker, and Bruce G. Johnston. "Moment–Rotation Characteristics of Column Anchorages," *Transactions*, ASCE, **122** (1957), 132–154.

13.97. John T. DeWolf and Edward F. Sarisley. "Column Base Plates with Axial Loads and Moments," *Journal of the Structural Division*, ASCE, **106**, ST11 (November 1980), 2176–2184.

13.98. David P. Thambiratnam and P. Paramasivam. "Base Plates under Axial Loads and Moments," *Journal of Structural Engineering*, ASCE, **112**, 5 (May 1986), 1166–1181.

Splices

13.99. Geoffrey L. Kulak and Deborah L. Green. "Design of Connectors in Web–Flange Beam or Girder Splices," *Engineering Journal*, AISC, **27**, 2 (Second Quarter 1990), 41–48.

Additional References

13.100. W. H. Yang, W. F. Chen, and M. D. Bowman. "The Behavior and Load-Carrying Capacity of Unstiffened Seated Beam Connections" *Engineering Journal*, AISC, **34**, 3 (Third Quarter 1997), 89–103.

13.101. C. J. Carter, W. A. Thornton, and T. M. Murray. Discussion of "The Behavior and Load-Carrying Capacity of Unstiffened Seated Beam Connections" by Yang, Chen, and Bowman (*Engineering Journal*, AISC, **34**, 4 (Fourth Quarter 1997, 151–156.

13.102. James A. Swanson. "Ultimate Strength Prying Models for Bolted T-stub Connections," *Engineering Journal*, AISC, **39**, 3 (Third Quarter 2002), 136–137.

13.103. Thomas M. Murray, W. Lee Shoemaker. *Flush and Extended Multiple-Row Moment End-Plate Connections*, Steel Design Guide Series 16. Chicago: American Institute of Steel Construction, Inc., 2002.

13.104. J. T. Borgsmiller and T. M. Murray. "Simplified Method for the Design of Moment End Plate Connections," Research Report CE/VPI-ST-95/19. Department of Civil Engineering, Virginia Polytechnic Institute and State University, Blacksburg, VA, November 1995.

PROBLEMS

All problems are to be done according to the AISC Load and Resistance Factor Design Method. The requirement of W section is intended to include W, S, and M sections. All given loads are service loads unless otherwise indicated. If needed, assume service loads are 2/3 of factored loads unless otherwise indicated. Assume lateral support consists of translational restraint but not moment (rotational) restraint, unless otherwise indicated. Assume all standard sections are equally readily available in the indicated grade of steel (even though actually they are not). Assume all contact surfaces are Class A and all holes are standard holes unless otherwise indicated. A design sketch to scale is required for all design problems.

13.1. Compute the maximum factored load (20% dead load and 80% gravity live load) reaction P_u that can be developed for the simple shear (framed beam) connection shown in the accompanying figure. The fasteners are $\frac{7}{8}$-in.-diam A325 bolts in bearing-type connections with threads excluded from the shear planes (A325-X).

(a) Neglect eccentricity as is customary on simple shear connections; compare detailed calculation results with *AISC Manual*.

(b) Consider eccentricity of loading; use the ultimate strength approach of Sec. 4.12 for eccentric shear, and use one of the combined shear and tension approaches of Sec. 4.15 for the connection to the column flange.

13.2. Repeat the analysis requested in Prob. 13.1 except the bolts are in slip-critical connections (A325-SC).

13.3. Design a double angle simple shear (framed beam) connection for a W16×50 beam having a reaction

Problems 13.1 and 13.2

of 20 kips dead load and 30 kips live load to connect to the column flange of a W8×67. Determine the angles to be used, and show the number and placement of fasteners. Use A36 steel. Use an *economical size* A325 bolt in bearing-type connections, with threads excluded from the shear planes (A325-X).

13.4. Redesign the double angle simple shear connection of Prob. 13.3 as a slip critical connection (A325-SC).

13.5. Design a double angle welded simple shear (framed beam) connection for a W33×118 beam connecting to a column flange which is $\frac{3}{4}$ in. thick. The beam reaction is 25 kips dead load and 100 kips live load. E70 electrodes are to be used with shielded metal arc welding (SMAW), and the base steel is A572 Grade 50. Design using basic principles; then compare with applicable *AISC Manual* tables.

13.6. Compute the maximum factored load (25% dead load and 75% gravity live load) reaction P_u the W14×30 beam may transmit to the seat angle *B*. What should be the size for angle *A* and what is its function? Use $\frac{7}{8}$-in.-diam A325 bolts in a bearing-type connection with threads excluded from the shear planes (A325-X).

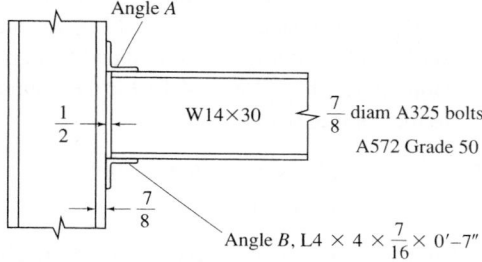

Problem 13.6

13.7. For the unstiffened seat of Prob. 13.6, what is the maximum factored load reaction P_u available when the connection is welded with E70 electrodes using shielded metal arc welding (SMAW). Will the 7-in. length of seat angle be satisfactory?

13.8. Assume a W16×57 beam is to have the maximum factored load (20% dead load and 75% live load) reaction P_u for a uniformly loaded 15-ft span. Design an unstiffened seat connection and specify its welded connection. Show detailed calculations and compare with any applicable *AISC Manual* tables. Use A36 steel, and E70 electrodes with shielded metal arc welding (SMAW).

13.9. Redesign the unstiffened seat connection of Prob. 13.8 using $\frac{3}{4}$-in.-diam A325 bolts in a *bearing-type* connection (A325-X).

13.10. Redesign the unstiffened seat connection of Prob. 13.8 using $\frac{3}{4}$ in.-diam A325 bolts in a *slip-critical* connection (A325-SC).

13.11. Design an unstiffened seat angle to support a W10×22 beam on the web of a W12×65 column. The beam reaction is 6 kips dead load and 16 kips live load. There is no beam on the opposite side of the web of the column. Use $\frac{3}{4}$-in.-diam A325 bolts in a bearing-type connection with no threads in the shear planes (A325-X) and A992 steel. Show detailed calculations for the design; then compare with applicable *AISC Manual* tables. Specify a size for the top clip angle.

13.12. Design a welded stiffened seat similar to that of Fig. 13.4.4 attached to a W14×74 column flange. The seat must support a W21×83 having a reaction of 30 kips dead load and 100 kips live load. Use E70 electrodes with shielded metal arc welding (SMAW) and A572 Grade 60 steel.

13.13. Design a welded stiffened seat to support a W18×65 beam having a reaction of 20 kips dead load and 70 kips live load. The beam lies at right angles to the stiffened seat as in Fig. 13.4.2. Use E70 electrodes, shielded metal arc welding (SMAW), and A572 Grade 60 steel.

13.14. Specify the weld size and triangular bracket plate thickness for the situation of the accompanying figure. Use $P = 11$ kips dead load and 35 kips live load, $e = 6$ in., and $L_w = 14$ in. Use E70 electrodes with shielded metal arc welding (SMAW), and A992 steel for base material.

13.15. Repeat Prob. 13.14 using $P = 21$ kips dead load and 42 kips live load, $e = 3$ in., and $L_w = 16$ in.

Problems 13.14 and 13.15

13.16. Design a stiffened seat (bracket plate) consisting of a structural tee that is bolted to the 1-in. flange of a column as in the accompanying figure. The seat plate is welded to the web of the tee. Use $\frac{7}{8}$-in.-diam A325 bolts in a bearing-type connection with threads excluded from the shear planes (A325-X) and A572 Grade 50 steel.

Problem 13.16

13.17. Investigate the adequacy of the connection shown in the accompanying figure. If the number of connectors is not adequate, determine the correct number. Then specify the dimensions of the plates, including their thickness. The load P is 5 kips dead load and 15 kips live load. Use $\frac{7}{8}$-in.-diam A325 bolts in a slip-critical connection (A325-SC).

Problem 13.17

13.18. Design a continuous beam-to-column connection for W21×55 beams to connect both flanges of a W14×84 column. Use the type of connection shown in Fig. 13.6.1b if stiffeners are required. Assume the nominal moment strength M_n at the connection is the plastic moment strength M_p of the beams and that elastic analysis has been used to obtain the factored load (20% dead load and 80% live load) internal forces. Use E70 electrodes with shielded metal arc welding and A36 steel for the elements being welded. The factored shear V_u in the beams adjacent to the column is 50 kips.

13.19. Design for the conditions of Prob. 13.18 but consider the W21×55 to frame in only on one side. Use the plate stiffeners as in Fig. 13.6.1b if necessary.

13.20. Design for the conditions of Prob. 13.18 except the beam frames in on only one side and the type of connection shown in Fig. 13.6.1d is to be used.

13.21. The split-tee beam connection shown in the accompanying figure is subject to factored moment M_u of 135 ft-kips and a factored end reaction P_u of 50 kips.

(a) Determine whether or not the tees are adequate, and if not, increase their size.

(b) Determine the number of $\frac{7}{8}$-in.-diam A325 bolts (A325-X) to connect the tees to the column flanges, the tees to the beam flanges, the clip angles to the column flange, and the clip angles to the beam web.

Problem 13.21

Cut from S24×105.9 × 1'–3"
Top and bottom

4"

WT8×28.5

W14×43

W14×159

WT10×31 × 3'–0"

4"

4" | 5@3=1'–3"

A325 bolts $\frac{7}{8}$ diam

$5\frac{1}{2}$

3 | 3

$\frac{5}{16}$

Problem 13.22

13.22. For the split-tee beam connection with the stub beam shown, determine the design moment strength ϕM_n and design beam reaction strength ϕP_n at the face of the column. Use $\frac{7}{8}$-in.-diam A325 bolts in a bearing-type connection with no threads in the shear planes (A325-X), and A992 steel.

13.23. Design a split-tee connection as shown in Fig. 13.6.2c to connect a W16×50 beam to a W14×132 column. Assume the nominal moment strength M_n to be developed is the plastic moment strength M_p of the beam. The factored shear V_u to be carried is 35 kips. Use $\frac{3}{4}$-in.-diam A325 bolts, or if it seems advantageous use A490 bolts, in a bearing-type connection with threads excluded from the shear planes (A325-X or A490-X).

13.24. Redesign the connection of Prob. 13.23 using an end-plate connection as shown in Fig. 13.6.2d.

13.25. Redesign the connection of Prob. 13.21 using an end-plate connection as shown in Fig. 13.6.2d. The factored moment M_u to be carried is 190 ft-kips, and the factored shear V_u is 70 kips.

13.26. Design a vertical tee stiffener (as in Figs. 13.6.3 and 13.6.9) connection to frame a W16×67 beam into the web of a W12×96 column. The beam must develop its plastic moment strength M_p and the factored shear V_u is 52 kips. Use A572 Grade 60 steel, and shielded metal arc welding (SMAW) with E80 electrodes.

13.27. Redesign the connection of Prob. 13.23 using top plate and seat connection similar to that of Fig. 13.6.12a. Use shielded metal arc welding (SWAW) with E70 electrodes.

13.28. Redesign the connection of Prob. 13.23 using a top and bottom plate connection similar to that of

Fig. 13.6.12b. Use shielded metal arc welding (SMAW) with E70 electrodes.

13.29. For the continuous beam connection of the accompanying figure, specify the plate size, weld size using E70 electrodes, and length of weld to develop full plastic moment strength M_p of the W10×39 beam. Consider only plate A and not other aspects of the connection. Use A992 steel and shielded metal arc welding (SMAW) with E70 electrodes.

Plate A

W10×39   W10×39

W21×62

Problem 13.29

13.30. Design a square knee connection (Fig. 13.8.1a or b) to join a W24×84 girder to a W30×108 column. Assume the plastic moment strength M_p must develop in the girder. Use A992 steel and shielded metal arc welding with E70 electrodes.

13.31. Design the column base plate for a W14×211 of A572 Grade 50 steel subject to factored axial load P_u of 1100 kips. The plate is to be supported on a 6-ft square concrete footing having $f'_c = 3000$ psi.

13.32. Design the column base plate for a W12×65 column to carry a factored axial load P_u of 370 kips. The base plate is supported on a 5-ft square concrete footing having $f'_c = 3500$ psi.

13.33. Design a four-plate beam splice for a W14×53 beam of A992 steel. Design for 80% of the moment strength and 60% of the shear strength of the section. Use $\frac{3}{4}$-in.-diam A325 bolts in a bearing-type connection with threads excluded from the shear planes (A325-X).

13.34. Redo Prob. 13.33 for W24×84 beam.

13.35. Redo Prob. 13.33 for W21×55 beam.

13.36. Design an eight-plate splice (plates inside and outside of the flanges and both sides of the web) for a welded girder consisting of $1\frac{7}{8}$×22 flanges and $\frac{7}{16}$×78 web. Use factored moment M_u of 6800 ft-kips and factored shear V_u of 510 kips. Use A36 steel and $\frac{7}{8}$-in.-diam A325 bolts in a slip-critical connection (A325-SC).

14

Frames—Braced and Unbraced

14.1 GENERAL

As discussed in Sec. 6.9, the effective length of the column members in frames is dependent on whether the frame is *braced* or *unbraced*. For the *braced* frame the effective length KL is equal to or less than the actual length. For the *unbraced* frame, the effective length KL always is greater than the actual length.

In order to understand frame behavior, consider in Fig. 14.1.1 the forces that arise in a column member of a frame as a result of lateral deflection due to a force such as wind. The moments M_Δ and shears Q_Δ are those portions of the moments and shears required to balance the moment $P\Delta$. In addition, there will be moments and shears caused by gravity loads at the particular floor level. Equilibrium in Fig. 14.1.1a requires

$$P\Delta = Q_\Delta h + 2M_\Delta \qquad (14.1.1)$$

The lateral deflection Δ is commonly called *drift* when it results from wind loading in multistory frames, as shown in Fig. 14.1.2. Drift consists of two parts; that resulting directly from horizontal load, and that arising from vertical load times the drift.

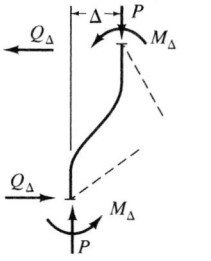

(a) Effects of lateral deflection

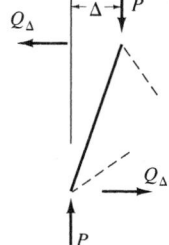

(b) Negligible moment restraint from columns and girders

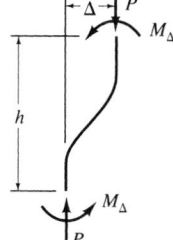

(c) Negligible shear restraint from diagonals or other bracing

Figure 14.1.1
Secondary bending moment due to $P\Delta$ in frames.

Welded unbraced rigid frame (Vierendeel truss). (Courtesy Bethlehem Steel Corporation)

A frame will deflect under lateral loading such as wind regardless of the pattern of its component members. However, the manner in which equilibrium is maintained against the moment $P\Delta$ differs depending on the restraint conditions. If the building were a vertical pin-jointed truss under lateral loading, there would be no continuity at the joints to allow

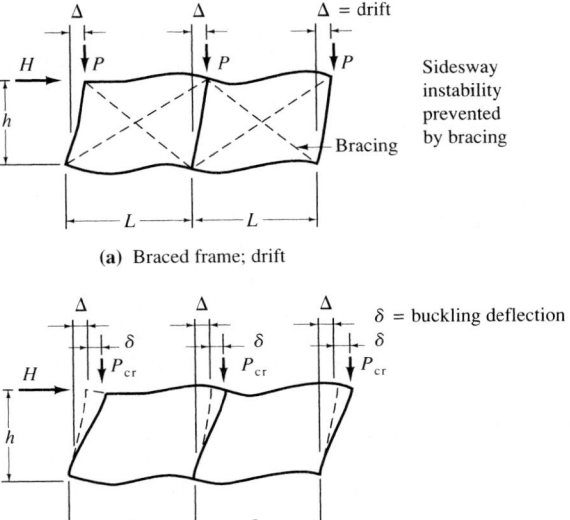

Figure 14.1.2
Comparison of braced and
unbraced frames.

the moment M_Δ to develop. In which case, as in Fig. 14.1.1b,

$$Q_\Delta = \frac{P\Delta}{h} \qquad (14.1.2)$$

Diagonal and horizontal members (web members of the truss) would have to carry the entire shear Q_Δ.

On the other hand, if the members are rigidly joined together but without diagonal members, there would be little shear resistance. Neglecting the shear resistance entirely would result in

$$M_\Delta = \frac{P\Delta}{2} \qquad (14.1.3)$$

as in Fig. 14.1.1c. In this case, the girders and columns would have to accommodate the moment M_Δ.

Braced Frame

A braced frame has relatively small moment resistance from columns and girders to counterbalance $P\Delta$, in comparison with the actual restraint from diagonals or other bracing. In other words, Eq. 14.1.2 is assumed to represent the braced frame in simplified design procedures. As is shown later in this chapter, there is both flexural resistance and shear resistance developed in the braced and the unbraced frame. It is the relative magnitudes of these resistances that make the difference between the braced and unbraced frame.

Basically, a braced frame is more appropriately defined as one in which *sidesway buckling is prevented* by bracing elements of the structure other than the structural frame itself. As will be seen in the next section, the theoretical elastic stability analysis of a braced frame assumes no relative joint displacements, which obviously could occur only with infinitely stiff bracing. However, it is practical for design and reasonably correct to assume negligible moment resistance as implied by Eq. 14.1.2, and to assume that for stability purposes the sidesway mode is thereby also prevented.

The term "sidesway" is used to refer to stable elastic lateral movement of a frame, usually due to lateral loads, such as wind. Sidesway buckling is the sudden lateral movement, such as δ of Fig. 14.1.2b, caused by axial loads reaching a certain critical value.

In conclusion, the braced frame accommodates the $P\Delta$ moments by developing shears Q_Δ in the bracing system.

Unbraced Frame

In the unbraced frame, as Fig. 14.1.2b, if the horizontal load H is maintained constant and the compressive loads P are increased sufficiently to cause failure, such failure will occur with a side lurch known as sidesway buckling. The lateral deflection will *suddenly* become greater than the drift as shown in Fig. 14.1.2b. For cases where there is no lateral loading H and, therefore, no initial deflection, the sudden sidesway will still occur when the vertical load reaches a critical value.

The practical design treatment of the unbraced frame assumes that, referring to Fig. 14.1.1c, no shears Q_Δ are capable of developing and Eq. 14.1.3 applies. Any $P\Delta$ effects are balanced by column and girder moments in the unbraced frame.

Classical methods for elastic buckling analysis of frames are widely available, particularly in Bleich [6.9], the *SSRC Guide* [6.8], and more recently in *Stability Design of Steel Frames* by Chen and Lui [14.1].

Detailed study of one-story frames is given by Goldberg [14.2], Galambos [14.3], Yura and Galambos [14.4], Lu [14.5], Zweig and Kahn [14.6], Lind [14.7], Schilling [14.8], Zweig [14.9], LeMessurier [12.50, 12.51], Scholz [14.10, 14.11], and Liu [14.12].

Multistory unbraced frames have been treated by Levi, Driscoll, and Lu [14.13], Korn and Galambos [14.14], Switzky and Wang [14.15], Liapunov [14.16], Cheong-Siat-Moy [14.17, 14.18], Cheong-Siat-Moy, Ozer, and Lu [14.19], Lu, Ozer, Daniels, Okten, and Morino [14.20], Haris [14.21], and Gaiotti and Smith [14.22]. Schultz [14.45] has reviewed methods of approximating lateral stiffness of stories in elastic frames. The space frame has been treated by Razzaq and Naim [14.23].

As a practical matter, unbraced multistory frames analysis is done using computer methods, at the very least using a first-order elastic analysis, wherein the original undeformed geometry is used. Computer capability has made possible sophisticated analyses. The link between research and practice has been discussed by White and Hajjar [14.24].

Matrix formulation using stiffness and/or flexibility coefficients is the standard procedure for frame analysis. Wang [14.25] and Halldorsson and Wang [14.26] illustrate the matrix formulation. Frame analysis methods are generally outside the scope of this text. Stiffness and flexibility coefficients are developed in Sec. 14.2 to provide introductory material; stiffness coefficients in the well-known slope-deflection method are used in explaining frame behavior in the remainder of this chapter. Elastic computer-aided analysis of multistory frames is generally available, for example, by Mehringer, Pierson, and Orbison [14.27].

Inelastic Buckling

Since some fibers of a cross-section usually yield prior to frame buckling occurring, inelastic buckling would usually govern the actual strength of a frame. Many studies of inelastic buckling have been conducted, including for braced frames the work of Ojalvo and Levi [14.28], and Levi, Driscoll, and Lu [14.29]. For unbraced frames, the reader is referred to Merchant [14.30], Yura and Galambos [14.4], Lu [14.5], Levi, Driscoll, and Lu [14.13], Korn and Galambos [14.14], Springfield and Adams [14.31], Daniels and Lu [14.32], Liapunov [14.16], Cheong-Siat-Moy [14.17–14.19, 14.33], Lu, Ozer, Daniels, Okten, and Morino [14.20], and Haris [14.21]. Inelastic buckling studies have been extended to the hybrid frame by Arnold, Adams, and Lu [14.34], and to space frames by McVinnie and Gaylord [14.35].

More recently, second-order inelastic analysis has been recommended by King, White, and Chen [14.46], and is recommended in Chap. 16 of the *Guide* [14.48].

14.2 ELASTIC BUCKLING OF FRAMES

The distinction between the braced and unbraced frame has been made in Sec. 14.1. In addition, there are two kinds of loading that may contribute to instability. For the *braced* frame of Fig. 14.2.1a, there are no primary bending moments; the only loading is axial compression. The critical load for such a situation is usually defined as a buckling load. No bending of members occurs until the buckling load is reached.

For Fig. 14.2.1b, bending moments exist when the structure is entirely stable. Due to the assumed "infinite rotational stiffness" of the rigid joints, moments are transmitted into column members. In addition to the primary bending moments in the column members, the axial compression induces secondary moments equal to P times the deflection. This was discussed in Chapter 12. Under certain combinations of axial compression and moment,

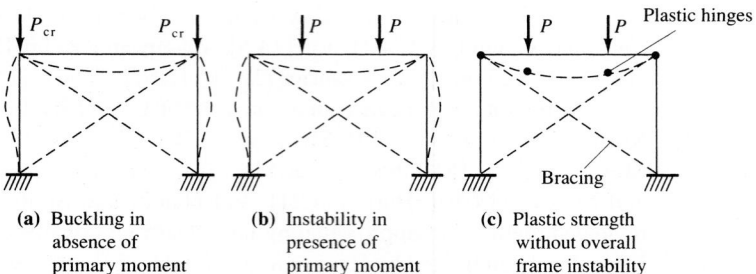

Figure 14.2.1
Strength of braced frames.

(a) Buckling in absence of primary moment

(b) Instability in presence of primary moment

(c) Plastic strength without overall frame instability

the lateral deflection of the column increases without achieving equilibrium; this is usually referred to as instability, or instability in the presence of primary bending moment.

When primary bending moments are present, enough plastic hinges may develop prior to achieving frame instability so that a mechanism forms, in which case for the braced frame the nominal strength is the plastic strength (Fig. 14.2.1c).

The strength of unbraced frames, as shown in Fig. 14.2.2, also may be separated into three categories; buckling in the absence of primary moment (Fig. 14.2.2a), instability in the presence of primary moment (Fig. 14.2.2b), and plastic strength (Fig. 14.2.2c). For the unbraced frame, achieving plastic strength frequently (though not always) means achieving a mechanism associated with overall geometric instability.

The remainder of this section treats the case of elastic buckling in the absence of primary bending moment, with the purpose of having the reader understand the difference in behavior of braced and unbraced frames.

To investigate the elastic stability of a rigid frame, it is first necessary to establish the relationships between end moments and end slopes for the individual frame members and then apply the compatibility of deformations requirement for rigid joints.

General Flexibility and Stiffness Coefficients for Beam-Columns

The reader is assumed to have some familiarity with the slope-deflection equations used in frame analysis when axial effect is not considered. For a prismatic section without axial load and without transverse load, as in Fig 14.2.3a,

$$M_a = \theta_a\left(\frac{4EI}{L}\right) + \theta_b\left(\frac{2EI}{L}\right) \tag{14.2.1a}$$

$$M_b = \theta_a\left(\frac{2EI}{L}\right) + \theta_b\left(\frac{4EI}{L}\right) \tag{14.2.1b}$$

Figure 14.2.2
Strength of unbraced frames.

(a) Buckling in absence of primary moment

(b) Instability in presence of primary moment

(c) Geometrically unstable at plastic strength

(a) Without axial load

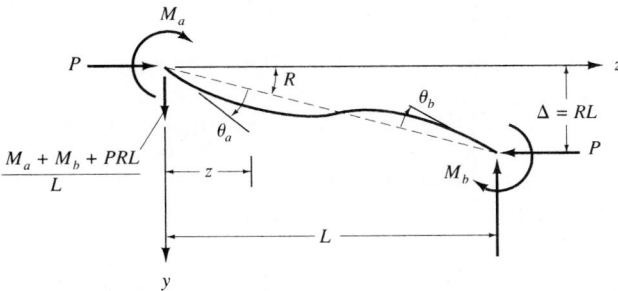

Figure 14.2.3
Definition of terms and
sign convention for slope-
deflection equations.

(b) Including axial compression

The following treatment, though similar to that for beam-columns in Sec. 12.2, is a more general approach that begins by expressing the moment at any section z of Fig. 14.2.3b:

$$-EI\frac{d^2y}{dz^2} = M_z = M_a + Py - \left(\frac{M_a + M_b + PRL}{L}\right)z \qquad (14.2.2)$$

$$\frac{d^2y}{dz^2} + \frac{P}{EI}y = -\frac{M_a}{EI} + \frac{M_a + M_b}{EIL}z + \frac{P}{EI}Rz \qquad (14.2.3)$$

Letting $k^2 = P/EI$, the solution for Eq. 14.2.3 is

$$y = A\sin kz + B\cos kz - \frac{M_a}{P} + \frac{M_a + M_b}{PL}z + Rz \qquad (14.2.4)$$

Applying the boundary conditions of zero deflection at $z = 0$ and $y = RL$ at $z = L$ gives

$$B = \frac{M_a}{P}$$

and

$$A = \frac{1}{P\sin kL}(M_a\cos kL + M_b)$$

Then

$$y = \frac{M_a}{P}\left[\frac{\sin k(L-z)}{\sin kL} - \frac{(L-z)}{L}\right] - \frac{M_b}{P}\left(\frac{\sin kz}{\sin kL} - \frac{z}{L}\right) + Rz \qquad (14.2.5)$$

Differentiating once to obtain the slope,

$$\frac{dy}{dz} = \frac{M_a}{P}\left[\frac{-k\cos k(L-z)}{\sin kL} + \frac{1}{L}\right] - \frac{M_b}{P}\left(\frac{k\cos kz}{\sin kL} - \frac{1}{L}\right) + R \qquad (14.2.6)$$

when $z = 0$, $\dfrac{dy}{dz} = \theta_a + R$ and when $z = L$, $\dfrac{dy}{dz} = \theta_b + R$.

Letting $\phi^2/L^2 = k^2 = P/EI$, θ_a and θ_b after some manipulation of terms may be expressed* as

$$\theta_a = \frac{M_a L}{EI}\left(\frac{\sin\phi - \phi\cos\phi}{\phi^2\sin\phi}\right) + \frac{M_b L}{EI}\left(\frac{\sin\phi - \phi}{\phi^2\sin\phi}\right) \qquad (14.2.7a)$$

$$\theta_b = \frac{M_a L}{EI}\left(\frac{\sin\phi - \phi}{\phi^2\sin\phi}\right) + \frac{M_b L}{EI}\left(\frac{\sin\phi - \phi\cos\phi}{\phi^2\sin\phi}\right) \qquad (14.2.7b)$$

where the ϕ functions (i.e., expressions within parentheses) are known as *flexibility coefficients*, f_{ii}, f_{ij}, f_{ji}, and f_{jj}. To obtain the beam-column counterparts to Eq. 14.2.1, solve Eqs. 14.2.7 (i.e., invert the matrix of coefficients) to obtain

$$M_a = \theta_a\frac{EI}{L}\left(\frac{\phi\sin\phi - \phi^2\cos\phi}{2 - 2\cos\phi - \phi\sin\phi}\right) + \theta_b\frac{EI}{L}\left(\frac{\phi^2 - \phi\sin\phi}{2 - 2\cos\phi - \phi\sin\phi}\right) \qquad (14.2.8a)$$

$$M_b = \theta_a\frac{EI}{L}\left(\frac{\phi^2 - \phi\sin\phi}{2 - 2\cos\phi - \phi\sin\phi}\right) + \theta_b\frac{EI}{L}\left(\frac{\phi\sin\phi - \phi^2\cos\phi}{2 - 2\cos\phi - \phi\sin\phi}\right) \qquad (14.2.8b)$$

where the ϕ functions are known as *stiffness coefficients*. The θ values are the end slopes measured with reference to the axis of the member.

Note that since $\phi^2 = PL^2/EI$, $\phi = 0$ means no axial compression and Eqs. 14.2.8 should become Eqs. 14.2.1. To verify the coefficient 4 when $\phi = 0$ in Eq. 14.2.8a, the numerator and denominator of the bracketed term must be differentiated four times in accordance with L'Hospital's Rule and then apply the $\phi = 0$ limit.

To simplify the use of Eqs. 14.2.8 for the slope-deflection solution of frame buckling problems, let the stiffness coefficients be referred to as S_{ii}, S_{ij}, S_{ji}, and S_{jj}. Because they are symmetrical $S_{ji} = S_{ij}$ and $S_{jj} = S_{ii}$. Thus Eqs. 14.2.8 become

$$M_a = \theta_a\frac{EI}{L}S_{ii} + \theta_b\frac{EI}{L}S_{ij} \qquad (14.2.9a)$$

$$M_b = \theta_a\frac{EI}{L}S_{ij} + \theta_b\frac{EI}{L}S_{ii} \qquad (14.2.9b)$$

Braced Frame—Slope-Deflection Method

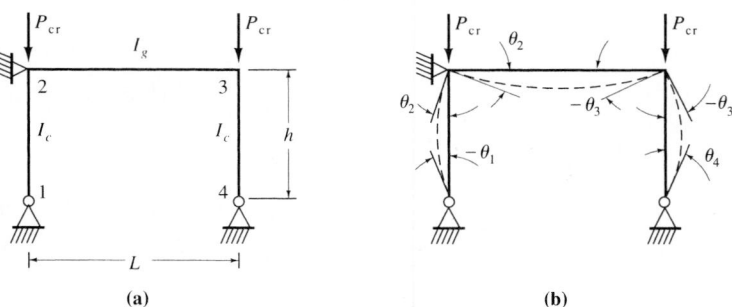

Figure 14.2.4
Braced frame—hinged base.

(a)   (b)

---

*The use of the symbol ϕ as the stability parameter throughout this chapter should not be confused with the resistance factor (strength reduction factor) ϕ used in Load and Resistance Factor Design.

The analysis of the rigid frame of Fig. 14.2.4 is presented using the slope-deflection method. Using clockwise rotations and rotational end moments as positive, the slope-deflection equations are as follows, using Eqs. 14.2.9:

$$M_{12} = \theta_1 \frac{EI_c}{h} S_{ii} + \theta_2 \frac{EI_c}{h} S_{ij} \tag{14.2.10}$$

$$M_{21} = \theta_1 \frac{EI_c}{h} S_{ij} + \theta_2 \frac{EI_c}{h} S_{ii} \tag{14.2.11}$$

$$M_{23} = \theta_2 \frac{EI_g}{L} S_{ii} + \theta_3 \frac{EI_g}{L} S_{ij} = \frac{2EI_g}{L} \theta_2 \tag{14.2.12}$$

Because no axial compression acts on member 2–3, $S_{ii} = 4$ and $S_{ij} = 2$ for Eq. 14.2.12. Use of symmetry reduces the number of moment equations from six (two for each frame member) to three, and $\theta_3 = -\theta_2$. The equilibrium equations for the joints are

$$M_{12} = 0 \tag{14.2.13}$$

$$M_{21} + M_{23} = 0 \tag{14.2.14}$$

Substitution of Eqs. 14.2.10 through 14.2.12 into Eqs. 14.2.13 and 14.2.14 gives

$$\left.\begin{array}{l} \theta_1 \dfrac{EI_c}{h} S_{ii} + \theta_2 \dfrac{EI_c}{h} S_{ij} = 0 \\[2ex] \theta_1 \dfrac{EI_c}{h} S_{ij} + \theta_2 \left(\dfrac{EI_c}{h} S_{ii} + \dfrac{2EI_g}{L} \right) = 0 \end{array}\right\} \tag{14.2.15}$$

Since the θ values cannot be zero when buckling occurs, the determinant of the coefficients of the θs must be zero. Thus the determinant, which is the stability equation, is

$$\left(\frac{EI_c}{h} \right)^2 \left(S_{ii}^2 + \frac{2I_g h}{I_c L} S_{ii} - S_{ij}^2 \right) = 0 \tag{14.2.16}$$

Since EI_c/h cannot be zero, the other term must be zero. Thus

$$S_{ii} - \frac{S_{ij}^2}{S_{ii}} = -\frac{2I_g h}{I_c L} \tag{14.2.17}$$

which in terms of ϕ becomes

$$\frac{\phi^2 \sin \phi}{\sin \phi - \phi \cos \phi} = -\frac{2 I_g h}{I_c L} \tag{14.2.18}$$

EXAMPLE 14.2.1

Determine the buckling load P_{cr} and effective length KL for a braced rigid frame, as in Fig. 14.2.4, which has $I_g = 2100$ in.^{4} (W24×76); $I_c = 796$ in.^{4} (W14×74); $L = 36$ ft; and $h = 14$ ft.

Solution:
The stability equation to be satisfied is Eq. 14.2.18, which inverted is

$$\frac{\sin \phi - \phi \cos \phi}{\phi^2 \sin \phi} = -\frac{I_c L}{2 I_g h} = \frac{-796(36)}{2(2100)14} = -0.487$$

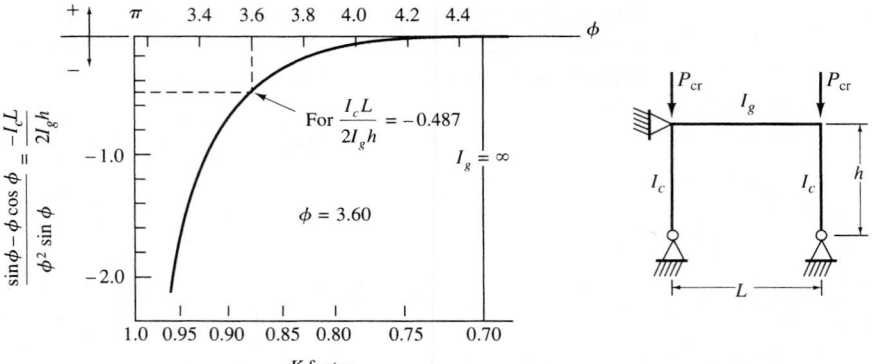

Figure 14.2.5
Braced frame—hinged base,
Example 14.2.1.

The smallest value of ϕ satisfying the buckling equation is the critical value; i.e., the one that governs. As I_g approaches zero, an isolated pinned column is indicated, with $\phi^2 = \pi^2$. As girder stiffness increases, ϕ should become greater than π. For an infinitely stiff girder, according to Fig. 6.9.1 which indicates $K = 0.7$, one would expect $\phi^2 = (\pi/0.7)^2 = (4.49)^2$.

The solution for $\phi = 3.60$ obtained by trial using values greater than π but less than 4.49 is shown in Fig. 14.2.5.

Comparing with the isolated pinned column,

$$\frac{\phi^2 EI}{h^2} = \frac{\pi^2 EI}{(Kh)^2} \tag{14.2.19}$$

it may be noted that the effective length factor K may be expressed

$$K = \frac{\pi}{\phi} \tag{14.2.20}$$

which for this problem means $K = \pi/3.60 = 0.87$. In other words, to account for frame buckling the column member could be designed using $0.87h$ as the pinned length of the isolated member. The K factor axis is also shown on Fig. 14.2.5 so that for various frame properties one could obtain the K factor directly. Note that a large increase in girder stiffness achieves only a small reduction in K.

The elastic buckling load is

$$P_{cr} = \frac{(3.60)^2 EI_c}{h^2} = \frac{(3.60)^2(29,000)796}{(14)^2(12)^2} = 10,600 \text{ kips} \quad\blacksquare$$

Unbraced Frame—Slope-Deflection Method

Next, the same frame that was treated as braced against sidesway at joint 2 is now to be analyzed as unbraced; i.e., with the horizontal support at joint 2 removed, as in Fig. 14.2.6. Even for the unbraced frame it is conceivable to consider a symmetrical buckling mode (Fig. 14.2.1a) exactly as when it was braced. It will be demonstrated that the sidesway buckling mode (Fig. 14.2.6b) will occur at a smaller load than the value obtained for the symmetrical case.

Figure 14.2.6
Unbraced frame—hinged
base. ($\theta' = \theta$ from Eqs.
14.2.9 and Fig. 14.2.3.)

Equations 14.2.9 are applied with the added factor that the axes of some members are tilted due to sidesway, so that letting θ represent *total* rotation Δ/h must be subtracted from it to get the end slope θ' measured with respect to the member axis.

$$M_{12} = \left(\theta_1 - \frac{\Delta}{h}\right)\frac{EI_c}{h}S_{ii} + \left(\theta_2 - \frac{\Delta}{h}\right)\frac{EI_c}{h}S_{ij} \tag{14.2.21}$$

$$M_{21} = \left(\theta_1 - \frac{\Delta}{h}\right)\frac{EI_c}{h}S_{ij} + \left(\theta_2 - \frac{\Delta}{h}\right)\frac{EI_c}{h}S_{ii} \tag{14.2.22}$$

$$M_{23} = \theta_2\frac{EI_g}{L}S_{ii} + \theta_3\frac{EI_g}{L}S_{ij} \tag{14.2.23}$$

With no axial compression considered on member 2–3, $S_{ii} = 4$ and $S_{ij} = 2$ for Eq. 14.2.23. This time if the structure is symmetrical, the sidesway buckling gives an antisymmetrical deflected curve; thus $\theta_3 = \theta_2$ and only three end moment equations are needed instead of six. The equilibrium equations are

$$M_{12} = 0 \tag{14.2.24}$$

$$M_{21} + M_{23} = 0 \tag{14.2.25}$$

which are the same as for the braced case. In addition, the sum of the base shears must be zero since no external horizontal force is acting. Because of antisymmetry $H_1 = -H_4$ so that only one column member needs to be considered. Referring to Fig. 14.2.6c,

$$H = \frac{M_{21} + P_{cr}\Delta}{h} = 0 \tag{14.2.26}$$

Equations 14.2.24, 14.2.25, and 14.2.26 then become

$$\left.\begin{array}{l}
\theta_1\dfrac{EI_c}{h}S_{ii} + \theta_2\dfrac{EI_c}{h}S_{ij} + \Delta\dfrac{EI_c}{h^2}(-S_{ii} - S_{ij}) = 0 \\[2ex]
\theta_1\dfrac{EI_c}{h}S_{ij} + \theta_2\left(\dfrac{EI_c}{h}S_{ii} + \dfrac{6EI_g}{L}\right) + \Delta\dfrac{EI_c}{h^2}(-S_{ii} - S_{ij}) = 0 \\[2ex]
\theta_1\dfrac{EI_c}{h}S_{ij} + \theta_2\dfrac{EI_c}{h}S_{ii} + \Delta\dfrac{EI_c}{h^2}(\phi^2 - S_{ii} - S_{ij}) = 0
\end{array}\right\} \tag{14.2.27}$$

Since θ_1, θ_2, and Δ cannot be zero (if buckling occurs), the determinate of the coefficients must be zero. Thus algebraic elimination of θ_1 and computation of the remaining four element determinant gives

$$\frac{EI_c}{h^2}(S_{ii}^2 - S_{ij}^2)\left[\frac{6EI_g}{L}\phi^2\frac{S_{ii}}{(S_{ii}^2 - S_{ij}^2)} + \frac{\phi^2 EI_c}{h} - \frac{6EI_g}{L}\right] = 0 \qquad (14.2.28)$$

Since $S_{ii} \neq S_{ij}$ the bracketed term must be zero. Substitution of the S_{ii} and S_{ij} expressions in terms of ϕ from Eqs. 14.2.8 gives the stability equation,

$$\frac{1}{\phi^2} - \frac{\sin\phi - \phi\cos\phi}{\phi^2\sin\phi} = \frac{I_c L}{6I_g h} \qquad (14.2.29)$$

or

$$\phi\tan\phi = \frac{6I_g h}{I_c L} \qquad (14.2.30)$$

EXAMPLE 14.2.2

Determine the buckling load P_{cr} and the effective length KL for the unbraced frame consisting of the same members and span as in Example 14.2.1.

Solution:

$$\phi\tan\phi = \frac{6I_g h}{I_c L} = \frac{6(2100)(14)}{796(36)} = +6.156$$

which is solved by trial to obtain $\phi = 1.354$, the smallest value satisfying the equation (see Fig. 14.2.7).

$$K = \frac{\pi}{\phi} = \frac{\pi}{1.354} = 2.32$$

Thus if the frame is unbraced it may be designed as an isolated member of effective length $2.32h$, whereas if it were braced the effective length would be $0.87h$. The range of K values may be studied for this frame in Fig. 14.2.7. With hinged bases, even with an infinitely rigid beam the effective length factor K is never less than 2 for an unbraced frame.

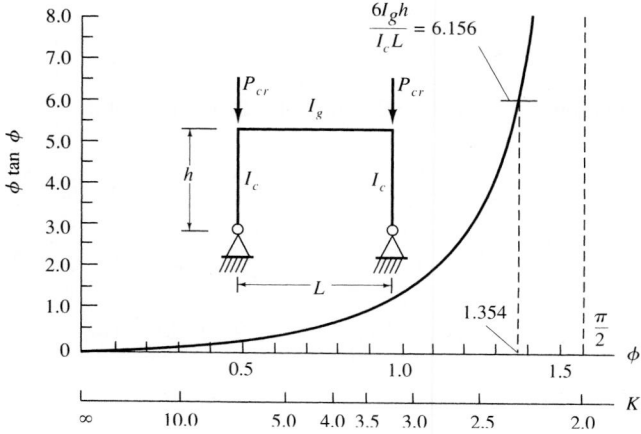

Figure 14.2.7
Unbraced frame—hinged base, Example 14.2.2.

The buckling load is

$$P_{cr} = \frac{(1.354)^2 EI_c}{h^2} = \frac{(1.354)^2 29,000(796)}{(14)^2\,(12)^2} = 1500 \text{ kips}$$

or about $\frac{1}{7}$ of what it was for the same frame when braced.   ∎

14.3   GENERAL PROCEDURE FOR EFFECTIVE LENGTH

For ordinary design, it is impractical to analyze an entire frame to determine its buckling strength and its effective length (equivalent pinned-end length). Thus it is desirable to have some general way of obtaining the K factor without the full analysis.

The usual procedure for obtaining the effective length factor K is to use the alignment charts of Julian and Lawrence [6.54]. These appear as Fig. 6.9.4 and their use is discussed in Sec. 6.9. The full mathematical development, including particularly the assumptions used, appears in Sec. 14.3 of the first and second editions of this book. Hajjar and White [14.47] have provided an excellent general treatment of effective length as related to unbraced frames. The practical application material from this section appears in Sec. 6.9. For unsymmetrical frames, Chu and Chow [14.36] have presented modifications to permit use of the alignment charts.

14.4   STABILITY OF FRAMES UNDER PRIMARY BENDING MOMENTS

Referring to Figs. 14.2.1 and 14.2.2, the distinction should be noted between (1) buckling in the absence of primary bending moments; i.e., *no* moments exist until buckling occurs; and (2) magnification of primary bending moments (which exist even without the presence of compressive loads) due to axial compression P times deflection Δ. The buckling of frames, which involved solving for the compressive loads P that make a determinant equal to zero, was treated in Sec. 14.2.

This section considers frame behavior when primary moments also exist. Consider a simple rectangular fixed base frame subject to primary moments as in Fig. 14.4.1a. If simple bending theory is used, and the stiffness of the girder is *not* considered reduced due to axial compression, any ordinary procedure of statically indeterminate analysis method will give the moments.

If compressive loads P are applied, there will be additional moments $P\Delta$, as in Fig. 14.4.1. The total effect may be treated as a magnification factor times the primary bending moments. This has been discussed in detail in Chapter 12 for members with no translation of joints (braced frames such as in Fig. 14.2.1b), and has been approximately presented for the unbraced frame beam-column in Sec. 12.5.

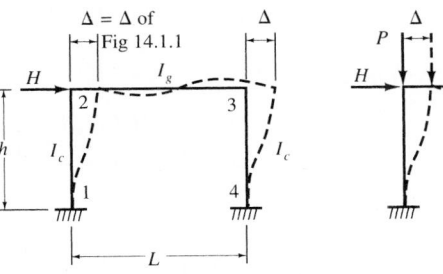

Figure 14.4.1
Frame with fixed bases.

(a) Primary bending

(b) Primary plus secondary bending

The following treatment is intended to provide an understanding of the mathematics and compare the actual magnification factor with the simple expression, Eq. 12.5.1, as used in AISC-C2.1b.

Primary Bending of Fixed-Base Frame

To determine magnification of primary moments, the primary moments must first be determined. While design practice could use any method of elastic frame analysis, the slope-deflection method is used here so that the reader may compare what follows with the modifications required to include the $P\Delta$ effect.

For the frame of Fig. 14.4.1a, use Eqs. 14.2.1 noting that θ is the angle measured from the axis of the member, and that for θ to represent the total rotation, Δ/h must be subtracted from it. The end moment equations are

$$\left.\begin{aligned}
M_{12} &= \left(\theta_1 - \frac{\Delta}{h}\right)\frac{4EI_c}{h} + \left(\theta_2 - \frac{\Delta}{h}\right)\frac{2EI_c}{h} \\
M_{21} &= \left(\theta_1 - \frac{\Delta}{h}\right)\frac{2EI_c}{h} + \left(\theta_2 - \frac{\Delta}{h}\right)\frac{4EI_c}{h} \\
M_{23} &= \theta_2\frac{4EL_g}{L} + \theta_3\frac{2EI_g}{L}
\end{aligned}\right\} \quad (14.4.1)$$

Because of symmetry in the structure,

$$\left.\begin{aligned}
M_{43} &= M_{12} \\
M_{34} &= M_{21} \\
\theta_2 &= \theta_3
\end{aligned}\right\} \quad (14.4.2)$$

Also, because of the fixed base, $\theta_1 = 0$.

For equilibrium, it is required that

$$M_{21} + M_{23} = 0 \quad (14.4.3)$$

and also the shears on the columns must equal H:

$$\frac{M_{12} + M_{21}}{h} + \frac{M_{43} + M_{34}}{h} + H = 0 \quad (14.4.4)$$

Substitution of Eqs. 14.4.1 into Eqs. 14.4.3 and 14.4.4 gives

$$\left.\begin{aligned}
\theta_2\left(\frac{4EI_c}{h} + \frac{6EI_g}{L}\right) - \frac{\Delta}{h}\left(\frac{6EI_c}{h}\right) &= 0 \\
\theta_2\left(\frac{12EI_c}{h}\right) - \frac{\Delta}{h}\left(\frac{24EI_c}{h}\right) &= -Hh
\end{aligned}\right\} \quad (14.4.5)$$

Solving for θ_2 and Δ/h gives

$$\theta_2 = \frac{Hh^2}{4EI_c}\left(\frac{I_cL/I_gh}{I_cL/I_gh + 6}\right) \quad (14.4.6)$$

and

$$\frac{\Delta}{h} = \frac{Hh^2}{12EI_c}\left(\frac{2I_cL/I_gh + 3}{I_cL/I_gh + 6}\right) \quad (14.4.7)$$

Substituting into the M_{12} and M_{21} Eqs. 14.4.1, gives

$$M_{12} = \frac{-Hh}{2}\left(\frac{I_cL/I_gh + 3}{I_cL/I_gh + 6}\right) \tag{14.4.8}$$

and

$$M_{21} = \frac{-Hh}{2}\left(\frac{3}{I_cL/I_gh + 6}\right) \tag{14.4.9}$$

M_{12} and M_{21} are the primary moments which are magnified when the axial loads P are applied.

Magnification Factor for Fixed-Base Frame

The frame of Fig. 14.4.1b is investigated next to determine the magnified moments M_{12} and M_{21}. Use the slope-deflection equations, Eqs. 14.2.9, that include effect of axial compression on stiffness. Again the rotation angle Δ/h must be subtracted from the full angle to obtain the value measured from the member axis:

$$\left.\begin{aligned}
M_{12} &= \left(\theta_1 - \frac{\Delta}{h}\right)\frac{EI_c}{h}S_{ii} + \left(\theta_2 - \frac{\Delta}{h}\right)\frac{EI_c}{h}S_{ij} \\
M_{21} &= \left(\theta_1 - \frac{\Delta}{h}\right)\frac{EI_c}{h}S_{ij} + \left(\theta_2 - \frac{\Delta}{h}\right)\frac{EI_c}{h}S_{ii} \\
M_{23} &= \theta_2\frac{4EI_g}{L} + \theta_3\frac{2EI_g}{L}
\end{aligned}\right\} \tag{14.4.10}$$

whereas for primary moment determination, the compression effect on the girder stiffness is not considered. The same symmetry and fixed-base condition ($\theta_1 = 0$) as in Eqs. 14.4.2 apply.

The equilibrium conditions are

$$M_{21} + M_{23} = 0 \tag{14.4.11}$$

the same as when the axial force P was not present; and

$$\frac{M_{12} + M_{21}}{h} + \frac{M_{43} + M_{34}}{h} + \frac{2P\Delta}{h} + H = 0 \tag{14.4.12}$$

where the term $2P\Delta/h$ is added to the previous condition of Eq. 14.4.4. Let $P = \phi^2EI_c/h^2$.

Substitution of Eqs. 14.4.10 into Eqs. 14.4.11 and 14.4.12, and recalling $M_{43} = M_{12}$ and $M_{34} = M_{21}$, gives

$$\left.\begin{aligned}
\theta_2\left(S_{ii} + \frac{6I_gh}{I_cL}\right) - \frac{\Delta}{h}(S_{ii} + S_{ij}) &= 0 \\
\theta_2[2(S_{ii} + S_{ij})] - \frac{\Delta}{h}[4(S_{ii} + S_{ij}) - 2\phi^2] &= \frac{-Hh^2}{EI_c}
\end{aligned}\right\} \tag{14.4.13}$$

Equations 14.4.13 compare with Eqs. 14.4.5 when Eqs. 14.4.5 are divided by EI_c/h.

Solving Eqs. 14.4.13 for θ_2 and Δ/h gives

$$\frac{\Delta}{h} = \left(\frac{S_{ii} + 6I_gh/I_cL}{S_{ii} + S_{ij}}\right)\theta_2 \tag{14.4.14}$$

and

$$\theta_2 = \frac{Hh^2}{2EI_c}\left(\frac{(S_{ii} + S_{ij})(I_cL/I_gh)}{(S_{ii}^2 - S_{ij}^2 - \phi^2 S_{ii})(I_cL/I_gh) + 6[2(S_{ii} + S_{ij}) - \phi^2]}\right) \quad (14.4.15)$$

Note that Eq. 14.4.15 becomes Eq. 14.4.6 when $P = 0$. When $P = 0$, $\phi^2 = Ph^2/EI_c = 0$, $S_{ii} = 4$, and $S_{ij} = 2$.

Substituting Eqs. 14.4.14 and 14.4.15 into Eqs. 14.4.10 gives

$$M_{12} = \theta_2 \frac{EI_c}{h}(S_{ij} - S_{ii} - 6I_gh/I_cL)$$

$$= \frac{-Hh}{2}\left\{\frac{(S_{ii} + S_{ij})(S_{ii} - S_{ij} + 6I_gh/I_cL)(I_cL/I_gh)}{(S_{ii}^2 - S_{ij}^2 - \phi^2 S_{ii})I_cL/I_gh + 6[2(S_{ii} + S_{ij}) - \phi^2]}\right\}$$

$$M_{12} = \frac{-Hh}{2}\left\{\frac{(S_{ii}^2 - S_{ij}^2)I_cL/I_gh + 6(S_{ii} + S_{ij})}{(S_{ii}^2 - S_{ij}^2 - \phi^2 S_{ii})I_cL/I_gh + 6[2(S_{ii} + S_{ij}) - \phi^2]}\right\} \quad (14.4.16)$$

and

$$M_{21} = \frac{-Hh}{2}\left\{\frac{6(S_{ii} + S_{ij})}{(S_{ii}^2 - S_{ij}^2 - \phi^2 S_{ii})I_cL/I_gh + 6[2(S_{ii} + S_{ij}) - \phi^2]}\right\} \quad (14.4.17)$$

which become the same as Eqs. 14.4.8 and 14.4.9 when $\phi = 0$; i.e., no axial compression.

The magnification factor, the *sway magnifier B_2* in Load and Resistance Factor terminology, is the ratio of the moment including the $P\Delta$ effect to the primary moment without $P\Delta$. Thus, dividing Eq. 14.4.16 by Eq. 14.4.8 gives the magnifier B_2 for the moment at the bottom of the column,

$$(B_2)_{\text{bottom}} = \frac{\text{Eq. } 14.4.16}{\text{Eq. } 14.4.8} \quad (14.4.18)$$

and the magnifier B_2 for the moment at the top of the column,

$$(B_2)_{\text{top}} = \frac{\text{Eq. } 14.4.17}{\text{Eq. } 14.4.9} \quad (14.4.19)$$

Note is made that the magnifier B_2 is applicable only in rectangular frames.

In actual practice it would be easier to do the second-order analysis to obtain the moments as obtained in this section than to compute correctly the magnification factor. The purpose here is to show the relative correctness of the sway magnification factor B_2 as given by AISC-C2.1b for the LRFD Method.

$$B_2 = \frac{1}{1 - \dfrac{\Sigma P_u}{\Sigma P_{e2}}} = \frac{1}{1 - \alpha} \quad (14.4.20)$$

where $\alpha = \Sigma P_u / \Sigma P_{e2}$

ΣP_u = sum of all factored loads in the columns resisting sway

ΣP_{e2} = sum of the Euler buckling loads in the columns resisting sway

$P_{e2} = \pi^2 EI/(KL)^2$

K = effective length factor in the plane of bending ≥ 1.0

Note that columns resisting sway in a story, such as the two columns in Fig. 14.4.1, cannot act independently; thus, α must include the effects on all columns resisting sway in the story. The term P_{e2} is the buckling load occurring in the absence of primary moment. Following the procedure of Sec. 14.3, P_{e2} may be determined for the frame of Fig. 14.4.1 by letting $H = 0$; then setting the determinant of the coefficients in Eqs. 14.4.13 equal to zero. Substitution of the S_{ii} and S_{ij} in terms of the stability parameter ϕ into that determinant will finally give the stability equation,

$$\phi \sin \phi \cos \phi \left(\frac{\cot \phi}{\phi} - \frac{I_c L}{6 I_g h} \right) = 1 \tag{14.4.21}$$

which for example, if $I_c L / I_g h = 1$, one obtains $\phi_{cr} = 0.865\pi$. For the fixed-base frame, the buckling load would be

$$P_{e2} = \frac{(0.865\pi)^2 \, EI_c}{h^2}$$

and

$$\alpha = \frac{P}{P_{e2}} = \left(\frac{\phi}{0.865\pi} \right)^2, \quad \text{for} \quad \frac{I_c L}{I_g h} = 1.0$$

The effective length factor K is

$$K = \frac{\pi}{\phi_{cr}} = \frac{1.0}{0.865} = 1.16$$

Table 14.4.1 provides a comparison of the theoretical magnification factor, Eq. 14.4.19 for the top of the column, with the AISC LRFD Eq. 14.4.20.

Comparison of the values for B_2 in Table 14.4.1 shows that AISC Eq. 14.4.20 compares favorably with the theoretical values.

Actual "unbraced frames" always have attachments which are designed as nonstructural elements; however, such items do actually contribute some bracing to the structure.

TABLE 14.4.1 Comparison of Theoretical Magnification Factor B_2 with $B_2 = 1/(1 - \alpha)$ for LRFD[†]

α	$\frac{I_c L}{I_g h} = 0.5$		$\frac{I_c L}{I_g h} = 1.0$		$\frac{I_c L}{I_g h} = 5.0$		AISC LRFD $\frac{1}{1-\alpha}$
	ϕ	Exact*	ϕ	Exact*	ϕ	Exact*	
0.10	0.92	1.102	0.86	1.107	0.66	1.114	1.111
0.20	1.30	1.230	1.22	1.242	0.94	1.257	1.250
0.30	1.59	1.394	1.49	1.415	1.15	1.441	1.429
0.40	1.84	1.612	1.72	1.645	1.32	1.687	1.667
0.50	2.05	1.918	1.92	1.968	1.48	2.031	2.000
0.60	2.25	2.377	2.10	2.453	1.62	2.548	2.500
0.70	2.43	3.143	2.27	3.261	1.75	3.410	3.333

[†] $\alpha = \Sigma P_u / \Sigma P_{e2}$

*B_2 from Eq. 14.4.19.

In other words, a real building can never be as flexible as the skeleton elastic frame. In addition, such items as exterior walls, partitions, and stairways, all tend to add to the overall stiffness.

For further study of the instability of frames in the presence of primary moment, the reader is referred to the work of Bleich [6.9], Lu [14.37], McGuire [3.11], Galambos [14.38], and the *SSRC Guide* [6.8].

14.5 BRACING REQUIREMENTS—BRACED FRAME

One of the decisions faced by the designer is the determination of whether a frame is braced or unbraced. Most efficient use of material in compression members is obtained when the frame is braced so that sidesway buckling or instability cannot occur. Some design guidelines are provided by Galambos [14.39]. As to what constitutes bracing, AISC-2.3a indicates that a frame is braced when lateral stability is provided by "diagonal bracing, shear walls or equivalent means."

The *amount* of stiffness required to prevent sidesway buckling has recently been defined in AISC-Appendix 6 of the *AISC Specification* [1.13]. It has been suggested [14.39] that "in many instances nonstructural building elements, curtain walls for example, can provide the necessary stiffness against sidesway buckling."

The following presentation from Galambos [14.38] provides some theoretical background on the strength requirements for diagonal braces. Before AISC-Appendix 6, this approach had been used by the authors to estimate the required strength of the diagonal brace. This discussion complements that in Sec. 9.13, where the emphasis was on beam and column bracing.

Stiffness Required from Bracing

It is the objective to use bracing to convert Fig. 14.2.2a into Fig. 14.2.1a. A simple and conservative procedure is to idealize the braced frame, as in Fig. 14.5.1. The following assumptions are made:

1. Columns do not participate in resisting sidesway.

2. Columns are hinged at ends.

3. Bracing acts independently as a spring at the top of columns.

Equilibrium using summation of moments about point 1 of Fig. 14.5.1 gives

$$P\Delta + P(\Delta + L) - \beta\Delta h - V_2 L = 0$$

$$V_2 = P + 2P\left(\frac{\Delta}{L}\right) - \beta\Delta\frac{h}{L} \qquad (14.5.1)$$

Figure 14.5.1
Idealized bracing
arrangements.

(a)   (b)

Similarly, summation of moments about point 4 gives

$$V_1 = P - 2P\left(\frac{\Delta}{L}\right) + \beta\Delta\frac{h}{L} \qquad (14.5.2)$$

Moments about the hinges at 2 and 3 for free-body diagrams of columns 1-2 and 4-3, respectively, give

$$H_1 = V_1\frac{\Delta}{h} \qquad (14.5.3)$$

$$H_2 = V_2\frac{\Delta}{h} \qquad (14.5.4)$$

Summation of horizontal forces gives

$$H_1 + H_2 - \beta\Delta = 0 \qquad (14.5.5)$$

Substitution of Eqs. 14.5.1 and 14.5.2 into Eqs. 14.5.3 and 14.5.4, and then into Eq. 14.5.5, gives

$$\frac{\Delta}{h}(2P - \beta h) = 0 \qquad (14.5.6)$$

where Eq. 14.5.6 is the approximate buckling equation. For a simple rectangular frame, as in Fig. 14.2.4, the exact buckling equation is given by Eq. 14.2.18.

From Eq. 14.5.6, substituting the nominal column strength P_n for P, gives for *two* columns

$$\beta = \frac{2P_n}{h} \qquad (14.5.7)$$

For a frame of several bays, the buckling load P (equal to nominal strength P_n) must include the sum of the loads carried on all columns resisting sway; thus Eq. 14.5.7 becomes

$$\text{Required } \beta = \frac{\Sigma P_n}{h} \qquad (14.5.8)$$

Equation 14.5.8 gives the stiffness (spring constant) β required for all bracing for the columns involved in ΣP_n. The ΣP_n includes the loads (factored loads $\Sigma P_u/\phi_c$) to be carried on all columns resisting sway.

Bracing Provided by Stiffer Members in an Overall Unbraced System

Overall sidesway buckling can occur only when the total lateral or sidesway resistance to horizontal movement is overcome. When the loading on individual members is less than their strength, the reserve strength can be utilized to provide bracing for other members. Yura [6.55] has presented the generally accepted design approach. More recently, Cheong-Siat-Moy [14.40], de Buen [14.41], Geschwindner [14.42], and Aristizabal-Ochoa [14.43] have treated the concept.

Examine the unbraced frame of Fig. 14.5.2a. Assuming the columns hinged at the junction with the beam, $K = 2.0$ for this cantilever situation. Members A, B, and C were proportioned for the axial loads 100, 300, and 400 kips, respectively. When sidesway occurs as in Fig. 14.5.2b, moments $P\Delta$ are produced at the base and the total load is the 800 kips.

If the system of Fig. 14.5.2 were a *braced* one, the practical effective length factor K would be 1.0 instead of 2.0, and the strengths of the members would be four times as great; 400, 1200, and 1600 kips, respectively, at columns A, B, and C.

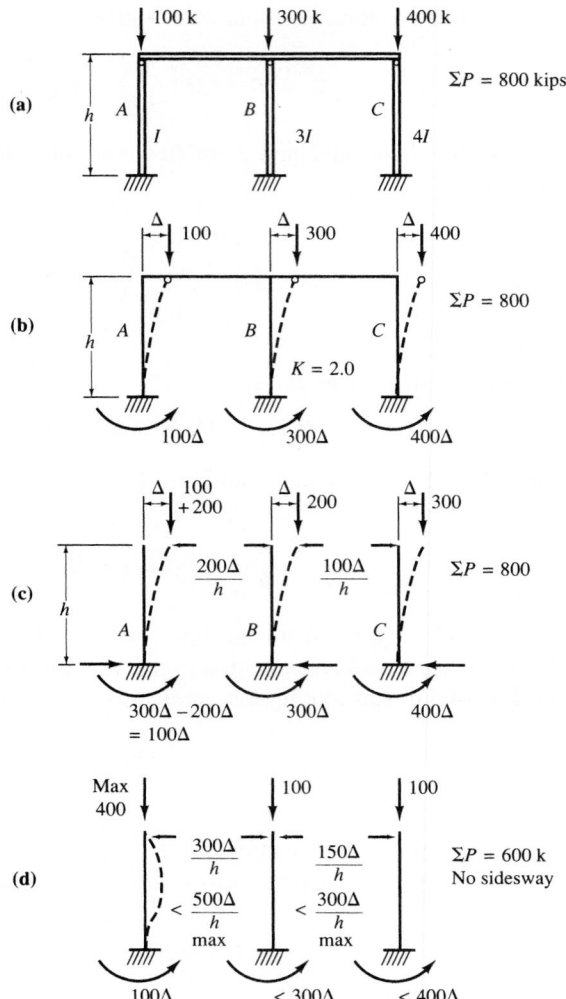

Figure 14.5.2
Stiffer members bracing less
stiff member in frame.

Suppose the loads applied to columns B and C are only 200 and 300 kips instead of the maximum strengths of 300 and 400 kips. As in Fig. 14.5.2c, columns B and C will not sidesway buckle until the moments developed at the bottom reach 300Δ and 400Δ, respectively. Those columns can therefore sustain horizontal forces at their tops sufficient to cause sidesway buckling. Column C and column B can each develop $100\Delta/h$ as a shear. These resistances are additive and act as horizontal restraint at the top of column A. Thus column A can carry the 100 kips original load, plus 200 kips as a result of the extra bracing from columns B and C. The total frame load is still 800 kips.

The maximum increase in strength that horizontal bracing can provide is the strength the member will have when its ends cannot translate with respect to one another. In other words, the shear developed at the top to prevent horizontal movement is the maximum resistance from other members that is usable. For instance, in Fig. 14.5.2d, the horizontal shear that can be developed based on columns B and C is $500\Delta/h$; however, when the shear at the top of column A is $300\Delta/h$ and the load on column A is 300 kips, no movement occurs and the member is fully braced. The maximum capacity for member A in a braced system is 400 kips and the resisting lateral force from columns B and C must be at least $400\Delta/h$, which is less than the maximum resistance of $500\Delta/h$.

As stated by Yura [6.55], "... the total gravity loads which produce sidesway can be distributed among the columns in a story in any manner. Sidesway will not occur until the total frame load on a story reaches the sum of the *potential* individual column loads for the *unbraced* frame." There still is the limitation that an individual column can carry no more than it could carry in a braced frame; i.e., with $K = 1$. Salem [14.6] has shown theoretically this procedure is valid regardless of the type of framing and ratio of member sizes.

The procedure described above will determine the load for which each column is to be designed; the columns having sway resistance carry more than their tributary compressive load. The modified loads are then used with effective length factors K determined in some "rational" manner, usually from the alignment chart, Fig. 6.9.4b (see discussion of effective length factors K in Sec. 6.9).

Alternative to adjusting the column loads to reflect the differences in sway resistance, the columns providing sway resistance could be designed for their actual loads, but use an *adjusted* effective length factor K. This approach has been presented by Geschwindner [14.42]. If the column supporting the "leaner" column (i.e., a column idealized as having zero rotational restraint at its ends) must carry its actual load P_u along with a load Q_u from the nonsway-resisting columns, then the Euler buckling load will be

$$P_u + Q_u = \frac{\pi^2 EI}{K_0^2 L^2} \tag{14.5.9}$$

where K_0 is the actual effective length factor based on actual frame members.

If one wishes the design of the column to be based on the actual column load P_u, then an adjusted effective length factor K_n must be used. The Euler equation becomes

$$P_u = \frac{\pi^2 EI}{K_n^2 L^2} \tag{14.5.10}$$

Solving Eq. 14.5.10 for $\pi^2 EI/L^2$ and substituting into Eq. 14.5.9 gives

$$K_n = K_0 \sqrt{\frac{P_u + Q_u}{P_u}} \tag{14.5.11}$$

where P_u is the actual factored gravity load to be carried by the frame sway-resisting column, and Q_u is the total factored gravity on all the "leaner" columns whose sway-resistance must be provided by the column carrying P_u.

Example 15.3.4 employing this concept is given in Chapter 15 on the design of rigid frames.

Diagonal Bracing

When cross-bracing is used, it is generally assumed that it can only act in tension; i.e., the diagonal which would be in compression buckles slightly and becomes inactive. Under a horizontal force F the diagonal brace in Fig. 14.5.3 must carry the force

$$\text{Brace force} = \frac{F}{\cos \alpha} \tag{14.5.12}$$

and the elongation of the brace, $\Delta \cos \alpha$, is

$$\text{Elongation} = \frac{(\text{brace force})(\text{brace length})}{(\text{area of brace})E} \tag{14.5.13}$$

Figure 14.5.3
Deformation of the bracing
member.

or

$$\Delta \cos \alpha = \frac{(F/\cos \alpha)\sqrt{h^2 + L^2}}{A_b E} \qquad (14.5.14)$$

Solving for F, using $\cos \alpha = L/\sqrt{h^2 + L^2}$, gives

$$F = \frac{A_b E L^2}{(h^2 + L^2)^{3/2}} \Delta \qquad (14.5.15)$$

Since $F = \beta \Delta$ according to Fig. 14.5.1b,

$$\beta = \frac{A_b E L^2}{(h^2 + L^2)^{3/2}} \qquad (14.5.16)$$

Substitution of $\beta = \Sigma P_n/h$ into Eq. 14.5.16 and solving for A_b gives

$$A_b = \frac{\left[1 + \left(\dfrac{L}{h}\right)^2\right]^{3/2} \Sigma P_n}{\left(\dfrac{L}{h}\right)^2 E} \qquad (14.5.17)$$

AISC Requirements for Frame Bracing*

It is always implied whenever the frame is defined as a braced that the AISC requirement for bracing is satisfied. AISC-Appendix 6 gives the requirements for the different types of column bracing that qualifies the overall frame system to be categorized as braced. There are two types of bracing: nodal bracing and relative bracing. Nodal bracing involves restraining the columns against a rigid abutment. Relative bracing utilizes diagonal braces and shear walls to reduce the relative lateral sway between nodes of rigid frames. Figure 14.5.4 shows both types of column/frame bracing.

The objective of either type of bracing, according to AISC-Appendix 6, is to allow for the design of members based on the length between brace points and using an effective length factor $K = 1.0$. Theoretically, it requires infinite stiffness to achieve a complete non-sway condition.

---

*In the design requirements sections of this chapter ϕ is the strength reduction (resistance) factor.

Figure 14.5.4
Types of column bracing.

(a) Nodal bracing   (b) Relative bracing

The design requirements for the relative (diagonal) bracing involves both stiffness and strength. Both requirements must be satisfied for adequate performance. These requirements are as follows.

1. Relative bracing

(a) Strength requirements:

The required brace strength is defined as 0.4% of the required axial compressive strength, using either LRFD or ASD. AISC-Appendix 6.2 gives this requirement as follows.

$$P_{\text{br}} = 0.004P_r \qquad (14.5.18)$$

(b) Stiffness requirements:

The brace stiffness β_{br} requirement was chosen to be twice the critical stiffness, when the relative sway between nodal points is at $0.002L$, and is given as follows

$$\beta_{\text{br}} = \frac{1}{\phi}\left(\frac{2P_r}{L_b}\right)(\text{LRFD}) \qquad (14.5.19a)$$

$$\beta_{\text{br}} = \Omega\left(\frac{2P_r}{L_b}\right)(\text{ASD}) \qquad (14.5.19b)$$

where $\phi = 0.75\,(\text{LRFD})$ $\Omega = 2.00\,(\text{ASD})$

2. Nodal bracing

(a) Strength requirements:

$$P_{\text{br}} = 0.01P_r \qquad (14.5.20)$$

(b) Stiffness requirements:

The brace stiffness β_{br} requirement was chosen to be the most severe case with multiple braces, and is given as follows

$$\beta_{\text{br}} = \frac{1}{\phi}\left(\frac{8P_r}{L_b}\right)(\text{LRFD}) \qquad (14.5.21a)$$

$$\beta_{\text{br}} = \Omega\left(\frac{8P_r}{L_b}\right)(\text{ASD}) \qquad (14.5.21b)$$

where $\phi = 0.75\,(\text{LRFD})$ $\Omega = 2.00\,(\text{ASD})$

EXAMPLE 14.5.1

Determine the area for the diagonal brace shown in Fig. 14.5.3 required to convert the unbraced frame of Example 14.2.2, Fig. 14.2.6a, into a braced frame. (a) Use Eq. 14.5.17 and (b) Use AISC-Appendix 6.

Solution:

(a) For a frame with $I_g = 2100$ in.^{4}, $I_c = 796$ in.^{4}, $L = 36$ ft, and $h = 14$ ft, the *braced* frame buckling load from Example 14.2.1 is

$$P_{cr} = \frac{(3.60)^2 E I_c}{h^2}$$

Since there are two columns, $\Sigma P_n = 2P_{cr}$. Substituting ΣP_n into Eq. 14.5.17 gives the required area A_b of the bracing as

$$A_b = \frac{\left[1 + \left(\dfrac{36}{14}\right)^2\right]^{3/2} 2(3.60)^2 E(796)}{\left(\dfrac{36}{14}\right)^2 E(14)^2(144)}$$

$$= \frac{2(7.6)^{3/2}(12.95)(796)}{6.6(196)(144)} = 2.3 \text{ sq in.}$$

(b) Use AISC-Appendix 6 for relative bracing requirements.

   (i) The required horizontal brace force

$$P_{br} \geq 0.004 P_r \qquad\qquad [14.5.22]$$

The required strength P_r is taken to be the design strength of both columns in the braced frame.

$$P_r = \Sigma \phi P_n = 2(0.9)(10,600) = 19,080 \text{ kips}$$

$$P_{br} = 0.004(19,080) = 76.3 \text{ kips}$$

$$P_{br(angle)} = 76.3(38.6/36) = 81.9 \text{ kips}$$

Assumming $F_v = 36$ ksi for the brace bar,

$$A_{br} = \frac{81.9}{0.9(36)} = 2.52 \text{ sq in.}$$

   (ii) The required horizontal stiffness for the brace

$$\beta_{br} \geq \frac{1}{\phi}\left(\frac{2P_r}{L_b}\right)$$

$$\beta_{br} = \frac{1}{0.75}\left(\frac{2(19,080)}{14}\right) = 3630 \text{ kips/ft}$$

$$\beta_{br} = \frac{A_{br} E}{L_{br}}\cos^2\theta = 3630 \text{ kips/ft} \qquad\qquad [14.5.23]$$

Rearranging gives the required brace area as follows,

$$A_{br} = \frac{3630(38.6)}{29,000\left(\dfrac{36}{38.6}\right)^2} = 5.55 \text{ sq in.}\quad \text{controls}$$

The brace area required is about 25% of the column cross-sectional area (21.8 in. for W14×74). It is noted that the example unbraced frame was very flexible with a higher than usual effective length factor K of 2.32; thus a some-what heavier than usual brace was required. ■

14.6 OVERALL STABILITY WHEN PLASTIC HINGES FORM

General

This chapter has focused primarily on basic concepts of frame stability, using the single-story frame as an example. AISC-C2 states. "Except as permitted in Section C2.2b, *required strengths* shall be determined using a *Second-order analysis* . . ." For one- and two-story frames, the second-order $P\Delta$ effect rarely affects the design, however. When plastic analysis is used as permitted under AISC-Appendix 1, consideration of second-order effects will be impractical and will offset any advantage of using plastic analysis. Traditionally, in one- and perhaps two-story lightly loaded frames, the $P\Delta$ effect has been neglected (seemingly without detrimental effects on the structure).

For multistory *braced* frames, the design of the bracing system must include the $P\Delta$ effect. For multistory *unbraced* frames, the $P\Delta$ effect must be included directly in the calculations for maximum strength.

Braced Frames

Braced frames are usually designed to cause any plastic hinges associated with the failure mechanism to form in the girders. For one- and two-story frames designed using plastic design, the $P\Delta$ effect must be included under AISC-C2. AISC-Appendix 1 requires the braces to remain elastic under the service load. Also, the required axial strength for columns and compression braces shall not exceed $\phi_c(0.85F_yA_g)$. The columns are then designed as beam-columns in accordance with concepts treated in Chapter 12. Design procedures for one- and two-story braced frames are presented in Chapter 15.

Multistory braced frames may be designed in accordance with *Plastic Design of Braced Multistory Steel Frames* [14.44].

Unbraced Frames

According to AISC-Appendix-1.5.2 for moment frames, the required axial strength of columns shall not exceed $\phi_c(0.75F_yA_g)$. This is a cautionary limit since little research has been done to explore the inelastic rotational capacity at high levels of axial force.

The treatment of plastic analysis for multistory unbraced frames is beyond the scope of this text; there is currently (2008) no generally recognized acceptable procedure for such an analysis. The reader is referred to work of Cheong-Siat-Moy, Ozer, and Lu [14.19], Springfield and Adams [14.31], Liapunov [14.16], Daniels and Lu [14.32], and LeMessurier [12.50, 12.51] for techniques of evaluating the strength of multistory unbraced frames.

SELECTED REFERENCES

14.1. W. F. Chen and E. M. Lui. *Stability Design of Steel Frames.* Boca Raton, FL:CRC Press, 1991.

14.2. John E. Goldberg. "Buckling of One-Story Frames and Buildings," *Journal of the Structural Division,* ASCE, **86,** ST10 (October 1960), 53–85. Also *Transactions,* **126** (1961), Part II, 482–515.

14.3. Theodore V. Galambos. "Influence of Partial Base Fixity on Frame Stability," *Journal of the Structural Division,* ASCE, **86,** ST5 (May 1960), 85–108. Also *Transactions,* **126** (1961), Part II, 929–969.

14.4. Joseph A. Yura and Theodore V. Galambos. "Strength of Single-Story Steel Frames," *Journal of the Structural Division,* ASCE, **91,** ST5 (October 1965), 81–101.

14.5. Le-Wu Lu. "Inelastic Buckling of Steel Frames," *Journal of the Structural Division,* ASCE, **91,** ST6 (December 1965), 185–214.

14.6. Alfred Zweig and Albert Kahn. "Buckling Analysis of One-Story Frames," *Journal of the Structural Division,* ASCE, **94,** ST9 (September 1968), 2107–2134. Disc. by Adel Helmy Salem, **95,** ST5 (May 1969), 1017–1029.

14.7. Niels C. Lind. "Simple Illustration of Frame Instability," *Journal of the Structural Division,* ASCE, **103,** ST1 (January 1977), 1–8.

14.8. C. G. Schilling. "Buckling of One-Story Frames," *Engineering Journal,* AISC, **20,** 2 (Second Quarter 1983), 49–57. Disc. by Dan S. Correnti, Alfred Zweig, John Springfield, and author, **21,** 4 (Fourth Quarter 1984), 207–215.

14.9. Alfred Zweig. "Force Method for Frame Buckling Analysis," *Journal of Structural Engineering,* ASCE, **110,** 8 (August 1984), 1893–1912.

14.10. H. Scholz. "*P*-Delta Effect in Elastic Analysis of Sway Frames," *Journal of Structural Engineering,* ASCE, **113,** 3 (March 1987), 534–545. Errata, **113,** 12 (December 1987), 2525. Also REF 12.54.

14.11. H. Scholz. "*P*-Delta Effect under Repeated Loading," *Journal of Structural Engineering,* ASCE, **116,** 8 (August 1990), 2070–2082.

14.12. Eric M. Lui. "A Practical *P*-Delta Analysis Method for Type FR and PR Frames," *Engineering Journal,* AISC, **25,** 3 (Third Quarter 1988), 85–98.

14.13. Victor Levi, George C. Driscoll, Jr., and Le-Wu Lu. "Analysis of Restrained Columns Permitted to Sway," *Journal of the Structural Division,* ASCE, **93,** ST1 (February 1967), 87–107.

14.14. Alfred Korn and Theodore V. Galambos. "Behavior of Elastic-Plastic Frames," *Journal of the Structural Division,* ASCE, **94,** ST5 (May 1968), 1119–1142.

14.15. Harold Switzky and Ping Chun Wang. "Design and Analysis of Frames for Stability," *Journal of the Structural Division,* ASCE, **95,** ST4 (April 1969), 695–713.

14.16. Sviatoslav Liapunov. "Ultimate Strength of Multistory Steel Rigid Frames," *Journal of the Structural Division,* ASCE, **100,** ST8 (August 1974), 1643–1655.

14.17. Francois Cheong-Siat-Moy. "Inelastic Sway Buckling of Multistory Frames," *Journal of the Structural Division,* ASCE, **102,** ST1 (January 1976), 66–75.

14.18. Francois Cheong-Siat-Moy. "Multistory Frame Design Using Story Stiffness Concept," *Journal of the Structural Division,* ASCE, **102,** ST6 (June 1976), 1197–1212.

14.19. Francois Cheong-Siat-Moy, Erkan Ozer, and Le-Wu Lu. "Strength of Steel Frames under Gravity Loads," *Journal of the Structural Division,* ASCE, **103,** ST6 (June 1977), 1223–1235.

14.20. Le-Wu Lu, Erkan Ozer, J. Hartley Daniels, Omer S. Okten, and Shosuke Morino. "Strength and Drift Characteristics of Steel Frames," *Journal of the Structural Division,* ASCE, **103,** ST11 (November 1977), 2225–2241.

14.21. Ali A. K. Haris. "Approximate Stiffness Analysis of High-Rise Buildings," *Journal of the Structural Division,* ASCE, **104,** ST4 (April 1978), 681–696.

14.22. Regina Gaiotti and Bryan Stafford Smith. "*P*-Delta Analysis of Building Structures," *Journal of Structural Engineering,* ASCE, **115,** 4 (April 1989), 755–770.

14.23. Zia Razzaq and Moossa M. Naim. "Elastic Instability of Unbraced Space Frames," *Journal of the Structural Division,* ASCE, **106,** ST7 (July 1980), 1389–1400.

14.24. Donald W. White and Jerome F. Hajjar. "Application of Second-Order Elastic Analysis in LRFD: Research to Practice," *Engineering Journal,* AISC, **28,** 1 (Fourth Quarter 1991), 133–148.

14.25. Chu-Kia Wang. "Stability of Rigid Frames with Nonuniform Members," *Journal of the Structural Division,* ASCE, **93,** ST1 (February 1967), 275–294.

14.26. Ottar P. Halldorsson and Chu-Kia Wang. "Stability Analysis of Frameworks by Matrix Methods," *Journal of the Structural Division,* ASCE, **94,** ST7 (July 1968), 1745–1760.

14.27. Vincent Mehringer, George Pierson, and James G. Orbison. "Computer-Aided Analysis and Design of Steel Frames," *Engineering Journal,* AISC, **22,** 3 (Third Quarter 1985), 143–149.

14.28. M. Ojalvo and V. Levi. "Columns in Planar Continuous Structures," *Journal of the Structural Division,* ASCE, **89,** ST1 (February 1963), 1–23.

14.29. Victor Levi, George C. Driscoll, Jr., and Le-Wu Lu. "Structural Subassemblages Prevented from Sway," *Journal of the Structural Division,* ASCE, **91,** ST5 (October 1965), 103–127.

14.30. W. Merchant. "The Failure Load of Rigid Jointed Frameworks as Influenced by Stability," *Structural Engineer,* **32** (July 1954), 185–190.

14.31. John Springfield and Peter F. Adams. "Aspects of Column Design in Tall Steel Buildings," *Journal of the Structural Division,* ASCE, **98,** ST5 (May 1972), 1069–1083.

14.32. J. Hartley Daniels and Le-Wu Lu. "Plastic Subassemblage Analysis for Unbraced Frames," *Journal of the Structural Division,* ASCE, **98,** ST8 (August 1972), 1769–1788.

14.33. Francois Cheong-Siat-Moy. "Consideration of Secondary Effects in Frame Design," *Journal of the Structural Division,* ASCE, **103,** ST10 (October 1977), 2005–2019.

14.34. Peter Arnold, Peter F. Adams, and Le-Wu Lu. "Strength and Behavior of an Inelastic Hybrid Frame," *Journal of the Structural Division,* ASCE, **94,** ST1 (January 1968), 243–266.

14.35. William W. McVinnie and Edwin H. Gaylord. "Inelastic Buckling of Unbraced Space Frames," *Journal of the Structural Division,* ASCE, **94,** ST8 (August 1968), 1863–1885.

14.36. Kuang-Han Chu and Hsueh-Lien Chow. "Effective Column Length in Unsymmetrical Frames," *Publications,* International Association for Bridge and Structural Engineering, **29–I,** 1969, 1–15.

14.37. Le-Wu Lu. "Stability of Frames Under Primary Bending Moments," *Journal of the Structural Division,* ASCE, **89,** ST3 (June 1963), 35–62.

14.38. Theodore V. Galambos. *Structural Members and Frames.* Englewood Cliffs, NJ: Prentice-Hall, Inc., 1968 (pp. 176–189).

14.39. Theodore V. Galambos. "Lateral Support for Tier Building Frames," *Engineering Journal,* AISC, **1,** 1 (January 1964), 16–19. Disc. by Ira Hooper, **1,** 4 (October 1964), 141.

14.40. Francois Cheong-Siat-Moy. "Column Design in Gravity-Loaded Frames," *Journal of Structural Engineering,* ASCE, **117,** 2 (May 1991), 1448–1461.

14.41. Oscar de Buen. "Column Design in Steel Frames Under Gravity Loads," *Journal of Structural Engineering,* ASCE, **118,** 10 (October 1992), 2928–2935.

14.42. Louis F. Geschwindner. "A Practical Approach to the 'Leaning' Column," *Engineering Journal,* AISC, **32,** 2 (Second Quarter 1995), 63–72.

14.43. J. Dario Aristizabal-Ochoa. "Slenderness *K* Factor for Leaning Columns," *Journal of Structural Engineering,* ASCE, **120,** 10 (October 1994), 2977–2991.

14.44. AISC. *Plastic Design of Braced Multi-Story Frames.* New York: American Institute of Steel Construction, 1968.

14.45. Arturo E. Schultz. "Approximating Lateral Stiffness of Stories in Elastic Frames," *Journal of Structural Engineering,* ASCE, **118,** 1 (January 1992), 243–263.

14.46. W. S. King, D. W. White, and W. F. Chen. "Second-Order Inelastic Analysis Methods for Steel-Frame Design," *Journal of Structural Engineering,* ASCE, **118,** 2 (February 1992), 408–428.

14.47. Jerome F. Hajjar and Donald W. White. "The Accuracy of Column Stability Calculations in Unbraced Frames and the Influence of Columns with Effective Length Factors Less Then One," *Engineering Journal,* AISC, **31,** 3 (3rd Quarter 1994), 81–97.

14.48. "Frame Stability," Chapter 16, *Guide to Stability Design Criteria for Metal Structures,* 5th ed. New York: John Wiley & Sons, 1998. (Note: Section 1.6 identifies Greg Deierlein and Don White, with assistance from members of Task Group 4, as the authors of Chapter 16.)

15

Design of Rigid Frames

15.1   INTRODUCTION

Concepts and procedures developed in previous chapters are combined in this chapter by the use of illustrative examples. Plastic analysis is extended from the continuous beam treatment in Chapter 10 to one-story frames. The reader is assumed to be familiar with elastic statically indeterminate analysis methods.

There are four approaches that can be used for designing one-story frames; (1) elastic analysis according to AISC LRFD Method; (2) plastic (inelastic) analysis according to AISC LRFD Method; (3) elastic analysis according to AISC ASD Method; and (4) plastic (inelastic) analysis according to AISC ASD Method.

Plastic design is likely the method of choice for one-story frames consisting of rolled W sections. When sections used have their limit states controlled by stability (lateral-torsional buckling, local flange buckling, or local web buckling), factored load elastic analysis will be the preferred method. Plastic analysis *requires* consideration of the $P\Delta$ effect, even in one-story frames.

15.2   PLASTIC ANALYSIS OF ONE-STORY FRAMES

As discussed in Chapter 10 for continuous beams, the plastic strength of a structure may be obtained either by using the equilibrium method, or the energy method. In braced frames, where joints cannot displace (i.e., no sidesway), the plastic strength may be obtained exactly as for continuous beams. For unbraced frames the sidesway mechanism creates a somewhat more complicated analysis than that for continuous beams. The following examples are intended to illustrate principles of plastic analysis for one-story unbraced frames.

A more extended treatment of plastic analysis is available in *Plastic Design in Steel* [15.1], in books by Beedle [15.2] and Massonnet and Save [15.3] devoted entirely to plastic analysis and design, and papers by Beedle [15.4] and Estes [15.5].

Welded multistory rigid frame. (Courtesy Bethlehem Steel Corporation)

Equilibrium Method

As discussed in Sec. 10.2, equilibrium must be satisfied at every stage of loading from a small load until the collapse mechanism has been achieved. When a sufficient number of plastic hinges have been developed to allow instantaneous hinge rotations without developing increased resistance, a mechanism is said to have occurred.

EXAMPLE 15.2.1

Determine the plastic strength for the frame of Fig. 15.2.1, using the equilibrium method.

Solution:
The elastic moment diagram shape is shown in Fig. 15.2.1b. For ease in solution it is often desirable to show the diagram with a horizontal baseline as shown in Fig. 15.2.1d. Assuming overall frame stability is adequate, the collapse mechanism is that shown in Fig. 15.2.1c.
The simple beam moment for the girder is

$$M_s = \frac{W_n L}{4} \tag{a}$$

which when superimposed on the moments at the ends of the girder $(H \times h)$ gives for equilibrium when the mechanism occurs,

$$M_p = M_s - M_p = \frac{W_n L}{4} - M_p \tag{b}$$

$$W_n = \frac{8 M_p}{L} \tag{c}$$

From the concepts of plastic analysis it might have been expected that only two plastic hinges should be required for a collapse mechanism, since the structure is statically indeterminate to the first degree. In this case both corner hinges form simultaneously, since the horizontal base reactions are equal and opposite. Depending on the ratio of h to L either the midspan plastic hinge, or the two corner plastic hinges form. If, however, h/L is large, the corner plastic hinges occur first and the structure has reached its collapse condition. If the beam and columns are of different cross-section the structure can be designed so the positive moment plastic hinge occurs first. This should be the objective. The occurrence first of the corner plastic hinges creates overall frame instability prior to utilizing the flexural strength of the girder; such a result should be avoided.

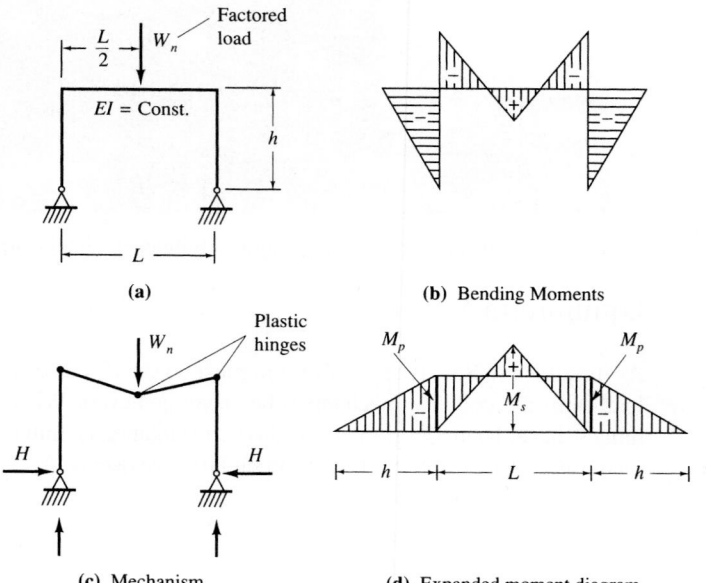

Figure 15.2.1
Example 15.2.1

(a)

(b) Bending Moments

(c) Mechanism

(d) Expanded moment diagram

EXAMPLE 15.2.2

Determine the plastic strength of the same frame as in Example 15.2.1 except in addition apply a horizontal load $0.5W_n$ at the top of the column (see Fig. 15.2.2a). Use the equilibrium method.

Solution:

Again, two plastic hinges should provide the collapse mechanism. This time, however, additional mechanisms are possible, as shown in Fig. 15.2.2. It is possible that (a) plastic hinges form at points 2 and 4 (Fig. 15.2.2b); (b) plastic hinges form at points 3 and 4 (Fig. 15.2.2c); and also (c) plastic hinges form at points 2, 3, and 4. The last situation would occur only if two of the hinges formed simultaneously.

(a) Plastic hinges at points 2 and 4,

$$M_s = \frac{W_n L}{4} \tag{a}$$

Equilibrium requires the positive moment M_p at point 2 $(0.5W_n h - Hh)$ to equal the negative moment M_p at point 4 (Hh):

$$M_p = 0.5W_n h - M_p \tag{b}$$

$$M_p = \frac{W_n h}{4}; \qquad W_n = \frac{4M_p}{h} \tag{c}$$

Further, in order that Eq. (c) is valid, the resulting moment at point 3 cannot exceed M_p:

$$M_3 = M_s + \tfrac{1}{2}(0.5W_n h) - M_p \tag{d}$$

$$= \frac{W_n L}{4} + \frac{W_n h}{4} - \frac{W_n h}{4} = \frac{W_n L}{4} \le M_p \tag{e}$$

Equation (e) requires $W_n L/4 < W_n h/4$, which means if $L < h$ plastic hinges form at points 2 and 4. If $L > h$ the plastic hinges will form at points 3 and 4.

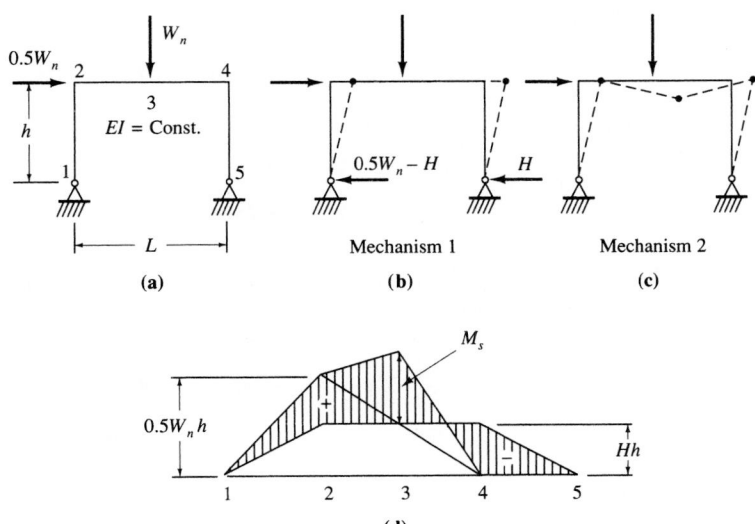

Figure 15.2.2
Example 15.2.2.

(b) Plastic hinges at points 3 and 4. For this, Eq. (d) is to be used letting $M_3 = M_p$,

$$M_p = M_s + \tfrac{1}{2}(0.5W_n h) - M_p \tag{f}$$

$$M_p = \frac{W_n}{8}(L + h) \tag{g}$$

$$W_n = \frac{8M_p}{L + h} \tag{h}$$

The above analysis has assumed the same moment of inertia (constant M_p) for both girder and column. If the girder and column are different, a combined mechanism with plastic hinges at points 2, 3, and 4 could be achieved. ∎

EXAMPLE 15.2.3

Determine the plastic strength W_n for the gabled frame of Fig. 15.2.3 using the equilibrium method. All elements of the frame are identical, having a plastic moment strength M_p of 200 ft-kips.

Solution:
Equilibrium requires a compatibility between the moment diagrams of the component loadings as shown in Fig. 15.2.3b. The moment due to H may be thought of as causing negative bending, while the other two components are causing positive bending. The moment diagrams for the components are shown on a horizontal baseline in Fig. 15.2.3c.

The maximum so-called positive moments, M_{s1} and M_{s2} are

$$M_{s1} = 7.5W_n$$
$$M_{s2} = 0.5W_n h = 7.5W_n$$

Several possible collapse mechanisms must be considered.

(a) Consider a sidesway mechanism (Fig. 15.2.3d) with plastic hinges at points 2 and 6. Equilibrium requires at point 2:

$$M_2 = M_p = M_{s2} - Hh \tag{a}$$

and at point 6:

$$M_6 = M_p = Hh = 15H \tag{b}$$

Thus

$$M_p = M_{s2} - M_p = 7.5W_n - M_p$$
$$M_p = 3.75W_n \tag{c}$$

(b) Consider a combination mechanism with plastic hinges at points 5 and 6 (Fig. 15.2.3e). At point 5,

$$M_5 = M_p = M_{s1} + \tfrac{1}{4}M_{s2} - H(h + 3.75)$$

$$M_p = 7.5W_n + \frac{7.5}{4}W_n - \frac{M_p}{15}(15 + 3.75)$$

$$M_p = 4.17W_n$$

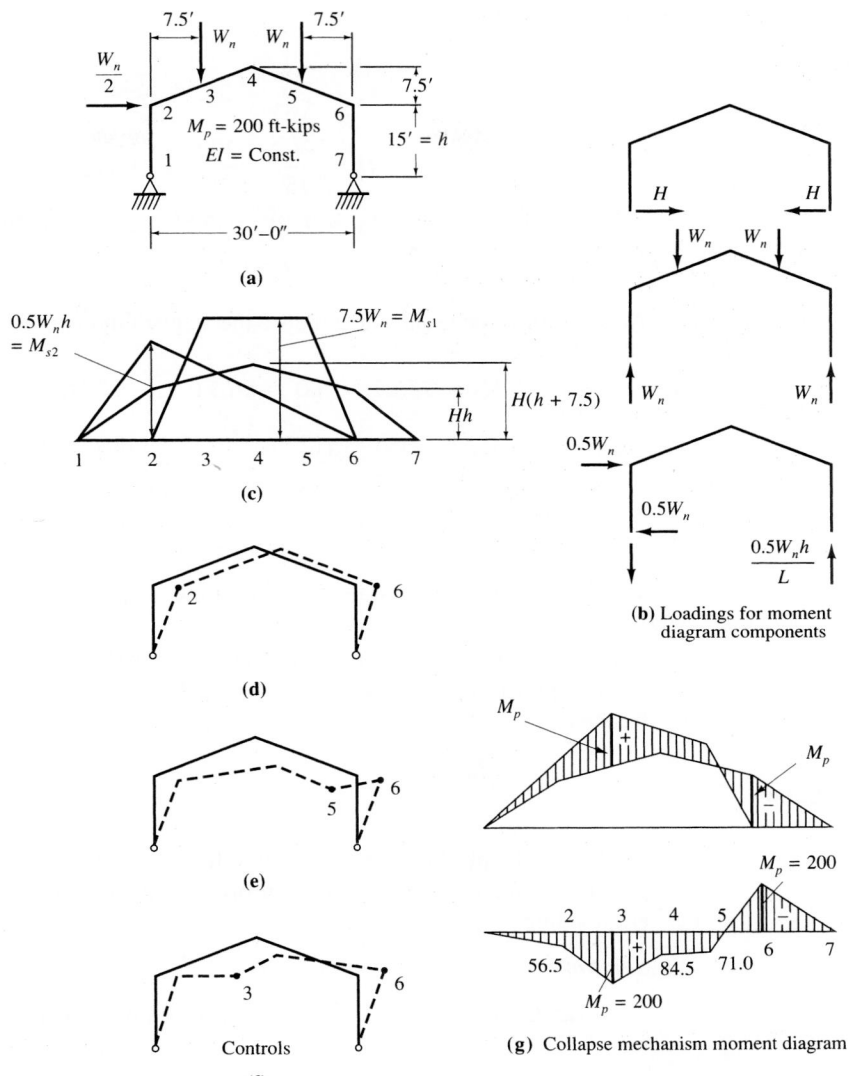

Figure 15.2.3
Gabled frame analysis—
Example 15.2.3.

Actually, the mechanism consisting of plastic hinges at points 5 and 6 could have been eliminated as a possibility by comparing the total positive moment at points 3 and 4 (Fig. 15.2.3c) which will indicate that point 3 will achieve a plastic hinge before point 5 can attain it.

(c) Consider the mechanism with plastic hinges at points 3 and 6:

$$M_3 = M_p = M_{s1} + \tfrac{3}{4}M_{s2} - H(h + 3.75)$$

$$M_p = 7.5W_n + \frac{3}{4}(7.5W_n) - \frac{M_p}{15}(15 + 3.75)$$

$$M_p = 5.83W_n \qquad \text{Governs}$$

The largest M_p, or the smallest W_n, for a mechanism to occur indicates the governing one. A check may be made by determining the collapse mechanism moment diagram assuming

$M_p = 200$ ft-kips. Thus

$$W_n = \frac{200}{5.83} = 34.2 \text{ kips}$$

$$H = \frac{M_p}{15} = \frac{200}{15} = 13.33 \text{ kips}$$

$$M_{s1} = 7.5W_n = 7.5(34.2) = 256.5 \text{ ft-kips}$$

$$M_{s2} = 256.5 \text{ ft-kips}$$

At the critical locations, the collapse mechanism moment is

$$M_2 = 256.5 - 200 = 56.5 \text{ ft-kips} < M_p \qquad\qquad \textbf{OK}$$

$$M_3 = 256.5 + \tfrac{3}{4}(256.5) - 13.33(18.75) = 200 \text{ ft-kips} = M_p$$

$$M_4 = 256.5 + \tfrac{1}{2}(256.5) - 13.33(22.5) = 84.5 \text{ ft-kips} < M_p \qquad\qquad \textbf{OK}$$

$$M_5 = 256.5 + \tfrac{1}{4}(256.5) - 13.33(18.75) = 71.0 \text{ ft-kips} < M_p \qquad\qquad \textbf{OK}$$

$$M_6 = 13.33(15) = 200 \text{ ft-kips} = M_p$$

The final collapse mechanism moment diagram is shown in Fig. 15.2.3g, both as it would be from graphical superposition and as it would be if the net moment is plotted on a horizontal baseline. ∎

EXAMPLE 15.2.4

Determine the plastic moment required for the frame of Fig. 15.2.4. Consider (a) uniform gravity loading w_n alone; and (b) uniform gravity loading in combination with uniform lateral loading.

Solution:

(a) Gravity loading only. The simple beam moment is

$$M_s = \frac{w_n L^2}{8} = \frac{w_n(75)^2}{8} = 704w_n \tag{a}$$

For a properly designed frame, the plastic hinges will occur at the ends and midspan of the girder:

$$M_p = M_s - M_p$$

$$M_p = \frac{1}{2}M_s = \frac{w_n L^2}{16} = 352w_n \tag{b}$$

and the superposition of the bending moments due to the loading components is shown in Fig. 15.2.4d.

(b) Combined gravity and lateral loading. There is an additional component of loading as seen in Fig. 15.2.4e which contributes an unsymmetrical effect of the same bending moment sign as caused by the gravity uniform loading. It may be observed from Fig. 15.2.4f that the maximum moments will occur slightly to the left of point 3 and at point 4. One could mathematically solve for the exact location and magnitude of the plastic

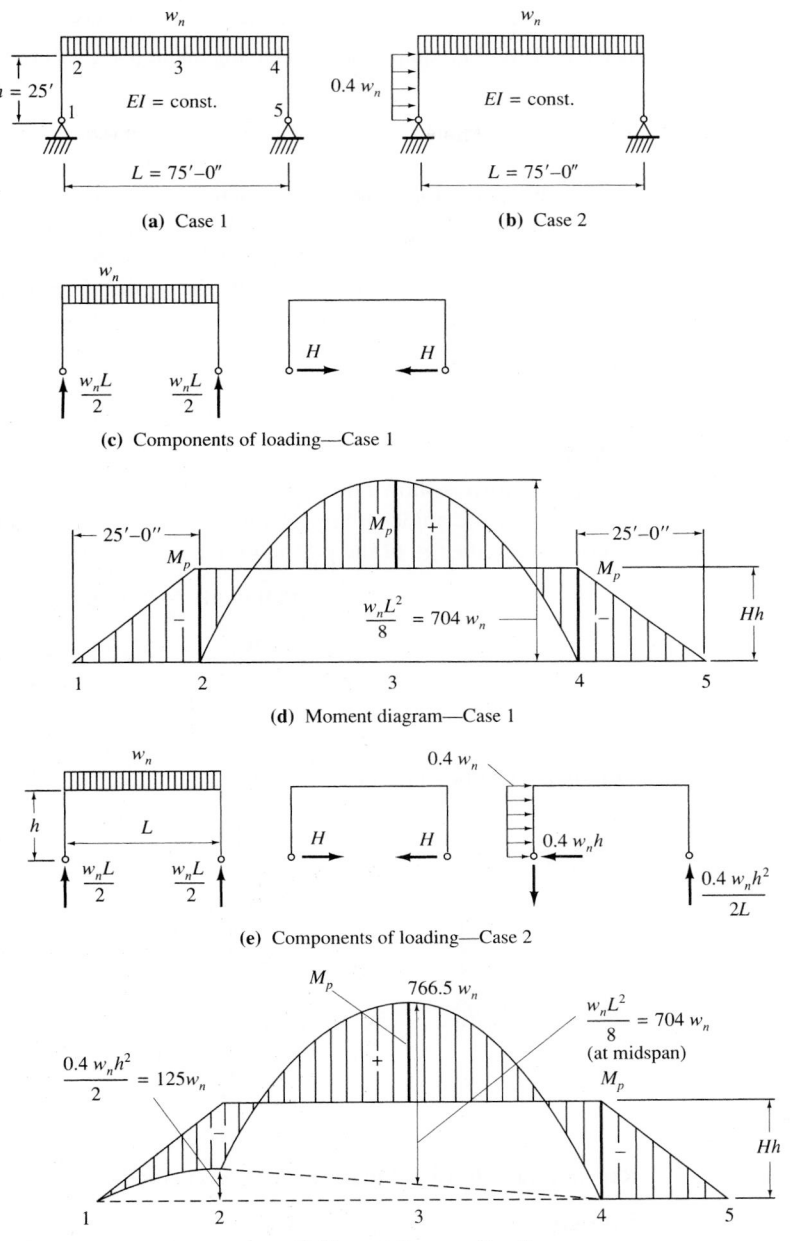

Figure 15.2.4
Example 15.2.4.

moment. However, for design purposes one may use a graphical construction and divide the maximum value of the parabolic curve by two to establish the horizontal line Hh. In other words, equalize the positive and negative moments. The moment at point 2 is obviously less than that at point 4 so that no plastic hinge will form at point 2.

By scaling, the maximum ordinate of the parabola near point 3 is $768w_n$. In which case,

$$M_p = \frac{768w_n}{2} = 384w_n \qquad (c)$$

The uniform lateral loading causes about 9% reduction in gravity carrying capacity, or alternatively it would require a section having about 9% greater moment strength. ∎

Energy Method

Just as discussed for beams in Chapter 10, the energy method may also be used for the plastic analysis of frames. For certain frame arrangements, the energy method will be found easier. The following examples show how the energy method may be applied to the same structures previously analyzed by the equilibrium method.

EXAMPLE 15.2.5

Repeat Example 15.2.1 except use the energy method.

Solution:
The mechanism is shown in Fig. 15.2.5. The plastic hinge locations are assumed and from the geometry the angles θ are established. The external work done by the load equals the internal strain energy of the plastic moments moving through their angles of rotation:

$$\text{External work} = \text{Internal work}$$

$$W_n \frac{\theta L}{2} = M_p(\theta + 2\theta + \theta)$$

$$W_n = \frac{8M_p}{L}$$

exactly as obtained in Example 15.2.1. ∎

Figure 15.2.5
Example 15.2.5.

EXAMPLE 15.2.6

Repeat Example 15.2.2 using the energy method.

Solution:
The possible mechanisms are shown in Fig. 15.2.6.

    (a) Mechanism 1

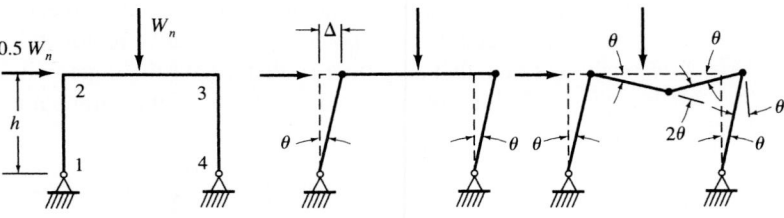

Figure 15.2.6
Example 15.2.6.

Mechanism 1   Mechanism 2

$$0.5W_n\theta h = M_p(\theta + \theta)$$

$$W_n = \frac{4M_p}{h}$$

(b) Mechanism 2

$$0.5W_n\theta h + W_n\theta\frac{L}{2} = M_p(2\theta + 2\theta)$$

$$W_n = \frac{8M_p}{L + h}$$

The result are identical to those of Example 15.2.2.   ∎

EXAMPLE 15.2.7

Repeat Example 15.2.3 using the energy method.

Solution:
The possible mechanisms are shown in Fig. 15.2.7.

(a) Mechanism 1

$$0.5W_n(15\theta) = M_p(2\theta)$$

$$M_p = 3.75W_n$$

(b) Mechanism 2. In order to treat this more complex mechanism, the concept of *instantaneous center* is used. When plastic hinges form at points 5 and 6 three rigid bodies remain which rotate as the structure moves. Segment 1–2–3–4–5 rotates about point 1; segment 6–7 rotates about point 7; segment 5–6 rotates and translates an amount which is controlled by the movement of points 5 and 6 on the adjacent rigid segments. If the segments are rigid, point 5′ is perpendicular to line 1–5, and point 6′ is perpendicular to line 6–7. Thus,

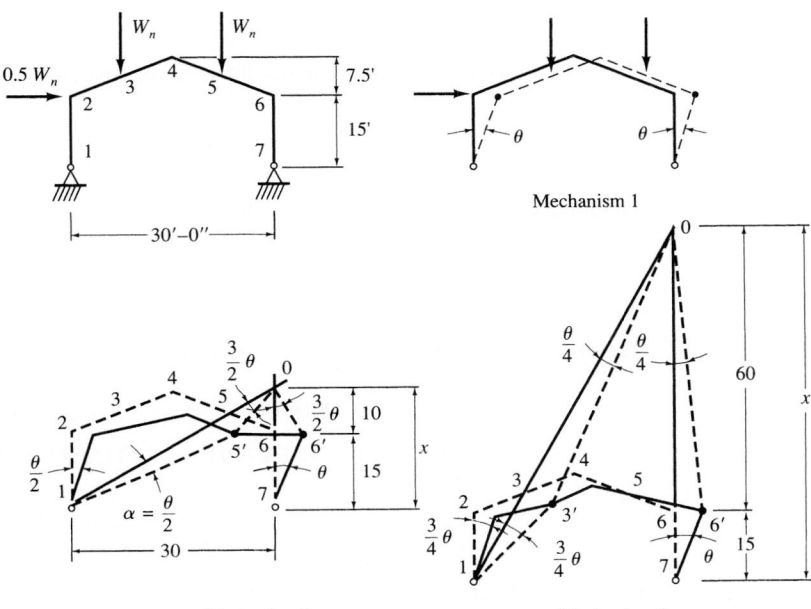

Figure 15.2.7
Example 15.2.7.

the points 5 and 6 may be thought of as rotating about point 0, the intersection of line 1–5 and line 6–7; i.e., the instantaneous center.

The first step in the energy method using the instantaneous center is to determine its location; since point 5 is 22.5 ft horizontally and 18.75 ft vertically from point 1, the vertical distance to point 0 from point 7 is

$$\frac{x}{30} = \frac{18.75}{22.5}, \qquad x = 25 \text{ ft}$$

Next, a reference angle θ is established arbitrarily as shown in Fig. 15.2.7. By proportion, the angle of rotation with respect to point 0 is $3\theta/2$. The rigid-body segment 5–6 rotates through this angle $3\theta/2$. By inverse proportion as distance 0–5 is to 1–5, the rotation of rigid-body 1–2–3–4–5 about point 1 is

$$\frac{\frac{1}{4}}{\frac{3}{4}} = \frac{\alpha}{3\theta/2}, \qquad \alpha = \frac{\theta}{2}$$

The relative plastic hinge rotation at point 5 is

$$\frac{\theta}{2} + \frac{3\theta}{2} = 2\theta$$

The relative plastic hinge rotation at point 6 is

$$\theta + \frac{3\theta}{2} = 2.5\theta$$

To compute external work done by the applied loads, the vertical distance moved due to rotation of points 3 and 5 and the horizontal distance moved at point 2 are required.

The vertical displacement of point 3 equals the angle of rotation times the horizontal projection from points 2 to 3. The load at point 3 moves vertically through a distance

$$\frac{\theta}{2}(7.5) = 3.75\theta$$

The load at point 5 moves vertically through the distance

$$\frac{\theta}{2}(22.5) = 11.25\theta$$

The load at point 2 moves horizontally through the distance

$$\frac{\theta}{2}(15) = 7.5\theta$$

The complete energy equation then becomes

$$\text{External work} = \text{Internal work}$$

$$0.5W_n(7.5\theta) + W_n(3.75\theta) + W_n(11.25\theta) = M_p(2\theta + 2.5\theta)$$

$$M_p = \frac{18.75}{4.5}W_n = 4.17W_n$$

the same as from the equilibrium method.

(c) Mechanism 3. The instantaneous center is found by intersecting the line 1–3 with line 6–7:

$$\frac{x}{30} = \frac{18.75}{7.5}, \qquad x = 75 \text{ ft}$$

If θ is defined by Fig. 15.2.7, then the angle of rotation with respect to 0 is $\theta/4$, since the distance 0–6 is four times the distance 6–7. Since the distance 0–3 is three times the distance 3–1, the angle 3–1–3' is $3\theta/4$ (3 times the rotation angle about 0).

The external work done by the various loads is

$$\text{Load at 2,} \qquad 0.5W_n\left(\frac{3\theta}{4}\right)(15) = \frac{22.5}{4}W_n\theta$$

$$\text{Load at 3,} \qquad W_n\left(\frac{3\theta}{4}\right)(7.5) = \frac{22.5}{4}W_n\theta$$

$$\text{Load at 5,} \qquad W_n\left(\frac{\theta}{4}\right)(7.5) = \frac{7.5}{4}W_n\theta$$

The internal strain energy is

$$\text{Moment at 3,} \qquad M_p\left(\frac{3\theta}{4} + \frac{\theta}{4}\right) = M_p\theta$$

$$\text{Moment at 6,} \qquad M_p\left(\theta + \frac{\theta}{4}\right) = M_p\frac{5\theta}{4}$$

External work = Internal work

$$W_n\theta\left(\frac{22.5}{4} + \frac{22.5}{4} + \frac{7.5}{4}\right) = M_p\theta\left(1 + \frac{5}{4}\right)$$

$$W_n\left(\frac{52.5}{4}\right) = M_p\left(\frac{9}{4}\right)$$

$$M_p = \frac{52.5}{9}W_n = 5.83W_n \quad \text{Governs}$$

The result is identical with Example 15.2.3.   ■

15.3   AISC LRFD METHOD—ONE-STORY FRAMES

There are two approaches that may be used under AISC Load and Resistance Factor Design: (1) elastic analysis using factored loads: and (2) plastic analysis (referred to as inelastic analysis in AISC-Appendix 1) using factored loads. These two approaches are compared in several examples.

Plastic analysis is permitted under AISC-B3.1 for steels having a yield stress F_y not exceeding 65 ksi (AISC-Appendix 1.4). The following other conditions must be satisfied:

1. In regions where plastic hinge rotation is necessary to reach the plastic strength collapse mechanism, local flange buckling must be precluded as a limit state by limiting $b_f/2t_f \le \lambda_p$ (AISC-Appendix 1.4), where for flanges of I-shaped sections

$$\lambda_p = 0.94\sqrt{\frac{E}{F_y}} = \frac{160}{\sqrt{F_y, \text{ksi}}} \tag{15.3.1}$$

2. The web local buckling limit state must be precluded in beams and beam-columns by limiting $h/t_w \leq \lambda_p$ (AISC-Appendix 1.4), where for webs in flexural and axial compression,

(a) For $P_u/(\phi_b P_y) \leq 0.125$:

$$\lambda_p = 3.76\sqrt{\frac{E}{F_y}}\left(1 - \frac{2.75P_u}{\phi_b P_y}\right) = \frac{640}{\sqrt{F_y,\text{ksi}}}\left(1 - \frac{2.75P_u}{\phi_b P_y}\right) \tag{15.3.2}$$

(b) For $P_u/(\phi_b P_y) > 0.125$:

$$\lambda_p = \left[1.12\sqrt{\frac{E}{F_y}}\left(2.33 - \frac{P_u}{\phi_b P_y}\right) = \frac{191}{\sqrt{F_y,\text{ksi}}}\left(2.33 - \frac{P_u}{\phi_b P_y}\right)\right] \tag{15.3.3}$$

$$\geq \left[1.49\sqrt{\frac{E}{F_y}} = \frac{253}{\sqrt{F_y,\text{ksi}}}\right]$$

3. The factored axial compression force in members of frames is restricted.

(a) In *braced* frames (AISC-Appendix 1.5),

$$P_u \leq 0.85\phi_c A_g F_y \tag{15.3.4}$$

(b) In *unbraced* frames (AISC-Appendix 1.5),

$$P_u \leq 0.75\phi_c A_g F_y \tag{15.3.5}$$

where P_u is the axial force in columns "caused by factored gravity plus factored horizontal loads."

4. The column slenderness ratio L/r may not exceed $800/\sqrt{F_y,\text{ksi}}$ (AISC-Appendix 1.6).

5. At plastic hinge locations associated with the plastic strength collapse mechanism, except in the region of the last plastic hinge to form, the lateral-torsional buckling limit state must be precluded by limiting $L_b \leq L_{pd}$ (AISC-Appendix 1.7), where for I-shaped sections, having compression flange equal to or larger than the tension flange, and bent about the major axis,

$$L_{pd} = \frac{3500 + 2200(M_1/M_p)}{F_y}r_y \tag{15.3.6}$$

where M_1/M_p is positive when moments cause reverse curvature, and negative for single curvature.

6. In the region of the last plastic hinge to form, and in regions not adjacent to a plastic hinge, the ordinary flexural strength requirements of AISC-F2.2 apply, as discussed in Chapters 7, 9, 10, and 12.

7. For beam-columns, the factored second-order moment M_u for use in the interaction equations of AISC-H1 must satisfy the $P\Delta$ effects (AISC-C2.1) and frame stability (AISC-C1.2). As previously stated, long practice with plastic design has specifically *neglected* the $P\Delta$ effect in one-story frames.

8. For flexural composite members (see Chapter 16), the nominal moment strength M_n must be computed using plastic stress distributions in accordance with AISC-I3.2. In accordance with AISC-I3.2, h/t_w cannot exceed $640/\sqrt{F_y,\text{ksi}}$.

EXAMPLE 15.3.1

Design a rectangular frame of 75-ft span and 25-ft height to carry 0.2 kip/ft dead load, 0.8 kip/ft snow load, 0.1 kip/ft gravity directional wind load, and horizontal uniform wind of 0.44 kip/ft. Lateral bracing from purlins occurs every 6 ft on the girder, and bracing is provided by wall girts on the columns every 5 ft. Use A992 steel and *plastic analysis* under Load and Resistance Factor Design.

Solution:
The frame is shown in Fig. 15.3.1.

(a) Factored load combination according to ASCE 7 Section 2.3

Case 1: Gravity load

$$w_u = 1.4D = 1.4(0.2) = 0.28 \text{ kips/ft}$$

Case 2: Gravity + wind(Case a)

$$w_u = 1.2D + 1.6S + 0.8W$$
$$= 1.2(0.2) + 1.6(0.8) + 0.8(0.1) = 1.6 \text{ kips/ft}$$
$$w_{uh} = 0.8W = 0.8(0.44) = 0.352 \text{ kip/ft} \quad \text{(horizontal)}$$

Case 3: Gravity + wind(Case b)

$$w_u = 1.2D + 1.6W + 0.5S$$
$$= 1.2(0.2) + 1.6(0.1) + 0.5(0.8) = 0.8 \text{ kips/ft}$$
$$w_{uh} = 1.6W = 1.6(0.44) = 0.704 \text{ kips/ft} \quad \text{(horizontal)}$$

It is clear that the gravity loading only load case is not critical and will not be considered further.

(b) Gravity and wind-Case 2. Using frame analysis as in Example 15.2.4, the location and value of the maximum moment is determined.

The moment variation in the girder along the horizontal line 2-4 (see Figure 15.2.4) can be expressed as

$$f(x) = \frac{w_u x}{2}(L - x) + \frac{w_{uh} h^2}{2}\left(1 - \frac{x}{L}\right)$$

Figure 15.3.1
Example 15.3.1.

and

$$M_u = \frac{f(x)_{max}}{2}$$

Location of the maximum value of $f(x)$ can be determined by obtaining the derivative of the expression above, equating it to zero and solving for x:

$$\frac{df(x)}{dx} = \frac{w_u L}{2} - w_u x - \frac{w_{uh} h^2}{2L} = 0$$

Maximum value occurs at $x = \dfrac{L}{2} - \dfrac{w_{uh}}{w_u} \dfrac{h^2}{2L}$

For this case:

$$x = \frac{75}{2} - \frac{0.352}{1.6} \frac{25^2}{2(75)} = 36.6 \text{ ft}$$

$$f(x)_{max} = \frac{1.6(36.6)}{2}(75 - 36.6) + \frac{0.352(25)^2}{2}\left(1 - \frac{36.6}{75}\right) = 1181 \text{ ft-kips}$$

Therefore,

$$\text{Required } \phi_b M_p = 1181/2 = 592 \text{ ft-kips}$$
$$\text{Required } M_p = 592/0.90 = 658 \text{ ft-kips}$$

(c) Gravity + wind—Case 3.

$$M_p = \frac{f(x)_{max}}{2}$$

$$x = \frac{75}{2} - \frac{0.8}{0.704} \frac{25^2}{2(75)} = 32.8 \text{ ft}$$

$$f(x)_{max} = \frac{0.8(32.8)}{2}(75 - 32.8) + \frac{0.704(25)^2}{2}\left(1 - \frac{32.8}{75}\right) = 678 \text{ ft-kips}$$

Therefore,

$$\text{Required } \phi_b M_p = 678/2 = 339 \text{ ft-kips}$$

$$\text{Required } M_p = 339/0.90 = 377 \text{ ft-kips} < 658 \text{ ft-kips} \quad \text{Hence Case 2 controls}$$

$$\text{Required } Z_x = \frac{658(12)}{50} = 158 \text{ in.}^3$$

Try W21×68. $Z_x = 160$ in.^{3}. $M_p = 50(160)/12 = 667$ ft-kips, or a deeper or heavier section if deflection is critical factor. Check the flange local buckling limits of AISC-B4

$$\left(\frac{b_f}{2t_f} = 6.04\right) \le \left(\frac{b}{t} = \frac{65}{\sqrt{F_y}, \text{ksi}} = 9.19\right) \qquad \textbf{OK}$$

(d) Preliminary section selected as a beam-column. Use Eq. 12.12.2 or 12.12.3 to select a satisfactory preliminary section for combined axial load and bending.

For a W21, estimate $p = 0.0005$, and $KL_{equiv} = 50/4.78 = 10.5$ ft
Check: $pP_u = 0.0005(61.5) = 0.031 < 0.2$ use Eq. 12.12.2

$$\frac{1}{2}\,pP_u + \frac{9}{8}M_{ux}b_x = \frac{0.031}{2} + \frac{9}{8}592b_x = 1.0$$

$$\text{Required } b_x \leq 0.00148$$

From AISC-Table 6-1, find W21×83 with $p = 0.00125$ and $b_x = 0.00134$ for $KL_{equiv} = 10.5$ ft

(e) Check lateral-torsional buckling for W21×83: $\phi_b M_p = 735$ ft-kips. In order to have used plastic plastic analysis L_b must not exceed L_{pd} (Eq. 15.3.4). The minimum L_{pd} will be when $M_1/M_p = -1$, giving

$$L_{pd} = \left[0.12 - 0.076\left(\frac{M_1}{M_2}\right)\right]\left(\frac{E}{F_y}\right)r_y$$

$$= [0.12 + 0.076(-1)]\left(\frac{29,000}{50}\right)\frac{1.83}{12} = 3.90 \text{ ft}$$

This means that the column lateral bracing of 3.5 ft is adequate, while the beam lateral bracing of 4.0 ft is slightly inadequate. The minimum r_y that satisfies this requirement can be determined as follows.

$$\text{Min } r_y = \frac{(L_{pd} = L_b)F_y}{3500 + 2200\,(M_1/M_p)} = \frac{4(12)(50)}{3500 + 2200(-1)} = 1.85 \text{ in.}$$

Try W24×84 with $r_y = 1.95$ in.

(f) Check the W27×84 as a beam-column according to AISC-H1.1. The effective length K_x for buckling in the plane of the frame must be determined by "structural analysis" according to AISC-C2.1, and may not be less than unity. Using the alignment charts, Fig. 6.9.4

$$G_A \approx 10 \text{ (estimated for hinge at bottom of column)}$$

$$G_B = \frac{I/L \text{ for column}}{I/L \text{ for girder}} = \frac{I/25}{I/75} = 3.0 \text{ (top of column)}$$

From Fig. 6.9.4b, unbraced frame find $K_x = 2.25$.

$$\text{Axis of bending } K_2L = K_xL_x = 2.25(25) = 56.3 \text{ ft}$$

At this point, the G factor may be adjusted as explained in Sec. 6.9 to account for inelastic behavior. Using AISC-Table 4-21, in which the stiffness reduction factor is based on the available strength of the column, the reduction factor τ_u is determined to be 1.0 for $P_u/A_g = 2.5$ ksi: no correction needed.

$$P_{e2} = \frac{\pi^2 EI}{(K_2L)^2} = \frac{\pi^2(29,000)2850}{(56.3 \times 12)^2} = 1787 \text{ kips}$$

Using the sway amplification factor B_2 on the entire moment as a conservative treatment of the $P\Delta$ effect,

$$B_2 = \frac{1}{1 - \dfrac{\sum P_{nt}}{\sum P_{e2}}} = \frac{1}{1 - \dfrac{1.6(75)}{2(1787)}} = 1.03$$

Examine the slenderness ratio,

$$\frac{K_y L_y}{r_y} = \frac{1.0(3.5)12}{2.07} = 20.3$$

$$\frac{K_x L_x}{r_x} = \frac{56.3(12)}{10.7} = 63 > 20.3 \qquad \text{controls}$$

From AISC-Table 4-22,

$$\text{For} \qquad \frac{K_x L_x}{r_x} = 63 \qquad \phi_t F_{cr} = 33.7 \text{ ksi}$$

Check the beam-column interaction equation (AISC-H1.1), Eqs. 12.10.1 or 12.10.2. According to AISC-C2.1b,

$$P_u = P_{nt} + B_2 P_{\ell t} = 60 + 1.03(1.5) = 61.6 \text{ kips}$$

and

$$\frac{P_u}{\phi_c P_n} = \frac{61.6}{(33.7)24.8} = 0.074 < 0.2$$

Use AISC Formula (H1-1b), for uniaxial bending,

$$\frac{P_u}{2\phi_c P_n} + \frac{M_{ux}}{\phi_b M_{nx}} \le 1 \qquad\qquad [12.10.2]$$

$$M_{nt} + M_{\ell t} = 370(1.60) = 592 \text{ ft-kips}$$

$$M_{ux} = B_2(M_{nt} + M_{\ell t}) = 1.03(592) = 613 \text{ ft-kips}$$

Note that in plastic analysis the combined gravity plus lateral loading is used: thus, M_{nt} and $M_{\ell t}$ are not separately computed. Magnifying the combined moment using the sway amplification factor is conservative, but permitted (see *User Note* in AISC-C2.1b). Thus Eq. 12.10.2 becomes

$$\frac{0.074}{2} + \frac{613(12)}{0.90(50)244} = 0.037 + 0.67 = 0.71 \qquad\qquad \textbf{OK}$$

If the shallower W24×84 is desired, $\phi M_p = 840$ ft-kips. The $K_x L_x/r_x = 68.9$, $P_{e2} = 1486$ kips, $B_2 = 1.042$, and $\phi_c F_{cr} = 31.8$ ksi. The strength criterion, Eq. 12.10.2, becomes

$$\frac{61.6}{2(31.8)24.7} + \frac{1.042(592)}{840} = 0.77 < 1.0 \qquad\qquad \textbf{OK}$$

Recheck Eq. 15.3.2 for h/t_w limit. For $P_u/(\phi_b P_n) = 61.6/[0.90(50)(24.7)] = 0.055 < 0.125$ limit.

$$\frac{h}{t_w} \le \frac{640}{\sqrt{F_y, \text{ksi}}}\left(1 - \frac{2.75 P_u}{\phi_b P_y}\right) = \frac{640}{\sqrt{50}}[1 - 2.75(0.055)] = 77$$

The actual h/t_w for W24×94 is 41.9; well below the limit.
Use W24×94.   ∎

It is clear that the $P\Delta$ effect is not significant (less than 4%) in this frame. To complete the design of the frame, the knee joint must be investigated, providing stiffeners if necessary, in a manner similar to Example 13.8.1.

EXAMPLE 15.3.2

Redesign the frame of Example 15.3.1 (Fig. 15.3.1) if the column cannot be deeper than 14.5 in. actual depth. Use A992 steel and the LRFD Method.

Solution:

For this solution the column and the beam will have different plastic moment strengths. The factored loads appear in Example 15.3.1, part (a).

(a) Select for plastic strength using loading Case 2 $(1.2D + 1.6S + 0.8W)$. There are many possible solutions using various sizes of columns. It has been determined in Example 15.3.1 that the inelastic design is governed by minimum r_y values: 1.65 in. for the girder and 1.85 in. for the column. So, try a W14×61 section having $\phi M_p = 383$ ft-kips and $r_y = 2.45$ in.

The plastic design moment $\phi_b M_p$ $(+M)$ required for the girder is (see Fig. 15.3.2)

$$\phi M_{p,required}(\text{girder}) = 1181 - 383 = 798 \text{ ft-kips}$$

$$Z_{x,required} = 798(12)/(0.9)50 = 213 \text{ in.}^3$$

Try W27×84 girder, with $\phi M_p = 915$ ft-kips and $r_y = 2.07$ in.

For making a corner connection, the two members should have approximately the same flange width. In this case both have 10-in. flange width. To fully utilize the $\phi_b M_p = 915$ ft-kips, perhaps the column can be reduced to W14×48 having $\phi_b M_p = 294$ ft-kips. The plastic strength for the girder will become

$$\text{Required } \phi_b M_p = 1181 - 294 = 887 \text{ ft-kips}$$

The most suitable is W27×84 girder, with $\phi M_p = 915$ ft-kips.

(b) Compute the column strength and the ratio $P_u/\phi P_n$. For this unbraced frame use the alignment chart, Fig. 6.9.4, to evaluate the effective length factor in the plane of the frame,

$$G_A \approx 10 \text{ (estimated for hinge at bottom of column)}$$

$$G_B = \frac{I/L \text{ for column}}{I/L \text{ for girder}} = \frac{484/25}{2850/75} = 0.51 \text{ (top of column)}$$

Figure 15.3.2
Factored load collapse
moment diagrams for plastic
analysis on Example 15.3.2,
assuming W14×61 columns
$(\phi_b M_p = 383$ ft-kips).

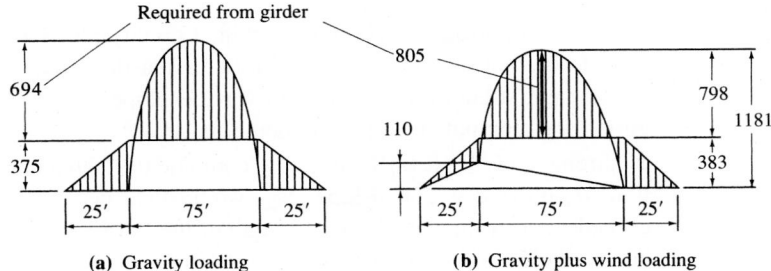

(a) Gravity loading

(b) Gravity plus wind loading

From Fig. 6.9.4b, unbraced frame, find $K_x = 1.8$.

$$\text{Axis of bending} \quad \frac{KL}{r} = \frac{K_x L_x}{r_x} = \frac{1.8(25)12}{5.85} = 92.3$$

Since the axial load on the column is low and the slenderness is high, there will be no adjustment in K_x for inelastic effect as discussed in Sect. 6.9.

$$\frac{K_y L_y}{r_y} = \frac{1.0(5)12}{1.91} = 31.4$$

Since $K_x L_x / r_x > K_y L_y / r_y$, the major axis controls $\phi_c F_{cr}$

$$K_x L_x / r_x = 92.3: \qquad \phi_c F_{cr} = 24.1 \text{ ksi}$$

The axial compression effect of the beam-column interaction equation is

$$\frac{P_u}{\phi_c P_n} = \frac{P_u}{\phi_c F_{cr} A_g} = \frac{61.6}{24.1(14.1)} = 0.181$$

Since this is less than 0.2, AISC Formula (H1-1b), Eq. 12.10.2 applies.

(c) Compute the sway amplification factor B_2. Again, as in Example 15.3.1, use the amplification factor on the total moment in the inelastic analysis of the frame. In evaluating P_{e2} use the unbraced length KL of $1.8(25) = 45$ ft

$$P_{e2} = \frac{\pi^2 EI}{(K_2 L)^2} = \frac{\pi^2 (29,000)484}{[45(12)]^2} = 476 \text{ kips}$$

Using the sway amplification factor B_2 on the entire moment as a conservative treatment of the $P\Delta$ effect,

$$B_2 = \frac{1}{1 - \dfrac{\sum P_{nt}}{\sum P_{e2}}} = \frac{1}{1 - \dfrac{1.6(75)}{2(476)}} = 1.14$$

(d) Check the beam-column interaction criterion. Use Eq. 12.10.2,

$$\frac{P_u}{2\phi_c P_n} + \left(\frac{M_{ux}}{\phi_b M_n}\right) \le 1.0 \qquad\qquad [12.10.2]$$

$$M_{ux} = B_2(\phi_b M_p) = 1.14(294) = 335 \text{ ft-kips}$$

Note that in plastic analysis the combined gravity plus lateral loading is used; thus, M_{nt} and $M_{\ell t}$ are not separately computed. In the plastic analysis, it was assumed the full plastic moment strength of the column was developed when the $\phi_b M_p$ requirement for the girder was computed. Since the actual $\phi_b M_p$ of 915 ft-kips for the girder exceeds the requirement of 887 ft-kips, one could compute the actual column $\phi_b M_p$ that is utilized and find M_u utilized is 266 ft-kips. M_{ux} would then become $1.14(266) = 303$ ft-kips. This exceeds 294: ft-kips the design strength of the W14×48. Increase the column to W14×61 having $\phi_b M_p = 383$ ft-kips and $b_f = 10$ in.

Recomputing for W14×61, $r_x = 5.98$ in., $K_x L_x / r_x = 90.8$, $\phi_c F_{cr} = 24.7$ ksi, $A_g = 17.9$ in., $I_x = 640$ in^4, $P_{e2} = 628$ kips, and $B_2 = 1.10$. Thus, Eq. 12.10.2 becomes

$$\frac{61.6}{2(24.7)17.9} + \frac{1.10(266)}{327} = 0.069 + 0.895 = 0.96 < 1 \qquad \textbf{OK}$$

The design is considered adequate since the hinged-base assumption certainly understimates the moment restraint and lateral stiffness of the frame.

(e) Check shear in the knee region using stiffeners as necessary, as illustrated in Example 13.8.1.

Use W27×84 for the girder and W14×61 for the column.   ∎

EXAMPLE 15.3.3

Redesign the rectangular frame of Example 15.3.1 using AISC LRFD Method *without using plastic analysis*. Use A992 steel.

Solution:

(a) Factored loads. The factored load combinations were computed in Example 15.3.1, part (a). The gravity plus wind (Case 2) was determined to control. Design here only for that case. The factored loads are

Case 2: Gravity + wind
$$w_u = 1.2D + 1.6S + 0.8W$$
$$= 1.2(0.2) + 1.6(0.8) + 0.8(0.1) = 1.60 \text{ kips/ft}$$
$$w_{uh} = 0.8W = 0.8(0.44) = 0.352 \text{ kip/ft} \quad \text{(horizontal)}$$

(b) Structural analysis. Elastic analysis of the structure under factored loads is needed; *assume the EI for the column and girder are equal*, as was assumed in Example 15.3.1. Statically indeterminate analysis is used to obtain the factored moments in Fig. 15.3.3. The details of statically indeterminate analysis are outside the scope of this text; many standard references are available.* According to AISC-C2.1b, two analyses are needed; one for the gravity part (1.60 kips/ft), and one for the sway part (horizontal 0.352 kip/ft). The alternative of using a second-order analysis to obtain M_u is not used here.

(c) Select a preliminary trial section. Neglecting the magnification effect, the maximum factored moment $(M_{nt} + M_{\ell t})$ is

$$M_{nt} + M_{\ell t} = 614 + 58 = 672 \text{ ft-kips}$$

and the factored axial compression P_u is

$$P_u = 1.60(75)/2 + (53 + 58)/75 = 61.5 \text{ kips}$$

---

*See for example, Chu-Kia Wang and Charles G. Salmon, *Introductory Structural Analysis*. Englewood Cliffs, NJ: Prentice-Hall, Inc., 1984; C. K. Wang, *Intermediate Structural Analysis*. New York: McGraw-Hill Book Company, 1983; Chu-Kia Wang, *Structural Analysis on Microcomputers*. New York: Macmillan Publishing Company, 1986.

Figure 15.3.3
Design under LRFD without
using plastic analysis;
Examples 15.3.3 and 15.3.4.

(a) Elastic factored moments M_{nt}
(gravity — nonsway part)
Load Case 2

(b) Elastic factored moments $M_{\ell t}$
(wind — sway part)
Load Case 2

Use Eq. 12.12.2 or 12.12.3 to select a satisfactory preliminary section for combined axial load and bending. From Table 12.12.1, and for W24 section; estimate $p = 0.0005$ and $KL_{equiv} = 2.25(25)/4.78 = 11.7$ ft.

Check: $pP_u = 0.0005(61.5) = 0.031 < 0.2$ use Eq. 12.12.2

$$\frac{1}{2}pP_u + \frac{9}{8}M_{ux}b_x \le 1 \qquad [12.12.2]$$

$$\frac{0.031}{2} + \frac{9}{8}672\,b_x = 1.0$$

$$\therefore \quad b_x \le 0.00130$$

From AISC-Table 6–1, find W24×84 with $p = 0.00133$ and $b_x = 0.00123$, based on an estimated $KL_{equiv} = 11.7$ ft.

$$\text{Try W24×84;} \qquad \phi_b M_p = 840 \text{ ft-kips.}$$

(d) Check column action. The effective length factor K_x in the plane of the unbraced frame is determined for $I_g = I_c$ exactly as in Example 15.3.1. Thus as before

$$K_x = 2.25 \text{ (unbraced)} \qquad K_y = 1.0 \text{ (braced)}$$

$$\text{Axis of bending } \frac{KL}{r} = \frac{K_x L_x}{r_x} = \frac{2.25(25)12}{9.79} = 69$$

$$\frac{K_y L_y}{r_y} = \frac{1.0(5)12}{1.95} = 31$$

Since $K_x L_x / r_x > K_y L_y / r_y$, the major axis controls $\phi_c F_{cr}$,

$$K_x L_x / r_x = 69 \qquad \phi_c F_{cr} = 31.8 \text{ ksi}$$

The axial compression effect of the beam-column interaction equation is

$$\frac{P_u}{\phi_c P_n} = \frac{P_u}{\phi_c F_{cr} A_g} = \frac{61.5}{31.8(24.7)} = 0.078$$

Since this is less than 0.2, AISC Formula (H1-1b) applies.

(e) Compute nonsway magnifier B_1. According to AISC-C2.1b, K_1 is set to equal 1.0. So, $K_x L_x = 25$ ft for the axis of bending.

$$P_{e1} = \frac{\pi^2 EI}{(K_1 L)^2} = \frac{\pi^2 (29,000) 2370}{[25(12)]^2} = 7540 \text{ kips}$$

$$B_1 = \frac{C_m}{1 - P_u / P_{e1}} \geq 1$$

For members subjected to axial compression B_1 may be calculated using first order estimate $P_u = P_{nt} + P_{\ell t} = 61.6$ kips

$$B_1 = \frac{1.0}{1 - 61.6/7540} = 1.01$$

Since $B_1 = 1.01$ it means the maximum moment is close to the end of the member.

(f) Compute the sway magnifier B_2. Using $K_x L_x = 56$ ft, for the axis of bending $K_x = 2.25$,

$$\sum P_{e2} = \sum \frac{\pi^2 EI}{(K_2 L)^2} = \frac{2\pi^2 (29,000) 2370}{[56(12)]^2} = 2970 \text{ kips}$$

$$B_2 = \frac{1}{1 - \dfrac{\sum P_{nt}}{\sum P_{e2}}} = \frac{1}{1 - \dfrac{120}{2970}} = 1.04$$

(g) Check the beam-column interaction equation, Eq. 12.10.2, AISC Formula (H1-1b),

$$\frac{P_u}{2\phi_c P_n} + \frac{M_{ux}}{\phi_b M_n} \leq 1.0$$

$$M_{ux} = B_1 M_{nt} + B_2 M_{lt} = 1.02(614) + 1.04(58) = 687 \text{ ft-kips}$$

The interaction criterion becomes

$$\frac{0.078}{2} + \frac{687}{840} = 0.039 + 0.82 = 0.86 < 1.0 \qquad \textbf{OK}$$

(h) Check local flange buckling, local web buckling, and lateral-torsional buckling limit states. The W24×84 is a compact section since $\lambda < \lambda_p$ of AISC-B4. For the lateral-torsional buckling, $L_b = 3.5$ ft $< L_p = 6.89$ ft and $\phi_b M_n = \phi_b M_p$ as assumed. Had the laterally unbraced length has been larger, the possibility of M_n less than M_p would have been considered earlier in the design process. Note that the girder with $L_b = 6$ ft also has $M_n = M_p$.

Use W24×84 (revised below).

(i) Reduction in end moment according to AISC-Appendix 1.7. When the conditions of AISC-Appendix 1.3 for using plastic analysis are satisfied *but plastic analysis is not used,* the beam-column (and the beam) may be proportioned for "nine-tenths of the negative moments produced by gravity loading at points of support, provided that the maximum positive moment is increased by one-tenth of the average negative moments."

In this design, the conditions for using plastic analysis *are* met. Thus, instead of M_{nt} of 614 ft-kips, $0.90(614) = 553$ ft-kips could have been used. The positive moment would then have increased to $511 + 0.10(614 + 614)/2 = 572$ ft-kips, which exceeds the reduced negative moment value 553 ft-kips. In this case, equalizing the positive and negative moments would be appropriate to give $M_{nt} = 563$ ft-kips.

If $M_{nt} = 563$ ft-kips were used in the design, the section could be reduced to W24×68. For that section $B_1 = 1.02$, $B_2 = 1.06$, $M_u = 630$ ft-kips, $P_u/\phi_c P_{cr} = 0.102$, and $\phi_b M_p = 664$ ft-kips. The beam-column interaction criterion, Eq. 12.10.2, AISC Formula (H1-1b) gives

$$\frac{0.102}{2} + \frac{630}{664} = 0.051 + 0.949 = 1.00 \qquad \textbf{OK}$$

(j) Examine alternate calculation of magnification factor B_2 using Eq. 12.11.12, AISC Formula (C2-6b). The lateral load deflection Δ_H at the top of the frame is computed as 1.02 in. for the loading in Fig. 15.3.3 using W24×68 section for both beam and column. The total factored load shear ΣH in the story is $0.352(25) = 8.8$ kips, and the total factored gravity load $\Sigma P_u = 1.60(75) = 120$ kips. Equation 12.11.12 then gives

$$B_2 = \frac{1}{1 - \dfrac{\Sigma P_{nt}}{\Sigma P_{e2}}} = \frac{1}{1 - \dfrac{\Sigma P_{nt}}{R_M \Sigma HL/\Delta_H}} = \frac{1}{1 - \dfrac{120}{0.85(8.8)(25)(12)/1.02}} = 1.058$$

which is the same value computed using the other method.

(k) Conclusion.

Use W24×68. The knee joint still must be checked for shear, and reinforced with stiffeners if necessary, as illustrated in Example 13.8.1. ∎

EXAMPLE 15.3.4

Verify the adequacy of W24×68 determined in Example 15.3.3, using the Direct Analysis Method of AISC-Appendix 7.

Solution:

(a) Governing load combination is the same as before; gravity plus wind. $(1.2D + 1.6S + 0.8W)$.

(b) According to AISC-Appendix 7.3(2), when the ratio of second order drift to first order drift is less than 1.5, notional loads are *not required* in conjunction with other lateral loads. Therefore no notional loads are to be used since the computed B_2 in the previous example is equal to 1.06 which is less than 1.5.

(c) Compute the reduced flexural stiffness EI^* according to AISC Formula (A-7-2).

Using the section W24×68 determined in the previous example, the values are

$$P_y = A_e F_y = 20.1(50) = 1005 \text{ kips}$$

$$\frac{P_u}{P_y} = \frac{61.6}{1005} = 0.06 < 0.5 \qquad \therefore \tau_b = 1.0$$

$$EI^* = 0.8\tau_b EI = 0.8(1.0)EI = 0.8EI$$

(d) Compute the magnification factors B_1 and B_2.

$$\frac{K_x L_x}{r_x} = \frac{1.0(25)12}{9.55} = 31$$

$$\frac{K_y L_y}{r_y} = \frac{1.0(5)12}{1.87} = 32 \qquad \text{governs}$$

and $\phi F_{cr} = 41.8$ ksi $\phi P_n = \phi F_{cr} A_g = 41.8(20.1) = 840$ kips

Reduced $P_{e2} = 0.8(P_{e2}) = 0.8(2200) = 1760$ kips

$$B_2 = \frac{1}{1 - \frac{\sum P_{nt}}{\sum P_{e2}}} = \frac{1}{1 - \frac{120}{1760}} = 1.07$$

$P_u = 61.6$ kips

$Pe_1 = 0.8(P_{e1}) = 0.8(5820) = 4656$ kips

$$B_1 = \frac{1}{1 - \frac{P_u}{P_{e1}}} = \frac{1}{1 - \frac{61.6}{4656}} = 1.01$$

$$M_{ux} = B_1 M_{nt} + B_2 M_{nt} = 1.01(563) + 1.07(58) = 633 \text{ ft-kips}$$

and the column-beam interaction criterion, Eq. 12.10.2, AISC Formula (HI-2b) becomes

$$\frac{61.6}{2(840)} + \frac{633}{664} = 0.99 \qquad\qquad \textbf{OK}$$

W24×68 is adequate.   ■

EXAMPLE 15.3.5

Select W sections for the column of the frame of Fig. 15.3.4a, containing rigid joints at the tops of columns A and D. The tops of columns B and C are to be considered pinned. The bases of columns A and D are restrained such that $G = (\Sigma I/L)_{col}/(\Sigma I/L)_{beam} = 3.5$. The structure is braced in the direction perpendicular to the given frame, with each column braced at the top and bottom. Use A992 steel. (This example illustrates the unbraced frame concept discussed in the latter part of Sec. 14.5.) Use the AISC LRFD Method.

Solution:

(a) Compute factored loads.

$$P_{u1} = 1.2(20) + 1.6(60) = 120 \text{ kips}$$
$$P_{u2} = 1.2(35) + 1.6(100) = 202 \text{ kips}$$

(a) Actual frame

Neglect moment here; assume as hinge

(b) Sidesway $P\Delta$ effects

(c) Equivalent sidesway restraint system

Figure 15.3.4
Frame of Example 15.3.5.

(b) Design of columns B and C. These columns are designed as pinned-end columns within a *braced* system; thus, $K = 1.0$. The bracing is provided by the rigid framing between the girder and column at points A and D.

$$K_y L_y = 15 \text{ ft}; \qquad P_u = 202 \text{ kips}$$

Select from the *AISC Manual* "Available Strength in Axial Compression" $\phi_c P_n$ tables,

$$\text{Find W8×31}; \qquad [\phi_c P_n = 230 \text{ kips}] > P_u \qquad \qquad \textbf{OK}$$

Check the section:

$$\frac{K_y L_y}{r_y} = \frac{1.0(15)12}{2.02} = 89 \qquad \phi_c F_{cr} = 25.2 \text{ ksi}$$

$$[\phi_c P_n = \phi_c F_{cr} A_g = 25.2(9.12) = 230 \text{ kips}] > [P_u = 202 \text{ kips}] \qquad \qquad \textbf{OK}$$

Use W8×31 for interior columns B and C.

(c) Design exterior columns A and D using Yura's method [6.55] as discussed in Sec. 14.5. These columns must brace the entire assemblage. In order to resist sidesway the interior columns B and C develop combined $P\Delta$ moments of $2(202)\Delta = 404\Delta$, as shown in Fig. 15.3.4b. To stabilize the system requires development of a horizontal bracing force $404\Delta/h$ at the top of the frame. Since *two* identical exterior columns are intended to provide the bracing, each exterior column must contribute $202\Delta/h$. Thus on columns A and D the total sidesway moment is

$$\text{Total sidesway moment} = \left(\frac{202\Delta}{h}\right)h + 120\Delta = 322\Delta$$

The above neglects end moments that may be developed by the partial restraint at the column bases. In other words, the extra bracing force required is the same as if the member itself were carrying 322 kips. Thus, when the exterior columns are designed to carry $P_u = 322$ kips those columns will brace the entire system.

The K_x factor to use for columns A and D in the frame will depend on the stiffness of those columns; assume to start that $K_x = 2.0$. Then $K_xL_x = 2(15) = 30$ ft; $K_yL_y = 1.0(15) = 15$ ft. Use *AISC Manual* tables "Available Strength in Axial Compression" where K_yL_y is tabulated vs $\phi_c P_n$. Since the $P_u = 322$ kips must be carried only for the frame action, K_xL_x *governs* for design to carry 322 kips. Since the "Available Strength in Axial Compression" tables are based on K_yL_y, one must enter the tables with an equivalent $K_yL_y = K_xL_x/(r_x/r_y)$. Using those tables, find

$$\text{Equivalent } K_yL_y \text{ for } W10 = \frac{30}{2.16} = 13.9 \text{ ft}; \qquad \text{Gives W10}\times45$$

$$\text{Equivalent } K_yL_y \text{ for } W12 = \frac{30}{2.64} = 11.4 \text{ ft}; \qquad \text{Gives W12}\times40$$

For the W12 section, the tables "Available Strength in Axial Compression" give only W12×40 which carries much more load than required; try a lighter W12×35 for which there is no tabulated load. Check that $b_f/2t_f \leq 95\sqrt{F_y}$ and $h/t_w \leq 253/\sqrt{F_y}$ are satisfied.

For the y-y axis (minor-axis buckling), the actual P_u of 76 kips must be carried.

$$\frac{K_yL_y}{r_y} = \frac{1.0(15)12}{1.54} = 117; \qquad \phi_c F_{cr} = 16.5 \text{ ksi}$$

$$\phi_c P_n = \phi_c F_{cr} A_g = 16.5(10.3) = 170 \text{ kips} > 120 \text{ kips} \qquad \textbf{OK}$$

For the x-x axis, the end restraint must be evaluated:

$$G_{\text{top}} = \frac{\Sigma I/L \text{ for column}}{\Sigma I/L \text{ for girder}} = \frac{285/15}{758/36} = 0.90$$

The alignment chart, Fig. 6.9.4b, is based on far ends of girders restrained. When the far end is not restrained, as assumed here, the G factor should be increased; multiplied by two when the far end is hinged. Thus,

$$G_{\text{top}} = 0.90(2) = 1.80$$

$$G_{\text{bottom}} = 3.5 \quad (\text{given})$$

From the alignment chart, Fig. 6.9.4b, find $K_x = 1.72$.

$$\frac{K_x L_x}{r_x} = \frac{1.72(15)12}{5.25} = 59; \qquad \phi_c F_{cr} = 34.9 \text{ ksi}$$

$$\phi_c P_n = \phi_c F_{cr} A_g = 34.5(10.3) = 355 \text{ kips} > [P_u = 332 \text{ kips}] \qquad \textbf{OK}$$

Check AISC-Appendix 1.5, Eq. 15.3.5,

$$[0.75\phi_c A_g F_y = 0.75(0.90)(10.3)50 = 348 \text{ kips}] > [P_u = 322 \text{ kips}]$$

This satisfies AISC-Appendix 1.5.2; note that the 322 kips is not the true column load but rather a pseudo load such that the columns A and D can have adequate lateral stiffness when interacting with the beam to provide the sway resistance for the entire assemblage. The design acceptable!

Use W12×35 for exterior column.

   (d) Repeat the design of columns A and D using Geschwindner's method [14.42]. In this method, the actual factored load $P_u = 120$ kips is used on columns A and D, but the effective length factor K_0 becomes K_n given by Eq. 14.5.11. In part (b) above, the effective length factor K (now called K_0) was found to be 1.72. That value is now adjusted, as follows:

$$K_n = K_0 \sqrt{\frac{P_u + Q_u}{P_u}} \qquad\qquad [14.5.11]$$

$$= 1.72 \sqrt{\frac{120 + 202}{120}} = 2.8$$

Try the W12×35 obtained in part (c).

$$\frac{K_x L_x}{r_x} = \frac{2.8(15)12}{5.25} = 97; \qquad \phi_c F_{cr} = 22.6 \text{ ksi}$$

$$[\phi_c P_n = \phi_c F_{cr} A_g = 22.6(10.30) = 233 \text{ kips}] > [P_u = 120 \text{ kips}]$$

Because of the low slenderness ratio (i.e., 59) of the original column, the answer using the increased K_n is less conservative than using the increased load for design.
   Try W12×30. Check local buckling limits,

$$\left(\frac{b_f}{2t_f} = 7.4\right) < \left(\frac{95}{\sqrt{F_y}} = 13.4\right) \qquad \textbf{OK}$$

$$\left(\frac{h}{t_w} = 41.8\right) > \left(\frac{253}{\sqrt{F_y}} = 35.8\right) \qquad \textbf{NG}$$

There is a potential web local buckling problem; Q may be less than 1.0. Initially assume $Q = 1.0$,

$$G_{\text{top}} = 2\frac{238/15}{758/36} = 0.75(2) = 1.51$$

$$G_{\text{bottom}} = 3.5 \text{ (given)}$$

From alignment chart, Fig. 6.9.4b, find $K_x = 1.68$, which leads to $K_n = 2.75$.

$$\frac{K_x L_x}{r_x} = \frac{2.75(15)12}{5.21} = 95$$

$$\frac{K_y L_y}{r_y} = \frac{1.0(15)12}{1.52} = 118 \text{ Controls!}$$

$$\phi_c F_{\text{cr}} = 16.1 \text{ ksi}$$

Check h/t_w according to AISC-E7.2 (Eqs. 15.3.1 to 15.3.3) using $f = 16.1$ ksi,

$$\frac{h}{t_w} < \left(1.49 \sqrt{\frac{E}{F}} = \frac{253}{\sqrt{f}} = \frac{253}{\sqrt{16.1}} = 63.1 \right) \qquad \textbf{OK}$$

Thus, $Q = 1.0$.

$$[\phi_c P_n = \phi_c F_{\text{cr}} A_g = 16.1(8.79) = 142 \text{ kips}] > [P_u = 120 \text{ kips}] \qquad \textbf{OK}$$

Check AISC-Appendix 1.5.2, Eq. 15.3.5,

$$[0.75\phi_c A_g F_y = 0.75(0.90)(8.79)50 = 297 \text{ kips}] > [P_u = 120 \text{ kips}] \qquad \textbf{OK}$$

This does satisfy AISC-Appendix 1.5.2; and here the check is against the actual P_u of 120 kips. The next lightest W12 will not work.

Use W12×30 for exterior column.   ■

When lateral load acts on a system such as the frame of Fig. 15.3.4, the same procedure is used as illustrated in Example 15.3.3. The factored gravity load elastic analysis is made for the nonsway situation (assuming the frame and loading are relatively symmetrical and cause insignificant sway). Then the factored lateral load elastic analysis is made without the gravity load. When the sway magnifier B_2 is computed, the total P_u participating in the $P\Delta$ effect is used in ΣP_u, including the loads on columns B and C. However, ΣP_e includes only the contributions of those columns actually resisting sway; i.e., columns A and D in this example.

15.4   MULTISTORY FRAMES

The design of multistory frames, either braced or unbraced, is outside the scope of this text. *Braced* multistory frames have been designed using Allowable Stress Design (Working Stress Method) since the earliest design of steel structures. Plastic Design of multistory braced frames was accepted by AISC with the adoption of the 1969 *AISC Specification*. The theoretical and experimental background, developed primarily at Lehigh University, was first presented to the engineering profession in 1965 [15.6]. Additional background material is available in the work of Driscoll and Beedle [15.7], Lu [15.8], Williams and Galambos [15.9], Goldberg [15.10], and Yura and Lu [15.11]. The design procedure is well-described in *Plastic Design of Braced Multistory Steel Frames* [14.37].

Regarding *unbraced* multistory frames, procedures are continuing to be developed. AISC LRFD and ASD Methods [1.15] do not restrict the number of stories, and the $P\Delta$ effect *must* be considered. Presently (2008), there seems to be no practical accepted procedure for using Plastic Design on multistory unbraced frames.

The *SSRC Guide* [6.8, Chap. 16] contains excellent treatment of methods to evaluate frame stability for multistory frames. Background on unbraced frame stability is available in the Lehigh Lecture Notes [15.6], Driscoll and Beedle [15.7], and Daniels [15.12]. Other studies on the stability of unbraced frames are given by Cheong-Siat-Moy [15.13, 15.15, 14.18], and Cheong-Siat-Moy and Lu [15.14].

SELECTED REFERENCES

15.1. AISC. *Plastic Design in Steel*. New York: American Institute of Steel Construction, Inc., 1959.

15.2. Lynn S. Beedle. *Plastic Design of Steel Frames*. New York: John Wiley & Sons, Inc., 1958.

15.3. C. E. Massonnet and M. A. Save. *Plastic Analysis and Design, Vol. I, Beams and Frames*. Waltham, MA: Blaisdell Publishing Company, 1965.

15.4. Lynn S. Beedle. "Plastic Strength of Steel Frames," *Proceedings,* ASCE, **81,** Paper No. 764 (August 1955). Also *Transactions,* **122** (1957), 1139–1168.

15.5. Edward R. Estes, Jr. "Design of Multi-span Rigid Frames by Plastic Analysis," *Proceedings of National Engineering Conference,* AISC, 1955, 27–39.

15.6. George C. Driscoll, Jr., et al. *Plastic Design of Multi-Story Frames,* Lecture Notes, Fritz Engineering Laboratory Report No. 273.20, Lehigh University, Bethlehem, PA, 1965.

15.7. George C. Driscoll and Lynn S. Beedle. "Research in Plastic Design of Multistory Frames," *Engineering Journal,* AISC, **1,** 3 (July 1964), 92–100.

15.8. Le-Wu Lu. "Design of Braced Multi-Story Frames by the Plastic Method," *Engineering Journal,* AISC, **4,** 1 (January 1967), 1–9.

15.9. James B. Williams and Theodore V. Galambos. "Economic Study of a Braced Multi-Story Steel Frame" *Engineering Journal,* AISC **5,** 1 (January 1968), 2–11.

15.10. John E. Goldberg. "Lateral Buckling of Braced Multistory Frames," *Journal of the Structural Division,* ASCE, **94,** ST12 (December 1968), 2963–2983.

15.11. Joseph A. Yura and Le-Wu Lu. "Ultimate Load Tests of Braced Multistory Frames," *Journal of the Structural Division,* ASCE, **95,** ST10 (October 1969), 2243–2263.

15.12. J. Hartley Daniels. "A Plastic Method for Unbraced Frame Design," *Engineering Journal,* AISC, **3,** 4 (October 1966), 141–149.

15.13. F. Cheong-Siat-Moy. "Stiffness Design of Unbraced Steel Frames," *Engineering Journal,* AISC, **13,** 1 (First Quarter 1976), 8–10.

15.14. F. Cheong-Siat-Moy and Le-Wu Lu. "Stiffness and Strength Design of Multistory Frames," *Publications,* International Association for Bridge and Structural Engineering, **36**-II, 1976, 31–47.

15.15. Francois Cheong-Siat-Moy. "Frame Design Without Using Effective Column Length," *Journal of the Structural Division,* ASCE, **104,** ST1 (January 1978), 23–33.

PROBLEMS

All problems are to be done in accordance using the AISC LRFD Method. Use the plastic analysis option under AISC-B3.1, if applicable, only when specifically assigned by the instructor; otherwise use elastic analysis. All given loads are service loads.

15.1. Design using W shapes an unbraced rectangular frame of 60-ft span as shown in the accompanying figure, to carry uniformly distributed gravity dead and live loads of 1.5 and 2.5 kips/ft, respectively, not including the weight of the girder. Use a lateral wind load of 0.8 kip/ft acting uniformly over the 18-ft height. Lateral bracing occurs every 5 ft on the girder and 4.5 ft on the columns. Include, if assigned by the instructor, the design of the knee region at the top of the column as discussed in Sec. 13.8. Use A992 steel.

Problem 15.1

15.2. Design for the conditions of Prob. 15.1 except assume the moment of inertia for the beam to be three times that of the column.

15.3. Design using W shapes the unbraced rectangular frame of the accompanying figure to carry uniformly distributed gravity dead and live loads of 1.4 and 2.2 kips/ft, respectively. Use a lateral wind load of 0.8 kip/ft acting uniformly over the 14-ft height. Assume the structure is braced perpendicular to the

plane of the frame such that columns are braced at the top and bottom with respect to the minor axis, and that $K_y = 1.0$. Assume all joints at the tops of columns are rigid; i.e., fully moment resisting. Use A992 steel.

15.4. Design as in Prob. 15.3 except use simple connections (assume as pinned) at the tops of columns BG, CH, and DI, with rigid moment resisting connections only at A and E.

Problem 15.3   4 @ 30′ = 120′–0″

16

Composite Steel-Concrete Construction

16.1 HISTORICAL BACKGROUND

Steel framing supporting cast-in-place reinforced concrete slab construction was historically designed on the assumption that the concrete slab acts independently of the steel in resisting loads. No consideration was given to the composite effect of the steel and concrete acting together. This neglect was justified on the basis that the bond between the concrete floor or deck and the top of the steel beam could not be depended upon. However, with the advent of welding, it became practical to provide mechanical shear connectors to resist the horizontal shear which develops during bending.

Steel beams encased in concrete were widely used from the early 1900s until the development of lightweight materials for fire protection in the past 50 years. Some such beams were designed compositely and some were not. In the early 1930s bridge construction began to use composite sections. Not until the early 1960s was it economical to use composite construction for buildings. However, current practice (2008) utilizes composite action in nearly all situations where concrete and steel are in contact, both on bridges and buildings.

Composite construction, as treated in this chapter, consists either of a solid cast-in-place concrete slab placed upon and interconnected to a steel rolled W section or welded I-shaped girder, as shown in Fig. 16.1.1, or most commonly, the concrete slab is cast upon

Figure 16.1.1
Conventional composite
steel-concrete beam.

804

Shear stud connectors on flanges of bridge girders to be embedded in the concrete slab in order to make steel section and concrete slab act as a unit (i.e., compositely). (Photo by C. G. Salmon)

cold-formed steel deck (Fig. 16.1.2), which itself is supported on a steel I-shaped section. The corrugations (ribs) may be either parallel to or perpendicular to the supporting beam. When the ribs are parallel to the beam, the behavior is essentially that of a variable thickness slab supported directly on the steel beam. When the ribs are perpendicular to the steel W section, special treatment is required. The many varieties of composite steel–concrete construction are discussed in the State-of-the Art Report [16.1].

The composite beam is one having a wide flange (concrete slab), typically spanning 8 to 15 ft between parallel beams. Ordinary beam theory, where the stress is assumed constant across the width of a beam at a given distance from the neutral axis, *does not apply*. Plate theory indicates the stress decreases the more distant a point is from the stiff part (steel section in this case) of the beam. Similarly to the treatment of T-sections in reinforced concrete, an *equivalent width* is used in place of the actual width, so that ordinary

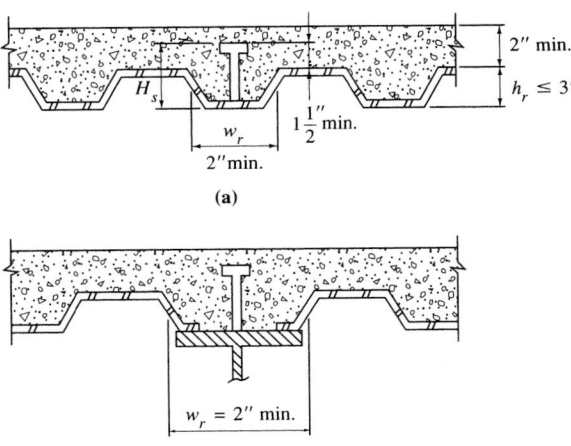

Figure 16.1.2
Composite section using formed steel deck. Steel beam supporting deck and slab may be parallel to ribs of formed deck (as in b.) or perpendicular to the ribs. (AISC-I3.2c)(Adapted from AISC Commentary [1.14])

beam theory can be used. An excellent summary of the factors involved in obtaining an effective width is given by Brendel [16.2] and Heins and Fan [16.3]. Vallenilla and Bjorhovde [16.4] have reviewed the effective width in the context of LRFD and the use of steel deck to support the slab.

Viest [16.5], in his 1960 review of research, notes that the important factor in composite action is that the bond between concrete and steel remain unbroken. As designers began to place slabs on top of supporting steel beams, investigators began to study the behavior of mechanical shear connectors. The shear connectors provided the interaction necessary for the concrete slab and steel beam to act as a unit; i.e., no slip between the concrete and steel beam parallel to the beam. For the earlier encased beams there had been sufficient contact area between concrete and steel so that friction provided the necessary interaction between the two materials.

The State-of-the-Art Report of 1974 [16.1] provides an overall survey of the subject of composite construction, including bibliography. Hansell, Galambos, Ravindra, and Viest [16.6] have provided the background for Load and Resistance Factor Design. Iyengar and Iqbal [16.7] have provided a modern review of composite construction in building design, and Lorenz and Stockwell [16.8] and Lorenz [16.9] have provided treatment of basic design concepts for Load and Resistance Factor Design.

A thorough treatment of steel-concrete composite construction in the context of *Eurocode 4* has been developed by IABSE [16.42].

16.2   COMPOSITE ACTION

Composite action is developed when two load-carrying structural members such as a concrete floor system and the supporting steel beam (Fig. 16.2.1a) are integrally connected and deflect as a single unit as in Fig. 16.2.1b. The extent to which composite action is developed depends on the provisions made to insure a single linear strain from the top of the concrete slab to the bottom of the steel section.

In developing the concept of composite behavior, consider first the noncomposite beam of Fig. 16.2.1a, wherein if friction between the slab and beam is neglected, the beam and slab each carry separately a part of the load. This is further shown in Fig. 16.2.2a. When the slab deforms under vertical load, its lower surface is in tension and elongates; while the upper surface of the beam is in compression and shortens. Thus a discontinuity will occur at the plane of contact. Since friction is neglected, only vertical internal forces act between the slab and beam.

Figure 16.2.1
Comparison of deflected beams with and without composite action.

(a) Deflected noncomposite beam

(b) Deflected composite beam

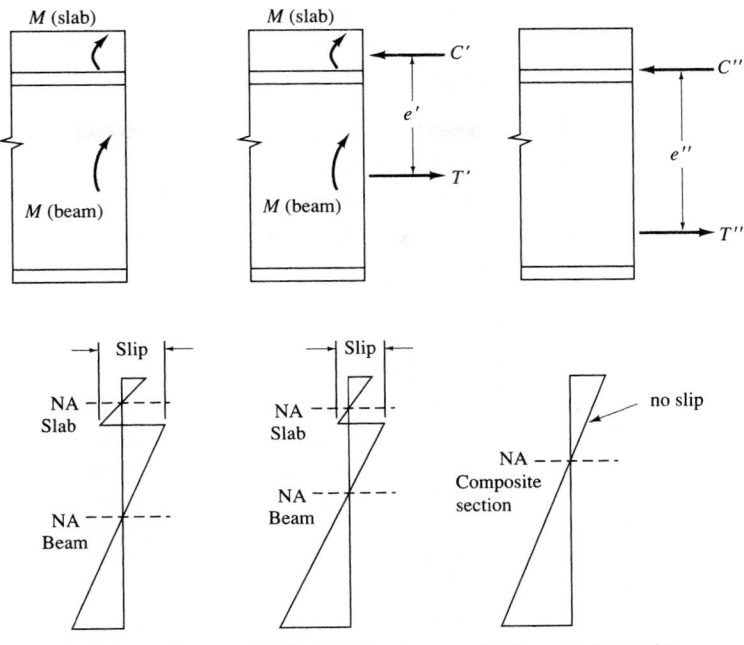

Figure 16.2.2
Strain variation in composite beams.

(a) No interaction   (b) Partial interaction   (c) Complete interaction

When a system acts compositely (Fig. 16.2.1b and 16.2.2c) no relative slip occurs between the slab and beam. Horizontal forces (shears) are developed that act on the lower surface of the slab to compress and shorten it, while simultaneously they act on the upper surface of the beam to elongate it.

By an examination of the strain distribution that occurs when there is no interaction between the concrete slab and the steel beam (Fig. 16.2.2a), it is seen that the total resisting moment is equal to

$$\Sigma M = M_{\text{slab}} + M_{\text{beam}} \qquad (16.2.1)$$

It is noted that for this case there are two neutral axes; one at the center of gravity of the slab and the other at the center of gravity of the beam. The horizontal slip resulting from the bottom of the slab in tension and the top of the beam in compression is also indicated.

Consider next the case where only partial interaction is present, Fig. 16.2.2b. The neutral axis of the slab is closer to the beam and that of the beam closer to the slab. Due to the partial interaction, the horizontal slip has now decreased. The result of the partial interaction is the partial development of the maximum compressive and tensile forces C' and T', in the concrete slab and steel beam, respectively. The resisting moment of the section would then be increased by the amount $T'e'$ or $C'e'$.

When complete interaction (known as full composite action) between the slab and the beam is developed, no slip occurs and the resulting strain diagram is shown in Fig. 16.2.2c. Under this condition, a single neutral axis occurs which lies below that of the slab and above that of the beam. In addition, the compressive and tensile forces C'' and T'', respectively, are larger than the C' and T' existing with partial interaction. The resisting moment of the fully developed composite section then becomes

$$\Sigma M = T''e'' \quad \text{or} \quad C''e'' \qquad (16.2.2)$$

16.3   ADVANTAGES AND DISADVANTAGES

The basic advantages resulting from composite design are

1. Reduction in the weight of steel
2. Shallower steel beams
3. Increased floor stiffness
4. Increased span length for a given number

A weight savings in steel of 20 to 30% is often possible by taking full advantage of a composite system. Such a weight reduction in the supporting steel beams usually permits the use of a shallower as well as a lighter member. This advantage may reduce the height of a multistoried building significantly so as to provide savings in other building materials such as outside walls and stairways. The overall economy of using composite construction when considering total building cost appears to be increasingly favorable [16.10, 16.11].

The stiffness of a composite floor is substantially greater than that of a concrete floor with its supporting beams acting independently. Normally the concrete slab acts as a one-way plate spanning between the supporting beams. In composite design, an additional use is made of the slab by its action in a direction parallel to and in combination with the supporting steel beams. The net effect is to greatly increase the moment of inertia of the floor system in the direction of the steel beams. The increased stiffness considerably reduces the live load deflections and, if shoring is provided during construction, also reduces dead load deflections. Assuming full composite action, the nominal strength of the section greatly exceeds the sum of the strengths of the slab and the beam considered separately, providing high overload capacity.

While there are no major disadvantages, some limitations should be recognized. In continuous construction, the negative moment region will have a different stiffness because the concrete slab in tension is expected to be cracked and not participating. In general, it is considered acceptable to assume the moment of inertia to be constant through both positive and negative moment regions, using the positive moment composite section moment of inertia AISC-I1. Tension in the concrete is neglected.

Long-term deflection caused by concrete creep and shrinkage could be important when the composite section resists a substantial part of the dead load, or when the live load is of long duration. This is discussed in Sec. 16.12.

16.4   EFFECTIVE WIDTH

The concept of *effective width* is useful in design when strength must be determined for an element subject to nonuniform distribution of stress. Referring to Fig. 16.4.1, the concrete slab of a composite section is considered to be infinitely wide. The intensity of extreme fiber stress f_c is a maximum over the steel beam and decreases nonlinearly as the distance from the supporting beam increases.

The effective width b_E of a flange for a composite member may be expressed

$$b_E = b_f + 2b' \tag{16.4.1}$$

where $2b'$ times the maximum stress f_c equals the area under the curves for f_c. Various investigators, including Timoshenko and Goodier [16.12] and von Kármán [16.13], have derived expressions for the effective width of homogeneous beams having wide flanges; and Johnson and Lewis [16.14] have shown such expressions are valid for beams in which the flange and web are of different materials.

The analysis for effective width involves theory of elasticity applied to plates, using an infinitely long continuous beam on equidistant supports, with an infinitely wide flange

Figure 16.4.1
Actual and equivalent stress
distribution over flange
width.

having a small thickness compared to the beam depth. The total compression force carried
by the equivalent system must be the same as that carried by the real system.

The practical simplifications for design purposes are given by AISC-I3.1a the *same*
for service load calculations as for nominal strength calculations when failure is imminent.

1. For an *interior girder*, referring to Fig. 16.4.2,

$$b_E \leq \frac{L}{4} \tag{16.4.2a}$$

$$b_E \leq b_0 \quad \text{(for equal beam spacing)} \tag{16.4.2b}$$

2. For an *exterior girder*,

$$b_E \leq \frac{L}{8} + \left(\begin{array}{l} \text{distance from beam center} \\ \text{to edge of slab} \end{array} \right) \tag{16.4.3a}$$

$$b_E \leq \tfrac{1}{2}b_0 + \left(\begin{array}{l} \text{distance from beam center} \\ \text{to edge of slab} \end{array} \right) \tag{16.4.3b}$$

Figure 16.4.2
Dimensions governing
effective width b_E on
composite steel-concrete
beams.

The American Concrete Institute (ACI) Code [16.15] has long used the following effective flange widths for *T*-sections:

1. For an *interior girder*, referring to Fig. 16.4.2,

$$b_E \leq \frac{L}{4} \tag{16.4.4a}$$

$$b_E \leq b_0 \qquad \text{(for equal beam spacing)} \tag{16.4.4b}$$

$$b_E \leq b_f + 16t_s \tag{16.4.4c}$$

2. For an *exterior girder*,

$$b_E \leq b_f + \frac{L}{12} \tag{16.4.5a}$$

$$b_E \leq b_f + 6t_s \tag{16.4.5b}$$

$$b_E \leq b_f + 0.5(\text{clear distance to next beam}) \tag{16.4.5c}$$

These 2008 *ACI Code* effective widths are identical to AISC effective widths used prior to the 1986 LRFD Specification. The present AISC rules are simpler, eliminating the beam flange width b_f and the slab thickness t_s as variables.

16.5 COMPUTATION OF ELASTIC SECTION PROPERTIES

The elastic section properties of a composite section can be computed by the transformed section method. In contrast to reinforced concrete, where the reinforcing bar steel is transformed into an equivalent concrete area, the concrete slab in the composite section is transformed into equivalent steel. As a result, the concrete area is reduced by using a slab width equal to b_E/n, where n is the modulus of elasticity ratio E_s/E_c. E_s is the modulus of elasticity of steel, taken as 29,000 ksi, and E_c in psi is given by the ACI Code [16.15], as follows:

$$E_c = 33(w^{1.5})\sqrt{f'_c}, \text{psi} \tag{16.5.1}*$$

where w is the density of concrete in pcf and f'_c is in psi. Since the *AISC Specification* uses stress in ksi for all formulas, AISC-I2.1b converts Eq. 16.5.1 approximately to the following for E_c in ksi:

$$E_c = w^{1.5}\sqrt{f'_c}, \text{ksi} \tag{16.5.2}*$$

Note that $\sqrt{1000}$ is 31.6; thus, Eq. 16.5.2 gives E_c about 4% lower than the ACI Code. For normal-weight concrete, weighing approximately 145 pcf, Eq. 16.5.2 gives E_c in ksi as

$$E_c = 1750\sqrt{f'_c}, \text{ksi} \tag{16.5.3}*$$

Within the accuracy that the modulus of elasticity of concrete may be predicted, either the *ACI Code* [16.15] value or the suggested value of AISC-I2.1b is acceptable. The modulus of elasticity ratio n is commonly taken to the nearest whole number. Table 16.5.1 indicates practical values usually used in computing elastic section properties.

---

*For SI units, giving E_c in MPa,

$$E_c = w^{1.5}(0.043)\sqrt{f'_c} \tag{16.5.1}$$

$$E_c = w^{1.5}(0.041)\sqrt{f'_c} \tag{16.5.2}$$

$$E_c = 4600\sqrt{f'_c} \tag{16.5.3}$$

where w is in kg/m³ and f'_c is in MPa.

TABLE 16.5.1 Practical Values for Modular Ratio n

f'_c (psi)	Modular ratio $n = E_s/E_c$	f'_c (MPa)
3000	9	21
3500	$8\frac{1}{2}$	24
4000	8	28
4500	$7\frac{1}{2}$	31
5000	7	35
6000	$6\frac{1}{2}$	42

Effective Elastic Section Modulus

A complete beam may be considered as a steel member to which has been added a cover plate on the top flange. This "cover plate" being concrete is considered to be effective only when the top flange is in compression. In continuous beams, the concrete slab is usually ignored in regions of negative moment. If the neutral axis falls within the concrete slab, present practice is to consider only that portion of the concrete slab which is in compression.

AISC-I3.2 permits reinforcement parallel to the steel beam and lying within the effective slab width to be included in computing properties of composite sections. These reinforcing bars usually make little difference to the composite section modulus in the positive moment region and their effect is frequently neglected.

EXAMPLE 16.5.1

Compute the elastic section properties of the composite section shown in Fig. 16.5.1 assuming $f'_c = 3000$ psi and $n = 9$. Use the effective flange width according to AISC-I3.1a.

Solution:
First, determine effective width (AISC-I3.1a).

$$b_E = L/4 = 0.25(30)12 = 90 \text{ in.} \quad \textit{controls}$$

$$b_E = b_0 = 8(12) = 96 \text{ in.}$$

The width of equivalent steel is $b_E/n = 10.0$ in. The computation of the moment of inertia I_x about the center of gravity of the W21×62 is shown, as follows:

Element	Transformed Area A (sq in.)	Moment Arm from Centroid y (in.)	Ay (in.³)	Ay^2 (in.⁴)	I_0 (in.⁴)
Slab	40.0	+12.495	+500	6245	53
W21×62	18.3	0	0	0	1330
Cover plate	7.0	−10.995	−77	846	1
	65.3		+423	7091	1384

$$I_x = I_0 + Ay^2 = 1384 + 7091 = 8475 \text{ in.}^4$$

$$\bar{y} = \frac{+423}{65.3} = +6.48 \text{ in.}$$

$$I_{tr} = I_x - A\bar{y}^2 = 8475 - 65.3\,(6.48)^2 = 5737 \text{ in.}^4$$

$$y_t = 10.50 - 6.48 + 4.0 = 8.02 \text{ in.}$$

$$y_b = 10.50 + 6.48 + 1.0 = 17.98 \text{ in.}$$

The symbol I_{tr} is used for the *fully composite uncracked transformed section* moment of inertia. The elastic section modulus S_{conc} referred to the top fiber of the concrete slab is

$$S_{conc} = I_{tr}/y_t = 5737/8.02 = 715 \text{ in.}^3$$

The elastic section modulus S_{tr} referred to the extreme fiber at the tension flange of the steel section (in this case the cover plate) is

$$S_{tr} = I_{tr}/y_b = 5737/17.98 = 319 \text{ in.}^3$$

The addition of a cover plate at the tension flange brings the neutral axis down and permits more economical use of the composite section. However, the cost of welding a cover plate to the rolled section usually exceeds any material saving; thus, a cover plate is rarely used. ∎

Figure 16.5.1
Composite section for
Example 16.5.1.

16.6   SERVICE LOAD STRESSES WITH AND WITHOUT SHORING

The actual stresses that result due to a given loading on a composite member are dependent upon the manner of construction.

The simplest construction occurs when the steel beams are placed first and used to support the concrete slab formwork. In this case the steel beam acting noncompositely (i.e., by itself) supports the weight of the forms, the wet concrete, and its own weight. Once forms are removed and concrete has cured, the section will act compositely to resist all dead and live loads placed after the curing of concrete. Such construction is said to be *without temporary shoring* (i.e., unshored).

Alternatively, to reduce the service load stresses, the steel beams may be supported on temporary shoring; in which case, the steel beam, forms, and wet concrete, are carried by the shores. After curing of the concrete, the shores are removed and the section acts compositely to resist all loads. This system is called *shored* construction.

The following example illustrates the difference in service load stresses under the two systems of construction.

EXAMPLE 16.6.1

For the steel W21×62 with the 1 by 7-in. plate of Fig. 16.5.1, determine the service load stresses considering that (a) construction is without temporary shoring, and (b) construction uses temporary shores. The dead- and live-load moment M_L to be superimposed on the system after the concrete has cured is 560 ft-kips.

Solution:
The composite section properties as computed in Example 16.5.2 are

$$S_{top} = 715 \text{ in.}^3 \quad (\text{top of concrete})$$

$$S_{bottom} = S_{tr} = 319 \text{ in.}^3 \quad (\text{bottom of steel})$$

The noncomposite properties for the steel section alone (see Fig. 16.6.1) are computed as follows:

$$\bar{y} = \frac{7.0(10.995)}{7.0 + 18.3} = 3.04 \text{ in.}$$

$$y_b = 10.495 - 3.04 + 1.00 = 8.45 \text{ in.}$$

Figure 16.6.1
Steel section for
Example 16.6.1.

$$I_s = I_0 \, (\text{W21} \times 62) + A_p y^2 - A \bar{y}^2$$

$$= 1330 + 7.0(10.995)^2 - 25.3(3.04)^2$$

$$= 1330 + 846 - 234 = 1942 \text{ in.}^4$$

$$S_{st} = \frac{1942}{13.55} = 143 \text{ in.}^3 \quad (\text{top})$$

$$S_{sb} = \frac{1942}{8.45} = 230 \text{ in.}^3 \quad (\text{bottom})$$

(a) *Without Temporary Shores.* Weight due to the concrete slab and steel beam,

$$w \text{ (concrete slab), } (4/12)(8)0.15 = 0.40$$

$$w \text{ (steel beam)} \qquad\qquad\qquad = \underline{0.06}$$

$$0.46 \text{ kips/ft}$$

$$M_D \text{ (DL on noncomposite)} = \tfrac{1}{8}(0.46)(30)^2 = 51.8 \text{ ft-kips}$$

$$f_{\text{top}} = \frac{M_D}{S_{st}(\text{steel section})} = \frac{51.8(12)}{143} = 4.3 \text{ ksi}$$

$$f_{\text{bottom}} = \frac{M_D}{S_{sb}(\text{steel section})} = \frac{51.8(12)}{230} = 2.7 \text{ ksi}$$

The additional stresses after the concrete has cured are

$$f_{\text{top}} = \frac{M_L}{nS_{\text{top}}(\text{composite})} = \frac{560(12)}{9(715)} = 1.04 \text{ ksi (concrete stress)}$$

where the stress in the concrete is $1/n$ times the stress on equivalent steel (transformed section).

$$f_{\text{bottom}} = \frac{M_L}{S_{\text{tr}}} = \frac{560(12)}{319} = 21.1 \text{ ksi}$$

The total maximum tensile stress in the steel is

$$f = f(\text{noncomposite}) + f(\text{composite}) = 2.7 + 21.1 = 23.8 \text{ ksi}$$

(b) *With Temporary Shores.* Under this condition all loads are resisted by the composite section.

$$f_{\text{top}} = \frac{M_D + M_L}{S_{\text{top}}(\text{composite})} = \frac{(560 + 51.8)12}{715(9)} = 1.14 \text{ ksi} \quad \text{on concrete}$$

$$f_{\text{bottom}} = \frac{M_D + M_L}{S_{\text{tr}}} = \frac{(560 + 51.8)12}{319} = 23.0 \text{ ksi}$$

Stress distributions for both with and without shores are given in Fig. 16.6.2. Since the dead load was small in this example, use of shores gave insignificant reduction in service load stress. Where thicker slabs are used, the dead load stresses may become as high as 30%, in which case using or not using shores will make a significant difference. ∎

Figure 16.6.2
Service load stresses for
Example 16.6.1.

(a) Without shores

(b) With shores

16.7 NOMINAL MOMENT STRENGTH OF FULLY COMPOSITE SECTIONS

The nominal strength M_n of a composite section having its slab in compression (positive moment) depends on the yield stress F_y and section properties (including slenderness $\lambda = h/t_w$ for the web) for the steel beam, the concrete slab strength f_c', and the strength of shear connectors providing the interface shear transfer between slab and beam.

The nominal strength (commonly called *ultimate strength*) concepts were first applied to design practice as recommended by the ASCE-ACI Joint Committee on Composite Construction [16.16], and further modified by Slutter and Driscoll [16.17]. Ultimate strength was reviewed in the State-of-the-Art Report [16.1], and treated in the context of Load and Resistance Factor Design by Hansell et al. [16.6].

Traditionally, since the Joint Committee Report [16.16] the design of composite beams has been based on nominal moment strength even though Allowable Stress Design was used. Load and Resistance Factor Design is particularly adapted to using composite flexural members since the concepts of strength are easier to understand without trying to convert them into a service load based Allowable Stress Design.

The nominal moment strength M_n when the slab is in compression (positive moment) is divided into two categories according to AISC-I3.2a, depending on web slenderness, as follows:

1. For $h/t_w \leq \left[3.76\sqrt{E/F_y} = 640/\sqrt{F_y}\right]$

M_n = nominal moment strength based on plastic stress distribution on the composite section (plastic moment)

$\phi_b = 0.90$

2. For $h/t_w > \left[3.76\sqrt{E/F_y} = 640/\sqrt{F_y}\right]$

M_n = nominal moment strength based on superposition of elastic stresses (shown in Sec. 16.6), considering the effects of shoring, for the limit state of yielding (yielding moment)

$\phi_b = 0.90$

Since the elastic properties and effects of shoring have been treated in Sec. 16.6, this section focuses on strength based on plastic stress distribution. It is noted that all current ASTM A6 W shapes satisfy the limit for Case 1.

The nominal strength M_n based on plastic stress distribution may be divided into two general categories: (1) the plastic neutral axis (PNA) occurs in the slab; and (2) the plastic neutral axis occurs in the steel section. When the PNA occurs in the steel section, the nominal strength M_n calculation will differ depending on whether the PNA is in the flange or the web.

The concrete is assumed to develop only compression forces. Although concrete is able to sustain a limited amount of tension, the tensile strength is negligible at the strains occurring when nominal strength is reached.

Case 1—Plastic Neutral Axis (PNA) in the Slab

Referring to Fig. 16.7.1b and assuming the Whitney rectangular stress distribution* (uniform stress of $0.85f'_c$ acting over a depth a), the compressive force C is

$$C = 0.85 f'_c \, ab_E \qquad (16.7.1)$$

The tensile force T is the yield stress on the beam times its area:

$$T = A_s F_y \qquad (16.7.2)$$

Equating the compressive force C to the tensile force T gives

$$a = \frac{A_s F_y}{0.85 f'_c b_E} \qquad (16.7.3)$$

According to the ACI-accepted [16.15, Sec. 10.2.7] rectangular stress distribution, the neutral axis distance x, as shown in Fig. 16.7.1d, equals $a/0.85$ for $f'_c \leq 4000$ psi. The nominal moment strength M_n, from Fig. 16.7.1b, becomes

$$M_n = Cd_1 \quad \text{or} \quad Td_1 \qquad (16.7.4)$$

When the slab is capable of developing a compressive force at least equal to the full yield strength of the steel beam, the PNA will be in the slab, the common situation for fully composite sections. Expressing the nominal strength in terms of the steel force gives

$$M_n = A_s F_y \left(\frac{d}{2} + t_s - \frac{a}{2} \right) \qquad (16.7.5)$$

The usual procedure for computing nominal strength is to assume the depth a for the rectangular stress distribution will not exceed t_s; i.e., use Eq. 16.7.3. If a is verified to not exceed t_s, Eq. 16.7.5 can be used to obtain nominal strength M_n.

In the past, Case 1 has been referred to as "slab adequate"; meaning that the slab is capable of developing in compression the full nominal strength of the steel beam in tension.

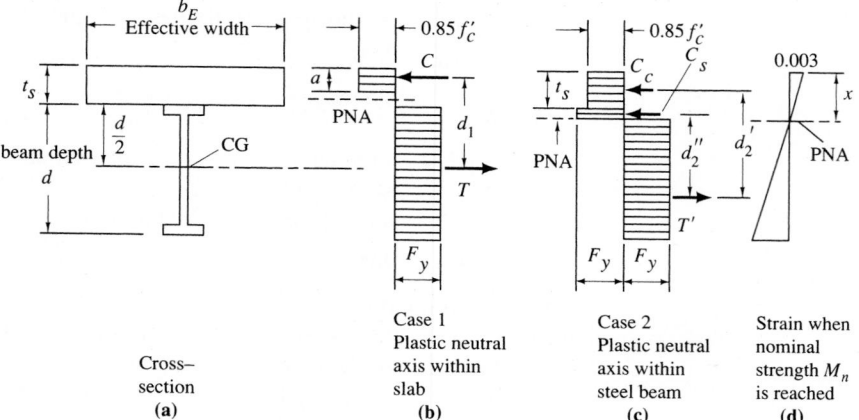

Figure 16.7.1
Plastic stress distribution at nominal moment strength M_n. (PNA = plastic neutral axis)

Cross–section (a)

Case 1
Plastic neutral axis within slab (b)

Case 2
Plastic neutral axis within steel beam (c)

Strain when nominal strength M_n is reached (d)

---

*For the development of the concept of replacing the true distribution of compressive stress by a rectangular stress distribution, see for example, Chu-Kia Wang, Charles G. Salmon, and José Pinchiera. *Reinforced Concrete Design*, 7th ed. (Wiley, 2006, Chap. 3).

Case 2—Plastic Neutral Axis (PNA) in the Steel Beam

If the depth a of the stress block as determined in Eq. 16.7.3 exceeds the slab thickness, the stress distribution will be as shown in Fig. 16.7.1c. The compressive force C_c in the slab is

$$C_c = 0.85 f'_c b_E t_s \qquad (16.7.6)$$

The compressive force in the steel beam resulting from the portion of the beam above the neutral axis is shown in Fig. 16.7.1c as C_s.

The tensile force T' which is now less than $A_s F_y$ must equal the sum of the compressive forces:

$$T' = C_c + C_s \qquad (16.7.7)$$

Also,

$$T' = A_s F_y - C_s \qquad (16.7.8)$$

Equating Eqs. 16.7.7 and 16.7.8, C_s becomes

$$C_s = \frac{A_s F_y - C_c}{2}$$

or

$$C_s = \frac{A_s F_y - 0.85 f'_c b_E t_s}{2} \qquad (16.7.9)$$

Considering the compressive forces C_c and C_s, the nominal moment strength M_n for Case 2 is

$$M_n = C_c d'_2 + C_s d''_2 \qquad (16.7.10)$$

where the moment arms d'_2 and d''_2 are as shown in Fig. 16.7.1c.

When the Case 2 situation occurs, the steel beam must be capable of accommodating plastic strain in both tension and compression to achieve the nominal strength condition. The lower the PNA occurs in the steel section the more local buckling may influence the behavior. As indicated earlier in this section, in order to use the plastic stress distribution at all, AISC-I3.2a requires the web $\lambda \le \lambda_p$.

When the flange of the steel section adjacent to the slab is in compression, there might be concern regarding flange local buckling. The combination of concrete bearing against the compression flange and the shear connectors used to attach the slab and steel beam together eliminates flange local buckling as well as lateral-torsional buckling as controlling limit states. AISC-I3.2 addresses only the issue of web local buckling; it is silent regarding flange local buckling.

EXAMPLE 16.7.1

Determine the nominal moment strength M_n of the composite section shown in Fig. 16.7.2. Use A992 steel, $f'_c = 4000$ psi, and $n = 8$.

Solution:
Assume the plastic neutral axis (PNA) is within the slab; i.e., that $a \le t_s$ (Case 1),

$$a = \frac{A_s F_y}{0.85 f'_c b_E} = \frac{10.6(50)}{0.85(4)60} = 2.60 \text{ in.} < t_s \qquad \textbf{OK}$$

$$C = 0.85 f'_c a b_E = 0.85(4)(2.6)60 = 530 \text{ kips}$$

Figure 16.7.2
Example 16.7.1.

$$T = A_s F_y = 10.6(50) = 530 \text{ kips} \quad \text{(checks)}$$

$$\text{Arm } d_1 = \frac{d}{2} + t - \frac{a}{2} = 7.925 + 4.0 - 1.30 = 10.63 \text{ in.}$$

The nominal moment strength M_n is then

$$M_n = Cd_1 = 530(10.63)\tfrac{1}{12} = 470 \text{ ft-kips} \qquad \blacksquare$$

EXAMPLE 16.7.2

Determine the nominal moment strength M_n of the composite section shown in Fig. 16.7.3. Use A992 steel, $f'_c = 4000$ psi, and $n = 8$.

Solution:
Referring to Fig. 16.7.3, assume the plastic neutral axis (PNA) is within the flange (i.e., Case 1),

$$a = \frac{A_s F_y}{0.85 f'_c b_E} = \frac{47.0(50)}{0.85(4)(72)} = 9.60 \text{ in.} > \left[t_s = 7.0 \text{ in.} \right] \qquad \textbf{NG}$$

Since the concrete slab is only 7 in. thick, the slab cannot develop enough strength to balance the tension force $A_s F_y$ capable of developing in the steel section; thus the PNA will be within the steel section; thus, Case 2 applies. Using Eq. 16.7.6,

$$C_c = 0.85 f'_c b_E t_s = 0.85(4)72(7) = 1714 \text{ kips}$$

Figure 16.7.3
Example 16.7.2.

Using Eq. 16.7.9.

$$C_s = \frac{A_s F_y - 0.85 f'_c b_E t_s}{2} = \frac{47.0(50) - 1714}{2} = 318 \text{ kips}$$

Assuming only the flange of the W36×160 ($b_f = 12.00$ in.) is in compression, the portion of the flange d_f to the neutral axis is

$$d_f = \frac{318}{50(12.00)} = 0.53 \text{ in.} < [t_f = 1.020 \text{ in.}]$$

Thus, the PNA is within the flange. The location of the centroid of the tension portion of the steel beam from the bottom is

$$\bar{y} = \frac{47.0(18) - 0.53(12)35.76}{47.0 - 0.53(12)} = 15.22 \text{ in.}$$

Referring to Fig. 16.7.3, the nominal composite moment strength M_n from Eq. 16.7.10 is

$$M_n = C_c d'_2 + C_s d''_2$$
$$= [1714(23.95) + 318(20.18)]/12 = 3960 \text{ ft-kips} \qquad ■$$

The nominal strength M_n has inherently assumed that shear connectors will provide sufficient shear transfer at the slab-to-flange interface to develop however much of the slab compressive strength that is required to balance the tension force developed in the steel beam. Shear connectors are treated in Sec. 16.8.

The nominal strength M_n is *independent* of whether or not the system is shored during construction. Even though service load stresses are different, as illustrated in Sec. 16.6, the nominal strength is the same, shored or unshored.

16.8   SHEAR CONNECTORS

The horizontal shear that develops between the concrete slab and the steel beam during loading must be resisted so that the slip shown in Fig. 16.2.2 will be restrained. A fully composite section will have no slip at the concrete-steel interface. Although some bond may develop between the steel and the concrete, it is not sufficiently predictable to provide the required interface shear strength. Neither can friction between the concrete slab and the steel beam develop such strength.

Instead, mechanical shear connectors are *required* (AISC-I3.2d), except for the totally concrete-encased steel beam. Some mechanical shear connectors are shown in Fig. 16.8.1. The only connectors specifically provided for in the *AISC Specification* are stud shear connectors [AISC-I3.2d(3)] and channel connectors [AISC-I3.2d(4)]. Currently (2008), nearly all shear connectors are headed studs.

Ideally, to obtain a *fully composite section* the shear connectors should be stiff enough to provide the complete interaction (i.e., no slip at the interface) shown in Fig. 16.2.2c. This, however, would require that the connectors be infinitely rigid. Also, by referring to the shear diagram for a uniformly loaded beam as shown in Fig. 16.8.2, it would be inferred, theoretically at least, that more shear connectors are required near the ends of the span where the shear is high, than near midspan where the shear is low.

Consider the shear stress distribution of Fig. 16.8.2b wherein the stress v_1 must be developed by the connection between the slab and beam. Under elastic conditions the shear stress at any point in the cross-section will vary from a maximum at the support to zero at

Figure 16.8.1
Shear connectors.

(a) Stud connectors

(b) Channel connectors

(c) Spiral connector

(d) Angle connectors

Figure 16.8.2
Shear variation for uniform
loading and distribution of
shear stress over the depth of
a steel-concrete composite
section.

midspan. Next, examine the equilibrium of an elemental slice of the beam, as in Fig. 16.8.3. The shear force per unit distance along the span is $dC/dx = v_1 b_E = V(\int y\, dA)/I$. (The $\int y\, dA$ is commonly given the symbol Q in elastic beam theory; this should not be confused with the nominal connector strength Q_n used below.) Thus, if a given connector has an allowable service load capacity q (kips), the maximum spacing p to provide the required strength is

$$p = \frac{q}{V(\int y\, dA)/I} \tag{16.8.1}$$

where $\int y\, dA$ is the statical moment of the transformed compressive concrete area (the slab) taken about the neutral axis of the composite section. Equation 16.8.1 is based on *elastic* beam theory and a fully composite section.

Figure 16.8.3
Force required from shear
connectors at service load
level.

Until recent years, Eq. 16.8.1 was used to space shear connectors. AASHTO-6.10.10.1.2 [1.3] requires using Eq. 16.8.1 to design for fatigue, a service load limit state related to the *range* of force applied, in this case the range of shear V_r resulting from live load (and impact). AASHTO also requires a strength limit state check.

According to the strength limit state, the shear connectors at nominal moment strength share equally in transmitting the shear at the interface between concrete slab and steel beam. This means, referring to Fig. 16.8.2a, that shear connectors are required to transfer the compressive force developed in the slab at midspan to the steel beam in the distance $L/2$, since no compressive force can exist in the slab at the end of the span where zero moment exists. The nominal shear transfer strength cannot exceed the maximum force the concrete can develop, namely

$$V' = C_{\max} = 0.85 f'_c b_E t_s \qquad (16.8.2)$$

where b_E is the effective slab width and t_s is the slab thickness. When the maximum force $T_{\max}$ that can develop in the steel is less than $C_{\max}$, the maximum shear transfer strength will be

$$V' = T_{\max} = A_s F_y \qquad (16.8.3)$$

where A_s is the cross-sectional area of the steel section.

Thus, when the nominal strength Q_n of one shear connector is known, the total number N of shear connectors required between points of maximum and zero bending moment is

$$N = \frac{C_{\max}}{Q_n} \quad \text{or} \quad \frac{T_{\max}}{Q_n}, \quad \text{whichever is } smaller \qquad (16.8.4)$$

Thus, the strength is achieved when the total number N of shear connectors is placed between the maximum moment and zero moment locations. Uniform spacing will be the simplest procedure, because the number of connectors rather than the spacings affects the strength.

The determination of the connector capacity analytically is complex, since the shear connector deforms under load and the concrete which surrounds it is also a deformable material. Moreover, the amount of deformation a shear connector undergoes is dependent upon factors such as its own shape and size, its location along the beam, the location of the maximum moment, and the manner in which it is attached to the top flange of the steel beam. In addition, any particular shear connector may yield sufficiently to cause slip between the beam and the slab. In the latter case the adjacent shear connectors pick up the additional shear.

As a result of the complex behavior of shear connectors, their capacities are not based solely on a theoretical analysis. In order to develop a rational approach, a number of research programs, summarized by Viest [16.1, 16.5], were undertaken to develop the strengths of the various types of shear connectors.

Investigators determined that shear connectors will not fail when the average load per connector is below that causing 0.003 in. (0.076 mm) residual slip between concrete and steel. The amount of slip is also a function of the strength of the concrete that surrounds the shear connector. Relating connector capacity to a specified slip may be realistic for bridge design where fatigue strength is important, but it is overly conservative with respect to failure loads. So-called "ultimate" capacities used prior to 1965 [16.17] were based on slip limitation, giving values about one-third of the strengths obtained when failure of a connector is the criterion.

When flexural strength of the composite section is the basis for design, the connectors must be adequate to satisfy equilibrium of the concrete slab between the points of maximum and zero moment, as discussed in the development of Eqs. 16.8.2, 16.8.3, and 16.8.4. Slip is not a criterion for this equilibrium requirement. As stated by Slutter and Driscoll [16.17], "the magnitude of slip will not reduce the ultimate moment provided that (1) the equilibrium condition is satisfied, and (2) the magnitude of slip is no greater than the lowest value of slip at which an individual connector might fail." Studies by Ollgaard, Slutter, and Fisher [16.19] and McGarraugh and Baldwin [16.20] included the effect of lightweight concrete on stud connector capacity.

All research work cited above was based on experimental work that used solid slabs or steel decks from flat steel plates. Most of composite steel floor decks used in buildings today are formed and have a stiffening rib in the middle of each deck flute. Recent research by Rambo-Roddenberry and others [16.47] concluded that shear stud strength equations in past AISC Specifications are unconservative. The stud strength whether the deck was perpendicular or parallel to the beams, is higher than those derived from either pushout or beam tests for studs embedded in modern steel decks. Also, because of the stiffener, studs must be welded off-center in the deck rib. Rambo-Roddenberry et al [16.47] have shown that shear studs behave differently depending upon their position within the deck rib. The "weak" (unfavorable) and "strong" (favorable) positions are shown in Figure 16.8.4.

Two currently accepted expressions for the nominal strength Q_n of shear connectors are as follows:

1. *Headed steel stud connectors welded to flange* (Fig. 16.8.1a). Load and Resistance Factor Design (AISC-I3.2) gives essentially the expression developed at Lehigh [16.19], and subsequently modified at Virginia Tech [16.47]

$$Q_n = 0.5A_{sc}\sqrt{f'_c E_c} \leq R_g R_p A_{sc} F_u \qquad (16.8.5)$$

where A_{sc} = cross-sectional area of stud shear connector, sq in.

F_u = specified minimum tensile strength of a stud shear connector, ksi

R_g = 1.0;

   (a) for one stud welded in a steel deck rib with the deck oriented perpendicular to the steel shape;

   (b) for any number of studs welded in a row directly to the steel shape;

   (c) for any number of studs welded in a row through steel deck with the deck oriented parallel to the steel shape and the ratio of the average rib width to rib depth $\geq$ 1.5

Figure 16.8.4
Weak and strong stud
positions [16.47]

$R_g = 0.85$;

    (a) for two studs welded in a steel deck rib with the deck oriented perpendicular to the steel shape;

    (b) for one stud welded through steel deck with the deck oriented parallel to the steel shape and the ratio of the average rib width to rib depth <1.5

$R_g = 0.7$ for three or more studs welded in a steel deck rib with the deck oriented perpendicular to the steel shape

$R_p = 1.0$ for studs welded directly to the steel shape (in other words, not through steel deck or sheet) and having a haunch detail with not more than 50 percent of the top flange covered by deck or sheet steel closures

$R_p = 0.75$;

    (a) for studs welded in a composite slab with the deck oriented perpendicular to the beam and $e_{mid\text{-}ht} \geq 2$ in.;

    (b) for studs welded through steel deck, or steel sheet used as girder filler material, and embedded in a composite slab with the deck oriented parallel to the beam

$R_p = 0.6$ for studs welded in a composite slab with deck oriented perpendicular to the beam and $e_{mid\text{-}ht} < 2$ in.

$e_{mid\text{-}ht}$ = distance from the edge of stud shank to the steel deck web, measured at mid-height of the deck rib, and in the load bearing direction of the stud (in other words, in the direction of maximum moment for a simply supported beam), in.

w_c = weight of concrete per unit volume (90 pcf $\leq w_c \leq$ 155 pcf)

E_c = modulus of elasticity of concrete, ksi

    = $(w^{1.5})\sqrt{f'_c}$, according to AISC-I2.1b, using f'_c in ksi. For normal-weight concrete having density $w = 145$ pcf, $E_c = 1746\sqrt{f'_c}$. Note that the ACI Code [16.15] gives slightly different values using $E_c = w^{1.5}33\sqrt{f'_c}$, with f'_c in psi instead of ksi.

2. *Channel connectors* (Fig. 16.8.1b). AISC-I3.2D(4) gives for the nominal connector strength Q_n,

$$Q_n = 0.3(t_f + 0.5t_w)L_c\sqrt{f'_c E_c} \qquad (16.8.6)$$

where Q_n = nominal strength of one channel, kips

    t_f = channel flange thickness (Fig. 16.8.1), in.

    t_w = channel web thickness, in.

    L_c = length of channel, in.

    f'_c = 28-day compressive strength of concrete, ksi

    E_c = modulus of elasticity of concrete (defined following Eq. 16.8.5), ksi

Connector Design—AISC LRFD Method

The nominal strength Q_n of the connectors is directly used in the AISC Design Methods. AISC-I3.2d requires "… the entire horizontal shear at the interface between the steel beam and the concrete slab shall be assumed to be transferred by shear connectors." For fully composite sections, the nominal horizontal shear strength V_{nh} to be provided by connectors is the smaller of Eqs. 16.8.2 and 16.8.3.

    The section may also be designed as *partially composite,* where the forces utilized of the internal couple are less than either the nominal compression strength available from the concrete, or the nominal tension strength available from the steel section. In partially composite sections, the strength ΣQ_n of the shear connectors determines the magnitude of the forces of the internal couple and nominal moment strength M_n, and correspondingly the required nominal horizontal shear strength V_{nh}. Lorenz and Stockwell [16.8] have discussed stresses in partial composite beams. Bradford and Gilbert [16.44] have provided recent work on partial interaction under sustained loads.

For the positive moment situations (i.e., compression in the concrete slab), the shear strength V_{nh} required is, therefore, the *smallest* of the following:

1. V_{nh} required $= 0.85\,f'_c b_E t_s$ [16.8.2]

2. V_{nh} required $= A_s F_y$ [16.8.3]

3. V_{nh} required $= \Sigma Q_n$ provided

When 3. applies, the number of connectors controls the nominal strength M_n of the section.

As stated earlier, and as specifically stated in AISC-I3.2d(6), the strength V_{nh} must be provided "... each side of the point of maximum moment ..." to the points of zero moment. Further, AISC-I3.2d(6) states that the connectors *shall* be distributed uniformly between the point of maximum moment and the point of zero moment. This is "unless otherwise specified", whatever that may mean. As long as adequate strength is provided, the spacing of the connectors is not important.

The nominal strengths Q_n for stud and channel connectors from AISC-I3.2d(3) and (4) are given by Eqs. 16.8.5 and 16.8.6; values for common stud diameters and some channels are given in Table 16.8.1.

When a formed steel deck is used (see Fig. 16.1.2) with shear studs embedded in the supported concrete slab, reduction factors must be applied to Q_n in accordance with AISC-I3.2d(3).

In the case of continuous beams (also see Sec. 16.13), the longitudinal reinforcing bar steel within the effective width of the concrete slab is permitted (AISC-I3.2d(2)) to be assumed to act compositely with the steel beam in the areas of negative moment. The total nominal horizontal strength V_{nh} needed from shear connectors between the interior support and each adjacent point of inflection (zero moment) equals the tension force available from the reinforcement (since the tension in the concrete is neglected),

$$T_{slab} = A_r F_{yr} \tag{16.8.7}$$

where A_r = total area of adequately developed longitudinal reinforcing steel within the effective width b_E of the concrete slab

    F_{yr} = minimum specified yield stress of the reinforcing steel

EXAMPLE 16.8.1

Determine the number of $\frac{3}{4}$-in.-diam $\times$ 3-in. shear stud connectors required to develop the fully composite section of Fig. 16.8.5. Assume the applied loading is uniform and the beam is simply supported. Use A992 steel, $f'_c = 4000$ psi, $n = 8$, and Load and Resistance Factor Design.

Solution:
Using Eqs. 16.8.2 and 16.8.3,

$$V_{nh} = C_{max} = 0.85 f'_c b_E t_s = 0.85(4.0)(72)7 = 1714 \text{ kips}$$

or

$$V_{nh} = T_{max} = A_s F_y = 47.0(50) = 2350 \text{ kips}$$

TABLE 16.8.1 Nominal Strength Q_n (kips) for Stud and Channel Shear Connectors Used with No Decking $(R_g=R_p=1.0)$ and Normal-Weight Concrete[†]

	Concrete strength f'_c (ksi)		
Connector	3.0	3.5	4.0
1/2" diam × 2" headed stud	9.4	10.5	11.6
5/8" diam × 2-1/2" headed stud	14.6	16.4	18.1
3/4" diam × 3" headed stud	21.0	23.6	26.1
7/8" diam × 3-1/2" headed stud	28.6	32.1	35.5
Channel C3×4.1	$10.2L_c$*	$11.5L_c$	$12.7L_c$
Channel C4×5.4	$11.1L_c$	$12.4L_c$	$13.8L_c$
Channel C5×6.7	$11.9L_c$	$13.3L_c$	$14.7L_c$

[†]AISC Formula (I3-3), Eq. 16.8.5, used for studs and AISC Formula (I3–4), Eq. 16.8.6, used for channels. Studs, A108 Type 2, F_u^b = 60 ksi.

*L_c = Length of channel, in.

Figure 16.8.5
Example 16.8.1.

As found from the analysis in Example 16.7.2, the neutral axis is located within the steel section; thus, $C_{max} < T_{max}$. The force in the concrete to be carried by shear connectors is 1714 kips.

The nominal strength Q_n per connector, from Eq. 16.8.5 or Table 16.8.1, is 26.1 kips. The number N of shear connectors required for each half span is

$$N = \frac{1714}{26.1} = 66$$

Use 66—$\frac{3}{4}$-in.-diam × 3-in. studs per half span.   ∎

Connector Design—Elastic Concept for Fatigue Strength

The 1992 AASHTO Specification [1.3] requirements for fatigue are based largely on the work of Slutter and Fisher [16.21]. For fatigue, the *range* of service load shear rather than strength under overload is the major concern. Fatigue strength may be expressed

$$\log N = A + BS_r \qquad (16.8.8)$$

where S_r is the range of service load horizontal shear; N is the number of cycles to failure; and A and B are empirical constants. The equation used for design is shown in Fig. 16.8.6.

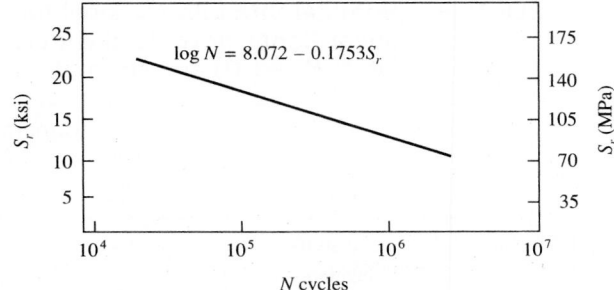

Figure 16.8.6
Fatigue strength of stud shear connectors. (From Ref. 16.21)

Since the magnitude of shear force transmitted by individual connectors when service loads act agrees well with prediction by elastic theory, the horizontal shear is calculated by the elastic relation VQ/I. Fatigue is critical under repeated application of service load; thus it is reasonable to determine variation in shear using elastic theory. The spacing of the connectors will vary along the span in accordance with V.

For *cyclical load,* Eq. 16.8.1 gives

$$V_{sr} = \frac{(V_{max} - V_{min})Q}{I} = \frac{\text{Allowable range } \Sigma Z_r}{p} \tag{16.8.9}$$

where p is the connector spacing. AASHTO-6.10.10.1.2 [1.3] gives Eq. 16.8.9 as

$$p \le \frac{\Sigma Z_r}{V_{sr}} \tag{16.8.10}$$

where V_{sr} = horizontal fatigue shear range per unit length, kip/in.

Z_r = allowable range of load per connector, lb/stud connector

$$= \alpha d^2 \ge \frac{5.5d^2}{2} \text{ (AASHTO 6.10.10.2)}$$

Σ = indicates the sum of Z_r for connectors at the section is to be used.

d = stud diameter, in.

$\alpha = 34.5 - 4.28 \log N$

EXAMPLE 16.8.2

Redesign the shear conncetors for the beam of Example 16.8.1 (Fig 16.8.5) using the service load stress fatigue requirement of AASHTO with $\frac{3}{4}$-in.-diam $\times$ 3-in. stud connectors. Design for 500,000 cycles of loading of live load. Whether or not the beam is shored, only the live load is the cylical load. Use uniform live load of 3.5 kips/ft, a spacing of 6 ft for beams, a beam span of 45 ft, $F_y = 50$ ksi, and $f'_c = 4$ ksi.

Solution:

(a) Loads and shears. For the fatigue requirement in AASHTO-6.10.10.1.2 only the *range* of service live load is needed. At the support with full span loaded,

$$V = \tfrac{1}{2}wL = 0.5(3.5)45 = 78.8 \text{ kips}$$

Using partial span loading of live load,

$$\text{Max } V\left(\text{at } \tfrac{1}{4} \text{ point}\right) = 3.5(45)(0.75)(0.375) = 44.3 \text{ kips}$$

$$\text{Max } V(\text{at midspan}) = \tfrac{1}{8}wL = \tfrac{1}{8}(3.5)45 = 19.7 \text{ kips}$$

Figure 16.8.7
Shear range diagram and stud
spacing according to elastic
fatigue theory used by
AASHTO-Example 16.8.2.

The envelope showing the *range* of live load shear is given in Fig. 16.8.7. Inclusion of dead load shear would change both V_{max} and V_{min} by the same amount at any section along the beam; however, $(V_{max} - V_{min})$, that is, the range V_r would not be affected.

   (b) Compute elastic composite section properties $(n = 8)$ (see Fig. 16.8.5).

$$\text{Effective slab width } b_E = b_0 = 72 \text{ in.}$$

Element	Effective area, A (sq in.)	Arm from CG of steel beam, y (in.)	Ay (sq in.)	Ay^2 (in.³)	I_0 (in.⁴)
Slab, 72(7)/8	63.0	21.5	1355	29,120	257
W36×160	47.0	—	—	—	9760
	110		1355	29,120	10,017

$$I_x = Ay^2 + I_0 = 29,120 + 10,017 = 39,100 \text{ in.}^4$$

$$\bar{y} = \frac{1355}{110} = 12.32 \text{ in.}$$

$$I_{tr} = 39,100 - 110(12.32)^2 = 22,400 \text{ in.}^4$$

$$y_t = 18.0 + 7 - 12.32 = 12.68 \text{ in.}$$

$$y_b = 18.0 + 12.32 = 30.32 \text{ in.}$$

$$S_t = \frac{22,400}{12.68} = 1767 \text{ in.}^3 \quad (\text{concrete at top})$$

$$S_b = S_{tr} = \frac{22,400}{30.32} = 739 \text{ in.}^3 \quad (\text{steel at bottom})$$

Determine the static moment of the effective concrete area about the centroid of the composite section,

$$Q = 63.0(y_t - 3.5) = 63.0(9.18) = 578 \text{ in.}^3$$

(c) Determine the allowable load for $\frac{3}{4}$-in.-diam $\times$ 3-in. stud connectors. AASHTO-6.10.10.1.2 gives an allowable service load range Z_r per connector based on fatigue for 500,000 cycles of loading as

$$\alpha = 34.5 - 4.28 \log N$$
$$= 34.5 - 4.28 \log(500,000) = 10.1$$

$$Z_r = \alpha d^2 = 10.1(0.75)^2 = 5.68 \text{ kips} > \left[\frac{5.5d^2}{2} = 1.55 \text{ kips} \right] \qquad \textbf{OK}$$

The AASHTO allowable values are based on a slip limitation.

(d) Determine spacing of connectors. Use 4 studs across the beam flange width at each location:

$$\Sigma Z_r \text{ for 4 studs} = 4(5.68) = 22.7 \text{ kips}$$

Using Eq. 16.8.10,

$$p = \frac{\Sigma Z_r}{V_{sr}} = \frac{\Sigma Z_r}{(V_{max} - V_{min})Q/I}$$

where $I/Q = 22{,}400/578 = 38.8$ in.

$$p = \frac{22.7(38.8)}{(V_{max} - V_{min})} = \frac{881}{(V_{max} - V_{min})(\text{kips})}$$

The values are computed in the table below and the spacing is determined graphically on the shear diagram of Fig. 16.8.7.

p (in.)	V_r (kips)	p (in.)	V_r (kips)
11	79	20	45
14	61	22	39
17	51		

$V_r = V_{max} - V_{min}$

The fatigue service load criterion requires 8% more connectors (68 vs 61 per half span) than the procedure based on strength. ∎

16.9 COMPOSITE FLEXURAL MEMBERS CONTAINING FORMED STEEL DECK

Composite flexural members may be made using formed steel deck, as shown in Fig. 16.1.2. The formed metal deck may be placed perpendicular to or parallel with the supporting beam. Furthermore, the beam may actually be an open web joist. Typically, the deck plate varies in thickness from 22 ga. (0.0336 in., 0.853 mm) to 12 ga. (0.1084 in., 2.75 mm). The deck rib height typically is $1\frac{1}{2}$, 2, and 3 in. for spans of, say, 8, 10, and 15 ft. As shown in Fig. 16.1.2, the thickness of the concrete slab above the top of the ribs must be at least 2 in. AISC-I3.2c and the embedment of the stud connectors into the concrete above the top of the ribs must be at least $1\frac{1}{2}$ in.

When the steel deck ribs are perpendicular to the steel beam, the stud strength Q_n may have to be reduced from that given by Eq. 16.8.5 by a reduction factor as explained earlier. Easterling, Gibbings, and Murray [16.43] provide a study of strength of shear studs in steel deck on composite beams.

Full treatment of formed steel deck supported slab composite beams is outside the scope of this chapter. The reader is referred to Grant, Fisher, and Slutter [16.23], and particularly with regard to LRFD design, to Vinnakota, Foley, and Vinnakota [16.24]. Composite open-web joists have been treated by Tide and Galambos [16.25] and Rongoe [16.26]. Two-way acting composite slabs with steel deck have been treated by Porter [16.29], and design *Specifications* and *Commentary* [16.27, 16.28] are available from ASCE. The special considerations regarding the design of "stubgirders" are treated by Buckner, Deville, and McKee [16.30].

16.10 DESIGN PROCEDURE—AISC LRFD AND ASD METHODS

The design of composite beams involves providing sufficient plastic strength ϕM_p of the composite section to equal the factored moment. Using rolled W shapes, local buckling ordinarily is not a controlling limit state although h/t_w should be checked when the PNA is in the web, and because the compression flange is attached to the concrete slab lateral-torsional buckling is precluded as a controlling limit state. Thus, it is required that

$$\phi_b M_p \geq M_u \qquad \text{(LRFD)} \qquad (16.10.1a)$$

$$M_p/\Omega_b \geq M_a \qquad \text{(ASD)} \qquad (16.10.1b)$$

where $\phi_b = 0.90$ and $\Omega_b = 1.67$ for a composite beam.

In general, the design should be started by assuming the plastic neutral axis (PNA) is within the slab (Case 1—Fig. 16.7.1b). Thus, using Eq. 16.7.5, the required area A_s for the steel section is

$$\text{Required } A_s = \frac{M_u}{\phi_b F_y \left(\dfrac{d}{2} + t_s - \dfrac{a}{2} \right)} \qquad \text{(LRFD)} \qquad (16.10.2a)$$

$$\text{Required } A_s = \frac{M_u}{\dfrac{F_y}{\Omega_b} \left(\dfrac{d}{2} + t_s - \dfrac{a}{2} \right)} \qquad \text{(ASD)} \qquad (16.10.2b)$$

Typically $a/2$ can be estimated as 1 in. for preliminary design.

In addition to the strength requirement under full dead and live load, AISC-I3.1c requires that when temporary shores are not used during construction, the steel section alone must have adequate strength "to support all loads applied prior to the concrete attaining 75% of its specified strength f'_c." For this condition, local buckling of the beam elements and lateral-torsional buckling must be considered.

16.11 AISC EXAMPLES—SIMPLY SUPPORTED BEAMS

EXAMPLE 16.11.1

Design an interior composite beam for the floor whose plan is shown in Fig. 16.11.1 assuming the beam is to be constructed without temporary shoring. Use 50 ksi, $f'_c = 4$ ksi ($n = 8$), a 4-in. slab, and the AISC LRFD Method.

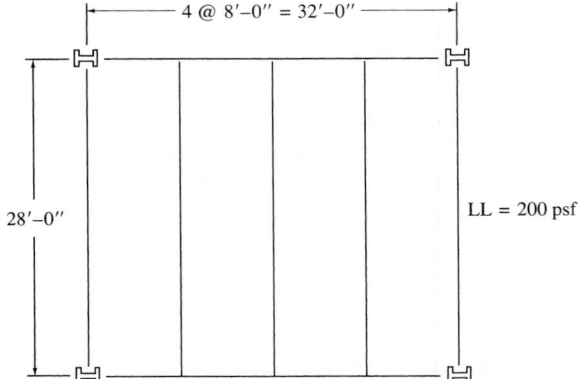

Figure 16.11.1
Beam framing plan for
Examples 16.11.1 and
16.12.1.

Solution:

(a) Compute factored and service loads.

Loads carried on steel beam:

$$\text{concrete slab, } \tfrac{4}{12}(0.15)8 = 0.40 \text{ kip/ft}$$
$$\text{beam weight (estimated)} = \underline{0.04} \text{ kip/ft}$$
$$\text{service dead load} = 0.44 \text{ kip/ft}$$
$$\text{factored dead load} = 0.44(12) = 0.53 \text{ kip/ft}$$

Load carried by composite action:

$$\text{service live load, } 0.2(8) = 1.6 \text{ kips/ft}$$
$$\text{factored live load} = 1.6(1.6) = 2.56 \text{ kips/ft}$$

(b) Compute service load and factored load moments.

$$M_D = \tfrac{1}{8}(0.44)(28)^2 = 43 \text{ ft-kips} \quad \text{(service load)}$$
$$M_L = \tfrac{1}{8}(1.60)(28)^2 = 157 \text{ ft-kips} \quad \text{(service load)}$$
$$M_u = \tfrac{1}{8}(0.53 + 2.56)(28)^2 = 303 \text{ ft-kips} \quad \text{(factored)}$$

(c) Select the section. Use Eq. 16.10.2 assuming the PNA (plastic neutral axis) is within the slab. Estimate $a \approx 1.0$ for preliminary selection.

$$\text{Required } A_s = \frac{M_u}{\phi_b F_y\left(\dfrac{d}{2} + t_s - \dfrac{a}{2}\right)} \qquad \text{[16.10.2a]}$$

From Eq. 16.10.2, the design strength $\phi_b M_n$ provided can be computed as A_s times the denominator. For a given value of $(t_s - a/2)$, $\phi_b M_n$ can be tabulated for a steel W section for any given yield stress; such tabulated information is given in the *AISC Manual*. Thus, for the 4-in. slab and estimated a of 1 in.,

$$t_s - \frac{a}{2} = 4 - 0.50 = 3.5 \text{ in.}$$

$$\text{Required } A_s = \frac{303(12)}{0.90(50)(7 + 3.5)} = 7.7 \text{ sq in.} \quad \text{(for W14)}$$

$$\text{Required } A_s = \frac{303(12)}{0.90(50)(8 + 3.5)} = 7.0 \text{ sq in.} \quad \text{(for W16)}$$

Using *AISC Manual*, Table 3-19 "Composite W Shapes-Available Strength in Flexure" entering with $Y2 = t_s - a/2 = 3.5$ in. and required $\phi M_n = 303$ ft-kips, find

W16×26	$\phi_b M_n = 327$ ft-kips	$A_s = 7.68$ sq in.
W14×26	$\phi_b M_n = 302$ ft-kips	$A_s = 7.69$ sq in.

The tabulated values selected are for the PNA within the slab (that is, $Y1 =$ distance from PNA to top of steel beam $= 0$ in.). When these tables are available, their use will be faster and more accurate than putting estimated d into Eqs. 16.10.2.

(d) Compute the plastic neutral axis location and check strength.

Try W16×26: Properties of the steel section alone are:

$$A_s = 7.68 \text{ sq in.} \qquad I_x = 301 \text{ in.}^4 \qquad b_f = 5.50 \text{ in.}$$

$$d = 15.70 \text{ in.}$$

Determine effective width of slab:

$$b_E = \tfrac{1}{4} \text{ of span} = 0.25(28)12 = 84 \text{ in.} \quad controls$$

$$\text{or} \quad b_E = \text{beam spacing} = 8(12) = 96 \text{ in.}$$

The compressive force in the concrete, assuming $a < t_s$, and the tension force in the steel section, are

$$C = 0.85 f_c' b_E a = 0.85(4)84a = 286a$$

$$T = A_s F_y = 7.68(50) = 384 \text{ kips}$$

Statics requires

$$C = T$$

$$a = 1.34 \text{ in.} < t_s \qquad\qquad \textbf{OK as assumed}$$

The nominal moment strength M_n is

$$M_n = T\left(\frac{d}{2} + t_s - \frac{a}{2}\right)$$

$$M_n = 384\left(\frac{15.7}{2} + 4.0 - \frac{1.34}{2}\right)\frac{1}{12} = 358 \text{ ft-kips}$$

$$\phi_b M_n = 0.90(358) = 322 \text{ ft-kips} > (M_u = 301 \text{ ft-kips}) \qquad \textbf{OK}$$

Note that M_u has been revised to include the correct beam weight.

(e) Check the strength of the steel section to support construction loads (AISC-I3.1c). This check is required when shores are not used. Assume adequate lateral support is provided during construction such that $L_b \le L_p$ and the section is compact for local buckling; therefore $\phi_b M_n = \phi_b M_p$, and $\phi_b = 0.90$ for the steel section acting noncompositely. There are no AISC-prescribed construction loads. It is prudent to consider that some of the wet concrete load should be treated as live load, say 50% of it (accomplished by using an average overload factor of 1.4). Further, other construction live load on the order of 20 to 25 psf should be included (20 psf used here).

$$\text{Slab} = 0.40(1.4) = 0.56 \quad \text{kip/ft}$$
$$\text{Construction} = 0.02(8)1.6 = 0.26 \quad \text{kip/ft}$$
$$\text{Steel section} = 0.026(1.2) = \underline{0.03} \quad \text{kip/ft}$$
$$w_u = 0.85 \quad \text{kip/ft}$$
$$M_u = \tfrac{1}{8}(0.85)(28)^2 = 83 \text{ ft-kips}$$
$$\phi_b M_p \text{ for W16×26} = 166 \text{ ft-kips} > 83 \text{ ft-kips} \qquad \textbf{OK}$$

(f) Design shear connectors. The compressive force in the slab must be carried by shear connectors,

$$C = 0.85 f'_c b_E a = 0.85(4)84a = 286a = 286(1.34) = 385 \text{ kips}$$

Since $a < t_s$, V_{nh} will be based on the 385 kips, which does equal $T_{max} = A_s F_y$. Using $\tfrac{3}{4}$-in.-diam × 3-in. headed studs, $Q_n = 26.1$ kips/stud from Table 16.8.1. The number N of connectors required to carry 385 kips is

$$N = \frac{V_{nh}}{Q_n} = \frac{385}{26.1} = 14.8, \quad \text{say 15}$$

which is the number of connectors required for the region between maximum moment and the support (zero moment location). Thus, 30 studs are needed for the entire span. Using a uniform spacing with two studs at each location, the spacing p required would be

$$p = \frac{L}{N} = \frac{28(12)}{15} = 22 \text{ in.}$$
$$\text{Maximum } p = 8t_s = 8(4) = 32 \text{ in.} \quad \text{(AISC-I3.2)}$$
$$\text{Minimum } p = 6(\text{diam}) = 6(0.75) = 4.5 \text{ in.} \quad \text{(AISC-I3.2)}$$

Use W16×26 section of A992 steel, along with 30—$\tfrac{3}{4}$-in.-diam × 3-in. headed stud connectors over the entire span, spaced at 22 in. The connectors are to be placed in pairs starting at the support. ∎

EXAMPLE 16.11.2

Design an interior composite beam to span 30 ft with a beam spacing of 8 ft, using the minimum number of $\tfrac{3}{4}$-in.-diam × 3-in. stud shear connectors. The slab is 5 in. thick. The beam is to be constructed without shores. The beam must support a ceiling of 7 psf, partitions and other dead load of 25 psf, and live load of 150 psf. Use A572 Grade 50 steel and $f'_c = 3$ ksi ($n = 9$) concrete. Use the AISC LRFD Design Method.

Solution:

(a) Compute factored loads and bending moments. The dead load and moment that must be carried by the steel beam alone during construction are

$$\text{5-in. slab, } \tfrac{5}{12}(8)0.15 = 0.50 \text{ kips/ft}$$
$$\text{Steel beam (assumed)} = \underline{0.03}$$
$$0.53 \text{ kips/ft}$$

$$M_D = \tfrac{1}{8}(0.53)(30)^2 = 60 \text{ ft-kips}$$

$$M_{u1} = 1.2(60) = 72 \text{ ft-kips}$$

The partition and ceiling dead loads, and the live load that must be carried by the composite section are

Live load 0.15(8)	= 1.2 kips/ft
Partitions 0.025(8)	= 0.2
Ceiling 0.007(8)	= 0.06
	1.46 kips/ft

$$M_L = \tfrac{1}{8}(1.46)(30)^2 = 164 \text{ ft-kips}$$

$$w_{u2} = 1.2(0.25) + 1.6(1.2) = 2.22 \text{ kips/ft}$$

$$M_{u2} = \tfrac{1}{8}(2.22)(30)^2 = 250 \text{ ft-kips}$$

$$M_u = M_{u1} + M_{u2} = 72 + 250 = 322 \text{ ft-kips}$$

(b) Select the section. One could use Eq. 16.10.2 assuming the PNA (plastic neutral axis) is within the slab and solve for required A_s as illustrated in Example 16.11.1 (part c). Alternatively, it will be simpler to use AISC Table 3-19, "Composite W Shapes-Available Strength in Flexure." Equation 16.10.2 is tabulated for the steel W shapes for various values of Y2. Estimate $a \approx 1.0$ for preliminary selection as in Example 16.12.1. For the 5-in. slab,

$$Y2 = t_s - \frac{a}{2} = 5 - 0.50 = 4.5 \text{ in.}$$

$$\text{Required } \phi M_n = 322 \text{ ft-kips}$$

Find:

W16×26	$\phi M_n = 356$ ft-kips
W14×26	$\phi M_n = 330$ ft-kips

The tabulated values selected are for the PNA within the slab (that is, $Y1 = $ distance from PNA to top of steel beam $= 0$ in.).

(c) Investigate the W16×26 further. For fully composite action, compute the plastic neutral axis location and check strength.

Try W16×26: Properties of the steel section alone are:

$$A_s = 7.68 \text{ sq in.} \qquad I_x = 301 \text{ in.}^4 \qquad b_f = 5.500 \text{ in.}$$

$$d = 15.69 \text{ in.}$$

Determine effective width of slab:

$$b_E = \tfrac{1}{4} \text{ of span} = 0.25(30)12 = 90 \text{ in.} \quad controls$$

$$\text{or} \quad b_E = \text{beam spacing} = 8(12) = 96 \text{ in.}$$

The compressive force in the concrete, assuming $a < t_s$, and the tension force in the steel section are

$$C = 0.85 f_c' b_E a = 0.85(3)90a = 229.5a$$

$$T = A_s F_y = 7.68(50) = 384 \text{ kips}$$

Statics requires

$$C = T$$

$$a = 1.67 \text{ in.} < t_s \qquad \textbf{OK as assumed}$$

The nominal moment strength M_n is

$$M_n = T\left(\frac{d}{2} + t_s - \frac{a}{2}\right)$$

$$M_n = 384\left(\frac{15.69}{2} + 5.0 - \frac{1.67}{2}\right)\frac{1}{12} = 384 \text{ ft-kips}$$

$$[\phi_b M_n = 0.90(384) = 346 \text{ ft-kips}] > [M_u = 322 \text{ ft-kips}] \qquad \textbf{OK}$$

The W16×26 section is adequate as a *fully composite section*. However, when a minimum number of shear connectors is desired and only partial composite action is used, the steel section usually must be heavier. Try W16×31 section.

(d) Minimum number of shear connectors required. The maximum spacing p along the span is

$$\text{Maximum } p = 8t_s = 8(5) = 40 \text{ in.} \quad \text{(AISC-I3.2)}$$

$$N = \frac{L}{p} = \frac{30(12)}{40} = 9 \text{ spaces}$$

The connectors would be in pairs which would mean 20 connectors for the 30-ft span, with 5 pairs (10 connectors) located between midspan and the end of the beam. When $\frac{3}{4}$-in.-diam studs are used, 10 connectors provide nominal strength ΣQ_n,

$$\Sigma Q_n = 10(21.0) = 210 \text{ kips}$$

Since the force in the slab based on connector strength is less than the maximum steel force,

$$T_{\max} = A_s F_y = 9.12(50) = 456 \text{ kips}$$

the plastic neutral axis (PNA) is within the steel section.

(e) Locate plastic neutral axis (PNA) and compute nominal strength. Check if PNA occurs within the flange,

$$\Sigma Q_n = 210 \text{ kips}$$

$$\text{Max force in flange} = t_f b_f F_y = 0.440(5.525)50 = 121.6 \text{ kips}$$

$$T_{\max} - 121.6 = 334.5 \text{ kips} > \Sigma Q_n$$

Thus, PNA is in the web. For equilibrium of internal forces, referring to Fig. 16.11.2, compute the compression force in the web,

$$\Sigma Q_n + C_f + C_w = T_{\max} - C_f - C_w$$

$$210 + 121.6 + C_w = 334.5 - C_w$$

$$2C_w = 2.9 \text{ kips}$$

$$\text{Depth to PNA from inside of flange} = \frac{C_w}{F_y t_w} = \frac{1.45}{50(0.275)} = 0.11 \text{ in.}$$

$$\text{PNA from top of slab} = t_s + t_f + 0.11$$
$$= 5 + 0.44 + 0.11 = 5.55 \text{ in.}$$

Figure 16.11.2
Example 16.11.2, showing
stress distribution to obtain
plastic neutral axïs.

Locate the centroid y_1 of the portion of the steel section in tension measured from the bottom of the steel section,

	Area, A	Arm, y	Ay
W section	9.12	7.94	72.41
Flange	−2.43	15.66	−38.07
Web	−0.03	15.39	−0.45
	6.66 sq in.		33.90 in.^{3}

$$y_1 = \frac{33.90}{6.66} = 5.09 \text{ in.}$$

(f) Compute the nominal moment strength M_n. Since the ΣQ_n representing the strength of the shear connectors used is less than the force in the concrete when there is fully composite action, force ΣQ_n is taken equivalent to $C_c = 0.85 f'_c b_E a$, the concrete force represented by the rectangular stress distribution in the concrete. That means

$$a = \frac{\Sigma Q_n}{0.85 f'_c b_E} = \frac{210}{0.85(3)90} = 0.91 \text{ in.}$$

Referring to Fig. 16.11.2, taking internal moments about the point of action of T_s gives

$$\Sigma Q_n: \quad M_{n1} = \Sigma Q_n(d - 5.09 + t_s - a/2)$$
$$= 210(15.88 - 5.09 + 5 - 0.91/2)\tfrac{1}{12}$$
$$= 210(15.33)\tfrac{1}{12} = 268.3 \text{ ft-kips}$$

$$C_f: \quad M_{n2} = C_f(d - 5.09 - t_f/2)$$
$$= 121.6(15.88 - 5.09 - 0.440/2)\tfrac{1}{12}$$
$$= 121.6(10.57)\tfrac{1}{12} = 107.1 \text{ ft-kips}$$

$$C_w: \quad M_{n3} = C_w(d - 5.09 - t_f - 0.11/2)\tfrac{1}{12}$$
$$= 1.45(15.88 - 5.09 - 0.440 - 0.055)\tfrac{1}{12}$$
$$= 1.45(10.30)\tfrac{1}{12} = 1.2 \text{ ft-kips}$$

$$M_n = M_{n1} = M_{n2} + M_{n3}$$
$$= 268.3 + 107.1 + 1.2 = 376.6 \text{ ft-kips}$$

$$\phi_b M_n = 0.90(376.6) = 340 \text{ ft-kips}$$

After correcting the dead load for the W16×31 section, the factored moment M_u becomes 321 ft-kips. Thus, $\phi_b M_n > M_u$ and the design is acceptable.

The designer should compare the economics of the W16×26 using connectors to develop a fully composite section with W16×31 using the minimum 20 connectors needed for this span length. To obtain a fully composite section the force to be carried by shear connectors would have been $T_{max} = A_s F_y = 384$ kips for the W16×26 section. The number of $\frac{3}{4}$-in.-diam studs needed would be

$$N = \frac{384}{21.0} = 18.3, \quad \text{say 20 for half the span}$$

Thus, the 40 connectors required for fully composite action can be reduced to 20 using partial composite action with the next heavier section.

(g) Check the strength of the W16×31 steel section to support construction loads (AISC-I3.1c). Refer to discussion in Example 16.14.1, part (e). Assume construction live load consists of 50% of the wet concrete (accomplished by using an average overload factor of 1.4), plus 20 psf for other construction loads.

$$\text{Slab} = 0.50(1.4) = 0.70 \quad \text{kip/ft}$$
$$\text{Construction} = 0.02(8)1.6 = 0.26 \quad \text{kip/ft}$$
$$\text{Steel section} = 0.031(1.2) = \underline{0.04} \quad \text{kip/ft}$$
$$w_u = 1.00 \quad \text{kip/ft}$$
$$M_u = \tfrac{1}{8}(1.00)(30)^2 = 113 \text{ ft-kips}$$
$$\phi_b M_p \text{ for W16×31} = 203 \text{ ft-kips} > 113 \text{ ft-kips} \qquad \textbf{OK}$$

Use W16×31 section ($F_y = 50$ ksi), with 20—$\frac{3}{4}$-in.-diam connectors over the entire span, spaced at 40 in. ∎

16.12 ASD EXAMPLE—SIMPLY SUPPORTED BEAM

EXAMPLE 16.12.1

Redesign the composite beam of Example 16.11.1 (see Fig. 16.11.1) using the AISC ASD Method. The materials are $F_y = 50$ ksi, $f'_c = 4$ ksi ($n = 8$), and a 4-in. slab.

Solution:

(a) Service load bending moments. From Example 16.11.1,

$$M_D = 43 \text{ ft-kips}$$
$$M_L = 157 \text{ ft-kips}$$

(b) Select steel section. Use Eq. 16.11.2b, assuming the PNA is within the slab. Estimate $a = 1.0$ in for preliminary selection.

The required allowable strength $M_a = 43 + 157 = 200$ ft-kips

$$\text{Required } A_s = \frac{M_a}{\dfrac{F_y}{\Omega_b}\left(\dfrac{d}{2} + t_n - \dfrac{a}{2}\right)} \qquad [16.11.2b]$$

$$\text{Required } A_s = \frac{200(12)}{\dfrac{50}{1.67}(7 + 3.5)} = 7.63 \text{ sq in.} \quad \text{(for W14)}$$

$$\text{Required } A_s = \frac{200(12)}{\dfrac{50}{1.67}(8 + 3.5)} = 6.97 \text{ sq in.} \quad \text{(for W16)}$$

Using AISC-Table 3-19, "Composite W Shapes-Available Strength in Flexure" with $Y2 = t_n - a/2 = 3.5$ in. and required $M_n/\Omega = 200$ ft-kips, find

W16×26	$M_n/\Omega = 217$ ft-kips	$A_s = 7.68$ sq in.
W14×26	$M_n/\Omega = 201$ ft-kips	$A_s = 7.69$ sq in.

This is identical to the sections determined by the LRFD Method.

(c) Compute the plastic neutral axis and check allowable strength.

From Example 16.11.1,

$$M_n = 358 \text{ ft-kips}$$

$$M_n/\Omega = 358/1.67 = 214 \text{ ft-kips} > (M_a = 200 \text{ ft-kips}) \qquad \textbf{OK}$$

(d) Check the allowable strength of the steel section to support construction loads. Loads values are obtained from Example 16.11.1.

Slab	$= 0.40$ kips/ft
Construction	$= 0.16$ kips/ft
Steel Section	$= \underline{0.026 \text{ kips/ft}}$
	$w_a = 0.586$ kips/ft

$$M_a = \tfrac{1}{8}(0.586)(28)^2 = 57.4 \text{ ft-kips}$$

$$M_p/\Omega_b \text{ for W16×26} = 110 \text{ ft-kips} > 57.4 \text{ ft-kips} \qquad \textbf{OK}$$

(e) Designing the shear connectors is identical to the LRFD Method.

Use W16×26 of A992 steel along with 30−3/4-in.-diameter × 3-in. headed stud connectors over the entire span, spaced at 22 in. The connectors are to be placed in pairs starting at support. ∎

16.13 DEFLECTIONS

The deflection of a composite beam will depend on whether it is shored or unshored during construction. Creep and shrinkage of the concrete in the slab also affect the result. Calculation of deflection requires obtaining the elastic cracked transformed section moment of inertia I_{tr} for the composite beam, and if unshored, also the elastic moment of inertia of the steel section alone.

When the steel beam is shored from below during the hardening of the concrete slab, the composite section will carry both the dead and live load. However, if the steel beam is *not* shored, the steel beam alone must carry the dead load.

When the construction is *without* shoring, the total deflection will be the sum of the dead load deflection of the steel beam alone and the live load deflection of the composite section.

When shoring provides the support during the hardening of the concrete slab, the composite section resists the entire load. Account should be taken to reflect the fact that concrete is subject to creep under long time load and that shrinkage will occur. This inelastic behavior may be approximated by multiplying the modulus of elasticity ratio n by a time-dependent factor such as two; thus reducing the effective width b_E/n. The result is a reduced moment of inertia I_{tr} to be used for computing the sustained load (dead load) deflection. The live load deflection would be computed using the elastic cracked transformed section moment of inertia.

Because the concrete slab in building construction is normally not too thick (say $t_s \leq 6$ in.) creep deflection is often not considered. The *AISC Specification* [1.15] gives no indication of any concern with creep of a concrete slab in composite construction. However, as discussed in Sec. 7.6, AISC-L3 states "Deflection in structural members and structural systems under appropriate service load combinations shall not impair the serviceability of the structure."

The ACI-ASCE Joint Committee [16.16] recommends using $E_c/2$ as the sustained concrete modulus of elasticity instead of E_c when computing sustained load creep deflection. AASHTO-6.10.1.1b [1.3] uses $E_c/3$ instead of E_c. Such arbitrary procedures can at best give an estimate of creep effects, probably no better than ±30%. The steel section, exhibiting no creep, and representing the principal carrying element, ensures that creep problems will usually be minimal.

More accurate procedures for computing deflections to account for creep and shrinkage on composite steel-concrete beams are available in a paper by Roll [16.31], and particularly in *Deformation of Concrete Structures* by Branson [16.32]. Lamport and Porter [16.45] have treated deflection prediction for concrete slabs reinforced with steel decking.

EXAMPLE 16.13.1

Compute the service dead and live load deflections for the composite beam consisting of W16×26 with 4-in. slab designed in Example 16.12.1 (see Fig. 16.13.1).

Solution:
Regardless of whether the selection of the steel section has been done by Load and Resistance Factor Design or by Allowable Strength Design, the deflections must be computed for *service* loads acting on the elastic section.

Figure 16.13.1
Beam cross-section for
Example 16.13.1.

(a) Compute the dead load deflection. From Example 16.11.1, part (a), the service dead load is 0.44 kip/ft, and all must be carried by the steel beam alone when the beam is unshored.

$$\Delta_{DL} = \frac{5wL^4}{384 E_s I_s} = \frac{5(0.44)(28)^4(12)^3}{384(29,000)448} = 0.47 \text{ in., } \quad \text{say } \tfrac{1}{2} \text{ in.}$$

The beam can be cambered or the slab can be thickened toward midspan so that this amount of deflection is compensated for during construction.

(b) Compute the live load deflection. From Example 16.11.1, part (a), the service live load is 1.6 kips/ft. This load must be carried by the composite section; thus, the elastic composite moment of inertia is required. Compute composite elastic section properties. Referring to Fig. 16.13.1 determine effective width b_E (ASCE-I3.1).

$$b_E = \tfrac{1}{4} \text{ of span } = 0.25(28)12 = 84 \text{ in.} \quad controls$$

or

$$b_E = \text{beam spacing } = 8(12) = 96 \text{ in.}$$

The width of equivalent steel is $b_E/n = 84/8 = 10.5$ in. The moment of inertia and elastic section modulus values are computed as follows:

Element	Transformed area A (sq in.)	Moment arm from centroid y (in.)	Ay (in.^{3})	Ay^2 (in.^{4})	I_0 (in.^{4})
Slab	42.0	9.85	413.7	4075	78.8
W16×36	7.68	0	0	0	301
	49.7		413.7	4075	379.8

$$I_x = I_0 + Ay^2 = 380 + 4075 = 4455 \text{ in}^4$$

$$\bar{y} = \frac{413.3}{49.7} = 8.32 \text{ in. above centroid of W16×26}$$

$$I_{tr} = I_x - A\bar{y}^2 = 4455 - 49.7(8.32)^2 = 1015 \text{ in.}^4$$

Often some of the dead load, such as partition loads and other items placed after the concrete slab has cured, acts on the composite section.

$$\Delta_{LL} = \frac{5wL^4}{384 E_b I_b} = \frac{5(1.6)(28)^4(12)^3}{384(29,000)1015} = 0.75 \text{ in.}$$

As discussed in Sec. 7.6, it has been traditional to consider that live load deflection exceeding $L/360$ may cause cracking of plaster. On the other hand, the *ACI Code* [16.15] restricts the live load plus creep and shrinkage deflection to a maximum of $L/480$. Thus, in the absence of any specific AISC limitation, a limit of approximately $L/400$ will likely give satisfactory serviceability for the floor system. In this case,

$$\Delta_{\text{limit}} = \frac{L}{400} = \frac{28(12)}{400} = 0.84 \text{ in.} > \Delta_{LL} \qquad \textbf{OK}$$

One may conclude that deflection will not cause concern. Note that $L/400$ is *not* an AISC limit. It is the designer's responsibility to establish any limit.

In the solution above, the full thickness of the slab was assumed to be contributing to the elastic stiffness of the composite beam. However, there is some uncertainty about the thickness of the slab that is actually contributing to the stiffness of the composite beam. The *AISC Manual* adopts a more conservative approach by assuming that the thickness of the concrete slab that is contributing to the strength is also defining the stiffness of the beam. So, regardless of whether the beam is fully composite or partially composite, the elastic stiffness of the composite beam is determined using the depth of the rectangular stress block a. The moment of inertia computed using a for the concrete thickness is defined as the lower bound moment of inertia I_{Lb}, because the actual moment of inertia would be always larger. I_{Lb} can be determined from AISC-Table 3-24 as follows:

$$\sum Q_n = 26.1(15) = 392 \text{ kips}$$

$$Y2 = t_s - \frac{a}{2} = 4 - \frac{1.34}{2} = 3.33 \text{ in.}$$

Enter AISC-Table-3.20 with these values.

By interpolation: $I_{Lb} = 781 \text{ in.}^4$

$$\Delta_{LL} = \frac{5wL^4}{384E_sI_{Lb}} = \frac{5(1.6)(28)^4(12)^3}{384(29,000)781} = 0.98 \text{ in.} > [\Delta_{\text{limit}} = 0.84 \text{ in.}] \quad \textbf{NG}$$

The authors believe the "exact" deflection probably falls near the $L/400$ limit. ∎

16.14 CONTINUOUS BEAMS

Traditionally on continuous beams the positive moment region has been designed as a composite section and the negative moment region where the concrete slab is in tension as a noncomposite section. However, some composite action has been known to exist in the negative moment region. Continuous composite beams have been studied by Barnard and Johnson [16.33], Johnson, Van Dalen, and Kemp [16.34], Daniels and Fisher [16.35], Hamada and Longworth [16.36, 16.37], and Kubo and Galambos [16.38]. Kubo and Galambos extended the treatment to plate girders (that is, beams having h/t_w exceeding $970/\sqrt{F_y}$).

According to AISC-I3.2b the negative moment strength is determined for the steel section alone. AISC allows for calculating the negative moment strength using the composite section, which accounts for the reinforcement contribution. However to analyze the beam as a composite section, the following conditions must be met:

1. The steel beam is compact and adequately braced.
2. Shear connectors are provided over the support region.
3. The slab reinforcement is within the effective width b_E and properly developed.

When the steel reinforcing bars in the concrete slab are utilized to contribute to composite action, the force developed by such bars must be transferred by shear connectors. The nominal strength developed would be

$$T_n(\text{for} -M \text{ region}) = A_{sr}F_{yr} \tag{16.16.1}$$

$$C_n(\text{for} +M \text{ region}) = A_s'F_{yr} \tag{16.16.2}$$

where A_{sr} = total area of longitudinal reinforcing steel at the interior support located within the effective flange width b_E

A'_s = total area of longitudinal compression steel acting with the concrete slab at the location of maximum positive moment and lying within the effective width b_E

F_{yr} = specified minimum yield stress of the longitudinal reinforcing steel

Thus, the nominal strength V_{nh} for which shear connectors must be provided in the *negative moment zone* is

$$V_{nh} = A_{sr}F_{yr} \qquad (16.16.3)$$

In the *positive moment zone,* when compression steel is included in the computation of composite section properties (plastic neutral axis), the nominal strength V_{nh2} from the compression steel is

$$V_{nh2} = A'_s F_{yr} \qquad (16.16.4)$$

The total horizontal shear force between the point of zero moment and the point of maximum moment is the smallest of $(0.85f'_c A_c + V_{nh2})$, $A_s F_y$, and ΣQ_n. AISC has no specific mention of the compression reinforcement in the positive moment zone; thus, inclusion of V_{nh2} is optional.

As discussed in Sec. 16.7, the usual limit state for composite sections in the positive moment zones is crushing of the concrete at the top of the slab. This assumes no shear connector failure, no longitudinal splitting because of inadequate reinforcing bar development, and no shear failure in the slab. In the negative moment region, the usual limit state is flange local buckling [16.37].

Regarding the lateral-torsional buckling limit state, the usual provisions for noncomposite steel sections apply to the negative moment regions of continuous composite beams. The limits on λ from AISC-B4 for the flange and web local buckling limit states must be applied in the negative moment zone.

EXAMPLE 16.14.1

Compute the plastic neutral axis (PNA) location and the nominal strength M_n for the section of Fig. 16.14.1 subject to negative bending moment. The W12×26 steel section is of A992 steel and the reinforcement in the slab has $F_{yr} = 60$ ksi.

Solution:

(a) Determine the plastic neutral axis location. The concrete slab will be in tension; therefore, none of the concrete is assumed to be effective. The reinforcing bars contribute

Figure 16.14.1
Composite section for negative bending of Example 16.14.1, including plastic stress distribution according to AISC-13.2b.

the nominal tension strength T_{sr},

$$T_{sr} = A_{sr}F_{yr} = 10(0.31)60 = 186 \text{ kips}$$

The maximum nominal compression force in the W12 section is

$$C_{max} = A_s F_y = 7.65(50) = 382.5 \text{ kips}$$

Since C_{max} exceeds T_{sr}, the PNA is within the steel W12 section. In which case the force equilibrium requirement may be expressed,

$$T_{sr} + T_s = C_{max} - T_s$$
$$2T_s = C_{max} - T_{sr} = 382.5 - 186 = 196.5$$
$$T_s = 98.3 \text{ kips}$$

Assuming that the PNA is within the flange of the W12,

$$\text{From top of flange to PNA} = \frac{T_s}{F_y b_f} = \frac{98.3}{50(6.49)} = 0.30 \text{ in.}$$

The assumption that PNA is within the flange is confirmed since $0.30 < (t_f = 0.38 \text{ in.})$. Thus, the distance PNA from top of slab is

$$\text{PNA} = t_s + 0.30 = 4.0 + 0.30 = 4.30 \text{ in.}$$

(b) Compute the nominal moment strength M_n. Locate the center of gravity y_1 of the compression force C_s in the steel section, measured from the bottom of the steel section,

	Area, A	Arm, y	Ay
W12 section	− 1.95	12.07	23.54
Flange	7.65		
	5.70 sq in.		23.54 in.^{3}

$$y_1 = \frac{23.54}{5.70} = 4.12 \text{ in.}$$

Referring to Fig. 16.14.1, taking internal moments about the point of action of C_s gives

$$T_{sr}: \quad M_{n1} = T_{sr} (d - 4.12 + t_s - 2.00)$$
$$= 186(12.22 - 4.12 + 4 - 2.00)\tfrac{1}{12}$$
$$= 186(10.1)\tfrac{1}{12} = 156.6 \text{ ft-kips}$$
$$T_s: \quad M_{n2} = T_s (d - 4.12 - 0.30/2)$$
$$= 98.3(12.22 - 4.12 - 0.30/2)\tfrac{1}{12}$$
$$= 98.3(7.80)\tfrac{1}{12} = 63.9 \text{ ft-kips}$$
$$M_n = M_{n1} + M_{n2}$$
$$= 156.6 + 63.9 = 221 \text{ ft-kips}$$
$$\phi_b M_n = 0.90(221) = 199 \text{ ft-kips}$$

Note that for composite action in the negative moment region, shear connectors must be used throughout the entire region. The required ΣQ_n equals the force T_{sr} in the reinforcement.

When partial composite action is used, ΣQ_n will be less than T_{sr}. In such a case, the PNA location and the nominal moment strength are computed using ΣQ_n instead of T_{sr}.   ■

16.15 COMPOSITE COLUMNS

A composite column can be defined as "a steel column fabricated from rolled or built-up steel shapes and encased in structural concrete or fabricated from steel pipe or tubing and filled with structural concrete." An example of the former is shown in Fig. 16.15.1, where a steel W section is encased in concrete; the concrete must contain longitudinal reinforcing bars and these must be surrounded by lateral ties in the manner of a reinforced concrete column.

The steel section must comprise at least 1% of the total cross-sectional area, otherwise the column must be designed as an ordinary reinforced concrete column. Research by Furlong [16.39, 16.40] and others was reviewed by Task Group 20 of the Structural Stability Research Council, chaired by Furlong [16.41]. This SSRC Task Group Report forms the basis for design of composite columns under AISC-I2.

Limitations

In order to qualify as a composite column, the limitations of AISC-I1.2 and I2 must be satisfied:

1.
$$A_s \geq 0.01A_g \qquad (16.15.1)$$

2. For a concrete encasement:

   (a) Longitudinal reinforcing bars must be used; load carrying bars must be continuous at framed levels (wherever a beam or slab frames to the column); other longitudinal bars used only to restrain concrete may be interrupted at framed levels.

   (b) Lateral ties must be used; spacing of ties may not exceed the smallest of 16 longitudinal bar diameter, 48 tie bar diameter, or 0.5 the least dimension of the composite section.

   (c) Area of lateral ties must be at least 0.009 sq in./in. of bar spacing.

   (d) The minimum required area of steel for continuous longitudinal reinforcement shall be $0.004A_g$.

   (e) Clear cover of at least 15 in. is required.

3. Concrete strength f_c':

   (a) Normal-weight concrete: $3 \text{ ksi} \leq f_c' \leq 10 \text{ ksi}$

   (b) Structural lightweight concrete: $3 \text{ ksi} \leq f_c' \leq 6 \text{ ksi}$

Figure 16.15.1
Composite column section; rolled steel shape encased in concrete.

Ties, commonly #3 bars

Longitudinal reinforcing bars

Rolled steel shape

4. Maximum yield stress of steel used in strength computations is 75 ksi for either structural steel or reinforcing bars.

5. Minimum wall thickness t for concrete-filled pipe or tubing:

(a) For each face width b in HSS rectangular sections:

$$\frac{b}{t} \leq 2.26\sqrt{\frac{E}{F_y}} \tag{16.15.2}$$

(b) For outside diameter D in circular sections:

$$\frac{D}{t} \leq 0.15\sqrt{\frac{E}{F_y}} \tag{16.15.3}$$

Nominal Strength

To account for slenderness effects, the AISC equations for composite columns are based on a modified form of the equations for steel columns in AISC-E. The yield strength becomes a modified strength P_e, and the elastic stiffness of the column is defined by an effective elastic stiffness EI_{eff} defined in what follows.

The resistance and safety factors adopted for composite columns are rather conservative in order to account for the uncertainty of composite columns and the use of ultimate strength of two different materials in defining the capacity. The factors are as follows:

$$\phi = 0.75 \,(\text{LRFD}) \qquad \Omega = 2.00 \,(\text{ASD})$$

The nominal compressive strength shall be determined according to AISC-I2.lb as follows

1. When $P_e \geq 0.44P_o$

$$P_n = P_o\left[0.658^{\left(\frac{P_o}{P_e}\right)}\right] \tag{16.15.4}$$

2. When $P_e < 0.44P_o$

$$P_n = 0.877P_e \tag{16.15.5}$$

where $\quad P_e = \pi^2(EI_{eff})/(KL)^2 \tag{16.15.6}$

For filled composite columns, AISC-I2.2b defines:

$$P_e = A_sF_y + A_{sr}F_{yt} + C_2A_cf'. \tag{16.15.7}$$

where $C_2 = 0.85$ for irregular sections and 0.95 for circular sections

$$EI_{eff} = E_sI_s + E_sI_{sr} + C_3E_cI_c \tag{16.15.8}$$

$$C_3 = 0.6 + 2\left(\frac{A_s}{A_c + A_r}\right) \leq 0.9 \tag{16.15.9}$$

For encased composite columns. AISC-I2.lb defines:

$$P_e = A_sF_y + A_{sr}F_{yt} + 0.85A_cf'_c \tag{16.15.10}$$

$$EI_{eff} = E_sI_s + 0.5E_sI_{sr} + C_1E_cI_c \tag{16.15.11}$$

$$C_1 = 0.1 + 2\left(\frac{A_s}{A_c + A_s}\right) \leq 0.3 \tag{16.15.12}$$

where A_c = area of concrete

A_r = area of longitudinal reinforcing bars

A_s = gross area of steel shape, pipe, or tube

E_c = modulus of elasticity of concrete, ksi

= $(w^{1.5})\sqrt{f'_c}$, where w is the density of concrete in pcf (i.e., 145 pcf for normal-weight concrete) and f'_c is in ksi

E_s = modulus of elasticity of steel

F_y = specified minimum yield stress of steel shape, pipe, or tube

F_{yr} = specified minimum yield stress of longitudinal reinforcing bars

f'_c = specified 28-day compressive strength of concrete

I_c = moment of inertia of concrete section

I_s = moment of inertia of steel shape

I_{sr} = moment of inertia of reinforcing bars

The *AISC Manual* contains tables for concrete filled HSS sections giving axial strengths ϕP_u and P_n/Ω. Note that $\phi_n = 0.75$ for composite columns.

Composite beam-column design has been treated by Uang, Wattar, and Leet [16.46]. AISC-I4 defines the method of treating composite beam-columns. An interaction curve similar to reinforced concrete needs to be developed, while accounting for the stability requirement of the column. The nominal strength of the section is to be determined using plastic stress distribution or strain compatibility. AISC-I2 is to be used to determine the nominal axial strength of the cross-section, using P_u as determined in AISC-I4.

SELECTED REFERENCES

16.1. Ivan M. Viest. Chairman, "Composite Steel-Concrete Construction," Report of the Subcommittee on the State-of-the-Art Survey of the Task Committee on Composite Construction of the Committee on Metals of the Structural Division, *Journal of the Structural Division,* ASCE, **100,** ST5 (May 1974), 1085–1139.

16.2. Gottfried Brendel. "Strength of the Compression Slab of T-Beams Subject to Simple Bending," *ACI Journal, Proceedings,* **61,** January 1964, 57–76.

16.3. Conrad P. Heins and Horn Ming Fan. "Effective Composite Beam Width at Ultimate Load," *Journal of the Structural Division,* ASCE, **102,** ST11 (November 1978), 2163–2179.

16.4. Cesar R. Vallenilla and Reidar Bjorhovde. "Effective Width Criteria for Composite Beams," *Engineering Journal,* AISC, **22,** 4 (Fourth Quarter 1985), 169–175.

16.5. Ivan M. Viest. "Review of Research on Composite Steel-Concrete Construction," *Journal of the Structural Division,* ASCE, **86,** ST6 (June 1960), 1–21; Also *Transactions,* ASCE, **126** (1961), Part II, 1101–1123.

16.6. William C. Hansell, Theodore V. Galambos, Mayasandra K. Ravindra, and Ivan M. Viest. "Composite Beam Criteria in LRFD," *Journal of the Structural Division,* ASCE, **104,** ST9 (September 1978), 1409–1426; Disc. **106,** ST2 (February 1980), 571–572.

16.7. Srinivasa H. Iyengar and Mohammad Iqbal. "Composite Construction," *Building Structural Design Handbook,* Chapter 23. New York: John Wiley & Sons, 1987.

16.8. Robert F. Lorenz and Frank W. Stockwell, Jr. "Concrete Slab Stresses in Partial Composite Beams and Girders," *Engineering Journal,* AISC, **21,** 3 (Third Quarter 1984), 185–188.

16.9. Robert F. Lorenz. "Understanding Composite Beam Design Methods Using LRFD," *Engineering Journal,* AISC, **25,** 1 (Second Quarter 1988), 35–38.

16.10. Robert F. Lorenz. "Some Economic Considerations for Composite Floor Beams," *Engineering Journal,* AISC, **20,** 2 (Second Quarter 1983), 78–81.

16.11. Mark C. Zahn. "The Economies of LRFD in Composite Floor Beams," *Engineering Journal,* AISC, **24,** 2 (Second Quarter 1987), 87–92.

16.12. S. Timoshenko and J. Goodier. *Theory of Elasticity.* New York: McGraw-Hill Book Company, Inc., 1959, Chap. 6.

16.13. Theodore von Kármán. "Die Mittragende Breitte," *August-Föppel-Festschrift,* 1924. (See also Collected Works of Theodore von Kármán, Volume II, p. 176.)

16.14. John E. Johnson and Albert D. M. Lewis. "Structural Behavior in a Gypsum Roof-Deck System," *Journal of the Structural Division,* ASCE, **92**, ST2 (April 1966), 283–296.

16.15. ACI Committee 318 (ACI 318-08). *Building Code Requirements for Reinforced Concrete.* Farmington Hills. MI: American Concrete Institute, 2008.

16.16. Joint ASCE-ACI Committee on Composite Construction. "Tentative Recommendations for the Design and Construction of Composite Beams and Girders for Buildings," *"Journal of the Structural Division,"* ASCE, **86**, ST12 (December 1960), 73–92.

16.17. Roger G. Slutter and George C. Driscoll. "Flexural Strength of Steel-Concrete Composite Beams," *Journal of the Structural Division,* ASCE, **91**, ST2 (April 1965), 71–99.

16.18. Peter Ansourian and Jack William Roderick. "Analysis of Composite Beams," *Journal of the Structural Division,* ASCE, **104**, ST10 (October 1978), 1631–1645.

16.19. Jorgen G. Ollgaard, Roger G. Slutter, and John W. Fisher. "Shear Strength of Stud Connectors in Lightweight and Normal-Weight Concrete," *Engineering Journal,* AISC, **8**, 2 (April 1971), 55–64.

16.20. Jay B. McGarraugh and J. W. Baldwin, Jr. "Lightweight Concrete-on-Steel Composite Beams," *Engineering Journal,* AISC, **8**, 3 (July 1971), 90–98.

16.21. Roger G. Slutter and John W. Fisher. "Fatigue Strength of Shear Connectors," *Highway Research Record No. 147,* Highway Research Board, 1966, pp. 65–88.

16.22. Charles G. Schilling. "Bending Behavior of Composite Hybrid Beams," *Journal of the Structural Division,* ASCE, **94**, ST8 (August 1968), 1945–1964.

16.23. John A. Grant, Jr., John W. Fisher, and Roger G. Slutter. "Composite Beams with Formed Steel Deck," *Engineering Journal,* AISC, **14**, 2 (First Quarter 1977), 24–43.

16.24. Sriramulu Vinnakota, Christopher M. Foley, and Murthy R. Vinnakota. "Design of Partially or Fully Composite Beams, with Ribbed Metal Deck, Using LRFD Specifications," *Engineering Journal,* AISC, **25**, 2 (Second Quarter 1988), 60–78.

16.25. R. H. R. Tide and T. V. Galambos. "Composite Open-Web Steel Joists," *Engineering Journal,* AISC, **7**, 1 (January 1970), 27–36.

16.26. James Rongoe. "A Composite Girder System for Joist Supported Slabs," *Engineering Journal,* AISC, **21**, 3 (Second Quarter 1984), 155–160.

16.27. ASCE. *Specifications for the Design and Construction of Composite Slabs.* New York: Technical Council on Codes and Standards, American Society of Civil Engineers, 1984.

16.28. ASCE. *Commentary on Specifications for the Design and Construction of Composite Slabs.* New York: Technical Council on Codes and Standards, American Society of Civil Engineers, 1984.

16.29. Max L. Porter. "Analysis of Two-Way Acting Composite Slabs," *Journal of Structural Engineering,* ASCE, **111**, 1 (January 1985), 1–17.

16.30. C. Dale Buckner, Danny J. Deville, and Dean C. McKee. "Shear Strength of Slabs in Stub Girders," *Journal of the Structural Division,* ASCE, **107**, ST2 (February 1981), 273–280.

16.31. Frederic Roll. "Effects of Differential Shrinkage and Creep on a Composite Steel-Concrete Structure," *Designing for Effects of Creep, Shrinkage, Temperature in Concrete Structures* (SP-27). Detroit, MI: American Concrete Institute, 1971 (pp. 187–214).

16.32. Dan E. Branson. *Deformation of Concrete Structures.* New York: McGraw-Hill Book Company, Inc., 1977.

16.33. P. R. Barnard and R. P. Johnson. "Plastic Behavior of Continuous Composite Beams," *Proceedings,* Institution of Civil Engineers, October 1965.

16.34. R. P. Johnson, K. Van Dalen, and A. R. Kemp, "Ultimate Strength of Continuous Composite Beams," *Proceedings of the Conference on Structural Steelwork,* British Constructional Steelwork Association, November 1967.

16.35. J. H. Daniels and J. W. Fisher. "Static Behavior of Continuous Composite Beams," *Fritz Engineering Laboratory Report* No. 324.2, Lehigh University, Bethlehem, PA, March 1967.

16.36. Sumio Hamada and Jack Longworth. "Buckling of Composite Beams in Negative Bending," *Journal of the Structural Division,* ASCE, **100**, ST11 (November 1974), 2205–2222.

16.37. Sumio Hamada and Jack Longworth. "Ultimate Strength of Continuous Composite Beams," *Journal of the Structural Division,* ASCE, **102**, ST7 (July 1976), 1463–1478.

16.38. Masahiro Kubo and Theodore V. Galambos. "Plastic Collapse Load of Continuous Composite Plate Girders," *Engineering Journal,* AISC, **25**, 4 (Fourth Quarter 1988), 145–155.

16.39. Richard W. Furlong. "Strength of Steel Encased Concrete Beam Columns," *Journal of the Structural Division,* ASCE, **94**, ST1 (January 1968), 267–281.

16.40. Richard W. Furlong. "AISC Column Design Logic Makes Sense for Composite Columns, Too," *Engineering Journal,* AISC, **13**, 1 (First Quarter 1976), 1–7.

16.41. Task Group 20, Structural Stability Research Council. "A Specification for the Design of Steel-Concrete Composite Columns," *Engineering Journal,* AISC, **16**, 4 (Fourth Quarter 1979), 101–115.

16.42. IABSE. *Composite Steel-Concrete Construction and Eurocode 4,* Short Course Notes, International Association for Bridge and Structural Engineering, Brussels, 1990, 191 pp.

16.43. W. Samuel Easterling, David R. Gibbings and Thomas M. Murray. "Strength of Shear Studs in Steel Deck on Composite Beams and Joists," *Engineering Journal,* AISC, **30**, 2 (2nd Quarter 1993), 44–55.

16.44. Mark Andrew Bradford and R. Ian Gilbert. "Composite Beams with Partial Interaction Under Sustained Loads," *Journal of Structural Engineering,* ASCE, **118,** 7 (July 1992), 1871–1883.

16.45. William B. Lamport and Max L. Porter. "Deflection Predictions for Concrete Slabs Reinforced with Steel Decking," *ACI Structural Journal,* **87,** September-October 1990, 546–570.

16.46. Chia-Ming Uang, Samer W. Wattar, and Kenneth M. Leet. "Proposed Revision of the Equivalent Axial Load Method for LRFD Steel and Composite Beam-Column Design," *Engineering Journal,* AISC, **27,** 4 (4th Quarter 1990), 150–157.

16.47. M. D. Rambo-Roddenberry, W. S. Easterling, and T. M. Murray. "Behavior and Strength of Welded Stud Shear Connectors-Data Report," Report No. CE/VPI-02/04, Virginia Polytechnic Institute and State University, Blackspurg, VA, 2002.

PROBLEMS

All problems are to be done according to the AISC LRFD Method or the ASD Method, as indicated by the instructor. All given loads are service loads unless otherwise indicated.

16.1. For the case (or cases) assigned by the instructor, compute the location of the transformed composite section neutral axis and moment of inertia I_{tr}. Refer to the accompanying figure.

Case	Steel section	F_y (ksi)	Slab t_s (in.)	b_E (in.)	f'_c (ksi)	
2	W12×26	60	4	72	3	($n = 9$)
4	W21×44	50	4	84	4	($n = 8$)
5	W24×55	50	4.5	90	4	($n = 8$)
6	W18×50	50	5	72	4	($n = 8$)
7	W18×50	50	4	72	3	($n = 9$)
8	W24×76	50	4.5	72	4	($n = 8$)
9	W24×94	50	4.5	72	4	($n = 8$)
10	W21×62	50	5	96	4	($n = 8$)
11	W21×147	50	4.5	96	4	($n = 8$)

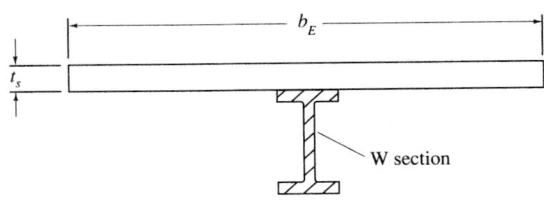

Problems 16.1, 16.2, and 16.3

16.2. For the case (or cases) listed for Prob. 16.1 and assigned by the instructor, compute the location of plastic neutral axis (PNA) measured from the top of the slab, as well as the nominal strength M_n. Assume the sections are *fully composite.* Refer to the accompanying figure.

16.3. For the case (or cases) listed for Prob. 16.1 and assigned by the instructor, select an economical size of headed stud shear connector from Table 16.8.1, determine the total number needed to develop a *fully composite* section for the beam, and specify the spacing. Assume the simply supported beam span equals $4b_E$.

16.4. For the case (or cases) assigned by the instructor, select a W section to design a *fully composite* section for span *BD* of the accompanying figure. Assume for simplicity that the slab is the only dead load to be considered. Also, select an economical size of headed stud shear connector from Table 16.8.1, determine the total number needed for the beam, and specify the spacing, to develop a fully composite beam. No shoring is to be used; therefore, assume that during construction wet concrete of 75 psf is live load and that an additional construction live load of 25 psf may act. The final composite beam may not have live load deflection exceeding $L/360$.

Case	Live Load (psf)	F_y (ksi)	Slab t_s (in.)	Span (ft)	Beam Spacing (ft)	f'_c (ksi)	
2	100	60	4	36	8	3	($n = 9$)
4	80	50	4	36	7	4	($n = 8$)
5	80	50	4	40	7	4	($n = 8$)
6	80	50	4	40	7	4	($n = 8$)
8	125	50	4.5	40	8	3	($n = 9$)
9	125	50	4.5	42	8	4	($n = 8$)
10	125	50	4.5	45	8	4	($n = 8$)
11	125	50	5	45	9	4	($n = 8$)
12	125	50	5.5	48	9	4	($n = 8$)

16.5. For the case (or cases) given in Prob. 16.4 assigned by the instructor, select a W section to design a *partially composite* section for span *BD* of the accompanying figure, using the minimum number of $\frac{3}{4}$-in.-diam headed stud shear connectors. Specify number of studs and the spacing. In addition to the slab dead load, using a ceiling load of 7 psf and partition load of 25 psf. No shoring is to be used; therefore, assume that during construction wet concrete of 75 psf is live load and that an additional construction live load of 25 psf may act. The dead load deflection before composite action is effective may not exceed $\frac{7}{8}$ in. and composite beam deflection resulting from superimposed dead load (i.e., ceiling and partitions) and live load may not exceed $L/360$.

16.6. Design a composite encased W shape column to resist a factored axial compression load P_u of 900 kips. The effective length $KL = 12$ ft, $F_y = 50$ ksi for structural steel, and $f'_c = 4.5$ ksi for concrete.

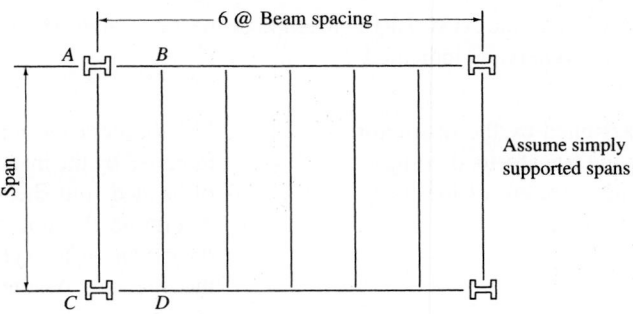

Problems 16.4 and 16.5
Framing plan.

Appendix

TABLE A1 Approximate Radius of Gyration

Column 1	Column 2	Column 3
$r_x = 0.29h$, $r_y = 0.29b$	$r_x = 0.42h$, $r_y = 0.42b$	$r_x = 0.31h$, $r_y = 0.48b$
$r_x = 0.40h$, $h = \text{mean } h$	$r_y = \text{same as for } 2L$	$r_x = 0.37h$, $r_y = 0.28b$
$r_x = 0.25h$	$r_x = 0.42h$, $r_y = \text{same as for } 2L$	$r_x = 0.31h$
$r = \sqrt{\dfrac{H^2 + h^2}{16}}$, $r = 0.35H_m$	$r_x = 0.39h$, $r_y = 0.21b$	$r_x = 0.31h$
$r_x = 0.31h$, $r_y = 0.31h$, $r_z = 0.197h$	$r_x = 0.45h$, $r_y = 0.235b$	$r_x = 0.40h$, $r_y = 0.21b$
$r_x = 0.29h$, $r_y = 0.32b$, $r_z = 0.18\dfrac{h+b}{2}$	$r_x = 0.36h$, $r_y = 0.45b$	$r_x = 0.38h$, $r_y = 0.22b$
$r_x = 0.31h$, $r_y = 0.215b = b(0.21 + 0.02s)$	$r_x = 0.36h$, $r_y = 0.60b$	$r_x = 0.39h$
$r_x = 0.32h$, $r_y = 0.21b = b(0.19 + 0.02s)$	$r_x = 0.36h$, $r_y = 0.53b$	$r_x = 0.35h$
$r_x = 0.29h$, $r_y = 0.24b = b(0.23 + 0.02s)$	$r_x = 0.39h$, $r_y = 0.55b$	$r_x = 0.435h$, $r_y = 0.25b$
$r_x = 0.30h$, $r_y = 0.17b$	$r_x = 0.42h$, $r_y = 0.32b$	$r_x = 0.42h$
$r_x = 0.25h$, $r_y = 0.21b$	$r_x = 0.44h$, $r_y = 0.28b$	$r_x = 0.42h$
$r_x = 0.21h$, $r_y = 0.21b$, $r_z = 0.19h$	$r_x = 0.50h$, $r_y = 0.28b$	$r_x = 0.285h$, $r_y = 0.37b$
$r_x = 0.38h$, $r_y = 0.19b$	$r_x = 0.39h$, $r_y = 0.21b$	$r_x = 0.42h$, $r_y = 0.23b$

* J.A.L. Waddel. *Bridge Engineering*, Vol. 1. New York: John Wiley & Sons, Inc., 1916, p. 504. Used by permission.

TABLE A2 Torsional Properties

O = shear center J = torsion constant, C_w = warping constant
G = centroid I_p = polar moment of inertia about shear center

$$J = \tfrac{1}{3}(2bt_f^3 + ht_w^3)$$

$$C_w = \frac{I_f h^2}{2} = \frac{t_f b^3 h^2}{24} = \frac{h^2 I_y}{4}$$

$$I_p = I_x + I_y$$

$$J = \tfrac{1}{3}(b_1 t_f^3 + b_2 t_f^3 + ht_w^3)$$

$$C_w = \frac{t_f h^2}{12}\left(\frac{b_1^3 b_2^3}{b_1^3 + b_2^3}\right)$$

$$e = h\frac{b_1^3}{b_1^3 + b_2^3}$$

$$I_p = I_y + I_x + Ay_0^2$$

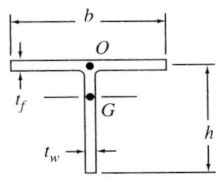

$$J = \tfrac{1}{3}(bt_f^3 + ht_w^3)$$

$$C_w = \frac{1}{36}\left(\frac{b^3 t_f^3}{4} + h^3 t_w^3\right)$$

$$\approx \text{zero for small } t$$

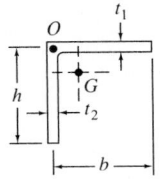

$$J = \tfrac{1}{3}(bt_1^3 + ht_2^3)$$

$$C_w = \frac{1}{36}(b^3 t_1^3 + h^3 t_2^3)$$

$$\approx \text{zero for small } t$$

$$J = \tfrac{1}{3}(2bt_f^3 + ht_w^3)$$

$$C_w = \frac{t_f b^3 h^2}{12}\left(\frac{3bt_f + 2ht_w}{6bt_f + ht_w}\right) = \frac{h^2}{4}(I_y + A\bar{x}^2 - q\bar{x}A)$$

$$q = \frac{th^2 b^2}{4I_x}$$

Index

N_i	=	notional lateral load, Sec. 12.11
p	=	pitch (spacing) of bolts; connector spacing (Chap. 16); $1/\phi_c P_n$, Eq.12.12.2
P_{bf}	=	design strength of column web to resist a concentrated factored load
P_{br}	=	required brace strength
P_{cr}	=	critical buckling load; compression force at buckling
P_e, P_{ex}, P_{ey}	=	Euler load $= \pi^2 E A_g/(KL/r)^2$ for axis of bending (using two subscripts for biaxial bending)
P_{e1}, P_{e2}	=	Euler load $= \pi^2 E A_g/(KL/r)^2$ for axis of bending, for use with magnification factors B_1 and B_2 according to the second subscript
P_n	=	nominal strength of an axially loaded compression member, $F_{cr} A_g$; nominal strength of weld configuration (Fig. 5.17.2)
P_u	=	factored axial load (Sec. 1.9); factored reaction or load
P_y	=	yield load, $F_y A_g$ (Chap. 12)
Q	=	form factor, $Q_a Q_s$ (Sec. 6.18); first moment of area (i.e., statical moment $\int y\, dA$) about the neutral axis from extreme fiber to section at which elastic shear stress is computed, (see Sec. 7.7)
Q_a	=	shape factor for stiffened compression element (Sec. 6.18)
Q_f	=	moment of area of one-half flange about y-axis (Sec. 8.5)
Q_s	=	shape factor for unstiffened compression element (Sec. 6.18)
r	=	radius of gyration, $\sqrt{I/A_g}$; radial distance from centroid to point of stress (Sec. 5.18)
r_i	=	distance from instantaneous center to a weld element (Fig. 5.17.2)
r_0	=	distance from instantaneous center to vertical weld line
r_{max}	=	distance to weld element farthest from instantaneous center
r_t	=	radius of gyration of a section comprising the compression flange plus one-third of the compression web area, taken about an axis in the plane of the web; used in ASD, Eq. 9.7.14
r_{ts}	=	effective radius of gyration used in determining L_r (Chap. 9)
r_x, r_y, r_z	=	radius of gyration about x-, y-, or z-axes, respectively
R_a	=	required strength (ASD)
R_e	=	moment strength reduction factor for hybrid girder (Secs. 11.7 and 16.9)
R_g	=	coefficient to account for group effect (Chap. 16)
R_i	=	resistance of a bolt at any deformation D (Chap. 4); strength of a fillet weld segment per unit length (Sec. 5.17)
$R_{i,ult}$	=	ultimate shear load on an element, Eq. 5.17.3
R_m	=	cross-section monosymmetry parameter (Chap. 9)
R_n	=	nominal strength of one fastener in tension, shear, or bearing; nominal reaction strength (Sec. 7.8)
R_{nt}	=	nominal strength of bolt in tension
R_{nv}	=	nominal strength of bolt in shear
R_{nw}	=	nominal strength of weld per inch of length
R_p	=	position effect factor for shear studs (Chap. 16)
R_{pg}	=	reduction factor for "bend-buckling" of the web, Eq. 11.4.3
R_s	=	direct shear component of bolt resistance
R_u	=	factored load per bolt; factored load per unit length of weld; factored reaction (Sec. 7.8)
R_{ult}	=	ultimate shear resistance in a bolt, $\tau_u A_b$
R_{us}	=	factored direct shear on bolt subject to eccentric load
R_{ut}	=	factored tension load on bolt
R_{uv}	=	factored direct shear component on bolt
R_{ux}, R_{uy}	=	factored shear on bolt, in x- or y-direction, respectively
R_v	=	direct shear component of bolt resistance; shear component of eccentric force on fillet welds; direct shear component of weld resistance/per unit length
R_x, R_y	=	x-or y-direction component of bolt resistance; x- or y-component of torsional moment force on fillet welds
s	=	stagger of bolt holes measured in the line of force (Chap. 3); distance from free edge along a thin wall section (Chap 8); band width for tension-field force T (Sec. 11.9)
S, S_x, S_y	=	elastic section modulus, $I/\bar{y}$ (Table 5.18.1), with respect to x- or y-axes $(I_x/c_y$ or $I_y/c_x)$, according to subscript

S_s = elastic section modulus of steel section alone, referred to its tension flange

S_{tr} = elastic section modulus of composite section, I_{tr}/y_b

S_{xc}, S_{xt} = section modulus S_x referred to the compression flange, S_{xc}, or the compression flange, S_{xt}

t = thickness; thickness of material against which bolt bears

t_e = effective throat dimension of a weld (Sec. 5.12)

t_f, t_{fb}, t_{fc} = flange thickness; for beam, t_{fb}; for column, t_{fc}

t_s = thickness of stiffener; slab thickness

t_w, t_{wb}, t_{wc} = web thickness; for beam, t_{wb}; for column, t_{wc}

T = tensile force; service load tensile force; torsional moment or torsional service load moment (Chap. 8); base metal thickness (Table 5.11.1)

T_0 = required tension strength (ASD)

T_b = initial force in bolt resulting from installation

= nominal strength of a tension member

T_u = factored tension load; factored torsional moment (required tension strength, required torsion strength) (Chap. 8)

u = displacement in the x-direction

u_f = lateral deflection of flange

U = reduction factor to account for shear lag (Sec. 3.9)

U_{bs} = stress reduction factor for nonuniform stress for block shear rupture

v = shear stress; displacement in the y-direction

v_s = St. Venant torsion shear stress (Chap. 8)

v_w = warping torsion shear stress (Chap. 8)

V = shear; service load shear force on a bolt

V_f = warping torsion shear force in flange

V_n = nominal shear strength

V'_n = nominal shear strength in the presence of bending moment

V_{nh} = nominal horizontal shear strength across interface between slab and steel section in a composite beam

V_r = range of service load shear force, Eq. 16.8.9

V_u = factored shear force

V_x, V_y = shear in the x- and y-directions, respectively

w = uniform loading; service uniformly distributed load on beam; displacement in z-direction (Fig. 6.14.2); width of stiffener plate (Chap. 11); density of concrete, Eq. 16.5.1

w_D, w_L = service uniform dead and live load, respectively

w_n = w_u/ϕ_b = required nominal uniform load causing collapse mechanism (Chap. 10)

w_u = factored uniform load

w_{uh} = factored uniform horizontal load

W = total service load on a span; concentrated load on beam; width of stiffener Chap. 11); seat width (Chap. 13)

W_n = W_u/ϕ_b = required nominal concentrated load causing collapse (Chap. 10)

W_u = factored concentrated load

x_0, y_0 = shear center distances from centroid measured along the x- and y-axes, respectively

y = deflection at a location z along axis of member

$\bar{y}$ = center of gravity (CG) of composite section measured from CG of gravity of steel W section

y_0 = $(V'_n/V_n)h$

y_1 = total deflection (including second-order deflection) of beam-column

y_b = distance to bottom of steel section from CG of composite section

y_c, y_t = distances from CG of the section to the compression and tension extreme fibers, respectively

Z, Z_x, Z_y = plastic modulus, $\int y\, dA$, with respect to the axes indicated by subscript

α = constant $GJ/(2EC_w)$, Eq. 9.4.7; , ratio of web yield stress to flange yield stress, F_{yw}/F_{yf}, (Sec. 11.7); P_u/P_e or $\sum P_u/\sum P_e$ (Chap. 12)

β = flexure analogy modification factor (Chap. 8); A_w/A_f, ratio of web cross-sectional area to cross-sectional area of the *compression* flange (Sec.11.7)

β_{br} = required brace stiffness (Chaps. 9 and 15)

β_s = E_t/E, Eq. 6.9.2 (Table 6.9.1)

β_w = selected ratio h/t_w for design (Sec. 11.14)